KNOWLEDGE ENTERPRISE: INTELLIGENT STRATEGIES IN PRODUCT DESIGN, MANUFACTURING, AND MANAGEMENT

IFIP – The International Federation for Information Processing

IFIP was founded in 1960 under the auspices of UNESCO, following the First World Computer Congress held in Paris the previous year. An umbrella organization for societies working in information processing, IFIP's aim is two-fold: to support information processing within its member countries and to encourage technology transfer to developing nations. As its mission statement clearly states,

> *IFIP's mission is to be the leading, truly international, apolitical organization which encourages and assists in the development, exploitation and application of information technology for the benefit of all people.*

IFIP is a non-profitmaking organization, run almost solely by 2500 volunteers. It operates through a number of technical committees, which organize events and publications. IFIP's events range from an international congress to local seminars, but the most important are:

- The IFIP World Computer Congress, held every second year;
- Open conferences;
- Working conferences.

The flagship event is the IFIP World Computer Congress, at which both invited and contributed papers are presented. Contributed papers are rigorously refereed and the rejection rate is high.

As with the Congress, participation in the open conferences is open to all and papers may be invited or submitted. Again, submitted papers are stringently refereed.

The working conferences are structured differently. They are usually run by a working group and attendance is small and by invitation only. Their purpose is to create an atmosphere conducive to innovation and development. Refereeing is less rigorous and papers are subjected to extensive group discussion.

Publications arising from IFIP events vary. The papers presented at the IFIP World Computer Congress and at open conferences are published as conference proceedings, while the results of the working conferences are often published as collections of selected and edited papers.

Any national society whose primary activity is in information may apply to become a full member of IFIP, although full membership is restricted to one society per country. Full members are entitled to vote at the annual General Assembly, National societies preferring a less committed involvement may apply for associate or corresponding membership. Associate members enjoy the same benefits as full members, but without voting rights. Corresponding members are not represented in IFIP bodies. Affiliated membership is open to non-national societies, and individual and honorary membership schemes are also offered.

KNOWLEDGE ENTERPRISE: INTELLIGENT STRATEGIES IN PRODUCT DESIGN, MANUFACTURING, AND MANAGEMENT

Proceedings of PROLAMAT 2006, IFIP TC5 International Conference, June 15-17, 2006, Shanghai, China

Edited by

Kesheng Wang
Department of Production and Quality Engineering, Norwegian University of Science and Technology Norway

George L. Kovacs
Computer and Automation Institute and Technical University of Budapest Hungary

Michael Wozny
Rensselaer Polytechnic Institute USA

Minglun Fang
Shanghai University China

Springer

Library of Congress Control Number: 2006925471

Knowledge Enterprise: Intelligent Strategies in Product Design, Manufacturing, and Management
Edited by K. Wang, G. Kovacs, M. Wozny, and M. Fang

p. cm. (IFIP International Federation for Information Processing, a Springer Series in Computer Science)

ISSN: 1571-5736 / 1861-2288 (Internet)
ISBN: 10: 0-387-34402-0
ISBN: 13: 9780-387-34402-0
eISBN: 10: 0-387-34403-9
Printed on acid-free paper

Printed in the United States of America.

9 8 7 6 5 4 3 2 1
springer.com

Organized and sponsored by

IFIP *(International Federation for Information Processing)*
NSFC *(National Natural Science Foundation of China)*
STCSM *(Science and Technology Committee of Shanghai Municipality)*
SHMEC *(Shanghai Municipal Education Commission)*
NTNU *(Norwegian University of Science and Technology)*
SHU *(Shanghai University)*

International Program Committee:

Chair: ***Kesheng Wang,*** Norway
Co-Chairs:
George L. Kovacs, Hungary
Michel Wozny, USA
Guy Doumeingt, France

Members:

Argentina, *G. Henning*
Australia, *C. Avran*
Australia, *P. Burnus*
Austria, *Man-Wook Han*
Canada, *D. Swayne*
China, *Renchu Gan*
Denmark, *P. Falster*
Finland, *R. Smeds*
France, *G. Doumeingts*
France, *R. Soenen*
Germany, *D. Kochan*
Germany, *E. Neuhold*
Germany, *R. Denzer*
Hungary, *G. L. Kovacs*
Israel, *G. Halevi*
Italy, *U. Cugini*
Japan, *E. F. Kimura*
Korea, *B. K. Choi*
Mexico, *N. Leon-Rovira*
Netherlands, *H. Afsarmanesh*
Norway, *K. Wang*
Norway, *Z. Deng*
Poland, *T. Czachorski*
Portugal, *L. M. Camarinha-Matos*
Slovenia, *I. Rozman*
Sweden, *T. Kjellberg*
Switzerland, *E. M. Gelle*
Switzerland, *N.Magnenat-Thalmann*
United Kingdom, *R. A. Earnshaw*
United Kingdom, *U. Bititci*
USA, G. *J. Olling*
USA, *H. Jamil*
USA, *M. B. Mc Grath*
USA, *M. Wozny*
USA, *R. Waxman*
USA, *J. Vlietstra*

National Program Committee:

Chair: ***Minglun Fang,*** Shanghai University
Co-Chairs: ***Xueyu Ruan,*** Shanghai Jiaotong University
Peigen Li, Huazhong University of Science and Technology
Zhihang Lin, Xian Jiaotong University

Members:

Egao Cai, Harbin Institute of Technology University
Mingyi Chen, Shanghai University
Weiming Cheng, Shanghai University
Minglun Fang, Shanghai University
Zhenbang Gong, Shanghai University
Peigen Li, Huazhong University of Science and Technology
Zhihang Lin, Xian Jiaotong University
Jinwu Qian, Shanghai University
Xueyu Ruan, Shanghai Jiaotong University
Min Tan, Institute of Automation Chinese Academy of Science
Xiaoqing Tang, Beihang University
Tao Yu, Shanghai University
Dongbiao Zhao, Nanjing University of Aeronautics and Astronautics
Qinzhi Zhou, Shanghai Mashine Factory Ltd. Co.
Songlin Zuang, University of Shanghai for Science and Technology

General Secretary:

Tao Yu**,** Shanghai University
Qingfeng Yuan, Shanghai University

Secretariat and Webmaster:

Cuilian Zhao, Shanghai University
Hongxia Cai, Shanghai University
Jiaying Ruan**,** Shanghai University
Jin Yuan**,** Shanghai University
Shuo Li**,** Shanghai University

Preface

PROLAMAT (**PRO**graming **LA**nguages for **MA**chine **T**ools) is a series of international conferences devoted to the filed of manufacturing - 1969 Roma, Italy; 1973, Budapest, Hungary; 1976, Stirling, Scotland; 1979 Ann Arbor, USA; 1982 Leningrad, USSR; 1992 Tokyo, Japan; 1995 Berlin, Germany; 1998, Trento, Italy; 2001, Budapest, Hungary; – organized by IFIP the International Federation for Information Processing. This triennial event has been a basic meeting for academic and industrial experts in manufacturing. Recently not only the original topics, but all other aspects of the computerized design and production of products and production systems from traditional to virtual ones have been covered and dealt with, including life-cycle issues.

This volume contains the edited version of the technical presentations of PROLMAT 2006, the IFIP TC5 international conference held on June 15-17, 2006 at the Shanghai University in China. The main theme of this conference is "Knowledge Enterprise". These proceedings focus on the issue of how to translate data and information into knowledge in manufacturing enterprises.

Profitability is no longer only a function of price, cost, and adequate quality. The way sustained competitive advantage relates to a firm is distinctive and it is difficult to replicate competencies. The basis for a firm's core competencies is its repository of organizational knowledge. This is highly-valued knowledge that provides opportunities for adding exclusive value to products and services of an enterprise. Knowledge strategy will be related to innovation, learning, and agility. Conversion of advanced research into advanced products; acquisition of knowledge from experience; and fast response-time to the dynamic market changes, are the important rules of international industrial competition at the moment. Innovation, learning, and agility are key factors for companies, which must shift in global markets rapidly and efficiently by delivering new products in the shortest time frame, while maintaining the highest quality and the lowest costs. The successful enterprise of our century will be characterized by an organizational structure that supports thinking process, experience transferring, knowledge discovery, and intelligence exploitation, all of which are based on data and information.

The success of a modern company is not based on its size but on its ability to adapt its operation to the changing environment. To answer to these constraints, the enterprise must develop a corporate culture that empowers employees at all levels and facilitates communication among related groups for constant improvement of its core competitiveness. Such an organizational structure requires a technological infrastructure that fully organizational structure requires a technological infrastructure that fully supports process improvement and integration, and has the flexibility to adjust to unexpected changes in corporate direction. To keep up with rapid developments in global manufacturing, the enterprise must first look at its organization and culture, and then at its supporting technologies. Too often, companies invest only in technology to compete in the global market and don't give enough attention to the training, the education, and the knowledge of their employees. This conference concentrates also on knowledge strategies in Product Life Cycle and brings together researchers and industrialists with the objectives to reach a mutual understanding of the scientific - industry dichotomy, while facilitating the transfer of core research knowledge to core industrial competencies.

These proceedings of PROLAMAT 2006 are published in the well-known IFIP Series with Springer Publishers. The conference will be regularly held in the future. We hope that this conference will lead to further IFIP TC5 activities in the area, and bring people from industries, research institutes, and educational organizations closer together.

On behalf of the chairs of PROLAMAT 2006, I would like to thank the authors for their proficient contribution, the program committees for their invaluable help in reviewing the papers and organize the conference, and all of the individuals who contributed to the editing of the conference proceedings. Finally, I would like to thank the National Natural Science Foundation of China (NNSFC), Science and Technology Committee of Shanghai Municipality (STCSM) and Shanghai Municipal Education Commission (SHMEC) for their financial support to the conference.

Kesheng Wang
15 March 2006
Trondheim

Table of Contents

Part 3. Data Mining in Manufacturing

Part 4. Manufacturing Logistics and Supply Chain Management

Table of Contents

DATA MINING IN MANUFACURING: THE NATURE AND IMPLICATIONS

Kesheng Wang
Department of Production and Quality Engineering, Norwegian University of Science and Technology, N-7491 Trondheim, Norway. Email:kesheng.wang@ntnu.no.

Abstract: Recent advances in computers and manufacturing techniques have made it easy to collect and store all kinds of data in manufacturing enterprises. Traditional data analysis methods are no longer the best alternative to be used. The problem is how to enable engineers and managers to understand large amount of data. Data mining approaches have created new intelligent tools for automated extracting, useful information and knowledge. All these changes will have a profound impact on current practices in manufacturing. In this paper the nature and trends of these changes and their implications on product design and manufacturing are discusses.

Key words: Knowledge discovery, data mining, design, manufacturing.

1. INTRODUCTION

The productivity and efficiency of manufacturing enterprises can be affected by product design, manufacturing decisions and production management. With recent advances in computers and manufacturing techniques, powered data acquisition systems are actively in use in manufacturing enterprises. This supports huge datasets which are related to bill of materials, product design, process planning and scheduling, process and systems, monitoring and diagnosis, and market forecasting, which are collected and stored in databases at various stages for easy manufacturing. The knowledge which is valuable for enterprises including rules, regulars, clusters, patterns, associations and dependencies, etc. is hidden in the

Please use the following format when citing this chapter:
Wang, Kesheng, 2006, in International Federation for Information Processing (IFIP), Volume 207, Knowledge Enterprise: Intelligent Strategies In Product Design, Manufacturing, and Management, eds. K. Wang, Kovacs G., Wozny M., Fang M., (Boston: Springer), pp. 1-10.

datasets. How to enable engineers and managers to extract, understand and use the knowledge from large amount of datasets is still a problem for most of companies. For example, Charts and tables are places on walls, but not read by all, reports and manuals generated, only to be stored on bookshelves, databases created, only to be archived away. Critical manufacturing-related knowledge may be hidden in the data. Example of such knowledge may include rules or regulations for finding defects of the quality of the products. Human operators may never find the rules by manually investigation. That means he may never discover such hidden knowledge from the data. Traditional data analysis methods are no longer the best alternative to be used. Data Mining (DM) approaches based on Computational Intelligence (e.g. Artificial Neural Networks, Fuzzy Logic Systems and Genetic Algorithms), Machines Learning (Decision Tress and Association Rules) and statistics have created new intelligent tools for automated extracting useful information and knowledge. These changes will have a profound impact on current practices in manufacturing.

The applications of DM primarily are related to the fields of business and health care. Applications of DM to manufacturing are still underutilized and infrequently used. In this paper the nature and trends of these changes and definition, functions, techniques and applications of Data Mining on manufacturing are discusses. A case study is used to demonstrate the overall DM process.

2. DATA MING

2.1 Definition

Data mining is an integration of analysis and modeling technologies developed over the last twenty years. Data mining is often defined as the process of extracting valid, previous unknown, comprehensible information from large data bases in order to improve and optimize business decision (Fayyad, et al, 1996; Last and Kandel, 2004).

DM techniques are at the core of DM process, and can have different functions (tasks) depending on the intended results of the process. In general, data mining functions can be divided into two broad categories: discovery data mining and predictive data mining.

(1). Discovery data mining

Discovery data mining is applied to a range of techniques, which find patterns inside your data without any prior knowledge of what patterns exist.

The following examples of discovery data mining: (1). Clustering; (2). Link analysis; and (3). Frequency analysis; etc.

(2). *Predictive data mining*

Predictive data mining is applied to a range of techniques that find relationships between a specific variable (called the target variable) and the other variables in your data. The following are examples of predictive data mining techniques: (1). Classification; (2). Value predication; and (3). Association rules; etc.

2.2 Techniques

A variety of techniques are available to enable the above functions. The most commonly used techniques can be categorized in the following groups (Kantardzic, 2003; Wang, 2005):

1. Classical statistical methods (e.g., linear, quadratic, and logistic discriminate analyses),
2. Modern statistical techniques (e.g., projection pursuit classification, density estimation, k-nearest neighbor, Bayes belief networks),
3. Artificial Neural Networks (ANNs),
4. Support Vector Machines (SVM),
5. Decision Trees (DT) and Rule Induction (RI) algorithms
6. Association Rules (AR),
7. Case Based Reasoning (CBS),
8. Fuzzy Logic Systems (FLS) and Rough Sets (RS),
9. Sequential Patterns (SP),
10. Genetic Algorithms (GAs),
11. Evolutionary Programming (EP), and
12. Visualization Methods (VM), etc.

Each class containing numerous algorithms, for example, there are more than 100 implementations of the decision tree algorithms. A hybrid DM system in which several techniques with different functions can be integrated to achieve a desired result are often more effective and efficient than a single one. For example, in order to identify the attributes that are significant in a manufacturing process, clustering can be used first to segment the process database into a given predefined number of categorical classes, then classification can be used to determine to which group a new data is belongs.

2.3 Procedures

The generic data mining procedures seen from IBM's viewpoint involves the following seven steps:

1. Defining the business issue in a precise statement,
2. Defining the data model and data requirements,
3. Sourcing data from all available repositories and preparing the data (the data could be relational or in flat files, stored in a data warehouse, computed, created on-site or bought from another party). They should be selected and filtered from redundant information,
4. Evaluating the data quality,
5. Choosing the mining function and techniques,
6. Interpreting the results and detecting new information, and
7. Deploying the results and the new knowledge into your business.

To understand how DM can overcome a variety of problems in manufacturing, we consider activities in a manufacturing company.

3. DM IN MANUFACTURING

3.1 Challenges

Application of data mining to several domain areas such as business, telecommunication and medicine has been investigated. Unfortunately, there has not been similar research interest and activity in the manufacturing domain, despite of the potential benefits. Different reasons can be explained:

1. The majority of researchers in the manufacturing domain area are not familiar with data mining algorithms and tools.
2. The majority of theoretical data mining researchers are not familiar with the manufacturing domain area.
3. The few researchers who are skilled in both data mining algorithms and manufacturing domain area are not able to access to often proprietary and sensitive manufacturing enterprise data.

The effort to explore the use of data mining in manufacturing enterprises started only a few years ago, mostly by the researchers in the manufacturing domain. The amount of interest in data mining has definitely grown in the manufacturing research community. However, most of manufacturing researchers are not familiar with DM algorithms and which domain in manufacturing is suitable for DM. The challenge will be where DM should be and what DM technique should be used.

3.2 Applications

Today's manufacturing systems and processes are very complex and complicated. There are many stages of operations and many variables related to each operation in each stage. Even most experienced engineers, charged with controlling variables to make quality consistent, time short and cost low, face problems that have unknown causes to defects and failures in the system and processes. These problems lead to product variability, reproduce, and rejects. System and process engineers always try to understand the relationship between variables using system and process models which are based on mathematics with much of assumptions. DM gives this burden to computers to quickly and exhaustively find those relationships which are useful for manufacturing systems and processes. With the knowledge, the system and process engineers can focus on improvement of the system and the process. (Braha, D., 2001)

Companies such as IBM, ORACLE, SPSS and SAS have developed DM software to large organizations for their data warehouses or data mart. Implementation of the DM systems requires the employment of people with sophisticated mathematical and computer skills. More recently, better users interface with advanced visualization have made DM accessible to a broader range of people. DM now can be applied to engineering and manufacturing easily, but experience and knowledge of how the tools work is still important. Users can encounter difficulty in getting their data loaded into the software, or in knowing which technique would work best for their particular problem. Once learned, however, DM can give the manufacturing engineer a powerful tool to uncover new truths and tackle the hardest problems.

The specific DM applications in advanced manufacturing include: (1). Manufacturing system modeling; (2). Manufacturing process control; (3) Quality control; (4) Monitoring and diagnosis; (5) Safety evaluation; and (6) Process planning and scheduling; (Chen, et al., 2005; Huang, et al., 2005; Hsu, et al., 2005; Tsironis, et al., 2005)

As we discussed above, collecting and cleaning of datasets are time consuming (80% of DM process) and critical to the success of data mining. Examples of variables in manufacturing datasets include:

1. Manufacturing process variables: machining, casting, forgings, extrusions, stampings, assembling, cleaning, etc.
2. Manufacturing process variables: Cutting speed, pressure, lubricants, coolants, voltage, current, forces, toques, vibration, etc.
3. Other variables: product materials, numbers, machines, fixtures, etc.
4. Environnement conditions variables: humidity, temperature, toxic, etc.
5. Working condition variables: duration, shift, injuries and accidents, etc.
6. Target variables: quality, yield, productivity, performance index, etc.

Successful implementations of data mining are not easy tasks. Some of the success factors are listed as the following: (Kasravi, 1997)

1. Datasets available – Availability of appropriate datasets is essential to the success of DM project.
2. Data preprocess – Cleanness of datasets is the major effort in DM.
3. Information content – not all data are suitable for data mining. Care must be taken to ensure that relevant variables and records are collected such that they could serve as the basis of any knowledge to be solved.
4. Correlation vs. causality – DM is a computer-based technique and it can only discover correlations, not causal relationships among data sets.
5. Privacy – Data sets are normally related to privacy of both personnel and companies which sometimes make it difficult to collect data and interpret the results.
6. Corporations – One of the important steps in data mining is interpretation. The patterns and rules obtained from data mining are domain-specific and the interpretation is a difficult task. Data mining experts should work with domain expert in order to make success of data mining.
7. Tools and software selection – Many data mining algorithms are standard, therefore a number of commercial data mining tools are currently available on the market. It will help manufacturing engineers to launch DM projects easily and quickly.

4. AN IMPLEMENTATION CASE

4.1 The scheduled maintenance data

Many examples of application of data mining in manufacturing can successfully include: Conceptual design, Design manufacturing systems, Fault diagnosis, Preventive machines maintenance, Manufacturing knowledge acquisition, Material requirements planning, Adaptive scheduling, Process planning, Manufacturing process control, Quality control, Optimization of yield, Assemble selection, and Learning in robotics. This section presents a case to show how to use data mining techniques to solve a manufacturing problem.

There are two goals for this data mining task: 1. Classification - to determine what subsystems or components are most responsible for downtime; 2. Predication - to predict when preventive maintenance would be most effective in reducing failures. This information can then be used in setting maintenance policy guidelines such as planed maintenance schedules.

Classification and predication can be done by the use of Decision Trees (DT). The advantages of DT are: 1. Robust performance, 2. Clear visualization of output, 3. Ease interpretation. Especially DT is a good choice when the number of classes are low.

The one of mining tasks for scheduled maintenance dataset is a classification of machine health. The target variable of DT is Equipment Availability (EA):

$$EA = \frac{ScheduledTime - DownTime}{ScheduledTime} \times 100$$

Because EA takes into account the amount of time needed to repair a machine, it is a good indicator of machine health. The target of the classification can be shown in the next section.

4.2 Machine health mining (Wang, et al., 2004)

The case study for machine health is run following the procedure and the data mining development is based on the commercial software, IBM Intelligent Miner. [Reinschmidt, et al, 1999]

4.2.1 The business issues

The machine health analysis is an important task for maintenance activities of a company because it permits it to find the most sensible machines, which are the most responsible to the low Equipment Availability of the whole company. The purpose of data mining is to use DT to detect a group of machines with low EA. It is useful in order to plan a maintenance action and we are able to focus on the group of machines in order to make the maintenance more effective and save time and cost.

4.2.2 The Data model

This datasheet has been created in order to obtain results in a particular way. Indeed, we would like here to classify the machine related to the importance of their downtime inside the company. Each generated node of the tree contains the different machines, which should be classified by the evaluation of the value of Equipment Availability. In the original datasheet, all downtime have been registered with respect to scheduling time. It is not suitable for the data mining process. Therefore we define each machine as a column of the datasheet with the values of the total downtime, which is the cumulative value (30+30 = 60). The modified datasheet is shown in Fig. 1.

4.2.3 Evaluating the data quality

The original dataset from the company is not quite suitable for our data mining purpose. It contains some errors, such as missing data, wrong data and other errors. A cleaning process is required. Afterwards a pure and modified datasheet can be ready for a data mining process.

As seen in Fig. 1, each record has the value of the total downtime for each machine. The last record contains the value of Equipment Availability of each machine

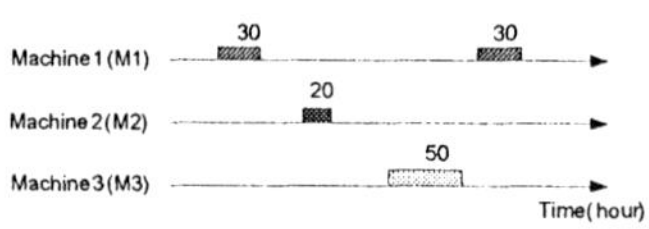

Machines	M1	M2	M3
Record1	30	0	0
Record2	30	20	0
Record3	30	20	50
Record4	60(30+30)	20	50

Figure 1. The datasheet for machine health.

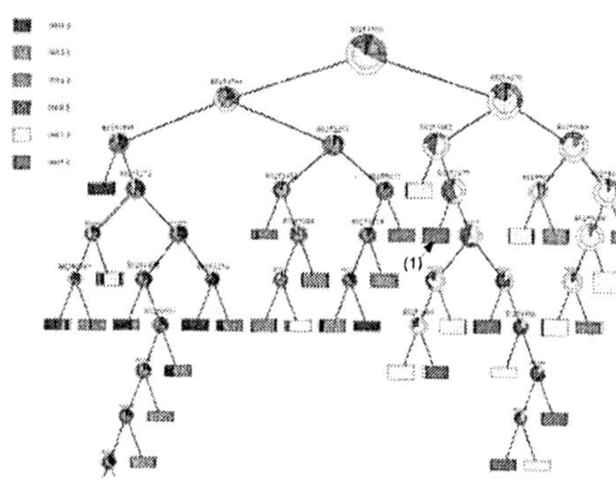

Figure 2. The tree structure of the machine health.

4.2.4 Selecting DT as a data mining approach

DT has been selected as a data mining algorithm for the task. We input the modified datasheet to IBM Intelligent Miner. We define six targets of classification, depicted from 0001,0 to 0006,0. The decision tree generated is shown in Fig. 2.

4.2.5 Interpretation of the tree

On the top left corner of the tree (Fig. 2), we can see the 6 main classes related to the variable of EA. 0001,0 corresponds to the highest availability and 0006,0 to the lowest one. Then the data are classified in the leaf nodes where you can see a repartition of colors. We have a repartition, a number of records and purity for each leaf node.

On each node of the tree there is a condition on the number of hours of downtime for each machine. For example, the root node corresponds to the Machine 80254580. And the condition is: "***If*** *80254580 < 322 hours of downtime,* ***Then*** *it goes to the left branch*".

In order to generate some rules, the principle is the same. We look at a leaf node. If the number of records and the purity are important enough, we can keep the rule and say that this rule is obviously relevant.

We can take an example of the leaf node (1) as shown in Fig. 3. We can see that label is 0005,0, Records is 145, Class Distribution is (1 0 0 1 2 141), Purity 97,2 %.

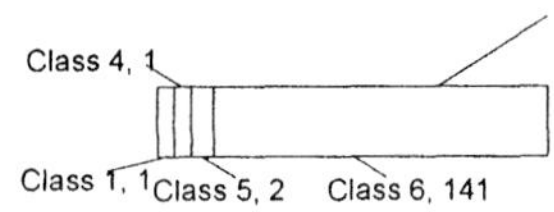

Figure 3. Zoom in the leaf node (1).

4.2.6 Generation of rules

Here we can say that this classification is meaningful because the number of records is important with a high purity so we can look at the rules:

IF (80254580 > 322)
And (80254270 < 22,5)
And (80253462 > 202,5)
And (80252477 < 36)
THEN Class is 0005,0

So for example the maintenance staff will be able to look at the machines 80254580 and 80253462 which seem to be the most responsible for the low availability in this part of the classification.

4.2.7 Deploying the results and the new knowledge into your business

When we look at the induced tree in Fig. 2, we clearly see that the emphasis on preventive maintenance should be on machines 80254580, 80254593, and 80254270 since these machines are close to the root node and are associated with low EA

So the tree structure is interesting for the maintenance management because it's pretty easy to detect the different part (high equipment availability, low equipment availability) with the colors on the charts and by the way that the machines are related to these results.

5. CONCLUSION

In the quest for a sustainable competitive advantage, manufacturing companies have finally come to realize that technology alone is not enough. What makes it sustainable is the knowledge residing in people's head, and process and experience in the companies. Modern manufacturing enterprises have collected and stored a huge amount of manufacturing datasets for easy access, such as product design data, manufacturing resource data, manufacturing system and process data, which are not fully used for increased competitive advantages in the global economy. Business Intelligence (BI), Data Mining (DM) and Knowledge Management (KM) will change the stakes. These techniques definitely are able to help the enterprises to collect, extract, create and delivery manufacturing knowledge in a competitive environment. In this paper we focus on DM techniques, which can be used to discover and extract previous unknown manufacturing knowledge from manufacturing data bases. The knowledge discovered can be used to improve the product manufacturing processes.

6. REFERENCES

Braha, D., (2001), Ed: Data mining for Design and Manufacturing – Methods and applications, Kluwer Academic Publishers, The Nederlands.

Chen, W. C., Tseng, S. S., and Wang, C. Y., (2005), A novel manufacturing defect detection methods using association rule mining techniques, *Expert Systems with Applications*, V. 29, No. 4, pp. 807-815.

Fayyad, U. Piatetsky-Shapiro, G., Smyth, (1996), From data mining to Knowledge discovery: An overview, in: *Advantaces in Knowledge Discovery and Data Mining,* Fayyad, U. Piatetsky-Shapiro, G., Smyth, P. and Uthurusamy, Eds., Cambridge, MA: MIT Press, pp. 1-36.

Hsu, C. H. and Wang, M. J. J., (2005), Using decision tree-based data mining to establish a sizing system for manufacture of garments, *Int J Adv Manuf Technol*, v. 26, pp. 669-674.

Huang, H., and Wu, D., (2005), *Lecture Notes in Artificial Intelligence* (sub series of Lecture Notes in Computer Science) V. 3614, No. Part II, *Fuzzy Systems and Knowledge Discovery: Second International Conference, FSKD, Proceedings*, pp.577-580.

Kantardzic, M., (2003), *Data mining – concepts, Models, Methods, and Algorithms*, Wiley-Interscience.

Karavi, K. (1997), Data mining and knowledge discovery in manufacturing, *AUTOFACT'97, Information Technologies for the Manufacturing Enterprise*, Michigan.

Last, M. and Kandel, A., (2004), Discovering useful and understandable patterns in manufacturing data, Robotics and Autonomous Systems, v. 49, pp. 137-152.

Reinschmidt, J., Gottschalk, H., Kim, H., and Zwietering, D., 1999, *Intelligent Miner for data: Enhance your business intelligence*, International technical support organization, IBM (http://www.redbooks.ibm.com).

Tsironis, L., Bilalis, N. and Moustakis, V., (2005), Using machine learning to support quality managm,ent – framework and experimental investigation, The TQM magazine, V. 17, No. 3, pp. 237-248.

Wang, K., (2005), *Applied Computational Intelligence in Intelligent Manufacturing Systems*, Advanced Knowledge International.

Wang, K., Fabien, C., Wang, Y., Yuan, Q. and Fang, M., (2004), The 6th International Conference on Frontiers of Design and Manufacturing, June 26-28, Xi'an, China.

MANAGEMENT AND PRODUCTION CONTROL ISSUES OF DISTRIBUTED ENTERPRISES

George L. Kovács
Technical University, Budapest & University of Pécs & Computer and Automation Research Institute, Kende u. 13, H-1111 Budapest, Hungary, E-mail: gkovacs@sztaki.hu

Abstract: Enterprises which are distributed in space and/or which are composed as a temporary joint venture of legally different units recently often called virtual (extended) enterprises. Planning, design and operation (management) goals and requirements of such firms are generally different from those of single, centralised enterprises. In this paper we suggest (software) solutions for design, planning and operation of complex, networked organisations represented as nodes of networks. First complex logistics flows of distributed SMEs are targeted, and then an European virtual institute for collaborative demand and supply network will be discussed. The third problem is a complex, web-based solution to manage large, expensive, multi-site, multi-company projects.

Key words: extended/virtual enterprise, supply/demand chain, ERP, management, SME.

1. INTRODUCTION

Recent manufacturing and service companies are creating "value", not "simply" products. *Supply Chain Management* (SCM) focuses on globalisation and information management tools, which integrate procurement, operations, and logistics from raw materials to customer satisfaction. Supply chain management is one of the leading business process re-engineering, cost-saving and revenue enhancement strategies in use today. Effective integration of a supply chain can save millions, and more if supply chain strategies are on the table early enough in the product development and design plans.

Please use the following format when citing this chapter:
Kovács, George, L., 2006, in International Federation for Information Processing (IFIP), Volume 207, Knowledge Enterprise: Intelligent Strategies In Product Design, Manufacturing, and Management, eds. K. Wang, Kovacs G., Wozny M., Fang M., (Boston: Springer), pp. 11-20.

There are some SCM solutions available in the recent market, most of them are for single, centralized organizations and they deal only with restricted tasks, i. e. with management of the supply chain itself. To be able to select the best from different possible solutions a deep analysis and appropriate simulation give a strong assistance.

Today the worldwide globalization and the appearance of virtual enterprises require more than only SCM for some tasks of a given enterprise. Due to the physically and logically distributed character of the co-operating units (workshops, plants, enterprises, etc.), taking advantage of the existence of Internet (www), web-based solutions are suggested.

There are 3 EU projects (FLUENT, WHALES and Co-DESNET, see (Esprit, 1998, 1999 and EU, 2004), which target proper goals. Some issues of these projects will be discussed in this paper.

As today more and more production and service activities are done by extended/virtual enterprises, the result of their activity, the "product" should be understood properly. As the term extended enterprise comprises more than just a single enterprise the term extended product (Jansson, 2002) should comprise more than just the core or tangible product. The EXPIDE (EU, 2003) project works on extending a tangible product. Different services in an intangible shell around the tangible product are defined by means of a layered model. This model shows a type of hierarchical, physical extension of (extended) product services. The concept is three rings around the tangible product. There is a *core product* that is closely related to the core functions of a product. The second ring describes the *packaging of the core functions*. The second ring includes only tangible features of the chosen product. The features of the tangible product are different from manufacturer to manufacturer (supplier to supplier). The third ring summarises all the *intangible* - often intelligent - *assets* of the product.

2. MANAGEMENT OF COMPLEX LOGISTIC FLOWS

New IT solutions are suggested for managing *complex logistic flows,* occurring in distributed *manufacturing networks* with multiple plants and co-operating firms. Networks of this kind are gaining relevance and diffusion, under the impulse of the following main factors:

- emerging virtual/extended enterprise paradigms
- pull-oriented production models, like Just-In-Time and Kanban
- lean/agile manufacturing models, based on horizontal, goal-oriented process chains (Hopf, 1994)
- evolving market conditions, calling for business globalisation and decentralisation of manufacturing facilities (Rolstadas, 1994).

In response to these changes, the newly developed solution and software provide manufacturing firms with advanced IT tools for logistics decision-making, thus enhancing their capability to operate in a distributed, networked (virtual, extended) production environment (Kovács, 2003).

Even in a traditional organization where such concepts are neither applied or envisaged, the complexity of logistics decision-making is now increased by factors such as: market globalization, decentralized manufacturing facilities, extended range of suppliers, highest emphasis on total quality issues and customer satisfaction (Hirsch, 1995; Bonfatti, 1996).

In these conditions, traditional logistics functions like sales and purchase are left alone to face problems far beyond their intended roles. Current Enterprise Resource Planning (ERP) systems can be of little help, due to some basic features, which are in conflict with the ERP foundations of:

- Hierarchic organisation,
- Embedding of business processes into the application code,
- Centralized data management based on company-specific standards;

Problems deriving from these evident limitations often induce large companies to assimilate their closest suppliers and sub-contractors, at least from the information system point of view. To overcome these limits, major ERP producers are developing supply-chain management add-ons on top of their production management solutions, often through partnerships with SCP producers. These often imply significant limitations, as:

- Centralized, static planning;
- Manufacturing vs. logistics orientation;
- Pre-defined, hard to adapt organisation model;
- High implementation costs, due to the complexity of SCP integration.

3. A NOVEL NETWOK/FLOW CONTROL SYSTEM

Traditional SCM implementations refer to a linear, standardised and relatively stable view of the supply chain: "Supply Chain Management is about managing the flow of products and services and the associated information, across the whole business system to maximise value to the end consumer" (PW, 1997).

The "whole business system" is a row of four to five actors (depending on whether electronic commerce issues are addressed or not) interacting with each other in pairs. The resulting SCM solutions are product suites including several independent tools, each designed to optimize a single link in this predefined sequence. For the whole picture to work, it is assumed that separate optimization of each link leads to overall performance improvements.

There is a potential failure behind this logic, especially where revenue increase is pursued instead of cost reduction. To look at the supply chain complexity as a competitive advantage, rather than as a source of costs, means a radical change of perspective in the organisation models supported by SCM tools: "For a start, the supply "chain" is really not a chain at all - it is a complicated web of relationships between demand and supply. The concurrent and multidimensional nature of these relationships creates a complex fabric woven step by step (Mirchandani, 1996).

Our solutions capture the inherent complexity of the supply network, allowing firms to manage the three fundamental barriers to supply chain performance improvement: visibility, velocity and variability. This is done by referring to a flexible, scalable and decentralized network model, based on the preservation of nodes' autonomy and on a case-by-case definition of links and dependencies between the nodes. There are *supplier-, company-, subcontractor- and customer-nodes* in the network. Management is done at knowledge-, planning- and control-level by a Flow Collector (input), Flow Processor (production management) and Flow Dispatcher (output).

Co-operation between nodes is realized through *links,* each representing a stable relationship for the exchange of a given product between a "supplier" node Flow Collector and a "receiver" node Flow Dispatcher. The Flow Processor is not directly involved in the link, since our flow control is based on a clear separation of logistics decision-making domains. Internal logistics are managed by each node on its own, and are perceived at the network level only through requirements, events and constraints on external logistics flows.

To support the above detailed organisation model, each node is provided with software tools designed to fulfil the requirements of a multi-site, multi-enterprise manufacturing network, as for example:

-To provide a unified and generalised representation of logistics flows
-To support decentralised and volatile organisation models
-To allow scalable and flexible network configuration
-To support decision-making at the tactical and operational level
-To manage and synchronise multiple decisional processes
-To integrate and distribute relevant information across the network;
-To integrate but not overlap with ERP and other internal-logistics management tools.

To fulfill all goals and requirements we provide an advanced IT infrastructure based on the following software components:

- a standard *Communication and Workflow Infrastructure*, for basic data interchange and message services.
- a high-level *Network Model,* on top of this basic layer to have an updated,

consistent representation of the network from the node point of view;

- an *Active Flows Control (AFC)* component, which monitors interaction with nodes in the Network Model to maintain updated information;
- a *Performance Measurement System (PMS),* acting in parallel with the AFC to keep historical records of the network activity;
- two *Decision Support Systems (DSS),* respectively for *input and output flows management,* that process internal and external demands; and allow flow planning based on AFC and PMS input,
- an *Interface with Enterprise Resource Planning (ERP)* that allows transparent interaction with the node local production management.

4. COLLABORATIVE DEMAND-SUPPLY NETWORKS

An advanced SCM means a demand network as well. Basically large-scale networks of production and service enterprises operating within a common industrial sector are taken into account. The relevant Co-DESNET project (EU, 2004) is devoted to organizing a European Virtual Institute: a permanent network linking all participating partners and other members in the future, based on a vertical web-portal. This portal supplies:

- a Virtual Library of documentations (academic and industrial articles, papers, presentations) concerning supply chains and related topics;
- a Virtual Laboratory of academic and industrial networks (of institutes, companies, SMEs) concerning supply chains and related topics;
- a Virtual Agency of information about conferences and meetings;
- a Discussion Forum for academic and industrial actors.

5. PROJECT MANAGEMENT ISSUES

5.1 Planning, deployment and monitoring

The objective was to provide a planning and management infrastructure for complex, distributed, multi-site, multi-enterprise organizations working on large scale engineering projects, characterized by huge investments in both materials and human resources and by concurrent, disparate activities (manufacturing, design and services) as well. Managing projects of this kind means dealing with several problems at the same time:

- Complexity of scope, in terms of time and resources employed, and variety

of activities to be planned, synchronized and monitored;
- Distributed organization, spanning through several companies and involving a multiplicity of actors and competencies;
- One-of-a-kind design, increasing planning complexity, hard to apply product and process standardization;
- Geographic distribution of sites and project activities;
- Strict time constraints, with dangerous critical-path dependencies;
- Contingency risks, due to the high planning uncertainty and difficult re-alignment of activities;
- Revenue-loss risks, due to difficulties in budgeting and high contingency costs.

Some other projects approach the problem in the following ways:
- Standards and systems are sought for product and process data modelling and interchange, and to support distributed design in concurrent and co-operative engineering environments, see (Esprit, 2000 and 2001).
- Virtual enterprises are studied for the creation of enterprise networks, see (IMS, 2000 and Esprit, 2002).

We focus on the *planning and day to day management* of (large) projects, assuming that technical departments and engineering functions have proper working tools and standards, and that the conditions for virtual enterprise creation have been fulfilled.

Several managers of the above type projects developed "in house" project databases and project management tools. These systems lack the distributed features and model sophistication. In response to these requirements, our system pursued two main objectives:
- To design and develop a set of software components supporting integrated planning, deployment and monitoring;
- To demonstrate the applicability and benefits of the software components through different pilot business cases.

We implemented the following general features for project management in complex and distributed organizations, the software:
- Provides a *unified and generalized representation* of project activities and related artefacts, comprising all material and immaterial work items (e.g., products, knowledge, design documents, etc.). Supports distributed organization models, crossing hierarchies and company boundaries.
- Provides a *scalable and flexible co-operation* environment. The system provides a project network infrastructure accessible to every node.
- Integrates and distributes *relevant information* across the project network, using a web-based environment.
- Supports *decision-making* in the project ideation, definition and deployment phases. Takes into account past performance, cost and capabilities to generate detailed plans considering both activities' timing,

equipment and materials availability, etc.
- Manages and synchronizes the *flow of decisions and events* in the project network.
- Integrates with local management and planning systems. It means to safeguard the nodes' autonomy and IT investments. The system shall not interfere with node internal procedures and management tools, as ERP, PPC, Human Resources, stand-alone Project Planning and Budgeting packages.

Prime contractors in large-scale projects are typically big companies with proper financial resources and assets. Nevertheless, this does not prevent our system to be extremely significant to SME users that can be involved as nodes (e.g. subcontractors for provision of services and components, to develop entire engineering packages, etc.) in a large project network.

Our project control/management software assists in project planning and deployment thanks to a software infrastructure producing the following measurable results on the end-users' business:
- Improved planning and budgeting; effective contingency management
- Improved monitoring, cost and risk assessment; higher flexibility and efficiency.

5.2 System software and innovation issues

To achieve the above improvements requires dealing with different enterprise functions and information sources, supported by heterogeneous and poorly integrated software applications, as:
- Enterprise Resources Planning systems (ERP) (as SAP, Baan, etc.),
- Production Planning and Control (PPC) and Warehousing systems,

Project planning tools provide graphical editing of GANNT and PERT diagrams, along with display of resources, workload and activities timing,
- Human Resources (HR) packages.

We have a distributed architecture that provides an integrated data and process infrastructure for different companies and actors participating in large projects' planning and execution, at the same time safeguarding each node's autonomy as regards local operations management and information system. These features match key requirements of the "virtual enterprise" organizations working on large one-of-a-kind projects, see (Esprit, 1993a).

The resulting system accommodates the needs of project networks independently of the industrial sector, thanks to its general and adaptable design, that comes from features like:
- Distributed project management environment; Project representation;

- Decentralized architecture and accountability structure;
- Powerful project and network data model; Flexible decision-support tools.

Three different layers are identified in the network logical architecture. A Network Data Model is closely connected to them:

The *Work Network Structure (WNS)* constitutes the bottom layer of the system infrastructure, crossing horizontally functional silos at companies co-operating on large scale projects.

The *Work Accountability Structure (WAS)* constitutes the intermediate layer of the system infrastructure, representing the temporary, multi-site and multi-company organization created to carry out one or more projects.

The *Work Breakdown Structure (WBS)* corresponds to the topmost layer of the system infrastructure, and represents a network-wise enhancement of WBS implementations supported by traditional Project Planning tools.

To support the outlined organization model the system network is provided with innovative software tools, as:

Web-based Project Environment (WPE). The system architecture is centered on a *WPE* – an integrated project management infrastructure,

Workflow Management System (WMS). The WPE embeds process automation features provided by a commercial WMS,

Network Data Model (NDM). The system organization model requires a common data infrastructure provided by the NDM – a distributed and decentralized database,

Local Applications Interfaces. Alignment between the new project network management and internal node activities is provided by a set of LAIs,

Decision Support Sub-systems (DSS). Two DSSs are developed and deployed as add-ons to the rest of the software, respectively for:

- *Project Planning & Budgeting to* support decision-makers in project ideation and bid preparation tasks,
- *Project Monitoring & Revenue Analysis to* support project and site managers in evaluating activities' progress, identifying risk elements and launching contingency actions.

To avoid replication of functionalities and waste of resources, our approach strongly relies on the integration with standard, and commercially available tools. The system is designed and implemented with common tools and principles; see (SCOR, 2001; Jacobson, 1993; Mezgár, 2002 and Horvath, 2004). We apply standards, if possible.

6. CONCLUSIONS

The implementation of our logistics flow management/supply chain approach represents a significant step forward on state-of-the-art logistics management techniques for the end-users. On the one side, in traditional enterprise practice the focus is on bilateral supply relations with each individual customer and supplier, with scant and informal co-operation possibilities and no supply-chain visibility. On the other side, multi-site planning extensions offered by major ERP and SCP vendors are still based on a centralised approach, lacking on-line integration and synchronisation with the other network actors. In this scenario we provide considerable benefits in terms of improved network visibility, better co-ordination and real-time control of materials flows.

Feasibility of the above improvements, along with the costs and time required for achieving them, are assessed through experimentation of the software system on selected user firms. The validation phase was successfully finished on four pilot cases in different industrial sectors: machine-tool industry, equipment production, textile industry and naval industry. Theoretical issues were discussed to give explanations to the extended products related to extended/virtual production. A new, virtual institute dealing with collaborative demand and supply network was outlined, too.

Experimentation was done on the basic *flow management* components, supporting network modelling, data-integration and workflow, in parallel with design and development of the decision-support components.

The web-based management software provides a *planning and management infrastructure* for complex distributed organizations working on large scale engineering projects, characterized by huge investments in both materials and human resources and by concurrent, disparate activities – manufacturing, design and services as well.

Four test business cases were co-ordinated dealing with application of common methodologies, metrics and best practices, and ensuring uniformity, comparison and joint evaluation of results produced by each business case:
- *shipbuilding* and ship-repairing services; - manufacturing *plant engineering*
- software *system engineering*; - *one-of-a-kind manufacturing* environment.

The first experiments at all the four different pilot sites have proven all advantages detailed in this paper.

7. ACKNOWLEDGEMENT

The author acknowledges the support of the European Union that provided the funds for the research projects FLUENT, Co-DESNET and WHALES the key results of which are presented in this paper. We are also thankful for all foreign and Hungarian academic and industrial partners for their cooperation.

8. REFERENCES

Bonfatti, F. et al, (1996), Co-ordination Functions in a SME Network. In: *Proc. of BASYS '96 International Conference*, Lisbon, Chapman & Hall.

Esprit (1998), FLUENT, Esprit IiM-1998-29088: Flow-oriented Logistics Upgrade for Enterprise NeTworks, *EU project documentations.*

Esprit (1999), WHALES, ESPRIT IST-1999-12538: Web-linking Heterogeneous Applications for Large-scale Engineering and Services, *EU project.*

Esprit (1999a), EP 20876 ELSEWISE, Esprit 20876: European LSE Wide Integration Support Effort.

Esprit (2000), EP 20377 OPAL, Esprit 20377, OPAL: Integrated Information and Process Management in Manufacturing Engineering.

Esprit (2001), EP 20408 VEGA, Esprit 20408, VEGA: Virtual Enterprise Using Groupware Tools and Distributed Architecture.

Esprit (2002), EP 26854 VIVE, Esprit 26854, VIVE: Virtual Vertical Enterprise.

EU (2003), EXPIDE, EU Project, http://www.expide.org (30.4.2003)

EU (2004), Co-DESNET, EU project, http://www.codesnet.polito.it/ (and DoW – not public)

Hirsch, B. E. et al., (1995), Decentralized and collaborative production management in distributed manufacturing environments, *Sharing CIME Solutions*, IOS Press.

Hopf, M.,(1994), Holonic manufacturing systems (HMS) - IMS Test Case 5, in J. H. K. Knudsen et al. (eds), *Sharing CIME Solutions*, IOS Press.

IMS (2000), *IMS GLOBEMAN21*, Globeman21 is an IMS project, http://www.ims.org/projects/project_info/globeman.html

Jacobson (1993), OOSE Methodology, Jacobson, I., et al. *Object Oriented Software Engineering: a Use Case driven approach*, Addison Wesley.

Jansson, K. (2002), Thoben, K-D.: The extended products paradigm, *DIISM 2002 - The 5th Int. Conf. on Design of Info. Infrastructure Syst. for Manuf. 2002.* Osaka, Japan

Horváth, L. and Rudas, I. J., (2004), *Modeling and Problem Solving Methods for Engineers*, ISBN 0-12-602250-X, Elsevier, Academic Press.

Kovács, G. L. and Paganelli, P. (2003), A planning and management infrastructure for large, complex distributed projects, *Computers in Industry*, 51, pp. 165-183.

Mezgár, I., (2002), Communication Infrastructures for Virtual Enterprises, *Proc. of the the IFIP World Congress,* 31 Aug.- 4 Sept. Vienna/ Budapest, pp. 432-434.

Mirchandani, V. and Block J. (1996), Supply Chain Management and the Back Office, *Gartner Group© Strategic Analysis Report*, September 1996.

PW (1997), Price Waterhouse©, Supply Chain Management Practice, *Supply Chain Planning for Global Supply Chain Management*, November 1997.

Rolstadas A. (1994), Beyond Year 2000 - Production Management in the Virtual Company, *IFIP Transactions B-19*: North Holland 1994.

SCOR (2001), SCOR Model, The Supply Chain Operations Reference-model (SCOR), http://www.supply-chain.org/

DIGITAL PROCESSES IN PRODUCT CREATION
An Extended Abstract

Dr. Gustav J. Olling
DaimlerChrysler Corp., 800 Chrysler Drive, Auburn Hills, MI 48326, USA
gjo@daimlerchrysler.com

Abstract: This paper underlines the importance of focusing on organization, culture, technology, and education to evolve an expanding global enterprise in today's complex and dynamic business climate.

Key words: product and component teams; cultural changes; improvement-oriented; CAD technology; well-integrated databases; information bridges; integrate expertise; global engineers; universities; conclusion.

1. INTRODUCTION

In an era of corporate expansion into global markets and mergers with international partners, many companies are discovering weaknesses in their organizational structures. Personnel, process, and technical solutions that worked well on a smaller scale do not necessarily succeed on a world stage. Which organizational strategies will support rapid product development and adaptation to market shifts while maintaining consistent productivity and quality levels?

In effecting large-scale organizational change, it is important to avoid overemphasizing the role of technology. Heavy investment in technology for technology's sake in the 1980s and 1990s demonstrated that changing an enterprise's technical framework without altering the ingrained attitudes and practices of the enterprise does not produce the expected gains in efficiency. Successful enterprise innovation results from broad organizational and cultural changes (80%) **supported by** suitable technological changes (20%).

Please use the following format when citing this chapter:
Olling, Gustav, J., 2006, in International Federation for Information Processing (IFIP), Volume 207, Knowledge Enterprise: Intelligent Strategies In Product Design, Manufacturing, and Management, eds. K. Wang, Kovacs G., Wozny M., Fang M., (Boston: Springer), pp. 21-25.

2. ORGANIZATION

One enterprise strategy that can be used to improve a company's infrastructure is to organize product development personnel into product and component teams. An automobile company may have previously partitioned its work into component groups, such as powertrain, chassis, body, and electrical, that worked on all vehicle types and operated separately from other groups, such as manufacturing, procurement, marketing, and finance. Under the new scheme, product teams divide the work by product type (Small Vehicle, Premium Vehicle, Family Vehicle, Activity Vehicle, Truck, and Powertrain), while the component teams continue to apply expertise and innovation to major components of the vehicle. The product and component teams work together on each new vehicle program. Moreover, these cross-functional teams include representatives from manufacturing, procurement, marketing, and finance. Hence, the collective effort becomes more collaborative and focused. Communication among groups improves considerably, and the closer cooperation enables flexibility and the ability to change direction quickly.

3. CULTURE

Organizational changes must be accompanied by a complementary set of cultural changes if the benefits of the structural changes are to be fully realized. For example, a manufacturing enterprise can establish a Product Development System to guide the product development process through a series of quality gates. It can launch a number of process improvement programs to analyze and enhance various manual, technical, and managerial activities, and it can establish Centers of Competence to encourage the exchange and capture of good ideas for key product areas. The net effects of these changes are twofold. First, they establish clear guidelines for how product development work is to be done, and, second, they provide procedures for doing the work and forums for discussing how to improve the work processes. Over time, these changes help to replace traditional results-oriented, "don't blame me" thinking with improvement-oriented, "let's get better" thinking. The cultural change releases an abundance of energy and fosters a climate of open communication.

4. TECHNOLOGY

While organizational and cultural changes are being made, technological changes continue to occur in the areas of computer-aided engineering, manufacturing automation, and management information systems. Many global enterprises today are trying to integrate the product development and manufacturing processes from end to end as much as possible. The consolidated product teams described above, insofar as they are integrative in nature, help to synchronize the activities of product development, process development, and resource planning. These efforts are supported by the evolution of design technology from manual and electronic drafting to CAD modeling to feature-based design to knowledge-based engineering.

Feature-based design possesses the virtue of storing rules, relationships, and attributes of the product-in-development in such as way as to support end-to-end integration efforts. Although knowledge-based engineering is still evolving, it will enable the automation of product and process design activities, associativity between the requirements of these activities, the simultaneous conduct of the activities, and eventually a seamless development cycle for product and process.

While intelligent CAD systems are essential for a growing global enterprise, they are not the only key technological element. On the back end, the enterprise needs well-integrated databases. The strategy in this area is to gather the sources of product geometry, bills of material, change notice information, engineering and manufacturing standards, resource capacity, long-range plans, and a myriad of other information types into an integrated data system. In a large international company that must function at full capacity every day, this is a daunting task. It comes as no surprise to say that most large enterprises have not achieved such corporate-wide integration.

Most companies can, however, make two important strides. They can adopt and customize a Product Data Management system to organize geometry, specify product configurations, and control changes. In addition, they can create in-house web applications that act as integration points between disparate databases. These measures allow companies to reference and exchange vital product information while evolving a more integrated set of databases.

Another important advancement that global enterprises can implement is the construction of information bridges, such as a Company Portal, that facilitates communication between corporate groups in different countries and between the company and its suppliers. The improved communication afforded by these bridges promotes part sharing and reuse, closer cooperation, and coordinated planning.

Current research is providing additional tools to assist expanding global organizations. The emergence of more intelligent process models and supporting systems, the evolution of decision-support, expert, and optimization systems, and the design of logical models that analyze the relationship of events and suggest alternative actions – all these will integrate expertise into the support structure of the enterprise. The thinking behind product creation will be increasingly automated.

5. EDUCATION

Besides the three factors so far discussed – organization, culture, and technology – one additional element is crucial to the viability of the expanding global enterprise: Education. The complex and changing environment of today's global businesses requires an extended set of job skills. Global engineers must be not only technical specialists but also product strategists, operation integrators, and ethical practitioners. They must solve unstructured, multi-solutional, and mutable problems. Unfortunately, educational institutions do not produce an abundance of people who have mastered the creative process and can communicate to all levels of an organization. Even fewer graduates appreciate the virtues of craftsmanship. To encourage global thinking, universities must expose students to a wide range of technical and liberal arts subjects, a wide range of computer-aided design, engineering, manufacturing, and customer-feedback processes, and a wide range of exercises in how to communicate with coworkers, management, and customers.

In addition to improving curricula, universities can hire faculty with broad educational backgrounds and diversified practical experience. They can support the efforts of faculty members to update their research skills and technical expertise. They can build laboratories, or work with independent laboratories, that realistically model current industrial practice, simulate variable manufacturing systems, and support quantitative analysis of processes and results. Finally, universities can collaborate with commercial enterprises to exchange ideas and methods pertaining to the latest technology.

6. CONCLUSION

To remain competitive, the expanding global corporation must make sure that it has the infrastructure to deal effectively with the speed and complexity of current industrial changes. The enterprise must be organized to focus on well-defined product areas with a comprehensive team of planners, developers, and implementers. It must establish a corporate culture with well-defined rules for all aspects of the product creation lifecycle, a cooperative climate of process improvement, and a community spirit that promotes the free exchange of ideas. The enterprise must continue to strive for improved integration of information and the automation of expert knowledge and decision-making. Finally, the enterprise must look for and encourage the education of creative, adaptive, honest, resilient individuals to deal with the difficult global issues of today and tomorrow.

COLLABORATIVE NETWORKS
Value creation in a knowledge society

Luis M. Camarinha-Matos [1], Hamideh Afsarmanesh [2]
[1] *New University of Lisbon, Q. Torre, 2829-516 Monte Caparica, Portugal – cam@uninova.pt*
[2] *University of Amsterdam, The Netherlands – hamideh@science.uva.nl*

Abstract: Collaborative networks show a high potential as drivers of value creation. The materialization of this potential however requires further progress in understanding these organizational forms and the underlying principles of this new paradigm. As a contribution in this direction, the notion of collaboration among organizations is clarified, a taxonomy of collaborative networks is proposed, and the basic elements of value creation are discussed, in the context of a holistic approach to collaborative networks.

Key words: Collaborative networks, virtual organizations, virtual enterprises, virtual communities, collaborative networks taxonomy, value creation

1. INTRODUCTION

Collaborative networks (CN) represent a promising paradigm in a knowledge-driven society. A large variety of collaborative networks have emerged during the last years as a result of the rapidly evolving challenges faced by business entities and the society in general [3]. For instance, the reduction of commercial barriers not only gave consumers wider access to goods, but also led to higher demands for quality and diversity as well as a substantial increase in the competition among suppliers. Therefore, highly integrated and dynamic supply chains, extended and virtual enterprises, virtual organizations, and professional virtual communities are just some manifestations of this trend that are indeed enabled by the advances in the information and communication technologies.

It is a common assumption that participation in a collaborative network has the potential of bringing benefits to the involved entities. These benefits

Please use the following format when citing this chapter:
Camarinha-Matos, Luis, M., Afsarmanesh, Hamideh, 2006, in International Federation for Information Processing (IFIP), Volume 207, Knowledge Enterprise: Intelligent Strategies In Product Design, Manufacturing, and Management, eds. K. Wang, Kovacs G., Wozny M., Fang M., (Boston: Springer), pp. 26-40.

include: an increase in "survivability" of organizations in a context of market turbulence, as well as the possibility of better achieving common goals by excelling the individual capabilities. On the basis of these expectations we can find, among others, the following factors: acquisition of a larger dimension, access to new/wider markets and new knowledge, sharing of risks and resources, joining of complementary skills and capacities which allow each entity to focus on its core competencies while keeping a high level of agility, etc. In addition to agility, the new organizational forms also induce innovation, and thus *creation of new value*, by confrontation of ideas and practices, combination of resources and technologies, and creation of synergies.

A particularly relevant issue in this context is to understand how value is created in a collaborative endeavor. Unlike the traditional and well-structured value chains, in new collaborative forms it is not easy to clearly identify the amount of "added value" contributed by each member. Consequently it is not easy either to devise general schemas for distribution of revenues and liabilities. The difficulty increases when the focus of value is moving from tangible goods to intangible, e.g. extended products, services, intellectual properties.

On the other hand, the notion of benefit depends on the underlying value system. In general, the structure of a value system, and therefore the drivers of the CN behavior, includes multiple variables or components. In a simplified view, the goal of a collaborative network can be seen as the maximization of some component of its value system, e.g. economic profit in a business context, or the amount of prestige and social recognition in altruist networks. Complementarily there are other factors that influence the behavior of a network and therefore its value generation capability such as the scheme of incentives, trust relationships, and management processes, ethical code, collaboration culture, and contracts and collaboration agreements.

Understanding these issues is fundamental for the sustainability of a collaborative network. It is important to find balanced ways of combining agility with some sense of stability (life maintenance support, knowing and trusting partners, having *fluid* interfaces, etc.). Therefore, a variety of organizational forms, including a mix of long-term strategic alliances and very dynamic short-term coalitions, shall coexist. As a contribution to better understand the role, behavior, and potential impact of collaborative networks in a knowledge-driven society, this paper first analyzes the main underlying concepts, as well as the specificities of the different organizational forms, and then addresses the issues of CN governance and value creation in a holistic perspective.

2. NOTION OF COLLABORATION

Although everybody has an intuitive notion of what collaboration is about, this concept is often confused with cooperation. For many people the two terms are indistinguishable. The ambiguities reach a higher level when other related terms are considered such as networking, communication, and coordination [6], [9]. Although each one of these concepts is an important component of collaboration, they are not of equal value and neither one is equivalent to another. In an attempt to clarify various concepts, the following working definitions can be proposed:

Networking – involves communication and information exchange for mutual benefit. A simple example of networking is the case in which a group of entities share information about their experience with the use of a specific tool. They can all benefit from the information made available / shared, but there is not necessarily any common goal or structure influencing the form and timing of individual contributions, and therefore there is no common generation of value.

Coordination – in addition to exchanging information, it involves aligning / altering activities so that more efficient results are achieved. Coordination, that is, the act of working together harmoniously, is one of the main components of collaboration. An example of coordinated activities happens when it is beneficial that a number of heterogeneous entities share some information and adjust the timing of for example their lobbying activities for a new subject, in order to maximize their impact. Nevertheless each entity might have a different goal and use its own resources and methods of impact creation; values are mostly created at individual level.

Cooperation – involves not only information exchange and adjustments of activities, but also sharing resources for achieving compatible goals. Cooperation is achieved by division of some labor (not extensive) among participants. In this case the aggregated value is the result of the addition of individual "components" of value generated by the various participants in a quasi independent manner. A traditional supply chain based on client-supplier relationships and pre-defined roles in the value chain, is an example of a cooperative process among its constituents. Each participant performs its part of the job, in a quasi-independent manner (although coordinated with others). There exists however, a common plan, which in most cases is not defined jointly but rather designed by a single entity, and that requires some low-level of co-working, at least at the points when one partner's results are delivered to the next partner. And yet their goals are compatible in the sense that their results can be added or composed in a value chain leading to the end-product or service.

Collaboration – a process in which entities share information, resources and responsibilities to jointly plan, implement, and evaluate a program of activities to achieve a common goal. This concept is derived from the Latin *collaborare* meaning "to work together" and can be seen as a process of shared creation; thus a process through which a group of entities enhance the capabilities of each other. It implies sharing risks, resources, responsibilities, and rewards, which if desired by the group can also give to an outside observer the image of a *joint* identity. Collaboration involves mutual engagement of participants to solve a problem together, which implies mutual trust and thus takes time, effort, and dedication. The individual contributions to the value creation are much more difficult to determine here.

A collaboration process happens for instance in concurrent engineering, when a team of experts jointly develop a new product. From this example it can be noticed that although some coordination is needed, collaboration, due to its joint creation facet, involves seeking divergent insights and spontaneity, and not simply a structured harmony.

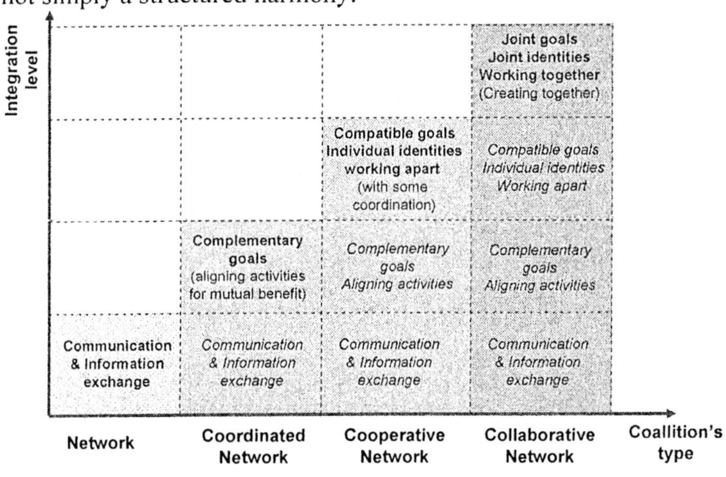

Figure 1. Examples of joint endeavor

As presented in the given definitions and depicted in Fig. 1, each of the above concepts constitutes a "building block" for the next definition. Coordination extends networking; cooperation extends coordination; and collaboration extends cooperation. As we move along the continuum from networking to collaboration, we increase the amounts of common goal-oriented risk taking, commitment, and resources that participants must invest into the joint endeavor. In the rest of this paper we focus on collaborative networks which subsume all other forms.

3. CLASSES OF COLLABORATIVE NETWORKS

A **collaborative network** (CN) is a network consisting of a variety of entities (e.g. organizations and people) that are largely autonomous, geographically distributed, and heterogeneous in terms of their operating environment, culture, social capital and goals, but that collaborate to better achieve common or compatible goals, and whose interactions are supported by computer network.

In today's society, collaborative networks manifest in a large variety of forms, including the production or service-oriented virtual organizations, virtual enterprises, dynamic supply chains, professional associations, industry clusters, professional virtual communities, collaborative virtual laboratories, etc. [3], [5].

Although not all, most forms of collaborative networks imply some kind of *organization* over the activities of their constituents, identifying roles for the participants, and some governance rules. Therefore, these can be called manifestations of **collaborative networked organizations (CNOs)** (Fig.2).

Other more spontaneous forms of collaboration in networks can also be foreseen. For instance, various **ad-hoc collaboration processes** can take place in virtual communities, namely those that are not business oriented – e.g. individual citizens contributions in case of a natural disaster, or simple gathering of individuals for a social cause. These are cases where people or organizations may volunteer to collaborate hoping to improve a general aim, with no pre-plan and/or structure on participants' roles and how their activities should proceed.

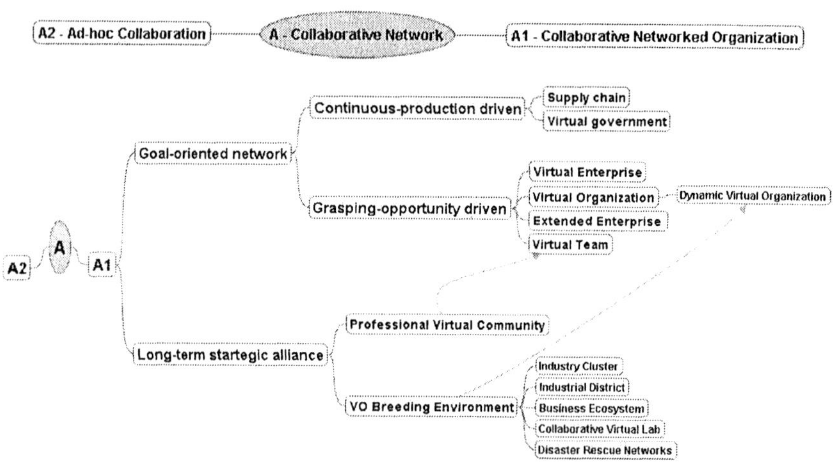

Figure 2. Examples of Collaborative Networks

Among the CNOs, some networks are goal-oriented in which intense **collaboration** (towards a common goal) is practiced among their **partners**, as opposed to longer term strategic alliances described below, where in fact not collaboration but **cooperation** is practiced among their **members**.

Goal-oriented networks can themselves be either driven by continuous production / service provision activities, or driven by the aim of grasping a single (collaboration) opportunity, as described below:

In Goal-oriented networks, the first case of CNOs labeled as *Continuous-production driven* in Fig. 2, includes those networks that have a long-term duration and remain relatively stable during that duration, with a clear definition of members' roles along the value chain. Typical examples include:

Supply chains – a stable long-term network of enterprises each having clear roles in the manufacturing value chain, covering all steps from initial product design and the procurement of raw materials, through production, shipping, distribution, and warehousing until a finished product is delivered to a customer.

Virtual government – an alliance of governmental organizations (e.g. city hall, tax office, cadastre office, and civil infrastructures office) that combine their services through the use of computer networks to provide integrated services to the citizen through a common front-end.

The second case of CNOs within the Goal-oriented networks are labeled as *Grasping-opportunity driven* CNOs in Figure 2, and are dynamically formed to answer a specific collaboration opportunity and will dissolve once their mission is accomplished. Examples in Fig. 2 and Fig. 3 include:

Virtual enterprise (VE) – represents a temporary alliance of enterprises that come together to share skills or core competencies and resources in order to better respond to business opportunities, and whose cooperation is supported by computer networks.

Virtual Organization (VO) – represents a concept similar to a virtual enterprise, comprising a set of (legally) independent organizations that share resources and skills to achieve its mission / goal, but that is not limited to an alliance of for profit enterprises. A virtual enterprise is therefore, a particular case of virtual organization.

Dynamic Virtual Organization – typically refers to a VO that is established in a short time to respond to a competitive market opportunity, and has a short life cycle, dissolving when the short-term purpose of the VO is accomplished

Extended Enterprise (EE) – represents a concept typically applied to an organization in which a dominant enterprise "extends" its boundaries to all or some of its suppliers. An extended enterprise can be seen as a particular case of a virtual enterprise.

Virtual team (VT) – is similar to a VE but formed by humans, not organizations, a virtual team is a temporary group of professionals that work together towards a common goal such as realizing a consultancy job, a joint project, etc, and that use computer networks as their main interaction environment.

The term "virtual" in the above organizations comes from the fact that these networks act or appear to act as a single entity, thanks to their organized communication and coordination mechanisms enabled by computer networks, although they are (usually) not a single legal entity, they may not have a physical headquarter, and are typically geographically distributed.

Besides the Goal-oriented networks, another class of CNOs is the *long-term strategic alliances* (see Fig. 2) aimed at offering the conditions and environment to support rapid and fluid configuration of collaboration networks, when opportunities arise. *VO breeding environments* and *professional virtual communities* exemplify these kinds of networks.

VO Breeding environment (VBE) – represents an association of organizations and their related supporting institutions, adhering to a base long term cooperation agreement, and adoption of common operating principles and infrastructures, with the main goal of increasing their preparedness towards rapid configuration of temporary alliances for collaboration in potential Virtual Organizations. Namely, when a business opportunity is identified by one member (acting as a broker), a subset of VBE organizations can be selected to form a VE/VO.

Earlier cases of VBEs were mostly focused on a regional basis, e.g. industry clusters, industry districts, and business ecosystem. Besides the production / services focus, a large number of more recent VBEs focus in new areas, e.g. science and virtual laboratories, crises management. Some examples include:

Industry cluster – is one of the earliest forms of VO breeding environments, consisting of a group of companies, typically located in the same geographic region and operating in a common business sector, that keep some "binds" with each other in order to increase their general competitiveness in the larger area. These binds may include sharing some buyer-supplier relationships, common technologies and tools, common buyers, distribution channels or common labor pools, all contributing to some form of cooperation or collaboration when business opportunities arise. Earlier forms of clusters did not require a strong ICT infrastructure but more and more collaboration resorts to such support.

Industrial district – is a term mostly used in Italy that represents a concept quite similar to an industry cluster. It can be focused on one single sector or cover a number of sectors in a given region.

Business ecosystems – is inspired by the mechanisms of the biological ecosystems, these networks try to preserve local specificities, tradition, and culture, and they frequently benefit from (local) government incentives. A business ecosystem, also sometimes called digital ecosystem, is similar to a cluster or industry district, although it is not limited to one sector but rather tends to cover the key sectors within the geographical region. In most aspects business ecosystems simply represents a renaming of the industrial district concept. Namely, differences are subtle and can perhaps be found only in a clearer emphasis on the involvement of a diversity of their actors – the *living forces* of a region – in addition to companies, and a more intense use of advanced ICT tools to support collaboration.

Virtual Laboratory (VL) networks – represent the alliance of autonomous research organizations, each having their own resources (equipments, tools, data and information related to their past experiments, etc.), enabling their researchers, located in different geographically-spread centers to be recognized and considered for taking part in potential opportunity based problem-solving collaborations (forming a kind of VO for each problem solving). During a problem-solving collaboration process, it is typical that some expensive lab equipments owned by one or more organizations is made available for (remote) use by the other collaboration partners.

Disaster rescue networks – a strategic alliance of governmental / non-governmental organizations specialized in rescue operations in case of disasters is another recent form of VBE aimed at facilitating a rapid and well-coordinated response in case of a disaster. This VBE could have a local / regional coverage or a global geographic span.

VBE is thus the more recent term that was coined to cover these cases and clearly extends their scope to both regional / global coverage, single / multi-specialty sector, and for-profit / non-profit organizations. Complementary views and coverage of these organizational forms are shown in Fig. 4.

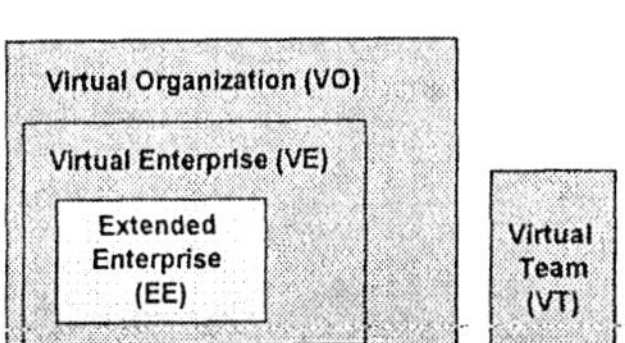

Figure 3. Single-opportunity CN

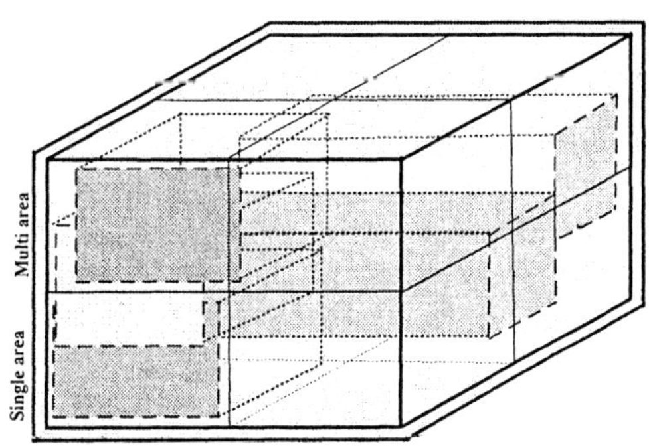

Figure 4. Long-term strategic alliances

Professional virtual community is an alliance of professional individuals, and provide an environment to facilitate the agile and fluid formation of Virtual Teams (VTs), similar to what VBE aims to provide for the VOs.

4. ISSUES ON COLLABORATIVE VALUE CREATION

What will my organization **benefit**, if embarking in a collaborative network? Will the benefits compensate for the extra overhead, loosing some control, and even taking the risks that collaboration implies? These are the main questions that many small and medium enterprise (SME) managers ask when the issue of collaboration is brought up. In fact, effective collaboration involves considerable preparation costs / time, in addition to the operational overheads and risks, which represent barriers to the rapid formation of dynamic coalitions in response to business opportunities. As a basic rule, in order to support rapid formation of collaborative networks, e.g. a VO, it is necessary that potential partners are ***ready in advance and prepared to participate*** in such collaboration. This readiness includes compliance with a common interoperable infrastructure, common operating rules, and common collaboration agreement, among others. Any collaboration also requires a base level of trust among the organizations. Therefore, the concept of **VO breeding environment** has emerged to provide the necessary context for the effective creation of dynamic virtual organizations.

The main aims for an organization's participation in a VBE would include: Agility for opportunity-based VO creation; effective common ICT infrastructure; mechanisms and guidelines for VO creation; general guidelines for collaboration; and increase chances of VO involvement.

Therefore, in this context, Table 1 illustrates some of the main reasons that can motivate an enterprise to join a VBE.

Table 1 – Examples of reasons to join a VBE

Market-related reasons	**Organizational reasons**
• Increase activities / profit • Coping with market turbulence • Increase chances of survival • More chances to compete with larger companies • Lobbying & market influence (branding / marketing) • Easier access to loans • Cheaper group insurance • Better negotiation power (e.g. Joint purchasing) • Prestige, reputation, reference • Access /explore new market /product (e.g. Multidisciplinary sector) • Expand geographical coverage • Increase potential for innovation • Economy of scale	• Management of competencies and resources • Approaches to build trust • Improve potential of risk taking • Support members through necessary re-organization • Learning & training • Shared bag of assets (e.g. shared tools, lessons learned, legal information, sample contracts) • Organize success stories & joint advertisement • Help in attaining clear focus / developing core competencies

• Develop branding • Achieve (global) diversity	• Sharing costs of branding / marketing

But another relevant question is "What keeps the current VBE members happy and loyal to the VBE?" Table 2 illustrates some of these potential reasons.

Table 2 – Reasons to stay as a VBE member

Motivation to stay
• Profit from businesses
• Benefiting from the existing infrastructure
• Better marketing possibilities (fairs, cheaper admission costs, better publicity/visibility (better location) ...)
• Better strategic position through the VBE
• Easy access to complementary skills
• Explore new market / new product (multi-disciplinary-sector), expand geographical coverage
• Potential for innovation
• Continue profiting from the opportunities only available through the VBE
• Fight against a common enemy
• Better negotiation power
• Existing success stories and advertising
• Gain higher rank (Gold, Silver, Bronze) for more opportunities

The initial attracting factors are not exactly the same that keep members happy in the long run! A company might have been attracted by the opportunity of getting access to new knowledge but after a while, once this objective has been achieved, different reasons are necessary to keep it involved in the VBE.

A number of other barriers have also been identified when interviewing running VBEs [1]. A challenge here is the identification and development of remedy measures to overcome these fears (see examples in Table 3).

Table 3 – Reasons to fear being a VBE member

Fear	**Possible remedy**
• Return on time / cost investment fear of not having ROI	- Proactive opportunity brokerage in the market / society Building success stories
• Loosing decision making power – a *collaboration side-effect*	- Creating flexibility in decision making - Transparent rules/regulations - Different levels of membership
• Trust and IPR problems	- Mechanisms to establish trust / forecast trustworthiness - Enhance trust in VBE establishment itself - transparent rules/regulations - Definition of incentives & sanctions - Enforcement of the defined rules to be a "good citizen" in VBE
• Fear of partner selection by a virtual system	- Neutral transparent definition of selection process
• Required high commitment level	- Different levels of membership

These are only some empiric examples collected from the interaction with several existing networks [5]. As the number and variety of networks increases, it is important to further and more systematically elaborate on these elements.

Another relevant issue is the creation of a **system of incentives** to motivate participants to pro-actively engage in collaborative value creation. Complementarily, **fairness** and the crucial issue of properly determining the individual organization's share are important; namely to identify both an organization's contributions towards, as well as benefits from, collaboration in the network. These are the other major issues for which transparent governance principles, whenever possible supported by objective indicators, are needed.

In fact it is frequently mentioned that the lack of objective measurements, clearly showing the benefits of collaboration, is an obstacle for a wider acceptance of these new organizational forms. Generally, it is difficult to prove the advantages of (dynamic) collaborative networks in contrast to more traditional organizational forms in terms of improved performance. Being able to measure the global performance of a collaborative network, as well as the individual performance of each of its members, could represent an important boosting element for the wide acceptance of the paradigm. However **performance indicators** tailored to CN are not yet available [2], [10].

A performance measurement depends on the premises of the measurement system used. Collaborative networks challenge the premises of the methods developed in the past [8], therefore the applicability of existing measurement systems in this area is questionable. First it is necessary to take into account that performance, and related generated values, can be seen from different points of view, e.g. from the individual participant perspective, from the network coordination perspective, and from the surrounding environment / society perspective.

Understanding and making the nature of collaboration benefits explicit, is also an important way to ensure that every member of the network understands the measurements in the same way (same "perception of value"). This is also a requirement for goals alignment in order to facilitate the coherence of members' goals with the measurements.

The actual meaning of a benefit depends on the underlying **value system** that is used in each context. It is commonly accepted that the behavior of an individual, society, or ecosystem is determined by its value system. It is also intuitively understood that the values considered in a business-oriented collaborative network are different from those in a non-profit context (e.g. disaster rescue network). In fact the business / economy oriented school tends to look at a value system in terms of the activity links between a company and its suppliers and customers, putting the emphasis on how much (monetary value), a product or service is worth to someone. On the other hand, the socio-psychological school considers a value system as the ordering and prioritization of a set of values that an actor or a society of actors holds. The two perspectives are not necessarily incompatible, but more research is needed in order to develop a general (abstract) theory of value systems that can then be instantiated to different application contexts.

In general, the structure of a value system, and therefore the drivers of the CNO behavior, might include multiple variables / aspects. Simultaneously other elements that may determine the behavior of the network and its members, include scheme of incentives, the existing level of trust, code of ethics, culture of collaboration, and collaboration agreements.

As a very preliminary step towards a theory of value systems and characterization of collaboration benefits, a number of indicators have been suggested in [4]. In this work the benefits received by a participant include both the benefits resulted directly from the activities performed by this participant and the benefits to this participant resulted by the activities performed by the other participants (external benefits). By combining base concepts from the transactions cost theory, social networks analysis, and game theory, a number of indicators is proposed (see summary in Table 4).

When combined with for instance a graphical representation, these indicators can provide the collaborative network participants a more objective view of their global and individual performance.

Although some experiments were carried out with real data from both a business-oriented network and a social-oriented network [4], much more research is needed in this area. Furthermore, in practice, collecting the base data from an operational network is rather a difficult task, which requires the development of new approaches and supporting infrastructures.

Table 4 Examples of indicators of benefits in collaborative networks

Social Contribution Benefits - sum of benefits contributed by an actor a_i to its partners as a result of its performance in the collaboration process.
External Benefits - sum of benefits received by an actor a_i as a result of the activity of the other actors involved in the collaboration process.
Total Individual Benefits - sum of external benefits plus self-benefits of an actor a_i
Individual Generated Benefits - sum of social contributed benefits plus self-benefits of actor a_i
Total Received Benefits - sum of external benefits achieved by a set of actors
Total Contributed Benefits - sum of social contributed benefits generated by a set of actors
Total Network Benefits - sum of benefits achieved by a set of actors in a specific collaboration process or over a period of time.
Progress Ratio - a macro indicator that represents the variation of the global benefits over a period of time.
Social Capital - density of the network benefits relation
Cooperative Development Ratio - measures the progress of cooperation benefits for a set of actors over a period of time.
Individual contribution index - Normalized contribution of an actor to the collaborative network
Apparent individual contribution index - based on the number of contribution links (i.e. the *out degree* of the actor in the graph representing the cooperation benefits), this index gives an apparent and simple to compute measure of the involvement of an actor in the collaboration process.
Individual external benefits index (IBI) - Normalized external benefits received by an actor, expressing the *popularity* or *prestige* of the actor.

Apparent individual benefits index - indicator based on the number of received contribution links (i.e. the *in degree* of the actor in the graph). Like IBI, this index also expresses the *popularity* or *prestige* of the actor.
Reciprocity index (RI) - balance between credit (sum of benefits contributed by an actor a_i to all its partners (or one specific partner)) and debit (sum of benefits received by an actor a_i as a result of the performance of all actors (or one specific partner) involved in the cooperation process). RI <0 – selfish behavior, RI = 0 – null balance, RI > 0 – altruist behavior.

5. PURSUING A HOLISTIC APPROACH

A large number of research projects are carried out worldwide and a growing number of practical cases on different forms of collaborative networks are being reported. This trend has so far led to an extensive amount of empirical base knowledge that now needs to be leveraged. In addition to the identification of many required components, tools, and the base infrastructure functionalities, awareness is being built and partially studied regarding the fundamental configuration and operational rules, as well as the behavioral patterns that emerge. It is urgent to consolidate and synthesize the existing knowledge, setting a sound foundation for the future research and development in this area. In this context, the ECOLEAD integrated project was launched in 2004 as a 4-year initiative, involving 20 industrial and academic organizations from Europe and Latin America. This project's vision is that in ten years, in response to fast changing market conditions, most enterprises and specially the SMEs will be part of some sustainable collaborative networks that will act as breeding environments for the formation of dynamic virtual organizations.

The fundamental assumption in this project is that a substantial increase in materializing networked collaborative business ecosystems requires a comprehensive holistic approach. Given the complexity of the area and the multiple inter-dependencies among the involved business entities, social actors, and technologic approaches, the substantial breakthrough cannot be achieved with only incremental innovation in isolated areas. Therefore, the project addresses three most fundamental and inter-related focus areas - constituting pillars - as the basis for dynamic and sustainable networked organizations including: VO Breeding Environments (VBE), Dynamic Virtual Organizations, and Professional Virtual Communities (PVC) and Virtual Teams.

As the main focus of work, the VBE pillar addresses the characterization of these networks, namely in terms of structure, life cycle, competencies, working and sharing principles, value systems and metrics, the governance principles and trust building processes, the VBE management, and VO creation. The dynamic VO area is mainly focused on the VO management and governance approaches, performance measurement, and VO inheritance.

PVC addresses the characterization of these communities in terms of the socio-economic context, governance principles, social and legal implications, value systems, metrics and business models, as well as the support platform for collaboration. Interactions and synergies among these three entities, namely in terms of business models, value creation and corresponding metrics, are a major issue of integration.

Implantation of any form of collaborative network depends on the existence of an ICT infrastructure. Therefore, in addition to specific services to support the three pillars mentioned above, a generic horizontal infrastructure for collaboration is also being developed. Sustainable development of collaborative networked organizations needs to be supported by stronger fundamental research leading to the establishment of Collaborative Networks as **a new scientific discipline**. Ad-hoc approaches and poor understanding of the behavior of the collaborative structures and processes mainly characterize the past developments in the area of collaborative networked organizations. The project includes the establishment of a sound theoretical foundation, and a reference architecture, as a pre-condition for the next generation of CNOs.

6. CONCLUSIONS

Collaborative networks are commonly recognized in society as a very important instrument for survival of organizations in a period of turbulent socio-economic changes. A growing number of varied CN forms are emerging as a result of the advances in ICT, the market and societal needs, and the progress achieved in a large number of international projects. Nevertheless most of the past initiatives have addressed only partial aspects, failing to properly understand and support the various business entities and their inter-relationships in complex and fast evolving business ecosystems. The ECOLEAD project is pursuing a more holistic approach considering both long-term and temporary organizations as well as networks of organizations and networks of people.

A major current research challenge is the elaboration of a sounder theoretical foundation towards a better understanding of value systems, benefits, and success factors for collaboration.

7. ACKNOWLEDGMENTS

This work was supported in part by the ECOLEAD project funded by the European Commission. The authors thank the valuable contributions of their partners in the consortium.

8. REFERENCES

1. Afsarmanesh, H.; Camarinha-Matos, L.M. – A Framework for Management of Virtual Organization Breeding Environments, in Proceedings of PRO-VE'05 – Collaborative Networks and their Breeding Environments, Springer, Valencia, Spain, 26-28 Sept 2005.
2. Alfaro Saiz, J.; Rodriguez, R.; Ortiz, A. – A performance measurement system for virtual and extended enterprises, in Proceedings of PRO-VE'05 – Collaborative Networks and their Breeding Environments, Springer, Valencia, Spain, 26-28 Sept 2005.
3. Camarinha-Matos, L. M.; Afsarmanesh, H. - Collaborative networks: A new scientific discipline, J. Intelligent Manufacturing, vol. 16, Nº 4-5, pp439-452, 2005.
4. Camarinha-Matos, L.M.; Abreu, A. - Performance indicators based on collaboration benefits, in Proceedings of PRO-VE'05, Collaborative Networks and their Breeding Environments, Springer, pp. 273-282, Valencia, Spain, 26-28 Sept 2005.
5. Camarinha-Matos, L.M.; Afsarmanesh, H.; Ollus, M. (Editors) – Virtual Organizations – Systems and Practices, Springer, 2005.
6. Denise, L. – Collaboration vs. C-Three (Cooperation, Coordination, and Communication), INNOVATING, vol. 7, Nº 3, Spring 1999.
7. Evans, S.; Roth, N.; Sturm, F. (2004). Performance measurement and added value of networks, in Collaborative Networked Organizations – A research agenda for emerging business models, Kluwer Academic Publishers, ISBN 1-4020-7823-4.
8. Folan, P.; Browne, J. (2005) – A review of performance measurement: Towards performance management. Computers in Industry, vol. 56, pp 663-680.
9. Himmelman, A. T. - On Coalitions and the Transformation of Power Relations: Collaborative Betterment and Collaborative Empowerment, American Journal of Community Psychology, Vol. 29, No. 2, 2001.
10. Kolakovic, M. (2003).VAVE – A model for measuring the business performance of virtual enterprise – Business Excellence I, performance measures, benchmarking and best practices in new economy, University of Minho, ISBN 972-8692-08-0.

KNOWLEDGE MANAGEMENT IN SMALL AND MEDIUM-SIZED ENTERPRISES

Wladimir Bodrow
University of Applied Sciences Berlin, Germany. Email: w.bodrow@fhtw-berlin.de

Abstract: In this paper the attempt to consolidate the diversity in theoretical aspects of knowledge management in Europe starting from 2004 (CEN/ISSS 2004) is briefly presented. Conform to this context some original results are discussed as well. In the thereon following part diverse knowledge management tools and systems – mainly developed using Information and Communication Technology (ICT) – are introduced. After it the European road map for implementing the knowledge management is described. In addition some aspects of the application of KM in small and medium-sized enterprises will be considered.

Key words: Knowledge management, knowledge-oriented enterprise, small and medium-sized enterprises

1. INTRODUCTION

The past decade activities in the global business *in general* and those in small and medium-sized enterprises (SMEs) in European countries *in particular* mark a paradigm shift. According to Peter Drucker "…the basic economic resource is no longer capital, nor natural resources, nor labor. It is and will be knowledge." (Drucker 1993) This message has now reached not only global players such as IBM, Hewlett Packard, Siemens, Cisco, McKinsey and Toyota but also a long list of small and medium-sized companies.

All together they accept knowledge as the most important factor of their business processes. During the last ten years many of them have started to implement a number of projects and solutions to improve the exploitation of

Please use the following format when citing this chapter:
Bodrow, Wladimir, 2006, in International Federation for Information Processing (IFIP), Volume 207, Knowledge Enterprise: Intelligent Strategies In Product Design, Manufacturing, and Management, eds. K. Wang, Kovacs G., Wozny M., Fang M., (Boston: Springer), pp. 41-53.

company knowledge in a wide range of business areas. A more thorough examination makes clear that until now there has not been one universal theory of knowledge management with a world wide acceptance by business and research institutions. In fact various concepts and approaches of knowledge management are currently in use in different European countries.

2. STATE OF THE ART IN THEORY – KM MODELS

Beginning with Ikujiro Nonaka and Hirotaka Takeuchi (Nonaka and Takeuchi 1995), Karl Erik Sveiby (Sveiby 1998) and other pioneers some dozens of knowledge management approaches were developed around the globe in the past decade. A part of them were implemented and approved in business practice worldwide. To name different knowledge managements approaches developed in Europe, here is a non complete list:

- Building blocks for knowledge management (Probst et al. 1999)
- Model of integrative knowledge managements (Reinhardt and Pawlowski 1997)
- Four steps toward knowledge management (Schüppel 1996)
- Lifecycle model of knowledge management (Rehäuser and Krcmar 1996)
- Company knowledge market model (North 2005)
- Process oriented model of Fraunhofer IPK (Mertins et al. 2003)
- Process-oriented knowledge management strategy (Maier 2004)

Brief consideration on these and other established approaches developed in the USA, Japan and other non-European countries shows a number of partially and completely agreeing aspects. This was the background for our investigation toward a common approach in knowledge management (Bodrow and Fuchs-Kittowski 2004, Fuchs-Kittowski and Bodrow 2004). The empirical analysis of 18 known approaches shows that the most important aspect of knowledge management actually is the *use* of knowledge. After the first presentation of the achieved results we extended the number of approaches to be investigated. This extension did not change the main tendencies observed before. Researchers and practitioners around the globe consider knowledge *using* ("application", "execution", "processing", "utilization" were used by different authors as well) as the most essential activity within knowledge management. On the second position with only a very small difference they see the *sharing* ("distribution", "dissemination", "communication", "exchange") of knowledge. The third and fourth place is occupied by the *generation* ("creation", "production", "discovery") and *integration* ("linkage", "interpretation", "adaptation") of knowledge into the structure of the resources available respectively. The *identification* ("filtering", "selecting") as well as the *acquisition* ("procuring", "collecting", "importing", "assimilating") and *development* ("derivation") achieved fifth places within the list of the 18 activities selected for the analysis. The *storage* ("saving", "retaining") of knowledge is on the sixth position. The most surprising point was the fact that all activities defined by Nonaka and Takeuchi – socialization, internalization, combination and externalization – hold their positions at the end of the hierarchy of activities in the analyzed knowledge management approaches. This is explainable by the knowledge management's strong orientation onto the exploitation of knowledge in enterprises and on the market likewise. These additional results derived from the investigation:

- This kind of empirical investigation can be considered a strategy toward a common or general approach to knowledge management. Further diversification of the existing approaches with their specialization according to the specifics of knowledge or (business) application is regarded as contra productive.
- The methodology we selected for our investigation delivered impulses for the development of a knowledge management ontology. Based on an empirical analysis of the different approaches it is possible to derive (or to sort out) the best accepted one and define its attributes and structure. Broad acceptance of an ontology build this way is the only factor indicating its quality!
- In the same investigation we changed the subject matter and analyzed the possibilities for building a knowledge ontology. Our conclusion is that the preferable basis to build such an ontology in general or in particular fields is an analysis of the activities. It allows to consolidate different perspectives and approaches and to establish a so-called activity ontology. For a correct analysis and satisfactory results in this field the investigation should be based on a well-founded linguistic knowledge. This is obligatory since the diversity of all the existing definitions has primarily been made by experts in special application areas. These definitions have to be communicated adequately among researchers and practitioners. While most research projects teams are mainly content or topic oriented they lack expertise in linguistic skills, which is the basis for a proper definition. As mentioned before the appropriate knowledge distribution and communication is one of the most important activities within knowledge management.

The importance of the ideas and results discussed were underlined by the workshop agreement published as a result of CEN/ISSS workshop in Madrid (CEN/ISSS 2004). Materials of this workshop with the title "European guide to good practice in knowledge management" (in following European guide) contain five parts:

1. KM Framework
2. Culture and KM
3. Implementing KM in Small and Medium-Sized Enterprises
4. Measuring KM
5. KM Terminology

and represent the state of the art in European research and practice in knowledge management. More than 100 researchers and practitioners from Europe, America and Asia participated in developing this European knowledge management agreement. Following the main ideas of this guide (except KM Terminology) will be briefly presented and discussed.

3. KM FRAMEWORK

The KM Framework proposes three layers for knowledge management:
a) The first and most important layer of the presented concept is oriented on business processes including all participants and their knowledge.
b) The second layer defines the core knowledge activities used by organizations in Europe – *identification, creation, storage, sharing* and *use* of knowledge.
c) The third layer represents the enabler of knowledge management. It consists of personal and organizational knowledge capabilities.

The only difference between this framework and our results (Bodrow and Fuchs-Kittowski 2004) is the absence of "storage" among the first five knowledge activities in our register. Instead of it we observed that the "*integration* of knowledge" into the existing structure is considered as more important. Possible explanation to these discrepancies can have semantic background. The mentioned before well-founded linguistic analysis is the way to clarify the situation.

4. CORPORATE CULTURE IS THE MOST SIGNIFICANT FACTOR OF KM

The survey of German TOP 1000 and European TOP 200 companies made by Fraunhofer IPK Berlin (Mertins et al. 2003) illustrated that the first five decisive factors of a successful knowledge management are corporate culture (47,1%), structural factors/external conditions (29,8%), information technology (27,9%), staff motivation and qualification (27,9%) and promotion by top management (26,9%). The importance of cultural aspects and their influence on a successful knowledge management was also confirmed in other surveys and projects (KPMG 2001, ME Survey 2003).

Starting from this crucial aspect the following factors of enterprise culture were defined and described in the European guide (CEN/ISSS 2004):

1. The relationship between knowledge and culture. In this part of the guide the barriers to a successful knowledge management are listed and the differences between traditional and knowledge-aware organizations are presented.

2. Individuals, groups and organizations. Beforehand authors defined enterprise culture and provided a short list of the corresponding facets. Subsequently they concentrated the examination on the different participants of culture, starting from a single person till the whole organization. Here different identities in respect to individual, functional, organizational or geographical characteristics of communities within and among enterprises are presented. Corporate culture acts as a mediator for the communication between the personal and organizational knowledge.

3. Trust and motivation. This chapter is devoted to the most important aspects of a successful knowledge management in the enterprise: *trust*, *motivation* and *leadership*. The results of the LexisNexis survey published in 2004 (LexisNexis 2004) are interesting in this context. They underline that 49% of all enterprises prefer humanists and social scientists and/or practitioners with a soft skill (educational) background for their knowledge manager position. Consequently various diplomas of Art are preferred to MBA graduates (only occupying the second place by 38%). A technical background, including computing, stands last in this line. The practice and development of actual knowledge management in Europe is looking for generalists, who can communicate all aspects of knowledge management and motivate the employees in the enterprise. According to the same survey 72% of employees are motivated by their own working achievements whereas only 10% are interested primarily in monetary rewards. The significant change of this trend from 2003 to 2004 emphasizes on the importance of cultural aspects for a successful knowledge management.

4. Competencies, learning and reflection. The crucial point of this chapter is a view on the process of knowledge creation in an enterprise. Authors subdivided the knowledge creation process into four parts: empathizing, articulating, connecting and embodying. The presented iterative cycle of knowledge creation differs significantly from the one by Nonaka and Takeuchi. The results of recent surveys (ME 2003, LexisNexis 2004) document a change in the knowledge generation as well. Actually they dealt with another object of change – the changing of the learning processes. The most important source of knowledge for the employee is the practical skill gained from team work. Moreover the majority of knowledge managers (85%) refuse to use books as an efficient source to improve their knowledge and prefer for that reason learning-by-doing directly in the project.

5. Tools to manage relationships that drive learning processes. The activities in this area have to follow the conviction that corporate-wide communication is the only basis for an efficiently organized knowledge exchange and production process. Correspondingly in the guide both formal and informal communities were suggested as important tools to improve the communication. At the beginning as well as accompanying knowledge management project activities, observation and questioning are considered mandatory. The knowledge audit is the proposed tool to identify the situation beforehand and during the project. Coaching and mentoring were accounted as well known and widely-established tools for knowledge transfer. Furthermore the authors examine the narrative, conversation and dialogue as additional possibilities for the knowledge exchange. In the guide the technology (ICT) is accepted as an enabler to connect people and support their communication a better way.

5. STATE OF THE ART IN TECHNOLOGY – KM SYSTEMS

The general goal of knowledge management is to provide each and every decision maker in all decision relevant areas with the right knowledge (according to his/her level of expertise) in the right form and quality, and at the right time and place. This goal is part of the common managerial vision in every single enterprise. A particular knowledge management system does not a priori imply information and/or communication technology. Today's wide consensus on ICT as an enabler to knowledge management has changed this position significantly. The number of approaches named before was implemented in various ICT-based knowledge management systems. They use different well-known and established information processing systems and technologies appropriate for the particular application. These systems support the already mentioned efficient providing of the decision makers with the relevant information of different kinds. Here is an incomplete list of technologies and systems in use: Groupware, Workflow, Document management system, Content management systems, Knowledge based or expert systems, Intelligent software agents, Data warehouse, Business intelligent systems, Collaboration systems, E-Learning systems, Customer relationship management, Intranet, Knowledge and enterprise portals, Human resources management.

With the help of these technologies various systems to support enterprise wide knowledge management were developed. They have many names from "knowledge management software" through "knowledge management support system" till "knowledge warehouse". A long structured list of ICT-based knowledge management tools is offered for instance by Ronald Maier (Maier 2004). In the analysis of the foundations of knowledge management systems Maier (Maier 2004, p.79) attempts to formulate a general definition for such systems. This seems a complicated endeavor just because of the variety of the existing views and definitions on knowledge, management and knowledge management.

In the investigation of knowledge management tools made by the research group of the Fraunhofer IPK (a knowledge management competence center in Germany) another standpoint is preferred (Mertins et al. 2003). Diverse established solutions are examined here in respect to the defined four core activities of knowledge management – *create* knowledge, *store* knowledge, *distribute* knowledge and *use* or *apply* knowledge. The investigated knowledge management tools are subdivided into eight groups:

1. Search engines/categorization tools/ intelligent agents.

According to Fraunhofer IPK, the tools in this group primarily assist the *application* of knowledge. Some other aspects could be added to this position: a) search engines can support the knowledge *creation* by providing the user with knowledge related to the topic from different perspectives. b) search engines and intelligent agents support the knowledge sharing and dissemination (target oriented knowledge *distribution*) as well. Among international providers like Alexa (www.alexa.com) or Magnifi (www.magnifi.com) listed in the paper one can find German products i.e. Knowledge Miner from USU AG (www.usu.de) and SchemaText from SchemaGmbH (www.schema.de).

2. Portals

Following Fraunhofer IPK, portals – including knowledge-, enterprise-, and information portals – are useful to *store*, *distribute* and *apply* knowledge. Apart from the RetrievalWare, Excalibur Technologies's tool (www.excalib.com), also AskMe offered by AskMe Corporation (www.askmecorp.com) is listed. The authors underscore that the most complete KM suites are implemented as portals. Consequently the *creation* of knowledge can be considered as an activity covered by portals as well.

3. Visualizing tools

These tools are to promote the presentation of knowledge adequately to the particular tasks and the individual user capabilities (e.g. skill and experience). This way they support the practical *application* of the knowledge presented. The German product Think Tool from Think Tools AG (www.thinktools.com) is listed here on the side of IBM tool KnowledgeX (www.ibm.com) and Eureka from InXight Software (www.inxight.com).

4. Skill management

Another kind of visualization is covered by skill management tools. They point to the location and quality of knowledge available in the enterprise.

Different solutions like knowledge maps, yellow pages, competence maps or skill maps can be realized using tools from this group. Some European products – Yellow Pages from altavier GmbH (www.altavier.de) – is listed here besides for instance Competence Manager from HR Hub.com (www.hrhub.com) or ScillView Enterprise from SkillView Technologies Inc. (www.skillview.com). These tools and the corresponding solutions are primarily responsible for the *storage* and *distribution* of knowledge in the enterprise.

5. Complete KM suites

Starting from the fact that the development of a complete KM suite is the goal of every provider, many European companies today offer a number of those complete tools. In the list presented one can find infonea from Comma-Soft AG (www.comma-soft.com), Knowledge Warehouse from SAP (www.sap.com), Business Process Management tools from IDS Scheer AG (www.ids-scheer.de) on the side of Aungate's suite from Aungate (www.aungate.com), Livelink from OpenText (www.opentext.com) or Hummingbird Enterprise KM from Hummingbird (www.hummingbird.com). All these and other listed tools support a complete life cycle of knowledge in the business solution.

6. Toolkits for developing individual solutions

These toolkits serve the development of individual solutions. Like the complete KM suites they cover all four activities within knowledge management: *creation, storage, distribution* and *application* of knowledge. Since their individuality a high effort is necessary to fit such tools to the company's individual needs. DynaSight from arcplan (www.arcplan.com) is listed here on the side of StoryServer from Vignette (www.vignette.com), Digital Dashboard from Microsoft (www.microsoft.com) and BackWeb solution of BackWeb (www.backweb.com).

7. Learn and teach

The first and foremost focus of MediaNaut from Chronomedia GmbH (www.medianaut.com), the only German tool in this group, is the *creation* of knowledge. Founded on an adequate presentation of the stored knowledge user obtains an appropriate assistance in learning and *creating* new knowledge.

8. Virtual teams/collaboration

The two tools in this group (eRoom from EMC Documentum (www.documentum.com) and HyperKnowledge from HyperKnowledge (www.hyperknowledge.com) provide a special support in the *creation, storage* and *distribution* of knowledge in the enterprise. Discussion groups and virtual teams primarily implemented within these tools are a result of the strong orientation on communication of these tools.

The main outcome of the investigation made by Fraunhofer IPK is the general adaptation of the knowledge management tools to the employees and their tasks within business processes.

6. APPLICATION OF KM IN SMES

The roadmap for the development and implementation of knowledge management solutions in the business praxis is summarized in the third chapter of the European guide (CEN/ESSS 2004). Following the European guide the general scheme for the implementation of knowledge management in small and medium-sized enterprises consists of five phases:

A Setting up a knowledge management project

In this phase the management defines and communicates in detail the mission, vision, strategy and the aims for the knowledge management project to *every* employee in the enterprise.

B Assessment

The activities in this phase support the understanding of the actual state and strategy of the knowledge management within the enterprise. The management is advised to apply an assessment tool for a better understanding. The authors give a short description of five assessment tools including diagnostic tools, knowledge audits and knowledge management audits.

C Development

The main results of this phase are a) a well defined concept for one kind of knowledge management solution and b) a suggestion for the tools and technologies to be applied. The tool classification according to their applicability presented in the same chapter involves more than just ICT-based tools as in both previously referred investigations. This classification is focused on five previously defined knowledge management activities.

D Implementation

Three factors were considered as crucial for the implementation of KM solutions – people, time and budgetary control. The authors formulate and give a detailed description of seven action steps within the implementation process. Different time restrictions with respect to the milestones and corresponding communicational facets were defined as well. As an important point of the implementation the adherence to a time limit was pointed out.

E Evaluation/Sustainability

In sum knowledge management's goal is to establish itself as an integral part of the business process. This way knowledge management becomes sustainable in the enterprise. The success of the project can be identified and measured based on it. A useful approach for a systematic evaluation of all stages of the project is the "lessons learned".

The milestones defined accordingly support an easy realization of each phase of the project implementation described.

7. MEASURING KM

Since knowledge has no originally quantitative attributes like other resources used in business processes, it is impossible to measure it directly in countable units. On the other hand knowledge has a value and consequently is the part of different business activities. A remarkable aspect is that the value of the same knowledge can vary from one situation to another and therefore this value is strong dependent on each particular business case. Outgoing from these ideas it has to be necessary to measure the knowledge (and following the knowledge management) to know and/or to improve the return of investment and other indicators of KM activities.

According to the European guide there are five dimensions directly related to intellectual capital (IC) for adding value through knowledge management:

1. Financial focus
2. Innovation focus
3. Process focus
4. Customer (client) focus
5. Human (employee) focus

On the basis of these dimensions three different types of intellectual capital are defined: *Human capital*, *Structural capital* and *Customer capital.* From this perspective the measurement of KM is the basis for the quantity of corporate intellectual capital. In that way the enterprise can control the development of IC. Different strategies and technologies to measure KM in the enterprise are also described in the European guide. In the same part several approaches for the measurement of the intangible assets are presented and partly described. Additionally the authors give a recommendation for the direct and indirect measures of KM. Some aspects of the German project in this area are the topic of the next chapter.

8. KM PRACTICE IN SMES

"Wissensbilanz – Made in Germany" (www.akwissensbilanz.org) is the current project in regard to measure knowledge management in small and medium-sized enterprises in Germany. The German expression Wissensbilanz (literally knowledge balance sheet) can rather be referred to as "intellectual capital report", "intellectual capital accounts" or "intellectual capital statement". It includes the evaluation of qualitative and quantitative outcomes of knowledge based and knowledge oriented business activities in a particular enterprise. Because of the unclear separation between knowledge and intangible assets as well as intellectual capital the literal transcription "knowledge balance sheet" is used in this chapter. The legal regulations for such a knowledge balance sheet is defined in the German Accounting Standards DRS 12 published in 2002 (www.drsc.de) and in the International Accounting Standards IAS from 2005 (www.iasb.org). According to DRS 12 enterprises can refer to different facets of intellectual capital in their accounting: human capital, customer capital, supplier capital, investor capital, process capital, location capital and innovation capital. To include the intangible asset into the accounting IAS (chapter 38) defines four critical attributes: identifiability, control, potential economic benefits and reliable acquisition and production costs. Examples of intangible assets listed in IAS include among others computer software, patents, copyrights, licenses, customer lists and marketing rights. "Wissensbilanz – Made in Germany" is part of a governmentally supported contest "Fit für den Wissenswettbewerb" (literally fit for knowledge competition) and initiative "Wissensmedia" (literally knowledge media). It represents the possible ways to establish knowledge management in a wide business practice. The roots of an intellectual capital analysis are to be found in pioneer publications of Scandinavian researchers Karl-Erik Sveiby (Sveiby1998), Leif Edvinsson (Edvinsson 1997). There are numerous approaches for knowledge evaluation based on monetary and non-monetary indicators (Bodrow and Bergmann 2003, Mertins et al. 2005, North 2005 p 219). They use the concept of Balanced Scorecard developed by Robert Kaplan and David Norton (Kaplan and Norton 1996) in various modifications. The idea to compose a regular balance sheet for corporate knowledge goes back to the Austrian Professor Günter Koch. Such knowledge balance sheets can be downloaded from the site of Austrian Research Centers since 1999 (www.arcs.ac.au). Many Austrian Universities have published their knowledge balance sheets beginning from 2002 (e.g. www.donau-uni.ac.au). In 2007 the Austrian administration will publish the first profound knowledge balance sheet for the whole country. Both the results achieved and the experiences gathered by Austrian colleagues provide a basis for German and other international projects. The main aim of the project "Wissensbilanz – Made in Germany" is different from the Austrian concept. It concentrates on the dissemination of knowledge management skills and technologies in small and medium-sized enterprises. As a result of the activities in this project several knowledge balance sheets were published last year on the project's site (www.akwissensbilanz.org). Currently similar projects are in the process of realization in other European countries as well e. g. Switzerland and Denmark. On the groundwork of the outcome of these projects the management will be able to identify and represent the knowledge available in the enterprise (remember knowledge identification is one of the five activities in the European guide). As a result the management of the enterprise receives instruments to control the development of its corporate knowledge and consequently its knowledge management.

9. OUTLOOK

Management change in large as well as in small and medium-sized enterprises around the globe today is a result of the broad acceptance of knowledge as the main business resource and the formation of knowledge economy in general. The gathered experiences in this field will lead to a generalization from the different approaches of KM toward one single universal approach. Parallel to this generalization SMEs will continue to integrate knowledge management in their business processes. Such specific favorable conditions for knowledge management like flat hierarchies, less organizational boundaries, efficient, traditional and already established informal knowledge exchange, clearly defined volume and reliability of knowledge are definitely advantageous to establish a successful knowledge management.

10. REFERENCES

Bodrow W, Bergmann P (2003) *Wissensbewertung in Unternehmen.* Erich Schmidt Verlag, Berlin

Bodrow W, Fuchs-Kittowski K (2004) *Wissensmanagement in Wirtschaft und Wissenschaft.* Wissens chafts fors chung.

Jahrbuch 2004 (in print) CEN/ISSS Knowledge management workshop (2004) European guide to good practice in knowledge management. http://www.cenorm.be/cenorm/businessdomains/businessdomains/isss/cwa/knowledge+management.asp, 9.2.2006

Drucker P (1993) *A post-capitalist society.* Econ, New York.

Edvinsson L (1997) Developing intellectual capital at Scandia. *Long Range Planning,* 30/6, pp 266–373

Fuchs-Kittowski K, Bodrow W (2004) Wissensmanagement für Wertschöpfung und Wissensschafung. In: Banse G, Reher EO (eds.) Fortschritte bei der Herausbildung der Allgemeinen Technologie. *Sitzungsberichte der Leibniz-Sozietät* Vol. 75, Verlag Irena Regener, Berlin, pp 81–104

Kaplan R, Norton D (1996) *The Balanced Scorecard.* Harvard Business School Press, Boston

KPMG (2001) Bedeutung und Entwicklung des multimediabasierten Wissensmanagements in der mittelständischen Wirtschaft. http://www.wissensmedia.de/studie_kpmg.pdf, 9.2.2006

LexisNexis (2004) Wissensmanagement in deutschen Unternehmen. http://www.lexisnexis.de/downloads/umfrage.pdf, 9.2.2006
Maier R (2004) *Knowledge Management Systems. 2nd edn.* Springer, Berlin Heidelberg New York
ME (2003) Wissensmanagement in deutschen Unternehmen. http://www.competencesite.de/wissensmanagement.nsf/0/497e5772f05ee450c1256e7e002ca3a8?OpenDocument, 9.2.2006
Mertins K, Alwert K, Heisig P (eds) (2005) *Wissensbilanzen –Intellektuelles Kapital erfolgreich nutzen und entwickeln.* Springer, Berlin Heidelberg New York
Mertins K, Heisig P, Vorbeck J (eds) (2003) *Knowledge management: concepts and best practices. 2nd. edn.* Springer, Berlin Heidelberg New York
Nonaka I, Takeuchi H (1995) *The knowledge creating company.* Oxford University Press, Oxford
North K (2005) *Wissensorientierte Unternehmensfürung.* 4th edn. Gabler, Wiesbaden
Probst G, Raub S, Romhardt K (1999) *Wissen Managen. 3d edn,* Gabler, Wiesbaden
Reinhardt R, Pawlowsky P (1997) *Wissensmanagement. In: Wieselhuber & Partner (eds) Handbuch Lernende Organisation.* Gabler, Wiesbaden
Rehäuser J, Krcmar H (1996) Wissensmanagement in Unternehmen. In Schreyögg G, Conrad P (Eds.) *Managementforschung* Vol.6 Wissensmanagement. Erich Schmidt Verlag, Berlin New York, pp 1–40
Schüppel J (1996) *Wissensmanagement: organisatorisches Lernen im Spannungsfeld von Wissens- und Lernbarrieren.* Gabler, Wiesbaden
Sveiby KE (1998) *Wissenskapital –das unentdeckte Vermögen.* Moderne Industrie, Landsberg (other publications at www.sveiby.com.au)

INTEGRATED QUALITY INFORMATION SYSTEM AND CHINA EXPERIENCE

Xiaoqing Tang, Guijiang Duan
School of Mechanical Engineering and Automation, Beihang University (BUAA), 37 Xueyuan Road, Haidian District, Beijing 100083, China, E-mail: tangxq@buaa.edu.cn

Abstract: China has been becoming one of the key members in global market, and being a "plant of the world" in the past two decades. Quality is one of the essential factors for Chinese manufacturing industry to success and information technology could assist manufacturing to improve company-wide quality systems. Quality Engineering Laboratory (QEL) of Beihang University (BUAA) has been focusing on research of related methodologies, and developing technologies, and modeling for integrated quality information process to support Chinese manufacturing industry for years. An Integrated Quality Information System (IQIS) – "*QQ-Enterprise*" has been developed by QEL for Chinese manufacturing industries, which is a platform with general-purpose quality functions and toolkits. A three-dimensional quality data integration model and a flexible architecture in *QQ-Enterprise* are proposed in this paper. Success in implementing of *QQ-Enterprise* proved that QEL technical approach is feasible, efficient and reliable.

Key words: Quality, Integrated quality information system (IQIS), Integrated model, Flexible system architecture, Chinese manufacturing industry.

1. INTRODUCTION

Quality management drew a great of interests in manufacturing, service, government and education, as well as in many countries around the world. There is no exception in China manufacturing industries. China has been opening door to global market since 1970's, and market-oriented economy encouraged Chinese manufacturers adopted technologies, managing experiences and know-how. The substantial actions have been taken in improving product quality, shifting focus from production operation to product process, transiting from quality control towards total quality

Please use the following format when citing this chapter:
Tang, Xiaoqing, Duan, Guijiang, 2006, in International Federation for Information Processing (IFIP), Volume 207, Knowledge Enterprise: Intelligent Strategies In Product Design, Manufacturing, and Management, eds. K. Wang, Kovacs G., Wozny M., Fang M., (Boston: Springer), pp. 54-61.

management (TQM), developing quality management, paying more attention in integrating ISO9000 into manufacturing process and enterprise business.

Quality data and information have been considering as an important resource to the success in quality management. Quality Engineering Laboratory (QEL) of Beihang University (original named Beijing University of Aeronautics and Astronautics) has been realized a need and focusing on the research of quality data integration in quality management since early 1990's.

To clear the concerns of Chinese manufacturing enterprises, QEL made an investigation in Shenzhen, an active manufacturing city in southern China, which included 35 questionnaires survey, covering state-owned large-size and middle-size manufacturing enterprises, joint-venture companies, and small-size or medium-size private-owned companies, and involving electronic products, electrical equipment, telecommunication, mechanical industries. The survey focused on three aspects, enterprises' attitudes and needs to QIS (Quality Information System), and current situation and practices of QIS, and problems and obstacles in implementing QIS. The Survey said: 100% were positive answers – manufacturer needs QIS technology.

On the basis of the result of Survey, QEL started research with methodology for IQIS (Integrated Quality Information System), and modeled a system framework, and developed an Integrated Quality Information System (IQIS) – "*QQ-Enterprise*" for Chinese manufacturing industries.

QQ-Enterprise endues to assist manufacturing enterprises to manage their quality data in more efficient way, and to strongly support the quality system implementation in manufacturing. QEL has been encouraging Chinese manufacturers to implementing *QQ-Enterprise* for years. *QQ-Enterprise* is playing an important role as a quality information processing platform in improving quality system operation in the enterprises. And it is proving QEL technological approaches feasible and successful. Based on *QQ-Enterprise* platform, more than 60 Chinese manufacturing companies have established their integrated quality information systems since the middle of 1990's. For example, a medium-size automobile accessories manufacturer in northern China had set up their IQIS in 3 months with 12 *QQ-Enterprise* modules including *QQ-QFD, QQ-CAIP, QQ-Inspection, QQ-Purchase, QQ-SPC, QQ-Analysis, QQ-Report, QQ- Improvement, QQ-Assessment, QQ-Service, QQ-Document, QQ-Gauge, QQ-Audit* (as in Table 1). And *QQ-Admin* was configuring to suit the company existing quality assurance system. The tailored *QQ-Enterprise* covered the major activities in company-wide quality assurance and management. The quality issues are managed in more efficient way and quality data was flowing in a more integrated way in the company.

2. INTEGRATED QUALITY MODEL IN MANUFACTURING ENTERPRISE

A common sense on quality management is covering the product lifecycle. It means management of quality data should be suited to company-wide functions and integrated all processes of product.

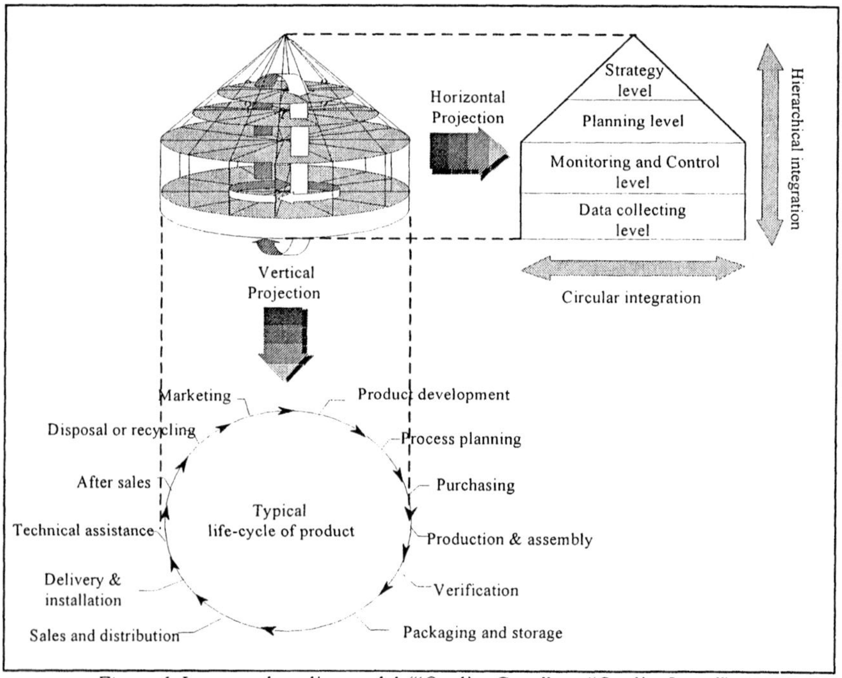

Figure 1. Integrated quality model ("Quality Cone" vs. "Quality Loop").

In fighting for the fiercely competitive global market and increasingly complex manufacturing environment, efficient and accurate information processing means more to quality than ever before. Ross pointed out that accurate and timely information permits purposeful planning and decision-making, and drives quality and technological resources (Ross 1991). Similarly, Chang considered that implementing TQC (total quality control) philosophy in a manufacturing organization would involve the management, analysis, interface and coordination of information for all phases of the manufacturing cycle. Information integration is vital to continuous quality improvements in TQC environment. Integrated information system is the challenges ahead of many manufacturing organizations as they continue to improve their competitiveness (Chang, 1990). Rzeznik presented a vivid evidence to show the benefits of quality information integration. In his example, owning to timely communication and quality data sharing between shop floor operators and manufacturing engineer through Internet, inferior products are prevented from being built. Company could thus save the money and reduces production time (Rzeznik 2000).

To present quality integration, a 3-dimension model named "Quality Cone" is constructed based on the 2-dimension model "Quality Loop" in ISO9000 (Figure 1). Two individuals but cross-related integration dimensions are defined in this model. The first one is "circular integration" which covers from product life-cycle, i.e. "Quality Loop". Another integration dimension is "hierarchical integration", which is a pyramidal architecture in manufacturing organizations.

2.1 Circular integration

Feigenbaum indicated, "the underlying principle of the total quality view, and its basic difference from all other concepts, is that to provide genuine effectiveness, control must start with identification of customer quality requirements and end only when the product has been placed in the bands of a customer who remains satisfied" (Feigenbaum 1983). In ISO9000, the concept of quality in life cycle is formalized as a "Quality Loop" which covers all activities affecting quality. In "Quality Loop", quality data keeps flowing along with the processes and activities, and quality data would be shared by the related processes and integrated in product lifecycle.

2.2 Hierarchical integration

Another integration dimension is the "hierarchical integration". Corresponding to "Quality Loop", there is also a vertical integration joining the vertical layers in manufacturing enterprises. The instructions and plans are relayed up-down, as well as the quality data are gathered along the quality loop, and reported bottom-up. As moving up hierarchically, quality data is aggregated, classified, filtered and processed. The hierarchical integration will endue a vertical data flow through quality management layers, and be bridged the "gap" between different layers.

The circular integration and hierarchical integration are not isolated but interrelated. The circular integration is inter-linked by hierarchical integration. Quality data integration at both circular and hierarchical dimensions will establish a tri-dimensional quality information framework.

"Quality Cone" model will ensure to send the right data to the right point at the right time.

3. FLEXIBLE ARCHITECTURE OF IQIS

As known, manufacturing enterprises are challenged by the changing global market and continuously business process reengineering (BPR). So, Integrated Quality Information System (IQIS) has to face the changing fact, and quickly reconfigured to keep pace with the dynamic environment and enterprise business process reengineering. Based on "Quality Cone" model, a flexible system infrastructure has been proposed, a general-purpose IQIS developing toolkit named *QQ-Enterprise* has been developed.

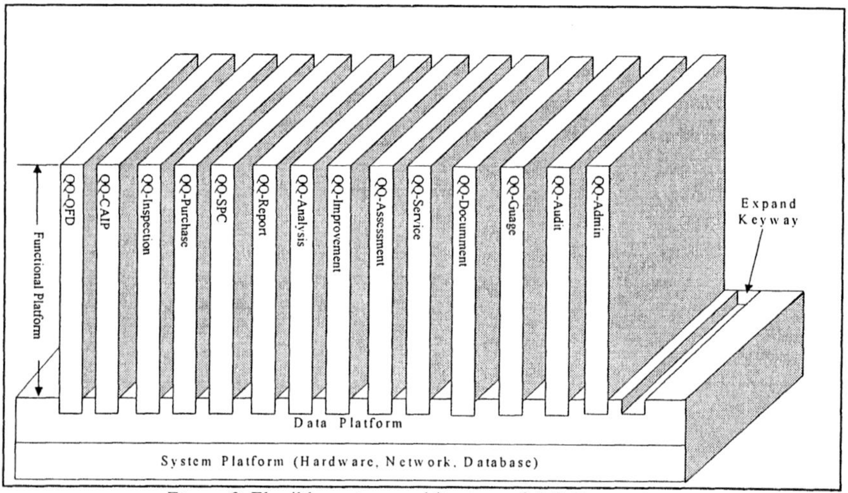

Figure 2. Flexible system architecture of *QQ-Enterprise*.

QQ-Enterprise is constructed in a flexible architecture. As shown in Figure 2, *QQ-Enterprise* is organized in three levels, system platform, data platform and functional platform. System Platform embodies network and DBMS (Database Management System), serving as a supporting platform. Data Platform provides service of *QQ-Enterprise* fundamental data, such as BOM, quality managing organization, process plans, standards and specifications, etc, which is in fact an integration platform of *QQ-Enterprise*. Functional Platform was constructed as a slot-structure, consisting of a set of functional modules (listed in Table 1), which cover all of quality activities in product lifecycle. In advantages of modularization, these functional modules can be "assembled" together, so as to ensure high flexibility.

Database in *QQ-Enterprise* is able to acquire, collect, categorize, store, analyze and manipulate quality data. The quality data in *QQ-Enterprise* is available at any time, any points, and to any person who has permit to access. *QQ-Enterprise* is a platform for quality data sharing in internal and external enterprise environment. *QQ-Enterprise* can communicate with CAD/CAM, ERP, CRM, SCM, etc. through reconfigurable interfaces. *QQ-Enterprise* can be reconfigured to suit BPR of enterprise. An open architecture with the development toolsets seems to be necessary.

Data Platform provides service to three levels – data collecting level, process monitoring and controlling level, and planning level. *QQ-Enterprise* Database was well organized to data logical relationship. Based on "Quality Cone" model, the quality data in different levels is effectively integrated and transferred bottom-up and top-down.

Table 1. List of functional modules in *QQ-Enterprise*

Function classification	Module name	Function description
Quality plan	*QQ-QFD*	Quality function deployment
	QQ-CAIP	Aiding to generate inspection plan
Quality control	*QQ-Inspection*	Capturing inspection data from production processes
	QQ-Purchase	Capturing inspection data of outsourcing goods
	QQ-SPC	A series of statistical process control tools
	QQ-Analysis	Analyzing tools to find out "quality bottleneck"
	QQ--Report	A flexible quality report generator
Quality Improve	*QQ- Improvement*	A set of quality improvement tools, such as 7 QM tools
Quality Assessment	*QQ-Assessment*	An index assessment system
Quality Management	*QQ-Service*	Tools for dealing with quality problems after-sale
	QQ-Document	Archive management and search the quality documents
	QQ-Gauge	Gauging management functions
	QQ-Audit	A toolkit to for quality system audit
System Management	*QQ-Admin*	Administrate and supervise the operation of *QQ-Enterprise*

4. IMPLEMENTING *QQ-ENTERPRISE*

Implementation of *QQ-Enterprise* is concerned with both technical and non-technical issues. *QQ-Enterprise* is employed to improve quality data processing but not to be able to solve all problems in enterprises. The success in implementing *QQ-Enterprise* requires Chinese manufacturers to adapt the existing quality system to an IQIS-featured quality system. Our experiences revealed that a thoroughly business process re-engineering (BPR) and a well-scheduled implementation guideline are effective and efficient.

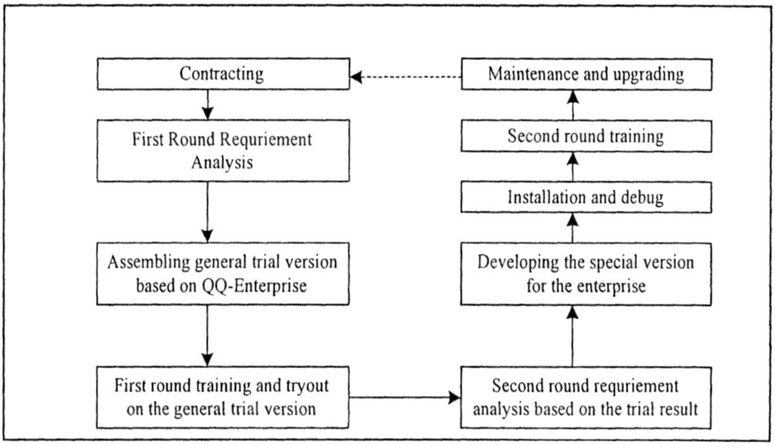

Figure 3. Implementing template for *QQ-Enterprise* in Chinese manufacturer

A *QQ-Enterprise* implementation template, as shown in Figure 3, has been planned by QEL according to China's manufacturing circumstance, which has been proved to be feasible and useful in pilot run of *QQ-Enterprise* in dozens of Chinese manufacturing companies. An initial system could be quickly configured according to first round requirements acquired from the user enterprise. The guideline will facilitate an effective discussion between *QQ-Enterprise* developers and users. With the thorough discussions, the second round requirements analysis becomes more accurate and actual. By expanding the initial system a customized system will be finally constructed. This template has been proved success in practical in China. The average project cycle can be reduced to 3-6 month; meanwhile the reliability of stability of *QQ-Enterprise* user systems would be improved.

5. CONCLUSION

IQIS approach can improve quality management for manufacturers. The survey and experience of authors reveals that, on the journey to merging into global economy for Chinese manufacturers, the needs for IQIS is increasing rapidly. Yet, their technical and managerial foundations for IQIS implementation have not been well prepared. Successfully implementing IQIS requires more efforts.

Quality data should flow in two-dimensional ways in a manufacturer, one is circular flow along with product lifecycle ("Quality Loop") and another is vertical flow across the hierarchical layers of an enterprise. A 3D quality information integration model ("Quality Cone") is aiming to endure the enterprise "3R" ability ---- "send the Right data to the Right point at the Right time".

Flexibility is important to IQIS. In global market competition, an enterprise has to improve its product and business process to suit a more dynamic environment, and improve their data accessing ability, dynamic data sharing ability and expanding ability. A flexible architecture of IQIS can support enterprises to reconfiguration and reengineering.

An integration model and flexible architecture are successfully applied in developing of *QQ-Enterprise*, which is a general purpose IQIS system designed for Chinese manufacturers. Implementation template for *QQ-Enterprise* practice was proved more efficient and reliable.

6. REFERENCES

1. Castle J A , (1998), New methodologies for integrated quality management, *TQM Magazine*, Vol.10, pp. 83-88.
2. Chang L C, (1990), Computer-Integrated Quality Information System, *Proceedings of 1990 IEEE International Engineering Management Conference, USA*, pp.85-88.
3. Duan G J and Tang X Q , (1999), Quality Information System In Dynamic Manufacturing Environment, *Computer Aided Manufacturing system (CIMS)*, Vol.5, pp.44-48.
4. Reimann M D and Sarkis J , (1999), An architecture for integrated automated quality control, *Journal of manufacturing systems*, Vol.12, pp.42-48.
5. Ross, D. F., (1991), Aligning the organization for world-class manufacturing, *Production and Inventory Management Journal, Second Quarter*, pp.22-26.
6. Rzeznik, T, (2000), Get more from collected data, *Quality Management, May 2000*, pp.76-81.
7. Tang X Q , (2001), Information System supporting Enterprise Quality System, *China Quality, 2001(2)*, pp. 35-37.
8. Tang X Q and LU Q L , (2002), Intranet/Extranet/Internet Based Quality Information Management System in Expanded Enterprise, *International Journal of Advanced Manufacturing Technology*, Vol.20, pp.853-858.
9. Wang X C and Tang X Q, (2001), Framework of Quality Management System for Virtual Enterprise, *Manufacturing and Automation*, Vol.23, pp.7-10.

AGENT-BASED FRAMEWORK FOR MODELLING OF ORGANIZATION AND PERSONAL KNOWLEDGE FROM KNOWLEDGE MANAGEMENT PERSPECTIVE

Janis Grundspenkis

Riga Technical University, Faculty of Computer Science and Information Technology, Department of Systems Theory and Design, 1 Kalku Street, LV 1658, Riga, Latvia, E-mail: Janis.Grundspenkis@cs.rtu.lv

Abstract: The paper tries to bridge gap between knowledge management and artificial intelligence approaches proposing agent-based framework for modelling organization and personal knowledge. The perspective of knowledge management is chosen to develop two conceptual models – one interprets organization as an intelligent agent, another models an agent-based environment of the knowledge worker for personal and organizational knowledge management support.

Key words: knowledge worker; organization's knowledge management system; personal knowledge management; intelligent agents; multi-agent system.

1. INTRODUCTION

Nowadays in the information age the organizational networks has become highly variable while their environments increasingly dynamic. As a consequence, organizations should react adequately, transform their infrastructure, interpret non-standardized information for problem solving and decision making, as well as change their management strategies in order to be competitive in the rapidly changing environment. Usually there are a lot of information and knowledge within organizations, but at the same time many organizations, in particular, service organizations are "information rich and knowledge poor". The information and knowledge assets, often called

Please use the following format when citing this chapter:

Grundspenkis, Janis, 2006, in International Federation for Information Processing (IFIP), Volume 207, Knowledge Enterprise: Intelligent Strategies In Product Design, Manufacturing, and Management, eds. K. Wang, Kovacs G., Wozny M., Fang M., (Boston: Springer), pp. 62-70.

an intellectual capital (knowledge that can be converted into value) make a great potential for organizations if utilized well (Apshvalka and Grundspenkis, 2003).

Knowledge management (KM) has become a new way of capturing and efficiently managing an organization's intellectual capital or, in other words, full past experience, skills and knowledge that is relevant for more effective performance in future. The main goal of knowledge management systems (KMS) is to provide knowledge creation (development, acquisition, inference, generation), storage (representation, preservation), aggregation (creation of meta-knowledge), use/reuse (access, analysis, application), and transfer (distribution, sharing), i.e., to support an effective knowledge flow.

KM tools and techniques afford an effective technological solution for acquisition, preservation and use of organization's knowledge. Typically it practices converting information to knowledge and connecting people. In principle, KM tools may be supported by information technology (IT) infrastructure and/or artificial intelligence (AI) techniques. Information management tools, for instance, data warehouses, data search engines, data modeling and visualization tools, etc., allow to generate, store, access, and analyze data. KM tools, such as, knowledge flow enablers, knowledge navigation systems, corporate memories, knowledge repositories and libraries, etc., in their turn, allow to develop, combine, distribute and secure knowledge. It is obvious, that a great part of functionality of both types of tools may be effectively supported by AI techniques. At the same time, it is needed to stress that the role of AI is underestimated and frequently even ignored by KM professionals. Analysis of a great number of publications shows that there is a difference of opinions even on fundamental terms, such as, "knowledge", "knowledge representation", "knowledge processing", etc. (Grundspenkis, 2001).

In this paper we try to bridge gaps between KM and AI. First, we suggest to interpret an organization in terms of intelligent agents. Second, we consider how intelligent agent paradigm may be integrated with the KMS, and develop a conceptual model of organization as multi-agent and knowledge management system. Finally, we describe a model of an agent-based environment of the knowledge worker that is considered to be a structure that can support the sharing and reuse of individual knowledge worker's knowledge (framework for personal KM), as well as to support an enterprise-wide knowledge, experience and lessons learnt.

The final goal of the research, presented in this paper, is to develop a model of an agent-based intelligent KMS and to implement this model into a working prototype.

2. ORGANIZATIONS AS INTELLIGENT AGENTS

From the systems theory point of view, any type of organization may be considered as a set of various objects together with relationships between them. In other words, organizations are systems which components are active or passive objects (Grundspenkis and Kirikova, 2005). Agents, in turn, may be natural, e.g., humans, or artificial ones, such as, software agents and robots. Artificial agents are acting within a real environment (robots) or within a virtual environment, that is, cyberspace (robots and software agents). All agents are called knowledge workers whose decisions effect their environment, which could consist of other agents and/or passive objects, for instance, other types of software and/or hardware that include also control devices. Environment entities can be local to the agent (the same platform or machine on which agent resides) or remote, if agent is connected via some type of network with other objects (Knapik and Johnson, 1998).

A wide variety of organizations considered as collections of active objects, i.e., agents or knowledge workers, and passive objects, allow to predict that it will be impossible to develop an effective general purpose KMS usable for all classes of organizations. At the same time, the role of KM is steadily growing, particularly for organizations operating in rapidly changing environments. Thus, new solutions appear, new technologies are introduced, and new methodologies are developed. In (Grundspenkis and Kirikova, 2005) organization's knowledge life cycle is represented as an organization's knowledge space that is organized form of data, information and knowledge captured in past, and used at present and in the future to get additional value out of them.

Each intelligent organization is trying to reach this goal autonomously making rational decisions and taking the best possible actions. So, the interpretation of an intelligent organization as a whole using the concept of an intelligent agent is quite obvious. For details of intelligent agents, their programs and architecture see (Russell and Norvig, 2003). Intelligent organization like an intelligent agent perceives the current state of the environment, using its detectors (sensors) for data, information and knowledge acquisition. The knowledge about the current state and the goal state is used to determine actions that through effectors will be applied in the organization's environment. This output is determined on the basis of percepts and knowledge captured in the knowledge space. The interpretation of an organization in terms of intelligent agents is shown in Fig. 1.

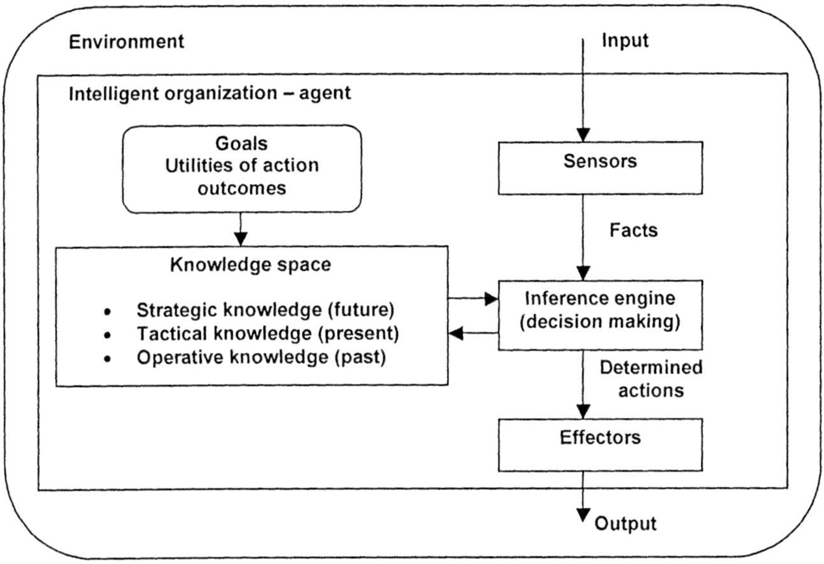

Figure 1. Organization as an intelligent agent

There are several main activities of organizations as intelligent agents that must be performed to build their intellectual capital. First, organizations must perceive and identify intellectual values, which are presented in the environment, as well as, inside the organizations themselves. Second, they must evaluate whether the identified intellectual values are sufficient for reaching the predefined goals, realizing business processes and rising the competitiveness. Third, the organizations must create an additional value from their intellectual capital by choosing more rational actions. The maintenance of the knowledge flow provided by the KMS is the vehicle for generation of new additional value from the organization's intellectual capital.

From intelligent organization – agent point of view KM means knowledge acquisition, processing, and use for rational decision making and choosing the best actions, and generation of new knowledge, i.e., systematic management of the intellectual capital of an organization. The main functions of the KMS from the agent viewpoint are the same that characterize intelligent agents in general:

- Detection of information and knowledge (the function of sensors)
- Storage of detected information and knowledge (the function of knowledge space that plays the role of memory)
- Decision making (the function of inference engine)

- Retrieval and visualization of knowledge

All organization's business processes are supported by intelligent organization – agent activities, such as decision making and acting. Intelligent organizations – agents generate alternatives and model possible solutions, which are the results of applying the chosen actions. Finally, they make decisions knowing goals and utilities of the predicted outcomes of actions.

3. ORGANIZATION AND PERSONAL KNOWLEDGE MANAGEMENT SUPPORTED BY MULTI-AGENT SYSTEM

More detailed examination of business process support from the inside reveals that managers, researchers and assistants, advisers, secretaries, etc. are employed as searchers, schedulers, and planners to do the diverse mundane tasks. Let consider perspectives of intelligent support of these activities by communities of intelligent agents, or so called multi-agent systems (Knapik and Johnson, 1998; Woodridge, 2002). This "inside look" on intelligent organization – agent is shown in Fig. 2.

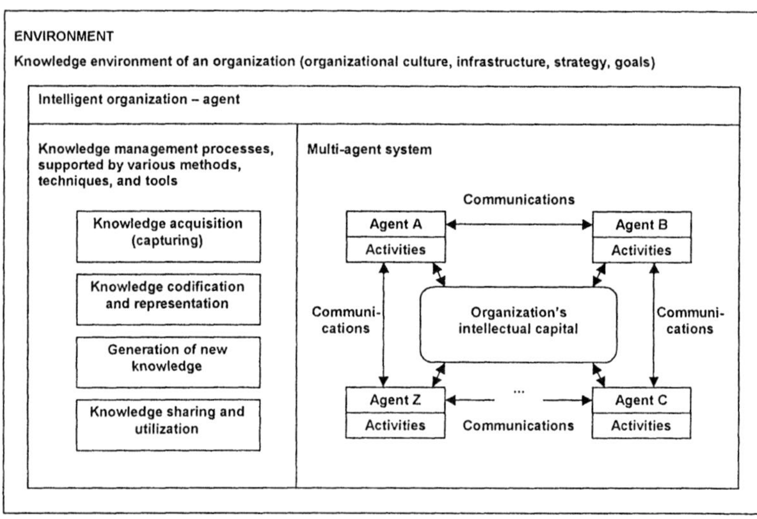

Figure 2. Organization's knowledge management supported by multi-agent system

The conceptual model of an intelligent organization's knowledge management system (OKMS) which is based on an intelligent agent paradigm and is under the development at the present moment is described in (Grundspenkis, 2003). The basic idea behind the conceptual model is that the OKMS should operate like the human brain and fulfill the following basic functions: knowledge acquisition through sensors, knowledge formalization, representation and storage in the knowledge space (memory), knowledge inference, sharing, and utilizing. As it is shown in Fig. 2, the conceptual model consists from two main parts: an intelligent organization as a multi-agent system for business process support and a KMS. The conceptual model has three layers called an "engine room", a structural layer, and a "cooperation platform". The "engine room" is an integrated set of technologies, hardware, and software to provide knowledge acquisition, storage, processing, retrieval, and representation. The purpose of the structural layer is to identify intellectual resources of the organization, and to organize knowledge to make it easily accessible and applicable. A "cooperation platform" is the physical and/or virtual environment where organization's intelligent agents may communicate with each other for effective knowledge sharing and distribution to achieve the business process goals. A "cooperation platform" maintains such components as video conferencing, chat rooms, electronic white boards, and other tools for cooperative work (groupware).

In (Grundspenkis, 2003) the potential already manifested by intelligent agents and multi-agent systems for KM is discussed and three groups of agents are marked. First, agents that may promote the KM and may be used as organization's common vehicle of the "engine room". Nowadays there already exist a number of agents that may be used as KM promoters, for example, network agents, database agents, connection and access agents, network software distribution agents, and intelligent Web agents (Knapik and Johnson, 1998; *Web Intelligence*, 2003). Second, among agents that provide communications, such agents as messaging agents, collaborative agents, cooperative agents, communication facilitators, team agents, and others may be listed (Knapik and Johnson, 1998; Ellis and Wainer, 2002). Third, so called personal agents are search, filtering, work-flow, and assistant agents (Knapik and Johnson, 1998). Their primary purpose is to support the knowledge work of each staff member of intelligent organization. Usually the term "knowledge worker" is used when the role and activities of staff members of modern intelligent organizations are discussed. The concept of the agent-based environment of the knowledge worker was proposed in (Grundspenkis, 2003). According to this concept the knowledge worker is embedded into a multi-agent system that consists of three circles of agents, namely personal agents, communication agents, and agents for access to external systems (network, databases, etc.) as it is shown in Fig. 3.

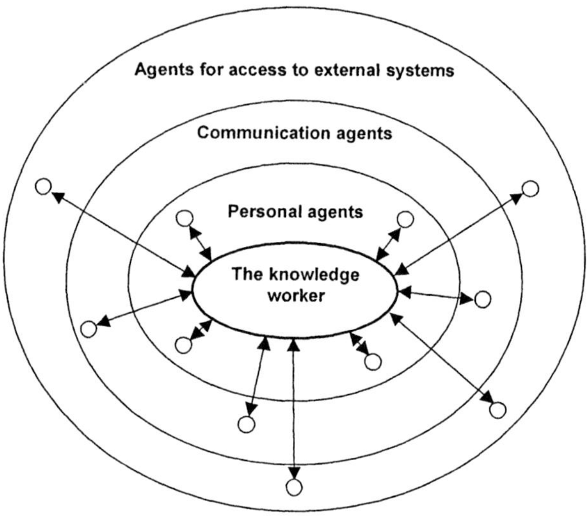

Figure 3. An agent-based environment of the knowledge worker

Knowledge management works best when knowledge workers take the initiative and responsibility for what they know, don't know, and need to know. Doing so not only makes the individual knowledge worker more valuable to the organization, but it also enhances the value of intellectual capital of an organization. The concept of personal knowledge management (PKM) emerges in this context (Apshvalka and Grundspenkis, 2005). The objectives of PKM extend further than giving employees access to intranets, systems, and standards. The final goal is to make knowledge workers better in capturing, sharing, and using knowledge, and maximizing their personal effectiveness in the social and relationship-building of their jobs (*KM Magazine*).

Concept of PKM is not a new concept, however, it is not notably popular in the field of KM. PKM is defined as a collection of processes that an individual needs to carry out in order to gather, classify, store, search, and retrieve knowledge in his/her daily activities (Tsui, 2002). PKM is considered from different perspectives, for instance, some authors focus on attempts how to utilize a computer to help the knowledge worker to manage his/her knowledge, others focus on problem-solving skills or arranging ideas (Apshvalka and Grundspenkis, 2005). To get a complete understanding of PKM, it is necessary to put all perspectives together and look at this kind of KM as a process of managing personal information and knowledge, and arranging ideas to be able to solve problems skillfully: to see problems, set goals, generate alternatives, make decisions, perform actions, learn from experience, etc. So, PKM is integrated discipline that integrates many aspects, in particular, psychological, social, and technological, and many perspectives from different fields.

Nowadays technologies can help individuals to make decisions and perform actions. In complex cases more sophisticated support is needed to make knowledge worker's activities really effective. This is where such technologies as intelligent agent and multi-agent systems should help to find and generate needed information and knowledge. Due to the importance of social aspects of PKM, multi-agent systems will be perspective for information and knowledge acquisition, storage, processing, and distribution.

In PKM the concept of knowledge worker is used twofold (Apshvalka and Grundspenkis, 2005). First, the knowledge worker possesses his/her individual knowledge that need to be managed because human agents possess tacit knowledge. Second, to provide effective knowledge management, the human agent's activities should be supported by software agents for management of explicit knowledge. In this case the concept of the agent-based environment described above and shown in Fig. 3 may be used, too.

4. CONCLUSIONS

This paper has identified the possible role of intelligent agents and multi-agent systems for organization and personal knowledge management support. Several conceptual models are discussed. First, organization as a whole is interpreted as an intelligent agent, which uses its own knowledge space for knowledge capturing. Second, some aspects how multi-agent systems may support organization and personal knowledge management are discussed.

Regardless that a lot of work should be done to achieve considerable results in implementation of the proposed framework, the potential of using intelligent agents and multi-agent systems to develop more intelligent KMS even now is rather high. We hope that conceptual model of an intelligent OKMS together with agent based environment of the knowledge worker will serve as a platform for researchers to investigate directions towards development of more and more intelligent systems for organization and personal knowledge management.

5. REFERENCES

1. Apshvalka, D. and Grundspenkis, J., (2003), Making organizations to act

more intelligently in the framework of the organizational knowledge management system, in: *Scientific Proceedings of Riga Technical University, 5th Series Computer Science, Applied Computer Systems,* Vol. 17, (RTU Publishing, Riga), pp. 72–82.
2. Apshvalka, D. and Grundspenkis, J., (2005), Personal knowledge management and intelligent agent perspective, in: *Proceedings of the 14th International Conference on Information Systems Development Pre-Conference-ISD 2005. Karlstad, Sweden, 14-17 August, 2005,* edited by A. G. Nilsson, et. at. (Karlstad University Studies, Karlstad, Sweden), pp. 219–230.
3. Ellis, C. and Wainer, J., (2002), Groupware and computer supported cooperative work, in: *Multiagent Systems. A Modern Approach to Distributed Artificial Intelligence,* G. Waiss, ed. (MIT Press, Massachusetts), pp. 425–458.
4. Grundspenkis, J., (2001), Concepts of organizations, intelligent agents, knowledge, learning and memories: towards an inter-disciplinary knowledge management, in: *Applied Computational Intelligence to Engineering and Business,* K. Wang, J. Grundspenkis, and A. Yeerofeev, eds. (Riga Technical University Publishing, Riga, Latvia), pp. 172–191.
5. Grundspenkis, J., (2003), Development of hybrid intelligent systems: integration of structural modelling, intelligent agents and knowledge management techniques, *Scientific Proceedings of Riga Technical University, 5th series Computer Science, Applied Computer Systems,* Vol. 17, (RTU Publishing, Riga), pp. 7–30.
6. Grundspenkis, J. and Kirikova, M., (2005), Impact of the intelligent agent paradigm on knowledge management, in: *Intelligent Knowledge-Based Systems*, C. T. Leondes, ed., Vol. 1: Knowledge-Based Systems (Kluwer Academic Publishers, Boston, Dordrecht, London).
7. Knapik, M. and Johnson, J., (1998), *Developing Intelligent Agents for Distributed Systems* (McGraw Hill, New York).
8. KM Magazine. Personal Knowledge Management, Vol. 7, Issue 7; www.kmmagazine.com.
9. Russell, S. J. and Norvig, P., (2003), *Artificial Intelligence. A Modern Approach,* (Pearson Education, Upper Saddle River, New Jersey).
10. Tsui, E., (2002), Technologies for Personal and Peer-to-Peer (P2P) Knowledge Management, CSC Leading Edge Forum Technology Grant Report.
11. *Web Intelligence,* (2003), edited by N. Zhong, J. Liu, and Y. Y. Yao, (Springer-Verlag, Berlin, Heidelberg).
12. Woodridge M., (2002), *An Introduction to Multiagent Systems* (John Wiley & Sons, Chichester, West Sussex, England).

EVALUATION MODEL OF MNES' KNOWLEDGE FLOW MANAGEMENT

Fengming Tao, Weidong Meng
College of Economics and Business Administration, Chongqing University, Chongqing, 400030. China, Email:taofengming@cqu.edu.cn

Abstract: With the brief introduction of knowledge and knowledge flow, this paper pointed out that knowledge flow management is essential for MNEs in such a global and information economy. Three elements affect the knowledge flow in MNEs and the index system is constructed based on these three elements. Evaluation model which is constructed with AHP and fuzzy set is built for MNEs to assess their knowledge flow management.

Key words: evaluation model; knowledge flow management; MNEs

1. INTRODUCTION

As Porter (1986: 17) observed, "we know more about the problems of becoming a multinational than about strategies for managing an established multinational", in today's global economy, if MNEs (multinational enterprises) want to be competitive as a whole, they must achieve a balance among the following four characteristics in knowledge management: the organizational ability to learn; the capacity to rapidly respond to the environmental changes; the ability to coordinate and integrate knowledge present in different locations; and the capacity to minimize costs in relation to competitors. Each one of the above requirements is essential but very difficult to complete. It is necessary to develop a knowledge strategy centrally managed, which leads the creation and application of strategic knowledge.

Please use the following format when citing this chapter:
Tao, Fengming, Meng, Weidong, 2006, in International Federation for Information Processing (IFIP), Volume 207, Knowledge Enterprise: Intelligent Strategies In Product Design, Manufacturing, and Management, eds. K. Wang, Kovacs G., Wozny M., Fang M., (Boston: Springer), pp. 71-78.

Globally distributed networks of subsidiaries constitute a potentially important source of competitive advantage for MNEs. By accessing the knowledge residing in these networks, MNEs can both exploit existing repositories of knowledge and combine these sources of knowledge to explore new issues. This argument, highlighting the potential importance of knowledge as a strategic resource, has brought the transfer of competence across units into focus as a central challenge for MNEs management. It has also triggered a considerable amount of research on factors influencing inter-unit knowledge transfer patterns within the differentiated MNEs. However, with certain notable exceptions (e.g., Gupta and Govindarajan, 2000), few efforts have been made to examine the influence of organizational mechanisms on knowledge flow within MNEs (Foss and Pedersen, 2002). In particular, there is a lack of research on the strategies that MNE headquarters may use to ensure that the competence of subsidiaries is transferred across different units. The question addressed in this paper is therefore: How do different factors impact on flows of knowledge within MNEs and how can we weight them and evaluate them?

2. KNOWLEDGE FLOWS IN MNES

2.1 Knowledge

Knowledge flows are understood as the aggregate volume of know-how and information transmitted per unit of time. Such a concept means to capture the overall amount of know-how and information transmitted between parent and subsidiaries and between subsidiaries themselves in all kinds of ways.

And there are two kinds of knowledge: explicit knowledge and tacit knowledge. Explicit knowledge is knowledge articulating and codifying in handbooks, computer programs, databases and training tools, among other elements, and this knowledge is transmissible. Tacit knowledge is personal, context specific and difficult to regularize. It includes cognitive element, which is mental patterns such as beliefs, points of views etc. that help individuals to perceive and define their environment. Organizations are considered to be a depositary of several types of knowledge (explicit and tacit) existing in different levels (individual, group, organizational and inter-organizational).

2.2 Knowledge flows

According to Mudambi (2002), this paper views knowledge flows through the source–target perspective. Each knowledge flow occurs between a source and a target along a channel. Knowledge flows are therefore taken to be node-specific and dyadic. Principally, four knowledge flows will be concerned.

1 Flows from the parent (and other MNE units) to the subsidiary: These flows from the parent to the subsidiary are the traditional flow, where the subsidiary exploits a home-base knowledge advantage.

2 Flows from subsidiary to parent (and other MNE units): High levels of these flows enable MNE headquarters to exploit local competencies and act as a knowledge intermediary or knowledge integrator.

3 Flows from host country to subsidiary: These flows consist of the subsidiary's learning, local competence exploitation, and local resource utilization.

4 Flows from subsidiary to host country: These flows are part of "spillovers". In the literature, spillovers have often been used to refer to flows both into and out of the firm. However, as our analysis is firm-centric rather than location- centric, here, spillovers only mean outflows from the subsidiary.

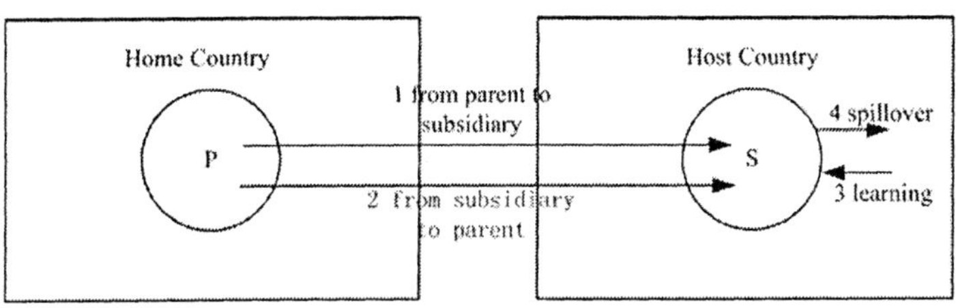

Figure 1 **Principal Knowledge Flows in MNEs**

3. IMPORTANCE OF KNOWLEDGE FLOW MANAGEMENT

Knowledge management is an organizational process that covers the creation or acquisition of knowledge, its combination, deployment, renovation, storage and transference of both intra- and inter- organizationally. Through adequate human resource management system architectures, the enterprise can support its knowledge management strategy. In this sense, especially the creation or acquisition of knowledge as well as its transfer are managed within the HR function.

The existence of a strong co-operative and collaborative culture is an important prerequisite for knowledge transfer between parent company and its subsidiaries. Without appropriate mechanisms to encourage co-operation, structured or technological interventions to facilitate knowledge transfer may not work. Establishing a collaborative and co-operative climate in an organization will improve knowledge transfer. Several works on international management and specifically on international human resource management suggest that an important aspect for MNEs is the degree in which management systems applied in the country of origin can answered back in a foreign country.

With all the discussion above, we can see that knowledge flow management is so important for MNEs that the directors of MNEs need to improve the efficiency of their knowledge flow management. Therefore, we should first of all build a scientific system to evaluate the knowledge flow management in MNE so that we can evaluate it and improve it. The following part is the brief introduction of the comprehensive evaluation model with AHP (Analysis Hierarchy Process) and fuzzy set.

4. INDEX SYSTEM OF THE EVALUATION MODEL

4.1 Hierarchies of Index

Since knowledge flow means knowledge flows between two or more units along a channel, we can evaluate knowledge flow management in MNEs through these three indexes: knowledge; units that send and receive knowledge; and the environment that knowledge flows happen. And the three indexes constitute the first hierarchy. That is, the first hierarchy is B_i, $i = 1, 2, 3$; and B_i is made up of the second hierarchies C_{ij}. $Bi = \{C_{ij}\}$. The following is the hierarchies of the index system.

4.2 Knowledge

Knowledge evolves through interactions between new knowledge and prior, related knowledge. Therefore, knowledge flows are kind of path dependence, and we ought to focus on knowledge stock and flow at the same time. Hereby, B_1 knowledge is composed of the following second hierarchies: size of MNE (C_{11}), history of MNE (C_{12}), relative economic level of MNE (C_{13}), and average annual investment in R&D in 5 years (C_{14})

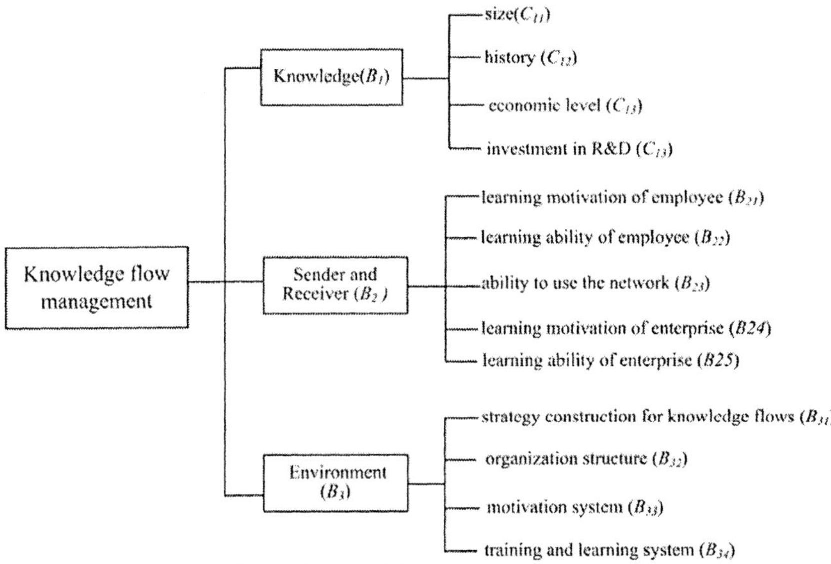

Figure 2 Hierarchies of the index system

4.3 Sender and receiver of knowledge

In nature, knowledge flow in enterprises is a process of mutual learning among individuals, groups and subsidiaries, hence, the learning abilities and motivation can somewhat decide the efficiency of learning. Hereby, B_2 sender and receiver is composed of the following second hierarchies: learning motivation of employee (B_{21}), learning ability of employee (B_{22}), ability to use the network (B_{23}), learning motivation of enterprise ($B24$), and learning ability of enterprise ($B25$).In nature, knowledge flow in enterprises is a process of mutual learning among individuals, groups and subsidiaries, hence, the learning abilities and motivation can somewhat decide the efficiency of learning.

4.4 Environment

Knowledge flow management is totally different from one enterprise to another, and of course different from one country to another. Therefore, MNEs can have an influence in the transferability of its knowledge through the design of the organization structure, business strategies and human resource management system with enough consideration of cultural difference. In this way, MNEs can create a better environment to have knowledge flow more efficiently. Hereby, B_3 environment is composed of the following second hierarchies: strategy construction for knowledge flows

(B_{31}), organization structure of MNE (B_{32}), motivation system (B_{33}), training and learning system (B_{34}).

5. EVALUATION MODEL WITH AHP AND FUZZY SET

5.1 Weight of index

Based on the index system above, we can fix the weight of indexes with AHP (Analysis Hierarchy Process). According to Seaty, we compare the indexes pairwise and quantify them on basis of stated scales and then construct the judge matrix *M*. This process should be implemented by experts. The Characteristic root of the judge matrix *M* is the weight coefficient of the indexes in the same hierarchy.

The weight of the second hierarchy *Wi* should be firstly found out. Take the four indexes belonged to Environment that knowledge flows happen (B_3) as the example, with several experts' suggestion, we can get the weight scores of the four indexes and then get the judge matrix.

Table 1 Example of weight scores of the second hierarchies

we / i / i	strategy construction	organization structure	motivation system	training and learning system
strategy construction	1	1/3	1/5	1/4
organization structure	3	1	4	4
motivation system	5	1/4	1	1/5
training & learning system	4	1/4	5	1

Therefore, the judge matrix is:

$$M=\begin{bmatrix} 1 & 1/3 & 1/5 & 1/4 \\ 3 & 1 & 4 & 4 \\ 5 & 1/4 & 1 & 1/5 \\ 4 & 1/4 & 5 & 1 \end{bmatrix}$$

5.2 Fuzzy comprehensive evaluations

For each index, there are five scales for experts to evaluate, that is,

V = {v_1, v_2, v_3 } = { very good, good, medium, bad, very bad}.

Then, based on experts' evaluation, we can get r_{ijk} which is the degree of membership. The implication of r_{ijk} is to what degree experts think index *Cij* can be judged as level. And *Ri* can be expressed as:

$$R_i = \begin{bmatrix} r_{i11} & r_{i12} & \cdots & r_{i1m} \\ r_{i21} & r_{i22} & \cdots & r_{i2m} \\ \cdots & \cdots & \cdots & \cdots \\ r_{in1} & r_{in1} & \cdots & r_{inm} \end{bmatrix}$$

n is the number of lines in the matrix and its implication is the number of indexes. And *m* is the number of rows in the matrix and its implication is the number of levels in *V*.

Finally, we can have the multi-level fuzzy comprehensive evaluation. According to the weight coefficient of the indexes *Wi* and the judge matrix *Ri*, we can calculate the fuzzy comprehensive evaluation sets of the second hierarchy: $Gi = Wi \cdot Ri = (g_{i1}, g_{i2}, g_{i3})$. In this expression, the implication of g_{ik} $(i = 1, 2, 3, 4, (5))$; $k = 1, 2, 3, 4, 5$) is the degree of membership of *Bi* to each level v_k ($k = 1, 2, 3, 4\ 5$) when assess the factors *Cij* belong to *Bi* in the second hierarchy.

With the result calculated from the above, we can have the fuzzy comprehensive evaluation of the first hierarchy based on the judge matrix of the second hierarchy. And the formula is as the following:

$$R = \begin{bmatrix} G_1 \\ G_2 \\ G_3 \\ G_4 \\ G_5 \end{bmatrix} = \begin{bmatrix} W_1 \cdot R_1 \\ W_2 \cdot R_2 \\ W_3 \cdot R_3 \\ W_4 \cdot R_4 \\ W_5 \cdot R_5 \end{bmatrix}$$

6. CONCLUSIONS

MNEs face important challenges. The enterprise must lean how to exploit its specific resources either acquired in the country of origin or in foreign markets; and they should know that the source of sustained competitive advantage takes place in the variety of skills and the diversity of knowledge, not in these resources homogeneity.

In MNEs, individual pieces of knowledge are embedded in an interconnected network of other pieces that provide an ecological context for changes in knowledge. Changes in some parts of the knowledge structure tend to induce changes in other, related or similar parts. Therefore, this paper builds a systematic evaluation model to test capacity of knowledge innovation and efficiency of knowledge transference in MNEs.

Through this model which is constructed with AHP and fuzzy set, MNEs can find out whether there exist any problems in knowledge flow management and dig out what lead to these problems. This evaluation model is comparatively scientific, but there are still some limits need to improve. Most of the indexes are subjective and might be totally different from one person to another. Therefore, a team of good and experience experts is the guarantee of the effect of this model.

7. REFERENCES

Foss, N.J. and Pedersen, T. 2002, Transferring knowledge in MNCs: the role of sources of subsidiary knowledge and organizational context, *Journal of International Management* 8(1): 1–19.

Frost, T. 2001, The geographic sources of foreign subsidiaries' innovations, *Strategic Management Journal* 22: 101–123.

Ghoshal, S.,&Nohria, N. 1989, Internal differentiation within multinational corporations. *Strategic Management Journal*. 10, 323-337.

Ghoshal, S., Khorine, H. and Szulanski, G. 1994, *Interunit communication in multinational corporations, Management Science*, 40(1): 96–110.

Gupta, A.K., & Govindarajan, V. 1991. Knowledge flows and the structure of control within multinational corporations. *Academy of Management Review*, in press.

Mudambi, R. 2002, Knowledge management in multinational firms, *Journal of International Management* 8(1): 1–9.

Nohria, N. and Ghoshal, S. 1994, Differentiated fit and shared values: alternatives for managing headquarters–subsidiary relations, *Strategic Management Journal* 15(6): 491–502.

Nonaka, I. and Takeuchi, H. 1995, *The Knowledge-Creating Company*, Oxford University Press: Oxford.

Osterloh, M. and Frey, B.S. 2000, Motivation, knowledge transfer and organizational forms, *Organization Science* 11(5): 538–550.

Porter, M.E. 1986, *Competition in global industries: a conceptual framework, competition in Global Industries*, Harvard Business School Press, Boston, MA: 15-20.

Torbiorn, I. 1985, The Structure of Managerial Roles in Cross-Cultural Settings, *International Studies of Management and Organization* 15(1): 52–74.

Van de Ven, A.H. and Polley, D. 1992, *Learning While Innovating*, Organization Science 3(1): 92–116.

White, R.E. and Poynter, T.A. 1985, The strategies of foreign subsidiaries, *International Studies of Management and Organization* 14(4): 91–106.

KNOWLEDGE BASED MANUFACTURINE SYSTEM

Gideon Halevi[1] and Kesheng Wang[2]
[1]Technion - Israel Institute of Technology; Technion City 32000, ISRAEL. [2]Department of Production and Quality Engineering, NTNU, N-7491, Norway. E-mail: halev@bezeqint.net and Kesheng.wang@ntnu.no

Abstract: Production management performance relies on data specified by engineers that are neither an economists nor production planners. Therefore, production planning and control becomes a complex task. This paper presents a method where engineer's task is not to make decisions but rather to prepare a "road map". Each user will generate routine that meets his needs at the time of need by using KBMS CAPP. Thereby increase dramatically manufacturing efficiency.

Key words: manufacturing; scheduling; capacity planning; shop floor control

1. INTRODUCTION

The manufacturing process is a dynamic; conditions are constantly changing and decisions have to be made within a short space in time. It is often preferable to have a decision on hand at the right moment than to seek the optimum decision without any time limit. The better the available relevant data at the right time, the better decision that will be reached. A computer is a tool that can be employed to narrow the gap between the conflictions demands of "time" and "decision". A computer system can store and manipulate large quantity of data in a short period of time. Therefore the computer was accepted enthusiastically by industry as a data processing center. Most computerized application used the methods and algorithms used by manual system, but performing them more often and quicker.

Please use the following format when citing this chapter:
Halevi, Gideon, Wang, Kesheng, 2006, in International Federation for Information Processing (IFIP), Volume 207, Knowledge Enterprise: Intelligent Strategies In Product Design, Manufacturing, and Management, eds. K. Wang, Kovacs G., Wozny M., Fang M., (Boston: Springer), pp. 79-88.

The real change in thinking and procedure in the manufacturing cycle was the MRP - material requiring planning / material resource planning. The objective of MRP is to plan the activities to be performed in order to meet the goals of the master production schedule.

The logic and mathematics upon which MRP is based is very simple. The gross requirements of the end product for each specific order is compared against on-hand and on-order quantities and then offset by the lead-time to generate information as to when assembly should be started. All items or subassemblies required for the assembly should be available on that date, in the required quantity. Thus, the above computation establishes the gross requirements for the lower level items. The same computation is repeated level by level throughout the entire product structure.

MRP [5] represents an integrated communication and decision support system that supports management total manufacturing business. The use of optimizing techniques drawn from operation research and management science was adapted by MRP. One significant reason why MRP was the technique that was adapted was that it made use of the computer's ability to centrally store and provide access to the large body of information that seemed necessary to run a company. It helped coordinate the activities of various functions in the manufacturing firm such as engineering, production and materials, purchasing, inventory, etc.

The attraction of MRP lay not only in its role as decision-making support, but more important in its integrative role within the manufacturing organization. Several areas of the plan are affected by MRP: production planning and scheduling, quality control, accounting, first line supervision, production labor, and the use to develop plant-wide scheduling.

A closed total working system can be based on MRP, starting with customer order, following by purchasing and sub-contractors order, shop floor scheduling and control, inventory management and control, bookkeeping and accounting. All are dependent activities such that each one is validated by status of the manufacturing system.

2. INTEGRATED MRP - IMRP

Integrated MRP is a working strategy where the computer is a working tool and not merely a data retrieval system. [3] It is an innovative strategy that covers all manufacturing stages to one comprehensive unique system. The logic of the system is that there is an objective, i.e. customer order that must be delivered. All activities of the manufacturing cycle are predicted and focused to meet that objective in the most efficient way and they can be performed automatically by using company databases, without human interventions.

IMRP prescribe what should be purchased, what should be subcontracted and what should be processed in shop. Using company databases a specification for purchasing the item on the list are prepared and sent to preferred suppliers for RFQ. Based on algorithm and suppliers rating data a decision is made.

The logic of the inventory system is that issue and receipt are all predictable. No components are received unless they were ordered, no items are issue unless they serve a specific order, or are needed for assembly or processing of a specific order. No shop-floor activity is performed unless it was scheduled by IMRP. No payment is done unless purchasing order was accepted and the payment terms are met.

All activities validate one another and action decisions are made automatically without manual introversions.

It sound as a perfect system, but unfortunately, IMRP was not a success story. Probably because it was introduced before the IT technology and the data collection systems were available to support its requirements.

Therefore manufacturing turned to other venues. One philosophy believes that production planning and control is very complex: therefore the only way to make such a system effective is to simplify it [4]. This trend proposed systems such as: Just in Time, Lean manufacturing, Agile manufacturing, kanabn, Kaizen, Grourp Technology and cellular manufacturing.

Another philosophy believes that production planning and control is very complex; therefore, there is no chance of developing a system that will solve production problems. Hence, the role of computers should be limited to supplying data while the decisions should be made by peole [1]. Several excellent systems are available. These manufacturing philosophies do not pretend to be integrated systems, and as people make decisions, it is not the SYSTEM fault if a disruption occurs, they are the decision making responsibility.

3. MANUFACTURING DATA

No matter what manufacturing strategy is employed it is based on data store in company computer databases. Databases such as: Order file; Product structure; Routing file; Inventory file; Resource file; Production feedback data file.

Data creation and storage is the responsibility of the appropriate discipline. For example: Product structure lists all items composing the product and the links between them. This data is created and stored by engineering department. Routing file specifies how each item and assembly is to be produced is created and stored by process planning department.

Each of the appropriate departments will probably consider many options and optimize its decision before introducing the data to the company database.

The data in those files is considered, by its various users, as a reliable, accurate and being what is stated beyond any possibility of doubt. It assumes that it is based on optimum values, validated in the process of inputting the data to the files.

Theoretical production planning and scheduling is actually very simple task. The plant gets orders which define the product, the quantity and delivery dates. The resources of the plants are known, the product bill of material is known. The task of production scheduling is to make sure that the orders will be ready on time, that's all. IMRP and other manufacturing strategy can generate a good operation list. The complexity appears when disruptions encounters, disruptions in the planning stage and in shop floor processing. In many cases disruptions appears due to the nature of the manufacturing data in the company databases.

Although the data is the optimum from the discipline that generated it, it must be remember that it is optimized by criteria of optimization of that of the specific discipline. Each discipline has different optimization criteria. For example, the design criteria are not the same as that of processing, and not that of shop floor control. The criteria of sales are not the same as those of the production management.

The probability of having overall company optimization, by using individual discipline optimization is very low and almost impossible. Especially when the company overall optimization is different than any discipline optimization, it is profit. But none of the individual disciplines, working with perfect local optimization assures company manufacturing meeting its objectives. Each discipline criteria of optimization must match the objective of the discipline. The main objectives are not negotiable they must be met; If they are not met the task solution is not acceptable. However there are several secondary objectives such as: cost, weight, appearance, maintainability, environmental issues etc. that the optimization criterion is a compromise of all of them.

Compromise means that there are several alternative solutions and the one that is in the company database is one of many that is selected by the engineer assigned to the job. Furthermore, in process planning, for example, there are three criterion of optimization: minimum cost; maximum production; and maximum profit. The data does not specify which criterion was selected for the recommended routing.

Process planning in spite of its importance to the manufacturing cycle, is predominantly labor intensive, depending on experience, skill and intuition of human labor. Dependence on such methodologies often precludes a thorough analysis and optimization of the process and nearly always results in higher than necessary production cost, delays, errors and non-standardization of processes.

There is time interval between the data creation and the data application. Optimized data is introduced into the databank a long time before its uses; therefore it cannot consider the present state of production. That is the main cause that production management becomes a complex task.

In production planning and scheduling resource overload are created and bottlenecks occur and must be resolved. As the data used was the optimum one, the only way to resolve the situation is to decide priority rules or develop a very complex capacity planning program.

4. KNOWLEDGE BASE MANUFACTURING - KBMS

The complexity is not a must [6], it occurs due to the strategy taken; a knowledge based manufacturing strategy will make manufacturing a simple straight forward system.

The objectives of the system are to increase productivity and reduce manufacturing costs. These objectives are the same as those of many other systems. The difference lies in the approach and strategy employed [2]. The Knowledge based manufacturing strategy approach makes use of the following notions:

- There are infinite ways of meeting design objectives.
- In any design about 75% of the dimensions (geometric shape) are nonfunctional (fillers). These dimensions can vary considerably without affecting the design performance.
- There are infinite ways of producing a product.
- The cost and lead time required to produce a component are functions of the process used.
- Transfer of knowledge between disciplines working to produces a product, should not be by transferring decisions, but rather by transferring alternatives, ideas, options considered, reasoning etc.
- Company database should be "open" and available to all disciplines.

These notions are what any of us does in personal life. For example: If you wish to go from point A to point B you, study the map and plan the optimum route to take. This is the present time decision. However, at another time when you have to move the same way, say, at night you might change the route, in winter you probably will look for a route with maximum shelter from the rain. In summer you might choose a route that protects you from the sun. In spring time you might choose a route with nice view. Despite the decision if you run into disruptions, such as a blocked road (bottleneck), red traffic light or with a crowded you might decide that instead of waiting it is better to consult the road map and change the route in order to find a path with no obstacles. Such change is done at each junction. It might be a longer route but it will be faster. The original decision must not prevent one from adapting the new route.

Similar strategy can be applied to production planning and management. The knowledge based manufacturing system proposes to supply each manager with a "road map" which is stored in the company databases, and allow him to deviate from the original processing route while accomplishing the production program assign for a period. Naturally the production program must be practical.

The problem is that by present day technology the manufacturing databases includes only decisions (without revealing why and how they were arrived at) and therefore the "road map" indicates only one path. Naturally the one that make the decision considered many options; alternatives, and optimization methods. To our sorrow all such considerations are lost, and merely decisions are transferred and stored in the company database. What should be done is to capture this available data and store it, instead of letting it go to waste.

The task of the process planner is to select out of the tremendous number of alternatives, the most economical process. The sequence of decisions affects the recommended process and thus can introduce many artificial constrains on the selected process planning.

However, the process planner is neither an economist nor a production planner. Therefore he should not make decisions that are beyond his field of expertise. His task should be to prepare a road map, covering all feasible routings, and let the expert in production planning decide which route to use, in light of order size and plant load at the moment a decision is needed. A product tree (network shown in Fig. 1) is used throughout all the planning steps.

The road map data can be achieved either by the process planner; instead of throwing away his scratch book computation, let him write it down in the spread sheet. Or list the selected routing in the spread sheet format, and use a computer program to fill in the operation time data for other resources. It takes a relative simple program to calculate the operation time based on the process planner computations. The time conversion can be based on resource specifications (such as power); assembly operations can be converted by ratios of manual operation, machine operation or robotics, etc.

The data ("road map") can be used in many applications: in resource planning, costing, cash flow planning, profit forecasting, budget and management control, master production plan, capacity planning and shop floor scheduling and control, performance measuring decision support system setting delivery date as a function of cost. Each application uses the data in a method fit its needs, but all use one main database.

Alternative routings can be generated by KBMS CAPP program that transformed process planning from a technological into a mathematics decision. Any user can generate a routing, without the assistance of a professional process planner, by setting his parameters, and the computer program will generate the routing.

Pro. Item	Op	PR	R1	R2	R3	R4	R5	R6	R7	R8	R9
BP2	10	00	3.12	3.17	3.68	99	4.02	3.27	99	99	99
	20	00	1.15	1.2	1.71	99	2.05	1.3	99	99	99
	30	20	1.49	1.53	2.05	99	99	1.64	99	99	99
	40	10	1.30	1.35	1.86	1.86	2.2	1.45	99	99	99
	50	40	1.28	1.33	1.84	99	2.18	1.43	99	99	99
	60	50	1.51	1.56	2.07	99	99	1.66	99	99	99
BP3	10	00	3.12	3.17	3.68	99	4.02	3.27	99	99	99
	20	10	1.15	1.2	1.71	99	2.05	1.3	99	99	99
	30	20	1.26	1.31	1.82	99	99	1.41	99	99	99
	40	10	1.57	1.62	2.13	2.53	2.47	1.72	99	99	99
	50	10	2.06	2.11	2.62	2.62	2.96	2.21	99	99	99
FB5	10	0	3.12	3.17	3.68	99	4.02	3.27	99	99	99
	20	0	1.15	1.2	1.71	99	2.05	1.3	99	99	99
	30	20	1.49	1.53	2.05	99	99	1.64	99	99	99
	40	10	1.30	1.35	1.86	1.86	2.2	1.45	99	99	99
	50	40	1.28	1.33	1.84	99	2.18	1.43	99	99	99
	60	50	1.51	1.56	2.07	99	99	1.66	99	99	99
	70	60	1.44	1.49	2.00	99	99	1.59	99	99	99
FB6	10	0	3.12	3.17	3.68	99	4.02	3.27	99	99	99
	20	10	1.15	1.2	1.71	99	2.05	1.3	99	99	99
	30	20	1.26	1.31	1.82	99	99	1.41	99	99	99
	40	10	2.06	2.11	2.62	2.62	2.96	2.21	99	99	99
P8	10	0	99	99	99	99	99	99	1.45	1.72	1.96
	20	10	99	99	99	99	99	99	1.45	1.72	99
P9	10	0	99	99	99	99	99	99	2.03	2.33	2.57
	20	10	99	99	99	99	99	99	1.91	1.85	99
C10	10	00	99	99	99	99	99	99	2.86	3.12	99

Where: Pro. Item = Product and Item; OP = Operation Number;
PR = Priority and constraint code; Ri = Resource Number

Figure 1. The road map time content.

The definition of the mathematical problem is as follows: Given a road map (Fig. 1) which lists operations to be performed; a list of available resources; and the time to process each operation on each resource. A decision is required as to which resource to use for each operation; in what sequence in order to generate an optimum routing. The operations in the list are not arranged according to any reasonable sequence. However, some operations must precede others or join with others. These constraints are stated in the road map in the PR (priority) column.

Extra expenses and time should be added to cover extra when moving from one resource to another. These extra expenses are called "penalty". Thus the penalty is for a batch, and is a function of the batch quantity to be produced. Naturally, for different quantities and constraint a different routing will result.

5. EXAMPLES OF USING THE ROAD MAP

The process planner recommends the best routing to produce a product. The question is: what is best process? Is it best from maximum production or for minimum cost? Is it independent of quantity? Is best to generate profit to the company?

To examine those questions we used the road map KBMS CAPP program to generate 22 process alternatives of one item. The alternatives are shown in Fig. 2. As can be seen there is no general "best" process, it depends on the objection and user of the routing. For sure you cannot leave such a crucial decision to be taken by the process planner.

Note: for max production use alt. #1; for min. cost use Alt. #22; for max. profit use alt. #9; for min. investment use alt. # 3; for No. of years to ROI use alt.13;

Alt No	Total Cost	Total Time	Max. Time	Resource costs $	Mach 1	Mach 2	Mach 3	Mach 4	Mach 5	Mach 6
1	23.76	5.94	5.94	400,000	5.94					
2	19.02	6.34	6.34	300,000		6.34				
3	15.54	11.10	11.10	70,000			11.10			
4	24.98	12.49	12.49	100,000						12.49
5	19.80	5.84	4.48	500,000	4.48					1.09
6	17.70	8.94	5.72	420,000	2.92				5.72	
7	22.80	7.10	5.00	405,000	5.00			1.50		
8	19.57	6.59	5.89	700,000	0.40	5.89				
9	16.97	6.22	4.83	400,000		4.83				1.09
10	15.53	9.19	5.72	320,000		3.17			5.72	
11	18.15	7.45	5.35	305,000		5.35		1.50		
12	15.18	9.99	7.73	370,000		1.16	7.73			
13	14.98	12.42	6.40	90,000			6.40		5.72	
14	15.19	10.47	9.08	170,000			9.08			1.09
15	24.90	13.50	11.40	105,000				1.50		11.40
16	16.09	13.34	9.99	120,000					9.99	2.75
17	14.94	12.98	6.40	95,000			6.4	1.50	3.88	
18	17.66	8.90	3.88	425,000	2.92			1.5	3.88	
19	14.90	10.09	6.23	375,000		1.16	6.23	1.5		
20	14.34	10.81	5.72	390,000		1.16	3.03		5.72	
21	14.83	9.36	6.85	470,000		0.82	6.85			1.09
22	14.30	10.77	3.88	395,000		1.16	3.03	1.5	3.88	

Figure 2. Alternate routing

To demonstrate the benefits of employing the KBMS method in production scheduling, let examine the throughput of processing a product with the data given in the company road map.

The objective of production management stage is to plan the activities on the shop floor in a manner that order delivery date will be met. Thus the planning must consider: the available resources, i.e. plan with finite capacity; the product structure. The use of the product structure is to make sure that all item required for assembly will be available when needed and not before. When an item is delayed, the planning of all other dependent items must be adjusted.

The KBMS planning method is with finite capacity and with the use of the product tree [3]. The detailed capacity plan is transformed into a real schedule for the purpose of job release to shop floor. Shop floor is still free to produce the product mix by any other routing which facilitates the solution of problems caused by disruption, with the restriction that the released product mix, as specified for the period, must be completed in time.

The KBMS planning steps are as follows:

1. Stock allocation.

   The allocation strategy is to assign stock to the critical order. A time base bill of materials is constructed using the CAPP program. The critical order is defined as the order whose low-level item has the start date at the earliest time. After each allocation the time base BOM is updated.
2. Adjust quantities by economic considerations
3. Capacity planning - resource loading

   Capacity planning strategy is to start with the critical low level item of the critical order, from its first operation. This item has priority. Loaded operations are recorded in a Gant chart format which indicated loading status. If there is no idle time or the preferred resource, or the waiting time is too long, the road map is used to scan that operation looking for an alternate resource.
4. Job release for execution
5. Shop floor scheduling and control.

   Shop floor strategy is that whenever a resource is free, it searches for a free operation. If there is a match the operation is loaded. If the resource is not the optimum for any of the free operations, a "look ahead" is used to determine the "waiting time" for a free operation. The waiting time (idle time) is used to compute a new optimum for a free operation, and load it. It is the efficiency of being in-efficient.

The decision is made by scanning the operation raw of the road map (see Fig. 1) and comparing the free resource time to the optimum resource time. This scheduling strategy was tested on: two orders; 12 items, 15 resources. The results are shown in Table 1. Fig. 3 show the two extreme detailed scheduling.

Table 1. comparison of scheduling strategies

Criteria of routing optimization	Number of periods to process orders	Unit cost
Maximum production	35	162
Minimum cost	32	76.2
Semi flexible	23	131
The KPMS strategy	21	102

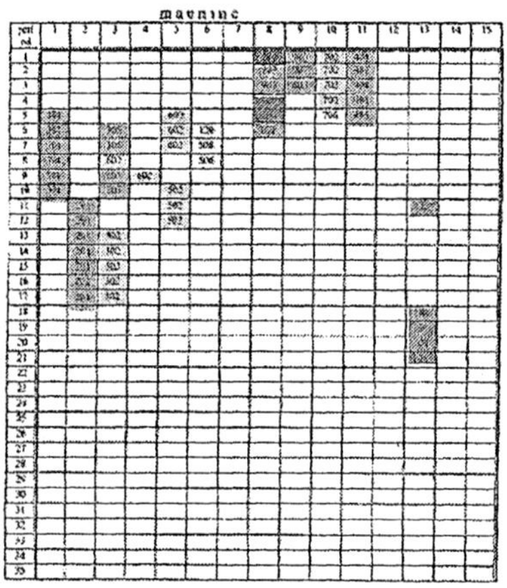

KPMS planning strategy

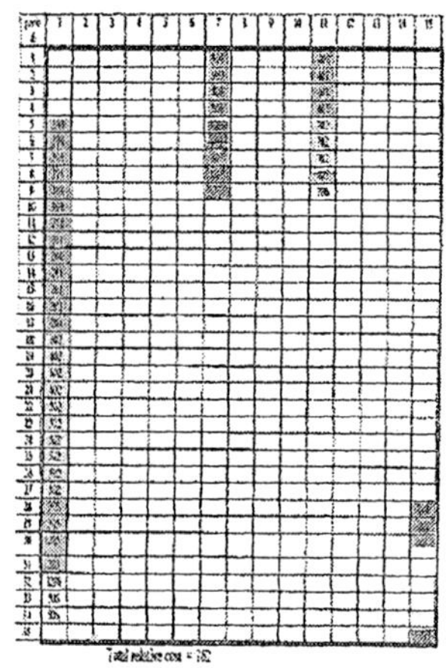

Max. production routing

Figure 3. Throughput as a function of scheduling strategy

6. REFERENCES:

1. Fitzgerald, A., (2000), Enterprise resource planning (ERP)-breakthrough or buzzword? In: *Third International Conference on Factory 2000. Competitive Performance Through Advanced Technology (Conf. Publ. No.359). IEE,* London, UK; pp.291-7
2. Guenov, M. D., (2002), Complexity and Cost Effectiveness Measures for Systems Design, *Second International Conference of the Manufacturing Complexity Network*, pp.. 455 - 466
3. Halevi, G., (1980); *The Role of Computers in the Manufacturing Processes,* Weily-Interscience.
4. Halevi, G., (1999): *Re-Structuring the Manufacturing Process – Applying the Matrix Method,* The St. Lucie/APICS Series on Resource Management.
5. Liu, J. and MacCarthy, B. L., (1996), The classification of FMS scheduling problems, *International journal of production research,* Vol. 34, No. 3, 647-656.
6. Tinham, B., (2000), What place MRP II in the new world? *Manufacturing Computer Solutions.* vol.6, no.1; p.14-18
7. Wiendahl, H. P. and Scholtissek, P., (1994), Management and Control of Complexity in Manufacturing, *Annals of the CIRP*, Vol. 43/2, pp.533-540.

KNOWLEDGE MANAGEMENT IN MANUFACTURING: THE SOFT SIDE OF KNOWLEDGE SYSTEMS

Ove R. Hjelmervik, Kesheng Wang
Department of Production and Quality Engineering, Norwegian University of Science and Technology, 7495 Trondheim, Norway

Abstract: Organizational learning is important for a firm's productivity growth. According to some literature, information and communication technology (ICT) do not support organizational learning and therefore not enhanced productivity. This research has demonstrated that in the case of manufacturing companies some computer-based systems do support organizational learning.

Key words: Knowledge Management, Knowledge Management System, Organizational learning, Manufacturing.

1. INTRODUCTION

There is much discussion about knowledge management (KM) and knowledge management systems (KMS), but mostly conceptually or consultatively. It is a common belief that KMS can be a source for strengthening an enterprise's knowledge, but little evidence is provided on the issue of information technology's impact on effectiveness or efficiency in business firms (Nonaka et al, 2001; Brynjolfsson and Hitt, 1998). Some literature (Brown and Duguid, 1991; Corley and Gioia, 2003) argues that learning takes place where practice is executed, in a master-apprentice relationship. We will in this research investigate if, and under which circumstances, a KMS can support a firm's routine development and thus enhance value creation (Huber, 1991). This is of primary importance for a firm in the way it does business. As the world is changing it is important for

Please use the following format when citing this chapter:
Hjelmervik, Ove, R., Wang, Kesheng, 2006, in International Federation for Information Processing (IFIP), Volume 207, Knowledge Enterprise: Intelligent Strategies In Product Design, Manufacturing, and Management, eds. K. Wang, Kovacs G., Wozny M., Fang M., (Boston: Springer), pp. 89-94.

firms to participate in its development. Knowledge management in a manufacturing context is a firm's ability to empower employees to enact reiterating processes between employees, technology and organization, leading to organizational learning. Our case study, in progress, describes the way two production sights within an aluminum company (ALU) apply its KMS, called Best Practice System (BPS). The study tries to identify why changing from the old to the new system made an influence on employees' support for organizational learning, as they apply computer technology in their everyday activities,. This paper will explore the use of ICT on organizational learning as the organization changes from one system to another. The data collection takes place in the period 2002 through 2005.

2. THE CASE

In year 2000 ALU's management decided to restructure the organization from a hierarchical to a processual structure. Today, the business process consists of three main activities: upstream activities (Primary Metal Sector), midstream activities (Metal Products Sector), and downstream activities (here represented by its Automotive Sector). Each Sector manager reports to the CEO. This restructuring resulted in fewer organizational layers. Management decided on the following strategy: more of the operational responsibilities to self-managed teams, and introduction of Best Practice to be aligned with Key Performance Indicators. The employee empowerment required changes to the information availability and structure. One of the important issues for management was an ICT capability allowing employees to transfer experience, process such experience and distribute it as new best practice. In 2003 the old system, SDOCS, was gradually replaced by the new system BPS, previously described in Wang et al (2001). The installation took place in the period 2003-2004. This longitudinal study reaches over a period of three years. The old system was a Lotus Note based document handling system, bringing up complete documents representing many separate routines. No supportive documentation, such as drawings, pictures or video could be attached, a capability the new system has. Furthermore, each document was restructured to bring up issue/task-specific routines linked into the value chain. At each task-specific routine one can encode and transfer an experience back to the process owner.

The Best Practice document structure (figure 1) installed in the global aluminum company has the following structure: User-interface business processes and system processes. Drill-down into the different process segments reveal sub-processes, flow chart and individual work activities with links to relevant documents. Furthermore, each work activity has an "experience transfer" button linking the relevant activity to the process owner. There is also a search engine for internal and external document search, based on "intelligent agents" technology. The technology structure for the business system can be seen from Fig. 2.

Management has provided the self-managed work teams with access to relevant information, routines tools and resources to carry out, and thus optimizing, the operating process. At any point along the production line a self-managed work team is responsible for a production segment. They also have to know what activity is upstream and downstream of the segment. By opening an electronic work activity page, the employee can see who is responsible for delivering what into the team's operating segment, its expected quality and timing of the delivery. Likewise, the electronic work activity page will tell the team who the customer is, what delivery responsibility the team has, the quality and timing of the delivery.

The business system is a KM system open to all employees. It is passive, except for the Intelligent Agent, and subsequently does not process data for intelligent analysis. From here you can reach all operational systems. These systems can be line-of-business application such as human resource systems, incident registration system "Synergy", enterprise resource planning system "SAP", or a document management system "Domino Doc" (Fig. 3). In order for employees to search on key words, concepts, etc., the system has an intelligent agent technology installed in the Oracle server. The agent is aiding the user by interacting with internal and external databases and the Oracle Knowledge Server. Figure 3 illustrates how interaction between databases and intelligent agent technology is aided by the Knowledge Server. The end result is served through the web server and out to the client.

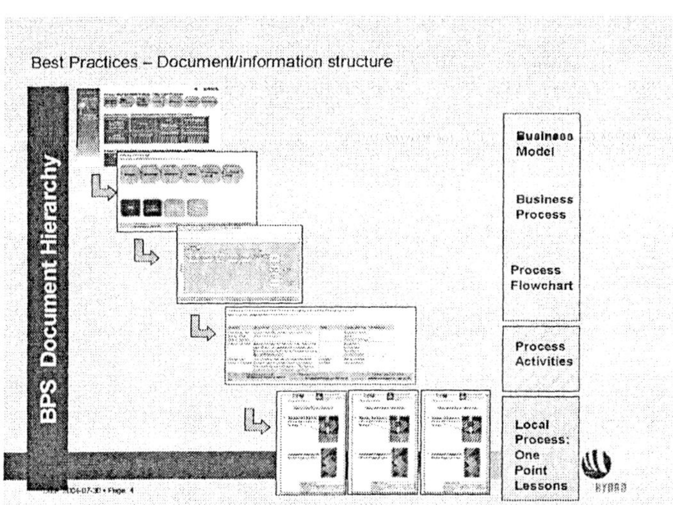

Figure 1 Best Practice document structure in BPS

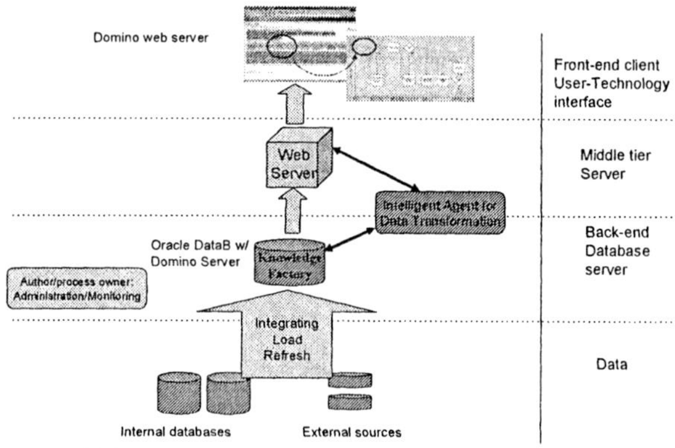

Figure 2 KM Business System

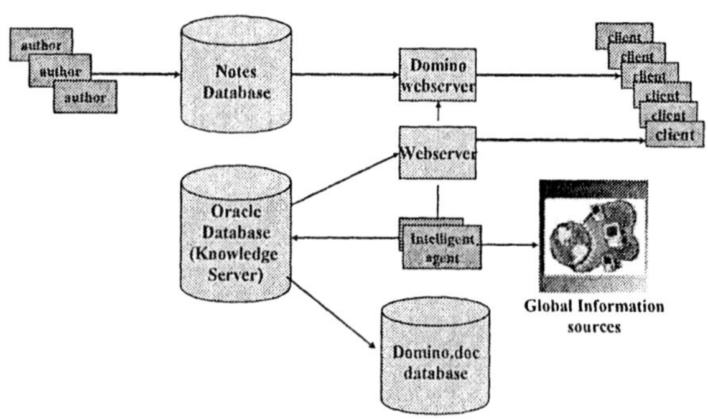

Figure 3 BPS Technical architecture

ALU's KMS is a directional and task related process system, supporting operational (production and staff) activities in mowing goods from raw materials to finished products, in a partly closed and partly open production.

3. REPORT FINDINGS

Knowledge management has something to do with learning, and since we associate management with organizations we consider knowledge management as part of organizational learning (Levitt and March, 1988). Furthermore, as learning is something you do to improve a current routine for the purpose of changing the institutional rules, values and processes, it is important for the firm to develop dynamic capabilities (Zollo and Winter, 2002). Such capabilities come from the organization's ability to learn, and develop, new routines in a reiterating process between employees, technology and organization. Such reiteration process can be understood in relation to structuration theory (Orlikowski, 1992).

We wanted to understand if a computer based business system could support organizational learning, and under which circumstances. Based on two plant location, and a set of nested levels of interviews from senior management to production operators, we found Zollo and Winters' knowledge evolution cycle to be reflecting the organization's dynamic capability, that is: variation, selection, replication and retention; and their three levels of learning mechanisms: experience accumulation, knowledge articulation and knowledge codification. Linking such dynamic capabilities to Orlikowski's (1992) expanded structuration theory, we found that in order for knowledge management to take place in the organization, all levels of the organization must participate. Management must allow dynamic capabilities to develop, employees must be encouraged to participate in organizational learning, and decision-making powers must be transferred to teams. Such strategy includes three elements: empowerment; a technology to represent organization knowledge; and employee participation in developing new or improved routines leading to enhanced productivity.

4. CONTRIBUTION

There is much discussion about knowledge management (KM) and knowledge management systems (KMS), but mostly conceptually or consultatively. Little evidence is provided on the issue of information technology's impact on effectiveness or efficiency in business firms. We found that employees, interacting with technology in their ongoing practices, enact structures which shape their emergent and situated use of that technology (Orlikowski, 2000), leading to participation in the development of product knowledge (Zhang, et al, 2005). This research investigated how KMS can support a firm's routines, process and product development, and thus enhance value creation. We found that organizations learn through empowered enactment, supported by KMS.

Based on this research we have defined KM in a manufacturing context as a firm's ability to empower employees to enact reiterating processes between employees, technology and organization. Furthermore we have defined a KMS as a push-pull technology providing employees with relevant information at point of decision, for the propose of execute current routines and processes, and support the development of new or improved routines and processes based on experience, trends and patterns in operating processes.

Further research includes organizational learning, supported by KMS, can lead to enhanced productivity.

5. CONCLUSION

We have in this research established that it is possible for a computer system to support organizational learning in manufacturing organizations. It is the way such computer system is structured which makes it valuable for the organization. Empowerment and access, for the purpose of organizational learning, are two important factors for a successful KMS.

6. REFERENCES

Brown, J.S., and P. Duguid, (1991), Organizational learning and communities of practice: toward a unified view of working, learning and innovation. *Organizational Science*, 2.

Brynjolfsson, E., and L.M. Hitt, (1998), Beyond the productivity paradox: computers are the catalyst for bigger changes. *Communication of the ACM.*

Corley, K.G., and D.A. Gioia, (2003), Semantic learning as change enabler: Relating organizational identity and organizational learning. *Handbook of Knowledge Management*, Blackwell.

Huber, G.H., 1991, Organizational learning: The contributing processes and the literatures, *Organization Science*, Vol., 2, No. 1, February.

Levitt, B., and J.G. March, (1988), Organizational Learning; *Annual Review of Sociology*, 14, pp. 319-340.

Nonaka, I., P. Reinmöller, and R. Toyama, (2001), *Integrated information technology systems for knowledge creation*, in Handbook of organizational learning and knowledge, eds.: M. Dierkes, A. B. Antal, J. Child, and I. Nonaka. Oxford University Press, Oxford.

Orlikowski, W. J., (2000) Using Technology and Constituting Structures: A Practice Lens for Studying Technology in Organizations; *Organization Science*, vol. 11, No. 4, pp. 404-428.

Orlikowski, W. J., (1992), The duality of technology: Rethinking the concept of technology in organizations. *Organization Science*, vol. 3, no. 3.

Wang, K, O.R. Hjelmervik, B. Bremdal, (2001), *Introduction to Knowledge Management, theory and practice*, Tapir Academic Publisher, Trondheim.

Zhang, J., Wang, Q., Wan, L. and Zhong Y., (2005), Configuration-oriented Product Modeling and Knowledge Management for Made-to-order Manufacturing Enterprises, Int. J Adv Manufacturing Technology, Vol. 25, pp. 41-52.

Zollo, M., and S.G. Winter; (2002), Deliberate Learning and the Evolution of Dynamic Capability, *Organization Science*, vol. 13, no. 3, p. 339.

DSM AS A KNOWLEDGE CAPTURE TOOL IN CODE ENVIRONMENT

Syed Ahsan Sharif, Berman Kayis
School of Mechanical and Manufacturing Engineering, The University of New South Wales, UNSW, Sydney, NSW 2052, Australia. [1]Email address: sasharif_unsw@yahoo.com

Abstract: A design structure matrix (DSM) provides a simple, compact, and visual representation of a complex system/process. This paper shows how DSM, a System Engineering tool, is applied as a knowledge capture (acquisition) tool in a Generic NPD process. The acquired knowledge (identified in the DSM) is then validated in an Australian manufacturing company. This acquired knowledge helps NPD teams, managers and stakeholders to benchmark their NPD efforts and select areas to focus their improvement efforts.

Key words: Design Structure Matrix (DSM); New Product Development (NPD); Customer Order Driven Engineering (CODE); Knowledge Management (KM)

1. 1. INTRODUCTION

In modern times we have focused on new manufacturing methods, shifting from mass to lean production, and are now at the next wave of manufacturing innovations – Customer Order Driven Engineering (CODE). Customers are demanding products that feature the latest in style and technology; offer utility, value and price; and meet quality and reliability expectations. In order to meet these customer needs CODE is the key concept for the manufacturing industry (Anderson, 2004; Chandra and Kamrani, 2004; Cheng and et al., 2002). CODE attempts to provide customized products for individual customers without losing many benefits of mass production – high productivity, low costs, consistent quality and fast response. To be successful, the CODE requires a major combination of

Please use the following format when citing this chapter:
Sharif, Syed Ahsan, Kayis, Berman, 2006, in International Federation for Information Processing (IFIP), Volume 207, Knowledge Enterprise: Intelligent Strategies In Product Design, Manufacturing, and Management, eds. K. Wang, Kovacs G., Wozny M., Fang M., (Boston: Springer), pp. 95-102.

efforts in all areas of business. Among the enablers of CODE, Knowledge Management (KM) is considered to be one of the most fundamental and challenging one, because the integration of activities and resources heavily relies on the integration and sharing of Knowledge (i.e. Knowledge, Information or Data) (Biesner and Brüggen, 2005; Davis et al., 2005). This paper shows how DSM, a System Engineering tool, is applied as a knowledge capture (acquisition) tool in a Generic NPD process. The acquired knowledge (identified in the DSM) is then validated in an Australian manufacturing company.

2. DESIGN STRUCTURE MATRIX (DSM)

DSM (variously expanded as the design structure matrix, problem solving matrix (PSM), dependency structure matrix and design precedence matrix) is a representation and analysis tool for system modeling, especially for decomposition and integration (Browning, 1999, 2001) purposes. A DSM displays the relationships between components of a system in a compact, visual, and analytical format. The advantages of DSMs in respect to other system representation and analysis techniques have led to their increasing use in a variety of context, including product development, project planning, project management, system engineering, and organization design.

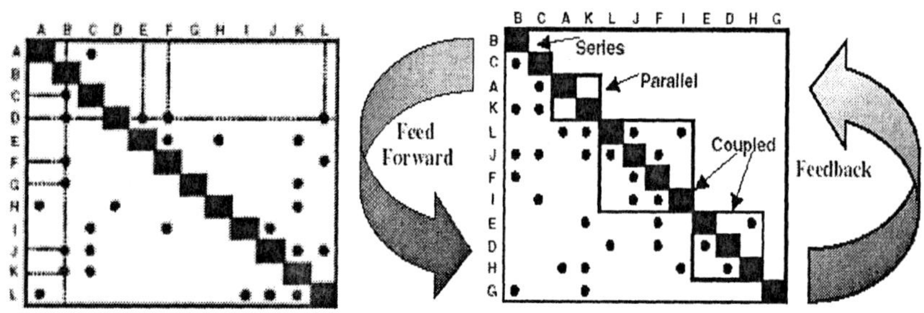

Figure 1. (a) Sample DSM, (b) Partitioned DSM matrix

DSM consists of an N-square matrix (Figure 1(a)) with one row and column per element and shows the interaction of each element with every other element in the model. DSMs elements are represented along the diagonal, usually represented by shaded area separate the upper and lower diagonal of the matrix. An off-diagonal mark signifies the dependency of one element on another. If the DSM elements represent tasks to be performed, then inspecting the row and column of a task reveals the inputs and outputs, respectively, for that task. For example, in Figure 1(a), B feeds

C, F, G, J, and K, while D is fed by E, F, and L. If there is a time sequence associated with the position of the elements in the matrix, then all marks above the diagonal are considered feedback marks. Feedback marks correspond to required inputs that are not available at the time of executing a task. In this case, the execution of the dependent task will be based on 'assumptions' regarding the status of the input tasks. As the project unfolds these assumptions are revised in light of new information, and the dependent task is re-executed if needed. It is worth noting how easy it is to determine feedback relationships in the DSM compared to the graph, which makes the DSM a powerful, but simple, graphic representation of a complex system or project.

DSMs partitioning algorithms (Steward, 1981; Eppinger at el., 1990; Yassine et al., 1999) re-sequence the elements in the matrix, which makes it a lower triangular matrix. Lower triangular matrix means each task begins only after it receives all the required information from its predecessors. This shows that there are no interdependent/coupled tasks. They are either in serial or in parallel. However, this sequencing analysis sometimes produces a block triangular matrix – a coupled block. Figure 2.15, shows the same sample matrix after partitioning. In Figure 1(b), task B and C are serial, i.e. C can be done after B is completed. Task A and K are completely independent of each other, and can be done in parallel. Elements involved in a coupled block must be done concurrently, for instance task E, D and H should be performed simultaneously by using iterations and/or negotiations. Partitioned matrix provides the first step and the mathematical result of optimization. However, Partitioning algorithms do not eliminate coupled blocks, rather they provide iteration sequences. Further study can be done after partitioning, including assigning dependency values, tearing, de-coupling and add-coupling, and other non-binary DSM techniques (Dong, 1999) to deal with the coupled blocks (Optimization of the DSMs).

3. DSM REPRESENTATION OF NPD PHASES

The construction of DSM in this research was undertaken by using the following steps:

Step 1: Define the NPD phases

Step 2: List all the elements of the phases (Task breakdown structure)

Step 3: Construct the DSM matrix to represent the interactions between different tasks

Step 4: Partition and Optimize the DSM matrix

Step 1 and 2: Since the DSM is a system engineering tool, it is important to define the boundary of the system in order to focus the research work. A Generic NPD process (with six phases) is defined, and then detailed into 34 tasks (Task Level 1). These higher level tasks then decomposed and detailed into lower level tasks (Level 2, 3 or lower – depending on the nature of the tasks). Like wise Task Level 1, these Task breakdown structure is also generic in nature. This definition of NPD phase and identification of tasks, tries to cover the whole NPD process (i.e. for entire Product Life Cycle) in CODE environment and under the philosophy of Concurrent Engineering (CE), which emphasize on collaboration, co-ordination and better communication among the cross-functional teams inside the organization as well as active involvement of the extended enterprise (Customers and suppliers).

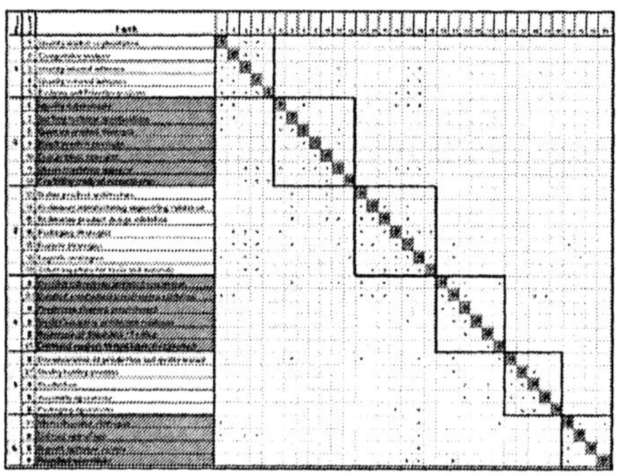

Figure 2(a). *Higher Order DSM for Generic NPD Process*

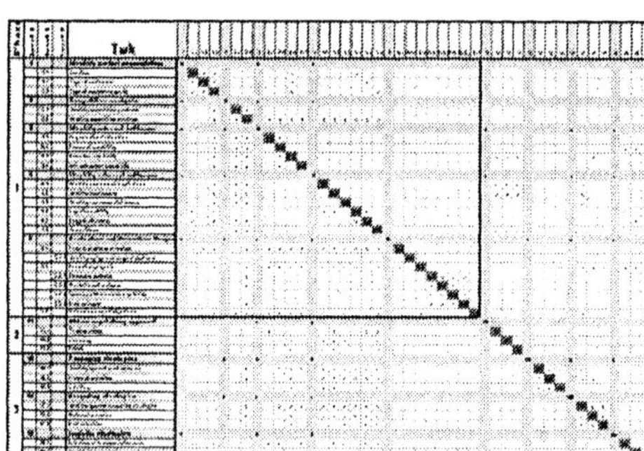

Figure 2(b). Decomposed DSM (Phase 1)

Step 3: The third step is to identify the interactions between different tasks and build the DSM matrix. The interactions between different tasks are identified by the "casual diagram" (Brian, 2006) and "surveying" the literatures related to NPD process. Figure 2(a) shows the DSM representation of a Generic Higher Order NPD process. Presenting very large models in a single matrix is challenging. When constructing models comprised of hundreds of tasks, the intuitiveness provided by the DSM representation diminishes. A very large DSM can be effectively structured into a hierarchy of smaller DSMs. This configuration avoids problems related to presenting extremely large matrices by shifting the focus to smaller ones, obtained through hierarchical decomposition. It also provides the flexibility to analyze the process at different levels of detail. This multi-tiered approach (developed by Dr. Grose (1994) at Boeing) is adopted to decompose the higher order DSM. Six more decomposed DSM (lower level) are developed from Figure 2(a) for the six phases of NPD. Figure 2(b) shows the decomposed DSM of Phase 1 only. Decomposed task breakdown structures are the source of these decomposed DSM.

Step 4: After constructing the DSMs, they are partitioned and optimized as per requirement as discussed in section 2. The constructed DSM matrixes show the task interactions not only between the same phases but also between other phases too, and with different task levels. They also indicate the complexities (coupled blocks), important interactions and loops (feedback and iterative) exist in the different NPD phases.

4. KNOWLEDGE REPRESENATAION

Knowledge items identified in the DSM (e.g. important interactions, feedback and iterative loops etc) are provided in the form of Questionnaires. An Assessment Model is developed, consisting of five performance indicators of the organization namely 'Marketing', 'Technical', 'Financial', 'Resource Management', and 'Project Management'. Around 150 questions are developed (in total), which are organized into these five categories for each of the six NPD phases. More than 40 questions developed for 'Phase 1' only, to tackle the 'fuzzy front end' of NPD. These Questionnaires provide rich enterprise knowledge and designed to assist in the improvement of a company's product development performance. Figure 3 shows some sample questionnaires which are scaled from 1 to 5. Generally "Scale 1" represents poor and "Scale 5" represents excellent.

5. VALIDATION OF THE ASSESSMENT MODEL

The validation of the assessment model developed has two fold objectives. The first objective is to check the sensitivity of categories used by means of comparing successful versus less successful product from the same product group of the same company. And the second objective is to observe whether the model can display the overall performance of the company in the five categories of NPD phases.

M5 How close is the link between your sales, marketing, planning and manufacturing functions?

Scale	1	2	3	4	5
Options	No link	Nominal link	Good link but major gaps exist	Strongly linked but some gaps exist	Very strongly linked (collaborative and cooperative environment)

T6 How well do you think you determine the project objectives (e.g., Quality, Reliability, Manufacturability, Sales target, etc)?

Scale	1	2	3	4	5
Options	Minimally determined	Partially determined	Determined but not very clear	Fully determined but some doubt exits	Fully and very clearly determined the project objectives

Figure 3. Sample Questionnaires from Phase 1

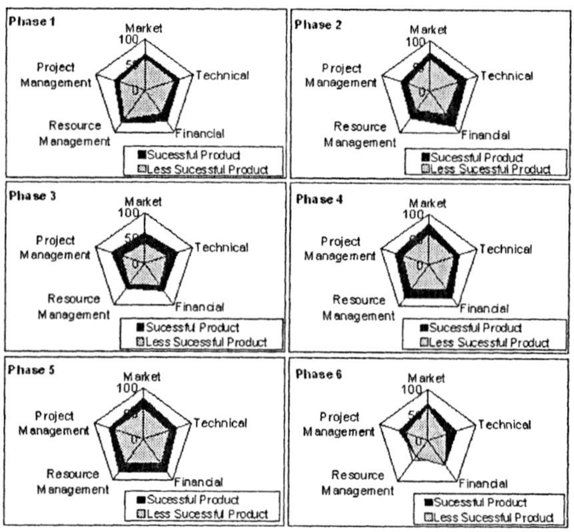

Figure 4. Sensitivity test

Figure 4 shows the sensitivity test of the Questionnaires, comparing Successful and Less Successful product. Both products are benchmarked against the acquired knowledge (100% score means the selected performance criteria are fully met). The results are consistent throughout the six NPD phases, i.e. less successful product have lesser percentage (or equal) than the successful product in each of five performance indicator axis. This sensitivity test (Figure 4) proves to be a powerful tool for the company to focus on areas that less successful product needs a remedial action.

The results of the successful product from previous "Sensitivity analysis" are considered again and the results are mapped (Figure 5) by "Performance Indicator" areas. This approach also assists with the benchmarking process where 100% score means the selected performance criteria are fully met. Figure 5 revealed the company's strong and weak areas in the product development process. For example, the company should give attention in 'Marketing' area of Phase 3, 'Technical' area of Phase 3, 4 and 6, and so on (as shown in circle in Figure 5). Although the data of the 'Project management' are steady for all of the six phases, it needs more consideration for all phases because of low scores which are in the order of 60%-70% only.

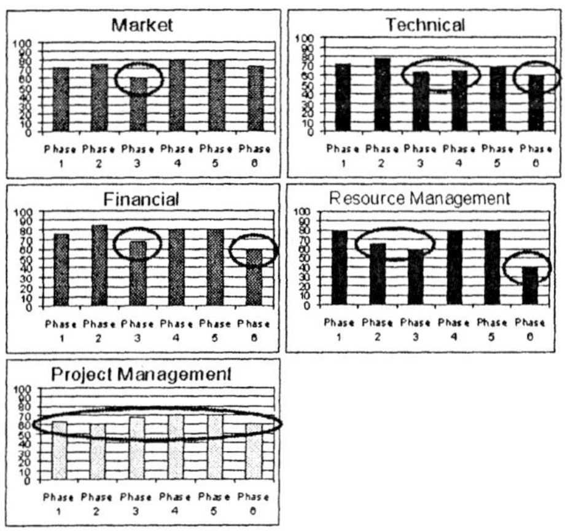

Figure 5. Total view

6. CONCLUSION

DSM is applied as a Knowledge capture (Knowledge acquisition) tool in the NPD process of CODE environment, which provides an essential knowledge to help NPD managers, teams and stakeholders in the complicated and uncertain nature of the NPD process. The industrial case used aims to validate the sensitivity of the acquired knowledge and more importantly shows the performance of the organization in five categories of each of NPD phases. It provides a means of assessing how well the company has met different product development performance criteria. Accordingly, companies would be able to identify the areas to focus for their improvement efforts.

7. REFERENCES

Anderson, D. M., 2004, *Build-to-Order and Mass Customization: The Ultimate Supply Chain Management and Lean Manufacturing Strategy for Low-Cost On-Demand Production without Forecasts or Inventory*, CIM Press, Cambria, California, pp. 1–9.

Biesner, J. and Brüggen, G., 2005, The Significance of Knowledge Management for a Larger Company, in: *Knowledge Management: Organizational and Technological Dimensions*, J. Davis, E. Subrahmanian, and A. Westerberg (Editors), Physica-Verlag Heidelberg, New York, pp. 159–167.

Brian, K.M.T., 2006, Ph.D. candidate, Research topic: *Implementing concurrent engineering with integration of intelligent agent based system in complex product*, School of Mechanical & Manufacturing Engineering, UNSW, Sydney, Australia.

Browning, T.R., 1999, The Design Structure Matrix, in: *Technology Management Handbook*, R.C. Dorf, Ed. Boca Raton, Fl: Chapman & Hall/CRCnet-BASE, pp. 103–111.

Browning, T.R., 2001, Applying the Design Structure Matrix to System Decomposition and Integration Problems: A Review and New Directions, *IEEE Transactions on Engineering Management*, **48**(3):292–306.

Chandra, C. and Kamrani, A., 2004, *Mass Customization: A Supply Chain approach*, Kluwer Academic / Plenum Publishers, New York, pp. v–vii.

Cheng, F., Ettl, M., Lin, G. and Yao, D. D., 2002, Inventory-service optimization in configure-to-order systems, *Manufacturing and Service Operations Management*, **4**:114–132.

Davis, J., Subrahmanian, E. and Westerberg, A. (Editors), 2005, *Knowledge Management: Organizational and Technological Dimensions*, Physica-Verlag Heidelberg, New York, pp. v–vii.

Dong, Q., 1999, *Representing Information flow and Knowledge Management in Product Design using the Design Structure Matrix*. Masters Thesis, M.I.T., pp.61–65.

Eppinger, S.D., Whitney, D.E., Smith, R.P., and Gebala, D., 1990, Organizing the Tasks in Complex Design Projects, *ASME Conference on Design Theory and Methodology*, Chicago, IL, September, pp. 39–46.

Grose, D. L., 1994, Reengineering the Aircraft Design Process, *Proceedings of the Fifth AIAA/USAF/NASA/ISSMO Symposium on Multidisciplinary Analysis and Optimization*, Panama City Beach, FL, Sept. 7–9.

Steward, D.V., 1981, The Design Structure System: A Method for Managing the Design of Complex Systems, *IEEE Transactions on Engineering Management*, **28**(3):71–74.

Yassine, A., Falkenburg, D. and Chelst, K., 1999, Engineering Design Management: An Information Structure Approach, *International Journal of Production Research*, **37**(19):2957–2975.

RESEARCH ON ACQUIRING DESIGN KNOWLEDGE BASED ON ASSOCIATION RULE

Jingmin Li[1], Yongmou Liu[1,2], Jiawei Yang[1], Jin Yao[1]
[1]*Department of Manufacturing Science and Engineering, Sichuan University, China; Email: hjljm@163.com.* [2]*IMAG Asia Ltd.*

Abstract: This study presents a method of acquiring product design knowledge. By analyzing the process and mathematics expression of product design, design attributes are taken as item sets and associated relationship among them are taken as analyzing target. Based on the expression and application for domain knowledge, design data are pretreated. Then the correlations knowledge among design attributes is obtained by association rules. A case study is given to illustrate that the method proposed is valid and practicable.

Key words: Product Design, Data Mining, Association Rule, Domain Knowledge.

1. INTRODUCTION

As we all known, product design consists of two kinds of knowledge: explicit knowledge and tacit knowledge[1]. Explicit knowledge means the standard knowledge which is like books and manuals and easy to express, to capture, to encode; tacit knowledge means the knowledge which is just experience derived from experts long-term working, such as how to choose materials or parameters, how to optimize the relationships among attributes, and how to make decision according to input conditions. This kind of knowledge can not be summed up easily since it is private, underlying and bitty. Gaining them is a challenging problem that we have to solve.

Relevant researches focus on the development of expert system and knowledge system[2-3], in which more attentions are paid to the expression and reasoning for domain knowledge. But the essential problem of design knowledge acquisition hasn't been solved and becomes the bottleneck of system further development. Data mining, as one of the promising technologies since 1990s, is to some extent a non-traditional data-driven

This project is supported by Sichuan Science and Technology department of China under the grant No.04JY029-003-2.

Please use the following format when citing this chapter:
Li, Jingmin, Liu, Yongmou, Yang, Jiawei, Yao, Jin, 2006, in International Federation for Information Processing (IFIP), Volume 207, Knowledge Enterprise: Intelligent Strategies In Product Design, Manufacturing, and Management, eds. K. Wang, Kovacs G., Wozny M., Fang M., (Boston: Springer), pp. 103-108.

method to discover novel, useful, hidden knowledge from massive data sets. Association rule mining is one of mining techniques and has attracted more and more attention of academic and practitioners with applications such as customer relation management[4], market baskets[5], fault diagnosis system[6], decision support system[7], time-series analysis[8] and so on, but to a less extent in engineering design. This paper tries to apply association rules to the field of product design for acquiring associated knowledge among design attributes, and the key technology is investigated.

2. PROBLEM STATEMENT AND DEFINITIONS

2.1 The graphic description of product design

The process of product design is an iterative analyzing-synthesizing-decision-making process. This kind of design problem can be expressed in the tree as shown in figure 1. P is function requirements, which can be divided into some sub-problems $\{P_1, P_2, ..., P_n\}$, each of which also can be divided into $\{F_1, F_2, ..., F_j\}$, until into $\{SP_1, SP_2, ..., SP_k\}$ which can not be divided any more. These sub-problems that can't be separated are called fontal-problems. Each fontal- problem corresponds with many solutions.

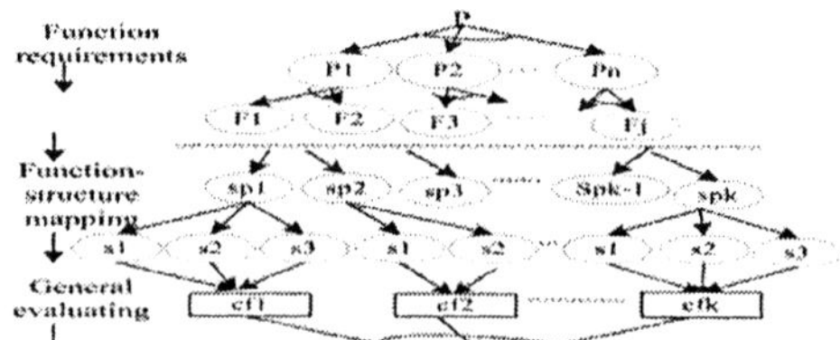

Figure 1. And & Or tree for product design problem

But since these fontal-problems are interrelated and interact on each other, corresponding solutions are various, similar, complicated, and sometimes contradictory on each other. Designer must make correct decision based on much knowledge to implement this mapping from problems to solutions.

2.2 The mathematic expression of product design

Product design problem can be described into the problem field and conclusion field in the form of attributes and attribute-values. The design case can be expressed in mathematics implication as follows:

$$P = < E, C, D, V, f > \quad (1)$$

The variable $E =< E_1, E_2,..., E_n >$ is the set of design case; $C \cup D = A$ is the set of design attributes, in which C expresses condition attributes including user requirements, design environment states and so on, and D expresses conclusion attributes resulted from condition attributes set; $V = E_{a\in A} V_a$ is the set of attribute values; $f : E_x A \rightarrow V$ is the mapping rules from attributes to attribute-values. During the design process, designer must perform the mapping from condition attributes C to conclusion attributes D on the constraints of f. Discovering this kind of strong correlations or independence among all attributes is the problem that this paper will solve.

3. THE METHOD AND IMPLEMENTATION

3.1 Association rule for design data

The basic principle of associations is shown as follows: let $I = (i_1, i_2,..., i_m)$ be a set of items. Let D be a set of transactions where each transaction T is a set of items such that $T \subseteq I$. A transaction T is said to contain A if and only if $A \subseteq T$. An association rule is an implication of the form $A \Rightarrow B$, where $A \subset I$, $B \subset I$, and $A \cap B = \phi$. The rule $A \Rightarrow B$ holds in the transaction set D with support s and confidence c, where s is the percentage of transactions in D that contain both A and B, and where c is the percentage of transactions in A that also contain B. The mining problem is to find all rules that satisfy both a minimum support threshold and a minimum confidence threshold. The process are two steps: (1) finding all frequent itemsets: each of these itemsets will occur at least as frequently as a pre-determined minimum support count; (2) generating strong association rules from frequent itemsets. It requires that data are expressed in the form of attributes and attribute-values, which is consistent with the expressing form of design case.

3.2 Domain knowledge for pretreatment

The essence of association rules is to search the correlations among "1(T)" values in a relational table in which all attributes are Boolean variable. Many discretization techniques, which can be used to reduce the number of values for a given continuous attribute by dividing the range of attribute into intervals, are required and investigated widely. Since design data has some domain meaning instead of purely numerical data, the determination of the interval number and the break point of a continuous attribute must depend on expert's domain knowledge which can come from domain expert, or can be refined from the result of data mining. The application process of domain knowledge are shown in figure 2, in which more attentions are paid to expressing the relations among design attributes or records and to constraining values of some attributes or records instead of being used for reasoning like in expert system

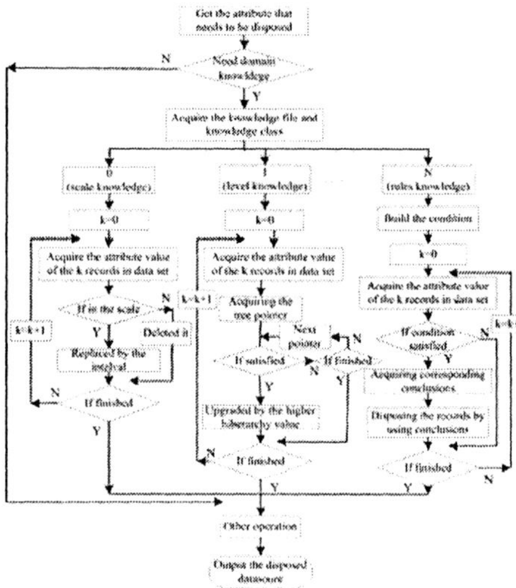

Figure 2. Domain knowledge application

4. A CASE APPLICATION IN DESIGN

Taking form-selection of gate valve as an example, each design case can be seen as a transaction T which includes many items such as working medium(WM), nominal pressure (PN), working temperature (WT), nominal diameter (DN), structure form and so on. These items have strong constraint relations. Acquiring these relations is the aim now. Firstly, the simple table D is set up shown in table 1. Since valve model has many different values that can't be conceptual, delete it; WT, PN and DN are all needed to be pretreated and corresponding domain knowledge is shown in figure 3.

Table 1. The simple table D

ID	Model	WM	PN	WT	DN	SS	WD	...
1	Z41T	water	1.0	100	50-450	B	W+S	...
2	Z65H	oil	2.5	425	15-50	A	W+S	...
3	Z40Y	acid	6.4	200	25-300	M	W+E	...
4	Z40H	oil	4.0	350	50-200	A	W+E	...
5	Z42W	gas	0.1	100	300-800	M	W+T	...
6	Z763F	petrol	6.4	30-80	700	M	P+S	...
7	Z43H	slop	0.6	120	50-350	M	P+S	...
8	Z44W	oil	1.0	100	50-400	M	P+T	...
...	...	...	...	...	...	...	...	...

The process of applying domain knowledge is as follows: (1) by searching for the relevant records according to “mining algorithm = association” and “Data sets = S-F”, find relevant knowledge file and knowledge class; (2) in .\asso1, according to the method of acquisition and application for rule knowledge, pretreat DN value into four values: small_dia, middle_dia, big_dia and super_dia. In.\asso2, process WT value into five values: super_low_T, low_T, normal_T, middle_T and High_T. In .\asso3, upgrade PN value according to its conceptual hiberarchy tree. Then, Let each $<x, l_k, u_k> \in split(x)$ map into logical attribute A, and each attribute value will be transformed into many Boolean variable. All these attributes compose the itemsets required by mining. The original simple table D can be changed into Boolean transaction base, shown as in table 2.

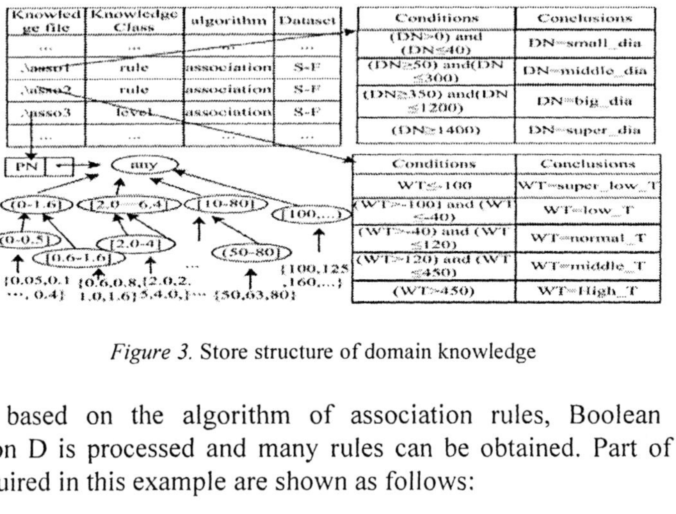

Figure 3. Store structure of domain knowledge

Last, based on the algorithm of association rules, Boolean design transaction D is processed and many rules can be obtained. Part of strong rules acquired in this example are shown as follows:

Table 2. Boolean design transactions table

ID	PN1	...	WM1	...	WT1	...	DN1	...	SS1	...	WD1	..
1	T	...	F	...	T	...	T	...	T	...	T	..
2	T	...	T	...	F	...	F	...	F	...	T	..
3	F	...	F	...	F	...	T	...	T	...	F	..
4	T	...	T	...	F	...	T	...	T	...	F	..
5	F	...	F	...	T	...	F	...	T	...	F	..
6	F	...	F	...	T	...	F	...	T	...	F	..
7	F	...	F	...	T	...	T	...	T	...	F	..
8	T	...	T	...	T	...	T	...	T	...	F	..
...	...	...	...	...	...	...	...	...	...	...	...	..

Rules1: $PN1 \wedge WT2 \Rightarrow WD3, \sup = 7.6\%, conf = 50.5\%$ .When PN is(0.5-6] and WT is high, the designs that the form of Wedge disc is parallel and single, are 7.6 percent in all design transactions. The designs that the form of wedge disc is parallel and single are 50.5 percent in all the design

transactions that PN is (0.5-6] and WT is high. This rule shows that parallel and single wedge disc fits the situation of high temperature and low pressure.

Rules2: $PN3 \wedge WM1 \Rightarrow WD1, sup = 8.3\%, conf = 65.2\%$. When PN is (10-80] and WM is water and steam, the designs that the form of wedge disc is wedge and single, are 8.3 percent in all design transactions. The designs that the form of wedge disc is wedge and single are 65.2 percent in all designs that PN is (10-80] and WM is water and steam. This rules shows wedge and single gate valve fits the situation of high pressure and low-viscidity medium.

These rules are the heuristic information with some supports and confidences, and through expert trimming, selecting, and evaluating, can be changed into useful and regular knowledge which are consistent with expert's thinking mode and provide important references for designer.

5. CONCLUSION

In this paper, based on domain knowledge discretization, association rule is introduced to the field of product design for acquiring the correlative knowledge among design attributes. This method integrates the computing ability of computer with expert domain knowledge. In the future, much more researches are needed to further improve its efficiency and accuracy in order to acquiring more complex design knowledge.

6. REFERENCES

1. Dienes Z,Perner J.A.,(1999), Theory of Implicit and Explicit Knowledge, *Behavioural and Brain Sciences*, vol. 22, pp. 735-808.
2. Tao Shanxin, Tang Wenxian, Li Limin,(2003), "Research and Implementation on Product Development System Based on Knowledge Driven", *Journal of Computer Engineering and application* ,vol.22, pp.129-131.
3. Liu Ying, Ni Zhonghua, (2005), The Expert System of Cooperative design Knowledge Based on machine tool products Full Life Cycle, *Group Technology & Production Modernization* ,Vol.22, pp.10-13.
4. Jiawei Han and Micheline Kamber, Data Ming: concepts and Techniques (China Machine Press, Beijing, 2001).
5. Haiming Chen, Wu qing, Chunhua Zhao, (2004), "Associated Rules Mining in DIAGNOSIS Based on Oil Monitoring", *Journal of WUHAN UNIVERSITY Of Technology (Transportation Science & Engineering)*, vol. 28, pp.729-731.
6. Show-Jane Yen, Arbee L.P. Chen Member, (2001), "A Graph-based Approach for Discovering Various Types, of Association Rules", *IEEE Trans.Knowledge and data Eng*, vol.13, pp.1354.
7. Hou Xue-bo, Tian Bin, (2005), "Application of association rules techniques in electric marketing analysis", *Proceedings of the CSU-EPSA*, vol. 17, pp.67-725.
8. Li Ming, (2005), "Application of Mining Association Rules with Multiple Minimum supports in Sales Data", *Computer Engineering,* vol.29, pp. 92-93.

ANALYSIS OF ISO 6983 NC DATA BASED ON ISO 14649 CNC DATA MODEL

Hiroshi Yamada, Fumiki Tanaka, Masahiko Onosato
Graduate School of Information Science and Technology, Hokkaido University, Japan; Email: yamada@dse.ssi.ist.hokudai.ac.jp.

Abstract: Retaining technical information is an important element for performing high quality machining. However, actual NC data (ISO 6983) contain implicit technical information which reflect the skilled worker's knowledge. In this research, for obtaining technical information, NC data are analyzed based on a next-generation CNC control language (ISO 14649) which contains technical information explicitly. In this paper, we demonstrate how ISO 6983 NC data generated by CAM software are applied to the test implemented system.

Key words: Data Analysis, Technical Information, ISO 14649, NC data

1. INTRODUCTION

Process planning is a time-consuming and highly skilled job, and t Retaining technical information is an important element for performing high quality machining. Manual collection of proprietary technical information is the most popular method of generating a database. Few research works have dealt with the generation of technical information databases automatically from NC data[1, 2]. However, actual NC data (ISO 6983)[3] contain implicit technical information which reflects the skilled worker's knowledge. In this research, as shown in Figure 1, in order to obtain technical information, NC data are analyzed based on a next-generation CNC control language (ISO 14649)[4] which contains technical information explicitly. The new ISO 14649 data model includes rich information such as machining features, machining operation, and machining conditions enabling feature-based programming. In this research, the machining features and process information are extracted from an NC program based on form-shaping function[5].

Please use the following format when citing this chapter:
Yamada, Hiroshi, Tanaka, Fumiki, Onosato, Masahiko, 2006, in International Federation for Information Processing (IFIP), Volume 207, Knowledge Enterprise: Intelligent Strategies In Product Design, Manufacturing, and Management, eds. K. Wang, Kovacs G., Wozny M., Fang M., (Boston: Springer), pp. 109-114.

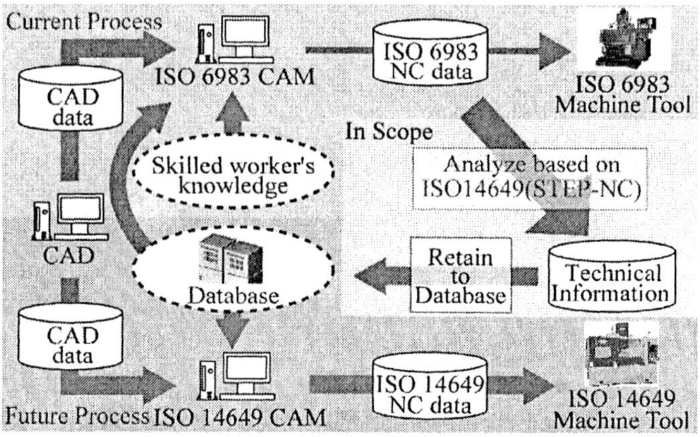

Figure 1. Objective of this research

2. ANALYSIS OF NC DATA BASED ON ISO 14649

2.1 Transformation procedure from ISO 6983 NC Data to ISO 14649 workingstep

The analysis procedure consists of five sub-procedures (activities) as shown in Figure 2. ISO 6983 NC data, cutting tool information and the workpiece blank model are used for the analysis.

The first sub-activity is the activity splitting ISO 6983 NC data into tool path elements for a cutting tool based on a tool change instruction, and classifying each tool path element as a bounded tool path. The classification of the tool path is based on the degree of freedom of the tool path element. The second sub-activity is generating the machining feature from the bounded tool path, the workpiece blank model, and the tool information. The details of this activity are described in **2.2**. The third sub-activity is generating the machining strategy and approach-retract strategy from the bounded tool path, the approach-retract path and the workpiece blank model. These are generated based on their classification and the mapping table between the bounded tool path and the machining strategy. The fourth sub-activity is generating the ISO 14649 workingstep from the tool information and the generated ISO 14649 machining feature, machining strategy, and approach-retract strategy. The fifth sub-activity is associating with the generated ISO 14649 workingsteps. The details of this activity are described in **2.3**. Figure 2 also shows the analysis example of slot machining.

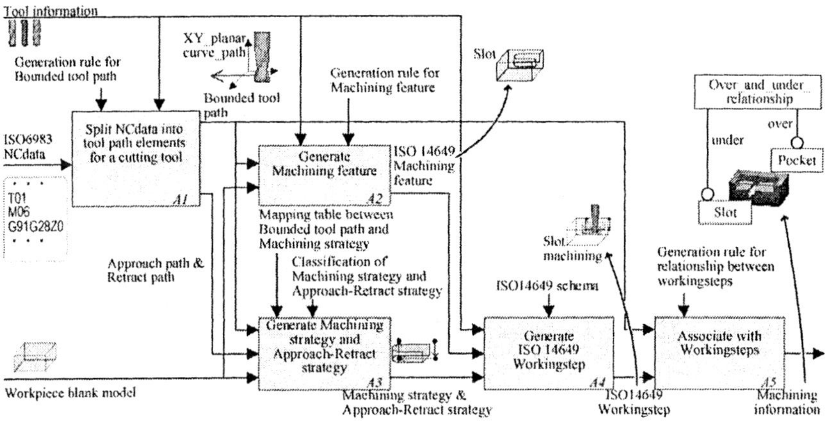

Figure 2. Procedure of analysis based on ISO 14649

2.2 Generating machining feature from Bounded tool path data

The activity of generating the machining feature from the bounded tool path, workpiece blank model, and tool information is shown in Figure 3 based on the form-shaping function[5]. In the first activity, the working tool is generated from the tool information and bounded tool path. The working tool is a functional model of the cutting edge generated from moving the macro tool that is generated by rotating the cutting edge around the spindle axis along the bounded tool path. The working tool is represented by the sweep profile which is the contact portion between the macro tool and the required shape and the sweep path which is the bounded tool path. The working edge is derived from the shape of the cutting tool. In the second activity, the removal volume is generated from the working tool. The removal volume is the subtracted volume of the working tool from the raw stock in the setup placement. To reflect the machining process information via the working tool, the removal volume inherits the information of the working tool and adds the boundary elements between the removal volume and the raw stock. In the third activity, the machining feature is generated from the removal volume. The machining feature is represented using the path and the profile in ISO 14649[4]. Therefore mapping between the machining feature and the removal volume which is represented as the path-profile is possible.

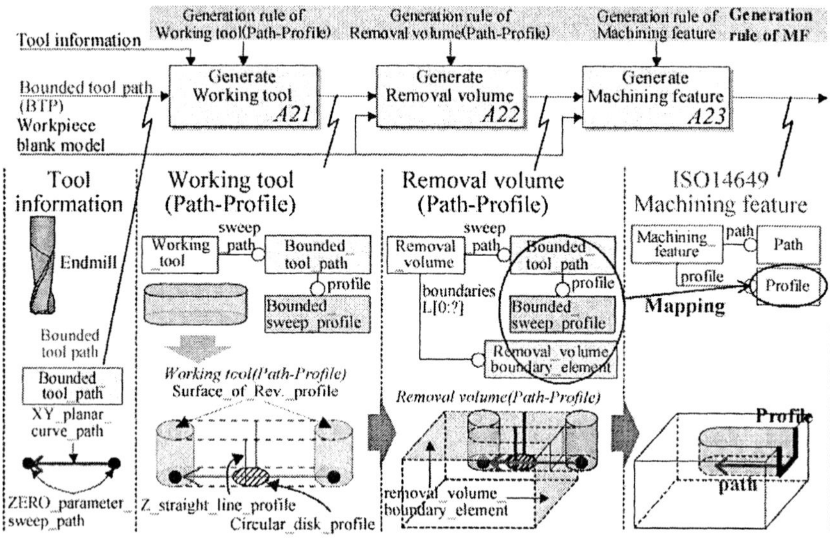

Figure 3. Generating machining feature from Bounded tool path

2.3 Representation of the Relationship between Workingsteps

In order to perform the process planning using the retained machining information, it is necessary to represent the relationship between workingsteps. However, in ISO 14649, there are few constructs which represent their relationships. Therefore, in this research, the new data model structure for describing the relationship between workingsteps is proposed as shown in Figure 4. In the proposed data structure, relationships consist of an operation relationship and a feature relationship.

The operation relationship describes precedence in machining processes. An example is the pocket which lays on the slot as shown in Figure 4 (c). In this case, the workingstep for pocket machining has priority over the one for slot machining. Such precedence is described by using Precedence_ relationship which is a subtype of Orientation_relationship. In addition, Roughing_and_finishing which is a subtype of Precedence_relationship is used for the relationship between roughing and finishing.

The feature relationship describes the geometric positional relationship between machining features. It is defined based on ISO 10303 Part 108[6]. As shown in Figures 4 (a), (b) and (c), respectively, Parallel_relationship is used for the machining features which are positioned parallel, Coaxial_relationship is used for the machining features which are coaxial, Over_and_under_relationship is used for the machining features which exist one above the other.

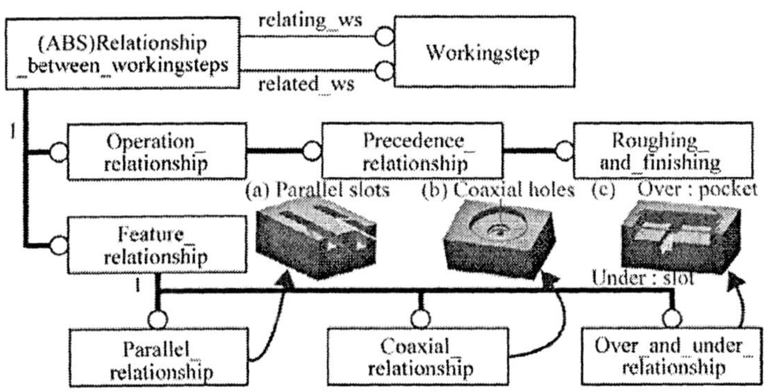

Figure 4. Data model structure for describing the relationship between workingsteps

3. EXAMPLE

In order to show the validity of the proposed system, we demonstrate how ISO 6983 NC data generated by CAM software are applied to the implemented test system as shown in Figure 5.

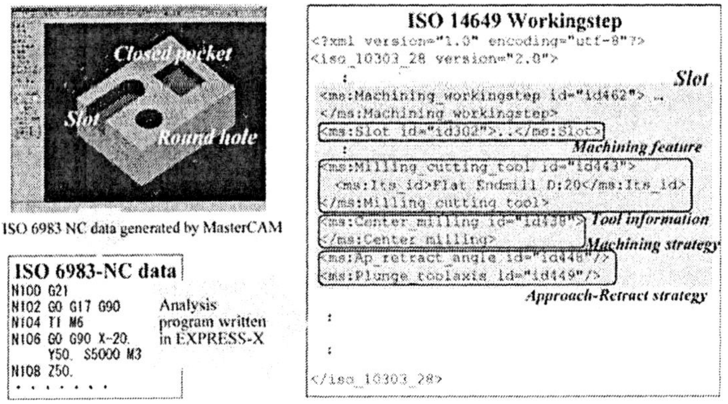

Figure 5. Data analysis example for slot, round hole, and closed pocket feature

The test ISO 6983 NC data for the slot, round hole, and closed pocket are generated by commercial CAM software (MasterCAM). As explained in previous sections, each activity in Figure 2 and Figure 3 is implemented by EXPRESS-X mapping language (EDMexpressX)[7], and EPM EXPRESS Data manager to execute the mapping process is adopted. From this example, the Machining_workingstep for the Slot, the Machining_workingstep for the Closed_pocket, and the Machining_workingstep for the Round_hole are obtained. These are the same as the contents of the ISO 6983 NC data. Therefore, valid ISO 14649 NC data are obtained.

4. CONCLUSIONS

In this research, technical information is extracted from the NC program, and is transformed into a next-generation CNC control language (ISO 14649) which contains the technical information. Conclusions of this research are as follows:

1. The method of extracting the machining feature and the process information from an NC program based on form-shaping function is proposed.
2. An analysis system for ISO 6983 NC data based on ISO 14649 is implemented using EXPRESS-X.
3. ISO 6983 NC data generated by CAM software are applied to the implemented test system, and valid ISO 14649 NC data are obtained.

REFERENCES

1. Yan X, Yamazaki K, and Liu J, Recognition of machining features and feature topologies from NC programs, *Computer-Aided Design* **32**(10), 605-616 (2000)
2. K.Nieda, H.Aoyama. Database Construction System of Cutting Techniques and Cutting Conditions Based on Skilled Worker's Operation and NC Data. Proc. of the International Conference on Leading Edge Manufacturing in 21st Century, 67-72 (2005)
3. ISO 6983-1 Numeric control of machines-Program format and definition of address words-Part 1: Data format for positioning, *line motion and contouring control systems* (1982)
4. ISO 14649-10: Industrial automation systems and integration- Physical device control-Data model for computerized numerical controllers-Part 10: *General process data* (2003)
5. Tanaka F., Takinami M., Kishinami T. , Removal Volume Modeling for Process Planning Based on form-shaping function. Proc. of the International Conference on Leading Edge Manufacturing in 21st Century, 747-752 (2003)
6. ISO10303-108: Industrial automation systems and integration-, Product data representation and exchange-, Part 108, *Parameterization and Constraints for Explicit Geometric Product Models* (2004)
7. EPM Technology, EDMassist Volume VI, EDMexpressX Language Reference (2002)

ENTERPRISE KNOWLEDGE MANAGEMENT BASED ON PLONE CONTENT MANAGEMENT SYSTEM

Chuanhong Zhou [1], Huilan Zeng [1]
[1]*CIMS & Robot Center of Shanghai University, Shanghai University, China; Email: chhzhou@staff.shu.edu.cn.*

Abstract: This paper represented a convenient and perfect Knowledge management Information System for enterprises by introducing Knowledge Management (KM) strategy into Content Management System (CMS). A Universal model was put forth on KM under the platform of CMS. In addition, an approach was researched for knowledge Acquisition by classifying under RSS in order to build a perfect Knowledge Process System (KPS). Various functions on KM were performed based on PLONE CMS, which imply the essential of Enterprise Knowledge Management (EKM). This research has a good prospect either in terms of economy or availability.

Key words: Knowledge Management; PLONE; knowledge Acquisition; Knowledge Process System

1. INTRODUCTION

Learning mode and knowledge reserved in enterprises reveal their performance and competitiveness in the market. Today more and more enterprises come to realize the importance of knowledge after suffering these troubles, such as core staff turnover, leak of key technique and loss of numberless important persons; as a result, they begin to build their own Enterprise Knowledge Management System (EKMS) to sharp the advantage. KM is an important conception in conducting Enterprises management and Information management, and aims at improving their overall abilities to acquire, develop, preserve, distribute and utility knowledge. Each enterprise, however, must combine its own feature with system.

Please use the following format when citing this chapter:
Zhou, Chuanhong, Zeng, Huilan, 2006, in International Federation for Information Processing (IFIP), Volume 207, Knowledge Enterprise: Intelligent Strategies In Product Design, Manufacturing, and Management, eds. K. Wang, Kovacs G., Wozny M., Fang M., (Boston: Springer), pp. 115-120.

Statistics shows, eighty percent of knowledge for many enterprises is hidden in the content directly or indirectly[1], so Content Management (CM) becomes a precondition in launching KM. CMS is regard as a support platform of KM with good feasibility and economy. In this paper a general model is presented on KM based on CM, and in which various functions are displayed to realize EKM in manufacturing sector.

PLONE is recognized as the most outstanding charge-free, open-source CMS with many merits, such as great convenience, use safety and reliability, internationalization and standard, expansibility etc. A new Knowledge Classifying and Acquisition method is explained based on RSS (RDF Site Summary) in the paper to perfect KPS, so as to extend the significance of KM in manufacturing firms.

2. UNIVERSAL MODEL OF KM BASED ON CM

2.1 Universal model of KM based on CMS

Literally CM means management of content, and KM means management of knowledge. The integration of these two becomes a rising trend with development. Statistics shows, eighty percent of knowledge for many enterprises is hidden in the content directly or indirectly, so content can be regarded as memories of enterprises, CM becomes a precondition in launching KM, meanwhile enterprises can also reach their goals by KM with help of content. A Universal model is put forth on KM under the platform of CMS with good feasibility and economy. CMS, as a support platform of KM, can collect fundamental information KM needs, and analyze and integrate KM by classifying and searching.

EKM Portal mainly consists of four big modules: CM module, Business process model, Communication Management and Project Collaboration, Knowledge Asset Management and Management of Employees with knowledge. The former comprises Documentary Management and official automation management (OA), Template Management, Workflow Management and Safe Management. The second one comprises Marketing Management (center in customer's relation management (CRM)), Logistic and Purchase Management (center in supply chain management (SCM)), Production and Manufacture management (combining the workflow management under PDM (Product Data Management), PLM (Product Lifecycle Management), and ERP (Enterprise Resource Planning)). Communication Management includes virtual meeting, right-time message, public white board, BBS, short message, online meeting, etc.

Knowledge in company helps take into shape group behavior, people can learn truth from their past experience and keep them in their common minds, working programs, rules or routine of company assets and activities[2] These four models penetrate flow of knowledge, which is formed by depositing, sharing and communicating to add great value to enterprises. It is needed to build a perfect KPS and workflow management system in establishing EKM based on CMS in manufacturing sector. See Fig 1.

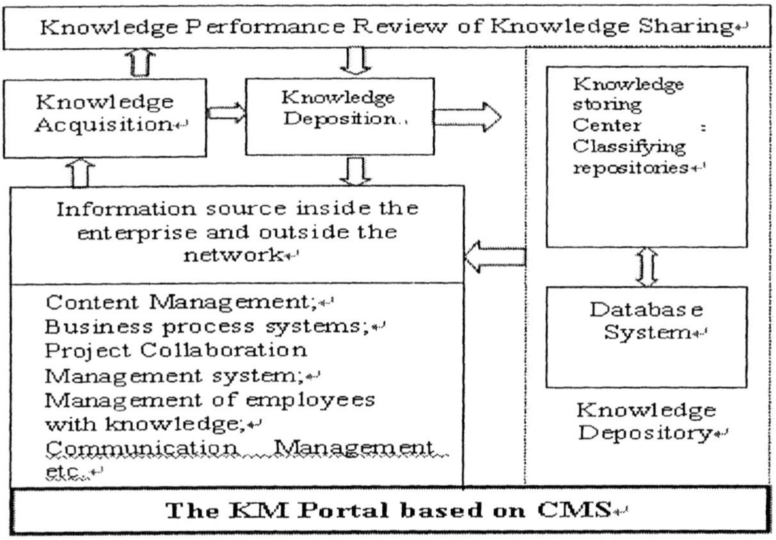

Figure 1. The Knowledge Process System model

2.2 Knowledge Acquisition

This paper presents two following crucial points in Knowledge Acquisition:

1. *Knowledge Classifying and Acquisition mechanism establishment based on RSS:* RSS[3] norms can be used to solve the issue like news publication. RSS is among of XML application, and can offer some optional, summarized web content to terminal customers, by which you can subscribe updated information offline as well. This paper proposed a method of Knowledge Classifying and Acquisition based on RSS, customers can get what they need through filling the specified condition. Tons of information and scattered data in company bring tougher difficulties to administrators, therefore, we introduce administrators a new approach to get necessary information timely in the light of specific requirements.

2. *Collecting information outside of intranet by RSS:* Users can read some refined materials and set latest time when not opening the web, giving that installing the third-party product BLOG (Weblog) in CMS, RSS can get BLOG information from outside timely, which is an elementary method to catch topics and latest information from BLOG.

2.3 Knowledge Deposition

Gathering all the knowledge is the first step to launch a campaign of KM in company, meanwhile collecting successful experiences and notorious lessons is necessary for enterprises so as to learn right from wrong. For this purpose, enterprises must come to change the model of handed Documentary Management to realize automation or semi-automation of business. Here three keys are pointed out to achieve this goal.

1) Knowledge classification system establishment: Based on current specified knowledge database, users must perfect their system consistently according to real situation, classify them respectively and issue them into the system timely.

2) A set of methodology put forth: Under different form, or knowledge, or environment, corresponding different measures should be taken.

The management of explicit knowledge: Make good use of current resources (such as instruction, various white papers, solution and proposal etc.) to neaten them and save into the relevant database in order to improve the efficiency and share of knowledge.

The management of tacit Knowledge: Motivate or force to realize the transition of knowledge from implicit to explicit. That valuable information kept in employee's minds, groups or special teams can be uncovered through BBS, knowledge motivation system, Knowledge Performance Review system, or specific review files or reports.

3) Knowledge Deposition by Knowledge Classifying and Acquisition mechanism: In this paper a new mechanism is established to acquire, deposit and retrieve knowledge[4] routinely by RSS.

3. KM PORTAL BASED ON PLONE

Various functions are designed to realize EKM in manufacturing sector so as to build enterprise KPS based on PLONE CMS. PLONE, giving birth at 2000, is advanced significantly and recognized as the most outstanding open-source CMS because of its strong function of ZOPE/CMF and the joint contribution from community leaders Limi and Runyan. PLONE CMS has many great virtues, such as convenience, safety and reliability, internationalization, accessibility, expansibility, strong searching power, customization etc, which make it successful to become the option for enterprises to set up EKM.

PLONE CMS offers overall function for Documentary Management and has perfect mechanism for privilege management. This paper proposes other functions in KM module by the following two aspects, 1.Utilize the colorful resources of the network, install the third plug-in (Products) of ZOPE or PLONE, in addition, you can also develop or create some requisite plug-in by PYTHON. 2. Strong workflow support is needed in KMS, the rule of workflow is built under OpenFlow, plug-in of ZOPE, which is a workflow management system based on activity and meets the criterion WFMC made. And what's more, visual workflow management system and workflow emulation system could be constructed based on PLONE so as to build perfect workflow management.

In the paper we take a case in a manufacturing company for example to explain KPS. Such items or processes are involved for most manufacturing enterprises, diagram papers, BOM (Bill Of Material) list, process planning, analysis model and result, cost structure, material specification, purchase price, training materials, knowledge forum, Email, message, and analysis report etc.

3.1 Knowledge Classifying and Acquisition mechanism establishment based on RSS

Knowledge Acquisition, a basis to deposition, includes in PLONE system, means of searching and filtering, RSS (RDF Site Summary), web navigation etc. In this paper, a classified searching system under RSS is initiated, which is a second-development package in Archetypes mechanism. Archetypes mechanism is a framework for developing new content types in PLONE.

Based on the above rule, customers can get what they need and create relevant dynamic RSS FEED through filling the specified condition. RSS Feed is a summary file about content of websites consisting of RSS elements. If typing a word like "machining" in "full text inquiry" of this Knowledge Classifying and Acquisition mechanism, the following result will be matched in RSS FEED format.

```
<?xml version="1.0" encoding="utf-8" ?>
-<rdf:RDF xmlns:rdf="http://www.w3.org/1999/02/22-rdf-syntaxns#"
        xmlns:dc="http://purl.org/dc/elements/1.1/"
        xmlns:sy="http://purl.org/rss/1.0/modules/syndication/"
        xmlns="http://purl.org/rss/1.0/">
  <channel rdf:about="...">
```

<image rdf:about=".../logo.jpg">
<item rdf:about="...">
<title>The method of improving machining accuracy</title>
...
<dc:creator>**designer3**</dc:creator>
<dc:rights />
<dc:date>**2005-06-22 22:48:10**</dc:date>
</item>

In addition, after installation of third-party product, BLOG for PLONE, RSS can get and filter sources outsides timely, which can be read and updated without opening web.

3.2 Knowledge Deposition

In this paper a new mechanism is established to acquire, deposit and retrieve knowledge routinely by RSS, system can get processing information regularly and timely and produce relative RSS FEED, which can be produced automatically by system, or also made manually. Then RSS FEED will be transformed into visual format and added into knowledge depositary.

4. CONCLUSION

In this paper a universal model of KM based on CM is put forward, and combined closely with business to build a EKMS based on PLONE CMS. This model can be introduced into various industries to establish related EKMS in order to run their business more efficiently. Furthermore, it also shows good accessibility to integrate with some prominent software for enterprise management, such as PLM, ERP.

5. REFERENCES

1. Thiruvathukal, G.K.; Laufer, K.; Plone and content management, Computing in Science & Engineering, Volume 06, Issue 4, July-Aug. (2004), pp. 88 – 95.
2. GAO Tianzhen, Model Framework of Knowledge Management System and the Analysis of the Key Problem in the Process of Knowledge Management Enforcement, SCIENCE OF SCIENCE AND MANAGEMENT OF S.& T, (2005), (3) ,pp. 69-72 (in Chinese).
3. Yuan Yongzhi, The Technology of Web Information Release and Integration Based on RSS, New technology of library and information service, (2004), (2),pp.60-62 .
4. Li Rong, A brief study on setting up an assessment mechanism of knowledge sharing, BEIJING TECHNOLOGY AND BUSINESS UNIVERSITY(SOCIAL SCIENCE), (2005), Vol.20, No.2, pp. 41-44 (in Chinese).

KNOWLEDGE MANAGEMENT FOR PROCESS PLANNING

Ketil Bø
IDI, NTNU, No-7490 Trondheim, Norway. E-mail: Ketil.Bo@idi.ntnu.no

Abstract: Knowledge Management offers enterprises a considerable competitive edge. Together with manufacturing features extracted from CAD models, the knowledge can be used for augmenting several manufacturing processes such as automatic process planning. Xplano is a Feature Recognition and Knowledge Management system using Manufacturing features, company resources and a knowledge base of rules containing "best practice" in the enterprise to automatically generate process plans.

Key words: Knowledge management, Process planning, Mechanical industry, Feature recognition, Resource database, Knowledge database, ISO STEP, 3D Visualization, Rule based reasoning

1. INTRODUCTION

Knowledge management is capturing, organizing, and storing knowledge and experiences of individual workers and groups within the company and making this information available to others in the organization.

Mechanical industry has invested heavily in software for computer aided design (CAD) and computer aided manufacturing (CAM). Increasing functionality for CAD/CAM is steadily reducing the amount of routine work connected to these activities.

On the other hand, the experts who decides WHERE and HOW to manufacture each product have limited access to modern tools.

To keep up with the designers and the manufacturing (more variants, shorter lifecycles etc.), the process planner is forced to look up his old plans from earlier work, and cut and paste from similar existing products. He is

Please use the following format when citing this chapter:
Bø, Ketil, 2006, in International Federation for Information Processing (IFIP), Volume 207, Knowledge Enterprise: Intelligent Strategies In Product Design, Manufacturing, and Management, eds. K. Wang, Kovacs G., Wozny M., Fang M., (Boston: Springer), pp. 121-126.

captured in routine work. His knowledge is not available for others, he has no time for developing new and more efficient solutions, and the knowledge will leave the company with him.

It is important for the competitiveness of the company to take care of "best practice" and make available the knowledge and the experience of the production process from those who know to those that need to know.

2. PROCESS PLANNING

The information associated with a machined mechanical part progresses through several steps, from its specification in a CAD system, through its interpretation in terms of manufacturing features, the associated sequence of machining operations to finally the tool paths for all operations. Each of these steps can be regarded as a transformation that produces a new representation of the object in another domain of discourse. Unlike most mathematical mappings, however, these transformations are interpretations of the object using knowledge about each successive domain to infer information that is not explicitly present in the model.

Process planning systems support the planning of the manufacturing procedure flow for a product. Here, both the manufacturing process and the manufacturing sequence must be determined. With basis in the geometry, the available resources and the company specific experience and knowledge, procedures for the production of work pieces are determined.

The planning of manufacturing sequences can be easier realized by using manufacturing features by means of which a tight coupling between geometry and manufacturing can be represented. For the interpretation of work piece data, a close connection between design and process planning must exist.

Recognition of manufacturing features, based on the CAD model, should be carried out as an integrated part of process planning to simplify the detection of tools and machines.

The process planning result mainly originate from the work piece, features, material, tools, machine hierarchies and the decision sequences for production planning. A high performance, object oriented form of representation is required, whereas the representation of decision sequences makes rule processing advantageous.

Databases should be employed for storage of resources and rules. The ability to modularize knowledge- and resource data bases is required for more complex planning systems in order to facilitate understanding of the program structure.

3. PRODUCTION KNOWLEDGE MANAGEMENT

The production knowledge consist of the resources available, the "best practice" of where, how and in what sequence the features of a product are produced in the company.

Xplano is a software system designed to take over this routine work. At the same time it gives the enterprise a modern tool to accumulate experience and ensure continuous process improvements.

The software is a visualization and knowledge management system, making all the production knowledge in the enterprise, available for those who need to know. In addition it also has a reasoning engine that uses the knowledge in the system to decide the best methods, tools and process data for any product feature. This makes Xplano capable of generating a detailed process plan for the products according to the knowledge and experience accumulated from the local manufacturing experts.

For chip removing production, the resources are systematized in a Resource database with Machine tools, Fixtures, Rotating tools, Turning tools and Raw material. The Knowledge is systematized in a Knowledge database with Manufacturing methods, Machine tools rules, Fixture rules and Operation order rules.

Some large international companies have been successful with rule based systems that are able to decide the manufacturing process for a limited family of similar products. The problem with these systems are that a wider range of products gives a complexity in rules that will easily create conflicts in certain areas, and/or lack of exact rules in other areas. To cover other product families, another expert system with many of the same elements is required. Some of the same data and the same rules will be created and maintained for many different applications which will absorb a considerable amount of resources from system people and process designers.

Xplano's goal is to make a generic software with enterprise specific Resource- and Knowledge data bases and a flexible reasoning engine able to decide the manufacturing process for all the products in the enterprise, still being both easier to use and easier to maintain than the existing systems based on conventional programming and exact rules. This approach lowers the "break even" point for the investment, and the software will be profitable also for small and medium size enterprises.

Knowledge management (KM) is the management of the company's individual and collective knowledge. The ultimate benefits are overall improved efficiency and reduced time from design to production, as well as cost reduction through reduced time in process planning and optimal production.

Promoting information sharing, motivating employees to stay with the firm, forging partnerships for knowledge acquisition is the benefit that makes industrial companies becoming increasingly aware of the need to manage individual and collective knowledge. The advantages deriving from knowledge management are not only explained by company size, specialization or research & development efforts. Knowledge management also stimulates innovation, a productivity factor. Preserving a company's skills and expertise, acquiring and disseminating new knowledge, is important factors to improve the competitiveness.

Different factors are promoting knowledge management:

Companies have to manage a more complex world because of changing technologies.

They are concerned about the loss of skills caused by today's extreme mobility of personnel. For all these reasons, companies are striving to motivate their employees to stay with the firm, alternatively take care of the expert knowledge if the expert leaves the company.

4. XPLANO

Xplano get its "expert knowledge" from the Knowledge Management System (KMS) built and maintained by the production engineer himself in the individual enterprise.

The KMS includes decision rules for all the steps in the design of a manufacturing process together with a Resource database containing information about the manufacturing resources available. The decision rules are expert knowledge, agreed by the manufacturing engineers in the company as the best known practice, and linked up to the preferred resources and process data for all the different conditions the experts can think of.

Automatic Process Design starts with a "feature extraction" process from the geometry model written in the international standard ISO STEP format (most advanced CAD systems are capable of storing the design model in ISO STEP format).

Output from the" feature extraction" process is a feature file, written in XML format, with all the information available for the manufacturing process. If this information is not complete, it is possible to fill in additional requirements, before the file is processed by the Reasoning Engine.

The Visualization Module visualize both the geometry and the extracted 3D feature model in a way optimized for Process planers, both for visual control and alternatively as a tool for manual or template based Process Planning.

The Reasoning Engine reads the feature file, consults the KMS and puts together a detailed description of the best possible production process for the component, using the best fit resources available.

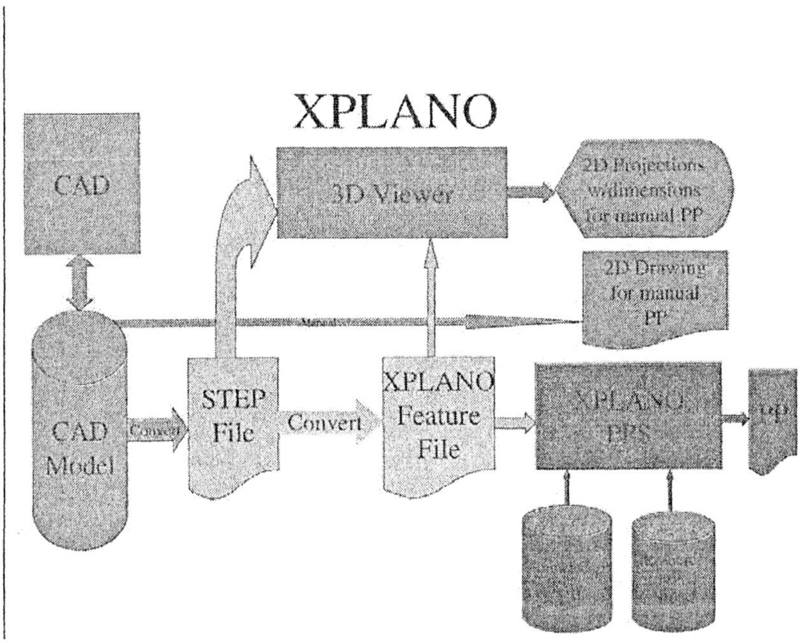

Figure 1. The integrated Xplano System

Artificial Intelligence techniques will assist the reasoning engine to take a decision when rules are in conflict or missing, and give the software the ability to learn from possible corrections made by the experts when the decision from the system is not acceptable.

The software is designed as a framework that must be filled in with the relevant rules and resources of the individual enterprise. The process specialist himself will enter the decision rules of the enterprise concerning methods, materials, machine tools, adapters etc..

The concept of Xplano is to collect as much product information as possible from the design file of the CAD system, and us it together with the decision rules from the company specific Knowledge database and the resources available from the Resource databases where the content are created and maintained by the local experts.

5. BENEFITS

Significant benefits can result from the production planning system only. In a survey of twenty-two large and small companies using generative-type CAPP systems, the following estimated cost savings were achieved [4]:

58% reduction in process planning effort.
10% saving in direct labor.
4% saving in material.
10% saving in scrap.
12% saving in tooling.
6% reduction in work-in-process.
In addition, there are a number of other intangible benefits.

6. CONCLUSION

Manufacturing Knowledge management (MKM) is an efficient way to take care of company specific knowledge and experience. In addition the knowledge can be used to support and augment manufacturing processes such as generation of process plans.

Xplano is a Manufacturing Knowledge Management (MKM) system for mechanical industries that collect and systemize resources and process knowledge in an enterprise and makes it available for everybody involved.

In addition the system also extracts and visualizes manufacturing features from the CAD model and automatically generates a process plan containing "best practice" of the company.

7. REFERENCES

1. Han, J.H., Pratt, M. And Regli, W.C. (2000). Manufacturing Feature recognition from Solid Models: A Status Report. IEEE Transactions on Robotics and Automation. Vol 16, No 6.
2. Bø K. (2006) Production Knowledge Management for the Mechanical Industry, *CAD User*, Vol 18 No 12.
3. Vandenbrande, J.H. and Requicha, A.G. (1994) Geometric computation for recognition of spatial interacting machining features *Advances in feature based manufacturing*, North Holland: Elsevier,
4. Crow, K. A. Computer Aided Process Planning. DRM Associates
5. Unigraphics: "Hybrid Modeling Fundamentals" Electronic Data Systems Corporation, Version 13.0, Unigraphics Division
6. Krause, F.L. and Schlingheider, J. (1995) Development and Design With Knowledge-Based Software Tools – An Overview. *Expert Systems with applications*. Vol 8 No. 2 pp233-248.

ADAPTING MANUFACTURING TO CUSTOMER BEHAVIOR

Lessons learned from trading goods on public market places

Stephan Kassel, Kay Grebenstein
University of Applied Sciences, Dr.-Friedrichs-Ring 2a, 08056 Zwickau, Germany,Email: Stephan.Kassel@fh-zwickau.de

Abstract: The globalization of competition urges manufacturers to reduce costs and simultaneously provide a better service to the customer. To achieve both goals, the prediction of customer behaviour plays a key role. This can be done by observation of the customers on public market places like eBay. These observations have to be combined with events influencing customer preferences. For this purpose, a decision support system for retailers was designed, combining an expert system with a data warehouse. The experiences of this project can be utilized for manufacturing companies as well.

Key words: event-driven customer behaviour, demand prognosis, knowledge management, data warehouse, public markets, decision support, customer-driven manufacturing

1. INTRODUCTION

Due to globalization and the emergence of customer-oriented markets the companies have to change their behaviour and provide additional value to their customers. This can be done in different ways, e.g. by providing a better service or by individualizing products to the customer. Customer-oriented or agile manufacturing (Kidd 1994, BüyüKözkan 2004) are the key words of this latter means of achieving customer satisfaction. But this leads to great problems for the manufacturing industry. With a higher degree of customer-orientation it becomes more and more impossible to produce goods in advance, leading to the advent of supply chain management (An and Fromm 2005, Becker 2005). This has been true for parts of the

---

Please use the following format when citing this chapter:
Kassel, Stephan, Grebenstein, Kay, 2006, in International Federation for Information Processing (IFIP), Volume 207, Knowledge Enterprise: Intelligent Strategies In Product Design, Manufacturing, and Management, eds. K. Wang, Kovacs G., Wozny M., Fang M., (Boston: Springer), pp. 127-133.

manufacturing industry, the special purpose machines industry, but it becomes more and more reality for consumer goods industry, too. For example, in the high-price car industry, the customers can choose from dozens of colours and a multitude of features. Even in the convenience goods industry, the number of products is ever growing.

On the other hand, there is the need to reduce stock capacities to achieve a minimized capital binding in products on stock. This can partially be solved by introducing lean production concepts (Womack and Jones 1996; Liker 2004). But often it leads to a dilemma, which is actually solved on the backs of the customers, who have to wait for high-price products very long, and/or on the back of the employees, who always have to stand by to solve customer needs quickly when they arise.

Some of the problems seem to be not solvable, because a multitude of manufacturing processes need time to be done. So every means to prognosticate the point of time, when the customer will buy some special goods, is increasingly important and can become a competitive advantage for the manufacturing companies.

There are two different ways to identify customer behavior successfully. One way to meet with general acceptance is to introduce customer relationship management (CRM). This approach tries to identify the needs and wishes of the individual customers to manage customer relationship and to provide a feedback channel to increase customer satisfaction (Kamakura et al. 2005). The second way is to analyze changes of demand on public market places and identify potential reasons for these changes (Liu et al. 2000). This is a generic approach and uses statistical procedures for the examination of the observed behavior. In the analysis of the reasons for the changes of customer behavior, one can identify different causes, as done by Verhoef et al. (2002). There are common and long-term trends, which are influenced by technological progress, fashions and culturally changing habits of the customers. Another class of reasons consists of special events, which can be periodic (like onset of winter, beginning of summer, Christmas or other holidays, or even week-ends or paydays), or acyclic (like Olympics, world championships, or big pop music events). This analysis and the utilization of the identified correlations is the core of our research activities.

2. A SELF-ADAPTING DECISION SUPPORT SYSTEM FOR TRADING ON PUBLIC MARKET PLACES

2.1 The initial research objective

Sales channels have changed dramatically in the past decade. After the e-commerce hype leading to big retailers selling via individually built e-commerce systems, public market places were built to provide a tool for selling goods for everyone. This trend will continue for the next years, as forecasted by „Jupiter Research" (Salcedo 2004): "The online commerce market in Europe will grow from €29 billion in 2003 to €117 billion in 2009, with 61 percent of European Internet users buying online and spending an average of €843 per buyer. Increased online tenure, growth in the online population, and improved broadband uptake will be the main drivers behind this solid growth". The same trend can be seen in each region of the world. Especially eBay was a big success story, starting with a platform to enable people everywhere selling things they didn't need any longer. But not only private trade was enhanced by eBay. Power sellers started up their businesses on this public market place and were successfully selling goods via this channel. Lately, the huge customer base of eBay attracted big retailers to make use of this trading channel, too.

To make business on these public market places, one doesn't need any complex e-commerce system. There are solution providers which deliver out-of-the box functionality for connecting inventory management systems with the market place. This can be done by using web services, a very flexible and light-weighted connection method for information systems.

So the connection to the public market place as a sales channel is neither a problem nor rather expensive. But for new participants on this market, there are a number of open issues which have to be solved to make optimal use of the sales channel. When should one place goods on the market? How many goods should be placed? How long should the goods be placed, or which auction duration is optimal? Which is the right (starting) price for the placed goods? Which sales campaign (fixed price and/or auction) should be chosen? The right answer to these questions depends on the goals of the seller, which can be a mixture of optimizing the selling volume or optimizing the sales price or both.

In the next sections, a solution for these issues is outlined, which has been prototypically implemented by the authors as part of an industry-funded research project.

2.2 Solution overview

The industrial partner AGETO has built a software system which interfaces between mail order systems and different market places like eBay. The product data from the sellers are transformed to offers and placed on the public market automatically; the orders on the market place are routed to the seller systems for fulfillment.

To be successful on the public market place, one has to learn from the participants which are already prospering. So the first step consisted of utilizing the knowledge of diverse power sellers. The knowledge of the power sellers should be shared by explicitly including it in a common knowledge base, as proposed by Sol (2002). An expert system was built, which determines the amount of goods to be placed on the market place as well as the exact point in time for the placement. The rules of the expert system reflect experience knowledge of power sellers. As an important influencing factor for the placement of offers on the market, facts and rules on external incidents like special events were included in the knowledge base from the beginning. The expert system was built using the open source shell Mandarax (Dietrich 2004)

In the next step, the actual sales data were collected and prepared to be utilized. A data warehouse has been built to hold internal data of the market place as well as external data of different data bases delivering information on events which were supposed to be influential for the sales success. This data warehouse, which has been built as a PostgreSQL database, was the first step towards business intelligence, enabling the enterprise to set up a successful customer relationship solution including the analysis of cross selling opportunities, as described by Vitt (2002). Basically, the development of prices over time, auction length, placement time, and external factors like seasons, weather, holidays, were included in the data base. From the basic data, classification numbers were derived to provide compact information on sales success from different perspectives. These classification numbers were related to the external events to determine possible correlations between the external events and the internal sales data.

After building the data warehouse and performing initial correlation analysis, the expert system and the data warehouse had been coupled to provide a closed loop application. This allowed for the automatic adjustment of the expert system rules to the feedback of the classification numbers leading to a higher conformance of the expert system with the changing behavior of the customers. Thus, recurring trends could be computed and used for a better prognosis of market activities.

2.3 Lessons learned from building the knowledge-based decision support system

In the project, a prototypical implementation of a knowledge based decision support system has been built. We have created an expert system containing a set of rules depicting the knowledge of power sellers for placing auctions and fixed-price bids on public market places. This expert system was coupled with the AGETO eBay web service system to provide sellers with assistance in placing their offers on the public market. This part of the work could be achieved well by using a public domain expert system shell.

The planning and implementation of the data warehouse has been challenging, caused by the vast amount of data we had from the user behavior on the market place. We needed to transform numerous log files to provide us with the data needed to measure the success factors of a bid. In extension, the external data, like weather forecasts, had to be condensed and classified. This classification was not easy, and has to be adjusted depending on the correlation quality which can be achieved with the classification.

The analysis of the data warehouse for finding correlations between external events and sales figures has been done manually. Classification numbers have been built and were used together with the external event data as input for some statistical measures. The found correlations were promising, but further analyses have to be done to find other connections as well. Maybe, we have to use data mining for raising the quality of the analysis.

Connecting the data warehouse with the expert system could be done easily; the expert system has a data base connection and could be enhanced to use some of the classification numbers as facts, leading to new results.

3. TRANSFERRING THE IDEAS TO MANUFACTURING

The principles of the presented solution can be utilized for a better prognosis of customer demand. The observation of special markets and the analysis on sales figures on these markets can be combined with external events influencing customer behaviour. A feedback system can be developed to predict demand for the manufactured good more precisely. Useful for this observation are especially the open public marketplaces, where you can achieve a multitude of data concerning preferences of customers. Additionally, the sales channel of public markets is rather fast. You can identify trends from the beginning, because the goods are rapidly traded, and the prices of auctions are market driven. This advantage of public market places is already utilized by big online retailers leveraging this sales channel as an ideal test market for new products, which are not announced with expensive marketing campaigns. This can be done anonymously, so the retailers get immediate response on customer affectations.

This advantage could be used by the manufacturers as well; they could sell prototypes via this channel and get immediate customer feedback.

By further enhancing the system with a direct connection to the production planning system, a demand driven manufacturing could be established, which is depending on the predicted customer demand. Together with the more direct interaction with the customers by using public markets, this behaviour can bear advantages in some industries, because the trade margin can be transformed to customer service and lower prices, further enhancing customer satisfaction and retention.

4. REFERENCES

1. An, C., Fromm, H., eds., (2005), *Supply Chain Management on Demand*, Springer, Berlin.
2. Becker, T., (2005), Prozesse in Produktion und Supply Chain optimieren, Springer, Berlin.
3. BüyüKözkan, G., Derelİ, T., and Baykasoğlu, A., (2004), A survey on the methods and tools of concurrent new product development and agile manufacturing, *J. Int. Manuf.* **15**(6): 731-751.
4. Dietrich, J., (2004), A Rule-Based System for eCommerce Applications, in: *Proceedings of Knowledge-Based Intelligent Information and Engineering Systems: 8th International Conference*, Springer LNCS 3213 / 2004, Heidelberg, 455-463.
5. Kamakura, W., Mela, C.F., Ansari, A., Bodapati, A., Fader, P., Iyengar, R., Naik, P., Neslin, S., Sun, B., Verhoef, P.C., Wedel, M., and Wilcox, R., (2005), Choice Models and Customer Relationship Management, *Marketing Letters* **16**(3-4): 279-291.
6. Kidd, P.T., (1994), *Agile Manufacturing*, Addison-Wesley, Reading.
7. Liker, J., (2004), *The Toyota Way*, McGraw-Hill, New York, New York.
8. Liu, B., Hsu, W., Han, H.-S., and Xia, Y., (2000), Mining Changes for Real-Life Applications, in: *Data Warehousing and Knowledge Discovery: Second International Conference, DaWaK 2000, London, UK, September 2000. Proceedings*, Y. Kambayashi, M. Mohania, A M. Tjoa, eds., LNCS 1874-2000, 337-346.
9. Salcedo, L., (2004), *Market Forecast Report European Commerce, 2003–2009*, JupiterResearch, Jupitermedia Corp.

10. Sol, H., (2002), *Expert Systems and Artificial Inteligence in Decision Support Systems*, Kluwer Academic Publishers.
11. Verhoef, P.C., Franses, P.H., and Donkers, B., (2002), Changing Perceptions and Changing Behavior in Customer Relationships, *Marketing Letters* **13**(2): 121-134.
12. Vitt, E., (2002), *Business Intelligence: Making Better Decisions Faster*, Microsoft Corporation, Redmond
13. Womack, J.P., and Jones, D.T., (1996), *Lean Thinking*, Simon & Schuster, New York, New York.

APPROACH FOR A RULE BASED SYSTEM FOR CAPTURING AND USAGE OF KNOWLEDGE IN THE MANUFACTURING INDUSTRY

Supporting manufacturing machines tasks with knowledge based services

Jivka Ovtcharova, Alexander Mahl, Robert Krikler

Institute for Applied Computer Science in Mechanical Engineering RPK, Universiy of Karlsruhe (TH), Kaiserstr. 12, D-76131 Karlsruhe, Germany, Email:ovtcharova@rpk.mach.uni-karlsruhe.de

Abstract: The support of manufacturing machines with IT solutions becomes state-of-the art in the production process. For this reason the European Chinese research project KoBaS (Knowledge Based Customized Services for Traditional Manufacturing Sectors Provided by a Network of High Tech SMEs) aims to create customized services for manufacturing machines providing a set of tools including advanced task and process planning machine configuration, maintenance, training and management support. This paper describes a software application for capturing and re-using rule based knowledge concerning manufacturing machine services listed before. The system is designed to support different kinds of manufacturing machines and manufacturing machine specific domains and furthermore to support an open and easy to implement data exchange and communication process.

Key words: Rule Based System, Knowledge Engineering, System Integration, Intelligent Manufacturing Machines

1. INTRODUCTION

Today the manufacturing companies have to cope with the increasing pressure from global marketplace. To compete successfully many industrial manufactured products need to be changed as described in [1], especially those that have common requirements for increased product customization

Please use the following format when citing this chapter:

Ovtcharova, Jivka, Mahl, Alexander, Krikler, Robert, 2006, in International Federation for Information Processing (IFIP), Volume 207, Knowledge Enterprise: Intelligent Strategies In Product Design, Manufacturing, and Management, eds. K. Wang, Kovacs G., Wozny M., Fang M., (Boston: Springer), pp. 134-143.

and improved competitiveness in terms of reduced cost, shorter delivery times and improved quality [2]. To fulfill the requirements coming from the market, the manufacturing industry needs a breakthrough regarding the current practices in the use of manufacturing machines.

In this context the European Chinese research project KoBaS (Knowledge Based Customized Services for Traditional Manufacturing Sectors Provided by a Network of High Tech SMEs) aims to support the manufacturing industry with a set of innovative software tools. These novel tools facilitate on one hand, an advanced task and process planning and, on the other hand, machine configuration, maintenance, training and management [2] for the manufacturing machine.

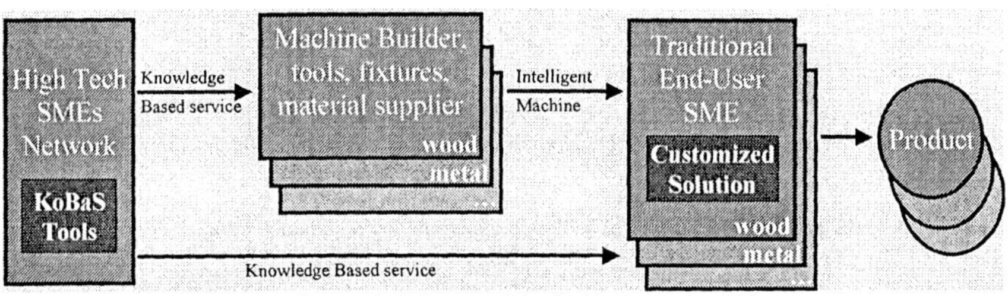

Figure 1. Schema of the KoBaS solution [2]

Since two manufacturing machines may be significant different in essential parameters like spindle RPM[1], maximal pressure, cutting speed etc. - even within one domain (wood, metal, plastic, stone etc.) - there is the need to provide a customized solution for each machine. The network of High Tech SMEs (to be established out of the KoBaS consortium) will install a customised solution of the software tools which take into account the requirements of the machine. In order to succeed in this task it is necessary to *re-use the knowledge* gained from former installations as well as to provide a *flexible, adaptable solution* which matches to the needs of the specific machine (and company). Thus the approach is to represent the knowledge using a rule based system which controls the flexible (customizable) part of the application logic of the software tools. This approach prevents to change/adapt the implementation of the software tools for each manufacturing machine.

Accordingly this paper focuses on the rule based system which builds the flexible knowledge core of the customized solution. The decisions of the rule based system on runtime have to take into account data from external IT-

[1] RPM: Revolution Per Minute

systems (e.g. PDM[2] or PPC[3] system). For example in order to provide a proper task planning the current status of the machine needs to be taken into account. Thus the rule based system must be flexible enough to be embedded into an existing heterogeneous system environment. Beside the description of the rule based system the paper will focus on an ontology based method for efficient realization of the setup phase of the customized solution. Within the setup phase the rule based system needs to adapt to the requirements coming from the involved IT-systems (e.g. data models).

2. RULE BASED SYSTEM WITH OPEN ARCHITECTURE TO SUPPORT MANUFACTURING MACHINE TASKS

The rule based system presented in this paper uses an object oriented approach for defining rules. Existing object oriented rule based systems like JESS[4][3], JRules[4], QuickRules[5], drools[6], JEOPS[5][7] or Rules4J[8] provide mechanisms to manage application logic of a software application. The integrated approach presented in this paper allows executing and defining rules on data objects (e.g. PDM data) as well as application specific objects. Thus the knowledge represented by rules could be used and applied on data and application objects.

Hence it is necessary to deal with several data models simultaneously, whereas the data models may change during runtime. Consequently the rule base operates on meta data model level. The paper will present the meta data model approach as well as the methodology of object oriented rule specification and execution within a heterogeneous software environment dealing with different data models in parallel.

2.1 Object oriented approach for rule representation

As stated before the paradigm of object oriented rule representation is implemented within several object oriented rule base engines. The prototype of the rule based system provides a generic rule description which may use any object oriented rule engine in its kernel. A rule base is defined through

- a data model,
- a set of rules and

[2] PDM-system: Product Data Management system
[3] PPC-system: Product Planning Control system
[4] JESS: Java Expert System Shell
[5] JEOPS: Java Embedded Object Production System

- instance data elements (e.g. objects, relations).

The set of rules as well as the instance data needs to be accordingly to the specified data model. Thus in case an object type will be deleted all instances (objects) will be deleted as well. Based on a request send to the rule base none, one or more rule(s) of the set of rules may be applied (this is called a rule "fires" [3]). The result will be send back.

Each rule consists of three sections: *declaration* of objects used within the rule, *condition* of the rule and *conclusion* (output of the rule). The condition is a logic expression (containing one or more logic operations (and, or etc.)) which decides whether the rule is applicable or not. The conclusion consists of one or more actions whereas an action is to create/modify an object, create/remove a relationship.

This paper introduces an exemplary a rule which may be used by the part program creation module (PCM) and access instance data of other IT-systems. The task of the part program creation module is to generate a CNC[6] program for a part to be manufactured on the machine. The knowledge how to generate the CNC code is modelled using the rule base. This will be done by specifying parameters which will later be used by the CNC code generator algorithm of the part program creation module. In case the rule "fires" there is a request with a part which fulfils the specified condition. The request refers to a part of a product structure administrated in the PDM system. The expected response of a request will be the information stating the possibility to manufacture the part on the machine and describing the CNC-program code which will be used by the part program creation module.

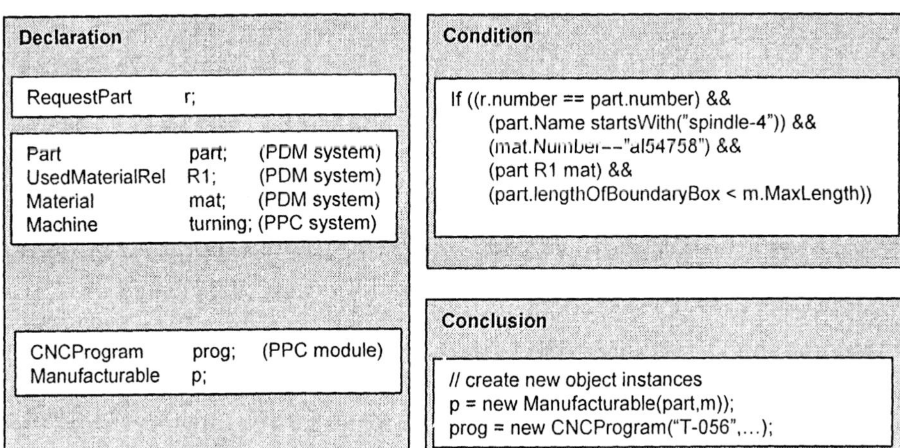

[6] CNC: computerized numerical control

Figure 2. Object oriented rule specification

Figure 2 presents an example for a Java based object oriented rule representation. The declaration section defines the objects which occur in the condition or conclusion section. The condition section is an (complex) if statement: The first "and-clause" searches for the requested part number in the rule base. If no part object with the specified requested number was found the rule will not fire. Otherwise it will checked, if the name of the part starts with "spindle-4", the specified material for the part has got the Id=″a154758″ and the Boundary box of the part is smaller than the maximum length of the working space specified in the machine.

In case the rule fires the (requested) part could be manufactured by this machine. This will be specified by the creation of a new instance of a "Manufacturable"-object. Furthermore the CNC-program will be specified using some internal parameters (e.g. CNC-module-Id=″Turning-056″ etc.). The rule will be applied in case the request send to the rule base contains a request part object which refers to a part object that fulfils the specified conditions.

2.2 Approach for generic data model description

To enable the rule based system to be embedded into an existing heterogeneous IT-system environment and adapted to the needs of a manufacturing machine, the rule based system needs to handle any data model. Hence the operations on instance data must be parameterized by the data model of the customized solution but act on meta model elements. A meta model describes the elements which are used to build a model. In case of an (object oriented) data model elements can be: object type, relationship type and attribute type as well as instance elements like object, attribute and relationship. The API[7] of the rule based system provides functionality for creating and manipulating these elements. Thus it is possible to modify the data model on runtime which will also effects the objects (internal facts) of the rule base. In case of deleting an object type all instances consequently will be deleted as well.

This approach bases on the results of the former European Chinese research project DRAGON [9] which was successfully finalized by RPK. The application case within DRAGON was limited to PDM data and didn't take in account (rule based) reasoning mechanisms nor object oriented rule specification. The approach in this paper enhances the DRAGON approach additionally by allowing the simultaneous work with several data models.

[7] Application Programming Interface

2.3 Exchange of instance data

Each rule based system contains one or more rule bases whereas each rule base consists of a data model, set of rules and instance data. The set of rules and instance data base on the data model. Thus each change of the data model might affect the set of rules and instance data. Modifications of the data model result in changes of the instance data (see 2.2). In order to avoid inconsistent rules a data model element must not be modified or deleted while it is used within the specification of the rule. This action is refused by the rule based system unless the rule is removed or modified first.

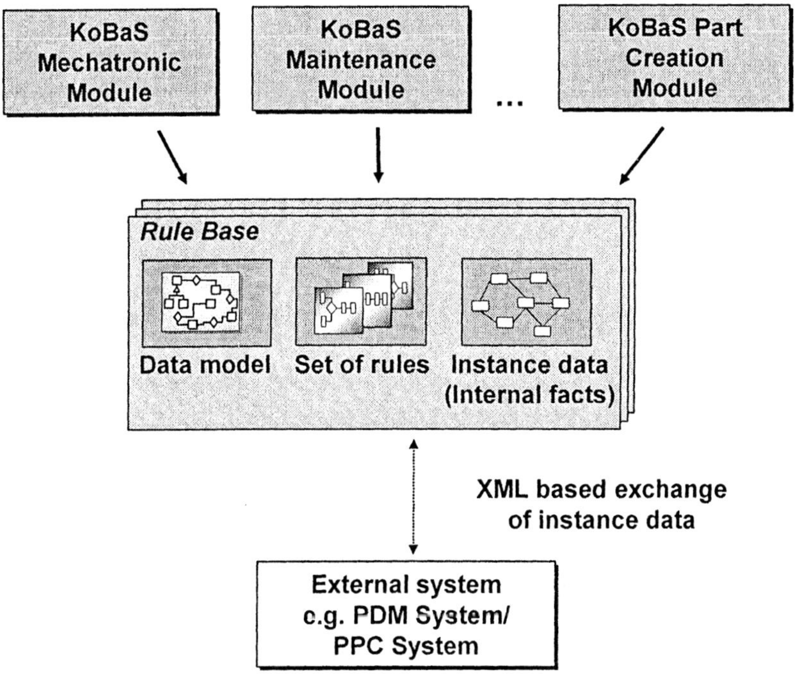

Figure 3. Exchange and access of instance data from heterogeneous data sources

The rule based system presented in this paper allows application to execute rules on instance data of application specific data objects and external systems like PDM or PPC systems (see *Figure 3*, furthermore see example presented in 2.1). Hence it is possible to use and modify product data within the execution of a rule.

The exchange of instance data is XML based whereas the correspondent XML document bases on a general description of instance data using the meta data model elements. E.g. each object has to specify its object type etc.

The instance data will be checked on runtime if the description is conform with the current data model of the rule base.

2.4 Setup phase of rule based system

The outcome of the whole KoBaS project will be a set of software modules which build the foundation for an intelligent manufacturing machine (see chapter 1). Within the setup phase these modules need to be adapted to the specific requirements of the manufacturing machine. Manufacturing machines of different domains fulfill different tasks and differ in design, operation mode and measurements. Thus the data model (and rules) for describing the machines will be different within machines of different domains. Within the research project the domains woodwork and metal are focused due to the presence of end-users while the approach is not limited to these domains. After project end it is expected to introduce the KoBaS solutions for various domains of manufacturing machines e.g. plastic, molding etc.

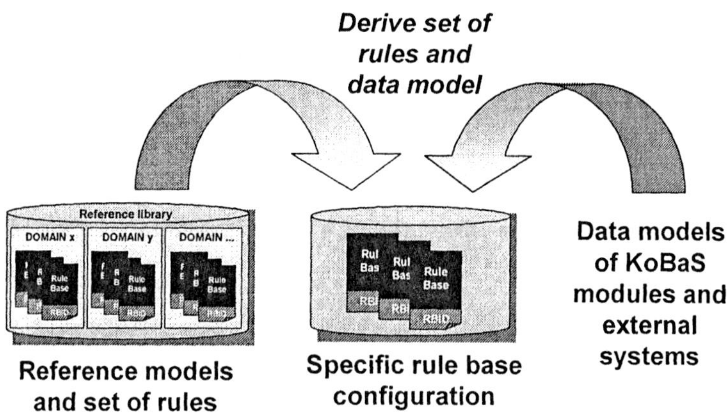

Figure 4. Exchange and access of instance data from heterogeneous data sources

In order to re-use the gained knowledge from former projects (installation of KoBaS software for other machines and enterprises) and speedup the setup phase a *reference library* for each domain will be created and incremental extended (e.g. after each KoBaS solution). This reference library contains for each domain one or more rule bases whereas (in general) one rule base is assigned to one KoBaS application.

Within the setup phase for each KoBaS application (e.g. part program creation module) the rule base (data model and set of rules) needs to be specified according to the manufacturing machine and enterprise specific processes. Therefore several different data models needs to be taken into account (see *Figure 4*): the reference data models (from reference library), data models of KoBaS applications (software modules) and data model(s) of external system(s) (e.g. PDM-system(s)) which manage(s) the data within an enterprise. The rule based system provides a *graphical rule base editor* which supports the setup process by providing views of several rule bases (and hence data models and instance data) as well as functionality for specifying the new data model based on these models. The rule based system provides an XMI[8][10] interface which allows importing the different data model specifications. Thus the data models may be specified using any XMI supporting UML tool like Borland Together, Enterprise Architect, Omondo or Rational Rose.

The specification of a rule rest upon on data model elements like object types, relationship types or attributes (see 2.1). Thus the reuse of existing rules is only possible if these elements exist within the new data model or if it is described which elements they correspond to. In order to specify the data model dependencies the rule base editor allows specifying mapping rules between data models.

The re-use of data model elements does not imply the re-use of the rules. This is only valid if the model elements are used with the same *semantics*. Therefore we will present an *ontology based method* which allows defining the semantic interpretation of the data model elements. An ontology is a formal, explicit specification of a shared conceptualization [11] and hence defines the semantic within a particular domain. The meta model of an ontology contains *concepts* and *semantic relations* between these concepts. A more formal definition could be found in [12]: *"An ontology is a tuple* $\Omega := (C, is\_a, R, \psi)$ *, whereas C is a set of concepts, „is_a" the inheritance relation on C, R a set of relationship names and* $\psi : R \rightarrow \wp(C \times C)$ *a function".*

While an object is assigned to exactly one object type, it could be linked to several concepts of the ontology. The link between object and concept describes the semantic interpretation of the object. Consequently the semantic (semantically interpretation) of an object type could be modeled by linking an object type with one or more concepts. It is possible to model synonyms translations of a concept. Thus the ontology based approach addresses the non-uniform naming of data model elements (synonyms, multi-lingual).

Figure 5 shows the necessary four steps within the setup phase. After the initial definition/update of the ontology the semantic interpretation of the *relevant* data model elements of the external system(s), KoBaS applications

[8] XMI: XML Metadata interchange (see: [10])

and reference library needs to be defined. The semantic interpretation of the reference data model will initially be defined and updated within each incremental update of the reference library. Therefore this step is limited to the data model of the external systems and KoBaS application. The last two steps are the derivation of the data model and initial set of rules. In a later stage of the implementation this could be done semi-automatically by using ontologies and data model matching algorithm of artificial intelligence.

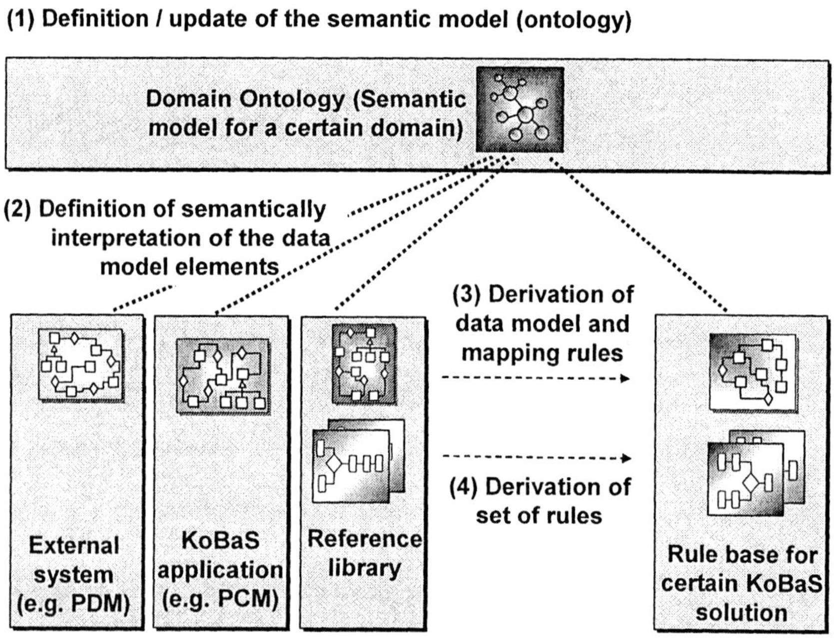

Figure 5. Setup phase of a rule base for a certain solution (machine)

3. CONCLUSION

In this paper the authors described a framework to create customized rule based system solutions in an object oriented way. The presented tool enhances the efficiency of the implementation, setup and management activities of rule based systems related to services for manufacturing machines. Moreover the advanced (ontology based) data integration utility of the system allows using distributed data and application objects during the rule definition (setup phase) and runtime (rule execution) phase of the customized solution. The open API (Application Programming Interface) of every customized solution allows in one hand the runtime management of the available rule bases, e.g. creation of new rule sets or rules, in other hand the deployment of new rule bases. Thus the software applications interfaced to the customized rule bases solution uses up-to-date data objects and knowledge. In order to proof the approach it will be validated within real industrial cases (manufacturing machines) in the KoBaS project.

4. REFERENCES

1. M. Rigamonti, T. Tolio; M. Tomasella, A Valente: Production requirements definition in a dynamic environment, CARV 2005, International Conference on Changeable, Agile, Reconfigurable and Virtual Production, Munich, Germany
2. KoBaS (Knowledge Based Customized Services for Traditional Manufacturing Sectors Provided by a Network of High Tech SMEs) project proposal, (2004)
3. E. Friedman-Hill: JESS in Action. Manning Publications Co., USA, 2003
4. N.N, JRules-Homepage: http://www.ilog.com, online at 12th January (2006)
5. QuickRules-Homepage: http://www.yasutech.com, online at 10th January (2006).
6. drools-Homepage: http://drools.org, online at 10th Januray (2006).
7. JEOPS-Homepage: http:// www.di.ufpe.br/~jeops/, online at 08th January (2006)
8. Rules4J-Homepage: www.rules4j.com, online at 08th January (2006)
9. A. Mahl, O. Hornberg, A. Ehrler : Concept and industrial use of platform for flexible cross enterprise collaboration, in: *product data journal* Vol. 2/11, (2004)
10. XMI-Homepage: http://www.omg.org/technology/documents/formal/xmi.htm
11. T. R. Gruber: A translation approach to portable ontology specifications. *Knowledge Acquisition.* Vol. 5. (1993).
12. Andreas Faatz, Ralf Steinmetz: *Ontology Enrichment with Texts from the WWW.* In Semantic Web Mining, WS02, Helsinki, Finland. (2002)

NEURAL NETWORK SYSTEM FOR KNOWLEDGE DISCOVERY IN DISTRIBUTED HETEROGENEOUS DATA

Timofeev A.V[1], Azaletskiy P.S.[1], Myshkov P.S[2], Kesheng Wang[3]
[1]Saint-Petersburg Institute for Informaion and Automation RAS, 14 line 39, St. Petersburg 199178 Russia; [2]Saint-Petersburg State University, 7-9 Univesrsitetskaya nab., St. Petersburg, 199034 Russia; [3]Norwegain Unversity of Science and Technology, N-7491, Trondheim, Norway. E-mail: Tav@iias.spb.su; kesheng.wang@ntnu.no.

Abstract: This paper proposed a distributed KDD system allowing remote usage of expert knowledge. Its implementation on the base of polynomial neural networks is described. The system is an universal KDD tool as it can build decision-making models in any subject field. Its implementation as a web-service will allow third-party software developers to create specialized applications, which focus on neural knowledge base usage.

Key words: knowledge discovery; polynomial neural networks; distributed KDD systems

1. INTRODUCTION

Modern computational systems and computer networks allow accumulation of huge amounts of data for solving problems in information analysis and optimal control. However, data in computer representation form contain information in a hidden way. To extract this information and knowledge, it is necessary to use special methods of data analysis and knowledge base synthesis.

Large amounts of data allow getting results that are more precise; but searching for solutions and knowledge synthesis is a very complicated task. As a result, a completely new class of systems serving as data analysis agents and knowledge engineers has emerged recently [2]. Such systems do

Please use the following format when citing this chapter:
Timofeev, A.V., Azaletskiy, P.S., Myshkov, P.S., Wang, Kensheng, 2006, in International Federation for Information Processing (IFIP), Volume 207, Knowledge Enterprise: Intelligent Strategies In Product Design, Manufacturing, and Management, eds. K. Wang, Kovacs G., Wozny M., Fang M., (Boston: Springer), pp. 144-151.

not simply analyze the data they contain, but also they are able to build decision-making models on the base of parallel data analysis. These systems are called Knowledge Discovery in Databases (KDD) systems [2].

The main task of KDD systems is analysis of data contained in databases with the purpose of discovering hidden, unobvious and unknown patterns and rules. The heterogeneous nature of modern distributed databases significantly complicates the task. Besides, each type of database requires different method of access to its data. Consequently, for each intellectual analysis system it is necessary to develop specific algorithms that take into account its features and characteristics. Modern databases are not only heterogeneous, but redundant as well. For a KDD synthesis, it is advisable to simplify the schema and the knowledge base structure as much as possible.

2. THE KDD TECHNOLOGY

Knowledge Discovery in Databases (KDD) is a process of search of meaningful knowledge in raw data [6]. The knowledge could be represented as a set of rules describing relations between data properties (decision trees), frequently encountered models (association rules), and classification results (neural networks) or data clusters (Kohonen maps), to name a few.

Regardless of knowledge representation model used, a KDD process consists of the following stages:

1. Data preparation. At this stage a data set is constructed, data consolidation is performed, and training and test sets are prepared.
2. Data preprocessing. Some tasks require input data be supplemented with certain information (e.g., expert information). Additionally, certain preprocessing algorithms, e.g. dimensionality reduction algorithms, are often required and executed at this step.
3. Data transformation and normalization. At this step the data is transformed into the form suitable for subsequent analysis.
4. Data mining. At the data mining step various knowledge discovery algorithms, such as neural networks, are executed.
5. Data postprocessing. Finally, different activities concerning interpretation of the results and their application to business are performed.

Knowledge Discovery in Databases does not determine what data analysis or processing algorithms should be used; instead, it defines the sequence of activities that should be performed in order to extract meaningful knowledge from source data. This approach is universal and does not depend on the application domain.

3. NEURAL NETWORK APPROACH FOR KDD

When using artificial neural networks (ANN) the choice of network architecture (the number of layers and the number of neurons in each layer) is of primary concern. The size and the structure of a network should correspond to the essence of the investigated problem (e.g., in terms of computational complexity). In the works [3-5] authors suggest three-tier polynomial and Diophantine neural networks, multi-tier "many-valued decision tree" neural network, as well as method for transformation of treelike neural networks into polynomial neural networks and for minimization of their complexity. Since at the beginning of the analysis the nature of the problem is often not well known, the choice of adequate ANN architecture becomes a rather complicated problem.

During the learning process a neural network uses input data to adjust its synaptic weights and to fine tune connections between neural elements. The resulting neural network expresses patterns present in the data and is able to make decisions on new data. The network serves as a functional equivalent for some data dependency model, i.e. a function of input variables, much like a one created using traditional modeling.

The principle advantage of neural network is that they are able to approximate any continuous function; therefore, there is no need for a researcher to assume any hypotheses regarding the model and sometimes even what variables are important.

A typical knowledge base construction system works that utilizes neural networks is shown on the Figure 1:

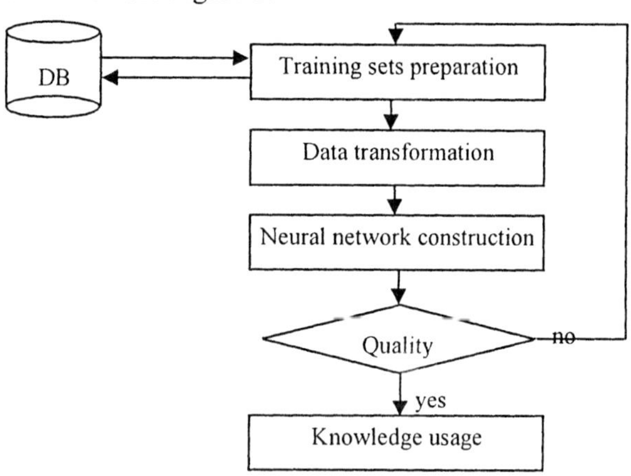

Figure 1. Sequence of operations for knowledge search with neural networks

3.1 Data sets preparation

At this stage qualified users or experts who possess knowledge in a certain application domain and want to automate the decision making process prepare a training set. This set $T = \langle Y, X \rangle$ is composed from known feature values $x_1, x_2, ..., x_n$ and decision function y values defined on an object set Ω:

$$\Omega = \{\omega_k : k = \overline{1,N}\}$$
$$Y = \{y(\omega_k) : k = \overline{1,N}\}$$
$$X = \{(x_1(\omega_k),...,x_n(\omega_k)) : k = \overline{1,N}\}$$

The set T can be easily represented as a table:

Table 1. Training set table representation

ω	Y	x_1	...	x_n
ω_1	$y(\omega_1)$	$x_1(\omega_1)$	...	$x_n(\omega_1)$
...	...	...	...	...
ω_N	$y(\omega_m)$	$x_1(\omega_m)$	...	$x_n(\omega_m)$

Here $y(\omega_k)$ serves as an expected result of the problem solution on the input set $\{(x_1(\omega_k), x_2(\omega_k)..., x_n(\omega_k)\}$. Training set is a database table with the expected values of decision function in the first column and input data feature values in the rest. The following constraints apply to this database:

- Validity: only qualified specialists take part in the creation.
- Completeness: all possible solutions are presented in the table.
- Consistency: there should not be two different values of decision function for the same input feature value.

These constraints allow to avoid noise influence, inconsistency and incompleteness in data representation.

3.2 Data transformation and normalization

If a KDD system uses neural networks, an expert (a user with strong knowledge in the application domain) has to fill in the training set. The system will then transform this input into numbers. Some systems use binary encoding of the set elements, i.e. {-1, +1}. If necessary, the training set could be sorted then.

3.3 Neural network construction and training

At this stage, a neural network is constructed and trained using various learning algorithms. As a result, a "neural knowledge base" is created.

3.4 Knowledge base quality control

After the neural network is trained, an expert tests its quality using training and test sets (these two sets should not intersect). The test set has the same structure as the training set. Based on the test results the expert makes decision on whether the constructed knowledge base is suitable for the given task or not. If not, the knowledge base construction process should be restarted from the step 1.

3.5 Knowledge base usage

To solve a task using the constructed knowledge base, a user prepares a table of feature values and sends it to the neural network input. The answer given represents the value of the decision function or the synthesized ANN.

3.6 Structure and organization of a distributed neural KDD system

The problem of development of a distributed system with remote knowledge base creation and usage capabilities is of high importance today. Such system often implements client-server architecture and uses Internet for information transmission.

A distributed neural KDD system operates in two main stages:

1. Neural knowledge base creation and control, implemented by an expert (the expert stage).
2. Knowledge base usage for the purpose of solving specific user tasks (the user stage).

A knowledge base system based on threshold polynomial neural networks has been developed [3-5]. It uses multiagent technology for processing and transmitting databases (DB) and knowledge bases (KB) through e-mail and HTTP protocols.

In the given model, experts producing a knowledge base and users working with it act as agents. The neural network type in this case can be described with arithmetic (Diophantine) polynomials. It is used for recognition of complicated (linearly non-separable) pattern types defined in a space of either binary or multi-valued feature-predicate. The scheme is illustrated on the Fig. 2.

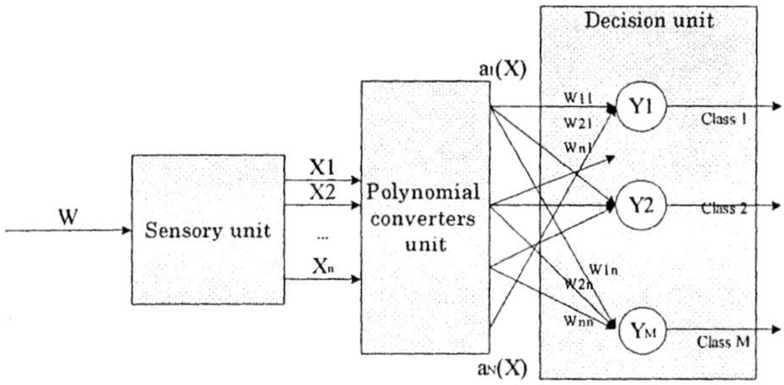

Figure 2. Structural scheme of a threshold polynomial neural network

At the input of a neural network there is a sensory unit consisting of threshold neuron-like elements, which encodes object features as binary codes $X = |x_1, x_2, ..., x_n|$.

The feature vector (X) is transmitted to a unit of polynomial converters (A-elements), which creates an m-dimensional vector of secondary (polynomial) features $Z = (a_1(X), a_2(X), ...a_N(X))$. These secondary features define a polynomial feature space $a_j(X), j = 1,2,...,n$, a so-called rectifying space. Explicit function form $a_j(X)$ is chosen in accordance with the given task and the training set, i.e. during the process of neural network construction and self-organization [3-5].

The output tier contains solution threshold neuron-like elements:

$$Y_i = sign(\sum_{j=1}^{N} w_j a_j(x)), \text{ where } i \in [1..M]$$

The recurrent learning algorithm used in threshold polynomial neural network is a supervised learning algorithm, which offers a number of advantages over the frequently used back-propagation of error (BPE) [1], including the following:

1. It is not necessary to determine network structure in advance, since the algorithm adjusts itself during learning process.
2. This is a single-pass algorithm, i.e. the third layer (decision layer) neurons' weights are adjusted in the first pass through the training set.
3. The algorithm guarantees error-free classification of elements in the training set.
4. The algorithm constructs a neural network with a high degree of extrapolation to data beyond the training set.

5. The synthesized neural network allows creation of a neural knowledge base based on the source database.

The structural scheme of the distributed KDD system is shown on the Figure 3:

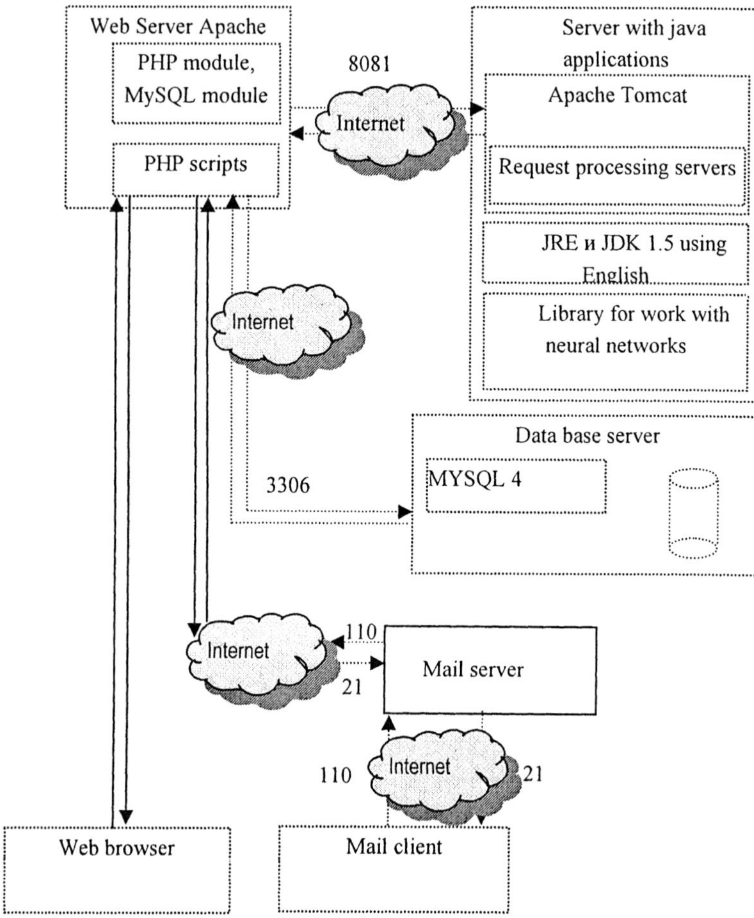

Figure 3. Structural scheme of the distributed neural KDD system.

At the first stage, an expert creates training and test sets in the form of database table. For this task, an expert query is constructed. The second stage concerns work with the knowledge base, at which a user query is constructed. As a reply, the system fills in the "Result" value corresponding to the certain feature cortege, which serves as an output generated by the system in response to the user query.

4. CONCLUSION

The distributed KDD system proposed in the article allows remote usage of experience of experts in a certain field, and its implementation as a neural knowledge base. The system is a universal KDD tool, since it makes it possible to build decision-making models in any subject field. Its improvement to a web-service will allow third-party software developers to create specialized applications oriented on neural knowledge base usage. The system can be applied in research as well: as a tool for in-depth study of different effects in ecology, economics and other fields. It serves as means to integrate problem solving experience of geographically distributed users. Apparently, the system could be useful in telemedicine area (for example, to help get the most probable diagnosis of complicated and rare illnesses), in the area of financial flows analysis and many others.

5. REFERENCES

1. Lugger, G. F. *Artificial Intelligence: Structures and Strategies for Complex Problem Solving*, 2003, p. 864.
2. Bagresan, A. A., Kuprianov, M. S., Stepanenko, V. V., Holod, I. I. *Methods and Models of Data Analyzing: OLAP and Data Mining.*
3. Simon Haykin, *Neural Networks a Comprehensive Foundation*, 2006, p. 1104.
4. Timofeev, A. V., Methods of Creation of Diophantine Neural Networks with Minimal Complexity – *RAS Report*, 1995, Vol. 345 No.1, pp. 32-35.
5. Timofeev, A. V., Parallelism and Self-Organization in Polynomial Neural Networks for Image Recognition – *Proceedings of the 7th International Conference on Pattern Recognition and Image Analysis: New Information Technologies* (18–23 October, 2004, St. Petersburg), pp. 97-100, 2004.

ASSESSMENT OF SURFACE ROUGHNESS MODEL FOR TURNING PROCESS

Yasir Hadi, Salah Gasim Ahmed
Department of Mechanical Engineering Technology, Yanbu Industrial College, Royal Commission for Jubai and Yanbu , Saudi Arabia. Email: yasirhadi@hotmail.com

Abstract: The quality surface of an engineering component plays an important role in its performance. This research focuses on developing an empirical model for prediction of surface roughness in finish turning. The model considers the following working parameters: feed, depth of cut, and spindle speed. Nonlinear regression analysis, with logarithmic data transformation is applied in developing the empirical model. Metal cutting experiments and statistical tests demonstrate that the model developed in this research produces smaller errors and has a satisfactory result.

Key words: surface roughness model, finish turning, prediction

1. INTRODUCTION

Surface roughness has received serious attentions for many years. It has formulated an important design feature in many situations such as parts subject to fatigue loads, precision fits, fastener holes and esthetic requirements. In addition to tolerances, surface roughness imposes one of the most critical constraints for selection of machines and cutting parameters in process planning. A considerable number of studies have investigated the general effects of the speed, feed, depth of cut, nose radius and others on the surface roughness.

A popularly used model for estimating the surface roughness value is as follows (Groover 1996) and (Boothroyd and Knight 1989):

Please use the following format when citing this chapter:
Hadi, Yasir, Ahmed, Salah Gasim, 2006, in International Federation for Information Processing (IFIP), Volume 207, Knowledge Enterprise: Intelligent Strategies In Product Design, Manufacturing, and Management, eds. K. Wang, Kovacs G., Wozny M., Fang M., (Boston: Springer), pp. 152-158.

$$R_i = \frac{f^2}{32r} \tag{1}$$

Where,
R_i: ideal arithmetic average (AA) surface roughness (mm)
f: feed (mm/rev)
r: cutter nose radius (mm)

This model assumed a none zero nose radius. The surface roughness models developed by Dickinson (1968), Fischer and Elrod (1971) considered the effect of feed rate and nose radius based on the motion geometry in a turning process, which provided the base for Equation (1). These models concluded that the effect of cutting speed is insignificant. However, different conclusions were presented in Shaw (1966), Hasegawa et al. (1976), Sundaram and Lambert (1979), Boothroyd and Knight (1989), Feng (2001), and Feng and Hu (2001). They demonstrated that cutting speed had a significant impact on surface roughness. The depth of cut was considered by Karmakar (1970), and Sundaram and Lambert (1981). Miller et al. (1983) considered the effect of cutting fluid on surface roughness. Although a qualitative analysis of machining variables of speed, feed and depth of cut on the surface roughness has been widely available in the literature, very few comprehensive predictive models have been developed. In this paper, an empirical surface roughness model, for high carbon steel, HRC40, will be developed based on metal cutting results from factorial experiments. The model will include the feed, depth of cut, and spindle speed. Regression analysis will be used to develop the models, and hypothesis testing will be conducted to validate the models.

2. SURFACE ROUGHNESS MODELS

L. Huang and J. Chen, 2001, applied a multiple regression modeling to express the surface roughness in the following form:

$$\begin{aligned} R = {} & \beta_0 + \beta_S S + \beta_F F + \beta_D D + \beta_V V + \beta_{SF} SF + \beta_{SD} SD \\ & + \beta_{SV} SV + \beta_{DF} DF + \beta_{DV} DV + \beta_{FV} FV + \beta_{SDF} SDF \\ & + \beta_{SDV} SDV + \beta_{SFV} SFV + \beta_{DFV} DFV + \beta_{SDFV} SDFV \end{aligned} \tag{2}$$

This formula is very complicated and cumbersome. In this work a simpler model is proposed in the form:

$$SR = \lambda S^{\alpha} F^{\beta} D^{\gamma} \tag{3}$$

Where: SR, surface roughness (micron).
λ, α, β, γ: constants.
S: spindle speed (rpm).
F: feed rate (mm/rpm).
D: depth of cut (mm).

3. EXPERIMENTAL DESIGN

A factorial experiment was carried out to estimate the values of the regression coefficients. A 3 factors full factorial experiment design was utilized to investigate the significant of the turning parameter and estimation of mathematical model coefficients. A carbide tool was used to machine a high carbon steel workpiece (HCR40) on a Colchester lathe (Master 2500) without any coolant. Table 1 shows the factorial design of the experimentation.

Table 1. Experimental Factors and levels

Run	Speed (rpm)	Feed (mm/rpm)	Depth of cut (mm)	Actual surface roughness (μm)	Predicted surface roughness (μm)
1	570	0.03	0.25	1.75	2.09
2	1030	0.05	0.50	2.41	1.74
3	770	0.03	1.00	2.16	2.11
4	770	0.05	0.25	2.25	1.94
5	770	0.04	0.50	2.51	2.05
6	1030	0.05	0.25	2.08	1.54
7	1030	0.04	0.25	2.08	1.44
8	1030	0.03	0.25	1.34	1.32
9	770	0.05	1.00	3.13	2.47
10	570	0.04	0.25	2.89	2.29
11	570	0.04	0.50	2.56	2.58
12	570	0.05	0.50	2.08	2.77
13	1030	0.04	0.50	2.14	1.63
14	770	0.05	0.50	2.20	2.19
15	570	0.03	1.00	2.80	2.67
16	770	0.03	0.25	2.41	1.66

17	770	0.04	1.00	2.44	2.31
18	570	0.05	1.00	4.57	3.12
19	770	0.03	0.50	2.23	1.87
20	570	0.03	0.50	2.20	2.36
21	1030	0.04	1.00	1.63	1.83
22	570	0.05	0.25	3.19	2.45
23	770	0.04	0.25	1.86	1.81
24	1030	0.03	0.50	1.67	1.49
25	1030	0.03	1.00	2.66	1.68
26	570	0.04	1.00	3.06	2.92
27	1030	0.05	1.00	3.49	1.96

After completing the experiments an analysis of variance (ANOVA) was conducted to discern whether differences in surface roughness between various runs were statistically significant.

Table 2 presents ANOVA results for experimental data generating during turning the high carbon steel workpiece. The Model F-value of 20.15 implies the model is significant. There is only a 0.01% chance that a "Model F-Value" this large could occur due to noise.

Values of "Prob > F" less than 0.0500 indicate model terms are significant. In this case A (speed), B (feed), and C (Depth of cut) are significant model terms. Values greater than 0.1000 indicate the model terms are not significant.

Table 2. Analysis of variance [Partial sum of squares]

Source	Sum of Squares	DF	Mean Square	F Value	Prob > F	
Model	1.33	3	0.44	20.15	< 0.0001	significant
A	0.95	1	0.95	43.29	< 0.0001	
B	0.11	1	0.11	5.10	0.0338	
C	0.27	1	0.27	12.05	0.0021	
Residual	0.51	23	0.022			
Cor Total	1.84	26				

Accordingly, the exponential model coefficients for surface roughness were given and the mathematical model will be as follows:

$$SR = 1113.5S^{-0.78} F^{0.3} D^{0.18} \tag{4}$$

Fig. 1 below shows how the prediction model match up to the actual experiments data.

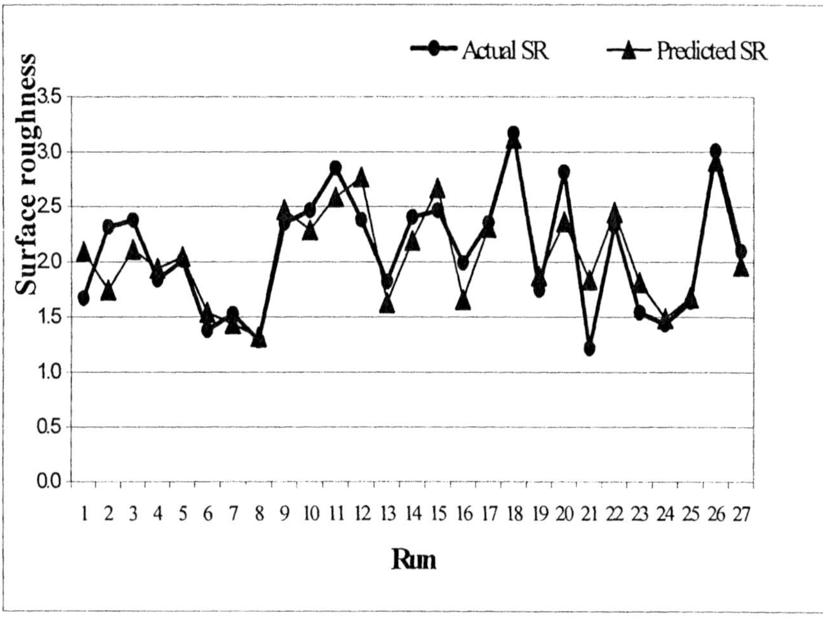

Figure (1) : Predicted versus actual surface roughness

4. CONCLUSION

1. The smallest values of surface roughness were produced when the material was machined with a smaller feed.
2. All cases showed that a higher speed would smooth the surface within the range of experiments.
3. A logarithmic transformation can be applied to convert the nonlinear form of Equation (3) into the additive (linear) form. This is one of the most popularly used data transformation methods in empirical model building (Box and Draper 1987).
4. The mathematical model, equation (4), can reasonably predicted the surface roughness within the range of machining parameters used in the experimentations

5. REFERENCES

Boothroyd, G. and Knight, W. A. (1989) *Fundamentals of Machining and Machine Tool*: Marcel Dekker, New York.

Dikinson, G. R. (1967/1968) Survey of Factors Affecting Surface Finish. *Proceedings of Conference on Properties and Metrology of Surfaces, the Institution of Mechanical Engineers (Great Britain)*, Vol. 182, Part 3K, pp. 135-147.

Feng, C-X. (2001) An Experimental Study of the Impact of Turning Parameters on Surface Roughness. *Proceedings of the 2001 Industrial Engineering Research Conference,* Paper #2036, Institute of Industrial Engineers, Norcross, GA.

Feng, C-X. and Hu, Z-J. (2001) A Comparative Study of the Ideal and actual Surface Roughness In Finish Turning. *Working Paper, Department of Industrial and Manufacturing Engineering,* College of Engineering, Bradley University, Peoria, Illinois 61625.

Fischer, H. L. and J. T. Elrod (1971) Surface Finish as a Function of Tool Geometry and Feed – A Theoretical Approach. *Microtecnic*, 25(3): 175-178.

Groover, M. P. (1996) *Fundamentals of Modern Manufacturing*. Prentice Hall, Upper Saddle River, NJ (now published John Wiley & Sons, New York).

Hasegawa, H., Seireg, A., and Lindberg, R. A. (1976) Surface Roughness Model for Turning. *Tribology*, December, 285-289.

*Karmakar, A. (1970) Factors Influencing Surface Finish During Fine Turning. *Proceedings of 4th All India Machine Tool Design and Research Conference,* India, 123-128.

Miller, J. C., De Vor, R. E., and Southerland, J. W. (1983) Surface Roughness Characteristics for Turning 380 and 390 Aluminum Casting Alloys. *Proceedings of North American Manufacturing Research Conference*: 282-288.

Shaw, Milton C. (1984) *Metal Cutting Principles*. Oxford University Press, Oxford.

Sundaram, R. M., and Lambert, B. K. (1981) Mathematical Models to Predict Surface Finish in Fine Turning of Steel, Parts I and II, *International Journal of Production Research,* 19: 547-564.

Sundaram, R. M. and B. K. Lambert (1979) Surface Roughness Variability of ANSI 4140 Steel in Fine Turning using Carbide Tools. *International Journal of Production Research,* 17(3):249-258.

6. APPENDIX

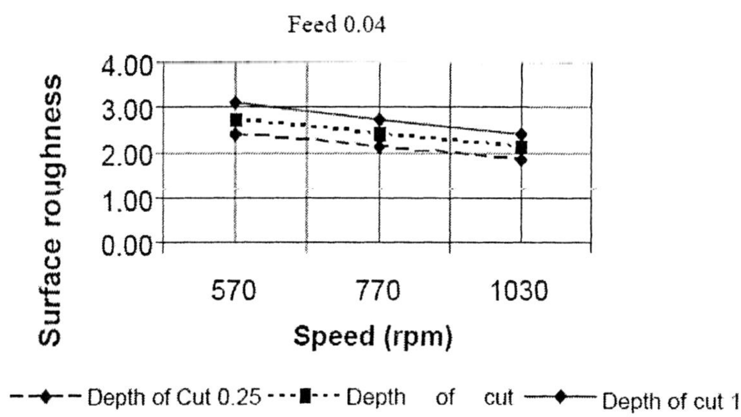

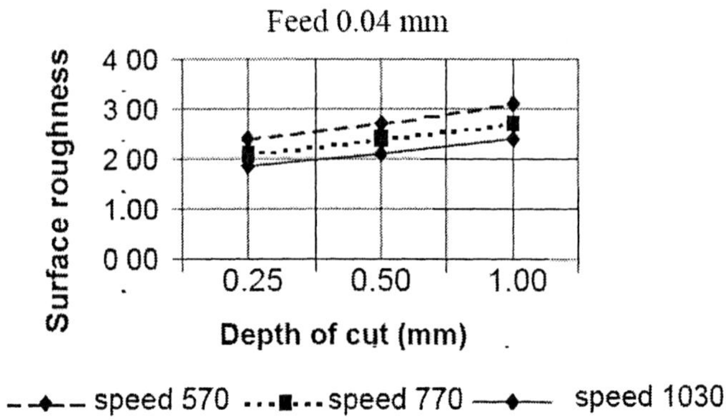
Feed 0.04 mm
Surface roughness
4 00
3 00
2 00
1.00
0 00
0.25
0.50
1.00
Depth of cut (mm)
speed 570
speed 770
speed 1030

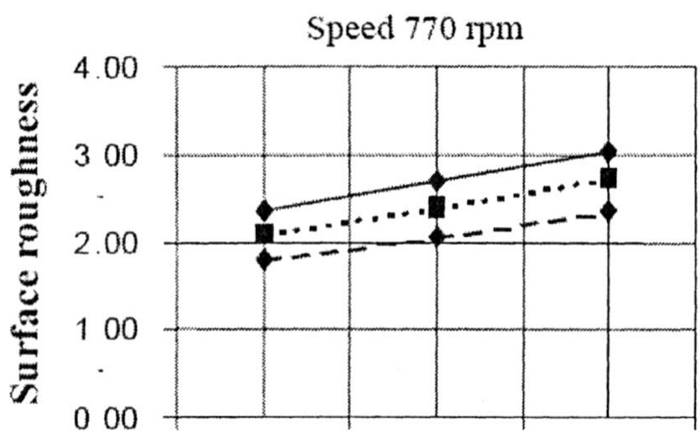
Speed 770 rpm
Surface roughness
4.00
3 00
2.00
1 00
0 00

A STUDY ON PRODUCT OPTIMIZATION DESIGN BASED ON GENETIC ALGORITHMS

Guiqin Li [1], Xiaojian Liu [2], Qinfeng Yuan [1], Minglun Fang[1]

[1]CIMS & Robot Center of Shanghai University, Shanghai University, China; Email: leeching@staff.shu.edu.cn. [2]2Industrial Design Institute, Northwestern Polytechnical University, China.

Abstract: Genetic algorithms (GAs) can be viewed as random searching algorithm that based on the ideas of natural evolution. It is quite suitable for the complicated non-structural problems, for it has some properties such as self-adaptability, parallelity, evaluation-based, probability and multi-solution. Its principle is similar to human designers' behavior, a trial-and-error process guided by the evaluation regarding the goal: product optimization schemes. The paper analyzes the comparability between product optimization design and GAs, and proposes a technical method with its algorithm structures based on GAs. The codified representation, evaluation function and optimize searching of product design was basically realized based on GAs.

Key words: Product Optimization Design, Genetic algorithms, codified representation.

1. INTRODUCTION

General evolution theory works up from the investigation of artificial intelligence and cognition science. It transposes the notions of evolution in Nature to complex social system, AI system and artificial product system. Evolving optimization design method based on GAs gained new optimum product plan through inheriting and improving successful design cases.

Evolution theory and GAs have got many achievements in the domain of products design in recent years. For example, M. A. Rosenman in Sydney University developed a non-routine design system.[1] The automatization department in Tsinghua University investigated soft-hard model based on generalized evolution. [2] Literature [3] gave an evolution model of requirements analysis. The study of J. Poon and M. L. Maher applied

This project is supported by the National Natural Science Foundation of China under the grant No.50375090, the Scientific Research Fund of Shanghai Education Commotion under the grant No.202624 and the Shanghai University Science and Technology Development Fund of China under the Grant No.202615.

Please use the following format when citing this chapter:
Li, Guiqin, Liu, Xiaojian, Yuan, Qinfeng, Fang, Minglun, 2006, in International Federation for Information Processing (IFIP), Volume 207, Knowledge Enterprise: Intelligent Strategies In Product Design, Manufacturing, and Management, eds. K. Wang, Kovacs G., Wozny M., Fang M., (Boston: Springer), pp. 159-164.

evolving principle to dynamic solution, taking into account the interaction of multi-factors during design process. [4] Literature [5] set up an evolution model for design thoughts.

This article analyses the implementation of product optimization design process in the design/manufacture integrative process based on GAs.

2. ANALYSIS OF THE PROBLEM OF PRODUCT OPTIMIZATION DESIGN

So far, many of the product design CAD method are mostly for generating shape or structure of products. It's far from enough for evaluation and automatic adjustment of a plan. This investigation provides solution to harmonize the relations among product design principles, diversification, and optimization and design freedom, and to deal with the fuzzy decisions problem, via systematic classification and formalization product structural optimization design problem.

The main task of formulization is construct reasonable coding scheme (also named "genotype") of design problem (also named as "represent type"), viz it should express all forms of availability design plans, and has feasibility of arithmetic operation. For example, we can treat with shape factors as follows:(1)classify factors according to the hierarchy and put them into difference design stages;(2)For the shape of certain rudiment required, increase fitness artificially in two aspects: the ratio of eliminate through selection or contest, and the offspring individual number. So it is a dominant force in evolving direction;(3)Deal with the authority of local code segment, to differentiate the relative importance;(4)Rigid shape restrictions are treated by Arithmetic;(5)Provide interactive interface of code forcible operation and plan artificial selection.

3. THE CODIFIED REPRESENTATION AND FITNESS FUNCTION OF PRODUCT OPTIMIZATION DESIGN

Firstly, it's necessary to define a solution space that contains all feasible solutions for the program as the search area. The solution space for GAs is constructed by the code of the target problem. To product, the code is a set of parameters that describe the product features and properties. Different parameters represent different products.

Fitness function is assessing rule that determines acceptance or rejection a design plan in throng of random generated. The structure of product function model includes three layers: physics function, physiology function and psychology function. Physics function used to accomplish input/output tasks of the substance, energy, information or data that were schemed out. Physiology function used to accomplish interactive tasks between enactment product and user on action, contact, language or sense. Psychology function indicates to users' sensation and cognition information, including aesthetics, semantics and individuation, etc. The work of constructing fitness function then bases on the quantity model of this three function constraints.

Product code is composed of parameters that denote its functions. The form of the product code is as following, in which M is function, whose value describes the function, such as length, weight, etc.

$$B = \{M_0, \quad M_1, \quad M_2, \quad \ldots \quad M_{n-1}\}, \quad M_i \in [M_{i0} \quad , \quad M_{i1}] \qquad (1)$$

All rational values of every M composed the solution space of the product, in which the best one lies.

According to the design flow, the search is based on the evaluation of the product code. There are different way of evaluating, depends on developer's purpose, such as technical evaluation, economical evaluation, aesthetics evaluation, etc. We adopted value engineering as the main evaluation method. We can get the value coefficient of each function:

$$ZI_i = \frac{MI_i}{S_i} \qquad (2)$$

where MI is function coefficient, $\sum MI_i = 1$, and S is cost coefficient, $\sum S_i = 1$. The total value of the product is calculated as:

$$Z = \sum ZI_i * M_{is} \qquad (3)$$

where M_{is} is the importance coefficient of function MI_i, and $\sum M_{is} = 1$.

The evaluation criterion discussed above is an integral standard, and is mostly often used. Sometimes, we can use different criterion to evaluate a product from several aspects, and the selection and decision work is left to the users.

4. REALIZATION OF EVOLVING OPTIMIZATION DESIGN ARITHMETIC

The conceit of arithmetic about main evolving concept and behavior in evolving calculation as follows:

a. Population: representation a series of coding plan to latency solution set of design problem

Original population generated with two methods: random method and object method. In which object method generates population with some design expectation, or some successful design cases through coding, or the optimization plans gained from previous round evolving calculation.

b. Inherit: maintenance in offspring design plans of part properties that are of some robust adaptability individual of the population

Inherit operation carry through reproduction and crossover.

c. Mutation: local changing in the coding segments

Mutation is main source of creativity in evolving design arithmetic. We plan traverse three mutation manners: innovation, adjustment and slighter shift. Innovation mostly used in conceptual design by way of biggish extent random mutation. In order to maintain creative content not eliminate through selection or contest, we can lower fitness function value artificially. Adjustment is to obtain diversity and highly fit individual through local modify code value in miniature. Slighter shift is to observe design results by continuance mutation of code segments to a certain direction based on some special requirements.

d. Selection: determination the selection of parent individual, survival of offspring, and the individual replaced by offspring in population based on the value of fitness function.

The product optimization design based on Genetic Algorithms is illustrated in Fig.1. The product optimization process is composed of four steps, which correspond to the five sub-processes of GAs——random search, evaluation, recombination / mutation, decoding and interactive selection. They are performed by ten basic operators as follows.

$$F = (M, N, O, P, Q, R, S, T, U, V) \tag{5}$$

where M—— individual construction operator, N—— constraints generating operator, O—— population maintenance operator, P—— individual evaluation operator, Q—— selection operator, R—— recombination operator, S——mutation operator, T——optimization control operator, U——decoding operator, V——interactive selection operator.

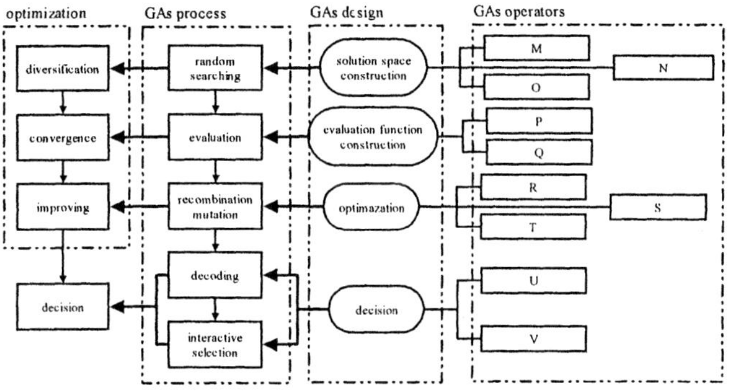

Figure 1. Product optimization design based on Genetic Algorithms

Fig.2 is the interface of the GAs based product design prototype we developed. Product optimization design flow can be divided into three stages: question conformation, automatic search and decision. Question conformation stage builds genotype and fitness function for design problem in interactive mode, and defines availability solution space. Automatic search stage is the operating process vs genotype in solution space. Decision stage is to select the ultimate plan through the interaction with user on the basis of work result, or decide next round search by user's determination till satisfied plan is obtained.

The system contains three functional parts: a) variable and parameter input, b) fitness function setting, c) search process setting. The design case in our prototype only take into account physics function, for other function is similarity. Computer is high efficient, but not versatile.

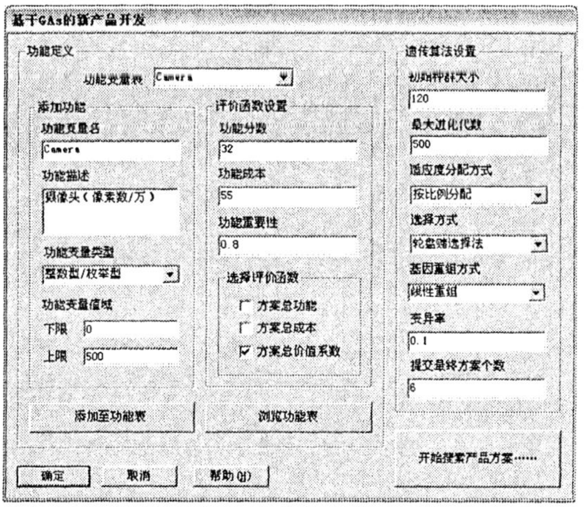

Figure 2. Interface of the GAs based product design system

5. CONCLUSION

The GAs-based product optimization design makes use of the advantage of computer in speed, and realized basic intelligence in the evaluation of complex product plans. At the same time, it lefts a free space for designer's decision making. GAs is a useful mathematical tool to the simulation of human designer's design process guided by the evaluation regarding the goal. Function analysis is a proper way to code the product model. It well describes the product's function feature, and is convenient for the construction of fitness function GAs need.

6. REFERENCES

1. M. A. Rosenman, *An exploration into evolutionary models for non-routine design, Artificial Intelligence in Engineering,* 11(3), 287-293(1997).
2. ZHAO Nanyuan, *Cognition Science and General Evolution Theory* (Tsinghua University Press, Peking, 2002) (in Chinese).
3. CHENG Jingping, An investigation of demand analysis modeling for evolutionary product conceptual design, *Hoisting and Conveying Machinery*, (11), 6-8(2002).
4. J. Poon and M. L. Maher, Co-evolution and emergence in design, *Artificial Intelligence in Engineering,* 11(3). 319-327 (1997).
5. ZHANG Zhiwei, YE Qingtai, and WANG Anlin, Thought and systems of evolutionary deign, *Journal of Shanghai Jiaotong University*, 34(10), 1449-1452(2000)

DESIGN OF THE BODY FRAME OF A RAIL DETECTOR

Jiangang Bao
North University of China Department of Mechanical Engineering, China; Email: baojiangang@nuc.edu.cn.

Abstract: According to the practical needs of the railroad engineering and the applying situations of the rail detectors, the assembling parts of the detector are designed and assembled with the SolidWorks toolkit to create the three-dimension model, and its reasonableness is analyzed with the help of the finite-element method.

Key words: rail detector, three-dimension modeling, SolidWorks.

1. INTRODUCTION

The development of the rail detector has undergone two stages: manual detecting and automatic detecting. In abroad, the application and development of the rail detector have a history of dozens of years. In developed countries today, the detection of the rail lines has been done with the detector, and manual detection is only an auxiliary way to check the defects. According to the now railroad situations, there are two kinds of working model for detection: discontinuous and continuous. The technology of the rail detector has been studied in the United States since the 1940s. After the 1970s, with the rapid development of computer technology, especially the application of the real-time data processing technique, the research of the rail detector advanced greatly. This technique enables the operation information detected by the car to be displayed real-time on the screen, the detected data to be stored and played back anytime to show the detected results. The development of the detector in China has taken two great strides, from tapping by hand to using ultrasonic rail detectors. The functions of the detectors have been updated, from the single resonance

Please use the following format when citing this chapter:
Bao, Jiangang, 2006, in International Federation for Information Processing (IFIP), Volume 207, Knowledge Enterprise: Intelligent Strategies In Product Design, Manufacturing, and Management, eds. K. Wang, Kovacs G., Wozny M., Fang M., (Boston: Springer), pp. 165-170.

detectors to micro-computer, semi-intelligence, and multi-channel ultrasonic ones. In order to raise the detecting efficiency of the rail defects to improve the railroad maintenance, large rail detectors are introduced, which have the advantages of rapid detecting speed, high reliability, and good repeatability. Now, there are many types of small detectors, produced home and abroad. The manually operated detector has a multi-probe structure, and the frame is made of light material. It is light and easy to handle to be used for moving on many types of rail and detecting. For example, the home-made Model MW37-GT-1C rail detector is a practical precision instrument. The latest Model SGT-2000 rail detector is a new, special digitalized rail detector, but it is inferior to the foreign products. It moves forwards only, and damages the probes if moving backwards. Its frame structure is complex and needs to be optimized, and its weight needs to be further reduced.

2. DESIGN OF THE DIGITALIZED THREE-DIMENSION MODEL

Here, the body frame of the rail detector is divided into six units: the probe support section, the bed, the box, the pulling section, the rack section, and the motion controlling section, with each made up of many parts.

The design of the three-dimension model of the detector and the assembly of the units and parts are carried out with the now popular three-dimension designing software, SolidWorks.

The bed of the detector is used to support the water box and the rack section, part of which is channeled. The channeled structure has many advantages, (1) high intensity, to bear very heavy pressure; (2) light weight, to be easily used for field work, and (3) material-saving, to reduce the cost greatly in mass production and thus improve economic benefits.

The bed is made up of 8 parts, the supporting arms of the body frame (2 pieces), the upright beams (2 pieces), the horizontal beams, the sliding rails for the probe support section (2 pieces), and the fixed base for the push bar.

3. DESIGN AND MODEL OF THE PROBE SUPPORT ARMS

There are two supporting arms for the probes, located at the front and rear ends of the detector, to be used for mounting the two probe stands and for mounting the rubber wheels. When the probe supporting arms are down, the clamping bar is linked with the bed to make the probes very close to the rail to detect; when the arms are up, the rubber wheels touch the land to support the detector to move on land. The arms are linked with the door-shape frame in the bed by its middle part.

The arms are designed to move forwards and backwards, without damaging the probes. The final arm is shown as in Figure 1.

Figure 1. Two-way motion probe arm

4. DESIGN OF THE RACK SECTION AND THE BOX

The rack section, the largest part of the detector, is mounted on the top of the body frame, whose shape is shown as in Figure 2. The box is mainly used to accommodate the ultrasonic controlling machine, Its three-dimension model is shown as in Figure 3.

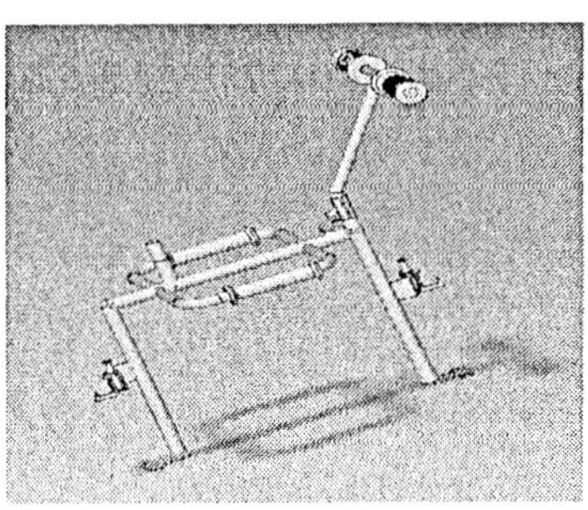

Figure 2 The rack section.

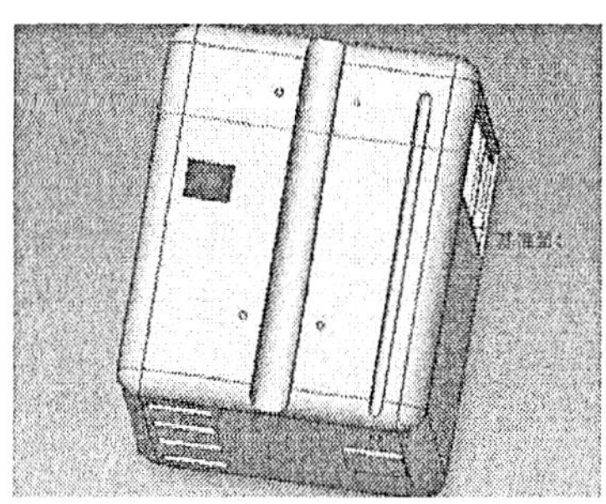

Figure 3 The three-dimension model of the box.

5. THREE-DIMENSION MODELS OF THE MOTION CONTROLLING AND PULLING SECTIONS

The motion controlling and pulling sections are primarily used to perform the motion functions of the detector, that is, to retract the wheels and to move back and forth on the rail. Their three-dimension models are shown as in Figure 4 and Figure 5.

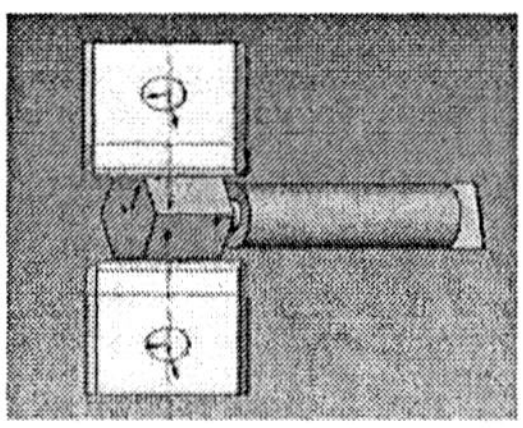

Figure 4 The motion controlling section.

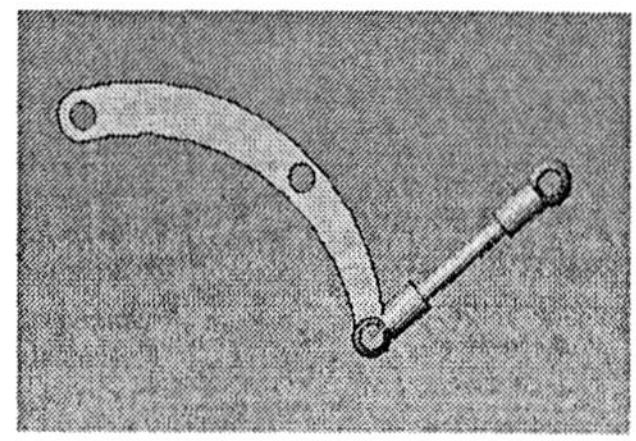

Figure 5 The pulling section.

6. GENERAL ASSEMBLY OF THE RAIL DETECTOR

The above parts are assembled in the SolidWorks software, and examined through interference[2] (supplied by SolidWorks, to detect the problems in time and solve them). Now the design of the model of the detector is accomplished, whose general assembly is shown as in Figure 6.

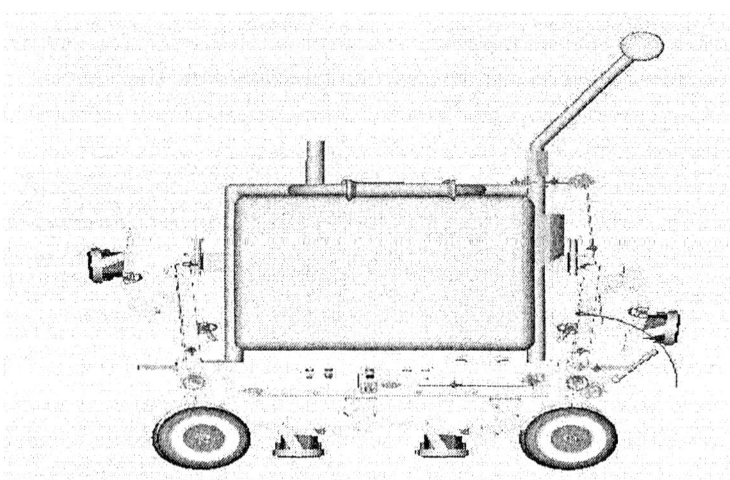

Figure 6 The general assembly of the new-type rail detector.

7. DYNAMIC SIMULATION

The first is to clarify the linking and driving relationships between the parts to determine which characteristics of the product is to be simulated and solved.

1) Determining which parts are added into the motion model

The pulling and relieving function of the InteliMotion explorer is used to point out which parts are motion parts, which are fixed and still, and which have closely fixed relationship with others. The COSMOSMotion tool is used to create the motion link for an assembled part in the motion model. Certainly, the user can manual create the motion link if there is not a suitable assembled part.

2) Imposing restrictions on the motion of the mechanical unit

The user can set the freedom of motion for the moving unit. The motion can be horizontal or rotary.

3) Adding additional load

The user can add to the parts in the mechanical unit additional load, external force or torque.

4) Simulating and checking the results

The user pushes the button to call the simulation function. The program will solve the problems with the motion of the mechanical unit: displacement, rate, accelerated velocity, and the acting forces between the moving parts in the unit.

After the simulation, the user can see the animated simulation results, which helps the user to know the details of the motion of the mechanical unit and exchange the results with others.

8. CONCLUSION

In this paper, the parts of the rail detector has been systematically classified and analyzed, and with the application of SolidWorks and the advanced modeling techniques --- parametric modeling and characteristic modeling, related parameters are determined and characteristics are classified, and the three-dimension model of the body frame of the rail detector is created. Finally, the motion simulation module is used to determine the form and parameters of the motion of the mechanical unit and create the spatial motion model in order to determine the reasonableness of the structure and the consistency of the motion with the theoretical analysis. The pilot rail detector practically works well.

9. REFERENCES

1. Liu Lijuan. Digitalized Simulation Design and development of Web-based Products. M.S. *Thesis of North University of China.* 2005.
2. Xie Hong & Shi Wei. Three-dimension Design of Assembled Units with SolidWorks. *Tool Techniques*, 2002.

THE SPATIAL SHAPE REPRESENTATION AND 3D MODELING OF SPECIAL-SHAPE SPRING

Xiangchen Ku[1,2], Runxiao Wang[1], Jishun Li[2], Dongbo Wang[1]
[1]School of Mechatronics Engineering,Northwestern Polytechnical University,Xi'an,China; Email:kuxch@sohu.com. [2]School of Mechatronics Engineering,Henan University of Science and Technology,Luoyang,China.

Abstract: The solid modeling of special-shape spring is important to improve production efficiency of spring coiler. However, how to describe and represent its spatial shape is difficult because there is various special-shape spring with complex spatial shape. Consequently, the 3D modeling and visualization of special-shape spring is not easy to implement. The paper presents a solution to this problem. First, we extract four kinds of elementary shape feature and define their data structure respectively using feature technology. The special-shape spring may regarded as the combination of different elementary shape features. The dynamic linked lists are employed to record the position relation between them. Then, this article studies the 3D modeling methods of special-shape spring based on OpenGL. The emphasis is the discretization of the spring solid surface. Last, the high reality solid modeling is created with C++ Builder.

Key words: special-shape spring; shape feature; representation; 3D modeling.

1. INTRODUCTION

Spring is one kind of basic mechanical parts used widely. The shapes of spring is quite different[1], thus it formed various special-shape spring with complex spatial shape. Along with the development of computer technology, the solid modeling technology is playing more and more important role in the product design and manufacture process. Currently there are some papers have discussed the solid modeling for compression spring with simple shape [2,3]. But all these 3D modeling of spring in these papers depends on some commercial 3D modeling software. Although these software can create solid

This project is supported by the Doctoral Programme Foundation of Education Ministry of China under the grant No.20050699033.

Please use the following format when citing this chapter:
Ku, Xiangchen, Wang, Runxiao, Li, Jishun, Wang, Dongbo, 2006, in International Federation for Information Processing (IFIP), Volume 207, Knowledge Enterprise: Intelligent Strategies In Product Design, Manufacturing, and Management, eds. K. Wang, Kovacs G., Wozny M., Fang M., (Boston: Springer), pp. 171-176.

modeling satisfactorily, but there are still some limitations in the application of these software, such as expensive cost, inconvenient operation and high specialization. Therefore, these software is unsuitable for the design and manufacture of special-shape spring.

In fact, it is a difficult task to express the spatial shape information of special-shape spring with uniform format especially for it's complex coil end. Consequently, there is no related papers which have discussed the solid modeling of special-shape spring at present. So the effective representation of special-shape is indispensable to its solid modeling. In this paper, a solution to the solid modeling of special-shape spring is proposed and implemented by feature-based technology. Firstly, four kinds of elementary shape feature are extracted in terms of spring shape feature. The spatial shape of the special-shape spring is described with these elementary shape features. Then using OpenGL and C++ Builder, the solid modeling for special-shape spring is researched.

2. REPRESENTATION OF SPATIAL SHAPE OF SPECIAL-SHAPE SPRING

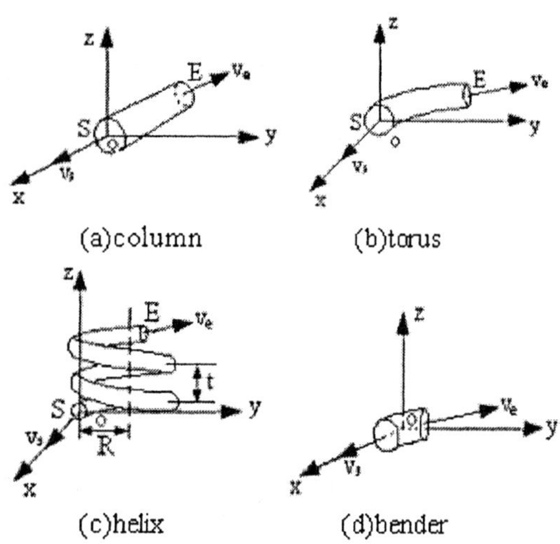

*Figure 1.*Definition of elementary shape feature

The shape feature is a basic unit to form mechanism parts. Therefore, the special-shape spring may regarded as the combination of different shape feature[4]. There are four kinds of elementary shapes abstracted from the special-shape spring, including helix, torus, bender and column. Figure 1 shows the definition and expression of these four elementary shape features.

The elementary shape feature is not only the abstraction of the shape but also the representation of the spring coiling, because each of these features correspond one coiling method. Each of these elementary shape features is defined by two vertexes and direction vector. The vertex is the connection points between two different elementary shape feature. The end vertex of the previous elementary shape feature is the start vertex of next one. By using this kind of connection relations, different elementary shape features can be assembled one by one and whole spring spatial shape information can be generated finally.

In order to describe and manage the spring shape information, the data structure of elementary shape feature is defined as follows:

```
①struct FeatureBase   //data structure of elementary shape feature
{
enum {line,circle,spring,bender} type;  //type of elementary shape feature
VERTEX *start_point;        //start vertex
VECTOR *start_vector;  // direction vector of  start vertex
VERTEX *end_point;    // end vertex
VECTOR *end_vector;   //direction vector of end vertex
VERTEX *start_position;//global coordinates of start vertex
VERTEX *end_position; //global coordinates of end vertex
HELIX *helixbody; // shape parameter of helix
float angle_plane;//the angle between the two planes which the
                  //elementary shape feature is located
struct FeatureBase *next; //pointer of next elementary shape feature
};
②struct VERTEX   //data structre of vertex position
{
float x,y,z;      //coordinates of vertex
};
③struct VECTOR   //data structure of vector
{
 float nx,ny,nz;    //coordinates of vector
};
④struct HELIX      //data structre of helix
{
char direction;    //direction of rotation
float num;         //number of coil
float bot_radius;  //bottom radius of helix
float top_ radius; //top radius of helix
float start_pitch; //start pitch of helix
float end_pitch;   //end pitch of helix
```

};

3. THE 3D MODELING OF SPECIAL-SHAPE SPRING

OpenGL used simple geometry unit to generate complex solid. Therefore, it is need to disperse the spring solid surface into a series of slices colored by OpenGL in the modeling of spring. One spring coil can be divided into small segment of *n*, then every small segment is dispersed as equal points. The triangle planes which OpenGL can filled and colored are constructed using these equality points. Many these triangle planes are put together according to the sequence and high reality solid modeling can be obtained finally.

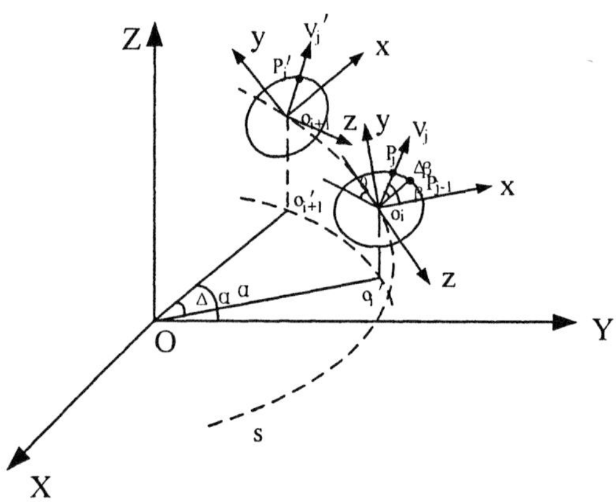

Figure 2. Discretization of spring solid surface

As shown in Fig. 2, in a global coordinate system $OXYZ$, s is a part of *k*th coil centric line of a right screw spring which spiral angle is θ . o_i and o_{i+1} is section center respectively, and their projection in the XOY plane is o_i' and o_{i+1}' .The part between o_i and o_{i+1} called *i*th segment when one coil is divided with step length of $\Delta\alpha$. Taking o_i as origin and establishing local coordinate system, x axis is parallel to Oo_i' and z axis is paraller to normal vector of section. The o_i global coordinate is: $x_i = R\cos\alpha$, $y_i = R\sin\alpha$, $z_i = (i/n+k)h$ (where, R is radius of spring, h is pitch, $\alpha = i * \Delta\alpha$, i=0,1,2,...).

In xoy plane, the circumference of section circle is divided into a series of points with step length $\Delta\beta$ started from point ($d/2$, 0)(where, d is diameter of wire). The relationship between the global coordinates (X,Y,Z) and local coordinates (x,y,z) of the points in the circumference is as follows:

$$\begin{pmatrix} X \\ Y \\ Z \end{pmatrix} = \begin{pmatrix} \cos Xx & \cos Xy & \cos Xz \\ \cos Yx & \cos Yy & \cos Yz \\ \cos Zx & \cos Zy & \cos Zz \end{pmatrix} \begin{pmatrix} x \\ y \\ z \end{pmatrix} + \begin{pmatrix} x_0 \\ y_0 \\ z_0 \end{pmatrix} \quad (1)$$

In the Eq.(1), (x_0, y_0, z_0) is the global coordinates of the origin, and the coefficient matrix is composed by direction cosine between the two coordinate system, the rest may be deduced by analogy.

As shown in Fig. 2, supposing P_j is jth point in the circumference, it's local coordinates is $((d/2)\cos\beta,(d/2)\sin\beta,0)$ (where, $\beta = j*\Delta\beta$). Using the Eq. (1), the global coordinates of this point (P_{jx}, P_{jy}, P_{jz}) may be obtain as:

$$P_{jx} = \text{-}(R+(d/2)\cos\beta)\sin\alpha + (d/2)\sin\beta\sin\theta\cos\alpha$$

$$P_{jy} = (R+(d/2)\cos\beta)\cos\alpha + (d/2)\sin\beta\sin\theta\sin\alpha$$

$$P_{jz} = (d/2)\sin\beta\cos\theta + (i/n+k)h$$

In order to obtain high sense of reality, it is necessary to calculate the normal vector of the disperse points. Because these points are located in the circumference of section circle, therefore the normal vector of the point equates the vector from the centre of a circle to this point. If the local unit normal vector of point P_j is $(\cos\beta, \sin\beta, 0)$, then the global unit normal vector $V_j(V_{jx}, V_{jy}, V_{jz})$ is:

$$V_{jx} = \sin\beta\cos\alpha\sin\theta - \cos\beta\sin\alpha$$

$$V_{jy} = \cos\beta\cos\alpha + \sin\beta\sin\alpha\sin\theta$$

$$V_{jz} = \sin\beta\cos\theta$$

Using the similar method, point P_j' and vector V_j' in another section circumference which correspond to the point P_j and vector V_j can be calculated. When these corresponding points is obtained, the helix curved surfaces are created using the continual small pieces of triangle surface, then the high reality solid modeling can generated by adding the illumination and material quality. The column and torus which compose special-shape spring also can be divided into small pieces of triangle surface. As a result, the solid modeling of whole spring is created.

4. IMPLEMENTATION AND EXAMPLES

After obtaining the shape parameter, the paper creates the solid modeling of special-shape spring using OpenGL under the C++ Builder 6.0 environment. Fig. 3 shows a solid modeling of spring.

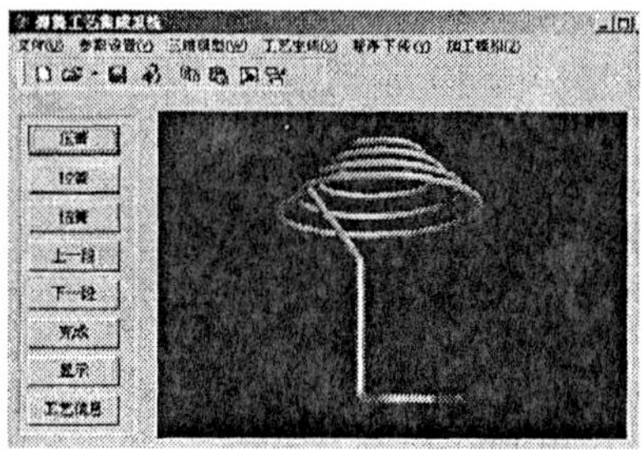

Figure 3. Solid modeling of special-shape spring

5. CONCLUSIONS

The advanced manufacture technology not only requires the blueprint, but also needs 3D model to describe products. This paper studies spatial shape representation and solid modeling of special-shape spring. The characteristics and contributions of the research includes (1) extracting four kinds of spring shape features and defining their data structure, (2) providing the complete shape information of spring with dynamic linked list which records the assemble process of different elementary shape features, (3)creating the solid modeling of spring, and (4)the system provides a sound foundation for the visual design and CNC coiling of special-shape spring.

6. REFERENCES

1. Michael Johnston,(2001),An engineer's guide to spring design, *Wire industry*, vol.68, pp.35-38.
2. Giuseppe Boschiero, (1999), Reducing setup times for critical compression springs, *Wire industry*, vol.66, pp 673-676.
3. YU Zhi-kun, WU Yan-ping, (2003), Study of Modeling Methods on Three Dimensional Spring Model, *Journal of Kunming University of Science and Technology*, vol.28, pp.46-48.
4. Ting-Kuo Peng, AMY J.C.Trappey, (1996), CAD-integrated engineering-data-management system for spring design, *Robotics and computer-integrated manufacturing*, vol.12, pp.271-281.

THREE-DIMENSIONAL MULTI-PIPE ROUTE OPTIMIZATION BASED ON GENETIC ALGORITHMS

Huanlong Wang[1],Cuilian Zhao[1] ,Weichun Yan[2] ,Xiaowei Feng[1]

[1]College of Mechanicals Engineering and Automation, Shanghai University, Shanghai 200072, China;Email:whllenny-2000@163.com; [2]Shanghai Turbine CO.,LTD. Shanghai 200240,China

Abstract: To optimize the design of three-dimensional multi-pipe, multi-constraint and multi-objective path planning, an approach based on Genetic Algorithms (GA) is presented in this paper, which includes definition of genes to deal with pipe routes, definition and application of fitness functions, and definition of punishing function set by constraints. An example and good simulation results are also presented to show the validity of this approach.

Key words: route optimization; GA; Encapsulated Oil Pipes (EOP).

1. INTRODUCTION

In the system of ships, aeroengines and turbines, which has complicated pipe system, the route optimization is always a research hotspot. The route optimization not only releases burden of manual design, but also enhances efficiency of design and reduces cost of design and manufacturing. Though having function of general pipe design, some commercial CAD softwares, such as UG, SOLIDWORKS and CATIA etc., don't have function of the route optimization for some special industries. And researchers mainly focused on the route optimization of two-dimensional pipes or three-dimensional and single one, but relatively less in the multi-pipe, multi-constraint and multi-objective route optimization previously. Taking the route optimization of the EOP[1] of turbine as an example, this paper will present an approach for the "three-multi" optimization problem, which may be useful to other route optimization.

Please use the following format when citing this chapter:
Wang, Huanlong, Zhao, Cuilian, Yan, Weichun, Feng, Xiaowei, 2006, in International Federation for Information Processing (IFIP), Volume 207, Knowledge Enterprise: Intelligent Strategies In Product Design, Manufacturing, and Management, eds. K. Wang, Kovacs G., Wozny M., Fang M., (Boston: Springer), pp. 177-183.

A lot of works have been done in the route optimization of three-dimensional pipes, and relative study is developing from the simple two-dimensional route optimization in the early days to the three-dimensional multi-pipe, multi-constraints and multi-objective route optimization now.

In 1976, sprout method adopted "Depth-First Search" to optimize route [2], which needed more information. In 1985, Dijstra adopted weighted graph (G=(V,E,W) where: V : a group of connecting points; E: cyclometer of connecting points; W; weighted value)to describe pipe space and solve the route optimization problem [3].And then some developed methods, such as adjoining nodes algorithm, the shortest route rapid algorithm and Dijkstra algorithm with evaluating function [4], were proposed, which were only to solve the single pipe problem, instead of considering the multi-pipe one just like the kind of route optimization problems of EOP of turbine. In 2003, Fan jiang proposed the path planning in the pipe system based on coevolution, which was only suitable to the two-dimensional pipe system [5].

Now the main algorithms of path planning are as follows:

- Lee Routing Algorithm [6]: it is a kind of maze algorithm, which is one of the oldest Lee Routing Algorithms. M.Dorigo proposed Ant Colony Algorithm [7] based on maze algorithm.
- Method based on AI and KBE: the process of pipe design is a kind of sequent one, which agglomerates designers' intelligence and experience. So the pipe planning and expert system can be developed to enhance the design efficiency by the technology of AI and KBE.
- Path planning based on Simulated Annealing (SA)[8].
- Path planning based on GA: it is quite useful to the route optimization, and can be applied to solve the multi-pipe, multi-constraint route optimization problem.

2. ALGORITHM OF THE ROUTE OPTIMIZATION UNDER KNOWN CONDITON

2.1 Condition setting

Supposed that the relative position of entrance and exit section of each pipe has been known. Data are transited to Excel as an initial data frame from an outer system by data transition (Fig.1). And then the route can be optimized by GA. Followed the flow chart (Fig. 2).

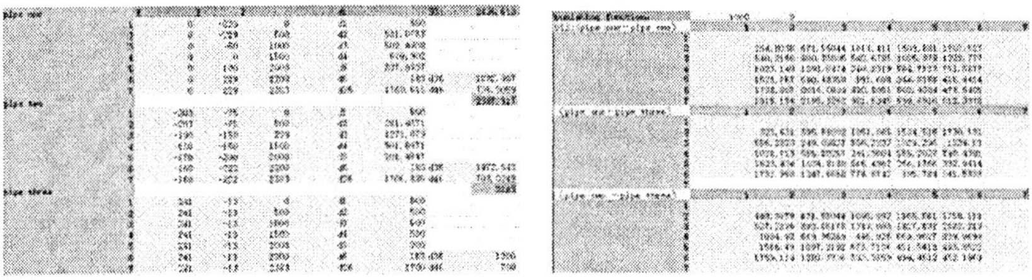

Figure 1. Data frame

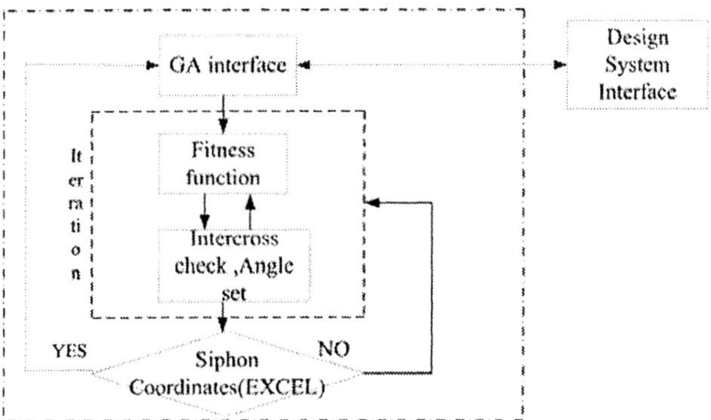

Figure 2. Flow chart

2.2 Coding method

Encoding is a chief aspect of designing and applying GA. Considering the three-dimensional and big data, the coordinate parameters are encoded by floating number, number of bend pipes by integer, the whole variable by coupling multi-parameter. And the code length of each individual "p" is changed with the variable "N" (the number of bend pipes). It is as follows:

$$x_{1,3}\cdots x_{1,m-2}x_{1,m-1}x_{2,3}\cdots x_{2,m-2}x_{2,m-1}x_{3,3}\cdots x_{3,m-2}x_{3,m-1}y_{1,3}\cdots y_{1,m-2}y_{1,m-1}y_{2,3}\cdots y_{2,m-2}y_{2,m-1}y_{3,3}\cdots y_{3,m-2}y_{3,m-1}$$

$$z_{1,2}\cdots z_{1,m-2}z_{1,m-1}z_{1,m}z_{2,2}\cdots z_{2,m-2}z_{2,m-1}z_{2,m}z_{3,2}\cdots z_{3,m-2}z_{3,m-1}z_{3,m}N_1N_2N_3$$

Where:

$x_{i,j}y_{i,j}z_{i,j}$: the coordinates of each pipe; $i=1$:oil pipe one; $i=2$:oil pipe two; $i=3$:oil pipe three; j: 1～m; and $z_{i,j+1}\geq z_{i,j}$;

N_k: the number of each bend pipe; $k=1$: oil pipe one; $k=2$: oil pipe two; $k=3$: oil pipe three; m: the number of strait pipe; m=N+1;

2.3 Selection of fitness function

The selection of fitness function is very important. For the optimization problem with constraints, it can be performance index, or be transformed from the performance index. This paper is about the shortest path, and constraint is that pipes can't be intersected, and bend pipe angle can't be over 60°. Although GA is the algorithm that deals with the maximum problem., the minimum problem with constraints can be transformed from the maximum problem with the punishing function.

According to pressure loss of fluid in the pipe, the shortest length of pipes and the least number of bend pipes are taken as performance index to form the fitness function F_i.(Each pipe has the same fitness function).

$$F_i = w_1 \times \left(1 + \frac{S_i}{\sum_{j=1}^{m} d_j} \right) + w_2 \times \left(\frac{N_{\max} - N_i}{N_{\max} - N_{\min}} \right) + R_p$$

Where:

i :suffix, its meaning is the same to the upper one;

S_i :the straight distance between the exit and the entrance section of each pipe;

f_i : the objective function of each pipe; F_i : the fitness function of each pipe; N_i : the number of bend pipes of each pipe;

w_1 : the weight value of pipe length in the fitness function; w_2 : the weight value of the number of bend pipes in the fitness function

d_i :the length of the single straight pipe; $N_{\min}$: the mini number of bend pipes; $N_{\max}$: the max number of bend pipes;

R_p : the punishing function, which is defined in case of the intersection between pipes and the bend pipe angle.

2.4 Selection of the punishing function R_p

It is a basic principle for the multiple pipes to avoid intersection and the bigger bend pipe angle. Hence, the intersection points must be deleted in order to the route optimization. And the bend pipe angle also needs restricting to reduce the searching space. For every pipe, the punishing function is added to the fitness function to reduce the fitness to delete the bad points when pipes intersect between each other and the bend pipe angle goes beyond the set one. Formula is as follows:

$$\sqrt{(x_{ij}-x_{kg})^2+(y_{ij}-y_{kg})^2+(z_{ij}-z_{kg})^2} > R_i + R_k$$

The bend pipe angle is limited between 0° and 60°, so the formula is as follows:

$$-1 \le \frac{d_{i,(j-1,j)}^2 + d_{i,(j,j+1)}^2 - d_{i,(j-1,j+1)}^2}{2 \cdot d_{i,(j-1,j)} \cdot d_{i,(j,j+1)}} \le -0.5$$

Where:

$$d_{i,(j-1,j)} = \sqrt{(x_{i,j}-x_{i,j-1})^2+(y_{i,j}-y_{i,j-1})^2+(z_{i,j}-z_{i,j-1})^2}$$

$$d_{i,(j,j+1)} = \sqrt{(x_{i,j+1}-x_{i,j})^2+(y_{i,j+1}-y_{i,j})^2+(z_{j+1}-z_j)^2}$$

$$d_{i,(j-1,j+1)} = \sqrt{(x_{i,j+1}-x_{i,j-1})^2+(y_{i,j+1}-y_{i,j-1})^2+(z_{i,j+1}-z_{i,j-1})^2}$$

i , k: $1 \le i, k \le 3$,and $i \ne k$, the symbol of pipe, its meaning is the same to the upper ones; R_1: radius of oil pipe one; R_2 : radius of oil pipe two; R_3 : radius of oil pipe three ;j, g : $1 \le j, g \le m$, symbol of the bend pipe coordinates .

For the punishing function, formula is as follows according to the upper rules:

$$R_p = \begin{cases} 0 \\ 1000 \end{cases} \quad \text{(if interference :0 ; otherwise: 1000)}$$

3. TRIAL ANALYSIS

The trial analysis of this algorithm is applied to one part of the EOP system, According to the design experience, the max number of the bend pipes is five. The pipe coordinates are as follows:

Table 1. The coordinates of entrance and exit

	Entrance X	Y	Z	Exit X	Y	Z	Diameter ¢
Sucker for bearing	-229.0	0.0	0.0	0.0	229.0	2383.0	219.0
Sucker for primary oil pump	-75.0	203.0	0.0	-160.0	-222.0	2383.0	168.0
Cream pipe for primary oil pipe	-13.0	-241.0	0.0	241.0	-13.0	2383.0	168.0

Set: population size M=100, cross rate Pc=0.9, mutation rate Pm=0.01, generation gap G=0.98, the weight value of length of the pipe w_1=0.3, the weight value of the number of the bend pipe w_2=0.7, the setting is as follows:

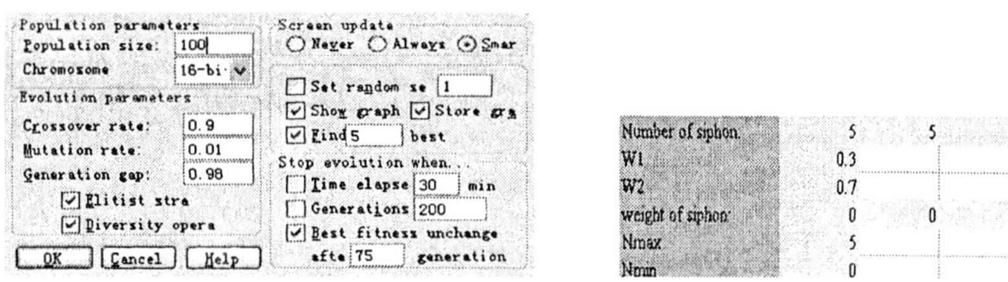

Figure 3. Parameter and weight value setting

Results are as follows after optimization.

Table 2. Optimization results

	X	Y	Z	Number of siphon
Sucker for bearing:	-229.00	0.00	812. 00	2
	0.00	229.00	1533.00	
Sucker for primary oil pump:	-75. 00	203. 00	299.00	4
	63. 00	122. 06	521.20	
	128. 12	84. 00	1138.74	
	-82. 88	-140. 09	1434.14	
	-160. 00	-222. 00	1629.00	
Cream pipe for primary oil Pump:	-13. 00	-241. 00	1021.24	5
	36. 33	-196. 72	1196.34	
	130. 32	-183. 48	1332.80	
	180. 55	-116. 79	1481.65	
	218. 57	-33. 19	1620.50	
	241. 00	-13. 00	1733.00	

The 3D modeling is as follows:

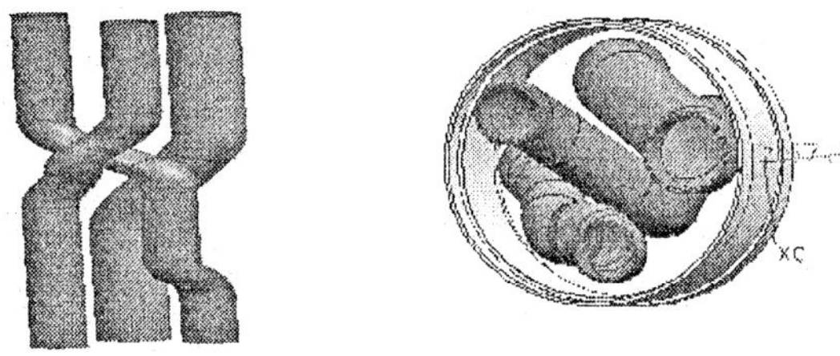

Figure 4. 3D modeling

4. CONCLUSION

Faced to the multi-pipe, multi-constraint and multi-objective optimization problem, a method based on GA is presented to deal with the pipe route planning. One part of the EOP system is also optimized and its result is quite satisfied. However, some further study is needed as following:

- The weight value is only set by the design experience, which may be thoughtless.
- There is only one part of the EOP system optimized, and the optimization of whole system needs improving.

5. REFERENCES

1. Wang Fufang, The quality control and check of the encapsulated oil pipes of large turbines , *Machinist (heat process)*, 3:62-63 (2002).
2. Van C and Koopmans J J G. The optimum routing of pipes in a ship's engine room, Proceedings, Computer *Applications in the Automation of Shipyard Operations and Ship Design II, ICCCS* 76, 333-341(1976).
3. Funaoka K and Wumeda H. new integrated ship design system MATES. Proceedings, Computer *Applications in the Automation of Shipyard Operations and Ship Design IV, ICCCS* 82, 257-268(1982).
4. Qing Li, Dingli Song and Shuangjiang Zhang. Two improved optimum path planning algorithms. *Journal of University of Science and Technology Beijing*,27(3),367-370(2005).
5. Jiang Fan, Mei Ma and Xiaoguang Yang. Path planning in pipe system based on coevloution. *Journal of Aerospace Power*. 19(5),593-597(2004).
6. Lee C Y. An algorithm for path connections and its applications. *Electronic Computres*.(9),346-364(1961).
7. Xiaobing Hu and Xiyue Huang,. Application of ant colony algorithm to maze problem. *Computer Emulation*, 114-116(1983).
8. Wenxun Xing and Jinxing Xie. *The modern optimization algorithms*,(Tsinghua Press, Beingjing,1999),pp.90-96.

AXIOMATIC DESIGN USING ONTOLOGY MODELING FOR INTEROPERABILITY IN SMALL AGRICULTURE MACHINERY PRODUCT DEVELOPMENT

Jiandong Jiang, Fang Xu, Xinrong Zhen, Xian Zhang,Yangyu Wang, Libin Zhang*

The MOE Key Laboratory of Mechanical manufacture and Automation, Zhejiang University of Technology, Hangzhou, China, 31003. Email:jiangjd@zjut.edu.cn

Abstract: Small Agricultural machinery (SAM) is applied widely in agriculture, forestry and stockbreeding. We focus the reconfiguration and Interoperability of the complex process of SAM product development, and propose a design architecture employing axiomatic design method and ontology information representation. In the proposed design process, product function decomposing and recomposing according to independent axiom of axiomatic design, all information of all design phases is expressed in ontology information model with XML language so that can be communicated among all engineers. A case study conducted in Small Agriculture Machinery (SAM) design helps to illustrate the proposed approach, it shows the proposed methods are feasible and reliable, lays a foundation of SAM product reconfigurable design system.

Key words: Small agricultural machinery, Axiomatic design, Ontology, Gearbox

1. INTRODUCTION

Small Agricultural machinery (SAM) is applied widely in agriculture, forestry and stockbreeding. With the ever drastically increment of market competition and the increase of requirement for product personality, Companies must solve numerous problems that have different origins. First of all, products are increasingly complex and should answer a great number of requirements imposed by customers, and manufactures have to improve quality of their products, to have more creativity than their competitors and to reduce costs and times of production, and the key to solve above-

*This project is supported by the Natural Science Foundation of China (Grant 50375146), National Basic Research Program of China (Grant 2004CCA05800), The Doctoral Program Foundation of Ministry of Education of China (Grant 20040337002), the department of international Science&Technology cooperation of China (Grant 2005DFA70380), Zhejiang Provincial department of international Science&Technology cooperation of China (Grant 2005C14005) and Zhejiang Provincial Natural Science Foundation of China (Grant Z503054).

Please use the following format when citing this chapter:

Jiang, Jiandong, Xu, Fang, Zhen, Xinrong, Zhang, Xian, Wang, Yangyu, Zhang, Libin, 2006, in International Federation for Information Processing (IFIP), Volume 207, Knowledge Enterprise: Intelligent Strategies In Product Design, Manufacturing, and Management, eds. K. Wang, Kovacs G., Wozny M., Fang M., (Boston: Springer), pp. 184-191.

mentioned problems lies in the product development which exhausts about 70% technical investment of enterprise. The difficulty to implement product development process can be explained by high number of different phases, and thus disciplines, that have to collaborate. Among all these phases, we can find (Ghodous, 1996, Prasad, 1996, Sriram, et al., 2002, Ullman, 1992, etc.): study of purposes, preliminary studies, planning, preliminary design, functional decomposition, detailed design of product, destruction and recycling. Due to all the phases (different objectives, constraints, manipulated information between functional design and manufacture, etc.) are difficult for them to communicate information.

We are interested in conceptual and virtual aspects of product development. Indeed, our objective is to find methods and data-processing tools, which can help experts to work in an individual and implement an automation design for product development. In this paper, we are involved in improvement of data communication from the first product development phases towards the most advanced one. More precisely, we focus on functional design and detailed design in small agricultural machinery (SAM) product development, and propose a method of axiomatic design using ontology modeling for Interoperability, as they are essential in the SAM product development process:

(1) The first one determines the main functions that are associated to an artifact and are mapped from customers requirements;

(2) The second one decomposes and recomposes these functions to generated the exact structural modal of the artifact employing axiomatic design theory and reconfigurable design methods;

(3) The last one builds a general representation model of design knowledge, experience and geometry information for all phases of product development.

These phases are difficult to relate because they involve different product artifact representations, Hence, our global objective is to find an artifact representation that would be common to these phases. To realize that, we decompose our work in two separate parts: we first consider form to physical structure mapping, which means that we try to find a way to improve interoperability between both functional design and detailed design and detailed design. The second part of our work consists to join the works that have been realized separately into a unique global approach for customer requirements-to-function mapping.

2. AXIOMATIC DESIGN

Axiomatic design provides a systematic and logical methodology for deriving, documenting and optimizing design processes. In the words of Professor Nam Suh of MIT, where the axiomatic design process was developed, axiomatic design methods work to make design more creative, reduce the random solution search process, minimize iterative trial–and–error processes and determine the best design among those proposed [Nam Pyo Suh, etc, 2001]. There are two axioms: (1) independence axiom; (2) information axiom.

Axiomatic design is a process in which engineers, designers and managers first think functionally and then follow through by the innovative creation of physical embodiment. This process provides a systematic way to satisfy many functional requirements (FRs) without introducing coupling of functions and creating integrated physical systems. It decomposes higher–level FRs and design parameters (DPs) into leaf–level FRs and DPs that can be implemented to construct the system according to the resulting design decision architecture. Figure 1 shows an axiomatic process schematically.

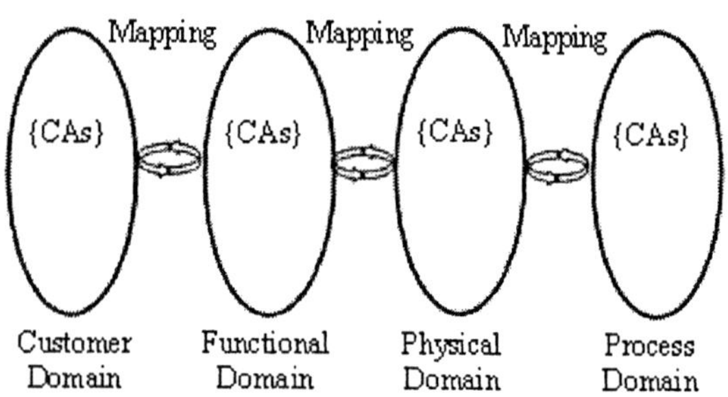

Functional requirements (FRs) *are a minimum set of independent requirements that completely characterize the functional needs of the design solution in the functional domain.*
Design parameters (DPs) *are the elements of the design solution in the physical domain that are chosen to satisfy the specified FRs.*
Constraints (Cs) *are bounds on acceptable solutions.*
Process variables (PVs) *are the elements of the process domain that characterize the process that satisfies the specified DPs.*

Figure 1. Fundamental concept of axiomatic design.

3. ONTOLOGY-BASED F-B-S INFORMATION MODEL

In the process of product design, redesign namely modified design is applied agilely, and reuse of information and knowledge becomes the key to advance design quality and efficiency. So we focus on how to establish design information model to organize and reuse corresponding design information and knowledge at the circumstance of collaborative design. The conception of ontology (Natalya; Michael; Liu, 2002; Li, 2002) in artificial intelligence is: it is a group of concepts or nomenclatures. It researches on classification and attributions of objects and the relationships among objects.

We focus on how to establish a design information model which organizes and reuses design information and knowledge obtained from collaborative design. This paper proposes an ontology-driven design information model and applies it to SAM to realize common use of information and knowledge in product development phases. On account of the three principles (uniqueness, completeness and minimum) constituted by Bogoni at three aspects of product description including function, behavior and structure, a information model satisfying different principles respectively is proposed, and the ontological information model presents all information in the process of design, simulation and production.

4. CASE STUDY

There is a case study conducted in SAM design which illustrates the XML frame of the ontological information model. The description of XML frame for F-B-S model is as following:

Table 1. Ontology representation for speed reducer of SAM

```
<?xml version="1.0" encoding="ISO-8859-1"?>
<rdf:ID = "SAM">
  ...
  <Object rdf:ID = "speed-reducer">
    <SBF:Function>
    Function
        Goal: [description]
        Input: {[flow]}
        Output: {[flow]}
        Sub-functions: {[convert-revert | magnify-diminish | veer-veer | couple-separate |
        conduct-insulate |sTtorage-withdraw]}
        Sub-function of: [Function]
        Referring-component: [Component]
    </SBF:Function>
    <SBF:Behavior>
    Behavior
        Input: {[flow]}
        Output: {[flow]}
        Relation: [convert | transmit]
        Referring- component: [Component]
    </SBF:Behavior>
    <SBF:Structure>
    Structure
```

```
            Main-connection: {[fixed connection | movement connection | transmitted
            connection]}
            Main-component: {[Generic-component-base]}
            Component-relations: {[Generic-component-constraint]}
            Number-component: [Number]
        </SBF:Structure>
        <SBF:Flows>
        Flow
            Category: [energy | signal | material]
        Energy
            Type: [Thermal energy, Hydro energy, Electrical energy, Mechanical
            energy ......]
            State: [Input, Output......]
            Efficiency: [percent]
        Signal
            Type: [Vision, Audition, Olfaction, Tactile sense ......]
            State: [Input, Output ......]
        Material
            Type: [Solid, Gas, Liquid, Mixture]
            State: [Input, Output ......]
            Capacity: [Intension, Tenacity, Elasticity ......]
        Source: {[Component]}
        Destination: {[Component]}
        Referring-Component: [Component]
        </SBF:Flows>
        ...
    </Object>
    ...
</rdf:ID = "SAM">
```

5. ARCHITECTURE OF PRODUCT DESIGN SYSTEM

Figure 2 illustrate the architecture of SAM product design system. It shows the SAM product design process according to four steps: the first transforms customer requirements into product functions employing product experts and sales through CRM tools and market investigation (Fig.2-A), the farther contents are referenced in our previous paper [Jiang Jiandong, et al., 2005,]. The second specifies how to decompose product functions according to independence axiom of axiomatic design method and how to recompose functions on principle of product module reconfiguration warranted by product library (Fig.2-B, C), detailed method is referred in interrelated literatures [Zhang Libin Jiang Jiandong, 2004, Zhang Libin, et al., 2005]. The third shows that a part, component and whole product will be expressed with functional ontology, behavioral ontology and structural ontology in individual phase, then a finished design is expressed in general ontology information model so that can be utilized by all engineers (Fig.2-D, E), it is referred in some literatures [Jiang jiandong, et al., 2005]. The last is that the purpose of designer is computed digitally, and translated into digital model in computer (Fig.2-F, G).

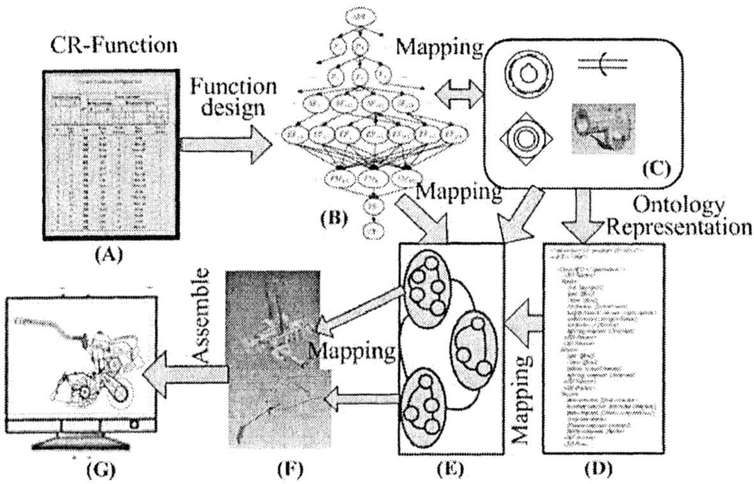

Figure 2. Design process of SAM product

The frame of product design system employs axiomatic design method, CRM, F-B-S ontology modeling reconfiguration design method and reconfigurable methods, integrates conceptual design, detailed design, intelligent design, artifact simulation and 3D product design, etc..

6. APPLICATION OF AXIOMATIC DESIGN FOR SAM GEARBOX

Axiomatic design theory adopts top-down decomposed method and designs function from abstract objects to detailed structure. A case study of SF series gearbox conducted in Small Agriculture Machinery (SAM) design helps to illustrate the proposed product function design and interrelated method. Firstly, the essential requirements include: low cost, small volume, agile manipulation and good adaptability, the mapped functions FRs and design parameters DPs may sort out: (1) FR_1-various speed, DP_1-gear design; (2) FR_2-low cost; DP_2-manufacture plan; (3) FR_3-integrated gear and case-frame, DP_3-case structural design; So the prime design matrix is shown in following

$$\begin{Bmatrix} FR_1 \\ FR_2 \\ FR_3 \end{Bmatrix} = \begin{bmatrix} X & 0 & 0 \\ 0 & X & 0 \\ X & X & X \end{bmatrix} \begin{Bmatrix} DP_1 \\ DP_2 \\ DP_3 \end{Bmatrix} \tag{1}$$

According to SAM task environmental requirements, we decide four various speed gearbox including four-speed, three-speed, two-speed and one-speed. So the design matrix for FR_1 shows:

$$\begin{Bmatrix} FR_{11} \\ FR_{12} \\ FR_{13} \\ FR_{14} \end{Bmatrix} = \begin{bmatrix} X & 0 & 0 & 0 \\ 0 & X & 0 & 0 \\ 0 & 0 & X & 0 \\ 0 & 0 & 0 & X \end{bmatrix} \begin{Bmatrix} DP_{11} \\ DP_{12} \\ DP_{13} \\ DP_{14} \end{Bmatrix} \tag{2}$$

Where, FR_{11} is four-speed gearbox structural design, FR_{12} is three-speed gearbox structural design, FR_{13} is two-speed gearbox structural design, FR_{14} is one-speed gearbox structural design, DP_{1i} (i=1~4) are design parameters for detailed structure of individual gearboxes.

All FRs and DPs are need decomposing and composing abided by independence axiom and information axiom, the final result includes almost eighty independent functional and one hundred design parameters, and these design parameters can be integrated into eighteen modules of SAM gearbox determined by forty-six structural parameters. The result is shown in Figure 3.

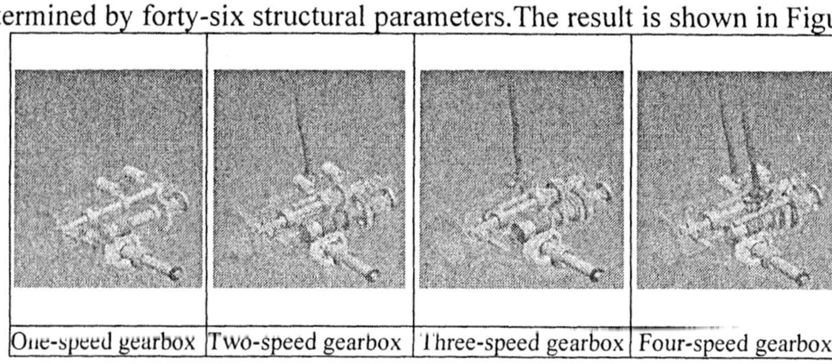

Figure 3. Artifacts for proposed design method

7. CONCLUSION

A design architecture employing axiomatic design method and ontology information representation is proposed. In the proposed design process, product function decomposing and recomposing according to independent axiom of axiomatic design, all information of all design phases is expressed in ontology information model with XML language so that can be communicated among all engineers. A case study of SF series gearbox is realized by axiomatic design method, and the functional, Behavioral and structural information for all design phases is expressed in ontology. This paper proves the feasibility and reliability by functions decomposed and mapping by actual SAM components. The proposed methods and design processes lay a foundation of SAM product reconfigurable design system.

8. REFERENCE

Ghodous, P. (1996), "Modélisation intégrée de données de Produit et de Processus de Conception", "Informatique", Lyon, Université Claude Bernard-Lyon 1.

Jiang Jiandong, Zhang Libin, Xu Fang, Zhang Xian, Ji Shiming, Lv Fangya. (2005), Study on the model of customer requirements for SAM product under MC. *Nongye Gongcheng Xuebao/Transactions of the Chinese Society of Agricultural Engineering,* v 21, n 9, September, p 98-102.

Jiang Jiandong, Zhang Libin, Xu Fang, Zhang Xian, Ji Shiming, Lv Fangya.(2004), Study on the model of customer requirements for SAM product under MC, *Soypsium of national cross-strait manufacture technology*, 15, Nov. p.80-85.

Liu Xijuan, Jiang Shouwei (2002), "Research on Reconfigurable Product Design Theory and Several Key Problems," thesis for doctor's degree, Shanghai Jiao Tong University, Shanghai, China. (In Chinese)

Li Shanping, Yin Qiwei, Hu Yujie (2004), "Overview of Researches on Ontology," *Journal of Computer Research and Development*, Vol. 41, No. 7, pp. 1041-1052. (In Chinese)

Michael Erdmann, Rudi Studer, "Ontologies as Conceptual Models for XML Documents," Institut für Angewandte Informatik und Formale Beschreibung sverfahren (AIFB) University of Karlsruhe (TH), D-76128 Karlsruhe (Germany).

Ullman, D. G. (1997), The mechanical design process, 2nd edition, McGraw-Hill.

Nam Pyo Suh, et al (2001) *Axiomatic design-advances and applications*, Oxford University Press, Inc.

Natalya F. Noy, Deborah L. McGuinness, *"Ontology Development 101: A Guide to Creating Your First Ontology,"* Stanford University, Stanford, CA.

Prasad, B. (1996), *Concurrent Engineering Fondamentals*, 2, 478.

Sriram, R. D. (2002), *Distributed and integrated collaborative engineering design*, 1, Sarven Publishers.

Zhang Libin, et al (2005) ."Scheme Design of the small agricultural machinery Based on reconfigurable Module Method". *Transactions of the Chinese society for agricultural machinery*, Vol.36(4),pp.78-81.

MAXIMUM ENTROPY OPTIMUM DESIGN FOR PRODUCT STRUCTURE

Xu Yang[1], Xingyuan Wang[2]
[1]*Shandong University of Architecture and Engineering, China; Email: yangxu@sdai.edu.cn.*
[2]*Shandong University, China.*

Abstract: Taking the product cost as goal function, a maximum entropy method is used to set up a mathematical model for mechanical structural optimization. By using MATLAB, the computing program is made. And then an example for a mechanical product is given out — a motorcycle frame optimum, this is an optimization with sizes, stresses and frequency constraints. Results show that the optimum plan is feasible and effective in saving cost.

Key words: Maximum Entropy; Structure Optimization; Motorcycle Frame; Cost Reduction.

1. INTRODUCTION

With the development of modern science and technology, the market competition has become more and more acute. How to talent show itself in many products with the same function and the quality, on the one hand, the mechanical product structure is required to be reasonable, the function is comprehensive and the quality be excellent, on the other hand, the relevant cost be lower. Accordingly, the designer must do their best to meet many demands in the product function, and so on, and at the same time, to reduce cost as fully as possible as. The contradictious problem can be described as a mathematics model to be solved by the maximum entropy optimal method (MEOM).

Based on the surrogate constraint and the information entropy concepts, the iteration for the surrogate multipliers was set up [1], and all constraints in different forms were united handled [2], replaced by a very simple formulas and the algorithm is effective in the static strength problems of a truss [3].

Please use the following format when citing this chapter:
Yang, Xu, Wang, Xingyuan, 2006, in International Federation for Information Processing (IFIP), Volume 207, Knowledge Enterprise: Intelligent Strategies In Product Design, Manufacturing, and Management, eds. K. Wang, Kovacs G., Wozny M., Fang M., (Boston: Springer), pp. 192-197.

The reference [4] introduces the convergence factors β and γ into the MEOM for the optimal design of rattled beam to force the design variables to get optimum. The convergence speed and the precision are improved obviously. In this paper, the author makes an optimal design example for a mechanical product—the lowest cost control of a motorcycle frame, which contains the safety and the comfortable constraint conditions.

2. PRODUCT STRUCTURE OPTIMAL DESIGN

To design a mechanical product which is composed of several parts with the same mechanical properties and the same price the target of the optimal design is to make the lowest cost, and satisfy the geometric size and the strength constraint conditions. Because the product has dynamic inputs, the natural frequency must escape the limiting frequency to prevent the resonance problem, by doing so, the safety and the economic properties of the product will be ensured.

Let: U denotes the cost; Fs denotes the surrogate constraint; X denotes the design variable. The optimization problem can be described as following:

$$Min \quad U = \sum_{i=1}^{n} C_i W_i(X) \tag{1}$$

$$s.t. \quad F_s = \sum_{j=1}^{m} \lambda_j F_j(x) \leq 0 \quad i=1, 2,\dots, n\,;\, j=1,2, \dots, m \tag{2}$$

Where: Ci, Wi are the unit price and the weight of the i-section separately, here the product is divided into n parts, each part has a design variable; λj, Fj(x) are the surrogate multiplier and the constraint function of the j-th constraint, there are m different constraints.

From the point of view of the probability [5], the λ satisfies:

$$\sum_{j=1}^{m} \lambda_j = 1, \lambda_j \geq 0, \quad j=1,2,\dots,m \tag{3}$$

Its value can be determined by: $\lambda_j = \exp[\theta F_j(X)] \Big/ \sum_{j=1}^{m} \exp[\theta F_j(X)]$ (4)

Where θ is a small positive number, let: $\theta_{j+1} = \theta_j + 10^{-3}$.

By introducing the Lagrange multiplier α for Fs, we structure the Lagrange function: $L_s(X,\alpha) = U(X) + \alpha \cdot F_s(X)$ (5)

From the stationary conditions, that is: $\partial L_s / \partial x_i = 0, \partial L_s / \partial \alpha = 0$.

We can obtain: $X_{k+1,i} = X[X_{k,i}, \lambda, \alpha]$ k=1,2,…,K (6)

Where K is the iteration times. It goes without saying that the possible paths exploring for the lowest cost is an iteration process, first, a group of initial value X_i^0 are given; then start to iterate and get a new group of design variable X_i^k, repeating this way, until the results satisfy the convergence criterions.

3. ENGINEERING EXAMPLE

3.1 Vibration test

Two-wheel-treadle motorcycle mainly drives in the city street, the speed of design in common use is the 30 Km/ h. driving in this speed, and the driver fell the vibration is rather greatly [6]. To explore the reasons of vibration, and adopt the measure, do two experiment researches as follows.

3.1.1 Road-driving vibration test

The measuring points of whole motorcycle are: Handlebar, Back-wheel axle, Engine in fixed-point of the frame, Back-vibration isolator in fixed-point of the frame, Treadle, Saddle in three directions. The results (take measuring point on Treadle as example) are shown in Fig. 1.

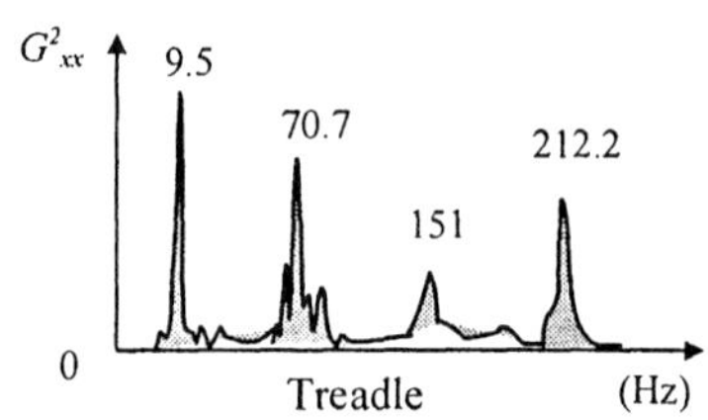

Figure 1. Vibratory frequency spectrum on line testing

Moreover, make the motorcycle rest in the laboratory, open the engine, test the vibration of each measuring point when the engine affect alone, that results and those on line test are similar, this test did not give.

3.1.2 Experimental modal analysis of a frame [7]

A motorcycle-frame structure is shown in Fig. 2. Its model parameters are shown in Tab.1.

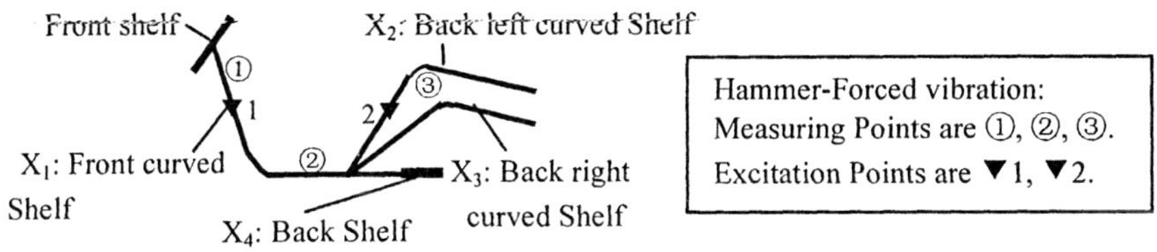

Figure 2. Motorcycle-frame Structure

Table 1. The frame model parameters

Order.No.	1	2	3	4	5
Model frequency(Hz)	66	151	185	258	284
Damping ratio (%)	1.515	0.662	0.541	0.386	0.335

From above results analysis, we know: That power spectrum curve has a resonance peak in low frequency 9.5 Hz or so, the peak is a vibration owing to road excitation to a frame, and it can be neglected. The engine is the most important vibration source of the motorcycle. Designing a frame structure again, make its natural frequency keep away excitation frequency, is one of the necessary measures for raising the driver comfort and safety.

3.2 Optimal model [8]

The frame is simplified to three-dimensional beam units, the quantity on each unit is divided equally in each node, and the moments of inertia also do the same processing. The Front shelf does not attend the optimal design, the finite element model: 40 units, 41 nodes. The finite element calculation results (the natural frequency) are: 66.8, 152.7, 192.6, 245.5, 285.9 Hz. There's a bit difference between the calculation values and test results.

A motorcycle frame is a space structure; the optimal control of the cost is that the cross acreage will be selected to lighten the weight of the frame, and satisfy the stress, the size and the frequency constraint conditions under the actual loads.

The optimization problem can be indicated as follows:

$$Min \quad U = \sum_{i=1}^{n} C_i \rho_i L_i A_i = \sum_{i=1}^{n} C_i \rho_i L_i x_i^{-1} \tag{8}$$

$$\text{s.t.} \quad F_1 = \left(\omega_1^2 / \underline{\omega}^2 - 1\right) \le 0; F_2 = \left(1 - \omega_2^2 / \overline{\omega}^2\right) \le 0 \tag{9}$$

$$F_{i+2} = \left(A^i{}_{\min} \cdot x_i - 1\right) \le 0 \tag{10}$$

$$F_s = \lambda_1 \cdot F_1 + \lambda_2 \cdot F_2 + \sum_{i=1}^{n} \lambda_{i+2} \cdot F_{i+2} \le 0 \quad , \quad i=1, 2, \ldots, n \tag{11}$$

Where ρ_i, L_i, A_i are the density, the length and the cross section of the i-th length acreage separately, using the reciprocal design variable x_i ($x_i=1/A_i$), the variable number is 4 (shown in Fig.2.), ω_1, ω_2, $\omega_{\underline{F}}$, $\omega_{\overline{F}}$, are the first, the second eigenvalue and its lower, the higher limit of the restricted bounds; A^i_{min} is the lower limit of i-th part cross section acreage constraints. The equation (8) satisfies not only the demand of the strength, but also the technology and the structure.

The frequency gradient formula:

$$\partial\omega_p^2 \Big/ x_i = \frac{\{U_p\}^T \frac{\partial K}{\partial x_i}\{U_p\} - \omega_p^2 \{U_p\}^T \frac{\partial M}{\partial x_i}\{U_p\}}{\{U_p\}^T M \{U_p\}} \quad (12)$$

Where M and K are the mass and stiffness matrices, let ψ denotes the quadratic moment of per unit acreage, then the moment of inertia of the cross section of the frame is: $I_i=\psi A_i^2$, which will be used in the finite element analysis of the structure. The optimal analysis steps are shown in figure 4. In Fig.4, β $(0<\beta<1)$ and γ $(0<\gamma<1)$ are the geometric and the dynamic convergence factors. Let: A^i_{min}= {0.27, 0.15, 0.15, 0.25} cm^2; Modulus of elasticity E=210GPa; C=$6000yuan/T$; ω_{down}=458rad/s, ω_{up}=955rad/s; The natural frequency f_1=67Hz, f_2=152Hz, $\omega=2\pi f$. The optimal results of a motorcycle frame are shown In Tab.2, and Tab.3.

Table 2. The optimal results

Plan.No.	Ai (cm2)	W (kg)	Reduced-W (kg)	f1, f2(Hz)
Original	{333.8,216.0,216.0,575.0}	4.443		66.83,152.7
Plan 1	{292.4,192.1,192.1,538.6}	3.886	0.557	71.24, 6.73
Plan 2	{275.1,200.3,200.3,538.6}	3.891	0.552	71.29, 6.83
Plan 3	{294.5,192.1,192.1,653.8}	3.929	0.514	70.86, 5.89
Plan 4	{310.2,200.1,200.1,596.4}	3.938	0.505	72.04,158.43
Plan 5	{276.4,200.2,200.2,596.4}	3.916	0.527	71.41, 56.63

3.3 Analysis

(1) In Each optimum process, the superior local value can be obtained through 4-6 iterations.

(2) The geometric convergence factor β affected the convergence speed, and the frequency convergence factor γ affected the precision, if its value is selected improper, the iteration will be difficult to converge even more.

Table 3. The saving cost per motorcycle

Plan. No.	1	2	3	4	5
Save cost (Yuan)	3.34	3.32	3.08	3.04	3.16
Rate of Reduced Cost	12.5%	12.4%	11.6%	11.3%	11.8%

4. CONCLUSIONS

(1) The frequency constraint is a non-linear function which it hidden the design variables. So, for obtaining the global optimum solution, we can select several representative groups of design variables as the initial values to calculate the optimal problem, generally speaking, each optimal result is the local solution, we may consider the there exists global optimization value among these. Above the process can be called the first-stage optimization process.

(2) The Second-stage optimization process [9]: The motorcycle frame uses the steel tube Structure, after determining the cross section acreage, many different standards can be selected, so the optimal design for the product structure will be obtained.

(3) Form table.3, we know: each motorcycle can be saved 1.6 Yuan on cost equally. In 2001, the manufacturer that the motorcycle discussed had obtained the highest sales volume (about a million motorcycle), if used the paper's design, they will save several millions Yuan; the economic benefit will rise greatly.

5. REFERENCES

1. E T Jaynes, (1957), Information Theory and Statistical Mechanics. *The Physical Review,* Vol. 106, pp. 620-630.
2. Templeman, A. B, Li Xingsi, (1987), A Maximum Entropy Approach to Constrained Non-Linear Programming. *Engineering Optimization,* No.3, pp. 191-205.
3. Li Xingsi, (1989), A Maximum Entropy Method for the Structure Optimization Design, *Computational Structural Mechanics and Application,* No.1, pp. 36-45.
4. Yang Xu, (1996), A Maximum Entropy Method for the Dynamical Structural Optimization Design, *Engineering Mechanics (Supplementary Issue),* Vol.2, pp. 598-601.
5. C E Shannon. (1948), *A Mathematical Theory of Communication,* Bell System *Technical Journal,* Vol.27, No.3, pp. 379-428.
6. Shuqian Cao. Etc. (2001), Vibration Structure Modal Analysis —Theory, *Experiment and Application,* Tianjing University Press, Tianjing.
7. K. H. Hueber, (1975), *The Finite Element Method for Engineers,* John Wiley.
8. Edward J. Haug, Jasbir S. Arora. (1985), *Applied Optimal Design,* Science Press. Peking.
9. Teng Haiwen, (2002), Two-Stage Frame Structure Optimum Design with Discrete Variables Based on Maximum Entropy, *Journal of Beijing Polytechnique University,* Vol28, No.3, pp. 78-84.

PRODUCT ROBUST DESIGN WITH A MIXTURE OF RANDOM AND INTERVAL FACTORS

Chuntao Liu, Zhihang Lin
CIM Institute, State Key Laboratory for Manufacturing Systems Engineering, Xi'an Jiaotong University, China; Email: chtliu@mailst.xjtu.edu.cn.

Abstract: To solve the product robust design problems involving a mixture of random and interval factors, a novel approach is proposed which combines an integrated metamodeling method with a percentile-based robust optimization model. The integrated metamodeling method embodies the integration of Response Surface Methodology (RSM) and Support Vector Regression (SVR), which is used to construct the metamodels of product performance efficiently. The percentile-based robust optimization model could bring both the design objective robustness and the feasibility robustness of the design constraint into account, which assures a product's reliability and quality robustness to the noise. A case study of a diaphragm spring in automobile clutch is described to show the effectiveness and practicability of the proposed approach.

Key words: robust design, random factors, interval factors, Response Surface Methodology, Support Vector Regression, percentile.

1. INTRODUCTION

Robust design is a method for improving the quality of a product through minimizing the effect of the causes of variation without eliminating the causes[1]. Under the concept of robust design, there are two broad categories of the causes of product performance variations: noise factors and control factors. In traditional robust design[2], uncertainties associated with noise factors and control factors usually imply for randomness and nondeterministic factors are assumed to follow certain distributions. However, in reality, some of the distributions related to random factors are often not precisely known. Moreover, uncertainties associated with other uncertain factors may not be the result of randomness[3]. Therefore, when

This project is supported by Chinese National Natural High-Tech. R&D Program for CIMS under Grant No. 2002AA413520 and National Fundamental Research Program (973) under Grant No. 2003CB716207.

Please use the following format when citing this chapter:
Liu, Chuntao, Lin, Zhihang, 2006, in International Federation for Information Processing (IFIP), Volume 207, Knowledge Enterprise: Intelligent Strategies In Product Design, Manufacturing, and Management, eds. K. Wang, Kovacs G., Wozny M., Fang M., (Boston: Springer), pp. 198-204.

nondeterministic factors are only known within intervals, robust design approach should treat them as interval factors. So far, there is still no method to deal with the robust design problems involving the combination of random and interval factors.

The purpose of this paper is to describe an efficient method for robust design to deal with problems under the mixture of random and interval factors. In the proposed approach, an integrated metamodeling method based on Response Surface Methodology (RSM) and Support Vector Regression (SVR) is combined with a percentile-based robust optimization model in which both robustness of the design objective and feasibility robustness of the design constraint are simultaneously assured.

2. A THREE-STEP ROBUST DESIGN APPROACH

A three-step robust design approach is proposed (see Figure 1) to deal with the problems under the mixture of random and interval factors:

Step 1 Build approximation models to associate each product performance with control and noise factors by using an integrated metamodeling method based on RSM and SVR.

Step 2 Model the robustness of the design objective and feasibility robustness of the design constraint by using percentile-based formulations.

Step 3 Use a percentile-based robust optimization model to find the robust design solution.

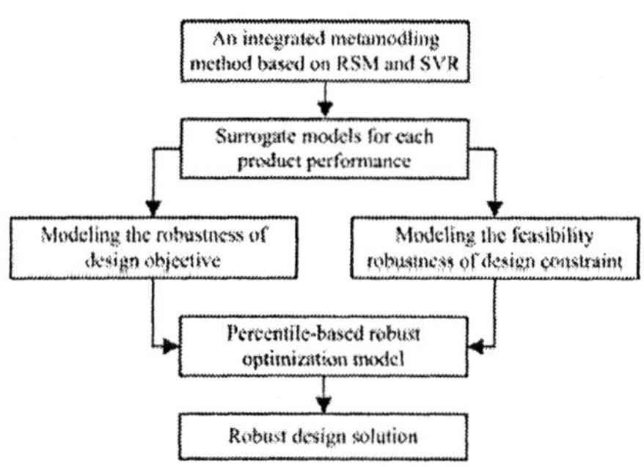

Figure 1. Flowchart of the three-step robust design approach

2.1 Integration of RSM and SVR

To take advantages of RSM[4] and SVR[5] and overcome their limitations, an integration strategy (as shown in Figure 2) is presented to build efficient surrogate models of product performance.

This strategy first employs the RSM techniques to generate a number of design experiments and create response surface models to represent the performance behaviors under study. If the models are not accurate enough, the SVR technique is implemented. Here, support vectors are trained on a set of data points which correspond to the design experiments generated previously based on RSM techniques. Additional points would be added to achieve the desired accuracy. When the results from either RSM or SVR techniques are accurate enough, they can be used as approximation models in robust optimization programs to determine the optimal values.

With the integration strategy, it is anticipated that the limitation of RSM in modeling highly nonlinear behaviors could be overcome by the SVR techniques. In the situation when the Response Surface Model is accurate enough, the complicated training process of the SVR can be avoided.

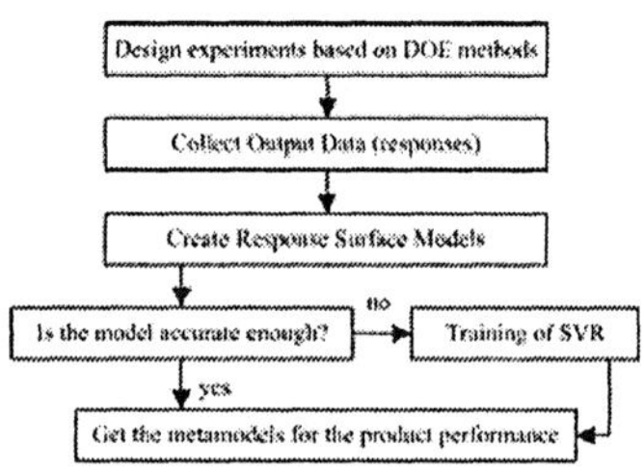

Figure 2. Flowchart of the integrated metamodeling method

2.2 Percentile-based robust optimization model

Under the mixture of random and interval factors, the percentile-based formulations are presented to model the robustness of design objective and the feasibility robustness of design constraint.

1) Modeling the robustness of design objective

Under the mixture of random and interval factors, the control factors $\boldsymbol{x}$ and noise factors z are classified into two types of factors: random factors $\boldsymbol{r}$ characterized by probability distributions; interval factors $\boldsymbol{d}$ within specified intervals $[d_L, d_U]$. As a result, the design objective $y = y(\boldsymbol{x}, z) = y(\boldsymbol{d}, \boldsymbol{r})$ is characterized by the mixture of distribution and interval as well. $y|\boldsymbol{d}$ is a random variable which is the function of random variables $\boldsymbol{r}$ and can be described by a marginal distribution, as $\boldsymbol{d}$ varies within the interval $[\boldsymbol{d}_L, \boldsymbol{d}_U]$, y performs as a sequence of probability distributions In this distribution sequence, there exists a maximum y^{p_2}, denoted as $y^{p_2}_{\max}$, and a minimum y^{p_1}, denoted as $y^{p_1}_{\min}$, where p_1 is a left-tail cumulative distribution functions (CDF), for example, 0.05 or 0.1, representing the product performance at the left tail of its probability distribution and y^{p_2} is a right-tail CDF, for example, 0.95 or 0.99. The percentile difference $\Delta y_{p_1}^{p_2}\Big|_{\max} = y_{\max}^{p_2} - y_{\min}^{p_1}$ reflects the variation range of a design objective. $\overline{y^{0.5}}$ is used to represent the variation location of a design objective.

Under the mixture of random and interval factors, for smaller-the-better type robust design, the robustness of design objective can be modeled as

$$\text{minimize} \quad [\overline{y^{0.5}}, \Delta y_{p_1}^{p_2}\Big|_{\max}] \tag{1}$$

2) Modeling the feasibility robustness of design constraint

Similarly, under the mixture of random factors $\boldsymbol{r}$ and interval factors $\boldsymbol{d}$, the constraint function $g_j(\boldsymbol{x}, z) = g_j(\boldsymbol{d}, \boldsymbol{r})$ performs as a sequence of probability distributions. The feasibility robustness of design constraint under the mixture of random and noise factors can be modeled as

$$g_j^{p_j}\Big|_{\max} \le 0 \quad j = 1, 2, \cdots, m \tag{2}$$

Where p_j is the user specified reliability.

In summary, through using the concept of percentile, under the mixture of random and interval factors, the unified robust optimization model for integrated design objective robustness and feasibility robustness of design constraint can be given as

$$\begin{aligned} \min \quad & [\overline{y^{0.5}}, \Delta y_{p_1}^{p_2}\Big|_{\max}] \\ \text{s.t.} \quad & g_j^{p_j}\Big|_{\max} \le 0 \quad j = 1, 2, \cdots, m \\ & x_L \le \bar{x} \le x_U \end{aligned} \tag{3}$$

3. A CASE STUDY

The design of a diaphragm spring in automobile clutch is used as a case study to verify the approach described in this paper. The purpose of this design is to find the dimensions of the spring as shown in Figure 3, which can meet the following design requirements:

1) Minimize the maximum stress σ in the diaphragm spring.

2) Produce sufficient load when the diaphragm spring clamps the pressure plate.

$$g_1 = 4T_{\max}(D^2 - d^2)/(D^3 - d^3) - P(\lambda_b) \le 0$$

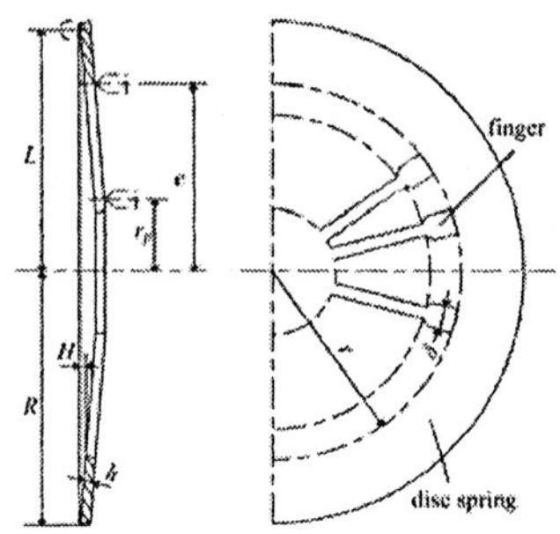

Figure 3. A diaphragm spring

3) Transmit the engine's power stably even when the clutch disc is worn.

$$g_2 = |P(\lambda_b - 1.2) - P(\lambda_b)| - 0.05P(\lambda_b) \le 0$$

Factors that are related to the requirements above are classified into eight control factors and three noise factors. In this case, the control factors are random variables and noise factors are interval variables (listed in Table 1).

Table 1. Information of random and interval factors

Random factors		
Factor	Standard deviation	Distribution
R /mm	0.04	Normal
r /mm	0.04	Normal
H /mm	0.05	Normal
h /mm	0.05	Normal
L /mm	0.02	Normal
e /mm	0.02	Normal
r_p /mm	0.02	Normal
δ /mm	0.02	Normal
Interval factors		
Factor	Lower Bound	Upper Bound
$T_{\max}$ /N•m	178.2	192.2
D /mm	239.6	240.4
d /mm	158.6	160.4

In this design problem, σ is treated as the design objective, while g_1 and g_2 are treated as the design constraint. To relieve the computational burden, the approximate models of σ , g_1 and g_2 are constructed by using the integrated metamodeling method aforementioned. Three metrics, named as, Max. Absolute Error (MAE), Average Absolute Error (AAE) and Root

Mean Square Error (RMSE) are used to measure the accuracy of metamodels. To be as a contrast, Artificial Neural Networks (ANN) models for the product performance on the same sample data are also built and tested. The accuracy of RSM, SVR and ANN models is shown in Table 2.

It could be noted from Table 2 that, for σ, which is highly nonlinear, the accuracy of SVR is the best while ANN is better than RSM. For g_1 and g_2, which are also nonlinear but not as much as σ, three methods have achieved a good accuracy – the accuracy of RSM is close to that of SVR while ANN is the worst. This indicates that, in most cases, SVR is better than RSM and ANN. The limitations of RSM and ANN, i.e., the defect of RSM in modeling highly nonlinear behaviors, and the necessity of a large set of training data for ANN, could be eliminated by SVR techniques. In the case of modeling lowly nonlinear behaviors, the use of RSM makes the complicated training process of SVR and ANN avoided.

Table 2. Accuracy of metamodels

Error	RSM			SVR			ANN		
	σ	g_1	g_2	σ	g_1	g_2	σ	g_1	g_2
MAE	2.739	0.485	0.178	0.524	0.473	0.165	1.724	0.959	0.175
AAE	0.457	0.098	0.029	0.073	0.095	0.026	0.287	0.161	0.028
RMSE	0.965	0.199	0.061	0.145	0.193	0.055	0.586	0.342	0.059

After constructing the metamodels of the design objective and design constraint, the percentile-based robust optimization model is given as

$$\min \ [\overline{\sigma^{0.5}}, \Delta\sigma_{0.01}^{0.99}\Big|_{\max}]$$
$$\text{s.t.} \quad g_j^{0.99}\Big|_{\max} \leq 0 \quad j = 1,2$$

Compromise Programming method[6] is used to solve this multi-objective model which allows the designer assign a weight factor w representing the relative importance of the two objectives. The optimal solutions under three different values of the weight factor are listed in Table 3.

For comparison, we also solve the problem by using the traditional deterministic optimization model ignoring the uncertainty in control and noise factors. The optimal solution is given in Table 4. To compare the effectiveness of the different methods, the variation range of stress, $\Delta\sigma_{0.01}^{0.99}\Big|_{\max}$, and the probability of constraint satisfaction at the optimum point from the conventional deterministic model are also calculated.

It can be noted from Table 3 and Table 4, the variation range of stress of the robust design solution is much less than that of the traditional deterministic optimization model, which means that the robust design solution is less sensitive to the variations in design conditions or operating conditions. Meanwhile, as predicted, the conventional deterministic model generates a low probability for constraint satisfaction, while the robust design solution produces the satisfying constraint feasibility.

Table 3. Robust design solutions of the diaphragm spring design

w	R /mm	r /mm	H /mm	h /mm	L /mm	e /mm	r_p /mm	δ /mm	$\overline{\sigma^{0.5}}$ /MPa	$\Delta\sigma_{0.01}^{0.99}\|_{max}$ /MPa
0.3	113	85	4.29	2.67	110	89.0	31	10	1439	21.01
0.5	116	90.1	4.32	2.70	112	92.4	30	9	1384	26.61
0.7	121	95.8	4.38	2.77	119	98.7	32	9	1341	33.34

Table 4. Optimal solution from the deterministic optimization method

R /mm	r /mm	H /mm	h /mm	L /mm	e /mm	r_p /mm	δ /mm	$\Delta\sigma_{0.01}^{0.99}\|_{max}$ /MPa	Probability of constraint satisfaction	
									g_1	g_2
123	100	4.19	2.66	121	103	30	11	61.21	0.58	0.791

4. CONCLUSION

When both the random factors and interval factors present in a product robust design problem, the product performance is also characterized with the mixture of randomness and interval. An integrated metamodeling method is combined with a percentile-based robust optimization model to solve this problem. The future work will target at how to improve the efficiency of solving the optimization model.

5. REFERENCES

1. M. S. Phadke, *Quality Engineering Using Robust Design* (Prentice Hall, New Jersey, 1989).
2. W. Chen, J. K. Allen, F. Mistree, and K. L. Tsui, A procedure for robust design, *ASME Journal of Mechanical Design* **118** (4), 478-485 (1996).
3. X. Du, A. Sudjianto, and B. Huang, Reliability-based design under the mixture of random and interval variables, *ASME Journal of Mechanical Design* **127**(6), 1068-1076 (2005).
4. R. H. Myers, and D. C. Montgomery, *Response Surface Methodology: Process and Product Optimization Using Designed Experiments* (John Wiley and Sons, New York, 1995).
5. N. Cristianni, and J. Shawe-Taylor, *An introduction to support vector machines and other kernel-based learning methods* (Cambridge University Press, Cambridge, 2000).
6. M. Zeleny, *Compromise programming, in Multiple Criteria Decision Making* (University of South Carolina Press, Columbia, 1973).

ROOT CAUSE ANALYSIS OF AN ARTILLERY BREECHBLOCK AND ITS IMPROVED

Wenyuan Song, Jianshe Kang, Lianyan Shi, Yabin Wang
6th department, Mechanics engineering college, Shijiazhuang, Hebei, 050003, China, Email: mumi13@163.com

Abstract: The breechblock failure is a repeatedly failure with high frequency on artillery. And it is unable to prevent effetely by Reliability Centered Maintenance. A Root Cause Analysis method, which is improved according to the feature of artillery, is used to analyze the breechblock failure. It was validated that one key root cause of this failure is the problem in pry-axis design of breechblock.

Key words: Root Cause Analysis; Artillery Breechblock; Improved method

1. INTRODUCTION

As an important unit of modern artillery, breechblock is used to stick into the shell pumer, closedown the chamber with powder flask when firing, and draw out the powder flask after firing. It is one of the highest frequency betake machine on artillery, especially in training. According to statistics, there is approximately 30 percent artillery failure caused by breechblock[1]. So frequency failure of the breechblock will disturb the normally training and foreclose the campaign effect of artillery. Although some of them can be prevented by traditional prevented maintenance measure based on Reliability Centered Maintenance(RCM) analysis, there are some reappearance failures, such as breechblock closedown failure, can't be reduced effectively. After analyzing, it was find that the breechblock is the lowest level FSI in RCM, and it was taken as the cause not the failure itself, the maintenance task is not prevented but repaired or replaced it. Clearly, It is necessary to find the root cause of breechblock exactly and timely, so relevant measure can be adopt to prevent the failure reappearance or reduce its probably. So, Root

Please use the following format when citing this chapter:
Song, Wenyuan, Kang, Jianshe, Shi, Lianyan, Wang, Yabin, 2006, in International Federation for Information Processing (IFIP), Volume 207, Knowledge Enterprise: Intelligent Strategies In Product Design, Manufacturing, and Management, eds. K. Wang, Kovacs G., Wozny M., Fang M., (Boston: Springer), pp. 205-210.

Cause Analysis (RCA) method is used in this paper.Considering the default of traditional RCA, A perfectly cycle root cause analysis method is present based on methods from references [2,3,4].

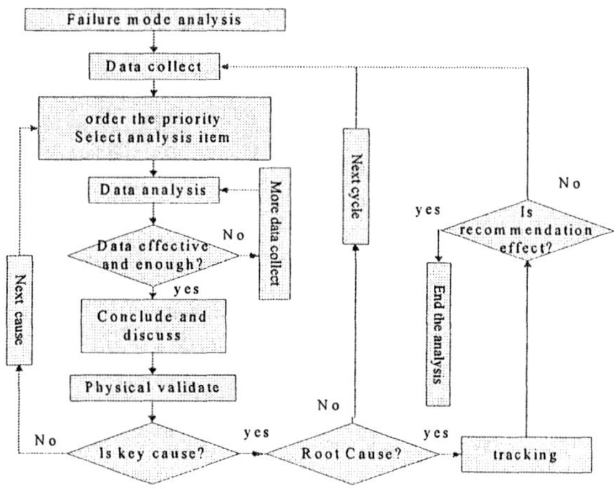

Figure 1. improved RCA analysis process

2. ROOT CAUSE ANALYSIS AND ITS IMPROVED

The originally root cause analysis method was presented by America navy nuclear department in 1970's. While until 1990's, root cause analysis method was gradually perfected and could be used in every walk of life, and attracted the eyes of America government, and military. And then there are many special website of RCA was founded and running, and tenth types commercial software about it were developed, such as Apollo、PROACT、Reason、TapROOT、Inhouse System. In 1992, DOE-NE- STD-1004-92 [5], a guide handbook of root cause analysis, was published by America Department of Energy (DOE). According to the result of 2001 RCA applying investigation by plant-maintenance Resource Center of Australia[6], RCA is used widely in nuclear energy, environment, oil, industry machine maintenance and body healthy with un-imaging profit.

As a general designation, RCA has several representations in practice, but all these representations have a similar essence step, it can be described in 5 steps:

- Preserving Failure Data

- Ordering the Analysis
- Analyzing the Data
- Communicating Findings & Recommendations
- Tracking for Success

Whereas, there are some defaults of traditional RCA in analyzing military materiel, such as none specific analysis end point, lacking select criteria of analysis direction, need operating data of long uninterrupted working period ——it is especially unsuitable for military materiel, etc.

According to the feature of military materiel, an improved RCA method is presented in this paper, limited by the space, figure 1 gave the analysis flow, and the detail analysis step is described with its application on artillery below.

3. ROOT CAUSE ANALYSIS OF BREECHBLOCK

3.1 Analyzing the failure modes of breechblock

The first step is describing the work principle detail, analyzing the key failure mode and cause by backfit FMEA. As an important part of artillery, breechblock has a multi-function. While firing, it is used to pin the shell pumer, burning the powder, and at the same time closedown the chamber with powder flask in order to forming a high temperature and high pressure gas in powder chamber so that the shell can be obtained a certain original velocity and revolving velocity. After firing, it is used to draw out the powder flask in order to backfill the next shell and blaze away. The failure mode and cause analysis see table 1.

Table 1. Failure Mode and Cause Analysis table

Describing Failure mode	Effected factors			Effect and probably of cause	Simulating result	note the key cause
	Components /design	Operating / maintenance	Environment factors			
	No.1					
	No.2					Key 1
	...					
Mode 1		Operating maintenance				Key 2
			Factor 1			
			Factor 2			Key 3
			...			
Mode 2						
...						

3.2 Cycle analyzing the cause-result chain

The second step is to build the failure cause-result(key) chain/tree of the objective breechblock failure mode by using RCA. Based on the cause-result chain, the cycle on the chain should be analyzed one by one until the key cause is found. Then all causes of the key cause in the chain one by one until the next key cause is found. And analyzing every level causes ordinal, until the root cause is found.

1. Building the cause-result chain of breechblock failure

For the breechblock, its failure cause-result chain can be protract according to system structure. All the physical relationship should be described in the cause-result chain. And the cause-result chain should extend to the first circle which is deviated the working environment of artillery, or the circle is no longer belong to maintenance measure, such as improve design technology, personnel training and so on. We call this point/circle the end point of whole cycle. Figure 2 show the cause-result chain of breechblock failure.

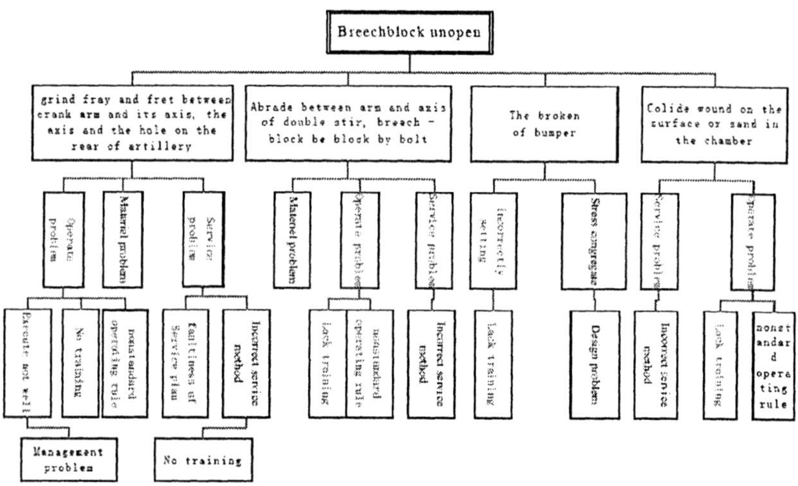

Figure 2. the cause-effect chain of breechblock

2. Classifying the cause

In general, cause of a materiel, such as artillery, can be classed in several types: components failed itself or son-components failed, operating rule, maintenance measure, environment cause, arts and crafts, design limitation and management problem and so on. All of these cause or part of them may be include in every layer.

3. Cycle flow

According to the analysis above, the flow of root cause analysis can be build as figure 1。

3.3 Determined root

In analysis, it is found that the main cause of insurance pry axes broken is the congregated of stress in oriented notch, which cause the crack appear and grown gradually in breechblock opening and closedown. When the crack extended to a limited point, the insurance pry axes broken subsequently. For this matter, there is no any suitable and effective maintenance task, such as 7 type preventive maintenance task based on RCM, CBM and even the proactive maintenance technology arisen recently, can be found by analysis. The only effective measure is redesign. By redesign, the structure can be improved. And the work of redesign is off the environment of breechblock operation and maintenance.

According to the guide line of end point judge, it can be concluded that the design limitation is the root cause of breechblock failure.

3.4 Recommendations and tracking

The recommendation is improve the structure and material of insurance pry axes. figure 3 to figure 5 show the scenario.

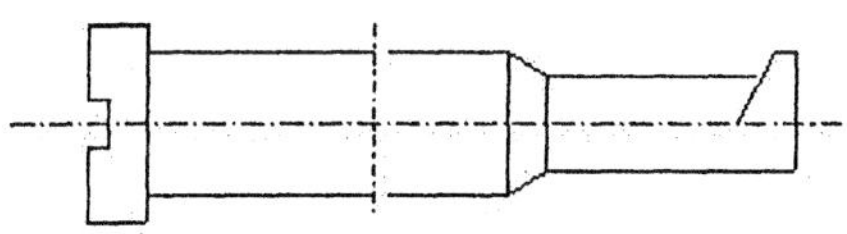

Figure 3. the originally insurance pry axes

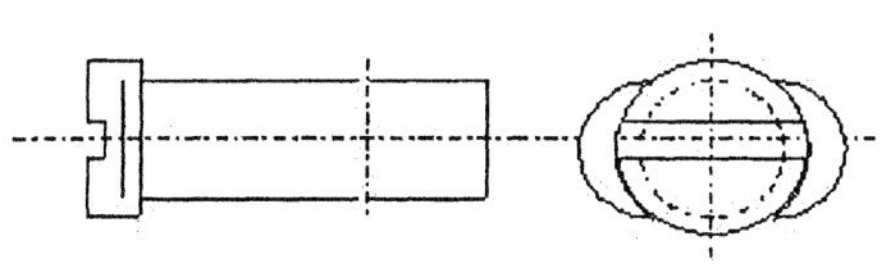

Figure 4. the shape of insurance pry axes improved

The aim of tracking is to checkout whether the recommenda -tion is effectiveness, the execution is correct and the result realized the object scheduled. By tracking, the execution of recommendation can be controlled and improved on real time, and fresh problem can be prevent on time. In this case, tracking work include several aspect as follow:

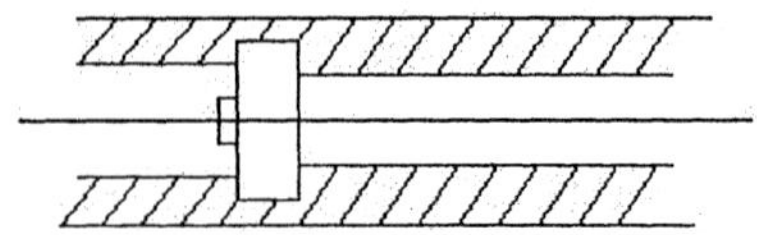

Figure 5. the hole shape of insurance pry axes

- Whether the circle load of insurance pry axes improved beyond the permitted limited;
- Whether the mean life of insurance pry axes improved longer enough;
- Whether the axes improved achieved the engage function in design;
- Whether insurance pry axes improved can avoided the similar broken.

According to the statistical data, the insurance pry axes improved has not broken in training since the refit. And the design function was achieved. So it can be concluded that the design limitation is the root cause of breechblock failure in deed. Thus, the analysis process can be finished, and the investigate report can be finalized.

4. CONCLUSION

Clearly, the improved RCA is effective in finding the root cause of breechblock closedown failure, and validating the legitimacy of recommendation. The cycle analysis and the rule of RCA end point judge, presented in this paper, is effectiveness in root cause analysis of breechblock failure despite its immature. And the lack of RCMA can be fetched up.

5. REFERENCES

1. FENG Xun-xin et. al. Study of cause-result relationship of common failure in self-propellant artillery, *vehicle & power Technology* 1999（03）: p32-34.
2. Ronald L. Hughes, Analysis: RCM Versus RCA, Maintenance Resources *On-Line Magazine*,May/Jun 2002.
3. Robert J.Latino, *Root Cause Analysis–Quality of Process?* Reliability Center Inc.May 1998.
4. Robert J. Latino, *The End of RCA, How, and Why,* TPMOnline.com Reliability Center, Inc - April/May 2001.
5. Doe-Ne-Std-1004-92 *Root Cause Analysis Guidance Document.* 2.1992.
6. Plant Maintenance Resource Center 2001 "Root Cause Analysis" Survey Result http://www.plant-maintenance.com/index.shtml.

A PROBLEM-SPECIFIC GENETIC ALGORITHM FOR PATH PLANNING OF MOBILE ROBOT IN GREENHOUSE

Xuemei Liu[1], Jin Yuan[2, 1], Kesheng Wang[3]
[1]School of Mechanical & Electronic Engineering, Shandong Agricultural University, China, Email:lxmywj@126.com; [2]CIMS & Robot Center of Shanghai University, Shanghai University, China; [3]Department of Production and Quality Engineering, Norwegian University of Science and Technology, Norway. Email:kesheng.wang@ntnu.no.

Abstract: This paper proposes a novel Genetic Algorithms (GAs) approach for a near-optimal path planning of a mobile robot in a greenhouse. The chromosome encoding features in inverse proportion between research spaces of GAs and complexity of obstacles. The fitness evaluation is designed for both incomplete and complete paths to guide the evolutional direction. The crossover and mutation operators are trimmed to the path planning problem. Two operators were presented to promote the effectiveness of evolution of problem-specific GAs. The simulation results obtained is satisfactory.

Key words: Genetic Algorithms; mobile robot; path planning.

1. INTRODUCTION

Use of robots is especially important when the duties performed are potentially harmful for the safety or the health of the workers, or when more conservative issues are granted by robotics. In order to utilize horticultural-related vehicles in a greenhouse as robots, it is necessary to plan the work path in advance. The typical task could be described as follows: for two arbitrary points in an area, finding the drivable path between these points minimizing the energy required.

There exist many approaches within Computational Intelligence to solve the problem of finding collision-free paths in the special working places[1-3],

Please use the following format when citing this chapter:
Liu, Xuemei, Yuan, Jin, Wang, Kesheng, 2006, in International Federation for Information Processing (IFIP), Volume 207, Knowledge Enterprise: Intelligent Strategies In Product Design, Manufacturing, and Management, eds. K. Wang, Kovacs G., Wozny M., Fang M., (Boston: Springer), pp. 211-216.

which considered the aspects of energy consumption and the length of the paths. However, what most of approaches have in common is that, much time is spent on computing whether the paths are feasible or not, especially the environment with more complicated obstacles. So an advisable approach of path planning with problem-specific GAs is presented.

2. PATH PLANNING BASED ON GAs

The proposed GA features its effective encoding, path evaluation and its problem-specific genetic operators. The planed paths are totally independent of the presence or distribution and shape of obstacles.

2.1 Problem description

The mobile robot working environment is represented by a 2-dimensional workspace subdivided into quadratic regions at the horizontal and vertical directions according to an appropriate resolution in representing contours of obstacles. Each region named grid is defined by a pair of integer-numbered coordinates. The obstacles are also represented by grids that covered the objects' vertexes. A potential robot path is formed by certain length grids connecting the start grid S to the target grid T, where S and T are represented by their natural coordinates.

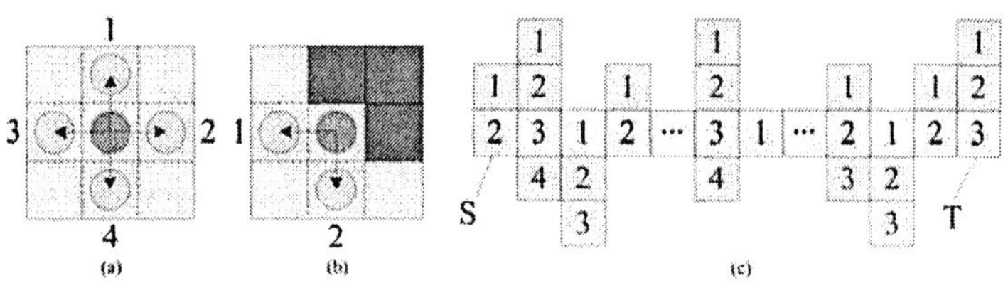

Figure 1. (a) The feasible space of one grid with four adjacent regions; (b) One grid with two adjacent regions; (c) The search space under chromosome encoding.

To overcome the huge time expenditure in determining whether a grid zone represents an obstacle or not, the feasible space of a grid was defined generally by 4 adjacent regions, as showed in Fig. 1a. That is, for the current grid (green one), there are four feasible next grids to cover. However, for the grid showed in Fig.1b, there are two feasible next grids to cover (the grids in black represents obstacles). Therefore, a potential feasible path could be represented by the variable length chromosome structure, where the sequence contains a number of feasible grids represented by real-valued coordinate. Such paths formed by the feasible grids are always collision-free. The length of paths would decrease or increase during the operation of GAs.

The feasible grids contain the research space of GAs, showed in Fig.1c, and correlate with the coverage rate of obstacles. Therefore, the chromosome encoding simplified the research space, and thus speeds up the computation, especially in a complex and congested greenhouse environment.

2.2 Path initialization and fitness evaluation

In the GAs, the fitness function guides the evolution direction. A well designed fitness function instructs the right evolution direction and high convergent speed[4]. A robot path was initialized from start grid along feasible space of next grids, so the paths may be either incomplete paths (inaccessible target grid) or complete paths. The evaluation should distinguish incomplete and complete paths and guides the different paths evolving with either category. The GA generates its initial solutions randomly, and evolves solutions starting from the initial population.

The evaluation of path here has two parts. The fitness defined to the incomplete path as F_i, and the complete path is defined as F_c:

$$F_I = 1000.0/(|X_s - X_T| + |Y_s - Y_T|)$$
$$F_C = 10000.0 \times (1 + 10.0 / sum) \qquad (1)$$
$$sum = sum + \begin{cases} 2 & \text{if } g_{i-1}, g_i, g_{i+1} \text{ in a line} \\ 1.5 & \text{else} \end{cases}$$

Where, X_S, X_T are the coordinates of start grid and target grid. The incomplete path closed to target grid has high fitness value in F_I, and the complete path with shorter sub-path has high fitness value in F_C. In addition, best-1-elitism was implemented to prevent the algorithm from losing the best path and accelerating the convergence of GAs.

2.3 Operator of Crossover

The algorithm adopts problem-specific One-Point Crossover. For a chosen two different individuals representing two paths, a feasible intersection points should be collected according to the same grid in two paths. After that, one intersection point is chosen at random, and the sub-paths following this point are exchanged (see Fig. 2a). Generally, one path (represented with red) fitness value increases while the other one (represented with black) decreases (see Fig. 2b).

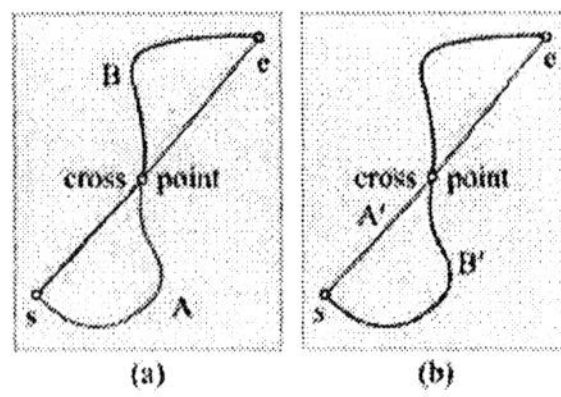

Figure 2. Operator crossover: (a) before crossover; (b) after crossover.

2.4 Operator of Mutation

There are two kinds of mutations due to mutating at different location: the mutation at mid-path and the mutation at end-path. The objects of the former are all paths, which optimize with the direction of shorter path, and the objects of the latter are just incomplete paths, which optimize with the direction of generating the complete paths. However, they are in the same operation mechanism.

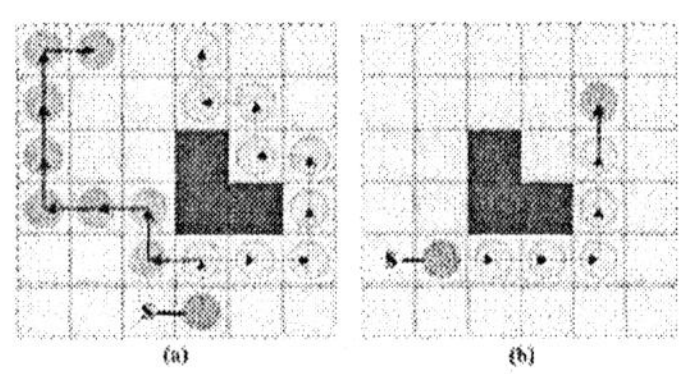

Figure 3. Operator mutation: (a) mutation at mid-path; (b) mutation at end-path.

In Fig. 3a, a grid in complete path was selected randomly. The next grid in its feasible space was selected randomly, and then repeat the process till the length of new path same as the old one. In the Fig. 3b, the grid was mutated only at the end of path.

2.5 Operator of Repair

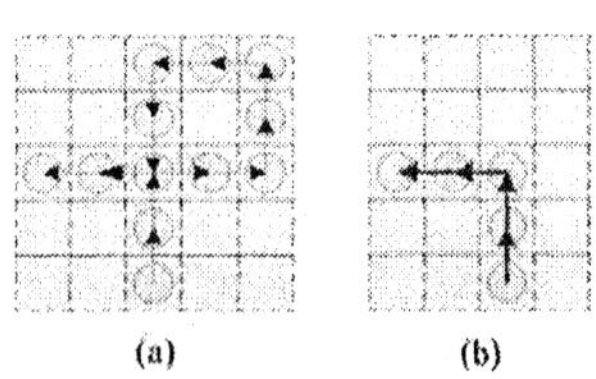

Figure 4. Operator of repair: (a) before repair; (b) after repair.

After the operations above have been applied to paths, some enclosed sub-paths may appear (Fig. 4a) , which does not contribute to the fitness value of paths, but decrease the mutation and crossover probabilities in the useful sub-paths. The operator of repair can scan the paths to find out the enclosed sub-paths and remove them according to 4 adjacent regions. The result is shown in Fig. 4b.

2.6 Operator of Tighten

There are still some undesired sub-paths as shown in Fig. 5b. When GAs reaches its global maximum, the sub-path would be straight, but it would consume much time and it will be difficult to converge with a longer path, making the tighten operation well compromised. First, judging the sub-path meet the condition that there are no obstacles between two arbitrary grids on the path line up in 4 adjacent directions, the sub-path between grids can be tightened. And then using straight grids substitute this sub-path. The sub-path does not meet the condition as shown in Fig. 5a. Fig. 5c shows the tighten result.

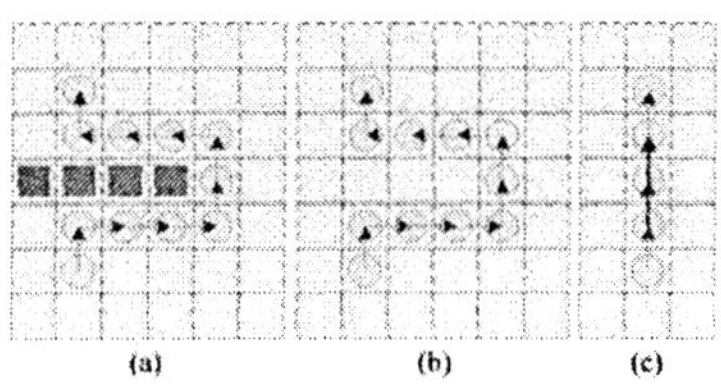

Figure 5. Operator of tighten: (a) path not meet the condition to tighten; (b) path can be tighten; (c) the result of tighten operation.

3. SIMULATION RESULTS

Several simulations were conducted about the effectiveness of the proposed GAs. In the simulations, parameters for the proposed GAs were set as follows: population size was 100; probabilities of mutation and crossover were 0.8 and 0.9; maximal generations were 100. Fig.6a shows the software interface with tackling the zigzag problem.

The proposed GAs can be easily applied to robot path planning in complicated environment, showed in Fig.6b, which contain 5 arbitrary shapes, and in Fig.6c, which simulating a workplace in a greenhouse. The results of this approach show satisfactory in the meaning of 4 adjacent regions. Fig. 6d shows a layout where some potential based path planning approaches and local path planning approaches may be trapped. The result also shows the ability of the proposed GAs to solve maze-like problem.

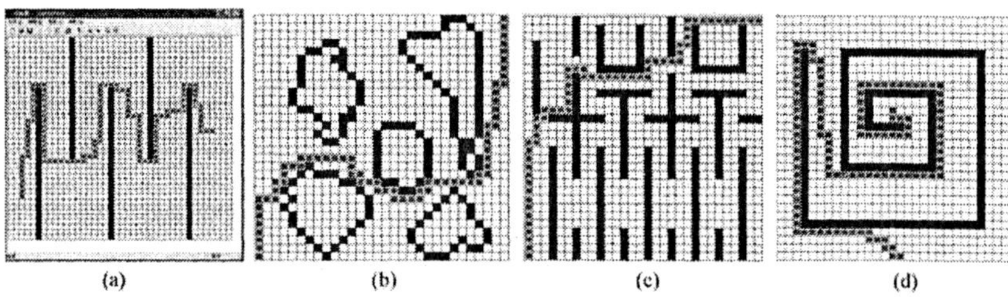

Figure 6. Simulation examples. (a) A zigzag layout. (b) An arbitrary shape layout. (c) The result in simulating greenhouse environment. (d) The result in maze-like layout.

4. CONCLUSIONS

To tackle the movements of mobile robots, a problem-specific GAs for path planning is proposed. Path representation based on feasible space of grids decreases the research space. The fitness evaluation is designed to fit incomplete and complete paths. The crossover and mutation operators are trimmed to the path planning problem. This paper also proposed two problem-specific genetic operators for robot path planning. The effectiveness of the problem-specific GAs is demonstrated by simulation studies. The efficiency in computation time makes the proposed genetic algorithm easily to be applied to real-life applications. The prospective work can be done as the following: (1) To tackle the difficult generation of complete path within the high resolution of environment representation. (2) To avoid the local minima using escaping technique.

5. REFERENCES

1. Sugihara, K. et al. (1997). Genetic Algorithms for Adaptive Motion Planning of an autonomous Mobile Robot, *Proceedings of the IEEE international Symposium on Computational Intelligence in Robotics and Automation*, pp. 138-146.
2. Gemeinder, M. et al. (2003), GA-based path planning for mobile robot systems employing an active search algorithm, *Applied Soft Computing Journal*, (3), p149-158.
3. Gihan, N. et al. (2004), Path planning for a mobile robot using genetic algorithms, *Proceedings of ICEEC'04*, p185-189.
4. Wang, K. (2005), *Applied Computational Intelligence to Intelligence Manufacturing Systems*, Advanced Knowledge International (ISBN 0-9751004-9-1).

RESEARCH ON THE DYNAMICS OF FLEXIBLE MANIPULATORS

Li Tu[1], Jinlian Deng[1], Huanya Cao[1], Pixuan Zhou[2]

[1]*Department of Mechanical Engieering, Zhejiang Mechanical&Electrical College, China; Email: tu_li22@yahoo.com.cn.* [2]*School of System and Industrial Engieering, University of Arizona, USA.*

Abstract: The dynamics of a Flexible Manipulator system are described by an infinite-dimensional mathematical model, since the model consists of partial differential equations. But to design a finite-dimensional controller, a finite-dimensional system model is needed. To achieve this goal, a finite dimensional approximation needs to be used to model a flexible manipulator, that is, to retain a finite number of modes, and to drop off the other, less significant modes based on the requirements of the controller. The scheme in developing a mathematical model is to use the Lagrangian method or Hamiltonian's Principle to the total kinetic energy, total potential energy and virtual work done by the torque actuated to the joint. This method will not introduce extra errors into system and will be used to obtain the state-space model for a flexible manipulator suggested in this paper. Dynamics of flexible manipulators with shear force and rotatory inertia are derived, and state-space equations with the integration of DC motor dynamics are developed as a theoretical base for mechatronic designs.

Key words: Dynamics, Flexible Manipulator; Mathematical Model.

1. INTRODUCTION

In last decades, researches on flexible manipulators have increased dramatically. From the perspective of control, recent researches were found in [1] [2] [3] [4] and references therein. These studies did control-intensive work to improve beam performance. But they ignored the impact applied from the beam mechanic shape. The resulting system was only local optimal. In the other hand, a better beam shape to move faster but for less vibration is another topic. One of the first papers on optimal shape for flexible

This project is supported by the National Natural Science Foundation of China under the grant No. 50275091.

Please use the following format when citing this chapter:
Tu, Li, Deng, Jinlian, Cao, Huanya, Zhou, Pixuan, 2006, in International Federation for Information Processing (IFIP), Volume 207, Knowledge Enterprise: Intelligent Strategies In Product Design, Manufacturing, and Management, eds. K. Wang, Kovacs G., Wozny M., Fang M., (Boston: Springer), pp. 217-223.

manipulators is [5] by Karihaloo and Niordson. Extensive studies in optimal shape design have been conducted recently by Wang's group.Wang (1991) [6] simplified substantially the scheme presented in [5]. A minimum weight design of flexible manipulators was developed by Wang and Russell (1992, 1993) [7] [8]. In 1995, they also investigated a new approach [9], the segmentized scheme of optimal design, which treats the flexible beam as a collection of small lengths of rigid beam constrained by each other's interfacial conditions. The new technique converts the optimal shape design problem into a matrix determinant problem. These studies focused on open-loop design. They only concerned the beam mechanic construction that is proper design the flexible beam shape can make it suffering less vibration. But in reality, all manipulators must be in closed-loop to get high performance. The couplings between the controller and construction were not considered at larger in these designs.

2. DYNAMICS OF FLEXIBLE MANIPULATORS

The dynamics of a Flexible Manipulator system are described by an infinite-dimensional mathematical model, since the model consists of partial differential equations. But to design a finite-dimensional controller, a finite-dimensional system model is needed. To achieve this goal, a finite dimensional approximation needs to be used to model a flexible manipulator, that is, to retain a finite number of modes, and to drop off the other, less significant modes based on the requirements of the controller. The N-mode expansion for the beam displacement $w(x,t)$ can be given by

$$w(x,t) = \sum_{i=1}^{N} \psi_i(x) q_i(\mathrm{t}) \quad (1)$$

The separability in this case refers to describing the displacement as a series of terms which are products of two separate functions, each of that is a function of a single variable, a spatial variable x, and time t, respectively. i is the $i-$th modal shape [10], or eigenfunction. q_i is the corresponded generalized modal coordinate describing the flexible deformation.

The scheme in developing a mathematical model is to use the Lagrangian method or Hamiltonian's Principle to the total kinetic energy, total potential energy and virtual work done by the torque actuated to the joint. This method will not introduce extra errors into system and will be used to obtain the state-space model for a flexible manipulator suggested in this paper.

2.1 State-space equations of flexible manipulators

The hub kinetics is written as,

$$T_h = \frac{1}{2} I_h [\dot{\theta}^2 + \dot{\omega}^2(0,t)'] \quad (2)$$

where $\omega = \partial\omega / \partial x = \omega'$. After applying Eq. 1 into Eq. 2,

$$T_h = \frac{1}{2} I_h \dot{\theta}^2 + \frac{1}{2} I_h \sum_{i=1}^{N} \sum_{j=1}^{N} \dot{q}_i \dot{q}_j \psi_i'(0) \psi_j'(0) \quad (3)$$

In the same way, the kinetics of the tip load is,

$$T_p = \frac{1}{2} \Gamma_1 \dot{\theta}^2 + \dot{\theta} \sum_{i=1}^{N} \dot{q} \Gamma_2(i) + \frac{1}{2} \sum_{i=1}^{N} \sum_{j=1}^{N} \dot{q}_i \dot{q}_j \Gamma_3(i,j) \quad (4)$$

where,

$$\Gamma_1 = (M_p L^2 + J_p + 2M_p L a_c,\ \Gamma_2(i) = (J_p + M_p L a_c)\psi_i'(L) + (L + a_c) M_p \psi_i(L), \quad (5)$$
$$\Gamma_3(i,j) = J_p \psi_i'(L) \psi_j'(L) + M_p \psi_i(L) \psi_j(L) + 2a_c \psi_i'(L) \psi_j(L)$$

The kinetic energy of the link,

$$T_L = \frac{1}{2} \Lambda_1 \dot{\theta}^2 + \dot{\theta} \sum_{i=1}^{N} \dot{q} \Lambda_2(i) + \frac{1}{2} \sum_{i=1}^{N} \sum_{j=1}^{N} \dot{q}_i \dot{q}_j \Lambda_3(i,j) \quad (6)$$

where

$$\Lambda_1 = \int_0^L (x^2 + S)\rho dx,\ \Lambda_2(i) = \int_0^L [x\psi_i(x) + S\psi_i'(x)]\rho dx,\ \Lambda_3(i,j) = \int_0^L [\psi_i(x)\psi_j(x) + S\psi_i'(x)\psi_j'(x)]\rho dx \quad (7)$$

S is the function of cross section area of the link. The total potential energy is

$$P = \frac{1}{2} \int_0^L EI(x) \omega^2(x)'' dx = \frac{1}{2} \sum_{i=1}^{N} \sum_{j=1}^{N} q_i q_j k(i,j) \quad (8)$$

where $k(i,j) = \int_0^L EI(x) \psi_i''(x) \psi_j''(x) dx$

The generalized virtual work is derived as,

$$W = \tau v'(0,t) = \tau\theta + \tau\omega'(0,t) = \tau\theta + \tau\sum_{i=1}^{N}\psi_i'(0)q_i(t) \quad (9)$$

In order to apply the Hamiltonian Principle, let us substitute Eqs. 3, 4, 6, 7, 8 for the following formula and group it in terms of θ, δq_i, and their derivatives,

$$\Xi = \frac{1}{2}\Omega_1\dot{\theta}^2 + \dot{\theta}\sum_{i=1}^{N}\dot{q}_i\Omega_2(i) + \frac{1}{2}\sum_{i=1}^{N}\sum_{j=1}^{N}\dot{q}_i\dot{q}_j\Omega_3(i,j) + \tau\theta + \tau\sum_{i=1}^{N}\psi_i'(0)q_i - \frac{1}{2}\sum_{i=1}^{N}\sum_{j=1}^{N}q_iq_jk(i,j) \quad (10)$$

where,

$$\Omega_1 = I_h + \Gamma_1 + \Lambda_1,\ \Omega_2(i) = \Gamma_2(i) + \Lambda_2(i), \Omega_3(i,j) = I_h\psi_i'(0)\psi_j'(0) + \Gamma_3(i,j) + \Lambda_3(i,j) \quad (11)$$

After applying the Hamiltonian method, we obtain,

$$\delta\Xi = \Omega_1\dot{\theta}\delta\dot{\theta} + \sum_{i=1}^{N}\Omega_2(i)\dot{q}_i\delta\dot{\theta} + \dot{\theta}\sum_{i=1}^{N}\Omega_2(i)\delta\dot{q}_i + \sum_{i=1}^{N}\sum_{j=1}^{N}\dot{q}_i\Omega_3(i,j)\delta\dot{q}_i + \tau\delta\theta + \tau\sum_{i=1}^{N}\psi_i'(0)\delta q_i - \sum_{i=1}^{N}\sum_{j=1}^{N}q_jk(i,j)\delta q_i \quad (12)$$

$$\int_0^{t}\dot{\theta}\delta\dot{\theta}dt = -\int_0^{t}\ddot{\theta}\delta\theta dt = -\int_0^{t}\ddot{\theta}\delta\theta dt,\ \int_0^{t}\dot{q}_i\delta\dot{\theta}dt = -\int_0^{t}\ddot{q}_i\delta\theta dt,\ \int_0^{t}\dot{\theta}\delta\dot{q}_i dt = -\int_0^{t}\ddot{\theta}\delta q_i dt,\ \int_0^{t}\dot{q}_i\delta\dot{q}_i dt = -\int_0^{t}\ddot{q}_i\delta q_i dt, \quad (13)$$

The Hamilton Extended Principle results in,

$$\int_{t_0}^{t_f}\delta\Xi dt = \int_{t_0}^{t_f}\{[-\Omega_1\ddot{\theta} - \sum_{i=1}^{N}\ddot{q}_i\Omega_2(i) + \tau]\delta\theta - [\ddot{\theta}\sum_{i=1}^{N}\Omega_2(i) + \sum_{i=1}^{N}\sum_{j=1}^{N}\ddot{q}_j\Omega_3(i,j) + \sum_{i=1}^{N}\sum_{j=1}^{N}q_jk(i,j) - \tau\sum_{i=1}^{N}\psi_i'(0)]\delta q_i\}dt = 0, \quad (14)$$

These coefficients of $\delta\theta$, δq_i , $1 \le i \le N$ must be zero, that is,

$$\Omega_1\ddot{\theta} + \sum_{i=1}^{N}\ddot{q}_i\Omega_2(i) - \tau = 0 \quad (15)$$

$$-\ddot{\theta}\Omega_2(i) - \sum_{j=1}^{N}\ddot{q}_j\Omega_3(i,j) - \sum_{j=1}^{N}q_jk(i,j) + \tau\psi_i'(0) = 0, \quad 0 \le i \le N , \quad (16)$$

or in matrix form, they are,

$$M\ddot{x} + kx = b\tau \quad (17)$$

where,

$$x=[\theta,q_1,q_2,q_3,...,q_N]^T_{(N+1,1)},\quad b=[1,\psi_1'(0),\psi_2'(0),\psi_3'(0),...,\psi_N'(0),]^T_{(N+1,1)},\quad (18)$$

$$M=\begin{bmatrix}\Omega_1 & \Omega_2(1) & \Omega_2(2) & . & \Omega_2(N)\\ \Omega_2(1) & \Omega_3(1,1) & \Omega_3(1,2) & . & \Omega_3(1,N)\\ . & . & . & . & .\\ \Omega_2(N) & \Omega_3(N,1) & \Omega_3(N,2) & . & \Omega_3(N,N)\end{bmatrix}$$

$$K=\begin{bmatrix}0 & 0 & 0 & . & 0\\ 0 & k(1,1) & k(1,2) & . & k(1,N)\\ . & . & . & . & .\\ 0 & k(N,1) & k(N,2) & . & k(N,N)\end{bmatrix}$$

As has been shown, $\Omega_3(i,j)=\Omega_3(j,i)$, $k(i,j)=k(j,i)$, thus the matrices M,K are symmetric and named mass and rigidity matrices respectively.

Next, the actuator dynamics need to be incorporated into the link system. It is assumed

that the arm is driven by a permanent magnet DC-motor. Therefore, the actuator dynamics can be described as,

$$-J_m\ddot{\theta}-(B_m+\frac{K_bK}{R})\dot{\theta}+\frac{K_m}{R}v_c=\tau \quad (19)$$

where J_m is the actuator inertia, B_m the friction coefficient, K_m the torque constant, K_b the back emf constant, R the armature resistance, and θ, υ_c the hub rotation and armature voltage, respectively. In general, all motor circuit parameters can be considered as design variables. The overall state variable is defined as,

$$q=\begin{bmatrix}x\\ \dot{x}\end{bmatrix} \quad (20)$$

Combining Eq. 17 with Eq. 18, we present the overall system state space equations as,

$$q=A\dot{q}+B,\quad u=v_c,\quad (21)$$

where

$$A=\begin{bmatrix}0 & I\\ -\overline{M}^{-1}K & -\overline{M}^{-1}\overline{B}\end{bmatrix},\quad B=\begin{bmatrix}0\\ \overline{M}^{-1}\overline{D}\end{bmatrix},\quad (22)$$

$$\overline{M}=(M+J_mb\underline{e_1})^{-1},\ \overline{B}=(B_m+\frac{K_bK_m}{R})b\underline{e_1},\ \overline{D}=b\frac{K_m}{R},\ \underline{e_1}=[1\ \ 0\ \ ...\ \ 0]_{1\times n}$$

A is the function of beam construction, thus any changes in beam mechanical shape will result in a different A, which provides the basis for simultaneously optimal construction and control based on the mechatronic formulation discussed early.

2.2 Output specifications

As can be seen from the system state-space equations, the state vector consists of θ, the generalized modal coordinates q_i, and their first order derivatives. So all controllers based on the state feedback are indirect, which means that the states need to be predicted. This will trade off with computation time and bring one more inaccuracy into a closed-loop, cutoff closed-loop bandwidth. This presents hindrances to real-time processing and high motion speed. Output feedback takes precedence over state space feedback as far as the mechatronic approach and objectives of this paper are concerned.

Tip deflection output, which is,

$$y = w(L,t) = \sum_{i=1}^{N} \psi_i(L) q_i = \begin{bmatrix} 0 & \psi_1(L) & \psi_2(L) & \dots & \psi_N(L) & \bar{0} \end{bmatrix} q, \quad (23)$$

where ψ_i is the ith eigenfunction, $\bar{0}_{1\times(n+1)}$ is zero vector. A CCD camera clipped on the hub can be used to measure the output.

The tip position output is,

$$y = v(L,t) = L\theta + \sum_{i=1}^{N} \psi_i(L) q_i = \begin{bmatrix} L & \psi_1(L) & \psi_2(L) & \dots & \psi_N(L) & \bar{0} \end{bmatrix} q, \quad (24)$$

The hub tangent angle output is,

$$y = \Theta = \theta + \sum_{i=1}^{N} \psi_i'(0) q_i = \begin{bmatrix} 1 & \psi_1'(0) & \psi_2'(0) & \dots & \psi_N'(0) & \underline{0} \end{bmatrix} q, \quad (25)$$

A potentiometer may be used for measuring Θ.

In this mechatronic design, output feedback maybe one of above outputs, or the mixture,

such as,

$$y = Cq = \begin{bmatrix} w(L,t) \\ v'(L,t) \end{bmatrix} q, \quad (26)$$

$$C = \begin{bmatrix} 0 & \psi_1(L) & \psi_2(L) & & \psi_N(L) & \bar{0} \\ 1 & \psi_1'(0) & \psi_2'(0) & & \psi_N'(0) & \bar{0} \end{bmatrix}$$

The corresponding state-variable equations are,

$$\dot{q} = Aq + bu, \qquad y = Cq. \qquad (27)$$

3. CONCLUSION

Mechatronic design is a global optimization of the overall system. For a flexible manipulator, the overall system is the integration of link dynamics, DC motor equation, measuring sensors, and the selected controller. As above, these output feedbacks can be used in control design to make the improvement of the suggested mechatronic approach and to make comparison between them .

4. REFERENCES

1. J. Doyle, K. Glover(1989), "State-Space Solution to Standard H2 and H Control Problem", IEEE AC, Vol. 34, No. 8 , pp. 831-847.
2. B. Francis, *A Course in H Control Theory,* (Springer-Verlag, Berlin-New York, 1998).
3. L. B. Guti 'errez, F. L. Lewis, J. A. Lowe (1998), *Implementation of a Neural Network tracking Controller for a Single Flexible link: Comparison with PD and PID Controller,* in IEEE Trans. on Industrial Electronics, Vol. 45, No. 2.
4. A. Jnifene, A. Fahim (1997), A Computed Torque/Time Delay Approach to the End-Point Control of a One-Link Flexible Manipulator, *Dynamics and Control*, Vol. 7, pp. 177-189.
5. M. Fujita, et al (1991), "Experimental Evaluation of H Control for a Flexible Beam Magnetic Suspension System," *S. Hosoe (ed) Robust Control*, Pro of a workshop held in Toyko, June 23-24.
6. F. L. Lewis, V. L. Syrmos (1995), *Optimal Control*, John Wiley and Sons, Inc., Second Edition.
7. K. Lezn, H. Ozbay, et al (1991), Frequency Domain Analysis and Robust Control Design for an Ideal Flexible Beam, *Automatica*, Vol. 27, pp. 947-961.
8. A. Tchernychev, A. Sideris, J. Yu (1997), *Constrained H Control of an Experimental Flexible Link*, Trans. o fthe ASME, June.
9. Fei-Yue Wang (1994), On the Extremal Fundamental Frequencies of One-Link Flexible Manipulators, *International Journal of Robotics Research*, Vol. 13, No. 2, pp. 162-170.
10. Fei-Yue Wang and J.L.Russell(1992), Optimum Shape Constructure of Flexible Manipulators with Tip Loads, *Proceedings of IEEE Intl,Conf. on Decision and Contral(Tucson,Arizona),* Vol. 1, pp. 311-316.

IMPLEMENTATION OF FPC CONCEPT IN ELECTRICAL WIRE EQUIPMENT DESIGN

Hui Zhao [1], Lixin Lu [2], Limin Li [1], Jun Chen [1]

[1]CIMS center, Shanghai University, Shanghai, 200072, China, Email of the first author::uu8490077@126.com.[2] Shanghai Feilo Co., LTD, Shanghai,200050, China.

Abstract: This paper introduces FPC concept, and exemplifies it with a full-automatic crimping machine developed recently, therefore proposed a new viewpoint---- assembly weight.

Key words: FPC, crimping machine, assembly weight.

1. INTRODUCTION

Modern manufacturing includes multiple advanced methods : Virtual Manufacturing, Concurrent Engineering, Lean Production, Clean Factory, Remanufacturing, etc, all of which are devoted to solving two great challenges faced with today's enterprises: depressing cost, and raising manufacturing capability. In this case, FPC （First Product Correct） is put forth, it meets the actuality of lack of resources, and raises a higher standard of time， quality, cost, service, and environment.

FPC can be regard as successfully making at the first try, or first design and success, it was initially proposed by American Integrated Manufacturing Technology Institute, and eventually becomes an important research orientation. Focus of FPC is to turn design concept into effective goods, or to turn single small batch production into multiple volume production, with particular emphasis on continuous production. It rejects drawbacks of volume mining, volume consuming, volume abandoning in conventional economy.

Please use the following format when citing this chapter:
Zhao, Hui, Lu, Lixin, Li, Limin, Chen, Jun, 2006, in International Federation for Information Processing (IFIP), Volume 207, Knowledge Enterprise: Intelligent Strategies In Product Design, Manufacturing, and Management, eds. K. Wang, Kovacs G., Wozny M., Fang M., (Boston: Springer), pp. 224-230.

Audiences of FPC vary from industry, government, to academy. Conventional concepts are also adopted by FPC, such as Integrated Manufacturing, Concurrent Corporation, CAD/CAPP/CAM， PDM, etc. FPC changes the ordinal manufacturing process which lacks of interactive communication. Once defect happens, the design will be transferred in time. With FPC, each department will feed back effectively before concept design, so that to achieve lower design alternation, shorter life cycle, lower cost, and first production correct.

2. MAINFRAME OF SYSTEM REALIZATION

Our project combines an electrical wire equipment (full-automatic crimping machine, processing electrical wire for automobile use), with embodied operational flow of FPC, and finally gains success from design to manufacturing at the first time.

As a modern managerial measure, FPC focuses on standardization of implementation, and clarity of responsibilities. Hence, we need to divide our members effectively. It is the viewpoint of FPC theory that the responsible team needn't too large, but must cross multiple departments. Corresponding teams are organized according to different production stages, and usually, that teams has a polar structure (see figure 1)

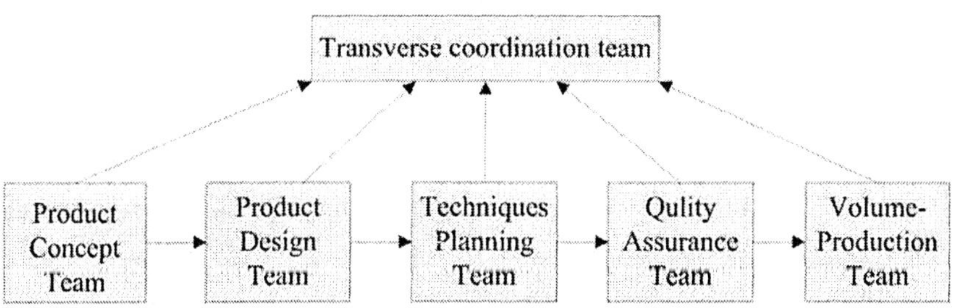

Figure 1. Structure of FPC Team

In order to optimize whole cycle of production, FPC needs to accurately define flow of output firstly, and then hand on each function branch to individual team. Each team doesn't only shoulder its own job, but also communicates immediately. Typical operation of FPC is depicted through operational flow of electrical wire manufacturing equipment in figure 2.

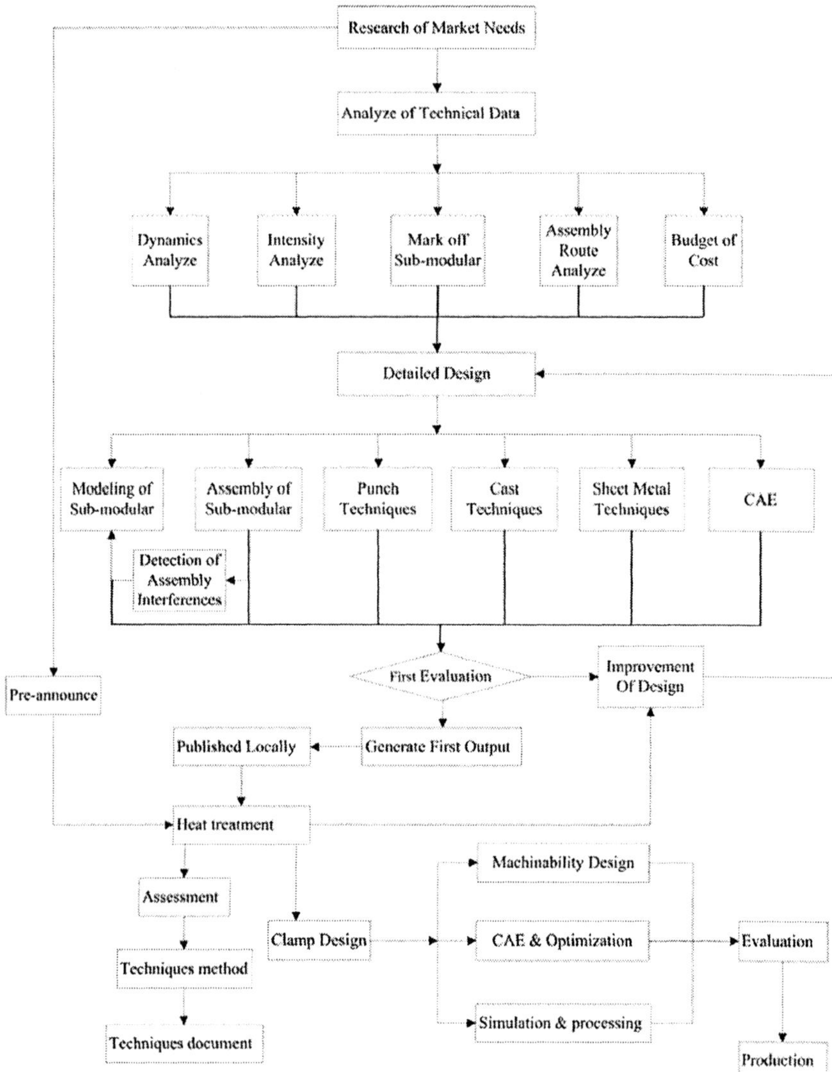

Figure 2. Operational Flow of FPC

3. DETAILED OPERATIONAL SCHEMES

At first, production concept team accurately orients market through fruitful marketing research, sketches out technical parameters, for example, cutting length of wire is 30-9999mm, stripping length of wire is 1.0-10mm, etc.Then conceives out rough assembly method, and divides relevant mechanism into modules, then plans assembly path. This full-automatic wire crimping machine can be divided into several mechanisms (see figure 3)

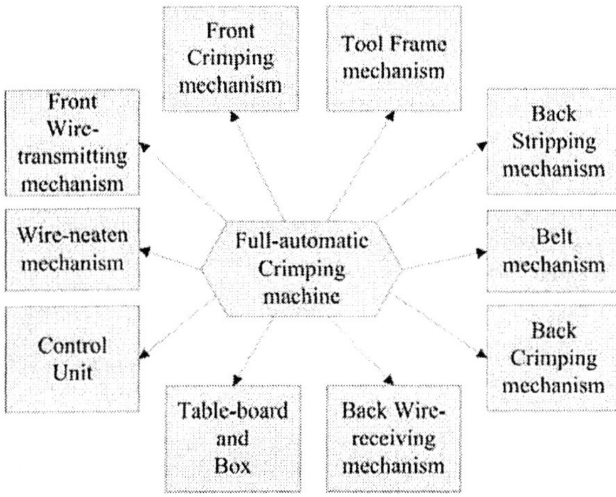

Figure 3. Mechanical Structure of Full-automatic Crimping Machine

Principle of this full-automatic crimping machine is as follows: wire-neaten mechanism strains wound wire straightly, for the convenience of processing. then the wire is sending to tool frame mechanism for stripping, after stripping it is time for front crimping mechanism to crimp one end of the wire, subsequently back Wire-receiving mechanism nips the wire, and waits for stripping of the other end and rotates to have this end crimped in the back crimping mechanism, finally processed wire is sent to belt mechanism which collects all the finishing wires.

Product design team members work collaterally to set up models through powerful virtual tool (UGNX) .With a information-sharing platform, they can communicate in time on material, heat treatment and so on that make product machinable and capable of assembly. Meanwhile, several design variables is defined, including intensity, rigidity, cost, ……, to compose a target function, which is finally optimized by convergent arithmetic, such as Genetic Algorithm, Ant Colony Optimization, etc. Finally, design rules come out. Each modular is made sure to be connected with whole assembly through low-cost standard elements that are easily found. Main modules are in figure 4 (Tool Frame & Back wire-receiving mechanism).

Afterwards, techniques planning team works out technique documents, and carries on CAE analysis. for example, whether moments of inertia in relevant rotating component match each other, whether mechanical jaws cause noises while working…….(As shown in figure 5)

At last, quality assurance team proceeds checking and revising, give all summarize to volume-production team for their convenience of establishing production planning.

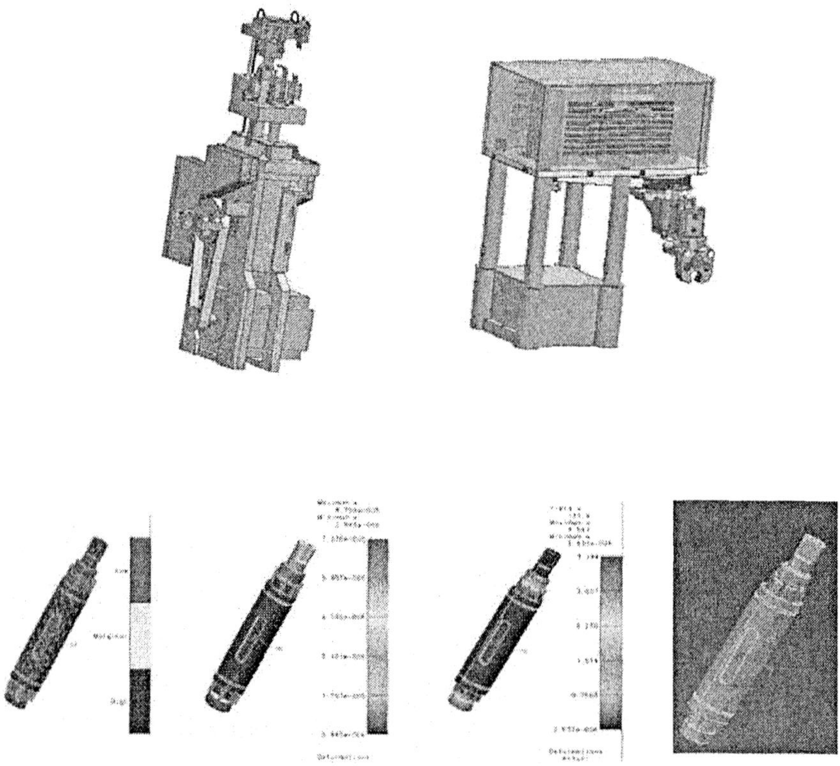

Figure 5. FEA meshing & Displacement & Stress & Structural Analysis

4. KEY TECHNOLOGY OF FPC— ASSEMBLY OF INTEGRATED PRODUCT

Assembly is the actual embodiment of machine's function, and is also the guideline of design and manufacturing, because requirements of assembly run through the entire designs of parts. All designs are devised on the behalf of implement of assembly function.

For that research of product in integrated manufacturing and assembly are key problems of FPC, it is necessary to consider constraints relevant to assembly manufacturing at initial stage. We should allocate different weights to different sub-assembly, place connections between sub-modules, feed back to designer of each modular to help them modify immediately, so that holistic design of machine links with every sub-modular friendly, while keeping reasonable remaining value to support reliability design.

Key weight of this full-automatic crimping machine lies in tool frame mechanism. Location of tool frame directly restricts relative positions of back stripping mechanism, front/back crimping mechanism, belt mechanism. Therefore we regard tool frame as the first weight (first consideration), and choose position of front/back crimping machine according to it. Back wire-receiving machine is the second weight for its movement area is only 58mm, 45° 。Function of belt mechanism is taking finished wire from back wire-receiving mechanism; we consider it as the third weight.

As for much larger scale mechanism, we can search for an all-around optimal solution, such as simulate anneal arithmetic. This arithmetic is suitable for scale combination optimization; it has an easy description and high efficiency, and is convenient to implement collateral computations. The main principle is to assume theory of solid annealing, which means heating the solid to desired high temperature, then cools slowly. When heating, particles inside the solid become disorder, inner energy increasing. By contrast, particles become ordered while cooling slowly, and get balance at each temperature, finally reaches ground state at normal temperature while its inner energy diminishes to minimum.

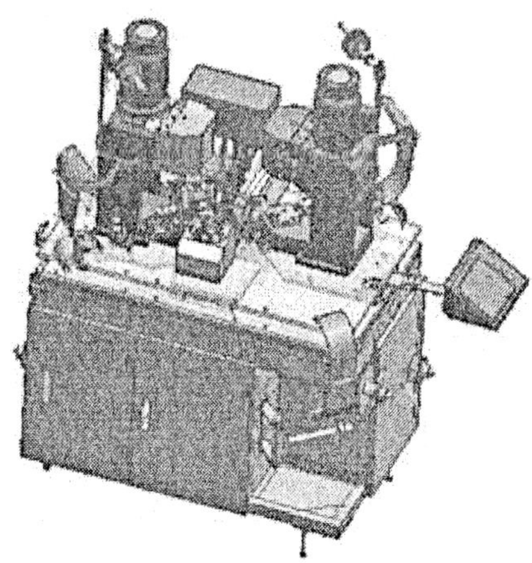

Figure 6. Whole Assembly

Assume that a machine has n mechanisms, which are represent by 1-n. association between mechanism i and j is d (i,j). We plan to define identity of the heaviest weight mechanism enjoys as our target function. We describe these mechanisms as (w_1, w2,⋯⋯.w_n), and $w_{n+1}=w_1$

Birth of new solution:

Randomly take k and m form 1-n, $k \neq m$. if k<m, then we change

($w_1,w_2,\cdots\cdots w_k,w_{k+1},\cdots\cdots, w_m ,\ldots, w_n$) into:

($w_1, w_2 ,\ldots, w_m , w_{m-1} ,\ldots, w_{k+1} , w_k ,\ldots, w_n$).

If k>m, we change ($w_1, w_2,\ldots, w_k , w_{k+1} ,\ldots, w_m ,\ldots, w_n$)

Into ($w_m, w_{m-1} ,\ldots, w_1 , w_{m+1} ,\ldots, w_{k-1} ,w_n , w_{n-1} ,\ldots, w_k$).

Through disparate substitution function, we can obtain margin of substitution function, eventually get ID of the most heavy weight mechanism.

Using FPC theory, our full-automatic crimping machine achieves direct success in designing and manufacturing. (See figure 6)

5. CONCLUSION

Development of our full-automatic crimping machine combines FPC with various latest bearing fruit in today's international society, with its feature being virtual, digital, network, integrated, intelligent, and green production, flexible manufacturing ,synergetic interaction.

Research of Integrated manufacturing/assembly is the distillate of this project. Main assembly designer mark off reasonable function space with advanced CAE analysis method at concept design Emphasis of development of integrated product in FPC lay in manufacturability (including synchronization between designer and technologist, and accept or reject design scheme according to machine cost), feasibility to assemble (in order to assemble easily, we need to use least assembly parts and optimize assembly path. Try to diminish connect constraints as least as possible on the premise that satisfy function requirements), feasibility to corporation (machine without reasonably programming cannot implement anticipated function. Designer of sub-mechanism doesn't only need to take charge of their own design, but also need to make conjunction parts facilitation, humanity.

6. REFERENCES

1. You Ying, "Research and Application of key technology in FPC", master thesis, mechanical and electrical engineering school, Wu Han technology university, China, 2004.

2 Integrated Manufacturing Technology Initiative, Inc. First Product Correct.America, 2000

THE DESIGNATION OF CAM'S CONTOUR FOR PISTON PUMP WITH STEADY FLUX

Zhongxu Tian [1], Xiaochuan Chen [2]
[1]*Internal Combustion Engine Institue, Shanghai Jiaotong University,200030,Shanghai, China; Email:zxtian@sjtu.edu.cn.* [2]*The Mechanical Engineering College, Donghua University, 201620, Shanghai, China*

Abstract: In order to get steady flux of the piston pump, an analytical method is given to push the driving cam's contour forward. The contour can be expressed by patch function analytically, which can be used conveniently as well as the study of the pump's performance is easy to be realized.

Key words: steady flux; analytical method; cam's contour; acceleration's continuity

1. ILLUSTRATION

Piston pump can provide high press and reliable work, so it is used in many fields such as mining oil, driving with hydraulic pressure and control system. In many cases, pumps' exporting steady flux is very important. Many people meet this requirement through the control of the driving motor, but this method needs high performance of driving motor and another control system to control the motor. In application, the precision of the flux is usually not steady enough. An analytical method is used for the designation of cam's contour, using which can drive pumps and give steady flux with high precision.

This project is supported by the National Natural Science Foundation of Shanghai under the grant No. 04ZR14053.

Please use the following format when citing this chapter:
Tian, Zhongxu, Chen, Xiaochuan, 2006, in International Federation for Information Processing (IFIP), Volume 207, Knowledge Enterprise: Intelligent Strategies In Product Design, Manufacturing, and Management, eds. K. Wang, Kovacs G., Wozny M., Fang M., (Boston: Springer), pp. 231-236.

2. CONSTRAIN OF PUMP'S STEADY FLUX

The export flux of the pump that contains a pair of pistons can be express by

$$Q = \frac{v_1 + |v_1|}{2} S + \frac{v_2 + |v_2|}{2} S \tag{1}$$

where S is the cross section area of the pistons, and v_1、 v_2 is the velocity of the two pistons respectively, whose positive direction is the direction when they push liquid out.

In order to give constant flux, the velocity of the pistons should meet

$$v_1 + | v_1 | + v_2 + | v_2 | = constant \tag{2}$$

Constrain (2) is expressed with the pistons' velocity, the designation of the cams should start from their velocity.

3. THE DESIGNATION OF THE PISTON'S VELOCITY

The phase difference between the two pistons usually is $180°$, and one type of the simplest velocity curve is shown in figure 1. The two piston pump out liquid alternatively, and meet the constant flux.

Figure 1. the Idealized velocity of the pistons

The velocity (figure 1) is not continuous, and the acceleration will become infinite sometimes, which means serious collisions between cams and pistons. So, adding a transition domain to make the velocity chances continually is very important.

Suppose the width of the transition domain is $2a$, and it's very easy to design the velocities change linearly in the domain (figure 2). Obviously, the pump can export liquid with steady flux. But the acceleration of this kind of cam is not continues, which means soft conflicts between the cam and the piston, and the cam can not run faster to avoid more vibrations.

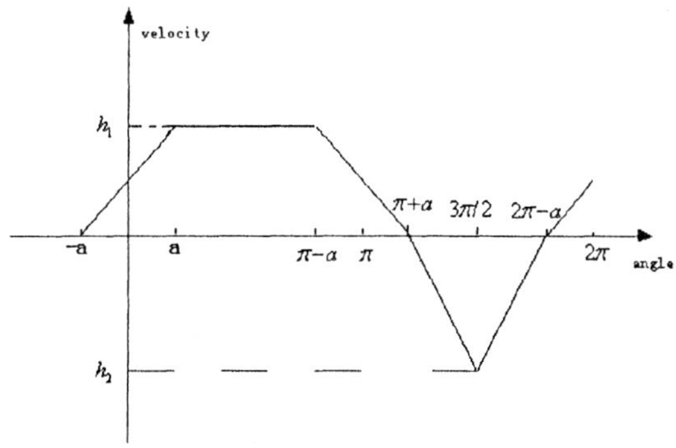

Figure.2 the velocity of the pistons with linear transition

Figure 3 shows one cam's velocity whose acceleration is continues, and the functions in the transition domain have to meet constrain expressed by equation (2).

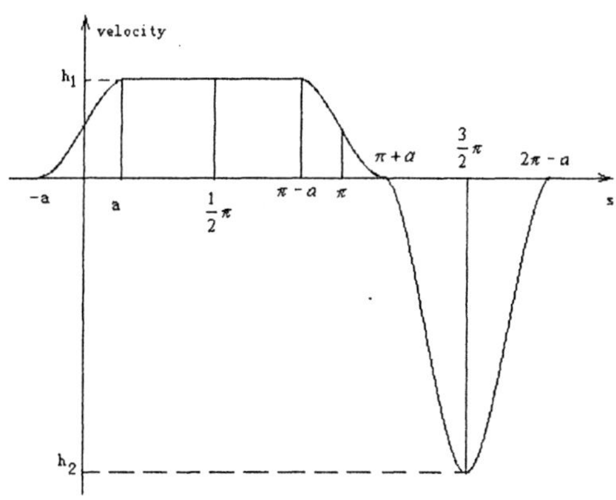

Figure 3 the piston's velocity with continues acceleration (s is the rotation angle of the cam)

The base function used in Hermite interpolations can just fit the former requirements, and the velocity of the cam can be expressed

$$v=\begin{cases} h_1 t^2(3-2t), \quad where,\ t=\dfrac{s+a}{2a} & s\in[-a,a] \\ h_1 \quad s\in[a,\pi-a] \\ h_1(1-3t^2+2t^3), \quad where,\ t=\dfrac{s-\pi+a}{2a} & s\in[\pi-a,\pi+a] \\ -h_2 t^2(3-2t), \quad where,\ t=\dfrac{s-\pi-a}{\pi/2-a} & s\in\left[\pi+a,\dfrac{3}{2}\pi\right] \\ -h_2(1-3t^2+2t^3), \quad where,\ t=\dfrac{s-3\pi/2}{\pi/2-a} & s\in\left[\dfrac{3}{2}\pi,2\pi-a\right] \end{cases} \tag{3}$$

It is very easy to verify that the flux will be steady without any bound.

4. THE COMPUTATION OF THE CAM'S CONTOUR

Integrate the velocity respect to time, and the displacement of the piston can be get:

$$f(s)=\begin{cases} f_1(s) & s\in[-a,a] \\ f_2(s) & s\in[a,\pi-a] \\ f_3(s) & s\in[\pi-a,\pi+a] \\ f_4(s) & s\in[\pi+a,\frac{3}{2}\pi] \\ f_5(s) & s\in[\frac{3}{2}\pi,2\pi-a] \end{cases} \tag{4}$$

where

$$f_1=\frac{-h_1s^4+6h_1s^2a^2+8h_1sa^3+16C_1a^3}{16a^3} \qquad f_2=h_1s+C_2$$

$$f_3=\frac{8h_1sa^3-6h_1s^2a^2+12h_1s\pi a^2+h_1s^4-4h_1s^3\pi+6h_1s^2\pi^2-4h_1s\pi^3}{16a^3}+C_3$$

$$f_4=(-4h_2s^4+20h_2s^3\pi+8h_2s^3a-36h_2s^2\pi^2-36h_2s^2\pi a+28h_2s\pi^3+48h_2s\pi^2a$$
$$+12h_2s\pi a^2-8h_2sa^3)/(-\pi+2a)^3+C_4$$

$$f_5=-h_2\left(s-\frac{(s-3\pi/2)^3}{(\pi/2-a)^2}+\frac{(s-3\pi/2)^4}{2(\pi/2-a)^3}\right)+C_5$$

and C_1、C_2、C_3、C_4、C_5 is the parameters to be determined. As the displacement is continue, and we will have

$$\begin{cases} f_1(a)=f_2(a) \\ f_2(\pi-a)=f_3(\pi-a) \\ f_3(\pi+a)=f_4(\pi+a) \\ f_4\left(\frac{3}{2}\pi\right)=f_5\left(\frac{3}{2}\pi\right) \\ f_5(2\pi-a)=f_1(-a) \end{cases} \tag{5}$$

Solving equation (4), and C_2, C_3, C_4, C_5, h_2 can be got as

$$h_2 = \frac{2h_1\pi}{\pi - 2a} \qquad C_2 = -\frac{3}{16}h_1 a + C_1$$

$$C_3 = \frac{-8h_1a^3\pi + 6h_1a^4 + 6h_1a^2\pi^2 - h_1\pi^4 - 16a^3C_1}{16a^3}$$

$$C_4 = \frac{h_1\left(-509\pi^4 a - 272\pi^5 - 224\pi a^4 + 712\pi^2 a^3 - 792\pi^3 a^2 + 48a^5\right)}{16\left(-\pi^3 + 6\pi^2 a - 12\pi a^2 + 8a^3\right)\left(-\pi + 2a\right)} + \frac{C_1\left(512a^3\pi - 256a^4 - 384\pi^2 a^2 - 16\pi^4 + 128\pi^3 a\right)}{16\left(-\pi^3 + 6\pi^2 a - 12\pi a^2 + 8a^3\right)\left(-\pi + 2a\right)}$$

$$C_5 = \frac{h_1\left(-19\pi a + 56\pi^2 + 6a^2\right) + C_1\left(16\pi - 32a\right)}{16\left(-\pi + 2a\right)} \tag{6}$$

By plugging (6), formula (4) becomes expanded curve of the cam, and in which, h_1 and C_1 parameters to be valued.

According to equation (3), the journey of the pistons is

$$S = h_1\pi \tag{7}$$

and the flux of the pimp is

$$Q = h_1 S\omega \tag{8}$$

The minimum and maximum size in radial direction of the cam are

$$r_{\min} = \frac{h_1\left(a^2 - 2C_1\right)}{4\alpha} \qquad r_{\max} = \frac{h_1\left(4\pi\alpha - \alpha^2 + 2C_1\right)}{4\alpha} \tag{9}$$

Application objects can be satisfied by giving h_1 and C_1 with reasonable values. Figure 4 shows an instance, the contour of the cam is calculated with expression (4).

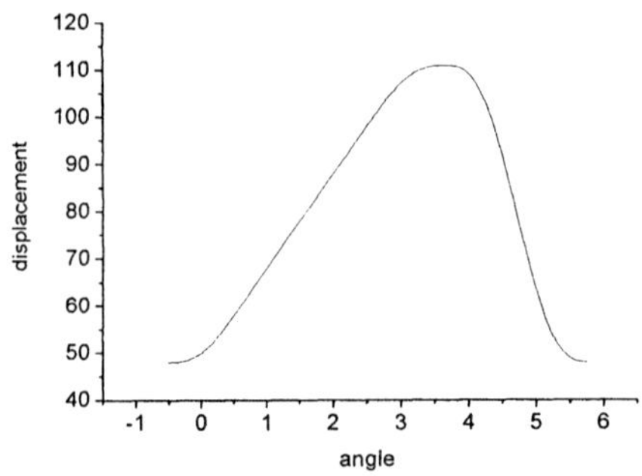

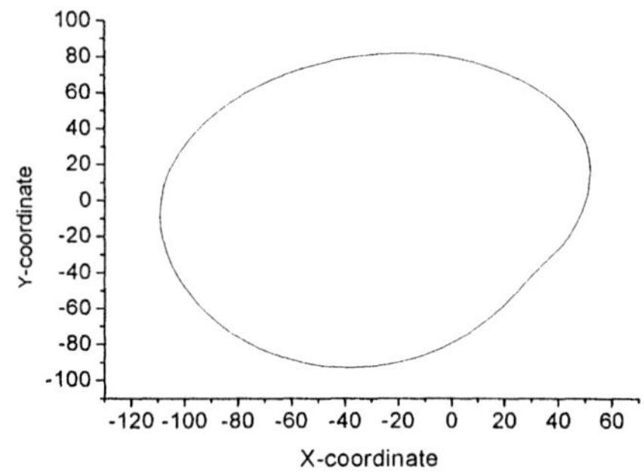

Figure 4 the calculated displacement of piston and the contour curve of the cam. $a = \pi / 6.0, h_1 = 20$ *and* $C_1 = 50$.

5. CONCLUSIONS

This paper has given the analytical function for the cams, which can drive piston pump and give steady flux. People can realize desired pump through given reasonable values for the parameters and the study of their performances is very conveniently.

6. REFERENCES

1. Zou, Huijun and Dong, Shiyu. *The modern design of the cam machine*. Publication of Shanghai Jiaotong University, 1991
2. Shi, Yonggang and Xu, Zhenhua. *The design of The cam machine*. Science technique publication of Shanghai, 1995
3. Yan, Guojun, Zhao, Junming and Dong, Yong, Study on Kinematics of Reciprocating Pump Valve, *China Mechanical Engineering* 2004(9):1617-1618

A STUDY OF SYSTEM BALANCE AND CYCLE OPTIMIZATION FOR MULTI-OPERATIONAL EQUIPMENT AND ITS IMPLEMENTATION

Huan You, Lixin Lu, Limin Li
[1]CIMS & Robot Center, Shanghai University, Shanghai, 200072, China; Email:alvaqqyh@hotmail.com. [2]Shanghai Feilo Co.,LTD., Shanghai, 200050, China.

Abstract: This paper introduces one-piece flow production concept and other related theories and methods to improve the balance of multi-operation mechanical system and optimize the operation cycle. An example of full-automatic crimping machine will be exemplified to prove the rationality and feasibility of this study on the last section of this paper.

Key words: one-piece flow (OPF) production; system balance; operation cycle; optimization.

1. INTRODUCTION

Equipment manufacture is a fundamental industry that has a great influence on the Comprehensive National Power. With the emergence of mechanical automation, automobile, light industry etc, the need of putting more emphasis on innovative and fast design as well as rapid manufacture is becoming more and more important for enterprises in the sharply changing market. Moreover, how to reduce product life cycle, find the solution to balance the multi-operational mechanical system and finally develop the potential capacity, attracts more and more focus at home and abroad.

To improve and strengthen the balance of multi-operational mechanical system and optimize the operation cycle, this paper introduces the theory of one-piece flow production into these systems to find out the bottle-neck working procedure and lower capacity resources whose mechanical structure

Please use the following format when citing this chapter:
You, Huan, Lu, Lixin, Li, Limin, 2006, in International Federation for Information Processing (IFIP), Volume 207, Knowledge Enterprise: Intelligent Strategies In Product Design, Manufacturing, and Management, eds. K. Wang, Kovacs G., Wozny M., Fang M., (Boston: Springer), pp. 237-243.

can be re-designed to coordinate the overall performances of the components. After analyzed and simulated by virtual assembly, new mechanical structure of these resources can be obtained, which can largely speed up the production efficiency. During this process, IE (Industry Engineering) related technique and methods will be used. An example of full-automatic crimping machine cooperatively developed by Shanghai Feilo Co., Ltd and Shanghai University will be exemplified to prove the rationality of this method.

2. SYSTEM BALANCE & OPERATION CYCLE

2.1 One-piece Flow Production Theory

One-piece flow (OPF) production is a market-oriented just-in-time (JIT) production system that strives to fulfill the requirements of total quality management (TQM), reduce product changeover time, increase productivity, reduce work-in-process inventory, and eliminate all sources of waste. Initially, the concept of one-piece flow production is closely related with the character of "one piece", which means collocating the working condition, workers, equipment to make sure that only one product stays at each workstation for the cycle time and none existence of buffers for semi-finished products. In this paper, one-piece flow production is extended into the multi-operational mechanical system, which means that the quantity of WIP (work in process) does not exceed the precedence process at any time between two neighboring processes, and that is a non-stop, non-exceed and non-lag process in the movement state. The production in each component carries on almost synchronously, making the WIP to be produced and passed on in a single piece. Once the product has been manufactured completely in one process, it will be put to next process immediately. The OPF production system is a hybrid of the best of all the traditional production systems that insists on waste reduction as the base for improvement.

2.2 System balance

Based on the theory of OPF production, the system balance requires coordinating the overall arrangement and adjusting the operation cycle that can make each component busy at its working period to make good use of its capacity.

Manufacturing a product on the multi-operational equipments requires partitioning the total amount of work into a set of elementary operation named task. Due to technological and organizational conditions, precedence constraints between the tasks have to be observed. If without the concept of system balance, i.e. all components (stations) operate at an individual speed, workpiece may have to wait before it can enter the next process and/or components may get idle when they have to wait for the next workpiece. Under the aforementioned restriction and difficulties, the system balance can be defined as making it easier to carry out the OPF production on the spot and to maximize the operation capacity that is a must to improve the working efficiency and raise the economic performance. The system balance rate can be assessed by the following formula:

$$\eta = \Sigma t/(s*C.T)*100\% \qquad (1)$$

In the Eq. (1), η denotes the system balance rate, S denotes number of components, Σt denotes overall processing time for one production, Cycle Time (C.T.) denotes Pitch Time that equals 3600/Q, and Q denotes the yield per hour.

2.3 Operation Cycle

The balance of production line depends on making each component's cycle time carry on simultaneously. So the time for each qualified output of each procedure can be defined as operation cycle, which can be obtained by the following formula:

$$C = T/R \qquad (2)$$

In the Eq. (2), C denotes operation cycle, T denotes efficient work time, and R denotes number of qualified output.

The efficient work time is the time actually applied into practical manufacture, which can be divided into two parts:

Basic working time, which is the actual time needed for accomplishing the necessary operations, i.e. manufacturing time.

Assistant working time, which is the time left for various assistant operations. These operations ensure the accomplishment of the basic operations.

From the Eq (2), we can know that the system balance relies on the efficient work time. Meanwhile, operation cycle time can be adjusted by the output produced in the efficient work time. So we can enhance the capability of mechanical systems by optimizing the efficient work time and the proportion of its sub-sections.

3. OPTIMIZATION AND EVALUATION

3.1 Optimization

Time-efficiency model is a commonly used model during improvement and optimization of mechanical and electronic products, of which the system balance and cycle optimization can be truly realized. The proposed model is given below:

$$\text{Time}=\sum_{i=1}^{n} k_i T_i c_i^{-\lambda_i}$$

$$\begin{cases} 0 < k_i \le 1; \\ c_i > 1; \\ \lambda_i = \begin{cases} 0(\mathit{improved}) \\ 1(\mathit{unimproved}) \end{cases} \end{cases} \tag{3}$$

In the Eq. (3), n denotes the number of components, Time denotes overall processing time for one production, T_i denotes processing time of each component with maximum capability, k_i denotes regulating factor for each component to get overall coordination, λ_i denotes improving factor for each component, and c_i denotes efficiency factor for each component after optimization.

There are two ways to realize the goal of optimization: 1) Improve the cooperation among each component on the basis of original system to realize optimal harmonicity, which is mainly to coordinate the value of k_i 2) Improve the overall efficiency (k_i) by restructuring each component (c_i). During re-designing, k_i and c_i are interactive and inter-restrict for finally optimization of the overall equipment.

Obviously, the first way just optimizes the original operation cycle, or only improves the processing ability instead of the original structure. Although it has some limitations, it is still useful in some simple cases. The second way is suitable for improving the system structure, based on which the optimization is carried on in this paper, i.e. finding out and increasing the efficiency of bottle-neck working procedure and other resources (with the lowest capacity) to improve the production ability of the whole system.

3.2 Evaluation

Certain assessment should be performed after optimization. By using the world-top analysis software of CAD and CAE, the whole structure will be improved and developed coordinately during which the detected interference and unreasonable structure will be modified and remodeled for virtual assemble and further assessment, accordingly, the efficiency of whole system can be evaluated by simulation.

4. CASE STUDY

With the fast development of electronic products and emerging application industry, especially automobile industry and consumer electronics, the need of wire harness is increasing dramatically. However, with the demand of high productivity, the problem of low efficiency is emerging.

Following the above concepts and methods, the optimization is carried out over structural improvement of the multi-operation equipment to enhance its efficiency and its market competitive ability

4.1 Flow Chart of Original System

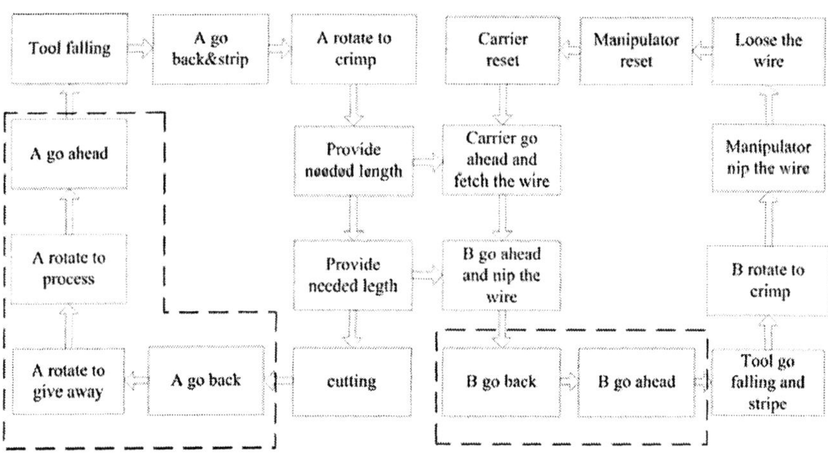

Figure 1. Flow Chart of Original System

As shown in figure 1, the system consists of eight main mechanisms: Front Wire neaten mechanism、Front Wire transmitting mechanism、Front Crimping mechanism 、Tool Rest mechanism 、Back Stripping mechanism、Back Crimping mechanism、Back Wire-receiving mechanism and belt mechanism. The principle of this full-automatic crimping machine is as follows: wire-neaten mechanism strains winding wire straight, which then sent to tool rest mechanism for stripping. After stripped, it is time for front crimping mechanism to crimp one end of the wire, while back crimping mechanism nips the wire to wait for its being cut off and rotates to have the other end crimped in the back crimping mechanism. Finally belt mechanism withdraws the finished wire.

Tool is a bottle-neck resource which should finish two tasks as cutting and stripping in original system. As repeatedly used in both ends A and B, especially when finishing crimping one end A and sending the wire to the other part for operation, according to the technical flow, the tool will be both needed in two ends A and B for consequent process. Thus unnecessary time results from the conflict using of tool resource because end A should give away during the operation process of end B and wait until B finishes.

4.2 Improving Scheme

According to the methods put forward in 3.1, we will innovatively substitute three tools for the original one tool in Tool Rest mechanism, which enables the operation of stripping and cutting finish simultaneously. It will alleviate the burden of bottle-neck resource and avoid the unnecessary time of giving away by rotating and waiting. Such an improvement optimizes each component's cooperation, accomplish parallel process and accelerate operation cycle so as to reach one-piece flow production and high efficiency. Two mechanisms are compared in Fig 2

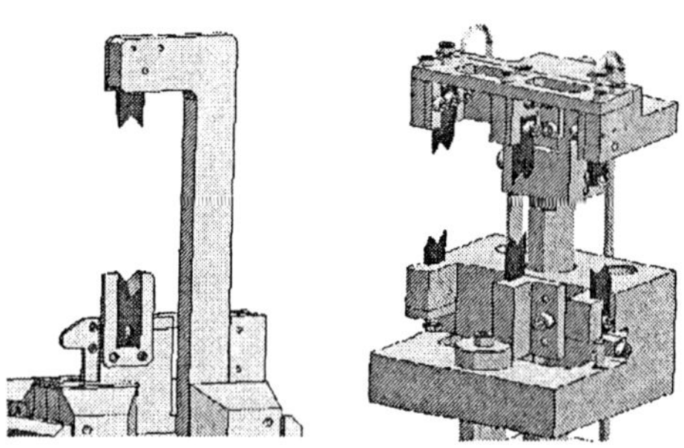

Figure 2. Comparison of the original mechanism with the new one

4.3 Simulation and Assessment

After the initially modeling, dimensions and parameters of each component, the virtual model and assembly will be established by powerful UGNX software for further assessment of collision, interference, etc. The design process will be repeated when meeting something unreasonable during evaluation. According to the Eq.(3), the optimal value of time for the improved model is 0.931s instead of 1.185s of the original one, or an increase of 31%. As to the Eq.(1) and Eq.(2), the optimal system balance rate is 80% instead of 67% of the original one, or an increase of 13%, which totally realized the original intention to redesign and optimization.

5. CONCLUSION

The related technique and methods of IE (Industry Engineering) is very important to reduce consumption and improve productivity. It will play a leading role once carried out in practical production. This paper introduces the theory of one-piece flow production into mechanical systems to find out the bottle-neck working procedure and related resources of low capacity. By re-building the mechanical structure of the bottle-neck resources, constructing the model of time-efficient to evaluate and using UG software to simulate, the system balance and cycle optimization of the machine can be optimized.

6. REFERENCES

1. Goldratt E. M., Cox J., (1992), *The Goal: A Process of Ongoing Improvement*, North River Press.
2. Matthias A., (2006), *Cost-oriented assembly line balancing: Model formulations, solution difficulty, upper and lower bounds*, European Journal of Operational Research, vol. 168, pp.747-770.
3. Tetsuo Y., Masayuki M., (2003), *A management design approach to assembly line systems,* Int. J. Production Economics, vol. 84, pp.193-204.
4. Pierre D.L, Alain D., (2003), Jean-Michel H., *An integrated approach for product family and assembly system design,* IEEE Transactions on Robotics and Automation, vol. 19, pp. 324-334.
5. Barnes C.J., Jared G.E.M., Swift K.G., (2004), *Decision support for sequence generation in an assembly oriented design environment,* Robotics and Computer-Integrated Manufacturing, vol. 20, pp.289-300.

ANALYSIS ON PRODUCT TECHNICAL RISK WITH BAYESIAN BELIEF NETWORKS

Ming Chen[1],Yun Chen[2], Bingsen Chen[1], Qun Wang[1]
[1]*Sino-German College of Applied Sciences, Tongji University, China; Email: cdhawcm@yahoo.com.cn.* [2]*School of Public Economy Administration, Shanghai University of finance & economics, China.*

Abstract: With the development of the science and technology, products are more and more complex, e.g. more functions integrated into one product, abilities of products are stronger and stronger, degrees of automatization and intelligentizing are higher and higher. In order to develop products successful, we must analyze product technical risk systematically and comprehensively. All products' behaviors that don't satisfy their functions are function failures, i.e. product technical risk. The sources of causing risk are the products' structures. We anticipate the behaviors of the target product, seek the potential function failures, and analyze the causes, and then set up BBN model. With this methodology, we have analyzed the product technical risk of a micro-electronic equipment.

Key words: Product developing, Product technical risk, Bayesian belief networks.

1. INTRODUCTION

With the development of the science and technology, people have been improving their living and working situations constantly. The requirements of products that serve people are more and more, e.g. multiple functions are integrated into a product, abilities of system are more and more powerful, degrees of automatization and intelligentizing are very highly. It makes required products more and more complex.

Frequently, Design indexes of are diversity and some indexes are not consistent with the others. The target product consists of massive elements.

Please use the following format when citing this chapter:
Chen, Ming, Chen, Yun, Chen, Bingsen, Wang, Qun, 2006, in International Federation for Information Processing (IFIP), Volume 207, Knowledge Enterprise: Intelligent Strategies In Product Design, Manufacturing, and Management, eds. K. Wang, Kovacs G., Wozny M., Fang M., (Boston: Springer), pp. 244-249.

And the product's interfaces are multitudinous and complex. Most products are involved into many fields, e.g. mechanical engineering, electronic engineering, optical engineering etc. Designers lack of the experiences of usage of the advanced technology that are used in product developing.

In order to develop products successful, we must analyze product technical risk.

There are some ways to assess the risk, for example, fuzzy mathematics and neuron network, but they can't solve the essential problem of risk assessment: reasoning integrated with experts' knowledge based on uncertain information.

Bayesian Belief Network is a kind of expert system that can describe uncertain information. It can reflect continuity and accumulation of risk assessment. The consistency of time can't be realized with non-memorable methods, for example, methods based on rules and neuron network. Because of reliability in mathematics, Bayesian Belief Network can be changed into standard model which can express human thinking and reasoning process, while the traditional expert system is entirely oriented to tasks, and different tasks have their own models and they lake of catholicity.

The Product Technical Risk is modeled with Bayesian Belief Networks, constitutes assessing arithmetic, simulates and analyzes the technical risk.

2. FUNCTION, STRUCTURE AND BEHAVIOR

2.1 Product system

Products are entities that provide customers utilities and interests. Large-scale products are systems composed of controls, units and interactions

Controls in product can be on product's level, on sub-systems' level or components' level. Units are non-control elements in product. Interactions are expressed that product's elements affect one other.

Any system has its own functions and shows its behaviors outside. System's behavior is process of realize relevant function, system's function is goal, results or constraints of behaviors. System's structure is total relations of elements in some system; it is a carrier of functions and behaviors.

2.2 Method of analysis on product technical risk

Owing to the complexity of systems in large-scale products, we must start analysis from the sources of product development: customers (Figure 1). All products' behaviors that don't satisfy their functions are function failure, i.e. product technical risk. The sources of causing risk are the products' structures. From the customers' requirements, the products' functions are acquired. According the functions, the structures are designed and after developing, the target product is manufactured. Analyzing the product's behaviors, we can judge whether or not the product satisfies the functions. We anticipate its behaviors, seek the potential function failure, i.e. product technical risk, and analyze the causes, i.e. its structure, and then set up Cause-and-Effect Chain, i.e. BBN model.

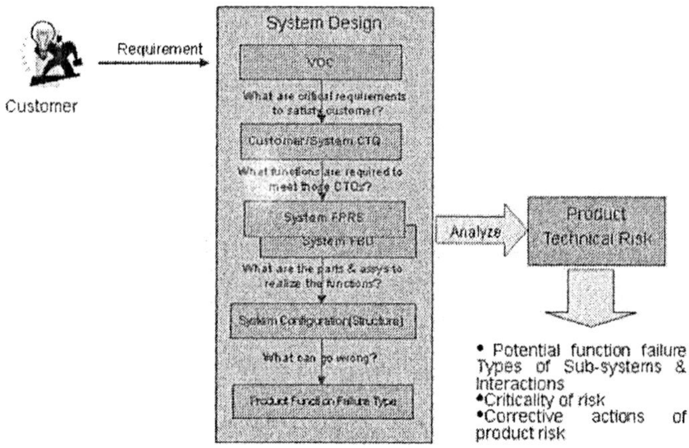

Figure 1. Method of product technical risk analysis

3. MODEL OF PRODUCT TECHNICAL RISK

3.1 VOC, CTQ, FPRS and FBD

The voice of the customer (VOC) is a process used to capture the requirements/feedback from the customer (internal or external) to provide the customers with the best in service/product quality. This process is all about being proactive and constantly innovative to capture the changing requirements of the customers with time.

Customer critical to quality (CTQ) is the key measurable characteristic of top level product whose performance standards or specification limits must be met in order to satisfy the customer. It aligns improvement or design efforts with customer requirements.

Starting with the customer's problem as outlined in the Customer/System CTQ, the Functional Performance Requirement Spec (FPRS) defines a product's functionality, usability, reliability, performance, supportability, availability, localizability, and other requirements. It describes the functions and performance levels that the proposed product should have to successfully address that problem.

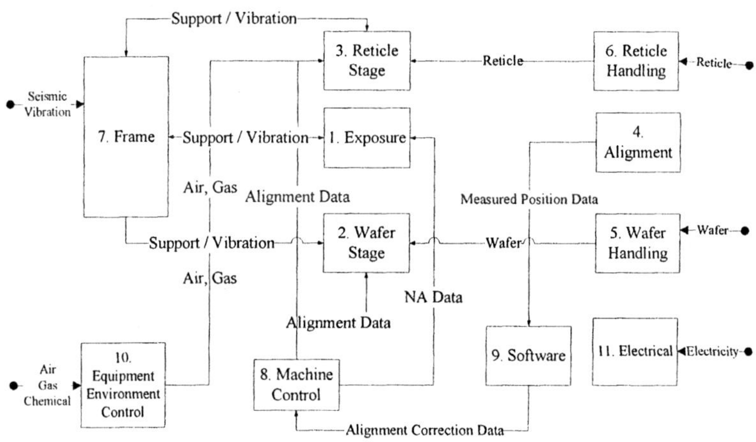

Figure 2. FBD of Step & Scanner

Functional Block Diagram (FBD) is used to show how the different sub-systems / parts interact with one another to verify the critical path. We prepare FBD in 3 different levels: system level FBD, sub-system level FBD and unit FBD. Figure 2 is a sample which shows the System Level of 300mm, 193 ArF, 130nm Step & Scanner.

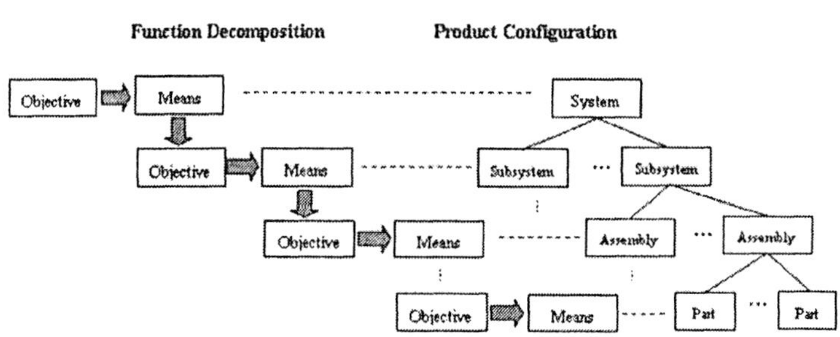

Figure 3. Function decomposition and product configuration

3.2 Product System Configuration

The functions of the target product decompose some objectives and means on the system level. We decide which structures on this level according to the means. These means on this level are the objectives on the next level. We break them to some means. Then the structures of subsystem are designed. In this way, all the system is developed (Figure 3).

3.3 BBN Model

A BBN model is a directed acyclic graph (DAG), which is defined to consist of the qualitative and quantitative relationships. Qualitative Relationships include a set of random or deterministic variables and a set of directed edges or arcs. Quantitative Relationships include a set of root and conditional probabilities.

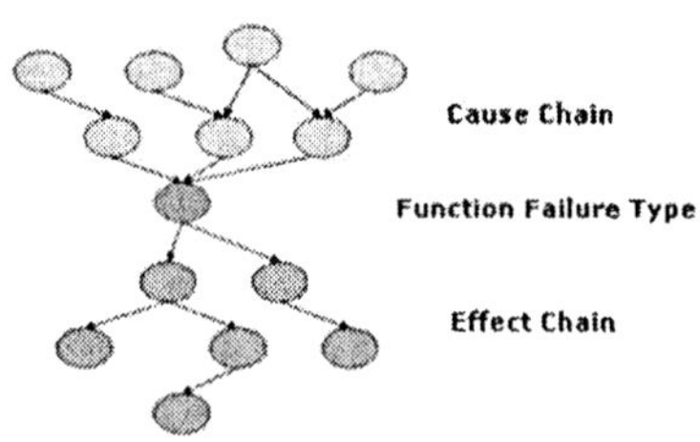

Figure 4. BBN model

Cause-and-Effect chain is a BBN model of the product risk analysis. Every function failure has its own model. It's easy to find causes that result in the function failure, and effect that the failure gives.

First we list the function failure, then list the phenomena occurs directly if the function failure appears, analyze the further effects and get effect chain. After that, we analyze the causes that result in the function failure, and analyze the further causes and get cause chain (Figure 4).

4. CASE STUDY

300mm, 193 ArF, 130nm Step & Scanner is a kind of large-scale complex product which is related to mechanical engineering, electronic engineering, optical engineering, control engineering etc. We must analyze its technical risk during developing. With this methodology, we have analyzed the potential function failures. There are 20 function failure type, for example, CD resolution failure, CD uniformity failure etc. With the BBN model, we calculated the subsystem technical risk (Figure 5). We took some measures to improve the design and the risk has been reduced.

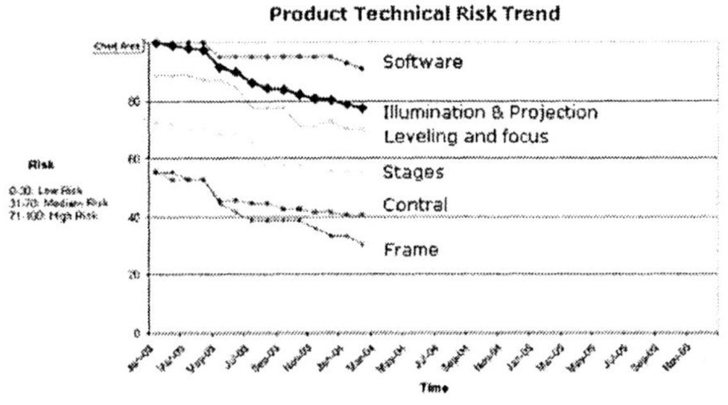

Figure 5. The product technical risk trend of the step & scanner

5. CONCLUSION

In order to develop large-scale complex products successful, we must analyze product technical risk systematically. Product technical risk is its function failure. We analyze the behaviors of the target product and anticipate the potential function, i.e. failure product technical risk, and then analyze the causes, i.e. structure. In this way, its BBN model will be set up. We calculate the risk, and then it can guide us to take corrective actions in order.

6. REFERENCES

1. Chen Ming, Chen Yun, Chen Bingsen. Management of technical risk in the development of complex products. *Manufacturing Automation* 2005; 10:pp. 6-8.
2. Wang Guohua, Chen Ming, Chen Yun. Fuzzy Synthetized Estimating of Technical Risk of Complex Products. *Mechatronic* 2006; 1:
3. Burton Hoyt Lee. Failure modes and effects analysis with Bayesian belief networks: bridging the design-diagnosis modeling gap. Dissertation of Stanford University 2002.
4. Elaine M.Hall. *Managing Risk——Methods for Software Systems Development.* Bosten: Addison-Wesley 1998
5. Wang Lianchen. *Engineering System Theory.* Beijing: China Astronautics publishing House, 2002
6. Eigenworx and ConneXwave. *Advanced FMEA.* SMEE 2003
7. ConneXwave . *Managing Complex Projects Plan.* SMEE 2003

THE APPLICATION OF INVERTER-DRIVEN TECHNOLOGY ON THE CRIMPING MACHINE

Runqing Zhu[1], Lixin Lu[2], Limin Li [1], Huan You[1]
[1]CIMS & Robot Center, Shanghai University, Shanghai, 200072, China; Email:wykyv_n@163.com. [2]Shanghai Feilo Co.,LTD, Shanghai, 200050, China.

Abstract: In this paper, an innovative design of crimping machine which is based on inverter-driven technology and adopted embedded PSOC system hardware platform is discussed. This improvement will abandon original disadvantages such as: bulky mechanical structure、low efficient、unsteady crimping quality、heavy noise etc, and make the inverter-driven technology available to control crimping machine.

Key words: Inverter-driven, Adjustable-speed Motor, Crimping Machine, Controller.

1. INTRODUCTION

With the fast development of electronic products and new emerging application industry, the requirement of wire harness is increasing dramatically. Therefore, the developing prospect of the wiring harness terminal processing machine-crimping machine is expansive and optimistical. Since 1980s, the action of crimping is driven by mechanical device. Its basic structure is as follows: driven mechanism、clutch and brake mechanism and crimping mechanism, which can be seen in Figure 1.

As shown in figure 1, the energy is stored by wheel which is driven by motor, and then crimping will be carried out by slide block which is driven by clutch mechanism, after that the equipment will be stopped by brake mechanism system. Because of the complex and heavy structure as well as unstable quality and heavy noise, etc, a new prototype will be brought up with the updated control method of inverter-driven technology rather than

Please use the following format when citing this chapter:
Zhu, Runqing, Lu, Lixin, Li, Limin, You, Huan, 2006, in International Federation For Information Processing (IFIP), Volume 207, Knowledge Enterprise: Intelligent Strategies In Product Design, Manufacturing, and Management, eds. K. Wang, Kovacs G., Wozny M., Fang M., (Boston: Springer), pp. 250-256.

bulky mechanical structure, which can abandon the original mechanical clutch and brake mechanism. This innovative design not only makes the overall equipment smooth in action, conservational in energy, high in efficiency as well as lets the machine compact in design, solid in structure and easy in maintenance.

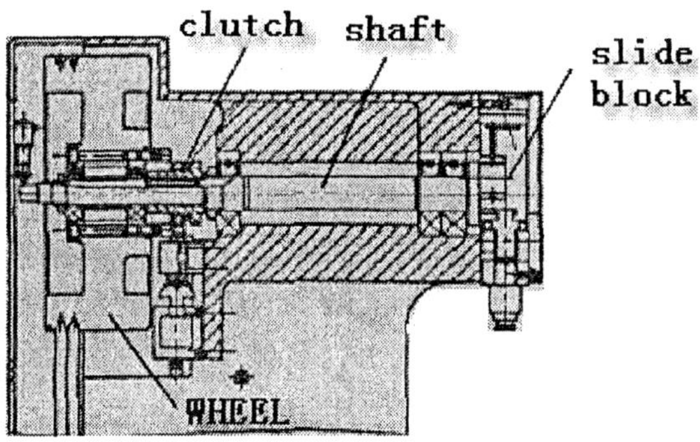

Figure 1. Mechanical structure of the crimping machine.

2. THE THEORY OF INVERTER-DRIVEN ADJUSTABLE-SPEED SYSTEM

Because the inverter directly drives the three-phase alternating current induction motor to drive the crimping machine, to study the theory, firstly we list the following formula that denotes the speed of the motor:

$$n=n_1(1-s)=60f_1/p(1-s) \quad (1)$$

Where "n" is the speed of motor rotor;
"f_1" is source frequency of the motor stator;
"p" is the number of magnetic pole pairs;
"n_1" is synchronous speed;
"s" is the slip;

The formula shows there are three methods to adjust the speed of induction motor.

1. Change the number of magnetic pole pairs;
2. Change the source frequency;
3. Change the slip;

The induction motor will run with a proportional and unsteady speed when changing the number of magnetic pole pairs; and have a low efficiency, unreliable mechanical characteristics when changing the source frequency; while if the slip is adjusted properly, the induction motor will has a wide speed range、 high efficiency, and could adjust for the desired speed with constant torque or constant power inverter-driven mode. With the development of Power Electronics Technology, inverter-driven adjustable-speed system is becoming the utmost important part in alternating current governor System.[1]But the source frequency will influence the other parameters of induction motor simultaneously, such as air-gap flux Φ_m、 field current、 power factor、 magnetic torque、 overload ability etc. So other factors must be considered to make full use of the induction motor.

Still we have formula as follows:

$$U_1 \approx E_1 = 4.44 f_1 N_1 K_{W1} \Phi m \qquad (2)$$

Where "U1" is the phase voltage of the motor stator;

"E_1" is phase electromotive force;

"N_1" is the total number of the motor stator winding turns;

"K_{W1}" is the fundamental turns coefficient;

"Φ_m" is the air-gap flux;

According to formula (1) and (2), the speed of alternating current induction motor will lower when reducing the source frequency, meanwhile in order to keep the phase voltage constant, air-gap flux must increase which will lead to some disadvantage such as: magnetic circuit saturation、 lower power factor、 core loss etc.[2]So the phase voltage and source frequency of the motor stator must reduce simultaneously, i.e. U_1/f_1= constant. And the mechanical characteristics of the induction motor can demonstrate as follows. The figure shows that it has mechanical characteristics in accord with the constant torque control mode.

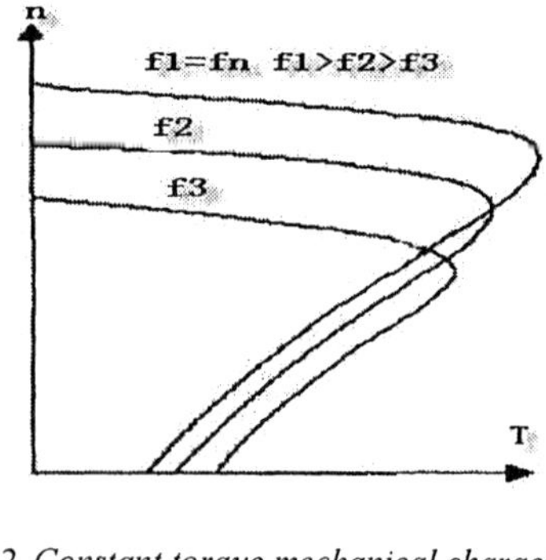

Figure 2. Constant torque mechanical characteristics.

If the source frequency is increased, the air-gap flux must reduce to keep the phase voltage constant, which will lead to the decrease of max electromagnetic torque 、 max start toque and overloading ability. And the mechanical characteristics of the induction motor can demonstrate as follows. The figure illustrates that it has mechanical characteristics in accord with the constant power control mode.

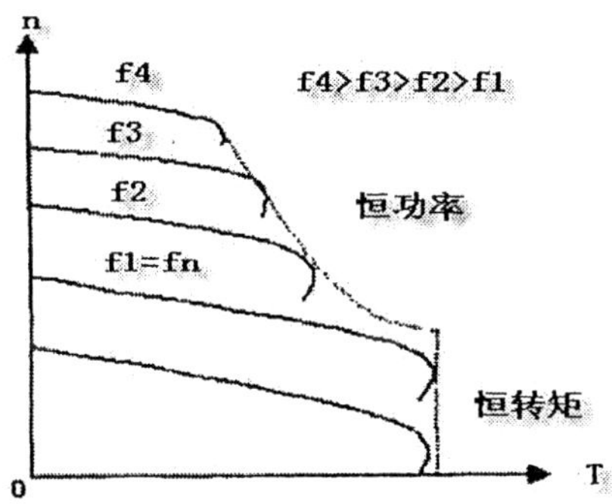

Figure 3. Constant power and constant torque mechanical characteristics.

So adopted inverter-driven technology, the induction motor not only provide better starting performance 、 adjustable speed 、 a wide speed range、 high efficiency, but also could run with constant power or torque control mode. Still the induction motor has a better performance such as: quick reflection、 forward/reverse control、 energy saving etc.[3]Constant torque control mode is recommended according to comprehensive analysis of the crimping machine.

3. INVERTER-DRIVEN SYSTEM CONTROLLER RESEARCH

3.1 Operation flow chart of the crimping machine

The system has three sensors, which transmit respective signal (accelerate、 decelerate、 stop) to inverter.

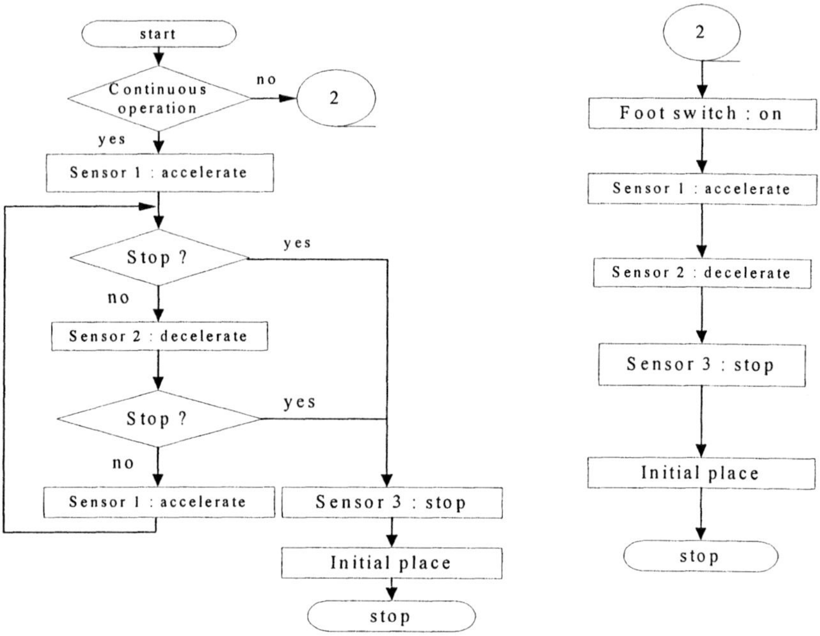

Figure 4. Operation flow chart of the crimping machine.

3.2 Develop the inverter-driven system controller of the crimping machine

Microprocessor is the core of the whole control system, how to choose a suitable one will have a crucial influence. Generally whether the microprocessor meets the requirement of the control system can be obtained by the following aspects: word length、 instruction sorts and quantity、 microprocessor speed、 interrupt handling etc. The PSoC device includes programmable logic blocks and a processor core. They are user-programmable devices, in which the designer can customize the configuration within development tools on their own computer. This microcontroller has digital and analog blocks (PSoC blocks) that can be configured into the desired peripherals. So all configurability can be handled through appropriate register settings that tell the specific PSoC block how to act, as well as controlling the traditional parameters such as input selection, compare values, and event actions etc.

3.2.1 The hardware module design of the controller

1) The whole framework of the hardware system is as follows:

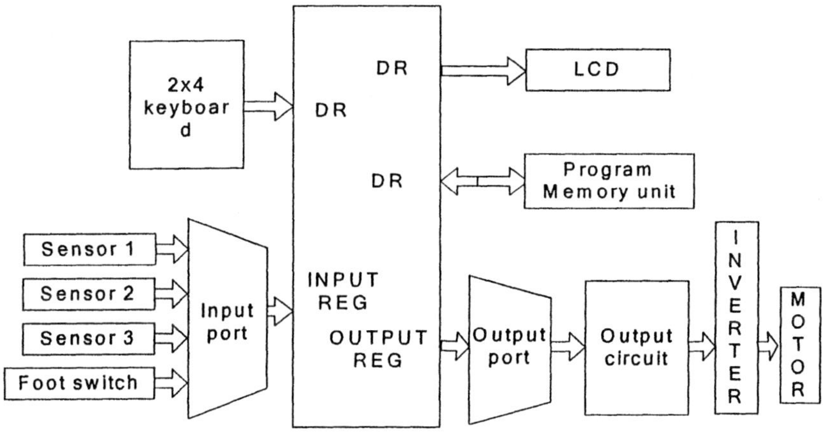

Figure 5. The framework of hardware system.

2) Inverter selection

Considering the requirement of quick start and efficient brake, as well as the high price versus performance ratio of the DELTA inverter, we take it into priority to drive the motor. The model of the motor is: YSJ8024,P=0.75KW,I_n=3.48A, n=1440r/min. the output current of the inverter should in accord with the formula as follows:

$I_b > I_{mn} = KI_n = 1.25I_n = 1.25x3.48 = 4.35A$

"I_n" : rated current of the motor;

"I_{mn}": max operation current of the motor;

"K": overloading ability, 1.25 is recommended;

So the inverter parameters are as follows: P=0.75KW, I_b=11.5A, P_{CN} =1.9KVA, model: VFD007M21A.

3.2.2 The software design of the controller

As we know the C programming language is widely used now, it's easy to interface with system devices, as well as compact、convenient、expressive、flexible and powerful, so software system is developed with the C programming language. The customer could not only set the number and mode of the operation, but read the number counted on the liquid crystal display. The framework of the software is as follows:

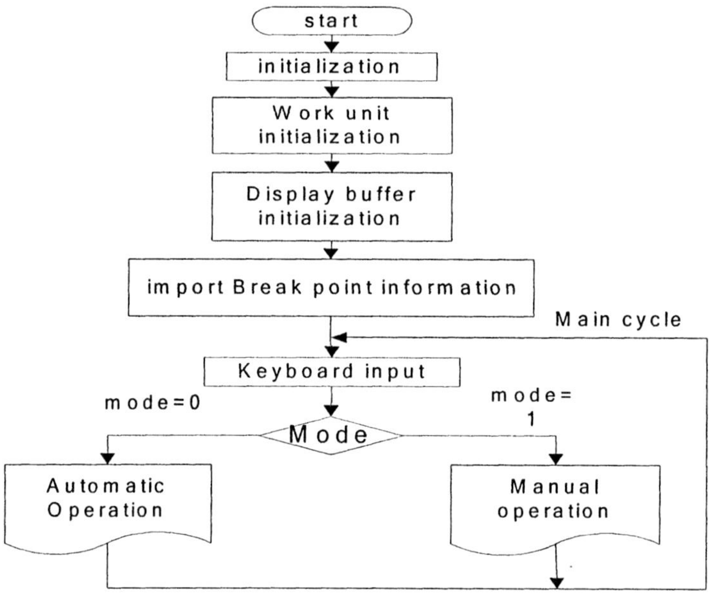

Figure 6. The framework of software design

4. CONCLUSION

Inverter technology is widely used for driving induction motors, in this paper the author has designed a new controller which allow the original crimping machine to abandon sophisticated mechanical device, reduce the mechanical impulsion, make the crimping machine noiseless、efficient and reliable. After innovative improvement, this machine must hit the market and obtain high economic returns for enterprises.

5. REFERENCES

[1]. Li Su. The direct torque control in asynchronous machine [M]. Beijing: China Machine Press, 2002.

[2] Bodson M, Chiasson J, Novotnak RT. A systematic approach to selecting optimal flux references in induction motors. *IEEE Trans Control System Tech.* 1999;3(4):388 - 97.

[3] Takahashi.I,Noguchi.T., A new quick-response and high efficiency control strategy of an induction motor. *IEEE Trans.ind.appl.* 1999, P820

AN EFFECTIVE ALGORITHM OF SHORTEST PATH PLANNING IN A STATIC ENVIRONMENT

Lingyu Sun[1], Xuemei Liu[2], Ming Leng[3]
[1]*Department of Computer Science, Jinggangshan College, Ji'an, China. Email:lmsly@263.net*
[2]*School of Mechanical & Electronic Engineering, Shandong Agricultural University, China*
[3]*School of Computer Engineering and Science, Shanghai University, Shanghai, China*

Abstract: Path Planning is generating a collision-free path in an environment with obstacles and optimizing it with respect to some criterion. Because of its importance in robotics, the problem has attracted a considerable amount of research interest. In this paper, we present an effective algorithm of shortest path planning for planar mobile robot whose time complexity is $O(4 \times n)$, n is the geometric complexity of the static planar environment. The success of our algorithm relies on exploiting both a tabu restriction and the greedy strategy of the *Dijkstra* algorithm. Our experimental evaluations on four different test cases show that our algorithm produces the shortest path very quickly.

Key words: path planning, greedy algorithm, *Dijkstra* algorithm

1. INTRODUCTION

During the last century, automation has become an extremely fast growing phenomenon impacting almost all facets of everyday life. Recently, robots have become a major part of this trend. Therefore, autonomously navigating robots have become increasing important [1]. The *path planning* problem is one of important problems in intelligent control of an autonomous mobile robot. It uses a prior information about a given environment and knowledge about robot motion capabilities to provide a path to be executed between two points in the robot workspace. The *path planning* problem is then an optimization problem where we want to find the best path under a set of constraints. Classical approaches to solve the *path*

Please use the following format when citing this chapter:
Sun, Lingyu, Liu, Xuemei, Leng, Ming, 2006, in International Federation for Information Processing (IFIP), Volume 207, Knowledge Enterprise: Intelligent Strategies In Product Design, Manufacturing, and Management, eds. K. Wang, Kovacs G., Wozny M., Fang M., (Boston: Springer), pp. 257-262.

planning problems include *Dijkstra* algorithm [2], *A* search* algorithm [3] and *dynamic programming* technique [4, 5].

The *Dijkstra* algorithm is an optimization algorithm that is mainly used for determining the shortest path. The *Dijkstra* algorithm is an uninformed search algorithm for finding shortest paths that relies purely on local path cost and provides a shortest path from a start node to a goal node in a graph. The *A* search* algorithm was developed by Hart et al. [6]. The algorithm uses a heuristic function $h(n)$ to estimate the cost of the lowest cost path from a start node to a goal node plus the path cost g(n), and therefore the search cost $f(n)=g(n)+h(n)$. Using f, g, and h values, the *A* search* algorithm will be directed towards the goal and will find it in the shortest possible route. The *dynamic programming* technique resorts to evaluating the recurrence in a bottom-up manner, saving intermediate results that are used later on to compute the desired solution. This technique applies to many combinatorial optimization problems to derive efficient algorithms and is also used to improve the time complexity of the brute-force methods [7].

In this paper, we present a Modified *Dijkstra* (MD) algorithm that is an $O(4\times n)$-time algorithm of shortest path planning for planar mobile robot in a static environment. Our work is motivated by the greedy strategy of the *Dijkstra* algorithm which consists of an iterative procedure that tries to find a local optimal solution. Furthermore, we integrate our algorithm with tabu search [8]. We test our algorithm on four test cases and our experiments show that our algorithm produces the shortest path very quickly.

The rest of the paper is organized as follows. Section 2 provides the problem formulation, describes the notation that is used throughout the paper. Section 3 presents an effective algorithm of shortest path planning. The proposed algorithm complexity is analyzed in section 4. Section 5 experimentally evaluates the algorithm.

2. PROBLEM FORMULATION

Our goal is to implement a shortest path planning system that uses the MD algorithm for planar mobile robot in a static environment. First, we use the traditional grid point representation to present a planar workspace for analysis. We consider the workspace for the planar robot that is subdivided into cells on a two-dimensional grid. A *grid node* which is located at a cell's centre-point is allocated to each cell as shown in Fig. 1(a). This method of space mapping generates a set of discrete nodes that cover the entire construction workspace domain, as shown in Fig. 1(b). Let *MaxRow* denote the total row of the two-dimensional grid. Let *MaxCol* be denoted as the total column of the two-dimensional grid. Naturally, the geometric complexity of the planar environment, denoted by n, is equal to *MaxCol*MaxRow*. Some dark areas indicate the presence of some obstacles that labeled as $O_j(x,y)$ where x and y are the coordinates of the obstacle j. The robot is considered as a punctual object and its position is given by $R(x, y)$. Next, a trajectory of the robot is composed of a set of the adjacent grid nodes which excludes the *grid node* occupied by some obstacles. The path cost evaluation function $F(p)$ is the number of the adjacent grid nodes residing between the start and the goal nodes where p is the entire path.

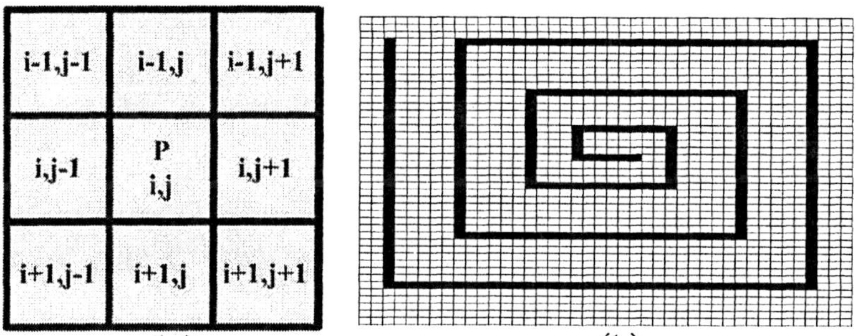

(a) (b)

Figure 1. (a) Grid point representation, (b) A sample of a discretized site layout

3. AN EFFECTIVE ALGORITHM OF SHORTEST PATH PLANNING

The *Dijkstra* algorithm can find optimal solutions to problems by systematically generating path nodes and testing them against a goal. Because the time complexity of the *Dijkstra* algorithm is $O(n^2)$, it becomes inefficient when it applies to large-scale problems. The reason of high time complexity is that it is designed to find the shortest path on graphs in which each edge has a nonnegative length.

However, as we see in Fig. 1, each edge that straddles two adjacent grid nodes has the unit length. Our MD algorithm is motivated by the greedy strategy behind the *Dijkstra* algorithm which has been proved that it does indeed find the shortest path. In the greedy strategy, the *Dijkstra* algorithm tries to find a node whose distance from the start node is the length of a shortest path in each step of expanding leaf node. Because the cost of expanding leaf node is unit length in Fig. 1, MD algorithm just adopts the breadth-first strategy to traverse the two-dimensional grid that needn't inspects every edge in each step. The time complexity of MD algorithms is small since only adjacent grid node is inspected exactly once by the algorithm. In the choice of adjacent *grid node*, the MD algorithm adopts simple tabu search strategy without aspiration criterion whose tabu restriction forbids expanding grid nodes which are designated as tabu status or traversed status. The pseudocode of the MD algorithm is given in Table. 1. The function *neighbors*(v) is used to iterate neighbors of grid node v in the two-dimensional grid.

Table 1. The pseudocode of the MD algorithm

INPUT: **1**.*MaxRow*; **2**.*MaxCol*;
3.the list of obstacles $O(x,y)$;
4.a start node s; **5**.a goal node g;

OUTPUT: **1**.the shortest path p;

```
1  set non-traversed status in the MaxRow*MaxCol matrix for all grid nodes;
2  set tabu status in the matrix for grid nodes of O(x,y);
3. initial empty queue and  insert s into queue;
4  while (the queue is non-empty) do {
5      v=pop(queue);
6      for (each u∈neighbors(v) ) do {
7          if (u's status==non-traversed status ) then
8              if (u==g) then {
9                  break;
10             } else {
11                 insert u into queue;
12                 set traversed status in the matrix for u;
13                 record v as the source of u;
14             }end if
15         }end if
16     }end for
17 }end while
18 if (u==g) then {
19     trace the source information to obtain the shortest path p
20 }end if
21 }
```

4. COMPLEXITY ANALYSIS OF MD ALGORITHM

The complexity of the algorithm is computed as follows. Step 2 costs $O(n)$ time. The while loop (Step 4-17) of MD is required to iterate n-1 times in the worst case and the for loop (Step 6-16) of MD is required to iterate four times at most because the planar mobile robot moves in 4-neighbour mode. As result, the total time complexity of the algorithm is $O(4 \times n)$. In the whole process, the MD algorithm uses a queue with the first-in first-out scheme and a two-dimensional matrix to store the status of all grid nodes whose size is n. In the worst case, the MD algorithm successively inserts $[n/2]$ grid nodes into the queue. Therefore, the total space complexity of the algorithm is $O(n*3/2)$.

5. EXPERIMENTAL RESULTS

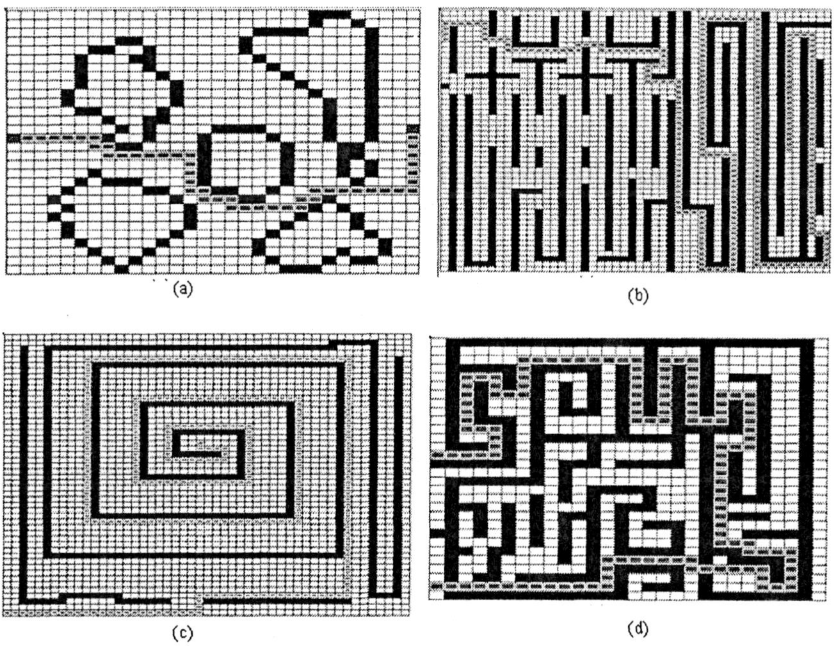

Figure 2. The result of our experimental evaluations on four different grids

The shortest path planning system is designed based on MFC programming technology. We use the four different grids in our experiments whose dimensional is 30 or 50 cells. The four evaluations are run on an 1800MHz AMD Athlon2200 with 128M memory. As expected, our experimental evaluations show that our algorithm produces the shortest path, as shown in Fig. 2, and its cost time is no more than 0.001 second.

6. CONCLUSIONS

In this paper, we have proposed an effective algorithm of shortest path planning whose time complexity is $O(4 \times n)$ and space complexity is $O(n * 3/2)$. The success of our algorithm relies on exploiting both a tabu restriction and the greedy strategy of the *Dijkstra* algorithm. Although it has the ability to find the shortest path very quickly, there are several ways in which this algorithm can be enhanced. This brings up two questions about possible improvement. The first question is how to find the shortest path with multi-constraints. Our algorithm just uses to produce the shortest path in static environment for planar mobile robot. Therefore, the second question is how to produce a new shortest path in response to environmental changes.

7. ACKNOWLEDGMENTS

This work was supported by the Foundation of Ji'an Municipal Commission of Science and Technology, grant No. 2005-28-7.

8. REFERENCES

1. Sugihara, K., Smith, J.: Genetic Algorithms for Adaptive Motion Planning of an autonomous Mobile Robot. *Proceedings of the IEEE International Symposium on Computational Intelligence in Robotics and Automation*, Monterey, CA(1997) 138-146
2. Dijkstra, E.: A note on two problems in connection with graphs. *Numerical Mathematic*, Vol. 1 (1959) 71-269
3. Russell, S., Norvig, P.: *Artificial intelligence: a modern approach.* Prentice-Hall Publications, New Jersey (1995) 230-247
4. Sedighi, K.H., Ashenayi, K.., Manikas, T.W., Wainwright, R.L., Tai, H.M.: Autonomous Local Path Planning for a Mobile Robot Using a Genetic Algorithm. *IEEE International Conference on Systems, Man and Cybernetics* (2004) 1338-1345
5. Xuemei Liu, Jin Yuan, Kesheng Wang. "A Problem-specific Genetic Algorithm for Path Planning of Mobile Robot in Greenhouse". *Proceeding of PROLAMAT 2006.* (In press)
6. Hart, P., Nilsson N., Raphael B.: Correction to 'A formal basis for the heuristic determination of minimum cost paths'. *SIGART Newslett*, Vol. 37 (1972) 9-28
7. Alsuwaiyel, M.H.: *Algorithms Design Techniques and Analysis.* World Scientific Publications, Singapore (1999) 203-205
8. Glover, F., Manuel, L.: Tabu search: *Modern heuristic Techniques for Combinatorial Problems.* Blackwell Scientific Publications, Oxford (1993) 70-150

PREDICTING ASSEMBLY QUALITY OF COMPLEX STRUCTURES USING DATA MINING

Predicting with Decision Tree Algorithm

Ekaterina S. Ponomareva, Kesheng Wang, Terje K. Lien
Department of Production and Quality Engieering, Norwegian University of Science and Technology, NTNU,, Norway. Email; ekaterina.s.ponomareva@ntnu.no.

Abstract: Our research aims at obtaining the relevant factors that cause the decrease in quality of assembly parts. Decision tree technique was employed to induce useful information hidden within a vast collection of data. The major objective of this study was to classify the existing data into certain types of segmentations and then predict the behaviour of a ball joint assembly. The intervals of the rolling time and achieved rolling force leading to occurrence of the high moment values of the ball joint during testing stage have been found.

Key words: Prediction, machine learning, decision trees, assembly.

1. INTRODUCTION

Many of automobile parts are being assembled from different components on the assembly lines. The components had to be machined with certain degree of precision before being assembled into the final product. To maintain the desired quality of the assembly product, the methods for electronic monitoring and data acquisition at the assembly stage have to be developed. The establishment of good models for the assembly processes will permit to make reliable performance predictions from observable data.

In order to make such performance predictions data collected from production process could be helpful. Automatic knowledge acquisition techniques have been developed to address this problem. Inductive learning is an automatic technique for knowledge acquisition. The inductive approach produces a structured representation of knowledge as the outcome of

Please use the following format when citing this chapter:
Ponomareva, Ekaterina, S., Wang, Kesheng, Lien, Terje, K., 2006, in International Federation for Information Processing (IFIP), Volume 207, Knowledge Enterprise: Intelligent Strategies In Product Design, Manufacturing, and Management, eds. K. Wang, Kovacs G., Wozny M., Fang M., (Boston: Springer), pp. 263-268.

learning. Induction involves generalising a set of examples to yield a selected representation that can be expressed in terms of a set of rules, concepts or logical inferences, or a decision tree.

In this paper, decision trees have been used to extract predictive intervals of roll forming operation parameters in order to reduce the possibility of appearance of high values of friction moment after testing the ball joint assembly part.

2. PROBLEM DESCRIPTION

As a case of quality control tasks in manufacturing by the use of data mining we will consider the control arm of a front wheel suspension of an automobile system. The control arm is made of aluminium. A critical part is the ball joint that is an assembly part. It is integrated in the housing on one of the ends of the control arm. The ball joint is connected to the nut part that supports the shaft of the wheel by a bearing. The ball joint consists of 8 parts, which need to be assembled together (Fig. 1).

The manufacturing process is automated, but there are still many potential sources of failures that lead to defective products. Inappropriate dimension matching, slight variations in surface roughness or variations in degree of deformation of the roll forming joining process can lead to out of tolerance friction forces in the ball joint.

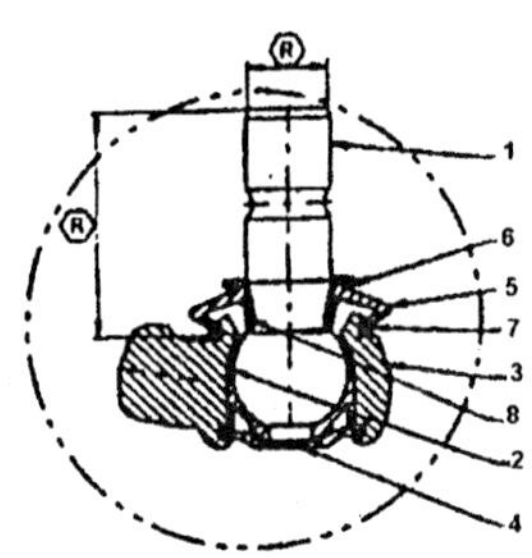

Figure 1. Ball joint assembly part: 1 – ball stud; 2 – plastic liner; 3 – housing; 4 – cap; 5 – seal; 6 – clamping ring;7 – clip ring; 8 – sleeve.

When ball stud – liner assembly is mounted in a housing **3** at the one of the ends of the control arm, the cap **4** of a special form is mounted on the top of the plastic liner **2** inside the top the housing **3.** The rim of the housing **3** is then rolled over the cap **4,** by applying force **F,** to fasten the housing. After this has been done, the test to check friction moment of the joint is carried out. In case, when friction moment has been registered higher or lower than the expected values, such parts considered being defective, the pallet is marked with special code 10, which means that these parts have a failure.

The marked pallets with those parts go through the assembling line without taking any further actions on them, until the end of the line, where these parts are discarded. The parts with satisfying results after the test are then assembled with other components: seal, clamping ring, clip ring, sleeve.

The quality control problem considered here is to reduce the possibility of appearance of high or low values of friction moment. The testing of the quality of assembled products occurs after the roll forming operation, when rim of the housing **3** is rolled over the cap **4,** by applying force **F,** to fasten the housing. Roll forming operation stage can be explained as follows: when the housing with other assembled components is ready for roll forming, rolling tool moves down until it meets the rim of the housing **3**, so the roll forming begins. Tool goes down applying force **F**, sensors register rolling time (RT) for how long tool is moving down, and achieved rolling force (RF).

If we assume that all the components of assembly part have been assembled in right order and without any flaws and defects, the only cause of appearance of high and low moment values of assembled components in testing stage can be unacceptable values of rolling time and achieved rolling force during roll forming operation.

Data-mining is one of the important techniques of information technology and known to be effective in dealing with the discovery of hidden knowledge, unexpected patterns and new rules from data (Adriaans & Zantinge, 1998). In the past few years, data-mining has also demonstrated enormous benefit in production (Milne, Drummond, & Renoux, 1998). Decision trees as one of the basic algorithms in Data Mining, can be used to build classification or regression models by recursive partitioning of data.

For data mining it is important to have good tool which combines advanced modeling technology with ease-of-use, helping to discover the interesting and valuable relationships within the data. One such tool is data mining tool *Clementine.*

3. DECISION TREES

A decision tree algorithm begins with the entire set of data, splits the data into two or more subsets according to the values of one or more attributes, and then repeatedly splits each subset into finer subsets until the split size reaches an appropriate level. The entire modeling process can be represented in a tree structure and the model generated can be summarized into a set of 'if–then' rules. Classification trees are similar to regression models in that they have a single dependent variable and multiple independent variables. Different form regression models, they can be translated directly into sets of logical if–then rules for predicting the conditional probability distribution of the dependent variable from the values of the independent variables. In addition to these, they can complement regression models by discovering additional useful patterns.

Ultimately, the advantages of decision trees include the following: (1) Non-parametric. (2) Easy to interpret. (3) Automatic interaction detection. (4) More informative outputs. Decision tree algorithms can be used to test conditional independence relations among variables in large data sets.

4. DATA ACQUISITION AND PROCEDURES

Data of the roll forming operation from the control arm assembly process have been collected. The file contains fields with Rolling Time (RT) and achieved Rolling Force (RF) for each control arm after roll forming operation with corresponding values of friction moment after the quality control test. Data have been cleaned up from noises and unwanted values. In addition the test procedure has been done on the 1782 instances.

The data contain the index number for every instance, time taken to roll over the rim of the aluminium housing **3** (Fig. 1), achieved rolling force, resulting friction moment, and a status of the quality control testing. The status is identified as follows: the acceptable limitations of the friction moment for the ball joint integrated into the aluminium housing are from 2 Nm to 8 Nm, in case, if values of friction moment are inside the limitation – the Status is "Normal"; in case when values of friction moment appear to be lower than 2 Nm – the Status is "Low"; in case when values of friction moment are higher than 8 Nm – the Status is "High".

There are hundreds of thousands data collected. Since the phase of data preparation takes a long time, it was decided to use first 1782 data points as a sample data set for the first iteration of data mining analysis. These samples contain only some records of friction moment with high moment values, and do not contain records of friction moment with low friction moment values.

4.1 Decision Trees Generation

To obtain the knowledge from the data, a decision tree is employed to determine a definition for each class.

The procedure consists of two phases for achieving our objective. Phase 1 is the tree-growing stage that is constructed by repeated splits of subsets of data into two descendant subsets. The fundamental concept relies on measuring for goodness of split (i.e. to select each split so that the data in each of the subsets are purer than those in the parent node). In this stage, the entropy function is used to measure each node's impurity (Breiman,

Friedman, Olshen, & Stone, 1984). The constructed tree in Phase 1 may be too complicated. Therefore, pruning this tree to remove branches with little statistical validity is the main work in Phase 2. In the tree pruning process, the cost-complexity pruning is used to reduce the size of the tree.

The entire procedure of generating decision tree has been done using data mining tool *Clementine 9.0*. For the construction of decision tree we used algorithm C4.5. It is an advanced version of famous algorithm ID3. In particular, C4.5 uses an improved criterion for the best attribute selection and a more sophisticated method of probability estimation. The system C4.5 includes an option that turns the tree into rules that are further tuned, which can bring further improvements. The tree is presented in Figure 2.

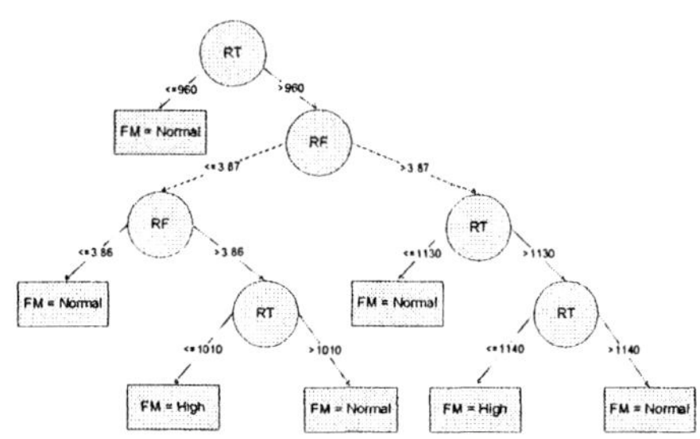

Figure 2. Decision tree for the prediction problem.

5. RESULTS

The particular interest is paid on to the values and rules associated with high moment values (FM = High). The set of rules for appearance of high moment values is shown below:

Rule 1: ***IF*** $960 < RT \leq 1010$ ***And*** $3.86 < RF \leq 3.87$ ***THEN*** ***FM = High***

Rule 2: ***IF*** $1130 < RT \leq 1140$ ***And*** $RF \leq 3.87$ ***THEN*** ***FM = High***

Graphically these rules can be presented as displayed in Figure 3.

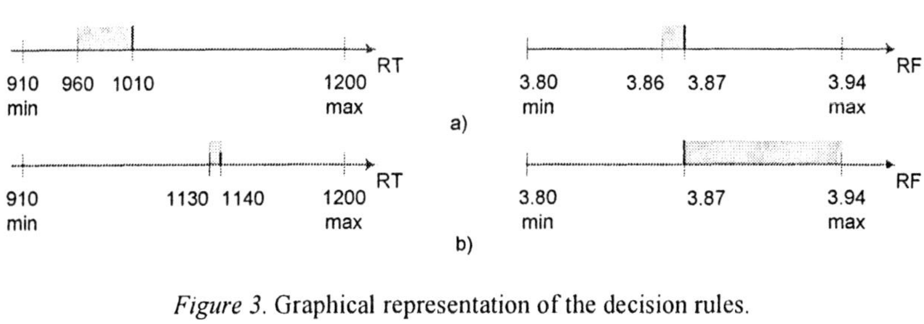

Figure 3. Graphical representation of the decision rules.

This is the most important result that has been extracted from decision tree. Rolling time and achieved rolling force are critical parameters, which cause the high values of friction moment of the ball joint. The main result of the decision tree induction is the identification of critical intervals for the rolling time and achieved rolling force causing the presence of high moment values.

Now it is possible to make a suggestion to avoid operation under following conditions: 1) if rolling time is in the interval (960 - 1010] and achieved rolling force is in the interval (3.86 - 3.87]; 2) if rolling time is in the interval (1130 - 1140] and achieved rolling force is in the interval [3.87 – 3.94).

6. CONCLUSIONS

In this paper, our research shows that we can obtain the relevant factors that cause the decrease in quality of assembly parts using Data Mining. Decision trees were employed to induce useful information and knowledge hidden in a vast amount of data. We have classified the existing data into certain types of segmentations and then predict behaviour of the ball joint assembly part. From practical views, results of this study show that it is desirable to avoid operation under certain conditions in order to improve the quality of the product. The most important factor is extraction of intervals of the critical parameters.

The intervals of the rolling time and achieved rolling force leading to occurrence of the high moment values of the ball joint during the testing stage have been found, further studies to find critical intervals of the same parameters for the occurrence of the low moment values have to be done.

7. REFERENCES

Adriaans, P., & Zantinge, D., (1998), *Data Mining*, Harlow: Addison-Wesley.

Breiman, L., Friedman, J., Olshen, R., & Stone, C., (1984), *Classification and regression trees*, Belmont, CA: Wadsworth International Group.

Chen, M.-S., Han, J., & Yu, P. S., (1996), Data mining: an overview from a database perspective. *IEEE Transactions on Knowledge and Data Engineering*, 8(6), 866–883.

Fayyad, U., & Stolorz, P., (1997), Data mining and KDD: Promise and challenge. *Future Generation Computer Systems*, 13(2), 99–115.

Li, X. B., Sweigart, J., Teng, J., Donohue, J., & Thombs, L., (2001), A dynamic programming based pruning method for decision trees. *Journal on Computing*, 13(4), 332–344.

Milne, R., Drummond, M., & Renoux, P., (1998), Predicting paper making defect on-line using data mining, *Knowledge-Based Systems*, 11, 331– 338.

THE DEVELOPMENT OF THE VIRTUAL DATA ACQUISITION AND ANALYSIS SYSTEM

Xiumin Yang , Weihua Tian, Ling Jiang, Yongshu Guo, Bo Wan
Dept. of Automatic Control, Shenyang Institute of Engineering, Shenyang, 110136,China;
Email: yxiumin@tom.com.

Abstract: A virtual multi-channel data acquisition system is developed based on computer technology and the virtual instrument development platform of LabWindows/CVI of NI Company of America. The PCI-6024E data acquisition card of NI Company of America is used as gateway device, and the Data Acquisition Library and data I/O function library (I/O for DAQ) provided by Lab Windows/CVI of NI company are used to abbreviate programs to realize data acquisition of the real physical signals and do frequency analysis. This paper introduces the technology of hardware configuration, data acquisition, waveforms display and so forth of the virtual data acquisition system. And an analysis is done about the time domain analysis module. Through testing proof, this virtual instrument system can realize the function of traditional data collection system completely, has more strong function even.

Key words: Data Acquisition, Waveforms Display, Virtual Instrument, Spectrum Analysis.

1. INTRODUCTION

The hardware platform of the virtual instrument is composed of two parts, the computer and the interface equipment of I/O. The interface equipment of I/O, mainly realize data acquisition, magnifying and A/D transformation of input signals. For a single-device virtual instrument, The interface equipment of I/O of hardware disposition is a data acquisition card. And for an measuring system composed of multiple virtual instruments, the interface equipment of I/O of hardware platform are buses including GPIB/GPIB488.2, RS232, VISA and VXI, etc.

This paper mainly proposes a design of single-device virtual data acquisition system, and a data acquisition card is adopted as interface

This project is supported by the Academic Science Research Fund of Liaoning Education Office of China No. 202263360.

Please use the following format when citing this chapter:
Yang, Xiumin, Tian, Weihua, Jiang, Ling, Guo, Yongshu, Wan, Bo, 2006, in International Federation for Information Processing (IFIP), Volume 207, Knowledge Enterprise: Intelligent Strategies In Product Design, Manufacturing, and Management, eds. K. Wang, Kovacs G., Wozny M., Fang M., (Boston: Springer), pp. 269-274.

equipment of I/O. The card is composed of 4 parts: multichannel switch, amplifier, sampling/holding element and A/D transformer. Besides these, some other circuits as well as timing counter and bus interface circuit are also involved to complete data acquisition, amplification and A/D transformation[1].

Based on the proper choice of data acquisition card, a virtual data acquisition and analysis system is developed by using the Data Acquisition Library and I/O function library (I/O for DAQ) of virtual instrument developing platform of LabWindows/CVI of NI Company of America. The system can realize acquisition of real physical signals, and display, analysis, storage and printing of the data curves.

2. DEVELOPMENT PLATFORM OF DATA ACQUISITION SYSTEM

2.1 Choice of data acquisition card

The proposed data acquisition system is mainly designed to realize analog signals acquisition of the output of secondary appearance. Considering the requirements of usual data acquisition system to sampling frequency, speed, resolution capacity and number of channels as well as comparison of performance and cost, we finally choose PCI-6024E data acquisition card of NI company. PCI-6024E has 68 I/O connecting foots and can provide high performance and reliable function of data acquisition. The specific performance indexes and parameters are as follows[2, 3]:

Analog inputs: 16 single inputs or 8 differential inputs
Input precision (bit): 12
Input range(V): ±10, ± 5
Input gain: 0.5,1,10,100
Sampling rate (S/s): 200K
Analog output: 2
Analog output precision (bit): 12
Digital I/O: 8
Counting timer: 2

2.2 Software development of the data acquisition system

There is two kinds of sampling technologies can be adopted: real time sampling and equivalent time sampling. Real time sampling is that the acquisitions of all sampling points of a signal are done during one acquisition process, and the sampled signal may be instant or periodic. Real time sampling is the most usual sampling mode of a data acquisition system. Under this mode, A/D transformer can once complete sampling of the whole input wave. This property is very fit for acquisition of single signal.

When doing data acquisition, DAQ (Data Acquisition) should be set first. LabWindows/CVI provides driving program libraries for various outside equipment [3], by which the distribution and initiation of each kind of device can be finished conveniently. When setting DAQ, naming numbers of devices and input channels, choosing types of input signals and limiting data values need to be done. Then, Data Acquisition Library functions should be called to realize triggering control, channels choosing control and time bases control.

Waveform display module is mainly to display the waveforms of sampled data. This data acquisition system can accomplish acquisition and display of 8 paths of signals and do frequency analysis at one time. Two channels A and B should be chosen at random, and the waveforms of the results of A+B or A-B as well as the corresponding frequency analysis can be displayed.

According to the difference of display modes, the programs are distinguished in design. In the part of waveform display, display modes can be divided into 2 sorts according to the realizing principle. The first sort is double channel, under which the 8 channels can be displayed at one time. The second sort is single channel, under which only one of A+B and A-B can be displayed. Because the graph controls can′ t support two formats of data, for the two different types of graphs, each uses a graph control to display. Some treatment is needed in users preface. The "Visible" property is adopted to control whether it is visible and ensure that every time there is only one graph displayed and the other is invisible.

Command controls include acquisition control button, pause button and waveform display button, etc. These controls are all located in Controls/Command. The acquisition control button controls data acquisition of the system. When it is pressed the acquisition module is called. And when the pause button is in use, the acquisition module is stopped working. After channel A and B are chosen and the button of A+B or A-B is pressed, then the waveform of A+B or A-B can be displayed. Besides, according to the catching function of the mouse, each acquisition point can be caught and its concrete value can be displayed by display control. It is two channels A and B addition waveform as well as its waveform of frequency spectrum to show as Fig.1.

The panel of the system is showed in Fig.2. Channel 0 is a sine wave as input of the data acquisition system produced by standard signal generator; channel 1 is a square wave as input of the signal produced by the standard signal generator. The others are waveforms built by the return values of the arrays to produce waveforms of signal generator.

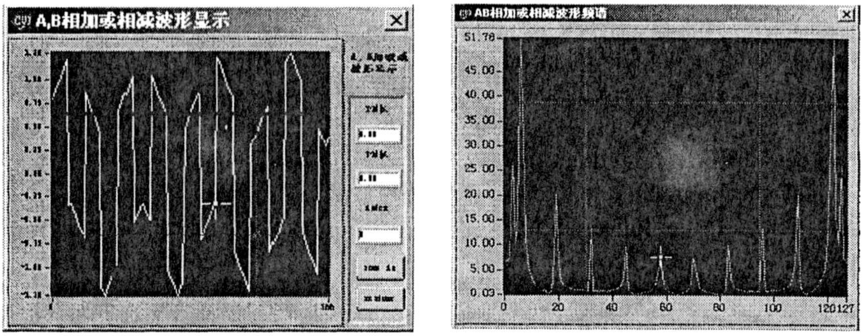

Figure 1. The superposition of square and sine waves and the frequency spectrum

The control parameters should be adjusted at suitable setpoints before the seed cleaner starts.

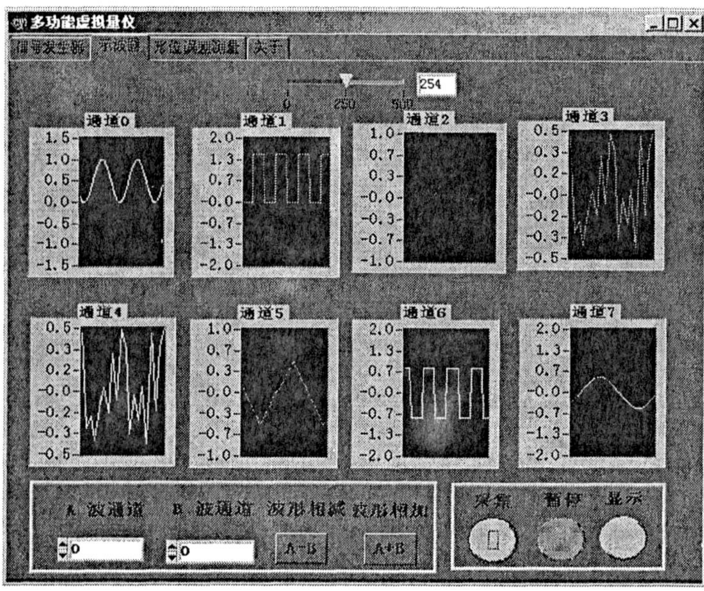

Figure 2. The panel of waveform displays

3. SPECTRAL ANALYSIS MODULE

There are two kinds of methods to observe signals. One is using oscillograph to do analysis in time domain. Here the oscillograph works at V-t mode, displaying the waveform of the measured signal itself. It is necessary to denote the relationship of time and phase of the signal in concept of time domain, but it can′ t show the whole characters of the circuit. The second method is using frequency analysis instrument, which can separate each frequency level when testing power levels. Here the amplitude is displayed in frequency domain, and each frequency component of the signal can be directly observed. The power and voltage as functions of frequency can be displayed in curves.

If the sampling value of the measured signal $f(t)$ is known as f_k, then the frequency spectrum of $f(t)$ can be worked out by the computer through Fast Fourier Transformation. Here, low-pass filters, sampling circuits, A/D transformers and memories build up the data collecting system, which can transform the measured signals into digitals. With these data, the frequency spectrum of the measured signal is worked out by FFT calculator through Fast Fourier Transformation, and is showed on display.

Different kinds of standard values in frequency domain and the frequency component information can be acquired from doing frequency-domain analysis to signals. The most important analysis methods are Fast Fourier Transformation and its inverse transformation. The frequency-domain analysis methods in LabWindows/CVI mainly involve Fast Fourier Transformation function FFT( ), real number serial FFT function ReFFT( ), the corresponding inverse transformation functions InvFFT( ) and InvReFFT( ), totally 8 ones. Here, FFT( ) is adopted to realize the performance of frequency-domain analysis. It employs the split-base algorithm, which is the most effective currently and needs the least computing amount.

The result got from FFT( ) is a Plural number. Let x and y denote the real part and imaginary part of the spectrum, then the expressions of amplitude spectrum and phase spectrum are showed as (1) and (2).

$$amp = \sqrt{x^2 + y^2} \qquad (1)$$

$$phase = \arctan(\frac{y}{x}) \qquad (2)$$

Seen from the expressions, the amplitude spectrum and phase spectrum are conformed after the Fourier frequency spectrum is worked out through FFT transformation.

The structure of a virtual frequency analysis system is similar to that of a virtual oscillograph system, but the data outlet of a frequency analysis system is mainly for spectrum analysis. The functions of a frequency analysis instrument must involve the data acquisition card to complete A/D sampling data input, the micro computer to do the consequent treatment and the proper display to realize functions of an oscillograph. Insertion or extraction to the input data at a certain multiple is needed to realize changing of the scan speed of consequent stage waveform. If it is required to change display amplitude, it is only needed to multiply the input data by corresponding coefficients.

To realize the functions of a frequency analysis instrument, the input data should be transformed from time domain to frequency domain first. The transforming method may be calling the corresponding digital signals processing modules such as the power-spectrum building module, and then the frequency analysis of the signal is done through programming. When calling the existing modules in Lab Windows/CVI, pay attention to the settings of some parameters.

The transformed data are already values in frequency domain. Take frequency as the axis of abscissas and make the data displayed, then the frequency spectrum graph is got. If it is together with local data insertion, extraction or being multiplied by corresponding coefficients, part of the spectrum graph can be magnified and deflated, and the information concerned in a certain frequency range can be analyzed in detail.

4. CONCLUSION

The proposed virtual data acquisition and analysis system can replace traditional measuring instruments and signal analysis and processing equipments. This system sufficiently embodies the flexibility of virtual instruments, and provides an instrument that has stronger functions and is simpler to operate. It is proved that the virtual system can completely realize the functions of a traditional data acquisition system, and has more strong function even.

Thanks for Jin Yinglian, graduate student of Northeast University, for her jobs in the development and debugging of the project.

5. REFERENCES

1. Yang Xiumin, Zhang Lei, Zhang Zhen, (2002), Calibration and analysis of the error for the forward channel of the measurement and control system , *Microprocessors* , vol. 93, pp. 53–55.
2. National Instruments Corporation, (1998), *Labwindows/CVI User Manual.*
3. National Instruments Corporation, (1998), *Labwindows/CVI Standard Libraries Reference Manual.*
4. Yang Xiumin, Qin Hon, Zhang Xiguo, (2001), The design of the soft-panel for virtual instrument, *Microprocessors*, vol. 92, pp. 24–26.
5. Zhang Lei, Zhang Yu, Sun Baoshou, (2004), Development of the virtual roundness error measuring instrument, *Machinery & Electronics*, vol. 134, pp. 60–62.

INTEGRATION OF DATA MINING WITH GAME THEORY

Yi Wang
Manufacturing Engineering center (MEC), Cardiff University, 24CF 3AA Cardiff, UK;
Email: sceyw2@cf.ac.uk

Abstract: Game Theory studies strategic situations where agents select different actions to maximize their returns. Game Theory has recently drawn attention from computer scientists because of its use in artificial intelligence and cybernetics. This paper presents a frame work of integrating Data Mining with Game Theory. Due to the reason of huge amount of data, it is hard for Game Theory alone to perform the modeling analysis. Data mining assists Game Theory to deal with the large amount of data and finds hidden rules to improve game analysis.

Key words: Classification; Clustering; Association; Data Mining; Game Theory; Strategy Selection.

1. INTRODUCTION

Classical Game Theory (GT) was pioneered by von Neumann, who formalized the concepts Game Theory (Neumann and Morgenstern, 1947). It was further improved by Nash, Selten, etc (Rasmussen, 1994), So far Artificial Intelligent (AI) have been incorporated in combinatorial Game Theory (Berlekamp, Conway and Guy, 2001), but it was mostly focus on computer game and board games. Combinatorial Game Theory tries to analyze the outcomes of "perfect-information" games such as poker game and chess.

In this paper, a new approach that integrates Data Mining (DM) with Game Theory attempts to discover and extract new knowledge from the

Please use the following format when citing this chapter:
Wang, Yi, 2006, in International Federation for Information Processing (IFIP), Volume 207, Knowledge Enterprise: Intelligent Strategies In Product Design, Manufacturing, and Management, eds. K. Wang, Kovacs G., Wozny M., Fang M., (Boston: Springer), pp. 275-280.

recorded data and information. These data are normally stored in databases, and can be all sorts of nature such as name and ages of the agent or payoff of return. The learned knowledge is represented in forms of rules, such as classification rules, prediction rules, association rules or clusters of rules. These results can be often used in a support for decision making for operational improvement.

In section 2 we briefly introduce the general framework combining Data Mining with Game Theory. Section 3 describes in detail how the combination could be accomplished and what has been previously done in the different area of Data Mining. The conclusion is drawn in section 4.

2. DATA MINING AND GAME THEORY

Data Mining is a combination of analysis, search, and modeling technologies. Data Mining is often defined as the process of extracting valid, previous unknown, comprehensible information from large data bases in order to improve and optimize business decision (Fayyad, et al, 1996). Data mining techniques involves different disciplines, such as databases, enterprise information systems, statistics, Machine Learning (ML), and Artificial Intelligence (AI).

Data Mining identifies new phenomena and uncovering patterns (J. Han and M. Kamber 2000), which are useful in enhancing our understanding of target systems represented by the datasets. In this paper, the functions of Data Mining are categorized as classification, clustering and association rules.

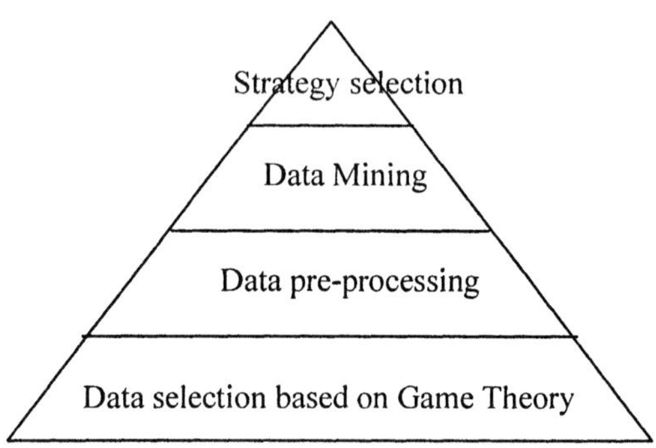

Figure 1. The process of integration of Data Mining with Game Theory.

Fig. 1 illustrates the knowledge discovery process of the integration of Data Mining with Game Theory. The process consists of 4 activities: (1). data selection based on Game Theory, (2). data pre-processing, (3). data mining and 4. Strategy selection. Knowledge discovery process can be iterative and each of these activities may be revisited multiple times in order to improve the knowledge discovery process. A good understanding of the whole process is important for any successful application. No matter how powerful the DM algorithm is, the resulting model will not be valid if the data based on GT are not selected and pre-processed correctly.

Game Theory (Rasmussen, 1994) is concerned with the formal analysis of situations called "*games*" where agents can choose different strategies that determine their actions under particular conditions. Conditions and outcomes unfold through the interactions of the agents' strategies.

There are many rules and patterns extracted by Data Mining which mirror the interaction of conflict between the agents, such as strategy of struggle in an environment of limited resource. Modelling these problems using the concepts of Game Theory improves the understanding of these rules. Classical Game Theory can offer strong mathematical proof techniques to the problem-solving approaches with high computational complexity. Adopting such strong proof techniques from Game Theory will lead to further advances in Data Mining.

In Game Theory due to the uncertainty with the opponent's action, to be on the safe side, thus leads to in a strategy selection matrix of a prisoner's dilemma, an agent will choose to confess. (Rasmussen, 1994) If both agents are rational, the solution to the *Prisoner's Dilemma* is that both should confess. In real life strategy selection is based on a agent's knowledge of the opponent. If there is perfect knowledge, then the agent has control of the *Game*. (Pham, Wang, and Dimov, 2004) This knowledge is gathered from past records by applying Data Mining methods.

For an agent following steps can be followed:

1. Observation. Data is observed and gathered. Only the relevant data will be taken into account.
2. Analysis. The relevant data is put into the Data Mining algorithm to get interesting rules, which is used for the game play.
3. Strategy selection. The rules provide a better understanding of the game environment, and the agent could make a better selection of the strategies available.
4. Testing. When the selected strategies are put into the real world, it will cause expected and unexpected feedback.
5. Return to step 1 and update the observation and continue the next steps.

3. MINING DESCRIPTION

3.1 Classification

Classification is the process of sub-dividing a dataset with regard to a number of specific outcomes. For example, we might want to classify our agents into 'winner' or 'loser' categories with regard to their payoff. The category or 'class' into which each agent is placed is the 'outcome' of classification process. A classification model is said to be 'trained' on historical data, for which the outcome is known for each record. It is then applied to a new, unclassified dataset in order to predict the outcome for each record.

Major researches done here is based on game mining, (Tveit and Tveit, 2002) Data Mining applied multi-agent games networking (Smed, et al., 2001) and game usage (Srivastava, et al., 2000). The main motivation for performing classification in computer games is to improve the game, so that the agents become more satisfied and stay longer, and to increase the revenue of the game service again. Game mining is defined as three main types: (1). game content mining; (2). game structure mining; and (3). game usage mining. It was inspired by well-known types of web mining, which are corresponding to: (1). web content mining; (2). web structure mining; and (3). web usage mining. (Cooley, et al. 1997).

The game mining approach (Tveit and Tveit, 2002) has been extended further here to find a relationship between different factors, such as previous selected strategy or the nature of agents. The nature includes age, income, etc. Thus datasets are divided into classes. It is important to find the proper training and target classes because real world classification problems usually involve many different types of classes.

3.2 Clustering

Clustering is usually achieved using statistical methods, e.g. a k-means algorithm. Each record is compared with a set of existing clusters. A record is assigned to the cluster it is nearest to, and this in turn changes the value that defines that cluster. Cluster analysis is the process of identifying the relationships on the basis of their targets similarity and dissimilarity. Unlike classification, clustering does not need pre-defined target variables.

Clusters are focused around a "centre" or "centres" which is initially defined and some measure must be used to estimates the similarity and their dissimilarity. The similarity and dissimilarity can be measured as the distance from each other and from the cluster centres.

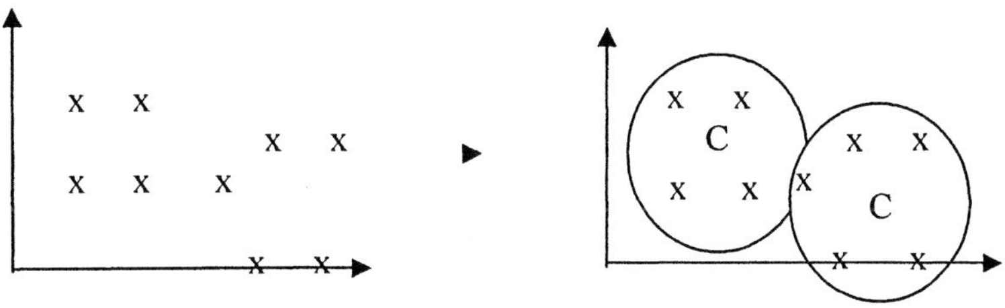

Figure 2. Clustering of data.

Huck, et al., (2000) mentioned that three clusters were identified with two equilibrium locations and one focal point. The observations are related to best-response in game dynamics, and to the fact that the agents rely on best-responses in particular when they are close to the equilibrium configuration.

Pavan and Pelillo (2003) used clustering for evolutionary game dynamics. The new framework for pair wise hierarchical clustering centred on the notion of a dominant set, a newly introduced graph-theoretic concept that has proven to be relevant in partitioned clustering and image segmentation problems. The approach is general and can be applied to solve a variety of problems in computer vision and pattern recognition. Its potential for the problem of image database organization has been demonstrated.

3.3 Association

Association Mining identifies links between attributes in a dataset. The task of association rule mining is to find certain association relationships among a set of items in a database.

A new way of extracting causal rules and making optimal selections has been proposed (Pham, et al. 2005). The method hinges on regarding a manufacturing strategy selection problem as a "*Game*" involving several agents. From a certain time point in the present to a certain time point in the future, any actions of any agents in the "*Game*" at any moment may influence on the actions of others. Each agent has a pattern of strategy selection, which is observable by other agents, and to which other agents have a counter pattern. A change in strategy pattern by an agent might cause other agents to change their selection pattern.

4. CONCLUSION

Game Theory studies strategic situations where agents select different actions to maximize their returns. Data Mining is the process of extracting valid, previous unknown, comprehensible knowledge from large data bases in order to improve and optimize business decision. In this paper, we have established a framework to integrate Data Mining with Game Theory in order to benefit from the two techniques. Some of the researches in this area are reviewed. We also classified Data Mining into three sub categories and showed how these two subjects can be successfully integrated together in detail.

5. REFERENCES

1. Berlekamp, E. R., Conway, J. H., and Guy R. K., (2001), *Winning Ways for Your Mathematical Plays*, A. K. Peters Ltd.
2. Cooley, R., Mobasher, B., and Srivastava, J., (1997), Web mining: Information and pattern discovery on the world wide web. *Proceedings of the 9th IEEE International Conference on Tools with Artificial Intelligence (ICTAI'97).* IEEE, November.
3. Pham, D.T., Wang, Y., and Dimov, S., (2005), Intelligent manufacturng strategy selection, *International Conference on Intelligent Production Machines and Systems 2005*, Cardiff. pp 363-369 july
4. Fayyad, U.,Piatetsky-Shapiro, G., Smyth, P., (1996), From data mining to knowledge discovery: An overview, in: *Advantaces Knowledge Discovery and Data Mining,* Fayyad, U. Piatetsky-Shapiro, G., Smyth, P. and Uthurusamy, Eds., Cambridge, MA: MIT Press, pp. 1-36.
5. Han, J. and Kamber, M., (2000), *Data Mining: Concepts and Techniques*, Morgan Kaufmann Publishers.
6. Huck, S., Müller, W. and Vriend, N., (2000), *The East End, the West End, and Kings Cross: On Clustering in the Four-Agent Hotelling Game,* Humboldt Universitaet, Berlin.
7. Neumann, V. J. , and Morgensten, O., (1947), *The Theory of Games and Economic Behavior*, Princeton, NJ: Princeton University Press.
8. Pavan, M. and Pelillo, M., (2003), Dominant sets and hierarchical clustering, *ICCV03*, pp. 362-369, WWW Version.
9. Rasmussen, E., (1989), *Games and Information: An Introduction to Game Theory*, Oxford: Basil Blackwell.
10. Smed, J., Kaukoranta, T. and Hakonen, H., (2001), Aspects of networking in multiagent computer games. *Proceedings of International Conference on Application and Development of Computer Games in the 21st Century*, pp. 74-81, November.
11. Srivastava, J., Cooley, R., Deshpande, M. and Tan, P. N., (2000), Web usage mining: Discovery and applications of usage patterns from web data. *SIGKDD Explorations*, 1(1):12-23, January.
12. Tveit, A. and Tveit, G. B., (2001) Game Usage Mining: Information Gathering for Knowledge Discovery Massive Multiagent Games. *Proceedings of International Conference on Application and Development of Computer Games in the 21st Century.*

DATA MINING AND CRITICAL SUCCESS FACTORS IN DATA MINING PROJECTS

Yun Chen [1], Dengfeng Hu [1, 2], Guozheng Zhang [1]
[1]Shanghai University of Finance & Economics, Shanghai, China. Email: hdengfeng0212@sohu.com. [2]Anhui University of Finance & Economics, Bengbu, China, 230135

Abstract: Utilization of historical data to support evidence-based decision making, is leading many organizations to recognize that the effective use of data is the key element in the next generation of client-server enterprise information technology. To deploy data mining projects successfully, organizations need to know the key factors for successful data mining. This article concentrates on studying the data mining and the key factors for successful data mining, pointing out the progress of the data mining and crucial success factors (CSFs) currently exist in China.

Key words: CSFs, progress of the data mining, organization data mining.

1. INTRODUCTION

The major reason that data mining has attracted a great deal of attention in the formation industry in recent years is due to the wide availability of huge amounts of data and the imminent need for turning such data into useful information and knowledge, and can be used for applications ranging from business management, production control, and market analysis, to engineering design and science exploration. Some applications include marketing, financial investment, fraud detection, manufacturing and production, and network management. Presently, the majority of ODM

This project is supported by the Shanghai Shuguang Project of China under the grant No.05SG38.

Please use the following format when citing this chapter:
Chen, Yun, Hu, Dengfeng, Zhang, Guozheng, 2006, in International Federation For Information Processing (IFIP), Volume 207, Knowledge Enterprise: Intelligent Strategies In Product Design, Manufacturing, and Management, eds. K. Wang, Kovacs G., Wozny M., Fang M., (Boston: Springer), pp. 281-287.

projects focus on customers and profitability, customer retention and attrition management, risk forecasting and profitability analysis.

What is attracting many analysts to data mining is the relative ease with which new insights can be gained in comparison to traditional statistical approaches. Data mining is always a hypothesis-free approach, whereas most popular statistical techniques require the development of a hypothesis in advance. Data mining algorithms typically produce a much wider set of data types and make fewer assumptions about their distribution – or no assumptions at all.

Data mining outcomes include classification, clustering, prediction, estimation, and affinity grouping. Berry and Linoff [1] used the term techniques for market based analysis, memory-based reasoning, cluster detection, link analysis, decision trees and rule induction, artificial neural networks, genetic algorithms, and on-line analytic processing (OLAP).

This paper is conducted via Chinese periodical and organization survey to find the progress in China, and this text into five sections. The first section introduces the concept, content and the progress of data mining. The second section discusses the research of data mining in China, focusing on published papers. The third section overviews previous literatures, pointing out the CSFs in organizations. The forth section concentrates in the application in organization. In this text, we use a questionnaire indirectly, for it is difficult in collecting data in organizations. Finally, the concluding section highlights the important of data mining to organizations.

2. RESEARCH OF DATA MINING IN CHINA

We can pass and search Chinese relevant periodicals, to understand China excavate the research situation of the respect in the data mining.

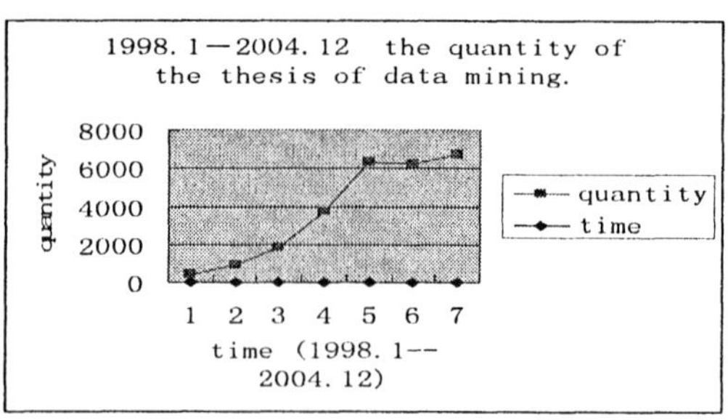

Figure 1. the quantity of relevant theses of research published.

Source of the materials: Chinese periodical network Chinese periodical full text date base

Now, we searched it from 1998 to 2004, figure 1 showing the quantity of relevant theses of research in the respect of the data mining.

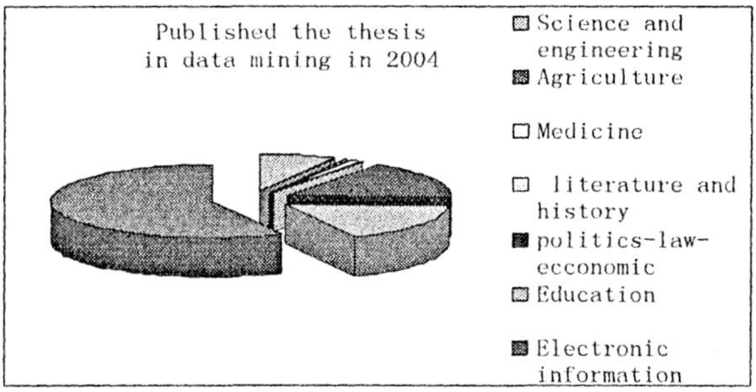

Figure 2. the quantity of relevant theses of research in each field Source of the materials: Chinese periodical network Chinese periodical full text data base

Based on "key words" and "abstract", We have carried on sort research .In 1998, 430 papers on data mining published, and from 1998 to 2004 the thesis published has been increasing progressively, 6448 published in 2004.In China, the research in this respect has been developed in each field, especially in the field of electronic information, secondly in political economy. About the theoretical research of this, have not represented and carry on relevant actual application in China.

We continue carrying on search to the articles published in journals in management from 1999 to 2003. Based on "key words" and "abstract", we have carried on sort research on data mining (table 1).

Table 1. The article published in data mining from 1998 to 2004

Journal name	quality
Chinese management science	35
Management engineering journal	39
System engineering theory and practice	33
Theory application of the system engineering	32
Theory application of the system engineering	29
Total	168

The research approach is a problem with most important research field, the mistake of the research approach will result in studying the mistake of the question directly. Based on Enrique Claver's achievement, the research approach is divided into two kinds: theory study and empirical study.

The theoretical study is mainly to the concept, structure and describing, but not to the realistic systemic observation, namely focal point studied on concept research, but not datum nor observe. The theoretical study can be divided into: the concept type studying, the proving type studying and the concept applied type studying .The concept type is mainly some definitions of frames, models or theories and provides relevant explanation. The proving type studying offers relevant action or in what kind of situation finishes a certain behavior mainly to study, and mainly as guidance of practice, what is focal point, how to do, but not why; the concept applied type studying is the combination that is above-mentioned two kinds of methods. The core of empirical study draws the meaningful conclusion of studying to observation of reality, or put forward new assumption or proposition to the new phenomenon, either examines or revises the existing theory. According to study course it may divide into positive research: case study, field study, field experiments study and laboratory experiments study.

We classify the study approach judging by 168 articles, get results shown table 2.

*Table 2.*the quantity of article of the research approach

Study approach	Theory study		Empirical study				
	Concept type	Proving type studying	Concept applied type studying	Case study	Field study	Field experim -ents study	Laboratory experiments study
Article quantity	102	48	16	2	0	0	0

From the above table2, the domestic study approach overweight on the research of the theory mainly, but empirical study little. The concept type study is generally introduction of new concept, theory and discussion, accounted for 60% of the total sample; At the same time some study overweight on the application in different fields of some theories, method and technology, accounted for 28%, which propose enterprise should do in data mining, and what is a key factor causing success of enterprise, and why should be done in this way, few people to study.

3. CRITICAL SUCCESS FACTORS (CSFS) IN DATA MINING PROJECTS IN ORGANIZATIONS

To deploy data mining projects successfully, organizations need to know the key factors for successful data mining [2]. The key factors can be critical success factors (CSFs), which is "the limited number of areas in which results, and if they are satisfactory, will ensure successful competitive performance for the organization [3]". " If these factors are crucial for the success of the system, then their failure would lead to the failure of the information system [4]". Pinto and Slevin [5] wrote of the diversity of reported project successes in the information technology area.

Now we review which data mining projects are planned and implemented. Chung and Gray (1999) suggest utilizing 9 steps in data mining. Fayyad, Piatetsky -Shapiro, and Smyth [6,7] suggested 7 steps for successful data mining projects. Han and Kamber [8] arranged data mining as 7 steps

Based on this previews successful research, the architecture of a typical data mining process involves of three major elements: 1) Data mining: data mining engine, techniques, applications, and interpretation and using discovered knowledge.2) Data management: database, knowledge base, data warehouse, or other information repository, data cleaning and data integration steps.3) Information system (IS) project: dealing with data mining projects based on IS environment.

While no strong consensus exists in the project management literature on how to define success, a number of models and techniques have been developed to aid in the definition and measurement process. There is also a link between choosing appropriate CSFs to engage management's attention, and their support of the project activity. Slevin and Pinto [9] developed supporting criteria for ten factors of project success; budget of these is more important in project.

They perceive this critical path layout to be important and relevant to successful implementation. A clear mission, top management support, a defined schedule or plan, client consultation, right personnel, appropriate technology, and client acceptance are defined as CSFs. Communication, monitoring, and feedback are simultaneous process steps, with troubleshooting occurring as needed throughout the implementation.

Outsourcing must confirm a creditable organization to implement. Because systematic maintenance time is longer, so how take precautions against agreement fulfilled, it determines the success of the project [18].

A number of principles emerged from the case study that may be relevant to applications of the CSF process in other contexts: (1) Top management support is essential; (2) Substantial group process is required; (3) Data collection is a significant problem; (4) Managers need to be continuously encouraged to consider the relevance of CSF for them; (5) The use of the factors must be pushed down through the organization; (6) Survey development will most likely be needed [10]; (7) Develop realistic project schedules; (8) Match skills to needs at the proper time; (9) Establish and control the performance of contractors.

4. QUESTIONS IN ORGANIZATIONS IN CHINA

We once did a questionnaire to 120 enterprises (annual sales revenue more than 100million), designing following 8 questions of various fields according to the principle of the above. In the end, we gained 86 shares of effective questionnaire.

Enterprises have already built information system to a certain degree, accounted for 91%, such as handling official business, online information obtained. Among these enterprises, 63% plans to set up the data warehouse, and only 20% has already been set up, but others are being looked around. In enterprises having already set up, only 40% of the enterprises think that there is some benefit brought to enterprises, 30% of the enterprises believe it is very useful, other 20% of the enterprises think it as useless.

Question and answer about the function of enterprise's historical data, 85% of the enterprises think they can find out the reason of the achievement in the past according to the report forms; 10% of the enterprises think that can find the question that existed in the past. 5% enterprise think they can find future chance from it, as for how to find the chance, how to use tools to analyses, but the answers are not very clear. So, we can draw the conclusion, a lot of enterprises do not know the function of the historical data at all in China, and do not know how to find these real values of historical data.

Who must advocate building database or warehouse? 80% enterprise think it is required by leaders, and 15% was promoted by other enterprises, only 5% by markets. This conclusion is similar with "Computer Sciences Corporation" analysis. But referring to data mining, 85% enterprise do not know the concept and 95% not how to use for their business, though the story of the relative between peer and nappy has been for a long time.

In our another questionnaire to graduates from science and engineering and IT, we find 60% knowing little about it, and 20% knowing principle, but no man use it in practice.

From the up questionnaires, it is easy for us to think the reason that the application level of IT is lower, not soul of touching enterprises in China.

5. CONCLUSION

Except crucial successful factor mentioned above, promoting factors of application of data mining has several factors in our country. Firstly, the domestic study on the data mining is at the introduction stage in each field at present, there is still relatively few practical application; secondly, the scarce of talents in the data mining has influenced the application of the data mining in enterprises directly, Especially the backwardness of our course offered has influenced the application with the data mining in enterprises directly; thirdly, how to build a data warehouse is key factor to the data mining in enterprise, but few of enterprises are engaged in the accumulation in this respect consciously. This will require organizations to further educate their employees if they want to narrow this knowledge gap, and to innovate not only by ways and means, but also through mechanism.

6. REFERENCES

1. Berry, M. and Linoff, G., *Data Mining Techniques: for marketing, sales, and customer support.* John Wiley & Sons, Inc., New York, NY, 1997
2. Jaesung Sim, B.P.A., M.P.A., M.S. Critical Success Factors in Data Mining Projects , Dissertation Prepared for the Degree of Doctor of Philosophy.
3. Rockart, J.F., Chief executives define their own needs, *Harvard Business Review*,57(2), March-April 1979, pp.81 - 93.
4. Zahedi, F., Reliability of information systems based on the critical success factors formulation, *MIS Quarterly* 11(2), June 1987, pp. 187 - 203.
5. Pinto, J. K. and Slevin, D. P. ,Project success: definitions and measurement techniques, *Project Management Journal* 19(1), February 1988, pp. 67-72.
6. Fayyad, U., Uthurusamy, R., Data mining and knowledge discovery in databases, *Communications of the ACM.* 39(11), November 1996, pp. 24-26.
7. Fayyad, U., Piatetsky Shapiro, G., and Smyth, P., The KDD process for extracting useful knowledge from volumes of data, *Communications of the ACM* 39(11), November 1996, pp. 27-34.
8. Han, J., and Kamber, M., *Data Mining: Concepts and Techniques*, Morgan Kaufmann Publishers, San Francisco, California, 2001
9. F.Warren McFarlan, Richard L. Nolan and Chen Guoqing, Seizing Strategic *IT Advantage in China*, 2003(4), Higher Education Press, P306-309.
10. Slevin, D. P., Stieman, P. A. and Boone, L.W. , Critical Success Factor Analysis for Information Systems Performance Measurement and Enhancement, *Information Management* 21(3), October 1991, pp. 161-174.

CUSTOMER SEGMENTATION IN CUSTOMER RELATIONSHIP MANAGEMENT BASED ON DATA MINING

Yun Chen, Guozheng Zhang, Dengfeng Hu, Shanshan Wang
School of Public Economy Administration of Shanghai University of finance & economics, Shanghai, 200433, China. E_mail:guozhengzhang@gmail.com

Abstract: Customer relationship management (CRM) is the new management principle that adapts the business enterprise strategy shift from product-centric to customer-centric. Customer segmentation is one of the core functions of customer relationship management (CRM).This paper will build customer segmentation function model based on data mining, and summarizes the advantages of customer segmentation function model based on data mining in customer relationship management(CRM).

Key words: Customer Relationship Management; Customer Segmentation; Data Mining; Customer Value;

1. INTRODUCTION

Over the past decade, there has been an explosion of interest in customer relationship management (CRM) by both academics and executives [1]. Organizations are realizing that customers have different economic value to the company, and they are subsequently adapting their customer offerings and communications strategy accordingly. Thus, organizations are, in essence, moving away from product- or brand-centric marketing toward a customer-centric approach.

Currently research demonstrates that the implementation of CRM activities generates better firm performance when managers focus on maximizing the value of the customer [2]. Customer segmentation is the

The project is supported by the Shanghai Shuguang Project of China under the grant No: 05SG38

Please use the following format when citing this chapter:
Chen, Yun, Zhang, Guozheng, Hu, Dengfeng, Wang, Shanshan, 2006, in International Federation for Information Processing (IFIP), Volume 207, Knowledge Enterprise: Intelligent Strategies In Product Design, Manufacturing, and Management, eds. K. Wang, Kovacs G., Wozny M., Fang M., (Boston: Springer), pp. 288-293.

base of how to maximize the value of customer. Again and again firms find that the Pareto principle holds true, with 20% of the customer base generating 80% of the profits. Both researchers and managers need to evaluate and select segmentations in order to design and establish different strategies to maximize the value of customer.

2. LIMITATION OF TRADITIONAL CUSTOMER SEGMENTATION METHODS AND ADVANTAGE OF DATA MINING METHOD

2.1 Traditional segmentation methods and limitation

Segmentation can be seen as a simplification of the messy complexity of dealing with numerous individual customers, each with distinct needs and potential value [4].Traditional customer segmentations methods commonly based on experiential classification methods or simple statistical methods. Traditional statistical methods segment customer according to simple behavior character or attribute character such as the product category purchased or the region resided in.

These segmentation methods couldn't do more complex analysis that what kind of customers has high potential value and what kind has high credit. With the extensive application of EC and CRM, enterprises have accumulated more and more customer data. Traditional technique such as multiple regressions cannot cope with this level of complexity. Consequently, the reliability and validity of the statistical functions used to generate segmentations or to build predictive models becomes a possible contributory factor to CRM user dissatisfaction [7].

2.2 Data mining and it's Advantage

Data mining can be considered a recently developed methodology and technology, coming into prominence in 1994.The SAS Institute defines data mining as the process of selecting, exploring and modeling large amounts of data to uncover previously unknown patterns of data [5]. Accordingly, data mining can be considered a process and a technology to detect the previously unknown in order to gain competitive advantage.

Data mining uses neural networks, decision trees, link analysis, and association analysis to discover useful trends and patterns from the extracted data [6]. Data mining can yield important insights including prediction models and associations that can help companies understand their customers better. Many large companies today have terabytes of data, within which they could probably find more information about their customers, markets, and competition than they would ever need. Data mining enables marketers to better extract valuable business information from the 'mountains of data' in a firm's systems. It is a potential solution to a big problem facing many companies: an overabundance of data and a relative dearth of staff, technology, and time to transform numbers and notes into meaningful information about existing and prospective customers. Data mining enables a firm to measure consumer behavior on the basis of 100 or more attributes, instead of the three or four associated with traditional statistical modeling [7]. The more attributes a firm uses, the greater the complexity of the data and the greater the need for data mining tools.

3. CUSTOMER SEGMENTATION MODEL BASED ON DATA MINING

3.1 Customer Segmentation model

As practitioners are enthusiastically seeking out groups of profitable customers whose loyalty is steady, some academics are beginning to question whether segments are actually stable entities and more fundamentally whether they really exist at all[7]. Segmentation method based on data mining put up by this paper can solve above problems because the model could study from new information that input afterward and get new rules. It provides completely support to the dynamic management process of customer acquiring, customer keeping and customer value increasing, customer satisfaction and customer loyalty promoting.

Building the mapping relationship between the conception attribute and the customer is the key step of segmentation method based on data mining. Customer data contain dispersive and continues attribute. Setting each customer attribute as a dimension and setting each customer as a particle, the whole customers in enterprise can form a multidimensional space, which has been defined as the attribute space of the customer.

The mapping relationship between customers attributes and conception category can be constructed by analytic method, or by sample learning method. Analytic method analyzes the attribute character of each conception category should have, subsequently constructs the mapping relationship between attributes space and conception space. But much mapping relationship between attributes space and conception space weren't clear, it need use sample learning method constructing the mapping relationship [11]. Sample learning method automatically generalize the mapping relationship between attributes space and conception space by applying data mining technology on known conception category in enterprise database. The data mining process is called sample learning.

Assume B={G_1, G_2, ..., G_n}, we can confirm a group conception category by B, L={L_1, L_2, ..., L_p}, $C' \subseteq C$, C' is known customer category.

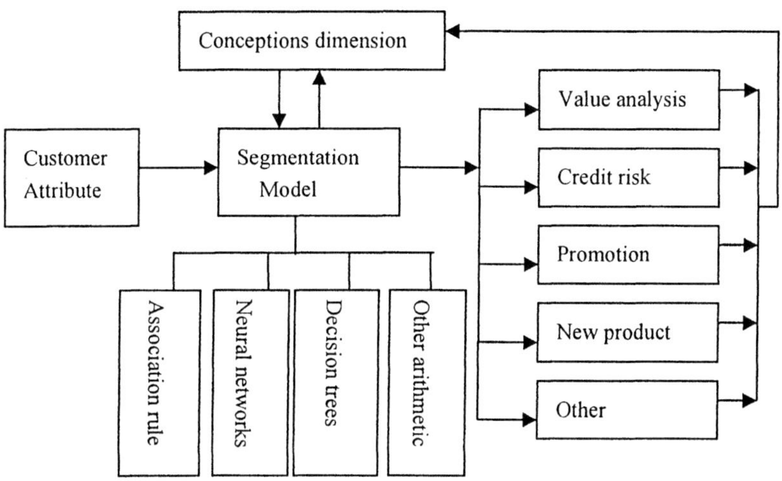

*Figure 1*Customer Segment Model based on data mining

The three steps of customer segmentation based on data mining:

(1) Ensuring mapping route. p: C→L, set $\forall c \in C'$, if $c \in L_i$, then $p(c)= L_i$.

(2) $\forall c \in C$, confirming which conception category c belong to by seeking the value of p(c).

(3) Rule creation and Function analysis. As showed in figure.1. The segmentation model applied data mining technology such as Association rule, Neural network, Decision tree, etc, as its method foundation. This segmentation model firstly segment customer according to the mapping relationship, subsequently processing various kind of business application.

1) Segmentation rule creation

Sort the customers with the Customer Segment Model and the Segment Function. After training the segment model, we get the segment rule or the network segment rule. We can effectively segment new customers based on the trained model.

2) Function Analysis

Function analysis includes customer value analysis, credit analysis and promotion analysis, etc, based on the foundation of the mapping relationship between customer and concept. Further on, new function requirement will be bring forward to CRM with the developing of the management practice. The new demand functions would be added to the conceptual dimension and reconstruct the mapping relationship with the customer characteristic.

3.2 Advantage of the customer segmentation model based on data mining

(1) Improving promotion effect

The customer segmentation based on the data mining can be helpful for the enterprises to make suitable promotion strategy, in suitable time, with suitable products and services, aiming at suitable customers.

(2) Analyzing customer value and customer loyalty

The customer value and the customer loyalty are important to enterpriser's stratagem and management tactics. Enterprises can confirm the rank of customer according to their expected value and loyalty analyzed by segmentation model based on data mining.

(3) Analyzing credit risk

Risk scoring is an effective way of evaluating certain specific types of customer risk, normally the risk of default.

(4) Instructing new products R&D

Enterprises can find out the preference of theirs customers by customer analyzing based on data mining, and make sure that various demand will be realized in the new design.

(5) Confirming target market

Customer segmentation base on data mining can make targeted customer group clear and locate the market explicitly.

4. CONCLUSIONS

A key role of marketing is to identify the customers or segments with the greatest value-creating potential and target them successfully with corresponding marketing strategies to reduce the risk of these high lifetime value customers defecting to competitors [10]. In this construction mode, segmenting customer is the basic work of data mining according to known historic segmentation information. The training data used to construct segment forecast mode can be historic data or exogenous data that gain from experience or survey.

Because customer behavior is uncertain and inconsistent, researchers and managers should construct dynamic customer segmentation model in order to objectively reflect the characteristic. In customer-centric era, customer segmentation result is concern with the establishment of enterprise's stratagem and tactics. Best practice demands that marketers develop their understanding of customer segmentation based on data mining techniques and use the output to develop marketing strategies creatively to maximize shareholder value.

5. REFERENCE

1. Werner Reinartz, Manfred Krafft, And Wayne D. Hoyer, The Customer Relationship Management Process: Its Measurement and Impact on Performance, *Journal of Marketing Research*, 293 Vol. XLI (August 2004), 293–305)
2. Gupta, Sunil, Donald R. Lehmann, and Jennifer A. Stuart, Valuing Customers, *Journal of Marketing Research*, 2004, 41 (February), 7–18.
3. Darrll Rigby, Frederick F. Reichheld, Avoid the Four Perils of CRM, *Harvard Business Review*, 2002, (1): 101-109
4. Hugh Wilson1,Elizabeth Daniel and Malcolm McDonald, Factors for Success in Customer Relationship Management (CRM) Systems, *Journal of Marketing Management*, 2002, 18, 193-219
5. SAS Institute, *From Data to Business Advantages: Data Mining, the SEMMA Methodology and SAS Software*, SAS Institute: Cary, North Carolina, 1998.
6. Jiawei Han, Michelline Kamber, *Data mining: conception and technology*, Beijing: Mechanic Industry Publish, 2002.
7. Agnes Nairn, and Paul Bottomley, Cluster analysis procedures in the CRM era, *International Journal of Market Research*, Vol. 45 Quarter 2 2003
8. Koh Hian Chye, Chan Kin Leong Gerry, Data mining and customer relationship marketing in the banking industry, *Singapore Management Review*, 2002; 24, 2;
9. Claudio Marcus, A practical yet meaningful approach to customer segmentation, *Journal of consumer marketing*, Vol. 15 No. 5 1998, pp. 494-504
10. Andrew Banasiewicz, Acquiring high value, retainable customers, *Database Marketing & Customer Strategy Management*, 2004, Vol. 12, 1, 21–31
11. Jon Kleinberg, Christos Papadimitriou, Prabhakar Raghavan, Segmentation Problems, *Journal of the ACM*, Vol. 51, No. 2, March 2004, pp. 263–280.

AN EFFECTIVE REFINEMENT ALGORITHM BASED ON MULTILEVEL PARADIGM FOR GRAPH BIPARTITIONING

Ming Leng, Songnian Yu, Yang Chen
School of Computer Engineering and Science,Shanghai University, Shanghai, PR China.
Email:lmsly@263.net

Abstract: The min-cut bipartitioning problem is a fundamental partitioning problem and is NP-Complete. It is also NP-Hard to find good approximate solutions for this problem. In this paper, we present a new effective refinement algorithm based on multilevel paradigm for graph bipartitioning. The success of our algorithm relies on exploiting both new Tabu search strategy and boundary refinement policy. Our experimental evaluations on 18 different graphs show that our algorithm produces excellent solutions compared with those produced by MeTiS that is a state-of-the-art partitioner in the literature.

Key words: min-cut, graph bipartitioning, multilevel paradigm, Tabu search

1. INTRODUCTION

Partitioning is a fundamental problem with extensive applications to many areas, including VLSI design [1], information retrieval [2], parallel processing [3], computational grids [4] and data mining [5]. The *min-cut bipartitioning* problem is a fundamental partitioning problem and is NP-Complete [6]. It is also NP-Hard [7] to find good approximate solutions for this problem. Because of its importance, the problem has attracted a considerable amount of research interest and a variety of algorithms have been developed over the last thirty years [8],[9]. The survey by Alpert and Kahng [1] provides a detailed description and comparison of various such

---

Please use the following format when citing this chapter:
Leng, Ming, Yu, Songnian, Chen, Yang, 2006, in International Federation for Information Processing (IFIP), Volume 207, Knowledge Enterprise: Intelligent Strategies In Product Design, Manufacturing, and Management, eds. K. Wang, Kovacs G., Wozny M., Fang M., (Boston: Springer), pp. 294-303.

schemes which include *move-based* approaches, *geometric representations*, *combinatorial* formulations, and *clustering* approaches.

As problem sizes reach new levels of complexity recently, a new class of graph partitioning algorithms have been developed that are based on the multilevel paradigm. The multilevel graph partitioning schemes include three phases [10], [11],[12], [13]. The *coarsening phase* is to reduce the size of the graph by collapsing vertex and edge until its size is smaller than a given threshold. The *initial partitioning phase* is to compute initial partition of the coarsest graph. The *uncoarsening phase* is to successively project the partition of the smaller graph back to the next level finer graph while applying an iterative refinement algorithm. In this paper, we present a new effective refinement algorithm based on multilevel paradigm which is integrated with Tabu search [14] for refining the partition. Our work is motivated by the multilevel partitioners of Saab [13] who promotes locked vertex to free by introducing two buckets for per side of the partition and Karypis [10], [11],[12] who proposes a boundary refinement algorithm and supplies **MeTiS** [10], distributed as open source software package for partitioning unstructured graphs. We test our algorithm on 18 graphs that are converted from the hypergraphs of the ISPD98 benchmark suite [15]. Our experiments show that our algorithm produces partitions that are better than those produced by **MeTiS** in a reasonable time.

The rest of the paper is organized as follows. Section 2 provides some definitions, describes the notation that is used throughout the paper and the *min-cut bipartitioning problem*. Section 3 briefly describes the motivation behind our algorithm. Section 4 presents an effective refinement algorithm based on multilevel paradigm for graph *bipartitioning*. Section 5 experimentally evaluates the algorithm and compares it with **MeTiS**. Finally, Section 6 provides some concluding remarks and indicates the directions for further research.

2. MATHEMATICAL DESCRIPTION

A graph $G=(V,E)$ consists of a set of vertices V and a set of edges E such that each edge is a subset of two vertices in V. Throughout this paper, *n* and *m* denote the number of vertices and edges respectively. The vertices are numbered from *1* to *n* and each vertex $v \in V$ has an integer weight $S(v)$. The edges are numbered from *1* to *m* and each edge $e \in E$ has an integer weight $W(e)$. A decomposition of a graph V into two disjoint subsets V_1 and V_2, such that $V_1 \cup V_2 = V$ and $V_1 \cap V_2 = \varnothing$, is called a *bipartitioning* of V. Let

$S(A)=\sum_{v\in A} S(v)$ denote the size of a subset $A \subseteq V$. Let $ID(v)$ be denoted as v's *internal degree* and is equal to the sum of the edge-weights of the adjacent vertices of v that are in the same side of the partition as v, and v's *external degree* denoted by $ED(v)$ is equal to the sum of edge-weights of the adjacent vertices of v that are in the different side of the partition. The *cut* of a *bipartitioning* $P=\{V_1,V_2\}$ is the sum of weights of edges which contain two vertices in V_1 and V_2 respectively, such that $cut(P)=\sum_{e\cap V_1 \neq \varnothing, e\cap V_2 \neq \varnothing} W(e)$. Naturally, vertex v belongs at boundary if and only if $ED(v)>0$ and the *cut* of P is also equal to $0.5\sum_{v\in V} ED(v)$.

Given a balance constraint *r*, the *min-cut bipartitioning problem* seeks a solution $P=\{V_1,V_2\}$ that minimizes $cut(P)$ subject to $(1-r)S(V)/2 \le S(V_1), S(V_2) \le (1+r)S(V)/2$. A *bipartitioning* is *bisection* if *r* is as small as possible. If the edges are ignored, the problem is the set partitioning problem which is NP-Complete [6]. Therefore, unless all vertices have unit sizes, it is NP-Hard to find just a bisection of arbitrary cost of graph. Most existing partitioning algorithms are heuristics in nature and they seek to obtain reasonably good solutions in a reasonable amount of time. Kernighan and Lin (KL) [8] proposed a heuristic algorithm for partitioning graphs, which requires $O(n^2 \log(n))$ computation time. The KL algorithm is an iterative improvement algorithm that consists of making several improvement passes. It starts with an initial bipartition $\{A,B\}$ and tries to improve it by every pass. A pass consists of the identification of two subsets of vertices $A' \subset A$ and $B' \subset B$ of equal size such that can lead to an improved partition if the vertices in the two subsets switch sides, that is to say, $\{(A-A')\cup B', (B-B')\cup A'\}$ is a better bipartition than $\{A,B\}$. Fiduccia and Mattheyses (FM) [9] proposed a fast heuristic algorithm for bisecting a weighted graph by introducing the concept of cell *gain* into KL algorithm. The previously unmoved vertex $v\in A$ (or $v\in B$) with highest *gain* that computed by $cut(\{A,B\})-cut(\{A-v,B+v\})$ is moved from A to B.

3. MOTIVATION

Tabu search has its antecedents in methods designed to cross boundaries of feasibility or local optimality standardly treated as barriers, and to systematically impose and release constraints to permit exploration of otherwise forbidden regions. Tabu search is a "higher level" heuristic procedure for solving optimization problems, designed to guide to other methods to escape the trap of local optimality [14].

During each pass of KL and FM, they have the same restriction that each vertex can move only once and the restriction may prevent the exploration of certain promising region of the search space. In the terminology of Tabu Search, the KL and FM strategy is a simple form of Tabu search without aspiration criterion whose prohibition period is fixed at *n*. In [13], Saab introduces the concept of *forward move step* and *restore balance step* for refining the partition and adopts aspiration criterion to allow a locked vertex to become free in the same passes of ALG2 algorithm. As a consequence of allowing locked vertices to move in an iterative pass, ALG2 algorithm can explore a certain promising regions of the search space. However, Saab limits the vertices *move-direction* of two kinds of steps in A→B and B→A respectively and is obliged to adopt two different aspiration criterions for two kinds of steps respectively to avoid cycling issue.

In both FM and ALG2 refinement algorithms, we have to assume that all vertices are free and insert the *gain* of all vertices in free bucket. However, most of *gain* computations are wasted since most of vertices moved by FM and ALG2 refinement algorithms are boundary vertices that straddle two sides of the partition. As opposed to the non-boundary refinement algorithms, the cost of performing multiple passes of the boundary refinement algorithms is small since only boundary vertices are inserted into the bucket as needed and no work is wasted. In [11], the boundary KL (BKL) refinement algorithm presented by Karypis swaps only boundary vertices and is much faster variation of the KL algorithm.

In this paper, we present a boundary Tabu search (BTS) refinement algorithm that combines the Tabu search theory with boundary refinement policy. It has three distinguishing features which are different from ALG2 algorithm. First, we initially insert into the free bucket the gains for only boundary vertices. Second, we remove the above limitation by introducing the conception of *step-status* and *move-direction*. Finally, we derive aspiration criterion from lots of experiments that is simpler than that of ALG2 algorithm.

4. BTS: A NEW EFFECTIVE REFINEMENT ALGORITHM

BTS introduces the conception of *move-direction* and *step-status* that consists of both *forward-move* and *restore-balance*. Given an input partition *Q* and balance tolerance *t*, if the input partition *Q* satisfies the balance tolerance *t*, current *step-status* is *forward-move* and we initially choose to move a vertex from the side whose bucket contains a highest *gain* vertex among all boundary vertices. Otherwise current *step-status* is *restore-balance* and we choose to move a vertex from the larger side of the partition. As BTS enters into the next step, if new partition satisfies the balance tolerance *t*, current *step-status* is *forward-move* and we choose the last *move-direction* to apply the current step, else current *step-status* is *restore-balance* and the *move-direction* of current step starts from the larger side of the partition. The strategy of BTS eliminates the limitation of *move-direction* of steps and its goal is to increase the chances of vertices that are closely connected migrating together from one side of the partition to the other by allowing moving sequences of vertices at one pass from one side of the partition to the other, although the highest *gain* maybe negative in a sequences of steps.

BTS uses free and tabu buckets to fast storage and retrieval the gains of free and tabu vertices respectively. At the beginning of a pass, all vertices are free and the internal and external degrees of all vertices are computed and two free buckets are inserted the gains of boundary vertices of two sides respectively that are computed by $ED(v) - ID(v)$. After we move a vertex v, the gains of itself and its neighboring vertices should be changed. First, we must lock the vertex v by deleting its original *gain* from bucket and insert its new *gain* (negative value of original *gain*) into the tabu bucket of the other side. Second, we update the gains of the neighboring vertices of vertex v. If any of these neighboring vertices becomes a boundary vertex due to the move of vertex v , we insert its *gain* into the free bucket of side in which it locates. If any of these neighboring vertices becomes a non-boundary vertex due to the move of vertex v, we delete its original *gain* from bucket that maybe free or tabu. If any of these neighboring vertices is already a boundary free vertex, we only update its *gain* in free bucket. If any of these neighboring vertices is a boundary locked vertex, we must delete its original *gain* from tabu bucket and insert its new *gain* into the free bucket of side in which it locates. In the terminology of Tabu Search, the tabu restriction forbids moving vertices which are designated as tabu status and the prohibition period of tabu vertex can be changed dynamically and decided by the above promotion rule. The purpose of promotion rule is to increase their chances of following neighbors to the other side of the partition and to allow the chance for the movement of a cluster from one side of the partition to the other.

The remains problem in BTS is how to select a vertex to move in current step. The vertex to move must be selected from free bucket or tabu bucket of the side that is start point of the *move-direction* of current step. When the current *step-status* is *forward-move*, the next vertex to move is the highest *gain* vertex in the free bucket if it is not empty. Otherwise, the highest *gain* vertex in tabu bucket is chosen to move. If the current *step-status* is *restore-*

balance, the next vertex to move is the highest *gain* vertex in both free and tabu buckets. It is not possible to run out of moves as long as tolerance t satisfies $0 < t < 1$. The choice rule of Tabu search is to select the highest *gain* vertex in free bucket and the aspiration criterion we have selected to override the tabu restriction is simple criterion that allows tabu vertex as candidate of vertex to move in the current step if current *step-status* is *restore-balance* or the free bucket is empty.

Table 1. The pseudocode of the BTS algorithm

```
BTS (initial partition Q, balance tolerance t, Total Moves k){
1  BEST=P;
2  COUNTER=0;
3  compute the internal and external degrees of all vertices;
4  compute the gains of boundary vertices of two sides;
5  insert the gains of boundary vertices in free bucket respectively;
6  while COUNTER <= k do {
7      decide the step-status and move-direction of the current step;
8      select the vertex to move by choice rule and aspiration criterion;
9      move the vertex and lock it;
10     original cut minus its original gain as the cut of new partition;
11     update the internal and external degrees of its neighboring vertices;
12     update the gains of its neighboring vertices by promotion rule;
13     if (the cut is minimum and satisfies balance constraints) then
14         BEST=P;
15     end if
16     COUNTER = COUNTER +1;
17   }end while
18 }
```

The pseudocode of the BTS algorithm is given in Table. 1. The while loop (lines 6-17) of BTS is iterated as long as improvements can be made and it is necessary in BTS that setting an upper limit on the parameter k. Because Tabu search aggressively selects the best admissible vertex based on the tabu restriction and aspiration criterion, it must examine and compare a number of boundary vertices by the bucket that allows to storage, retrieval and update the gains of vertices very quickly. It is important to obtain the efficiency of BTS by using the bucket with the last-in first-out (LIFO) scheme, as Tabu search memory structure, can enforce the "locality" in the choice of vertices to move. The internal and external degrees of all vertices, as complementary Tabu search memory structures, help BTS to facilitate computation of vertex *gain* and judgement of boundary vertex.

5. EXPERIMENTAL RESULTS

We use the 18 graphs in our experiments that are converted from the hypergraphs of the ISPD98 benchmark suite [15] and rang from 13,000 to 210,000 vertices. Each benchmark comes with 3 files, a .net file, a .are file and a .netD file. We convert hyperedges of .netD file into edges by the rule that every subset of two vertices in hyperedge can be seemed as edge. We create the edge with unit weight if the edge that connects two vertices didn't exist, else add unit weight to the weight of the edge. Next, we get the weights of vertices from .are file. Finally, we store 18 edge-weighted and vertex-weighted graphs in graph format of **MeTiS** [10]. The characteristics of these graphs are shown in Table 2.

Table 2. The characteristics of 18 graphs to evaluate our algorithm

benchmark	vertices	hyperedges	edges
ibm01	12752	14111	109183
ibm02	19601	19584	343409
ibm03	23136	27401	206069
ibm04	27507	31970	220423
ibm05	29347	28446	349676
ibm06	32498	34826	321308
ibm07	45926	48117	373328
ibm08	51309	50513	732550
ibm09	53395	60902	478777
ibm10	69429	75196	707969
ibm11	70558	81454	508442
ibm12	71076	77240	748371
ibm13	84199	99666	744500
ibm14	147605	152772	1125147
ibm15	161570	186608	1751474
ibm16	183484	190048	1923995
ibm17	185495	189581	2235716
ibm18	210613	201920	2221860

We implement the BTS refinement algorithm in ANSI C and integrate it with the leading edge partitioner **MeTiS**. In the evaluation of BTS refinement algorithm, we must make sure that the results produced by BTS refinement algorithm can be easily compared against those produced by **MeTiS**. First, we use the same balance constraint r and random seed in every comparison. Second, we select the sorted heavy-edge matching (SHEM) algorithm during *coarsening phase* because of its consistently good behavior in **MeTiS**. Third, we adopt the greedy graph growing partition (GGGP) algorithm during *initial partitioning phase* that consistently finds smaller edge-cuts than others algorithms. Finally, we select the BKL algorithm to compare with BTS algorithm during *uncoarsening and refinement phase* because BKL can produce smaller edge-cuts when coupled with SHEM algorithm. These measures are sufficient to guarantee that our experimental evaluations are not biased in any way.

The quality of partitions produced by our BTS refinement algorithm and those produced by **MeTiS** are evaluated by looking at two different quality measures, which are the minimum *cut* (Mincut) and the average *cut* (AveCut). To ensure the statistical significance of our experimental result, two measures are obtained in twenty runs whose random seed is different with each other. For all experiments, we use a 49-51 *bipartitioning* balance constraint by setting r to 0.02. Next, we pass the same *initial partition*, result of *initial partitioning* phase, to BTS and BKL algorithms in every comparison that guarantees a fair evaluation. Furthermore, we set the number of vertices of the current level graph as the value of parameter k and 5% as the value of parameter t.

Table 3. Min-cut bipartitioning results with up to 2% deviation from exact bisection

benchmark	Metis		BTS		BTS/Metis	
	Mincut	AveCut	Mincut	AveCut	Mincut	AveCut
ibm01	517	1091	506	1081	0.979	0.991
ibm02	4268	11076	4184	8410	0.980	0.759
ibm03	10190	12353	7548	10299	0.741	0.834
ibm04	2273	5716	2336	2789	1.028	0.488
ibm05	12093	15058	12074	14597	0.998	0.969
ibm06	7408	13586	2470	11692	0.333	0.861
ibm07	3219	4140	2867	3449	0.891	0.833
ibm08	11980	38180	12845	18470	1.072	0.484
ibm09	2888	4772	2917	3982	1.010	0.834
ibm10	10066	17747	6359	9743	0.632	0.549
ibm11	2452	5095	2406	3593	0.981	0.705
ibm12	12911	27691	12383	19301	0.959	0.697
ibm13	6395	13469	4582	8702	0.716	0.646
ibm14	8142	12903	7608	10359	0.934	0.803
ibm15	22525	46187	12112	36590	0.538	0.792
ibm16	11534	22156	10111	14787	0.877	0.667
ibm17	16146	26202	14792	20620	0.916	0.787
ibm18	15470	20018	14903	19199	0.963	0.959
average	---	---	---	---	0.863	0.758

Table 3 presents *min-cut bipartitioning* results allowing up to 2% deviation from exact bisection and Fig. 1 illustrates the Mincut and AveCut comparisons of two refinement algorithms on 18 graphs. As expected, the BTS refinement algorithm reduces the AveCut by 1% to 52% and reaches 24% average AveCut improvement. Although the BTS refinement algorithm produces partition whose Mincut is up to 7.2% worse than that of **MeTiS** on some benchmarks, we still obtain 14% average Mincut improvement and between -7% and 67% improvement in Mincut. All evaluations that twenty runs of two algorithms on 18 graphs are run on an 1800MHz AMD Athlon2200 with 512M memory and can be done in half an hour.

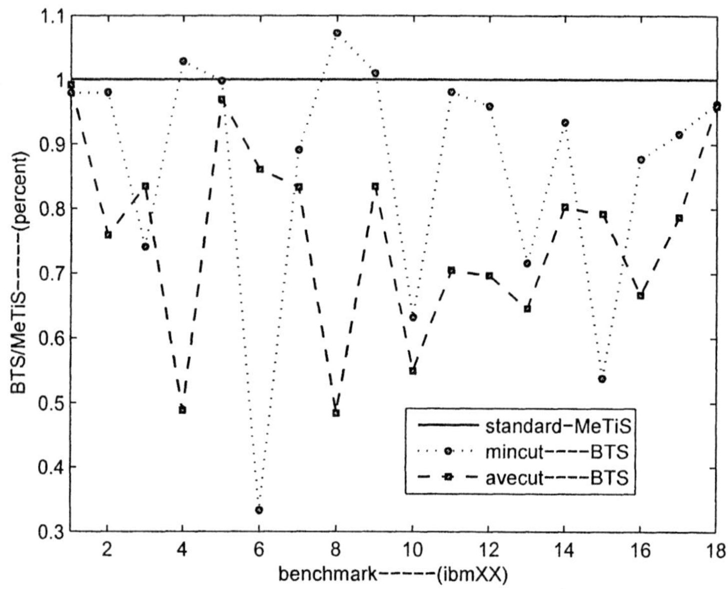

Figure 1. The Mincut and AveCut comparisons of two refinement algorithms on 18 graphs

6. CONCLUSIONS

In this paper, we have presented a new refinement algorithm based on multilevel paradigm. The success of our BTS algorithm relies on exploiting both new Tabu search strategy and boundary refinement policy. We obtain excellent *bipartitioning* result compared with those produced by **MeTiS**. Although it has the ability to find cuts that are lower than the result of **MeTiS** in a reasonable time, there are several ways in which this algorithm can be improved. We raise three questions about possible improvement below. The first question is how to find an optimal value for balance tolerance t. The second question, how to find an optimal value for k, is similar with Saab's question about r [13] because we observe that larger values of k lead to better partitions at the expense of a proportional increase in running time and the improvement in the quality of partitions in not linearly related to k. In the Mincut evaluation of benchmark ibm04, ibm08 and ibm09, the BTS refinement algorithm is 2.8%, 7.2%, 1% worse than **MeTiS** respectively. This shows that a good initial partition has been lost in the *uncoarsening phase* before it can yield a good bisection for the original graph. Therefore, the third question is how to guarantee find good approximate solutions by setting appropriate value for k and t.

7. ACKNOWLEDGMENTS

This work was supported by the Science Foundation of Shanghai Municipal Commission of Science and Technology, grant No. 00JC14052, and by "SEC E-Institute: Shanghai High Institutions Grid" project.

8. REFERENCES

1. Alpert, C.J., Kahng, A.B.: Recent Directions in Netlist Partitioning. Integration, *the VLSI Journal*, Vol. 19 (1995) 1-18
2. Zha, H., He, X., Ding, C., Simon, H., Gu, M.: Bipartite graph partitioning and data clustering. *Proc. ACM Conf Information and Knowledge Management* (2001)
3. Hendrickson, B., Leland, R.: An Improved Spectral Graph Partitioning Algorithm for Mapping Parallel Computations. *SIAM Journal on Scientific Computing*, Vol. 16 (1995) 452-469
4. Wanschoor, R., Aubanel, E.: Mesh Partitioning for Computational Grids. *Proc. 2nd Annual Conf on Communication Networks and Services Research* (2004)
5. Ding, C., Xiaofeng, H., Hongyuan, Z., Ming, G., Simon, H.: A Min-Max Cut Algorithm for Graph Partitioning and Data Clustering. *Proc. IEEE Conf Data Mining* (2001) 107-114
6. Garey, M.R., Johnson, D.S.: *Computers and Intractability: A Guide to the Theory of NP-Completeness*. WH Freeman New York (1979)
7. Bui, T., Leland, C.: Finding Good Approximate Vertex and Edge Partitions Is NP-Hard. *Information Processing Letters*, Vol. 42 (1992) 153-159
8. Kernighan, B.W., Lin, S.: An Efficient Heuristic Procedure for Partitioning Graphs. *Bell System Technical Journal*, Vol. 49 (1970) 291-307
9. Fiduccia, C., Mattheyses, R.: A Linear-Time Heuristics for Improving Network Partitions. *Proc. 19th Design Automation Conf* (1982) 175-181
10. Karypis, G., Kumar, V.: MeTiS 4.0: Unstructured Graphs partitioning and sparse matrix ordering system. *Technical Report, Department of Computer Science, University of Minnesota* (1998) Available on the WWW at URL http://www.cs.umn.edu/~metis
11. Karypis, G., Kumar, V.: A fast and highly quality multilevel scheme for partitioning irregular graphs. *SIAM Journal on Scientific Computing*, Vol. 20 (1999) 12.
12. Selvakkumaran, N., Karypis, G.: Multi-Objective Hypergraph Partitioning Algorithms for Cut and Maximum Subdomain Degree Minimization. *IEEE Trans. Computer Aided Design*, Vol. 25 (2006) 504-517
13. Saab, Y.G.: An Effective Multilevel Algorithm for Bisecting Graphs and Hypergraphs. *IEEE Trans. Computers*, Vol. 53 (2004) 641-653
14. Glover, F., Manuel, L.: *Tabu search: Modern heuristic Techniques for Combinatorial Problems*. Blackwell Scientific Publications Oxford (1993) 70-150
15. Alpert, C.J.: The ISPD98 Circuit benchmark suite. *Proc. Intel Symposium of Physical Design* (1998) 80-85

A FRAMEWORK OF ANALYSIS: APPROACHES IN THE APPLICATIONS OF E-CRM

Dengfeng Hu [1,2], Yun Chen [1], Guozheng Zhang [1]
1 Shanghai University of Finance & Economics, Shanghai, China, 2. Anhui University of Finance& Economics, Bengbu, China. Email: hdengfeng0212@sohu.com;

Abstract: Data mining techniques provide a popular and powerful tool set to generate various data-driven classification systems. Asim M Ansari and Carl F. Mela(2003) studied a statistical and optimization approach for customization of information on the Internet., and Balaji Padmanabhan (2004) explored the application of optimization and data mining. In this paper, based previous literatures we illustrate how optimization and data mining interact in e-CRM settings, and expound a framework of analysis in e-CRM.

Key words: approaches, data mining, optimization, e-CRM

1. INTRODUCTION

According to the professional and trade literature, more companies are using data mining as the foundation for strategies that help them outsmart competitors, identify new customers and lower cost [1]. In particular, data mining is widely used in marking, risk management and fraud control [2].

A successful e-business strategy integrates both e-commerce and customer intelligence to provide a complete view "e-customer intelligence". A lucrative opportunity exits for vendor to provide an e-customer intelligence solution that can market, sell, and service customers across e-commerce and customer intelligence technologies such as OLAP, data mining, data warehousing, and campaign management. The Patricia Seybold

This project is supported by the Shanghai Shuguang Project of China under the grant No.05SG38.

Please use the following format when citing this chapter:
Hu, Dengfeng, Chen, Yun, Zhang, Guozheng, 2006, in International Federation for Information Processing (IFIP), Volume 207, Knowledge Enterprise: Intelligent Strategies In Product Design, Manufacturing, and Management, eds. K. Wang, Kovacs G., Wozny M., Fang M., (Boston: Springer), pp. 304-310.

Group has dubbed this opportunistic void between e-commerce and customer intelligence the "e-Gap'.

We systematically explore how optimization and data mining can help one another for certain customer relationship management (CRM) applications in e-commerce, termed analytical e-CRM.

The remainder of the paper is organization into four sections. The first section introduces analytical e-CRM in general. The second section discusses the optimization opportunities in e-CRM, customer interaction in e-CRM and the data mining literature. The third section illustrates possible approaches to solve traditional DM problems applications and examples of data mining in the literatures. Finally, the concluding section highlights the limitations of data mining and suggests other possible.

2. DATA MING AND OPTIMIZATION OPPORTUNITIES IN E-CRM

While the tremendous business value of customer-centric marketing and management strategies is intuitive, implementing CRM initiatives have data storage capabilities, data warehousing applications, and data mining techniques (Berry and Linoff, 1997) , it is not just about computer software and hardware. Data mining uses sophisticated statistical processing or artificial intelligence algorithms to discover useful trends patterns from the extracted data. Data mining can yield important insights including predication models and association that can help companies understand their customers better.Swift [3] describes analytically e-CRM as a four-step iterative process consisting of (1) collecting and integrating online customer data, (2) analyzing this data, (3) building interactions with customers based on this analysis such that certain performance metrics such as LTV are optimized, and (4) measuring the effectiveness of these interactions in terms of these performance metrics.

2.1 Maximizing Lifetime Value

Customized products and communications attract customer attention and foster customer loyalty and lock-in. A typical performance metric used in many CRM applications is the LTV of a customer. One of the key questions in CRM is how to develop proactive customer interaction strategies that maximize LTV. Substantial work has been done on modeling LTV, mainly in the marketing literature [4]. The bulk of previous research in CLV has mainly focused on measuring the lifetime value customers or the value of entire customer base.

In contrast, in measuring the value of customers using individual customers of a firm, or limited customer acquisition in the form of switching from some known alternatives. The DM community studied the LTV problem in the presence of large volumes of customer data [5]. Rosset et al. (2002) compute LTV based on large volumes of customer data by focusing on using DM to estimate customer churn and future revenues. Because DM methods are used for computing LTV values, and optimization methods determine incentives resulting in highest LTV values, such applications provide examples of how DM methods can help solve optimization problems.

2.2 Customer Analysis in e-CRM

Customer analysis includes two main steps in the e-CRM context: (1) preprocessing data that tracks various online activities of the customers—this involves starting with individual user clicks on a site and constructing logical user "sessions" and summary variables; and (2) building customer profiles from this and other data to learn these patterns from data. DM is used Customer profiles are then built from these results. As argued by Zheng et al.[6], data preprocessing is a critical step of the knowledge discovery process in e-CRM, and the success of most DM methods to a large extent depends on this step.

Some popular session identification methods include session-level characterization [7] that aggregates user clicks into sessions, a fixed-length sliding window method that breaks a session into several sliding windows, and different types of clipping methods that break a session into windows of different sizes using various splitting methods.

2.3 Customer Interaction in e-CRM

Current DM approaches are based on mining Web log files for path traversal patterns and restructuring sites based on the analysis. In Perkowitz and Etzioni (1998), the problem is defined in terms of automatic construction of index pages based on log file data, with the goal of constructing index pages that provide users access to information that they are likely to view. The algorithm proposed has four steps [8]: (1) processing the log files into user visits, (2) computing co-occurrence frequencies between pages and creating a similarity matrix, (3) creating a graph from this matrix and finding cliques in this graph, and (4) creating index pages corresponding to each clique in the graph.

The website is structured as per the solution to the optimization problem. DM can, thus, identify additional constraints using patterns identified by this data. For example, DM can build that most users access the sports pages.

The optimization problem is then solved incorporating the additional constraints. This process can be iteratively done until some user-defined stopping criterion. However, DM will play a key role in the specification of the optimization space, and will also play a part in the specification of the constraints (Padmanabhan, 2004).

2.4 Personalization and Chance

Most of the approaches developed for solving this type of a recommendation problem use statistical and DM approaches, and usually try to determine "good" products to recommend to the customer. Various existing recommendation methods were classified by Balabanovic and Shoham (1997) into content-based, collaborative and hybrid approaches and have been reviewed in surveys, such as the collaborative filtering method — k 'nearest neighbors'. Many proposals have been made for defining distances and weights to compute the weighted ratings. As argued in Adomavicius and Tuzhilin (2003), the recommendation problem can also be formulated as an optimization problem that selects the best items to users.

3. OPTIMIZATION——THE APPROACH TO SOLVE TRADITIONAL DM PROBLEMS

Optimization can contribute to DM in one of two ways: (1) optimization can be a component of a larger DM process, or (2) new DM techniques can be built using entirely optimization-based methods. Support Vector Machines (SVM) [9] fall into the second category, because selection of an optimal separating a surface constitutes the DM method. So, we consider the following DM problems: feature selection, active learning, DM model optimization, selection of the best DM model or pattern, classification using mathematical programming, rule and constraint discovery, clustering and visualization toolkit.

3.1 Feature Selection and Active Learning

When optimization is used to preprocess click-stream data, as described above, optimization is used before DM. In general, optimization can be useful during the preprocessing stage of the DM process. Good feature selection techniques can significantly improve DM algorithms in two ways. First, these techniques can contribute to dimensionality reduction and faster running times. Second, they can contribute to building more accurate models from data.

Mannino and Mookerjee (1999) address a related problem where they optimize the cost of information acquisition for expert systems. By selecting only the data with high utility to the model, active learning aims to minimize the data needed for model building. The different active learning methods fall into two categories: (1) heuristic based, and (2) optimization based. The data points chosen are those in which there is maximum disagreement among the models. Optimization approaches employ an objective function, and those data points that optimize this objective function are selected.

3.2 DM Model Optimization

Many DM methods can be directly formulated as optimization problems or can have an optimization component that is a part of the DM problem (Hand et al. 2001). As another example of the heuristic search for an optimal solution to a DM problem, consider the problem of selecting the best decision tree using the Minimum Description Length (MDL) principle.

3.3 Selection of the Best DM Model or Pattern

Fu et al. [6] describe how a genetic algorithm selects the best decision tree from a set of decision trees generated by tree induction methods. This selection process is performed in the post processing stage after the DM algorithm generates decision trees. Kennedy et al. (1997) present genetic algorithm encoding decision trees as chromosomes, using a variation of one of the standard tree traversal methods. Fu et al. (2003) build multiple decision trees from the data by subsampling from the data and building individual decision trees for each sample.

3.4 Classification, Rule and Constraint Discovery

Using Integer Programming.Logical Analysis of Data (LAD) is method for classification proposed in the optimization literature (Crama et al. 1988). It builds a classifier for a binary target variable based on learning a logical expression that can distinguish between positive and negative examples in a data set. In this sense, the problem is similar to the concept learning studied in machine learning (Mitchell 1997). The solutions to the problem are developed by formulating and solving optimization. Early work in this area was targeted at binary domains. Recent work has extended this target to include numeric predictor attributes by using a process known as "binarization" . In DM, a challenge for optimization researchers building DM algorithms is to explicitly address this issue and to study how optimization over training sets produces optimal models for unseen data.

The opportunity for optimization researchers is to develop new optimization-based search methods that can focus on learning rules faster than traditional DM search methods. Related work in machine learning presents an initial approach to exploring this opportunity.

There are opportunities for DM to help solve problems that are addressed traditionally using an optimization framework. For example, an opportunity for using DM with optimization is to use DM to learn patterns in data that can, in turn, be used as constraints in a traditional optimization framework. This can ensure that the solutions derived from the optimization approach are consistent with the dominant patterns that emerge from the data and, hence, the solution quality can be improved. Further, using better constraints can help solve the optimization problem faster by appropriately restricting the search space. Consider the content management problem of placing advertisements on Web pages, such as done by Double Click. This can be formulated as an optimization problem pointed out by Geoffrion and Krishnan (2001).

3.5 Clustering and Visualization Toolkit

When data are limited to two or three dimensions, the most powerful tool for judging cluster quality is usually the human eye. Clusion, our CLUSter visualization toolkit, allows us to convert high-dimensional data into a perceptually more suitable format, and employ the human vision system to explore the relationships in the data, guide the clustering process, and verify the quality of the results.

The visualization toolkit Clusion allows even non-specialists to get an intuitive visual impression of the grouping nature of objects that may be originally defined in it can be extended to other domains, as illustrated by our results on document clustering and grouping web-logs. Alexander Strehl(2003) spent two years in studying the Dell customer data, and pointed out that Clusion effective for getting clusters balanced the number of customers or net dollar ($) amount, and even more so for conveying the results to marketing management.

4. CONCLUSIONS

The paper has studied the approaches in e-CRM, simultaneity, addressing e-CRM problems by utilizing the strengths of both DM and optimization can, therefore, be a useful area for future research. A better understanding of the problems in e-CRM and of the possible ways in which DM and optimization can interact, can help exploit the opportunities for synergistically using both approaches in the solution of e-CRM problems.

Specifically, we describe an optimization algorithm based on the assignment algorithm to optimize the design and content of e-CRM. We believe that such a general approach (combining choice models with optimization models) has utility beyond customization and can be used in the design of tailored services, customer designed catalogs, and bundling of goods

Optimization can help DM in e-CRM applications by developing more systematic methods of preprocessing click-stream data, in a manner that optimizes the performance of a DM model subsequently built on the preprocessed data. As (Balaji Padmanabhan, 2004) pointed out there are opportunities for researchers to formulate and solve this optimization problem.

5. REFERENCE

1. Dver, A. (2003), *Customer relationship management: The good, the bad, the future.* Bus. Week (April 28).
2. Kuykendall, L. (1999), The data-mining toolbox, *Credit Card Management* 12(6): 30-40.
3. Swift, R. (2002), *Analytical CRM Powers Portable Relationships: Creating Success by Letting Customers Guide You.* DM Rev.
4. Berger C (1999), *Data Mining to Reduce Churn,* Target Marking, 22(8).
5. Rosset, S., E. Neumann, U. Eick, N. Vatnik, Y. Idan (2002), Customer lifetime value modeling and its use for customer retention planning, *Proc-ACM Internat Conf. Knowledge Discovery Data Mining Edmonton.*
6. Zheng, Z., B.Padmanabhan, S.Kimbrough(2003), On Data Preprocessing Biases in Web Usage Mining, *INFORMS J. Computer*, 15(2) 148 - 170.
7. Srikant, R., Y. Yang (2001), Mining Web Logs to Improve Website Organization. *Proc. Tenth World Wide Web Conf., Hong Kong.*
8. Balaji Padmanabhan, Lexander Tuzhilin, on the Use of Optimization for Data Mining: Theoretical Interactions and e-CRM Opportunities Management Science. 2003 *INFORMS* Vol. 49, No. 10, October 2003.
9. Balabanovic M., Y. Shoham (1997), Content-based, Collaborative Recommendation. *Comm. ACM* 40(3) 66 - 72.

PARALLEL SEQUENCE ALIGNMENT ALGORITHM FOR CLUSTERING SYSTEM

Yang Chen, Songnian Yu, Ming Leng
School of Computer Engineering and Science, Shanghai University, Shanghai, PRC. Emai: chenyang2000@163.com

Abstract: Sequence alignment is one of the most important fundamental operations in bioinformatics. It has been successfully applied to predict the function, structure and evolution of biological sequences. In this paper, the sequence alignment algorithms based on dynamic programming are analyzed and compared. We present a parallel algorithm for pairwise alignment and implement it on a clustering system with MPI. The experimental results demonstrate the effectiveness in performance promotion. We encapsulate the algorithm into a grid service for practical use.

Key words: Bioinformatics; sequence alignment; parallel algorithm; clustering system; MPI

1. INTRODUCTION

With the genome project being carried out, new biological molecular information has emerged in large numbers, in which some valuable knowledge does exists. However, it is really hard for us to discover them. To fulfill such a demanding task, a new cross-disciplinary subject called bioinformatics has been rapidly developed.

In bioinformatics, the similarity comparison among biological sequences called *sequence alignment* is a fundamental method of information processing, which plays a significant role in revealing the function, structure and evolution of biological sequences. Sequence alignment finds out the maximal matched bases or residues by means of some specific mathematical model or algorithm whose outcome to some extent tells the relationship

Please use the following format when citing this chapter:
Chen, Yang, Yu, Songnian, Leng, Ming, 2006, in International Federation for Information Processing (IFIP), Volume 207, Knowledge Enterprise: Intelligent Strategies In Product Design, Manufacturing, and Management, eds. K. Wang, Kovacs G., Wozny M., Fang M., (Boston: Springer), pp. 311-321.

among sequences and their biological characteristics. Therefore, the design of a reasonable and efficient algorithm for sequence alignment has become a very important research issue in this field.

For the alignment of sequence pairs whose lengths are n, there are totally

$$\binom{2n}{n} = \frac{(2n)!}{(n!)(n!)} \cong \frac{2^{2n}}{\sqrt{\pi n}}$$

different cases to be considered. That seems a computational problem with exponential complexity.

Needleman and Wunsch [10] presented the first pairwise sequence alignment algorithm based on dynamic programming (DP), on which Gotoh [9] made some further modifications. An algorithm with linear space complexity was proposed by Hirschberg [6], which was later improved by Mayers and Miller [7]. Still based on DP technique, Smith and Waterman [11] presented an algorithm for local alignments.

2. BIOLOGICAL SEQUENCE ALIGNMENT ALGORITHM

The data of sequence are partitioned into DNA sequences and protein sequences. Each DNA sequence consists of four types of base A, T, C and G and each protein sequence is made up of 20 types of amino acid, hence any sequence can be represented as a string over specific alphabet. The similarity of sequences is described in quantitative values or qualitative description. The *degree of similarity* and the *edit distance* are used to quantify the similarity of sequences. The *dot plot* indicates in an intuitive and clear way the area where conspicuous similarity exists between two sequences, and then helps to locate the possible alignment.

2.1 Sequence alignment

Given a *score function*, the sequence alignment computes the optimal alignment among two or more sequences, ending up with the alignment of maximal similarity among sequences via matching the corresponding symbols or inserting gaps for insertion or deletion within sequences. Score function is a key parameter involved in sequence alignment algorithms, which is used to evaluate the *edit operations* applied to sequences (shown in Fig. 1).

$$p(a,a) = 1$$
$$p(a,b) = 0 \qquad (a \neq b)$$
$$p(a,-) = p(-,b) = -1$$

Figure 1. Evaluate edit operation with score function

Several algorithms are presented for sequence alignment. Most of them are based on dynamic programming, upon which various improvements are made. According to the number of sequences to be aligned, the algorithms are classified into pairwise alignment algorithms and multiple alignment algorithms. In this work, we mainly focus on the former ones.

There are two main criteria for the performance evaluation of sequence alignment algorithm. One is the running speed of algorithm; the other is the result sensitivity for optimal alignment. We care more about the execution time of algorithm which reflects the running speed.

2.2 Pairwise sequence alignment algorithm

The typical algorithm for global alignment is the Needleman-Wunsch algorithm, which is designed for sequence pairs with a high degree of similarity on the global level. The degree of similarity for sequence pairs is computed using an iterative method (shown in Fig. 2) and stored in the *score matrix* from which the optimal alignment is found through the backtracking procedure based on dynamic programming.

$$d_{i,j} = \max\begin{cases} d_{i-1,j-1} + p(s_i, t_j) \\ d_{i-1,j} + p(s_i, -) \\ d_{i,j-1} + p(-, t_j) \end{cases}$$

Figure 2. Compute degree of similarity iteratively

The basic algorithm for local alignment is the Smith-Waterman algorithm. Different from the global one, the local alignment algorithm is applied to sequence pairs that are distantly related with some similarity in local areas rather than on the whole. With regard to this, some slight modifications are made so as to seek the subsequence pairs with maximum similarity.

The complexity of algorithms mentioned above is determined by the scale of score matrix; in other words, it is closely related to the product of sequence lengths. The time complexity of computing score matrix is $O(n^2)$

and that of the backtracking procedure is $O(n)$, where n is the length of sequence. Therefore, the overall time complexity comes to $O(n^2)$.

2.3 Visualized implementation of algorithms

For the convenience of comparison among various sequence alignment algorithms, we implement and join them into a same graphical user interface. The program is designed based on MFC programming technology, running on Windows platform as shown in Fig. 3.

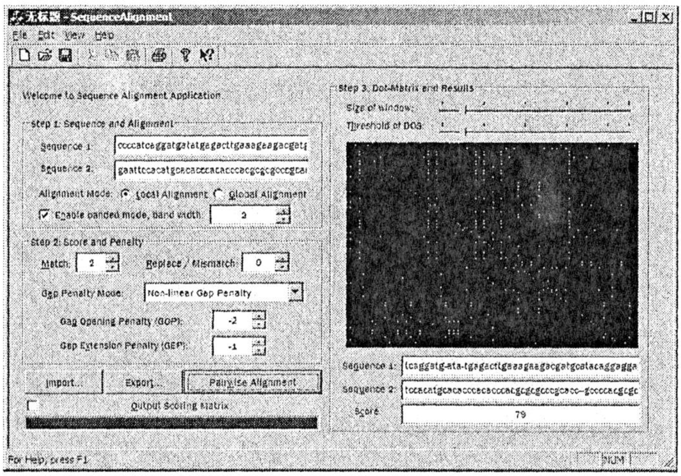

Figure 3. Visualized implementation of sequence alignment algorithms

Here, different cases of pairwise sequence alignment are integrated and the environmental parameters such as alignment mode, score function, gap penalty, etc. can be configured and adjusted freely. Meanwhile, the visualized results including score, dot-plot and optimal alignment can be instantly observed from the window.

3. PARALLEL SEQUENCE ALIGNMENT ALGORITHM

3.1 Feasibility analysis

As previously stated, the time consumption of sequential algorithm mainly lies in the computation of score matrix. The recursive formula of score matrix indicates that the computation of $d(i, j)$ can be started only if $d(i\text{-}1, j\text{-}1)$, $d(i\text{-}1, j)$ and $d(i, j\text{-}1)$ obtain their values, which is shown in Fig. 4.

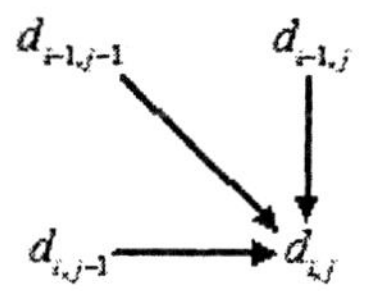

Figure 4. Data dependency in computing elements of score matrix

As a result, the computation of score matrix can be conducted successively in order of anti-diagonals. Therefore, the elements in same anti-diagonal marked by the symbol → can be computed simultaneously (Fig. 5).

		A	T	G	C
	0	1	2	3	→
A	1	0	1	→	
T	2	1	→		
G	3	→			
C	→				

Figure 5. Compute score matrix in a parallel manner

3.2 Model for problem solving

Driven by the data dependency in computing score matrix, we parallelize the pairwise alignment algorithm to improve its performance for multi-processor systems. The major models for parallel solving are listed below.

Pipeline model: As a basic unit, each row of the score matrix is computed sequentially by a processor, which blocks itself till the required elements in the row above are computed. This forms a continuous pipeline (Fig. 6).

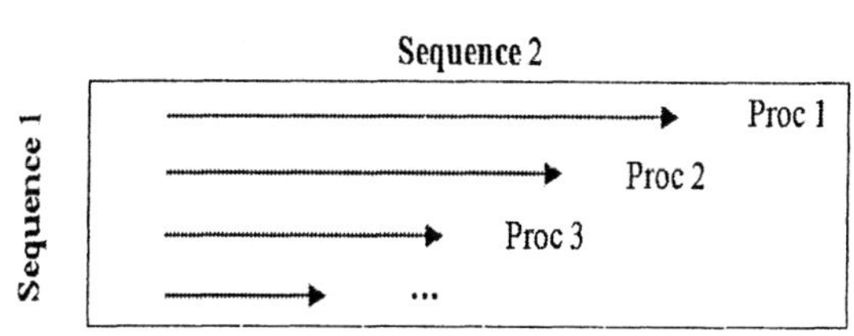

Figure 6. Pipeline model for parallel solving

Anti-diagonal model: All processors compute simultaneously along an anti-diagonal of score matrix, from the left-top corner to the right-bottom corner. Each idle processor selects an element that is not computed from current anti-diagonal. When all elements in current anti-diagonal are processed, the computation moves on to next anti-diagonal (Fig. 7).

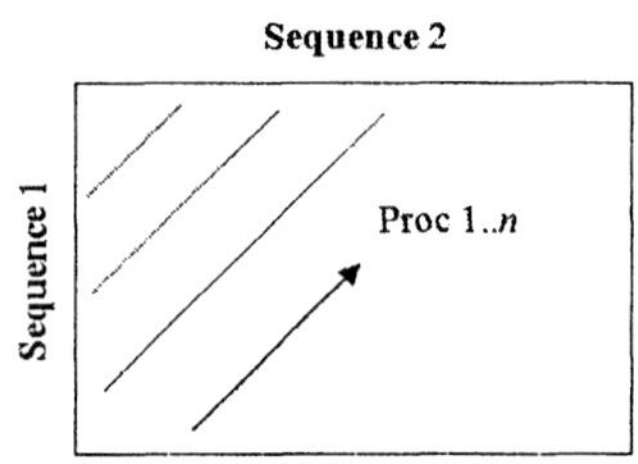

Figure 7. Anti-diagonal model for parallel solving

3.3 Design and implementation

In accord with the pipeline model above, we design a synchronous medium-grained parallel algorithm for distributed memory systems, by using single program multiple data (SPMD) technology.

The algorithm is based on the wave-front method. The score matrix is partitioned into several *bands* by row and several *blocks* by column. All the bands are distributed to multiple processors via a balanced allocation. A typical example is shown in Fig. 8. In line with the order of anti-diagonals, each processor computes the block in its own band concurrently. Due to the data dependencies, the communication between processors is required to transfer data on the boundary of bands. The height of bands and the width of blocks can affect the performance of the parallel algorithm.

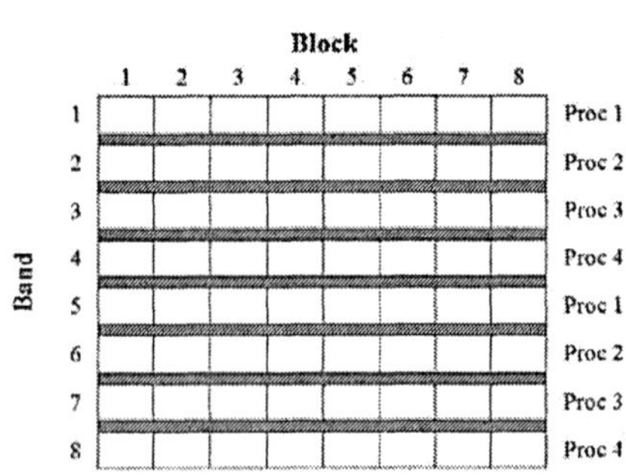

Figure 8. A 8*8 partition for 4 processors

By applying this algorithm, the time complexity is reduced to $O(n)$ when n processors are used. Specifically, if the sequence pairs to be aligned are closely related, only parts of DP matrix are worth computing. In this case, while the sequential algorithm has $O(cn)$ time complexity, the parallel counterpart can compute in $O(n)$ time with $O(c)$ processors, where c is a constant factor of restriction.

The algorithm is implemented on a clustering system with message passing interface (MPI) and it works properly on Linux platform. The outputs of algorithm for different length-specific inputs are verified, which to a certain extent guarantees the correctness of algorithm.

4. EXPERIMENTAL RESULTS

To evaluate the performance of the parallel algorithm, we design a experimental scheme to measure the execution time of algorithm under various lengths of sequences and various numbers of processors which are related to the complexity of algorithm.

The biological sequences we use for experiment are retrieved from the SRS database provided by research centre of bioinformatics, Peking University. The testing environment is the laboratory of high performance computing in Shanghai University, which is a PC cluster constituted by 70 nodes (Intel Pentium 4 CPU 2.4 GHz with 256 MB memory) running on Linux operating system (Red Hat Enterprise Linux WS 3, gcc 3.2.3, MPICH 1.2.5). The experimental results are shown in Table 1 as follows.

Table 1. Comparison in execution time of algorithms

sequence length	sequential algorithm	parallel algorithm, *np* processors					
		$np = 1$	$np = 2$	$np = 4$	$np = 8$	$np = 16$	$np = 32$
500	0.244804	0.365445	0.247524	0.165411	**0.126663**	0.570836	1.30005
1000	1.79986	2.71574	1.59836	0.959565	**0.654181**	1.14644	1.53744
1500	5.9714	8.83867	4.86467	2.95273	1.8195	**1.78117**	2.14013
2000	13.9534	20.3237	10.9388	6.30774	3.79232	**3.22145**	3.47226
2500	26.2785	39.5335	21.014	11.6055	6.9633	5.1999	**4.56656**
3000	45.5543	67.2282	35.4311	19.5344	11.4641	8.196	**6.54519**
3500	71.6498	106.583	55.7057	30.1991	17.4998	11.809	**8.95839**
4000	107.255	158.605	82.4512	44.275	25.0185	16.3415	**12.1245**
4500	152.542	225.746	118.905	62.4093	34.7657	21.5546	**15.9286**
5000	208.827	310.181	158.51	86.4587	46.905	28.465	**20.2812**

On one hand, given specific number of processors, the execution time of sequential algorithm and parallel algorithm with only one processor ($np = 1$) rises sharply with the increase of sequence length, which is shown in Fig. 9. It is highly imperative to take the advantage of parallel processing.

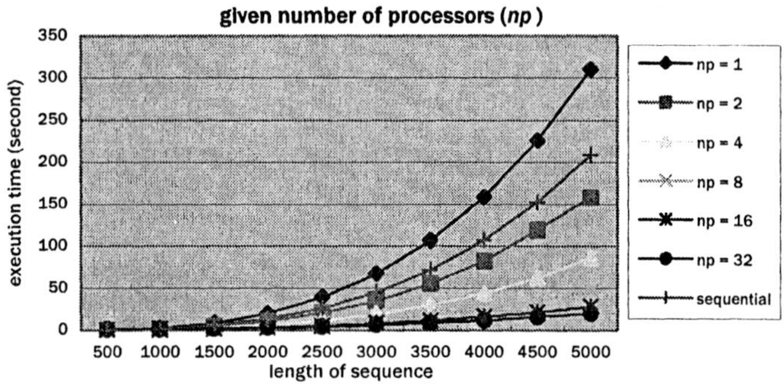

Figure 9. Execution time for various lengths of sequence

On the other hand, given specific length of sequence, the execution time of parallel algorithm tends to fall as the number of parallel processors increases, which is shown in Fig. 10. Obviously, our algorithm successfully captures the parallelism of the problem itself and unleashes the power of PC clustering systems, consequently improving the computational performance effectively. However, the parallel efficiency is found in a fairly low level partly because of the high proportion of communication for data transmission among processors.

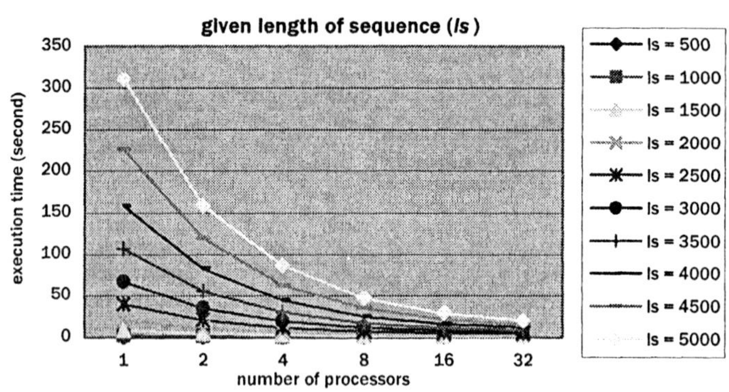

Figure 10. Execution time for various numbers of processors

5. ENCAPSULATION OF GRID SERVICE

Grid is an integrated environment for shared resources and services, which aims to establish a virtual supercomputer by organizing the geographically distributed computers all over the world via Internet.

5.1 Grid development environment

We attempt to encapsulate the core algorithm of sequence alignment into a grid service for the convenient access to end users. In doing so, we use the Globus Toolkit 3.0 (GT3), a software toolkit for building grid applications based on OGSI 1.0. Before all, a development environment is set up through the installation and configuration of JDK, Ant and GT3 Core.

5.2 Programming and deployment

We follow the proxy-stub model, a general programming model for distributed computing. It comprises two parts: the *server end* and the *client end*, which are weakly-coupled via service description.

The server end describes the service information in a file written in WSDL that includes service interface, invoking method and its binding low-level communication protocol. We rewrite the previously implemented algorithm and deploy it into a Web service container provided by GT3 itself. The client end generates the stub for service invoking according to the WSDL file from the server end. It receives the request data submitted by end users, invokes the core algorithm deployed on the server end via predefined interface and eventually returns the processed results.

In the respect of a friendly user interface, we embellish the client end using HTML and JSP technology such that users are allowed to gain entrance to the grid service of sequence alignment by means of web browsers and submit their specific tasks through online forms; the results are organized in the standard format and a concise report with statistical and analytical information is presented (shown in Fig. 11).

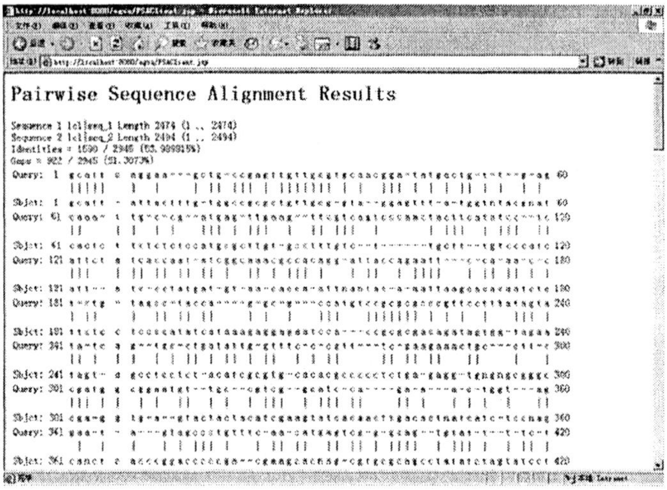

Figure 11. Obtain results of sequence alignment via grid service

6. CONCLUSION

In this work, several algorithms for sequence alignment are analyzed and a visualized program is implemented for further research. We focus on the pairwise alignment algorithm based on dynamic programming and have it parallelized for distributed memory systems based on pipeline model and wave-front method. The algorithm is implemented on a PC cluster with MPI and the experimental results indicate a remarkable enhancement of algorithm performance in comparison with the sequential algorithm. Furthermore, we encapsulate the algorithm into a grid service for practical use.

We are currently trying to discover more about the potential parallelism of algorithm especially in the backtracking procedure and also the hardware characteristics of clustering systems such as its network topology. The future work will involve transplanting the algorithm into a grid computing environment, and direct to the research of multiple alignment algorithm. Our goal is to find an efficient and scalable solution for biological sequence alignment.

7. ACKNOWLEDGMENTS

This work was supported by the Science Foundation of Shanghai Municipal Commission of Science and Technology, grant No. 00JC14052, and by "SEC E-Institute: Shanghai High Institutions Grid" project.

8. REFERENCES

1. Andreas D. Baxevanis, B. F. Francis Ouellette. "Bioinformatics: the practical guide for analysis of genomics and proteomics". Tsinghua University Press. 2000
2. Borja Sotomayor. "The Globus Toolkit 3 Programmer's Tutorial". 2004.
3. Cyntbia Gibas, Per Jambeck. "The computer technology in bioinformatics". China Dianli Press. 2002
4. C. Xavier, S.S. Iyengar. "Introduction to the parallel algorithm". China Machine Press. 2004
5. D. S. Hirschberg. "A Linear Space Algorithm for Computing Maximal Common Subsequences". Comm. ACM, vol. 18, no. 6, pp. 341-343, 1975
6. E. W. Mayers, W. Miller. "Optimal Alignments in Linear Space". Computer Applications in the Biosciences, vol. 4, no. 1, pp. 11-17, 1988
7. Joshy Joseph, Craig Fellenstein. "Grid Computing". Tsinghua University Press. 2005
8. O. Gotoh. "An Improved Algorithm for Matching Biological Sequences". J. Molecular Biology, vol. 162, pp. 705-708, 1982
9. S. B. Needleman, C. D. Wunsch. "A General Method Applicable to the Search for Similarities in the Amino Acid Sequence of Two Proteins". J. Molecular Biology, vol. 48, pp. 443-453, 1970
10. T. F. Smith, M. S. Waterman. "Identification of Common Molecular Subsequences". J. Molecular Biology, vol. 147, pp. 195-197, 1981
11. Zhihui Du. "High-performance computing and parallel programming technology: MPI parallel programming". Tsinghua University Press. 2001

MANAGING THE SUPPLY AND DEMAND UNCERTAINTY IN ASSEMBLY SYSTEMS

Hongze Ma[1] and Kesheng Wang[2]

[1]*Logistics, Turku School of Economics and Business Administration, Turku, Finland; E-mail: Hongze.Ma@tukkk.fi.* [2]*Norwegian University of Science and Technology, Trondheim, Norway*

Abstract: To react to changing market rapidly, some companies outsource all components, and then assemble them into end product to fulfill customer demand. For such assembly systems, the marketing environment at both supply and demand sides is quite volatile. So, agility is one of the key factors for them to win the competition. In this paper, a simulation model is built for an assembly system, in which both supply and demand uncertainties are considered. Then, by this model, the effect of supply and demand uncertainty on the performance of the system is studied, and the approach on how to manage such uncertainties is provided. To achieve agility, managers of the assembly systems can utilize such results to adjust their assembly schedule to adapt to the different levels of supply and demand uncertainty.

Key words: Assembly system, inventory management, supply and demand uncertainty, simulation.

1. INTRODUCTION

In recent years, more and more companies, especially electronic companies, implement pure assembly system, i.e. outsource all the components, and assemble them into end products to fulfill customer demand. Such strategy possesses advantages of easy to start up and easy to achieve agility. But, at the same time, a company implementing pure assembly system also faces challenges of volatile component supply and customer demand. At the supply side, frequently updated product makes it difficult to organize the component procurement. At the demand side, shorter product life cycle makes the company difficult to forecast the customer demand. To meet these challenges, the operation managers in an

Please use the following format when citing this chapter:

Ma, Hongze, Wang, Kesheng, 2006, in International Federation for Information Processing (IFIP), Volume 207, Knowledge Enterprise: Intelligent Strategies In Product Design, Manufacturing, and Management, eds. K. Wang, Kovacs G., Wozny M., Fang M., (Boston: Springer), pp. 322-327.

assembly system need to position the safety stock levels wisely, and make the assembly schedule carefully. Especially, when the supply and demand variability changes, the managers need to adjust the safety stock and assembly schedule to adapt to the new situation.

In this paper, to study how to manage the supply and demand uncertainty, a simulation model for the assembly system is built, in which the supply and demand variability are considered. By this simulation model, the effect of supply and demand variability on the system performance is assessed, and the managerial implication on how to adjust safety stock to cope with different levels of variability is provided.

2. LITERATURE REVIEW

Inventory problems have been widely studied since the seminal work by Clark and Scarf.[1] There are mainly two types of inventory control policies: continuous review policy[4, 6] and periodic review policy.[2, 3, 4, 5] Most of these models utilized mathematics to determine the optimal inventory control parameters. To model the complicated assembly systems, simulation techniques had been adopted.[7, 8, 9] In most of the existing literatures, internal management for the system was taken as the main concern. But, the issues on how to react to the changing market environment were not well addressed. For example, when the variability of demand or supply changes, how will the company change its internal management to adapt to the new situation? In this paper, we will try to analyze the sensitivity of system performance to the changing variability in supply and demand, and then, based on this analysis, the managerial levers on how to adapt to the changed situation are provided.

3. MODELING AND SIMULATION OF AN ASSEMBLY SYSTEM

3.1 The assembly system to be modeled

In this paper, a two-echelon assembly system is considered. The BOM (Bill of Material) of the product to be assembled at the company is shown in Figure 1. The company purchases component A, C, and D, and then assemble them into end product. Both supply and demand for this assembly system are uncertain. The customer demand at the end product is a random process with normal distribution. The uncertainty of component supply is expressed as random lead time. To deal with these uncertainties, the company needs to keep some safety stock. To manage this assembly system properly, the company needs to determine an optimal safety inventory level. Moreover, this safety stock level must be adjusted to adapt to the changing variability of supply and demand.

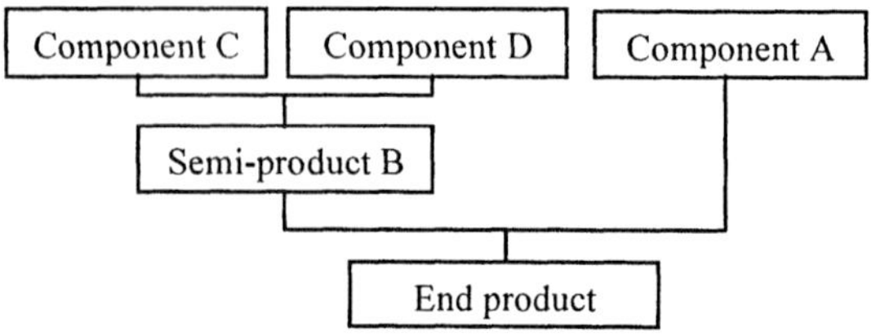

Figure 1. BOM of the product to be assembled

3.2 The simulation model

In this paper, we decided to use simulation technology to study this system. Vensim is selected to model and simulate the assembly system mentioned above. The entire model is divided into three parts: demand generating and fulfilling, product assembling, and component supply.

3.2.1 Demand generating and fulfilling

Demand is generated randomly with normal distribution. The variability level of demand is indicated by *Mean Absolute Percentage Deviation (MAPD)*. After demand is generated, the end product inventory is checked. If there are enough products in the store, the demand is shipped; otherwise, it is backlogged. Concerning customer service, the performance of the system is indicated by *fill rate*. In the later part of this paper, the issues on how the system performance (i.e. *fill rate)* reacts to the different levels of demand variability (i.e. *MAPD of demand*) will be addressed, and the corresponding managerial levers will be given.

3.2.2 Product assembling

The assembly process is organized according to the product's BOM. First, the inventory for components is checked, then, based on inventory status and assembly capacity, the production schedule is made. After a lead time, the end product is assembled. This lead time is random with normal distribution.

3.2.3 Component supply

When the at hand inventory (including the end product inventory and work-in-process) for a component is less than or equal to the reorder point, an order is placed at its supplier. The reorder point consists of two parts: lead time demand and safety stock. Safety stock is used to cover the uncertainty in both supply and demand. It has direct effect on customer service level, i.e. *fill rate*. In the later part of this paper, we will discuss: when the variability of demand and/or supply changes, how the safety stock will be adjusted to keep the same *fill rate*.

The order placed at the supplier will arrive at the assembly system after a lead time. This lead time is random with exponential distribution. Similar to customer demand, the level of variability for component supply is expressed by its *MAPD of lead time*.

4. ANALYSIS OF SIMULATION RESULT

4.1 Managing the demand uncertainty

The objective for this part is to find how to manage the assembly system when the demand variability changes. Generally, when safety stock increases, fill rate is improved. For a fixed safety stock, when demand variability increases, the fill rate declines, but their relationship is not linear. The larger the demand variability is, the more sensitive the fill rate is to the demand variability. When MAPD of demand is more than 0.5, the backlogged demand increases remarkably as demand variability increases.

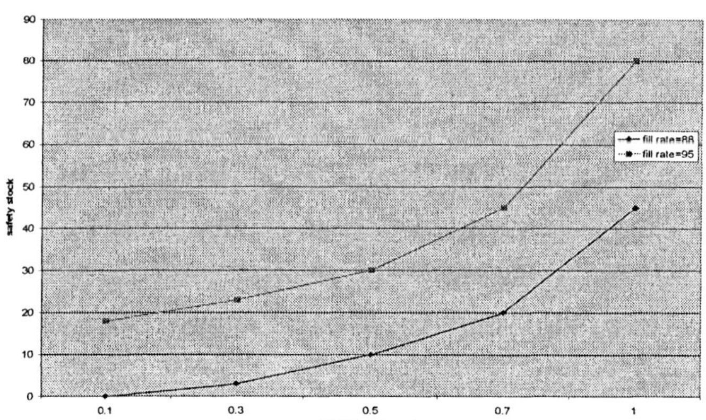

Figure 2. The needed safety stock to keep a given fill rate for different demand

In practice, when demand variability increases, we need to raise the safety stock to keep the given fill rate. The question is, how much to increase?

Or in other words, for a given fill rate, which safety stock is needed for a demand variability? Figure 2 answers this question.

As shown in Figure 2, if we want to keep the fill rate more than 95%, safety stock must be more than 18 when MAPD of demand is 0.1, 30 when MAPD is 0.5, and 80 when MAPD of demand is around 1.0. By Figure 2, we may also get the following conclusion: when MAPD of demand is less than 0.5, it is relatively easy to manage the demand uncertainty. We do not need to increase safety stock so much to compensate the increase in demand variability. When MAPD of demand is more than 1.0, it becomes very difficult to keep the fill rate more than 95%.

4.2 Managing the supply uncertainty

In an assembly system, another uncertainty comes from supply side. Figure 3 shows how to adjust safety stock when supply MAPD changes. For example, if we want to keep fill rate more than 95%, safety stock must be more than 15 when MAPD of supply is 0.1, and 40 when MAPD of supply is 0.5. It is more difficult to manage supply uncertainty when it is larger. For example, if we want to keep fill rate more than 95%, safety stock must be more than 90 when MAPD of supply is 1.0. Moreover, for this level of supply uncertainty, fill rate does not increase significantly when we increase safety stock above 90.

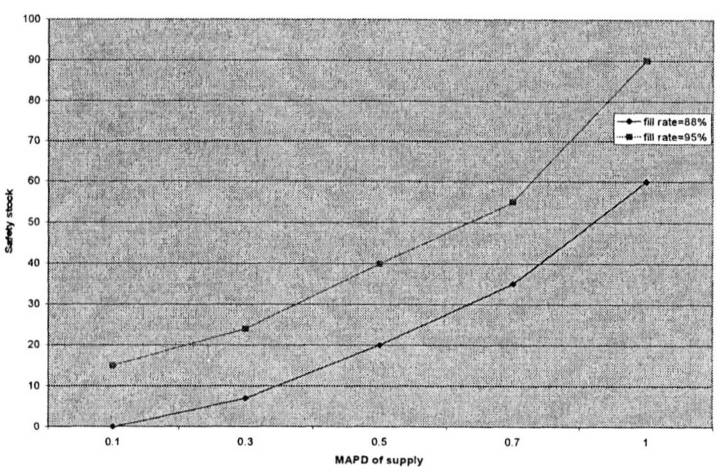

Figure 3. The needed safety stock to keep a given fill rate for different supply variability

In this section, based on the simulation results, the issues on how to manage the supply and demand uncertainty are discussed, and the results are summarized in Figure 2 and Figure 3. To apply these results, first, a reasonable fill rate must be determined. Normally, if the cost in losing demand is high, a high fill rate needs to be kept; otherwise, fill rate can be relatively low to avoid high inventory. Chopra[6] provided some formulae for determining optimal fill rate.

5. CONCLUSION

In managing an assembly system, one of the main difficulties is to deal with the uncertain environment. The uncertain delivery in component supply side and the variable order at the customer demand side are the main sources for this difficulty. In this paper, to find the way on how to manage the supply and demand uncertainty, a simulation model is built for an assembly system. In this model, the level of uncertainty for supply and demand is expressed as their Mean Absolute Percentage Deviation respectively, and the performance of the assembly system is measured by the fill rate. The reason for this arrangement is that, the uncertainty in supply and demand has direct influence on the fill rate for the customer demand. By this simulation model, the effect of supply and demand uncertainty on the performance of the system (i.e. fill rate) is assessed. Then, the approach on how to deal with the supply and demand uncertainty is provided. By these results, managers can adjust the safety stock level according to the observed supply and demand variability to keep a reasonable fill rate.

6. REFERENCES

1. Clark, A., Scarf, H., (1960), Optimal policies for multi-echelon inventory problem, *Management Science* 6, 475-490.
2. Diks, E.B., Kok, A.G., (1998), Optimal control of a divergent multi-echelon inventory system, *European Journal of Operational Research.*
3. Heijden, M., (1999), Inventory control in multi-echelon divergent systems with random lead times, *Springer Verlag.*
4. Graves, S.C., Willems, S.P., (2000), "Optimizing strategic safety stock placement in supply chains", *Manufacturing Service Operation Management 2*, 68-83.
5. Cachon, G.P., (2001), Exact evaluation of batch ordering inventory policies in two-echelon supply chains with periodic review, *Operation Research*, Vol. 49, No. 1.
6. Chopra, S., Meindl, P., (2004), *Supply chain management: strategy, planning and operation*, Prentice Hall, 250-340.
7. Bertolini, M., Rizzi, A., (2002), A simulation approach to manage finished goods inventory replenishment economically in a mixed push/pull environment, *Logistics Information Management*, Vol. 15, No. 4, pp. 281-293.
8. Bollapragada, R., Rao, U.S., Zhang, J., (2004), Managing inventory and supply performance in assembly systems with random supply capacity and demand, *Management Science*, Vol. 50, No., 12, pp. 1729-1743.
9. Decroix, G.A., Zipkin, P.H., (2005), Inventory management for an assembly system with product or component returns, *Management Science*, Vol. 51, Issues 8, pp. 1250-1266.

STUDY OF DATA MINING TECHNIQUE IN COLLIERY EQUIPMENTS FAULT DIAGNOSIS

Hongxia Pan, Jinying Huang, Zonglong Xu

College of mechanical engineering and automatizition, North University of China, China; Email: panhx1015@yahoo.com.cn.

Abstract: In this paper, aiming at the fault phenomenon of many colliery electromechanic equipments can't be expressed with the structural data, and the traditional expert system that based on the rule reasoning is very difficult to extracting the rule, put forward a kind of method that regard the numerous diagnosis case examples in the past and the faults that possibly occur and the elimination project as the knowledge source. It set up a structure frame of the data mining system that was adopted in the fault diagnosis of the colliery equipments, which based on the association rule, and discuss the data mining based on the association rules of the single layer fault and the multilayer fault in the colliery equipments system.

Key words: Data Mining, Fault Diagnosis, Colliery Equipments, Association Rule.

1. INTRODUCTION

Along with the flying development of science and technology, the open-air excavate coal equipments have gained a fast renewal. Adopting the efficient importing equipments largely, heightened the automation degree and the production efficiency of the excavate coal. At the same time, because of the structure complexity and function perfectness of the excavate coal electromechanic equipments, after they are used several years, the multifarious faults begin to occur continually. Therefore, to carry on the efficient and fast fault diagnosis for these importing electromechanic equipments, and keep its good usage performance, is an important assurance to heighten the production efficiency and reduce the cost. However, to carry

This project is supported by the National Natural Science Foundation of China under the grant No.50575214.

Please use the following format when citing this chapter:

Pan, Hongxia, Huang, Jinying, Xu, Zonglong, 2006, in International Federation for Information Processing (IFIP), Volume 207, Knowledge Enterprise: Intelligent Strategies In Product Design, Manufacturing, and Management, eds. K. Wang, Kovacs G., Wozny M., Fang M., (Boston: Springer), pp. 328-333.

on the sufficient exposure of fault source and ascertaining the fault position for the numerous and such complicated equipments, only depending on the diagnosis and maintenance man who used the traditional methods to carry on the fault diagnosis and exclusion, which has more localization generally. To introduce the artificial intelligence into the fault diagnosis field of the excavate coal equipments, which provides a new method for the fault diagnosis of the excavate coal equipments.

In this paper, aiming at the characteristic of the excavate coal equipments, which the numerous parts of which the performance is complicated and each constitute a complicated system, researched to make use of the data mining technique to obtain the association rule between fault and reason. It makes the decision maker easily get the fault relating degree among each subsystem of the excavate coal equipments. It provides the important reference information for fleetly and thoroughly eliminating faults..

2. DATA MINING TECHNIQUE IN EQUIPMENTS FAULT DIAGNOSIS

The data mining technique is one of the important researched contents of the intelligence system theory and technique nowadays [1]. It refers multi-domain — machine learning, pattern recognition, inductive reasoning, statistics, database, data viewable, high performance calculation, etc. The data mining technique aims at discovering the knowledge concealed in the numerous data, and will be applied to resolving the problem that is "data rich but knowledge poor". It has been applied to some trades—industry, business, finance, medicine, administration management, communicate network, etc. It has played a pole in some ways, such as the modeling of the fuzzy controller, the modeling and forecasting of fault diagnosis, etc.

In the fault diagnosis of equipments, it is the key step to build the diagnosis knowledge model. The knowledge model of the diagnosed equipments is expressed by 4 kinds of modes, such as the equipments running model, the diagnosis experience rule, the diagnosis decision model and the diagnosis case. According to the diagnosis knowledge sequence from shallow to deep and from general to special, they can be divided into four layers to be organized. The first layer knowledge describes the elementary diagnosis knowledge, providing the essential elucidation for the diagnosis. The second layer knowledge describes the general diagnosis knowledge, applying to the diagnosis that based on rules. The third layer knowledge is the diagnosis decision model that was formed according to the similar diagnosis case via training and structuring. The fourth layer knowledge is consisted of the case database that based on the binary tree, applying to the diagnosis that based on cases [2].

2.1 Build self-learning model of fault diagnosis knowledge

The strategy of the modeling is: ①Learn the diagnosis rule between the symptom and the diagnosis result from dataset. ② Carry on testing and processing, every data sample that can be explained by mined diagnosis rule should be deleted from dataset. Regard the dataset as the training sample set, applying to build the diagnosis decision model; ③After structuring the decision model via the training and leaning, carry through testing and processing for the dataset, every data sample that the error of diagnosis result is less than a certain threshold should be deleted from the dataset. ④A small quantity of data that haven't been deleted in the dataset, after getting rid of the noise data via verifying mutually by users, are deposited in the case base. These cases express the diagnosis knowledge of special fault that is different with generic diagnosis knowledge model, and these cases carry on the organization of binary tree according to its exceptional characteristics, in order to carry on the fault diagnosis that based on cases.

2.2 Constructing fault diagnosis model based artificial intelligence knowledge

To constitute the fault diagnosis model of the fault diagnosis example set of diagnosis equipments via training, the form of model lies on the structure of model and its learning rule. They can be selected according to the instance of diagnosis equipments. It may be the Nerve Net model, or may be the Fuzzy Clustering model or other diagnosis models. After completing model training, can use the diagnosis models that have been built to compute the corresponding diagnosis result according the actual symptoms of diagnosed equipments.

2.3 Fault diagnosis reasoning based on mixed knowledge model

After building fault diagnosis knowledge models of the diagnosed equipments via self-training, can apply mixed knowledge model (function model, rule, decision model, case) to carry on fault diagnosis reasoning. Its strategy is: First, confirm the symptom according to the function model of diagnosed equipment, and carry on the diagnosis reasoning matching the corresponding fault diagnosis rule. If without the promise that fault diagnosis matched with the symptom, use the diagnosis decision model to carry on fault diagnosis. If the precision of the diagnosis result of decision model is under the given threshold, then carry on the fault diagnosis that based on cases according to the symptom of diagnosed equipments matching with the diagnosis cases in case base.

3. EQUIPMENT FAULT DIAGNOSING ANALYSIS BASED ASSOCIATION RULE

3.1 Building database of colliery equipments fault [3]

Colliery equipments fault has layer, correlation and synthesis. Layer is that low-layer fault must induce high-layer fault, but high-layer fault can be induced by low-layer fault. Correlation is that after a certain structure cell has occurred faults, certainly will induce the correlative structure cell or correlative state with it to be changed, further will be likely to induce the correlative structure cell or correlative state to happen faults too. Synthesis is that any of primary faults has multifarious potential evocable faults. This shows, the colliery equipment is a system that many faults occur simultaneously. In this paper, the colliery electromechanic equipments are divided into four great systems — machinery and hydraulic subsystem, lubricating and cooling subsystem, electric and control subsystem, power drive subsystem. Such as the power drive subsystem, to build the fault database of power drive subsystem, as Table 1 showed.

Table 1. Fault database of power drive subsystem

Engine assembly	Fault 1	Fault 2	...	Fault k
Turbine supercharger	Fault 1	Fault 2	...	Fault m
...	...	...	...	...
Gear-box assembly	Fault 1	Fault 2	...	Fault n

For every element of the subsystem database, can build its each database respectively as well as. Take example for engine assembly, can build its own fault database as Table 2 showed.

Table 2. Fault database of Engine assembly

Fault 1	Cause 1	Cause 2	...	Cause k
Fault 2	Cause 1	Cause 2	...	Cause m
...	...	...	...	...
Fault k	Cause 1	Cause 2	...	Cause n

3.2 Mining of association rule in single-layer fault

The purpose of data mining rests with finding out the reliable and significative rules. Reliability predicates the reliable degree of rules. Support degree represents the important degree in all data. As a result, the association rule which reliability is small (that is the weak association rule) has a lower application value generally. After carrying through the mining of single-layer association rule for the database of Table 1, can acquire such as association rules of the faults that occur in every constitute component of power drive subsystem. These association rules can afford the reference information that is worthily regarded for the elimination of relativity faults in single-layer equipments.

3.3 Mining of association rule in multi-layer fault

In the numerous faults that occur in every subsystem of colliery electromechanic equipments, many faults are often interactional and inter-restricted. The rules of relational faults are concealed in the different abstract hierarchies of database, for example, the fault of clutch can cause all devices which were used in the subsystem of transmission back-end not to work normally. The fault logger of current equipments can't show the correlation of these faults, and numerous traditional fault diagnosis expert systems haven't the capability of finding out the correlation of multi-layer faults, the result can influence the comprehensive diagnosis of faults, therefore, it is considerably necessary and valuable for fault diagnosis to finding out the association rule of multi-layer faults.

Actually, all kinds of faults that occur in colliery equipments aren't often in the same layer, for example, engine system fault and clutch fault aren't in the same layer, but in fact clutch fault sometimes can cause the output of engine to change drastically, thereby induce engine system fault.

4. STRUCTURE FRAME OF DATA MINING SYSTEM IN COLLIERY EQUIPMENTS FAULT DIAGNOSIS

Basing on the analysis of the above application of association rule in colliery equipments fault diagnosis, in the colliery equipments fault diagnosis data mining system, its concept model of structural frame can be designed to be mainly composed of user query interface, query coordinator, database management module, knowledge base management module, model base management module, knowledge discover pretreatment module, knowledge evaluate module and conclusion interpret module etc. The structural frame of system is showed as Fig1.

5. CONCLUSION

Colliery equipment is a sort of system whose multi-fault is concurrent, and its fault has some characteristics, such as layer, relativity and syntheses etc., the traditional fault diagnosis expert system is not easy to extract the rule from the data structured difficultly. The data mining technique can mine the association rule on the base of the numerous fault data. It also can gain the valid association rule between the fault and the reason, as well as between faults. It provides some very important reference information for obviating the fault fast and thoroughly.

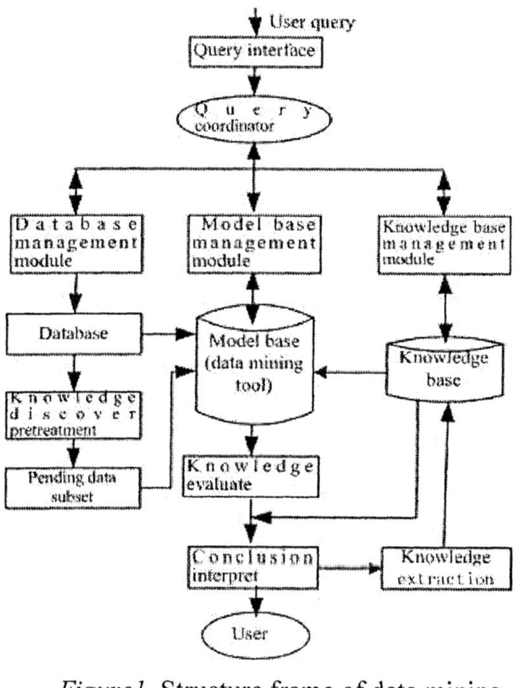

Figure1. Structure frame of data mining

6. REFERENCES

1. Fyyad U M, Piatetsky-shapiro G, Smyth P. *Advances in knowledge discovery and data mining. California*: AAAI/MIT, press, 1996.
2. Yang Jie, Huang Xin, LüYong. Data mining technique and its applications in modeling optimization and fault diagnosis [J]. *Infrared and Laser Engineering*, 2000,29(3):23-27.
3. Dong Doudou, Li Dengfeng, Chen Yuwen. Data Mining System Frame of Ship Fault Diagnoses based on Association Rules[J]. *Ship Engineering*, 2001,(4):61-64

A MULTI-AGENT SYSTEM FRAMEWORK FOR SUPPLY CHAIN MANAGEMENT UNDER DISTRICT ALLIANCE SETTING

Fangzhong Qi[1], Gengui Zhou[1], Yongzhu Qi[2]

[1]*College of Business & Administration, Zhejiang University of Technology, China; Email: fz_qi@zjut.edu.cn.* [2]*Institute of Advanced Manufacturing Technology, Lanzhou University of Technology, Lanzhou 730050, China.*

Abstract: Multi-agent system (MAS) is regarded as an efficient approach to solving dynamic and distributed problem in complicated operation process of supply chain. The MAS organization architecture for supply chain of district alliance is discussed based on virtual agent federation. A MAS architecture model for supply chain system is created and the functions of facilitator agent and Workflow engine have been emphasized in the model. Multiple upper-layer agent federations are designed to execute the complex system tasks and to implement the system functions. A MAS system framework which has features of high reactivity, reusability, flexibility and reliability is proposed based on CORBA standards. Application example is illustrated in a synergic business prototype system under district alliance setting.

Key words: MAS, Supply Chain Management, District Alliance, System Framework.

1. INTRODUCTION

District alliance, a developing manufacturing mode widely practiced in Southeast China, has contributed a lot to Chinese economy during the last decade. In such alliance, production system is a dynamic and complex supply chain system which is organized by several large, almost dominant enterprises and their tightly collaborative partners such as suppliers, manufacturers, distributors. The combination of these independent entities and the number of collaborative partners in a supply chain often change according to market opportunity and business strategic. Highly dynamic environment requires synergic work in supply chain management that provides more accurate allocation of costs to products or activities by

This project is supported by the National High Technology Research and Development Program of China under the grant No. 2002AA414070.

Please use the following format when citing this chapter:

Qi, Fangzhong, Zhou, Gengui, Qi, Yongzhu, 2006, in International Federation for Information Processing (IFIP), Volume 207, Knowledge Enterprise: Intelligent Strategies In Product Design, Manufacturing, and Management, eds. K. Wang, Kovacs G., Wozny M., Fang M., (Boston: Springer), pp. 334-343.

optimizing inter-organization resources. Therefore, the challenge of today's district alliance is to find novel ways to run their businesses to become more flexible, reliable, and responsive. Agent has emerged as an effective technology to support the management and operation of supply chain [1, 2]. Considering the features in district alliance, it is reasonable to abstract these independent entities as autonomous agents which deal with business transactions. The integrated supply chain system can be viewed as a multi-agent system formed by agents that subject to different organizations.

The structure of any MAS is naturally a distributed one, and the connection between agents is a point-to-point relationship. Such fully distributed structure makes the system to be highly agile, but every agent in the system is required to own more knowledge about its environment. Supply chain faces dynamic and fluctuant market place, and the scope of resource optimization extends to even wider district, this will cause the structure of agent to be more complicated. Along with the number of involved enterprises and entities in supply chain increases, the cost of system communication in such point-to-point network connection will become a serious problem. We adopt the concept of agent federation [3, 4]: the whole supply chain system in a district alliance can be designed as a multi-agent system constructed by cohesive and dynamic agent federations for all business functions of partner enterprises participated in the supply chain.

The organization of agent federation is not unchangeable, but can be dynamically reconstructed according to the changing business process. Combined with the viewpoint of dynamic alliance, we further introduce the concept of virtual agent federation: before a production task or a customer order is put forward, each distributed autonomous agent is on the state of loosely coupling; As soon as system task is confirmed, all agents send their own intention to other agents by their status and interest based on the mechanism of contract-net. For a particular system task, these agents will come into being a virtual agent federation to accomplish the task according to coordination protocols. When system task is completed, the virtual agent federation will disperse automatically and resource will be released waiting for next task. The existence presupposition of virtual agent federation is that each agent owns its superiority and domain knowledge, and can execute determinate business functions.

2. A MULTI-AGENT SYSTEM ARCHITECTURE MODEL

The whole supply chain system can be viewed as an autonomous multi-agent system which is composed of agents belong to different enterprises, so any node enterprise or organization on a supply chain can be abstracted to a set of agent federations – the lower layer agents in a agent federation can be supervised by the upper layer agents and by contraries, the upper layer agents can specify tasks to the lower layer agents. The agent federation can process nesting according to actual requirements, namely the agent at the middle layer of an agent federation will be the administrator of their lower layer agent and will be supervised by their upper layer agent. Based on above discussion, we created a multi-agent system architecture model for supply chain under district alliance setting, as shown in Figure 1.

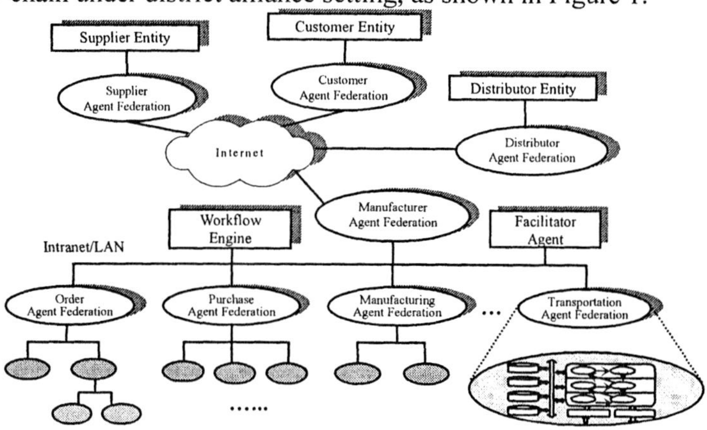

Figure 1. A Multi-Agent System Architecture Model in Supply Chain.

Generally, supply chain is a production mode that aims at a certain market opportunity and is driven by customer orders, so the top layer agent in the federation is usually an order agent. The tasks of an order agent are to administrate the inter-organization cooperative works and to receive, send or monitor customer orders at enterprise level. After an order agent has received an order message from a customer, it queries other agents about the production circumstantiality or capacity to make sure if the order could be promised. As soon as the order has been confirmed, it specifies the task according to the different interest and capability of its lower layer agents. During this process, the work of task distribution could be implemented by adopting one-to-one negotiation or one-to-many auction mechanics. Finally, a feasible task implement plan is set down and the planning result is submitted to work flow engine. Actually, the plan set here is an order execution plan.

While plan is being executed, the work flow engine sends the work task to the work list of every function agent based on order execution plan. Each function agent carries out the practical function operation about the task assigned by the work flow engine. After the task has been accomplished, the relevant information is written into process database and the work flow engine does not startup the next task of the function agent accordingly until the order has been fulfilled. Actually, the work flow engine acts as the role of monitoring the process of plan executing within the whole multi-agent system. For exceptional situation such as order postponed or plan readjusted, the plan or activity will be set to the state of suspension. The work flow engine sends request message to the order agent who will negotiate with other functional agent, and this leads to the next negotiation process.

During the process of task distribution, sometimes conflicts between local and overall benefit could occur probably. Under such circumstances, the lower layer agent must submit to the upper layer agent. In regard to the number of layers in an agent federation, it relates to the scale of the system and the granularity of each agent de-signed. Commonly, we think that 2 to 4 layers will be appropriate in practice.

Considering the peculiarities in the operation process of district alliance, the multi-agent system of supply chain we designed is composed of several upper layer agent federations. We define their core functions as followings:

- Order Agent Federation: in charge of accepting customer orders and negotiating with other functional agent about the order production task. It also creates or adjusts order execution plan, deals with exceptional event, customer information maintenance, order auction and negotiation. This federation is the core part of the whole supply chain system.
- Resource Agent Federation: in charge of the dynamic management of all available inner- and inter-organization resource, gathering the capability and status information of all production resource.
- Purchase Agent Federation: in charge of selecting supplier, maintaining the supplier information, creating purchase plan, monitoring the purchase process, and managing account payable. Other functions of this federation include investigating the material and parts or components supplied, inquiring price and inviting public bidding for the material demanding.
- Transportation Agent Federation: in charge of managing the transfer and convey of material within the organization or among collaborative enterprises, functions also include transportation route planning, coalition of transportation task, creating the transportation plan and maintaining the transportation equipment.
- Finance Agent Federation: in charge of inner- or inter-organization cost control and management, including cost accounting, cost planning, cost forecasting and cost analytic Statistics.

- Manufacturing Agent Federation: in charge of implementing the production task in each workshop or manufacture cell, mainly including the creating of workshop synergic operation plans, scheduling the production task in workshop or work center.

These agent federations are classified by executing functions in the business process of supply chain management. Several children agents could be employed cooperatively to fulfill a certain task under a certain circumstance. For example, the order agent federation could be decomposed to such children agents as customer ser-vice agent, task auction agent, order planning agent. The magnitude of agent granularity to be designed is a problem noteworthy. If an agent granularity is thick enough, the level of system reusability will be getting worse, and will lessen the distributing and autonomous capability of the multi-agent system. Reversely, if an agent granularity is too thin, the complexity of the system will increase accordingly, and the cost of collaborative communication among multiple agents will also increase greatly. During the process of our research, we make detailed plan and system analysis for the supply chain system applied in the district manufacture alliance. Based on this necessary work in advance, the functional agents with appropriate granularity are created to constitute the whole supply chain multi-agent system.

When the collaborative enterprises participated in a supply chain for a certain market opportunity is too many, there is a serious problem about site allocation and capacity matching between service demander and service provider. In the multi-agent system of supply chain, we introduce a third part agent called facilitator or match-maker [5, 6] whose main task is to play the role of bridge over the service provider and service demander by abstracting and describing the capability of information service it owned. The service provider registers its capacity to the facilitator agent. As any service demander makes a query, the facilitator will match these two parts, and the facilitator's task is finished after sending the matching information to the service demander. The multi-agent system is configured with several facilitator agents to solve the site allocation and capacity matching problems among multiple agents. The introducing of facilitator provides an agile and effective interactive channel in the communication between agents.

3. MULTI-AGENT SYSTEM SOFTWARE FRAMEWORK

From the view of system design, the action of agent is to implement a system function call. As there are many Legacy systems running in different collaborative enterprises inevitably, we encapsulate the modules to be called by each agent by employing CORBA standard. The whole software system is composed of Legacy system, CORBA middleware, CORBA objects and agent/agent federation.

Figure 2 illustrates a multi-agent software framework in the supply chain of district alliance. The structure of the whole system is a distributed architecture based on CORBA mechanism. Because the application service based on CORBA represents to the outside as an object-oriented interface definition which employs the interface definition language (IDL) to define, the co-operation between application services can be independent from location and environment. These application services will be convenient to implement the action of agents. Application to be developed should act in accord with CORBA standards, and the final application designed will be either a CORBA object or an application server which can provide remote call service. As for Legacy system, we introduce a system encapsulation method called Adapter to encapsulate the Legacy system forming the CORBA object or application server with the remote call function.

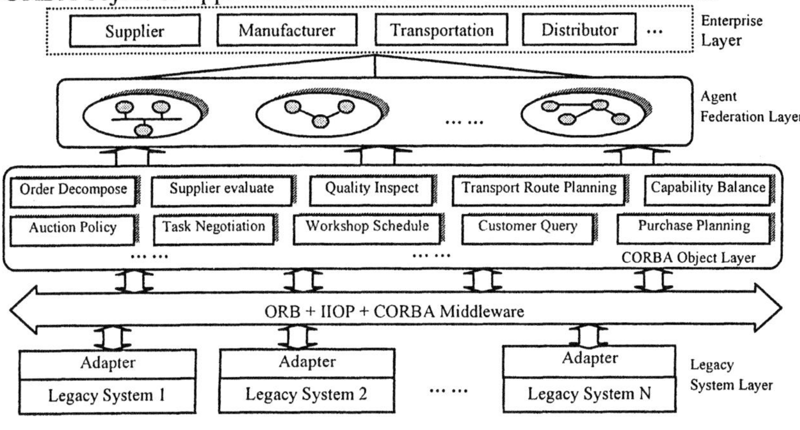

Figure 2. System Software Framework of Multi-Agent System in Supply Chain.

The supply chain multi-view models created at our initial system analysis phase has been used to guide the designing of the CORBA application layer and agent/agent federation application layer of the multi-agent system by using information encapsulation techniques. By adopting such hierarchy framework and integrated multi-view models, it will accelerate the establishment of application and operation environment in the supply chain system design [7]. The modeling idea of the supply chain system runs through the creating of system application objects and the designing of agent – any changing about the multi-view model merely impacts on the changing of the action or function of the agent. Above the CORBA object is the agent federation layer – they can call or communicate with each other. Several relevant agents compose agent federation, and the agent federations represent the enterprise functions which participate in the implementation and operation of the supply chain system.

The upper layer agent federations which are designed based on dynamic alliance concepts, can be created, adjusted and reconstructed in response to the market demand and to the optimization of operation process and organization. It can plan, schedule and manage the main process of collaborative work in the operation of supply chain management. By negotiating and adjusting adequately among these upper layer agents, organizations in the supply chain system could contact suppliers or customers as earlier as possible. To optimize every process in the supply chain system, they can also provide the chance to discover problems and take remediation action as soon as possible. The steps in process and subroutines that involves fewer steps are dominated by lower layer agent federations, so the functions of lower agents are relatively steady. This will increase the response speed to market, increase the system reliability and capability of reuse. When the market opportunity is fulfilled and the supply chain is decomposed, the reengineering of the supply chain can be realized quickly by re-configuring the agents' or agent federations' functions according to new business process. The implementation of agent employs CORBA middleware techniques, so each heterology agent is provided with the feature of plug and play. Ideally, if all processes in the supply chain system employ the same techniques, information in the supply chain system will realize real seamless integration, what needs to do is just adjusting the agents' functions to reconstruct the supply chain based on the actual requirement.

4. CASE STUDY—AN EXAMPLE

Based on the above discussion, we have designed a district alliance-oriented supply chain system model and developed a prototype system. The system includes sales agents ES_Agent1 and ES_Agent2 which represent the suppliers in the system, purchase agent P_Agent, manufacture agent M_Agent, sales agent S_Agent, transportation agent T_Agent and the enterprise Workflow management system (WfMS). The architecture of the distributed multi-agent system is shown in Figure 3.

Each agent in the system is programmed in Java, and the connection between Java and database employs JDBC_ODBC Bridge. Among these agents we designed, P_Agent from the purchase department and ES_Agent1, ES_Agent2 from suppliers apply the technology of mobile agent. These mobile agents have been developed on the Aglets development platform which is provided by IBM Corporation. The main process of the multi-agent system is divided into two phases:

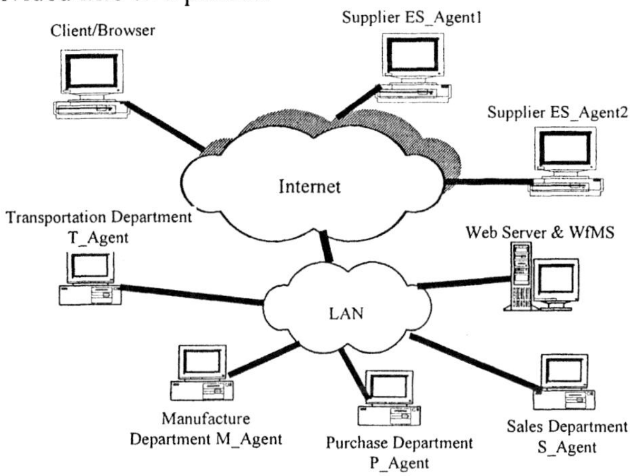

Figure 3. Distributed System Architecture of MAS in Supply Chain.

1. The confirmation of customer order and the process of negotiation;
2. The actual production process.

Customers will obtain interrelated data through their interaction with the web server firstly. If a new order is created, the request message of new order is sent and the sales department's S_Agent will be activated. Then S_Agent will query the inventory about finished product based on the material code of the product, and if the quantity on hand is enough, it will send transportation message to the T_Agent. Its task of this new order will be finished as soon as its data item is updated. If the quantity is not enough, S_Agent will calculate the net amount of requirement, and then it will send the product style, quantity and date of delivery to the manufacture department's M_Agent. M_Agent will then confirm whether it can finish the task based on its capability. If there is any conflict, a feasible solution will be found by sufficient negotiation between M_Agent and S_Agent at this stage. Subsequently, the result of negotiation will be sent to purchase department's P_Agent. After P_Agent gets the production message from M_Agent, it will decompose the task based on BOM and combine the raw and processed materials to produce a purchase bill. Then the P_Agent will verify the requirements to see if they could be satisfied and if there is any conflict, it will solve the problem by negotiating with M_Agent. By contraries, it will send confirmation message to M_Agent. As soon as M_Agent gets the confirmation message, it will send confirmation message to S_Agent. Thus, the whole order confirmation process is finished, and S_Agent will mark the order confirmation as ready to the Workflow management system (WfMS).

The production process is driven by the Workflow management system [8]. Each work process and the assigning of executive role to each process have been predefined in WfMS, and here we realize this by the method of database table. The production process of the order is started up by system administrator at an appointed time. The basic process is: purchase, quality inspect, in and out of inventory, parts manufacture, components assemble and finished product in stock.

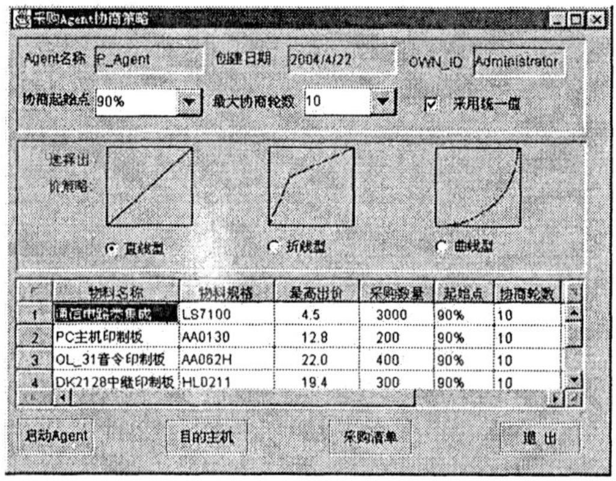

Figure 4. Negotiation Policy of the Mobile Agent P_Agent.

During the process of purchasing, we introduce the mode of mobile agent. The mobile agent can move from local host to the suppliers' host according to the suppliers' IP addresses which the purchase staff has provided. The mobile agent communicates with the suppliers' ES_Agent1 and ES_Agent2 to get the quoted price of raw materials. After comparing the quoted price of same material from the two suppliers, the quotation will be fetched to the local host. Then the purchase staff is able to fix on a supplier and execute the further actions in the purchasing process.

After the purchase process of raw and processed materials has finished, P_Agent marks the purchase accomplishment as ready to the WfMS. The WfMS then sends messages to M_Agent and T_Agent sequentially according to the predefined process. Figure 4 shows the confirmation process about the negotiation policy of the mobile agent P_Agent.

5. CONCLUSION

The MAS framework we designed provides a supply chain system with the feature of high reconstruction. While outside environment of the system is changing, the only work we need to do is to adjust the functions of the relevant agent or to let the agent join in or leave from the system automatically according to the actual requirement, and this will not affect the natural work of any other agent within the system. Furthermore, these agents could exchange information directly with the Legacy system at the interface layer through the interface definition of Legacy system based on CORBA standards. Such practice shortens the cycle of system development because we need not to redesign the whole system. As the agents are designed in conformity to the intelligent models of different functions within the different enterprises' problem domains, they are able to respond continuously to the ever-changing environment while they are executing a certain tasks the system designers have defined under the dynamic circumstances, if the conditions are satisfied. By employing the agent technology, the integrated scope of enterprise application system could be extended to either upper or downstream cooperative partners. At the same time, these agents have the characteristic of high reactivity, so multi-agent system can answer incidents such as cancelled order or client system breakdown – these will reduce the enterprise's losing farthest.

6. REFERENCES

1. Jin, D., Yueting, C. and Jiaben, Y. Coordination and Optimization in Agile Supply Chain Management Based on Multiple Agent System. *System Engineering – Theory & Practice,* Vol. 20, No. 8, Aug. (2000), pp. 7-11
2. Teigen, R. Information Flow in a Supply Chain Management System. *Diploma Thesis. Univer-sity of Toronto,* May (1997)
3. Hai, Z., Jian, C., Yulin, F. and Xiaoqing, S. A Federation–Agent–Workflow Simulation Framework for Virtual Organisation Development. *Information & Management,* Vol. 39, No. 4, Jan. (2002), pp. 325-336
4. Khedro, T. A Distributed Problem-Solving Approach to Collaborative Facility Engineering. *Advances in Engineering Software,* Vol. 25, No. 2-3, March-April (1996), pp. 243-252
5. Fischer, K. The Design of an Intelligent Manufacturing System. In Proceedings of the 2nd International Working Conference on Cooperating Knowledge-based Systems, University of Keele, 1994, pp. 83-99
6. Finin, T., Fritzon, R. McKay, D. and McEntire, R. KQML – A Language and Protocol for Knowledge and Information Exchange. Tech. Report, University of Maryland, Baltimore, 1993
7. Zhigang, Z., Xiaohong, P., Ming, G. and Yaodong, C. Research on Modeling Technology of Agile Supply System Supporting Dynamic Alliance. *China Mechanical Engineering,* Vol. 12, Suppl. 1, May (2001), pp. 64-67
8. Fangzhong, Q., Zhengxiao, W., Xiaohong, P., Yaodong, C. and Zhigang, Z. Research on Workflow Management System (WfMS) Framework in Agile Supply Chain. *China Mechanical Engineering,* Vol. 13, No. 9, May (2002), pp. 750-755

RESEARCH INFO "BULLWHIP EFFECT"IN SUPPLY CHAIN MANAGEMENT SYSTEM BASED ON WSRF IN GRID

Yan Kang , Shiying Kang
Institute of Computer Science and Information Engineering,Chongqing Technology and Business University, Chongqing, China;Email:kangsy@ctbu.edu.cn

Abstract: Strengthening the information share extent and establishing the supply chain resources share system based on network is the important measure to weaken "bullwhip effect". The reason that causes the low efficiency for supply chain information share platform based on traditional Internet and Web Services to weaken "bullwhip effect" is expounded. Through the analysis on supply chain management platform designed with the WSRF.NET technology in Grid environment, the effect of the platform on reducing "bullwhip effect" is also demonstrated theoretically and technologically.

Key words: Grid; WSRF (Web Service Resource Framework); Supply Chain Management; Bullwhip Effect.

1. INTRODUCTION

"The Bullwhip Effect" is the phenomenon that the change extent of the order quantity that was from lower reaches to upper reaches of supply chain increase step-by-step, or even greatly contort. It is a hot issue that is focused by supply chain enterprises and their managers. To "the Bullwhip Effect", many scholars have done their research. The famous "Beer game"[1] carried out by Steman of Sloan School of Management in the American Massachusetts Institute of Technology is the research model on Bullwhip Effect, which is based on the system dynamics. This experiment shows that in the linearity cost structure, the order quantity increases when it moves to the upper reaches of the supply chain. Because the information is delayed,

Please use the following format when citing this chapter:
Kang, Yan, Kang, Shiying, 2006, in International Federation for Information Processing (IFIP), Volume 207, Knowledge Enterprise: Intelligent Strategies In Product Design, Manufacturing, and Management, eds. K. Wang, Kovacs G., Wozny M., Fang M., (Boston: Springer), pp. 344-349.

order form will not reflect the present practical beer demand, and it is difficult for the decider to forecast the demand. Researchers of this experiment believe that the Bullwhip Effect in supply chain system is decided by the structure of this system. System structure affects every part's behavior in system, so the appearance of Bullwhip Effect is inevitable.

Nowadays, to weaken "the Bullwhip Effect" by setting up Internet supply chain share system has become the common understanding of many domestic and foreign scholars. Especially the system which is open, can well combine with information system of every enterprise in supply chain, and can make every enterprise smoothly and perfectly realize the direct information communication and resource communication, is the ideal system in people's heart. So, we can find out that the problem through realization of information share and supply chain system structure change to weaken "The Bullwhip Effect" has been solved. However, the practice in recent years cannot be all that could be desired, which is showed in the following facts.

2. REDUCTION OF "THE BULLWHIP EFFECT" THROUGH SUPPLY CHAIN INFORMATION SHARE SYSTEM ON TRADITION INTERNET

In recent years, scholars are devoted to set up supply chain information resource share system on traditional Internet. The presented system can be divided into two kinds: (1) Supply chain information share system based on Internet, (2) Supply chain information share system based on Web Service

2.1 Supply chain share system based on Internet

In 2002, Gong Bengang put forward a information share system model based on Internet; Pan Yigang et al further put forward the technology for setting up supply chain management system applied XML and SOAP technology on Internet], to provide interface for different manufacturers on different platforms of supply chain members, so that the reduction of "The Bullwhip Effect" through information share in supply chain is realized. But this method doesn't work well; because:(1)Different software in asynchronous platform actually doesn't have a uniform description language and uniform format. Uniform information transfer cannot be realized among platforms.(2) Concurrency Control and asynchronous control in information transfer can not be solved. We can see that though an information share system has been connected in Internet form, actually information share is hard to realize. So the reduction of "Bullwhip Effect" is very limited.

2.2 Supply chain share system based on Web Service

The appearance of Web services that are open and span-platform realizes a good mechanism for supply chain to set up asynchronous platform information system and "share resource". In 2002 Xu Wei et al present "the workflow technology integrate in logistics based on Web Service"[4], and from 2004 to 2005, Xian Feng, Jin Jiwen et al carried out the supply chain system research and realization based on Web Services, and the supply chain information share is made great progress.

But supply chain doesn't realize the real "resource share" in technology, because:(1) Many important "Non share part" resources which are closely relevant to supply chain separately distribute on every network website, they cannot be effectively found and used. "Information isolated island" and "resource isolated island" cannot be removed. (2) The speed and efficiency of Web services are always severely restricted by non-share software or hardware and network, and even long time wait or computer system crash happens. (3) What the especially serious situation is that when the connected software and hardware are too many, every enterprise information system of every supply chain can not be well combined, or either information communication and resource communication can smoothly carried out. Now we can see that though the theoretical research on "Bullwhip Effect" has achieved great progress, when Grid isn't referred, any effort for people to apply share platform to reduce "Bullwhip effect" fails to meet the demand.

3. REDUCTION OF SUPPLY CHAIN PLATFORM ON BULLWHIP EFFECT BASED ON WSRF

Grid brings new hope for supply chain share platform that can truly reduce "The Bullwhip Effect". But the building process is tough and hard.

3.1 Web Service based on Grid

Grid is a new technology constructed in the new generation Internet. It will combine high speed Internet, PC of different geography locations, high capability computer, large database, sense organ, long distance equipment, peripheral equipment and so on together, so the whole Internet becomes a huge super computer that is comparatively transparent to clients. The resource share and cooperation among computer resource, storage resource, communication resource, software resource, information resource, knowledge resource and so on are realized, and information isolated island and resource isolated island are eliminated.

Since both the Grid and Web Services have such advantages, it is natural for people to consider bringing Web Service in Grid. "Web Service based on Grid"—Grid Service is formed to realize the combination of Grid and Web Service. So through Grid, every application and resource on "asynchronous structure" network is connected. "Huge super computer" is formed. It not only solves the "bottleneck" problem of speed and efficiency of Web service, but also provides every kind of resource for the whole Grid. So in the recent 1 or 2 years, the supply chain information share system based on Grid technology is put forward.

3.2 The presentation of supply chain information share system based on Grid technology

In 2003, Liu Wei et al put forward the assumption of "supply chain management system based on Grid calculation"; Zhao Xinjuan et al put forward the "supply chain management system model based on Grid calculation" and the concept of "based on facing to grid service".

However at that time, Grid and Web Services were not truly combined in technology. The former Grid system structure based on the OGSA/OGSI (Open Grid Service Architecture/Open Grid Service Infrastructure) that has some shortages and disadvantages: (1) The criterion and explanation are too complicated to support the future extension and development; (2) When used, serious problems exist among OGSI and the present Web service, WML tool, and the expansion of WSDL is difficult; (3) The concept and method of "face to object" are over used. These shortages form a certain obstacle to prevent the further development of Grid technology. Every grand computer company of the whole world puzzles its brain about how to make full use of the advantages of Grid and Web Services. The research on the combination of Grid and Web Services is on the initial stage.

3.3 Reduction influence of supply chain management platform based on WSRF on the Bullwhip Effect

The real combination of Grid and Web Service was after IBM, Globus union and HP put forward the WSRF in 2004.

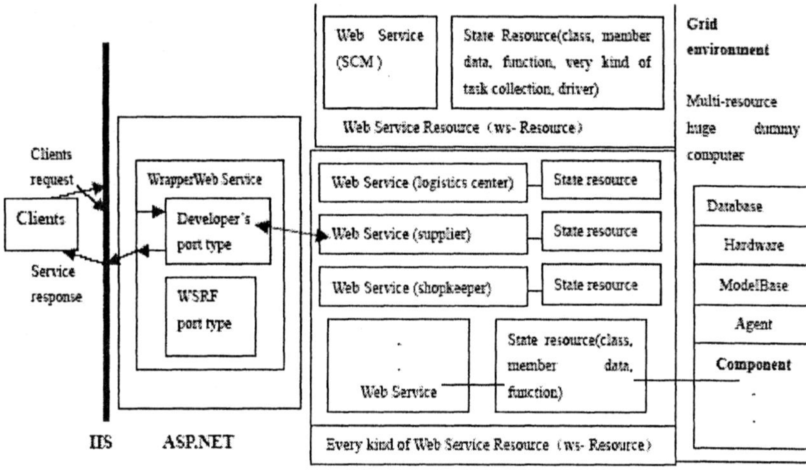

Figure 1. Supply Chain Management platform Model based on WSRF in grid

WSRF is a new method to express the relation between Web Service and state resource, and is the realization of combination of Grid technology and Web Service. It adopts different structures to modeling state resource and its related Web Service, while OGSI adopts one structure to modeling one state resource as one Web Service. Only one to one mapping relationship is allowed between Web Service and its related state resource in OGSI. WSRF can form many to many mapping relationship. So as a interface and management technology of calling WS-Resource, WSRF replace OGSI to provide the realization conditions for the supply chain information share system to achieve real resource share and cooperation, and to get rid of information isolated island and resource isolated island.

Figure 1 is the Model of the supply chain management platform that is designed through applying WSRF.NET2.0, and using related distributed system based on WSRF.NET technology for reference by us [9].

In this platform, the "Wrapper Web Service" is generated by WSRF.NET service (ASP.NET service) tooling. This Service is from the web service written by service developer port types and certain WSRF specification-defined port types that the developer chooses to include. This Service can run as a normal Web Service in ASP.NET worker process. IIS dispatches HTTP requests to the service, which internally invokes either a method on a port type written by service developer or a port type defined by WSRF.

Before the method is invoked. The Wrapper Web Service uses the value of the EPR (EndpointReference) in the <To> header of the invocation SOAP message to interact with a particular WS-Resource and retrieve state values.

WSRF.NET automatically resolves the execution context presented in the EPR and provides a programmatic interface to access the specified WS-Resource (class, member data, function, every kind of task collection, driver and so on). Although there are many different resolution mechanisms and types of WS-Resource, WSRF.NET implements WS-Resource by executing method on developer's port type or WSRF port type. These methods include visiting database, driving hardware and processing information for every Web service and client side.

WSRF.NET2.0 can provide interface for developer to design the function he needs, and can carry through the building, producing, loading, storing and so on of every kind of WS-resource" [9][10][11]]. So this platform can use WSRF to integrate every member heterogeneous information system in supply chain, and to realize the share of smoothly direct information communication and resource communication and every software and hardware support platform; meanwhile, recurring to Grid technology environment, high capability resource can be chosen to meet the special demand of clients in high speed and efficiency (such as decision-making analysis, high speed calculation, intelligent calculation, and high speed flow media transfer). So not only the "information isolated island" and "resource isolated island" phenomenon can be eliminated, but also every enterprise member in supply chain can make use of technology condition to set up close cooperation between each other in supply chain.

The appearance of WSRF is the sign that the Web Service integration technology based on Grid has developed to a new stage. From this stage on, supply chain management platform can possess the realization conditions to realize the information share and resource share span time and space in technology.

4. REFERENCES

1 Sterman J D. The beer distribution game [A]. In Heieke J, Meile L, Eds. Games and Exercises for Operations Management [M] .New Jersey: Prentice Hall, (1995). 101—112
2. Wei Xu, Zong-Kai Yang, Jing Xia, Application of Workflow Based on Web Services in Logistics System Integration [J], *Logistics Technology*, (2002).2:16:19
3. Wei Liu, Supply Chain Management System based on Grid Computing [J], *Group Technology and Production Modernization* Vol.20, No.3 (2003):28-31
4. Xin-juan Zhao, Guo-zhen Tan, Xun-y1J Wang, Research on the Model of Supply Chain Management System Based on Grid Computing [J], May (2004):82-84
5. Glenn Wasson,Marty Humphrey, Exploiting WSRF and WSRF.NET for Remote Job Execution in Grid, *Appears in the Proceedings of the 19th IEEE International Parallel and Distributed Processing Symposium(IPDPS 2005).*April 3-8.2005.Denver.CO.
6. *WSRF.NET 2.0 Programmer's Reference*
7. *WSRF.NET Developer Tutorial*

A FRAMEWORK FOR HOLISTIC GREENING OF VALUE CHAINS

Wei Deng Solvang, Elisabeth Román, Ziqiong Deng, Bjørn Solvang
Virtual Manufacturing and Supply Chain Management Research Group, Narvik University College, Lodve Langes gate 2, 8505 Narvik, Norway,Email:wds@hin.no

Abstract: Protecting and sustaining nature environment are two of the most important conditions regarding economic and social development. While the resources are extracted from and wastes are disposed to the same mother Earth, managing wastes of a value chain has been segmented and treated isolated. This segmentation leads to the partial optimization in waste minimization. This paper proposes a framework that facilitating holistic greening of a value chain so that total waste minimization can be achieved from entire chain's perspective.

Key words: value chain, waste minimization, holistic greening

1. INTRODUCTION

Effective waste management and environment protection are essential for the quality of life for current and future generations. Never before have these being more challenged by rapid economic development. The global ecosystem is facing a severe challenge of its economic sub-system as its capacity of energy as well as capability of waste disposal is reaching the limit (Goodland, 1991).

One of the driven forces to this economic development is manufacturing industry. By continuous improvement of material transformation process, a manufacturing system aims to constantly reduce costs and increase value-added to its products and services. This transformation can usually be viewed in value chains as it often appears as chains of value-added activities.

Please use the following format when citing this chapter:
Solvang, Wei Deng, Roman, Elisabeth, Deng, Ziqiong, Solvang, Bjørn, 2006, in International Federation for Information Processing (IFIP), Volume 207, Knowledge Enterprise: Intelligent Strategies In Product Design, Manufacturing, and Management, eds. K. Wang, Kovacs G., Wozny M., Fang M., (Boston: Springer), pp. 350-355.

A conventional definition of value is rather as narrow as profit or productivity. However, as current increasingly intensified focus on sustainable development and pressure from government regulation, the connotation of value needs to be refined and extended to embed environmental aspects. Further, as Handfield et al. (1997) declaimed that, in order to be successful with environmentally-friendly practices, environmental management strategies must be integrated into all states of the value chain.

In the recent years, manufacturing industry, being accused as one of the major enemies to environment, has already started with recycling and reusing. Topics cover remanufacturing (Shu & Flowers, 1999), green disassembly (e.g., Kuo, 2000; Lambert, 2002), cleaner production (Vlavianos-Arvanitis, 1998), reverse logistics (Hirsch et al., 1998), green supply chain management (Handfield et al., 1997), to mention a few. Comparing with the recycle and reuse termed in waste management, the definitions and applications of these two terms in manufacturing industry are rather straight forward in sense that there is only mechanical rather than chemical or biological processes engaged.

Currently, waste management has been treated differently in different knowledge disciplines. The lack of interdisciplinary cooperation and joint venture in integrated waste reduction and minimization leads to local rather than overall optimized waste management. A methodology for enabling overall waste minimization is therefore needed for dealing with following challenges:

- implementing upfront pollution prevention and minimization rather than end-of-pipe pollution control
- eliminating hazardous waste at the source; and
- optimizing overall waste reduction at entire value chain.

The rest of the paper is organized as follows. In section 2 we define a general model of green value chain. Section 3 briefly describes waste management and Environmental Protection Agency (EPA) approaches for pollution prevention and waste management. In Section 4, we propose a framework for holistic greening of value chains. Finally, section 5 provides some concluding remarks and suggestions for future work.

2. GREEN VALUE CHAIN MANAGEMENT

Conventionally, a value chain can be defined as the set of activities spanning the entire customer order cycle, including design, procurement, manufacturing and assembly, packaging, logistics, and distribution (Handfield et al., 1997). A green value chain incorporates a new dimension of value into the traditional value chain, namely, environment. As showed in

Figure 1, the primary material flow which starts from raw material supplier and ends at consumer has been extended to cover a 'from birth to graveyard' perspective. The products that have no utility value will be decomposed or dissembled and the utilizable material/parts go back to the chain. The rest of the products go further to indirect treatment where methods as chemical (i.e., incineration) and biological transformation are applied (Shah, 2000). Examples of outputs of this stage are renewed material and/or energy. The residuals after this stage are transported to landfill and, from there, further transformed to harmless and reusable materials.

However, during each of these value chain stages, wastes are generated in the same processes where value is created and added. These wastes can be hazardous to environment. Among which those end up at landfill are the

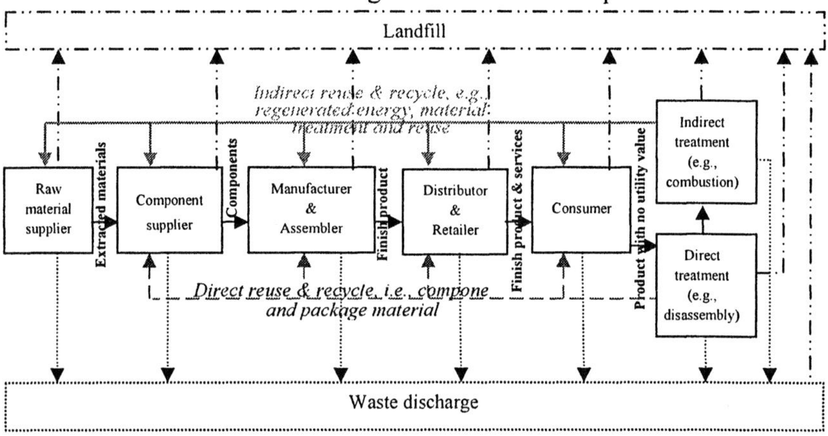

Figure 1. A General Model of Green Value Chain

most hazardous ones. As disposal at landfill is the least preferred form of wastes, efforts should be devoted to minimize total portion (Shah, 2000).

3. WASTE MANAGEMENT

Waste management can be generally defined as processes of dealing with waste (DEQ[a]). It is a sustainable process for reducing the environmental impact of the disposal of all types of materials used by businesses. Over the last decade wastes that end up at landfills are the most hazardous ones (Lagerkvist, 2001). There is a common principle that the pollution from landfills – emissions as gasses and leachate reflects, for example, categories of waste being put into landfills as follows:

[a] http://www.deq.state.or.us/wmc/solwaste/cwrc/glossary/glossary1.html

- Organic waste from households produces gasses as methane, hydrogen and ammonia;
- Plastic material are broken down very slowly;
- Heavy metals and environmental poisons such as dioxine, polyaromatich hydrocarbons (PAH) are dissolved in the leachate. Small concentration can make much harm when coming into contact with ecosystems since they are persistent.

In order to reduce the portion of wastes which ends up at landfill, U.S. Environmental Protection Agency (EPA) has proposed a general hierarchy of recommended approaches to pollution prevention (Figure 2) together with a preferred approach to hazardous waste management (Figure 3).

These approaches to pollution prevention and hazardous waste management stem from the environmental desirability of source reduction as the preferred means of minimizing waste. Recycling techniques allow hazardous materials to be put to a beneficial use and treatment options are considered only after acceptable waste minimization techniques have been identified (Shah, 2000).

To sum up, waste management need to purpose waste minimization that being implemented through following steps:

- source reduction
- resource recovery and recycling
- treatment
- ultimate waste disposal

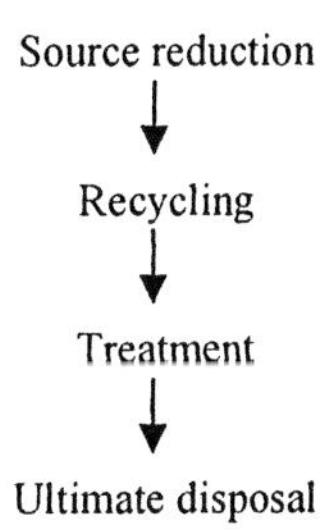

Figure 2. U.S. EPA hierarchy of recommended approaches to pollution prevention

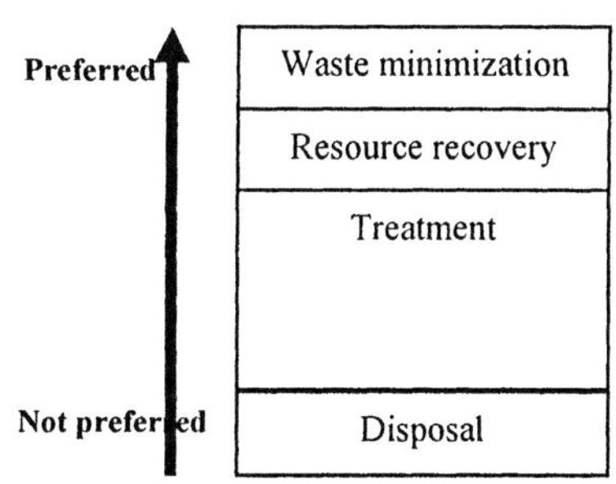

Figure 3. U.S. EPA preferred approach to hazardous waste management

4. A FRAMEWORK FOR HOLISTIC GREEN SUPPLY CHAIN MANAGEMENT

It is stated that there is an urgent need to shift from anthropocentric to biocentric values in society, and to place environmental appreciation at the core of every action and thought (Vlavianos-Arvanitis, 1998). In line with this statement, we propose a framework for holistic greening of value chains as shown in Figure 4.

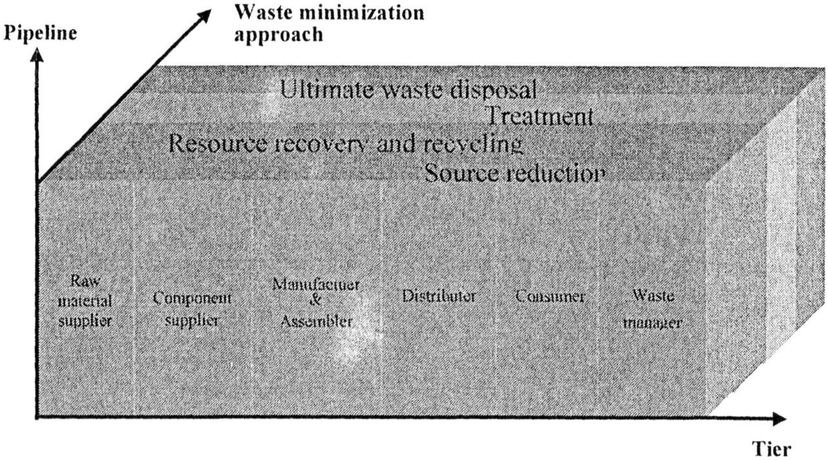

Figure 4. A Methodological Framework for Holistic Greening of Value Chains

This framework is three-dimensional. The tier dimension categorizes six types of actors corresponding to six stage of a product life cycle. A pipeline within a supply chain refers to the stream of materials, components, and assemblies that are associated with one particular product or tight family of products (Bhaskaran, 1998). While a pipeline illustrates partial horizontal view of a supply chain, a tier represents a vertical segment of the chain. The third dimension categorizes four approaches to achieve waste minimization, namely, source reduction, resource recovery and recycling, treatment, and ultimate waste disposal.

This framework suggests that, no matter where an actor situates or what function it has on its pipeline, it needs to conduct all four approaches at its strategic, tactic, and operational levels in order to achieve holistic greening. This is from an individual actor's perspective. From a pipeline perspective, in order to achieve optimized overall waste minimization, actors in that pipeline needs to coordinate their waste reduction activities in terms of four categories. From waste minimization perspective, our ultimate goal in holistic greening is to minimize the portion of waste (especially hazardous waste) which ends up at landfill. This can only be achieved by conducting all four approaches at all actors on all pipelines.

5. SUMMARY AND FUTURE WORKS

This paper proposes a methodological framework for achieving holistic greening of value chains. Future research needs to be devoted to model, measure, and evaluate current value chain greening performance.

6. REFERENCES

Bhaskaran, S., (1998), Simulation analysis of a manufacturing supply chain, *Decision Science* 29(3):663-657.

Goodland, R., (1991), Det som taler for at verden har nådd grenser – mer presist at nåværende produksjonsvekst i den globale økonomien ikke kan opprettholdes, in *'Økonomisk Politikk for en Bærekraftig Utvikling: oppfølging av Brundtlandkommisjonen'*, Goodland, R., Daly, H., Serafy, S.E, von Droste, B. ed., J.W. Cappelen Forlag, pp. 1-14.

Handfield, R.B., Waltor, S.V., Seegers, L.K., Melnyk, S.A., (1997), 'Green' value chain practices in the furniture industry, *Journal of Operations Management*, 15:293-315.

Hirsch, B.E., Kuhlmann, T., Schumacher, J., 1998, Logistics simulation of recycling networks, *Compters in Industry*, 36:31-38.

Kuo, T.C., (2000), Disassembly sequence and cost analysis for electromechanical products, *Robotics and Computer Integrated Manufacturing*, 16:43-54.

Lambert, A.J.D., (2002), Determining optimum disassembly sequences in electronic equipment, *Computer & Industrial Engineering*, 43:553-575.

Lagerkvist, A., (2001), Landfill Technology, Luleå university of technology, department of environmental engineering, *division of waste science & technology*.

Shah, K.L., (2000), Basics of Solid and Hazardous Waste Management Technology, *Prentice Hall*.

Shu, L.H., Flowers, W.C., (1999), Application of a design-forremanufacturing framework to the selection of product life-cycle fastening and joining methods, *Robotics and Computer-integrated Manufacturing*, 15:179-190.

Vlavianos-arvanitis, A., (1998), Cleaner production: profit for future generations, *Journal of Cleaner Production*, 6:381-385.

STUDY ON VALUE-BASED MODEL OF PRODUCTS PACKAGE MANAGEMENT

Ying Liu[1], Fengming Tao[2] and Fei Liu[1]

[1]College of Mechanical Engineering, Chongqing University,China, 400044; Email: lyjfb@xinhuanet.com. [2]College of Economics & Business Administration Chongqing University, China, 400044

Abstract: Package management is an important part of supply chain management. This paper analyzed the different levels of the value of package and the structure of the value chains in products package management. A value-based model of products package management was brought forth and the key technologies of package management were discussed.

Key words: value; package management; supply chains; green manufacturing

Today the importance of products package has been recognized by the society and firms. In some industries, the package can even decide whether the product can be popular [1].Package management is the important chain in supply chains. Even though product packaging improves marketing a lot, package management is comparatively new in China and very few researches on the value weight of package management in supply chain and on the control model of package management. Problems of package have made huge economic loss and environment pollution. The value-based package management is the optimization of value chain and human-centered.

1. PACKAGE AND PACKAGE MANAGEMENT

- Concept and classification of package

The function of package is to protect and beatify products so to increase their add value. Besides, it can act as advertisement to trigger consumers' interest and motivate them to buy products.

According to the shape of package, it can be classified as single package, inside package and outside package; according to the location of supply

---

Please use the following format when citing this chapter:

Liu, Ying, Tao, Fengming, Liu, Fei, 2006, in International Federation for Information Processing (IFIP), Volume 207, Knowledge Enterprise: Intelligent Strategies In Product Design, Manufacturing, and Management, eds. K. Wang, Kovacs G., Wozny M., Fang M., (Boston: Springer), pp. 356-362.

chain, it can be classified as business package and industrial package; according to the material, it can be classified as wooden box, paper box, and glass container, and package of plastic, cloth, hemp rope, metal, plant and etc.; according to the ways of package, it can be classified as water-proof, wet-proof, rust-proof, vacuum, decompression, fire-proof, frozen, gas-proof, explosion-proof and etc.

- Package management

Package management is the sum of several act including design, manufacturing, delivery and use of package so that consumers can be guaranteed to totally, conveniently and accurately get and take the value of the product. In a broad sense, package management does not only mean the manufacturing, encapsulation and usage of package itself, instead, it concerns the management of design, logistics, marketing and recycling of package.

2. THE VALUE SYSTEM OF PACKAGE

There are three levels in the value system of package(figure 1).

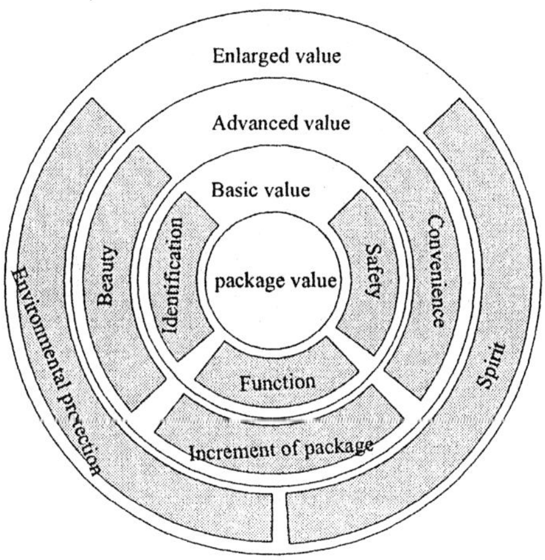

Figure 1. Levels of package value

2.1 Basic value

It is the core of package:

- Function of package: package protect production from leakage, stealing, fresh keeping and so on.

- Identification of package: good package can explain the function of products and trigger consumers' interest to buy.
- Safety of package: package itself must be safe, which protect products from damage and protect users safe as well.

2.2 Advanced value

It is essential to complete and increase the value of package so that value added can be gained.

- Beauty of package: beautiful and practical package can attract consumers' attention and help products gain better benefit.
- Increment of package: if cost and wastage of package is reduced, recycle rate is increased and management flow is amended, the value of products package can be increased, better economic benefit can be obtained.
- Convenience of package: some kind of convenience can be get from package so that the convenience of product storage, transportation, sale and usage can be realized.

2.3 Enlarged value

It is essential to expand the value of package so that the responsibility of product package toward consumers and society can be completed.

- Environmental protection of package: products package and the industry must be green, they must deal the recycling and wastes in a right way.
- Spirit of package: as the clothes of products, package can be the important carrier of brand value. It can describe the concepts of products and show the taste of products. Some of them even become one part of the firm culture, for instance, the bottle of Coca Cola.

3. VALUE CHAIN OF PACKAGE MANAGEMENT

The value of package exists in every unit of the "supply chain" of package. Such a chain consists of package design, package manufacturing, package encapsulation, package transportation, marketing, after sale service and so on. All the firms, organizations and individuals concerned can benefit from this supply chain of products package, hence, they shape the value chain of products package.According to the degree of cooperation among package users, suppliers and sellers and the influence on the value chain, we can divide the units of the chain into three groups as follows.

- **Core value units.** This units main refer to the firms that use products package. Two kinds of firms belong to this group: firms manufacturing products; and sellers re-packaging products. They are responsible for the organizing and management of the supply chain, resolving conflicts among units and maximizing the value of the chain.

- **Alliance value units.** These units include the firms responsible for design, manufacturing, supply and transportation of package. They provide service to the core value units and get benefit from it.
- **After sale value units.** These units are responsible for the recycling and waste dealing of products package. This kind of firms are very special because most of them are under government's control and only a very few firms that use products package involve in the recycling. However, with the popularization of environmental protection and the advance of waste dealing, package recycling and re-use is becoming a new industry with great potential value.

4. VALUE-BASED MODEL OF PACKAGE MANAGEMENT

4.1 Puzzles

- Integration of elements

In value system of products package, eight elements are relative and at the same time contradictory. For example, the value increment of package and environmental protection of products package are always contradictory. Therefore, when considering a certain product package, it is necessary to compare all the eight elements and determine the weight of each one according to actual situation, and hence receive an optimal plan.

- Management and balance of the value chain

Along the value chain, core value units, alliance value units and after sale value units have respectively their own value and benefits. All the value and benefits might be conflict, so it is essential to establish a system to manage and balance their interests and benefits.

4.2 Value-based model of package management

Aimed to the puzzles, this paper put forward the Value-based model of package management (figure 2). The model is composed of three levels.

4.2.1 Package value system level

The concrete contents in this level consist of four parts.

- **Principles of package value management.** They include principles of standard generalization of package, principles of innovation, principles of moderation, principles of intensivism, principles of economy and principles of recycling and re-use.

- **Weight evaluation and strategies of package value management.** When considering a certain product package with limiting factors, it is necessary to compare all the eight elements and determine the weight of each one, and hence receive an optimal plan. Based on the optimal plan, a suitable strategy can be obtained. The weight evaluation and strategies of package value management should be completed by the core value units who must quickly feedback and adjust the plan and strategies when limiting factors change.
- **Balance mechanism for different value units.** Along the value chain, core value units, alliance value units and after sale value units have respectively their own value and benefits. All the value and benefits might be conflict, so for the core value units, it is essential to establish a system to manage and balance their interests and benefits. With such a mechanism, the construct of the value chain and modes of process; interests distribution and coordination along the chain; and control mechanism in package management can be settled down.
- **Standards and policies of products package.** There are a series of national and international standards and policies, such as international environment standards in ISO14000. If products package did not meet the standards or policies, the manufacturing and sale of products would be limited and great economic loss would happen. Therefore, the package management must have the direction of standards and principles.

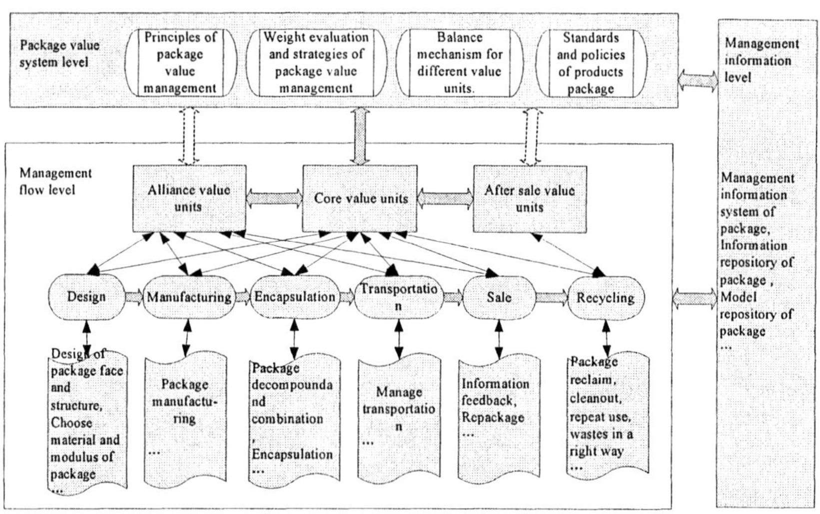

Figure 2. Value-based model of package management

4.2.2 Management flow level

Management flow level is the process of concrete management and control of the supply chain of products package. It is a course including design, manufacturing, encapsulation, transportation, sale and recycling. In this flow, core value units, alliance value units and after sale value units separately complete their duties and gain their benefits under the direction of the strategies and plans of products package. The key points are as following.

- The core value units must hold the overall situation, obtain dynamic information, constitute the flow model and guild the process of the flow.
- Core value units, alliance value units and after sale value units are closely tied up and they action and information are tightly related.
- There are sub-circles in the management flow such as multi-package and disassembly. For instance, disassembling packages in transportation and repackage in business promotion.

4.2.3 Management information level

This level is made up of management information system of package, information repository of package and model repository of package. By the ways of internet and other network, information coding management, information management of management flow, management of information repository, consumer management and computer modeling, the information in the supply chain of package can be collected, summarized and managed. Consequently, the value units and consumers can get and communicate information and consultation and solve plan can be obtained.

In the three levels above, package value level provides the value-based plan, management flow level implements the plan, and management information level offers information and service to the two levels. The three levels influence each other and constitute the management model together.

5. KEY TECHNOLOGIES

- The value chain of package management and its control strategies. In order to obtain the optimal strategies, the value units must analyze the relationship among them and build up the cooperation mechanism and control mechanism.
- The value-based package design. Package design is so important that more than 40% successful packages resulted from good designs. The value-based package design considers the structure design, reasonable

economic materials, beautiful and practical appearance design, function design and storage design [2], and etc.

- Innovation of material and structure of package. The usage and development of new material and new structure of package can improve the function and reduce cost, and attract customers to buy.
- Green package and its management technologies. Green package and its management technologies constitute the suitable layer management, simplifying sectors of packaging and integration, green design, usage and development of green material, recycling and re-use of package and degradation technologies of wastes[3].
- Information management and tracing. It constitutes the classification and encoding technologies of information, extraction technologies of information, information transportation and integration, information tracing technologies and information repository and model repository and so on[4].

6. CONCLUSION

Aimed to increase the comprehensive value of the "supply chain" of products package, the value-based package management integrated multi-knowledge, value chain and value units network, green package and information management together. And the kernel of the value-based package management is the optimization of value chain and human-centered. This kind of package management can increase the value of packages and has a bright future. The development of package industry will promote the value-based package management to a better improvement.

7. REFERENCES

1. C. F.Dong, Y. He, How Chinese package firms improve their international competitive capacity, *Package Engineering.* (2004), 32 (6),165-167
2. F. Liu, H. Zhang, H.H. Yue, Green manufacturing: The model of sustainable development for modern manufacturing, *Chinese Mechanical Engineering.* (1998), 9(6): 76-78
3. H. M. Dai, Theory, standard and manufacturing model of green package, *Chinese Mechanical Engineering.* (2002), 23(13): 2048-2052
4. G. B. Jin, Intelligent package technologies and the development, *Chinese Package.* (2002), (5): 37-41.

PROJECT MANAGEMENT STANDARDIZATION AND R&D PROJECT PERFORMANCE: A CROSS-NATIONAL COMPARISON

Xiaohong Zhou[1, 2]

[1] *School of Management, Zhejiang University, China; Email: zhouxiaohong666@yahoo.com.cn.*

[2] *Department of Management, Anhui University of Technology and Science, China.*

Abstract: R&D project management is a very complex and difficult work. Many companies attempt to implement standardized project management (SPM) defined as a standardized set of project management (PM) practices. Previous empirical studies of US researchers show that SPM tools, leadership skills, and process may impact project success. Using data collected from R&D project in the China high-velocity industry, we conduct a comparative study to test the impact of PM standardization on R&D project success. The results of the testing indicate that standardized PM process, tools, culture, and leadership are of higher interest, and that standardized metrics, the information management system, and project organization may be of lower interest.

Key words: R&D project, Project management, Standardization.

1. INTRODUCTION

To be successful in today's competitive marketplace, companies must continuously development new products. Unfortunately, many new products fail and never enter the launch phase at all[1]. Of the nearly 16 000 new products introduced in 1991, almost 90% did not reach their business objectives[2]. Many studies have found that the many projects failed are due to PM improperly to some extent[3,4,5,6].To solve the problem, some companies have taken SPM as the strategy for R&D project management, which can be defined as a standardized set of PM practices[7,8]. It suggests that SPM may increase project performance. Meanwhile, some companies perceive research and development as somewhat fuzzy, involving high uncertainty[9]. The literatures we reviewed did not provide adequate evidence that PM

Please use the following format when citing this chapter:

Zhou, Xiaohong, 2006, in International Federation for Information Processing (IFIP), Volume 207, Knowledge Enterprise: Intelligent Strategies In Product Design, Manufacturing, and Management, eds. K. Wang, Kovacs G., Wozny M., Fang M., (Boston: Springer), pp. 363-369.

standardization may increase R&D project performance. SPM refers to a process of managing projects composed of standardized practices. It was reported that 85% of the Fortune 500 use standardized approaches and procedures[10]. The Project Management Institute (2004) issued a new PM standard which suggest that SPM as a major strategy[11].Specifically, Dragan Milosevic and Peerasit Patanakul (2005)[8] have studied the impact of SPM on project performance, and proposed that standardized PM process, tools, and leadership are of higher interest, the standardized PM organization, information management systems, metrics, and culture are of lower interest.

The purpose of this study is, therefore, to test the impact of PM standardization on R&D project performance. We test the results in Dragan Milosevic and Peerasit Patanakul(2005) in the China high-velocity industry. We begin this study by defining the variables and hypotheses from Dragan Milosevic and Peerasit Patanakul (2005) [8]. Next the methodology for collecting and analyzing the China data is discussed, and the results are reported. Finally, the implications of the study are presented.

2. RESEARCH HYPOTHESES

Corresponding to the theoretical hypotheses proposed in Dragan Milosevic and Peerasit Patanakul (2005)[8], we test the following seven hypotheses using the China data.

H1:A higher degree of standardized PM process tends to increase the success of the R&D projects in high-velocity industries.

H2:R&D projects in high-velocity industries organized by more standardized practices of the project organization are more successful.

H3:Using a more standardized OPM-level information management system leads to higher success of projects in high-velocity industries.

H4:R&D projects in high-velocity industries that use more standardized PM tools tend to increase their project success.

H5:R&D projects in high-velocity industries using a more standardized system of metrics to measure and monitor project performance will have higher project success.

H6:R&D projects in high-velocity industries where cultural values are more standardized tend to have increased project success.

H7:R&D projects in high-velocity industries managed by managers with more standardized skill sets tend to have improved project success.

3. RESEARCH METHOD

3.1 Measures

The purpose of this study requires that we follow the same measure evaluation and scale development process in the original US study[8].

The dependent variable is the degree of project success. The degree of project success can be operationalized as the degree to which the projects accomplished their schedule, cost, quality, and customer satisfaction goals. To capture the numerical responses of the respondents as to the degree of project success, we used a 5-point Likert scale (5 being the highest degree, 1 being the lowest degree). The seven independent variables are PM process, organization, information management systems, tools, metrics, culture, and leadership. According to Dragan Milosevic and Peerasit Patanakul (2005) [8], we defined "standardized" as "the degree of uniformity or consistency applied in implementing project management process". Thus, the highest degree of uniformity, i.e., standardization, is when the PM process is implemented by all project managers in the same way. The lowest degree of uniformity, or standardization, is when the PM process is inconsistently used by all project managers. To measure the numerical responses of the respondents as to the degree of PM process standardization, we again used a 5-Likert scale. Using the same way, we measure the degree of standardization of project organization, information management systems, tools, metrics, culture, and leadership.

On account of only one question for each independent variable in original US study[8], the single item construct are less effective than multi-item constructs. Thus, we redesigned anther questionnaire with two or three questions for each variable on the basis of the previous questionnaire again. We found that Cronbach's alphas were higher than the minimum value of 0.70 recommended by Nunnally (1978)[12]. It shows that the scale is reliable and acceptable level of internal consistency for the study.

3.2 Samples and data collection

The data for this study were obtained from the firms in China. To enhance to the validity of the data, we mailed the questionnaires to the project directors only. Meanwhile, to ensure the external validity of the findings, the sample included projects varying in size. The final qualifying sample included 65 R&D projects. Of the samples, 26 were in computer industries; 39 were in electronics industries. Approximately 48 of these firms stated that they had an emphasis on the SPM.

3.3 Data analysis

The purpose of this study requires that we follow the same analysis method in the original US study[8]: two methods of bivariate data analysis along with one multivariate method. The bivariate methods were Pearson product-moment correlation between each independent and dependent variable, and t-test, which assesses the significant difference of the project performance in the terms of project success. The assumption here is that higher project success will have the higher degree of SPM factors. If so, the t-test will show significant differences for each factor. Finally, we use stepwise multiple regression analysis to validate the previous analysis.

4. RESEARCH RESULTS

The summery statistics of the bivariate analysis and stepwise regression results of testing the hypotheses are presented in Table1 and 2.

4.1 SPM tools, leadership, process, and culture analysis are of interest

Table 1. Impact of SPM factors on project performance (bivariate analysis)

Factors	Performance		
	Pearson Correlation	Sig.(2-tailed)	N
Process	.536(**)	.000	65
Organization	.231	.064	65
Information	.185	.141	65
Tools	.562(**)	.000	65
Metrics	.167	.184	65
Culture	.437	.000	65
Leadership	.428(**)	.000	65

** Correlation is significant at the 0.01 level (2-tailed).* Correlation at the 0.05 level.

Table 2. Multiple regression analysis of PM performance Versus standardization (multiple analysis)

Model		Unstandardized Coefficients		Standardized Coefficients	t	Sig.
		B	Std. Error	Beta		
1	Constant	-.240	.371		-.647	.519
	Tools	.891	.095	.762	9.333	.000
2	Constant	-.754	.416		-1.811	.075
	Tools	.815	.097	.696	8.363	.000
	Leadership	.259	.107	.200	2.406	.019

The correlation coefficients of 0.536, 0.562, 0.437, and 0.428 show a significant relationship between the standardized PM process, tools, culture, and leadership. T-tests confirmed that there are significant differences in the standardization of these variables between these projects. This indicates that higher standardization of PM process, tools, culture, and leadership will lead to higher project success.

From the results of the stepwise multiple regression (Table 2), we found that only two factors (tools, leadership) entered into the equation. The two predicted variable may capture the explained variance of the dependent variable by using its correlated factors. As a result, the correlated factors may not enter the equation. Since some SPM factors were strongly correlated with leadership and tools. For example, the correlation coefficient between leadership and culture is 0.283 and 0.627 between tools and process (at the 0.05 level). Thus, it is quite possible to get a short list of factors in the equation. In summary, Hypotheses 1, 4, 6, and 7 were support.

4.2 The three SPM factors of lower interest

Table 3. Results for US and China

Factors of interest		Factors of lower interest	
US	China	US	China
Process	Process(H1)	Organization	Organization
Tools	Tools(H4)	Information system	Information system
Leadership	Culture(H6)	Metrics	Metrics
	Leadership(H7)	Culture	

The three factors with little or not impact on project success in the statistical analysis were standardized metrics, the information management system, and project organization. The results of the regression also indicate this finding (Table 2). Because they have lower impact on the project success, hypotheses 2, 3, and 5 are not supported.

To facilitate comparison of the US and China findings, we report the results of the testing (Table 3). From table 3, we can find that the results from the China data are some different from the results from US study.

5. DISCUSSION AND MANAGERIAL IMPLICATIONS

The objective of the study is to verify the findings in US study[8]. The findings inconsistent with that in original study [8] show that increasing the level of the standardization of some PM factors will lead to higher project performance.

A standardized PM process may increase project success. PM process can improve project performance. Standardized PM process can result in less rework, fewer mistakes, fewer delays and snags, and better use of time.

When a standardized PM process is in place, the level of repeatability is higher. Thus, those projects using a standardized process will minimize variation in how they are executed, and will be improved in performance.

PM tools are structural factors in determining PM success[13]. Standardized PM tools may impact project success. The companies should create their standardized PM toolbox (e.g., templates, WBS, Gant Chart) in order to help project team to accomplish project goals.

Formal project leadership play a key integrative role in innovation projects both within the project team and between the team and outsides[14]. The project managers with standardized leadership skill sets may be a factor in project success[8]. The standardized leadership will reduce uncertainty for data transmission between the team and top management. The reliance on standardized leadership is likely to be more crucial for performance for projects involving high uncertainty. In this condition, project managers are more likely to deliver on cycle time, customer satisfaction, quality, and cost goals when given a standardized skill set that can address all sorts of project challenges than they are when left to randomly develop such skills[8].

The project team with standardized project culture may help project team to accomplish project goals. The efficiency of project teams are influenced by culture diversity [15], because certain culture values make team members resist teamwork. The cultural standardization might help shape standardizing value and beliefs of employees that reducing variation in their behavior especially in China. The culture standards of different countries' will have different impact on teamwork[16].

6. CONCLUSION

Some companies have tried to introduce SPM for increasing project success, because the R&D PM is very complex. Some researchers have studied the impact of PM standardization on R&D project performance. An important study of US has shown that the SPM would increase project success. Compared to the US, the PM standardization is still immature in developing countries like China. Thinking about the cultural difference between US and China, the PM practices may be also different. Therefore, we test the findings in US using the China data. The results of the testing are not consistent with findings in US study[8]. The testing of the seven hypotheses indicated that standardized PM process, tools, culture, and leadership are of higher interest, and that standardized metrics, the information system, and project organization may be of lower interest.

Since the China survey was designed to replicate the previous US study, we did not redesign another research framework. The research model may be viewed as simple. It does not state the correlation between each of the different critical standardization factors affecting project success. Another limitation is that this research is based on a final qualifying sample including 65 projects only. To be broadly generalizable, the results need to be replicated in a broader set of sample. Future research should attempt to address these questions.

7. REFERENCES

1. Mark A.A.M. Leender,Berend Wierenga,(2002),The effectiveness of different mechanisms for integrating marking and R&D,*Journal of Product Innovation Management,*19,pp.305-317.
2. R.Balachandra,(1997), Factors for Success in R&D Projects and New Product Innovation: A Contextual Framework. IEEE Transactions on Engineering Managemnet,44(3),pp.276-287.
3. Jeffrey K. Liker,et al,(1999),Flexibility and Standardization: Test of a Contingency Model of Product Design–Manufacturing Integration. *Journal of Product Innovation Management*,16,pp.248-267.
4. Jonathan West,Marco Iansiti,(2003),Experience,experimentation,,and the accumulation of knowledge:the evolution of R&D in the semiconductor industry,Research Policy;32,pp..809-825.
5. John Davis,Alan Fusfeld et al,(2001),Determining a project's probability of success,Industrial Research Insitute inc, May-June ,pp.51-56.
6. Jeffrey K.Pinto and Dennis P.Slevin,(1989),Critical success factors in R&D projects,*Research technology management*,32(Jan-Feb),pp.31-35.
7. Shona L Brown, Kathleen M Eisenhardt,(1997),The art of continuous change:Linking complexity theory and time-paced evolution in relentlessly shifting organizations, *Administrative Science Quarterly,* 42(1),pp.1-34.
8. Dragan Milosevic， Peerasit Patanakul,(2005),Standardized project management may increase development projects success,*International Journal of Project Management*,23,pp.181-192.
9. D. Nobelius,(2004),Towards the sixth generation of R&D management. *International Journal of project management,* 2(2),pp.369-375.
10. Dragan Milosevic,et al,(2001), Impact of project management standardization on project effectiveness, *Engineering Management Journal,*13(4),pp.9-16.
11. PMI,(2004),Organizational Project Management Maturity Model(OPM3),Newtown Square: Project Management Institute.
12. Nunnally JC,(1978), Psychometric theory, New York: McGraw-Hill.
13. Might RJ,Fisher WA,(1985),The role of structural factors in determining project management success,*IEEE Trans Management*,32(2),pp.71-77.
14. Helene Sicotte,Ann Langley,(2000),Integration mechanism and R&D project performance, *Journal of Engineering and Technology Management,*17,pp.1-37.
15. Tomas,(1999)Cultural diversity and work group effectiveness-An experimental study,*Journal of Cross Cultural Psychology*,30(2),pp.242-263.
16. Amanda Dunkel,Sylvia Meierewert,(2004),Culture Standards and their impact on teamwork-An empirical analysis of Austrian,German,Hungarian and Spanish culture differences,*Journal for East European Management Studies*,9(2),pp.147-174.

A BPR METHODOLOGY FOR PRODUCT DEVELOPMENT PROCESSES

M. Bertoni[1] , U. Cugini[1], D. Regazzoni[2], C. Rizzi[2], M. Ugolotti[1]
[1]*Dipartimento di Meccanica, Politecnico di Milano, Milano, Italy.*
[2]*Dipartimento di Ingegneria Industriale, Università degli Studi di Bergamo, Dalmine (BG), Italy, Email: caterina.rizzi@unibg.it*

Abstract: This paper presents methodology and guidelines about the use of process modeling languages to support BPR activities in relation with *Product Development* process. The methodology is based on the complementary adoption of different modeling techniques such as IDEF, UML and ARIS. Starting from the analysis of different models, we have selected and associated the most appropriate modeling suite to each phase of the re-engineering process. At this purpose, diagrams coming from IDEF, UML and ARIS families have been compared with respect to following issues: degree of formality, completeness, simplicity vs. detail, capability to describe "business goal" and support to costing operations.

Key words: Product development process, Business process re-engineering, Process modeling, IDEF, UML, ARIS.

1. INTRODUCTION

Those companies that consider *product development* a core competency, cannot ignore that during everyday work a big amount of information, in many different forms, is produced. This *knowledge* has to be captured, stored and correctly classified. At this purpose, PLM/PDM systems have been introduced in the last few years in order to manage the evolution of a product during its life cycle from the conception to the disposal [1]. When a PLM/PDM solution is designed and implemented, it has to be customized to best satisfy enterprise needs: main features are set and adapted in order to best fulfill company's process requirements. Thus, the design and implementation of a PLM solution requires an adequate analysis of the

---

Please use the following format when citing this chapter:
Bertoni, M., Cugini, U., Regazzoni, D., Rizzi C., Ugolotti, M., 2006, in International Federation for Information Processing (IFIP), Volume 207, Knowledge Enterprise: Intelligent Strategies In Product Design, Manufacturing, and Management, eds. K. Wang, Kovacs G., Wozny M., Fang M., (Boston: Springer), pp. 370-378.

involved business processes. Often, such implementation is coupled with a Business Process Re-engineering (BPR) activity to better deploy technologies and/or methodologies [1][2]. A BPR project consists on a "fundamental rethinking and radical redesign of business processes to achieve dramatic improvement in critical, contemporary measures of performance such as cost, quality, service and speed" [3]. It can be seen as an iterative activity. After performing the initial AS-IS and TO-BE analysis of the company's process (step1), a new solution is physically implemented (step 2) and then, once the new system is working, it needs to be monitored and controlled (step 3). Detailed process description are used in order to guide and support all the activities performed along the "Re-engineering cycle". In such context, process modelling plays a very important role: models are used to understand how the current process is working (AS-IS models) and to give a preview about how process will behave once it will be re-engineered (TO-BE models). In order to increase BPR project's rate of success, modeling languages have to be carefully selected [4]. This choice is crucial, and not easy. The world of business modeling languages is characterized, in fact, by the lack of one leading business modelling technique. Models are heterogeneous and no official accepted standards exist in this area [5]. Many languages, like IDEF [6] and ARIS [7] are not equipped to support the description of software and IT systems. On the other hand, UML [8] is not specifically targeted for business-oriented models [9].

Starting from the concept of the *Re-engineering cycle*, main goal of this paper is to defined a methodology, named *Integrated Methodology*, and guidelines to be applied for the use of process modeling techniques in the context of BPR. We decided to focus on *Product Development* activity, basing on our previous experiences in this field [1][10]. The described methodology has been experimented within an industrial context in order to recognize and evaluate benefits and drawbacks. The study case we considered refers to the Product Development process of an Italian SME producing hydraulic systems [1]. Main goal was to implement a PLM/PDM system: decision taken in order to grant a better coordination and data exchange among company's department, facilitate testing and prototype activities and improve the product development lead-time. To have a complete overview of the entire design process, we considered all design process typologies described in the Total Quality Management procedures, where each type of design process is subdivided into four main phases: *Concept generation*, *Preliminary design*, *Detailed design*, *Product engineering*.

2. PROCESS VIEWS & MODELLING TECHNIQUES

Starting from the KAEMaRT methodology [10], we tried to define a new methodology for BPR project evaluating the possibility to use, in addition to IDEF models, different types of diagrams like those provided by ARIS and UML. The introduction of these modeling techniques suggested to enlarge AS-IS and TO-BE views, in order to represent with an higher level of completeness several aspects of enterprise processes. The modeling suites have been analyzed and compared realizing the AS-IS and TO-BE model of the considered Product Development process. Having the possibility to choose among different views allows the process analyst to work with the ad *hoc* instrument in relation with the context, the objective and re-engineering steps.

Five different views have been considered (Fig.1):

- **PROCESS VISION**. Models of this view are characterized by the smallest granularity in process description. From this perspective, an abstract process representation focused on macro-activities is given.
- **WHAT**. This view lets the analyst to describe WHAT the process does. It collects functional models, which represent information flows, considering all the other aspects of the process (time, data and products) that concur to function definitions.
- **HOW**. Diagrams related to this view permit to the user to describe HOW the process works. They provide a dynamic picture of system behavior providing a time sequenced description of the activity flows, in terms of parallelism and sequence relations.
- **DATA/OBJECTS**: This view concerns data, objects and all elements that contribute to data/objects definition. Models here are used to describe logical data structure, their state and their modification along the process flow.
- **INTERACTIONS**. These diagrams show patterns of interaction among people and objects that populate the system.

Figure 1 shows the diagrams we selected from the three modeling families for each process view. We could observe many similarities among models populating WHAT and HOW, both in terms of information collected and in terms of the graphical structure. IDEF0, Eriksson Penker BE, and ARIS Function View have been chosen in order to represent information's flows characterizing the WHAT row. At the same time, IDEF3 PF (Process Flow), UML Activity Diagrams, and ARIS EPC have been used to describe dynamic system's behavior. Data and objects were represented with IDEF 3 OSTN, with UML State and Class Diagrams and with ARIS ERM. UML Use Case Diagrams and ARIS Control View were elaborated in order to describe company's activities with minimum level of detail. Finally, thanks to UML Collaboration and Sequence Diagrams, we could represent message exchange between people and objects in the system.

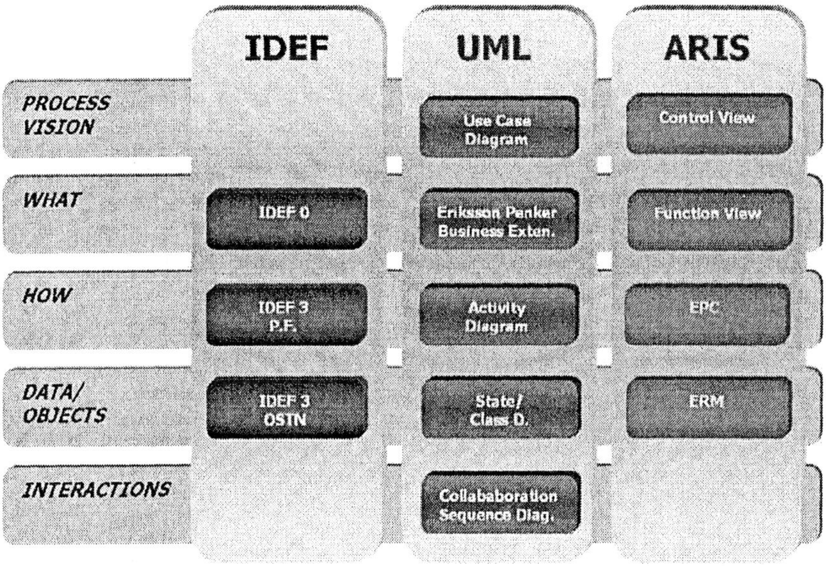

Figure 1. Process Modelling views and related modeling techniques.

3. METHODOLGY FOR OPTIMAL CONTROL

Both IDEF, UML and ARIS provide valuable tools in order to support BPR activity. Using diagrams from different methodologies we can horizontally enlarge our vision of the process, describing companies from five different points of view, and go deeper into the description of information, objects and activities flow. To define what kind of model best fulfill the requirements of each BPR step, we evaluated diagrams considering: Formality extent of the modeling language, Easiness of understanding vs. level of detail, Ability to describe business goals, Approach completeness, and Support to costing operations.

♦ *Formality extent of the modeling language*

A language is considered to be formal if its syntax and semantics can be precisely defined. When the semantics is formally defined, sentences in the language have a unique interpretation. Although these basic principles are common to most modeling techniques, the meaning of graphical symbols and the nature of their relations can be defined with a different level of precision. We can say that, in general, any business model should be easy and unambiguously readable by a person.

Selected IDEF techniques have a very precise syntax and semantics definitions [11]. Moreover, they are very simple and concise, easily readable also by non professionals of the modeling technique. On the other side, IDEF diagrams suffer from semantic limitations that narrow their use especially in the IT description field [12]. Syntax and semantics of UML diagrams are concisely defined [5]. However, due to its origin, UML is based on a semantic and syntax more software design oriented than business process oriented. ARIS suite contains an enormous quantity of graphical symbols, but its semantics is not very precisely defined [5]. For example, it's not so clear, what is the meaning of various connection types between org-units in org-charts or connection types between functions and org-units in eEPC.

- *Easiness of understanding vs. level of detail*

When choosing a modeling technique, we have to keep well in mind who will read and use the models. For instance, during the AS-IS and TO-BE analysis models should be easily understandable by people with different competences and backgrounds; this facilitates communication and share of process knowledge. On the contrary, during the implementation step, the representations must be complete and well detailed to facilitate domain experts' work.

In general, IDEF notations are well suited to describe, with an high level of abstraction, the behavior of a very complex system thanks to the hierarchical decomposition of its diagrams [6]. IDEF models show to be inadequate when a detailed representation of enterprise processes, at the Implementation Description level is needed [12]. UML suite provides a very complete view of the company processes, particularly focalized on the IT aspects [13]. On the other side, newcomers can be disoriented in front of the formality and the intrinsic complexity of UML constructs. ARIS suite is composed by more or less 50 different diagrams [14], that can be used separately depending on the scope of the description. Therefore it can be used in many different ways, both by experts for a detailed analysis of the process or newcomers for a rough view of enterprise system.

- *Ability to describe business goals*

High level business goals as well as constraints and limitation influencing objectives achievement should be represented within the process diagrams to better address process analysis, especially during the evaluation of the status quo and the TO-BE definition.

IDEF modeling techniques do not contain specific constructs for the description of business system goals [15]. IDEF5 diagrams could be used at this purpose however they are poorly used [11]. UML suite comprises the specific Goal Model but, in practice, it's scarcely utilized [15]. Goals are anyway expressed inside UML Eriksson-Penker Business Extension providing a representation of business process completed with the description of "goal objects". High level goals description is best developed in ARIS. In fact, it proposes the Objective Diagram, which permit to describe goals, critical factors influencing the objectives and a list of functions supporting their achievement. However we cannot link goals list to the activity flows representation and Object Diagrams nevertheless give a clear description of the process situation and are not utilized in the context of enterprise BPR activities [11].

- *Approach completeness*

This aspect concerns the ability of modeling languages to provide an holistic view of the entire business process to be re-engineered.

IDEF modeling approach is largely function oriented [15]. It provides useful means to describe enterprise activity flows; in fact, IDEF notations have been merely designed with the aim to model company's system from a defined viewpoint, such as a ''function viewpoint'' or an ''information viewpoint''. This is both a strength and a weakness of this technique. Also considering that IDEF models are very useful to quickly represent processes behavior, on the other side, no effective means are provided in order to represent some key constituents of the process. UML modeling methodology provides about ten diagrams, which can be used to describe business process, choosing the level of granularity desired [15]. Depending on the scope we can decide to represent only aspects related to a precise perspective, selecting only the best suited set of models [16]. UML mainly provides suitable and widely-used modeling constructs for developing multi-perspective descriptions of IT systems. ARIS house, as already said, comprises more than 50 diagrams [7][17], which enable users to investigate the process from many different points of view, choosing the detail extent best fitting the analysis scope. Anyway we could be disoriented in front of this big amount of tools.

- *Support provided to costing operations*

When dealing with product or process problem solving, another important issue is metrics: we must define and measure what we mean by improvement considering techniques and tools that permit to evaluate and report on productivity and innovation. For example, during the Conceptual step of the re-engineering cycle, analysts need to evaluate two or more alternatives in the field of cost effectiveness, in order to identify the solution with the highest input/output ratio.

IDEF language supports the ABC (Activity Based Costing) costing technique. ABC feature is typically defined only for IDEF0 diagrams but not for IDEF3, although this last is the most used business process models inside

IDEF suite [15]. Due to its origin, UML suffers a lack of extensions dedicated to the cost effectiveness analysis. Parameters necessary to perform gap analysis cannot be correctly defined inside UML language. ARIS methodology is supported by a number of software components which support cost analysis, such as ARIS Process Cost Analyzer used to plan and analyse prices and allocation rates [7]. It support ABC costing technique and each function has a very rich list of cost attributes, which can be used to obtain total values either via static analysis or via simulation [11][18].
Figure 2 summarises previous considerations also in relation with process re engineering life-cycle.

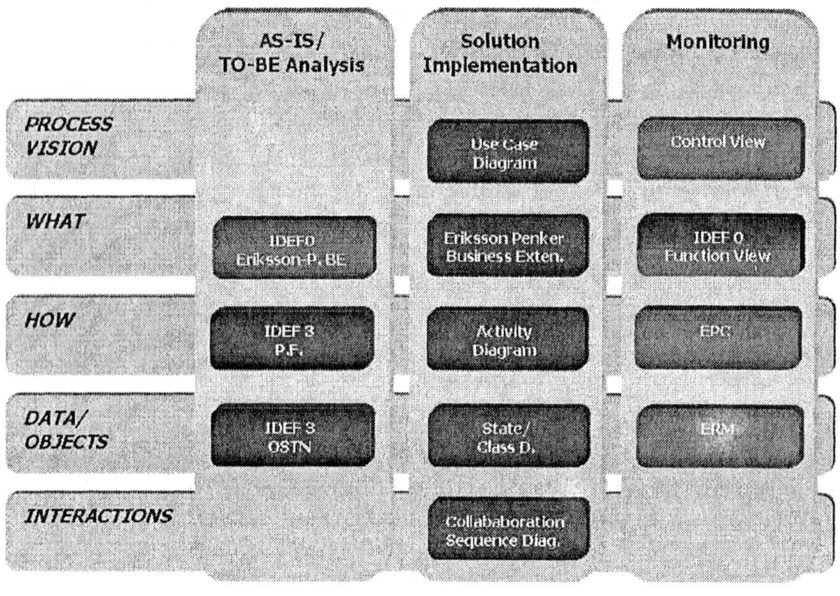

Figure 2. Proposed Integrated Methodology.

Conceptual analysis requires a clear and understandable representation of the enterprise process. Many people are involved coming from different company functions and with different backgrounds. Diagrams should be easily understandable, many process flows alternatives have be evaluated and representations should few detailed to allow an easy reading and a fast realization. IDEF techniques seem to be the perfect tools to provide a brief and concise description of the AS-IS and TO-BE models of the process to be re-engineered. However, in order to better share BPR objectives among different people and to motivate personnel, UML Eriksson-Penker BE can be used to represent the information flows during the *Conceptual* step.

During the *Implementation* activity, process representations will be utilized by domain experts to implement the solution set in the previous step. Therefore models should describe in detail what the new system will do, how its different components will be combined and how they will work together. This BPR step is perfectly supported by UML diagrams.

In the *Monitoring* step, diagrams are mainly used by business people, not experts in the modeling techniques. The diagrams should be simple, to be easily understood also without a technical background, but, at the same time they should contain specific constructs to support gap analysis and costing operations. ARIS diagrams and IDEF0 seems to be the correct choice in order to support process description at this step.

4. CONCLUSION

Main scope of this work has been to develop a methodology and guidelines for the use of modeling languages during Product Development process re-engineering. In order to obtain a more realistic and reliable description of the process under analysis, we considered five different perspectives (views) and three different modeling suites. Choosing among different diagrams and depending on the scope of the description, one can focus the attention on data, objects, interactions, activity flows. Thanks to the *Integrated Methodology* it is possible to explicit the need of an holistic view of the company structure to abandon the narrow vision provided by the traditional approach, and to enlarge our understanding of the process, analyzing relationships and behaviors between processes inside the global enterprise.

Classifying IDEF, UML and ARIS diagrams inside the five different views, depending on the information they allow to represent, and comparing modeling techniques on the basis of the five aspects previously introduced, it was possible to detect most appropriate technique to be utilized in the three steps of the *Re-engineering cycle*.

Finally, we tried to realize a mapping among the information contained in models populating the same row of the table (Fig. 2). We evaluated the possibility to export information from a language to another, in order automate or semi-automate diagrams building up, with the scope to make feasible the adoption of the *Integrated Methodology*. It is concretely possible, making use of "neutral format", like .CSV files, opportunely translated through simple application written in C++ code. Moreover, some special application has been recently developed in order to support, for instance, the unidirectional conversion from ARIS models to UML diagrams [19].

5. REFERENCES

1. Cugini, U., Ramelli, A., Rizzi, C., Ugolotti, M., (2004), *PLM Paradigm in a SME: Total Quality Management and Process Modeling Techniques,* in Proceeding 14th International CIRP Design Seminar 2004 – Design in the Global Village (CD-ROM), Cairo (Egitto) 16-18 May 2004.
2. Scarbrough, H., (1997), *Business Process Re-design: The knowledge dimension*, Warwick Business School, University of Warwick, UK.
3. Hammer, M. and Champy, J., (1993*), Reengineering the Corporation: A Manifesto for Business Revolution*, Nicholas Brealey Publishing, London.
4. Childe, S., Maull R., Mills B., (1997), *UK Experiences in Business Process Re engineering*, University of Plymouth, UK.
5. Kalnis, A., *Business Modelling. Languages and Tools*, Institute of Mathematics and Computer Science, University of Latvia.
6. http://www.idef.com
7. ARIS toolset, 2000, www.a-cci.it
8. Bennett, B., Skelton, J., Lunn, K., (2001), *Introduzione a UML*, Ed. McGraw-Hill, UK.
9. Rittgen, P., *Business Process in UML*, Information System Group II, Technical University Darmstadt.
10. Rizzi, C., Ruozi, D., Gherri, N., (2004), *A methodology to acquire and formalise process knowldege for technology innovation: an industrial application*, Knowledge Intensive Design Technology, Eds J.C. Borg, P.J. Farrugia, K.P. Camilleri, 2004, Kluwer Academic Publishers, pp. 65-80.
11. Kalnins, A., Kalnina, D., Askolds Kalis, (1999), *Comparison of Tools and Languages for Business Process Reengineering*, Institute of Mathematics and Computer Science, University of Latvia, Latvia.
12. Cheol-Han, K., Weston, R.H. , Hodgson, A., Kyung-Huy L., (2001), *The complementary use of IDEF and UML modelling approaches,* Ed. Elsevier, UK.
13. Eriksson, H.E., Penker, M., (2000), *Business modeling with UML*, Ed . John Wiley and sons, INC, USA.
14. Scheer, A.W., Abolhassan, F., Jost, W., Kirchmer, M., (2002), *Business process excellence, Aris in practice*, Ed. Springer, GER.
15. *Business modeling: UML vs. IDEF*, Griffith University School of computing and information technology, © Ovidiu S. Noran.
16. Loos, P., Allweyer, T., (1998), *Process orientation and Object orientation – An approach for integrating UML and Event Driven Process Chain (EPC),* Publication of the institute für Wirtscheftsinformatik, University of Saarland, Saarbrücken.
17. ARIS Methods Manual Version 4.1, AG Sheer, ED. IDS, July 1999.
18. Lewis, R.J., (1993), *Activity Based Costing for Marketing and Manufacturing,* Ed. Quorum Books.
19. ArcStyler MDA, Business Transformation Tutorial, ww.omg.org/mda.

DATA MANAGEMENT IN GROUP PURCHASING SYSTEM

Zongfeng Zou, Tao Yu
CIMS&Robot Center, Shanghai UniversityShanghai,China, Email:zfzou,tyu@mail.shu.edu.cn

Abstract: Introduces a method to solve group purchasing and Price Comparison System framework. It stresses on data loading and data management including data model, database, algorithm and buffer Data Structure. The framework offers a efficient method for loading large and complex scientific databases.

Key words: Group purchasing, Price Comparison System, Data management

Modern group-companies are often made of many subsidiary companies. Stocking centralization can economize much fund for its large-scale purchasing action. Subordinate companies hope to choose the best provider according to reliable data in most possible area. Advanced data in variety units is massive amounts of data that must be post-processed and organized to support provider evaluation actions. Increasing data volumes from all of the companies challenge to state-of-the-art database system and data-loading techniques. Price Comparison System consists of an algorithm for data loading, data structure to support data integrity, and counterpoising system. The system will do great benefit to group economy.

1. INTRODUCTION

With the development of technologies, purchase action research has taken place great changes. Advanced data collection technologies can collect many data. They have been built to contain the data and to be used as valuable resources for economy analyses and other purpose. The characteristics are reflected in several demands that must be addressed when loading such data. Firstly, data loading speed must keep up with data

Please use the following format when citing this chapter:
Zou, Zongfeng, Yu, Tao, 2006, in International Federation for Information Processing (IFIP), Volume 207, Knowledge Enterprise: Intelligent Strategies In Product Design, Manufacturing, and Management, eds. K. Wang, Kovacs G., Wozny M., Fang M., (Boston: Springer), pp. 379-385.

acquirement speed. Secondly, it must be possible to put to different database tables from one source file. Thirdly, it often have to perform complex data transformations and computations during the loading process. At last, automatic error recovery is required during the lengthy data-loading process.

The Price Comparison System is a collaborative system among the Shanghai Electric Group Co. Price Comparison is a multi-time, multi-item category purchase recorder at the Shanghai Electric Group Co. located in all area of Shanghai city. The purchase recorder consists of 18 kinds of information and can give evaluation of providers. In contrast to traditional purchases, all of the indent must be recorded in the system and examined and approved by experts of group purchase department. System allows users to statistically analyze the price and trend contributes to the data of the whole group.

2. DATA LOADING REQUIREMENT AND METHOD

The large data-collection rates and volumes noted in the previous part give the necessity for a fast data repository loading process that is capable of keeping up over time with the speed of data get. Much need contribute to the difficulty of achieving this goal. Collected raw data and computed catalog data are usually archived in a mass storage system that is separated from the database server. The catalog data that should be transferred from the mass storage system to load the database repository typically saturates the available network bandwidth, introducing the network as the first bottleneck to fast data loading.

Purchase action data encompasses information of many different types, from different information level specifications to the purchase details of goods. This variety of information is interleaved in the catalog data set that is generated when the raw data is processed. During the data-loading process the complex catalog data must be listed, the correct destination tables must be identified, and the data must be loaded into multiple target tables in the repository. Loading data into multiple tables is further complicated by the presence of multiple relationships among tables. Relationships that must be maintained by obey the primary and outer key constraints during the loading process.

Additional operations are also performed during the data loading process. These operations include transformations to convert data types and change precision, validation to filter out errors, and calculation of values such as the Hierarchical Triangular Mesh ID (HTMID). All such intensive operations place an additional burden on the loading process. Finally, because data loading is typically a lengthy process, a mechanism of automatic recovery from errors is a basic requirement.

To solve the problems in building the repository, we designed framework, called Price Comparison System, which is made up of (1) a logical network to join the different data position of all of the companies. (2)an efficient algorithm to perform bulk data loading, (3)an effective data structure to maintain table relationships and allow proper error handling, (4)active database and system tuning to achieve the best data-loading performance. Using this framework we can bulk load data, insert data into multiple database tables at the same time without locking and constraints, and recover the loading process from errors. The Price Comparison System framework has improved the ability of data loading.

3. THE PRICE COMPARISON SYSTEM FRAMEWORK

3.1 Data Model and Price Comparison System Mission

The original data created by the company in the past are archived in mass storage system. A program is arranged on the data to extract catalog data, which includes a wide range of information. Typically, the catalog data includes information on the user's position, the amount they wanted, the parameters described, the providers presented, the frames generated, and the feedback written. The catalog information is first written to an ASCII file, which is saved in the mass storage system and then uploaded to a repository database. The format of the catalog file varies depending on the extraction program used. Commonly, different aspects of the catalog information are interleaved in the file. For example, a row of frame information is followed by four rows of frame information, and a row of object information is followed by four rows of finger information. Each row in the catalog data file usually has a tag or a keyword that can be used to determine the destination table in the database.

Oracle 10g is the best choice to host the data repository. The repository database has been designed to store the catalog data and support data analysis. Each table stores a unique aspect of the purchase information. For example, metadata related to a purchase action such as company position, state in use, and created time. Metadata related to the company such as company id and attribute goes into the table company_columns. Detailed information related to purchase goods goes into the merchandise table.

3.2 Data Loading Algorithm of Price Comparison System

Data loading allows multiple operations to be inserted into a single area and waiting for one database call, minimizing network communication and disk I/O. It is straightforward to perform bulk loading to a single table. If the data belonging to a child table is loaded before the corresponding parent keys, a foreign key constraint is disarranged. To avoid this problem is to first buffer the data into separate arrays designated for different tables, and then to follow the parent-child relationship order when performing the bulk inserts. The table-loading order is illustrated in Figure 1.

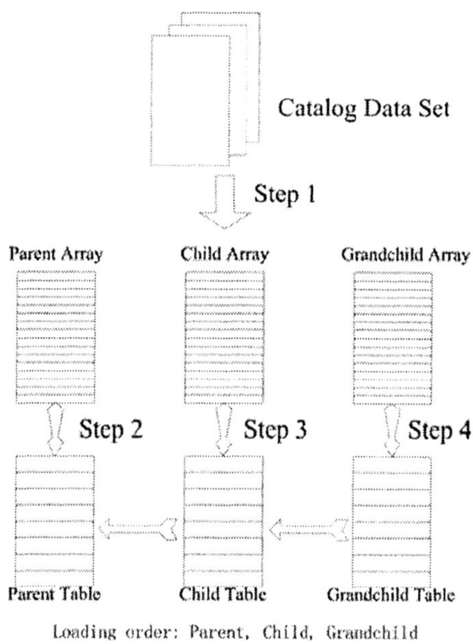

Figure 1. Data Loading Order with Multiple Tables

Another difficulty in data loading is recoverability in a lengthy data-loading process. The catalog data set to be loaded sometimes contains errors such as missing and/or invalid values. The algorithm not only speeds up data loading by a factor of 7 to 9, but also maintains the relationships of multiple tables and enables the system to recover from errors during data loading. The algorithm is introduced in Figure 2.

When any data array reaches array-size (Line 5), the batch_row procedure is called (Line 10) for each array based on the parent-child relationship. The array for the parent table is processed first, followed by the child tables. This processing sequence depends entirely on the data model.

Our approach does not cover circular parent-child relationships, as a good database design does not have circular dependencies between tables.

```
Input: a series of input data files
Output: populated database tables
int array-size /* the size of an array */
int batch-size /* the size of a batch; typically << array-size */
Procedure data_loading
(1) for each data file {
(2) open the file
(3) for each row {
(4) parse the row, do validation, transformation and
computation, and buffer it in a designated array based on the
destination table;
(5) if (any array.size >= array-size) {
(6) for each array ordered by parent-child relationship {
(7) first_idx = 0;
(8) last_idx = array.size;
(9) while (first_idx <= last_idx)
(10) first_idx = batch_row(array, destination_table,
first_idx, last_idx)
(11) } /* for each array */
(12) } /* if reach array-size */
(13) } /* for each row */
(14) } /* for each data file */
Function batch_row (array, destination_table, first_idx,
last_idx)
(15) while (first_idx <= last_idx) {
(16) prepare SQL statement;
(17) add to batch;
(18) if (batch-size reached) { /* time to insert */
(19) insert batch into the destination table;
(20) if (successful insert) {
(21) first_idx += batch-size;
(22) } else { /* if an error occurred skip that row */
(23) skip_one_row;
(24) return (the_next_index);
(25) }
(26) } else if ( first_idx == last_idx) { /* array done */
(27) insert batch into the destination table;
(28) if (successful insert) {
(29) return (last_idx + 1 );
(30) } else { /* if an error occurred skip that row */
(31) skip_one_row;
(32) return (the_next_index);
(33) }
(34) }
(35) } /* while there are more rows to process */
```

Figure 2. data loading Algorithm

The data loading algorithm has been implemented using the JDBC core API. In the best case, that is when the data set is error-free, the algorithm will generate database calls and result in database I/Os. In the worst case, for example primary key violations on every row caused by repeatedly loading duplicate rows, data loading will deteriorate to a series of singleton insert operations which make database calls and perform database I/Os. This behavior results from the algorithm breaking up the problematic batch, skipping the error row, and repacking the batch to continue each time that an error is encountered.

3.3 Buffer Data Structure

The interleaving of data for multiple target tables, combined with the presence of multiple relationships among tables, relationships that must be maintained by obeying with the primary and foreign key constraints during the loading process—makes bulk loading especially challenging. To manage the crossing data and complex table relationships, and to improve quick recovery when an error is detected during the data-loading process, an effective data structure, array-set, is necessary.

The number of arrays in the array-set at a given time during data loading depends on the degree to which the data in the catalog data set is interleaved. As the input catalog data set is processed, the framework creates a new array in array-set whenever it reads an input row targeted for a database table for which no array is currently maintained. When any of the arrays in array-set are fully moved, data loading finished. At the end of the data loading cycle, the arrays in array-set are destroyed and their memory released. The framework resumes reading the input catalog data and creates new arrays as required to buffer the incoming table rows. In order to load the catalog data items into different destination tables and keep the proper relationships, array-set can buffer the data and execute the data loading in the order of parent-child sequence.

3.4 Active Database and System Adjusting

In Price Comparison System framework, active database and system adjusting are set to achieve the best configuration. The performance adjusting is very important to achieve the fast loading of massive data volumes required for repository. Since the Price Comparison System is a multi-year continuous effort, the price repository must serve two purposes at the same time. First, it must be a warehouse to store increasingly loaded data. Second, it must act as a query engine to support economic analyzing research. For example, query performance is necessary to create indices on database tables. However, indices usually make data loading slower because every insert requires an update of all index entries.

Allocating a smaller database data cache actually improves the data-loading performance. Since a database writer needs to scan the entire data cache when writing new data from data cache to disk, the reduced data cache size minimizes the work that the database writer has to do each time . This reduced cache configuration should be adjusted after the intensive data-loading phase is entirely because a larger data cache often performs better for user queries.

4. CONCLUSIONS

The Price Comparison System is a project to collect, archive, process, and distribute trade data for economics collaborations. The repository being built at Shanghai Electric Group Co. is to hold catalog data for the complex provider analyzing. The first significant was to load the catalog data into the repository database in a in time fashion. Data loading with array buffering is a good method to accomplish the task. System framework consists of an efficient algorithm for data loading, an effective data structure to support data integrity and proper error handling during the loading process and guidelines for database and system tuning. This framework can help to decrease much of the loading time.

5. REFERENCES

1. Du Zhihui, Chenyu, Liupeng. *Grid Computing.* Tsinghua University Publishing Press. Apr 2002.
2. Gabrielle Allen, Thomas Dramlitsch, Ian Foster, Nick Karonis, Matei Ripeanu, Ed Seidel, Brian Toonen, Supporting Efficient Execution in Heterogeneous Distributed Computing Environments with Cactus and Globus, Supercomputing 2001. [Winning Paper for Gordon Bell Prize (Special Category)]
3. Donald McMullen, Randall Bramley, et al. The Xport Collaboratory for High-Brilliance X-ray Crystallography, http://www.cs.indiana.edu/ngi/sc2000/index.html
4. http://www.iumsc.indiana.edu/
5. Y. Dora Cai, Ruth Aydt, Robert J. Brunner. " Optimized Data Loading for aMulti-Terabyte Sky Survey Repository". National Center for Supercomputing Applications (NCSA),Nov, 2005
6. S. Amer-yahia and S. Cluet. "A Declarative Approach to Optimize Bulk Loading into Databases". ACM Transactions on Database Systems, Vol. 29, Issue 2, June 2004.
7. G. Reese. Database Programming with JDBC and Java. O'Reilly. 2nd Edition, Aug. 2000.

SUPPLY CHAIN MANAGEMENT IN SMALL AND MEDIUM-SIZED ENTERPRISES

Peter Nyhuis, Katja Hasenfuß
Institut of Production Systems and Logistics (IFA), University of Hannover. Email: nyhuis@ifa.uni-hannover.de.

Abstract: In the production logistics context, the success of supply chains corresponds to their ability to meet the logistic performance expectations of customers. This ability depends on the contributions of all supply chain members to the logistical performance of the entire chain. Moreover the overall performance is affected by the dynamic interrelationships between the specific logistic performances of the separate supply chain members. This includes small and medium-sized companies (SMEs). The paper mainly focuses on current supply chain management methods and supply chain management concepts to improve logistical performance and their present application in today's industries especially SMEs.

Key words: supply chain management (SCM), logistical performance, SCM-methods, small and medium-sized companies (SMEs)

1. INTRODUCTION

Confronted with constant structural changes in society and the manufacturing industry, enterprises have undertaken several innovative activities to optimize their processes and to improve their logistical performance. There is hardly any enterprises still regard and treat logistics as a function of simply transportation and stock-keeping which may not influence their competitiveness. Rather, most enterprises see that their logistics significantly supports their business strategy. In some enterprises their business strategy even arises, at least partly, out of their logistic strengths. In continuation of this trend enterprises cooperate closely with their suppliers as well as with their customers in order to improve business

Please use the following format when citing this chapter:
Nyhuis, Peter, Hasenfuß, Katja, 2006, in International Federation for Information Processing (IFIP), Volume 207, Knowledge Enterprise: Intelligent Strategies In Product Design, Manufacturing, and Management, eds. K. Wang, Kovacs G., Wozny M., Fang M., (Boston: Springer), pp. 386-392.

performance. This enables them to further focus on core competencies. Originally, this led to the forming of supply chains, which are characterized by the share of information between suppliers and producers. An intensive flow of data helps the enterprises of a supply chain to plan more accurately and to adjust their capacities more effectively, since information is provided both earlier and in more detail. Thus, supply chains facilitate the avoidance of large deviations of demand along the value added chain as described by Forrester (1961).

2. SUPPLY CHAIN MANAGEMENT

The understanding of SCM is ever-changing. Still SCM is intensively and controversial discussed in respect of its content regarding practical experiences in a fast developing business environment. Nevertheless, current definitions of SCM emphasize the collaborative aspects, that this management paradigm incorporates – on a strategic (Schönsleben, 2000) as well as on an operational level (Wiendahl et al., 1998). Hieber and Alard (2000) state that there is still a gap between the conceptual ideas of supply chain management and its application in industry. This is especially true for the case of SMEs. There are still disparities between the best practices of SCM and their implementation in practice: enterprises in supply chains pursue individual rather than common objectives. Nevertheless SCM promises more effective processes, better customer orientation and lower costs.

2.1 Concepts and Methods

SCM is internationally accepted by research and industrial practice as one element of the business strategy. Within the scope of SCM it is essential to operationalize the objectives which have been defined within the supply chain strategy. For this purpose relevant functions and tasks can be differentiated by their planning horizons and their objective targets. Therefore, a number of different SCM-concepts and SCM-methods as well as Best Practice examples exist to realize the possible potentials in supply chains (Hieber 2002). As a part of a research project on network-oriented production management accomplished at the Institute of Production Systems and Logistics (IFA), a number of theoretical methods from the fields of production planning and control, such as the funnel model, throughput diagrams and logistic operating curves, have been adopted for production management in a network of several partners (Wiendahl et. al., 1998). Based on the supply chain planning matrix (Meyr 2002) and the SCM-function model of the RWTH Aachen University a categorization scheme was developed by the IFA as a part of a current research project on methods for cooperative business processes. The new categorisation scheme is shown in figure 1. It includes the functions SC-Configuration, SC-Planning, SC-Execution and SC-Controlling. In this scheme the four SCM-functions are arranged accordingly to their planning horizon.

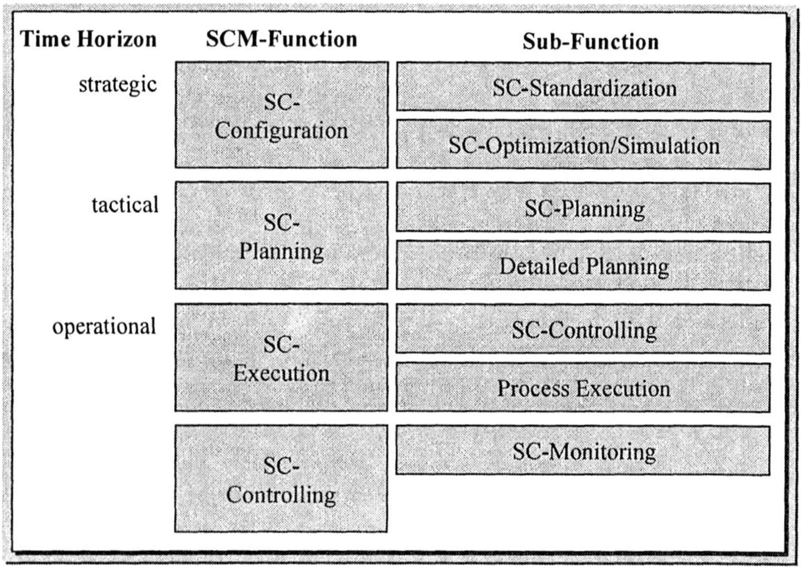

Figure 1. SCM categorization scheme

2.2 Application of Supply Chain Management

To analyze the actual status of application and implementation of SCM best practices in industry a survey in German speaking countries was conducted by the ETH-Zurich and the IFA in autumn 2005. Some 51 questionnaires have been analyzed on the subject „supply chain management and business success" (Schnetzler et al. 2006). Enterprises from different branches and of different sizes took part in the survey. About 30% of the enterprises are SMEs. The European SME definition set the upper limit of employees to 250 and the annual turnover at a maximum of €50 million or the annual balance sheet total at a maximum of €43 million. 20% of all enterprises of the conducted survey have 250 to 500 employees. Figure 2 shows a fairly heterogeneous distribution of different best practices. The analysis of the survey showed that only 30% of the enterprises apply for example order promising, meaning available/capable to promise and that it is also seldom planned to implement it. This is astonishing especially as current SCM-/ERP-Software provides this functions. With this best practice enterprises would be able to offer a more realistic order confirmation. This would lead to an increase of their delivery reliability and consequently, to an improvement of their logistical performance. It is especially relevant if enterprises produce by the principle of make to order - which most of the questioned enterprises do.

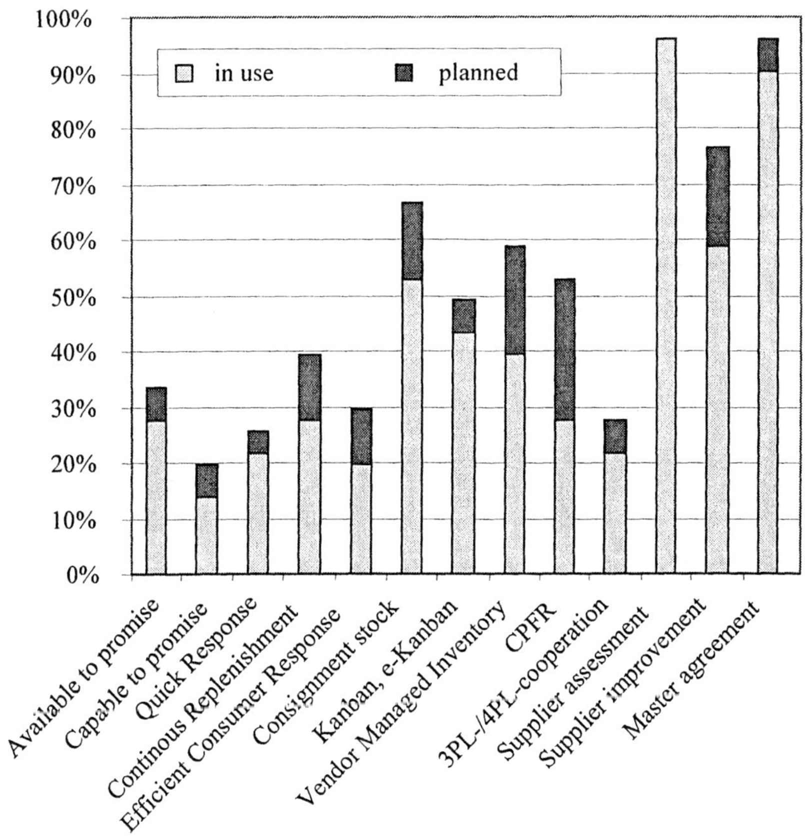

Figure 2. Extract of best practices in use and planned

Furthermore, the following supply concepts where analyzed in the survey: quick response (QR), continuous replenishment (CR), efficient consumer response (ECR), consignment stock, kanban und vendor managed inventory (VMI). The three first supply concepts are most commonly in use. This may be due to the fact that QR, CR and ECR were originally developed for trading. In many cases enterprises plan to implement monitoring, controlling or benchmarking concepts to record, evaluate and compare the performance of their SCM. The analysis of the survey showed, that the implementation of best practices in the German-speaking countries is still not widely spread. The main obstacle seams to be the scarcity of resources. This comes along with the incompatibility of the present information systems. Furthermore SMEs denominate the name recognition of SCM-methods and SCM-concepts as rather low. Therefore, the selection of suitable concepts and methods obviously becomes extremely challenging for SMEs. The planning and implementation of suitable SCM-concepts and SCM-methods becomes even more difficult due to the fact that they do not have the required know-how and responsibilities for implementation are not defined.

3. SUPPLY CHAIN MANAGEMENT IN SMEs

A differentiated analysis of the goals of SMEs shows that cooperating SMEs seldom possess one consistent strategy. In addition if individual SMEs do have a SCM-strategy, they often do not derivate their individual objectives from this strategy. Their concentration on core competences interferes with strategic and policy. It also explains the multiplicity of varying SCM-objectives. Most of those objectives focus on operational functions. Those operational objectives may be structured by the SCOR processes source, make and deliver. Out of all objectives the logistical objectives can be consolidated to the logistical objectives of supply chains. These are short delivery time, high service level, even machine utilization and low work in progress and stock. Therefore they are looking for simple but effective and efficient methods or concepts to conduct the vitally important SCM-functions, which shall improve their logistic performance.

3.1 Restrictions and Requirements of SMEs

Since the application of SCM-concepts and SCM-methods is not that widespread, it was investigated, which restrictions and requirements of SMEs are existent. The determined restrictions in the requirement analysis may be split up into the categories transparency, cooperation, realization and costs. In the category transparency it was for example discovered that beside the fact that different SCM-concepts and SCM-methods are not well known like mentioned before companies do quite often neither know the field of application of a known SCM-concept or SCM-method nor do they know which benefit they may generate out of the employment of a specific concept or method. One reason among others might be that different management approaches are subsumed in SCM-concepts. Therefore it is difficult to get a general conspectus on existing concepts and methods.

4. CONCLUSION

Concluding one may state that from a logistical point of view the advantages of SCM are obvious and demonstrated by the results of SCM in industrial practice. However there is still a need for further implementation into practice and further development which will be executed in research and in industrial practice. The demand for SME oriented SCM-concepts and SCM-methods may be explained by the existent inherent difference between major enterprises and SMEs. Furthermore, the characteristic comprehension and definition of SCM differs as well as the particular goals of these enterprises. Usually, SMEs state that they need to focus on their core competencies. Consequently, many SMEs expect to get operational assistance in their day to day business. Both realization and application of SCM-concepts and SCM-methods demands the acceptance of the employees, therefore and because of the lack of resources SMEs prefer simple methods and concepts which require less specialized knowledge. Thus a guideline to support cooperative business processes of SMEs shall be developed within the current research project at the IFA. This guideline will help SMEs to select suitable SCM-concepts and SCM-methods and to pre-estimate the benefit of the selected concept or method.

5. REFERENCES

1. Forrester, J. W., (1961), Industrial Dynamics, John Wiley & Sons, New York.
2. Hieber, R., (2002): Supply chain management. Zürich.
3. Hieber, R., Alard, R., (2000), Supply Chain Management - A Survey of Practices and Trends in Swiss Industry. In: Strandhagen, J.O.; Alfnes, E. (Eds.): Information and Communication technology (ICT) in Logistics and Production Management. Proceedings of the Annual IFIP WG 5.7 Working Conference, Conference Proceedings 2000. Tromsö, Norway, 28-30 June 2000.
4. Meyr, H., Wagner, M., Rohde, J., (2002), Structure of Advanced Planning Systems. In: Stadtler, H., Kilger, C., Supply Chain Management and Advanced Planning – Concepts, Models, Software and Case Studies. Berlin, Heidelberg, New York.

5. Schönsleben, P., 2000, Integral Logistics Management – Planning and Control of Comprehensive Business Processes. St. Lucie Press, Boca Raton et al.
6. Wiendahl, H.-P., Hoebig, M., Kuhn, A., et al., 1998, Kennzahlengestützte Prozesse im Supply Chain Management. In: Industrie Management, Vol. 14, No. 6, pp. 18-24.
7. Schnetzler, M., Nölle, A., Hasenfuss, K., et al., 2006, SCM und Unternehmenserfolg. Zürich.

AN OPERATIONS MODEL TO SUPPORT MANUFACTURING LOGISTICS PROCESSES IN THE SUPPLY CHAIN

Jan Ola Strandhagen, Heidi Dreyer
Department of Production and Quality Engineering, Norwegian University of Science and Technology, Norway.E-mail: ola.strandhagen@ntnu.no

Abstract: This paper presents an Operations Model, including a Control Model, developed to support reengineering of manufacturing logistic processes. The reengineering process is supported by a methodology. The Operations Model concept was first designed to support the enterprise perspective, and this paper explains the expansion of the model and methodology to supply chains. The methodology has been developed through several case studies and supports the realisation of Supply Chain Control dashboards and studio.

Key words: Manufacturing Logistics, Operations Model, Control Model, Reengineering

1. INTRODUCTION AND METHODOLOGY

In today's global economy operational excellence is required for manufacturing companies. Competitiveness can only be obtained and maintained if operations management is developed in order to support market requirements and customer needs. It becomes essential to develop innovative demand oriented business models which is adapted to the products and manufacturing processes in the company, and to provide for a holistic supply chain orientation towards suppliers and customers. ICT is an enabler leading to increased competition between companies, and at the same time being an instrument which supports companies in gaining competitive advantages. The ICT-based operational processes must be constantly reviewed and changed paving the way for new control concepts to

Please use the following format when citing this chapter:
Ola Strandhagen, Jan, Dreyer, Heidi, 2006, in International Federation for Information Processing (IFIP), Volume 207, Knowledge Enterprise: Intelligent Strategies In Product Design, Manufacturing, and Management, eds. K. Wang, Kovacs G., Wozny M., Fang M., (Boston: Springer), pp. 393-398.

be applied. Implementation of new control concepts are best supported by a set of enterprise models that represent the operations processes and resources involved. These models should be unified along the supply chain, allowing a process view rather that a functional view of the operation. The model supporting the reengineering process and the Operations Model itself must be neutral and hence generic to the actual control principles both in the as-is and to-be state of the operation. Also the models applied in the reengineering process should support the actual operation itself, allowing online decisions support functionality.

The aim of this paper is to describe the Operations Model and its applicability for supply chains. The research issues addressed is what the key characteristics of a supply chain Operations Model are, and what the main elements of a model supporting the reengineering process are.

The research strategy followed is a combination of a theoretical discussion, exploratory research and empirical data use. The empirical data is collected from a number of cases. The case study approach is preferred due to the need for a deep and extensive study about thoughts and reflections of operations models in supply chains [11][6]. The data sets are analyzed through traditional quantitative and qualitative approaches as statistical processing and textual interpretations of field notes.

2. OPERATION MANAGEMENT

A traditional definition of operations management is *"the design, operation, and improvement of the systems that create and deliver the firm's primary products and services"* [4]. In this paper we adopt this definition but broaden it to focus on and include the supply chain system.

2.1 Extended enterprise – supply chain

Today one of the most significant competitive enablers is innovative supply chain systems based on collaborative models between companies which see the needs of coordinating the supply, production and delivery processes. Supply chain management, which is *"a set of approaches utilised to efficiently integrate suppliers, manufacturers, warehouses, and stores, so that merchandise is produces and distributed at the right quantities, to the right locations, and the right time, in order to minimize system wide cost while satisfying service level requirements"*, is both a management philosophy and a strategy for active operation and integration of the company's market and supply relations [10]. Here active operation means that the company together with suppliers and customers forms holistic operation supply chain system between them.

Collaboration in the supply chain has a wide range of forms with one common goal: to gain information and to create a transparent, visible demand pattern that paces the entire supply chain [8]. Information and visibility is considered as *"The Holy Grail"* leading to higher predictability and insight into the demand situation which is an essential element in an Operations Model. Several studies have identified the problems caused by a lack of information and to what extent competitive advantages can be gained from a seamless supply chain [7][9][5].

2.2 Modelling and operations model

Enterprise models are used for a range of purposes and can differ from some coarse sketches to detailed numerical models. This paper focuses on the conceptual type of models, which usually are developed to make sense of aspects of an enterprise and communicate with other actors. Such models are normally descriptive, visual, and built on high level modelling entities.

Reengineering an enterprise requires a clear understanding and definition of customer requirements and the operation's capabilities [1]. To gain this understanding requires the collection of data from the enterprise's information systems, interviews with key personnel, examination of documents etc. This data is then used to create a textual and visual description of the Operations Model.

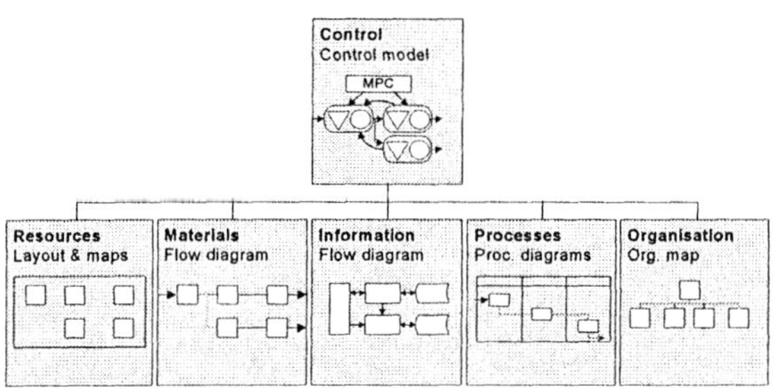

Figure 1. The operations model set.

A Control Model is a representation of how operations are organized and controlled in manufacturing [12][3][2]. A Control Model is normally developed by the following building blocks: customer order decoupling point, control principles and methods, main operations processes and the related information's flows and ICT systems. The Control Model is the key model in the comprehensive Operations Model. The Operations Model proposes six views that should be modeled as shown in Figure 1.

3. THE SUPPLY CHAIN OPERATIONS MODEL

A simplified example of a Control Model for the manufacturing enterprise HÅG is shown in Figure 2. This model is the result of a reengineering process to improve the logistics performance of the enterprise.

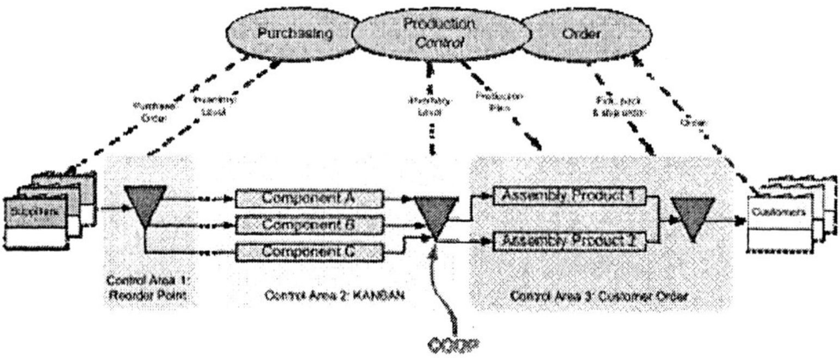

Figure 2. Control Model for Manufacturing Enterprise.

The focus in this model is on control of the internal processes of the enterprise. To expand from an enterprise view to a supply chain view the supply chain Operations Model has been developed to include the aspects given in Table 1.

Table 1. The expanded supply chain view

Resources	Processes	Material	Information	Organisation	Control
Suppliers	Forecasts	Demand pattern between actors	Informations flow between actors	Roles of different actors	CODP placing (several in VC)
Transporters	Order	Distribution principles	ICT systems of actors	Contract principles	Control principles
Wholesalers	Call offs				Control areas combining actors
Customers	Procurements				KPI measure in customer-supplier relations
	Wholesalers control				
DATA SET (documentation)					

The expanded supply chain view is applied in a reengineering process to create an Operations Model for a pharmaceutical supply chain. The TO-BE supply chain Operations Model is based on the concepts of automatic replenishment and it resulted in the following Operations Model in Figure 3. Again the reengineering process methodology explained in section 4 was used. The figure shows a holistic view of the production, distribution and demand system in the pharmaceutical supply chain. In the AUTOMED project such a model is developed between a global manufacturer, a

Norwegian wholesaler, and a Norwegian pharmacy retail chain in order to allow automated replenishment of pharmaceuticals based on point-of-sale-data (POS-data). The essential elements in this model are the continuously replenishment of the pharmacies based on POS-data, the placing of the customer order decoupling point (CODP) and that the control process is integrated and coordinated in and between the companies.

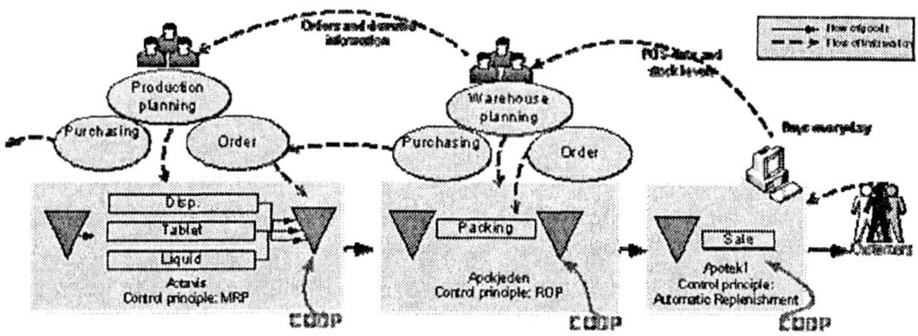

Figure 3. Supply chain operations model for AUTOMED.

4. THE REENGINEERING PROCESS

Process reengineering requires an in-depth understanding of the enterprise's current manufacturing logistics processes. The Enterprise Reengineering Methodology organises the reengineering process in six steps; see Figure 4. The purpose of the process is to create an Operations Model that describes the AS-IS and TO-BE logistics processes. The process purpose and preliminary problem hypothesis are defined in the initial stage. These guide the choice of methods for the mapping and analysis stages.

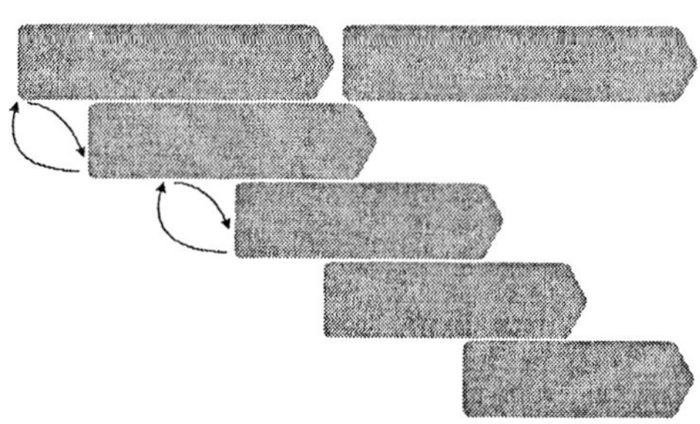

Figure 4. Enterprise Reengineering Methodology.

Mapping consists of creating a description of the AS-IS Operations Model and identifying potential improvement areas. Subsequently a selection of general and specific analyses is performed on chosen aspects of the enterprise's operations. This should identify the key characteristics of the supply chain, and based on the findings, the TO-BE Operations Model and solution elements are designed, and implemented in the enterprise.

5. CONCLUSION

This paper has presented an Operations Model, including a Control Model, developed to support reengineering of manufacturing logistic processes. The reengineering process is supported by a methodology. The Model and Methodology has been applied in reengineering of more than twenty manufacturing enterprises. The Model has now been expanded into a Supply Chain view, based on the same Reengineering Methodology, and has been applied to several cases. Further research is focused on developing concepts and solutions for Supply Chain Dashboards and Supply Chain Control Studios based on the Operations Model.

6. REFERENCES

1. Alfnes, E. (2005) *"Enterprise Reengineering; A Strategic Framework and Methodology"*, Doctoral Thesis at NTNU, 2005:153
2. Alfnes, E., Strandhagen, J. O. (2000) Enterprise Design for Mass Customization – The Control Model Meth., *International Journal of Logistics*, Vol. 3, No. 2, July
3. Andersen, B., Strandhagen, J. O., Haavardtun, L. J. (1998) *Material- og produksjonsstyring*, Cappelen Akademisk Forlag, Oslo
4. Chase, R.B., Jacobs, F.R., Aquilano, N.J (2004) *Operations management for competitive advantage*, tenth edition, McGraw-Hill
5. Chen, F. Drezner, Z., Ryan, J., Simichi-Levi, D. (2000) Quantifying the Bullwhip Effect in a Simple Supply Chain: The Impact of Forecasting, Lead Times, and Information, *Management Science*, Vol. 46, No. 3, pp. 436-443
6. Eisenhardt, K. M. (1989) Building Theories from Case Study Research, *Academy of Management Review*, Vol. 14., No. 4, pp. 532-550
7. Forrester, J. W. (1961) *Industrial Dynamics*, MIT Press, Cambridge, MA
8. Holweg, M., Disney, S, Holmström, J, Småros, J. (2005) Supply Chain Collaboration: Making Sense of the Strategy Continuum, *European Management Journal*, Vol. 23, No. 2, pp. 170-181
9. Lee, H. L., Padmanabhan, V., Whang, S. (1997) Information Distortion in a Supply Chain: The Bullwhip Effect, *Management Science*, Vol. 43, No. 4, pp. 546-558
10. Simchi-Levi, D., Kaminsky, P., Simchi-Levi, E. (2003) *Designing and managing the supply chain: concepts, strategies & case studies*. McGraw-Hill.
11. Yin, R. K. (1994) *Case Study Research: Design and Methods*, Thousand Oaks, CA, Sage Publications, Inc.
12. Quistgaard, T., Grøtte, P. B., Hauknes, B. A., Engh, Ø., Persson, G. (1998) *Håndbok for flytorientert MPS.* Tapir Forlag et. al.

DEVELOPMENT OF A DECENTRALIZED LOGISTICS CONTROLLING CONCEPT

Peter Nyhuis, Felix Wriggers, Andreas Fischer
Institute of Production Systems and Logistics, University of Hannover, Germany, e-mail: wriggers@ifa.uni-hannover.de

Abstract: Nowadays, companies have to achieve a high logistics performance combined with low costs and high quality. Conventional, resource-oriented production planning and control (PPC) is unable to meet these demands. Both research publications and industrial projects show that logistics controlling based on logistics principles is an effective instrument for improving PPC. A newly developed controlling concept that increases the quality of production planning and control in a significant way is presented in this paper.

Key words: logistic controlling; performance measurement

1. INITIAL SITUATION

In today's consumer markets, enterprises are increasingly being forced into adapting their products to changing customer demands, promoting technological developments and fulfilling the high requirements placed on delivery lead time and reliability, quality and costs [Wil04].

Therefore, enterprises have to redesign their products and processes in such a way that profitable production will be possible under present and future boundary conditions [Schu97]. Only goal-oriented reproduction, evaluation and interpretation of production processes expose weak spots and help detect means of improvement [Wie02, Sc.Wo99].

Methods of reproduction and analysis that can be generally applied to the order flow are especially scarce for manufacturing processes. The conventional, mere resource-oriented reproduction of manufacturing processes will not be able to cope with the anticipated demands due to lack

Please use the following format when citing this chapter:
Nyhuis, Peter, Wriggers, Felix, Fischer, Andreas, 2006, in International Federation for Information Processing (IFIP), Volume 207, Knowledge Enterprise: Intelligent Strategies In Product Design, Manufacturing, and Management, eds. K. Wang, Kovacs G., Wozny M., Fang M., (Boston: Springer), pp. 399-405.

of information. Production planning and control can be effectively supported by suitable controlling concepts that are based on reproduction of both, the resource view and the order view. An appropriate controlling concept allows the analysis, evaluation and design of a manufacturing process and helps to manage the lack of information. Thus, the objective of a research project at the Institute of Production Systems and Logistics was to develop a decentralized, participative logistics controlling concept utilizing "intelligent" transponders. Further, employees should be integrated into the logistics controlling. Based on key figures which are calculated decentralized, they shall be qualified to take actions independently to improve processes logistically.

2. DEFINITION OF THE CONTROLLING OBJECT

Experience in designing company departments has shown that it makes sense to create, preferably, simple images of the observed departments to begin with. This decreases the complexity of the controlling object and thus increases transparency. Therefore, the controlling object consists of four work systems which are interrelated by so-called "1:1 relationships (as shown in Fig. 1).

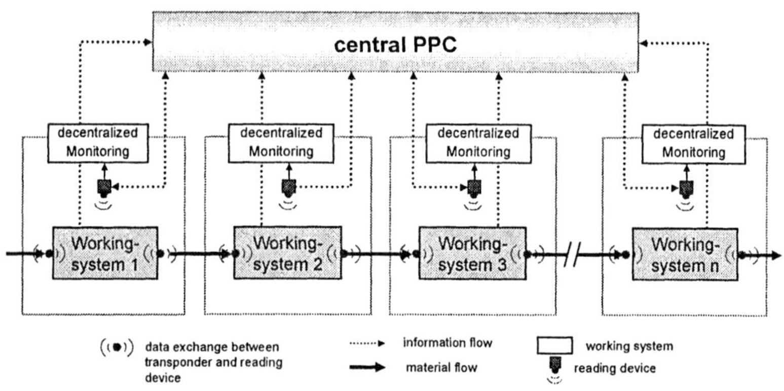

Figure 1. Controlling object of the decentralized participative controlling concept

The work systems of the controlling object are to be equipped with a computer unit that allows decentralized monitoring. The data required will be captured and passed on by "intelligent" transponders. In this context, "intelligent" transponder refers to a pre-processing label (PPL), providing functions for local calculation of order-specific key data, which has been developed by the Institute for Transport and Automation Technology.

Key figures derived from business operations form a fundamental basis for controlling. Their task is the focused registration of those facts and trends that can be quantified and thus enable reliable statements. The intention of this specific form of information is to describe complicated facts in a simple way. The significance of key figures is not only determined by the kind of values used for their generation but also by their form of presentation.

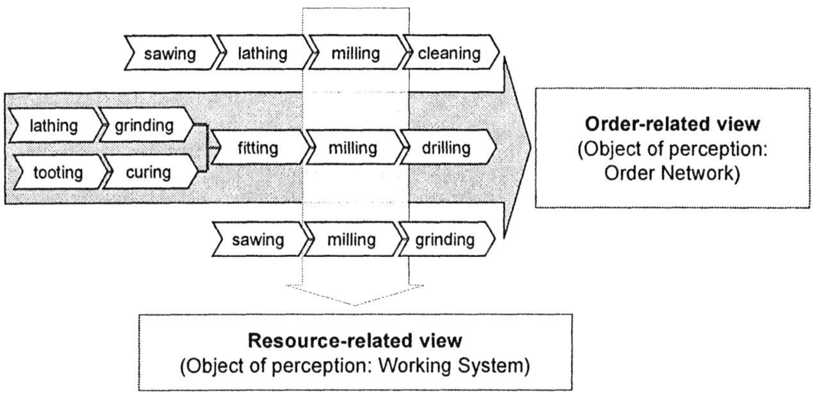

Figure 2. Differentiation between resource-related and order-related view

The performance measurement system of the innovative controlling concept is intended to form the basis for the order- and resource-oriented analysis. The two general perspectives of controlling are related through the individual work processes (see Fig. 2). The individual work process therefore represents a unified basis for all perspectives, which establishes the fundamental requirements for a holistic controlling concept.

3. RESOURCE-RELATED MONITORING

Within the monitoring system, the results should be displayed in a clear and straightforward way. Like for the performance measurement system, we distinguish between a resource-related and an order-related representation.

The funnel model describes manufacturing processes by means of the input and output behavior of the participating systems [Wie97]. It facilitates the qualitative and exact-time representation of the dynamic behavior of manufacturing systems and represents the interrelationships between the logistics key figures for throughput time, schedule variance, mean inventory and capacity utilization in the form of characteristic operating curves and makes them accessible for mathematical description [Ny.Wi03].The funnel model and the theory of logistics operating curves have not only proved themselves in science, they also have been successfully applied in industry projects to improve logistics performance. Moreover, the model has been incorporated into commercial ERP software and has found applications in controlling. For this reason, the representation of logistical interrelations by means of characteristic operating curves is used in resource-related monitoring. The characteristic operating curves for delivery reliability and throughput time represent the logistics performance (Fig. 3, left), those for performance and costs reflect the logistics expense situation (Fig. 3, right).

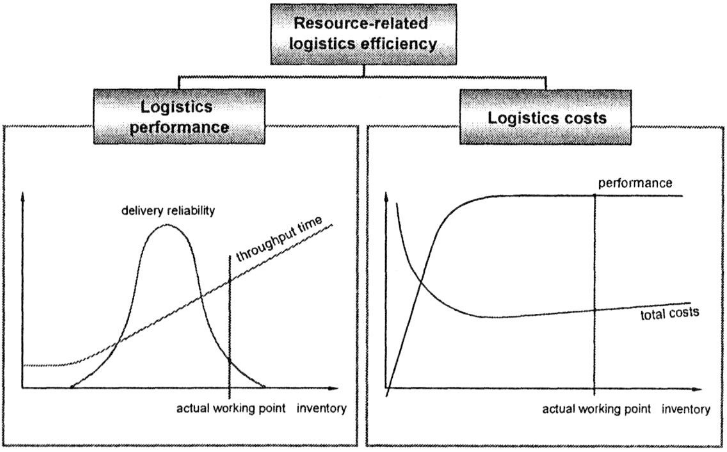

Figure 3. Resource-related logistics monitoring

4. ORDER-RELATED MONITORING

The aggregated, order-related key figures for logistics performance and logistics costs are displayed in a different way, because the order reference does not permit a direct connection between these key figures and the inventory level.

The order-related logistics performance can be reproduced by a progress graph (Fig. 4, left). The actual time is plotted as completion confirmation data at defined measuring points and is then compared with the target. Based on this, the scheduling delay can be displayed. This form of representation facilitates the analysis of where and to what extent the delay occurs. This helps users to respond quickly with appropriate countermeasures. The order-related logistics costs are illustrated in a cost–time diagram (Fig. 4, right). This diagram displays manufacturing costs and capital tie-up costs, with actual and target values in direct comparison.

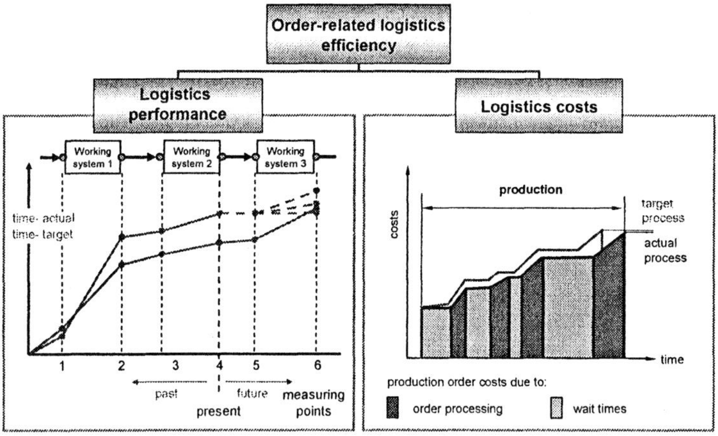

Figure 4. Order-related logistics monitoring

5. DEFINITION OF MEASURES

It is necessary to discuss at what time which measures have to be taken and who has to initiate them. The intention is to adjust the load or the capacity in such a way that the target values can be achieved as far as possible. Basic measures for adjusting load and capacity have been subsumed under the concept of capacity adjustment in Fig. 5.

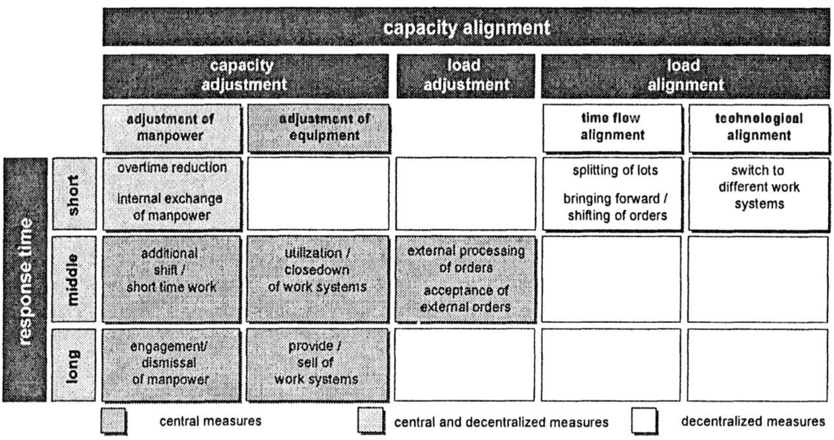

Figure 5. Differentiation of measures for capacity alignment [Büchel]

The response time needed for the realization of measures is an important criterion for the decision which measures should be taken. A short response time necessitates decisions at plant level. Thus, they can be implemented faster and more flexibly and the employees are integrated into the process. Employees can only participate if measures come under their sphere of competence and influence. This is true for short-term measures only. Medium-term or long-term measures require an interdisciplinary perspective that a production employee usually does not sport. Consequently, decisions regarding long-term and medium-term actions have to be taken centrally.

6. CONCLUSIONS

To meet the high demands placed on delivery lead time and reliability, quality and costs, the reproduction, analysis and interpretation of logistics processes are becoming increasingly important. To fulfill these requirements, PPC can be supported by suitable controlling concepts. However, analyses have shown that current controlling concepts possess deficits. For this reason, the Institute of Production Systems and Logistics has developed a new decentralized participative controlling concept based on "intelligent" transponders that decreases these deficits and so increases PPC quality. In the next step the results of this research work will be transferred to assembly orders with integral buffers, finally to whole supply chains.

The authors would like to thank the German Research Foundation for sponsoring this research project.

7. REFERENCES

1. [Ny.Wi03] Nyhuis, P. and Wiendahl, H.-P., (2003), *Logistische Kennlinien. Grundlagen, Werkzeuge und Anwendung*. Springer-Verlag, Berlin,.
2. [Sc.Wo99] Scholz-Reiter, B. and Wolf, H., (1999), Logistik-Controlling: Mobilisierung von Optimierungspotenzialen durch Transparenz in der Logistik. *Industrie Management* 15.
3. [Schu97] Schuh, G. and Wiendahl, H.-P. (Hrsg.), (1997), *Komplexität und Agilität*; Springer Verlag, Berlin,.
4. [Wie02] Wiendahl, H.-P., (2002), Erfolgsfaktor Logistikqualität: *Vorgehen, Methoden und Werkzeuge zur Verbesserung der Logistikleistung.* 2. Aufl. Hans-Peter Wiendahl (Hrsg.), Springer-Verlag, Berlin,.

5. [Wie97] Wiendahl, H.-P., (1997), *Fertigungsregelung. Logistische Beherrschung von Fertigungsabläufen auf Basis des Trichtermodells*. Carl Hanser Verlag, München, Wien,.
6. [Wil04] Entwicklungstrends in der Automobil- und Zulieferindustrie. TCW Transfer-Centrum GmbH & Co. KG, ISBN 3-934155-85-5, München, 2004

MODELING THE VARIABILITY WITH UML FOR TRIZ BASED CAI SYSTEM

Jianhong Ma[1], Runhua Tan[2]
[1]School of Computer Science and Software Engineering, Hebei University of Technology, TianJin, 300130, China, Eamil:jh_ma@eyou.com.; [2]School of Mechanical Engineering, Hebei University of Technology, TianJin, 300130, China

Abstract: Going with the inventive knowledge applied broadly in different science fields, special Computer Aided Invention(CAI) software product customized for different science fields should be developed with low cost and short time. In order to minimize the number of unexpected adaptations and features within application engineering projects, variability is analyzed not only for current requirements but also for future requirements. Based on analyzing CAI software domain, the development trend of CAI family products is predicted, and the variability of this domain is modeled with extended UML in this paper. The feature-based object-oriented domain analysis approach is adopted to model the variability of CAI family products.

Key words: Computer Aided Invention (CAI); TRIZ; modeling variability; UML; software product line

1. INTRODUCTION

TRIZ, the theory of solving inventive problem, is proved to be useful in invention in practice. Many CAI products have been completed and applied in industries. Going with the inventive theory developed thoroughly and generalized broadly and the requirements of market changefully, the power of related products should be enhanced rapidly. The characteristics of CAI family tools are:

1. Evolving

   The CAI family tools are evolving and becoming powerful with new modules gradually being added to the previous version. For example,

This project is supported by the Project of Natural Science Foundation of China under Grant No.50375045, the Project of Natural Science Foundation of Tianjin under Grant No.043802211 and the Project of Educational Foundation of HeiBei Province under Grant No.2004412

Please use the following format when citing this chapter:
Ma, Jianhong, Tan, Runhua, 2006, in International Federation for Information Processing (IFIP), Volume 207, Knowledge Enterprise: Intelligent Strategies In Product Design, Manufacturing, and Management, eds. K. Wang, Kovacs G., Wozny M., Fang M., (Boston: Springer), pp. 406-411.

CAI tool 3.0 version is developed by integrated TMMS module and 76 standard solutions module into 2.0 version.

2. Integrative
   In inventive design domain, it is trend that CAI tool will be a synthetical tool with many different inventive theories, such as QFD, TRIZ and AD etc[1].
3. Flexibility and Customizability
   The CAI tool has been applied in different domains. So, the special CAI tool for special domain must be developed with flexibility and customizability.
4. Extensibility
   The knowledge of enterprises is playing a more and more important role in activities of enterprises culture and inventive design. It is the customer's requirements to mix the databases of the CAI tool with the knowledge of inner enterprises to form the repository with own intellectual property rights.

So, efficient method to model and manage the variability of family products for reuse turns out to be necessary, both from single product process and economic viewpoint. Lots of researches on exploiting an efficient method to model the commonality and variability have been undertaken.

2. SOFTWARE PRODUCT LINE

2.1 Commonality and Variability

A product family consists of a group of related software sharing some common features.

2.1.1 Commonality

Commonalities of a product family are characteristics that all the family products own. They serve to characterize the domain. The determination of whether a characteristic is a commonality or variability is often a strategic decision rather than an inherent property of the product family.

2.1.2 Variability

The common definition of variability is given as: Software variability is the ability of a software system or artifact to be changed, customized or configured for use in a particular context.

Variability in software systems can be identified as the functional and non-functional variability. Functional variability means that the system can provide different functionalities in different contexts.

2.2 Software product line

The approach of software product line aims in decreasing the costs and lifecycle required to produce a customer specific product. It includes two processes: Application engineering and Domain engineering. In domain engineering the communality and the variability of the product family are defined and developed as core asserts for reusability. The variability in such a domain (or several domains) is explicitly modeled and separated from the common parts. During application engineering a customer specific application will be defined and ideally developed by selecting and configuring core assets resulted form the domain engineering. Domain engineering core assets are evolved based on the feedback from the application engineering.

3. MODELING VARIABILITY WITH UML FOR CAI FAMILY PRODUCTS

An effective representation of the variability not only identifies and models alternatives among the products in a product line, but also defines what characteristics are associated with what products, as well as what dependencies and interrelationships exist among variability.

The representative approaches are classified into three catalogues: feature-oriented method2, object-oriented method3 and integrative method4. In the paper, the feature-based object-oriented method, a typical integrative method, is applied to modeling the variability for CAI product family, and variability is described at feature model, object model and subsystem model with extended UML.

3.1 Modeling feature with extension UML

Feature means the attribute and character belong to the systems. Feature model captures the end user's understanding of the general capabilities of applications in a domain. It represents the common and the variable features of concept instances and the interdependencies between the variable features.

Features in feature model are classified into following categories:

1. Mandatory feature: means features must be selected, it represents the core attributes and characters of the domain.

2. Optional feature: means that there are 1 to n feature to be selected. In this case, some of them may not belong to several family products.
3. Alternative feature: at least one feature should be selected. In this case, the different products will take different methods at the same attributes and characters.
4. External feature: a feature realized by the underlying platform, not by the system itself. It is useful for describing relations to external (e.g. platform) requirements.
5. Extension feature: features that will be extended in future.

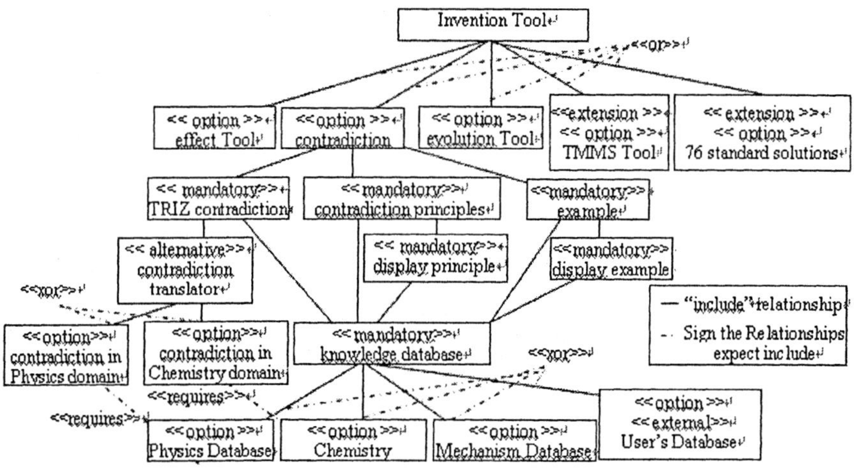

Figure 1. Feature model for CAI family products with UML

The features are organized in a tree with the modeled concept as the root constructed from composition or generalization relationships. The relationships between features include: (the relationships are modeled with their UML notation – filled diamond and generalization-arrow)

1. composition or generalization relationships: expressed as <<include>> or omitted .
2. dependency relationship: expressed as << requires>>.
3. exclusive and non-exclusive alternatives: expressed as << xor>> and <<or >>.

The feature model of CAI product family is shown as figure 1 above. It supplies some modules to be selected according to the customer' requirements, some have been finished, such as effect tool, contradiction tool, evolution tool, and some will be developed or finished in future, such as TMMS tool and 76 standard solution tool. So, the relationship of these features is expressed as <<or>>. Take the contradiction tool as the example, contradiction tool must have the following features: TRIZ general contradiction, contradiction principles and examples, and these features are the result of querying the TRIZ contradiction database. If the being developed product is applied in special domain, such as physics, etc, the "contradiction translator "feature must be selected, which the special domain contradictions will be translated into TRIZ general contradictions. Of course, the feature of special contradiction databases depends on the selection of the special application domain, so the relationships between "contradiction special domain" and "special databases"expressed as <<requires>>.

3.2 Object model

Object model represents the domain structure in terms of objects and their relationships. There are three steps to model objects: identify object from feature model and their relationships; handle the variability through analyzing the relationships between objects, such as aggregation, generalization and association, and assigning the variant to object model.

3.2.1 Identifying the object from feature model

The guideline presented by K. Lee[4] is:

1. Modeling capability features as service-state hiding objects or user role objects.
2. Modeling operating environment features as interface objects.
3. Modeling domain technology features as objects that encapsulate requirement decisions.
4. Modeling implementation technique features as objects that encapsulate communication methods, design decisions and implementation methods.

The relationships between objects fall into generalization and aggregation relationships in object-oriented programming. The aggregation and generalization relationships can be respectively derived from the same types of relationships in the feature model and interactions between the identified objects.

3.2.2 Handling the variability in object model

To describe the variability in object model, the notion of variation points is applied into the object model. The extension for variation points consists of two parts: explicitly marking the location of a variation and the distinguishing the different ways of binding this variation.

Variant points exist in the objects which identified from extended features, optional features, alternative features and extension features, also including the feature relationships signed as {xor} {or}. The variation point determines when (binding time) and how many variants (multiplicity) can be bound (i.e. selected for the desired product). In addition, constraints for the variants can be modeled.

4. CONCLUSION

The approach of feature-based object oriented domain analysis has been applied to the CAI domain. With this method, the feature model and object model explicitly distinguish the commonality and variability. A new special application developed based on the CAI software product line will be completed quickly through tailoring or binding the core asserts.

5. REFERENCES

1. Tan Rnnhua, Kraft Dieter. A conceptual design methodology for variety using TRIZ and QFD [A]. *Proceedings of the ASME Design Engineering Technical Conference* [C]. 2002. V4. 377-384
2. Kang KC, Kim S, Lee J, Kim K, Shin E, Huh M. FORM: A feature-oriented reuse method with domain-specific reference architectures. *Annals of Software Engineering*, 1998, 5:143~168.
3. Griss ML, Favaro J, d'Alessandro M. Integrating feature modeling with the RSEB. In: Devanbu P, Poulin J, eds. *Proceedings of the 15th International Conference on Software Reuse. Victoria: IEEE Computer Society*, 1998. 76~85.
4. Kwanwoo Lee, Kyo C. Kang, Wonsuk Chael and ByoungWook Choi Feature-based approach to object-oriented engineering of applications for reuse *Software—Practice And Experience*. 2000; 30:1025–1046
5. Hassan Gomaa, *An Object-Oriented domain ananlysis and modeling method for software reuse*. 0073-1129-1/92 1992 IEEE.

EFFECT AND EFFECT CHAIN IN FUNCTIONAL DESIGN

Guozhong Cao[1], Haixia Guo[2], Runhua Tan[1]
[1]*School of Mechanical Engineering, Hebei University of Technology, Tianjin, China. Email: cgzghx@163.com;* [2]*School of Management, Hebei University of Technology, Tianjin, China.*

Abstract: Functional design is a key step in the product design process, whether original or redesign. To push functional design into the realm of repeatable and computational, this paper proposes an automated functional design approach. The realization of function is supported by effect chain which is produce by effects using six effect modes. This approach can produce numerous feasible principle solutions in conceptual design process. A design example for functional design of a Chinese medicine mechanism is presented to demonstrate the proposed functional design methodology.

Key words: Effect, TRIZ, Functional Design

1. INTRODUCTION

Functional design in engineering design research theory is a well-researched and active field of engineering study[1]. Functional design plays the central role in ensuring design quality and product innovation.

Theory of Inventive Problem Solving (TRIZ) is an integration of concepts, principles, methods and cases which are abstracted by analyzing patents of the world wide. According to Altshuller's patent search, for any given problem, there is more than a 90% of chance that a similar problem has already been addressed somewhere, at some time[2]. Tan[3] proposed a process of two stages analogy-based design employing TRIZ, in which the process of determining TRIZ special solution from a domain problem is called the first stage analogy process which can perform automatically. The

Please use the following format when citing this chapter:
Cao, Guozhong, Guo, Haixia, Tan, Runhua, 2006, in International Federation For Information Processing (IFIP), Volume 207, Knowledge Enterprise: Intelligent Strategies In Product Design, Manufacturing, and Management, eds. K. Wang, Kovacs G., Wozny M., Fang M., (Boston: Springer), pp. 412-420.

question now becomes: how the appropriate information can be accessed and organized?

Although there are now some general methodologies dealing with functions or reasoning about functions[1, 4-7], few computational tools exist to assist designers during the functional design, and functional design success is often heavily weighted on personal experience and innate ability. To push functional design into the realm of repeatable and computational, this paper proposes an automated functional design approach that utilizes six effect modes and the existing effect knowledge base to support functional realization.

2. FUNCTIONAL DESIGN BASED ON BEHAVIOR AND EFFECT

Cao[8] proposes a functional design approach based on behavior and effect.

Function is a statement to describe the transformation of input and output flows, aiming to achieve the designer's purpose. Function is expressed as verb. The 30 functions in TRIZ are expanded and reclassified and the standard set of functions is presented. The object moving between functions is called a flow, which is divided into material, energy and parameter flows based upon the work by Pahl and Beitz[4]. Flow is expressed as noun. The standard set of flows is presented. Each flow contains a set of attributes to describe its state; two flows are different if their attributes are not the same. The standardized sets of functions and flows are intended to comprehensively describe the design space.

Behavior is a causal relationship between input and output flow. The transformation of input and output flows performs function. Function is an abstracted and subjective representation of behavior, and behavior is a physical interpretation of function. The difference in function and behavior only lies in the identification of their input and output. From this observation, it is shown that function and behavior are different views of the same thing.

The function types derived from the relation between input and output flows identify the behavior. The behaviors, which characterize the implementation of function, are called external behaviors. As with most complex system, it is generally good to break large external behaviors down into smaller and more easily sub-behaviors. An internal behavior is a sequence of alternating sub-behaviors and sub-behaviors transitions, which represents the way of external behavioral achievement. The causal behavioral reasoning based on effect in TRIZ is to support functional modeling. The mapping from function to structure is realized by relationship between effects and examples based on the analysis of patents. An example contains structural feature, such as what elements the design is composed of, what the attributes of the elements are and how they are related.

The functional design approach based on behavior and effect provides an automatic computational solution to complex conceptual design systems. Due to the mappings among function, behavior, effect and example are many-to-many relationship, many principle solutions are usually produced. For the multi-solutions, it is needed to determine one or a few domain solutions through evaluation. The result of detailed design is design files (drawings or data file) that can be used for manufacture.

3. EFFECT AND EFFECT CHAIN IN FUNCTIONAL DESIGN

3.1 Effect

Effect is one of the knowledge base tools in TRIZ. By the analysis of hundreds of thousands patents, effects are emerged from the relevance between behaviors delivered by a design product described in a patent and a principle used in the product[5]. An Effects is an input and output relationship that combines the laws of science including physics, mathematics, chemistry and geometry, and their corresponding engineering applications, which helps to bridge the gap between science and engineering. According to the included information effects can be divided into physical, chemical, geometric effects, etc.

Effects can be characterized by its input, output relations[9]. Generally speaking, an effect has an input and output flow, which is called as basic effect, thus the effect model has two poles, as shown in Figure 1(a). Most transitions from input to output with effect are controlled by auxiliary flow, so the controllable effect should be denoted with three poles, as shown in Figure 1(b). The control flow specifies the factors that can be manipulated to change the output intensity of an effect.

An effect may have multiple input flows, output flows or control flows, so the effect has multiple input poles, output poles or control poles (two, three or more). This paper only discusses effects with less than three input flows, output flows, or control flows.

Effects can fulfill transition from inputs to output, namely, the happening of sub-behaviors depends upon effects. When an effect is said to fulfill a given sub-behavior, it implies that there exists at least an effect, such that the sub-behavior can be achieved.

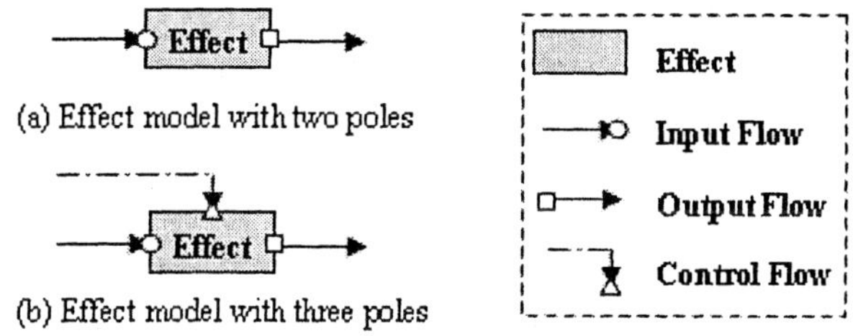

Figure 1. Effect

3.2 Effect Mode

Effects can be connected to one another through its input or output ports and compatible relationships among adjacent effects, which confirm causal relation and structural relation of sub-behaviors. The internal behavior can be achieved by the following effect modes, in which the directed link represents one or several flows.

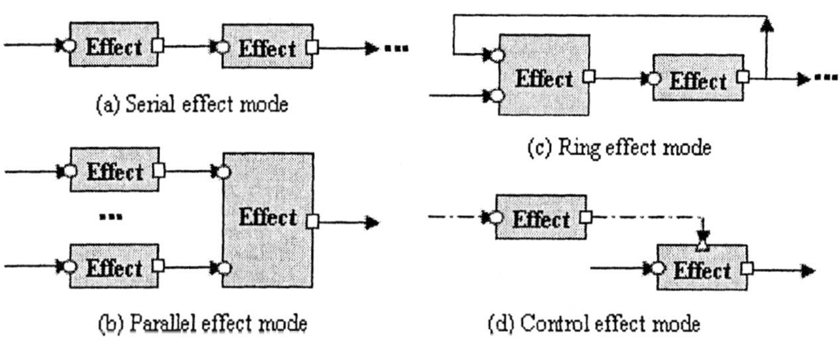

Figure 2. Effect modes

- Single effect mode: achieve an internal behavior by an effect, namely, the internal behavior only includes single sub-behavior, as shown in Figure 1(a). An effect can fulfill several behaviours. A behavior can be fulfilled by several effects respectively.
- Serial effect mode: achieve an internal behavior by a set of effects occurring in sequence, as shown in Figure 2(a).
- Parallel effect mode: achieve an internal behavior by a set of effects occurring at same time, as shown in Figure 2(b).
- Ring effect mode: achieve an internal behavior by a set of effects, as shown in Figure 2(c).

- Control effect mode: the internal characteristic of an effect can be controlled by other effects in order to control the achievement mode of internal behavior, as shown in Figure 2(d).
- Combined effect mode: achieve an internal behavior by several above effects modes.

3.3 Effect Chain

The effects can be linked into an effect chain by using effect modes. This is useful when you do not know how to produce the input required by an effect you want to use. Multiple effect chains that produce the same output can be generated, allowing you to then select the chain that best fits your available resources and interrelated constraints. The effect chain can help engineers achieve breakthrough innovation by proposing new and unexpected variations in producing a specific output.

To generate an effect chain, only those effects where the output flow matches the input flow of the next one can be linked. The consistency between input of the first effect and output of the next one can be denoted by D_c (Degree of consistency). Supposed there are *m* attributes in required output flow, and there are *n* attributes in produced output flow which satisfy the attributes in budget output flow ($n \leqslant m$). D_c can be shown as:

$$D_c = \frac{n}{m} \times 100\% \qquad (1)$$

The effects chosen for an effect chain must be compatible with each other, that is, the name of output flows of the first effect must be same as that of input flow of the next one and D_c =100%.

During the transformation from input flow to output flow by effect modes, there are three methods, namely, method of exhaustion, method of minimal path length and method of consistent degree.

- Method of exhaustion

If times of transformation from input flow to out flow are unlimited, there will be theoretically uncountable effect chains which produce the same output. Figure 3 shows two reasoning model for effect chain: forward direction reasoning model and backward direction reasoning model.

In forward direction reasoning model, first of all identify the input flow (f_i) and search for the effects whose input is consistent with the input flow (f_i). Then according to each effect the output flow (f_{po1}) is produced. If the name of produced output flow is the same as required output flow (Name: f_{po1}=f_q) and D_c =100%, an effect chain is generated and the path length is evaluate as one (p=1). When the first reasoning is over (t_r=1) or all the produced output flows don't match required output flow, take the produced output flow (f_{po1}) as new input flow and carry through the second reasoning (t_r=2). Repeat the above process until the reasoning times (t_r=n) designer permits.

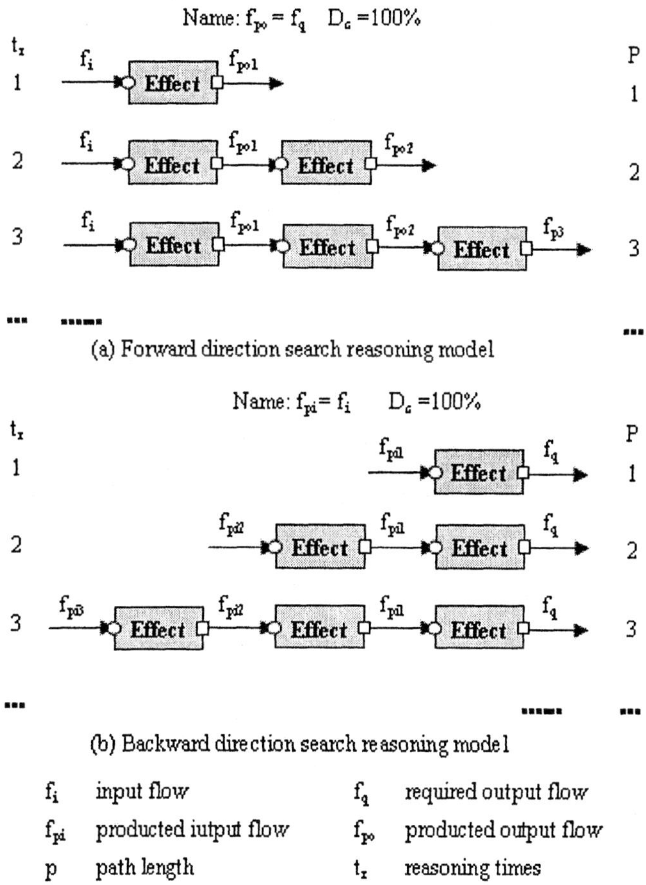

Figure 3. Reasoning model for effect chain

In backward direction reasoning model, first of all identify the required output flow (f_q) and search for the effects whose output is consistent with the required output flow (f_q). Then according to each effect the input flow (f_{i1}) is produced. If the name of produced input flow is the same as input flow (Name: f_{pi1}=f_i) and D_c =100%, an effect chain is generated and the path length is evaluate as one (p=1). When the first reasoning is over (t_r=1) or all the produced input flows don't match input flow, take the produced input flow (f_{pi1}) as new output flow and carry through the second reasoning (t_r=2). Repeat the above process until the reasoning times (t_r=n) designer permits.

Method of exhaustion can produce multiple effect chains in large-scale effects, which offer various potential principle solutions and are helpful for product innovative design, but the computational of method of exhaustion is complex and inefficient.

- Method of minimal path length

In principle, the numbers of effects that consist of effects chains can be chosen according to user discretion. But the effects chains should be as short as possible in order to obtain a simple system. The path of between two flows can be identified by method of exhaustion. The minimal path length (p_m) is the minimum value of path length of transformation between two flows. The knowledge of minimal path length (k_p) is presented by input flow (f_i), output flow (f_o) and minimal path length (p_m), as shown in the following:

$$k_p = \{f_i, f_o, p_m\} \qquad (2)$$

For example, {E, Δ l, 1} denotes that the minimal path length from electric field (E) to length change (Δ l) is one (by electrostriction effect).

Designers can rapidly achieve transformation from input flow to required output flow by the knowledge of minimal path length (k_p). The method is high performance, but maybe the effect chain couldn't satisfy the requirement.

- Method of consistent degree

In the method of exhaustion, each reasoning process produces various effects. If degree of consistency (D_c) between produced output flow of an effect and required output flow or between produced input flow of an effect and input flow is figured out, the produced output/input flow whose D_c is maximum can be identified and used as new input/output flow and carry through the next reasoning. Thus the method of consistent degree is more effective than the method of exhaustion.

4. CASE STUDY

Pill is a kind of good form of Chinese traditional medicine, but it can not be produced by Western medicine facility for its process and physics characteristic. The present condition is long process, high energy consume and great labor intension, so it is important to develop continue forming and shorten process to meet the need of modern times.

The granulator system can be initially modeled as a black-box in Figure 4, whose inputs are powder (medicinal powder) and liquid (cementing liquid), and whose outputs are sphericity and particle (pill).

According to the known inputs and outputs, search for the effects by forward/backward direction reasoning models. In order to shorten manufacturing process of pill, the reasoning times is set as three (t_r=4). The effects can be automatically linked into effect chains by using effect modes. Figure 5 shows the part of effect chains of granulator system, which are mainly based on fluidized bed effect, vibration effect, plastic deformation effect, shear effect, friction effect and Pascal's effect. The solution structure of granulator system based on Figure 5(d) is shown in Figure 6.

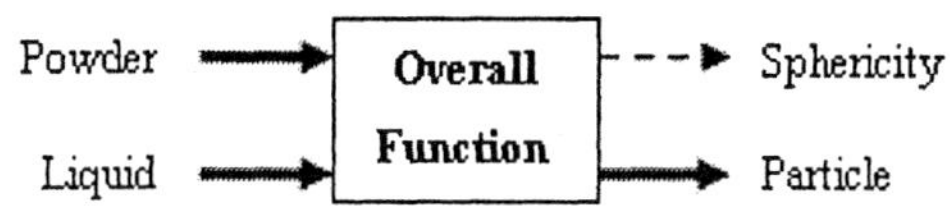

Figure 4. Black-box model of granulator system

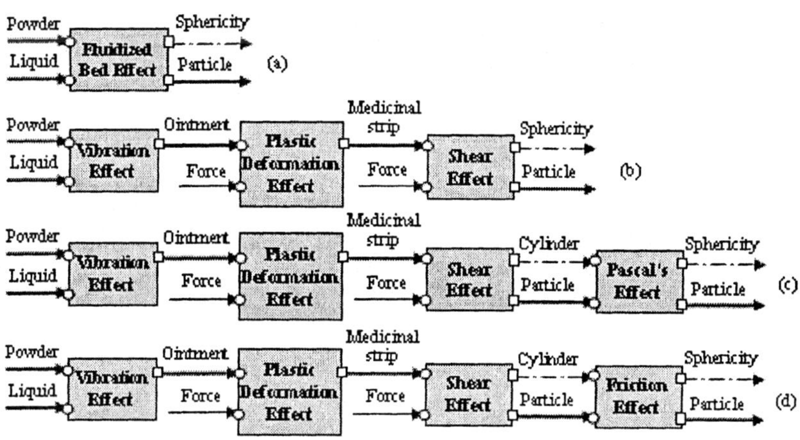

Figure 5. Effect chains of granulator system

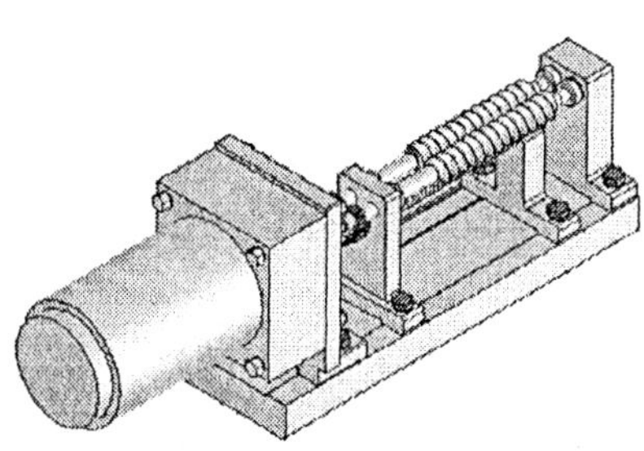

Figure 6. Principle solutions of granulator system

5. CONCLUSION

Conceptual design confirms the principle solution that contains the required function. Functional design is a process in conceptual design that impacts the key features of the design result.

This paper proposes an automated functional design approach that utilizes six effect modes and the existing effect knowledge base to support functional realization. Based on effect modes, the Effect Module in computer-aided innovation design software InventionTool 2.0 has been developed. This approach can produce numerous feasible principle solutions in conceptual design process.

A design example for functional design of a Chinese medicine mechanism is presented to demonstrate the proposed functional design methodology and prove that the method is feasible.

ACKNOWLEDGEMENTS

The research is under the support of Natural Science Foundation of China (No. 50375045) and Natural Science Foundation of Tianjin (No. 043802211).

REFERENCE

1. Stone, R. B. and Wood, K. L., (1998), Development of a functional basis for design, *Transactions of the ASME, Journal of Mechanical Design,* vol.122, pp.359-370.
2. Altshuller, G., (1999), *The Innovation Algorithm, TRIZ, Systematic Innovation and Technical Creativity,* Technical Innovation Center, INC, Worcester.
3. Tan, R., (2005), *Process of two stage analogy-based design employing TRIZ,* The 1st IFIP Working Conference on Computer Aided Innovation, ULM, Germany.
4. Pahl, G. and Beitz, W., (1996), *Engineering Design – A Systematic approach*, The 2nd Edition, Springer-Verlag London.
5. Altshuller, G., (1984), *Creativity as an Exact Science*, Gordon and Branch Publishers, Luxembourg.
6. Umeda, Y., Ishii, M., Yoshioka, M., et al., (1996), Supporting conceptual design based on the function-behavior-state modeler, *Artificial Intelligence for Engineering Design, Analysis and Manufacturing: Aiedam*, vol.10, pp. 275-288.
7. Qian, L. and Gero, J. S., (1996), Function-Behavior-Structure Paths and their Role in Analogy-based Design, *AIEDAM,* vol10, pp. 289-312.
8. Cao G., Tan R., Zhang R., et al., (2005), Study on Conceptual Design Process Model Based on Effects, *China Mechanical Engineering*, vol.16, pp.823-827. (in Chinese)
9. Tan R., (2002), *Innovation Design—TRIZ: Theory of Innovative Problem Solving,* China Mechanic Press. (in Chinese)

STUDY ON INTEGRATING APPLICATION METHOD FOR AD AND TRIZ

Zishun Chen, Runhua Tan
School of Mechanical Engineering, Hebei University of Technology, Tianjin, 300130, China; Email:chenzishun@126.com.

Abstract: The AD and TRIZ are the important methods of conceptual design. A planeness measurement equipment is analyzed and the finding problems are solved through using the AD and TRIZ. In order to satisfy the new requirements presented by production line to planeness measurement equipment, the redesign of planeness measurement equipment is carried out by using AD and TRIZ. The new conceptual design possesses many excellencies such as higher measurement efficiency, lower work strength etc., and satisfied the requirements of production line.

Key words: AD, TRIZ, Conceptual Design, Contradiction Matrix.

1. INTRODUCTION

For the manufacturing industry enterprises, the quality and efficiency of production line in manufacturing product is very important for competition in the current market. In order to control the varying tendency of product quality, it is a usually method that adopts special measurement equipments to carry out on-line measurement for work-piece of production line. However, in many small and medium-sized enterprises, there exist a problem that results in the time of measuring one work-piece is much longer than one manufacturing cycle when using on-line measurement equipments. Therefore the operation of work-piece measurement is a bottle-neck for the production line, which leads to not only not quickly to know the quality state of work-piece, but also affect the efficiency of production.

For the planeness measurement equipment of work-piece, the total time of measurement one work-piece includes adjusting work-piece position time and measuring time. But most of the time is for the adjusting work-piece

This project is supported by the National Natural Science Foundation of China under the grant No.50375045

Please use the following format when citing this chapter:
Chen, Zishun, Tan, Runhua, 2006, in International Federation for Information Processing (IFIP), Volume 207, Knowledge Enterprise: Intelligent Strategies In Product Design, Manufacturing, and Management, eds. K. Wang, Kovacs G., Wozny M., Fang M., (Boston: Springer), pp. 421-432.

position, therefore, the problem is the main reason that affects the quality control and production efficiency.

In this paper, the Axiomatic Design (AD) [1, 2] and TRIZ theory [3] are used to analyze and redesign the measurement equipment of work-piece planeness.

2. THE IMPROVEMENT DESIGN OF PLANENESS SPECIAL MEASUREMENT EQUIPMENT

In many small and medium-sized enterprises, the tools of measuring work-piece's planeness are usually standard plate, height gauge, support screw poles, and the measurement work is carried out by manual methods. Due to the measurement method possesses some excellencies such as simple in tool's structure, low cost, so it is used widely. When measuring the planeness of work-piece, the first step is to place three support screw poles (P1、P2 and P3) on the standard plate, and then place the work-piece on three support screw poles and lets the waiting measurement plane A face to up. Before using the height gauge to measure the plane A, the position of work-piece must be adjusted, namely through adjusting P1、P2 and P3 of contacting with the bottom plane B of work-piece, the three measurement benchmark points (M1, M2 and M3) on plane A have to be on the same height relative to the standard plane, as shown in Fig.1.

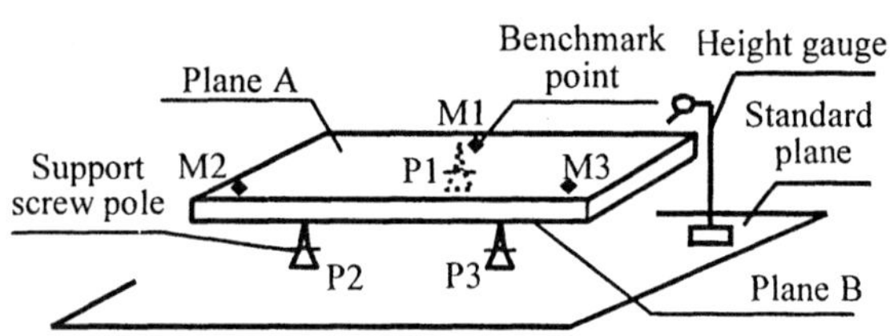

Figure 1. The measurement method of planeness

2.1 The problems description

In order to cause the three benchmark points have the same heights by adjusting the plane A, then need to adjust the three support screw poles. But because adjusts anyone vertical position of support screw pole, it will affect the vertical positions of three benchmark points simultaneously, thus must have to adjust the three support screw poles repeatedly, and result in the preparative time of adjusting workpiece's position occupies mostly the time of entire measurement process, so the efficiency of measurement is very low.

2.2 Using the AD to analyze the measurement equipment

According to the AD theory, In order to obtain the leaf level FRs and DPs of hierarchy, the zigzagging mapping and decomposition must be carried out between the FRs and the DPs of upper level [2].

On the level 1, the FRs, Cs and DPs of the planeness measurement equipment are described respectively as:

FR_0: Measure the planeness of plane A.

C_{11}: Speedily adjust the plane A to parallel with the standard plate.

C_{12}: Apply the manual adjustment equipment.

DP_0: The measurement equipment of plane A.

According to the AD, the level 1 design equation can be expressed as:

$$\{FR_0\} = [A]\{DP_0\} \qquad (1)$$

The FRs and DPs of level 2 are described as:

FR_1: Adjust the vertical position of point M1 on the plane A.

FR_2: Adjust the vertical position of point M2 on the plane A.

FR_3: Adjust the vertical position of point M3 on the plane A.

C_{21}: In order to enhance measurement precision, the benchmark points M1, M2 and M3 must be dispersed and closed the brim of plane A.

C_{22}: The three support screw poles (P1、P2 and P3) can be moved only in the area F, as shown in Fig.2.

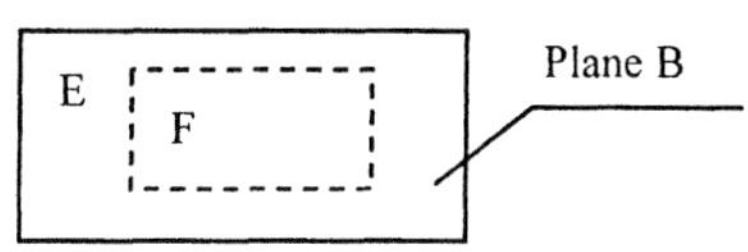

Figure 2. The situation of plane B

DP_1: Using support screw pole P1 of contacting with plane B.

DP_2: Using support screw pole P2 of contacting with plane B.

DP_3: Using support screw pole P3 of contacting with plane B.

The design equation of level 2 can be written as:

$$\begin{Bmatrix} FR_1 \\ FR_2 \\ FR_3 \end{Bmatrix} = \begin{bmatrix} A_{11} & A_{12} & A_{13} \\ A_{21} & A_{22} & A_{23} \\ A_{31} & A_{32} & A_{33} \end{bmatrix} \begin{Bmatrix} DP_1 \\ DP_2 \\ DP_3 \end{Bmatrix} \qquad (2)$$

For equation (2), the design matrix [A] can be expressed as:

$$[A]=\begin{bmatrix} A_{11} & A_{12} & A_{13} \\ A_{21} & A_{22} & A_{23} \\ A_{31} & A_{32} & A_{33} \end{bmatrix} \quad (3)$$

When using the measurement equipment to measure workpiece, the relations between positions of three measurement benchmark points of plane A and the positions of three support screw poles of contacting with plane B are shown in Fig.3. In Fig.3, the letter symbols are indicated as:

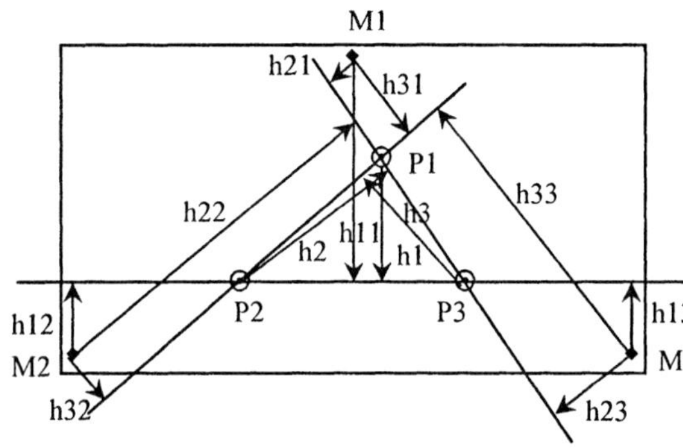

Figure 3. The position relations of three benchmark points and three support screw poles

h_1 is the distance of P1 reaching the P2 and P3 line.
h_2 is the distance of P2 reaching the P1 and P3 line.
h_3 is the distance of P3 reaching the P1 and P2 line.

h_{11}, h_{12} and h_{13} are the distances of benchmark points M1, M2 and M3 reaching P2 and P3 line respectively when adjust the support screw pole P1.

h_{21}, h_{22} and h_{23} are the distances of benchmark points M1, M2 and M3 reaching P1 and P3 line respectively when adjust the support screw pole P2.

h_{31}, h_{32} and h_{33} are the distances of benchmark points M1, M2 and M3 reaching P1 and P2 line respectively when adjust the support screw pole P3.

According to the design matrix's elements formula $A_{ij} = \partial FR_i / \partial FR_j$ [2], the elements of design matrix $[A]$ may be determined as:

$$\begin{array}{lll} A_{11} = h_{11} / h_1 & A_{12} = -h_{21} / h_2 & A_{13} = -h_{31} / h_3 \\ A_{21} = -h_{12} / h_1 & A_{22} = h_{22} / h_2 & A_{23} = -h_{23} / h_3 \\ A_{31} = -h_{13} / h_1 & A_{32} = -h_{23} / h_2 & A_{33} = h_{33} / h_3 \end{array}$$

In terms of the design matrix's elements above, the design equation of level 2 can be expressed as:

$$\begin{Bmatrix} FR_1 \\ FR_2 \\ FR_3 \end{Bmatrix} = \begin{bmatrix} h_{11}/h_1 & -h_{21}/h_2 & -h_{31}/h_3 \\ -h_{12}/h_1 & h_{22}/h_2 & -h_{32}/h_3 \\ -h_{13}/h_1 & -h_{23}/h_2 & h_{33}/h_3 \end{bmatrix} \begin{Bmatrix} DP_1 \\ DP_2 \\ DP_3 \end{Bmatrix} \quad (4)$$

Because the design matrix's elements A_{ij} of design equation (4) are all nonzero elements, so that the design of level 2 is a coupled design, it does not satisfy the independence axiom [1, 2] and need to decouple.

2.3 The coupled design of measurement equipment decoupling and improvement

For the design matrix's elements A_{ij} of equations (2), if lets the $A_{ij}=0$ except i=j, namely $h_{12}=h_{13}=h_{21}=h_{23}=h_{31}=h_{32}=0$, then the design matrix $[A]$ is become a diagonal matrix, and satisfies the independence axiom. For this example, i.e. must lets the positions of three support screw poles superposing with the positions of three benchmark points respectively.

Due to the design of measurement equipment exists C_{21} and C_{22}, so these positions can not be superposed between the three support screw poles and the three benchmark points, which exists a contradiction in the decoupling process. For the coupled design can not be decoupled directly, therefore it needs to use the TRIZ to eliminate the contradiction.

Because exists a contradiction, so causes the positions of three measurement benchmark points can not be adjusted one to one with three support screw poles, and results in adjusting repeatedly and need a long time. Thus the first characteristic is described as 'Loss of time' by using 39 characteristics of TRIZ[4]. The distributive positions of support screw poles are restricted by C_{22}, and then using the 39 characteristics to describe the second characteristic is 'Length of a stationary object'. Because the contradiction relates to two different characteristics, so the contradiction belongs to technical contradiction [5].

For solving this technical contradiction, the inventive principles No.30, 24, 14 and 5 can be obtained by using the contradiction matrix[6, 7] of TRIZ. Though analyzing, the inventive principle No. 24 can be used, namely No.24 is a principle solution. The content of No. 24 is 'Mediator'.

In this real example, the 'Mediator' is a piece of segregation plate, and placed between three support screw poles and plane B when the workpiece is measured. Thus the distributive positions of three support screw poles can be superposed with three benchmark points (here $A_{11}=A_{22}=A_{33}=1$), and the area E of plane B can not be destroyed simultaneously, as shown in Fig.4.

The additive segregation plate is a field solution of using the principle solution 'Mediator'. In the scheme of improving design, when adjusts the vertical position of anyone support screw pole, then only affects the vertical

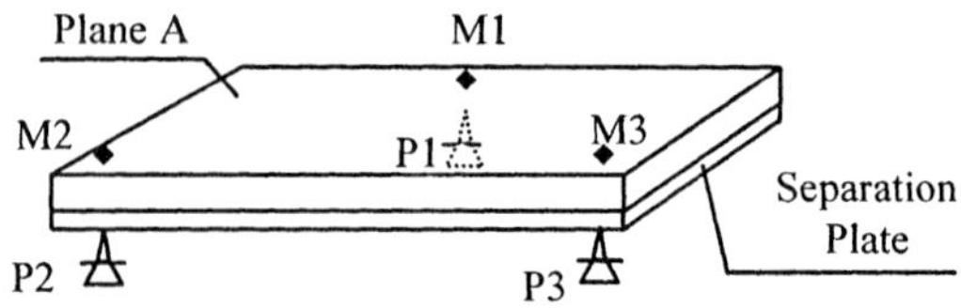

Figure 4. The conceptual design of measurement equipment after improving

position of corresponding measurement benchmark point, therefore the time of adjusting workpiece's position can be shortened.

The improving design of the measurement equipment is an uncoupled design, and the design equation is written as:

$$\begin{Bmatrix} FR_1 \\ FR_2 \\ FR_3 \end{Bmatrix} = \begin{bmatrix} 1 & 0 & 0 \\ 0 & 1 & 0 \\ 0 & 0 & 1 \end{bmatrix} \begin{Bmatrix} DP_1 \\ DP_2 \\ DP_3 \end{Bmatrix}$$

3. THE NEW DESIGN OF SPECIAL PLANENESS MEASUREMENT EQUIPMENT

When measures the workpiece, because the work of putting in and taking out the workpiece from measurement equipment is performed by manpower, so the planeness measurement equipment improved design above is suitable for the production line that only produces pint-sized or light weight products. But for the production line that produces the good-sized or heavier workpieces, it will cause the inspector to be borne heavy load and the measurement time will be protracted. So the measurement equipment improved must be redesign in order that the measurement equipment suitable for heavy workpiece production line.

3.1 The framework model of new planeness measurement equipment

Conveying the heavy workpieces when measurement is not suitable for manpower. Therefore, there is a contradiction between the operation method of using measurement equipment improved and the design constraint that for the heavier workpiece must avoid conveying by manpower, and this contradiction can be described by 39 characteristics of TRIZ as:

The worsening characteristic is No. 2: 'weight of a stationary object';

The improving characteristic is No. 37: 'difficulty of detecting and measuring'.

Clearly the contradiction belongs to a technical contradiction, and can obtain 4 commending inventive principles from contradiction matrix, i.e. No.6, 13, 28 and 1.The inventive principles No.13 and No.1 can be used by analyzing. The content of inventive principle No.13 is 'The other way round', and the content of inventive principle No.1 is 'Segmentation'.

For this real example, the principle solution No.13 'The other way round' can be understood and described as: when measures the planeness of workpiece, the operational method is to convey the measurement equipment instead of workpiece; the principle solution No.1 'Segmentation' can be described as: when measure the workpiece' planeness, the measuring tool is a multi-gauge system instead of a height gauge.

According to the inventive principles No.13 and No.1, the framework model of new planeness measurement equipment can be illustrated in Fig.5.

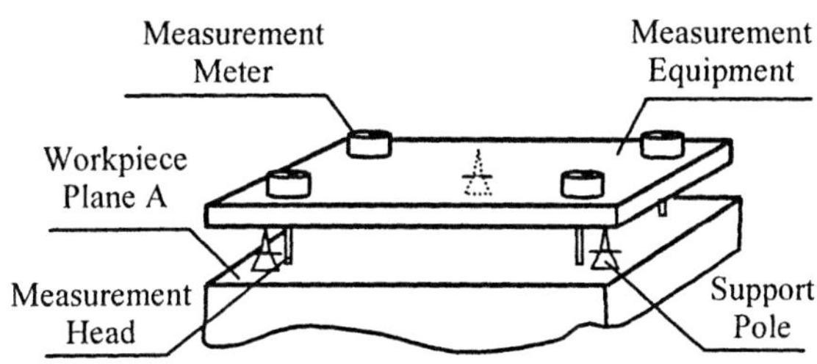

Figure 5. The framework model of planeness measurement equipment

3.2 Describing for the FRs, Cs and DPs of new measurement equipment

In order to suit the production line of producing good-sized or heavy weight workpiece, the FRs, Cs and DPs of measurement equipment need to redefine and need decompose between FRs and DPs.

At level 1, the FRs, Cs and corresponding DPs are expressed as:

FR_1: Measure the planeness of plane A.

FR_2: Measurement process can be carried out advantageously.

C_{11}: Use the manual method to measure workpiece.

C_{12}: Only convey the measurement equipment.

C_{13}: The work strength is low in measurement process.

DP_1: Using the special planeness measurement equipment.

DP_2: Using the maneuverable special measurement equipment.

According to the AD theory, the design equation of level 1 is:

$$\begin{Bmatrix} FR_1 \\ FR_2 \end{Bmatrix} = \begin{bmatrix} x & 0 \\ 0 & x \end{bmatrix} \begin{Bmatrix} DP_1 \\ DP_2 \end{Bmatrix}$$

At level 2, the FRs, Cs and corresponding DPs are expressed as:

FR_{11}: According to the shape and size of plane A determine the quantity and position of collecting measurement data.

FR_{12}: The structure of measurement equipment possesses high stability.

FR_{21}: Obtain the measurement data easily.

FR_{22}: Calibrate the measurement equipment easily.

FR_{23}: Convey the measurement equipment easily.

Cs: Equal to C_{11}, C_{12} and C_{13}.

DP_{11}: The measurement equipment that distributes the measurement points reasonably at the proper positions of plane A.

DP_{12}: Using the rigid structure to prevent distortion.

DP_{21}: The measurement equipment can display measurement data.

DP_{22}: Special calibration device.

DP_{23}: Using a pair of handles structure.

The design equation of level 2 is:

$$\begin{Bmatrix} FR_{11} \\ FR_{12} \end{Bmatrix} = \begin{bmatrix} x & 0 \\ 0 & x \end{bmatrix} \begin{Bmatrix} DP_{11} \\ DP_{12} \end{Bmatrix}$$

$$\begin{Bmatrix} FR_{21} \\ FR_{22} \\ FR_{23} \end{Bmatrix} = \begin{bmatrix} x & 0 & 0 \\ 0 & x & 0 \\ 0 & 0 & x \end{bmatrix} \begin{Bmatrix} DP_{21} \\ DP_{22} \\ DP_{23} \end{Bmatrix}$$

At level 3, the FRs, Cs and corresponding DPs are expresses as (in here hypothesize the new measurement equipment distributing 10 waiting measurement points on the reasonable positions of plane A):

FR_{1101}: Collect the measurement data at reasonable position 1.

FR_{1102}: Collect the measurement data at reasonable position 2.

... ...

FR_{1110}: Collect the measurement data at reasonable position 10.

Cs: Equal to C_{11}, C_{12} and C_{13}.

DP_{1101}: Installing a special meter to display measuring data at position 1.

DP_{1102}: Installing a special meter to display measuring data at position 2.

... ...

DP_{1110}: Installing a special meter to display measuring data at position 10.

Through decomposing FR_{11} and DP_{11}, the design equation can be expressed as:

$$\begin{Bmatrix} FR_{1101} \\ FR_{1102} \\ FR_{1103} \\ FR_{1104} \\ FR_{1105} \\ FR_{1106} \\ FR_{1107} \\ FR_{1108} \\ FR_{1109} \\ FR_{1110} \end{Bmatrix} = \begin{bmatrix} x & 0 & 0 & 0 & 0 & 0 & 0 & 0 & 0 & 0 \\ 0 & x & 0 & 0 & 0 & 0 & 0 & 0 & 0 & 0 \\ 0 & 0 & x & 0 & 0 & 0 & 0 & 0 & 0 & 0 \\ 0 & 0 & 0 & x & 0 & 0 & 0 & 0 & 0 & 0 \\ 0 & 0 & 0 & 0 & x & 0 & 0 & 0 & 0 & 0 \\ 0 & 0 & 0 & 0 & 0 & x & 0 & 0 & 0 & 0 \\ 0 & 0 & 0 & 0 & 0 & 0 & x & 0 & 0 & 0 \\ 0 & 0 & 0 & 0 & 0 & 0 & 0 & x & 0 & 0 \\ 0 & 0 & 0 & 0 & 0 & 0 & 0 & 0 & x & 0 \\ 0 & 0 & 0 & 0 & 0 & 0 & 0 & 0 & 0 & x \end{bmatrix} \begin{Bmatrix} DP_{1101} \\ DP_{1102} \\ DP_{1103} \\ DP_{1104} \\ DP_{1105} \\ DP_{1106} \\ DP_{1107} \\ DP_{1108} \\ DP_{1109} \\ DP_{1110} \end{Bmatrix}$$

The results of decomposing FR_{12} and DP_{12} are as:

FR_{121}: The main body of measurement equipment has no distortion.

FR_{122}: When measure, the measurement equipment's position keep invariability relative to the plane A

FR_{123}: Can install the measuring meters.

DP_{121}: The main body of measurement equipment can be made of high strength material as carbon steel or aluminium alloy etc.

DP_{122}: Using three rigid and fixed support poles.

DP_{123}: Using the structure that can install the measuring meters.

By decomposing FR_{12} and DP_{12}, the design equation can be obtained:

$$\begin{Bmatrix} FR_{121} \\ FR_{122} \\ FR_{123} \end{Bmatrix} = \begin{bmatrix} x & 0 & 0 \\ 0 & x & 0 \\ 0 & 0 & x \end{bmatrix} \begin{Bmatrix} DP_{121} \\ DP_{122} \\ DP_{123} \end{Bmatrix}$$

The results of decomposing FR_{21} and DP_{21} are as:

FR_{211}: Can display the measurement data.

FR_{212}: Can read the measurement data easily.

DP_{211}: Special measuring meters.

DP_{212}: The dial of measurement meter face to up (see Fig.6).

By decomposing FR_{21} and DP_{21}, the design equation is expressed as:

$$\begin{Bmatrix} FR_{211} \\ FR_{212} \end{Bmatrix} = \begin{bmatrix} x & 0 \\ 0 & x \end{bmatrix} \begin{Bmatrix} DP_{211} \\ DP_{212} \end{Bmatrix}$$

Figure 6. Special measuring meter

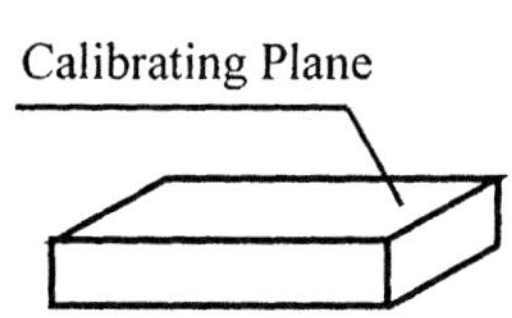

Figure 7. The calibrating plate

Using the same method to decompose the FR_{22} and DP_{22} as:
FR_{221}: Can provide the benchmark data.
FR_{222}: The 10 measuring meters can be calibrated synchronously.
DP_{221}: Using the calibrating plate (see Fig.7).
DP_{222}: The calibrating plane that possesses proper shape and size.
Via decomposing FR_{22} and DP_{22}, the design equation is expressed as:

$$\begin{Bmatrix} FR_{221} \\ FR_{222} \end{Bmatrix} = \begin{bmatrix} x & 0 \\ 0 & x \end{bmatrix} \begin{Bmatrix} DP_{221} \\ DP_{222} \end{Bmatrix}$$

Because of the FR_{23} and DP_{23} are at the leaf level already through decomposing above, so they will not be decomposed continually.

The function of calibrating plate is to calibrate 10 measuring meters synchronously before measuring planeness of plane A. The process of measurement planeness is first putting the measurement equipment on the calibrating plane to regulate the 10 meters 'zero', then places the calibrated measurement equipment on the plane A of workpiece, the maximum difference value of the showing data of 10 measuring meters is the planeness value of plane A.

The new conceptual design of measurement equipment has 18 leaf level FRs and 18 leaf level DPs through decomposing. The corresponding relations of these leaf level FRs and DPs are shown in Table 1.

Table 1. The relations of leaf level FRs and leaf level DPs.

FR \ DP	DP1101	DP1102	DP1103	DP1104	DP1105	DP1106	DP1107	DP1108	DP1109	DP1110	DP121	DP122	DP123	DP211	DP212	DP221	DP222	DP23
FR1101	x	0	0	0	0	0	0	0	0	0	0^1	0	0	0	0	0	0	0
FR1102	0	x	0	0	0	0	0	0	0	0	0^1	0	0	0	0	0	0	0
FR1103	0	0	x	0	0	0	0	0	0	0	0^1	0	0	0	0	0	0	0
FR1104	0	0	0	x	0	0	0	0	0	0	0^1	0	0	0	0	0	0	0
FR1105	0	0	0	0	x	0	0	0	0	0	0^1	0	0	0	0	0	0	0
FR1106	0	0	0	0	0	x	0	0	0	0	0^1	0	0	0	0	0	0	0
FR1107	0	0	0	0	0	0	x	0	0	0	0^1	0	0	0	0	0	0	0
FR1108	0	0	0	0	0	0	0	x	0	0	0^1	0	0	0	0	0	0	0
FR1109	0	0	0	0	0	0	0	0	x	0	0^1	0	0	0	0	0	0	0
FR1110	0	0	0	0	0	0	0	0	0	x	0^1	0	0	0	0	0	0	0
FR121	0	0	0	0	0	0	0	0	0	0	x	0	0	0	0	0	0	0
FR122	0	0	0	0	0	0	0	0	0	0	0	x	0	0	0	0	0	0
FR123	0	0	0	0	0	0	0	0	0	0	0	0	x	0	0	0	0	0
FR211	0	0	0	0	0	0	0	0	0	0	0	0	0	x	0	0	0	0
FR212	0	0	0	0	0	0	0	0	0	0	0	0	0	0	x	0	0	0
FR221	0	0	0	0	0	0	0	0	0	0	0	0	0	0	0	x	0	0
FR222	0^2	0^2	0^2	0^2	0^2	0^2	0^2	0^2	0^2	0^2	0	0	0	0	0	0	x	0
FR23	0	0	0	0	0	0	0	0	0	0	0	0	0	0	0	0	0	x

Note:

1. The design of measurement equipment's rigidity should be carried out after the positions and assembly structure of 10 measurement meters were determined.
2. Before determining the shape and size of calibrating plate, the installing positions of 10 meters on measurement equipment body must be considered.

In the Table 1, the shadow parts indicate the corresponding relations of the leaf level FRs and leaf level DPs that obtained from the same one design equation; no shadow parts indicate the corresponding relations of the leaf level FRs and leaf level DPs that obtained from the different design equations. If there exist correlations between the leaf level FRs and leaf level DPs in the no shadow parts, then the existing correlations can be considered as design constraints [2].

For measuring heavy weight or good-sized workpiece, in order to assure measurement precision, measurement time and decrease work strength, on the foundation of original design, several new FRs and Cs are added, and a new conceptual design of measurement equipment is obtained.

The added FRs and Cs in the new conceptual design are as:

FR_2: Measurement process can be carried out advantageously.

C_{12}: Only convey the measurement equipment.

C_{13}: The work strength is low in measurement process.

The new conceptual design of measurement equipment is illustrated in Fig.8, and its characteristics are:

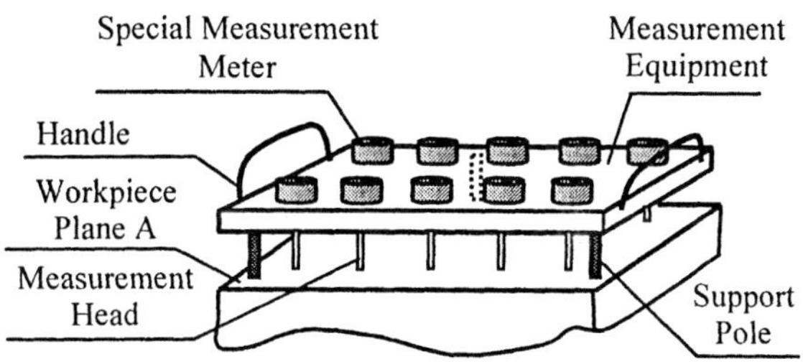

Figure 8. The new conceptual design of measurement equipment

(1) The measurement process is rapid and the efficiency high.

The new measurement equipment is a multi-meters device that may measure many points on the plane A synchronously.

(2) To calibrate the 10 meters is easy.

When calibrating the 10 meters of measurement equipment to 'zero', it only need one time through using the calibrating plate.

(3) The strength of measurement work is low.

In the phase of detail design, the measurement equipment's designing must satisfy the requirement of proper weight.

(4) The Applying scope is wide.

The new measurement equipment is not only suited for measuring the planeness of heavy weight workpiece, but also for measuring the planeness of light weight workpiece. The new conceptual design of measurement equipment is an uncoupled design, which solved the bottle-neck problem of measurement process as well as satisfied the requirements of production line.

4. CONCLUSION

(1) The AD and TRIZ are important methods for conceptual design. Using the design equations of AD can survey the conceptual design whether existing 'coupling' problem or not, so the AD can make the designer rationally judge the design, and provides a powerful tool for designer to find the good design.

The TRIZ provides an effective tool for designer to solve the contradictions of design process, and designer can obtain the anticipant design quickly by using the TRIZ to solve the coupling problems in AD process. So the powerful design method is formed by integrating the AD and TRIZ.

(2) An improved design of workpiece planeness measurement equipment is carried out by using the AD and TRIZ. In order to expand the applying scope of measurement equipment, the contradictions among the new FRs, Cs and DPs must be solved. A new conceptual design of measurement equipment is obtained by using the AD and TRIZ and can Satisfy the requirements of production line.

5. REFERENCES

1. Nam P. Suh, *The Principles of Design* (Oxford University Press, 1990).
2. Nam P. Suh, *Axiomatic Design-Advances And Applications* (Oxford University Press, 2001)
3. Tan runhua, *The Innovation Design* (Mechanical Industrial of china Press, 2002).
4. Ellen Domb, Joe Miller, Ellen MacGran, and Michael Slocum (1998/11); http://www.triz-joumal.com/archives/
5. Michael S. Slocum, K.O.St.I., Ellen Domb, Catherine Lundberg (2003/01); http://www.triz-journal.com/archives/
6. Zinovy Royzen (1996/11); http://members.aol.com:/zroyzen/triz.html
7. Todd Williams, Ellen Domb (1998/05); http://www.triz-journal.com/archives/

RESEARCH ON PRODUCT INNOVATIVE DESIGN METHOD DRIVEN BY CLIENT DEMANDS

Limin Li, Guiqin Li, Shi'an Huang
Center of CIMS and Robotics,Shanghai University,Shanghai 200072 P.R.China,Email: leeching@staff.shu.edu.cn

Abstract: Based on the research on acquiring methods of client demands for product innovation development, apply system cluster analyses in modeling demands. In study of foundation of QFD and TRIZ, an integration process model is put forward for product innovation design of QFD and TRIZ. The model is developed taking in the advantages of both, which respond to client demands information effectively. The whole process is set up in structure and regularization, according with general process of human innovative thinking. And it has strong maneuverability as well.

Key words: client demands, innovative design, QFD, TRIZ, integrated model

1. INTRODUCTION

Nowadays, "to treat the customers as the core, rapid response to the markets, and to satisfy the client's requirements with might and main" has been the main intention to enhance the competition ability of the enterprises. Since product innovation has been hardly indifferent for the survival and developing of the manufacture enterprises, to seek an effective design theory which is suitable for our nation's status and with it to design the products which can satisfy the client's demands as well as to realize the technique and product innovation are popular in design field. Therefore, based on the principle that client's demand is the motivation of the new products innovation and with the advantages in products design of the QFD (Quality

Please use the following format when citing this chapter:
Li, Limin, Li, Guiqin, Huang, Shi'an, 2006, in International Federation for Information Processing (IFIP), Volume 207, Knowledge Enterprise: Intelligent Strategies In Product Design, Manufacturing, and Management, eds. K. Wang, Kovacs G., Wozny M., Fang M., (Boston: Springer), pp. 433-439.

Function Deployment) and TRIZ (Theory of Inventive Problem Solving), a product innovation model driven by client's demands is put forward.

2. ACQUIREMENT AND ANALYSIS METHODS OF CLIENT'S DEMANDS

2.1 Methods to acquire client's demand

No matter the idea of a new project come from the exterior or interior of an enterprise, the enterprise should take market survey and analyses it with the consideration of its competitiveness and its products ability. The person related with the new project should be clear on two things: who were the clients and which kind of the client's demands we should satisfy.

There are several methods to confirm client's demands: Designers themselves act as users; consumer survey; Organize different users to take part in some special discussion; Talk to the users face to face; Inspect the products in real life; research the competitor's products. Summarize the user's feedback and suggestions. As the web technology is developing, to survey on the internet having a good performance as well.

The most difficulty of the market research is how to dig the client's potential demands. For technique evolution theory can not only forecast technology development, but also exhibit the realization of conceivable product structure. It takes on directive effect for product innovation design. Directed Evolution is the branch of the technique revolution theory which core is the management with control process and technology evolution. The theory has been figured it out there are 10 modes for the product evolution. Under each mode there are many evolution ways. With the application of the modes and the evolution ways, the designers can generate the initial idea of product innovation. Apply the technique evolution theory to forecast the development directions of product, and then translate those directions into client's words, farther to dig up and induce client's potential demands combined with the market survey methods.

2.2 Analysis methods of client demands

Applying the Hierarchical clustering method to neaten of the client demands, we could acquire a relatively objective quantitative tool for the personnel participate in product development project to sort the scattered client demands.

1. Set up a demands matrix with the data from the market research

$$X=\begin{bmatrix} x_{11} & x_{12} & \cdots & x_{1p} \\ x_{21} & x_{22} & \cdots & x_{2p} \\ \vdots & \vdots & \vdots & \vdots \\ x_{n1} & x_{n2} & \cdots & x_{np} \end{bmatrix} \tag{1}$$

To eliminate the influence of the dimension, normally transform the row vectors, that is

$$x_{ij}'=(x_{ij}-\bar{x}_j)/S_j \tag{2}$$

in which: $\bar{x}_j=\frac{1}{n}\sum_{i=1}^{n}x_{ij}$; $S_j^2=\frac{1}{n-1}\sum_{i=1}^{n}(x_{ij}-\bar{x}_j)^2$

2.Comparability measurement. Measure the distance between client demands with Euclidian distance.

$$d_{ij}=\left[\sum_{k=1}^{p}(x_{ik}-x_{jk})^2\right]^{1/2} \tag{3}$$

3.To use square-sum subtraction, the basic idea root in squared difference analyses. If sort n items of client demands into k groups, $G_1,G_2,\cdots,G_k,x_{it}$ denote the i item of client demands in G_t, n_t denotes the number of the items in $G_t,\bar{x}_t$ is the cg of G_t, then the square-sum subtraction in cluster G_t is

$$S_t=\sum_{i=1}^{n_t}(x_{it}-\bar{x}_t)'(x_{it}-\bar{x}_t) \tag{4}$$

The entire square-sum in groups is

$$S=\sum_{t=1}^{k}\sum_{i=1}^{n_t}(x_{it}-\bar{x}_t)'(x_{it}-\bar{x}_t)=\sum_{t=1}^{k}S_t \tag{5}$$

First put n items of demands as one group respectively, and reduce one cluster every time, each time the square-sum subtraction will augment, combine two groups to make augment of S least, till all groups turn to one group.

4. Draw a dendriform pedigree chart of above cluster operation result. Thus the congregate process and distance of each group can be visually made out. Ascertain the groups collective based on the need of analyses and disposal next step.

3. INNOVATIVE DESIGN ORIENTED QFD AND TRIZ INTEGRATED PROCESS MODEL

QFD can effectively transform the client demands into technical requirements of engineering designers via the quality house, and it has great advantages in acquiring information. The quantitative description and evaluation for functions, will administer to widen the ideas of plan and to grab the key and the essential of the problems. TRIZ theory has a set of comparatively integrant system info in solving the conflicts in product design, and it has a powerful logic strictness and operation procedure. Furthermore, TRIZ has regularity cognition at the human's inventive process, which is coherent with the human's learning process that is studied in cognition psychology. Thus, based on the advantages of TRIZ in eliminating conflict, and integrate the theories and methods of QFD, and then combined with substance field analyses, educe the process model for innovative design as shown in the fig.2.

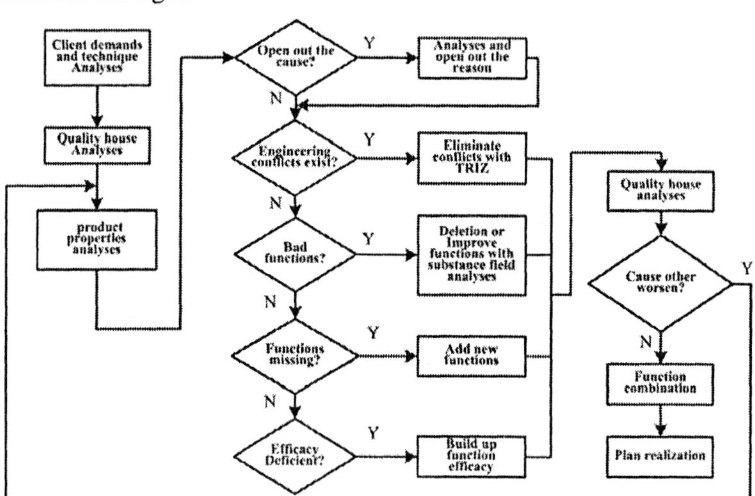

Fig.2 the process model of solving problem using integrated QFD and TRIZ

We can accomplish the innovative design with this process model in several steps:

1) Based on the analysis of client demands and product functions, we should firstly set up the quality house from the requirement to product properties and functions, transform "languages of client demands" to "languages of engineering design", and, according to the analyses tools DFD provided and analysis product properties, find out the negative correlative guideline, namely design conflicts. Choose the negative correlative guideline and issue function as the key to be solved next step.
2) Classify the design conflicts, make sure the negative correlative are caused by the engineering conflicts or the disability of self-function, and then utilize the corresponding tools showed in fig.1 to consummate the functions, eliminate the technical or physical conflicts.
3) During eliminating the conflicts and consummating, the basic functions may simultaneously worsen other functions, which will cause new conflicts, so re-analyses with the quality house should be taken after step 2, to judge whether new problems have been caused till designer satisfaction.
4) Synthesize the improved accession function with the other functional structures, and then the feasible collective design plan can be produced.

4. APPLICATION EXAMPLES

The development of the vacuum infuse equipments is a greatly benefit from the method in this investigation. The quality house constructed by the requirements and technical properties is showed in fig. 3.

Then, apply the method of eliminating conflict to solving the conflict between airproof and unlock of the airproof gate.

There is two main functions for the airproof gate of the vacuum infuse equipments: providing alleyway for mould raw material; airproofing to form vacuum field. It has a high frequency of open and close, and it should have a press distortion beforehand if the elasticity rubber airproof action based on airproof principles. If the distortion is strong, the airproof effect will has a good performance, but a strong distortion with a high pressure will decrease the open close performance. If the distortion is weak, the elasticity airproof piece distort after press, thus the gate and ream pivot will bear the force, the gate body will be distorted or a period of time action to decrease the performance of the airproof. Substance field model is showed in fig.4. The technical conflicts described by the 39 standard parameters of TRIZ as follows:

- The predicted parameter improvement: the reliability of equipment.
- Technical attrition parameter: maneuverability.

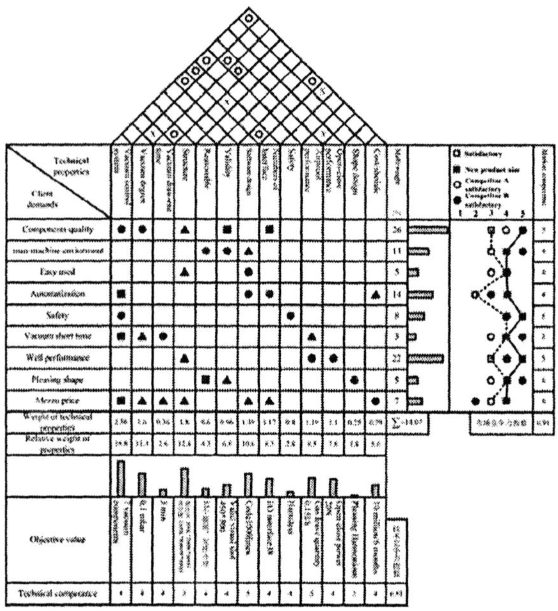

Fig.3 the quality house constructed by requirement and technique properties

Checked from the conflicts resolve matrix, the usable principle to solve the conflicts constituted with reliability and maneuverability of equipment are the principles of 27th, 17th and 40th. Finally we find the first cue from principle of 17th (change dimension): change the moved or rest object from one dimension space to two dimension space, and change the object in two dimension space to three dimension space. It provided a remarkable idea for the improvement of the airproof gate.

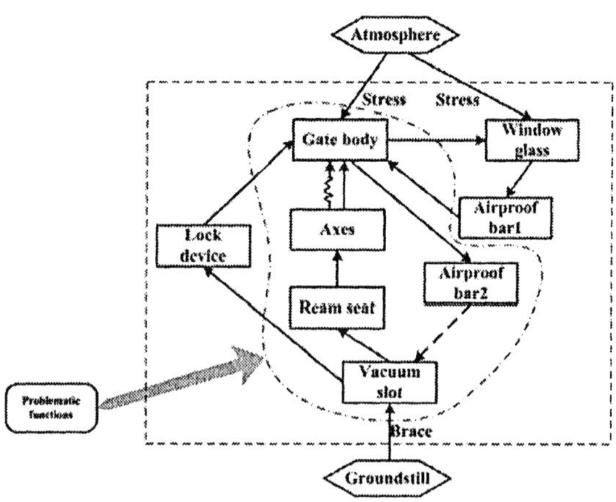

Fig.4 substance field model of airproof gate system

According to the cue, adopt the first principle alter the gate originally in the space only rotate the reamer pivot with the circular rotation. But being transformed to be the whole gate body rotates through the reamer pivot along the force direction after being pressed. In the real life application, change the reamer pivot structure, as a clearance conjugate during rotating, and keeping a 3mm abundant capacity between the hole and the pivot (fig.8 shows the conjugate instance after gate close) to ensure that move with the gate when being pressed. The gate will not be affected by torque action.

The practice result proves the design eliminate the conflicts between the performance of airproof and open-close effectively.

5. CONCLUSIONS

With the client demands as the driven force of the product innovation, we can guarantee the client satisfaction during product design process and reduce the risks of products development. Utilizing the structural model integrated by QFD and TRIZ, we settle the malpractice of lack in solving problems by QDF, and avoid the "two layer questions" caused by simply using TRIZ, and effectively transform the client demands into engineering realization.

6. REFERENCE

1. Rogers Beth, *Creating Product Strategies* (Northeast Finance and Economics University Press, 2003)
2. Tan Runhua, Progress of some problems in product design for innovation, Chinese *journal of mechanical engineering*, 39(9), 11-16(2003)
3. Mei Changlin, Zhou Jialiang, *Practical Statistics Technique* (Science Press, 2002)
4. Hu Qinxi, Yu Tao, Fang Minlun, *Principle and Application of Parallel Engineering* (Shanghai University Press, 2001)
5. Cristiano J, Liker J K,White Chelsea C. Customer-Driven Product Development Through Quality Function Deployment in the U.S. and Japan. *J PROD INNOV MANAG*, (17), P286-308(2000)
6. Sushkov V V, .Mars N J, Wognum P. Introduction to TIPS: a theory for creative design. *Artificial Intelligence in Engineering*, (9): P177-189(1995)

THE INNOVATIVE DESIGN OF RECIPROCATING SEAL FOR HYDRAULIC CYLINDER BASED ON TRIZ EVOLUTION THEORY

Fuying Zhang, Hui Liu, Linjing Zhang
Department of Mechanical Engineering, Tianjin University of Science & Technology, Tianjin 300222, P.R.Chain;Email:zhfuying@tust.edu.cn

Abstract: The design of reciprocating seal is extra important to the hydraulic cylinder's performances at lower speed, sensitive, dynamic and static rigidity. Technology evolution of TRIZ provides the laws or prevailing lines of technology system evolution which a designer should follow to find the inventive solutions. To help a designer to get the innovative ideas in a short time by using TRIZ evolution theory and to make the innovative design process operable, a CAI software tool for gauging and positioning of the state of a system in relation to each law or line of technology system evolution is developed. It can draw the evolutionary potential radar plot of studied system automatically on the basis of evolution analysis. The innovative design of reciprocating seal for hydraulic cylinder is detailed according to the future possible development opportunities that evolutionary potential plot recommends.

Key words: Hydraulic cylinder; TRIZ technology evolution; reciprocating seal; evolutionary potential; CAI software tool

1. INTRODUCTION

The reciprocating seal is a key technology of hydraulic cylinder. The hydraulic cylinder's performances at lower speed, sensitive, dynamic and static rigidity depend more on the capability of the hydraulic reciprocating seal.

The reciprocating seal is built upon the theories of lubrication, tribology, wear, heat transfer, material properties, and structure design of sealing

This project is supported by The Science Research Fund of Tianjin University of Science & Technology under the grant No.20050217.

Please use the following format when citing this chapter:
Zhang, Fuying, Liu, Hui, Zhang, Linjing, 2006, in International Federation for Information Processing (IFIP), Volume 207, Knowledge Enterprise: Intelligent Strategies In Product Design, Manufacturing, and Management, eds. K. Wang, Kovacs G., Wozny M., Fang M., (Boston: Springer), pp. 440-449.

interface[1]. It is the most complex application situation and the most severe demand for dynamic seal in the hydraulic equipments. Therefore, it is important to develop new reciprocating seal technology by analyzing and evaluating the development potential of current sealing technologies in the aspect of structure, material, shape, and consequently to predict their possible future structure stage.

Technology evolution theory, which is special in technology predicting, is an important branch of TRIZ[2] (The Theory of Inventive Problem Solving). Laws of technology system evolution are the theoretical fundament of TRIZ evolution theory. Presently, TRIZ technology evolution has different versions: evolution of technique[3] (ET), guided technology evolution[4], directed evolution[5] (DE). TRIZ technology evolution does not predict the future of technology; it also forces the system to its most probable future development by inventing it before it would occur naturally. Designers can systematically generate the creative ideas necessary for developing a next-generation product or process by using technology evolution theory.

In this paper, the technology evolution of TRIZ is first reviewed. Then, using the CAI software tools in which TRIZ evolution theory forms the very basis, we present the future development directions of the reciprocating seal by analyzing its evolutionary course and the lines of evolution. Finally, the innovation design of reciprocating seal for hydraulic cylinder is carried out according to the recommendation of evolutionary potential plot.

2. TRIZ TECHNOLOGY EVOLUTION THEORY

Laws of technical systems evolution form the very basis of TRIZ. By the term law, Altshuller means a trend that governs the development of any technical system. However, it is not easy to use them in the perspective of product development. To make these laws operational in the design process, patterns/lines of technology evolution are summarized, and evolutionary potential of technical system are presented. These patterns/lines, and evolutionary potential plot can be easily understood and used by designers to get innovative product[6].

2.1 The evolution s-curve of technology system

According to Altshuller, technology is always in evolution. As they evolve, most technologies follow predetermined patterns rather than representing a collection of random events. The evolution process of technology systems can be represented by the classic s-curve, which illustrates the life cycle stages of infancy, growth, maturity and decline.

TRIZ researchers discover that technological systems always tend to evolve in the direction of increased ideality. Fig.1 provides a generic graphical description of technology system evolution toward ideality through a family of S-curves. With this chart, we are able to forecast that there always will be a future technology development that will replace current technology - a very powerful and insightful knowledge indeed. We should make current technology jump to the next-generation S-curve by eliminating the inherent conflicts of technology systems, when it reaches the limitation of its performance.

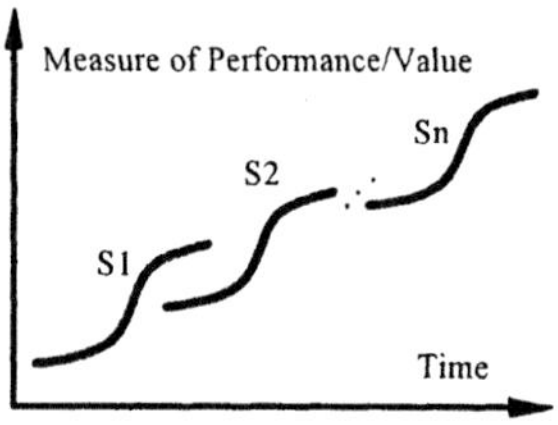

Figure 1. Evolution S-curve of technology systems

2.2 Patterns/ lines of technology system evolution

The patterns of evolution represent a compilation of trends that document strong historically recurring tendencies in the development of manmade or natural systems. There were three different versions about the patterns of evolution since the work was first published, such as Altshuller's ten set of patterns[2], Zusman's eight set of patterns[7] and Darrell's eleven set of patterns [6]. In our software tool, Darrell's eleven set of patterns of technological evolution are used to summarize the evolution of technology systems. These trends describe evolution towards:

- Pattern.1: Systems with increasing benefits and decreasing cost and harm.
- Pattern.2: Increased dynamization within systems.
- Pattern.3: Increased system segmentation.
- Pattern.4: Increased space segmentation.
- Pattern.5: Increased surface segmentation.
- Pattern.6: Increased controllability.
- Pattern.7: Increased complexity followed by reduced complexity.
- Pattern.8: Use of all available physical dimensions within a system.
- Pattern.9: Decreased number of energy conversions.
- Pattern.10: Increased rhythm co-ordination.
- Pattern.11: Increased action co-ordination.

Every pattern of technology evolution includes many lines of evolution[8]. In each line of evolution, researchers have identified a number of generic evolution steps up to and including a final level of evolutionary potential. The study of lines of technology evolution can provide ideas for technology systems' jumping to next-generation. The current state of the art now brings the total number of technical evolution lines to over 300[8]. It is significant for product innovation to study the evolution lines.

2.3 Evolutionary potential of technical system

The patterns/lines of evolution and their incorporation into a design method generate the new concept of evolutionary potential[6]. Evolutionary potential radar plot is illustrated in Fig.2. Each spoke in the plot represents one of the TRIZ patterns relevant to the given component .The outside perimeter of the plot represents evolutionary limit, and the shaded area represents how far along each pattern the current system has evolved. Thus the area difference between shaded area and perimeter is a measure of evolutionary potential. The counts of evolutionary patterns covered in evolutionary potential radar plot are different for different products. Evolutionary potential radar plot is used as a way of describing how far along each of the TRIZ patterns a given system has evolved, and is as powerful guides to help determining the future development opportunities and limits for a wide variety of technical systems.

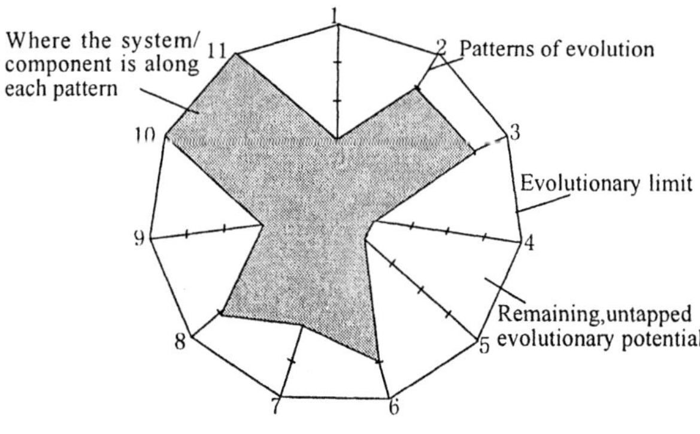

Figure 2. Evolutionary potential radar plot

3. THE EVOLUTION ANALYSIS OF RECIPROCATING SEAL FOR HYDRAULIC CYLINDER

Reciprocating seals have various kinds with the difference in pressure and speed of liquid, dimensions of radial and axis, and the property of dynamic and static of hydraulic cylinder. In this paper, the reciprocating seal we will discuss is focused on the development of seal, and is not a restriction of detail hydraulic elements.

3.1 Evolution S-curve of reciprocating seal

Lack of space dictates, the reciprocating seal's evolution s-curve given by CAI software tool is shown in Fig. 3. It is the result of the quantitative analysis of reciprocating seal's performances at material, configuration, controllable, wear and scuffing. Obviously, the reciprocating sealing is still positioned at the growth stage of its evolution curve and the future development opportunity of hydraulic reciprocating sealing technology will increase the seal, wear and scuffing characteristics by improving the configuration, material and shape.

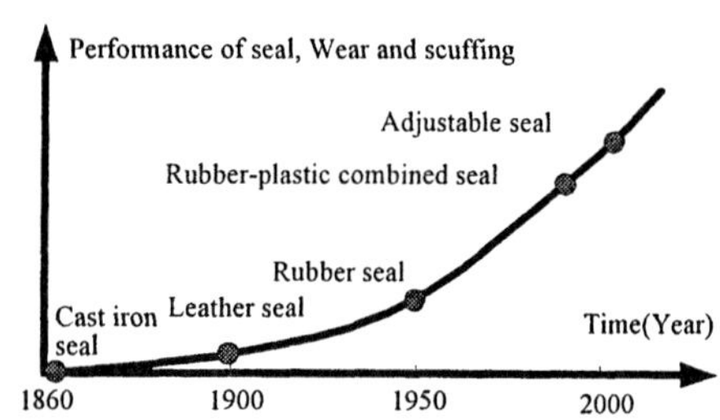

Figure 3. The evolution S-curve of reciprocating seal

3.2 Evolution lines of hydraulic reciprocating seal

In order to find a point along the trend that best describes the current evolutionary state of the reciprocating seal design, we compare reciprocating seal design with each of the Darrell's evolution patterns, and summarize its evolution lines. The evolution states of existing reciprocating seal are shown in Fig.4. Table 1 presents the general evolution lines which are discovered by TRIZ researcher based on the study and application of the patterns of evolution in various systems and are closely relevant to the evolution of reciprocating seal.

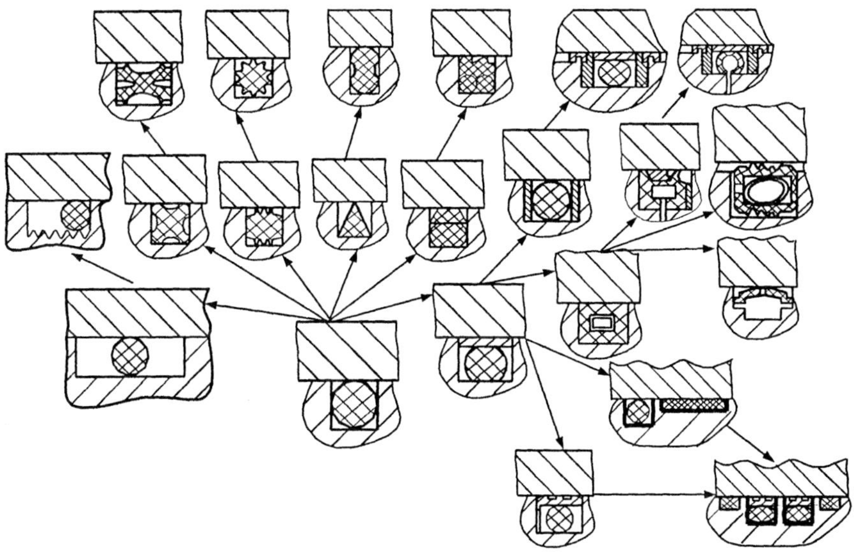

Figure 4. Evolution states of reciprocating seal

Table 1. The general evolution lines closely relevant to the evolution of reciprocating seal

Darrell's evolution patterns	The general lines of evolution	The lines of evolution of reciprocating seal
Pattern1	Dear material → general material → compound material → function material.	Rubber seal →synthetic rubber seal → synthetic rubber seal with wearable additive.
Pattern 2	Immobile system → joint → many joints →completely elastic →liquid, gas →field.	O-ring seal →magnetic liquid seals → magnetic seals.
	Mono-system → Bi-system →Three system → poly-system →combined poly-system.	Combined coaxial seal → O-ring seal with backup-rings →combined coaxial seal with backup-ring and bearings.
	Mechanical field → hydraulic field→ electromagnetic field → chemic field → biologic field.	Elasticity seal → spring elasticity seal → liquid elasticity seal.
	Single field → combined field with two-field →combined field with multi-field.	Rubber seal → plastic and spring seal →plastic, spring and liquid seal.
Pattern 3	System → segmental monolith → liquid or powder → gas → field.	Single O-ring seal → two-ring seal → three-ring seal.

Darrell's evolution patterns	The general lines of evolution	The lines of evolution of reciprocating seal
Pattern 4	Mono-system →system with a cavity →system with multiple cavity →capillary and porous system →system with active capillary	Combined coaxial seal → coaxial seal with oil grooves to form hydrodynamic lubrication.
Pattern 5	Flat surface → surface with single protrusion → surface with two-protrusion → rough surface → combined surface.	O-ring seal → multi-grooved surface seal →grooved surface seal.
Pattern 6	Direct control action → action though intermediary → addition of feedback → intelligent feedback.	Elasticity unalterable combined seal with PTFE and spring →PTFE seal with liquid-loaded elasticity →coaxial seal with back-up rings, bearings and adjustable elasticity by external liquid
Pattern 7	Single function →multi-function →integrated function.	O-ring seal → Tandem seal with different components to deliver functions of seal and wearable →Tandem seal with different components to deliver functions of seal and wearable, while grooves form hydrodynamic lubrication.
Pattern 8	Line → 2D-curve → 2D-curve → compound 3D-curve.	O-ring seal → anti-twist X-seal.
	Plane → single curvature of surface →double curvature of surface → combined curvature of surface.	O-ring seal → triangle-ring seal → abnormity-ring seal.
Pattern 9	Sliding friction → combined friction with sliding and rolling→ rolling friction→ fluid friction	O-ring seal → O-ring seal with a free seal chamber → O-ring seal with grooved in the seal chamber
Pattern 10	Continuous action→pulsation action→pulsation in the resonance mode → several actions → travelling wave	O-ring seal → continuous seal in bidirectional →one direction seal discontinuous
Pattern 11	Non-coordinated actions → partially coordinated actions → coordinated actions → actions during intervals	Tandem seal of PTFE and O-ring →Combined coaxial seal

In all the evolution lines being presented, TRIZ depicts an evolutionary progression from left to right across each trend, in which benefits increase as a design travels further rightwards.

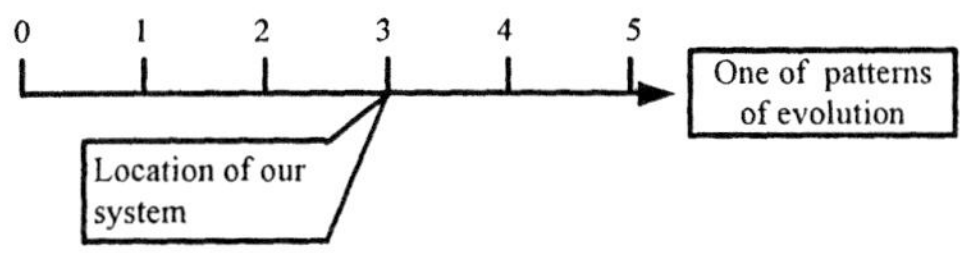

Figure 5. The process of positioning a system in relation to the patterns of evolution

4. THE INNOVATIVE DESIGN OF RECIPROCATING SEAL FOR HYDRAULIC CYLINDER USING CAI SOFTWARE TOOLS

4.1 The method of gauging and positioning the state of a system in relation to patterns of evolution

The main difficult encountered in procedure of making the patterns or lines of evolution operational is the gauging and positioning of the state of a system in relation to each pattern or lines of evolution[9]. In our CAI software tools, we use the fuzzy sets to gauge the relative evolutionary position of system. Fig.5 shows the process of positioning the system in relation to a pattern of evolution. The meaning of fuzzy value 0, 1, 2, 3, 4 and 5 is presented in Table 2.

Table 2. Positioning a system in relation to the patterns

Fuzzy value	The meaning of value
0	The pattern is not applied at all, it is necessary to focus on its development
1	The pattern is applied to a small extent, the future development of the system probably depends on developing this pattern
3	In most cases, the pattern is applied; it would nonetheless be useful to observe a development in this direction if no other pattern is really deficient.
5	The pattern is applied to the latter, a development in this direction is probably not what is most urgently needed in the system's development overall
2,4	Intervenient of 0 and 1, 3 and 5

4.2 Positioning the reciprocating seal in relation to the patterns of evolution

According to table 2, we can position the reciprocating seal in relation to the patterns of evolution. Lack of space dictates the details positioning process of the reciprocating seal in relation to each pattern of Darrell's evolution, the contents is omitted in this paper. The detail process can reference paper 10.

4.3 Evolution potential plots of hydraulic reciprocating seal

Fig.6 presents the evolutionary potential plot. The plot clearly shows that there to be considerable amounts of untapped potential in the reciprocating design, such as the potential to improve the ability of adjustable of the system and to increase the segments of configuration, shape, highlighting the utility of systems resource. Therefore, there are consequently significant improvements for reciprocating seal design, if we explore all above-mentioned untapped potential.

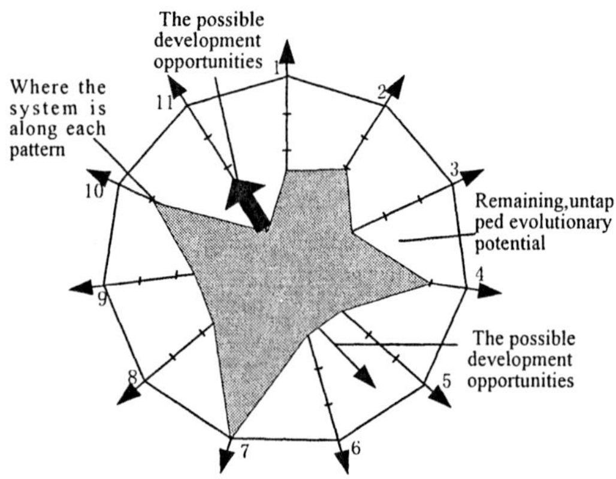

Figure 6. Evolutionary potential radar plot of hydraulic reciprocating seal technology

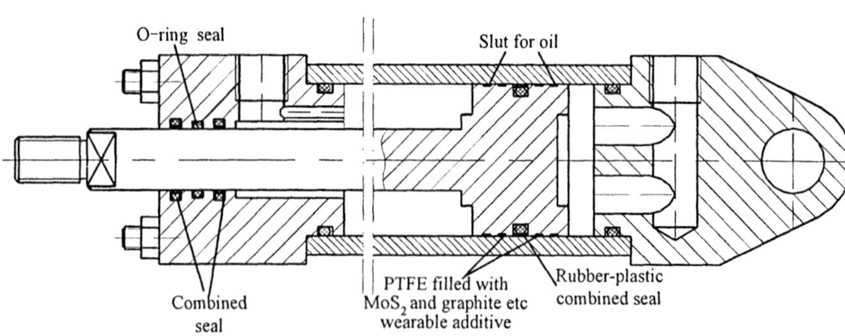

Figure 7. Reciprocating seal design for hydraulic cylinder

4.4 The innovative design of reciprocating seal using TRIZ

On the Basis of our work and the result of evolutionary potential radar plot, we present the detail design of reciprocating seal for hydraulic cylinder in Fig.7.

5. CONCLUSIONS

Traditional product development is based on the principle that everything is feasible in terms of the thought process, and encourages reflections from every angle and potentially leads people in a broad spectrum of directions. CAI software tools based on TRIZ evolution theory provide a powerful structure methodology to guide designers closer to the right solution. In this paper, by using CAI software tool, we position the evolution stage of reciprocating seal in its evolution S-curve and draw its evolutionary potential radar plot. Making use of the possible development opportunities that are recommended by evolutionary potential radar plot, we detail the innovative design of reciprocating seal for hydraulic cylinder.

6. REFERENCES

1. Mnler,H.K. and Nau,B.S., *Fluid sealing technology-principles and applications* (China Machine Press, Beijing, 2002).
2. G.S.Altshuller, *Creativity as an Exact Science*(Gordon & Breach , New York, 1988).
3. Savransky S D, *Engineering of creativity* (CRC Press , New York, 2000).
4. Fey V R, Rivin EI, Guided technology evolution-TRIZ technology forecasting (1999); http://www.triz-journal.com/article2.html.
5. Zusman A, Zlotin B and Zainiev G, An application of directed evolution (2003); http://www.ideationtriz.com/ Endoscopic_case_study.asp.
6. Mann, D.L, Better technology forecasting using systematic innovation methods, *Technological Forecasting and Social Change*.Vol.70, pp.779-795(2003).
7. B.Zlotin, A.Zusman: *Directed Evolution: Philosophy, Theory and Practice* (Ideation International Inc, USA, 2001).
8. S.R.Luke, *A Conceptual Design Tool for Engineers: An Amalgamation of Theory of Constraints, theory of Inventive Problem Solving and Logic* (Old Dominion University ,U.K.2002).
9. Cavallucci D Integrating Altshuller's development laws for technical systems into the design process, *Annals of the CIRP*, Vol.52(1),(2002)
10. Zhang Fuying, Xu Yanshen, Evolution study of hydrodynamic reciprocating sealing for hydraulic cylinder based on TRIZ, *Run Hua Yu Mi Feng/ Lubrication Engineering*, Vol.163,pp.14-17 (2004).

INNOVATIVE DESIGN OF THE SEAL STRUCTURE OF BUTTERFLY VALVE BASED ON TRIZ

Jianhui Zhang [1], Runhua Tan [1], Ping Jiang [1], Jinling Dai [2]
[1]Institute of Design for Innovation, Hebei University of Technology, Tianjin, 300130, PRC; Email: zjh2031@sina.com. [2]Institute of Pharmaceutical Research, Tianjin, 300073, PRC.

Abstract: Innovative ideas generation is a key step in the innovative design of the products. Computer-aided innovation (CAI) systems based on TRIZ (Theory of inventive problem solving) support new ideas generation of innovation. The patterns and lines of technology evolution in TRIZ provide a powerful structured methodology for a directed development of new products. The concept of 'evolutionary potential' and the method to construct an evolutionary potential plot are introduced. The Innovative ideas for new product development can be generated by the analysis of the evolutionary potential. A general process of the innovative ideas generation is then developed. A butterfly valve's seal structure case study is presented to demonstrate the process.

Key words: TRIZ; patterns and lines of technology evolution; evolutionary potential; innovative idea generation; butterfly valve.

1. INTRODUCTION

Butterfly valve is a controlling valve used in large-size pipe, which has function of opening or closing pipe, regulating flux and check, etc. With the development of the industry, the demand for the valve capability becomes stricter and stricter. So the design of the seal structure of the butterfly valve has been an important research task at all times.

The theory of technology evolution in TRIZ (Theory of inventive problem solving) provides the powerful tools of technological forecasting—patterns of evolution and lines of evolution, which can forecast the

Please use the following format when citing this chapter:
Zhang, Jianhui, Tan, Runhua, Jiang, Ping, Dai, Jinling, 2006, in International Federation for Information Processing (IFIP), Volume 207, Knowledge Enterprise: Intelligent Strategies In Product Design, Manufacturing, and Management, eds. K. Wang, Kovacs G., Wozny M., Fang M., (Boston: Springer), pp. 450-456.

probabilities of certain design parameters falling within particular confidence intervals at some future time (Alla, et al. 2002). Evolutionary potential can be found when patterns and lines of technological evolution, which match the historical evolution of the core technology of the products, are applied to analyze the products. And the innovative ideas for future technological R&D plans can be generated by analysis of the evolutionary potential. The computer-aided innovation (CAI) systems based on TRIZ are practical tools to support this process.

This paper introduces a general process of the innovative ideas generation supported by computer-aided innovation (CAI) systems. A butterfly valve's seal structure case study is presented to demonstrate the process step by step.

2. TECHNOLOGY EVOLUTION AND EVOLUTIONARY POTENTIAL

Patterns of technical system evolution are revealed by decades of analysis of hundreds of thousand of the global inventive patent. The patterns define general technological evolutional directions and each pattern includes multiple lines of evolution that specify particular evolutional stages or scenarios along these directions (Altshuller, 1984; Tan, 2002).

By comparing the products under study with each of the lines and making a connection with one line stage, the state of the lines that can best describe the current evolutionary state of the design defined as 'current state', and the unused states are defined as 'potential states'. The 'evolutionary potential' concept is a means of observing the relevant evolution lines of product together (Darrell, 2002). Fig. 1 illustrates the structure of evolutionary potential plot. An evolution line known as 'space segmentation' in the figure describes the evolution of the use of space inside structures. There are five states shown in this line, the distance between the ending state and the current state in the line exists the 'evolutionary potential'.

If a system is under evaluation relative to the 'space segmentation' line, a state of the art design will be at stage 3 of 5 possible stages. Repeating this process for each line in turn, then arranging each line as one spoke on a radar plot reveal a picture of the overall current evolutionary state of the product. As shown in Fig. 1, the shaded area of the radar plot represents how far along each line the current system has evolved, the area difference between shaded area and perimeter is a measure of untapped 'evolutionary potential'. The evolution lines and the untapped potential act as signposts indicating the directions that other successful systems have evolved in. So the new ideas generated by analysis of the potential states are more reliable than that generated by traditional methods.

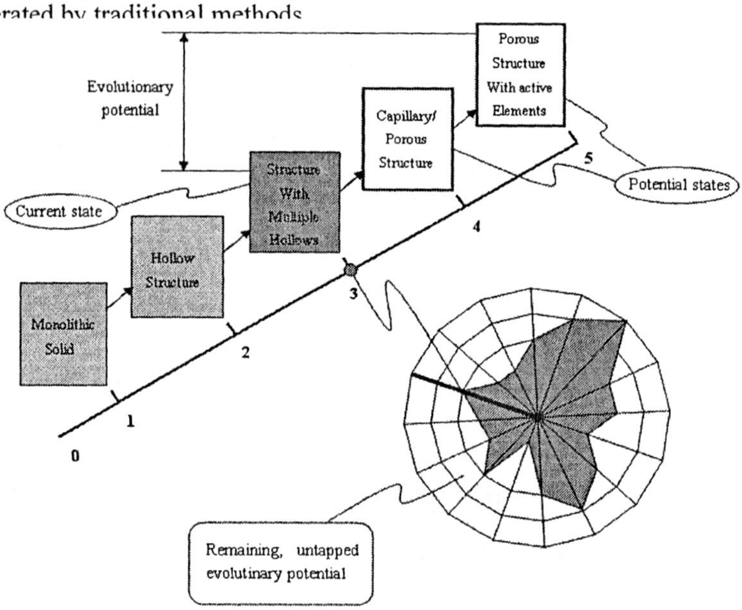

Figure 1. Evolutionary potential radar plot structure

3. A GENERAL PROCESS OF INNOVATIVE IDEAS GENERATION USING CAI SYSTEM

The technological system evolution is propelled by core technical innovation, and the key patents about a certain product are the track record of core technical innovation of the product and can represent the history of product evolution. So the patent information of the product is treated as the main source of product evolution analysis. The general process of the innovative ideas generation includes the following six steps:

Step 1: Analyze the history of product evolution

By searching and analyzing the relevant key patent of the products, valuable technical and business information regarding target or key technology of the product can be available.

Step 2: Apply the relevant patterns and lines to analyze the product

By comparing the product under study with each of the selected patterns and lines of CAI system, and making a connection with one evolution state, the current state and the potential states for the product are determined. The

CAI systems, such as Goldfire Innovator of Invention Machine (USA) and InventionTool (China) support this step.

Step 3: Draw the evolutionary potential radar plot

On the basis of the analysis of step 2, trace out the product's evolutionary potential radar plot.

Step 4: Search evolutionary potential

Evolutionary potential can be found from one or more lines selected in the evolutionary potential radar plot. InventionTool (China) supports step 3 and step 4.

Step 5: Generate innovative ideas

Under analyzing the potential line in the evolutionary potential radar plot, one or more innovative ideas can be generated.

4. INNOVATIVE IDEA GENERATION AND APPLICATION FOR BUTTERFLY VALVE'S SEAL STRUCTURE

The innovative idea generation and application for the butterfly valve is carried out using the general process step by step.

Step 1: Analyze the history of product evolution

Search the butterfly valve patents of different industry in Chinese patent databases and study tends of the technology evolution for the main structure of the valves. A butterfly valve is divided into three modules: the controlling module (includes drive and transmission section), executing module (as shown in Fig. 2, ④, includes valve disc, valve shaft, valve seat, and seal ring) and valve body. The evolution of the executing module shows the core technology changes of this kind of valve. The evolutional tree of the executing module of butterfly valve is shown in Fig. 2 (Zhang, 2004).

In order to clearly show the evolution of various generations of butterfly valve systems, we define the generations of butterfly valve as follows: the first generations –①, the second generations –②, ③, ④, the third generations –⑤,⑥, ⑦, ⑧ and the fourth generations –⑨, ⑩. The same generation products share some common features. For example, the offset between shaft and disc has transformed from zero to single, double and triple, as shown in fig. 2 (①→③→⑦→⑩).

Step 2: Apply the relevant patterns and lines to analyze the product

Applying the patterns and lines that match the evolution of butterfly valve to analyze the main structure of butterfly valve, there are eleven lines has been used in the evolution of the butterfly valve. The current state and the potential states for the product are determined. As shown in Fig. 3. These lines are selected from Goldfire Innovator 4.0 of Invention Machine (USA)

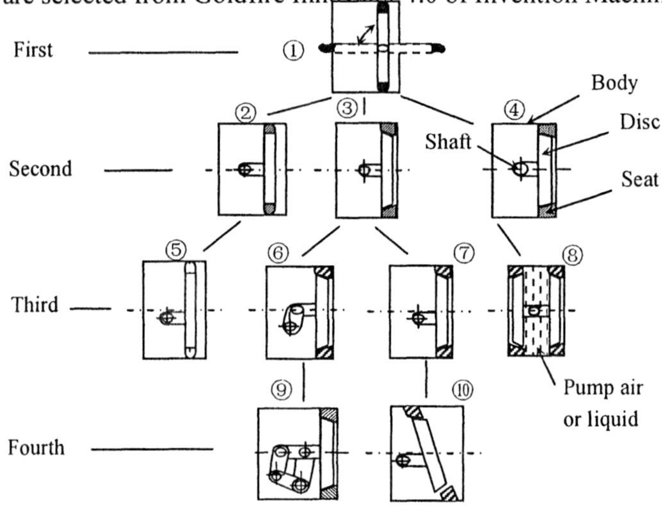

Figure 2. The evolution tree of executing module structure of butterfly valve

and InventionTool 3.0 (China).

Step 3: Draw the evolutionary potential radar plot

In order to see how quickly the evolutionary potential is being used by the various generations of design of butterfly valve systems, overlaying the evolutionary potential plots for each generation, a composite picture highlighting the evolution process that has taken place from the first generation butterfly valve to the most recent is illustrated in Fig. 3. The plot labeled '1', '2', '3' and '4' describes the evolutionary state of the four generations respectively. InventionTool 3.0 (China) supports this step and the next step.

Step 4: Search evolutionary potential

As shown in Fig. 2, undoubtedly butterfly valve technology may be seen to be mature in terms of dimension and geometric structure evolution-because all of the available dimensional freedoms have been used. Beyond that, however, the plot clearly shows there is still considerable potential in the area difference between shaded area and perimeter. Therefore, there are consequently significant improvements that can be developed.

Many butterfly valve patents have related to sealed issues, in order to improve the integral leakage index. On the basis of the above analysis, evolution line of 'increased dynamics' was selected to find the evolutionary potential in relation to the seal technology of the butterfly valve. There are six states in this line:

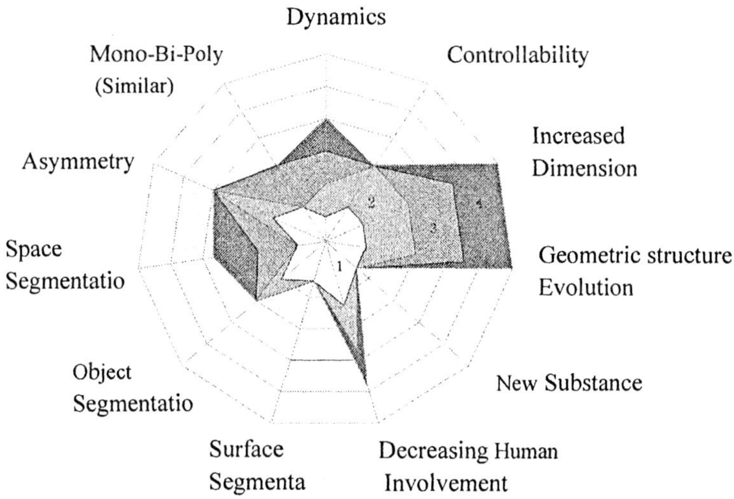

Figure 3. Evolutionary history of butterfly valve technology

stiff system→a joint→many joints→elastic→molecular (liquid, gas) →field

The current state is "elastic", such as the application of elastic material of seal ring. The states of "molecular" and "field" are potential states.

Step 5: Generate innovative ideas

The key technique of the valve is to compensate for the wear of the seal ring in order to reduce leakage. The action of the compensation can be achieved by making the valve disc to move forward in a small displacement after the rotation of the valve disc when closing the valve, or making the valve disc to move backward in a small displacement before the rotation of the valve disc when opening the valve.

The application of the elastic elements in current state can implement the function above-mentioned. For example, the elasticity of the spring can implement that function from the current state "elastic", the compressibility of liquid or gas can implement that function from the first potential state "molecular", the second potential state is also applicable. For example, a permanent magnet can implement the function of the compensation. So, some innovative ideas are generated.

Innovative ideas: Apply elasticity of spring, or compressibility of liquid or gas, or a permanent magnet to implement the function of the seal ring compensation.

These three innovative ideas should be possible because there are successful applications in industry or life. These new ideas are generated from the 'evolutionary potential' that other successful industries have already made, so the uncertainty of the new ideas is reduced. The engineers of the butterfly valve industry should also make evaluation for the ideas and implement them by the conditions of management, design and manufacture capacity of their firm, they should make an adaptive and specific design for the existing products that need to be improved. Accordingly, the pace of the products innovation is accelerated.

5. SUMMARY

To generate innovative ideas is a key step in the innovative design of the products. CAI systems based on TRIZ are applicable to support this process. A general process of the innovative ideas generation supported by computer-aided innovation (CAI) systems is developed.

Innovative design of seal structure of butterfly valve is carried out and shows the general process of the innovative generation and application step by step. Although perhaps thought of as 'mature' technologies, the valve still have considerable levels of untapped evolutionary potential remaining, and that there are consequently significant opportunities for future development of it.

6. ACKNOWLEDGMENT

The research is supported in part by Natural Science Foundation of China under Grant Numbers 50375045 and TianJin Natural Science Foundation Key Projects under Grant Number 043802211.

7. REFERENCES

1. Altshuller, G.S., 1984, Creativity As An Exact Science: The Theory of the Solution of Inventive Problems, The United Kingdom: Gordon and Breach Science Publishers Inc., London, pp. 205-216.
2. Tan Runhua, 2002, Innovation Design—TRIZ: Theory of Inventive Problem Solving, China Machine Press, Beijing, pp. 58-60 (in Chinese).
3. Alla, Z., Boris, Z., and Gafur, Z., (June 2001); An Application of Directed Evolution, http://www.Ideation-triz.com.
4. Darrell Mann, 2002, Evolutionary-PotentialTM in Technical and Business Systems, TRIZ Journal, 6, http://www.triz-journal.com.
5. Zhang, H.G., Tan, R.H., and Zhao, W.Y., 2004, The technology maturity of product and its mapping, ASME Proceedings of IMECE'04, IMECE2004-59260. Anaheim, USA.

THE DETERMINATION METHOD AND RESOLVING PROCEDURE OF DESIGN CONFLICT BASED ON EVOLUTION PATTERN AND PREREQUISITE TREE

Lihui Ma,Runhua Tan
Hebei University of Technology, Tianjin, 300130, China. Email: malh198@yahoo.com.cn

Abstract: In this paper, a determination model of design conflict based on the evolution theory of TRIZ and the Prerequisite Tree (PRT) is proposed. The evolution patterns are used to identify evolving objectives and the PRT is used to identify all evolving obstacles. A new conflict determining diagram and template are set up to identify design conflict. Then the separation principles or the conflict matrix in TRIZ are applied to solve the design conflicts. A case of engineering has verified the application of the process model.

Key words: TRIZ, PRT, design conflict or contradiction, evolution pattern

1. INTRODUCTION

The key of product innovation lies in finding and solving the conflict or contradiction in design. Conflict is basic to the definition of a technical problem in TRIZ. TRIZ (Theory of Inventive Problem Solving) is a method for solving inventive problems based on the knowledge where there is no obvious solution and where there is at least one major contradiction that appears to be irresolvable. TRIZ provides a wide-ranging series of techniques and tools for solving technical problems of various complexities, which can help designers and inventors to avoid trial-and-error approach in design process and to solve problem in creative and powerful ways[1]. However, TRIZ has not provided a method with stronger effectiveness to

Please use the following format when citing this chapter:
Ma, Lihui, Tan, Runhua, 2006, in International Federation for Information Processing (IFIP), Volume 207, Knowledge Enterprise: Intelligent Strategies In Product Design, Manufacturing, and Management, eds. K. Wang, Kovacs G., Wozny M., Fang M., (Boston: Springer), pp. 457-464.

confirm the conflict of design[2]. How to confirm the design conflict has already become a focus of the research on TRIZ in recent years[3-7]. In some literatures[8-10], TRIZ and Thinking Process from TOC are integrated to identify conflict. But just as Darrell MANN and Roy STRATTON stated in their thesis[10] that the area of greatest common ground between CRD and TRIZ lies in the way in which CRD helps to define physical contradictions.

In this paper, a determining model of design conflict based on the evolution patterns of TRIZ and the Prerequisite Tree (PRT) of TOC (Theory of Constraints) is proposed, which can be applied to determine physical and technical conflicts. The synergy of TRIZ and Thinking Process will form the systemic process for identifying and solving design conflicts.

2. CONTRADICTIONS IN DESIGN

G.S.Alshuller divides contradictions into administrative contradictions, physical contradictions and technical contradictions[7,11]. Administrative contradictions don't belong to the contents of TRIZ.

Technical contradiction: an interaction between system's parts when the useful function/action causes simultaneously a harmful effect, or introduction (intensification) of the useful function/action, or elimination or reduction of the harmful function/action causes deterioration or unacceptable complication of one of the parts or of the whole system. Technical contradiction is usually shown as the conflict between two subsystems in a system.

Physical contradiction: a situation when the same component must satisfy mutually exclusive demands to its physical state, e.g., be hot and cold, electrically conductive and insulative, etc. For physical contradictions, there exist two opposing requests simultaneously on one aspect of the system.

3. EVOLUTION PATTERNS AND PRT

3.1 Evolution patterns of TRIZ

Evolution procedure of a product is a self-sustaining process that resolves the profound conflict constantly and pushes the product ever closer to its ideal state.

The TRIZ technology evolution developed by Altshuller[12] is a set of patterns and paths, which show the trends of technological systems evolution in structures.

The TRIZ technology evolution has different versions now[7]. DE[13] is an important branch of the Evolution theory in TRIZ. The core of DE is the management and control of technical evolution process. DE has many patterns and paths, which are suitable for the innovation of technological system and production process. In this paper the patterns of DE are applied.

According to DE, there are eight patterns of evolution as follows.

- Stages of evolutions.
- Increasing ideality.
- Uneven development of subsystems.
- Increasing dynamism and controllability.
- Increasing complexity, followed by simplicity.
- Matching and mismatching of parts.
- Transition toward micro-level and increasing use of fields.
- Decreasing human involvement.

3.2 Prerequisite Tree (PRT) from TOC

The Prerequisite Tree from TOC (Theory of Constraints)[14] is a necessity based logic structure. Its purpose is to help us identify all obstacles and the intermediate steps (action, objective, or injection) that we need, to overcome them in realizing an objective.

4. DETERMINATION METHOD AND RESOLVING PROCEDURE OF DESIGN CONFLICT

Product design or development is the evolution process of overcoming contradictions constantly. Finding and determining conflict is the first step of innovation.

Identify the core problem (or root causes) is the key of the whole design process. PRT shows the designer how to identify all obstacles and the required sequence of actions needed to realize a desired course of action. TRIZ is strong in guiding designers how to identify objectives and solve the problems. It is very useful to integrate the Prerequisite Tree (PRT) from TOC and TRIZ for determining and resolving contradictions.

4.1 Method to confirm conflict based on evolution patterns and PRT

During the evolution of product, improvement of any part of the product, which can make the system reach its pinnacle of functional performance, will lead to conflict with another part of the product. Design of product is the process that overcomes the obstacle and resolves the corresponding conflict and pushes the product ever closer to its ideal state.

Figure 1 shows the determining procedure of contradictions, which includes 3 steps as below.

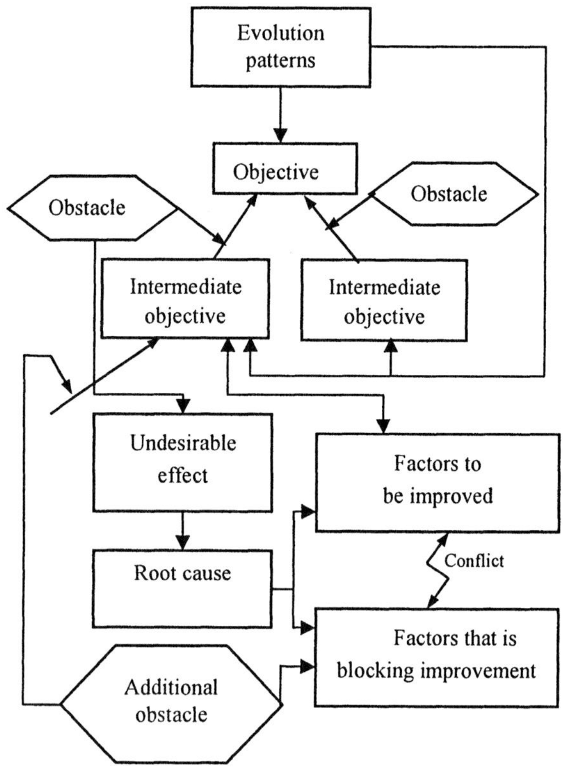

Figure 1. Contradiction identifying diagram

Step 1 Set up objective using evolution patterns of TRIZ. Because the result of evolution will cause the improvement of the performance or the parameter of product, the evolution objective can thus be determined directly.

Step 2 Build the Prerequisite Tree

List all the obstacles at first. Then, for each obstacle, identify a condition that overcomes the obstacle. This is usually a mutually exclusive condition, which implies the existence of contradiction.

By focusing on individual obstacles, we identify the smaller intermediate objectives (IOs) that lead us to our more ambitious goal.

Step 3 Identify contradictions through obstacles. A new conflict determining diagram is set up as shown in Figure 1.

The determination of conflict can adopt the following template:

According to evolution objective, a useful function or a parameter will exist or be improved, but because of the existence of some obstacles, it won't exist or will cause the deterioration of other functions or parameters.

Physics conflict can be determined according to the template and conflict determining diagram directly.

When determining technical conflict, the improved parameter can be confirmed according to the template and conflict determining diagram. The worse parameter usually comes from an additional obstacle, which belongs to another subsystem. Converting them into 39 standard engineering parameters of TRIZ, the standardization of technical conflict is accomplished.

4.2 Resolve design conflict

For eliminating physical contradictions, four generic principles (separation principles) are used.

For technical contradictions, the 39 standard engineering parameters, contradiction matrix and 40 inventive principles in TRIZ should be used to solve them.

5. ENGINEERING EXAMPLE

Dropping pill is a rapidly developing form of Chinese traditional medicine. Dripping the mixture of liquid medicine into a cooling liquid under thermal condition, the process of pill formation results in a solid-dispersoid in a pill shape[15, 16].

In the traditional machine, the dropping pills are acquired through the freedom dropping way. Because of the uncontrolled state , it is very difficult to keep the dropping pills dimensional uniformity during the dropping course. Figure 2 is the conflicts confirming procedure for the design of dropping pill machine.

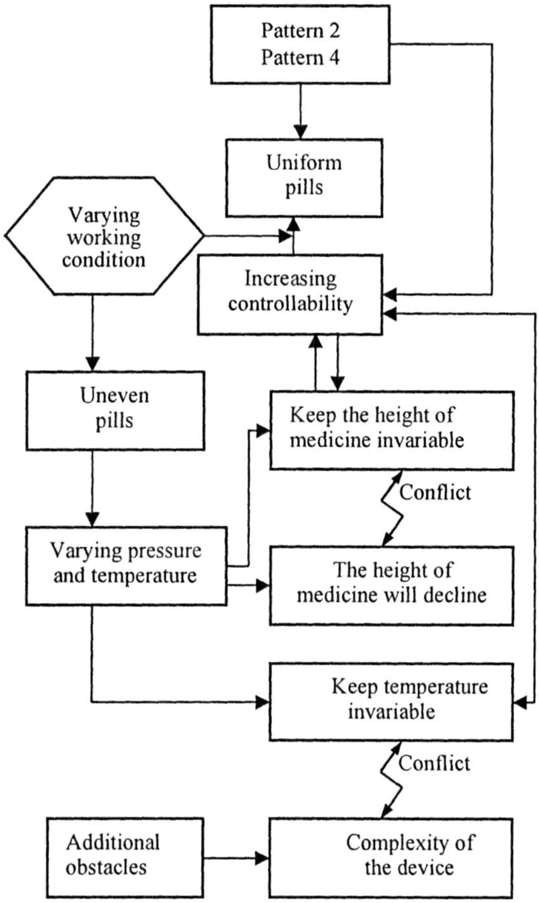

Figure 2. Confirm conflicts for design of dropping pill machine

Select pattern 2 and pattern 4 for the evolution direction of dropping pill machine, i.e., increasing ideality, dynamism and controllability. The ambitious goal, intermediate objective and obstacle can be determined. There are physical conflict and technical conflict as shown in Figure 2. For the physical conflict , the first generic principle can be applied and the corresponding inventive principles are No.9,No.10,No.11,No.15,No.16, No.18, No.19, No.20, No.21, No.29, No.34 and No.37. The usable inventive principles are No.11(Beforehand cushioning), No.15 (Dynamics) and No.29(Pneumatics and hydraulics).

The technical conflict can be expressed as the one between No.17 (Temperature) and No.36 (Complexity of a device) or No.38 (Level of automation) and No.36 (Complexity of a device) of the standard engineering parameters through standardization of technical conflict.

According to contradiction matrix of TRIZ, for the technical conflict between the standard engineering parameters No.17 and No.36 or No.38 and No.36, the corresponding inventive principles are No.2 (Extraction), No16 (Partial or excessive action) and No.17 (Another dimension) or No.10 (Prior action), No.15 (Dynamics) and No.24 (Mediator). No.2, No.10, No.15 and No.24 can be adopted here.

Figure 3 is the design of dropping pill machine for Chinese traditional medicine.

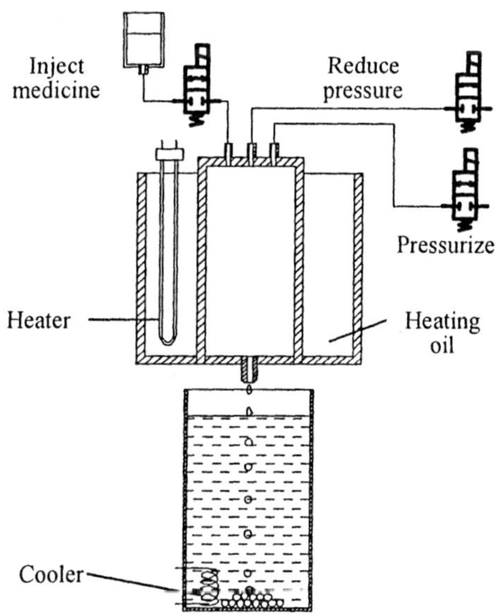

Figure 3. Innovation design of dropping pill machine

Before dropping, dropping nozzles are heated up by heating oil in advance. During dropping, the temperature of medicine can be fixed to stability constant by heating oil. The pressure of liquid medicine can be cushioned by compress air when it's falling. In this system, the dimensional uniformity of dropping pills can be kept.

6. CONCLUSIONS

A determination model of design conflict based on the evolution theory of TRIZ and the Prerequisite Tree (PRT) is proposed, which can be used to identify physical and technical conflict. In this model, the evolution patterns are used to identify objectives and the PRT is used to identify all obstacles. Then the physical or technical contradiction can be found by the conflict determining flow and template. The separation principles or the conflict matrix in TRIZ are applied to solve them. The design example of dropping pill machine has verified the application of the process model.

7. REFERENCES

1. Kai Yang and Hongwei Zhang, A comparison of TRIZ and Axiomatic Design, *TRIZ Journal* (August, 2000); http://www.triz-journal.com.
2. Cavallucci D, Integrating Altshuller's development laws for technical systems into the design process, *Annals of the CIRP*, 2001, 50(1): pp.115-120.
3. Lee Noel, A new model of the conceptual design process using QFD/FA/TRIZ, *TRIZ Journal*, TRIZ 9 to Create the Ideal Product Design, http://www.triz-journal.com.
4. Terninko J, Selecting the best direction to create the ideal product design, *TRIZ Journal*, http//:www.triz-journal.com.
5. Runhua Tan et al., Study on the conceptual design process using QFD and TRIZ, *Chinese Journal of Mechanical Design*, Vol.9,pp.1-4 (2002).
6. Ellen Domb, Finding the zones of conflict: tutorial, *TRIZ Journal* (June, 1997); http://www.triz-journal.com.
7. Runhua Tan, *Theory of Inventive Problem Solving* (Science Press, Beijing, China, 2004).
8. Eduardo C. Moura, TOC trees help TRIZ, *TRIZ Journal* (Sep, 1999); http:// www.triz-journal.com.
9. Ellen Domb and H.William Dettmer, Breakthrough innovation in conflict resolution marrying TRIZ and the Thinking Process, *TRIZ Journal* (May, 1999); http:// www.triz-journal.com.
10. Darrell Mann, Roy Stratton, Physical contradictions and evaporating clouds, *TRIZ Journal* (April, 2000); http:// www.triz-journal.com.
11. Semyon D. Savransky, Elementary Contradictions in TRIZ, International Conference on Engineering Design, ICED 99 MUNICH, August 24-26,1999.
12. Altshuller G, The Innovation Algorithm, TRIZ, Systematic Innovation and Technical Creativity, Technical Innovation Center, INC, Worcester, 1999.
13. Zusman A, Zlotin B, Zainiev G, An application of directed evolution, http://www.ideationtriz.com/Endoscopic_Case_Study.htm.
14. H.William.Dettmer, *Goldratt's Theory of Constraints* (ASQC Quality press, 1997).
15. Chunlin Cao, *Chinese Medicine Pharmacy* (Shanghai Science Press, China, 1999).
16. Zhaowang Zhang, *Chinese Medicine Pharmacy* (Chinese Medicine Press, China, 2003).

RESEARCH ON DEVELOPMENT OF COLOR SORTER USING TRIZ

Lihong He, Zhanwen Niu, Dongliang Chen
School of Mechanical Engineering, Tianjin University, 300072, China. Email:lihong-he@126.com

Abstract: Color Sorter has been widely used in crops further processing industries. And it has a broad market in China. Improving product quality or developing new products is very important for enterprises to keep technical advantages in market competitions. Technology evolution theory can forecast the future technology in products. It can guide a designer along the right direction of developing a new product. The solutions from contradictions solving give starting points to solve a problem. According to these solutions, the designer can detail the design using professional knowledge. This paper is organized as follows. Firstly, it proposes to integrate technology evolution and the design of Color Sorter and forecast the future technologies of the Color Sorter parts, such as the optical signal-collecting system and grooves. Secondly, using the contradiction and inventive principle, this paper proposes several solutions for improving the adjust method of background board.

Key words: Color Sorter, Technology evolution, TRIZ, Contradiction, Inventive principle

1. INTRODUCTION

Color Sorter[1-3] is further processing equipment which can separate materials based on the color aberration between falling particles and background. As is shown in Figure.1, particles are sliding down through grooves to working area from feeder. Two sets of lamps, high sensitive photoelectric detectors, and background boards are fixed on the front and back of the working area. When all particles pass through the work area, photoelectric detectors compare the detected reflex and transmission

Please use the following format when citing this chapter:
He, Lihong, Niu, Zhanwen, Chen, Dongliang, 2006, in International Federation for Information Processing (IFIP), Volume 207, Knowledge Enterprise: Intelligent Strategies In Product Design, Manufacturing, and Management, eds. K. Wang, Kovacs G., Wozny M., Fang M., (Boston: Springer), pp. 465-470.

luminous flux with benchmark, and then identify those that have different colors. The signal of color aberration is magnified to drive the aerodynamic ejector, which blows away the rejected particles.

Technology is always in evolution. TRIZ[4,5] was developed in Russia by Genrich Altshuller and his followers in 1940s. It includes many solution tools, such as technology evolution, contradiction, effect and ARIZ[6], etc. The technology evolution is very helpful for technology forecasting since they can identify the most promising directions in the development of a system[7,8].

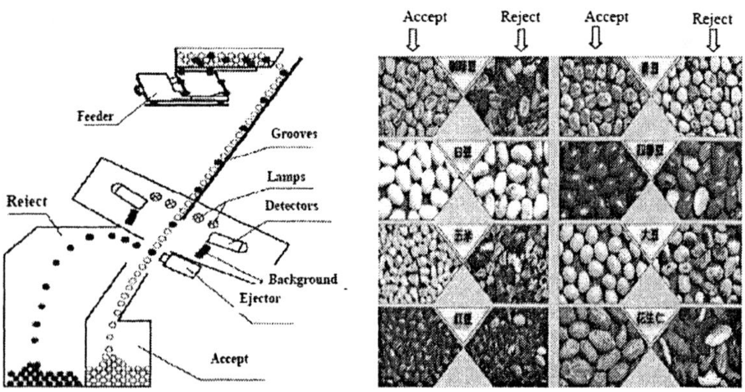

Figure 1. Color Sorter work principle and color sorting efficiency

An inventive situation is usually inherent in the technical and/or physical contradictions in the technique. The technical contradiction represents the confliction between two subsystems of a system. It is the main contradiction in the product evolution. In TRIZ, 39 features[9] can be used to describe the technical contradictions and 40 inventive principles[10] are given to address these contradictions.

2. ANALYSIS OF COLOR SORTER DEVELOPMENT

2.1 Technology Evolution Analysis and Forecast of Optical Signal Collecting System

The optical signal-collecting system plays the most important role in color sorter. Its upgrade means the upgrade of the color sorter. The first color sorter was made in Britain in 1960s, and now such equipment has evolved into its fourth generation. Based on the evolution laws and principles of TRIZ, the evolution analysis was performed, and the fifth generation was predicted, as is shown in Figure. 2.

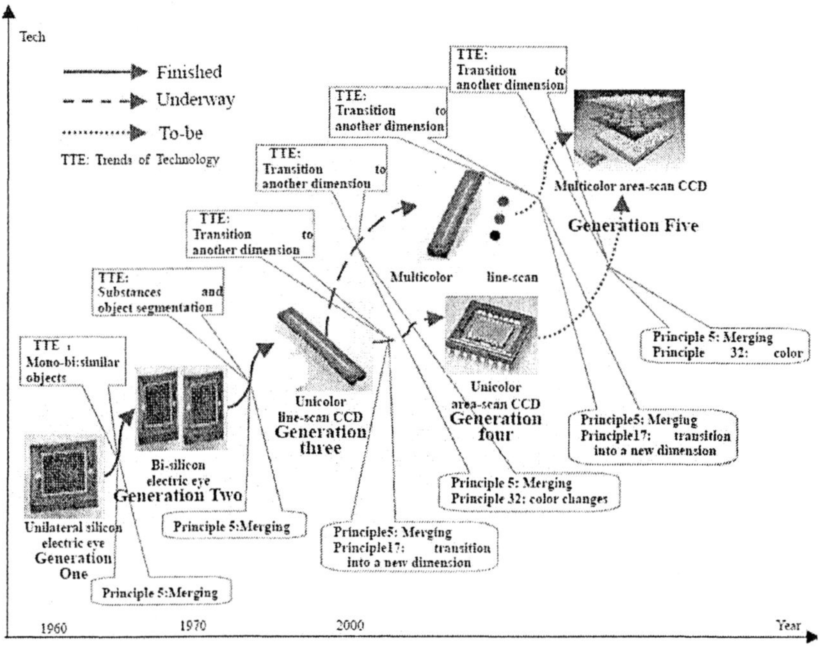

Figure 2. Evolution of optical signal collecting system

Because the unilateral silicon electric eyes that were fixed on the first generation could only catch the unilateral information, color-sorting precision was not enough. According to the laws of increasing flexibility, the eyes were fixed on the front and back of the particles flow in the second generation. More information could be got. The precision of color-sorting was improved. The silicon electric eyes could hardly detect tiny differences because of their large ray-receiving area and low sensitivity. So the area should be subdivided into smaller ones to increase the sensitivity. Based on the law of transition to micro-level, a new photosensitive component called CCD (Charge Coupled Device) was introduced to its' third generation. The unicolor line-scan Camera was first put into use, it worked as many mini eyes worked together. The information collected by CCD system was one hundred times more than before. However, a higher information processing speed was required because of more information to be processed. According to the principle of transition into a new dimension, there were two evolution directions in which the fourth generation, one was multicolor line-scan Camera; the other was unicolor area-scan Camera. Unicolor CCD could only perform gray processing while missing large amount of valuable information In addition; the line-scan camera had a poor function of processing multi-dimension signals. Multicolor line-scan camera based on the RGB-mixture principles could collect all the valuable information contained in the visible light. That broadened the processed spectrum. The application of multicolor line-scan camera and the unicolor area-scan camera not only improved the color-sorting precision, but also broadened the particle sorts.

So, the optical signal collecting system is a processor aiming at getting as more valuable information as possible given that the information processing speed is high enough. According to the laws and principles of TRIZ, we predict that multicolor CCD would be used in the fifth generation of the color sorter.

2.2 Technology Evolution Forecast of the Grooves

Corresponding to the optical signal collecting system, grooves were used in color sorter as the main part of particles transmitting system. But it made the area between every two grooves bordered on unoccupied, then the work area utilization ratio was reduced. After the application of CCD technique, the planar signals could be processed; grooves could be replaced by plate. Here we introduced the law of transition to micro-level to separate the relative mass material flow into smaller ones. The work area utilization ratio was enhanced greatly, and the work efficiency of a single Color Sorter

Figure 3. Evolution of the grooves

would also be improved. When using plate, signal interference caused by particle overlap was prevented. Figure 3 shows the improved design of grooves.

2.3 Improvement of the background board adjust method using contradictions

For the most popular particles were milk-white rice, the background board was generally milk-white, and it was able to reflect the beam from lamps to the optical signal collecting system, at the same time, the standard background lightness could be controlled by changing the rotating the reflection angle of the board. As we know, when specific particles were passing, there must be optimal standard background lightness. With this lightness, the color-sorting precision was the highest theoretically.

The evolution of adjusting method has been evolving through three phases, manual adjust, ordinary motors adjust with feedback, and servomotors adjust, see figure 4.The servomotor has been widely used in the background lightness adjustment till now.

- Problem description: aiming at improving the applicability but the driver complication must be enhanced.
- Describing the problem with Contradiction matrix is equivalent to:

How to improve the applicability without enhancing the driver complication?

- The Contradiction Matrix suggests the following Inventive Principles:

15—Dynamics (A. Change the object (or outside environment) for optimal performance at every stage of operation. B. Divide an object into parts capable of movement relative to each other. C. Change an object from immobile to mobile or adaptive)

29—Pneumatics and hydraulics (A. Use gas and liquid parts of an object instead of solid parts, e.g. inflatable, filled with liquids, air cushion, hydrostatic, hydro-reactive)

37—Thermal expansion (A. Use thermal expansion, or contraction, or materials. B. Use multiple materials with different coefficients of thermal expansion)

28—Mechanics substitution (A. Replace a mechanical means with a sensory (optical, acoustic, taste or smell) means. B. Use electric, magnetic and electromagnetic fields to interact with the object, etc.)

Principle 15 suggested dividing the system into movable parts, the rotating angle fixed to the optimal and obtaining standard lightness by changing the movable background board.

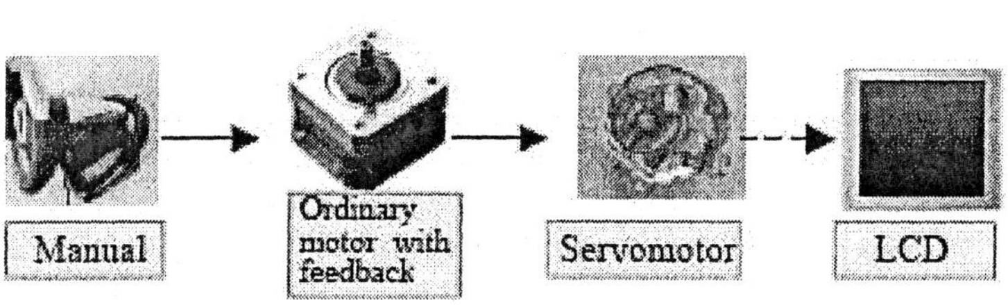

Figure 4. Improvement of the background board adjust method

Principle 28 recommended a starting point to solving the problem, we can replace mechanics with electric. We replaced mechanics with LCD lightness adjust system.

Though there were many advantages in servomotor adjust, the ultimate adjusting action was finished by mechanical movement, which was constrained inevitably by machining, material, and the installation precision. With LCD, we could get variable lightness by adjusting input voltage. It would be a breakthrough for the development of the background lightness adjust.

3. CONCLUSION

Through studying Color Sorter, using the patterns and principles of technology evolution, the future technologies of the optical signals collecting system and the grooves were given. The new states were the sticking points of new products. Then using the contradiction and inventive principle, this paper proposed several solutions for improving the color-sorting precision.

TRIZ has proved to be a very strong tool in helping to solve difficult technical problems that required inventive thinking. A competitive company pays much attention to the development of new products. Currently, there has been no structured methodology to perform this crucial activity. TRIZ provides a powerful structured methodology to achieve improved products.

4. REFERENCES

1.M.C.Pasikatan, F.E.Dowell, High-Speed NIR Segregation of High-and Low-Protein Single Wheat Seeds, *Cereal Chem.*Vol.81, No.1, 2004, pp.145-150

2.Tom Pearson, Machine Visio System for Automated Detection of Stained Pistachio Nuts, *Lebensm. -Wiss.u. -Technolo.*, 29, 1996,pp.203-209

3.Melchor C.Pasikatan, Floyd E.Dowell, Evaluation of a High-Speed Color Sorter for Segregation of Red and White Wheat; written for presentation at the 2002 ASAE Annual International Meeting/CIGR XVth World Congress, Chicago Illinois, USA, July28-July31, 2002

4.Tan Runhua, *The Theory of Inventive Problems Solving,* (China Machine PRESS, 2002)

5.Niu Zhanwen, Xu Yanshen, Lin Yue, The Scientific Methodology-TRIZ. *China Mechanical Engineering,* vol.10 (1), 1999. pp: 84-89

6.Altshuller G. *The Innovation Algorithm, TRIZ, Systematic Innovation and Technical Creativity.* Technical Innovation Center, INC., Worcester, 1999

7. Petrov V.M. & Zlotina E.S., *Laws of System Evolution,* (Leningrad). 1990

8. Zhang Qinghua, *Researching on the Theory of Technology Evolution based on TRIZ and the Engineering Application,* MS thesis, Hebei university of technology, 2003

9. Domb Ellen, The 39 Features of Altshuller's Contradiction Matrix(1998); Http: // www.triz-journal.com

10.Tate Karen, Domb Ellen, 40 Principles with Examples(1997); Http: //www.triz-journal.com

COMPUTER AIDED INNOVATON OF CRANKSHAFTS USING GENETIC ALGORITHMS

Humberto Aguayo Téllez[1], Noel León Rovira[2]
[1]*Center for Innovation in Design & Technology (CIDT), ITESM; Email: haguayo@itesm.mx.*
[2]*Professor CIDT, Director Research Program Creativity and Innovation in Engineering, ITESM, Campus Monterrey, Chairperson WG5.4 Computer Aided Innovation IFIP T;. Email: noel.leon@itesm.mx. Phone: +52 81 81582012.*

Abstract: In this paper a method is described for determining the design unbalance of crankshafts and also the recommended procedure for a balanced design strategy. The use of a search tool for solutions is suggested based on Genetic Algorithms (GA). GAs have been used in different applications, one of them is the optimization of geometric shapes, a relatively recent area with high research potential. The interest towards this field is growing, and it is anticipated that in the future mechanical engineering will be an area where many applications of shape optimization will be widely applied.

Key words: Genetic Algorithms, Splines, Unbalance

1. INTRODUCTION

The objective of the balance analysis of engine crankshafts is to develop its geometry so that a specified previous unbalance is obtained on two correction planes. These two planes lie where the outermost counterbalances are located. This previous unbalance represent a value of mass by radial position which effect will be eliminated by means of drilled holes around such counterbalances (at the correction planes, for instance) to values near zero, taking in count the effect of pistons and rods weights (these last commonly replaced for equivalent "fictitious" weights). This intentional previous unbalance is designed just for the drilling to take out mass

Please use the following format when citing this chapter:
Téllez, Humberto Aguayo, Rovira, Noel León, 2006, in International Federation for Information Processing (IFIP), Volume 207, Knowledge Enterprise: Intelligent Strategies In Product Design, Manufacturing, and Management, eds. K. Wang, Kovacs G., Wozny M., Fang M., (Boston: Springer), pp. 471-476.

(unbalance mass) from the external profile surface of the counterweight, at a limited zone, restricted because of the engine geometry and manufacturing process. Figure 1 shows a schematic sketch of a crankshaft.

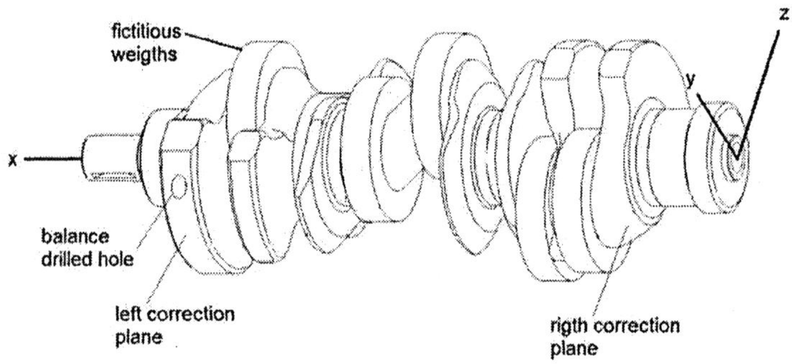

Figure 1. Combustion Engine Crankshaft

In the pivoted crankshaft turning with constant angular velocity the sum of moments produced by the differentials of mass on its location is equal to a bending moment acting on the crankshaft. The equations that govern the dynamic behavior of rotating crankshafts are:

$\Sigma M_x = 0$
$\Sigma M_y = I_{xz}\omega$^2
$\Sigma M_z = -I_{xy}\omega$^2
Restricted to
$\sum F=0;$

The sum of the inertia forces produced by each differential of the crankshaft mass and the mass from the pistons and rods is equal to the reactions on the supports at the rotation fixations [1]. Those equations allow finding the equivalent unbalance. The procedure for doing so is described next.

The sum of moments of the equation is equal to the resultant of the moment caused by the centrifugal force due to the crankshaft mass at the center of gravity, and the moment caused by the unbalance masses at the correction planes. It is a common practice to take this sum of moments about the left correction plane (into the counterweight close to the flange) in order to eliminate the moment caused by the left unbalance mass and finding the unbalance on the right plane:

$$m_R r_{Ry} = ((l/2)\ m_g\ r_{gy}\ + I_{xy})/l$$
$$m_R r_{Rz} = ((l/2)\ m_g\ r_{gz}\ + I_{xz})/l$$

where l is the distance between the two correction planes, m_g is the mass of the crankshaft, r_{gy} and r_{gz} are the coordinates y and z of the center of gravity. The same way moments are taken about the right correction plane to find the unbalance on the left plane:

$$m_L r_{Ly} = ((l/2)\ m_g\ r_{gy} - I_{xy})/l$$
$$m_L r_{Lz} = ((l/2)\ m_g\ r_{gz} - I_{xz})/l$$

The mass properties of interest (m_g, r_{gy}, r_{gz} and the inertia products I_{xy} and I_{xz}) of the crankshaft model can be obtained from the CAD software, orienting the geometry respect to a system of coordinates with its x axis towards the flange end upon the rotation axis and the z axis in the proper vertical orientation, using the proper info commands of the commercial CAD software. It is worth noticing that these equations do not depend on the angular velocity of the crankshaft.

If it is specified a target for unbalance ($m_L r_L$ y $m_R r_R$) the difference between the target and the current unbalance is a set of vectors that is going to be minimized by making modifications to the crankshaft geometry.

These vectors are dependent of the counterweight shape. The search for an optimal shape can be extensively time consuming but different strategies can be used to reduce the search time and effort for obtaining an optimal geometry which consider the constrains of the most important parameters involved in the crankshaft performance and manufacturing process such as weight and the filling of cavities during the forging process.

2. STRATEGY TO OPTIMIZE THE UNBALANCE

Next the general method is described for developing a balance optimization procedure using Genetic Algorithms (GA) [2]. GAs have been used in different applications [3], one of them is the optimization of geometric shapes, a relatively recent area with high research potential. The first step is to have a 3D CAD parametric model of the crankshaft. From this model the mass properties required to calculate the unbalance are obtained (using the equations shown earlier).

Now that the difference between the unbalance target and the CAD model unbalance is known. The next step is to modify the CAD geometry for minimizing this difference. This implies two restrictions:

- the weight of the crankshaft should not be increased

- the filling process of die cavities during forging should remain unaffected or even improved.

Because unbalance depends on location and amount of mass, it is not straightforward to modify the counterweights. On the other side, only certain locations of the counterweights are allowed to be modified to adjust balance.

It was decided to substitute with splines the profile of the counterweights close to the initial design. This process has been called "splinization". A procedure that was developed at the Center for Innovation on Design and Technology has been used for this substitution [4]. From this initial with splines adjusted profiles, variations are made by changing the positions of the control points. Figure 2 shows the "splinization" strategy to modify the profile of a counterweight for one of the possible profiles represented.

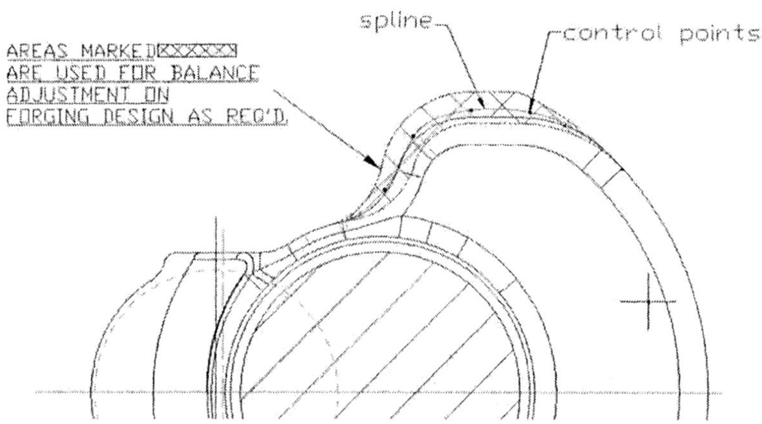

Figure2. Profile of counterweight represented by a spline

The two main reasons for using splines instead of the traditional arcs are:

1. arcs contain fewer degrees of freedom for manipulating geometry (just the center, radius and angle) and

2. the second derivative of a set of arcs contains sharp edges, something not so good for material fluency in the forging dies.

The splinization of the counterweights needs a codification of the control points to form the chromosomes of the Genetic Algorithm. This codification was achieved as follows: From each control point only the Y coordinate will be used as the X coordinate may remain fixed in proper locations. For each Y coordinate an upper and a lower value are defined, those limits are then codified as +1 and -1 and these define the phenotype. Table 1 shows these first three values with the upper and lower limits and its codification as +1, 0 and -1.

Table 1. First four Y coordinates of the CW1 spline

	CW1 TOP				CW1 TOP CODIFIED			
upper	51	66	73	78	1	1	1	1
	48.33	62.47	71.21	76.2	0	0	0	0
lower	46	60	68	74	-1	-1	-1	-1

Because splines with at least four free control points are used, i.e. on top and bottom side profiles of counterweights number 1,2, 8 & 9 the chromosome must have at least 4x2x4=32 "genes". The genetic algorithm is going to be run with an initial population representing different counterweight shapes for a crankshaft.

Each resulting crankshaft unbalance is evaluated according to the equations derived before and those that have the smallest difference vectors are combined to form a new population until the unbalance target within a certain tolerance is reached. Fig 3. Shows the proposed algorithm.

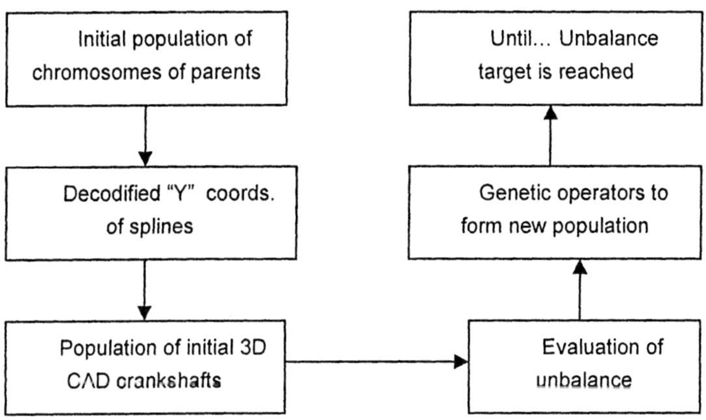

Figure 3. Proposed algorithms for balance optimization using Genetic operators

The use of commercial software I-SIGTH™ is required to integrate the genetic algorithms programmed in MATLAB™ with the modeling of CAD crankshafts in PRO/ENGINEER™

3. CONCLUSIONS

It is possible to optimize the balance on forging crankshafts in order to take the performance of the combustion engine far beyond the limits reached by conventional methods, using strategies of evolutionary computation. The use of Genetic Algoritms on crankshaft balance is based on the premise that the geometry of counterweights can be modified automatically by algorithms that optimize critical parameters of the engine performance

Further work is required to find the best approach on taking the new counterweights profiles under the filling of die cavities because it is perceived that these new profiles are better than conventional rounded profiles

As sketches are controlled through its parametric structure using control points through the splinization approach, a solution is presented for substituting the initial parametric structure by a new size controlling structure that allows parametric similar control of the size of parts and features being analyzed as the genetic algorithms provide.

On the other side, as parameters are used as dimensioning basis for drawings further analysis is required on how to achieve a dimensioning structure (design for manufacturing) of parts that have been "splinized".

4. REFERENCES

1. Rao. *Mechanical Vibrations*. Wesley
2. Goldberg DE. *Genetic algorithms in search, optimization and machine learning*. Reading, MA: Addison-Wesley; 1989. ISBN 0-201-15767-5.
3. J. Lampinen. Cam shape optimization by genetic algorithm. *Computer Aided Design* 35 (2003) 727–737.
4. Noel Leon, José María Cueva, Jorge Gutiérrez, David Silva. Automatic Shape Variations in 3D CAD Environments. 1ST IFIP-TC5 Working Conference on Computer Aided Innovation. November 14-15 2005, Ulm, Germany

DEVELOPMENT OF A CAI SYSTEM OF STANDARD SOLUTIONS BASED ON TRIZ

Bojun Yang, Runhua Tan , Yumei Tian
School of Mechanical Engineering, Hebei University of Technology, TianJin, 300130, China; Email: ybj@hebut.edu.cn

Abstract: Standard Solutions is a powerful tool based knowledge for quickly finding effective solutions to complex product design and production problems. As an independent part of TRIZ, it can be used to solve level three inventive problems in product design for innovation. Standard Solutions is usually used with Su-F Model at first step. The CAI system InventionTool was developed using an algorithm concerning Standards, different Standards were selected according various constraints. The development process of software supported by Object-Oriented method, VC++ language and XML.

Key words: Standard Solutions; Theory of Inventive Problem Solving; Object-Oriented Method; XML; Computer Aided Innovation

1. INTRODUCTION

TRIZ is a useful tool and an important the instructional theory for product innovation design. As an independent part of TRIZ, 76 Standard Solutions can be used to solve relatively common optimization problems, and more particular to solve level three inventive problems[1]. With the help of software designer can make more effective and more doable product design scheme, and that is the aim of Computer Aided Innovation. Many CAI software based TRIZ have been developed, such as TRISolver, IWB, Goldfire and so on. But to using Standard Solutions, no software has a maturity implement. In the CAI software Invention Tool, Standard Solutions

Supported by the Project of Natural Science Foundation of China under Grant No.50375045, the Project of Natural Science Foundation of Tianjin under Grant No.043802211

Please use the following format when citing this chapter:
Yang, Bojun, Tan, Runhua, Tian, Yumei, 2006, in International Federation for Information Processing (IFIP), Volume 207, Knowledge Enterprise: Intelligent Strategies In Product Design, Manufacturing, and Management, eds. K. Wang, Kovacs G., Wozny M., Fang M., (Boston: Springer), pp. 477-482.

had been developed a module of software through Object-Oriented method and XML software design method.

2. STANDARD SOLUTIONS

2.1 TRIZ

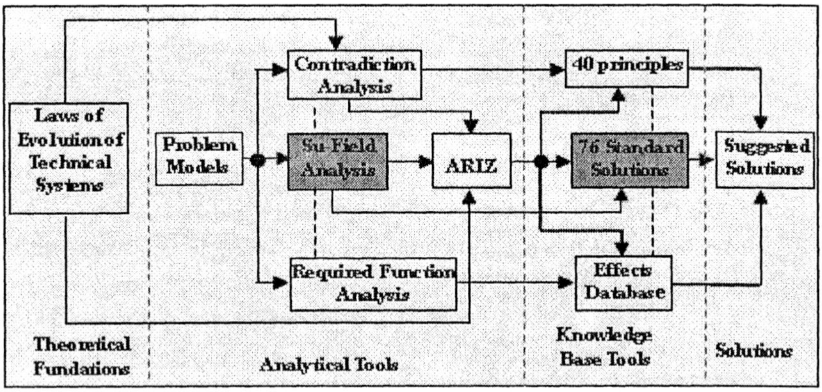

Figure 1. Structure of TRIZ Methodology

TRIZ[2], the methodology of inventive problem solving, is developed by G. S. Altshuler from 1946. TRIZ specialists have taken about 3500 man-years and analyzed approximately 2.5 million patents of all over the world for the development of the theory. After more than 50 years development, TRIZ become a powerful methodology for technical problem solving that leads to enhancement of existing technique and strong acceleration of progress. TRIZ has been defined a human-oriented knowledge-based systematic methodology of inventive problem solving. According to the process of inventive problem solving, TRIZ system constructed from the theoretical foundation, the analytical tools and knowledge base tools (Figure 1). The key step of design for innovation is to produce new concepts in conceptual design stage. The knowledge is needed to support designers at that stage. Standard Solutions is a kind of TRIZ knowledge [3](Table 1).

Table 1. The knowledge to support inventive problem solving

Knowledge	Search engine	Levels of solutions
40 principles	39 problem parameters and Matrix	2
76 Standard Solutions	Su-Field Model	2, 3, 4
Effects	30 TRIZ standard functions	2, 3, 4

Knowledge	Search engine	Levels of solutions
TE patterns and paths	TE patterns	2, 3, 4

2.2 Substance-Field Model

Substance-Field, or Su-F, is a graphical model of a minimal working technique in TRIZ[4]. Figure 2 is a simple description of Su-F Model. As a rule, a Su-F model comprises three components: two substances and a field usually. The identification of substances depends on the application, often S1 is a product or a raw object and S2 is a tool. Su-F Models and Su-F Analysis, created by G. S. Altshuller, provide a fast, simple description of subsystems and their interaction in a technical system via a well-formulated model of the technique in which all subsystems, inputs, and outputs are known or can be quite easily determined. Su-F Model is one of the most analysis tools in TRIZ, usually used in ARIZ with Standard Solutions just as Figure 1 shown. Figure 3 shows a complete description of solving a problem in Su-F model.

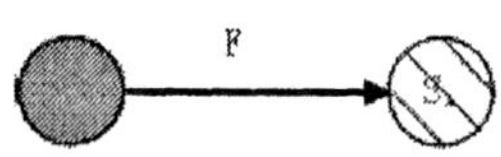

Figure 2. a simple Su-F Model

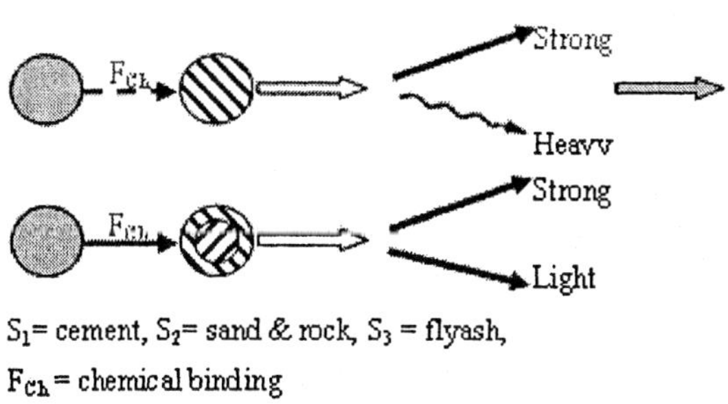

Figure 3. complete process of solving a problem in Su-F Model

2.3 76 Standard Solutions

Altshuller discovered various engineering problem from different fields can be solved by generic conceptual approaches. He formulated and systematized these approaches, known as TRIZ Standard Solutions or Standards. The Standards is a powerful tool based knowledge for quickly finding effective solutions to complex product design and production problems. They are grouped into 5 large categories as Table 2:

Table 2. The Standard Solutions categorization

1. Improving the system with no or little change	13 standard solutions
2. Improving the system by changing the system	23 standard solutions
3. System transitions	6 standard solutions
4. Detection and measurement	17 standard solutions
5. Strategies for simplification and improvement	17 standard solutions
Total:	76 standard solutions

The 76 Standard Solutions are used to solve relatively common optimization problems. Further more, they are useful for level three inventive problems. Level 3 inventions significantly improve existing systems, and represent 18% of the patents. Typically, the 76 Standard Solutions are used as a step in ARIZ, after the Su-F Model has been developed and any constraints on the solution have been identified. Standards and Su-F are linking tightly each other. Before use Standards, a Su-F Model of technical problem must be constructed mostly.

Standards are the precepts of synthesis and transformation with the aim to overcome or circumvent technical and physical contradictions.[4] When a problem is solved with the help of Standards, there is no need to formulate a contradiction. For easy to orient in Standards, some algorithms for choosing the right Standards had been designed. In an algorithm which is base for development of the CAI system concerning Standards, different Standards were selected according various constraints.

3. THE CAI BASED TRIZ

After more than 50 years development and many successful applications in superior companies in the world, TRIZ is regarded as a useful innovation methodology in practice. Computer Aided Innovation software based on innovational theory is a kind of CAD that is suitable for conceptual design stage[5]. TRIZ-based software as the main content of CAI developed rapidly in recent years.

TRISolver 2.1™ is the early English-language version of the CAI software from TRISolver Consulting Group in Germany. The Invention Machine in America had issued TechOptimizer, and than upgrade to Goldfire Innovator. Another America firm Ideation International developed the Innovation WorkBench (IWB) and so on. Those English-language software of CAI integrate the concepts, principles and tools of TRIZ. Those CAI software were applied in practice and made great success. But to use of Standards, some of them mention little or no, some is just a database or a guide to Standards knowledge. How to use the Standards in CAI software need develop.

In china, some kinds of English-language version software such as Goldfire Innovator were introduced. For the limitation of language and thinking custom, that kind of software is hard success completely. Many Chinese-language version CAI software were developed by university or institute organization. But most of them are experiment. InventionTool1.0 and 2.0 based on TRIZ had been completed by Hebei University of Technology and were in commercial. [6]

4. THE DEVELOPMENT OF STANDARDS SYSTEM

According to the algorithm for using 76 Standard Solutions, a Standard Solutions module of InventionTool CAI software is developed. In one CAI software it is realized not only to support technical knowledge but to solve problem using in 76 Standard Solutions systemically.

A software development process is fallen into 6 steps in term of the lifecycle of software: analysis by Fusion method, design, implementation, module test, system test and executing. The Standard Solution module follows the same process. Object-Oriented method[7] is the mainstream method to develop software. Fusion method integrated and expended the advanced characteristics of some other Object-Oriented methods. The module analysis obeys Fusion method, and implement with VC++ language and XML.

Several object models have been created during the software development process. Each object model presents one aspect of the software. The total of object models make up of the object model of Standards System. A life-cycle model of the Standards module includes operators occurring orders, operations is then established.

XML (eXtensible Markup Language) is used to develop the interface of the software. XML Standard by W3C (World Wide Web Consortium). For its extensible, structured and self-describe data, XML is more impactful then HTML. This software use a mode of XML+XSL→HTML. On its strong suit of data transfer and data store, XML also can be used to update the CAI software.

5. CONCLUSION

With minutely analyzing on TRIZ method, Su-F model and 76 Standard Solutions, according to the algorithm for using Standard Solutions, the

Standard Solutions module software is completed. The development process supported by Object-Oriented method, VC++ language and XML. Compared with others CAI software based TRIZ in the world, InventionTool and its Standard Solutions module have their own characters and value.

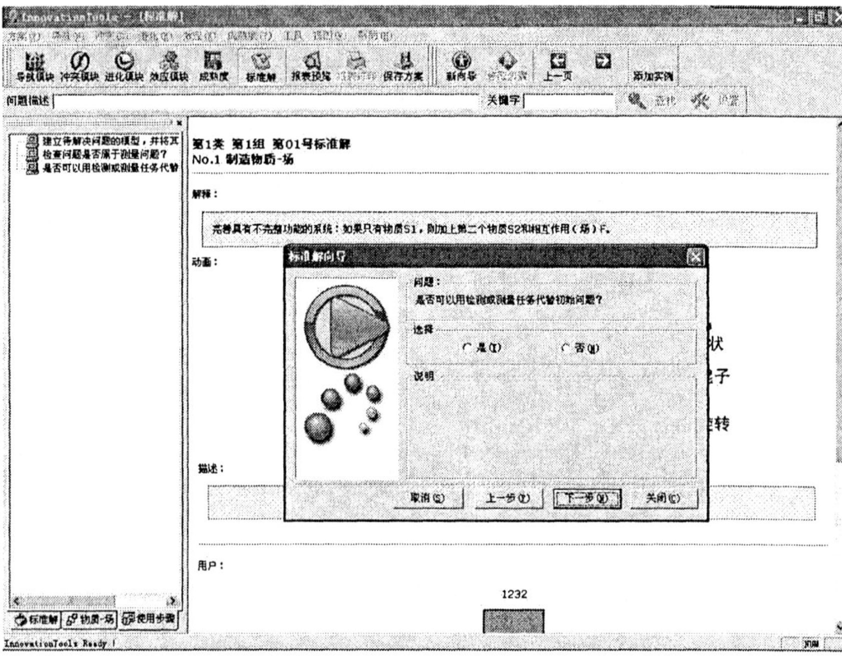

Figure 4. interface of Stanards Module

6. REFERENCES

1. John Terninko, Ellen Domb, Joe Miller, The Seventy-six Standard Solutions with Examples, *The TRIZ journal*, February, 2000, http://www.triz-journal.com.
2. Altshuller G., The Innovation Algorithm, TRIZ, systematic innovation and technical vreativity, *Technical Innovation Center*, INC, Worcester, 1999
3. Kalevi Rantanen, Levels of solutions, *The TRIZ Journal*, Dec. 1997, http://www.TRIZ-Journal.com
4. Savransky S D, Engineering of Creativity, New York: CRC Press, 2000
5. Tan Runhua, Progress of some problems in product design for innovation, *Chinese Journal of Mechanical Engineering*, Sep., 2003
6. Computer Aided Inventive Problem Solving System, InventionTool1.0, The number of computer software register: 2003SR2694
7. Hassan Gomaa, An Object-Oriented domain ananlysis and modeling method for software reuse, 0073-1129-1/92, 1992 IEEE.

STUDY ON OPTIMIZATION OF STIR HEAD FOR FSW BASED ON GENETIC ALGORITHM

Zhaoxin Meng[1], Hui Chen[1], Xihua Yue[2]
[1]School of Mechatronics Engineering, [2]College of Foreign Languages, Northeast Forestry University, China; Email:mzhxmail@163.com.

Abstract: The shape selection of stir head is processed to Friction Stir Welding (FSW), and the mathematical model and constraints for the optimization of stir head are established, what is more, the multi-target optimization algorithm of stir head dimension based on genetic algorithm is presented. Utilizing this algorithm, an example of optimization is analyzed, and its optimized finite element modeling is given, at last the reasonable dimension parameters are obtained.

Key words: Friction Stir Welding (FSW), Stir head, Optimization, Genetic algorithm

1. INTRODUCTION

Friction Stir Welding (FSW), which is a new solid phase welding process, has been largely used in aluminum alloys welding[1]. To date, FSW has attracted considerable interest because of its better weld ability of aluminum alloy weld joint, and considerable research on FSW has been carried out. However, research is lacking on the optimization procedure of FSW, it is well known that stir head is a critical component of the friction welding machine, and the quality of the weld joint is dependent on its property. So, this situation has led to the study on optimization procedure of stir head for FSW in order to be superior in property to any other stir head.

Please use the following format when citing this chapter:
Meng, Zhaoxin, Chen, Hui, Yue, Xihua, 2006, in International Federation for Information Processing (IFIP), Volume 207, Knowledge Enterprise: Intelligent Strategies In Product Design, Manufacturing, and Management, eds. K. Wang, Kovacs G., Wozny M., Fang M., (Boston: Springer), pp. 483-491.

2. SELECTION OF STIR HEAD SHAPE

2.1 The shape of shaft shoulder

The function of shaft shoulder is to provide friction contact points and sealed welding condition so that plastic metal cannot be extruded through it[1]. In the middle of the shaft shoulder end, the general shape is concave, especially the juncture of the stir needle and the shaft shoulder. Therefore, the general shape of shaft shoulder, the end of which is concave, has been selected[2,3].

2.2 The shape of stir needle

With a cone screw stir head, it has been proved that weld joint has higher tensile strength. And what is more, with a tapered thread stir head, tensile strength is more stable with the welding speed changing. Material in heat-affected zone of cone screw is harder, and the hardness of onward side of cone screw weld joint is clearly larger than usual.

In the process of welding, plastic material receives a kind of shear force from screw thread, and receives a kind of downward pressure from conical surface at the same time, since tapered thread has cone angle. The affection of these two forces can make the crystal grains in the heat-affected zone elongated, where slender grains that should be parallel with conical surface come into being. The temperature duration of receding side of weld joint is slightly longer, so the hardness of receding side is lower with the larger size of grains[4].

With a tapered thread stir head, weld joint has higher tensile strength. And what is more, with a tapered thread stir head, tensile strength is more stable with the welding speed changing, therefore, a tapered thread stir head has been selected in the research.

3. MATHEMATICAL MODEL FOR THE OPTIMIZATION OF STIR HEAD DIMENSION

3.1 Mathematical model for the optimization of stir head dimension

It is supposed that R_0 is the middle diameter of stir needle, L is the length, R_2 is the lower extreme diameter of it, and R_1 is the upper extreme diameter of it, and $R_2 \leq R_1$. In the process of welding, we suppose that the welding speed is v, the cone angle of stir needle is α, and rotating speed is ω.

In the process of welding, the material ahead of stir needle has been under high plastic state, which brings to bear pressure intensity to stir needle, P is referred to the pressure intensity.

As the stir needle moves forward, welding materials engender plastic deformation in response to mechanical forces, the volume of which is ΔS_u when stir needle has moved ΔS after Δt, and the volume of the plastic metal that stir needle extrude is ΔJ, from the given information, we find that the thread any infinitesimal rotates is $\Delta t \cdot \omega \cdot r$, using this equation, the generated heat quantity is calculated as:

$$\Delta Q = F \cdot \Delta t \cdot \omega \cdot r = \frac{P_s \cdot f \cdot r \cdot \Delta r \cdot \Delta \theta}{\sin \alpha} \cdot \Delta t \cdot \omega \cdot r$$

After Δt, we find that the friction heat generated from the front of the stir needle is:

$$\Delta Q_Z = \int_{R2}^{R_1} \int_0^{\pi} \frac{P_s \cdot f \cdot r^2 \cdot \omega}{\sin \alpha} dr \cdot d\theta = \frac{\pi \cdot P_s \cdot f \cdot \omega \cdot \Delta t}{3 \sin \alpha} (R_1^3 - R_2^3)$$

and the heat comes from shaft shoulder is calculated as:

$$\Delta Q_J = \frac{4\pi\omega \cdot f \cdot F \cdot (R^3 - R_0^3)}{3 \cdot R^3} \Delta t$$

where F is the axial pressure from shaft shoulder and R is the diameter of shaft shoulder.

Thus, solving for the heat output coefficient $\Delta Q / \Delta t$ and substituting the values gives:

$$\frac{\Delta Q}{\Delta t} = \frac{\Delta Q_Z + \Delta Q_J}{\Delta t} = \pi P_s \cdot f \cdot \omega \cdot \left(\frac{L^3 \cdot tg\alpha}{12\cos\alpha} + \frac{R_0^2 \cdot L}{\cos\alpha}\right) + \frac{4\pi\omega \cdot f \cdot F \cdot (R^3 - R_0^3)}{3 \cdot R^3}$$

The parameters of the heat input coefficient with relation to the stir head are L, R, R_0 and α.

Since L is always thinner than welding material thickness, it is not likely to change. It is appreciable that R and R_0 should be optimized, for they are the decisive parameters of stir head and the heat output coefficient of stir needle. From the given numerical model, we find that the value of the cone angle of stir needle α can influent the heat input of stir needle, in practice, too large α will not be considered because of its decreasing the capacity of undergoing impact forces, and too small α will also not be considered because the proximate cylindrical shape of stir head is not ideal. According to the above analysis, it is necessary to optimize the diameter of shaft shoulder and the diameter of stir needle in this study.

In the process of FSW, the strength of weld joint is connect with the shape of the stir head, the shape parameters of this frustum stir head are considered to be the diameter of shaft shoulder and the diameter of stir needle. It is important recognize, however, the relation between them is not functional, but statistical. In this study, the optimization of stir head for FSW has considered the diameter of shaft shoulder and the diameter of stir needle as independent variable, which are selected within numeral value. At this point, in establishing mathematical model for the optimization of stir head, the diameter of shaft shoulder R and the diameter of stir needle R_0 should be restricted: $R = 17 \sim 27$mm, $R_0 = 3 \sim 5$mm.

3.2 Mathematical model for the heat input coefficient of stir head

We have known the heat output coefficient $\Delta Q / \Delta t$ is calculated as:

$$\frac{\Delta Q}{\Delta t} = \frac{\Delta Q_Z + \Delta Q_J}{\Delta t} = \pi P_s \cdot f \cdot \omega \cdot \left(\frac{L^3 \cdot tg\alpha}{12\cos\alpha} + \frac{R_0^2 \cdot L}{\cos\alpha}\right) + \frac{4\pi\omega \cdot f \cdot F \cdot (R^3 - R_0^3)}{3 \cdot R^3}$$

The parameters of the heat input coefficient with relation to the stir head are L , R , R_0 and α .

4. DESIGN OF THE ALGORITHM FOR THE OPTIMIZATION OF STIR HEAD DIMENSION

4.1 Multi-target optimization based on genetic algorithm

The most typic method to solve problems of multi-target optimization is to integrate several objective functions in the way of weighted average so that one objective function, which may be solved easily, can be achieved. However, the achieved objective function has no definite physical meaning, though separate objective functions have different physical meaning. Moreover, since order of magnitude of different objective functions should be different, the weighted value cannot be used to substitute the proportion of the integrated objective function. And it is a fact that the weighted value, which is selected in term of the designer's experience, will tamper with the objectivity.

Considering the given reason, a new multi-target optimization with constraint based on genetic algorithm is utilized for the optimization of stir head dimension.

When genetic algorithm is selected to optimize multi-target function, this study will not optimize the integrated objective function only, but all objective functions at the same time. A optimum solution to the multi-target optimization is on the lower left boundary of the doable range of values, therefore, if the lower left boundary of the doable range of values can be found, we can select the optimum solution directly according to the actual need.

4.2 The operation process of algorithm

There are several processes in the operation of algorithm. First of all, a certain number initial chromosomes are created at random, which make up of a population; then, a degree of adaptability function will confirm the degree of adaptability of each individual, which is to circumstance. The next thing to do is to select the chromosomes to obtain a new population, and this new population will perform chiasm operation and variation operation. The repeating selection operation, chiasm operation, and variation operation will be performed until the given iterations are fulfilled. Fig.1 shows the operation process of algorithm, according to the principium and process of genetic algorithm.

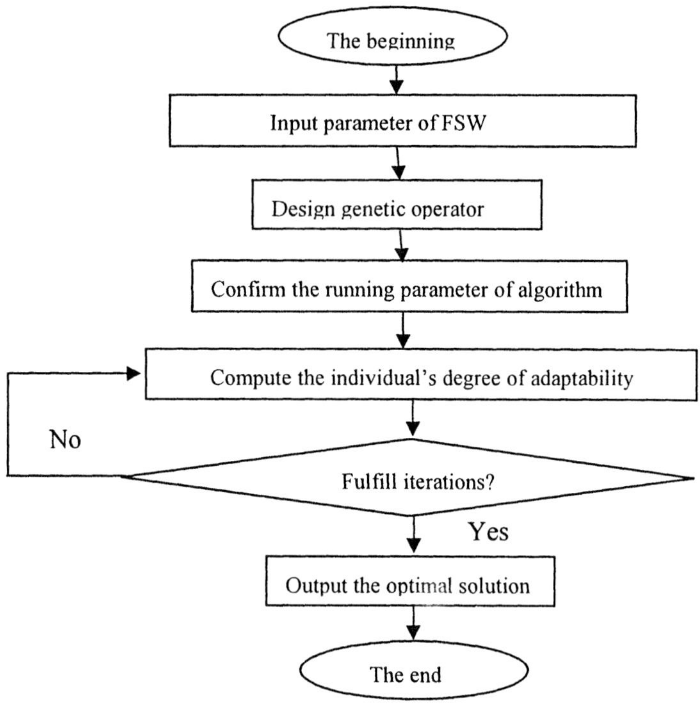

Figure 1. The operation process of algorithm for the optimization of stir head dimension

4.3 The retention strategy of algorithm and the analysis of astringency

The fundamental genetic algorithm can be described as an odd capital Markov chain, the reason is that the selected operation, chiasm operation, and variation operation should be performed autonomously and stochastically, new populations are connected with their father's population, but are not the former populations, that is, populations have no after working. And the transition probability between every two populations has nothing with the initial point of time. Therefore, the probability that the fundamental genetic algorithm can converge to the optimal solution is less than 1.

However, in theory, the genetic algorithm in this study must be ameliorated so that the genetic algorithm can converge to the optimal solution. So, optimal retention strategy is used in this study, which is helpful to converging the optimal solution, and its probability will be 1.

5. CASE ANALYSIS

5.1 Computation of case

EXAMPLE: Aluminum-lithium alloy is new material used in the modern aerospace field because of its excellent properties; however, it is always a difficulty to connect with each other[5]. This FSW joining method can largely promote the joining property of Aluminum-Lithium alloy. Which kinds of shapes of stir head will be selected has been the key point of Friction Stir Welding, something in common is the concave shape in the middle of the shaft shoulder end, especially the juncture of the stir needle and the shaft shoulder, and the stir needle is frustum. So, a stir head for opposite joint of Aluminum-Lithium alloys with straight stitch, of which the serviceable thickness is 5mm, has been optimized as follows[6].

This study threats the given mathematical model as the objective function, the value of the length of coding string: l=50, the value of the size of population: M=80, the value of the chiasm probability: p=0.5, and the terminate algebra: T=50.

After optimizing the stir head, the study has calculated: R=13.4mm, R_0=4.2mm. This optimized finite element modeling is given.Fig.2 shows the optimized finite element modeling of stir head.

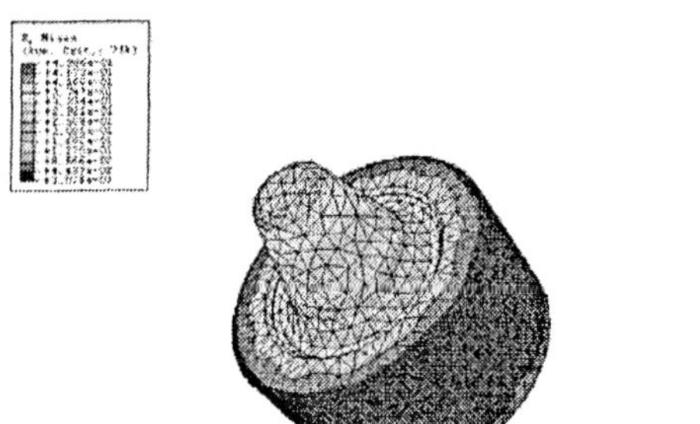

Figure 2. The optimized finite element modeling of stir head

5.2 The mechanical property analysis of the weldment with optimal stir head

The test result of weld strength has showed that the weld joint strength with cone screw, of which R is 13.4mm, R_0 is 4.2mm, can reach to 370 MPa, the highest value of the strength, this is consonant with the result of the optimized mathematical model of stir head size. Fig.3 has clearly showed that weld strength will go up with the growth of weld speed. It is delightful that this stir head has the highest strength in the experiment; moreover, it is most likely that the highest strength can be increased. Therefore, if the diameter of shaft shoulder R is 13.4mm and the diameter of stir needle R_0 is 4.2mm, the cone screw stir head can be considered as the optimal stir head.

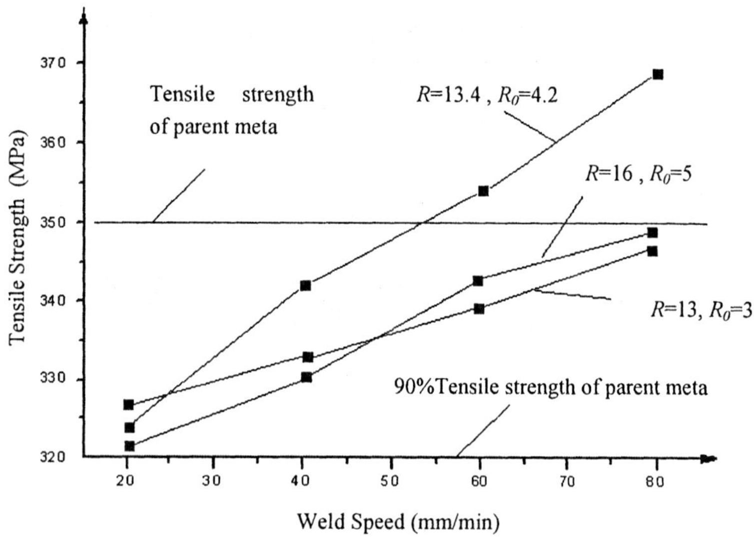

Figure 3. Effect of weld speed on the strength of the optimal stir head

6. CONCLUSION

- The cone screw stir head has been selected according to the analysis to its advantages.
- The mathematical model for the optimization of stir head is established, which has been treated as the optimized objective function.
- The optimization algorithm and its running parameters of the stir head dimension based on genetic algorithm are presented by analyzing the convergent of the algorithm.
- It has been proved that the stir head in this study is the optimal stir head after optimization. The character of genetic algorithm such as easy

fulfilling, unrelated problems and overall convergent is making its practicability and application potential extensive.

- In the process of running, considering different processing conditions, different running parameters of algorithm can be selected, and more reasonable stir head can be obtained by several pilot calculations to the running parameters.
- The optimization of stir head dimension can be regarded as a problem of multi-target optimization, of which the way of weighted average has intrinsic limitation. A new multi-target optimization with constraint based on genetic algorithm can be utilized for the optimization of multi-target function, with the application expedient and rapid, and the optimized result can also reflect the internal relation between the objective functions.

7. REFERENCES

1. Lohwasser D. Application of friction stir welding for aircraft industry. *The 2nd International Symposium on Friction Stir Welding.Sweden:* Gothenburg, 2000
2. Thomas MW, Nicholas ED, Needham JC, etal. Friction stir welding. *GB Patent Application,* 9125978. 1991-12-08
3. Lohwasser D. Welding of airframe by friction stir welding. *The 3rd International Symposium on Friction Stir Welding.* Japan: Kobe, 2001
4. Talwar R, Bolser D, Lederich R, etal. Friction stir welding of air frame structures .*The 2nd International Symposium on Friction Stir Welding.* Sweden: Gothenburg, 2000
5. Marie F. Development of friction stir welding of 7000 series extrusions for central wing box application. *The 3rd International Sym 2posiumon Friction Stir Welding. Japan:* Kobe, 2001
6. Pedwell R, Davis H, Jefferson A. The application of friction stir welding to aircraft wing.*The1st International Symposium on Fric2tion Stir Welding.* USA: Thousand Oaks, California, 1999

RESEARCH ON CNC TECHNOLOGY BASED ON CAD&CAPP

Tao Yu[1], Tan Liu[1], Shuzhen Yang[2], Wenbin Wang[1]
[1]*CIMS & Robot Center of Shanghai University, Shanghai University, China; Email:* *alanltnew@citiz.net.*[2] *Electromechanical Engineering Institute, Shanghai Second Polytechnic University, Shanghai, China.*

Abstract: This paper is based on a new philosophy of CNC (Computer Numerical Control), which is named as INC (Integrated Numerical Control or Intelligent Numerical Control). First it has a brief introduction of INC and describes important characteristics of INC. Second it introduces and analyses the one of INC's three key technologies: CNC technology based on CAD&CAPP, which has mended general CNC system to fit CAD&CAPP. It mainly includes interface technology with a STEP (Standard for the Exchange of Product Model Data) extension, embedded system and microcontroller technology, real-time Linux technology and so on. It discusses the principle and application of them also.

Key words: INC, CNC, Integrated, Intelligent.

1. INTRODUCTION

INC [1], proposed by Manufacturing Grid Resource Group of CIMS & Robot Centre of Shanghai University, in a narrow sense is the abbreviation of Integrated Numerical Control. It means that abstracting some functions of traditional CIMS (Computer Integrated Manufacturing System), such as CAD, CAPP, CAM, into a series of modules, which makes the application be divided into a group of tasks frame. The tasks frames could be decomposed into detailed sub-functions. Every function module is a practical module oriented to manufacturing, such as code explaining, interpolation and PLC control. The function modules are organized and integrated into the bigger parts modules with the tasks frame. And these

This research work is financially supported by the Science and Technology Development Foundation of Shanghai Science and Technology Committee (036505008 and 05DZ11001).

Please use the following format when citing this chapter:
Yu, Tao, Liu, Tan, Yang, Shuzhen, Wang, Wenbin, 2006, in International Federation for Information Processing (IFIP), Volume 207, Knowledge Enterprise: Intelligent Strategies In Product Design, Manufacturing, and Management, eds. K. Wang, Kovacs G., Wozny M., Fang M., (Boston: Springer), pp. 492-497.

parts modules are integrated into a concrete application CNC system finally with a unified user interface. The flow chart of INC is as shown in figure 1.

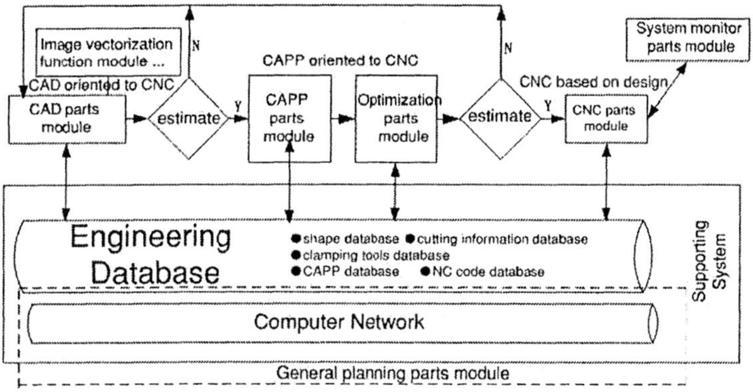

Figure 1. Flow chart of INC.

Different with some general CAD/CAPP/CAM integrated system, the focus of INC is the CNC module. Every foregoing module is the preparation of the CNC module. The other modules work for the kernel module - CNC module. For example, CAD parts module, which is oriented to CNC module, is not same as the general CAD system. It excludes some advanced and complex functions of computer aided design, such as three-dimensional design and three-dimensional modeling, but includes some useful functions oriented to CNC, such as distilling of image outline, image binarization, image vectorization and so on.

In a broad sense, INC is the abbreviation of Intelligent Numerical Control. It not only has the functions of INC in a narrow sense, but also has the function of collecting system information, monitoring system conditions to do self-adapting adjustment, recording historical information to analyse and synthetize, making decision, enlarging it own expert database and etc. It has extra expert system and information collecting system than INC in a narrow sense. It also has the function of remote control.

2. KEY POINTS OF INC

From the concept of INC, we can see the differences between INC and DNC. DNC [2] integrates communication, control, plan and management together, maybe includes CAD, CAPP and CAM in some degree. But it aims at controlling a group of NC machines effective, or further controlling a whole factory. It can be seen as a kind of Distributed Manufacturing. Instead, the heart of INC is NC. It aims at manufacturing. All its work is for its last task: NC manufacturing. Its CAD parts module and its CAPP module, which are oriented to CNC module, are not same as the general CAD system. It has a good CSCW (Computer Supported Collaborative Work) performance and can be seen as a good application case of parallel engineering. Contrasted to Distributed Manufacturing, it is a kind of Collaborative Manufacturing. And it is a more cost-effective approach to realize CIMS than DNC. Theoretically, it can realize CAD/CAPP/CAM in a CNC machine. INC also absorbed the open attribute of ONC. For INC has adopted embedded system as its hardware platform, it is integrated and is open as ONC [3] in the same time.

Because INC is an integrated and intelligent CNC, its key technologies not only include the technique of CNC, but also include extra integration and intelligentization technology. And we conclude there are three kernel technologies should be considered: CAD technology oriented CNC (CADoCNC), CAPP technology oriented to CNC (CAPPoCNC) and CNC technology based on CAD&CAPP (CNCbCAD&CAPP).

In the next pages, we emphasize on discussion and study of CNCbCAD&CAPP (CNC technology based on CAD&CAPP) used in INC.

3. CNC TECHNOLOGY BASED ON CAD&CAPP

In the traditional CAD/CAPP/CAM manufacturing process, CNC can only get data from CAD&CAPP, but it can't feedback data to CAD&CAPP. There is no two-way communication between CNC and CAD&CAPP. For general CNC controller is made of SCM (Single Chip Micyoco), it hasn't a good extensible and configurable characteristic. Now we amend INC with newly STEP technology and embedded system technology, which is discussed as follows.

3.1 Interface technology with a STEP extension

In recent years, Occident has proposed and advocated a new CAM/CNC interface specification, which is named as STEP-NC [4]. STEP-NC data model is formed on the base of STEP (ISO6893) about three-dimensional geometric model. Manufacturing technology and information is added to it to substitute the traditional G-code (ISO 6893).

At present, the international standard ISO14649 comprises general principles (ISO 14649-1), universal data model (ISO 14649-10), CNC milling model (ISO 14649-11) and milling tools (ISO 14649-111), etc. The standard adopts EXPRESS (ISO 10303-11) language to define the data interface of milling and cutting operations of CNC in detail.

We referred its milling and turning machine part. We can express manufacturing characters such as a cutting path with a STEP extension as follows. The operating object is got by milling/cutting tool, which goes from start point to end point. And the tool may work in one depth. So the motion can be expressed in depth and path. The path is made up of some line segments, which are closed or not closed. The process parameter includes rotating speed basic shaft, cutting speed, and cutting path, direction (clockwise or anti-clockwise), etc. The full operation includes some special parameters, such as parts motion and rotation, arc forward and back cutting, cutter compensation, retract strategy and so on. In the lower definition, we adopt the profile style of ISO14649-10 in path description, and we adopt the approach retract strategy of ISO14649-11 in the retracting step. The following is the motion wrote in EXPRESS language [5].

```
ENTITY motion_feature_M
dep th: length_measure;
path_curves: p rofile;
start_point: cartesian_point;
end_point: cartesian_point;
path_direction: OPTIONAL rot_direction;
its_cutting_speed: rot_speed_measure;
rotation_direction: rot_direction;
its_feedrate: length_measure;
approach_and_lift: app roach_retract_strategy; retract_height: length_measure;
END_ENTITY
```

3.2 Embedded system technology

SCM is widely used in traditional CNC. But with the development of science and technology of computer, SCM has been found that it can not meet the requirement of more and more integrated and multifunctional and modularized CNC system. And embedded controller has been the trend of CNC development in the future.

Here we talk about a distinguished one of embedded systems: embedded Linux. Embedded Linux is a small type of operating system designed towards the requirement of embedded system. It comprises a kernel and some system modules customized. A small type of embedded Linux includes threes basic elements of bootloader, microkernel, and initializing process. CPU which supported embedded Linux comprises x86, Alpha, Sparc, Mips, PPC, Arm and so on. And the memory device of embedded Linux is not traditional soft disk or hard disk, but Rom, compact flash, disk on chip of M-system or Memory stick of Sony. And the inner memory can use general memory or special RAM.

Compared to other embedded systems, Linux is not only free, but also is an open-source project. As a customizable platform Linux is supposed to be a popular system of the embedded product by many companies. Considering embedded Linux is open-source, good network-supporting, steady and safe and some other merits, we prefer to adopt embedded Linux as the embedded system solution in INC.

We have divided tasks to two categories: real-time task and nonreal-time task. Nonreal-time task interact with kernel through OS API (Application Programming Interface). And we use an embedded GUI between kernel and nonreal-time application to facilitate users to operate. There are three popular embedded GUI solutions in the present: Qt-embedded, Microwindows and MiniGUI. We choose MiniGUI for its small kernel and good Chinese support attributes. And for real-time tasks, they have different interplay style with kernel. We talk about it in next paragraph in detail.

3.3 Real-time Linux technology

The rapid development of Linux is for its applications have good compatibility of POSIX, but the architecture of Linux is a typical time-share design.

Although Linux is a single-kernel time-share OS, it has absolute advantage than other real-time embedded system for it has a better development environment and running environment. And some research organizations has developed different Linux real-time edition to make Linux used conveniently in real-time environment.

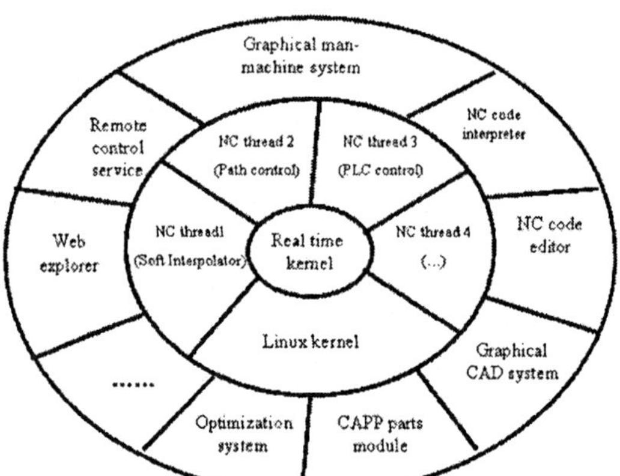

Figure 2. Software structure of CNC module

At the present, the real-time Linux solutions can be divided two categories. The first is to modify the original kernel adding or correcting some codes of original edition to make the kernel has a foreseeing capability. The second is to use a hybrid structure. It needs to write a new extra real-time OS kernel to handle real-time tasks. It is generally called dual-kernel architecture [6].

We can choose the first manner to modify the kernel. But if response time need to be less-than 1ms, hard real-time is recommended. We need to use the dual-kernel architecture in the condition. When using the real-time kernel, the CNC function modules could be divided into two layers as figure 2.

The NC threads functions module such as PLC control and Interpolation have a high priority level, and the others have a lower priority level.

And CADbCAD&CAPP includes some other technologies also, such as programming of device drive, kinetic control technology and so on. But they are not as distinguished as the above three, so we don't introduce them here.

4. CONCLUSIONS AND PERSPECTIVES

This paper proposes a new type of CNC - INC. It is a new conception in CNC field which gives an integrated theory oriented to manufacturing. And paper emphasizes on discussion and analysis of one of its key technology: CNC technology based on CAD&CAPP (CNCbCAD&CAPP).

INC is a system replacing Distributed Manufacturing with Collaborative Manufacturing that it is a novel philosophy using a cost-effective means to directly facilitate the manufacturing to enhance productivity. It is an integrated manufacturing or intelligent manufacturing.

As far as CNCbCAD&CAPP is concerned, it solves the problem how to setup a two-way communication between CNC and CAD&CAPP. And it gives a solution using embedded system and real-time Linux kernel to substitute traditional PC or SCM to make INC have a unified external standard controller interface and have extensible and configurable modules.

5. REFERENCES

1. Tan Liu, Tao Yu, Shuzhen Yang and Wenbin Wang, "INC: A New Philosophy of CNC", *Proceedings of 2005 International Conference on Machine Learning and Cybernetics,* Guangzhou, China, August. 17-21, 2005, pp 1338-1344
2. P. Stout, and R. Leonard,"The introduction of DNC technology as a partial approach to achieving the objectives of CIM", *Computer-Aided Engineering Journal*, China, 1999.2, pp. 17-20.
3. Yulong Wang, "Open architecture for CNC controller and their developing prospect", *CNC & Software*, China, 2004.4, pp. 75-76
4. Official STEP-NC website: www.step-nc.org
5. Weiyue Xiao, *Intelligent Information Porcessing Technology in CAPP*, National Science Publisher, Changsha, China, 2002.
6. Kanfu Wang, *Technological Base of Numerical Control*, Machinery Industry Press, Beijing, China, 2001.4

DEVELOPMENT OF ENTERPRISE PORTAL FOR INTELLIGENT MOLD SHOP

B.K. Choi, H.Y. Lee, D.W. Lee, and T.D. Kim
VMS Lab, Department of Industrial Engineering, Korea Advanced Institute of Science and Technology (KAIST), 373-1, Guseong-dong, Yuseong-gu, Daejeon, Republic of Korea;
Email: bkchoi@vmslab.kaist.ac.kr

Abstract: Presented in the paper is an enterprise portal for *intelligent mold shop* (IMS) called *IMS-portal* to support the *IMS system* proposed earlier by the authors that consists of six software modules enabling to transform a skill-based mold-shop into an intelligence-based one. The proposed IMS-portal is an *enterprise BPM portal* in which the BPM (business process management) functionalities are embedded in the existing enterprise portal framework.

Key words: Intelligent mold shop; Enterprise portal; BPM (business process management) system; Enterprise BPM portal; IMS-portal.

1. INTRODUCTION

Mold shops are faced with a number of difficulties including severe global competition, increasing labor cost, increasing demand for lead-time reduction, and shortage of skilled manpower. Recently, a framework of IMS (intelligent mold shop) software system was proposed to help cope with these difficulties[1]. The proposed IMS system, which aims to substitute operators' skills and cognitive abilities by software modules and hardware components as much as possible so that the mold-machining shop can be run by fewer and less skilled operators, consists of six functional modules as depicted in Figure 1: *NC-data filtering* module for high-performance and error-free machining; *Tooling plan* module for robust and optimized tool-selection; *Setup plan* module for standardized and robust setup; *Virtual inspection* module for machining quality assurance; *Real-time monitoring* module for unattended NC machining; *Loading simulation* module for optimized job-progress scheduling.

Please use the following format when citing this chapter:
Choi, B.K., Lee, D.W., Kim, T.D., 2006, in International Federation for Information Processing (IFIP), Volume 207, Knowledge Enterprise: Intelligent Strategies In Product Design, Manufacturing, and Management, eds. K. Wang, Kovacs G., Wozny M., Fang M., (Boston: Springer), pp. 498-504.

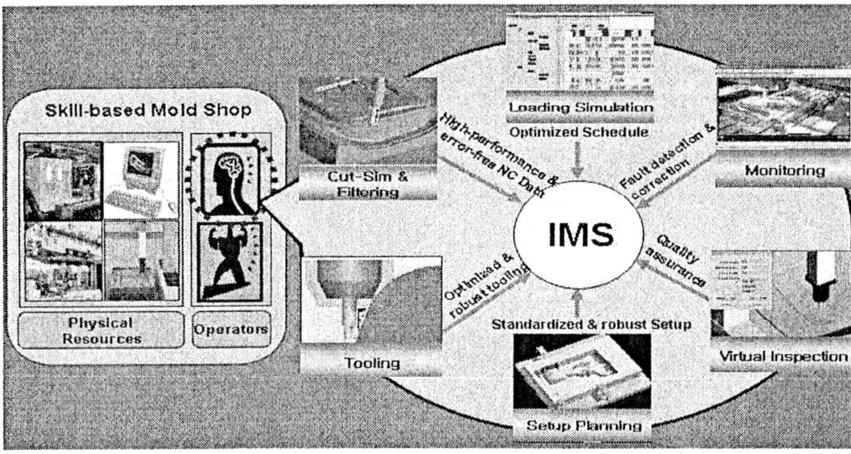

Figure 1. IMS (Intelligent Mold Shop) Framework[1]

A difficulty with the introduction of the IMS system is that it adds complexity in managing the use of diverse types of software systems and other resources. It is also required that the activities related to the IMS software system need to be integrated with other business processes and decision-making activities. Thus, we need a *higher level information system* framework that can provide a single-point gateway to all stakeholders while orchestrating diverse business processes involving the six IMS modules as well as other software systems in the mold shop.

2. CONCEPTS OF ENTERPRISE PORTAL

A ***search engine*** whose main goal is to facilitate access to information contained in documents spread throughout the internet has evolved to the concept of ***enterprise portal*** by integrating other functions such as virtual communities and by adding the ability to personalize interfaces and to access specialized contents[2]. As depicted in Figure 2, the search engine has evolved into portals along two paths: 1) it has evolved into a knowledge management system (KMS) and then to EKP (enterprise knowledge portal) for production, acquisition, transmission and management of knowledge focused on enterprise business processes[3]; 2) it has evolved into EDMS and DW (data warehouse) and then into EIP (enterprise information portal) enabling a company to provide users with a single, personalized gateway to the information they need to make informed business decisions[4].

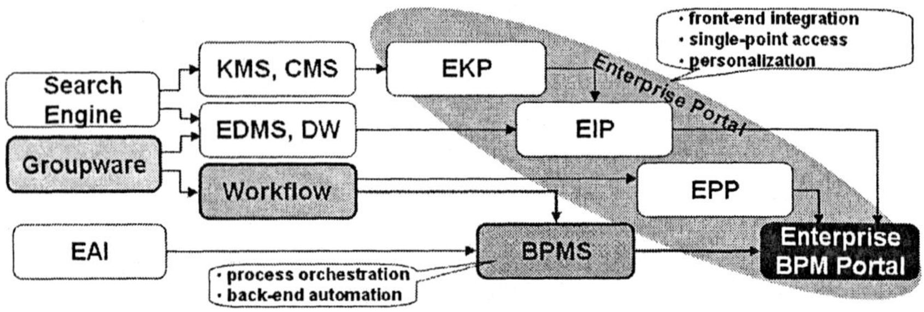

Figure 2. Enterprise Portal Concepts

Another line of development started from the concept of ***groupware*** that had evolved into ***workflow*** management systems (WfMS). A WfMS that provides intelligent user interfaces for workflow operations and augments with relevant process information is called an EPP[5] (enterprise process portal). ***BPM*** (business process management) system is the evolution of WfMS as well as of EAI (enterprise application integration)[6]; it is a software system that is to orchestrate *operational business processes* that interact with people and systems within and across organizations and are automatically driven by explicit process designs[7,8].

Key words of BPMS are process orchestration and flexible automation of back-end processes, while existing enterprise portals focus on single-point access, front-end integration and personalized interfaces. By incorporating the functionalities of BPMS into the exiting framework of enterprise portal, we propose a new type of enterprise portal called ***Enterprise BPM Portal***.

3. IMS PORTAL

In this paper, an *enterprise BPM portal* for a mold shop to orchestrate diverse business processes involving the six *IMS modules* as well as other software systems in the mold shop is called ***IMS-portal***. An abstract framework of IMS-portal is depicted in Figure 3. The main functionalities of the IMS-portal are 1) orchestration of *process-oriented services* such as technical data processing and job progress management, 2) real-time support for *decision-making services* such as CTP (capable-to-promise) and out-sourcing decision, 3) process-oriented management of *information resources* such as mold design data and manufacturing instruction data, and 4) proactive management of *physical resources* such as running NC machines and die-spotting machines. On top of these, it has to provide a single-point gateway via portlet[9] that can be personalized for each of the stakeholders.

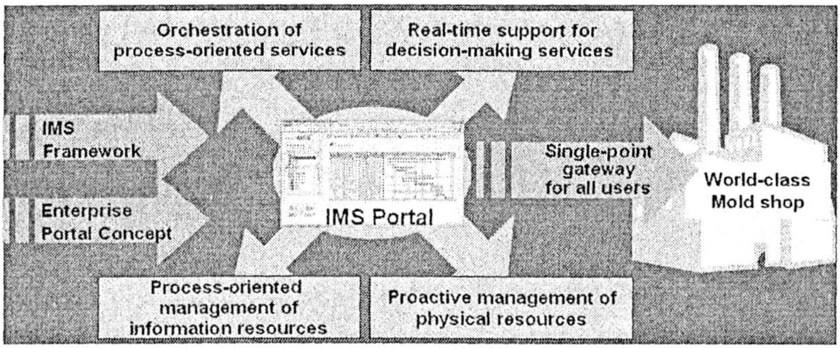

Figure 3. IMS-portal Framework

Shown in Figure 4 is a system architecture for the IMS-portal framework of Figure 3. The IMS-portal architecture consists of four layers: 1) *resource layer* containing all databases and business process repository, 2) *application layer* with all application software systems including the six IMS modules of Figure 1 (enclosed in the dashed rectangle), 3) *orchestration layer* providing milestone management and deliverable management functions and various process enactment services, 4) *portal-service layer* for providing searching and notification services to the stakeholders via the personalized portlet.

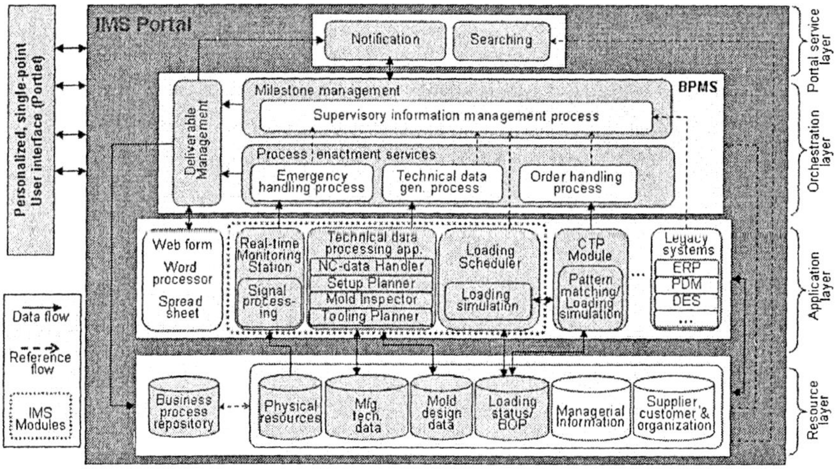

Figure 4. IMS-portal Architecture

Comparing Figure 4 with Figure 3, the *deliverable management* module is in charge of 'process-oriented management of information resources', and the *emergency handling process* and *real-time monitoring station* modules are in charge of 'proactive management of physical resources'. The *order handling process* module supporting the *CTP Module* is an example of 'real-time support for decision-making services'. There are two types of 'process-oriented service': technical data-processing and job-progress management. The *technical data-generation process* module supporting *NC-data Handler* etc. is in charge of technical data processing services, and Loading Scheduler is in charge of (one type of) job-progress management services.

4. IMPLEMENTATION OF IMS PORTAL

The six IMS modules shown in Figure 1 had been implemented for a mold shop in Korea, and some of the modules have been used in the mold shop since 2004. On top of the IMS modules, an ad hoc prototype of the IMS-portal was implemented as depicted in Figure 5. Major modules of the ad hoc IMS-portal include Work-list handler, RTM (real-time monitoring) interface, Milestone planner for the loading schedule, and NC-data handler interface.

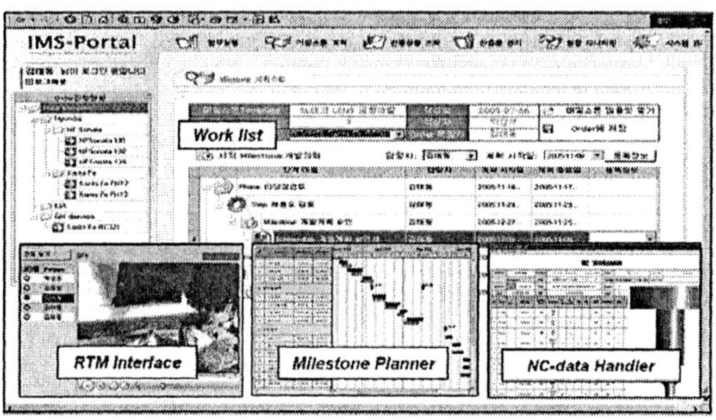

Figure 5. An Ad Hoc IMS-portal

A formal IMS-portal reflecting the architecture given in Figure 4 is under development. The modules in the orchestration layer are realized using an in-house BPM-engine named Harmony®. The IMS-portal is a Web-based system employing the concept of portlet, and there are five main menus as shown in Figure 6:

1. *Work-list* menu is used to activate the 'Progress-oriented services work-list portlet' which provides the user with a list of pending workitems related to the *Process enactment services* (emergency handling, tech. data-generation and order handling) in the Orchestration layer of Figure 4.

2. *Mold-shop management* menu is for the 'Job-progress monitoring portlet' which provides the user with a customized view of *Milestone management* in the Orchestration layer. It also controls the 'Physical resource monitoring portlet' for monitoring the current state of NC machines.
3. *Information management* menu is for the 'Information resources management portlet' which provides the user with a customized view of the *Deliverables management* in the Orchestration layer of Figure 4.
4. *Application execution* menu allows the user to directly invoke the various application software modules in the Application layer of Figure 4.
5. *My information* menu is for the 'Groupware portlet' which prides the user with personalized information services including notifying messages, e-mails, personal schedules, and portlet-configuration information.

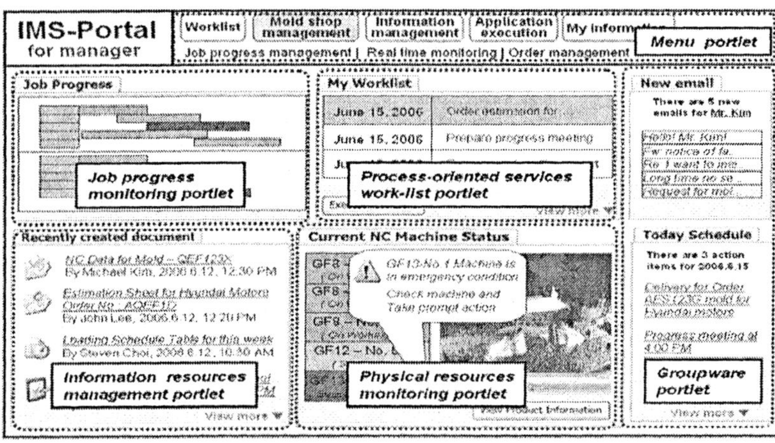

Figure 6. User Interface Design for IMS-portal

5. CONCLUSION

In this paper, a new type of enterprise portal called Enterprise BPM Portal is proposed and it is applied to an intelligent mold shop (IMS) to come up with a framework of IMS-portal. Software architecture of the IMS-portal is also proposed that can provide a single-point gateway to all users while orchestrating diverse business processes involving various types of application software systems in the mold shop. An ad hoc IMS-portal was developed, and it is being upgraded to a 'formal' portal conforming to the proposed IMS-portal architecture. If fully developed and deployed, the proposed IMS-portal is expected to provide an unprecedented 'IT leverage' to mold-making companies.

6. REFERENCE

1. B. K. Choi, K. Ko and B. H. Kim, Development of intelligent mold shop, *Computer-Aided Design & Applications*, 2005, Vol. 2, No. 5, pp. 619-626
2. Claudia Dias, Corporate portals: a literature review of a new concept, *International Journal of Information Management* 21 (2001) 269–287
3. Firestone, J. M., DKMS Brief No. Eight: Enterprise information portals and enterprise knowledge portals, April 2000. [http://www.dkms.com/EKPandEIP.html].
4. Aneja, A., Rowan, C. & Brooksby, B., Corporate Portal Framework for Transforming Content Chaos on Intranets, *Intel Technology Journal*, 2000, Q1, pp. 1-7.
5. Chi-Tsai Yang, Bin-Shiang Liang, Shung-Bin Yan, Feng-Jian Wang., An Intelligent and Personalized Enterprise Process Portal, *Workflow Handbook* 2004, pp207-220
6. Howard Smith and Peter Fingar: *BPtrends, January 2004.* [http://www.bpmi.org]
7. What does business process management mean for the IT industry?, *KMworld Magazine,* 2002, Sempember, Vol 11, Issue8, Ovum
8. BPM 2003 *Market Milestone Report*, Delphi, 2003
9. Sun Microsystems, Introduction to JSR 168-The Portlet Specification, http://developers.sun.com/prodtech/prtalserver/reference/techart/jsr168

REALIZATION OF INTELLIGENT CONTROL FOR ROLL SHAPE IN ROLL GRINDER NC MACHINE

Liwen Yan[1,2], Tao Yu [1] ,Yi Pan[2]
[1]CIMS & Robot Center of Shanghai University, Shanghai University, China; Email: ylw@cslg.edu.cn. [2] Mechanical Engineering Department , Changshu Institute of Technology , China.

Abstract: During the roll grinding process, it is difficult to attain a high rolling shapes and surface quantities for several factors such as the emery wheel abrasion. This paper puts forward a new Neural Networks intelligent error compensation algorithm to measure round error, which is based on the analysis of the causation of rolling shape. And it has an experiment to verify the algorithm further.

Key words: Round, Neural Networks, Intelligent Compensation, Roll Grinder

1. INTRODUCTION

Rolls are used as tools themselves. They flatten size, shape or impart a finish to other materials like Steel, Aluminum, Paper, etc. In recent years, because of the metallurgy, deckle, rubber, plastics and print to dye the quick development of the industry economy of etc, there appear various rolls with different shapes of rolls, such as CIR, LINE, SINE, FREE and CVC curve roll. To meet the requirement of the market, the higher request is put forward to ensure the surface quantity and roll form accuracies of the different form for roll. For example, while whetting to pare the work roll, the target value with actual value deviation should be less than ±1.0μm, and while whetting to pare to prop up the roll, the target value with actual value deviation should be less than ±1.5μms. The roll cylinder degree should be less than 0.005 mm, and the radius direction deviation (roundness tolerance) should be less than

Please use the following format when citing this chapter:
Yan, Liwen, Yu, Tao, Pan, Yi, 2006, in International Federation for Information Processing (IFIP), Volume 207, Knowledge Enterprise: Intelligent Strategies In Product Design, Manufacturing, and Management, eds. K. Wang, Kovacs G., Wozny M., Fang M., (Boston: Springer), pp. 505-510.

0.005mm, disallowing to grinding trace or chafe, etc, in the roll circumference direction.

Compared to other machine manufacturing methods, there are more and complex factors that affect the precision of manufacturing result for the complicated characteristics of grinding. In addition to the static state accuracy of the tool machine, there still has the e wear away of emery wheel, grinding force, the flexibility transform of craft system and some other factors, etc. In practical manufacturing process, while grinding, the influence of these factors is to be reflected in the roll form accuracy and the size error of the dimension. The actual result which we got indicates that characteristics as follows:

(1) The size of the roll has the trend of the gradual aggrandizement.

(2) The variety of the size of roll is not linear, even is not monotonous augmentative.

(3) On the same grinding machine, the size and the form varying regulations of the roll processed is also incompletely in phase.

From above, we can see that the higher request has been demanded to the numerical control roll grinder to meet the requirement of high-quality roll grinding. To make the roll grinding to be more accurate, more efficient, more automatic and more intelligent and to meet the requirements of the society and customers, we had developed a numerical control system of the roll grinder to accomplish the intelligent compensate control based on neural networks(NNs). The approaches of intelligent roundness error compensation of grinding roll realized as follow.

1.1 The former feedback compensation control system of NC roll grinder

In order to compensate the roundness error in the grinding, the methods used applied currently are showed in figure 1. Because of the production of roundness error of roll is nonlinear, this kind of simple way of error compensate can't offset the accidental factor and decrease the effect of roundness variety of roll by nonlinear factor, so we focus on the intelligent compensation system of shapes error of roll (including roundness error) based on neural networks.

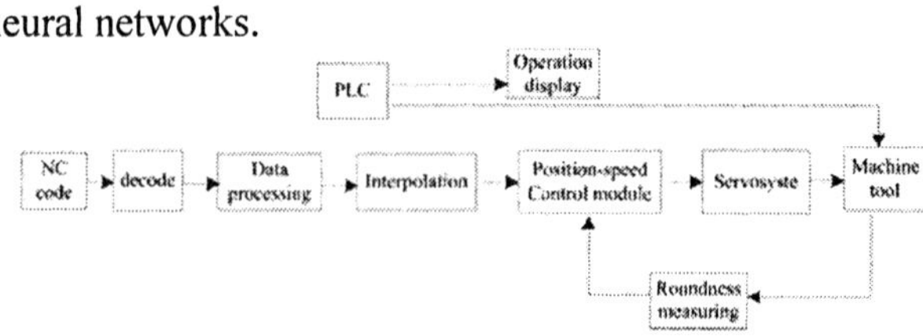

Fig. 1. Control system workflow of NC roll grinder

1.2 The adoption intelligence compensates system using NNs

Artificial neural networks (ANN) comes into being the model from the microstructural and functional simulation of nervous system of human brain, which simulate the part capability of imagery thinking. Its features contain nonlinear characteristic, learning capability and self-adaptive, which is the most important way to simulate the intelligence of human being. For meeting the high performances of roll grinder, when the self-adaptive control is applied to real control process, where has many uncertain environment and manufacturing procedure, there are still some problem, for example, the complex structure of self-adaptive controller and the model reference adaptive control system (MRACS) can't insure zero steady-state error with certainty interference and etc.

In order to make full using of the self-adaptive control technology to improve the capability of robust, real-time, fault-tolerant control and the self-adaptive and learning ability for control parameters, and even more effectively control the process with many uncertainty and difficult to describe the nonlinear complex procedure, the self-adaptive control and NNs were combined together. Only the NNs directly act as controller in the control system can it utilize the advantage to construct perfect novel control system. [1]

1.3 The working principle of the intelligent compensating system[2][3][4]

The neural network intelligent control system which we adopted is an error compensating system. As is shown in figure 2, the architecture of neural-network intelligent error-compensating system of roll roundness, y_d, y are the standard roundness value and output roundness value of the roll in the system, u is the control value. Normalized factor K is added before and after the neural network, (K=kyd, k is the fine turning coefficient, generally span 0.5~1), which made that the input value of the system is not limited, and made the system practical. The import values of the neural network are the normalized input value y' d and the output values of three different time y' (k),y'(k-1),y'(k-2), which show the phase of the system.

The system remains the basic structure of neural network controller. The controller is a neural network, whose parameter, the connecting weight of neural network, can be adjusted online, and is not limited by the controlled object model structure, while the controlled object can be a non-linear, time-variant and unknown model. We can use the neural network as the compensating controller directly without a advance training, and the initial weight value get randomly from [-0.1, +0.1]. The system can adjust the weight value of the neural system, which made itself have a good adaptable capability.

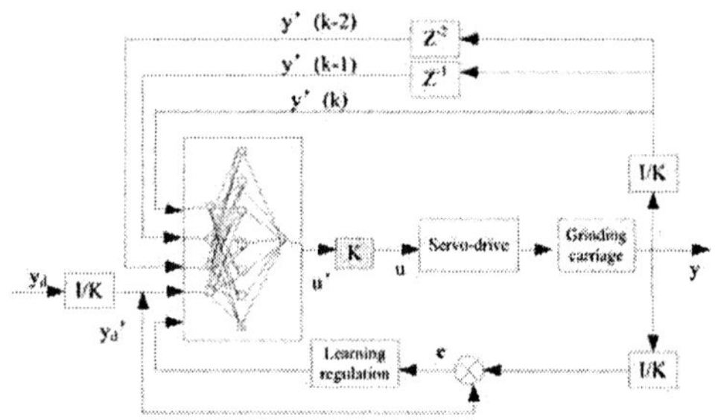

Fig. 2. Intelligent roundness error compensation based on NNs

1.4 The roundness error compensating algorithm based on neural network

1.4.1 The formulation of nonlinear compensating control system

The difference equations of nonlinear compensating control system are given as follows:

$$y(k+1)=f\{y(k),y(k-1),\ldots,y(k-n+1)\}+\sum_{i=0}^{m} g_i \times u(k-i) \qquad (1)$$

Where $m \leq n$; The g_i is constant coefficient; The outputs of system $y(k) \in Y \subset R$; inputs of system is $u(k) \in U \subset R$; And have $f\{0\}=0$.

The output of NNs compensating controller is given by:

$$u(k)=K\left(\sum_{j=1}^{N} V_j \times F\left(\sum_{j=1}^{N} P_{ij} \times X_i\right)-\theta\right)$$

$$F(x)=1/(1+e^{-x})$$

$$\theta=\frac{1}{2}\sum_{j=1}^{N} V_j$$

Where F(x) is transfer function of hidden layers. θ is the threshold of output layer node which insure the stability of control system.

1.4.2 The learning algorithm of the NNs

Define the quadratic function of error system e as value function J :

$$J(K+1) = e(k+1)^2/2$$
$$e(k+1) = (y(k+1) - y_d)/k$$

The weight matrix of NNs is defined by:

$$W = \{W_{i,j}\} = \begin{bmatrix} V_1 & P_{11} & L & P_{M1} \\ V_2 & P_{12} & L & P_{M2} \\ M & M & O & M \\ V_N & P_{1N} & L & P_{MN} \end{bmatrix}$$

Regulate weights of NNs using the gradient descent algorithm, that is, adjust the descending direction of at the J gradient:

$$W_{i,i}(k+1) = W_{i,i}(k) - \beta\, \partial J(k+1)/\partial W_{i,i}$$

Where β is constant coefficient. Assumption:

$$\Delta W_{i,j}(k+1) = -\beta\, \partial J(k+1)/\partial W_{i,j}$$

$$\Delta W_{i,j}(k+1) = -\beta \frac{\partial J(k+1)}{\partial u(k)} \cdot \frac{\partial u(k)}{\partial W_{i,j}} =$$

$$-\frac{\beta}{K} \cdot e(k+1) \cdot \frac{\partial y(k+1)}{\partial u(k)} \cdot \frac{\partial u(k)}{\partial W_{i,j}} = -\eta \cdot e(k+1) \cdot \frac{\partial u(k)}{\partial W_{i,j}} \qquad (2)$$

The learning algorithm of the nerve network can get from the type (2):

$$\Delta V_j = -\eta \cdot e \cdot k(X_j - \frac{1}{2})$$
$$\Delta P_{ij} = -\eta \cdot e \cdot V_j \cdot (1 - X_j) \cdot X_j \cdot X_i, i = 1, \cdots, M \quad , \quad j = 1, \cdots, N$$

In order not to produce paralysis phenomenon of network, when NNs compensation controller works, the system error e is less than $\pm 5\%$ and the control parameter u overcome to the biggest threshold, the neural network would stop learning.

2. EXPERIMENTAL PARAMETERS AND ANALYSIS

Machine for experiment, a roll grinder machine: MKZD84125
Specimen for grinding, length 1600mm, diameter 1000mm
The parameters of NNs compensation controller were set as:
The efficiency of study, G=0.1
Nerve unit in the hidden layer, N=6
Initialized value was selected randomly form [-0.1,+0.1], and the coefficient of fine tuning (k) was fixed at 0.65
Analysis on the experimental result:

The experiment on error compensation by NNs method was carried on the roll "SYZG02". When the intelligent error compensation was chosen, the outline of the roll fluctuated around the perfect outline and the stability of algorithm was nice. Since the error compensation just on roundness of the roll, there were several other parameters such as roundness error, size error, height error, roundness error and roughness error which should be considered in the real grinding process because the interrelationship exist among them. If the intelligent compensation by NNs was applied on all of the parameters that have been mentioned above, input more and output more could be achieved. The result from the experimental work showed the error margin shrink and the effectiveness on compensation was obvious.

3. CONCLUSION

The method of nerve network intelligent compensation could be applied on roll NC. After the nerve network was used, the ability of roll NC on handling the unknown nonlinear controlled device and self-adapting was improved. Such method could also be used in the control on other non-linear machines.

4. REFERENCES

1. Yang shuzi, Ding hong. Development and research of intelligence control technique and intelligences manufacturing system [J]. *China Mechanical Engineering*, 1992,3:(2) 15-18.
2. Li shiyong. Fuzzy control • neural control and Intelligent Cybernetics [M]. Harbin: *Harbin Institute of Technology press*,1996. 169-178.
3. Zhang Ying, Zuo Li. Application of Neural Network in NC Machine Intelligent Control System [J]. *Journal of Wuhan Transportation University*, 1999, 23(3):226-229
4. Tian Xincheng CNC machining technique and control system of ceramic tools in: *the Tongji University Proc of IEEE IC2IT'96 Shanghai: the Tongji University*, 1996 34-36

APPLICATION OF GREY SYSTEM MODEL TO THERMAL ERROR MODELING ON MACHINE TOOLS

Yongxiang Li[1], Jianguo Yang, Hongtao Zhang and Hengchao Tong
[1]*School of Mechanical Engineering, Shanghai JiaoTong University, China; Email: liyongxiang@sjtu.edu.cn.*

Abstract: The basic principle and method of grey system model and its application in thermal error modeling on machine tools has been presented. How to utilize thermal error sequence surveyed to model and predict is discussed by adopting total data GM (1,1) model, new information GM(1,1) model and metabolic GM(1,1) model after function transform is brought forward to improve degree of smoothness of primary data sequence. Finally prediction precision of metabolic GM (1,1) model is 7.2% and 15.46% higher than those of new information GM (1,1) model and total data GM (1,1) model respectively. The comparison of the experiment's results indicates the accuracy and robustness advantages of metabolic GM (1,1) model over other two models.

Key words: Machine tool, Thermal error, Metabolic GM (1,1) model, Modeling

1. INTRODUCTION

In the precision machinery processing, machine tool thermal error is the major factor to cause workpiece's error, how to control machine tool thermal error becomes essential technique to increase their processing precision. These errors could be decreased by amending machine tools' self-structure through advanced design and manufacture technique, but with precision's improvement of machine tools, corresponding expense would be increased. Error compensation technique is regarded as a effective and economical way, it could make workpiece more precise than machine tool without improving machine tools' themselves precision. So the kind of way has already received more and more widespread value in the mechanical processing industry.

This project is supported by A Foundation for the Author of National Excellent Doctoral Dissertation of China under the grant No. 200131

Please use the following format when citing this chapter:
Li, Yongxiang, Yang, Jianguo, Zhang, Hongtao, Tong, Hengchao, 2006, in International Federation for Information Processing (IFIP), Volume 207, Knowledge Enterprise: Intelligent Strategies In Product Design, Manufacturing, and Management, eds. K. Wang, Kovacs G., Wozny M., Fang M., (Boston: Springer), pp. 511-518.

If compensation technique is used for machine tool thermal error, three problems must be solved: first, the detection of thermal error; second, the establishment of prediction model of thermal error; third, the implementation of real-time compensation of thermal error. Establishment of prediction model is the most essential among them. For thermal error is in great degree decided by many kinds of factors such as the processing condition, the processing cycle, the use of coolant as well as the environment, moreover, machine tool thermal error presents the non-linearity and interaction, therefore, it is quite difficult to precisely establish mathematical model of thermal error with theoretical analysis only [1]. Grey system theory was founded by Professor *Deng Julong [2-3]* of Huazhong University of Science and Technology. For grey system model includes content of general linear regression and power function, it could make up the deficiency of traditional difference and disperse model and present a strong tendency, especially for the system with both the known information and the unknown information, grey system model has very good fitting, prediction effect. Moreover, grey system model has quick speed of modeling and computation, has not high requirement on primary data and does not depend on massive data information as modeling foundation, so grey system model is a simple and convenient model which has lower request on gathered data information [4]. The work of this article is to apply improved grey system model to modeling of machine tool thermal error, the results show that metabolic GM (1,1) model is the most satisfied among three kinds of GM (1,1) model.

2. GREY SYSTEM MODEL

2.1 Establishment of GM (1,1) model

Assumption $\left\{x^{(0)}(i)\right\}$ $(i=1,2,\cdots,n)$ is a primary data sequence, after once accumulating generation, it becomes a new sequence as follows:

$$x^{(1)}(k)=\sum_{j=1}^{k}x^{(0)}(j);\quad k=1,2,\cdots,n \qquad (1)$$

Then corresponding differential equation of GM (1,1) model is expressed as:

$$\frac{dx^{(1)}}{dt}+ax^{(1)}=u \qquad (2)$$

where a is development grey number, u is endogenetic control grey number.

$\hat{a}$ is defined as estimating parameter vector, $\hat{a} = \begin{bmatrix} a \\ u \end{bmatrix}$, according to least square method, using following formula

$$\hat{a} = \left(B^T B\right)^{-1} B^T y_N \qquad (3)$$

where $B = \begin{bmatrix} -\frac{1}{2}\left[x^{(1)}(1) + x^{(1)}(2)\right] & 1 \\ -\frac{1}{2}\left[x^{(1)}(2) + x^{(1)}(3)\right] & 1 \\ \cdots & \cdots \\ -\frac{1}{2}\left[x^{(1)}(n-1) + x^{(1)}(n)\right] & 1 \end{bmatrix}$,

$$Y_n = \left(x^{(0)}(2), x^{(0)}(3), \cdots, x^{(0)}(n)\right)^T$$

to resolve differential equation (2), the solution is expressed as

$$\hat{x}(k+1) = \left[X^{(0)}(1) - \frac{u}{a}\right] e^{-ak} + \frac{u}{a} \qquad (4)$$

Equation (4) is just GM (1,1) and it is carried on once inverse accumulating generation, the reductive model is got by

$$\hat{x}^{(0)}(k+1) = \hat{x}^{(1)}(k+1) - \hat{x}^{(1)}(k) \; or \; \hat{x}^{(0)}(k+1) = (-a)\left[x^{(0)}(1) - \frac{u}{a}\right] e^{-ak} \qquad (5)$$

2.2 Function transform modeling of grey system

Considering grey system model's requirement of degree of smoothness for disperse data sequence, for non-minus primary data $\left\{x^{(0)}(i)\right\}$ $(i = 1, 2, \cdots, n)$, $f(i)$, as a non-minus and monotonously degressive sequence, would be chosen to transform $x^{(0)}(i)$, a new sequence is given by: $F\left(x^{(0)}(i)\right) = x^{(0)}(i) \times f(i)$, which would evidently improve degree of smoothness of primary data $x^{(0)}(i)$; then $F\left(x^{(0)}(i)\right)$ would be modeled by grey system model and be reverted by contrary transform, finally, modeling precision of system would be heightened markedly [5].

2.3 Improvement of GM (1,1) model [6]

In practical modeling, not all data of the primary data sequence is used in modeling, part of data could also be used to establish a model. Generally speaking, different data could establish different model, even if the same kind of GM (1,1) model is established, the values of parameter a , b are different with different selected data. This kind of instance precisely reflects the influence of different situation and condition on system characteristic in the model.

Assumption $X^{(0)}=\left(x^{(0)}(1),x^{(0)}(2),\cdots,x^{(0)}(n)\right)$, $x^{(0)}(n)$ is chosen as origin of time axis, then $t<n$ is considered as past, $t=n$ is present and $t>n$ is future.

1. GM (1,1) model, established by $X^{(0)}=\left(x^{(0)}(1),x^{(0)}(2),\cdots,x^{(0)}(n)\right)$, is called total data GM (1,1) model;
2. GM (1,1) model, established by $X^{(0)}=\left(x^{(0)}(k_0),x^{(0)}(k_0+1),\cdots,x^{(0)}(n)\right)$, $\forall k_0>1$, is called partial-data GM(1,1) model;
3. Assumption $x^{(0)}(n+1)$ is latest information, which is placed in $X^{(0)}$, then GM (1,1) model, established by $X^{(0)}=\left(x^{(0)}(1),x^{(0)}(2),\cdots,x^{(0)}(n),\right.$ $\left.x^{(0)}(n+1)\right)$ is called new information GM(1,1) model;
4. If latest information $x^{(0)}(n+1)$ is placed and the oldest information $x^{(0)}(1)$ is taken out, then GM (1,1) model, established by $X^{(0)}=\left(x^{(0)}(2),\cdots,x^{(0)}(n),x^{(0)}(n+1)\right)$, is called metabolic GM(1,1) model.

3. APPLICATION AND ANALYSIS OF GREY SYSTEM MODEL

3.1 Measurement and analysis of thermal errors of machine tool

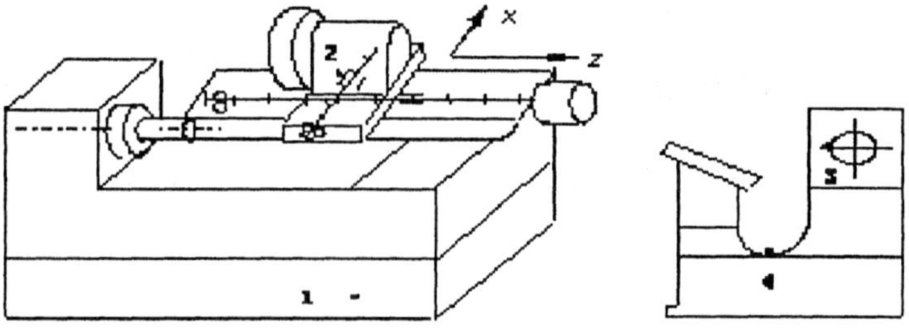

Figure. 1. Sketch map of turning center

The method is applied to a CNC turning center, as shown in Fig. 1, and is used in study of thermal error's compensation [7]. Sketch map of thermal error's measurement is shown in Fig. 2.

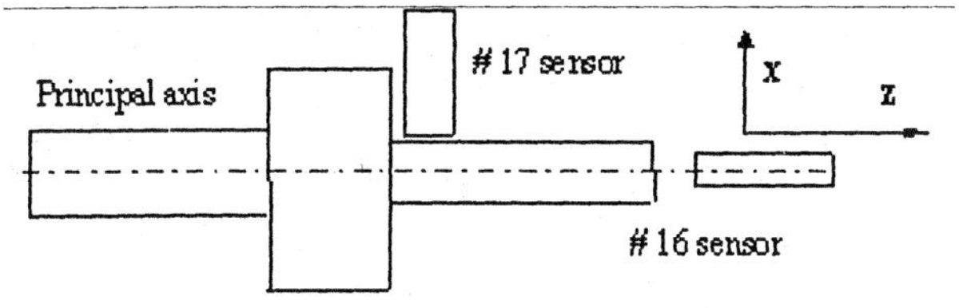

Figure. 2. Measurement of thermal errors of principal axis

Two displacement sensors fixed on knife rest measure thermal errors between principal axis of X direction (17) and Z direction (16) and knife rest. For workpiece is very short, the error of leaning angle is neglected.

3.2 Prediction analysis of grey system model

Grey prediction model is to predict self-development according to dynamic differential equation established by its own data [8]. The pretreatment of grey varible is generally accumulating or inverse accumulating primary data, in order to weaken even eliminate randomicity information of original sequence and generate data sequence with strong regularity [4]. The text chooses radial thermal errors in the process of turning as researchful objects (for workpieces' precision of radial dimension has rather high request, radial thermal error of machine tool is only considered), in practice, researchful objects are workpieces' errors of radial dimension changed along with the machine tool's temperature or the time [9]. Change figure of thermal errors of turning center is shown in Fig. 3.

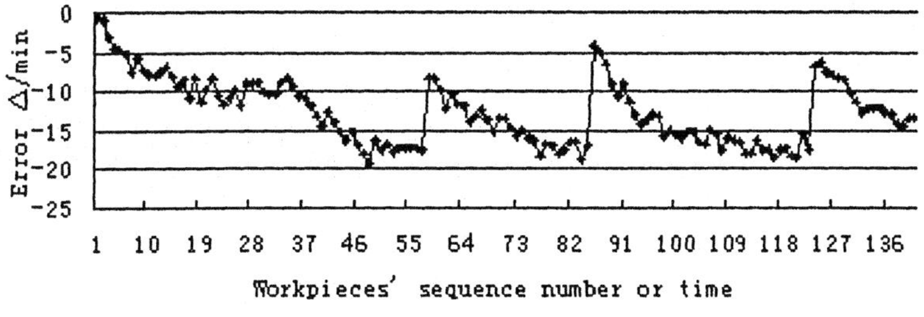

Figure. 3. Change figure of thermal errors of turning center

According to above mentioned grey system theory, the most basic GM (1,1) model, namely total data GM (1,1) model, is firstly used in thermal error sequence modeling of turning center. Considering the fact that grey system model requires that data sequence must have high degree of smoothness, first, primary data sequence, $x^{(0)}(i)$ $(i = 1, 2, \cdots, n)$, would be transformed into positive data sequence by absolute value operation due to the negative values of thermal errors of machine tool shown in Fig. 3; second, positive data sequence would be transformed by a non-minus and monotonously degressive sequence $f(i) = 10/i$ $(\mathrm{i} = 1, 2, \cdots, \mathrm{n})$ again, then get new data sequence, $F\left(x^{(0)}(i)\right) = x^{(0)}(i) \times f(i)$; third, the former 100 data of new data sequence are used to establish GM (1,1) model according to (1)- (5); fourth, GM (1,1) model would be reverted into forecasting GM (1,1) model by contrary function transform; fifth, forecasting GM (1,1) model could forecast the latter 41 data of primary data sequence, then forecasting value of thermal error for machine tool would be obtained, compared with primary data sequence, absolute error and relative error between forecasting values and practical values are also obtained. Finally, both new information GM (1,1) model and metabolic GM (1,1) model are also used to model and predict thermal error sequence.

The comparisons of predicted value and practical value of all models are drawn in Fig. 4-Fig. 6 according to above predicted results. It can be seen from 3 figures clearly, prediction precision of metabolic GM (1,1) model is highest, new information GM (1,1) model is next, total data GM (1,1) model is worst. Prediction precision of metabolic GM (1,1) model is 7.2% and 15.46% higher than those of new information GM (1,1) model and total data GM (1,1) model respectively. Thermal error of turning center can be kept in 7.58 μm from primary about 20 μm through error compensation using metabolic GM (1,1) model. The reason was that along with passing time, some random vibration or driving factor would enter into the system constantly in the developing course of any grey system, and then affect systemic development, however, metabolic GM (1,1) model just renews new information constantly and removes old information duly, so that modeling sequence could reflect systemic present characteristic and obtain higher prediction precision. The comparison of error curve of three grey prediction models is shown in Fig. 7.

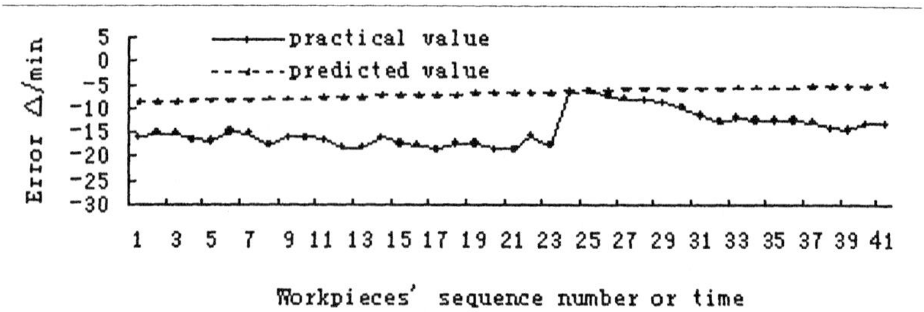

Figure. 4. Comparison of predicted curve and practical curve of total data GM (1,1) model

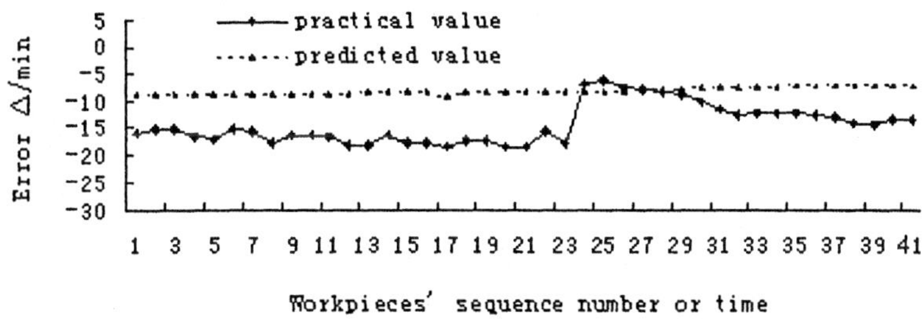

Figure. 5. Comparison of predicted curve and practical curve of new information GM (1,1) model

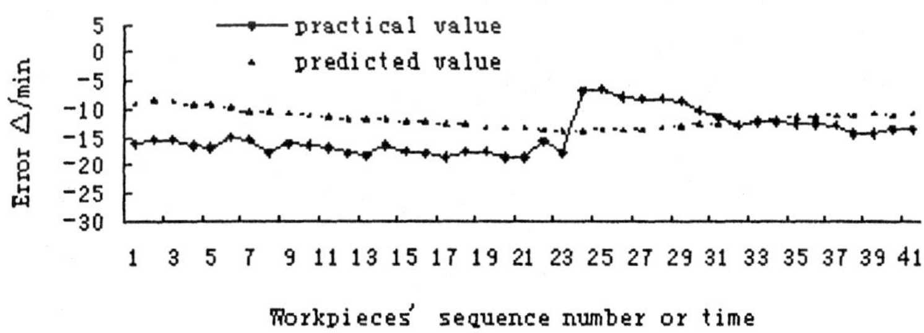

Figure. 6. Comparison of predicted curve and practical curve of metabolic GM (1,1) model

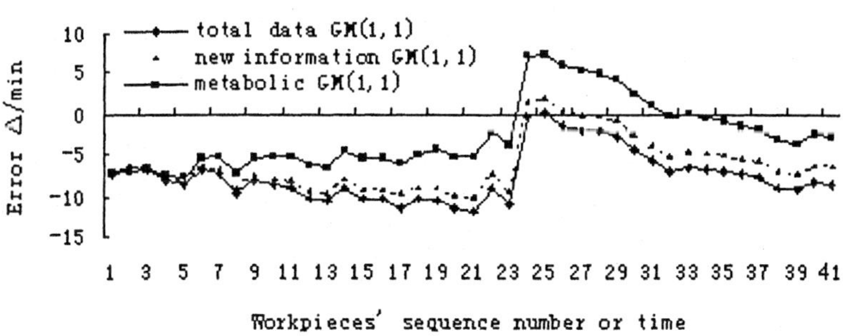

Figure. 7. Comparison of errors curve of three kinds of GM (1,1) model

4. CONCLUSION

Grey system model, a dynamic model, weakens randomicity and shows systemic assured developing trendency [10]. The text just regards machine tool thermal error as a grey variable according to generation mechanism and anfractuosity of machine tool thermal error, and adopts grey system model to model. At the same time, after degree of smoothness of primary data sequence is improved by function transform, based on total data GM (1,1) model, the text brings forward new information GM (1,1) model and metabolic GM (1,1) model to model for machine tool thermal error.

It could be seen that both new information GM (1,1) model and metabolic GM (1,1) model have higher predicted precision than total data GM (1,1) model from predicting result of machine tool thermal error, metabolic GM (1,1) model is optimal model among three kinds of GM (1,1) model, for the function of the old data would decrease with the development of system, it is obviously reasonable to remove those old data which could not reflect systemic present characteristics, moreover, metabolic GM (1,1) model could avoid the difficult caused by continual increase of computing quantity with the increase of information quantity.

5. REFERENCES

1. Y. Q. Ren, (2004), *Applied research on high-efficiency measurement, modeling and compensation for errors on NC machine tools*, Shanghai: Shanghai JiaoTong University.
2. J. L. Deng, (1989), Introduction to grey system theory, *The Journal of Grey System,* vol. 1, pp. 1-24.
3. J. L. Deng, (2002), *Basis of grey theory*, Wuhan: Huazhong University of Science and Technology Press.
4. Y. L. Pan, S. X. Zheng, (2002), Improvement for the precision of instruments by a grey system model, *Cryogenics and Superconductivity,* vol. 30, pp. 65-67.
5. Y. X. Luo, L. T. Zhang and M. Li, (2001), *Grey system theory and application in the mechanical engineering,* Changsha: National University of Defense Technology Press.
6. S. F. Liu, T. B. Guo and Y. G. Dang, (1999), *Grey system theory & Application,* Beijing: Science Press.
7. J. G. Yang, J. X. Yuan and J. Ni, (1999), Thermal error mode analysis and robust modeling for error compensation on a CNC turning center, *International Journal of Machine Tools and Manufacture,* vol. 39, pp. 1367-1381.
8. Z. T. Wu, Y. Li and Q. Li, (1998), Grey prediction of running status trendency of complex mechanical equipment, *Mechanical Research & Application,* vol. 1, pp. 44-45.
9. J. G. Yang, (1998), *Error synthetic compensation technique and application for NC machine tools*, Shanghai: Shanghai JiaoTong University.
10. P. Gong, Y. T. Fei, (2001), Grey evaluation method of dynamic measurement errors based on error propagation theory, *Tool Engineering,* vol. 5, pp. 31-34.

NOZZLE DESIGN AND ASSISTANT-GAS FLOW ANALYSIS IN THE CO2 LASER CUTTING

Jun Chen [1], Lixin Lu [2], Limin Li [1], and Hui Zhao[1]
[1]*CIMS center, Shanghai University, Shanghai, 200072, China, Email:pedrorichter@163.com*
[2]*Shanghai Feilo Co.,LTD,Shanghai,200050, China*

Abstract: This paper investigates the nozzle design and assistant-gas analysis. By some experiments, we present important date and some conclusions about the significant parameters of the laser.

Key words: laser cutting, nozzle, kerf.

1. INTRODUCTION

Lasers are widely used in industry as cutting tools due to ultra flexibility of the cutting conditions, obtaining high quality end product, quick set up, non-mechanical contact between the workpiece and the tool, and small size of the heat affected zone.

In china, various lasers products are now widely used for material processing such as cutting & drilling, welding, surface treatment, ablation, and so on. CO_2 lasers and their applications to the fine cutting of the steel plate seems to have reached completion for meeting the user's demands. However, CO_2 lasers are continuously being improved so that they may become applicable to more effective and more precise material processing. High laser power, high laser-beam quality, and highly controlled power-mode are the essential items for attaining the purser.

In a laser system, laser beam focuses on the surface of workpiece and molten it though lens; meanwhile the co-axially compressed gas puffs away the dross in this process. The laser beam moves above the workpiece relatively and forms a cutting kerf on it.

Please use the following format when citing this chapter:
Chen, Jun, Lu, Lixin, Li, Limin, Zhao, Hui, 2006, in International Federation for Information Processing (IFIP), Volume 207, Knowledge Enterprise: Intelligent Strategies In Product Design, Manufacturing, and Management, eds. K. Wang, Kovacs G., Wozny M., Fang M., (Boston: Springer), pp. 519-524.

The nozzle is utmost important part of a laser machine. Besides focus the laser beam on the workpiece, it provides the assistant-gas for cutting process. The composition, pressure, speed of the assistant-gas is all the top important parameters to a cutting system.

2. NOZZLE DESIGN AND ANALYSIS

As laser systems are becoming more industry demanding, the need for developments in the area of monitoring, diagnosis, regulation and modeling becomes essential to achieve and maintain a high-quality cutting process. The design of nozzle in a laser cutting system, is significant to a laser machine, Especially it's assistant-gas will greatly affect cutting process and cutting quality. Assisting gas protects the surface from high temperature oxidation reactions while high temperature exothermic reactions take place when oxygen is used as assisting gas.

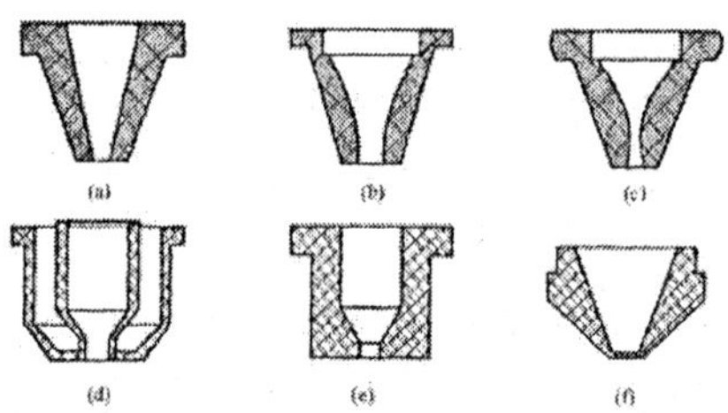

Figure 1. Different co-axial nozzle

The nozzle could be designed in different styles; fig1 presents some common co-axial nozzles. The diameter of the nozzle outlet should let the laser beam pass successfully and avoid touching the inner surface of the nozzle. In addition, the jet assistant-gas of the nozzle should couple to blow off the dross in the kerf and strength cutting effectively. Different styles of nozzle bring different pressure, velocity of the assistant-gas, and different cutting quality. If the diameter is too large, kerf width will be too wide, while it is too small, it will make the value of kerf inaccurate, diameter of the nozzle outlet is about 1~4mm.

3. ASSISTANT-GAS ANALYSIS

In general, it is impractical to investigate all of the factors at one time. The identification of the dominant factors for which variation could lead to large effects on the cut quality is important. The gas-composition variation, the gas pressure and the gas speed are the dominant factors; their effects on the cut quality need to be investigated.

As a co-axial nozzle system is used in laser cutting to assist the cutting process, the gas flow velocity at the nozzle exit is an important factor for the process. Because of the rapid change of pressure and temperature in nozzle, the wall surface is insufficient to transfer any appreciable amount of heat to the fluid. Therefore, the flow is considered adiabatic. Because of the short length of the nozzle and the smooth variation of its dimensions, the eddy effect and skin friction are neglected. Thus, the flow is considered as reversible and adiabatic, which means that the flow is isentropic.

The analysis associated with the influence of assisting gas including cooling and exothermic reaction contribution on the cutting process was examined in the previous study; therefore, the analysis will be described briefly. Assisting gas forms a boundary layer flow over the liquid surface (molten metal surface). The heat transfer from liquid surface to boundary laser flow occurs, since the assisting gas is at room temperature. Since the cooling effect of the assisting gas as well as melting and evaporation taking place at the laser irradiated surface are introduced in the scaling laws, only the exothermic reaction contribution of the assisting gas is introduced. Moreover, high temperature oxidation reaction (exothermic reaction during which metal oxides are formed) at the melt surface provides excess energy to the laser-irradiated region during the machining operation.

The species formed after the chemical reactions (product of exothermic reactions) contribute to the heat transfer-taking place at gas–liquid interface. In this case, heat transfer at the gas–liquid interface due to high temperature oxidation reaction can be described through the ratio of dimensionless heat transfer coefficients due to diffusion (CHd) and chemical reaction (CHc), which was presented as (1):

$$\frac{C_{Hd}}{C_{He}} = \frac{[1 + B_1(u_L/u_e)]^{Pr-1}}{[1 + B_1(u_L/u_e)]^{Sr-1}}, \quad B_1 = \frac{2(\rho u_e)_g}{\rho_e u_e C_f} \tag{1}$$

And Pr and Sc are turbulent Prandtl and Schmidt numbers, respectively. Ue is the gas jet velocity at the edge of the boundary layer. UL the liquid velocity, and Cf the skin friction coefficient. The rate of heat transfer (q˙LJC) per unit area (Kerf width ×Kerf depth) of the molten metal excluding the cooling effect while including chemical reaction contribution of the assisting gas can be written as:

$$\dot{q}\, LJC = \rho_e u_e C_{H_c}[(\frac{C_{Hd}}{C_{He}} - 1)h_c] \qquad (2)$$

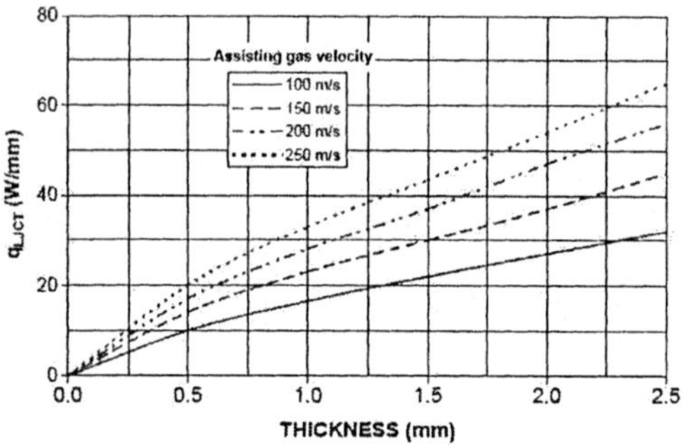

Figure 2.digram of different velocity

The quantity $\dot{q}$ LJC was computed for various Kerf depths as shown in Fig2; therefore, the values of q LJC computed in the previous study are accommodated in the analysis. Moreover, the quantity q LJC can be integrated over the cut width to obtain the rate of chemical energy contribution during the cutting process. Fig2 shows graphs of four assisting gas velocity, as we can see, the q LJC in a certain thickness is increasing with the rising of velocity.

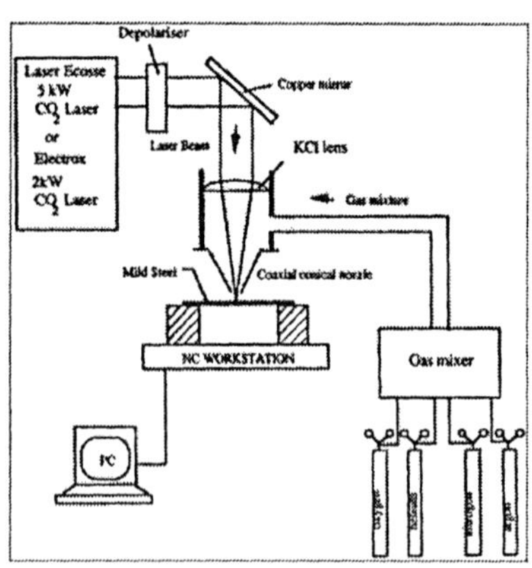

Figure 3.model of different gas composition

Through investigation of how the composition of the assistant-gas affecting the cutting process, a model can be set up in Fig3.to exemplify this.

If the gas flow condition inside the kerf during the cutting process is considered similar to pope flow, the gas flow from the nozzle is uniform with a free stream velocity at the slot entrance. As it moves down the slot, the friction between the cut wall and the flow stream element will generate a shear force to retard the flow motion and cause the boundary layer to grow until it is fully developed. For laminar flow, the Length (entrance length) required for the velocity to be unchanged with axial position can be approximated to (0.05)*Re*D, where Re is the Reynolds number and D is the diameter of the pipe.

From the table1, it is easy to see that the shear force on the cutting front is influenced largely by the gas flow velocity and the density. Increasing the gas flow velocity will significantly increase the shear drag force on the cutting front. High density has higher shear drag force, thus from table1 argon may be the most promising gas for CO_2 lasers cutting and helium the least promising.

Table1:date of different gas parameters

Parameter	Gas			
	Oxygen	Argon	Helium	Nitrogen
ρ:Density (kg m^{-3})	1.2997	1.8435	0.1779	1.1367
μ:dynamic viscosity (106 N s m^{-2})	2.0854	2.2693	1.9618	1.8003
ν:kinematics' viscosity (10^5 m^2 s^{-1})	1.5498	1.2734	11.1007	1.5498
C_P: specific heat of constant pressure (kJ K^{-1} kg^{-1})	0.9213	0.5198	5.2015	1.0436
k*: thermal conductivity (10^2 $W^{-1}K^{-1}m^{-1}$)	2.6735	1.7637	15.3879	2.5768

We also obtained table2 and Fig4 by investigating the pressure of the assistant-gas. These experiments were performed with 2.0 kW CO2 laser, the laser mode structure being TEM01. The new cutting head was used in these high-pressure cutting experiments. Four different assistant gases used in this study, being pure oxygen, pure Argon, pure nitrogen, and air (78% nitrogen and 22% oxygen). The first three types of gases are commercial standard gases, whilst the air used in the experiments is supplied by an air compressor through an air cleaner. The cutting experiments were carried out primarily on 3mm thick mild steel plate.

The influence of the cutting speed by the gas pressure can be seen by table2, for a certain assistant-gas, increase the pressure can improve the cutting speed obviously. The Fig4 tells us the different cutting speed and pressure is significant to the kerf quality.

Table2:experiment date of gas in different pressure

mm s^{-1}	2 bar	4 bar	6 bar	10 bar
Oxygen	52.5 ± 2.5	57.5 ± 2.5	60.5 ± 2.5	110 ± 2.5
Nitrogen	17.5 ± 2.5	17.5 ± 2.5	17.5 ± 2.5	22.5 ± 2.5
Argon	17.5 ± 2.5	17.5 ± 2.5	17.5 ± 2.5	27.5 ± 2.5
Air	22.5 ± 2.5	27.5 ± 2.5	37.5 ± 2.5	

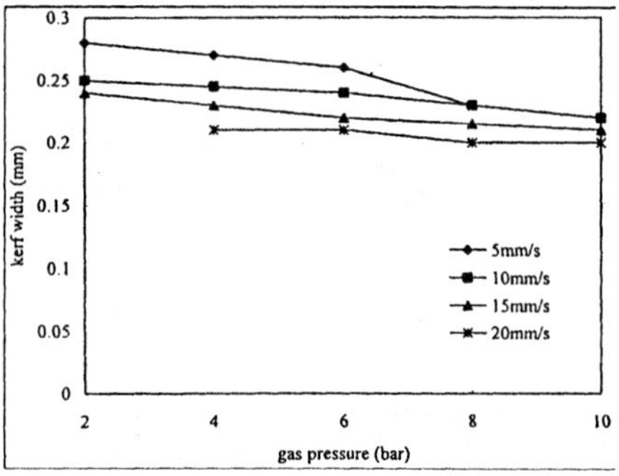

Figure 4. kerf in different gas presure

4. CONCLUSIONS

We investigate nozzle design and gas analysis by some experiments. These research show, the design of nozzle, composition, velocity, pressure of assistant-gas, will affect greatly in the cutting process.

5. REFERENCE

[1] K.A. Bunting, G. Cornfield, Toward a general theory of cutting: a relationship between the incident power density and the cut speed, *ASME J. Heat Transfer* 97 (1975) 116–121.

[2] K. Danisman, B.S. Yilbas, Z. Yilbas, Study of some characteristics of plasma generated during CO2 laser beam cutting process, *Opt. Laser Technol.* 24 (1992) 33–38.

[3] S.L. Chen, The effects of gas composition on the CO2 laser cutting of mild steel, *J. Mater. Process. Technol.* 73 (1998) 234–242. 89–95.

[4] M.J. Hsu, P.A. Molian, Off-axis, gas-jet-assisted, laser cutting of 6.35 mm thick stainless steel, *ASME J. Eng. Ind.* 117 (1995) 272– 276.

[5] B.S. Yilbas, A.Z. Sahin, Turbulent boundary layer approach allowing chemical reactions for CO2 laser oxygen-assisted cutting process, *Proc. Inst. Mech. Eng.* C 208 (1994) 275–284.

[6] B.S. Yilbas, The analysis of CO2 laser cutting, *Proc. Inst. Mech. Eng.* B 211 (1997)

[7] B.S. Yilbas, A study into laser cutting process, *Heat Mass Transfer* 32 (1997) 175–180.

INTELLIGENT METAL POWDER LASER FORMING SYSTEM

Fei Xing[1, 2], Weijun Liu[1], Kai Zhang[1,2], Xiaofeng Shang[1],Tianran Wang[1]
[1]*Advanced Manufacture Lab, Shenyang Institute of Automation, CSA, Shenyang, Liaoning Province, China; E-mail: xingfei@sia.cn.* [2]*Graduate School, Chinese Academy of Sciences, Beijing, China.*

Abstract: Metal Power Laser Forming (MPLF) is a high precision and complicated process. The quality of the formed parts is affected by many parameters and may vary significantly during the laser forming process. In order to guarantee the stabilities of the process, a state-of-the-art intelligent MPLF system was developed in this paper. The intelligent system includes CNC system, energy supply and control system, powder delivering and control system, and real-time feedback control system. The established state-of-the-art apparatus and several metal parts formed by the MPLF system are presented in this paper.

Key words: Metal Power Laser Forming, monitor, control, laser cladding, apparatus.

1. INTRODUCTION

Metal Power Laser Forming (MPLF) is an advanced laser materials processing technique, which is widely used in manufacturing, part reparation, surface modification, and Solid Freeform Fabrication (SFF). In the MPLF process, the metal powder sprayed form the powder delivery nozzle and injected into the laser focal zone, together with a thin layer of the moving substrate, is melted by a laser beam and deposited a layer on the substrate [1]. A great variety of materials can be deposited on a substrate to form a layer with a thickness ranging from 0.1 to 2 mm. This technique can produce a much better coating, with minimal dilution, minimal distortion, full-density and good surface quality than other techniques such as arc welding and plasma spraying. These advantages have recently been found attractive in industries for tools manufacturing and metallic rapid prototyping [2]. This

This project was financially supported by the National "863" High-technology Research Project (Grants No. 2002AA420060) and the Knowledge Innovation Project of the Chinese Academy of Science (Grants No.F040208).

Please use the following format when citing this chapter:
Xing, Fei, Liu, Weijun, Zhang, Kai, Shang, Xiaofeng, Wang, Tianran, 2006, in International Federation for Information Processing (IFIP), Volume 207, Knowledge Enterprise: Intelligent Strategies In Product Design, Manufacturing, and Management, eds. K. Wang, Kovacs G., Wozny M., Fang M., (Boston: Springer), pp. 525-535.

technique is also considered as the best way for manufacturing of functionally-graded parts.

Similar to other rapid prototyping techniques, in rapid prototyping using laser forming, a three-dimensional (3D) CAD solid model is used to produce a part without intermediate steps. This approach to produce a mechanical component in a layer-by-layer fashion allows us to fabricate a part with features that may be unique to laser cladding prototyping. These features include a homogeneous structure, enhanced mechanical properties, and production of complex geometries. However, the clad quality may vary significantly during a laser cladding process. Variations of the quality may even be observed between processing cycles performed using the same operating conditions. This poor reproducibility arises from the high sensitivity of laser cladding to small changes in the operating parameters such as laser power, beam velocity and powder delivery rate, as well as to process disturbances such as variations in absorptivity [3]. As a result, development of an intelligent MPLF system is essential for overcoming the effects of disturbances in the process.

At present an advanced intelligent MPLF system is successfully established at Shenyang Institute of Automation Chinese Academy of Sciences. In this paper, the components of the intelligent MPLF system are explained in detail, and several metal parts formed by the LMDS system are presented.

2. THE INTELLIGENT MPLF SYSTEM

The intelligent MPLF system includes CNC system, energy supply and control system, powder delivery and control system, and real-time feedback control system. It can fabricate fully dense, highly accurate complex parts and tools directly from a CAD solid model. Firstly a 3D CAD solid model is sliced into many layers and produces a CLI (Common Layer Interface) files by the system application software. The software also provides a graphical user interface (GUI) for the control of CAD data processing, the motion of the stage, and the powder regulation, as well as the controls of the safety and processing shutters. The laser head and the worktable governed by the CNC system move together to insure that the injected powder is melted onto the substrate as the laser scans over the part. Sensing and control systems are developed to guarantee the stabilities of the process, such as the stability of the melt pool's temperature and shape, the stability of the powder delivery rate, the stability of the single cladding's height. The real photograph and schematic illustration of the MPLF system are presented respectively in Fig.

1(a) and 1(b). The detail explanation of the developed MPLF apparatus will be individually presented as follows.

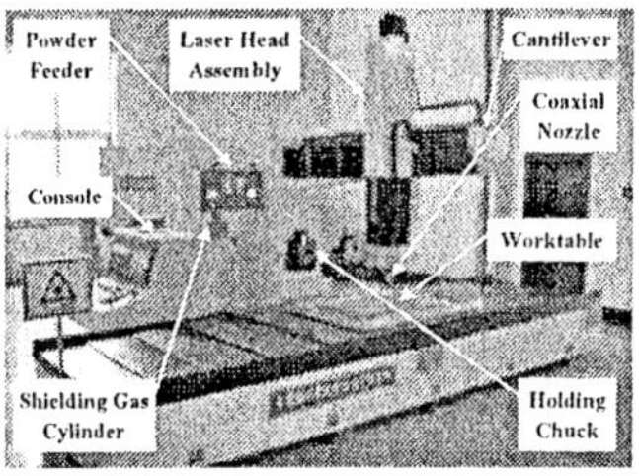

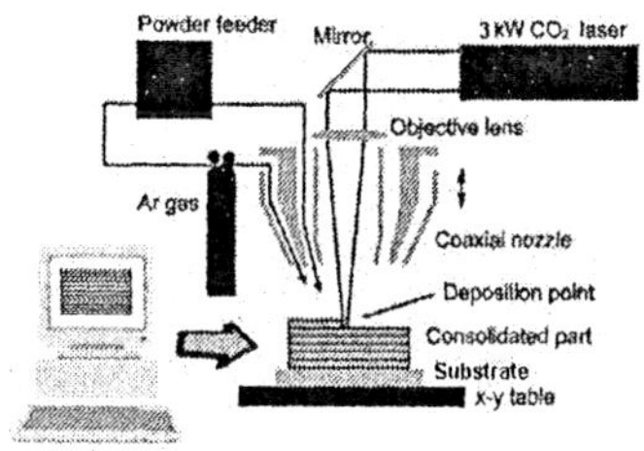

(a) Real photograph of MPLF system. (b) Schematic of the MPLF system.
Figure 1. The intelligent Metal Powder Laser Forming System.

2.1 CNC system

The CNC system consists of industrial computer, several motors, motion control cards, NC worktable, system application software, and corresponding circuits. It mainly includes a six-axis NC machine tool, namely three precise controlled linear axes (i.e. X, Y, Z) and three free rotation axes that rotate around the X, Y, and Z axes individually. The machine tool can achieve four-axis linked motion by the control of the industrial computer.

Two sets of PCI-1240 motion control cards are installed in the computer. The NC worktable is communicated with the industrial computer through one control card. Two motors of the powder feeders are governed by the other control card. During the MPLF process, the X-Y plane worktable achieves the horizontal motion of substrate and deposited material. The powder nozzle head is mounted on a vertical linear Z-axis of the NC worktable. The Z-axis provides the vertical motion of the nozzle head during the deposition process. The positions and velocities of X, Y, and Z axes are regulated via the PCI-1240 motion card by the industrial computer. Each linear axis is equipped with two limit switches which restrict its own motion range in order to ensure the safety and stability of the MPLF system. From Table.1 we can see that the system has good accuracy and repeatability.

Table 1. The performance of the CNC system

Motion range of the worktable	2000 mm × 1000 mm × 500 mm
Accuracy of X, Y, Z axis	± 0.08 mm, ± 0.08 mm, ± 0.05 mm
Repeatability	± 0.05 mm

However, the exact degree of repeatability in this MPLF equipment will vary from table to table but in comparison with the variability associated with the laser and the material it is unlikely to be a significant factor. At the same time a laser forming system based entirely on predictive and theoretical work would be impractical. A better approach to achieve precision laser forming would be to introduce on-line measurement to continually monitor the process and control key parameters rather than rely entirely on predictive information.

2.2 Energy supply and control system

The laser power supply system is a console mounted energy supply with a cooling system. A crosscurrent, pipe sheet, multimode, 3 kW CO_2 LASER, whose wavelength is 10.6μm, is employed in the MPLF system as the power source.

The parameters of the energy supply system are the laser power, beam diameter, beam mode structure, and beam focal point. The effective energy (the amount of energy delivered to the substrate) depends on the laser power, CNC scanning speed, and beam diameter and is expressed as $E = P/(Dv)$ where E is the specific energy (J/mm^2), v is the CNC traverse speed (mm/s), P is the laser power (W), and D is the spot size (mm). Thus, in order to regulate the amount of energy delivered to the substrate, the laser power, beam diameter, and traverse speed have to be adjusted. A power collecting sensor is installed to monitor the laser power, which is displayed simultaneously. A feedback hardware system is developed to adjust the laser power. The feedback instruction sent from the computer via a PC RS232 serial port and an adjusting circuit is used to ensure optimum working electric current. Besides the high-energy LASER, the laser energy supply system includes light gate, calibration laser, reflector, plano-convex lens, and cooling system, as shown in Fig. 2 [4]. During deposition, the light gate, the calibration laser and reflector move together. A semiconductor laser, whose wavelength is 0.633μm, emits the red calibration laser. The calibration laser and the CO_2 laser are concentric. By real-time control of the laser power, many other process parameters, such as the temperature and geometry of the melt pool, can be optimized.

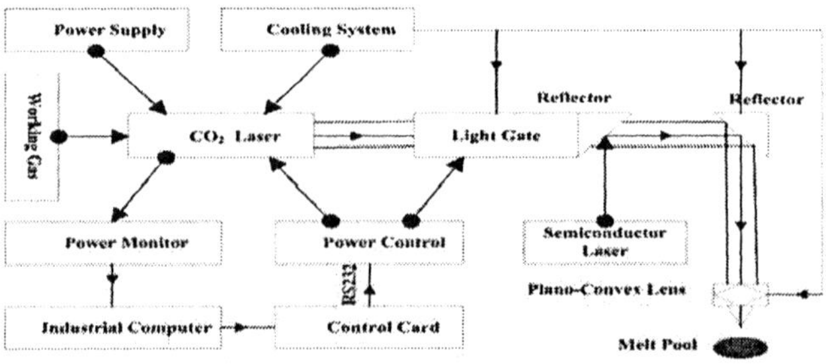

Figure 2. Schematic of the energy supply and control system

2.3 Powder delivery and control system

The stability of the powder delivery speed is one of the most important parameters affecting the quality of the deposited parts. The controllable powder delivery can greatly improve the performance of 3D laser cladding. It also provides the feasibility to build a functional graded material or alloy.

There are two types of powder supply patterns, which are coaxial and lateral powder nozzle. However, when lateral powder nozzle is used, the metal powder can only be injected in one direction. 3D metal powder laser forming mainly depends on the directions of the powder delivery and the worktable movement. In order to solve the problem, a coaxial nozzle is invented to insure the metal powder is injected in the direction of the laser beam, and has been applied into the MPLF system. The real photo and sectional view of the coaxial powder nozzle is individually shown in Figs. 3(a) and 3(b). The powder is fed from four sides of the nozzle and uniformly distributed between powder taper sleeve and cooling water taper sleeve. The nozzle end has been specifically designed to enable the powder to be delivered out at a specific angle to the vertical so that the focal point of the powder coincides with the focal point of the laser beam. This allows maximum powder utilization during deposition. Moreover, the powder flow angle can be regulated by varying the shielding gas flow rates. Indispensably, water-cooling system has been well designed to keep the appropriate operating temperature for nozzle. Since the powder is injected directly to the molten pool, the powder nozzle can move in any directions on the surface. It makes the MPLF system process a high degree of flexibility and efficiency.

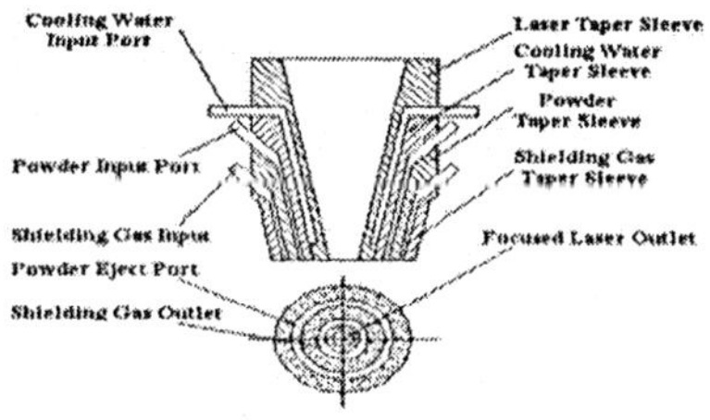

(a) Real photo of coaxial powder nozzle. (b) Sectional view of coaxial powder nozzle.
Figure 3. Coaxial powder nozzle of the MPLF system

Besides the coaxial powder nozzle, two powder feeding chambers is designed, with two individual powder hopper units assembled together, as shown in Fig.4(a). The powder nozzle and the powder feeding chambers together make up of the powder delivery system. Each powder hopper can hold a significant portion of the powder. During the deposition, the powder flows from the hopper outlet down to an annular groove of a rotary table in a chamber, as shown in Fig. 4(b). With the rotation of the disk, the powder in the annular groove is transported to a nearby suction nozzle, which is fixed over the annular groove, but a little space is left between them. By virtue of the air pressure of carrier gas, the powder is forced into the suction nozzle, and then is delivered to the nozzle head along the powder feeding tube. Each powder feeder is driven by a motor, and also has its own feeding tube. The two feeding tubes are connected into a mixing channel through a three-way coupling joint. The converged tube is subsequently divided into four parts by splitters, which uniformly distribute the powder into four tubes leading to the nozzle end. Then the powder is injected into the laser and generated melt pool. When two different kinds of powders are delivered, they can be mixed on line after they pass through the three-way coupling joint. By regularly adjusting the rotational speed of each motor, the functionally graded material can be fabricated freely.

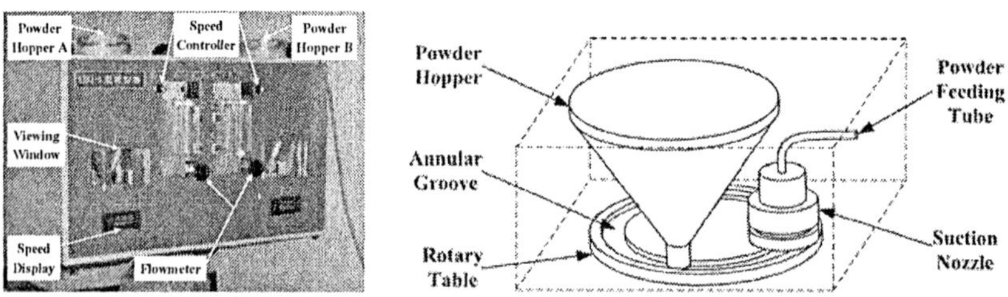

(a) (b)

Figure 4. Powder feeding chambers of the MPLF system. (a) Real photo of powder feeding chambers. (b) Schematic of inner structure of powder feeding chamber.

In order to adjust the delivery speed of the metal powder, an optoelectronic sensor is developed to sense the real-time powder rate, as shown in Fig. 5(a). The sensor consists of a laser diode, a photo diode, and a glass window. The components are installed in such a way that the laser beam emitted from the laser diode passes through the powder stream flowing inside the glass chamber and is received by the photo diode. The carrier gas and the metal powder are mixed well, so the powder particles distribute uniformly in the carrier gas. Because of the diffusion, absorption, and reflection of the powder particles to the laser beam, the laser energy received by the photo diode decreases if the powder delivery rate increases, which means there is a higher percentage of powder particles in the carrier gas [5].

An on-line electronic scale is used to calibrate the sensor. The averaged powder delivery rate is calibrated by measuring the time period and the weight of the powder delivered during the time period. The corresponding sensing signal from the sensor is also acquired by a data acquisition card installed on the industrial computer. A linear relationship between the powder delivery rate and the averaged sensor output is established, as shown in Fig. 5(b). Then we can acquire the delivery rate according to the sensor output. A PID controller is developed to control the powder feed motor based on the delivery rate and the cladding height. The delay time of the powder delivery is measured beforehand in the control system [6].

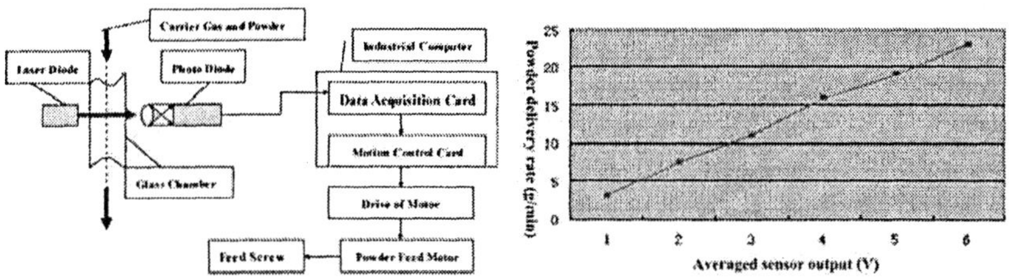

(a) Schematic of the feedback control system. (b) The performance of the sensor.
Figure 5. Metal powder monitoring and control system

2.4 Real-time feedback control system

As addressed in above sections, close monitoring and control of the variables of these systems are required to enhance the parts quality. Feedback controllers have to be designed to regulate these variables mainly to control melt pool size, temperature distribution in the melt pool, and cladding height and width. These variables may vary during the operation due to fluctuations in system parameters such as powder delivery rate, beam position and diameter, output power, and scanning speed, and presetting these operating parameters are not appropriate. They have to be monitored and optimized continuously to obtain the desired conditions.

2.4.1 Feedback control system of the melt pool

Temperature is a critical factor that requires continuous monitoring and control. It determines the melt pool dimensions and, hence, the dilution. If the temperature is too low, the resulting melt pool catches little powder and if the temperature is too high, it may melt back the substrate.

During the MPLF process, the temperature of the melt pool can reach 2500K while high melting-point powder is injected, so non-contact measurement technique should be chosen to monitor the features of the melt pool. Colorimetric technique is developed to obtain the 2D temperature distributions of the melt pool [7]. The melt pool is sensed by a high frame-rate camera installed at the side of laser head, which takes grey images with a 640×480 resolution. The infrared radiation from melt pool produced by the laser beam passes through the lens and forms an image on the CCD chip of the camera.

In terms of the light path structure, the sensor is installed in the form of single passage, which signifies only one CCD sensor is used in the measuring system, as shown in Fig. 6(a). Two kinds of optical filters, which respectively capture infrared ray with specific wavelengths (790nm and 921nm), are fixed on a turnplate. With the rotation of the turnplate, the filters are put into the way of the light alternately. Since a step motor is used as driver and a photoelectric switch is used to ensure the position of the filters, the turnplate keeps fast rotary speed and accurate fix position. Considering the thermal inertial of the melt pool, two images of the melt pool are obtained in so little time by the high frame-rate camera that the two images can be regarded as being taken at the same time. Then we can acquire the temperature field and the shape of the melt pool through the colorimetric technique. As mentioned above, the features of the melt pool are greatly judged by the laser power and the scanning speed. Fig. 6(b) shows a fuzzy logic controller which is designed to maintain the stability of the melt pool by adjusting the inputs of the scanning speed and laser power based on the feedbacks [8].

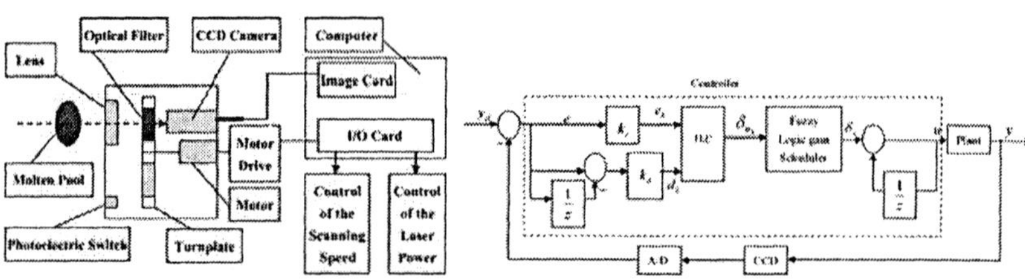

(a) Schematic of the feedback control system of melt pool. (b) Fuzzy logic controller.

Figure 6. Feedback control system of melt pool.

2.4.2 Feedback control system of the cladding

The dimensional accuracy of the parts depends on the uniformity and repeatability of the clad height and width being deposited. The cladding geometry is another important factor that requires continuous sensing and control. The feedback control system can not only improve the precision of the formed parts, but also save costs for the post treatment of the parts.

The monitoring and control system is constructed by an industrial CCD camera, a line-structure LASER, an industrial computer, an image card, and system software. Fig. 7(a) shows the schematic diagram of the monitoring system of the cladding. The CCD camera and the line-structure laser are installed at different sides of the laser head. A line laser emitting from the line-structure LASER forms a line on the surface of the metal part. At the same time the camera takes photos of the part, which are grabbed by the image card and transmitted to the computer. The program designed beforehand in the computer would manipulate these images, calculate the three-dimensional coordinates of the image points based on the world reference frame, and obtain the feedback of the height. The computer program on calculation of the cladding height includes four primary steps: (1) calibration of the camera and the line laser plane [9]; (2) calculation of the three-dimensional coordinates of the image points based on the camera reference frame; (3) calculation of the relational matrix between the camera reference frame and the world reference frame; (4) acquisition of the corresponding points of different claddings and calculation of the height. A feedback is acquired by comparing the actual height with the presetting height of the cladding layer. Besides the height of the cladding, the width can also be acquired by the CCD camera. As mentioned above, A PID controller is designed to adjust the powder feed motor in order to guaranteeing the stability of the cladding height. Fig.7 (b) shows the schematic of the feedback control system of the cladding.

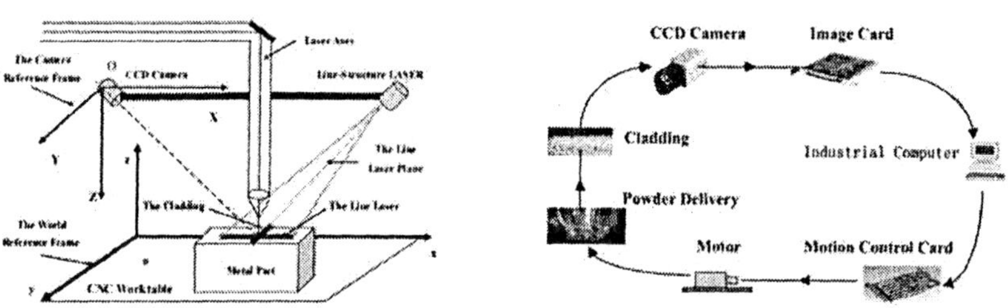

(a) Schematic of the monitoring system. (b) Schematic of the feedback control system

Figure 7. Monitoring and control system of the cladding

3. METAL PARTS FORMED BY THE INTELLIGENT MPLF SYSTEM

Base on intelligent MPLF system mentioned above, multilayered claddings were successively made to explore the feasibility of the fabrication of metal parts. With the optimal processing parameters ascertained by previous experiments and listed in Table 3, several fully-dense nickel-based metal parts were formed by the MPLF system, four of which are shown in Fig. 8. All of them demonstrate fine appearances and good surface quality. Besides, they are free of porosity, fully-dense, and near-net shaped. The density of these metal parts is expected to be the same as the parts fabricated by conventional manufacturing processes, because each layer was fused with metallurgical bonding and is free of porosity. In addition, rapid solidification during deposition ensures the fine microstructure and perfect properties of the fabricated metal parts.

Table 2. The optimal processing parameters of the MPLF system

Laser power (W)	Scanning velocity (mm/s)	Spot diameter (mm)	Powder delivery rate (g/mm)	Scanning space (mm)
1000	5	2	10	1.3

(a) Thin-wall sample. (b) Complex polygon sample. (c) Rotary blade sample. (d) Acronymic emblem of Shenyang Institute of Automation.
Figure 8. Nickel-based metal parts formed by intelligent MPLF system

4. CONCLUSIONS

In this paper, an intelligent MPLF system is introduced. The detailed design and components of the MPLF system is presented, and its performance is evaluated. A CNC system is established to guarantee the accuracy and repeatability of the MPLF system. From experiments we find the performance of the system is good. A 3 kW CO_2 LASER is employed as the power source. The laser power is acquired and displayed by a power collecting sensor, and can be adjusted by a control card and an adjusting circuit. Powder supply and control system plays an important role in MPLF process. The design of the coaxial nozzle makes metal parts 3D fabrication possible, and the design of a state-of-the-art powder feeding apparatus makes that the functionally graded material parts can be fabricated freely. An optoelectronic sensor is developed to monitor the metal powder speed, which can be adjusted via the powder feed motor by the industrial computer and motion control card. The real-time sensing and control systems of the powder delivery speed and the laser power indicate feasible methods to guarantee the stability of the important process parameters, such as the temperature and geometry of the melt pool, the height and width of the cladding.

The process parameters need to be monitored and optimized continuously. Two feedback control systems of the melt pool and cladding, which are two critical factors of the MPLF process, are developed in this paper. Machine vision systems, based on industrial CCD cameras and computer, are used to sense the forming process. Colorimetric technique, a non-contact measurement technique, is developed to obtain the 2D temperature distributions of the melt pool. A stereovision technique is developed to acquire the height of the cladding. Based on the feedbacks, the scanning speed (worktable feeding speed), laser power, and powder delivery rate are regulated instantly. Based on many experiments and analyses, a number of metal parts have been formed by the intelligent MPLF system. These parts illustrate the capability of the system to fabricate fully-dense and near-net shaped parts.

All of the above-mentioned achievements pave the road towards an intelligent MPLF system. However, this multidisciplinary technology needs to be developed in many different aspects of technique to arrive at a fully functional intelligent MPLF machine. These techniques include the state-of-the-art apparatus, advanced and knowledge-based controller, classic fabricating technics, and abundant experiment experience.

5. REFERENCES

[1] Zhang Kai, Liu Weiju, Shang Xiaofeng, (2005), Research on the processing experiments of laser metal deposition shaping, *Optics & Laser Technology*, In press.

[2] J. Mazurnder, A. Scifferer, and J. Choi, (1999), Direct materials deposition: designed macro and microstucture, *Material Research Innovation,* vol. 3, no. 3, pp. 118-131.

[3] Radstok E, (1999), Rapid tooling, *Rapid Prototyping Journal,* Vol.5, No.4, pp. 164-169.

[4] Shang Xiaofeng, Liu Weijun, Wang Tianran,(2004), Laser Engineered Net Shaping Technology, *Journal of Shanghai University*, 8 (A02), pp. 26-29.

[5] Dongming Hu, Radovan Kovacevic, (2003), Sensing, modelling and control for laser-based additive manufacturing, *Machine Tools and Manufacture,*vol.43, pp.51-60.

[6] Guoqing Ning, et al, (2002), Research about close-loop control system during laser direct manufacturing metallic components, *Applied Laser,* vol.22, no.2.

[7] Yuichiro Tago, Funmie Akimoto, Kuniyuki Kitagawa, Norio Arai, (2004), Measurements of surface temperature and emissivity by two-dimensional four-color thermometry with narrow bandwith, *ENERGY.*

[8] J. Yen and R. Langarani, (1998), Fuzzy logic intelligence, control, and information, *Prentice Hall,* pp. 194-210.

[9] Liu Weijun, Dong Zaili, Hao Yingming, Zhu Feng, (2001), A robot self-location method based on stereovision system, *High Technology Letters*, pp.91-94, no.10.

STUDY ON DYNAMIC CHARACTERISTICS OF A NEW TYPE OF TRANSMISSION

Changping Zou
College of Mechanical and Packing Engineering, Chongqing Technology and Business University,China;Email:cpzou@ctbu.edu.cn

Abstract: The new type of transmission is a kind of swaying mechanism specially developed for certain aircraft, having low rotational speed and great torque. Utilizing criterions of mass equivalent and bending stiffness equivalent, this paper firstly applies simplification to the transmission, getting a piece of continuous beam with varying circular sections. According to the mounting condition of transmission, then establishes the finite element dynamic analyzing model, using I-DEAS software, selecting method of synchronous vector iteration, finishes normal modal analyzing. Next, carries Hertzian contact resonance analyses to ball straight-rotary pair, ball spline pair and ball unload pair, calculates contact resonance frequencies. Considering the resistance torque being moving still concludes formulae of quantity of contact resonance frequencies. Lastly, based on load spectra, makes a conclusion that antiresonance behavior of the transmission is well, showing that the design plan is feasible. Experimentation of prototype indicates that the above theoretical analyses are correct.

Key words: Ttransmission; Hertzian contact; Dynamic analysis; Finite element method

1. INTRODUCTION

The new type of transmission is a new type of ball helical transmission with helical lift angle of 72° specially developed for certain aircraft,As shown in Fig.1. As shown in Fig.2, the new type of transmission is composed of two parts, one is high-pressure oil cylinder with piston style, another is swaying perform mechanism, including ball straight-rotary pair, ball spline pair and ball unload pair. It transforms linear movement of piston

Please use the following format when citing this chapter:
Zou, Changping, 2006, in International Federation for Information Processing (IFIP), Volume 207, Knowledge Enterprise: Intelligent Strategies In Product Design, Manufacturing, and Management, eds. K. Wang, Kovacs G., Wozny M., Fang M., (Boston: Springer), pp. 536-543.

into sway of leading edge of vane using pure scroll of working balls. Thus, ball straight-rotary pair, ball spline pair and ball unload pair are important assemblies for transferring dynamic force and supporting loads.

There is no doubt that dynamic question is very important to aircraft. Hence, analyses on Dynamic characteristics of the new type of transmission are significative. This paper intends to compute natural characters of this transmission by its dynamic analytical simplified beam model, and apply Hertzian contact resonance analysis to ball straight-rotary pair, ball spline pair and ball unload pair.

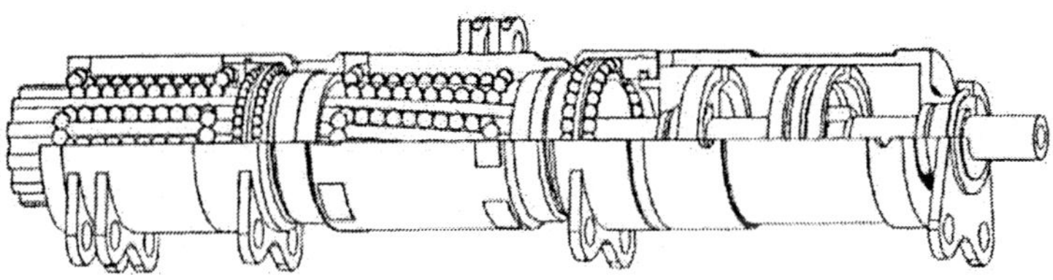

Figure 1. Configuration of the new type of transmission

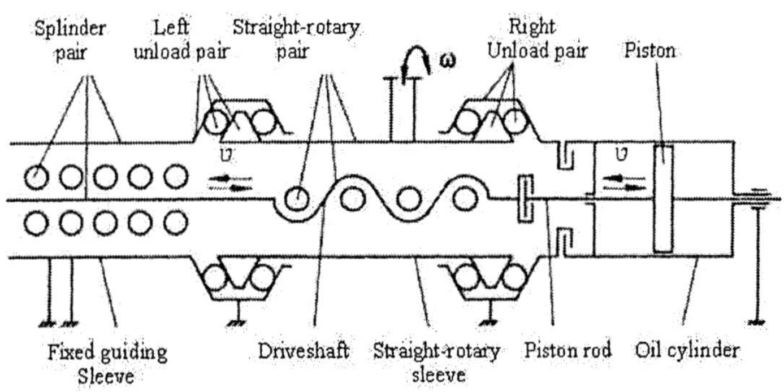

Figure 2. Working principle of the new type of transmission

2. CONFIGURATION AND WORKING PRINCIPLE OF THE NEW TYPE OF TRANSMISSION

2.1 Configuration of the new type of transmission

As shown in Fig.1, main body of the new type of transmission is cylinder; its top and bottom flanks are ears. During installation, five fixed ears of the transmission, sited on bottom flank, were hinged to the base; two moving ears, sited on top flank, were hinged to leading edge of vane needing rotary motion, can sway.

2.2 Working principle of the new type of transmission

The high-pressure oil cylinder is power source;it transforms pressure of high-pressure oil into axial thrust. Ball straight-rotary pair transforms linear movement of driveshaft into sway of straight-rotary sleeve, at the same time, transforms axial force into driving torque, as showed in Fig.2.

3. COMPUTATION ON NATURAL CHARACTERS OF THE NEW TYPE OF TRANSMISSION

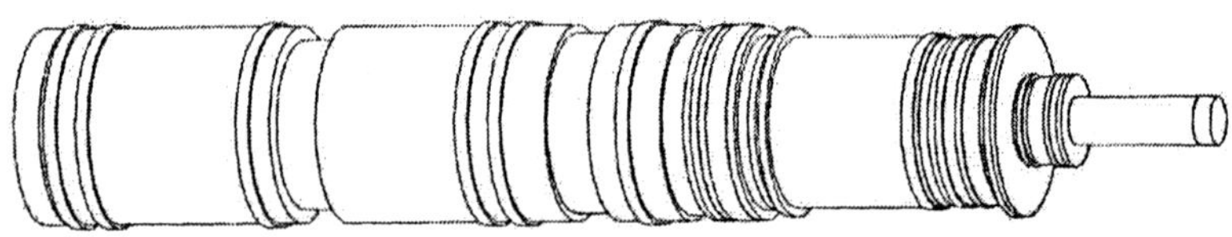

Figure 3. Simplified model of the ball straight-rotary actuator

According to the criterions of mass equivalent and bending stiffness equivalent, the new type of transmission is simplified to a piece of continuous beam with varying circular sections (Fig.3). Using I-DEAS software, according to the mounting condition of transmission, select 2D-beam element, finite element dynamic analytical model of the new type of transmission can be got easily (Fig.4). In the abstract, weather synchronous vector iteration (i.e. SVI method) or Guyan reduction can be selected for dynamic calculating. In fact, according as load spectra, since only modes falling into 0~10000 Hz need to be computed, it is better to select SVI method.

Total eight modes are got; each order natural frequencies are listed in Table 1,their corresponding modal shape are shown in Fig.5.From Table 1 and Fig.5, it can be seen that natural frequencies and modal shapes of orders 1~2, 3~4,5~6,7~8 are equal respectively. This is because that modes of order 1,3,5 and 7 denote all orders of transversal vibration within XY plane of the new type of transmission, whenas modes of order 2,4,6 and 8 denote all orders of transversal vibration within XZ plane of the new type of transmission. Otherwise, since when applying boundary condition, the X, Y and Z degree of freedom of all nodes on inner cylindrical face of five bottom ears of the transmission are restrained to death, there is no vibration along axis X of the transmission.

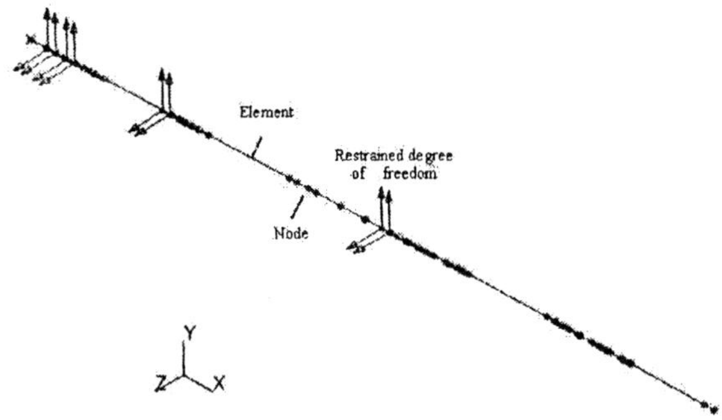

Figure 4. Finite element dynamic analytical model of the new type of transmission

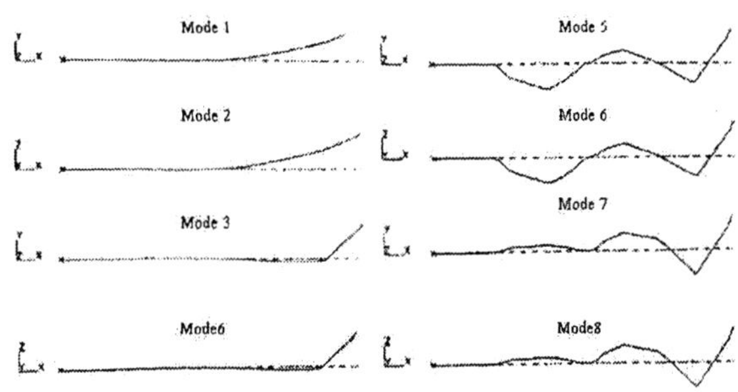

Figure 5. Preceding eight orders natural modal shapes of the new type of transmission

Table 1. **Natural frequencies of the new type of transmission (Unit: Hz)**

Order of mode	1	2	3	4	5	6	7	8
Natural frequency	1892	1892	3461	3461	4738	473	7889	7889

4. HERTZIAN CONTACT RESONANCE ANALYSES

4.1 Conception of Hertzian contact and basic calculating formulae

When the new type of transmission is operating, balls are pressed closely to the inner and outer rolling ways. Material rounding the contact point deforms under the normal pressure; incipient contact point is changed into a small plane. Profile of this contact plane is generally an ellipse. Stress and deformation acted on contact zone can be determined according to Hertzian method. Such contact is name as Hertzian contact.

Loaded under pressure *P*,the two contacted elastomers are compressed along normal direction with elasticity,coming near with a quantity of δ.If materials of the elastomers are steel,taking E_1=E_2=2.07×$10^{11}$$P_a$, υ_1= υ_2=0.3, simplified formula of *a*,*b* and δ are:

$$a = 0.0236 m_a (P / \sum \rho)^{1/3} \qquad (1)$$

$$b = 0.0236 m_b (P / \sum \rho)^{1/3} \qquad (2)$$

$$\delta = 2.79 \times 10^{-4} (\frac{2K}{\pi m_a})(P^2 \sum \rho)^{1/3} \qquad (3)$$

where a is long radius of contact ellipse,mm;b is short radius of contact ellipse,mm; δ is decrease of distance between the two contacted elastomers,mm; P is normal pressure,N;. Σ ρ = ρ 11+ ρ 12+ ρ 21+ ρ 22, ρ ij is master curvature,1/mm, ρ ij =1/Rij; moreover,Rij is master radius of curvature of elastomers surface at contact point,mm,i,j=1,2,i or j denotes one of the two elastomers;ma,mb and 2K/(π ma) are corresponding to function F(ρ) of master curvature, can be looked up from Table 5-1 [4];for ball straight-rotary pair, ball spline pair and ball unload pair,F(ρ) is given by Eq.(4) as

$$F(\rho) = (\rho_{11} - \rho_{12} + \rho_{21} - \rho_{22}) / (\rho_{11} + \rho_{12} + \rho_{21} + \rho_{22}) \qquad (4)$$

4.2 Conception of Hertzian contact resonance and basic calculating formulae

Hertzian contact can be regarded as a spring because of elastic, it relates to masses of two contacted elastomers, may bring vibration. If frequency of duty is the same as resonance frequency, resonance will create, which is named as Hertzian contact resonance.

For ball straight-rotary pair, ball spline pair and ball unload pair, contact of ball with inner or outer rolling ways or collars can be regarded as a spring, thus each contact pair can be considered as compound of two springs (Fig.6a). Since mass of ball is smaller than inner and outer rolling ways or collars, ball is omitted, as shown in Fig.6 (b). In Fig.6,mo,mb mi denotes masses of outer rolling way or collar, ball and inner rolling way or collar respectively. Ki and Ko denote stiffness of spring connecting mb with mi or mo, N/m.

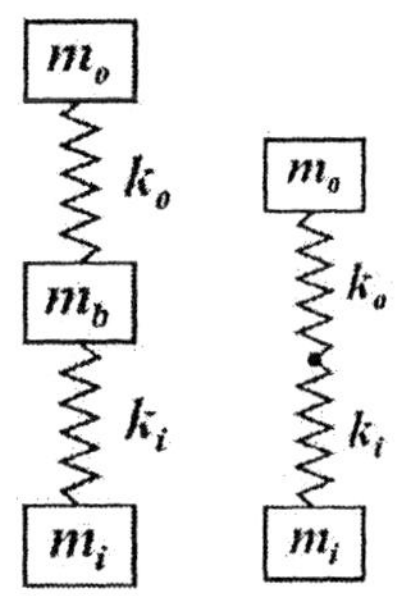

(a) (b)

Figure 6. Vibrating systems with masses and springs of Hertzian contact

Assume one body be fixed, Hertzian contact resonance can be determined by Eq. (5) as

$$m\ddot{x} + kx = 0 \qquad (5)$$

Solving Eq. (5), we find that resonance frequency is

$$f = \sqrt{k/m}\,/(2\pi) \qquad (6)$$

where

$$k = 1/(1/k_i + 1/k_o) \qquad (7)$$

4.3 Calculating formulae or methods of master radii of curvature and master curvatures

The key is to determine two master planes correctly. For ball spline pair, the first master plane is coinciding with the rolling way, i.e. longitudinal; the second plane is perpendicular to the first plane, i.e. transversal. For ball straight-rotary pair, the first master plane is the normal section of helical rolling way at contact point, being perpendicular to tangent line passing contact point, master radius of curvature is computed by program compiled according to differential geometry; the second plane is perpendicular to the first plane. For left or right ball unload pairs, refer to centripetal thrust ball bearing, the first master plane is an axial section passing contact point; the second plane is perpendicular to the first plane passing contact point.

4.4 Computation of contact stiffness and resonance frequencies

4.4.1 Computing formulae of contact stiffness

In Eq. (7), K is total stiffness; formulae of K_i and K_o are given by Eq. (8) as

$$k_i = P_i / \delta_i , k_o = P_o / \delta_o \tag{8}$$

where P_i and P_o is normal pressing force, N.

4.4.2 Computing method of resonance frequencies

For each contact pair, meanings of m in Eq. (6) are different.

1. For ball straight-rotary pair, assume driveshaft be fixed, only straight-rotary sleeve is studied; there are six rows of balls, one of which is studied; m equals to one-sixth of mass of straight-rotary sleeve.
2. For ball spline pair, fixed guiding sleeve is fixed, one-sixth of driveshaft is studied, m equals to one-sixth of mass of driveshaft.
3. For left ball unload pair, left bearing base (i.e. outer collar) is fixed, left bearing housing (i.e. inner collar) is studied, m is mass of left bearing housing.
4. For right ball unload pair, right bearing base is fixed, right bearing housing is studied, m is mass of right bearing housing.

4.4.3 Result of computation

For ball straight-rotary pair, ball spline pair, left ball unload pair and right ball unload pair, Hertzian resonance frequencies are 2638,3061,1984 and 2731 Hz, respectively. Since the four Hertzian resonance frequencies approach upper limit of the frequency of load spectra, we can make a conclusion that antiresonance behavior of the transmission is well, and the design plan is feasible.

4.5 Calculating formulae of contact resonance frequencies

In fact, considering the resistance torque acted on transmission is moving, accordingly Hertzian contact pressure P is variable, thereby, above result is just corresponding to maximal control torque. Calculating formulae of contact resonance frequencies of four contact pairs are given by Eq. (9) as

$$\begin{aligned} f_Z &= 654.19P^{1/6}, f_H = 758.98P^{1/6}, \\ f_{ZX} &= 672.66P^{1/6}, f_{YX} = 751.90P^{1/6} \end{aligned} \tag{9}$$

where f_Z,f_H,f_{ZX} and f_{YX} denote Hertzian contact resonance frequencies of straight-rotary pair, spline pair, left unload pair and right unload pair,Hz.

5. CONCLUSIONS

According to the structural feature of the new type of transmission, this paper has built the dynamic analytical simplified beam model,and finished the theoretical calculation of natural frequencies and normal modes using I-DEAS software. This paper has also completed the Hertzian contact resonance analysis of the four ball contact pairs. The above analyses and computations prove that dynamic characteristics of the new type of transmission satisfy the demands of certain advanced aircraft.

6. REFERENCES

1.Zou Changping.Three-dimensional full parameterized modeling methods on the key parts of a new type of swinging device (*in Chinese*).*Machine Tool & Hydraulics*, 2005, 12: 172-173,182.
2.Xu Yanzhong,Jiang Shuyun.Dynamic characteristics of high speed angular-contact ceramic ball bearing.*Journal of Southeast University*,2004,Vol.20,No.3:319-323
3.Wu Zongze.Advanced *machine design(in Chinese)*. Beijing: the publishing company of Tsinghua University, 1991
4.Rao Zhengang,Wang Yongwei. *Ball screwed rod pair and locking equipment(in Chinese)*. Beijing: the publishing company of national defense industry,1990
5.Cheng Guangren,etc. *Design elements of ball screws transmission (in Chinese)*. Beijing: the publishing company of mechanical industry,1987

THE APPLICATION OF EMBEDDED TECHNOLOGY IN NC SYSTEM

Yong Yin[1], Zude Zhou[1], Quan Liu[1], Fangmin Li[1], Yihong Long[1]
[1]*School of Information Engineering, Wuhan University of Technology; Email: yiyng_hust@126.com*

Abstract: The paper analyzes the inherent drawbacks of the traditional NC system based IPC, puts forward and then provides the embedded NC system with the architecture of ARM and DSP, in which the ARM takes charge of the G code input, compiling and interpretation, the simulation display, the user's interface and the network interface etc, while the DSP is mainly in charge of real-time tasks such as interpolation computation and path control. The paper also introduces the hardware and software design in detail. Experiments show that NC system built on the embedded technology can reduce the system hardware scale, facilitate application development, cut down the costs and promote the system's reliability and real-time performance.

Key words: NC system, Embedded technology, ARM, DSP.

1. INTRODUCTION

Embedded system is the production integrated with advanced technologies of computer, communication, semiconductor, microelectronics and sensor etc. Its hardware and software can be tailed and configured readily, and it is especially applicable to systems with specific requirements in functionality, reliability, cost, volume and power consumption etc [1].

Present NC systems are mainly based on the IPC (Industrial Personal Computer) architecture, which have their inherent defects on both the structure and the performance [2]. First, the IPC structure is not defined for NC processing. As the IPC is on the basis of the universal structure, it dose not take special account for the NC features from the point of both the bottom hardware design and the operation system environment, thus it is difficult to construct NC system. Second, the NC system of IPC structure

This project is supported by the National Natural Science Foundation of China under the grant No. 50335020.

Please use the following format when citing this chapter:
Yin, Yong, Zhou, Zude, Liu, Quan, Li, Fangmin, Long, Yihong, 2006, in International Federation for Information Processing (IFIP), Volume 207, Knowledge Enterprise: Intelligent Strategies In Product Design, Manufacturing, and Management, eds. K. Wang, Kovacs G., Wozny M., Fang M., (Boston: Springer), pp. 544-549.

can't ensure the real-time property and reliability. As the IPC with the universal operation system is occupied by many other tasks unrelated with NC processing, which consume large amount of system resource, interfering with the NC's foreground priority. The irrelevant tasks also result into the instability of the NC system. Third, the cost of the NC system on the ground of IPC structure is very high. To equip with the hardware and software of an IPC system will take thousands of dollars, while a piece of embedded microcontroller is no more than fifty dollars, which reduces the cost of the whole system dramatically [3].

In the paper, the application of embedded technology base on ARM7 and DSP to NC system is introduced. The embedded NC system has been developed successfully by Hubei Digital Manufacturing Key Laboratory and Embedded System Research Laboratory in Wuhan University of Technology. The system is based on the S3C44B0 microcontroller of ARM7 series of the Samsung Corporation and the TMS329F2812 microprocessor of the TI Company. The following sections of the paper are to introduce the architecture, the hardware and software design of the system based on embedded technology.

2. SYSTEM ARCHITECTURE

The embedded NC system of the paper is not built on IPC, but on the embedded architecture of ARM and DSP. Other accessorial units such as servo control unit, PLC unit, communication unit, display unit and data input unit etc, are also constructed on the basis of embedded technology. The embedded NC system is composed of G code interpretation module, interpolation module, tools' compensation module, servo control module, operation interface module, network interface module, system management module, simulation and display module and signal detection and fault diagnosis module etc, as shown in Fig 1. The executive workflow of the system is as follows. The G code interpreter translates the user's codes into instructions to control each motor. The instructions include G, T, S, F and M. For M instruction, the system sends it to the interpolation module to control the feed motors. The T, M, S and F instructions are accessory, with which the NC system controls the assistant devices such as the cooling liquid controller, the clamps and the lubricant controller etc. As the NC system executes the accessory instructions, it transfers them to the PLC to control the peripheral devices. In the system, the PLC module is designed with soft-core, which drastically decreases the cost and reduce the volume of the hardware [4]. Besides executing these accessory instructions, the PLC also accepts the I/O signals from the control panel and the alarm signals from the system. The interpolation module, tools' compensation module and the position control module are also the essential unit of the system, whose accuracy and real-time property are the vital indices of the embedded NC system.

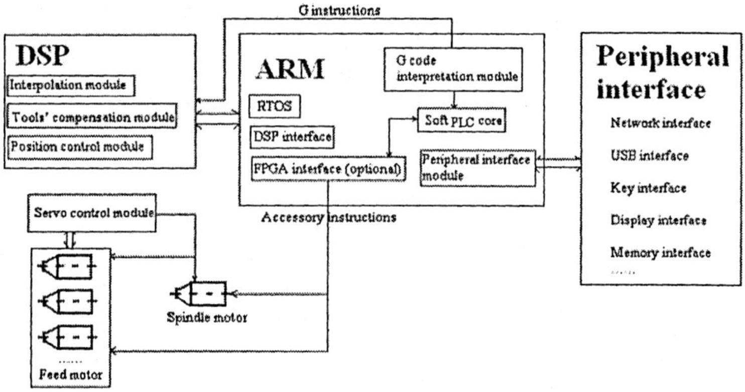

Figure 1. NC system with the architecture of ARM and DSP.

The G code interpretation module, the soft PLC module, the processing simulation module and the operation interface module are implemented in the ARM. So the ARM takes charge of the functions of G code input, compiling and interpretation, the simulation display, the user's interface and the network interface etc. The DSP is mainly in charge of real-time tasks such as interpolation computation and path control.

3. SYSTEM HARDWARE

The main control CPU of the system is S3C44B0 chip. This chip has abundant and powerful peripheral resources and communication ability with very high operation and data processing speed, so it can realize functions of data collection, disposing and process control conveniently. The circuit board is four layers with 5v DC power support and its highest main system frequency is 66MHz. The system is configured with 32M Flash, 16M SDRAM, two USB interface of USB2.0 protocol, two Ethernet interface with RTL8019AS chip, one 256 color STN LCD with 640*480 pixels and one PS2 keyboard interface, as shown in Fig2

As there is no control and driving core of CAN bus protocol inside S3C44B0, CAN bus controller and driver are indispensable in the system. In here we choose SJA1000CAN CAN bus controller and 82C250 bus driver of Philips Company to implement the CAN bus interface with S3C44B0 chip, as shown in Fig3.

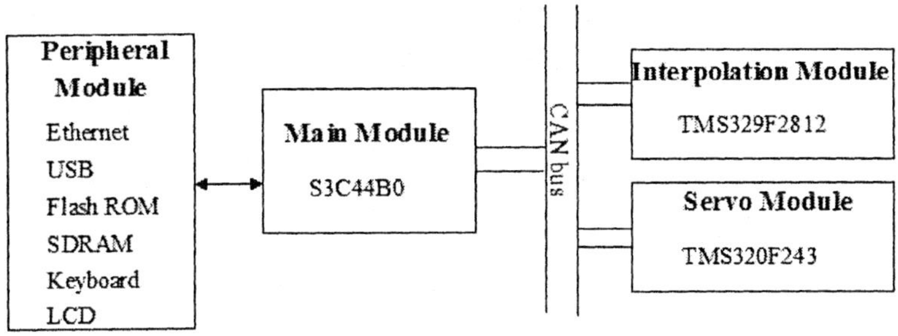

Figure2. System hardware framework

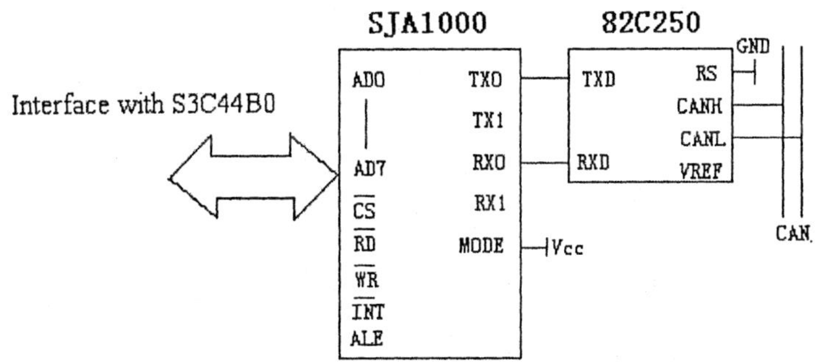

Figure3. CAN bus interface with S3C44B0

Real-time tasks such as interpolation computation and path control are in the charge of DSP. In the system, the DSP is adopted of TMS329F2812 microprocessor of the TI Company, with its highest main frequency 150MHz and its clock cycle 6.67ns, which meets the requirement of the system's real-time performance absolutely. The chip TMS329F2812 microprocessor communicates with S3C44B0 chip through CAN bus. As a CAN bus controller is integrated into TMS329F2812 microprocessor, that is, all functions of the data link layer and application layer of CAN bus protocol is realized inside the TMS320F2812 chip, only a CAN bus drivers 82C250 is needed in the system to implement CAN bus linkage.

As DC motor, AC motor or step motor can all be used in the NC system, considering the generality and reconfiguration of the system, the servo module is also linked to the system through CAN bus. Since a clock frequency of 20MHz is enough for servo control, taking cost into account, the TMS320F243 chip of TI Company is adopted as servo CPU, not TMS329F2812 chip. TMS320F243 is one of the signal processing chips of the Fixed-point series, which is designed specifically for digital motor control. Because a CAN bus controller is also integrated into TMS320F243 chip, it is easy to fulfill communication functions through CAN bus between the main control module of S3C44B0 and the servo module of TMS320F243 and to achieve the reconfiguration functions of the system.

4. SYSTEM SOFTWARE

The software architecture of the embedded NC system can be separated into three parts, namely the operation system environment, embedded NC core and application program.

The RTOS in the embedded NC system is based on U-cosii, which is transplanted into the ARM microcontroller, providing file system, network interface, memory management and multi-task schedule etc to the NC system. The embedded NC core software, including the PLC unit, I/O control unit, position control unit, G code interpretation unit and tool's compensation unit etc, are supplied in the form of API to the user's application program. On the upper layer, users can develop various kinds of application program according to specific numerical controlled processing.

As U-cosii is a high performance, deterministic, real-time, scaleable, reliable and modularized RTOS, it can be embedded into the NC system based on ARM readily, and users can extend or delete functional modules of both hardware and software easily [5].

In the system, closed-loop adaptive motion control is introduced, so the tool path can be adjusted adaptively in a pre-specified manner with the online feedback from the processing. A fuzzy-PID control algorithm is implemented in DSP, which improves the control accuracy of the system. The system adopts a two-step interpolation algorithm, namely first rough interpolation and second fine interpolation. The first step is in the charge of ARM processor, which outputs not control pulses, but many line sections to the DSP processor. DSP takes charge of the second step, namely it marks out the motion paths of each feed axis, and then outputs fine and real-time motion control pulses.

5. CONCLUSIONS AND FUTURE WORK

An economical embedded NC system prototype shown in Fig4 has been developed successfully in the Embedded Lab of Wuhan University of Technology.

Figure4. Fig6 Embedded NC system prototype

Experiments over the prototype demonstrate that NC system built on the embedded technology can reduce the system hardware scale, facilitate application development, cut down the cost and enhance the system's reliability and real-time performance. Members in the embedded lab of WHUT are bending themselves to realize core NC system functions on SOC (System On Chip) and try to realize user interface based on MiniGUI. We hope that products based on this technology have a perspective market in the future.

6. REFERENCES

1. Zhang X, (2003), *Embedded System and Embedded Operation System*, Modern Electronic Technique, vol.8, pp.20-24
2. Zhou. Z.D, Liu, Q, Long. Y.H, Li. F.M, (2005), *Embedded Technology & Digital Manufacturing.* Digital Manufacturing Science, vol.3. pp.1-10
3. Sun. K, Wang. T.M, Wei. H.X, (2003), *Design and Implement of Embedded and Reconfigurable CNC System Based on ARM,* Machine Tool & Hydraulics, vol.6, pp.302-304
4. Zhou. Z.D, Chen. Y.P, Yu. W.Y, (2001) *Concept and Related Scientific Problems of Digital Manufacturing.* China Mechanical Engineering, vol.12, pp.100-104
5. Han. Q.K, Deng. Q.X, Wen. B.C, (2005), *Applications of Embedded Technology to Large Intelligent Equipmennts.* Digital Manufacturing Science, vol.3, pp.1-17

THE RESEARCH OF FUZZY PID CONTROL STRATEGY BASED ON NEURAL NETWORK IN THE TENSION SYSTEM

Yongyi He, Shuai Guo, Qihong Zhou
School of Mechanical Engineering and Automation, Shanghai University, Shanghai, China.
Email: richie_gs@sina.com.

Abstract: Based on back-propagation (BP) arithmetic, a neural network Fuzzy PID control strategy is presented for the tension systems, the parameters of which are of deep coupling, nonlinearity, time-variability, and indeterminacy. In this control strategy, the parameters such as proportion, integration and differentiation of PID controller are adjusted on-line through self-learning of a network based on classical PID control algorithm. The decoupling control is carried out in the process. The performance of control strategy has been performed on the tension tested-bed under the varying load conditions. The experimental results show that the almost dynamic decoupling and completely static decoupling are obtained.

Key words: Tension System; Decoupling Control; Neural Network

1. INTRODUCTION

Tension control technology is one of the common technologies exclusively used in the engineering industry such as wire saw, IC manufacture, printing equipment, fibre wind and etc. Along with the modern wind machine developing to higher speed and more precision, tension control technology becomes more important. Due to the fact that tension system can be affected by many factors, its structures are complex and its parameters are of deep coupling, nonlinearity, time-variability, and indeterminacy. Traditional control method is hard to carry out effective control. Some papers[1,2] proposed the fixed-gain and variable-gain PID control of web tension in the winding section. But the effect is not satisfactory. Other engineers have tried to derive a mathematical model of

Please use the following format when citing this chapter:
He, Yongyi, Guo, Shuai, Zhou, Qihong, 2006, in International Federation for Information Processing (IFIP), Volume 207, Knowledge Enterprise: Intelligent Strategies In Product Design, Manufacturing, and Management, eds. K. Wang, Kovacs G., Wozny M., Fang M., (Boston: Springer), pp. 550-556.

the tension system used in a model-based controller[3], which results a fifth-degree differential equation described the mechanical behavior. In theory, it could work, but in fact it does not. For the complex electromchanical tension control system there are still many problems on its basal theoretics, check and measure technology and control technology to be solved. It is one of the problems of great significance for the technician to study.

Taking "Fuzzy logic PID decoupling control strategies based on neural network" as a main topic, this paper established a new kind of MIMO fuzzy PID decoupling controller based on BP neural network for the complex electromchanical tension systems. Compared with other control strategies, the structure and learning algorithm of the strategies presented in this paper are relatively simple and physical meanings of neuron in input and output layer are determinate. The parameters of PID controller on-line are adjusted according to a certain control law and through self-study of neural network. Then the manipulated variable gotten through classical PID control algorithm is employed to the control and the decoupling is realized in this course while it need not use the given sampling signal to study offline. By the way the structure of controller is defined in the light of the number of input and output variable. The structure of control object needs to be known in advance. The weight coefficient of network connection and values of parameters in PID controller can be taken by course of fuzzy logic control.

2. STRUCTURE OF NEURAL NETWORK FUZZY LOGIC PID CONTROLLER OF TENSION CONTROL SYSTEM

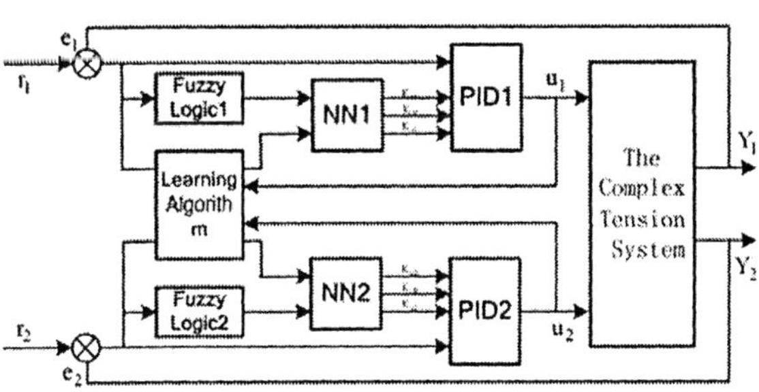

Fig. 1 Basic Tension Controller Architect[4]

Tension systems are couple input and couple output system. The structure of tension control system is made up of neural network NN and classical fuzzy logic PID controller. Structure diagram is shown in figure l, in which r_1 and y_1 correspond with the given value and output value of practical system tension respectively, while r_2 and y_2 correspond with the given value and output value of system speed, respectively. Close loop control is carried out to control object by classical PID controller. While the three parameters K_P, K_i, and K_d are adjusted by neural network, they includes learning algorithm, fuzzy logic and neural network NN. Parameters of PID controller can be adjusted on line through the learning algorithm based on the operation state. So performance figures can get to optimization.

Neural network NN is a sub-network including three layers. The number of input layer nodes is M =3, hidden layer nodes Q = 6, output layer nodes L = 3. The three nodes in input layer accept error $E(k)$, error sum $\sum E(k)$ and error increment $\Delta E(k)$ respectively; The three nodes in output layer correspond with the parameters of PID controller, K_P, K_i, and K_d. Synthesis of all the sub-network N=2 is carried out in control study algorithm. And MIMO control system is formed through combining it with fuzzy logic PID control.

3. DESIGN SCHEME OF FUZZY PID DECOUPLING CONTROLLER BASED ON NEURAL NETWORK FOR THE TENSION SYSTEM

The computational method of fuzzy PID controller based on the neural network is made of three sectors: fuzzy logic, classical PID control algorithm and multi-variable learning algorithm based on neural network.

Fuzzy logic algorithm: As fuzzy logic tuning mechanism is used for the on-line adjusted BP neural network, we use standard terminology from fuzzy set theory. Associated with the map F is a collection of linguistic values, each membership function is a map from the real line to the interval [0,1]. The membership functions used are of the triangular type. The height of the membership functions in this case is 1, which occurs at the points -1, -0.66,-0.34, 0, 0.34, 0.66, 1 respectively.

The realization of the function $r(k) = F[e(k), \Delta e(k)]$, based on the standard fuzzy method, consists of three stages: fuzzification, Fuzzy rule inference, and defuzzification. We describe each of these stages in turn.

Fuzzification: The process of fuzzification transforms the inputs $e(k)$ and $\Delta e(k)$ into the setting of linguistic values. This consists of scaling the inputs $e(k)$ and $\Delta e(k)$ appropriately and then converts them into fuzzy sets. We use the symbol C_e for the scaling constant for the input $e(k)$, and the symbol $C_{\Delta e}$ for the scaling constant for the input $\Delta e(k)$. For each linguistic value $l \in L$, we assign a pair of numbers $n_e(l)$ and $n_{\Delta e}(l)$ to the inputs $e(k)$ and $\Delta e(k)$ via the associated membership function u_l, by :

$$n_e(l) = u_l(C_e e(k)) \qquad (1)$$

$$n_{\Delta e}(l) = u_l(C_{\Delta e} \Delta e(k)) \qquad (2)$$

The numbers $n_e(l)$ and $n_{\Delta e}(l)$, $l \in L$ are used in the computation of $F[e(k), \Delta e(k)]$, which we will describe when discussing the defuzzification stage.

Fuzzy rule Inference: Associated with the decision making process is a set of fuzzy rules $R = \{R1, R2, R3, \cdots Rr\}$ where r is the total number of rules. Our rules derived by using a combination of experience "trial and error" and our knowledge of the response of the system. Suppose that the command signal is a constant y_m, the error $e(k)$ is zero and the change of error $\Delta e(k)$ is a negative number, which means that the output $e(k) = y_m(k) - y_p(k)$ is increasing, i.e., heading in the direction of overshoot. To compensate for this, we decrease the command signal. This corresponds to applying a correction term $r(k)$ which is negative. Hence, we get the rules "if error is zero and change-of-error is negative- medium then output is a negative-big correction term". The other rules are obtained in the similar manner.

Defuzzification: The defuzzification process maps the result of the fuzzy logic rule stage to a real number output $F[e(k), \Delta e(k)]$. We use the height defuzzification method, which is simple to implement and comes to relatively good results. To describe this method, let $p(l)$, $l \in L$, be the location of peak of the membership function μl. Recall that locations are given by $p(NB) = -1$, $p(NM) = -0.66$, $p(NS) = -0.34$, $p(ZO) = 0$, $p(PS) = 0.34$, $p(PM) = 0.66$, $p(PB) = 1$, define:

$$a_i = n_e(l_i^e) \wedge n_{\Delta e}(l_i^{\Delta e}) \qquad (3)$$

Where Δ represents the min operator; Then, the output of the defizzification process is given by

$$F[e(k), \Delta e(k)] = C_F \frac{\sum_{i=1}^{27} p(l_i) a_i}{\sum_{i=1}^{27} a_i} \qquad (4)$$

Where C_F represents the scaling factor, this completes our description of the term $r(k) = F[e(k), \Delta e(k)]$.

Multi-variable study algorithm based on BP neural network resembles other BP algorithm[5]. It is made up of two sectors: forward and back propagation algorithm. The three neurons in input layer correspond with the parameters of PID controller K_P, K_i, and K_d respectively and they are not to be negative number. So activation function in output layer neural network can take sigmoid function of positive and negative symmetric. Alphabet p stands for sequence number of sub-network.

Neural network algorithm: In any sampling time the computation formula of forward propagation algorithm of neural network is given as following:

Input of input layer of neural network NN is:

$$\begin{cases} x'_{p1}(k) = e_p(k) \\ x'_{p2}(k) = \sum_{i=0}^{k} e_p(i) \\ x'_{p3}(k) = e_p(k) - e_p(k-1) \end{cases} \tag{5}$$

Input and output of hidden layer is:

$$\begin{cases} net''_{pi}(k) = \sum_{j=1}^{M} \omega''_{pij} x'_{pj}(k) - \theta''_{pi} \\ x''_{pi}(k) = f\left[net''_{pi}(k)\right] \end{cases} \tag{6}$$

In expression (6), M is neuron number of input layer, ω''_{pij} is connection weight coefficient in sub-network of hidden layer; θ''_{pi} is threshold value of neuron in hidden layer; $f(\bullet)$ is activation function in hidden layer $f(\bullet) = \tanh(x)$;

Input and output of output layer is:

$$\begin{cases} net''_{pi}(k) = \sum_{j=1}^{Q} \omega''_{pij} x'_{pj}(k) - \theta''_{pi} \\ x''_{pi}(k) = g\left[net''_{pi}(k)\right] \end{cases} \tag{7}$$

In expression (7), Q is neuron number of hidden layer; ω''_{pij} is connection weight coefficient in output layer; θ''_{pi} is corresponding threshold value; $g(\bullet)$ is activation function in output layer $f(\bullet) = 0.5[1 + \tanh(x)]$.

4. DESIGN OF TENSION CONTROLLER BASED ON DSP

In order to proof the performance of the tension controller with fuzzy PID decoupling control strategy based on neural network, we designed the tension control board based on DSP controller. DSP controllers available today are able to perform the computation for high performance DMC structures for different motor technologies and motion control configurations.

The core part is based on a novel 16-bit TMS320LF2407 DSP module. It has a 30MIP's 16 fixed point DSP core with up to 30MHz CPU frequency. It also integrates the following power electronics peripherals—two event manager modules (EVA&EVB) with two general-purpose timers (GP Timers), three capture units and eight PWM outputs for each module. It also has other features such as a 3.3V power supply, a built-in quadrature encoder pulse (QEP) circuit, a high speed A/D converter, as well as a large number of digital I/O. The characteristics mentioned above make it quite flexible especially for digital motion control applications.

In the software design, the real time control is a very serious problem in the tension control system. Now the interrupt technology is very popular in real time control. In general sense, the real time is implemented by interruption of the real time clock. The real time clock is in charge of managing control process, collecting the related signals, inspiring all kinds of control task, and sending the related control signal. So the real time clock is very important in the real time control[6].

The performance test of the tension control board with PID control algorithm based on DSP has been performed on the tension tested-bed under the varying load conditions. Fig. 2 shows the basic system architecture of the tension tested-bed. Fig. 3 shows the plots of tension and speed responses in a speed control mode when there is a change in the speed from 0 to 100 RPS with 25% parameter error. Fig. 3 (a) shows the case with the conventional PID controller, while Fig. 3 (b) shows the case with proposed fuzzy PID control strategy based on neural network controller. The speed variation in Fig. 3 (a) tells us the imperfect decoupling, and Fig. 3 (b) shows its advantages of higher speed response and stronger robustness.

5. CONCLUSIONS

PID decoupling control algorithm based on neural network is proposed in this paper. It makes classical PID control law induct in BP neural network. The structure of controller is simple and its principle is definite. So it is easy to be designed and to be carried out. The ability of dealing with dynamic problems is very good. We can control the system under the circumstance of unknown state perfectly and search and correct parameters of the controller on line. Experimental study is carried out on the tension tested-bed under the varying load conditions. Results show that the system has strong robustness, adaptive faculty and good control quality even when the model has changed.

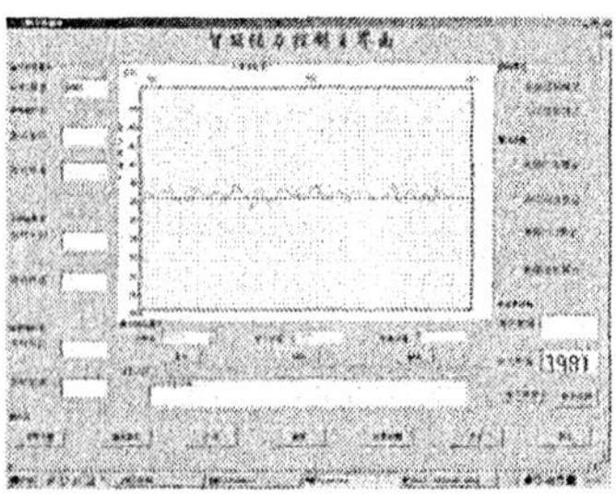

Fig.3a Results of Experiment

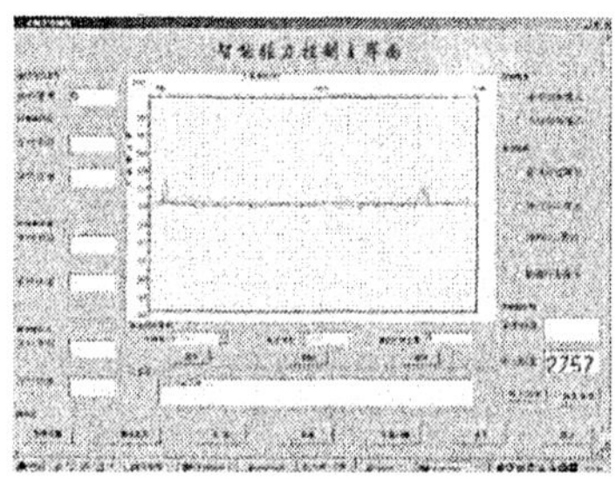

Fig. 3b Results of Experiment

6. REFERENCE

1. Y. Hou, Z. Gao, F. Jiang and B. T. Boulter, "Active Disturbance Rejection Control for Web Tension", Decision and Control, 2001. *Proceedings of the 40th IEEE Conference* , Volume: 5, 4-7 Dec. 2001 Pages: 4974 - 4979 vol.5 Regulation
2. Berger Larhr, *Technical Document*, November 1998, pp.1-3
3. J. C. Basilio and S. R. Matos, "Design of PI and PID Controllers with Transient Performance Specification", *IEEE Transactions on education,* VOL.45, NO.4, November 2002, pp.364-366
4. Jong-Hwan Kim, "Fuzzy Precompensation of PID Controllers", *Second IEEE Conference on Control Applications*, pp.184-186
5. Karl N. Reid, Kee-Hyun Shin, and Ku-Chin Lin, "Variable-Gain Control Of Longitudinal Tension In A Web Transport System", AMD-Vol. 149, *Web Handling, ASME* 1992, pp87-100.
6. TAN shi-zhi, Yang Ru-qing, "Design of Industrial Robotic Controller Based on DSP", *Journal of Donghua University*(Eng. ED.), VOL.20, No.1, 2003, pp.61-65

AN EFFICIENCY MODEL OF MULTISTAGE ELECTROMECHANICAL MACHINES

Lixin Lu, Limin Li, Yanjun Huang
CIMS & Robot Center of Shanghai University, Shanghai University, China; Email: lulixin@feilo.com.cn.

Abstract: This paper presents an efficiency model of multistage electromechanical machines design. The aim of design and redesign can be got by analyzing the rhythm of work with the model.

Key words: efficiency model; multistage; electromechanical machines.

1. INTRODUCTION

With the development of science and technology, multistage electromechanical machines become more and more favorite in market. How to harmonize these machines' working rhythm and optimize the structure becomes the focus of the design. This paper presents an efficiency model and put it into practice.

2. SYSTEM BALANCE & OPERATION CYCLE

2.1 One-piece Flow Production Theory

Supposing an equipment is composed of n sub-components which contain one or several operation cycles for each one, and make an part needs time: Time; the theory time is Ti of every sub-component. The efficiency model:

Please use the following format when citing this chapter:
Lu, Lixin, Li, Limin, Huang, Yanjun, 2006, in International Federation for Information Processing (IFIP), Volume 207, Knowledge Enterprise: Intelligent Strategies In Product Design, Manufacturing, and Management, eds. K. Wang, Kovacs G., Wozny M., Fang M., (Boston: Springer), pp. 557-562.

$$\text{Time}=\sum_{i=1}^{n} k_i T_i c_i^{-\lambda_i}$$

$$\begin{cases} 0 < k_i \leq 1; \\ c_i > 1; \\ \lambda_i = \begin{cases} 0(\textit{improved}) \\ 1(\textit{unimproved}) \end{cases} \end{cases} \quad (1)$$

The k_i is rhythm parameter and stand for the harmony of sub-components' operation cycle; λ_i is the developing parameter of sub-department; c_i is the developed sub-component efficiency parameter, which shows performance of the sub-component.

There are two ways to raise the working efficiency of a machine according to the model: 1. Developing k_i. We can optimize the rhyme basing on the old machine, which is seldom happen in practice, or developing one or several sub-components' structure. 2. Parameter c_i. This parameter can be raised by redesign the sub-components. Parameter k_i cares of the macroscopical development of machine, while c_i pay attention to some sub-component's development. We use k_i to develop a machine as soon as c_i will show the changing tendency, which can direct our redesign. The value of c_i should be tested by k_i, and make sure whether it is appropriate. All of these will affect the machine development. The two parameters have a close relationship, and have a influence on the other, which lead to the optimization of the system.

3. APPLICATION

"Fully Automatic Crimping Machine" is used to process wire, just as picture 1 shows. The description of process can be subscribed: crimping the both ends exodermis of wire, then stamping the both ends. According to this process and the other situations, the machine should include 7 sub-components: Neatened Component, Sending Wire Component, Tool Post, Stamping Component (2), Back Crimping Component, and Taking Products Component. Every one of them has about one or several operation cycle but Neatened Component. The picture 2 shows the machine's operation cycle. The rhythm of these operation cycles is key factor of successful design, which is the main mission of redesign. The paper will analyze parameter k_i

at first in order to get preliminary optimizing the system; then computer parameter c_i, At last, we can get the aim of this design and redesign.

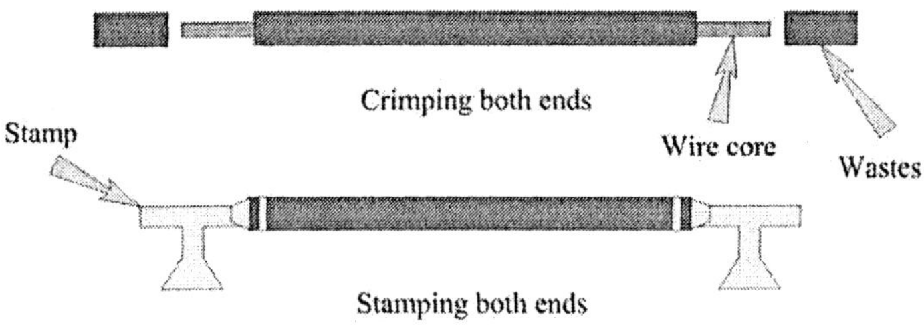

Figure 1. processed wire

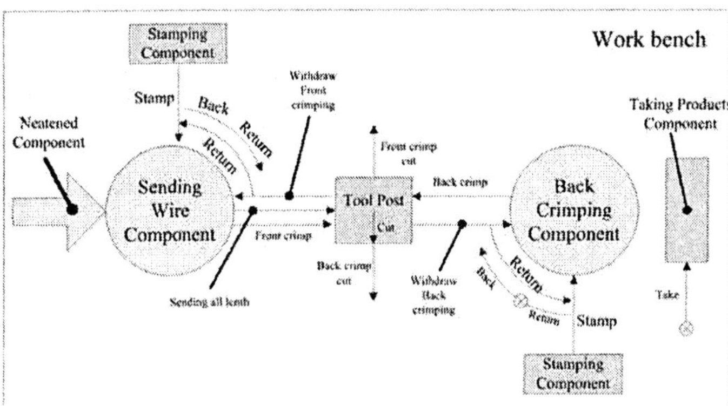

Figure 2. the old Fully Automatic Crimping Machine's operation cycles

3.1 k_i in products' optimization

Picture 2 shows the machine's operation cycle. This machine can be seen as front and back parts according to its' structure and the wire processed. The picture 3 shows the synthetical rhythm that can be got after analyzing the machine's operation cycle. It's clear that processing the first wire end time (this process happens in Sending Wire Component, for short: FC, Front Crimping) is overlap with the second wire's top (this process happens in Back Crimping Component, for short: BC, Back Crimping), and the end must wait for the top for a while form the picture. The Sending Wire Component has lots of potential to use after analyzing it's work ability. This problem can be easily resolved if crimping the first wire's end and crimping the second wire's top are done at the same time. So we should study the Tool

Post and try to give it a development. If the Tool Post contains three tools, other than one, this situation will happen and the Tool Post running ability has not a raise. The machine's rhythm just as picture 4 and the development is cycled by long and two short dash line now. The description of developed operation cycles in detail: the old Tool Post's top crimping and end crimping can be processed by one operation cycle; the sending operation of Sending Wire Component and Back Crimping Component can be completed at the same time (this operation cycle is named Sending wire); the draw back wire operation cycles of Sending Wire Component and Back Crimping Component happen simultaneously (named as: Withdrawing wire). The synthetize of operation cycles is showed in picture 5. Now the running burthen of Sending Wire Component is same as Back Crimping Component, which optimizes the machine's process rhythm and leads to efficiency rising.

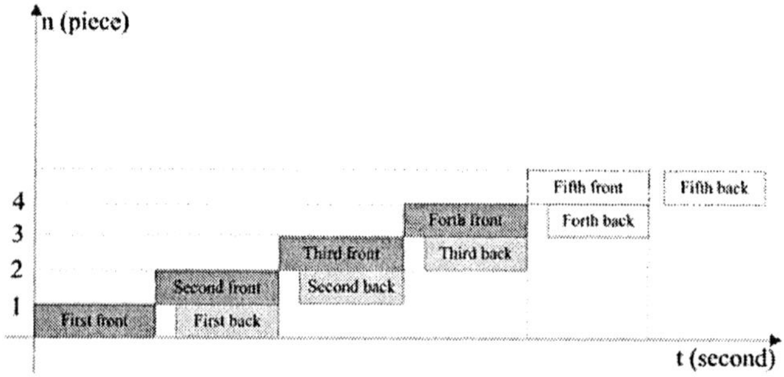

Figure 3. the preliminary synthetical rhythm

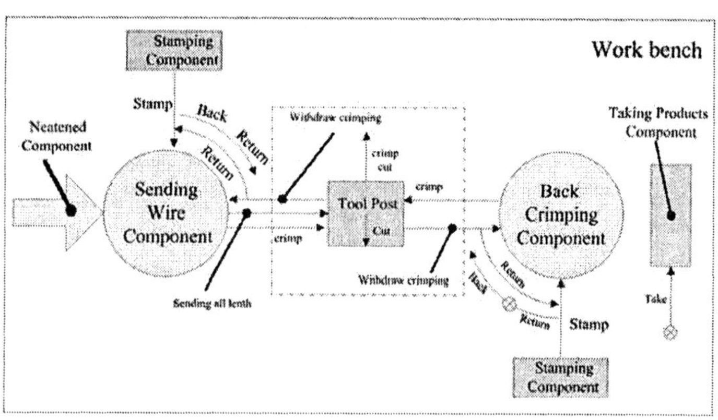

Figure 4. the developed Fully Automatic Crimping Machine's operation cycles

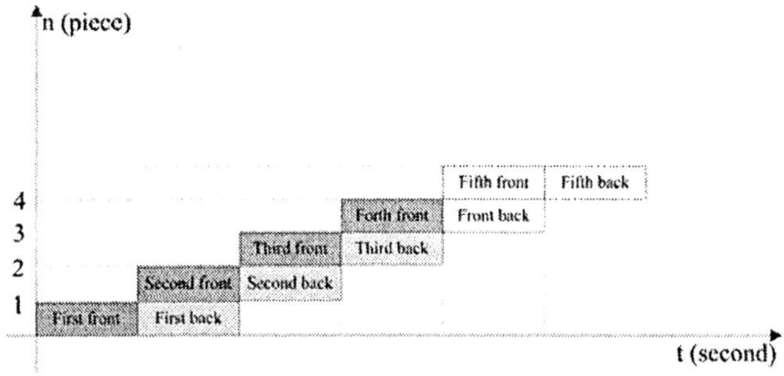

Figure 5. the developed synthetical rhythm

3.2 Parameter c_i analyzing

Analyzing the performances of every sub-components, we can get their detail stage-time bar, bar 6(Note: ① the same color belongs to the same component; ② Stamping Component and Tool Post's half operation cycle is useful to process). Based on Sending Wire Component, study the Stamping Component and Tool Post's operation cycles, and compare their relation with the system.

Stamping Component:

Process ability of theory:

$$2T_1=2/3s/cycle$$

The other sub-components' time:

$t=0.010+0.286+0.015+0.045+0.015+0.045+0.010=0.426s$

Compare:

$2T_1>t$

Tool Post:

Process ability of theory:

$2T_2=0.027s/cycle$

The other sub-components' time:

$t1=0.045s$;

$t2=0.045+0.010+0.335+0.010+0.286=0.686s$

Compare:

$2T_2<t1$， $t2$

It can be get that Stamping Component affect the machine's process efficiency from the above result. So the next step: developing Stamping Component, increasing parameter c_i, and optimize parameter k_i again, in order to harmony system's process rhythm.

Every parameter of the Model has a close relationship between them, which shows in the above study. When one parameter changes, the other parameters will show their trends of changing, that gets the design better.

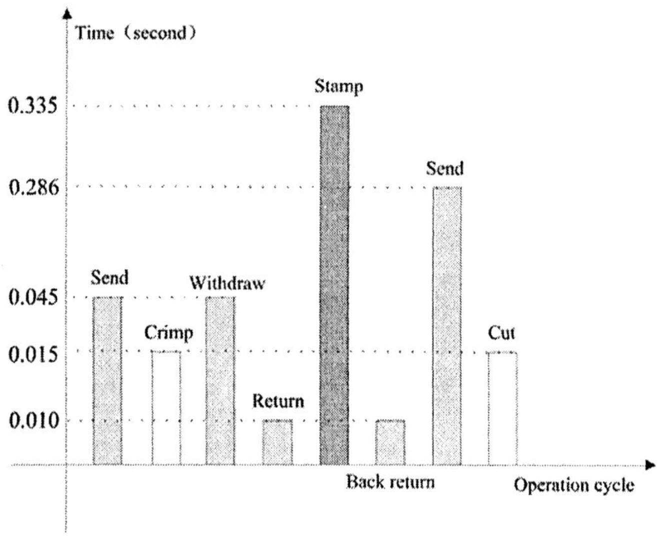

Figure 6. the detail stage-time bar Case Study

4. CONCLUSION

The project of Fully Automatic Crimping Machine Design has been completed by using this Efficiency Model, which direct this design and redesign.

5. REFERENCES

1. Huang Yanjun. The research and Application of DFA based on Systems Engineering [D]. Shanghai University. 2006.02:p23

PREDICTION OF THE CUTTING DEPTH OF ABRASIVE SUSPENSION JET USING A BP ARTIFICIAL NEURAL NETWORK

Xiaojian Liu[1,2], Tao Yu[1], Wenbin Wang[1]
[1]CIMS &Robot Center of Shanghai University ,Shanghai 200072, China. [2]Shandong Institute *of Light Industry Shandong 250100, China. Email: lxj@sdili.edu.cn.*

Abstract: Abrasive suspension jet is a new embranchment of abrasive jet. In this paper, the abrasive suspension technology is first used in cutting process in domestic market. The abrasive grain concentration in suspension is constant, so the cutting quality is more stable. In this paper, a prediction model based on a back-propagation (BP) artificial neural network is presented for predicting the cutting depth generated by abrasive suspension jet. In the application of the BP neural network, the mean error of the output in the model training is 0.01, the relatively discrepancy is below 8.70%. The modeling method based on the BP neural network is much more convenient and exact compared with traditional methods, and can always achieve a much better prediction effect. It is verified with experiments to be reasonable and feasible, and it is the better foundation for the future study of abrasive suspension jet.

Key words: neural network; abrasive suspension jet; cutting depth

1. INTRODUCTION

Abrasive water jet (AWJ) machining has been accepted in the industry due to several factors, among which are its versatility for use on a wide range of materials and the high-quality surfaces produced. These surfaces are free from heat affected zones(HAZ)and mechanical distortions. however, AWJ machining has some limitations:

Please use the following format when citing this chapter:
Liu, Xiaojian, Yu, Tao, Wang, Wenbin, 2006, in International Federation for Information Processing (IFIP), Volume 207, Knowledge Enterprise: Intelligent Strategies In Product Design, Manufacturing, and Management, eds. K. Wang, Kovacs G., Wozny M., Fang M., (Boston: Springer), pp. 563-569.

- Relatively slow cutting rates when compared to laser(when HAZ is not critical).
- Limited in cutting thin kerfs.
- When used with garnet abrasives, not efficient for cutting carbides, ceramics, and hard materials.

As the need for improved cutting and machining capabilities increases, more attention is being focused on improving the above limitations of the AWJ. Work is underway to introduce an alternative technique to complement the AWJ process capabilities without sacrificing any of its advantages ,so abrasive suspension is proposed. It's an embranchment of abrasive jet in theory. In this paper, abrasive suspension jet (ASJ) is concerned. The advantages of this cutting technology is:

- The energy transfer to particles is more efficient
- Nozzles can be miniaturized to fit in tight spaces.
- The ASJ can machine thinner kerfs and smaller holes than AWJ.

The influence factors on cutting ability of low-pressure abrasive suspension jet are complex, they often have nonlinear relations with each other. It's difficult to propose a mathematics model of cutting depth exactly, and a model can be set based on neural network. Unlike the traditional method, the artificial neural network (ANN) does not require a mathematical model. ANN can perform not only a highly accurate non-linear, but also retains memory of data. Further, the ANN is powerful in handling dispersed data, and moreover it can also adjust the state of the network on the base of the original network in order to adapt to new data- sets through training with these new data- sets. In this paper, an ANN is used in predicting the cutting depth and the optimum structure of the ANN is also proposed.

2. THE BASIC PRINCIPLE OF ARTIFICIAL NEURAL NETWORK

An artificial neural network has multi-layers. There is an input layer where data are presented to the neural network, and an output layer that holds the response of the network to the input. It is the intermediate layers, also known as hidden layers, that enable these network to represent and compute complicated associations between patterns. Neuron is the basic element. Each neuron is connected to all the neurons in the next layer.

For a training pattern $(X_k,Y_k)(k=1,2,\ldots N)$ in which X_k is the input pattern and can be expressed as an no-dimensional vector $(x_{k1},x_{k2},\ldots x_{kn})^T$ and Y_k is the corresponding expected solution and can be expressed as an no-dimensional vector $(y_{k1},y_{k2},\ldots y_{km})^T$, the net input to a neuron is usually computed as the sum of all inputs to it:

$$I_{kj}^{l} = \sum_{i} W_{ji}^{l} O_{ki}^{l-1} + \theta_{j}^{l}$$

where W_{ji}^{l} is the weight of the link between neuron u_j^l in layer l and the neuron u_i^{l-1} in layer l-1, O_{ki}^{l-1} is the output of the neuron u_i^{l-1}. θ_j^l is the threshold of the neuron u_j^l.

The net input to a neuron is used by its activation function, here the most commonly used sigmoid function, to produce an output

$$O_{kj}^{l} = f(I_{kj}^{l}) = \left[1 + \exp(-I_{kj}^{l})\right]^{-1}$$

where f(.) is the transfer function. The transfer function has several forms, such as the hyperbolic tangent, step, or sigmoid function. However ,the sigmoid function is the most common in practical application and is used here.

The BP neural network is characterized by gaining possession of hidden layers. Generally, a BP network with one hidden layer is enough for most applications, and one more hidden layers tends to make the network too complicated, a potential that results in more minimums and causes lower convergence speeds and larger errors. Theoretically, BP with two hidden layers is enough to explain the complicated system with more general classic boundaries and has faster convergence speeds.

In the BP learning scheme, the calculated outputs in the output layer, y'_{kj} are compared with the desired outputs y_{kj}, to find the error, before the error signals are propagated backward through the network. The error function E k can be defined as

$$E_k = \frac{1}{2} \sum_{j=1}^{m} \left(y_{kj} - y'_{kj}\right)^2$$

where m is the total number of the output neurons.

In the BP algorithm, each training pattern is presented once and the weight correction is calculated, but the weights are not actually adjusted; calculated weight corrections for each weight are added together for all the patterns and then weights are adjusted only once using the cumulative correction. The momentum strategy implements a variable learning rate coefficient implicitly by adding a fraction of movement in the weight change to the current direction of movement in the weight apace. The new equation for weight change is given by

$$W^{l}_{ji}[T+1]=W^{l}_{ji}-\eta\left(\frac{\partial E}{\partial W^{l}_{ji}[T]}\right)+\alpha\times\left(W^{l}_{ji}[T]-W^{l}_{ji}[T-1]\right)$$

where η is the step size or the learning rate coefficient, the index T labels the iteration number in the learning process, and α is a momentum coefficient that can have a value between 0 and 1.

3. THE PREDICTION OF CUTTING DEPTH BY NEURAL NETWORK

3.1 Experiments of cutting depth

There are many factors that influence the cutting depth of abrasive suspension jet. The dominating factors are cutting pressure, feed rate, additive concentration and standoff distance. So in this paper, four factors are in considered. The experiment conditions are: grain mesh: 80,grain type :garnet ,cutting material: A3 steel, cutting angle: 90°.

3.2 Determination of the structure of the BP network

In order to establish the process model, we first need to decide how many layers and how many neurons per layer of the network are utilized to describe the behavior of cutting. It is easy to determine the number of neurons in the input layer and the output layer during applications. There are four neurons in the input layer denoting the content of cutting pressure, feed rate, abrasive flow rate, additive concentration, respectively, and one neuron in the output layer representing the corresponding the cutting depth. However it is not so easy to choose the appropriate number of neurons in the hidden layer for there is currently no definite rule to determining it except through experimentation. Using too few neurons impairs the neural network and prevents the correct mapping of input to output. Using too many neurons impedes generalization and increases training time. In our study, ten hidden neurons are finally employed in the BP network and the structure of this neural network is shown in Figure 1.

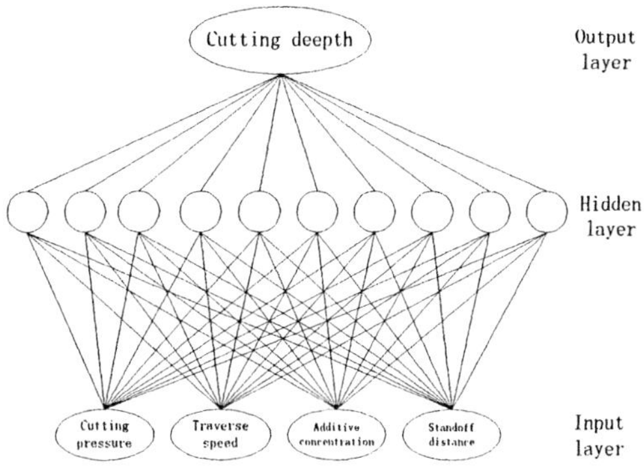

Figure 1. The depth prediction model of BP neural network

3.3 Choice of training precision

In the BP algorithm, training is carried out by repeatedly presenting the entire set of training patterns (with the weights updated at the end of each cycle) until the energy functions over all the training patterns are minimized and within the preset training precision. In principle, the BP network can always remember all the training patterns after enough large iterations. But what should the training precision be set to terminate the training course? in actual engineering systems the training data are usually erroneous, so the too high training precision will over fit the training patterns and then impede generalization of the BP network. The training precision of this paper is determined with 0.01.

3.4 Network training and verification

Data of 20 typical samples are gathered for use in training and testing the BP network, 16 of which are randomly chosen as training patterns and the others are chosen as testing patterns, shown in Table1. These data basically contain the various distinct characteristics of the problem the BP network is likely to encounter in the finished application. Preprocessing of the data is usually required before presenting the patterns to the BP network. This is necessary because the sigmoid activation function modulates the output of each neuron to values between 0 and 1. The prediction results shown in Table2.

Table 1 The samples of the BP neural network mode

serial number	Cutting pressure (MPa)	Feed rate (mm/min)	Additive concentration	Standoff distance (mm)	Cutting depth (mm)
1	15	15	1:10	2	12.1
2	15	30	1:10	2	8.6
3	15	60	1:10	2	4.5
4	15	120	1:10	2	2.1
5	20	15	1:12	4	14.8
6	20	30	1:12	4	10.3
7	20	60	1:12	4	6.6
8	20	120	1:12	4	3.2
9	25	15	1:14	6	20.1
10	25	30	1:14	6	12.5
11	25	60	1:14	6	7.4
12	25	120	1:14	6	5.3
13	30	15	1:16	8	23.2
14	30	30	1:16	8	15.5
15	30	60	1:16	8	9.4
16	30	120	1:16	8	7.3
17	15	15	1:16	2	13.1
18	20	30	1:14	4	9.8
19	25	60	1:12	6	7.1
20	30	120	1:10	8	6.9

Table 2 prediction results of the BP neural network model

serial number	Measured depth (mm)	Predicted depth (mm)	Absolute discrepancy(mm)	Relative discrepancy (%)
17	13.1	13.7	0.6	4.58
18	9.8	9.6	-0.2	2.04
19	7.1	7.4	0.3	4.23
20	6.9	6.3	-0.6	8.70

4. CONCLUSIONS

A BP neural network can predict the cutting depth generated by abrasive suspension jet, and the prediction results show greater accuracy than those from the traditional statistical models. The modeling process based on the BP neural network is much easier and direct, for there is no necessity to specify a mathematical relationship between the input and output variables. The BP neural network can be effective for analyzing a system containing a number of variables, to establish patterns and characteristics not previously known. However, there are some problems need to be well discussed in the applications of BP neural network, such as the choice of the preset training precision and the proper number of hidden neurons etc.

5. ACKNOWLEDGEMENTS

The authors wish to acknowledge the funding by the Science Technology Committee of Shanghai .Acknowledgement is also due to Professor Liu Linsheng who provided help in the test. The authors are indebted to Dr. Wang Wenbin, Hu guihua for their help in the investigation.

6. REFERENCES

1.Xiao-jian Liu Tao Yu ,Study on the Theory of A New Kind of Abrasive Jet Cutting Technology and the Test Analyses, *Machinery* ,2005.2.

2. Liu Xiaojian, Yu Tao, Liu Linsheng Wang Wenbin "The study on abrasive suspension jet cutting technology", *Lubrication Engineering,* 2005,169(3), 31–33,

3. Cybenko, G., *Mathematics of Control Signal and Systems,*2003, 2, 303.

4. J.M. Cabrera, A. Alomar, J.J. Jonas, J.M. Prado, Metall. Trans. A 28 (1997) 2233-2244.

5. Taylor. J.G.(1998) Neural Networks :an overview in L.J. Landau, &J.F. Taylor. *Concepts for neural networks – a Survey* London :Springer.

7. G. A. Carpenter and S. Huang, The ART of adaptive pattern recognition by self-organizing neural networks, *Computer* 2004 ,22(6),77-88.

8. L. Monostori and D. Barschdorff, Artificial neural networks in intelligent manufacturing, *Robotics and Computer Integrated Manufacturing* 2002,9(2),412-436.

9. Refenes A. N. Stock performance modeling using neural networks : A Comparative study with regression models, *Neural Networks,*1994,17 (2).36-48.

RESEARCH AND IMPLEMENT OF QUALITY MANAGEMENT SYSTEM IN ERP

Yibing Li
Coll. Of Mech., Wuhan University of Technology, China; Email:ahlyb@tom.com

Abstract: Quality is crucial for the enterprise. The structure of quality management system is discussed in this paper. And the modules of function are set up under the environment of ERP. Meanwhile, this paper studies statistical process control (SPC) technology and implements the combination of SPC and ERP according to an enterprise practice. Finally, this paper gives an example to illustrate by using C++ Builder6.0 program and SQL Server2000 database.

Key words: Enterprise Resource Planning; Quality Control; Quality Management, SPC

1. INTRODUCTION

Quality is crucial for the enterprise. In order to improve the quality, the enterprise must rely on not only the application of technology but also a strong quality management system. Enterprise Resources Planning (ERP) is the embodiment of the most latest and advanced enterprise management ideas. Quality management system (QMS) under the environment of ERP combines the theory of total quality management (TQM), the ideas of ISO9000, and the advantages of information integration and multi-angle data analysis. And it is a power tools for the quality continuous improvement, and is indispensable to ERP [1].

Now, ERP system has been used in more and more enterprise. There produces a great deal of data. The new facing problem is how to make use of these data for decision [2]. This paper studies the structure of QMS and sets up the modules of function based on the enterprise's practice, and the combination of SPC technology and ERP. Finally, this paper gives an example to illustrate by using C++ builder6.0 programs and SQL server 2000 database.

Please use the following format when citing this chapter:
Li, Yibing, 2006, in International Federation for Information Processing (IFIP), Volume 207, Knowledge Enterprise: Intelligent Strategies In Product Design, Manufacturing, and Management, eds. K. Wang, Kovacs G., Wozny M., Fang M., (Boston: Springer), pp. 570-575.

2. STRUCTURE AND MODULES OF QUALITY MANAGEMENT

Setting up a valid quality management system is indispensable to the enterprise. According to the characteristic of the quality management, the quality management system has three layers [3,4] , including plan layer, administrator (control) layer and execution layer. The next layer receives the plan or control orders from the previous layer, and transmits its own control orders to the under layer and returns the feedback of the quality information to super layer. The advantage of this structure is that each layer can fully make use of its respective information and accomplish its own function. So this systematic structure of quality control has the advantage of quickly response speed, real-time control. Also it is easy to recompose and extend, and satisfy the demanding of plasticity and correctness.

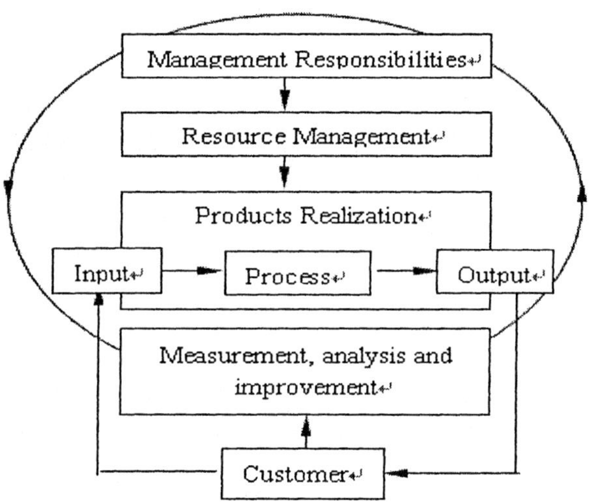

*Figure1. F*unction of quality management

Quality management mainly includes ‘management responsibilities’, ‘Resources management’, ‘products realization’, ‘measurement, analysis and improvement’ [5], shown as figure 1. The main target of the system contacts outside is the customer. It reflects the basic principle of regarding customer as the centre.

According to the quality management characteristics and functions [6,7,8], the system can compartmentalize several important modules, as shown as fig.2. This system can help the enterprises improve the dependability and quality of their products, and satisfy the demands of the customer, and realize the profits of the enterprise.

(1) Quality standard management

The standard is a guide to acquire the best quality for the enterprise. This module manages the quality standard, the enterprise quality goal and the quality authentication, such as ISO9000 etc. So, the result of realizing standardization would be greatly improved the quality, it is also the target of all enterprise as a conscientious effort toward an end.

(2) Quality inspection and control

According to the product lifecycle quality management ideas, the main sub-modules include the inspection and control of the raw materials, semi-finished product, finished product, key procedure spots and the after sale quality etc. And the main function is to collect and record all daily quality information, such as production quantities, waster quantities. And the symptom of inspection and the formative cause also must be recorded. All these are the foundation of statistic and analyses.

(3) Quality statistic and analyses

Quality statistic and analyses are important significance for increasing the economic benefits, reducing the cost, guaranteeing and improving production quality. Quality statistic and analyses are actually the data analyses, and include the fuzzy and multi-selection query and form report, the analysis of quality cost, the appraisement of provider, the management of quality advices, the disposes of quality problem etc. And the approach to analyze the quality data is the full use of statistical process control (SPC), including the data collection form, stratification, scatter, pareto, histogram, cause-effect diagram, and control chart.

(4) Quality improvement

The quality management target is the continual quality improvement. This module includes the improvement plan of customers' satisfaction, the encouraging measures of employee values' actions, quality planning, quality train and quality education, the management and selection of supply chain partnership and partners' supervising, the management of the quality control groups etc.

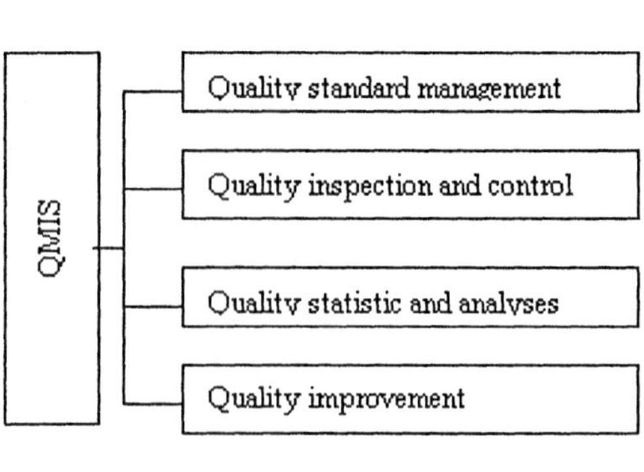

Figure 2. Modules of QMS

3. DESIGN AND APPLICATION OF QMS

Quality management system is an information disposal system with computer and information technology, and can improve the quality from the point view of management and control. There is no control, no management, the opposition is right. The key of management is for valid control. So the ideas of quality management are whole participate, total control and priority of preventive measures. Now, this paper gives an example to illustrate according to an enterprise practice based on C++ Builder6.0 program and SQL 2000 database.

(1) Data collecting and maintenance

Quality data, especially the data of key procedure is the basis of quality statistics and analyses. In generally, the approach of data acquirement includes the production record, periodical and selective examination, key procedure record etc. Of course, in the process of data collecting, the enterprise usually gives attention to two or more approaches for collecting the impersonal data. And a good software system should have convenient but powerful function. So that the user can be easy to add, delete, modify and browse the data.

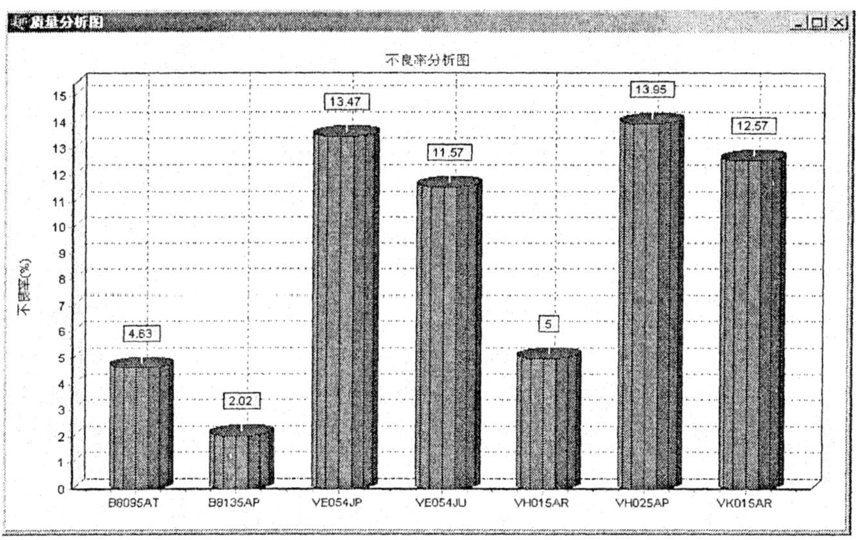

Figure 3. Histogram of ratio

(2) Quality statistics and analyses

The target of quality management is to help the manager to understand the current quality status and environment so that they can make correct decision. Because the quantities of quality data are very large, even some data are redundant and at random, and at the same time, different manager wants to acquire different data, the system should be able to induce and analyze these data according to the respective demands. Here, the author gives the histogram to illustrate, shown as figure 3. From this figure, it is easy to know that the high ratios of bad production are the production VHO25AP, VEO54JP, Vk015AR and WEO54JU. And the manager must be laid a strong emphasis on these productions later. And the figure4 shows the concrete symptom of the production VHO25AP daily. From this form report, it is easy to find out the quantities of every symptom, and compare the quantities of every symptom and every day each other in these period of time.

项目	07-20	07-21	07-22	07-23	07-24	07-25	07-26	07-27	07-28	07-29	合计
检查数	45902	32620	43679	45162	34437	27018	27104	39071	22411	20193	337597
不良点数	3074	2447	3434	3000	2275	2098	1982	2190	1122	1001	22623
不良率	6.7	7.5	7.86	6.64	6.61	7.77	7.31	5.61	5.01	4.96	6.70
IC 浮起	2	3		46		1	3	1	1	4	61
IC 欠品	1	1	1	3	3		1	1			11
IC短路	101	34	298	243	144	194	136	111	96	35	1392
IC位移	3	12	12	8	18	23	4	6		12	98
错件	1		5	1	2	3		2	1		15
短路	256	191	116	120	167	226	70	116	51	68	1381
二极管反贴	2	4	8	31	2	1	1	13		11	73
二极管欠品	69	9	26	26	63	26	5	3	5		232
反贴	4	8	11	18	7	1	17	10	1		77
浮起	435	498	747	543	321	140	360	302	73	105	3524
烂件	92	115	90	102	82	86	126	111	21	25	850
其他						1					1
欠品	379	321	390	368	234	312	204	142	165	153	2668
竖起	512	329	460	500	324	208	298	437	230	208	3506
脱件	6	23	26	59	30	7	12	9	18	1	191
位移	947	830	1016	890	802	750	707	856	423	352	7573
无锡	264	69	228	42	76	119	38	70	37	27	970

Figure 4. Quality form report of VHO25AP

(3) Quality improvement

The aim of quality statistics and analyses based on SPC are to estimate the process under the condition of control whether or not, namely estimate the process stability. After calculating the process capability index, the manger should adjust and control the process according to the values of the process capability index so as to ensure the continual improvement. Now, the author uses the production VHO25AP to illustrate the process capability.

Here, we calculate the result: C_p =1.165. According to the judgement and disposal table of process capability, we know the process degree is secondary, and the capability of process is passable. So the continual improvement disposals include the full use of the control chart to avoid foreign fluctuation, and insisting on the sample inspection etc.

4. CONCLUSION

This paper gives the structure of the quality management system under the environment of ERP, and sets up the modules and describes its functions in detail. Meanwhile, the author tries to combine the ideas of ERP and SPC, and discuss the approach of data collecting, data analyses and process capability. Finally, the paper gives an example to illustrate according to an enterprise practice. Now, the system has been well used in this enterprise, and reached the anticipative effects.

5. REFERENCES

1. Luo Hong. *Implementation and Design of ERP theory [M].* Beijing: Electron technology Press.2002.
2. Li Yong, Gao Jianming, Chen Fuming. The combination approach of Statistical process control and Quality System of ERP. *Computer applications [J].* (2004), 12: 149~152
3. Yang Xu, Wang Dongpeng, Ma Yulin. Study on quality management information system based on CIMS. *High technology communication [J].* (2000), 8:70~75
4. Yuan Feng, Xie Liyang, Yi Gang. CIMS-oriented integrated quality management system in workshop. *Computer engineering and application* [J]. (2003),31:23~25
5. ISO/FDIS9001:2000Quality management system – requirement.
6. Liu Wenhui, Ma Ke, Zhao Xiaohui. Research of the quality analysis and control in manufacturing based on enterprise resources planning. *Light Industry Machinery [J].* (2004), 4:155~160
7. Miao Rui, Wang Dongpeng, Yao Yingxue. The study of quality management information system for large-piece OKP based on CIMS. *Computer engineering and application [J].* (2002),8:27~29
8. Kong Zaojie. Comparative research for quality information system frame model under the CIMS environment. *CIMS [J].* (2000), 4:85~88
9. Wang Xueming, Guan Xiaohong, Song Guoji. Design and implementation of parts products quality control management system in MIS environment. Proceeding of 3th World Congress On Intelligent Control and Automation. June 28-July 2, 2000, HeFei, P.R.China
10. Ermer. D. S. A. Total Quality Process for design of a performance appraisal system. Transactions of Nanjing University of Aeronautics & Astronautics. (1998),6
11. Senn J A. Info System in management. Wadsworth Publishing Company, (1987)
12. Ermer. D. S. A Total Quality Process for design of a performance appraisal system. Transactions of Nanjing University of Aeronautics & Astronautics. (1998),6

A MONITORING SYSTEM FOR PLC CONTROLLED MANUFACTURING SYSTEM BASED ON FIELDBUS

Xiaofeng Song[1], Shili Tan[2] and Junjian Ding[3]
[1]Department of Precise Machinery, Shanghai University, Shanghai, 200072. Email:mail2song@163.com. [2]Department of Precise Machinery, Shanghai University, Shanghai. [3] Department of Precise Machinery, Shanghai University, Shanghai.

Abstract: In this paper, we propose a monitoring system for Programmable Logic Controllers (PLCs) controlled manufacturing system based on CC-Link fieldbus in the process monitoring and control for film plating. The PLCs are mainly utilized for collecting data as well as realizing auto-tuning PID and sequence control strategies. In addition, Mitsubishi GT Designer software with a Human-Machine Interface (HMI) is used to monitor the dynamical process.

Key words: PLC; fieldbus; CC-Link; HMI; PID; monitoring system; winding system.

1. INTRODUCTION

The winding vacuum systems for plating metal such as zinc and aluminium on thin film are very common in the capacitance industry. The main goal is to increase as much as possible the film transport velocity while controlling the tension of the film. The tension and the speed of the film should be maintained at a reference value. The main concern is the coupling existing between film velocity and tension. There exist many sources of disturbances on the velocity (roller non-circularity, film sliding). If the delivery speed fluctuates, then the treatment on the film becomes non-uniform. On the other hand inappropriate tension may cause wrinkles or even a film break. Once film break takes place, the operator needs open the winding chamber again. Obviously the pressure of winding chamber comes back to standard atmosphere pressure. And it will take about 20~30 minutes for winding chamber to reach the required pressure (about 1.3×10-1～2.67

Please use the following format when citing this chapter:
Song, Xiaofeng, Tan, Shili, Ding, Junjian, 2006, in International Federation for Information Processing (IFIP), Volume 207, Knowledge Enterprise: Intelligent Strategies In Product Design, Manufacturing, and Management, eds. K. Wang, Kovacs G., Wozny M., Fang M., (Boston: Springer), pp. 576-581.

×10-2Pa). Accordingly the output will decrease evidently. So a monitoring system to monitor tension fluctuation in order to avoid film break is very needful.

CC-Link is a field network system that processes both control and information data at high speed, to provide efficient, integrated factory and process automation. Using CC-Link enables users to reduce the amount of control and power wiring required in the complicated production line. Since users can select the suitable devices from more than 354 field devices supported CC-Link it is easy to expand multi-vendor environment.

As for the communication of speed, this enables the communication with the sensor input which requires a high speed response and intelligent devices which requires a large amount of data communication.

As one of CC-Links features, RAS functions such as, stand-by master function, detaching slave station function and auto return function enable recovery from the communication fault and the system debug at start up time. As for test and monitoring function, confirmation of the data link status, hardware test and circuit test are available. The CC-Link network diagram is shown as Fig. 1. The CC-Link monitoring system is composed of master station、 slave stations、 personal computer、 inverter、 servo motors、 remote I/O units, ect.

In order to meet the quest for automation and flexibility, many complex manufacturing systems are controlled by Programmable Logical Controllers (PLC). This is because that PLC's are adaptable, modular, user-friendly and acquired at low cost. And that winding vacuum film plating system is just such a complex manufacturing system.

The paper presents an intelligent monitoring system based on CC-Link fieldbus which consists of upper personal computer and the lower Programmable Logic Controllers for the aim of control and information management. A new kind of technology using CC Link fieldbus is adopted in this monitoring system.

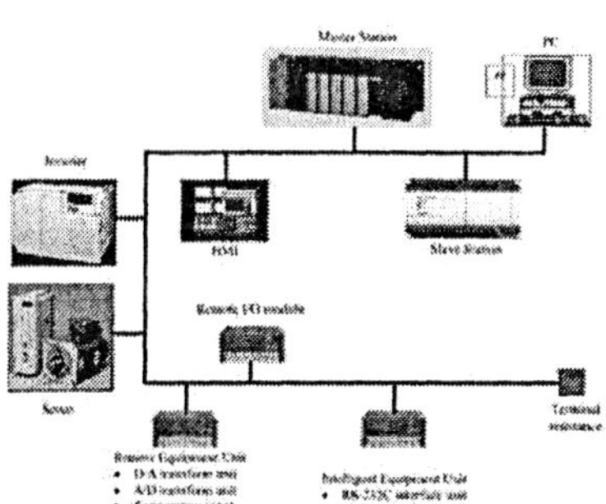

Figure 1. CC-Link network diagram

2. STRUCTURE OF THE MONITORING SYSTEM

This fieldbus-based monitoring system has two arrangements including monitoring units in field and information management and control unit in control room. All intelligent monitoring units located nearby the winding vacuum machine have microprocessors, which have the following function as respective signal sampling, A/D conversion, data calculation, ect. Fieldbus is the most important connection between each discrete monitoring unit in slave stations and master station. Analog signal is superseded by digital signal for bi-directional communication. It is beneficial for an operator to realize control, verification and parameter setting in control room. The kind of structure, on one hand, increases the precision and anti-interference capability of the monitoring system. On the other hand, it saves the investment cost.

The quantities of intelligent monitoring units should be determined by practical needs. This system has five kinds of monitoring units: temperature, vacuum degree, roll diameter, film tension and transfer velocity.

Whether the monitoring system could work normally depends on the performance of the sensors to a considerable degree. Their precision and stability has a direct influence on the monitoring system. Data obtained from these sensors are processed by PLC and displayed on HMI as Fig. 2 shown.

Intelligent monitoring module has two functions: on one hand it filters, amplifies and adjusts the output signal got from sensors to appropriate value which can be processed by PLC (FX series). On the other hand it carries through signal sampling, A/D conversion, data processing, software anti-interference, parameter calculation and transmits data to central control unit by virtue of communication interface of a PLC.

Central control unit (CCU) is the center of this monitoring system which consists of PC and RS232. It takes control of working condition of each intelligent monitoring unit, gets data and abnormal information and saves them. In addition, CCU can be visited by upper user computers. Furthermore, it has been practicable to analyze data and give a diagnosis for winding system by taking advantage of data management and fault diagnosis software.

The monitoring system makes use of CC-Link fieldbus which consists of communication module and twisted-pair wire acting as communication medium. Sixty-four stations could be connected to this kind of fieldbus at most. The communication module is connected with monitoring units and respective stations through CC-Link fieldbus. Transmission distance even reaches 1.2km while communication rate of fieldbus is 156kpbs.

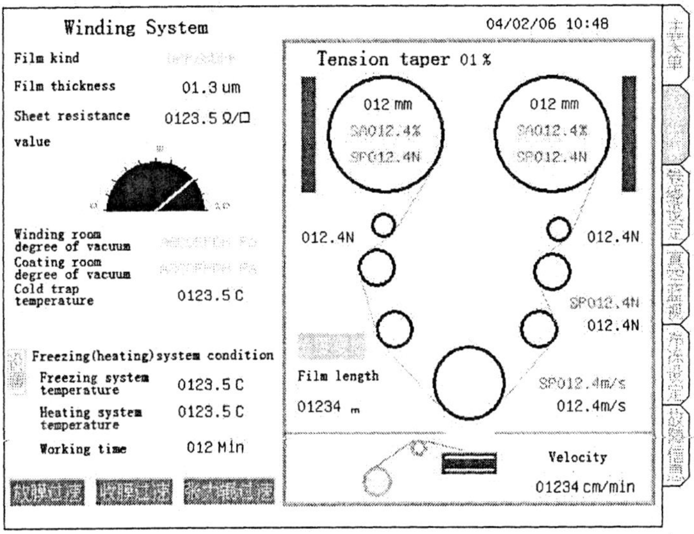

Figure 2. Monitoring display on HMI

3. PLC PROGRAMMING

In this winding system, we use Mitsubishi A series PLC as master station PLC because it has the characteristic of quick response and great ability of information processing. It is used to control the behaviors of the total winding system together with FX series PLCs of winding and unwinding system. The operating actions of the system and the sequence of these actions were edited beforehand into the control program by the designer. The control program sets a series of operations of the winding system, which tells the PLCs how to control a system. The current states of all sensors or actuators are saved as an array of input, output or flag signals in the PLC memory. Therefore, the PLC program is the basis of monitoring in a PLC controlled manufacturing system.

The programming method used is the ladder diagram method. The PLC system provides a design environment in the form of software tools running on a host computer terminal which allows ladder diagrams to be developed, verified, tested, and diagnosed. First, the high-level program is written in diagrams. Then, the ladder diagram is converted into binary instruction codes so that they can be stored in random-access memory (RAM) or erasable programmable read-only memory (EPROM). Each successive instruction is decoded and executed by the CPU. The function of the CPU is to control the operation of memory and I/O devices and to process data according to the program. Each input and output connection point on a PLC has an address used to identify the I/O bit. The method for the direct representation of data associated with the inputs, outputs, and memory is based on the fact that the PLC memory is organized into three regions: input image memory, output image memory, and internal memory[4].

The PLC program uses a cyclic scan in the main program loop such that periodic checks are made to the input variables. The program loop starts by scanning the inputs to the system and storing their states in fixed memory locations (input image memory). The ladder program is then executed rung-by-rung. Scanning the program and solving the logic of the various ladder rungs determine the output states. The updated output states are stored in fixed memory locations (output image memory). The output values held in memory are then used to set and reset the physical outputs of the PLC simultaneously at the end of the program scan[4].

As we know, logic control is a remarkable property of PLC, and it can be applied to efficiently handle analog data[2].

1) Analog data acquisition and transformation

Analog inputs and outputs such as pressure and temperature need to be measured on-line. For instance, temperature is first acquired by the Platinum resistances. A signal transformation module then converts the measured resistance value into the voltage range of 1-5V. The output of this transformation module is finally collected and transmitted back to the above PLC.

2) PLC control algorithms

The analog variables could be any controlled variable, e.g., temperature or pressure in our winding system. As a matter of fact, there exist two control modes, auto and manual. In the manual mode, operators modify the output values according to the desired levels; while in the auto mode, output values are regulated by the pre-designed control algorithms. It is worth mentioning that the outputs of PLC are always incremental values. Although auto-tuning PID controllers can be employed here, manual tuning is always used at the beginning of the production for routine initialization, and after that the system is switched to the auto mode. We emphasize that user's experiences oriented fuzzy logic controllers are utilized here to further improve the production performance as well. A film tension control system using fuzzy logic controller (FLC) on PLC to cope with the winding system uncertainty is proposed. Fuzzy logic algorithm is realized by basic PLC instructions.

To deal with the system with fixed model and parameter conventional PID is enough, but fixture is relative, in fact, the parameters of the winding system are variable in a definite range, therefore the controller must has robustness to obtain high performance. For this purpose, fuzzy reasoning is used to adjust the factors of PID, by this way the PID factors can be adjusted according to the changes of system state and plant parameters.

3) Servo motors control

At the beginning the film is transferred at a low speed, and then the servo motors are accelerated to set speed to wind at high speed at about 8 miles per second. The film tension is detected by load cells which control servo motors velocity to keep tension constant. The objective of control is not only to keep film tension constant but also to make winding velocity arrive at set value at least time. So the main problem is how to control servo motors accurately.

4. CONCLUSION

A monitoring system for PLC controlled manufacturing System Based on fieldbus for film plating production process is presented in this paper. It has been successfully on operation in practical production since 2003. Promising results have demonstrated its great advantage in practice as well as significant profit improvement.

5. ACKNOWLEDGMENT

The research has been supported by Shanghai Educational Committee and Shanghai Kelai mechatronics engineering Co., Ltd..

6. REFERENCES

1. J. C. Xing, P. Wang, B. Y. Cheng, S. Q. Wang, and Q. L. Yang, "Fieldbus Based Distributed Control System," 8th International Conference on Control, Automation, Robotics and Vision Kunming, China, 6-9th December 2004, pp. 1690-1694.
2. Q. Sheng, X. Z. Gao, and X. Y. Zhuang, "PLC-based Control Systems for Industrial Production of Fuel Alcohol" IEEE ICIT' 02, Bangkok, THAILAND, pp. 827-832.
3. W. Hu, M. Schroeder, and A. G. Starr, "A Knowledge-Based Real-Time Diagnostic System for PLC Controlled Manufacturing Systems," 1999 IEEE, pp. 499-504.
4. Maria G. Ioannides, "Design and Implementation of PLC-Based Monitoring Control System for Induction Motor," IEEE Transactions On Energy Conversion, VOL.19, NO.3, September 2004, pp. 469-476.
5. Y. M. Jia, F. H. Su, J. Liu, "An On-line Insulation Monitoring System Based On Fieldbus" , pp. 769-772.

RESEARCH OF PARAMETER MONITORING SYSTEM IN SUGAR BOILING PROCESS

Jianjun Su[1], Yanmei Meng[2], Huaqiang Qin[2], Xu Kai[1]
[1]College of Mechanical Engineering, Guangxi University, China; E-mail: gxu_mengyun@163.com. [2]School of Light Industry and Food Engineering, Guangxi University, China.

Abstract: Sugar Boiling is a most important link in producing sugar. The quality of it direct influence product. Its production line is a complicated chemistry and physics processes. The super saturation and Brix of confection in sugar boiling process, sugar boiling bowl's temperature, press...etc. Various factor effects mutually and hard to reasonable control. This paper makes use of the technique of virtual apparatus to set up on-line detection and control system of sugar boiling process, making use of the high definition of the virtual apparatus and the characteristics of easy to progress data handling, making the monitor system to have strong data amendment and analyzed processing abilities. At the same time the super saturation of confection is used as the detection and the control objects in the article, which to research the method to eliminate Brix and temperature effect of confection when measures the super saturation with method of conductance in sugar boiling process. Adopt the artificial neural network method to deal with the output of super saturation sensor by data fusion method. Using the measured value of electrical-conductivity sensor, measured value of Brix and temperatures as the inputs, the network is trained by improvement BP algorithm to achieve the comparatively precision and stabilization output. The results of test shows that the data fusion method based on the artificial neural network could effectively eliminate the change of Brix and temperatures has influenced on the super saturation. Combine virtual technique of apparatus and data fusion techniques can useful exaltation the on-line detection and control techniques in sugar boiling process, carry out its process optimization control.

Key words: Virtual apparatus; super saturation; data fusion; neural network.

The research is supported by Guangxi Province Nanning Key Technology Research Project

Please use the following format when citing this chapter:
Su, Jianjun, Meng, Yanmei, Qin, Huaqiang, Kai, Xu, 2006, in International Federation for Information Processing (IFIP), Volume 207, Knowledge Enterprise: Intelligent Strategies In Product Design, Manufacturing, and Management, eds. K. Wang, Kovacs G., Wozny M., Fang M., (Boston: Springer), pp. 582-590.

1. INTRODUCTION

The craft of Sugar Boiling is a most important link in producing sugar. The quality of it direct influence product. Its production line is a complicated process. This process include heat transfer and mass transfer, is a complicated chemistry and the physics processes, include various factor effects mutually. The core of sugar boiling craft is to control the super saturation of syrup. And adjust the parameters, such as temperature, press of sugar boiling process and the liquid potential...etc., to carry out its process optimization control. The boil sugar automatic system has already had the more applications abroad, but its equipment devotion is huge. The automatic control boil sugar system that is manufacture by domestic didn't yet see up to now. The great majority plant still depends on extracting sampling bar to get material, observing and judging the crystal dimension on the sheet glass and the content of sweet in the mother liquor, judging the degree of the further absorption and the value of sugar liquid plumb degree, and judging the operation of the next move on these grounds. The technical of boiling sugar is learning by teaching, this kind of method although go its availability, but can't get away from the man-made factor always. And can't carry on the large-scale production of automation forever, can't compare to with abroad on size and stable product quality, even can't compete. So the research of automatic detection and control system in keeping with domestic sugar production line has the realistic and important meaning. Because the control parameters in boil sugar production line are too many and mutual influence of this parameters is greatly, there are bigger limit to carry out the whole control of boiling sugar process with the traditional detection and control meter. Virtual apparatus is an all-new concept apparatus, which are deep combines computer technology with technique of apparatus. It is an effective coalescent of computer resource (processor 、 memory 、 display) 、 Hardware of apparatus (D/A、 A/D、 the digital input and output timing and signal processing etc) and the software using for data analysis、 the process communication and graphic user interface. It made use of the storage and the calculation functions of the calculator well, passing the software to carry out analyzing and treating input signal. The virtual apparatus software's programming environment provided a full space for the user's capacity and imagination force. You can bases the conceive and request of your to design、 sets up own apparatus system get across programming. After establish hardware platform, the virtual apparatus can pass the change of software's mode to adapt different request. Its function is vivid and opens, easy to connect with other establish outside、 the network, constituting a stronger system. It can upgrade of apparatus and system along with the development of the computer technology and the requirement of consumer, has advantage in capability maintenance and vivid configuration etc that the traditional apparatus can't compare to, and throw in little, yield results greatly. The main research of this paper adopts the technique of virtual apparatus and data fusion to development detection and control system of sugar boiling process.

2. TECHNIQUE OF MAIN PARAMETER MONITOR IN SUGAR BOILING PROCESS.

2.1 Main control parameter in sugar boiling process.

The main mission of boiling sugar is that using hot function which produce by heating steam in sugar bowl to condense the syrup (or molasses) to the certain super saturation, separate out muster sugar crystal and keep it greatly, boil the good confection. Pass through again help crystal and use the centrifuge removal to eliminate mother liquor, acquire good quality of granulated sugar. The key operation of boiling good confection is to control the super saturation of mother liquor in boiling confection, make the sedimentation velocity of sugar phase equilibrium. Integrate the method of the boiling sugar, can induce for 4 main factors of following: 1、Hold vacuum、the press of steam、the concentration of input and the temperature stabilizations; 2、Insure the good backward flow of confection which are boiling; 3、Control the super saturation adequate; 4、Boil the confection becomes tight. To carry out the automatic control of boiling sugar process, must monitor this a few main parameters: the super saturation of confection 、temperature of boiling sugar bowl、liquid place, press of steam, degree of vacuum…etc.

2.2 The main parameter detection and control system in boiling sugar process.

It can adopt the control method which use control super saturation as the core in boiling sugar process, and attain to the request control as follows: (1) Measure the liquid level of the syrup or confection continuously, for the purpose of the serve of automatic feed; (2) Control super saturation depend measure the conductivity of conductance or the value of not conduct electricity continuous; (3) Continuous control the absolute pressure in sugar bowl; (4) Continuous control the serve of steam in the whole boiling sugar process; (5) Continuous control the temperature of confection; (6)Measure the power dissipation of the agitate machinery automatically, on these grounds let it condense to the most suitable value of sugar liquid plumb degree before put it in sugar bowl further. The main parameter monitoring system of boiling sugar process is shown in Figure 1.

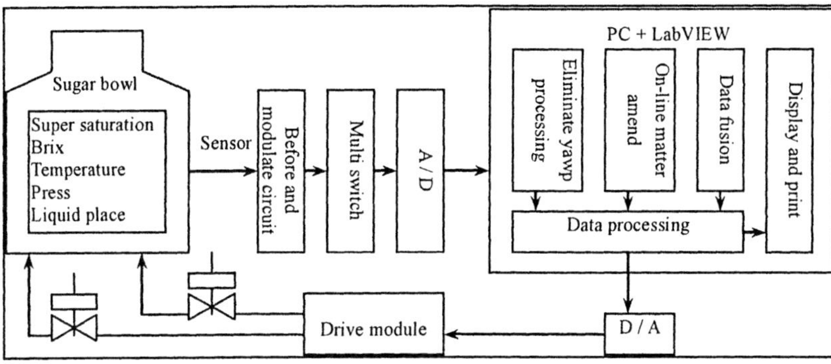

Figure 1. Main parameter on-line detection and control system

Detection and control system include inspect measure and control. The main parameters of boiling sugar process such as super saturation, Brix, temperature, press and liquid place etc., those are converted as electricity passes homologous sensor. It then pass beforehand modulate circuit and multiplex select switch, make use of the A/ D circuit transform it as digital signals, enter computer treating system to go along data handling. Result can display on computing and print. Then the signals, which have been dispose, pass D/ A conversion and drive circuit to adjust the valve switch. Thus carrying out to the automatic control of boiling sugar process. Detection and control system make use of the high definition of the virtual apparatus and the characteristics of easy to processing data well, make it have the function of strong data analyzed processing and amend. It can eliminate yawp in meterage and accurate amend the date which have error. For example it can compensate the measure cold end of thermoelectric couple that use for measure temperature and amend the nonlinear of measure, it can do date fusion with super saturation which effect by temperature and Brix bigger. It raised the accuracy and the reliability of detection data consumedly.

3. ON-LINE DETECTION TECHNIQUE OF SUPER SATURATION

Boil sugar this production link is very important in refine sugar industry produce. The syrup through compression、 vaporize and other chemistry treating, complete fluid- solid convert in the crystal pot finally. The syrup is placed in metastable section of super saturation and born finished product sugar by crystal gradually. Boil sugar process change constantly because of the change of material degree and concentrations. The craft condition of boiling sugar such as Brix of confection、wave motion of temperature or steam press, this factor affected the super saturation of syrup directly. From the craft of refine sugar we can know, the key of keeping the quality of boiling sugar is to control super saturation of syrup and keeping in adequacy range. The sensors of conductance can on-line measure the super saturation. But use the conductance method measure the super saturation is under the influence of the Brix of confection and the temperature of boiling sugar. But we can use method of data fusion that is base on neural network to eliminate this influence.

3.1 Method of data fusion about super saturation

The electrical-conductivity sensor is placed in the boiling sugar bowl to measure super saturation of syrup, while a Brix sensor and a thermocouple that respectively measured the Brix and temperature of syrup are added to eliminate the influence. The super saturating is regarded as object parameter and the Brix and temperature of syrup are interferential parameters. Then the three parameters would be dealt with by data fusion to achieve the objective output. The frame of data fusion system is shown in Figure 2. In Figure 2, the outputs of super saturation sensor, Brix sensor and temperature sensor are U_1, U_2 and U_3 respectively, B shows super saturation with no disposal, Bs shows the super saturation dealt with the method of data fusion.

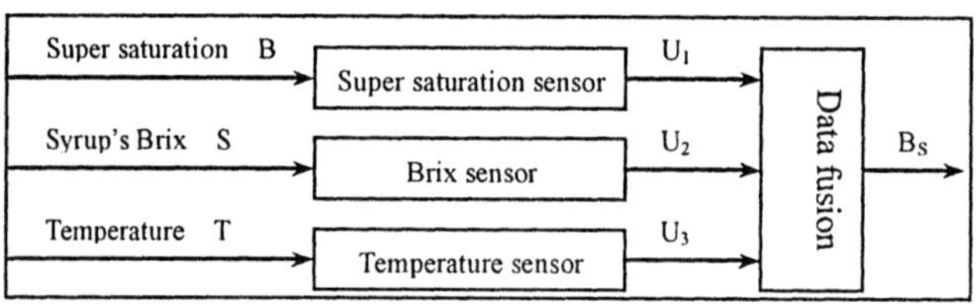

Figure 2. The model of date fusion system

3.2 BP forward neural network algorithm

Back Propagation network (BP network) is used widely in the models of neural network. The BP network is made up of three layers included input layer, hidden processing layer and output layer. The layers is interlinked all, and inside a layer, each neuron is not connected one another. The BP network has the great capability of nonlinear mapping and paradigm function; furthermore, any continuous function or mapping could be implemented by three-layer BP network. According to the research, the output of network is linear combination of a set of weight value、 threshold value and input value. The network could approach any continuous function when the function of hidden processing layer has differential coefficient with any step. Then the multi-sensor data could be fused by using a three-layer neural network which would be trained by optimized BP algorithm. BP algorithm that belongs to the δ algorithm is a surveillant type algorithm and its main theory is: for the input swatches that will be learned (p^1, p^2, …, p^q), corresponding output swatches (T_1, T_2, …, T_q) is known, then, the learning objective of network is to modify the weight value by using the error between the actual output (A_1,A_2,…,A_q) and target vector (T_1, T_2, …, T_q) in order to achieve the least sum of errors' square. The BP algorithm approaches the objective by continuously computing the weight value of network and the change of error along the decreased direction of error function's slope. Each change of weight value and error become the direct proportion with the influence of the network error. And it transfers to each layer in the way of reverse spread. BP algorithm is composed of two parts, forward transfer of information and reverse spread of error. In the forward transfer, the information of input is transferred passing the hidden processing layer from the input layer to the output layer. The outputs of each layer's node act on the inputs of next layer's node. If the output layer couldn't get the anticipant value, BP would compute the change value of output layer and turn to back propagation and then modify the node's weight value of each layer to achieve the anticipant objective according to the error signal's feedback which has been back from quondam connected gateway. The algorithm of reverse transfer is used widely but its long training time tends to plunge the partial infinitesimal because it is limited by using the gradient-descent method. The training of BP network substantively is an optimized problem about nonlinear objective function. And if the weight values of BP network could be trained by using the algorithm based on the method of numerical optimization, a faster convergent speed than using the gradient-descent method would be attached. Levenberg-Marquardt algorithm (LM) has been adopted to train the network at the same time. The search direction of LM method is given as follows.

$$X^{(k+1)} = X^{(k)} + \eta^{(k)} S(X^{(k)}) \qquad (1)$$

$$S(X^{(k+1)}) = (H^{(k)} + \lambda^{(k)} I)^{-1} \nabla f(X^{(k)}) \qquad (2)$$

Where $X^{(k)}$ is a vector merging the all power and offset in the network, $S(X^{(k)})$ is the search direction of X vector, $\eta^{(k)}$ is the min step of $f(X^{(k+1)})$ with the direction $f(X^{(k+1)})$.

If $\eta^{(k)}=1$, then $X^{(k+1)} = X^{(k)} + S(X^{(k)})$. The strongpoint of LM is that the network converges quite quickly as less weight value.

3. 3 Test Data Fusion and Analysis of the Conclusion

At first the demarcated test is taken to attain the output (U_1) of super saturation sensor with different super saturation in the specifically Brix and temperature, the output (U_2) with different syrup's Brix in the specifically super saturation and temperature and the output (U_3) with different temperature in the specifically syrup's Brix and super saturation. And then the demarcation value will be used to train the network and enforce data fusion so as to achieve the exacter information of super saturation.

According above theory, a three-layer BP network has been created to achieve the data fusion by using the Neural Network Toolbox of MATLAB and the fused result would be transferred to the foregrounding of measurement by the interface of LabVIEW and Matlab. There are three nodes on the input layer and one node in the output. In the network the *tansig* function is taken to activate from the input layer to hidden processing and the *purelin* function is used to activate from the hidden processing layer to the input layer. Firstly the input data of network must be done with unitary method to ensure the network's convergence. In this way, the three inputs are U1, U2 and U3 with unitary disposal and the output is Bs that has dealt with data fusion. Having done with the data, the network could be trained by the date of swatch database. It couldn't be convergent after 10000 steps using the non-optimized gradient-descent method, while it only need 1804 steps to attain the convergent precision quickly by using the LM algorithm. The measured value of super saturation sensor and fused results used the network with different Brix in the especially temperature (375k) are shown in Table 1.

Table 1. The measured value and fused value of super saturation with different Brix

Brix (^{0}Bx)	Demarcated Value=1.100		Demarcated value=1.150		Demarcated value=1.20		Demarcated value=1.25		Demarcated value=1.30	
	Meas-ured value	Fused value	Meas-ured value	Fused value	Meas-ured value	Fused value	Meas-ured value	Fused value	Meas-ured value	Fused value
75	0.870	1.090	1.031	1.147	1.021	1.199	1.063	1.236	1.213	1.295
75.5	1.031	1.085	1.121	1.138	1.153	1.201	1.231	1.251	1.291	1.298
76	1.131	1.110	1.231	1.149	1.230	1.205	1.206	1.249	1.352	1.306
76.5	1.153	1.103	1.364	1.151	1.362	1.203	1.248	1.252	1.483	1.310
77	1.136	1.105	1.302	1.156	1.445	1.202	1.339	1.251	1.537	1.312
77.5	1.270	1.102	1.231	1.159	1.311	1.205	1.265	1.249	1.491	1.301
78	1.126	1.109	1.186	1.499	1.243	1.219	1.232	1.248	1.387	1.306
Max ΔB	0.4		0.333		0.424		0.302		0.324	
Max ΔBs		0.025		0.021		0.020		0.016		0.017

Table 2 shows the measured value of super saturation sensor and fused results using the network with different temperature in the specially Brix (76°Bx).

Table 2. The measured value and fused value of super saturation with different temperature

Temp-erature (K)	Demarcated Value=1.100		Demarcated value=1.150		Demarcated value=1.200		Demarcated value=1.25		Demarcated value=1.300	
	Meas-ured value	Fused value	Meas-ured value	Fused value	Meas-ured value	Fused value	Meas-ured value	Fused value	Meas-ured value	Fused value
350	0.878	1.093	1.021	1.131	1.026	1.189	1.073	1.246	1.213	1.292
355	1.030	1.082	1.161	1.140	1.153	1.201	1.231	1.251	1.298	1.304
360	1.212	1.109	1.241	1.148	1.230	1.205	1.206	1.249	1.362	1.309
365	1.156	1.106	1.379	1.151	1.362	1.203	1.248	1.252	1.417	1.317
370	1.141	1.105	1.392	1.157	1.453	1.202	1.349	1.251	1.569	1.312
375	1.232	1.112	1.215	1.149	1.311	1.205	1.449	1.249	1.458	1.310
380	1.126	1.109	1.182	1.141	1.243	1.213	1.212	1.248	1.410	1.303
Max ΔB	0.354		0.371		0.427		0.376		0.356	
Max ΔBs		0.030		0.026		0.024		0.022		0.025

Before it has been fused, the relative value of sensor's fluctuant output that is affected by the change of syrup's Brix is given as Eq (3),

$$\delta = \frac{\max|\Delta B|}{B_{标}} \qquad (3)$$

Where δ is the relative value of output's fluctuant electrical pressure, max|ΔB| is the max absolute value of sensor's output, $B_{标}$ is the value of demarcating super saturation.

Having been fused, the relative value of sensor's fluctuant output is affected by the change of syrup's Brix is given as Eq (4).

$$\delta' = \frac{\max|\Delta B_s|}{B_{标}} \qquad (4)$$

Where δ is the relative value of output's fluctuant electrical pressure, max|ΔB_s| is the max absolute value of sensor's fluctuant output, B_d is the demarcated super saturation.

From the demarcation data, the relative value of sensor's fluctuant output is 35% when the demarcated super saturation is 1.20% and the Brix changes from $75^0 B_x$ to $78^0 B_x$; therefore, the Brix of syrup affected its super saturation strongly. Having been fused, the relative value of fluctuant output is decreased to 1.7%.

In the same way, the relative value of sensor's fluctuant output is 36% when the demarcated super saturation is 1.20% and the temperature changes from 350K to 380K; therefore, the temperature affected the syrup's super saturation strongly. Having been fused, the relative value of fluctuant output is decreased to 2.0%. It shows that the effect of Neural Network data fusion is distinct. The neural network method is quite availability in eliminating the effect by temperature and Brix to sensor at work.

4. CONCLUSION

It is convenience, fast, good quality of transplant and reuse that the on-line detection and control system in boiling sugar process use the open and graphics mode form of programme method to set up the system. Detection and control system also make the best of characteristics of virtual apparatus: high precision and easy to progress data handling, make it have the function of strong data amend and analyses treating. Hang together the technique of virtual apparatus and data fusion organic when measure the super saturation degree of confection organic. The method of date fusion, which is base on the artificial neural network, eliminate the influence of the change of confection's Brix and temperatures, thus get the precision、stable-measured value. The fluctuant value of super saturation sensor's fluctuant output is 30～40% without disposal, which compared with the value is less then 2.5% with data fusion; therefore, the effect of data fusion is distinct. This experimentation proves that the automatic monitoring system of boiling sugar process achieve the anticipant effect.

5. REFERENCES

1. Liang Zhou, *Instrument and Automation of Refining Sugar Industry*, Beijing: Light Industry-publishing Company, 1988.
2. Shen Yi etc. *Study on Data Fusion of Bedload's Content with Capacitive Sensor [J]*, Journal of Huazhong Agricultural University, 2004, vol23，No4：459～462.
3. Cong Shuang, *Application of Nerve Network and Fuzzy System on Movement Control [M]*, Beijing: Publication of University of Science and Technology of China, 2001.
4. Qingdong Du, Lingyu Xu, Zhao Hai, *Differentiate and Analyse of Pressure and Indraught System of Water and Electricity Factory Based on Neural Fusion Arithmetic [J]*, Control and Decision-making, 2001,16 (Subjoined): 787~789.
5. Beiping Hou, Lu Pei, *Modeling and System Simulation of BP Nerve Network Based on MATLAB [J]*, Automation and Instrument, 2001, 16 (1): 34~36.

A DECISION METHOD OF MATERIEL MAINTENANCE BASED ON FAILURE LIFE CYCLE

Wen Yuan Song, Lian Yan Shi, Jian She Kang
6th department of shijiazhuang mechanics engineering college, Shijiazhuang, Hebei, 050003, China, Email: mumi13@163.com.

Abstract: For a complex materiel, the failure of significant item with higher reliability require can not be satisfied by Reliability Centered Maintenance Analysis alone, nor by Root Cause Analysis and Proactive Maintenance. According to the feature of failure come into being, a concept: failure life cycle, was presented. And it was divide into several parts according to each phase feature. The incidents occurred on each phase can be resolved by this three technics. Integrated them based on failure life cycle, a perfect dynamic maintenance decision technology can be formed in solved any failure on complex materiel.

Key words: maintenance decision, life cycle, decision optimized, reliability centered maintenance, proactive maintenance, root cause analysis.

1. INTRODUCTION

In the implementing of Reliability Centered Maintenance (RCM), there are many items which failure can not be prevented effectively. After analyzing on three modern maintenance decision technology, RCM, Root Cause Analysis(RCA) and Proactive Maintenance(PaM), it is found that there is not a single method of them able to solve these problems perfectly or econamically and meet the require of high reliability or availability level[1,2,3]. It is necessary to find a systematic method focus on the whole life cycle of an evident failure in order to determine the best maintenance task. To the important materiel, it is worthy of finding any multifarious preventive maintenance policy for its significant. Based on that, an item, failure life cycle(FLC), was presented. And then, FLC was divided into several parts

Please use the following format when citing this chapter:
Song, Wen Yuan, Shi, Lian Yan, Kang, Jian She, 2006, in International Federation for Information Processing (IFIP), Volume 207, Knowledge Enterprise: Intelligent Strategies In Product Design, Manufacturing, and Management, eds. K. Wang, Kovacs G., Wozny M., Fang M., (Boston: Springer), pp. 591-596.

according to its feature. Different maintenance decision technology can be taken for the different parts of FLC. By integrated these three methods, a perfect dynamic maintenance decision technology can be formed finally.

2. DIVIDING THE FLC

For complex materiel, the failure forming process of significant function item is always an event chain or tree with a causal relationship (causal chain or causal tree), which can describe from the initial root cause[3] to the aimed function failure. FLC was divided into five parts in this paper, see figure I .

1. Pregnant Phase: Pregnant phase is a phase from FSI begin working to the root cause occur, not a necessary phase. There is not a pregnant phase if the root cause existed at the very begging, such as fixing problems.
2. Root Cause Action Phase: This phase begins at the root cause occur and end at the next causal factors occur, a necessity phase. Root cause can be a fault element, a operating/maintenance error, design problem, environment problem, and it can also just the normal work station, such as vibrating. Root cause may be not just one, and it can be combined by several event or factors. These event or factors may occur in different time, while, there is always a key cause, which directly effected whether the causal chain or tree can be formed.

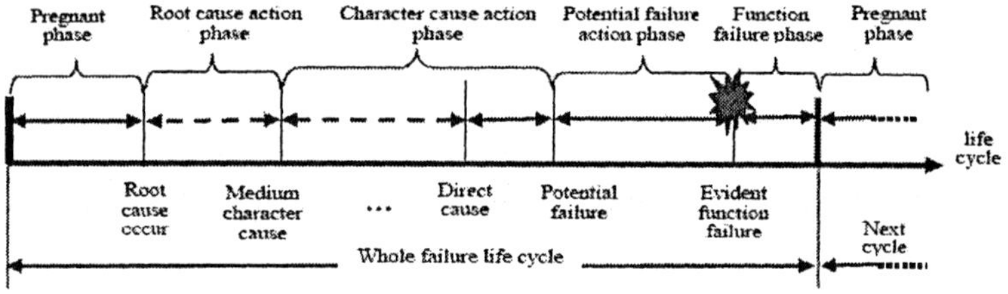

Figure 1. The divided of failure life cycle

3. Medium character causes action phase: This phase began at the next cause occurred after root cause and ended at the potential or function failure occurred, not a necessity phase. These causes prefer the defined identifiable condition that indicates that a degradation/failure process is taking place that will lead to a next item functional failure or the aim item-FSI potential/function failure. It may have several events or have none when the root cause is the direct cause of the aim function failure.
4. Potential Failure Action Phase: This phase begins at the potential failure appearing and end at the evident function occurred. And the potential failure is a defined identifiable condition that indicates that a degradation process is taking place that will lead to the evident function failure.

5. Evident Function Failure Phase: This phase is from the aim item evident function failure occurs until it is repaired.

3. MODERN MAINTENANCE DECISION TECHNICS

RCM, PaM and RCA are the three methods used commenly in plant maintenance planning currently.

In RCM, it is by answered the 7 questions[5] in turn to determine all the potential failure modes or direct cause of the analysis item, and develop a maintenance policy in order to reduce the failure effect by using logic decision[4]. And the policy is focus on these potential failure mode or direct cause. While in fact, most of the failures occurred in working are the repeatedly failure which caused only by a root cause in generally, only few of the failure caused by happenstance. Obviously, RCM can not prevented the failure itself, but reduced its effect. So, reference [6] considered that the RCM is driven by maintenance policy not by security policy.

RCA is a normalized causal analysis approach, which research and trace the accident and the cause relationship, until to find the root cause of accident, and then adopt correspond measure in solving the problem thoroughly[3]. It is focus on the root cause on materiel of failure, not directly cause. The corrective measures of RCA are far beyond traditional maintenance work types, which include all works able to improve the materiel work condition, such as training, improving operation type and design etc. In general, the beginning level of element in RCA is the ending level in RCMA[6]. The object of RCA is the item with high reliability require or frequently failure.

PaM is a method which identify the "root cause" of the evident function failure systematically, and then determined maintenance policy before the object performance or material degradation, that is preposition maintenance by dint of modern detection technology in order to reduce the whole maintenance require and extend the service life of materiel[7].

Contrast with RCM and RCA: First, it analysis object should be the high reliability require item with the evident condition that must be identified and controlled by relative advanced detection technology, and there still be some suitable, economic and effectiveness PaM measure; Second, the "root cause" is any surround factors effected the lowest analysis item of RCM, not the "root cause" used in RCA. In this paper, it is taken as the directly cause or medium character cause; The last not the least, PaM is not a independent maintenance work type, but a maintenance measure based on condition monitoring aim at the failure root cause, it contain the failure station monitoring and elimination.

4. FLOW OF MAINTENANCE DECISION BASED ON FAILURE LIFE CYCLE RESULTS

Traditional preventive maintenance aimed at the phase from directly cause to function failure and the key point is potential failure phase. It is considered that the closer the maintenance task to function failure, the more economics of the work premising the function not occur. While, the principle of RCA is prevented the aimed failure occurred at root by avoiding the root cause or medium characteristic cause occur. And the focus of PaM are the directly cause or the cause of directly cause, such as material performing, the polluting of liquid uncommonly, etc. In PaM, the modern detected technology, such as automatic particle counting, oil and water analyzer, etc, is necessary. And the figure II show the suitable phases of the 3 modern maintenance decision method. Figure III show the flow of integrated maintenance decision method, and it is described in detail as follow.

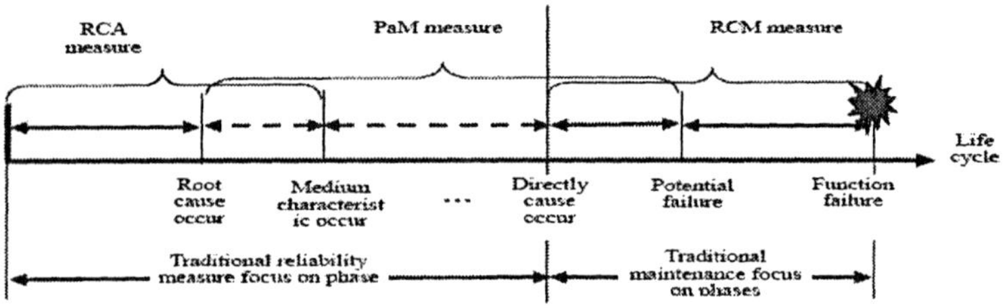

Figure II. 3 maintenance decision technics active phases and failure life cycle.

4.1 Determined the analysis item

The analysis items must be the significant function item(FSI), just as FSI in RCM. And divided these item according to its reliability require and failure feature that obtained by FMEA. The selected criteria as below:

- the FSI with high reliability requirement, and serious failure effect need RCA and/or PaM analysis;
- the FSI with frequency failure feature, and due to some root cause or semi-root cause need RCA and/or PaM analysis.

4.2 Build the causal chain/tree

In this sub-process, RCA is used to resolve two matters: one is determined whether a root cause exist or not; the other is build the key causal chain. If there is a root cause indeed, relative measure should determine according to the root cause character, if there is not a real root cause, a causal chain should build. It should be noted that[8]:

- The data collected include all effective information, such as test data, history data, even the operating information. While the veracity, accuracy of data should be verified. The data collecting direction should be modified according to the real information in time.
- The estimating needs be verified. The process is analyzing one level events, and evaluating, and then verifying, until the root cause was find or the key factor un-convergence.
- Tracking result is most important in analysis processing, it is an effective method with long period and semi-simulation property. In normal analysis, it is unable to practice, but marked to do that after analysis.

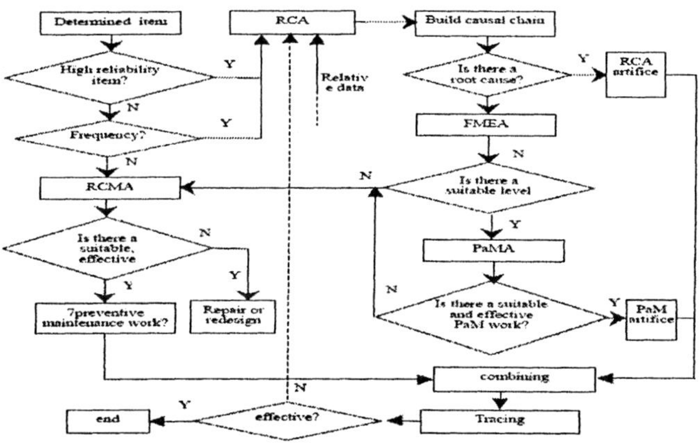

Figure III. maintenance decision flow based on FLC

4.3 Proactive maintenance analyzing

PaMA is based on RCA, which focus on the item without real root cause or effective RCA measure. General, there is only one key failure mode/event in on link/level of causal chain/tree with special property. Considering the require of item in PaM, not all the key failure/event are suitable for PaM. So, it is need analysis one by one in order to determine the suitable link. It should be noted that:

- The analysis item must be key factor in the causal chain/tree.
- The location of analysis item in material must be suitable to be monitoring or/and detecting by test devices on line.

- There must be a suitable test devices or test technology to monitoring or/and detecting the item.
- The interval of the analysis item acting phase must be reasonable and constant in order to ensure the worthy of the PaM measure.

4.4 Determined and Combine the Preventive work

The item need RCMA in here is items which failure will cause direct adverse effect on operating capability or economic effect. Otherwise, those items without real root cause or relative measures is still need be analysis by traditional RCM logic decision. And then combing all the preventive work/artifice determined according to the item or the interval of maintenance work and finally come into a maintenance plan or program.

4.5 Result Tracing

Result tracing is an important step in this analysis, the result of tracing will validate the accuracy of maintenance plan or program and modify the RCA process.

5. CONCLUSION

Above all, this method can analysis the items with repeatedly failure or/and high reliability requirement on complex materiel preferably which cann't be resolved effectively by traditional RCM. The maintenance plan developed by this method is more perfectly. Limited to space, the optimizing analysis of the decision precess and a case study are not present in this paper.

6. REFERENCE

1. Moubray J, (1997) *Reliability Centered Maintenance*, Butterworth - Heinemann, oxford.
2. E.C.Fitch, (2002), Proactive Maintenance for Mechanical Systems, *Magazine,* May/Jun.
3. W.Y. Song, (2005), RCA——A Normalized Causal-Result Analysis Method, *Materiel Quality* 2005.6, pp. 34-39.
4. *Requirement and Procedure of Developing Preventive Maintenance Program for Materiel*, Milltary standard Publisher (GJB1378-1992).
5. SAE JA1011, Evaluation Criteria for Reliability Centered Maintenance (RCM) Processes.
6. Ronald L. Hughes, Analysis: RCM Versus RCA, Maintenance Resources On-Line, *http://mt_online.com.*
7. Z.H.Cheng, X.SH. Jia, etc (2003). Active Maintenance, Proactive Maintenanc, and RCM, *Jurnal of Ordnance Engineering College* Vol.14 No.11,p23-28.
8. *Root Cause Analysis Guidance Document*, (DOE-NE-STD-1004-92) 2,1992.

DISCUSSION ON CSCW METHODS FOR EMBEDDED SYSTEMS

Tao Yu[1], Tan Liu[1], Shuzhen Yang[2], Wenbin Wang[1]
[1]CIMS & Robot Center of Shanghai University, Shanghai University, China; Email: alanltnew@citiz.net.[2] Electromechanical Engineering Institute, Shanghai Second Polytechnic University, Shanghai, China.

Abstract: This paper discusses the Computer Supported Collaborative Work (CSCW) methods that can facilitate the work of embedded systems. The focus is on how to design and build effective embedded network architecture, and which CSCW tools should be used in embedded system. A brief conception of CSCW and embedded system is presented, followed by designing and building an embedded network system currently available, and CSCW tools used in embedded systems are discussed additionally. Finally, predictions are made on what CSCW for embedded systems might expect to be in future years.

Key words: CSCW, Embedded System.

1. INTRODUCTION

Doing business over Palms is cheap and convenient, and a commercial conference can be hold just through Palm communications. C4P (CAD, CAPP, CAM, CAE and PDM) [1] can be collaboratively worked in the microcontrollers such as SCM or Soc (System on chip). A project can be carried out from general engineer to partial managers, and to every executor without a desktop computer. A work-flow can be finished smoothly through your GSM. You can control robots or machine tools just with your mobile phone through the internet, and so on.

All of above are embedded products in our daily lives. The CSCW research for general computers has been presented for a long history, but the CSCW study for embedded systems is done with even less people. For the reason that embedded systems are too many and they are designed for all kinds of different special purposes, the research is complicated to carry out.

This research work is financially supported by the Science and Technology Development Foundation of Shanghai Science and Technology Committee (036505008 and 05DZ11001).

Please use the following format when citing this chapter:
Yu, Tao, Liu, Tan, Yang, Shuzhen, Wang, Wenbin, 2006, in International Federation for Information Processing (IFIP), Volume 207, Knowledge Enterprise: Intelligent Strategies In Product Design, Manufacturing, and Management, eds. K. Wang, Kovacs G., Wozny M., Fang M., (Boston: Springer), pp. 597-603.

This paper focuses on how to build an effective and preferable embedded system network and select some suitable tools to finish their different works collaboratively. And, here we give a general project to solve the CSCW problem of embedded systems in final.

2. EMBEDDED SYSTEM AND CSCW

Embedded system [2] is a special type of computer system that is centered on application, adjustable in software and hardware and suitable to the strict application requirements in function, reliability, cost, size and energy consumption. The embedded computer well surpasses the general computer in the extent of application. They also are widely used in manufacturing industry, process control, communications, instruments and meters, automobiles, ships, aeronautics and astronautics, military equipment, and consumer goods. The strong points of embedded system are of good real-time performance, low-cost, miniaturized, specialized, high reliability, extensible and configurable.

Talking to CSCW, we must say something about the birth of CSCW-Office automation. In the mid 1970's, single user applications such as spreadsheets and word processors ran amok on mini-computers. Office automation was attempt to integrate those applications, but without any real idea of the user requirements or of the sociology of sharing the source documents. CSCW then began as an attempt by software engineers to incorporate into new software tools know on how people work in groups. This knowledge was obtained form practitioners in the fields of, for example, social psychology and organizational theory [3].

3. COMMUNICATION OF EMBEDDED SYSTEMS

3.1 Connection to the Intranet/Internet

Considering the performance and the cost, you must estimate the resource demands earlier. It includes the memory demands of protocol stacks. Using more protocols will make you system bigger. And you must consider what hardware is needed in your system, maybe an Ethernet controller, or a UART asynchronous adopter and a serial interface, or even a wireless receiver. These factors should be considered in your project [4].

And you must consider the additional consumption to your CPU added by the protocol stacks, because adding a network support are likely to make your system sluggish. If you bought the protocol stacks from a RTOS developer, the stacks are integrated in the operation system. The developer will afford the technical support.

There are many models that connect embedded systems to the Intranet/Internet for different network demands and construction purposes. But generally speaking, we may just consider how to implement the TCP/IP network protocol. The two questions of connecting embedded systems in the Intranet/Internet are the media selection of information transmission and the protocol selection. Oriented to the two questions, the most common connecting method is: using the Ethernet and TCP/IP protocol, achieve physical layer and link layer protocol through network interface controller (NIC), and at the same time the micro-processor run the TCP/IP protocol communication module to complete the connection with Intranet/Internet.

3.2 Preferable bus architecture

Many communication styles can be used in the embedded systems based on micro-processor. It is different with the speeds and times in communication processes. The most common transmission is the communication between micro-processor and memory, and that the low-speed communications between micro-processor and peripheral equipments are less. All communications can be done in one high-speed bus. But it has the following disadvantages: First, every peripheral device must have high-speed bus interface, which will add logic switches, power waste and cost; Second, generally high-speed bus are particularly designed for special processor, so the peripheral devices with the interface above are not easy to be transplanted; Third, the speed may be greatly reduced when the peripheral devices in the bus are too many.

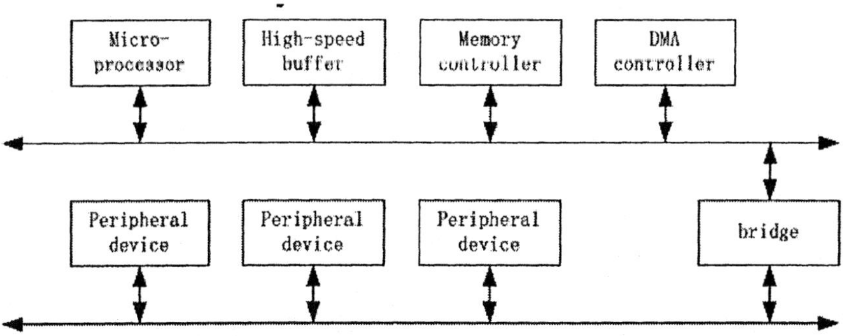

Figure 1. Two layers bus architecture.

For the reasons above, the network can be designed to be with two layers (Fig.1). It is the high-speed local bus and the low-speed peripheral bus. The high speed local bus is used to connect micro-processor, high-speed buffer, and memory controller and high-speed co-processor. It is a special bus and it is generally as wide as the width of a word of the memory.

The peripheral bus should be designed to have a good transplanting performance. It should have a low power-waste, and have less logic switches. It is generally compatible with the standard industry bus, such as ISA and PCI, so it can be transplanted easily. It is narrow and low-speed compared with the local bus, so it doesn't need many logical switches, and doesn't consume much energy. And now there is some other peripheral bus currently used, such as CAN, SPI, which has a higher speed [5].

The bridge is a single-purpose processor; which can transfer the communication with this bus standard to the communication with another bus standard.

Sometimes, the architecture can be built with three layers. The first layer is the micro-processor local bus, the second is system bus, and the third is the peripheral device bus. The system bus is high-speed bus. It undertakes some communication tasks in the local bus. This three-layer architecture is beneficial to build some complex systems with multi-coprocessors.

3.3 Wireless communication

Besides the traditionally serial communication and parallel communication, wireless communication can avoid the physical contact with the networked devices. The physical layer used in short-range wireless communication is IR (infrared ray) channel or RF (radio frequency) channel. There are three famous wireless protocol: IrDA (Infrared Data Association), Bluetooth, and IEEE802.11. They are widely used in embedded communication devices. In a long-distance communication, GSM/GPRS is the selection.

4. DISCUSSION OF CSCW METHODS FOR EMBEDDED SYSTEMS

4.1 CSCW tools for embedded systems

To date, the technology used in CSCW can be broken down into several areas. By examining the functionality of the available tools in each area, it is apparent that the distinction between each is blurring towards a few hypothetical, homogeneous "super tools". The "super tools" would cover E-mail/Messaging, Groupware, Conferencing, Software Development and Virtual Rooms. The common theme in what is available is the underlying use of the Internet, and in particular the Web, as an inter-personal computer connection transport.

Table 1 shows the diversity of CSCW categories as they appear today. Comparisons are made between embedded systems and general computers. For each category there is an estimate of how widely it can be used by the embedded system, and some example tool names given.

Table 1. CSCW categories can be used in embedded systems

Category	Can be used in embedded systems	Example projects and tools in PC
Bulletin Boards	Yes, Easy	WebLines
Chat	Yes, Easy	WebChat, NetMeeting
Collaborative Design	Yes, Difficult	n-dim
Collaborative Drawing and Writing	Yes, Difficult	GroupSketch, Conversation Board, Emacs with
Collaborative Software Management	Yes, Difficult	CVS, RCS, SCCS
Collaborative Learning	Yes, Difficult	MUNIN
Email	Yes, ubiquitous	Numerous
Groupware toolkits	Yes, Difficult	Lotus Notes, Microsoft Exchange,
MUDs	Yes, Not difficult	XMX, XTV
Newsgroups	Yes, Easy	Navigator, Internet Explorer
Virtual Labs and Rooms	Unkown, Very difficult	TeamWave Workplace, UARC, AAEM
Share Windows	Yes, Difficult	NCSA Xcollage(defunct)
Videoconferencing	Yes, Difficult	Vic, vat, rat, nv, CU, NetMeeting, WebPhone, VidCall, FreeVue, Vfone, etc.
Virtual Reality	Yes, Very difficult	VENUS, CAVE, DIVE
Web-based conferencing	Yes, Difficult	WebNotes, NetMeeting
Web-based calendars	Yes, Easy	WebCal

Pure videoconferencing is technology that allows two or more users to interact using audio, video or both. It is covered by a set of standards relating to CODEC, ISDN and packet-based conferencing, which are more or less supported by the available tools. And Virtual labs and rooms contains tools that allow users to enter virtual spaces, manipulate objects within them, and interact with other users "present" in the space. Meeting ware is for tools which integrate videoconferencing as part of a suite of subtools to support meetings. Because these tools need high bandwidths and abundant resource, it is very difficult to be transplanted to embedded systems.

Workflow tools or groupware help in the management of project schedules. They allow the sending, reception and annotation of electronic documents, and the triggering of messages to be sent when certain conditions have been met. They can be easily to be transplanted to embedded systems for their flow-style mechanism. But they should be cut down to meet the embedded systems.

Email and messaging is the most widely used and platform-independent groupware system. It is the lowest common collaborative computer tool. Embedded system can make heavy use of Email, and it is not further treated here, and some embedded systems have included it. Usenet news can also be widely used in embedded system. Most embedded Web browsers have built in News readers. Bulletin board systems were mainly designed in the late '70s as dial-up systems dedicated to the exchange of files between people. There are Web-based BBS systems, such as WebLines, available. In general, BBS systems can also be used in embedded systems. Multi-user dungeons (MUDs) are also referred to as Multi-User Dimensions. The basic idea is that, using a text-based system, the user enters an environment that contains objects and other users, and then interacts with the environment by typing commands. The main use of MUDs is in game playing coupled with social intercourse between the participants. And it might be a synchronous tool can be used in embedded systems [6].

4.2 Synchronous or Asynchronous

Do CSCW tasks intrinsically need high bandwidths? To answer this question, divide CSCW applications into synchronous and asynchronous. A synchronous CSCW requires communications facilities as the system is being used; an asynchronous CSCW requires communications facilities between users. Thus synchronous CSCW permits users to work on the same task simultaneously; whereas tasks in an asynchronous CSCW must be interleaved. There are clearly advantages in both approaches, depending on the exact nature of the tasks being supported. It is clear, however, that the communications media for the alternatives need not be the same, in particular, synchronous and asynchronous CSCW have very different maximum and average bandwidth requirements considered as a function of the number of active participants in any task. For embedded systems, considering the limited bandwidth, MUDs might be a good synchronous tool, and Email, Bulletin board systems might be good asynchronous tools.

5. FUTURES FOR CSCW IN EMBEDDED SYSTEMS

What is driving the market for the paper talked above? As embedded systems become more and more networked, small groups of workers become potential new customers for groupware tools. Mature single user embedded applications can be embellished with groupware features. Telecom companies seek to increase demand for networks by promoting multimedia tools that make use of them. Commercially, the platform being targeted to the embedded systems are rapidly developed, and the corresponding CSCW tools are unavoidable to be put on the table.

Apart from commercial factors, there is considerable interest in the academic community in wide area project collaboration, and the need for tools that support collaborative work at a distance is evident when considering the size of new generation industrial embedded systems collaborations and their geographical spread, such as meeting-ware and shared applications. Transplanting existing CSCW toolkits to embedded systems to support wider area collaboration looks promising for the future [7].

6. CONCLUSIONS

This paper discusses CSCW methods that can facilitate the work of embedded systems. It focuses on how to design and build embedded network architecture, and which CSCW tools should be selected in embedded systems. Generally, TCP/IP protocol and two-layer bus architecture are preferable. The CSCW tools such as MUDs, Email, Bulletin Boards and so on can be used in embedded system. It is evident that the CSCW for embedded systems has a promising future.

7. REFERENCES

1. John V Farr. Dennis M Buede, "Systems engineering and engineering management: Keys to the effient development of products and serives.", *Engineering Management Journal*, Vol 15, No. 3, pp. 3, Sep. 2003
2. Rick Grehank, Rober Moote, Ingo Cyliax, *Real-Time Programming*, Addison-Wesley. USA. 2001.
3. Junzhong Gu, *An Introduction to Computer Supported Cooperative Work*, Tsinghua university publisher, Beijing China, 2002.
4. Frank Vahid & Tony Givargis, *Embedded System Design: A Unified Hardware/Softwre Introduction*, John Wiley & Sons, USA, 2002.
5. Hassan Gomaa, *Software Design Methods for Concurrent and Real-Time System*, Addison-Wesley, USA, 2003
6. Yulong Wang, "Open architecture for CNC controller and their developing prospect", *CNC & Software*, pp. 75-76, Apr 2004
7. Shuzi Yang, "Trend in the development of advanced manufacturing technology", *Chinese Journal of Mechanical Engineering*, Vol 5, No. 1, pp. 73-77, Dec. 2003.

A SYSTEM OF WINE BLENDING BASED ON NEURAL NETWORK

Jianping Ren, Zhimin li
Department of Mechanical Engineering, North University of China, China; Email: renjp@nuc.edu.cn.

Abstract: According to the investigation of some liquor company, this paper built a physics and chemistry information database of wine bodies and a management information system, and achieved wine auto-blending through some training to physics and chemistry information of wine bodies by exercising BP network. At the same time, a satisfactory obligatory condition (the best feeling, the balance harmony of a variety of slight ingredient) was produced by making use of the linear programming, and the wine auto-blending was achieved.

Key words: wine bodies, BP network, wine auto-blending, linear-programming, Computer-aided wine blending.

1. INTRODUCTION

In practical working course, the quality of liquor is not stability because the quality of wine blending totally depends on the personal feeling of the person who is very knowledgeable about liquor (gustation and smell), and yet the personal ability has some ambiguity, and is subject to being effected by outside factors. For example, the person who is very knowledgeable sometimes feel good about the liquor, but its result of the chemical examination is not passed the standards of the state; or its result of the chemical examination is passed the standards of the state, but the person who is very knowledgeable feel not good about the liquor. Now the key problem should be solved at once that how to rebuild the traditional process of making-liquor and the experiential techniques of making-liquor to managing by computers and adopting mathematical method and neural network algorithm.

Please use the following format when citing this chapter:
Ren, Jianping, li, Zhimin, 2006, in International Federation for Information Processing (IFIP), Volume 207, Knowledge Enterprise: Intelligent Strategies In Product Design, Manufacturing, and Management, eds. K. Wang, Kovacs G., Wozny M., Fang M., (Boston: Springer), pp. 604-609.

2. THE ENSEMBLE PROJECT DESIGN OF THE SYSTEM

According to the practical and process analysis of some liquor company, we fixed on the process of making-liquor (is shown as Figure 1). Management information system module of wine bodies is shown as Figure 2. The managing process of system work is shown as Figure 3.

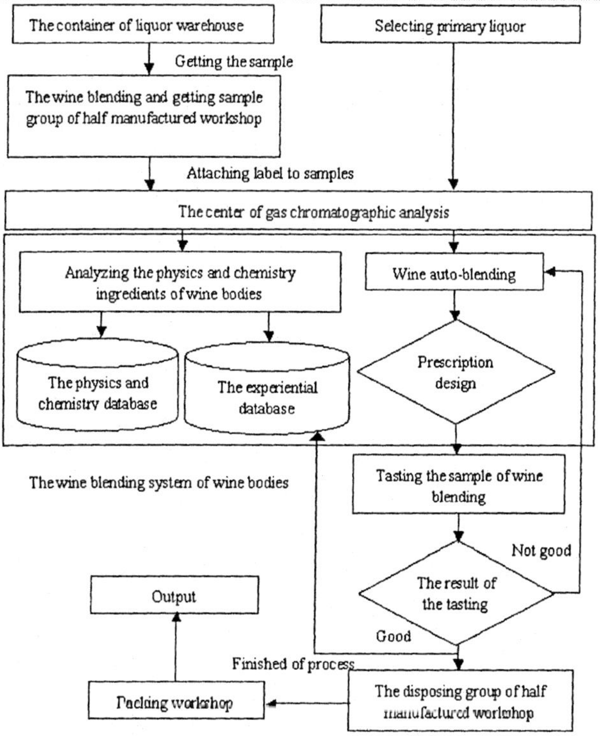

Figure 1. The process of making-liquor.

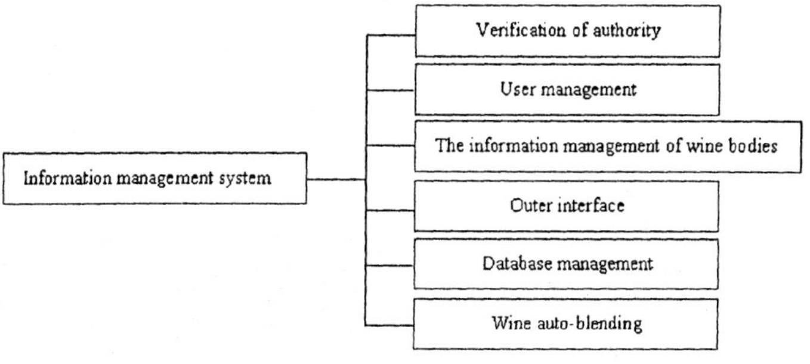

Figure 2. The management information system module of wine bodies.

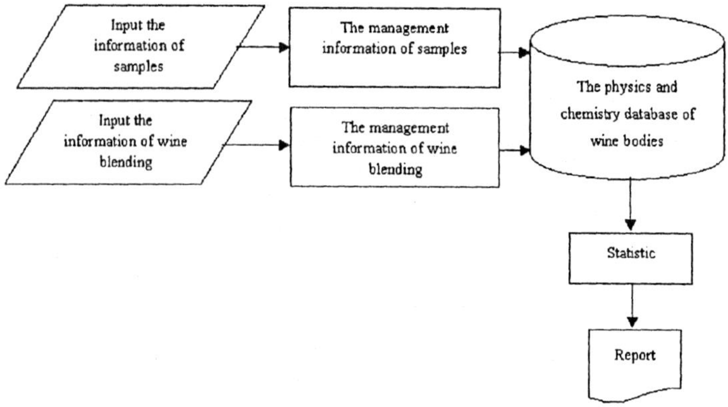

Figure 3. The managing process of system.

3. THE IMPLEMENT OF WINE AUTO-BLENDING SUB MODULE

Wine auto-blending was reformed tradition techniques to that: experimenters get small sample of different level liquor to test, and input the result into the system. Then the system automatically fills a prescription in accordance with the standards of the state, and its prescription adopts linear programming algorithm. For this reason, the wine blending will be with the standards of the state. In the end, the person who is very knowledgeable tastes the wine blending and adjusts blending parameters in accordance with the tasted feeling. So the whole process was under the standards, and it could improve the speed and the quality of the wine blending.

The designed user interface sub module of wine auto-blending is shown as Figure 4. Firstly, the range of the liquor index is input into the system, then it designs the suitable arithmetic to calculate the best method of mixing in terms of the range and information of the liquor. The linear programming and the object programming are important branches of the operational research, and it was studied and applied in earlier time and developed correspondingly mature. Introducing linear programming on the wine auto-mixing sub module, we could make the wine blending more pertinence, veracity, maneuverability, and reducing the fluctuation of the quality.

The designed user interface sub module of wine auto-blending is shown as Figure 4. Firstly, the range of the liquor index is input into the system, then it designs the suitable arithmetic to calculate the best method of mixing in terms of the range and information of the liquor. The linear programming and the object programming are important branches of the operational research, and it was studied and applied in earlier time and developed correspondingly mature. Introducing linear programming on the wine auto-mixing sub module, we could make the wine blending more pertinence, veracity, maneuverability, and reducing the fluctuation of the quality.

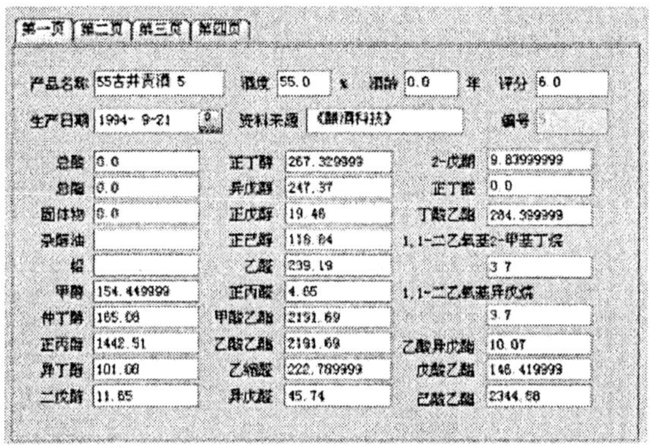

Figure 4. The input interface of physics and chemistry information of wine bodies.

When the wine blending, the indexes of physics and chemistry ingredients principally rest on the standard of the state and the corporations, and are compatible with the demand of foodstuff security and taste. The information of the physics and chemistry ingredients is dredged up by sub module of wine bodies, and data mining is good to test and verify if the standards of corporations are maturity, and could improve the product quality.

4. THE IMPLEMENT OF ANALYZING PHYSICS AND CHEMISTRY INGREDIENTS OF WINE BODIES

The sub module of analyzing physics and chemistry ingredients of wine bodies is established on the physics and chemistry indexes of wine bodies offered by a liquor manufactory. Except for the solids and the plumbum, 120 physics and chemistry ingredients of every wine body are all organic. More than ten of these ingredients are skeleton ingredients. That is what is known as important to the quality of liquor; the rest are not skeleton ingredients, and their effect is unknown too. The person who is very knowledgeable marks all samples of liquor. "The analysis of physics and chemistry ingredients of wine bodies" should get below knowledge: (1) the relation of physics and chemistry ingredients and mark; (2) the weighting of physics and chemistry ingredient in the network. At last, this module could mark for the sample of new type liquor.

There are three steps in the analyzing process. Firstly, with the liquor samples offered by users, we will design the database management system and provide some functions (including adding, deleting, changing, searching) for the information, in order to insure the validity of the data and the robustness of the program. Besides little additive information, there are about 120 floating-point types fields whose veracity decides the success. After building the winy sample information database, dactylogram atlas of the different liquor samples should be established to figure physics and chemistry information; secondly, we will design the interface which pick DBMS up MATLAB, and then input the physics and chemistry data of liquor samples to MATLAB to support the next step calculation; Lastly in the MATLAB, we will design proper neural network algorithm by making use of the toolbox of neural network to train the liquor samples data and dredge up the connotative information among them.

5. DRAWING THE DACTYLOGRAM OF LIQUOR

Just as everyone has his dactylogram and the dactylogram of everyone has its peculiarity, every kind of liquor sample has difference in its peculiarity and principal ingredient. With the help of computes and modem analyzing technology, the liquor sample peculiarity and its effective ingredients could be depicted in the way of atlas to make every kind of liquor has a standard atlas as the dactylogram of human. This is the dactylogram atlas of liquor.

When establishing the dactylogram atlas, we will firstly chose the typical sample to analyze, then draw the graph with different peak value in terms of different physics and chemistry ingredients of the sample. Then this atlas is the dactylogram atlas of this sample. The dactylogram atlas of Fenjiu-flavor liquor (A) and compared liquor sample (B) is shown as Figure 5.

When the quality of new produced liquor is tested, we could draw its physics and chemistry ingredients into dactylogram atlas, and analyze its Gas Chromatographic method. Then, compare it to the standard atlas in the database. For different the quality of liquor, we can find the factors that affect the quality by contrasting the dactylogram atlas. We also could introduce some mathematical tools, such as the cluster analysis algorithm, and as the assistance to estimate the quality.

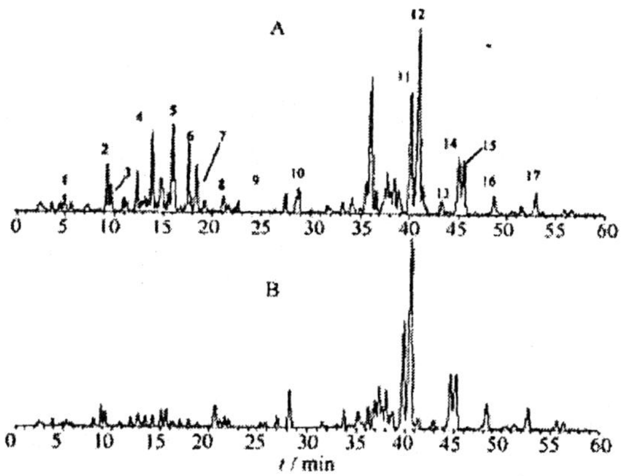

Figure 5. The dactylogram atlas of Fenjiu-flavor liquor (A) and compared liquor sample (B).

6. CONCLUSION

According to some liquor company, this paper was making use of VC++ and SQL Server, and have designed a physics and chemistry information database of wine bodies and the management information system, and then has automatically marked for the quality of liquor from the information of physics and chemistry analysis. A satisfactory obligatory condition (the best feeling, the balance harmony of a variety of slight ingredient) was researched by making use of the linear layout, and achieved the wine auto-blending.

7. REFERENCES

1. Zhai Xulong. The research and exploitation of characteristic ingredients dactylogram atlas system on Fen-flavor Liquor with Daqu. *Project application report.* 2004.8
2. Huang Hanying&Wei Mingxin&Yao Jicheng. The application of linear programming method in the technology of wine blending. *Journal of Huazhong Agricultured University*, 2001(4): 400-403.

FAMILY CARS' LIFE CYCLE COST (LCC) ESTIMATION MODEL BASED ON THE NEURAL NETWORK ENSEMBLE

Xiaochuan Chen[1],Jun Shao[1],Zhongxu Tain[2]

[1]The Mechanical Engineering College, Donghua University, Shanghai 200051, China Email:xcchen@dhu.edu.cn ;[2]Internal Combustion Ebgin Institute, Shanghai Jiaotong University, 200030

Abstract: Design for cost (DFC) is a method that reduces life cycle cost (LCC) at design stage. From the angle of DFC, the design features of family cars were obtained, such as all dimensions, engine power and displacement. At conceptual design stage, cars' LCC were estimated using back propagation (BP) artificial neural networks (ANN) method based on the features. An example was given. Levenberg-Marquardt（LM）and Genetic algorithm (GA) were used to train BP ANN' s weights. The results obtained through the adoptions of neural network ensemble is better than simply use GA or LM algorithm.

Key words: life cycle cost (LCC); artificial neural networks (ANN) ensemble; family cars

1. INTRODUCTION

Now family cars have become popular among some areas in China as a symbol of growing rich from a well-off standard. The demand of family cars has increased strongly, but market competition is also aggravated increasingly. As the continuous reduction of the sales prices, the profits of cars' production enterprises were also depressed. As a result, to increase benefits, the key is to reduce cars' cost. Generally, about 60%-70% of production cost is decided implicitly during the design stage[1]. So the cost estimation in the design stage is crucial.

With the crisis of petroleum, customers consider not only the purchase price of the car but also the expense in use. Therefore in this paper, through a

Please use the following format when citing this chapter:

Chen, Xiaochuan, Shao, Jun, Tian, Zhongxu, 2006, in International Federation for Information Processing (IFIP), Volume 207, Knowledge Enterprise: Intelligent Strategies In Product Design, Manufacturing, and Management, eds. K. Wang, Kovacs G., Wozny M., Fang M., (Boston: Springer), pp. 610-618.

review of the car's design, the artificial neural networks (ANN) method of back-propagation model based on feature is selected to establish a model to estimate life cycle cost (LCC) and some performances of cars at the conceptual design stage from the viewpoint of design for cost (DFC). The estimation results can help designers enhance the accuracy of the LCC estimation.

2. BRIEF DISCUSSION OF DFC

Design for Cost (DFC) is a design method which analyzed and evaluated the product's life cycle cost (include manufacturing cost, sale cost, use cost, maintenance cost, recycle cost, etc.), then modified the design to reduce the life cycle cost[2]. DFC need confirming parameters of manufacturing, usage, maintenance phases, for example, assembly cost percent unit, usage cost percent unit. Designer should balance performance, schedule, reliability, LCC and so on. In DFC, LCC serve as a critical parameter for design and provide support tools for designers to analyze and evaluate cost. For more details about DFC, please refer to Methodology and technology of design for cost (DFC)[3].

The cost in DFC refers to LCC, which consists of the total expense of research, design, development, production, usage, maintenance and disuse used in a large range from plan, argumentation, research, design, development, production, usage, maintenance to the final disuse phases[4,5]. The concept of LCC is first presented and then used by DoD (Department of Defense). In a typical weapon system, the cost of usage and maintenance occupy about 75%, so the research for LCC must be conducted then. But the technique developed by DoD aim at purchase instead of design. LCC include plan cost, manufacturing cost, sale cost, maintenance cost, use cost, recycle and disuse cost, while design cost occupies about 10%-15%, manufacturing cost 30%-35%, use and maintenance cost 50%-60%, others less than 5%[5].

3. ANALYSIS OF LCC FEATURES

3.1 Selection of LCC Estimation Methods

In order to reduce the LCC of a product, the key is to adopt a proper method to estimate it. In the conceptual design stage, design information is not integrated and is uncertain highly. For example, material information is not decided basically; machining process is considered roughly and is not include detail information. In this case it is difficult for us to translate highly uncertain design feature into detail cost feature. Therefore in this paper the ANN method based on feature is selected according to the character of conceptual design.

1. Identify cost-related features, such as: material, process, product structure, tolerance etc.
2. Classify and quantify the identified features. As the value of feature input into the neural network is usually between 0-1, a method called quantification must be conducted to deal with the value in practice.
3. Construct and train neural network.
4. Train and adjust the weight of neural network continuously in practice.

The advantages of this method are as follows:

1. No need of detailed machining time.
2. No need of cost function in practice, due to the self-study ability to real cost statistics of ANN.
3. The cost estimation among different designs at the conceptual stage can help to improve design.
4. From the viewpoint of DFC, in order to conduct cost estimation in the design stage, the design feature of family cars need to be transformed into cost features, namely, feature mapping[6]. Using neural network method, mapping can be realized automatically and then decrease the work of people consequently.

3.2 The presentation and application of neural network ensemble

Hansen and Salamon first presented the trailblazing method of neural network ensemble in 1990.They proved that the generalization ability can be prominently improved simply by training several neural networks and combining their results in some way. Sollich and Krogh defined neural network ensemble as a collection of a (finite) number of neural networks or other type of predictors that are trained for the same task[7]. Now, this definition has been widely acknowledged.

Perrone and Cooper proved in 1993 that, compared with the mean value of each network's generalization error, ensemble generalization error decrease by N times when merge regression estimation with neural network ensemble, if simply mean is adopted and error of each network is independent random variable and their likelihoods are zero. Here, N refers to the number of network included[8].

It's an easy trap for some common-used neural networks to fall into local minimum during its study, which is considered to be one of the biggest weaknesses for neural networks. But Perrone and Cooper argue that this characteristic play an important role for neural network in the promotion of its generalization ability. Since these neural networks are uncorrelated to each other, they will easily be trapped by local minimum. Thus the variance of neural network ensemble should be pretty big and then reduce generalization error. In other words, the negative effect of each local minimum counteracts in the end. All the networks included in the neural network ensemble give the same or approximate outputs to the same input. At this moment the diversity factor of the ensemble is close to zero while its generalization error is close to the weighted average of each network's generalization error. On the contrary, if the network included in the ensemble is independent, then the diversity factor of the ensemble will be pretty big, and its generalization error will be far smaller than the weighted average of each network's generalization error. Consequently, to enhance the generalization ability of neural network, the errors in the ensemble should be as uncorrelated to each other as possible.

3.3 Obtain the design features of family cars

Vehicle design process included conceptual design, technology design, trial-manufacture and sales stage[9]. In this paper we research the earlier stage of conceptual design. At this stage the following features can be confirmed in a car: general dimension/size, wheelbase, engine and so on. Considering DFC theory and some facts, we pick volume (length, width, and height), engine's parameter, car's weight, feature of electronic equipment and actuating feature as design characters. But in consideration of the difficulty in data-collecting and the characteristics to the conceptual design stage, we select these parameters: length, width, high, wheelbase, maximum power, maximum twisting moment/ torque and exhaust quantity.

Because it is difficult to collect LCC data, we compute use cost by oil/100KM , 100KM/day and 10 years in this paper. Then we get LCC adding sale price to usage expense. The data of this paper are obtained through Internet, which may not be fully accurate, but that's definitely enough to illustrate the feasibility of our method.

4. A CASE ABOUT FAMILY CAR'S LCC ESTIMATION BASED ON NEURAL NETWORK ENSEMBLE

4.1 Structure of Back Propagation ANN

Pilot calculation has been conducted first so as to compare the parameter selected. 3-layer structure is applied as the structure of Back Propagation (BP) ANN. The number of input units is the same with that of input features, and the number of hidden units is 1 more than that of input features while there is only 1 output number: LCC. Namely, input units number are 7, hidden units number are 8. The training error is 0.001. To facilitate programming, the neural network tool box of MATLAB 6.5 is chosen to carry out some relevant calculations.

4.2 Collection and choice of samples

We collected 19 groups' data of economic family cars on Internet. In this paper we adopt 12 groups of data as train sample of ANN, 7 groups as inspection sample of ANN. The output value of train samples are:

T=[11.132 15.514 16.122 15.488 18.572 20.076 24.124 24.198 37.29 35.788 21.133 40.074] Unit: 10thou. Yuan

ANN value of inspection samples are :

TT=[35.388 10.112 14.618 17.096 19.887 20.29 24.514] Unit: 10thou. Yuan

4.3 The training and LCC estimation of neural network

The train algorithm of traditional BP ANN adopts Gradient Descent. Its rapidity of convergence is slow and it is easy to fall into local minimum value. We have applied improved train algorithm, namely, Levenberg-Marquardt (LM) algorithm and genetic algorithm (GA). The advantage of LM is its swift convergence ability when the quantity of network weights is not too much. It has combined the advantage of Gradient Descent and Newton method. We need point out that there exists local minimum value in this algorithm. Due to the problem of local minimum for all the algorithms applied in this article, repeated train was performed toward the network with different initials. The final calculations were given in Table 1. Through actual application, we discovered that LM and GA methods can get comparatively better results than traditional train algorithm of ANN.

GA algorithm has a characteristic to convergence at the global optimum, so to optimize the weight of ANN with it might acquire global optimum result.

Table 1. results on LCC estimation with different train methods

NO.	LCC computational value	Appended momentum technique		Momentum technique combined with self-adapting learning rate		LM method		ANN's weight selection method based on GA	
		Train step:21150 Train error: 1.653		Train step 11881 Train error:0.01		Train step: 30 Train error:0.001		Evolution generation number:25 Mean train error:13.0%	
		predictive value	relative error(%)	predictive value	relative error(%)	predictive value	relative error(%)	predictive value	relative error(%)
1	35.3880	39.236	10.8740	42.1587	19.1326	39.3300	11.1395	27.8739	-21.2335
2	10.1120	8.0722	-20.1721	11.1669	10.4321	11.2599	11.3516	10.8568	7.3652
3	14.6180	17.951	22.8014	15.2406	4.2590	14.1796	-2.9992	16.2908	-11.4435
4	17.0960	18.192	6.4102	15.0595	-11.9121	14.5095	-15.1295	16.1443	-5.5670
5	19.8866	22.732	14.3060	26.9708	35.6228	25.3349	27.3966	21.5937	8.5842
6	20.2900	21.135	4.1639	24.1018	18.7867	23.6370	16.4956	19.8623	-2.1079
7	24.5140	36.541	49.0629	21.7315	-11.3506	24.2586	-1.0417	21.1299	-13.8047
mean relative error(%)		18.256		15.928		12.222		10.0151	

We get the LCC estimation result through the application of LM train method and GA train method in Table 1 to apply Neural Network Ensemble. The estimation result is showed in Table 2; we can find mean relative error is obviously smaller than simply use LM method or GA method.

Table 2. Results on LCC estimation of neural network ensemble based on LM train method and GA train method

NO.	LCC computational value	Estimation results of neural network ensemble	
		predictive value	relative error(%)
1	35.3880	33.60195	-5.04705
2	10.1120	11.05835	9.358683
3	14.6180	15.2352	4.222192
4	17.0960	15.3269	-10.348
5	19.8866	23.4643	17.99051
6	20.2900	21.74965	7.193938
7	24.5140	22.69425	-7.42331

NO.	LCC computational value	Estimation results of neural network ensemble	
		predictive value	relative error(%)
mean relative error(%)		8.797673	

The content discussed before treat LCC as a whole to estimate, so in the rest part we will estimate usage expense (oil consumption cost) and purchase expense respectively, and then add the latter to the former to obtain the estimation result. Because the effect of LM train method and GA train method is pretty good, the following estimation use them respectively. From the comparison of Table 3, Table 4 and Table 1, we can find some errors of estimation value increase while some others decrease without regularity, but all the overall mean errors decrease. The final calculations are given in Table 5. It can be observed that its mean relative error is the smallest and actually that's the result of neural network ensemble, namely, estimate LCC by neural network: estimates every part of LCC respectively, and then add all of them to obtain LCC which is feasible. Though this method add to the computation quantity, but the information obtained are more abundant than simply estimate the total quantity of LCC.

Table 3. purchase expense and usage expense obtained through LM train method

NO.	LCC computational value	Purchase expense train step: 40 train error: 0.001			Usage expense train step: 24 train error: 0.001			LCC estimation	
		actual value	predictive value	relative error(%)	actual value	predictive value	relative error(%)	predictive value	relative error(%)
1	35.3880	24.000	31.558	31.490	11.388	12.581	10.4775	44.138	24.727
2	10.1120	3.9800	5.0449	26.756	6.1320	5.9085	-3.6446	10.953	8.3208
3	14.6180	6.8800	7.1125	3.3797	7.7380	7.0825	-8.4709	14.195	-2.8937
4	17.0960	10.380	6.5844	-36.567	6.7160	8.8224	31.3644	15.407	-9.8806
5	19.8866	10.820	12.004	10.940	9.0666	8.3671	-7.7149	19.719	-0.8448
6	20.2900	10.800	11.977	10.898	9.4900	9.0846	-4.2721	21.062	3.8023
7	24.5140	12.980	11.977	-7.6677	11.534	11.566	0.2759	23.543	-3.9622
mean relative error(%)			18.2426			9.4601		7.7758	

Table 4. purchase expense and usage expense obtained through GA train method

NO.	LCC computational value	Purchase expense Evolution generation number:25 train error: 28.7%			Usage expense Evolution generation number:25 train error: 14.6%			LCC estimation	
		actual value	predictive value	relative error(%)	actual value	predictive value	relative error(%)	predictive value	'elative error(%)
1	35.3880	24.000	23.1705	-3.4561	11.3880	12.5943	10.5931	35.7648	1.06477

NO.	LCC computational value	Purchase expense Evolution generation number:25 train error: 28.7%			Usage expense Evolution generation number:25 train error: 14.6%			LCC estimation	
		actual value	predictive value	relative error(%)	actual value	predictive value	relative error(%)	predictive value	'elative error(%)
2	10.1120	3.9800	3.5479	-10.8571	6.1320	5.4205	-11.6034	8.9684	-11.3093
3	14.6180	6.8800	10.9628	59.3432	7.7380	7.1751	-7.2747	18.1379	24.0792
4	17.0960	10.380	9.5297	-8.1917	6.7160	7.7368	15.2002	17.2665	0.99731
5	19.8866	10.820	13.2279	22.2542	9.0666	0.8987	20.2067	24.1266	21.3209
6	20.2900	10.800	11.1698	3.4244	9.4900	9.5213	0.3302	20.6911	1.97684
7	24.5140	12.980	13.0058	0.1986	11.5340	11.1505	-3.3251	24.1563	-1.45917
mean relative error(%)		15.3893			9.7905			8.8868	

Table 5. Results on LCC estimation of neural network ensemble based on purchase expense and usage expense

NO.	LCC computational value	LCC value estimated through LM method		LCC value estimated through GA method		LCC value estimated through neural network ensemble	
		Predictive value	relative error(%)	Predictive value	relative error(%)	Predictive value	relative error(%)
1	35.3880	44.1382	24.7265	35.7648	1.06477	39.9515	12.89561
2	10.1120	10.9534	8.3208	8.9684	-11.30934	9.9609	-1.49426
3	14.6180	14.1950	-2.8937	18.1379	24.0792	16.16645	10.59276
4	17.0960	15.4068	-9.8806	17.2665	0.99731	16.33665	-4.44168
5	19.8866	19.7186	-0.8448	24.1266	21.3209	21.9226	10.23805
6	20.2900	21.0615	3.8023	20.6911	1.97684	20.8763	2.889601
7	24.5140	23.5427	-3.9622	24.1563	-1.459166	23.8495	-2.7107
mean relative error(%)		7.7758		8.8868		6.466096	

Reference 10 hold the view that in the data acquisition process, requirement change in the various phases and product information keep enriched. Because information are unavailable and incomplete during conceptual design stage, the accuracy of cost estimation range from -30% to +50%; When design information are further enriched, and the historical data resembling to current design are available, the accuracy of estimation can achieve -15%～+30%; According to the results of LCC estimation in this paper, error is well controlled between -5% to +13%.

5. CONCLUSIONS

This paper adopt self-adapting learning rate, LM algorithm and genetic algorithm respectively to train Back Propagation neural networks, LCC estimation was made for economically family cars through the application of neural network ensemble method. The results obtained through the adoptions of neural network ensemble is better than simply use genetic algorithm or LM algorithm, and we can easily find that its computational results are pretty stable, with mean error between 7% to 9% while other methods such as LM is not quite stable.

In order to better estimation LCC and function index to conformed them to the real practice, Something more can be research in the fields of enterprise size, repair and maintenance etc. We are fully convinced that the application of these methods must lead to the reduction of family car's LCC and actively enhance the level of design.

6. REFERENCES

1. S. Dowlatshahi, "Product design in a concurrent engineering environments an optimisation approach", *Journal of Production Research*, 30(8), 1803–1818(1992).
2. Xiao-chuan Chen, ZhangBao-bao,Feng Xin-an, Key technology and conceptual model about design for cost(DFC),*Journal of Dalian University of Technology*,39(6),775-780(1999). (in Chinese)
3. Chen Xiaochuan, Yang Jianguo, Li Beizhi, Feng Xin-an. Methodology and technology of design for cost (DFC), *The 5th World Congress on Intelligent Control and Automation*, (WCICA'04) Hangzhou, China, June 14-18, (2004) pp. 3129-3134.
4. Benjamin S. Blanchard. Life cycle costing-A review, *Terotechnica*, 1,9-15(1979).
5. Chen Xiao-chaun, "Research on theories and methods of Design For Cost(DFC) in concurrent engineering", *Dalian University of Technology, Ph. D. Dissertation.* (2000). (in Chinese)
6. Chen Xiaochuan, Fang Minglun, Feng Xinan. Application of multiple domain feature mapping in Design For Cost(DFC), *CE2002*, Cranfield University, U.K. July, 27-31, (2002).
7. P. Sollich, A. Krogh. Learning with ensembles: How over fitting can be useful. In: *Advances in Neural Information Processing Systems*, edited by D Touretzky, M Mozer , M Hasselmo. (Cambridge, MA: MIT Press, 1996), pp.190- 196.
8. M. P. Perrone, L. N. Cooper. When networks disagree: Ensemble method for neural networks. In: *Artificial Neural Networks for Speech and Vision*, edited by Mammone R J (New York: Chapman & Hall, 1993), pp. 126- 142.
9. Wangyu Wang, *Automobile design* (fourth edition), (Beijing: China Machine Press.2004).
10. R. C. Creese, L. T. Moore ,Cost modeling for concurrent engineering, *Cost Engineering* ,32(6),23-27(1990).

A XML-BASED RESEARCH ON INTEGRATION TECHNOLOGY BETWEEN PDM AND ENTERPRISE CONTENT MANAGEMENT SYSTEM

Chuanhong Zhou, Jigan Zhan
CIMS & Robot Center of Shanghai University, Shanghai University, China; Email: chhzhou@staffe.shu.edu.cn

Abstract : The advantage of XML in Enterprise Application Integration(EAI) had been investigated , putting stress on the exchange of non-structural data between PDM and Enterprise Content Management System(ECMS) ,giving the integration prototype system model based on XML and introducing the algorithm of transforming the structure tree and the method of serializing XML document object to realize the core block—parser . The results show that the technologies solve the information "isolated island" among enterprise application systems, which has the meaning at large.

Key words: PDM; ECMS; XML; integration; XML Object Serialization

1. INTRODUCTION

With the rapid development of network technology, network scope are expanding, information resources appear different isomer, distributing, loose characteristic of coupling. Meanwhile, with the globalization of the market, the increasing aggravation of the market competition, the connection among modern enterprises is becoming more frequent, cooperative in forming alliances, and the working way has the characteristic of colony 、 interaction 、 distributing and collaboration .Enterprise Content Management System (ECMS) can gather, manage , store, save and submit textural and multimedia content across and between enterprises, it is the important conjunction between enterprise's customer-oriented application and deal

Please use the following format when citing this chapter:
Zhou, Chuanhong, Zhan, Jigan, 2006, in International Federation for Information Processing (IFIP), Volume 207, Knowledge Enterprise: Intelligent Strategies In Product Design, Manufacturing, and Management, eds. K. Wang, Kovacs G., Wozny M., Fang M., (Boston: Springer), pp. 619-624.

within the enterprise So, ECMS may be as the information integration platform while structuring the Enterprise Information Portal (EIP), which realizes the information share between various kinds of enterprise information system(for instance: PDM , ERP , CAX).

As a part of structuring EIP, our paper has proposed a kind of integration framework between PDM and ECM based on XML, and have introduced relevant technologies.

2. SYSTEM INTEGRATION FRAMEWORKS

PDM manages all information correlated with product development (Including project norm, electronic documents, scan the picture , CAD/CAE/CAM file , product structure, products order , supplier state, etc.) and some related process (including Work flow, Approve / Release process, Engineering Change) [1]. In order to make all in products life cycle participants including customer, supplier participate into the design process from the beginning of product design, two kinds of information need transmitting from PDM system to the enterprise content management platform First some generic document can be generally imported by hand, second exists some non-structural data relying mainly on designing BOM (D-BOM) information. In our paper we mainly resolve how to import design BOM into ECMS accurately and timely.

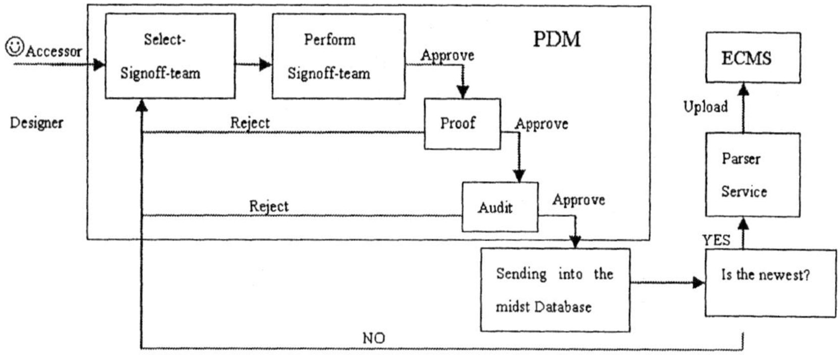

Figure 1 an Overview of Integrated Framework

Among this integrated framework, designers release these design information (D-BOM) in the form of enclosure through the native workflow engine of PDM. If the auditor approve, these information will be sent into the midst Data Base in XML-format[2] and transfers its special trigger to judge whether this information is the latest edition. If YES, transfer outside web service – parser to realize datum format-transformation, extract the useful information from D-BOM and upload it into ECMS's Data Base .At this moment, utilizing ECMS's embedded workflow template, ECMS system manager can release related design information into the EIP.

3. CORE FUNCTION MODULE---- PARSER

The Parser module fully takes the advantage of XML in data exchange, adopting XSLT template and XML Object Serialization[3] technology, so that we can realize conveniently the format transformation of structural and non-structural data exported from PDM system.

3.1 Parser Model

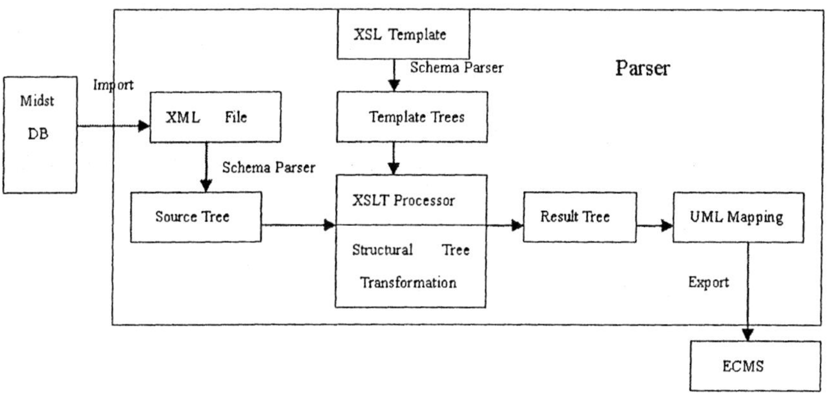

Figure 2 Parser Model

First of all, the Parser reads the D-BOM information that is expressed with XML format from the midst database , transferring corresponding XML Schema parser to parse the input file and XSL template file, showing for tree structure - source tree and template tree; Then XSLT transfer corresponding template rules in the template tree to carry on structure transformation for source tree , reject those sensitive information to supplier and customer, draw a XML file with " standard " pattern; According to the particularity of two systematic data interchange finally (it is necessary to share synchronously the outside non-structural datum linked with "locationRef" element in result tree's Dataset node),serialize the result tree by using UML Mapping and export the result into ECMS .

3.2 Algorithm of the Structure Tree Transformation

Begin
1. Input XML file, transfer schema parser and parse it into the tree-like structure
2. According to the designated source schema , carry on the legitimacy check of the source xml file.
   If false
   Print " XML file pattern is wrong "
   Return
   Else
3. Transfer XSL template and transfer corresponding schema parser to parse it into the template tree
4. Read the root node of the source tree, look for a template rule to agree with nodal name to deal with in the template tree
   If template rule number 〉 1
   Determine to suitable template rule to deal with the root node according to the value of priority attribute
   Else if template rule number=1
   This template is needed.
   Else if template rule number=0
   Transfer inside template rule
5. Carry out the template rule
   If "select" attribute exists
   Enumerate and Carry out specific node and son's nodal corresponding template rule that "select" attribute appoint
   Else
   Enumerate and Carry out the present node and sub nodes
6. Output the result tree
End

3.3 XML Object Serialization

Utilizing XSLT format-transformation technology can conveniently convert XML format file into HTML, text file and PDF,etc[4]. However, in result tree there is a "locationRef " element behind Dataset node which can link with some outside dataset (Word document、CAD drawings .etc). At this moment, utilizing XSLT[5] is unable to send synchronously these outside datasets and other structural datum into ECMS. So we utilize XML Object Serialization technology to map XML document into a normal class, and "locationRef " is a attribute of this class. Using Object-Oriented (OO) method can conveniently satisfy the demand of two systematic information integration.

Fig. 3 has explained how to map XML document into a UML object's instance.

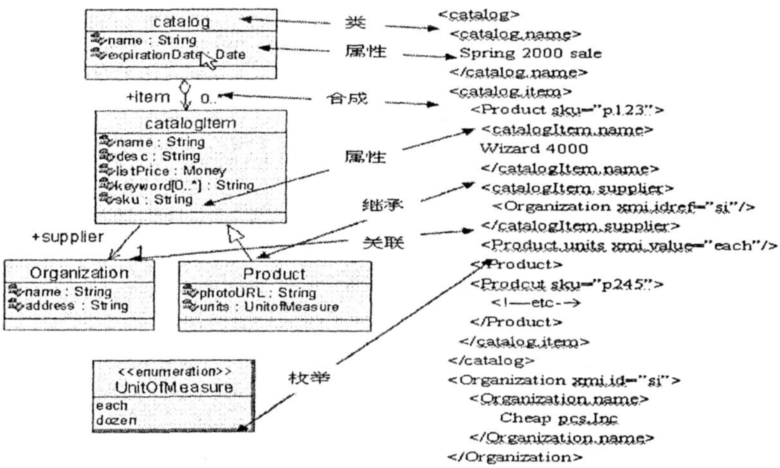

Figure 3: XML Document Mapping into UML Object's Instance

4. AN IMPLEMENTATION

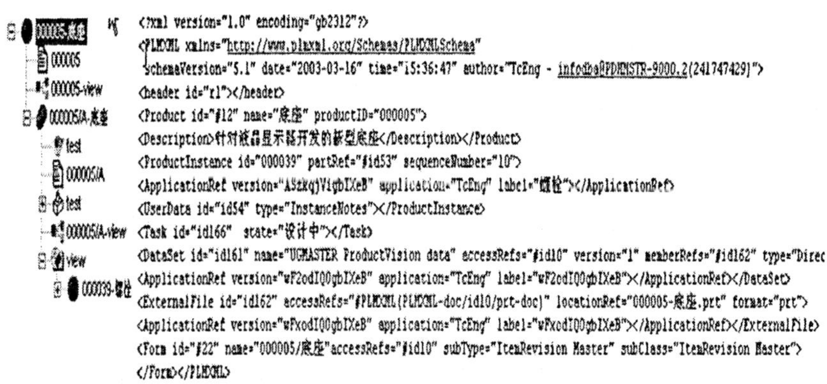

Figure 4 Design BOM (D-BOM)

We have adopted TeamCenter Engineering of UGS. and Plone while implementing the information-based project for FEIYUE Group, China. Through using the system integration framework mentioned above, we have released smoothly the D-BOM (Fig.4) in TeamCenter Engineering into Plone systematic platform.

5. CONCLUSION AND FUTURE WORK

In conclusion, through an actual project, the integration model which this paper adopted we have succeeded in solving the structural and non-structural data interchange between PDM system and Enterprise Content Management System (ECMS). Because XML language is an international industry standard, presently most corporations support this standard, the successful implementation of this integration model will offer strong technical support for enterprise application integration (EAI) in the future which is based on Enterprise Information Portal using ECMS.

At the same time, present research still concentrates on the single-direction data integration mainly. In the future, integrating enterprise application systems in double-direction., making the information of ECMS feedback into enterprise's other information system accurately and automatically will be our research emphases.

6. REFERENCES

1. Li Haifeng, Wang XianKui, Wu Dan . Research on Product Data Exchange Among Heterogeneous PDM Systems Based on XML[J] *Aviation Precision Manufacturing Technology* 2003，39（1）：p.20—24 (in Chinese)
2. Xu Xiangzhong, Wang Jingye ,Jiang Haibin . XML-based DIF and Challenges It Faces[J] .*Computer engineering and application* 2003,(01): p.27—30 (in Chinese)
3. David Carlson .Modeling XML Applications with UML [M] . Tsinghua University Press , 2003 (in Chinese)
4. David Carlson .*Modeling XML Application with UML[M]*： Rational Software Corporation，2003
5. Chen Jianxuan. *Introduction to Mastery for XSLT [M]* . China Railway Publication House ,2003 (in Chinese)

APPLICATION RESEARCH OF WIRELESS EQUIPMENTS IN PLM

ChuanHong Zhou[1], Yong Shi[1], Qiang Wang[1]
[1]*CIMS & Robot Center of Shanghai University, Shanghai University, China; Email: chhzhou@staffe.shu.edu.cn*

Abstract: The approach of the extend to wireless terminal equipments from enterprise's PLM system was investigated , using Peer-to-Peer technique which is based on java, putting stress on solving the technical framework and key technologies of wireless equipments in application to PLM system, giving its apply form and integration model in PLM system. The results shows that, the WPLM system that using the technical framework meets enterprise's need of cooperation design, extends enterprise's cooperation working environment to wireless field, carves out a new space for PLM.

Key words: JXME (Java2 mobile edition); Peer-to-Peer; WPLM (Wireless Product Lifecycle Management)

1. THE PRESENT SITUATION

1.1 The general situation of PLM

The PLM is under the environment of Web, based on the product dada integration within whole life cycle, investigating product's management and cooperation from product layout, design, manufacture to distribution and so on within its life cycle, aiming at reduce the time which products come into the market and lower the cost, providing integrated product's cooperation and manufacture system which sustain product rapid design and optimal scheduling of manufacturing for enterprises. It is a strategically thought way[1].

Thanks to the fast development of advanced wireless technique and wireless equipments presently, it provides possibilities of adding wireless accessing functions for PLM system.

Please use the following format when citing this chapter:
Zhou, ChuanHong, Shi, Yong, Wang, Qiang, 2006, in International Federation for Information Processing (IFIP), Volume 207, Knowledge Enterprise: Intelligent Strategies In Product Design, Manufacturing, and Management, eds. K. Wang, Kovacs G., Wozny M., Fang M., (Boston: Springer), pp. 625-630.

1.2 The general situation of wireless equipments

Applying wireless equipments to PLM system , the significance lies in no matter whether the employee is using wireless local area network in factory area or the employee in other places can all accessing electronic product data and enterprise information in time and conveniently, it contributes to the improvement of employee's work efficiency and enterprise's information degree. Besides, as the wireless network technology become more and more mature and popular, its price is far lower than traditional wire network, .it has more advantages to enlarge PLM technology's extent.

2. WIRELESS EQUIPMENTS' TECHNOLOGY FRAMEWORK AND KEY TECHNOLOGIES IN APPLICATION OF PLM SYSTEM

2.1 Technology framework

The system structure of applying wireless equipments to PLM system is shown as Fig. 1:

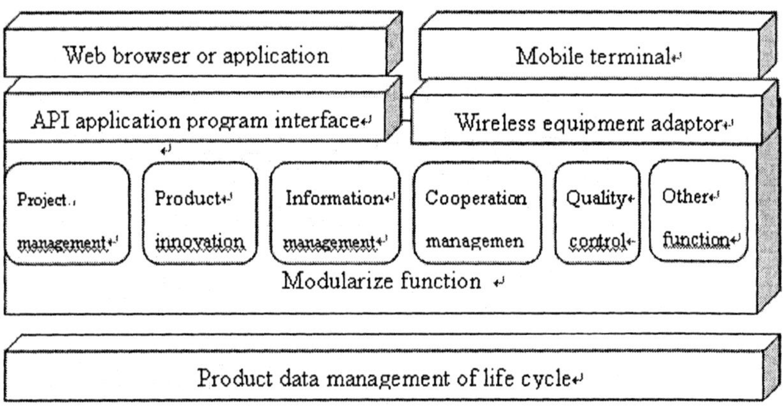

Figure 1 the system structure of wireless equipments in PLM system

Based on traditional PLM system, the PLM system which accessed to wireless local area network has added wireless equipment adapter and increased mobile terminal at client.

According to the description of whole system structure above, the following physical structure of wireless equipments in PLM system is put forward.

Accessing way can make those customers using mobile terminal such as PDA, mobile telephones, pocket computers accessed into wire power grids under mobile environment, so as to achieve WPLM (Wireless Product Lifecycle Management) system operation environment.

2.2 Key technologies

In hardware aspect, present wireless network equipments mostly use 802.11 protocol, compared with other protocols, it has some advantages such as high data transmission efficiency, far transmission distance, performance stabilization, According to Palm's practical application environment, selecting appropriate wireless equipments is the key of successfully making PLM system wireless.

In software aspect, the software applying to mobile terminal mainly adopts JXME used for mobile equipments——the J2ME platform based on JXTA technology.

J2ME is developed for those equipments using limited energy, limited network connection (often wireless connection) and limited Graphics User Interface faculties. There are great differences between hardware platforms which sustain J2ME, Among them there are more high- tech equipments, for instance: TV STB,IPTV and so on; there are also low-tech equipments, such as: mobile telephone, pager, PDA and so on. J2ME aimed at hardware's qualifications such as data processing capability, memory capability, network connection capability decided CLDC (Connected Limited Device Configuration),CDC(Connected Device Configuration)configuration and MIDP（Mobile Information Device Profile）description.

JXTA technique apply itself to create a universal platform, set up special service and application of peer-to-peer and distribution in simple and effective ways, making developers avoiding the necessity of considering how to solve the technique problems of P2P computing excessively, then they can absorbed in how to achieve and perfect high level application of extendable, better interoperability and usability[2]. JXTA makes use of some protocols, each protocol can achieve and integrated into P2P service and application easily, thus between different P2P systems ,it is easy to communicate with each other, work cooperatively and provide service for the opposite. JXTA was designed to which independent of program languages such as C or JAVA, platforms[4] such as WINDOWS or UNIX and network platforms such as TCP/IP and Bluetooth. Furthermore, JXTA was designed to which could be realized on any digital equipment, including sensor, mobile telephone, PDA, network router, desktop computer, server and memory equipment[3].The network structure of JXME is shown as Fig. 2:

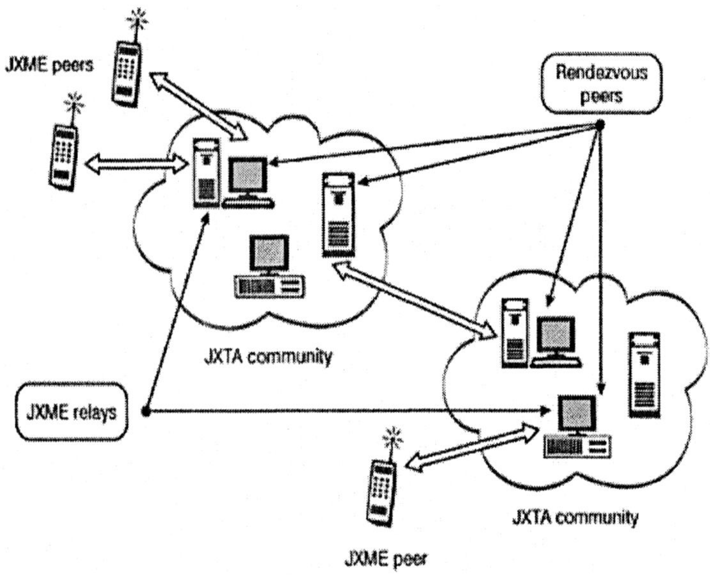

Figure 2 JXME's network structure

By using JXME technique to achieve mutual transfer of mobile peer information, the load of wireless network can be reduced greatly, the access speed of data can be improved, and accordingly guarantee is provided for the abroad application of wireless network equipments.

3. THE APPLICATION MODE AND ISSUES REQUIRED TO BE SOLVED OF WIRELESS EQUIPMENTS IN PLM SYSTEM

3.1 Application forms

The wireless equipments in PLM system are mainly digital terminal (mobile telephone, PDA, pocket computer etc.), wireless network interconnection equipments (wireless access point, wireless network card, wireless network bridge and wireless router etc.), these equipments have many application styles.

1) Flow's examine and approve

In WAN or LAN, by accessing operation interface of PLM system based on web, using mobile terminal such as PDA, mobile telephone to process flow's examine and approve timely and accurately, personnel's work efficiency can be increased, the workflow can be processed successfully.

2) Close integration with PLM's manufacture

PLM's effect run through product's every step from layout, design, manufacture, distribution to discard. Among them, wireless equipments in application to manufacture step is of realistic significance. In manufacture environment, via mobile terminal it could access upstream product design data real time, for example, design BOM, product change information, the problems occurred in production can be changed and solved in time. At the same time, in bigger workshop, every machine's work information can be seen rapidly by using wireless equipments.

3) Close integration with PLM's product distribution and the after sales support

Applying wireless equipments to PLM system can provide product's correlative information and technical supports for customers at any time and anywhere, reflecting customers' requirements to product design department through mobile terminal, thus it can rapidly response customers' requirements, increase customers' satisfaction. In the same way it's also helpful to serve staff to improve work, shorten serve turnover time, and make enterprises increase its after-sales service income.

3.2 Problems to be solved

1) The security problem of wireless LAN

JXTA Version 1.0 provides the following security primitive:

A simple bank supporting Hashfunction (such as MD5) ,symmetric encrypt algorithms (such as RC4) and asymmetric encrypt algorithms (such as RSA) ; a distinguish framework; a simple entry strategy based on password; a simple access control mechanism, this is based on peer group, the member of a group automatically authorized accessing other member's share data; security transmission strategy based on SSL/TLS[5].

2) The conversion of photograph format

The photograph of mobile terminal such as PDA, mobile telephone mostly adopts BMP, JPG and GIF formats. As for PLM system based on web, the transmission of Common JPG and GIF photographs haven't much problems, but for three-dimension CAD photographs of special formats ,it still require to change them into BMP,JPG and GIF formats that can be received by PDA, mobile telephone at server end through photograph format transmission.

3) Data compression:

As for mobile terminal requiring mass access to all kinds of product data, the present transfer rate of wireless network data is difficult to meet the requirement. So better data compression technique requires to be developed to meet the increasing data visiting quantity.

4. EXAMPLE OF FLOW'S EXAMINE AND APPROVE

In the example of flow's examine and approve, we can create and start a new flow from mobile terminal such as PDA, mobile telephone and so on or general computer, besides, task flow can be successfully processed by decisions such as permit, veto, unsettled etc. between different mobile terminal even fixed terminal , consequently increase personnel's work efficiency.

For example, starting flow's revise, examine and approve of some part in computer, when the personnel who hold mobile equipment received the examine and approve task, he will make out decision such as permit, veto, unsettled, consequently ensure flow could be processed immediately and successfully.

5. REFERENCES

1. Budiarto, Shojiro Nishio , Masahiko Tsukamoto, Data management issues in mobile and peer-to-peer environments,*Data & Knowledge Engineering* 41 (2002) 183–204
2. Li Zupeng, Huang Daoyin, Zhuang Lei etc. Research of JXTA Technology Based on Peer-to-Peer Network. *Computer Engineering and Application*,2003.11
3. Akhil Arora,Carl Haywood,Kuldip Singh Pabla. *JXTA for J2ME– Extending the Reach of Wireless With JXTA Technology*,Sun Microsystems, Inc,2002.3
4. Michael J. Yuan. *Develop universial P2P network's Mobile Extension.P2P* China (PPcn.net),2003.9
5. Zhang Zhiqiang. *The Introduction of JXTA Technique and Prototype's Implement,* developerWorks China Web, 2001.7.

METHODS FOR CALCULATING THE TRANSMISSION TORQUE OF MAGNETIC TRANSMISSION MECHANISMS

Shunqi Mei [1], Zhiming Zhang[1], Renbin Xiao[2]
[1]*Wuhan University of Sci.& Eng.,China; Email: meishunqi@vip.sina.com.* [2]*Huazhong University of Sc. & Tech., China.*

Abstract: The methods of analyzing and calculating the magnetic field and the main problems of designing magnetic mechanisms are reviewed and discussed. The comparatively precise engineering methods for calculating the transmission torque of the plane axial and coaxial radial magnetic transmission mechanisms based on the magnetic theory are introduced, which can be used for optimization of designing the magnetic transmission mechanism. Two application instances of magnetic transmission mechanisms used in the Three-For-One and Two-For-One textile twisting machine are introduced, and the transmission torque of the mechanisms are measured by experiment and calculated by means of the engineering methods.
Key words: Magnetic mechanism, Torque, Calculation, Design.

1. INTRODUCTION

The magnetic coupling transmission mechanism is one of the main application of magnetics. The main task for designing the magnetic coupling transmission mechanism is determination of the reasonable mechanical electromagnetism structural style and the parameters, so as to realize the transformation of magnetic energy into mechanical energy, accomplish a certain movement and drive requirement. Simplifying and transforming the problem of magnetic field as the problem of magnetic circuit is a general design method used at present[1, 2, 3]. Generally, the following factors must be considered: (1) the transmission torque ought to be able to meet the operation needs; (2) the structure size ought to be as far as possible compact, the magnetic material used ought to be as far as possible little; (3) the dynamic performance of the mechanism(vibration, noise and so on) ought to

This project is supported by the Science & Technology Foundation of Hubei Province 2005 (NO. 2005ABA230)

Please use the following format when citing this chapter:
Mei, Shunqi, Zhang, Zhiming, Xiao, Renbin, 2006, in International Federation for Information Processing (IFIP), Volume 207, Knowledge Enterprise: Intelligent Strategies In Product Design, Manufacturing, and Management, eds. K. Wang, Kovacs G., Wozny M., Fang M., (Boston: Springer), pp. 631-636.

be satisfy the requirements. At present, there is still no more complete satisfactory method for calculating the transmission torque of the magnetic mechanism, it is still relied on the empirical data and the empirical formulas.

2. THE METHODS OF ANALYSIS AND DESIGN OF MAGNETIC MECHANISMS

The method for analyzing and calculating the magnetic field and the transmission torque of magnetic mechanism is the key problem of designing magnetic mechanism, which is attracted great attention in the academic and engineering circles. So far, some methods for calculating the magnetic field and the transmission torque have been given as follows[1, 2, 3, 4, 5].

(1) The test-calculation method together with experience or experiment based on the magnetic circuit concept. The main limitation of this method is that the calculating formula is generally made very great simplification, and the values of some parameters of the formula have to be given based on experience. Thus, the calculated results by using this method have low precision, on the other hand the calculation and design are blindness in some degree, and the operation load is great.

(2) The Finite Element Method (FEM). The method FEM used in the electromagnetic field analysis is generally based on the Maxwell equation. For analyzing the permanent magnetic field with the FEM method, the permanent magnet is processed as boundary electric current model, this brings on massive mathematic analysis and calculation. Until now the effective mathematic analysis model for the permanent magnetic field hasn't been established.

(3) Analytic and approximate analytic methods. In the process of these methods, the processes of variable separation, effective equipotential surface, and even magnetization are always used. Among them, the variable separation is the basic analysis method for the problem with boundary value. For the effective equipotential surface method the ferromagnetic surface is modified as equipotential surface. For even magnetization method it is assumed that the magnetic material is magnetized evenly under the action of external field, then analyze and calculate the interelectrode field distribution.

In recent years the domestic scholars attempt to develop the equivalent magnetic charge method[6]. It is considered that the equivalent magnetic charge of permanent magnet is distributed on the two surfaces which are vertical to the direction of magnetization, and then the Coulomb force of the equivalent magnetic charge acted on the two surfaces, the torque between the two magnetic charge, the torque of magnetic field can be calculated.

3. ENGINEERING METHODS FOR CALCULATING THE TRANSMISSION TORQUE

There is much progress in the above-mentioned researches about the analysis and calculation of magnetic transmission torque, however, at present the research results are still immature and too complex, and almost impossible to instruct the engineering design. For instance, the paper [6] obtained the four multiple integrals expressions of calculating the magnetic torque for the plane axial magnetic transmission mechanism. But according this method and its expression the values of some parameters must be measured in advance, and this is almost impossible for the first development and design of a magnetic mechanism. Thus, here we attempt to give effective engineering methods.

Figure 1 shows the schematic drawings of the plane axial (a) and coaxial radial (b) magnetic transmission mechanisms. The transmission torque can be described as follows[4, 7]:

$$M = \frac{1}{2}\frac{V_m B_r^2}{\mu_0}\frac{r}{L_g}\frac{L_m}{L_g}\frac{K_f}{K_r}\frac{\sin\phi\cos^2\phi}{[1+(K_f/K_r)\bullet(L_m/L_g)\cos\phi]^2} \quad (1)$$

Where V_m---the sum of magnets volume in the drive rotor and driven rotor of magnetic mechanism, m3 ; $V_m = 2A_mL_m$; L_m---the average axial length of permanent magnets , m; A_m---the axial cross-sectional area of the permanent magnet, m^2 ; L_g---the axial space between the drive and the driven rotors, namely the air gap length, m; r---the mean radius of the permanent magnets distribution, m ; μ_0---magnetism constant, $\mu_0 = 4\pi\times10^{-7}$, H/m; B_r---the remanence of the permanent magnet, T; K_f , K_r---the coefficients of the magnetism leakage and magnetism resistance respectively while the drive and driven rotors are rotating; $tg\,\phi = r\theta / L_g$, θ—relative rotation angle between the drive and driven rotors, rad.

The equation (1) indicates that the transmission torque M is the function of angle ϕ, the relative rotation angle between the drive and driven rotors, while the other parameters are determined. For getting the limited value of the transmission torque from the equation (1), let $\partial M/\partial\phi = 0$, we have:

$$P\cos^3\phi + 3\cos^2\phi - 2 = 0 \quad (2)$$

where $P = (K_f / K_r)/(L_m / L_g)$.

For the magnetic mechanism with determined structure parameters, the value K_f/K_r can be determined by means of magnetic circle analysis[5], then the value of P is determined, and the value of ϕ can be obtained by solving the equation (2), and then solving the equation (1) the value of maxim transmission torque M is determined. Thus in the process of designing magnetic mechanism, we can in advance suitably give the value of parameters L_m , L_g , r , V_m according the structure and dimension limit, calculate the value of maxim transmission torque M until it is equal or close to the required design value by means of above mentioned method.

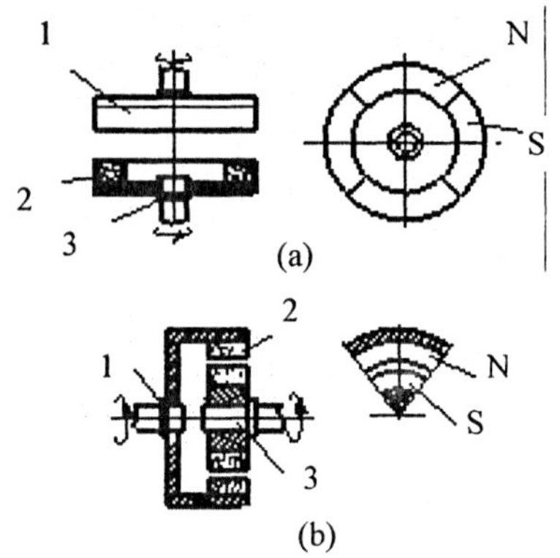

Figure 1 Schematic drawings of magnetic mechanism
1: drive rotor, 2:magnet, 3:driven rotor.

At present, for the coaxial radial magnetic mechanism there is also no really effective method for calculating the transmission torque, which is researched not adequately. For the purpose of engineering design the transmission torque M can be determined by following equation[4, 8]:

$$M = n \bullet B^2 \bullet S \bullet R /(2\mu_0) \tag{3}$$

where n—the number of magnets; B—the magnetic induction intensity, T; S—the area of the air gap in the direction of transmission, mm^2; R—the mean radius of the air gap, m.

4. APPLICATION OF MAGNETIC MECHANISMS IN TEXTILE MACHINERY

Nowadays along with the progress of the technology of textile machinery and magnetic material production, the magnetic transmission mechanism is gotten wider and wider application. In the Three-For–One and Four-For-One twisting machine the plane axial magnetic transmission mechanism is used for driving spindle, in the Two-For-One twisting machine the coaxial radial magnetic transmission mechanism is used for fixing the yarn package and balloon limitor.

For instance, in the Three-For-One twisting spindle developed by us (Figure 2) the parameters of the magnetic transmission mechanism are as follows: r=0.037m， L_g=0.012m， B_r=4000Gs=0.4T, the drive and driven rotors are installed with 16 magnets, the total volume of which V_m =1.584× 10-6 × 32=5.07 × $10^{-5}m^3$. The measured maxim transmission torque by experiment is 1.17Nm. On the other hand, by means of magnetic circle analysis, we can get the coefficients of the magnetism leakage and magnetism resistance: K_f =2.14, K_r =1.25; solving the equations (2), (1) in terms of above mentioned method we get the maxim transmission torque with the value of 1.2274Nm. Meanwhile, if the influence of magnetism leakage is neglected, i.e. $K_f / K_r = 1$, the maxim transmission torque is solved from the equation (1) with the value of 1.2333Nm. Obviously, the two calculated results are close. In fact, in the case of little air gap with the little value of L_m /L_g(but it must be L_m /L_g > 1) for the plane axial magnetic transmission mechanism, it will have no serious influence on the calculated result that the magnetism leakage is neglected, but the analyzing and calculating process will be simplified to a great extent. The experiment shows that the magnetic transmission mechanism in the spindle runs well. In the Two-For-One spindle manufactured by Zhejiang Xinya textile machinery Co. (Figure 3) the relative parameters of the coaxial radial magnetic transmission mechanism are as follows: material of magnet is ferrite, B_r=0.4T(4.0kGs), H_{cb}=175kA/m(2.2kOe), number of magnets n =12, L_m=0.017m, L_g=0.013m, R =0.0745m, S =1.3×10-4m^2. By means of magnetic circuit analysis we have B=0.0886T(886Gs), substituting it in the equation (3) we obtain the maxim transmission torque M=0.358Nm. In the experiment we measured the maxim transmission torque M with the value of 0.333Nm, which is comparatively close to the calculated result, the difference between them is 6.5%, and the magnetic mechanism fulfill its own task well.

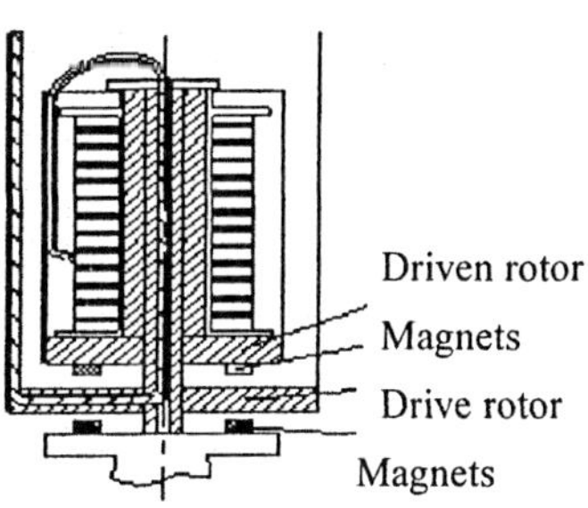

Figure2. Three-For-One spindle.

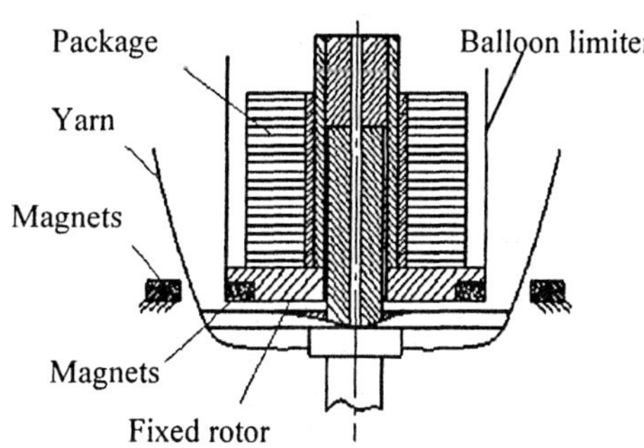

Figure 3. Two-For-One spindle.

5. CONCLUSIONS

Analysis and calculation of transmission torque is the key problem for designing magnetic transmission mechanism. The magnetic circle method and analytical method can describe directly and approximately the relationship between the transmission torque and parameters of the magnetic mechanism, but can not obtain precise solution. The numerical method such as Finite Element Method (FEM) has advantage of precisely including the influence of magnetism leakage and bow wave on the solution, but the analysis model and process are rather complex, and difficult to be applied in engineering design. In this paper an improved method of calculating magnetic transmission torque based on the theory of magnetic filed is provided, which has comparatively precise solution and is convenient for engineering design. It is clear that establishing effective analysis model and method for designing magnetic transmission mechanism is a important direction in this filed.

6. REFERENCES

1. Zhao Han, etc, (2003), Review of study of magnet machine and Mechanism, *Chinese Journal of Mechanical Engineering*, 39(12), pp. 31--36.
2. S.M. Huang, (2002), Analytical of magnetic couplings with parallelepiped magnets. *Journal of Magnetism and Materials*, 239, pp. 614--616.
3. Jean Frediric Charpentier, (1999), Optimal design of Cylindrical Air Synchronous Permanent Couplings. *IEEE Trans. Mag*, 35(2), pp. 1037.
4. Zhong Wending, (1987), *Magnetics of Magnets,* Science Press, Beijing, China.
5. Lin Qiren,etc, (1987), *The Principle of Magnetic Circuit Design*, Machinery Industry Press, Beijing, China.
6. Tan Qingchang, (1993), Study on the Drive Torque of Plane Permanent Magnetic Coupling, *Chinese Journal of Mechanical Engineering*, China, 27(1), pp. 12--16.
7. Mei Shunqi, etc, (1999), Study on the Calculation Method of Magnetic Driver Torque, *Machinery Design & Manufacturing,* China, (3), pp. 18-19.
8. Wang Biyun, (1985), The Design and Calculation of Magnetic Coupling Transmission, *Journal of Magnetic Materials and Devices*, China, 16(3), pp. 1--5.

SUSTAINABLE INDUSTRIAL SYSTEM (SIS)

Odd Myklebust
SINTEF, Trondheim, Norway, Email: Odd.Myklebust@sintef.no

Abstract: Sustainability is a topic which is important both for national and international research. Sustainability can both be a competitive factor for manufacturing industry but also reduce the competitiveness with cost driving environmental legislation for a single region e.g. EU only. Approaches for sustainability claim to have both a holistic and international views. They need to manage the product life cycle and keep track of the information flow within of all the lifecycle phases of the product. A special attention is kept on End of Life Management (EOL)
Key words: Sustainability, Product Life Cycle, International, Industry, Standardization, End of Life Management

1. INTRODUCTION TO SUSTAINABLE INDUSTRIAL SYSTEM (SIS)

The Sustainable Industrial Systems (SIS) approach brings together the different challenges and dimensions of sustainability into a common platform comprising a system oriented approach for industry. As a result, this will lead to improved methodologies, easier and clearer standards and legislations. One important factor that has been almost absent in research projects is the consumer and market relation, which has been called the consumption paradigm.

SIS's overall goal is to obtain a holistic view of product cycles in the manufacturing industry and optimize the life-cycle of industrial systems, products and services. Methodologies and tools to support the manufacturing of products and production must be increasingly life-cycle and service

Please use the following format when citing this chapter:
Myklebust, Odd, 2006, in International Federation for Information Processing (IFIP), Volume 207, Knowledge Enterprise: Intelligent Strategies In Product Design, Manufacturing, and Management, eds. K. Wang, Kovacs G., Wozny M., Fang M., (Boston: Springer), pp. 637-642.

oriented, in addition to the requirements of intelligence, cost-effectiveness, safety and cleanliness. The objectives are to:

- Combine the environmental dimension of product development and manufacturing and plant operations
- Develop viable and sustainable industrial systems that support customer satisfaction in a sustainable consumption paradigm.
- Develop corporate social responsibility (CSR) principles
- Develop a framework for business organizations to measure progress of commitment and results regarding environmental, economic and social criteria.

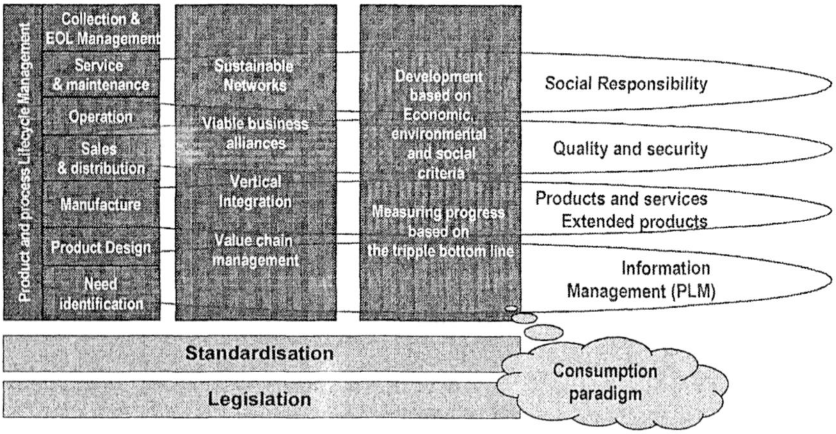

Figure 1. Illustration of The Sustainable Industrial System framework.

The modules in SIS, as they are shown in Figure 1 above, can shortly be described as follows:

- Product and Process Lifecycle Management
- Sustainable Network creation (sustainable extended enterprise)
- Social- economical- ecological combined assessment and performance measurement.

This will lead to a Sustainable Industrial Development Model where:

- Non-renewable resources can be used only at the rate where alternatives are developed or loss of opportunity compensated
- Waste and emissions should not be generated at rates that exceed the ecosystem capacity to assimilate them
- Renewable resources should not be used at rates that exceed the systems ability to replenish itself

1.1 Product - service

Product Service System strategies where products and services are seen as mutually interrelated systems, may contribute to separate economic growth from growth in material and energy throughput, and thereby contribute to the required dematerialization goals. There is a growing understanding among industrialists and policymakers that environmental initiatives need to be seen in relation to each other rather than in isolation.

Technological challenges involve understanding and improving the effectiveness of the material and energy flows. Social challenges are linked to how to redirect society at large, companies and individuals towards a more cyclic- and systems-thinking on production and consumption. The relationships between the different disciplines are vital to understand the action-effect mechanisms and interpret incentives to obtain desired results.

1.2 Product lifecycle information management (PLM)

PLM can be described an a strategic business approach that applies a consistent set of business solutions in support of the collaborative creation, management, dissemination, and use of product definition information across the extended enterprise from concept to end of life.

PLM is for the manufacturing industry looking to improve business performance. PLM should consolidate business activities that create, modify and use data to support all phases of a product's life cycle. PLM technology with seamless access to product model structures with data and information on the one hand and distributed knowledge repositories.

Functionality for handling product information of Middle of life (operational phases) and End Of Life (recycling phases) are limited conceived up to now. In a end of life perspective the product model supported by the PLM system also need to "reincarnate" product or material data for reuse of material and/or parts (closing the information cycle).

2. INDUSTRIAL ECOLOGY CHALLENGES

2.1 Improvements within an organization

Within a company the challenge is to make environmental improvements in accordance with industrial ecology principles along the internal value chain. In product development central aspects are design and material choices that secure good quality, long life for the product, and reuse and recycling. In production processes the challenges are to optimize the use of energy and materials. In the final link, the distribution processes, the challenges are to distribute and redistribute the products for use and reuse/end-of-life treatment with a minimum of environmental impact.

2.2 Value chain management

Industrial ecology goes beyond conventional environmental work in industry by expanding the system borders of a company. These transactions and actions have environmental impacts, and by organizing them together and looking for synergies, major environmental improvements at a system level can be achieved. The customer relationship between manufacturers gives opportunities for demanding environmental performance to be the preferred supplier. A manufacturer has a lot of suppliers and dealers, and by putting forward environmental demands one single company can create large environmental chain reactions far beyond its own organizational borders. The challenge is to understand how green chain reactions are started and managed.

2.3 End-of-Life of Product Systems

The EU project AEOLOS (An End-of-Life of Product Systems) developed an integrated methodology and a toolset that examine the issues relating to the overall sustainability of the End-Of-Life (EOL) treatment of waste products. The AEOLOS methodology enables users to systematically define EOL products and processes to a level of detail that will allow meaningful economic, environmental and societal costs and values of different EOL scenarios. Targeted users include resource recovery and recycling companies, original equipment manufacturers (OEM), and policy makers.

2.3.1 Background and Project Definition

The AEOLOS EOL methodology is designed to enable users to define product and process to a level of detail that will allow meaningful economic, environmental and social costs and values of different EOL scenarios to be calculated, analyzed and compared. In so doing, the methodology will assist the user to determine the most sustainable treatment of EOL products [1].

2.3.2 The Step-By-step Methodology for End of Life Management

The methodology comprises four main steps, preceded by user and facility / product set-up steps [2].

- Step 1 Scenario definition

Select a product and EOL option(s), which together make up an EOL scenario. The definition of the scenarios is unlimited by any constraints, such as local availability of facilities – the refinement of scenarios occurs later in the "loop". The EOL options included in the AEOLOS methodology and software are re-manufacturing, component reclamation, material recycling, incineration with energy recovery, incineration without energy recovery, and landfill. The minimum amount of information required for a product definition is the weight of that product.

- Step 2 Scenario assessment

Select which environmental, social and economic indicators are to be calculated. A description of each indicator will be provided by the system. Once the indicators have been selected, data are entered for the selected indicators. In some cases data is supplied by the software, for instance, the environmental inputs and outputs associated with consumption of energy.

- Step 3 Scenario Analyses

Analyze results (and compare scenarios). The user can choose the method of analysis from a "toolbox" of methods, which range from a simple comparison of indicator results, though to more complex multi-criteria decision analysis techniques.

- Step 4 Scenario Refinement

Checklist of questions to ensure that the preferred scenario is feasible is:

- Check system boundaries and allow adjustment
- Collect more data in order to be able to calculate more indicators
- Seek expert opinion on feasibility of scenarios
- Check local capacity for implementation of a scenario.

3. CONCLUSION, SUSTAINABILITY IN AN INTERNATIONAL DIMENSION:

Sustainability within a industrial international society is overall important. This should concern both future international R&D framework programs and national programs. Unless research within environment and sustainability does not obtain a global perspective the main challenges will never be met. Therefore we need to find international activities and agreements that try to include the whole world like the Rio- and Kyoto agreements. The programs can be made a driver to focus on important international topics such as sustainability. This is overall important when we se the growth in the global industrialization and the total growing consume in material and energy.

3.1 Methods for End Of Life treatment

In Europe the regulations for recycling products has been developed for quite a while. EU has the e.g. WEEE regulative and the automotive recycling regulative. This is probably just a start of to treat product in middle of life and end of life in the future. This is a typical topic that can be made global. The global waist streams could be made optimal and the methods and regulations for recycling of the products could be standardized.

3.2 Green logistics

Industry is global and they work in networks and chains. Products and parts are distributed all over the world. For every mile of transport CO2 gas is sent out in the air. A more balanced assessment models between transport, cost and environmental factors should be developed. This is important both for logistics in manufacturing and distribution and the return logistics after end of life treatment.

3.3 Standardization

At present, national and international standardization and legislation act as guidelines and frameworks for direct regulations and normative value settings. A significant contribution to further development and enhancement of national and international standards and legislations within environment and sustainability is an important activity that must continue.

4. REFERENCES

1 Adda, S., Bruce, W.,_Goggin, K., Ray, E. and Lamvik, T., (2001), *Goal and Scope*, AEOLOS internal deliverable 1.1.

2 Adda, S., Snaddon, K., Goggin, K., Thurley, A., Lamvik, T., Miljeteig, G., Schnatmeyer, M. and Bufardi, A., (2002), *The AEOLOS Methodology*. AEOLOS internal deliverable 5.2.

MODELING TECHNOLOGY AND APPLICATION OF REPAIRING BONE DEFECTS BASED ON RAPID PROTOTYPING

Xusheng Yuan, Qingxi Hu, Hanqiang Liu, Chunxiang Dai, Minglun Fang

Rapid Manufacturing Engineeing Center, Shanghai University, Shanghai 200444,P. R.China; Email:huqingxi@mail.shu.edu.cn

Abstract: Now some people suffer from the bone defects for various reasons, and how to solve this problem has become a hot topic internationally in the field of tissue engineering. By the analysis of all kinds of the bone defects, an all-purpose approach using bionic bone scaffold to repair the bone defects is proposed in this paper to help the doctors to diagnose the state of the bone defect illness with the real models manufactured by the rapid prototyping technology. Meanwhile, a clinic application is given to prove the feasibility of this approach.

Key words: bone defects; surface reconstruction; rapid prototyping; tissue engineering; bionic scaffold.

1. INTRODUCTION

Because of the traffic accidents, production accidents, skeletal diseases and the like, the number of persons suffered from the bone defects or the bone injured is in excess of 3,000,000 annually in China.[1] Many of them need to make a bone graft surgery to repair the defects, which includes autogenous bone grafts, bone allograft[2], bionic bone grafts, etc. Due to the former two methods with the deficiencies like bones insufficiency, immunobiologic reaction, the bionic bone grafts can avoid the above deficiencies, which is a new approach to repair the bone defects.

The graft implanted to the human body is a scaffold using bionic bone grafts, which is seeded by the cells. The scaffold, depending on the tissue or

This project was supported by Shanghai Education Key Foundation 03AZ01.

Please use the following format when citing this chapter:

Yuan, Xusheng, Hu, Qingxi, Liu, Hanqiang, Dai, Chunxiang, Fang, Minglun, 2006, in International Federation for Information Processing (IFIP), Volume 207, Knowledge Enterprise: Intelligent Strategies In Product Design, Manufacturing, and Management, eds. K. Wang, Kovacs G., Wozny M., Fang M., (Boston: Springer), pp. 643-649.

organ in production, can be anything from a matrix of collagen, a structural protein, to synthetic biodegradable plastic laced with chemicals that stimulate cell growth and multiplication. The "seeded" cells that initiate regeneration come from laboratory cultures or from the patient's own body.[3]

How to repair the irregular bone defects is a key problem in the whole process (see Fig. 1). The process of repairing the bone defects will be proposed detailedly in this paper.

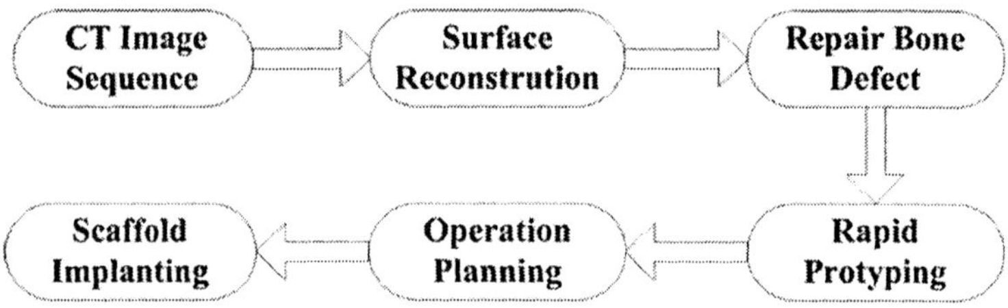

Figure 1. Repairing process

2. SURFACE RECONSTRUCTION

Many clinical situations ask for the visualization of anatomical surfaces. Common examples include hip replacement surgery, intra-operative visualization of surgical instruments or probes, visualization of planning information, and implanting surgery. The visualization of anatomical information with mesh-surfaces is useful for treatment planning and surgery.[4]

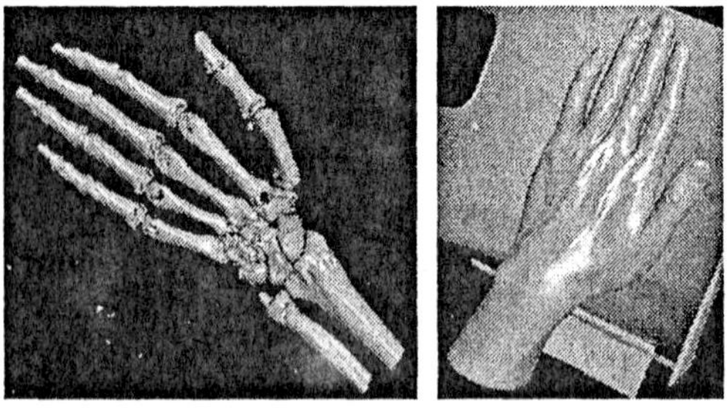

Figure 2. Reconstruction with different threshold

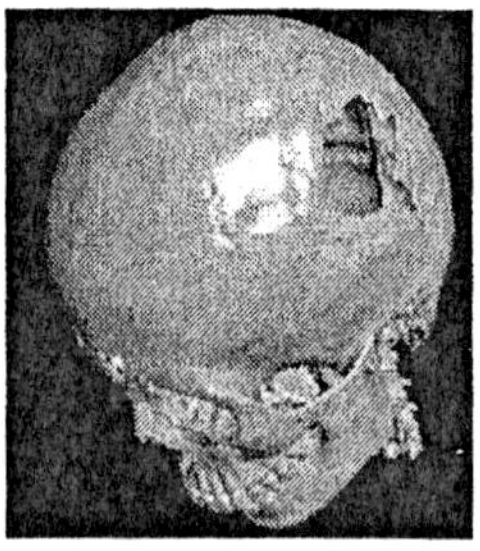

Figure 3. Model of bone defect

In order to display the model on the screen of computer, a 3D reconstruction algorithm called "Marching Cubes"(MC in short) was proposed by W. Lorensen and H. Cline in 1987.[5] The iso-surface can be reconstructed from the irregular and nonlinear volume datasets especially the medical volume datasets via the MC algorithm. And a threshold is needed to extract iso-surface, which stands for density of the substance we want to extract (see Fig. 2).

3. REPAIRING THE BONE DEFECT

There are some methods to repair the bone defects such as: mirror-image method, surface interpolation method and neural network method[6], etc. Mirror-image method is a common method to repair the bone defects, but it is based on the hypothesis that the repaired object is symmetrical. In fact, the bone is not always naturally symmetric. The results of the surface interpolation method depend on the selection of control vertices and basis function, which form a smooth profile. Now an all-purpose approach is proposed to repair bone defects by constructing free-form surfaces in this paper, which can meet the individuation of different bone defects.

For the sake of interpreting the process of repairing the bone defect, a 3D model reconstructed from CT images (see Fig. 3) is taken for example. The model is composed of so many spatial discrete points, which can be organized to a point cloud data to be dealt with via Imageware software. The point cloud can be handled as you like by the strong operation of Imageware.

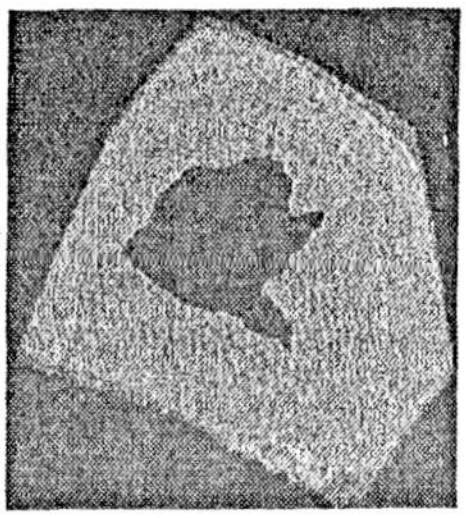

Figure 4. Region of interest

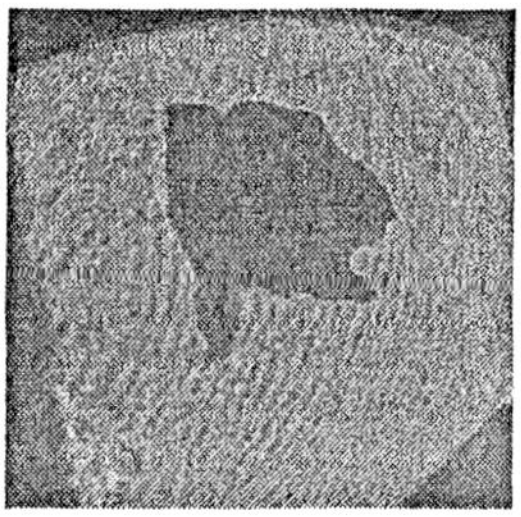

Figure 5. 3D B-Splines

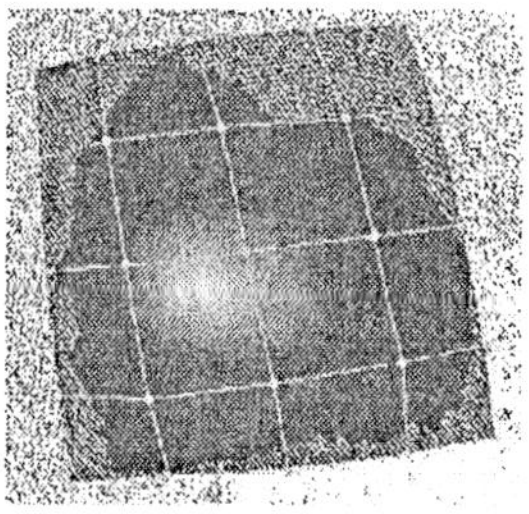

Figure 6. Free-form surface with control points

Fist of all, the point cloud of 3D bone model is imported to the software, and the quantum of the point cloud should be reduced to leave the region of interest (like the defect as shown in Fig. 4) in order to increase the velocity of each operation, because the whole data of the model is so big that the runtime of each operation is much longer.

For the exterior surface, curves can be created through several points chosen from the points along the margin of the defect on the exterior surface, which are 3D B-Splines in the 3D space. Generally, 4 B-Splines (see Fig. 5) are enough to enclose the defect fully. If the defect is too complex to fit the outline of the defect with 4 B-Splines, more 3D B-Splines are needed to fit the outline.

A free-form surface on the exterior surface is swept through these 3D B-Splines. It also can be adjusted to fit the trend of the whole curved surface through the control points (see Fig. 6), and its boundaries also should be adjusted to cover the defect. As for the internal surface, do the above process again to generate another free-form surface (see Fig. 7).

Figure 7. Upper and lower surface

Figure 8. Solid block

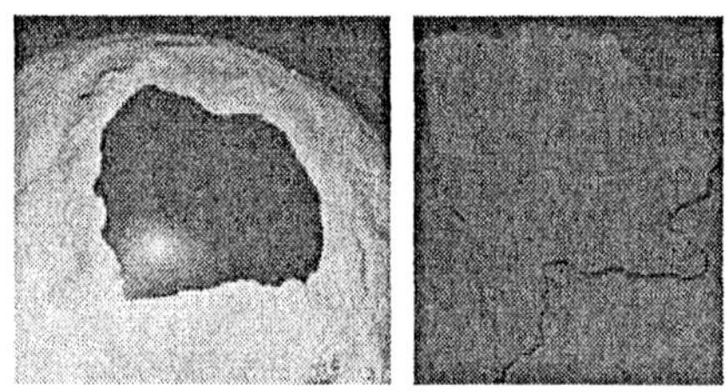

Figure 9. Subtraction operation

Afterwards, the other 4 surfaces can be blended to occlude the whole surface through the corresponding boundaries between upper surface and the lower surface. The 6 surfaces can be sewed together to form a solid block (see Fig. 8). In the end, the bone model is subtracted from the block, and the remnant of the block is the model fitting the defect we want (see Fig. 9). Sometimes, there are some edges and corners around the brim of the remnant, which are not necessary commonly for the clinic application. No matter how complex the bone defect is, it can be solved through this method. It is an all-purpose method to deal with all kinds of bone defects.

4. MANUFACTURING THE MODELS

The models of the bone and the substitute are so helpful that the doctor can make a pre-simulation of the implanting surgery and make an operation planning. On the basis of the analysis of models, the doctor can get the reasonable shape of implant and manufacture the scaffold with porous structure. But how to manufacture the model of he bone and the scaffold is very difficult, because the structure of the bone and the scaffold is quite complex and the surface is too irregular and the shape differs from each other. Using the traditional manufacturing technology, the production cycle will be longer and the shape could not accord to the real shape absolutely. In the field of medicine and engineering, an engineering method is being explored to manufacture the scaffold, which helps to solve the repair of the bone defects.

However, a new advanced manufacturing technology called rapid prototyping emerged in the past few years, which is capable of directly generating physical objects from CAD models. The prototype is produced by adding materials rather than removing materials. The model of the prototype is mathematically sliced into a series of parallel cross-section pieces. For each piece, the curing or binding paths are generated, which are directly used to instruct the machine for producing the part by solidifying a line of material. After a layer is built, a new layer is built on the previous one in the same way. Thus, the model is built layer by layer from the bottom to the top.[7] The models of the bone and the scaffold are quite complex, but they are can be manufactured by rapid prototyping technology. The production cycle will be shorter and the shape will accord to the real model (see Fig. 10).

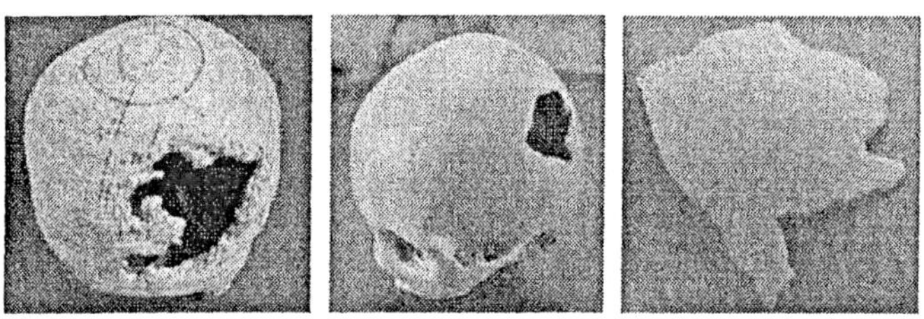

Figure 10. Models manufactured by rapid prototyping technology

5. CLINICAL CASE

For a male patient, there is a defect in the up left corner of the skull (see Fig. 11), which results from cell tumor. In the beginning, the model of skull should be reconstructed and repaired, and then, the models (see Fig. 12) should be manufactured by rapid prototyping for the doctors to simulate the surgery and make an operation planning. The models also supply references to the shape of the implant. Sometimes there is no need to keep the edges and corners along the boundary of the implant (see Fig. 13), which can make the surgery easier to carry out.

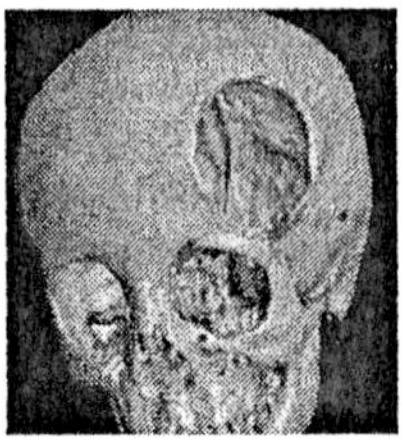

Figure 11. Defect

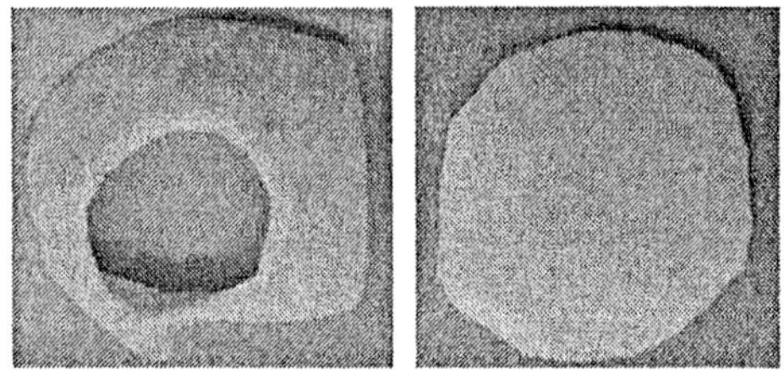

Figure 12. Models

When the shape of the implant is identified after surgery pre-simulation, the doctor divides the implant into several parts, and the parts are manufactured using biologic material, which structure is porous. The reason why doing so is that it could avoid the dead bone, because the implant needs to be seeded the cells in order to make the implant and the bone growing together.

According to the return visit with the doctors, there is a wonderful concrescence between the implant and the bone, and the patient recovers gradually (see Fig. 14 and Fig. 15).

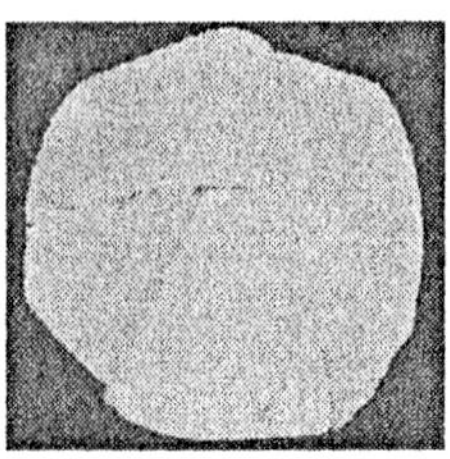

Figure 13. Implant

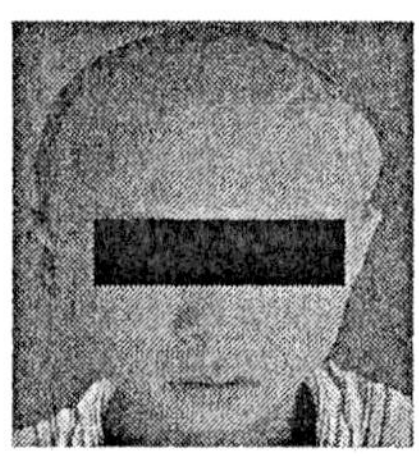

Figure 14. Ante-operative photograph

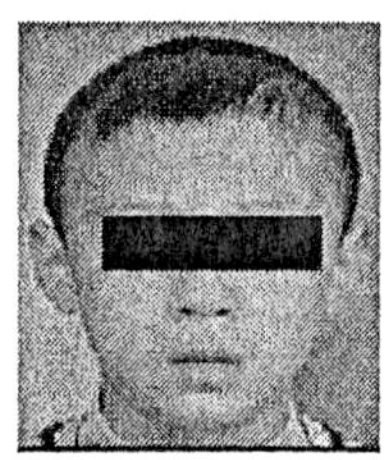

Figure 15. Post-operative photograph

6. CONCLUSION

In this paper, we have presented a new approach to repair the bone defects, and the repairing process is depicted in detail. This approach is proved to be feasible by the successful clinic application. Through this method, the diagnosis of bone defects will no longer depend upon the experience of the doctors mostly, and the success rate of cure will be increased greatly. It can be predicted that the rapid prototyping technology will be widely applied in the field of biomedicine with its strong superiorities. The combination of rapid prototyping with material, biology, medicine and other interrelated subjects, will make it possible that the bone defects are fully cured, which result from disease, accident, war and so on.

7. REFERENCE

1. Yu Yaoting, Wang Lianyong and Wang Shengqi, The development and countermeasure of biomedical material, www.cas.ac.cn/html/Books/O61BG/a1/2002/2/2.3_1.htm
2. Weng Xisheng, Application of bone grafts and its substitute in spine surgery, *China Contemporary Medicine,* 8 (1), 33—39(2002).
3. Ho Saey Tuana, Dietmar W. Hutmacher, Application of micro CT and computation modeling in bone tissue engineering, *Computer-Aid Design* 37, , 1151—1161 (2005).
4. S. van Foreest-Timp, Iso-surface volume rendering for implant surgery. *International Congress Series* 1230, 733—738 (2001).
5. Lorensen W, Cline H. Marching cubes: a high resolution 3D surface construction algorithm. *ACM Computer Graphics* 21(4), 163—170 (1987).
6. Jung-HsiangHua, Ching-ShiowTseng. Application of three-dimensional orthogonal neural network to craniomaxillary reconstruction. *Computerized Medical Imaging and Graphics* 25, 477—482 (2001).
7. Tang Yiping, *Advanced manufacturing technology* (China Machine Press, Beijing, 2004).

A STUDY OF THE METHOD OF RECONSTRUCTING THE BIONIC SCAFFOLD FOR REPAIRING DEFECTIVE BONE BASED ON TISSUE ENGINEERING

Hanqiang Liu, Qingxi Hu, Limin Li, Minglun Fang
Rapid Manufacturing Engineering Center, Shanghai University, Shanghai 200444, P.R.China; Email: huqingxi@mail.shu.edu.cn.

Abstract: Bionic modeling technology of bone defect repair plays a key role in the process of repairing defective bone based on tissue engineering. In this paper, via studying the structure of bone defect repair, new concept of bionic scaffold is put forward and a new method of integrating medical image processing, 3D reconstruction and CAD technology to construct the bionic scaffold of repairing bone defect is presented. Using the system implemented by this method, the bionic scaffold of repairing a skull is constructed. The evaluation of porosity of scaffold is also analyzed.

Key words: bone defect repair, bionic scaffold, modeling, tissue engineering.

1. INTRODUCTION

Repairing bone defect based on tissue engineering is a research hotspot for the moment. It comes down to many aspects such as bionic modeling of repair, materials of repair, processing of repair and tests of osteoblasts seeding etc. Among these, bionic modeling of repair is one of the key technologies.

At present, for bionic modeling of bone repair, researchers mostly adopt conventional CAD methods[1-2]. Woodfield,T. in Canterbury University of

This project was supported by Shanghai Education Key Foundation 03AZ01

Please use the following format when citing this chapter:
Liu, Hanqiang, Hu, Qingxi, Li, Limin, Fang, Minglun, 2006, in International Federation for Information Processing (IFIP), Volume 207, Knowledge Enterprise: Intelligent Strategies In Product Design, Manufacturing, and Management, eds. K. Wang, Kovacs G., Wozny M., Fang M., (Boston: Springer), pp. 650-657.

New Zealand and Manjubala, I. in Max Planck Research Community of Germany both have utilized conventional CAD technology combining processing technology to obtain the porestructure of bone repair[3–4]. Professor LU Bingheng and his team in Xi'An JIAOTONG University of China have obtained bioceramics artificial bone which possesses microstructure of 200~600μm in aperture size by using CAD software combining rapid-prototyping (RP)[5]. MAO Ya in WUHAN University of Technology of China presents a design method of bionic bone possessing similar 3D structures as real bone by utilizing fractal technology[6].

Constructing model of repair via conventional CAD technology needs to be accomplished by professional technicians holding CAD modeling technology, which is a bottleneck to implement the industrialization of repairing bone defect. Hence, based on the study of medical image processing, 3D reconstruction and CAD modeling technology, a new method is established to obtain the repair bionic scaffold from bone medical tomography images in this paper, and meanwhile a professional system of constructing repair bionic scaffold is developed.

2. CONCEFT OF THE FEATURES OF REPAIR BIONIC SCAFFOLD

Repair bionic scaffold is the carrier suitable for histiocytes to grow which must meet the requirements of tissue engineering. According to the requirements of tissue engineering and rapid manufacturing, in this paper, the features of repair bionic scaffold are classified as external model feature (external feature) and internal model features (internal features), and the internal features are classified as macrostructure feature (macrostructure) and microstructure feature (microstructure), Figure 1 shows the classification.

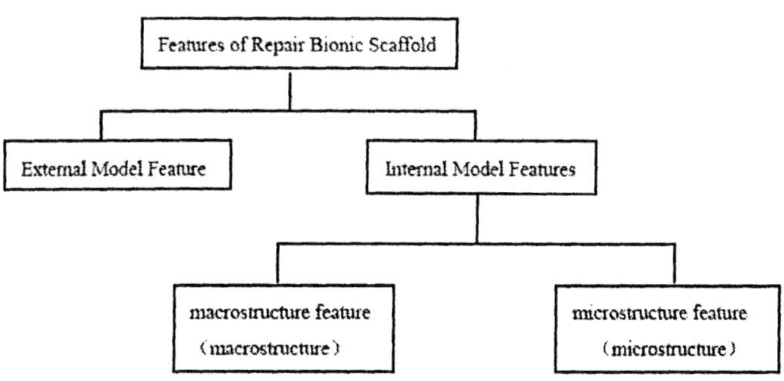

Figure 1. Classification of the features of repair bionic scaffold

External feature is the external shape feature of repair bionic scaffold, which is the shape of repair. The shape of scaffold should fit the shape of bone defect regions and should have continuous curvatures as possible as the surface of fine bone sections.

Internal features are the internal porestructures of scaffold that offer the environment of osteoblasts growing, and they are the passages through which osteoblasts can get nurture and oxygen. Macrostructure is the porestructure that is constructed in scaffold through modeling technology and it should be constructed according to the requirements of processing. Microstructure is the porestructure formed in scaffold through forming technology, which is regulated by materials and processing. Constructing macrostructure and microstructure is to guarantee that scaffold has favorable porosity and interconnectivity meeting the requirements of tissue engineering.

The modeling methods presented in this paper are used to construct the external feature and macrostructure of scaffold.

3. METHOD OF CONSTRUCTING THE EXTERNAL MODEL FEATURE OF REPAIR BIONIC SCAFFOLD

3D reconstruction from tomography images is one of the key steps in the process of constructing repair bionic scaffold from bone tomography images. At the present time, there are mainly three ways of 3D reconstruction from images:(1) Generating fitting surface through contours of every layer of image (surface reconstruction based on contours). (2) Generating isosurface directly from 3D volume data (surface reconstruction based on volume data) (3) Enduing color and opacity for every voxel data to render model directly, not to constructing surface[7]. Because scaffold needs to be processed in solid, and generally bones have complex features and the requirements of interaction and calculation amount must be counted, the method of surface reconstruction on volume data is adopted in this paper to construct the model of whole bone, external feature and macrostructure of scaffold.

By the method of surface reconstruction based on volume data, a serial of CT images of a skull with defect from Shanghai X People's Hospital have been reconstructed into 3D model, as shown in Figure 2:

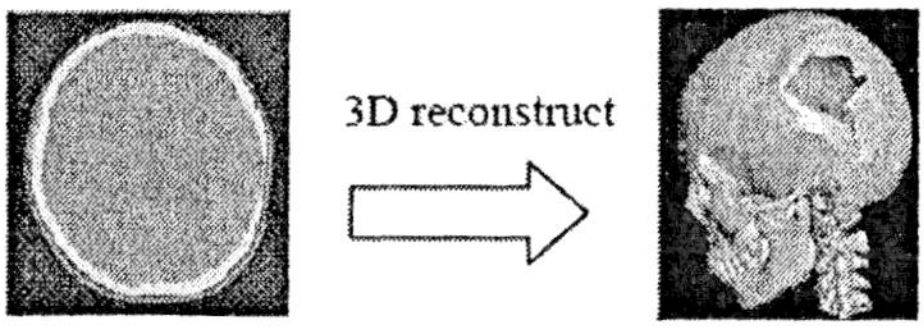

Figure 2. 3D reconstruction of a whole skull

Here, a method of constructing the external feature of repair bionic scaffold is presented as followed: firstly, by programming, pixels are filled up in defect regions of bone images; secondly, Boolean operation is proceeded between the pixels of filling up sections and the pixels of original bone images; finally all filling up sections are 3-dimensionally (3D) reconstructed to get the favorable external feature of scaffold.

Figure 3 shows this process of constructing the external feature of the skull repair bionic scaffold through the method above.

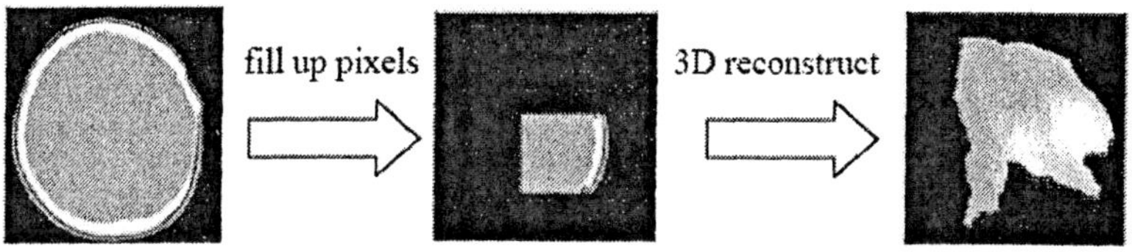

Figure 3. Process of constructing the external feature of repair bionic scaffold

4. CONSTRUCTING THE MACROSTRUCTURE OF REPAIR BIONIC SCAFFOLD

4.1 Requirements of constructing macrostructure

In the view of tissue engineering, the internal features of repair bionic scaffold should possess appropriate porosity to plant ample osteoblasts in it, appropriate aperture to facilitate attachment and growth of osteoblasts. On the other hand, the structures of internal features should have as high interconnectivity as possible to guarantee that osteoblasts can obtain as ample nurture and oxygen supply as possible. Hence, the requirements of constructing macrostructure are that it had best be fully interconnected, its apertures should range from 200μm to 400μm and the whole porosity of repair bionic scaffold should exceed 75%[8].

4.2 Method of constructing macrostructure

At the present time, there are mainly two methods of constructing internal porestruture of repair bionic scaffold as followed: one is to construct porestruture through the porosity of bionic material (namely microstructure in this paper), this method emphasizes on processing technology to control pores, which has been applied successfully to tissue engineering of cartilage, the most obvious shortcoming of this method is that it is difficult to control porosity; and it is difficult to implement the industrialization of repair bionic scaffold; another method is to construct macrostructure through conventional CAD modeling technology, favorable porosity can be obtained by this method, but, this method is difficult to be implemented.

Here a new method as followed is put forward to construct the macrostructure of repair bionic scaffold:

(1) Choosing the negative of macrostructure (spatial scaffold structure)

(2) Utilizing programming to obtain the crossing sections from intersecting between the negative of macrostructure and the plane that every bone tomography image exists in.

(3) Proceeding Boolean operation between the pixels in filling up sections of every bone image (seen as above) and corresponding cross sections of the negative of macrostructure derived from the step (2).

(4) 3D reconstructing all filling up sections after processed in the step (3) to obtain the macrostructure of scaffold.

There are two key points with this method: one is the calculation of cross sections of the negative of macrostructure; another is to proceed Boolean operation between the pixels in the filling up sections and corresponding cross sections of the negative of macrostructure. The negative of macrostructure is a kind of spatial structure. The law that the negative of macrostructure varies along the increasing axis (generally z axis) of all layers of images is studied, and then the crossing sections at corresponding position of every image are calculated. This process is implemented though programming. The process of Boolean operation between filling up sections and cross sections of negative of macrostructure is within pixels which sets the gray scale values of pixels in cross sections out of the ranges of the gray scale value of bone pixels, thus macrostructure in the scaffold can be formed when all filling up sections are 3D reconstructed.

There are mainly two methods in constructing the shape of macrostructure: One is that the repair bionic scaffold with constructing macrostructure in it is has regular structure, by this method larger porosity

(>80%) can be obtained, for example cylinder can be chosen as the shape of solid not as the shape of pore, and the scaffold with macrostructure constructed in it by this method has lower mechanical strength, and it is more suitable to repair cartilaginous tissue; another method is to choose regular structure as the shape of pore. The scaffold with macrostructure constructed through the second method has higher mechanical strength but lower porosity than the scaffold by the first method, and it is more applicable to repairing bones. According to practical requirements, the regular shape can be cylinder, square column, prism and etc. the intersecting angles between columns can be right angle and bevel. A zooming-in view of the negative of macrostructure adopting intersecting cylinder in spatial right-angle is shown in Figure 4:

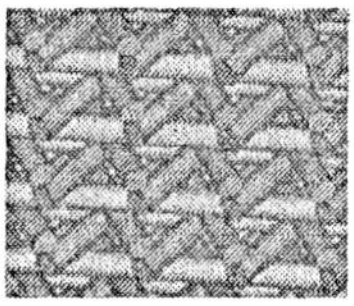

Figure 4. A zooming-in view of the negative of macrostructure

The negative of macrostructure obtained by programming is a spatial structure that is derived from intersecting between many columns three-dimensionally. The negative of macrostructure is fully interconnected (as shown in Figure 4 and Figure 5), so the macrostructure of repair bionic scaffold is also fully interconnected.

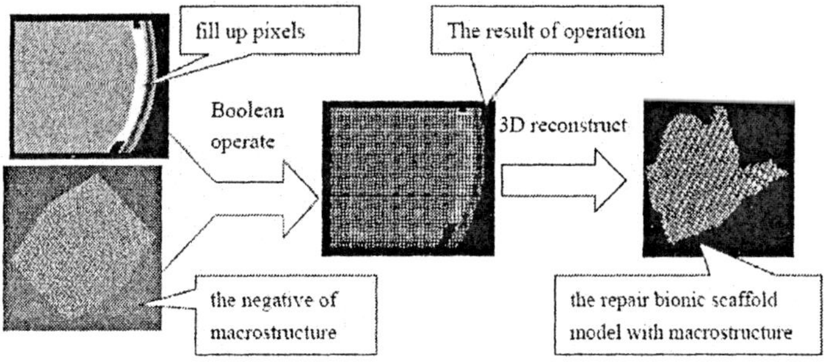

Figure 5. Process of constructing microstructure of repair bionic scaffold

Repair bionic scaffold is a kind of individual model whose external shape and macrostructure both have complex features, and it is difficult to be manufactured in solid by conventional processing methods. RP has excellent processing flexibility, which can manufacture models with complex features in high precision; so, the 3D model of scaffold is saved as STL format applicable to RP methods.

Through setting some parameters of macrostructure such as aperture and interval of pores and adjusting the porosity of scaffold, the macrostructure of repair bionic scaffold can be obtained. Favorable porosity and interconnectivity of repair bionic scaffold can be obtained through constructing macrostructure via this method and constructing microstructure via processing. The process of constructing repair bionic scaffold through this method is shown in Figure 5:

4.3 Evaluation of porosity of repair bionic scaffold

The porosity of repair bionic scaffold includes two kinds: the porosity of macrostructure derived from modeling and that of microstructure formed through processing. Here, P represents the whole porosity, P_1 the porosity of macrostructure, P_2 the porosity of microstructure, then, P can be obtained by Eq. (1):

$$P = P_1 + P_2 \quad (1)$$

P_1 can be derived through measuring the volumes of repair bionic scaffold. The volume V_1 of repair bionic scaffold without macrostructure and the volume V_2 of repair bionic scaffold with macrostructure are respectively measured, then, P_1 can be obtained by Eq. (2):

$$P_1 = (1 - V_2 / V_1) \times 100\% \quad (2)$$

Here, RP processing is adopted to obtain the solid of repair bionic scaffold; the material of repair bionic scaffold is mixed powder of β-tricalcium phosphate (β-TCP) and adhesive (mixture of epoxy resin and nylon). Porestructure (microstructure) can be formed when scaffold solid is burned to hard ceramic body under high temperature because of volatilization of the adhesive. Supposed that the volume ratio of β-TCP and adhesive is 1: β, the porosity P_2 can be derived by Eq. (3):

$$P_2 = V_2 / V_1 \times (\beta \div (1 + \beta)) \times 100\% \quad (3)$$

Synthesizing Eq. (1), Eq. (2) and Eq. (3) to obtain the whole porosity of repair bionic scaffold as shown in Eq. (4):

$$P = ((1 - V_2 / V_1) + V_2 / V_1 \times (\beta \div (1 + \beta))) \times 100\% \quad (4)$$

The methods mentioned above is adopted to construct the external feature and macrostructure of the repair bionic scaffold of the skull above, the shape of the pores of macrostructure is cylinder, the intersecting angle is right-angle the aperture of pore is 320μm, the interval of pores is 400μm, and the volume ratio of β-TCP and adhesive is 1:1, finally the whole porosity P=80.57% is obtained by Eq. (4). As can be seen, repair bionic scaffold with favorable porosity to plant osteoblasts can be obtained by using the methods presented in this paper combining processing technology.

5. CONCLUSION

This paper presents the concept and classifications of the features of repair bionic scaffold. The calculation method of porosity and the modeling method of scaffold for repairing bone defect are established. A professional system for repairing bone defect is also developed. It will promote the development of biomanufacturing technology.

6. REFERENCES

1. SHI Zhen-sheng, HU Qing-xi, WU Mao-liang, DAI Chun-xiang, FANG Ming-lun, (2004), Modeling Technology of Artificial Bone Substitute Based on Reverse Engineering Journal OF Shanghai University (English Edition) ,vol. 8, pp. 28–30.
2. LI Limin, GUO Guifang, HU Qingxi, WU Maoling, (2005), Fabircation of 3D Porous Scaffolds for the Bone Tsissue Engineering *Manufacturing Information Engineering of China5*, vol. 34, pp. 86–88.
3. Woodfield T, et al. (2005), Polymer scaffolds fabricated with pore-size gradients as a model for studying the zonal organization within tissue–engineered cartilage constructs *Tissue Engineering* vol. 11, pp. 1297–1311.
4. Manjubala I, et al. (2005), Biomimetic mineral-organic composite scaffolds with controlled internal architecture, *Journal of Materials Science: Materials in Medicine*, vol. 16, pp. 1111–1119
5. Li Xiang, Li Dichen, Su Yanping, Lu Bingheng, (2005), Structure design and fabrication of β-tricalcium phosphate artificial bone based on rapid prototyping technique, *Journal of Xi'an Jiaotong University*, vol. 39, pp. 9–12.
6. Mao Ya, Cao Zuobing, Yu Xinming, Rao Gang, (2004), 3D Bionics Artif icial Bone Design Based on Medical Images, *Chinese Journal of Stereologyand Image Analysis*, vol. 9 pp. 160–163.
7. Tang Guo, Zhao Xiaodong, Wang Yuanmei, (1998), Three–dimension medical image segmentation and visualization, *Chinese Computers*, vol. 20, pp. 204–209.
8. Laurence Marcelle Mathieua, et al. (2006), Architecture and properties of anisotropic polymer composite scaffolds for bone tissue engineering, *Biomaterials*, vol. 27, pp. 905–916.

HYDRAULIC CYLINDER CAD DESIGN SYSTEM BASED ON PRODUCT DISPOSITION

Yigang Hu, Yonggang Shen
Shanghai University of Engineering Science, Shanghai, China;Email:hgg163@sohu.com

Abstract: To adapt the requirements of the mass product-made production model, the three-dimension parametric design software about hydraulic cylinder CAD is developed. This software possesses the product disposition function for the product series design with product data management (PDM) basic idea. This design system is more useful to the mass product-made production for the customer requirements (MC, Mass Customization).

Key words: Hydraulic Cylinder, Parametric Design, Product Disposition, MC.

1. INTRODUCTION

Since the requirements of the persons for the products are going forward the variety and the individuality, the tasks of the enterprise research and developing are more and more heavy. According to the developing experiments of the enterprise new products, for one kind of new product there are 80% parts and components that are carried on composing and promoting. The enterprise should change its design concept, do its best to use the ripe technology of the old products and to use the machining plants, and do its best to use now available parts to compose more new products. The enterprise should convert the production model of many kinds and small quantity to the production model of many kinds and mass quantity, i.e. the mass product-made production (MC, Mass Customization).The mass product-made production importantly adapts the product disposition design model. This design method is forward product series, can lower the design cost and the production cost, and can short the design period.

This project is supported by Shanghai education committee science and technology development fund.No.03FK17

Please use the following format when citing this chapter:
Hu, Yigang, Shen, Yonggang, 2006, in International Federation for Information Processing (IFIP), Volume 207, Knowledge Enterprise: Intelligent Strategies In Product Design, Manufacturing, and Management, eds. K. Wang, Kovacs G., Wozny M., Fang M., (Boston: Springer), pp. 658-663.

At present, the old CAD software does not possess the function of product disposition, lacks the means of repeatedly using design information sources, and can not adapt the enterprise requirements of the mass product-made production model. Although using PDM technology can satisfy the requirements of product disposition, implementing PDM, the period is long, the cost is high and the risk is big. As one kind of standard PDM system is implemented one year and more in the enterprise, the expenses are up to several 100000 dollars of RMB-1000000 dollars of RMB apart from the expenses of the software and the solid ware. Considering the present cases of Chinese enterprises, particularly the middle enterprises and the small enterprises can not undertake these expenses. Hence facing to these enterprises, developing the CAD software that possesses PDM feature for the product series design possesses the important real significance.

2. GENERAL SCHEME OF HYDRAULIC CYLINDER CAD SYSTEM

Hydraulic cylinder is the most usual component of executing in hydraulic system. It is widely applied in common machinery, engineering machinery, lifting and transport trades etc. The chief parts and the chief components of hydraulic cylinder include cylinder lid, cylinder body, cylinder base, piston, piston rod, piston ring, guide sleeve etc. There are many kinds of the structure shape for every kind of part. There are many kinds of series dimension for the part of same structure shape. And there are many kinds of type in the link fashion and the seal fashion.

Firstly for hydraulic cylinder series products of the enterprise, according with the design method facing to product series, we sum up and conclude the product basic prototype (BP), construct the product databases, pick up the change-pattern parameters as the three-dimension templates of series product design and change-pattern design. On the parametric base, the series products are formed. Then according with the requirements of the customer, go on disposing of the product, so that form the new products under satisfying the definite constraint conditions.

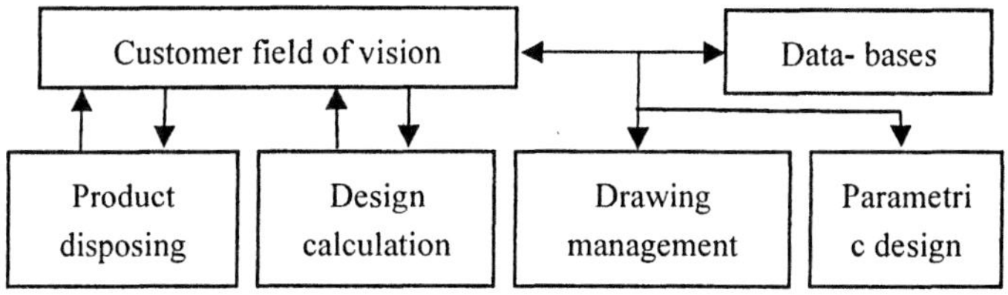

Figure1. General scheme flow diagram.

3. GENERAL SCHEME PROTOTYPE SELECTING OF HYDRAULIC CYLINDER

Under the production model of the mass product-made, from the point of view of constructing prototype of hydraulic cylinder product series, we analyze, sum and conclude the series products of the hydraulic cylinder enterprises, design three kinds of basic structure shape, i.e. clamping key link, flange link and crew link. Through system mutual effect between the person and the computer, select the link between the cylinder lid and the cylinder body, the link between the piston rod and the piston edge, the link between the cylinder base and cylinder body, and select the basic structures of the seal fashion of hydraulic cylinder parts and hydraulic cylinder components. We also should input the requirements of the customer, e.g. the stroke of hydraulic cylinder, the installation dimension, the piston force and the rated press of hydraulic cylinder. Then the general scheme design of hydraulic cylinder is completed.

4. PARTS AND COMPONENTS STRUCTURE DIMENSION DESIGN OF HYDRAULIC CYLINDER

According with the structure information of hydraulic cylinder parts and hydraulic cylinder components, under satisfying these conditions of the strength, the stability, the machining technology and the fixing requirement, this system can automatically calculate the structure dimensions of the structure parts, and can store the design parameters of hydraulic cylinder product into the product engineering databases. In the design process of hydraulic cylinder parts and hydraulic cylinder components, this system simultaneously goes on function calibrating about the strength and the stability, checks how the design scheme satisfies the design requirements. If the design scheme can not satisfy the design requirements, this system sends out a warning, and may return and again goes on mending for the structure shape and the structure dimension. Design calculation flow is as follow, first, from customer land mould into strength check mould, if it satisfies the requirement, then into whole scheme select mould and stability check mould, or back into strength check mould. If strength check and stability check are satisfied, then into part component select mould, or back into whole scheme select mould. All of these data are sent into hydraulic cylinder databases. Databases management is controlled by this CAD design system.

5. THREE-DIMENSION PARAMETRIC PROTOTYPES DATABASES

According with the basic shapes of clamping key link, flange link and crew link of hydraulic cylinder, we construct the three-dimension parametric prototypes databases of facing to hydraulic cylinder product series, include and sum the design important parameters of hydraulic cylinder parts and hydraulic cylinder components. Using the design important parameters controls the geometry dimensions of the parts and the developing relations. This system realizes the three- dimension parametric design of hydraulic cylinder through picking up the important parameters of the three-dimension parametric prototypes, and again vests with the values.

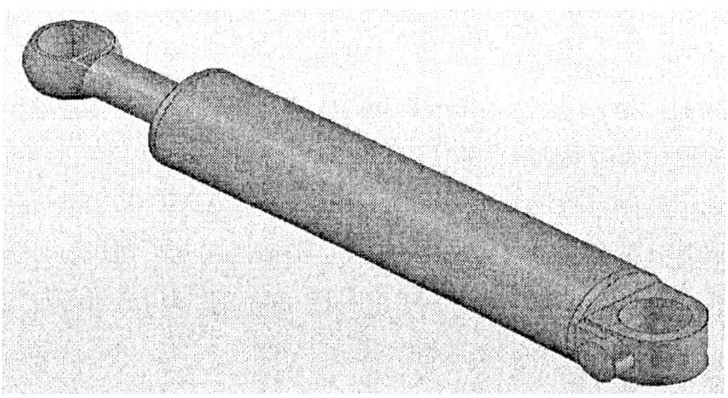

Figure 2. Hydraulic cylinder fixed drawing.

6. FILES MANAGEMENT

After completing the product design of hydraulic cylinder, this system automatically places the design parameter data files of the product and the three-dimension drawing data files into the product engineering databases. This system can dimly inquire to the product engineering databases, or set inquiring conditions to look for hydraulic cylinder series product of conforming to the requirements.

File management not only can inquire the parameter data of the product, but also can inquire the three-dimension drawing data of the product using OLE (objective link embed) technology.

7. PRODUCT DISPOSITION

Considering the drawing numbers of the product parts and the product components as the index, forms the layers structure tree of the product. The joint of the tree root represents the product. The joints of the second layer represent the parts and the components. The joints of the third layer represent the standard parts and the seal parts. This method with drawing view reflects the layers relations between the product parts and the product components. Each joint of the structure tree links the related properties of the part and the component, it becomes the shining relation with the alphabetic string of the product databases. Through the joints of the structure tree, we can inquire the design file of the related parts and the related components. According with disposing regulation, we construct the structure tree of the product in the definite scope of the design parameters, complete the product disposition through disposing for the joints of the structure tree, and for disposing product may be out shape design using OLE technology under UG surroundings.

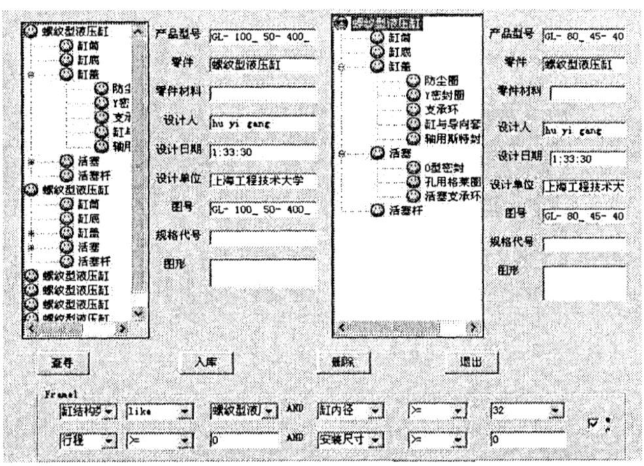

Figure 3. Disposition field of vision.

8. ENGINEERING DATABASES MANAGEMENT

Hydraulic cylinder is a product of more standardization and more regularization. In design process, we index the structure dimensions and the specifications numbers about O pattern seal ring, V pattern and Y pattern seal ring, anti-dust ring, spindle and hole composing seal, support hoop and guide hoop. We place various marks of the materials and related mechanics function targets, and the design parameters of standardization and regularization into the engineering databases. This system can inquire, index, mend, erase and add information to the engineering databases.

9. CONCLUSION

1. Realize CAD parametric design method facing to the product series.

Analyze hydraulic cylinder series products, go on product forecasting for the requirements of the available customers, design the basic prototype (BP) of hydraulic cylinder series products, then pick up the shape parameters considering the three-dimension constraints and the geometry constraints of the product parts and the product components. Based on the mould driving technology, construct the three-dimension parametric mould databases, and realize CAD parametric design and shape design facing to the product series.

2. Support MC product disposition.

Through product disposition, the enterprise does its best of using the product information sources, controls the part kinds and the part quantity into full limitation to dispose more product kinds with less part quantity, forms MC production model, increases the design efficiency, decreases the production cost.

3. Embed file management and engineering databases management in CAD design.

This system can store and inquire to design parameter files and drawing data files. This system also can inquire, index, mend, erase and add to the engineering databases.

4. Design and property checking calculation simultaneously treating.

At the same time of design, this system can automatically go on calculating and checking about the strength and the stability to judge if the design scheme and the part structure dimension satisfy the property regulations. If they can not, this system puts forward the warn information.

10. REFERENCES

1. Ren Jian-ping, Li Ju, (2004),Research on Heterogeneous Environment of Engine CAD Engineering Databases, *Journal of Shanghai University*,Vol. 8,Suppl.,pp.149-151.
2. Zhuang Pin, Wang Ning-sheng,(2003), Research on Supply Chain Management for Mass Customization, *Aeronautical Manufacturing Technology*,Vol. 7,pp.46-47.
3. Hu Dian-wei,Zhang Xu,Zhang Xu-hui(2005), Management of Standard Part Library Based on PDM, *Aeronautical Manufacturing Technology*,Vol. 2,pp.85-88.
4. Gu Yong,Chen Wei et al,(2005), Parametric Design of Drawing Die Structure for Auto Cover Panel Based on UG, *Aeronautical Manufacturing Technology*,Vol. 3,pp.75-77.
5. Lu Ji-ping,Tang Shui-yuan,Yan Yan, (2005),PDM-Based Integrating Technology, *Aeronautical Manufacturing Technology*,Vol. 4,pp.96-97.

THE REUSE STUDY OF SMALL AGRICULTURAL MACHINERY'S PRODUCT INFORMATION MODEL

Xuejun Sun, Jiandong Jiang, Fang Xu, Libin Zhang
The MOE Key Laboratory of Mechanical manufacture and Automation, Zhejiang University of Technology Email:sunxuejun@zjut.edu.cn.

Abstract: It is estimated that more than 90% of industrial design activity is based on variant design. How to use the product information model of similar products and to improve design efficiency and the designer's creativity is a subject that is worth studying. This paper has set up the structure frame of small agricultural machinery product information model, and uses ontology to describe product information model. The described mechanism of product information model has been set up. The revise mechanism of product information model is researched. The results of study are used in the reused system of small agricultural product information model.

Key words: Product information model; ontology, Modification of product information, Small agricultural machinery

1. INTRODUCTION

The development of product is the intersection of market requests, customer requests and the knowledge, experience, creativeness, artistic quality of designer and all kinds of standards. How to unite the three items is a subject that is worth studying. In the methods of traditional product development, designers concentrate their energy on the unification of

This project is supported by the Natural Science Foundation of China No,50375146, National Basic Research Program of China No,2004CCA05800, The Doctoral Program Foundation of Ministry of Education of China No,20040337002, the department of international Science&Technology cooperation of China No,2005DFA70380, Zhejiang Provincial department of international Science & Technology cooperation of China No,2005C14005, and Zhejiang Provincial Natural Science Foundation of China No,Z503054)

Please use the following format when citing this chapter:
Sun, Xuejun, Jiang, Jiandong, Xu, Fang, Zhang, Libin, 2006, in International Federation for Information Processing (IFIP), Volume 207, Knowledge Enterprise: Intelligent Strategies In Product Design, Manufacturing, and Management, eds. K. Wang, Kovacs G., Wozny M., Fang M., (Boston: Springer), pp. 664-672.

requests and knowledge, experience, and making the product accord with all kinds of standards. This method limited the exertion of creativeness and artistic quality. Designers face on choose of being in a dilemma between efficiency and creativeness, artistic quality. It is estimated that more than 90% of industrial design activity is based on variant design.[1] So the reuse of product information model is an attractive product development method. The reuse of product information model can avoid the repetition of design effort. The reuse of product information can improve the exertion of designer's creativeness and artistic quality. The expressions, reasoning, choose, modification of design and appraisement of product information are the key technology during the reuse of product information. This paper sets up the reuse system structure of product information model, then expresses the product information model, and finally studies the design modification of product information.

2. THE SYSTEM STRUCTURE

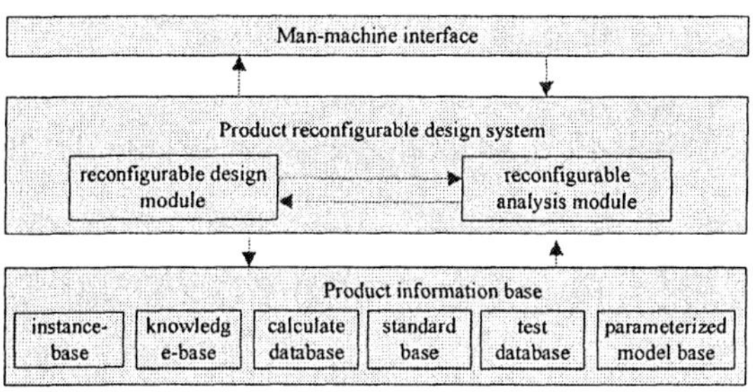

Figure 1. system structure chart of small agricultural machinery reconfigurable design system

Fig.1 is the system structure chart of small agricultural machinery reconfigurable design system. It mainly includes three parts: man-machine interface, production reconfigurable design system and production information base. Production reconfigurable design system is composed of by reconfigurable design module and reconfigurable analysis module. The reconfigurable design module receives the user requests from man-machine interface, then picks up the corresponding instance and parameterization model to have quickly and individuation design, which is according to some rules. The design results of reconfigurable design module are handed in reconfigurable analysis module to have some performance analysis, such as vibration analysis, work performance analysis and so on. The production information model-base includes instance-base, knowledge-base, calculate database, standard base, test database and parameterized model base. The instance-base mainly be used stored the instance that has successfully developed, during a new product development these instances can be reused or consulted. The product development experience of workers and the knowledge that is used during product development is stored in knowledge-base. When a new product is developed, designer and performance analysis engineer can study and use this knowledge to solve new problems. Calculate base mainly stores these data that need be calculated and will be used in the follow-up product development, for example the data of natural calculate frequency etc.. The national standard and trade standard of product development, performance analysis and the product performance will be stored in standard base. Test database mainly store some test results such as the data of production performance test and vibration test data etc.. Parameterized model base is used store the parameterized model of product parts and components. These bases compose of the production information base, and how to organize these data and make them become an organic whole, can be convenient used is the content of next part.

3. PRODUCTION INFORMATION MODEL

Fig.1 shows that the data structure of product information model includes parameterized geometry model, all kinds of calculated data and curve, mature theory knowledge, the practical experience of product developer, all kinds of test data and curve and all kinds of statutory standard. How to logically express and organize these data, and provide to reconfigurable design module and reconfigurable analysis module using is the key of modeling the product information model. Here we use the "ontology "to express and organize the product information model.

"Ontology" is a philosophy term, and is a description about the concept and relation of external being.[2,3] This paper uses ontology to express product information model with "task, problem solve method and domain" three aspects and " meta-ontology, ontology, and application ontology" three levels and "intranet connector, intranet connector" two tools.

Task (T) is the problem that this system will solve. Problem solve method (PSM) is the method that uses to solve problem, a problem can have one or many methods to solve, and every sub-problem also have one or many methods to resolve. Domain (D) is the knowledge when resolve problem, it can be separate into the domain knowledge of every sub-problem. Intranet connector (intra-C) mainly expresses the interrelation of the same aspect of one problem, for example: the interrelation between T1 and T2, the interrelation between PSM1 and PSM2 and so on. Internet connector (inter-C) is used to express the interrelation of different aspects of one problem, for example: the interrelation between T and PSM, the interrelation between T and D and so on. T, PSM, D, intra-C and inter-C are expressed with meta-ontology, ontology and application ontology.

Definition1 Meta-ontology meta-ontology is the cell ontology that can not be decomposed, and it is expressed by MOi.

$$DefMO_i = (Atr_i - name,\ MOb - name,\ Atr_{ij}, Val) \quad i = (1,2,\cdots,n);\ j = (1,2,\cdots,m)$$

Where: Atri-name—the description of attribute, is corresponding with one attribute of meta-ontology;

MOb-name—the name of meta-ontology;

Atrij—the explanation of the jth attribute;

Val—the value of Atrij.

Definition2 Ontology ontology can divide into meta-ontology, and it is the aggregation of multi-meta-ontology. One ontology should integrally describe the smallest subclass or the smallest connector of one aspect of problem. It is described with O. Ontology is the aggregation of meta-ontology, but it is not the simple element combination, and it is the composing with some constraint condition, it can be described as following:

$$O = (O - name,\ \Sigma\, DefMO_i,\ \text{int}\, ra - C_{ij},\ Atr_{ij},\ Val)$$

Where: O-name denotes the name of ontology;

$$\Sigma\, DefMO_i = \sum (DefMO_1, DefMO_2, \cdots DefMO_n),$$

the aggregation of metaontology attributes;

intra-Cij—the intranet connector of between meta-ontology;

Atrij—the explanation of the jth concrete attribute, it is the attribute that ontology has after the aggregation of meta-ontology;

Val—the value of Atrij.

Difination3 Application ontology application ontology is the smallest resolve cell of one problem; it is expressed with Apl-O. Application ontology is a aggregation of ontology and meta-ontology that aims at one concrete problem. It is expressed as following:

$$Apl - O = (Apl - O - name,\ \sum DefMO_i,\ \text{int}\, ra - C_{ij},\ \sum DefO_i,\ \text{int}\, er - C_{ij},\ Atr_{ij},\ Val_i)$$

Where: Apl-O—name—the name of application ontology;

$\Sigma DefMO_i = \sum(DefMO_1, DefMO_2, \cdots DefMO_n)$, — the attribute aggregation of ontology who compose the application ontology;

intra-Cij—the intranet connector between meta-ontology;

$\Sigma DefO_i = \sum(DefO_1, DefO_2, \cdots DefO_n)$,—the attribute aggregation of meta-ontology who compose the application ontology;

inter-Cij — the internet connector between meta-ontology, between ontology and between meta-ontology and ontology;

Atrij—the explanation of the jth concrete attribute of solved results.

Val—the value of Atrij.

From the description we can know that ontology is an extensible knowledge expression method, and it includes and describes the data, data structure, data type and document structure with the structure of tree. Besides that, ontology can include semantic to describe the knowledge that has good structure. In knowledge database, first order predicate logic, production rule, frame, table and so on all these expression methods have good structure at format, and can be directly described by ontology. Ontology also can include hyperlink in document, and can make one node with many nodes, and realize the relation of one to many and many to many, and can directly describe the relation of all kinds of graph structure. Ontology provides a uniform format to describe first order predicate logic, production rule, frame, table, process, semantic network and graph etc. many kinds of type knowledge expression methods, can inosculate different kinds of knowledge in a whole knowledge database. Ontology takes the extensible tree structure to describe knowledge, which can expediently extend, modify and delete knowledge of knowledge database, and can expediently manage and maintenance knowledge database. These merits of ontology can realize the description, expression, management and maintenance etc. function of engineering database.

4. THE MODIFICATION OF PRODUCT INFORMATION

The reuse will be concerned on the search of data and knowledge and the modification of product information model and components [4]. Here we will study the modification of task model and components. The process of redesign will be expatiated, and how to modify the product information model to satisfy the new design request will be studied. Fig.2 is the control flow chart of product information. The reused model divided into three parts: design request, the description of design object and the harmonization of design process. The four components that related to design request are the following. The first component is modification, this component mainly analyses current design requests, comes into being the modification advices. These advices will be compared with each other. The second component is guided mend, which according to the design theory of knowledge base modify the design request with guided mend. The third component is the updating of current description, this component stores and modify the current design requests. The fourth component is the updating of modification history, this component stores and modifies the modification history of design requests. The four components that related to design object description also includes modification, guided mend, updating of current description and updating of modification history.

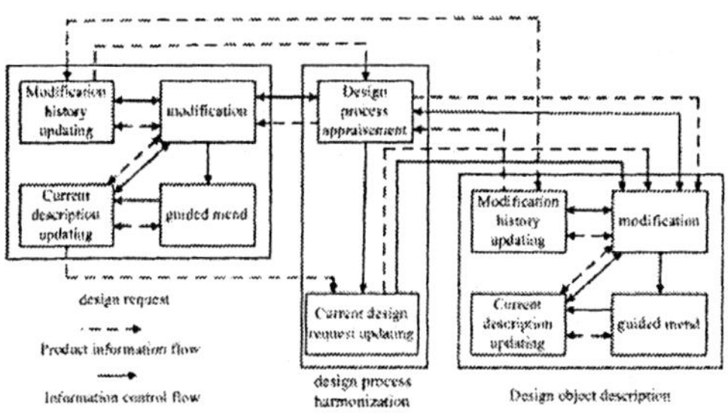

Figure 2. product information modification control flow chart

The two components that related to the harmonization of design process are the following. The first is the appraisement of design process. The status of design process will be appraised and controlled. This component mainly appraises the description of design object whether or not softies the design requests. The second is the updating of current design requests. This component stores and modifies the current design requests

While redesign the product based on the product information model, the sub-components of design requests and design object description that is redesigned will be modified. Every sub-component has the following operations. The first is the analysis of current description. This phase will study and analyze the problem that need solve in current design request constrain set and design object description. The second is the confirmation of modification center focus. This phase will confirm the part that must be modified in current design requests and design object description, so that to resolve the problems. The third is the conformation of modification method. This phase will confirm the modification method that be used to modify the part. The modification method is obtained from knowledge base or is provided by designer or expert. The fourth is modifying component according to the modification method.

The component and sub-component will be described by ontology. Following is an example of product information modification.

The initial design requests are: RQS_0 = { RQa, RQb, RQc, RQd}

Where: RQS—the product design request set;

RQ—the product design request.

The initial design object description is: DOD_0 = { PF, HD, HE, ID }

Where: DOD—the description of design object;

PF—the formatting disposal of product information;

HD—the confirmation of precondition;

HE—the appraisement of precondition;

ID—the description of product information.

The design process will be appraised by the harmonization component of design process. The result of appraisement shows that DOD_0 can satisfy RQS_0. The system restarts the design request and the design will continue. Knowledge engineer (KE) brings forward two new design requests RQ1 and RQ2. New design request is analyzed. New design request RQ1 makes the function more perfect, and it must be satisfied. RQ2 is a secondary design request. The design object description is analyzed and to check if it satisfies the design request. The result shows that the design request of RQ1 is not satisfied. In order to solve this problem, the design object description is modified.

The updating of modification history:

RQS_0 = { RQa, RQb, RQc, RQd}

RQS_1 = { RQa, RQb, RQc, RQd, RQ1, RQ2}

Rationale ($<RQS_0, DOD_0>$, $< RQS_1, DOD_0>$, method(KE)).

The modification of DOD:

1) the analysis of current description, current DOD can not satisfy the all the current design requests.
2) The confirmation of modification center focus, setting the modification focus as {HD}.
3) The conformation of modification method, the select of modification method is operated based on knowledge base.
4) Modification, HD is replaced by component – libstructD that can have a disposal of structurization for precondition. This component can produce general sub-component – libG, contrastive sub-component – libC and select sub-component – libS. These operations will be renamed.

The updating of current description of DOD:

The current description is updated as: {PF, HE, IO, HD*, HD*:HG, HD*:HC, HD*:HS}

The updating of modification history:

The modification history is updated as: DOD_0 = { PF, HD, HE, ID }

$DOD_1 = \{PF, HE, ID, HD^*, HD^*:HG, HD^*:HC, HD^*:HS\}$

Rationale(<RQS_1, DOD_0>, <RQS_1, DOD_1>, replaced_by (HD, HD*))

Rationale(<RQS_1, DOD_0>, <RQS_1, DOD_1>, meant_to_satisfy(HD*, {R1}))

Rationale(<RQS_4, DOD_0>, <RQS_4, DOD_1>, method(library_consutation))

Rationale(<RQS_4, DOD_0>, <RQS_4, DOD_1>, is_based_on(HD* , libstrucD))

Rationale(<RQS_4, DOD_0>, <RQS_4, DOD_1>, is_based_on(HD*:HG , libstrucD:libG))

Rationale(<RQS_4, DOD_0>, <RQS_4, DOD_1>, is_based_on(HD*:HC , libstrucD:libC))

Rationale(<RQS_4, DOD_0>, <RQS_4, DOD_1>, is_based_on(HD*:HS , libstrucD:libS))

The appraisement of design process:

The product object description satisfies the RQ1, and does not satisfy the RQ2. DOD will be continued operation to satisfy RQ2.

5. APPLICATION INSTANCE

The product information of small agricultural machinery is described by ontology. During the design the product information model of small agricultural machinery is reused and modified by the method of section3. Fig.3 is the frame of small agricultural machinery reused module. Through this system designer can expediently reuse the product information model of small agricultural machinery, and it also can make the designer concentrate their energy on the creativeness and artistic quality of product.

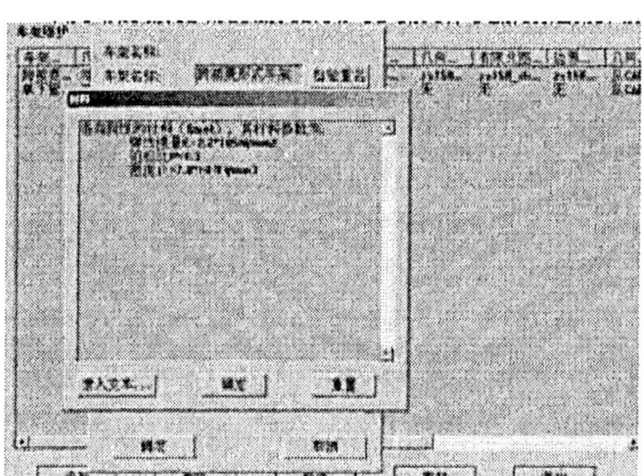

Figure 3. The reuse model of frame

6. CONCLUSION

This paper sets up the system structure of small agricultural machinery product information reused design. The description of product information model and the reused mechanism is emphases researched. The product information model is set up by ontology. The idea of component is used to set up the product information model mechanism. This paper brings forward the method of product information model selecting, modification and reuse. The results of research are primitively used in the small agricultural machinery configurable design system.

7. REFERENCES

1、 Gao Yongsheng, Zeid Ibrahim; Bardasz Theodore. Characteristics of an effective design plan system to support reuse in case-based mechanical design, *Knowledge Based System*, 1998(10): 337~350

2、 Lovett P. J., Ingram A., Bancroft C. N. Knowledge-based engineering for SMEs — a methodology, *Journal of materials Processing Technology*, 2000 (1~3): 384~389

3、 Oussalah M., Reuse in KBS: a components approach, Expert Systems with Application, 2003 (24): 173~181

4、 Brazier Frances M. T., van Langen Pieter H. G., Treur Jan, Wijngaards Niek J. E.. Redesign and reuse in compositional knowledge-based system, *Knowledge Based System*, 1996(9): 105~118

STUDY OF CAD-INTEGRATED ANALYSIS FOR COMPLEX STRUCTURES

Hirpa L. Gelgele
Dept. of Mechanical/Structural Engineering and Material Technology, University of Stavanger, Norway
Email: Hirpa.G.Lemu@uis.no

Abstract: In the classical product development process, design of a product is verified by testing physical prototypes. This approach falls short of fulfilling the demands for providing products at lower cost and short delivery time. This article illustrates the use of the digital data from CAD model in analysis and simulation of complex structures. Two industrial examples are shortly discussed to demonstrate the advantages of integrating analysis in the early phase of the product development process.

Key words: Product development, finite element analysis, integrated CAD/FEA, CAE

1. INTRODUCTION

Providing products to the market at lower prices, shorter time and better quality have been the means of gaining better competitive advantage for manufacturing industries within the last 4 – 5 decades. For products that require mechanical design, in particular, being able to use computers in the design process was the first step towards design automation that resulted in today's CAD (Computer-Aided Design) systems. For example, the early computer aided drafting system like CADAM from Lockheed Martin were developed to reduce time to market by replacing the conventional drawing board in early 1970s. Manufacturing engineers have also attempted to automate the machining process through developing numerically controlled tools such as APT (Automated Programming Tool) from MIT. However, the standalone Computer-aided technologies faced short of fulfilling the required goal due to constraints of lack of common language between the CA-techniques. Research in late 80s and early 90s has focused on

Please use the following format when citing this chapter:
Gelgele, Hirpa, L., 2006, in International Federation for Information Processing (IFIP), Volume 207, Knowledge Enterprise: Intelligent Strategies In Product Design, Manufacturing, and Management, eds. K. Wang, Kovacs G., Wozny M., Fang M., (Boston: Springer), pp. 673-678.

establishing an integrated system between CAD/CAM systems including the automation of process planning, production control and the like.

In the traditional design methodology, verification of the design process is based on testing physical prototypes that have similar performance to the required product or machine with certain dimensional scaling. For complex machines and structures, however, this approach is not only difficult, but also time and resource consuming.

This article illustrates an integrated approach to early analysis of complex products at the design phase using Finite Element Analysis (FEA) as a replacement for simulations and verifications using physical prototypes.

2. WHY ANALYSIS AT EARLY DESIGN PHASE?

As stated above, verification of the design for complex products through prototype testing is a time-consuming process. Studies show that 85% of the total time and cost of product development are committed in the early stages of the product development phase when only 5% of the project resources are expended. This is because in the early conceptual stage, necessary decisions are made concerning basic geometry, material type, system configuration and manufacturing processes. Further, mistakes or wrong decisions made at the early stage of product development phase can be modified with none or minimal increment in the product cost. In other words, the time and cost to correct problems increases about ten-fold with each step of the product development cycle: concept definition, detailed design, prototype development, prototype testing, and manufacturing of final product.

Industries have attempted to improve these and other drawbacks of the traditional product development process through extensive use of computers in various phases of the process, concurrent engineering, computer integrated design and manufacturing systems, and other product development compression techniques. These methods have shown better results for products of few components, simpler geometry and configuration.

3. AN INTERACTIVE DESIGN AND SIMULATION

Today, several computer-based tools have been available to assist a product's life cycle. Solid modeling has enabled designers to define parts and assemblies in 3D space and utilize the geometric data in several of downstream applications such as process planning, prototyping, simulation and analysis. It has been possible to convert CAD models into physical prototype parts in order to visualize, fit together and evaluate the appearance of components. Using recently developed knowledge-based systems; we can capture technical standards, experiential know-how, defined procedures and other design and manufacturing information to automate design routines.

Most recently, the term Computer-aided engineering (CAE) appeared in the engineering field to imply the interactive work of CAD, CAM and FEA (Finite Element Analysis) systems. This concatenation of FEA in CAD/CAM systems to give birth to CAE systems indicates the necessity to study stress, deformation, vibration, temperature distribution and other behavior of products or structures using the CAD data. As far as these tools are computer-based, being able to share the digitalized information among these tools truly contributes to the compression of the product's time to market.

3.1 Existing implementation challenges

The primary precondition for an effective application of numerical analysis tool in engineering to detect potential areas having excessive stress, deformation, vibration and/or other problems in complex structures is existence of a 3D model. An interactive design and analysis through integration of CAD and FEA tools allows designers to quickly iterate the proposed design solution and evaluate the merits and demerits of different ideas, compare alternatives, and identify design weaknesses before a costly production has been done.

However, the real advantages of this tool cannot be achieved without problems. Primarily, the powerful graphics capabilities, and specially the non-standardized user interfaces require considerable time and expertise of a dedicated analyst having the basic knowledge of finite element methods (FEM) and FEA theory. As FEA is an approximation method, the level of the analysis error depends on the skill, training and background of the user to apply proper mesh densities, element types and boundary conditions. Furthermore, the analyst must have sufficient knowledge to translate CAD geometry into proper format for building the FEA model, as well as appropriately interpret results, plots and other output information.

In many manufacturing companies, this knowledge gap between the designer and the analyst is so high that FEA is typically reserved for critical components. To have a competitive advantage, however, efforts are being done to re-orient the product development process itself so that analysis is performed much earlier in product development [Roth, 2000].

The other issue is that the embedded analysis systems within the CAD environment often do not contain the full capabilities that a standalone FEA system would have, and this has to be addressed in future developments.

3.2 FEA can replace physical testing

The trend in the manufacturing industry shows that the cost of materials and manpower to build prototypes is ever increasing. Conversely, computer hardware costs decreased significantly and computing power is increasing, which has led to a wider use of virtual prototyping using CAE tools.

Application of FEA on solid models of products benefits manufacturers in that they don't rely on expensive and time-consuming prototypes. While physical tests can still be required for certain product certifications, simulation using 3D models is more cost-effective and repeatable than physical tests. The other advantage of integrating FEA in early design phase is that it reduces the knowledge gap by allowing the test and redesign performed on software prototypes than on physical prototypes (Adams and Askenazi, 1999).

3.3 Industrial examples

3.3.1 Analysis of a robot arm

In this section, the rationale for virtual prototyping at the early product development phase or as an integrated part of the whole design process for an industrial robot is illustrated. The studied robot has motion about 6 axes. The load in the components is not so high but very complex during the motion in those different axes.

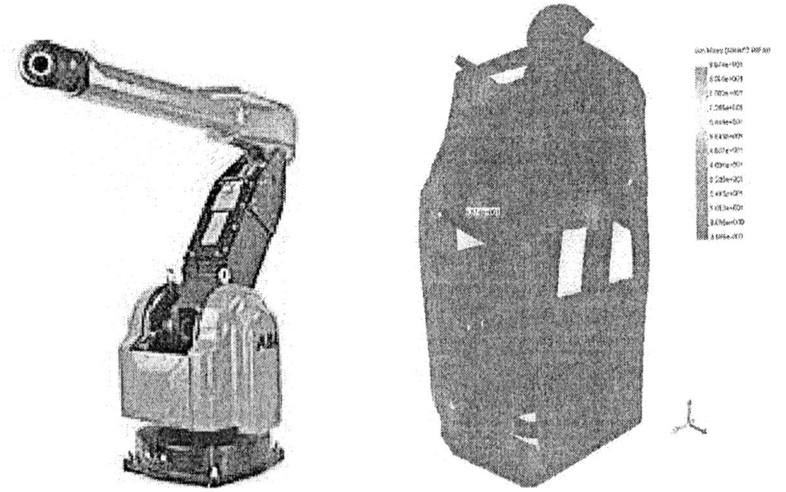

Figure 1. Analysis of robot arm

Because of the precision required, robots are traditionally designed for high stiffness. On the other hand, the vertical arm of the robot is exposed to various forms of loading, and at the same time stress concentration due to access holes for electronic components makes this part of the robot the weakest link. This weak part also influences the programmed precision.
The robot arm has two parts:

- Upper part - of Aluminum alloy AlSi10Mg, yield strength 200 MPa and
- Lower part - of nodular graphite iron SjK-400, yield strength 250 MPa (analysis model not shown).

The modeling was done in SolidWorks, a 3D CAD software. The fully embedded analysis software COSMOSworks within the 3D CAD environment facilitates the smooth transfer of the geometry to the analysis model. Based on the geometry data, the analysis model describes the product in terms of finite element meshes, loads and constraints. Using this approach, it has been possible to make many "what-if" scenarios and identify the weakest portion of the structure for various loading and motions while it is still in digital form.

Nowadays, many manufacturing industries are re-orienting their design procedures so that a single digital file (i.e. the solid model) contains all necessary information of the product's life cycle. Studies in the local industry show also that virtual analysis at early phase of the product development process has, among others, the following advantages:

- Reduced product development time 40 – 50 %
- Reduced design error 30 – 35 %
- Improved customer satisfaction
- Simplified product configuration
- Improved product quality and safety.

3.3.2 Stress concentration factor in a pressure vessel

The effect of stress concentration in pressure vessels is mostly estimated using hand calculations based on empirical relations. The results of such calculations are very unreliable particularly in the localized high stress areas due to holes, welded joints and dimensional variations. Studies (Sang *et al.*, 2002) show that a reliable and accurate design method for pressure vessel intersections with diameter ratio $d/D \geq 0,5$ is necessary, where d = nozzle diameter and D = shell diameter.

In this study, a cylindrical pressure vessel with spherical ends and mean diameter D = 800 mm was modeled in the CAE tool I-DEAS 10 NX series. The cylindrical shell with a cylindrical nozzle at the middle is positioned horizontally and supported at two bottom points.

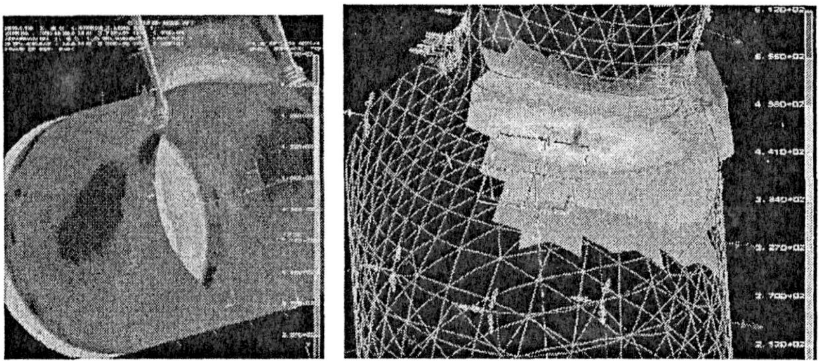

Figure 2. Analysis of stress concentration factor in a pressure vessel

The main objective of this study is to identify how local stresses vary with dimensions of a pressure vessel, where making physical tests are not amenable due to certain practical reasons. The results show that the stress concentration factor for a given D/d ratio decreases with increasing fillet radius (Gelgele, 2005). At the same time it has been observed that smaller D/d ratio has higher stress concentration factor.

4. CONCLUSIONS

The article demonstrated the possibilities and advantages of executing analysis of complex structures at an early design phase. An integrated CAD/FEA approach enables to study the performance of structures by varying design parameters in a cost effective way using the digital data of the CAD model. This signifies the transition from the traditional *design by formula* to *design by analysis* approach. However, there are still many practical issues that have to be addressed to get the real implementation.

5. REFERENCES

Adama, V. and Askenazi, A. (1999), *Building better products with Finite Element Analysis*, OnWord Press, 1. ed.

Gelgele, H. L. (2005), Stress concentration in pressure vessels using FEM, Proc. of 9th Int. conf. on *Trends in development of machinery and related technologies*, Antalya, Turkey.

Roth, G. (2000), The value of early analysis, *Benchmark, Int. magazine for engineering design and analysis*, NAFEMS.

Sang, Z. F. *at el.*, (2002) Limit and burst pressures for cylindrical shell intersection with intermediate diameter ratio, *Int. Journal of Pressure Vessels and Piping*, **79**, 342-349.

CAD/CAM RESEARCH ON RADIAL NON-CIRCULAR GEARS

Jiang Han [1] , Dongdong Zhang [2], Lian Xia [3]
CIMS Insitute, Hefei University of Technology , China; Email: peterzdd_2002@hotmail.com

Abstract: Based on the comparison with the traditional normal non-circular gears, a new type of radial non-circular gears and its generation method are discussed, and the thought for tooth-by-tooth modification is introduced. In this paper, the whole process from the mathematical analysis to CAD/CAM Integration is discussed, which has an extensive application in the field of flow meters.

Key words: Radial Non-circular Gears; Tooth-by-Tooth Modification; Injection Mold; CAD/CAM Integration.

1. INTRODUCTION

The non-circular gear, widely applied in automation and instrumentation, is a kind of driving component used to transmit variable motions. Non-circular gears can substitute for the linkages and cams in many fields because of the following merits[1-3]: (1) Non-circular gears are more compact and work well; (2) It is easier to guarantee the change drive with a regular pattern by non-circular gears than by other mechanisms.

In the positive-displacement flow meters, non-circular gears are always used as the central parts to measure the expensive liquid medium, such as oil and natural gas. Because of the accuracy of this kind of flow meter, it is extensively used in various industrial fields, such as petroleum, chemical industry, medicine and energy. And the design and manufacture of non-circular gears has being studied. Tong S H[4] investigated the generation of identical non-circular pitch curves, Chien-Fa Chen[5] and Biing-Wen Bair[6] studied the elliptical gears with circular-arc teeth and Chang S L[7] focused on the tooth profiles generated by rack cutter. In this paper, a particular type of radial gears is introduced, which has shown excellent advantages in the

Please use the following format when citing this chapter:
Han, Jiang, Zhang, Dongdong, Xia, Lian, 2006, in International Federation for Information Processing (IFIP), Volume 207, Knowledge Enterprise: Intelligent Strategies In Product Design, Manufacturing, and Management, eds. K. Wang, Kovacs G., Wozny M., Fang M., (Boston: Springer), pp. 679-684.

minute-flux meters and other occasions. Hence, the CAD/CAM research on radial non-circular gears is of central importance.

2. THE ANALYSIS OF RADIAL NON-CIRCULAR GEARS

Traditional non-circular gears are basically normal gears[1, 2], that is, all teeth are arranged in the normal direction of the pitch curve, as displayed in Figure 1. However, the novel radial non-circular gears are quite different in that the teeth are arranged in the radial direction of the pitch curve, as shown in Figure 1. Both of these types of gears meet the demand of engagement, and their pitch curves are coincident. The difference lies in the generation mechanism and the arrangement method of the gear teeth. Compared with traditional normal gears, the improved radial gears have some excellent characteristics:

1. The accuracy and the sensitivity of the flow meters are greatly enhanced. In positive displacement flow meters, under the pressure difference between the entrance and the outlet, the torque is produced to make the gears rotate continually[8]. Because certain pressure acting on the teeth of radial gears could produce more force moment than that on the teeth of normal gears, the reaction to the minute flux is quit sensitive.
2. The security and the stability of the flow meters are improved. The contact ratio of a couple of radial gears is higher than that of normal gears, which minimizes the probability of latch and improves the ability to adapt to the variation of temperature.

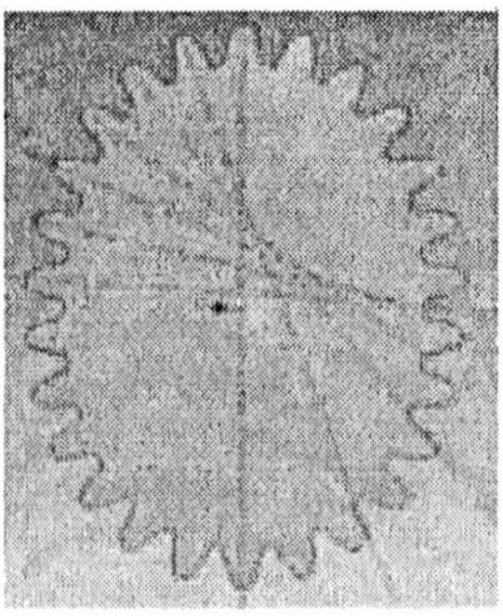

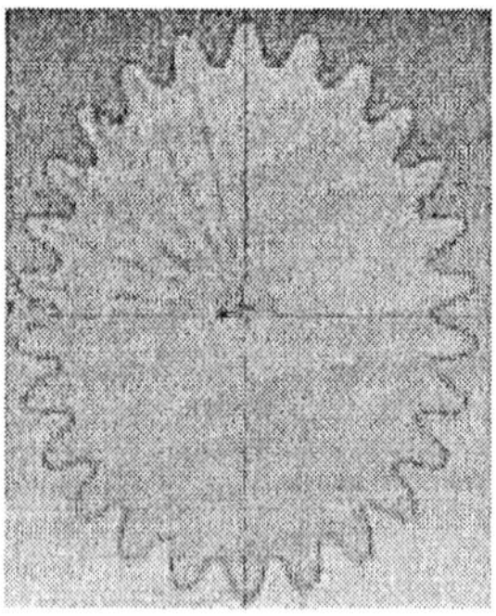

Figure 1. Normal Gears (Left) and Radial Gears (Right)

3. DESIGN OF RADIAL GEARS

The toughest part of the design of radial non-circular gears lies in the generation of the tooth profile, the involute between the addendum circle and the dedendum circle. In the paper, the method of equivalent pitch radius and tooth-by-tooth modification is introduced to develop the mathematical model of the tooth profiles.

3.1 Method of Equivalent Pitch Radius

This method is that the radius vector of every equant point* on the pitch curve is regarded as the pitch radius, called equivalent pitch radius[1, 9]. Therefore, according to the different radius vector, the corresponding tooth profile is generated. As in Figure 2, Cartesian coordinate system is established, in which the origin is set in the geometric center of the gear, and axis y is in the direction of radius vector of every equant point. Thus, the standard tooth profile is generated by the following equation (1):

$$\begin{cases} x = \dfrac{R_J}{\cos a_2} \sin\left[\dfrac{0.25\pi m}{R_d} + (\tan a_1 - a_1) - (\tan a_2 - a_2)\right] \\ y = \dfrac{R_J}{\cos a_2} \cos\left[\dfrac{0.25\pi m}{R_d} + (\tan a_1 - a_1) - (\tan a_2 - a_2)\right] \end{cases} \quad (1)$$

In Eq.(1), parameter m is the module, R_j the base radius, R_d the equivalent pitch radius, a_1 the pressure angle of the crossing point of involute and pitch curve and a_2 the pressure angle of any point required.

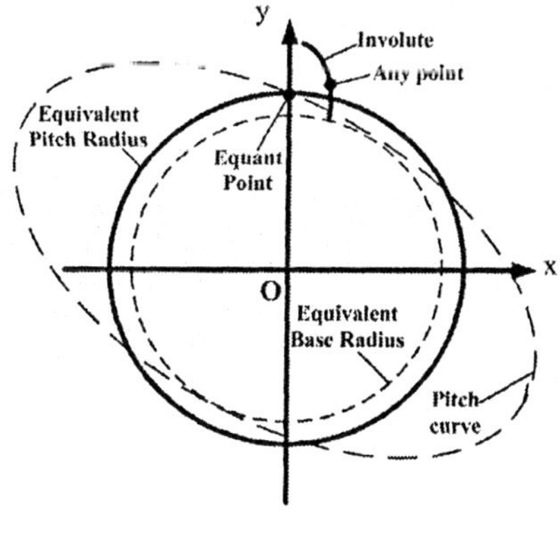

Figure 2. Method of Equivalent Pitch Radius

*equant point: On the pitch curve, the arc length between the gear teeth is identical, and the crossing point of the center line of every gear teeth and the pitch curve is called equant point.

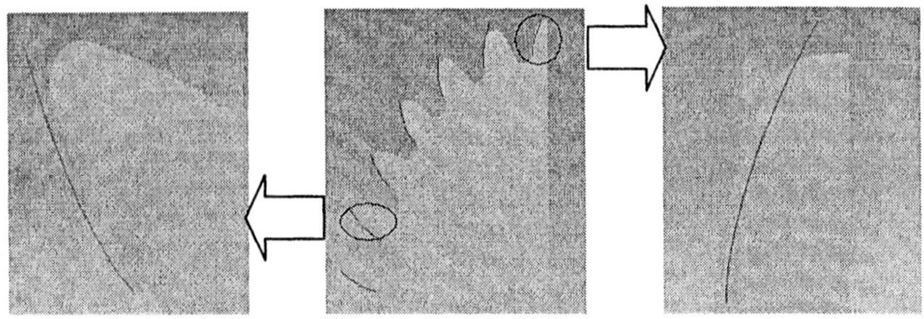

(1) (2) (3)

Figure 3. Tooth-by-tooth Modification

3.2 Tooth-by-Tooth Modification

Tooth-by-tooth modification is defined as the gear shift with different coefficients which are determined by the equivalent pitch radius of every gear tooth. The standard of comparison is $\pi m/2$, in which m stand for the module of the gear. When the equivalent pitch radius is greater than $\pi m/2$, the positive-shifted modification is applied, as shown in Figure 3(1). On the contrary, if the equivalent pitch radius is less than $\pi m/2$, the negative-shifted modification is used, as shown in Figure 3(3). After the modification, the used part of the involute is changed, but the center distance and the anticipated drive ratio of the gear pair are invariable[9]. The thought for tooth-by-tooth modification is put forward to eliminate the meshing interference, a phenomenon that the tooth face of one gear exceeds the specified kinematic boundary and interferes with the tooth face of the mating gear. But via tooth-by-tooth modification, the gear teeth near the major axis are wider and those near the minor axis are narrower, which guarantees the exact engagement of the couple of gears.

4. CAD/CAM INTEGRATION

Based on the design method above, the parameterized 3-dimensional model of radial non-circular gear can be established using Pro/E[10]. And the drive process can be simulated[11], as shown in Figure 4.

The traditional manufacture method of gear shaping and hobbing is unlikely to be utilized to produce radial gears. In this paper, the injection model is introduced, which guarantees not only the quality of products, but also the mass-production. And the injection model can be cut by transferring the CNC code obtained automatically via the NC module of Pro/E to the linear cutting machine directly.

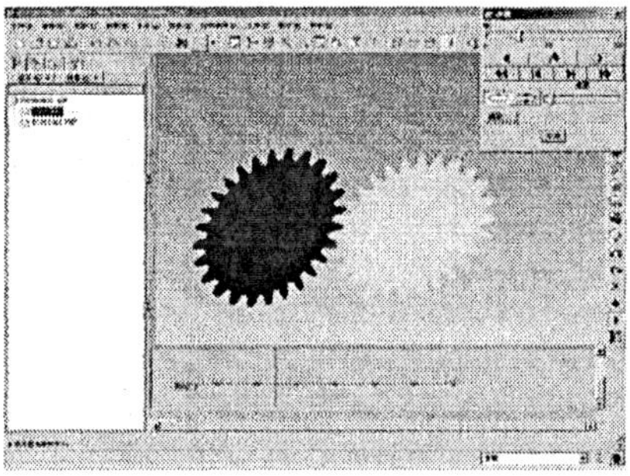

Figure 4. Dynamic Simulation of Radial Gears

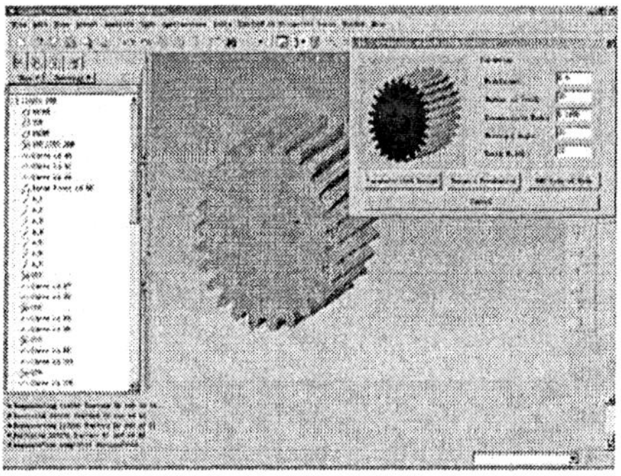

Figure 5. Interface of CAD/CAM Integration

Through Pro/TOOLKIT[12, 13] and VC++, the compiled application code, shown in Table 1 can be linked to the library files of Pro/TOOLKIT, and the executable program, so-called the exclusive design module will be started when Pro/E is opened. So the developed module can meet the demand of special design and manufacture of radial non-circular gears efficiently. The mutual interface of the integration system of radical non-circular gears is shown as Figure 5.

Table 1. The Compiled Application Code (part)

BOOL CDlg8_2::OnInitDialog()	if (status==PRO_TK_NO_ERROR)
{ CDialog::OnInitDialog();	{ProMdlToModelitem(model,
MoveWindow(600,50,450,300,TRUE);	&modelitem);
ProMdl model;	....
ProModelitem modelitem;	if (status==PRO_TK_NO_ERROR)
ProName ParamName;	{ProParameterValueGet(&param, &value);
ProParameter param;	m_B=value.value.d_val; }
ProParamvalue value;	}
ProError status;	UpdateData(false);
status=ProMdlCurrentGet(&model);	return TRUE;
	}

5. CONCLUSIONS

The concise modeling and the manufacture of gears have been the tough issue in CAD/CAM field, not to mention non-circular gears. In this paper, the following conclusions are drawn.

1. In the field of flow meters, normal non-circular gears can be replaced by radial gears, which have an extensive application. The model of modified tooth profiles complies with the theory of gearing, and the method of establishing and simulating the model is clear.
2. Tooth-by-tooth modification is a feasible method to eliminate the meshing interference. This method can be used not only in radial gears but also in normal gears in order to improve the condition of gear drive.
3. The radial gears can be produced by the injection model, which can be cut according to the CNC code generated automatically.
4. Secondary development of Pro/E can be achieved by Pro/TOOLKIT. And the developed CAD/CAM integration module of radial non-circular gears can make design and manufacture more accurate and efficient.

6. REFERENCES

1. Li Fusheng, *Non-circular Gears* (Mechanical Industry Press, Beijing, 1975), PP.117-145.
2. Wu Xutang and Wang Haigui, *Non-circular Gears and Nonuniform-speed-ratio Drive* (Mechanical Industry Press, Beijing, 1997), PP.1-56.
3. Wu Xutang, *The Principle of Gear Engagement* (Mechanical Industry Press, Beijing, 1984), PP.1-86.
4. Tong S H, Yang D H, Generation of Identical Noncircular Pitch Curves, *ASME Journal of Mech Trans Des*, 1998,120 (2), PP.337-341.
5. Chien-Fa Chen, Chung-Biau Tsay, Computerized Tooth Profile Generation and Analysis of Characteristics of Elliptical Gears with Circular-arc Teeth, *Journal of Materials Processing Technology* 148 (2004),pp.226–234.
6. Biing-Wen Bair, Computer Aided Design of Elliptical Gears with Circular-arc Teeth, *Mechanism and Machine Theory* 39 (2004),pp.153–168.
7. Chang S L, Tsay C B. Mathematical Model and Undercutting Analysis of Elliptical Gears Generated by Rack Cutters, *Mechanism and Machine Theory*, 1996, 31 (7),pp.879-890.
8. Positive Displacement Flow Meter, http://www.chinaflow.com.cn/basic/rongji13-1.htm.
9. Zheng Wenwei and Wu Kejian, *The Theory of Mechanism 7tth ed.* (Higher Education Press, Beijing, 2000.8).
10. Fang Xing, Xia Lian, and Han Jiang, Development of A Parametric Function Design System of Involute Gears Based on Pro/Engineer, *Journal of Hefei University of Technology*, 2004,27(8),pp.915-918.
11. Han Yulong, *The Unit Design and Dynamic Simulation on Pro/Engineer Wildfire*, (Tsinghua University Press, Beijing, 2004.11).
12. Liu Wenjian, Chang Wei, Jin Tianguo and Bai Heming, *Technology of CAD/CAM Integration* (Harbin Institute of Technology Press, Harbin, 2000.12).
13. Parametric Technology Corporation, *Pro/TOOLKIT User's Guide*, Waltham, USA.

SPECIAL CNC BASED ON ADVANCED CONTROLLER

Junxi Bi[1,2],Tao Yu[1],Qiang Li[2]
[1]CIMS & Robot Center of Shanghai University, Shanghai University, China; Email: junxibi163@163.com. [2]College of Mechanical, Inner Mongolia University of Technology, Hohhot, Inner Mongolia , China.

Abstract: This paper introduces a new type of CNC, which is based on Sinumerik 840D and STEP-NC. This is a kernel of I^2MS (Intelligent integrated Manufacturing System), which includes compute-aided design, computer aided process plan, computer aided manufacturing, and CNC. It provides them a unified platform oriented to CNC manufacturing, which make their work turned from distributed work to collaborative work. In this paper, we describe and analyze the conceptual framework and key technologies such as AI, KM, and etc systematically. We also discuss STEP-NC approaches using in CNC.

Key words: CNC; STEP-NC; 840D; KM; AI

1. INTRODUCTION

Roll grinder is a special machining equipment in the manufacturing industry, it mainly be used to grind some special roll such as CVC roll shape in the steel rolling field. Because of special accuracy requirement of plate shape, thus the grinding accuracy of roll which decided by roll grinder is more and more high. In order to solve this issue, we must adapt higher accuracy CNC roll grinder. With the development of artificial intelligence, modern CNC machine tools have great changes. But with the continuously increase of the manufacturing system information amount and the lasting improvement of flexibility and automation requirement, the deflection of computer numerical control(CNC) system served in the bottom layer of factory has been shown. It only loyally executed preplanning machining program, the effective apperceive ability and self-control ability are

This project is supported by the Science and Technology Department Foundation of Shanghai Science and Technology Committee under the grant No. 05DZ11C04.

Please use the following format when citing this chapter:
Bi, Junxi, Yu, Tao, Li, Qiang, 2006, in International Federation for Information Processing (IFIP), Volume 207, Knowledge Enterprise: Intelligent Strategies In Product Design, Manufacturing, and Management, eds. K. Wang, Kovacs G., Wozny M., Fang M., (Boston: Springer), pp. 685-690.

insufficient. There is a trend that advanced manufacturing technologies are to be developed to include numeralization, precision, uttermost, automatization, integration, latticing, intelligentization and green production technologies [1]. Manufacturing enterprises are facing serious challenges and pressures from the growing globalization of the economy and the market as well as from the rapid developments of science and technology [2]. Small and medium-sized enterprises especially, have to reform their traditional manufacturing methods by using advanced technologies, such as Artificial Intelligence (AI), Knowledge Management (KM), etc, in order to succeed in the increasingly intense competition for markets. Thus, it is of great importance for them to accept the concept of the special computer numerical control based on advanced controller. As the brain for industrial machinery, Computer Numerical Control (CNC) is the core element in modern intelligent integrated manufacturing system (I^2MS). The structure of I^2MS can be shown in figure.1.

In a traditional CNC manufacturing environment, the operation flow usually starts from product design via geometric models created in CAD system, and ends with the generation of machining instructions required to convert raw material into a finished product, based on geometric model. In fact this process includes CAD/CAPP/CAM/CNC/MT. It is an executing mechanism without intelligent and it is a single direction signal workflow in current industrial manufacturing system.

In this paper, a brief conception of special CNC is presented first, followed by introducing the framework of it based on STEP-NC. Successful implementation in application in roll grinding is given to illustrate how CNC based on 840D controller works.

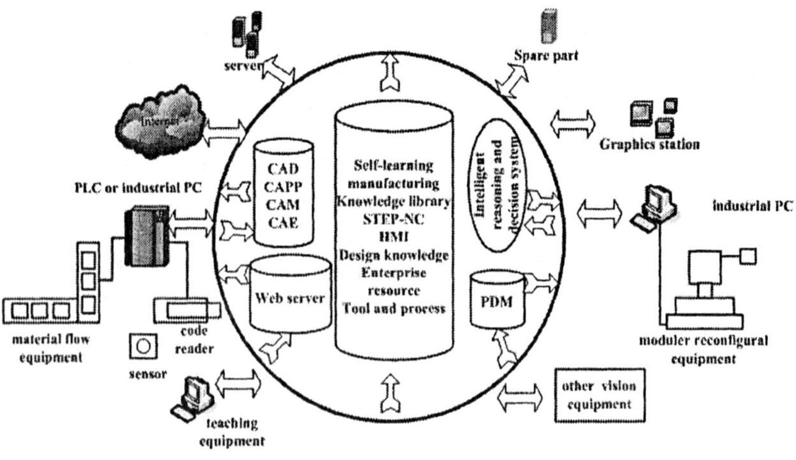

Figure 1. The new model for an I^2MS

2. CONNOTATION OF SPECIAL CNC

CNC is the abbreviation of Computer Numerical Control. Here intelligent and integration are two key facts in special CNC system. AI based on the operating principle of the human brain and natural biological evolution, have been received considerable and increasing interest over the past decade and have been applied to CNC machining. Most of the developments are concentrated on an intelligent control algorithm and its application to a traditional CNC machining system, so they have not achieved good results [3, 4]. Various part of system being simply condensed is the idea of intelligent manufacturing model, the system is composed of these condensed intelligent body. Each condensed intelligent body is relatively isolated cell, the system is composed of these cells which can divide and collaborate to work. In order to realize the total intelligence of manufacturing system, it is essential for the intelligence of each cell device. For the intelligence of CNC system, the direction and expression of intelligence is firstly decided, for the integration of CNC system, the method of integration is secondly considered. The intelligence of CNC system is shown as follow [5].

1) The intelligence of job plan

2) The ability of self-learning and improvement

3) The self-adapted man machine interface

4) The intelligent control of manufacturing process

5) The intelligent failure diagnosis

Because the G and M code are employed in the traditional CNC technology, this has brought some critical problems such as incompatibility between various systems, data sharing and integration, and programming, and so on. Moreover it restricted the open ability and intelligence of CNC system, and led bottle-neck phenomena between the CNC and CAX technology. In order to solve the above-mentioned some reasons, the Europe Union passed a planning of OPTIMAL and provided the concept of STEP-NC. STEP-NC (14694) expands STEP into the field of CNC. It redefined a kind of interface between technology planning system of CAM and the control system of NC. This is helpful in decreasing machining time of traditional CAD/CAM and in eliminating post processor required in the manufacturing process, and it supports the coming intelligent manufacturing equipments. In fact STEP-NC is a CNC language which has covered the total life cycle of product. That is to say, it is a higher-level programming language for seamless integration of the CAD-CAM-CNC chain. At the present time the research and development of the standard of STEP-NC is still in the laboratory. 840D is an open advanced platform in HMI ,NCK and PLC, thus it is primary task to study and develop open CNC system based on advanced controller and STEP-NC.

Here CNC means that abstracting some function of traditional CIMS, such as CAD, CAPP, CAM, into a series of modules, which makes the application be divided into a group of tasks frame. The tasks frames could be decomposed into detailed sub-functions. Every function modules is a practical module oriented to manufacturing. The function modules are organized and integrated into the bigger parts modules with the task frame, and these parts modules are integrated into a concrete application CNC system finally with a unified user interface. CNC is also a kind of new philosophy and method in I^2MS.

3. THE FRAMEWORK OF SPECIAL CNC

The basic architecture of CNC is composed of:

1) Human-machine interface (HMI)

2) Control function

3) Data processing within CNC [6]

This holds true for CNC in the architecture presented, the user requirements are implemented in basic architecture in an extended fashion. As shown in figure2, it includes the following main modules:

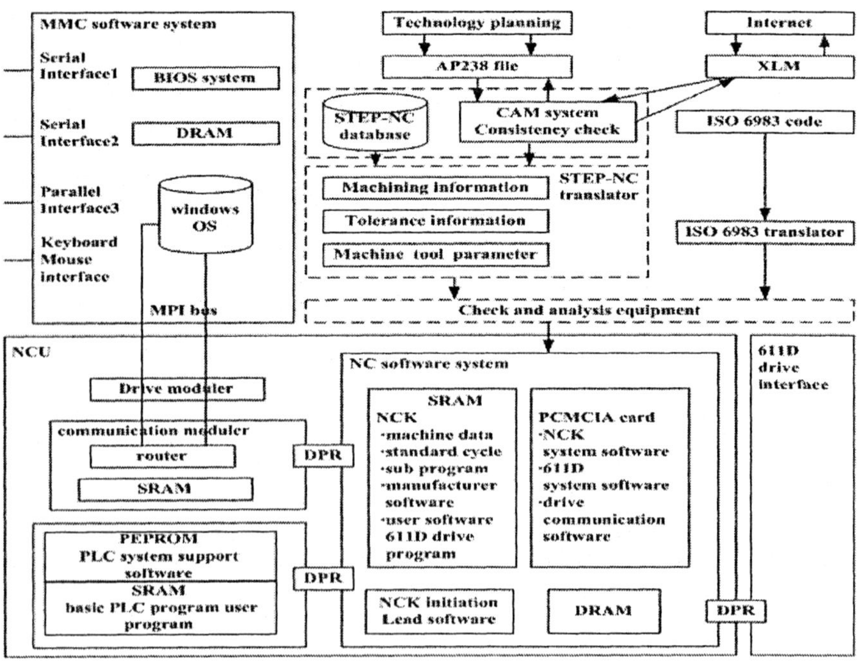

Figure 2. A conceptual architecture of special CNC

CAD/CAPP/CAM module, which are extended HMI comprehensively covering part programming and tool path generation based on a STEP-NC data model. These modules incorporate the CAD/CAPP/CAM function into the programming system. The standard CAD data (STEP, AP203) can be converted into built-geometric modeling kernel data, by recognizing the machining features and processing these feature information. Output is stored in the machining feature database. CAPP determines the processing sequence and other processing information by using the machining knowledge. Sequentially, CAM extracts NC instruction based on STEP-NC (ISO14649).

NCK/PLC control module which is related to the activities of CNC to carry out successfully the instruction given to it. NCK interprets the tool path commands and executes them by activating the servo mechanism, and PLC executes machinery commands, such as tool change and workpiece loading or unloading. The principal function is to control the machine tool so that the desired shape is produced accurately by the machining operation. Various intelligent control functions are covered here, such as adaptive control, human interaction, quality control, resource management, and self-learning.

Engineering Database based on KM, which provides comprehensive data for the part programming and Tool Path Generation. This database include machining feature database, machining resource database, machining process database, machining knowledge, diagnosis database and so on.

4. IMPLEMENTATION OF SPECIAL CNC

We have done researches on special CNC upon Windows NT4.0, Sinumerik 840D platform. Here we describe working process of special CNC under the windows-based environment as an example. All the researches and development of Windows NT4.0 based CNC system were based on RG (Roll Grinder).RG is a product of Shanghai Machine Tool Co ltd. RG is used to grind various roll shape in the steel industries. The hardware section of machine is composed of SINUMERIK 840D, industrial PC, MPI (Multiple Point Interface), ProfiBus, and so on.

The software section has an open and modularized structure, which has a standard module interface and each module can be independent with each other and be easily integrated together. One of the important modules is CAD parts module. Self-design CAD of roll shape has been embedded in the software, this module is based on feature, we can perform all the CAD feature function via that interface. Another important module is CAPP intelligent expert process planning system based on knowledge. Here we can obtain grinding process parameters easily and we can generate machining instructions based on STEP-NC for the next step via CAM module.

From that illustration above, we can see that special CNC has integrated some special function of CAD/CAPP/CAM, and all the information are for machining features based on STEP-NC.

5. CONCLUSION

This paper discusses a new type of CNC. It provides a unified platform that makes the workflow of project easy to be done. It focuses on some key modules namely CAD/CAPP/CAM/CNC.ISO STEP-NC standard is also introduced and applied in CNC based on sinumerik 840D.

With the development and implementation of STEP-NC and many advanced AI, KM etc. technologies in I^2MS, the research and application of CNC based on advanced controller must be the trend in future I^2MS.

6. REFERENCES

1. Shuzi Yang, "Trend in the development of advanced manufacturing technology",*Chinese Journal of Mechanical Engineering*,Vol 5,No.1,pp.73-77,Dec.2003.
2. Tao Cheng, Jie Zhang, Chunhua Hu, etc. Intelligent Machine Tools in a Distributed Network Manufacturing Mode Environment. *International Journal of Advanced Manufacturing Technology*,2001,17,pp.221-232.
3. S. Tarng, T. Hwang and Y. S. Wang, "A neural network controller for constant turning force", *International Journal of Machine Toolsand Manufacturing*, 34(4), pp.453–460, 1994.
4. M. K. Kim, M. W. Cho and K. Kim, "Application of the fuzzy control strategy to adaptive force control of on-minimum phase end milling
5. Shi Xiaojuan, Wang Xiaochun. The modeling Analysis of the Intelligent Numerical Control System, *CHINESE JOURNAL OF SCIENTIFIC INSTRUMENT*,2003,24(6)
6. SUH S H, CHEON S U.A framework for an intelligent CNC and data model. *International Journal of Advanced Manufacturing Technology*,2002,19,pp.727-735.

A UNIFIED DECISION MODEL FOR EVALUATION AND SELECTION OF MES SOFTWARE

Liang Chao[1], Qing Li[2]
[1]*Department of Automation, Tsinghua University, Beijing, China; Email: liangc04@mails.tsinghua.edu.cn.* [2]*Department of Automation, TsinghuaUniversity, Beijing, China.*

Abstract: In this paper, we explore the potential of applying the analytic network process (ANP) to evaluate and select MES software. Different from AHP, ANP allows feedback and interdependence between factors. In our decision model, we consider MES' unique attribute as an information hub related to nearly all of the other information systems. In addition, the relationships among functionalities which MES software performs are also taken into consideration. Finally, we validate the decision model through a test.

Key words: Analytic Network Process (ANP), Manufacturing Execution Systems (MES), Decision Model, Evaluation, Selection.

1. INTRODUCTION

Since introduced in 1990, Manufacturing Execution Systems (MES), a category of industrial software for the manufacturing environment, has been adopted in a number of industries.[1] As increasing manufacturers realize that MES is a key technology for improving manufacturing operation and financial performance in the competitive global marketplace, and want to implement it, how to select the most suitable software inevitably becomes a critical problem encountered by the IT managers or executives

The paper provides a decision model for MES software evaluation and selection based on the analytical network process (ANP). Since MES often

Sponsored by China 863 Program, No. 2001AA415340; Sponsored by China Natural Science Fund, No. 60474060

Please use the following format when citing this chapter:
Chao, Liang, Li, Qing, 2006, in International Federation for Information Processing (IFIP), Volume 207, Knowledge Enterprise: Intelligent Strategies In Product Design, Manufacturing, and Management, eds. K. Wang, Kovacs G., Wozny M., Fang M., (Boston: Springer), pp. 691-696.

act as an information hub, the decision model not only evaluates the software itself, but also its broad organizational impact.

In the following part of the paper, Section 2 presents the decision model. Introduction of the ANP methodology and its application are provided in sections 3.Finally, we complete the paper with a conclusion.

2. THE DECISION MODEL FOR MES SOFTWARE SELECTION

The decision model of this study is shown in Fig.1. The factors considered and relationships among them are described following in detail.

2.1 Factors considered

As we know, when we evaluate and select software, we not only need to compare the software performance, but also need to compare the vendor performance, because besides the factors related to software performance, other factors requiring critical consideration, such as cost, implementation time and support, vary dramatically from different vendors. So we divide factors considered into two categories: *Software Requirements (SOR)* and *Vendor Requirements (VER).* [7]

For the *SOR*, we consider two clusters of factors:

Functionalities (FUN): MESA International has identified eleven principal functions of MES[4]. Because an individual customer may have individual needs, we can not define the constant functions in the decision model. Here we construct the cluster *FUN* based on the assumption that a certain customer requires the seven functions as showed in decision model. Of course, we can change the functions in the cluster according to practical needs.

It is the functions that serve plant's needs. Therefore, during evaluating process, we must identify whether and to which extent functionalities of the MES software match the production needs firstly.

Information Technology Requirements (ITR): This cluster relates to technical factors pertaining to the MES.

- *Flexibility* refers to the extent to which MES can be modified for use in applications or environments. In the fiercely competitive and ever-changing society, plants may be required to produce different products to survive, so *flexibility* is very important for a software system.
- *Reliability* refers the ability of MES to perform its required functions under stated conditions during the time when it is running
- *Compatibility* indicates how the MES accommodates with the existing

and future systems with which it is interacting. Because MES touches nearly all of the other information systems, integration between MES and the other systems is a key to gaining full benefits not only from MES but also from other information systems.[4]

• *Security* entails the protection of the MES' data and processes.[2]

• *Ease of use* is defined as the extent to which a prospective user can operate without having to overcome many difficulties. Since MES is mainly operated by workers in plant rather than the professionals, the factor is also considered critically.

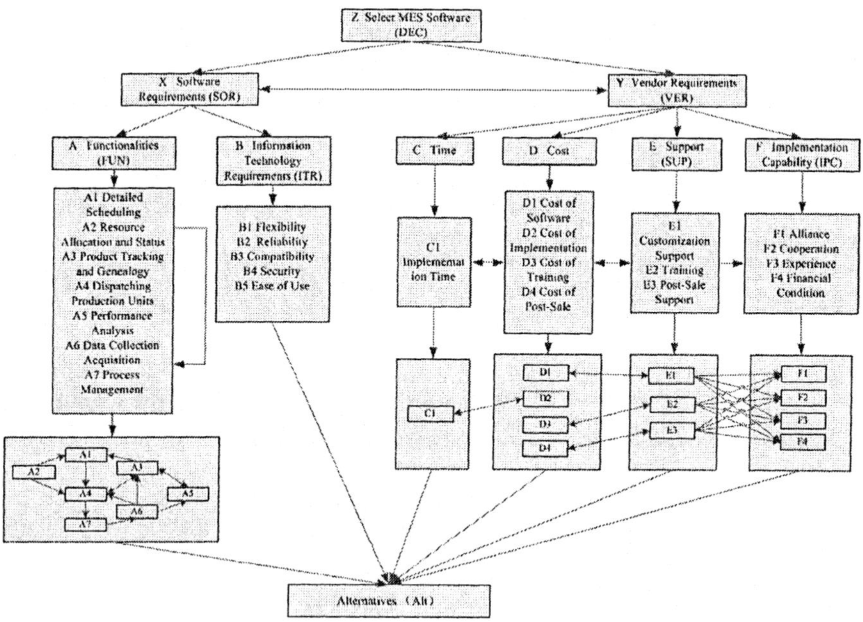

Figure 1. Decision Model for MES Software Selection

For the *VER*, we consider four clusters of factors:

1) *Time*: in this cluster we include only one factor

•*Implementation Time*: since the implementation time for different vendors varies greatly, the factor needs to be considered.

2) *Cost*: We should consider the total cost rather than only *cost of software*, because *cost of software* is only a part or even a small part of the total cost, and other types of cost such as *cost of post-sale* may account for more portions.

3) *Support (SUP)*: this cluster involves three types of related services provided by vendors.

•*Customization Support* refers to the ability of MES vendors to redesign or adjust the software according to customers' practical needs.

•*Training* indicates the activity provided by vendors to help end-users familiar with and operate the system quickly.

•*Post-sale Support* involves a series of services during MES whole life cycle, such as maintenance and software upgrade. Since the relationship between customers and vendors is not a short one, and MES affects not only manufacturing operation but also financial performance, the supports would have a potential impact on the company strategy in the long term.

4) *Implementation capability (IPC)*: *Cooperation* attitude, rich *experience,* powerful and helpful *alliance*, and favorable *financial condition* would help vendors fulfill their promise.

2.2 Relationships among factors

Fig.1 gives the relationships not just among the clusters but also among the factors (if the relationships exist), which is showed in the Fig .1 just below the relationship among clusters.

The hierarchical relationships between *SOR and FUN, ITR* clusters are easy to justify. So are the hierarchical relationships between the *VER and Time, Cost, SUP, IPC* clusters.

In addition to these hierarchical relationships, there are three types of interdependent (two-way)relationships between different clusters, one type of one-way relationship between different clusters, and one type of intradependent (two-way) relationship in the same cluster that require explanation.

Among the three types of interdependent relationships, the first is a two-way relationship between *SOR* and *VER* .This relationship is substantiated in that different vendors supply different MES software, and different software requirements lead to different vendors who satisfy the requirements .Our second interrelationship is between *Time* and *Cost* clusters, which is due to the interrelationship between *C1* and *D2* factors, and the interrelationship is easy to prove. The third interrelationship is between *Cost* and *SUP*. The relationship indicates the fact that the *Supports* provided by vendors and the corresponding *Cost* the *Supports* charge influence each other.

The one-way relationship is from *SUP* to *IPC* cluster. The explanation is that each factor of *IPC* affects each factor of *SUP,* in other words, the factors that influence the *Implementation Capability (IPC)*, are also the ones that influence *Supports (SUP)* the vendor can provide.

The intradependent relationships are among factors of the *FUN* cluster. Many of these functions logically contribute to each other. For example, *Data Collection/Acquisition* can provide data to *Product Tracking*, while

Performance Analysis feed *Product Genealogy*. In this model, we depict all the possible relationships among the functions.[4,8]

3. APPLICATION OF ANP

ANP is proposed by Thomas L. Saaty[6], and has been used as a practical method for evaluation and selection in many fields. In this study, to validate the decision model, we invited Mr. Chen to help us to select the best MES software from three alternatives: $Alt1$, $Alt2$ and $Alt3$. Mr. Chen was an IT manager in a big plant, and he was with over ten years experience as a consultant for enterprise software implementation and integration. The plant was planning to implement MES. After identifying not only the theory and procedure of ANP, but also the factors in each cluster and the relationships among them, under our directions, Mr. Chen made the paired-comparisons in both the cluster and factor level[2]. Based on the preference from Mr. Chen, we compute the priority vectors, and form the initial supermatrix.[2,5,6] *Table 1* provides the overall cluster priorities, each column represents the priorities of the clusters on the left with respect to the controlling cluster on the top of the column .*Table 2* shows partial data of the initial supermatrix.

Table 1. Overall cluster priorities

	A	B	C	D	E	F	X	Y	Z	Alt
A	0.75	0.00	0.00	0.00	0.00	0.00	0.60	0.00	0.00	0.00
B	0.00	0.00	0.00	0.00	0.00	0.00	0.20	0.00	0.00	0.00
C	0.00	0.00	0.00	0.14	0.00	0.00	0.00	0.09	0.00	0.00
D	0.00	0.00	0.33	0.00	0.20	0.00	0.00	0.13	0.00	0.00
E	0.00	0.00	0.00	0.43	0.00	0.00	0.00	0.22	0.00	0.00
F	0.00	0.00	0.00	0.00	0.40	0.00	0.00	0.19	0.00	0.00
X	0.00	0.00	0.00	0.00	0.00	0.00	0.00	0.37	0.67	0.00
Y	0.00	0.00	0.00	0.00	0.00	0.00	0.20	0.00	0.33	0.00
Z	0.00	0.00	0.00	0.00	0.00	0.00	0.00	0.00	0.00	0.00
Alt	0.25	1.00	0.67	0.43	0.40	1.00	0.00	0.00	0.00	0.00

Table 2. Initial supermatrix

	A1	A2	A3	…	B4	B5	C1	D1	…	X	Y	…	Alt3
A1	0.00	0.50	0.54	…	0.00	0.00	0.00	0.00	…	0.28	0.00	…	0.00
A2	0.00	0.00	0.00	…	0.00	0.00	0.00	0.00	…	0.11	0.00	…	0.00
A3	0.00	0.00	0.00	…	0.00	0.00	0.00	0.00	…	0.11	0.00	…	0.00
…	…	…	…	…	…	…	…	…	…	…	…	…	…
B4	0.00	0.00	0.00	…	0.00	0.00	0.00	0.00	…	0.12	0.00	…	0.00
B5	0.00	0.00	0.00	…	0.00	0.00	0.00	0.00	…	0.07	0.00	…	0.00
C1	0.00	0.00	0.00	…	0.00	0.00	0.00	0.00	…	0.00	1.00	…	0.00
D1	0.00	0.00	0.00	…	0.00	0.00	0.00	0.00	…	0.00	0.29	…	0.00
…	…	…	…	…	…	…	…	…	…	…	…	…	…

	A1	A2	A3	…	B4	B5	C1	D1	…	X	Y	…	Alt3
X	0.00	0.00	0.00	…	0.00	0.00	0.00	0.00	…	0.00	1.00	…	0.00
Y	0.00	0.00	0.00	…	0.00	0.00	0.00	0.00	…	1.00	0.00	…	0.00
…	…	…	…	…	…	…	…	…	…	…	…	…	…
Alt3	0.21	0.50	0.21	…	0.40	0.44	0.16	0.40	…	0.00	0.00	…	0.00

After normalizing the initial supermatrix by multiplying it by the corresponding elements in *Table 1,* and raising the multiplicative resulting to a high power,[2,5,6] we obtain the final priorities of the alternatives as following: $Alt1 = 0.469$, $Alt2 = 0.232$, $Alt3 = 0.299$

Obviously, $Alt1$ is the MES software recommended for selection in our study. Mr. Chen told us that $Alt1$ was just the one which he considered to be most favorable before the test, and the results agreed with his former assessments by and large.

According to feedbacks from Mr. Chen, the decision model is helpful for potential customers to make their selection of MES, and it is a practical way to apply ANP in this model, though the process is somewhat tiring.

4. CONCLUSION AND SUMMARY

In this paper, we explore the potential of applying the ANP technique to evaluate and select the MES software. The evaluation and decision model provides a systematic view of the alternatives. Based on the ANP technique, the decision-making process is lucid.

However, we must point out that the underlying assumption in this paper is that manufacturers have decided to invest in MES, if it is not the case, more consideration would be required.

5. REFERENCE

1. The Benefits of MES: A Report from the Field, MESA international, White Paper No.1, May, 1997.
2. J. Sarkis and R.P. Sundarraj, Evaluation of Enterprise Information Technologies: A Decision Model for High-Level Consideration of Strategic and Operational Issues, *IEEE Trans. on Systems, Man, and Cybernetics-Part C Volume*, Issue 99,2005 pp.1 – 14.
3. Controls Definition & MES to Controls Data Flow Possibilities, MESA international, White Paper No.3, February, 2000.
4. MES Explained: A High Level Vision, MESA international, White Paper No.6,9, 1997.
5. T. L. Saaty, Fundamentals of Decision Making and Priority Theory with the Analytic Hierarchy Process, Pittsburgh, PA: RWS, 1994.
6. T. L. Saaty, Decision Making With Dependence and Feedback: The Analytic Network Process, Pittsburgh, PA: RWS, 1996.
7. MES Software Evaluation / Selection, MESA international, White Paper No.4, 1996.
8. International Standard IEC/WD 62264-3, Enterprise-control system integration -Part 3: Activity models of manufacturing operations management.

3D MODELING AND CAPACITY CALCULATION SYSTEM IN TANK CALIBRATION

Cuilian Zhao [1], Minglun Fang [1], Zhongxu Tian [1], Yongsui Lin [2]
[1]*Cims center of shanghai University, China; Email:clzhao@mail.shu.edu.cn* [2]*Shanghai Merchant Ship Design & Research Institute*

Abstract: This paper introduces a new system to calculate tank capacity based on 3D CAD modeling and then to get capacity forms. The whole system platform and framework are introduced, and some key technologies are also described in detailed.

Key words: 3D modeling, Capacity Calculation, Tank Calibration

1. INTRODUCTION

Tank calibration is used to calibrate a new built or a rebuilt ship for liquid cargo trades. Tanks have double kinds of functions, on the one side, itself a storage, where a liquid cargo can be stored inside; on the other side, a meter of measurement, by checking a special document for conversing the current measured height to a volume of oil. In Australia, China, United Kingdom and some other countries, these tanks should be calibrated by an organization or company certified by the national organization of testing authorities or by the local state government trade measurement department. Obviously, the accuracy of tank calibration is important. According to the calibration regulation in China, uncertainty for oil tanker should be less than plus or minus 0.3%. According to OIML Recommendation[1]: The relative error of the volumes stated in the document issued do not exceed: plus or minus 0.5 % of the stated volume for tanks of irregular shape that cannot be calibrated using the volumetric method.

In fact, there are a various tank layout, a different cross-section, and complex inside components among tankers. Generally, tanks near tanker's

Please use the following format when citing this chapter:
Zhao, Cuilian, Fang, Minglun, Tian, Zhongxu, Lin, Yongsui, 2006, in International Federation for Information Processing (IFIP), Volume 207, Knowledge Enterprise: Intelligent Strategies In Product Design, Manufacturing, and Management, eds. K. Wang, Kovacs G., Wozny M., Fang M., (Boston: Springer), pp. 697-703.

fore are irregular, but in most of rebuilt tanks and some new built ship, there are more and more irregular tanks because of the trend to get the maximum of tank capacity. The question is how to develop a common system, which is suitable for most types of ships' situation, easy to use, make a better calibration' accuracy and take less time. With the technical development in the field of Compute Aided Design (CAD), its open platform can make full use of powerful functions. CAD is strong in describing 3D structure; its advantages in free-form feature modeling, visualization and simulation are widely accepted.

So, in this paper, a new method to calculate the irregular tank capacity faster and more accurately will be introduced. The whole system framework will be described in detailed, including some key technologies: the modification of theoretical tanker model, the way to create full oil model and etc. But in our research two following factors aren't considered:

- temperature differences between the time of calibration and the practical use.
- Pressure and deformation difference between the empty tank in calibration and the full loaded tank in the practical situation.

2. TANK CAPACITY CALCULATION SYSTEM

The architecture of the proposed computer-aided capacity calculation system for tank calibration is developed as shown in Figure1. Generally, the calibration operation will begin from analyzing tanker's design drawing and then measuring the tanker on site. The physical measurement can be done by a classical method - a means of tape measures, or by an optical method. So there are two kinds of input information, one is the design data, the other is the measured data. Hence, there are two data flows, which is necessary to be unified into the practical model from the model based on the design data.

Main parameters include ship's name and width, cross-section type, tanks' layout type, bottom's type, deck's type and etc. Generally, in a drawing ribs' number is used to locate the position of ship's structure, which should be transformed to absolute coordinate system in order to fulfill CAD modeling.

The oil ship is classified to three different types according to its cross-section and four types according to Bulkhead layout after we analysis a lot of ship drawings. The type of cross-section of tanker: curve type, fold type and user-defined type (a special cross-section, not included in the another two types), which decides automatically modeling function's callout for tanker's hull. Modeling with Curve type need use three degree of NURBS free-form feature modeling. Modeling with Fold type use one degree's modeling.

According to Bulkhead layout's character, some named ways for bulkhead and longitudinal bulkhead are gives.

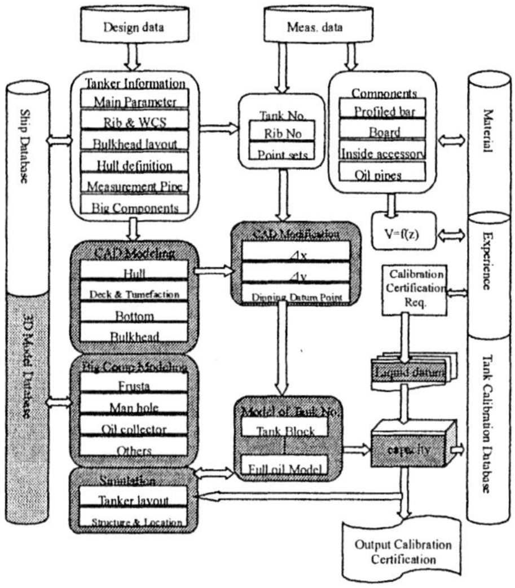

Figure 1. Architecture of 3D modeling and capacity calculation system

The above information, including 3D basic models and the expert experience, is managed by a database. We use Visual Basic 6.0 to develop the user interface, and use ACCESS 2000 as the system database, which can be accessed by a CAD open platform - UG/OPEN[2] through the engine of ODBC.

Using CAD to do modeling for the hull, bulkhead, deck and some components (note: not all of components), which we consider is serious to influence the accuracy of capacity calculation. Assembly structure of a tanker will be automatically created by the developed programs. The hull and deck will be created first. With WAVE geometry linker, the hull is associated with all of other CAD components.

The output format is stipulated by the national organization of testing authorities. Automatically generate tank capacity form by the template of XLS format with the content of Ullage height, Innage height, tinl, and etc.

Figure 2 gives an example of a tank partial capacity form. Since all of data can be accessed by the tanker's database, it is easier to get a tank' conversion from a measurement height to the current capacity .

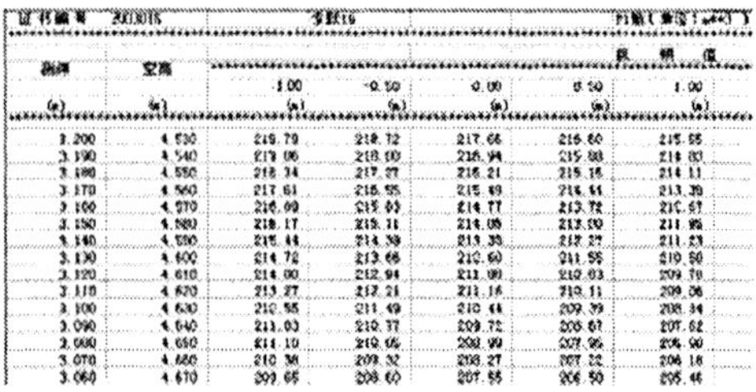

测深	空高	纵倾值				
		-1.00	-0.50	0.00	0.50	1.00
(m)	(m)	(m)	(m)	(m)	(m)	(m)
3.200	4.530	219.79	218.72	217.66	216.60	215.55
3.190	4.540	219.06	218.00	216.94	215.88	214.83
3.180	4.550	218.34	217.27	216.21	215.16	214.11
3.170	4.560	217.61	216.55	215.49	214.44	213.39
3.160	4.570	216.89	215.83	214.77	213.72	212.67
3.150	4.580	216.17	215.11	214.05	213.00	211.95
3.140	4.590	215.44	214.39	213.33	212.27	211.23
3.130	4.600	214.72	213.66	212.60	211.55	210.50
3.120	4.610	214.00	212.94	211.88	210.83	209.78
3.110	4.620	213.27	212.21	211.16	210.11	209.06
3.100	4.630	212.55	211.49	210.44	209.39	208.34
3.090	4.640	211.83	210.77	209.72	208.67	207.62
3.080	4.650	211.10	210.05	208.99	207.95	206.90
3.070	4.660	210.38	209.32	208.27	207.22	206.18
3.060	4.670	209.65	208.60	207.55	206.50	205.46

Figure 2. an example of a tank partial capacity form

3. MODIFICATION OF TANKER MODEL

Since all of data come from the ship design drawings, the tanker model is called theoretical model. With the developed programs, the modeling can be done automatically. in Figure 3 shows the assembly tree of a tanker and simulation of whole layout[3]. All the modeling based on the input data is about the shape of left side. Reference tool is used to transform the view. So it is easy to simulate the tanks' layout in a tanker, to check every component's data and to inquire the information of some key points and to know if the shape of model is right[4].

However, the above model is only based on the design data. Tank Measured data is based on the tank's bottom, which is a set of points, represented by Xp,Yp, Zpb. Xp is got from rib's number and offset from it, where

Xp=Fn+N.

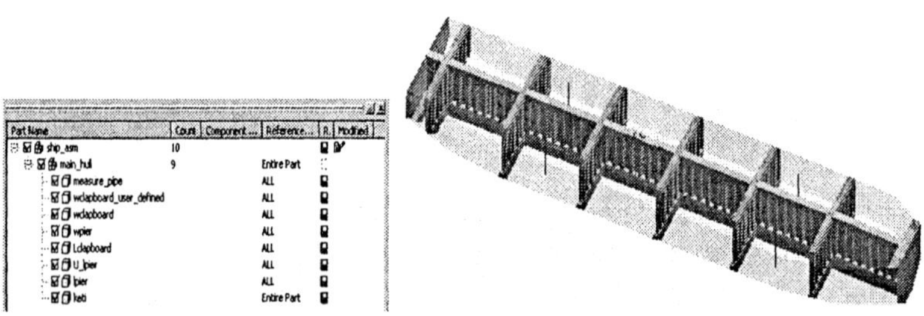

Figure 3. the assembly tree and 3D modeling of a tanker

Zpb is based on the bottom of tank. Firstly, need to get each tank' theoretical base point, which is the lowest point (Z direction) in a tank. Then transform the measured data of each tank into absolute coordinate system,

Zp. The free-form surface of hull, which has been gotten by modeling of the design data, is interpolated by the following:

Yt=F(Xp,Zp)

Yt represents the theoretical value.

So, we can get the modification:

$$\Delta x=\Sigma(Yp-Yt)/n$$

$$\Delta y=\Sigma(Xp-Xt)/m$$

where, m, n represent the number of measured points. Similarly, there is the third important datum point in each tank – Dipping Datum Point, which decides the liquid alltitute along the vertical measurement axis. Interpolate the ship hull with each location of measurement pipe:

Zj=F(Xj,Yj)

where, Xj,Yj represents a measurement pipe location in a tank.

4. FULL OIL MODEL AND TANK CAPACITY CALCULATION

The full oil model of a tank means the oil full of the space enveloped by six sides which are free-form features or regular form features. Also the volumes of inside big components are deducted from the full oil model, such like a man hole, strong crossbeam, and etc. We made modeling of a block big enough. The six sides will cut the block which like a rough material being manufactured. Figure 4 shows the method. And then the big components will be as tools to cut the rough material continually. Using this method avoids the strict requirement for calculating a volume enveloped by sheets from CAD software. Finally the result is the full oil model in the tank.

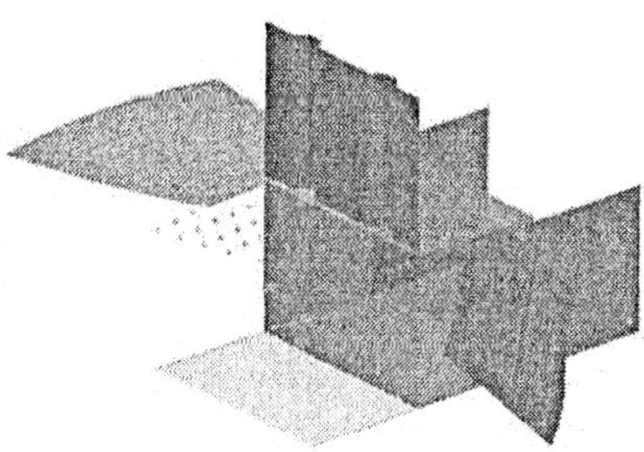

Figure 4. using a block to cut rough material using

Simulating the model, the specialist can check each tank's 3D model again. Thus, the mistake can be avoided in advance.

After each of 3D model with full oil is ready, a liquid plane, which pass the vertical point offset from the datum point, will be created to cut off the full oil model, the left is a new model in the current measurement height and tilt. The tilt represents the difference between the height of bow and stern. The new model volume is generated by CAD volume calculation function.

For example, according to the requirement of calibration certification, the interval of heights is 10 mm, and the interval of tilts is 0.5m. Hence, there will be (6000/10)*12=7200 times of volume calculation only for a tank with 6M high, this is tough thing by hand. But easier to use computerizable programs.

5. CERTIFICATE THE ACCURACY OF A TANK CAPACITY CALCULATION BASED ON CAD MODEL

According to UG document, for simple solids with analytic faces (i.e., planar, cylindrical), the default accuracy value is sufficient. For more complicated solids with B-Surface faces, accuracy values closer to 1.0 may be necessary. The maximum accuracy value used by the system is 0.999999 (UG ref). As calibration requirement the accuracy should reach less than 0.3%, according to the above, its accuracy seems to be satisfied. However, it should be certified. Since no code of CAD system is openly released. We need to prove the result of using 3D model calculation by mathematical theory.

Table 1. capacity results based CAD model and mathematical model

Altitude (mm)	UG Volume	Math Result	Absolute Error	Relative Error
268	0.5323803	0.532639	0.000259	0.0005
735	1.561296	1.560722	-0.00057	0.0004
738	1.568156	1.567582	-0.00057	0.0004
825	1.768082	1.76752	-0.00056	0.0003
1337	3.011928	3.01129	-0.00064	0.0002
1340	3.019673	3.019045	-0.00063	0.0002
1341	3.022256	3.021909	-0.00035	0.0001
1417	3.220681	3.220838	0.000157	0.0000
1420	3.228597	3.228728	0.000131	0.0000
1456	3.323487	3.323766	0.000279	0.0001
1459	3.331465	3.331715	0.00025	0.0001

A mathematical model of the simplest tank is created to make comparison with UG result. We create mathematical model and CAD model for computing a test tank, which is a standard tank by volumetrically calibration. Using grid with 2500 x 2500, the result is satisfied. See the table 1. This proves the higher accuracy using CAD model to calculate volume. So tank capacity calculation can be based on CAD model.

6. SUMMARY

The overall of framework of tank capacity calculation has been developed and introduced in this paper. The objective of the research is to provide a computerized tool for rapid and accurate calibration of tank capacity with simulation of 3D modeling. In the system, tank information is organized by the tank layout and cross section, tanker modeling based assembly makes the simulation for whole tanker or a special tank easily. And the modification for a practical manufacturing error can be achieved through interpolating the theoretical free-form features in CAD model. Finally, the result of tank capacity can be exported into XLS according to the regular templates. This system has already been used by Shanghai Merchant Ship Design & Research Institute. The result proves it is accuracy enough to using CAD model for tank's capacity.

7. ACKNOWLEDGEMENTS

This paper is partially supported by Shanghai Merchant Ship Design & Research Institute, authors would like to thank Yonggang, Lu, Qinwen Sun, and two graduate students from Shanghai University, Hong Zhang, Hualong Yang for their contribution in the system development.

8. REFERENCES

1. http://www.oiml.org First Committee Draft Revision OIML International Recommendation R71 21 – 22 April 2004, Vienna, Austria
2. *V18.0 Unigraphics Help*
3. Zhang Hong, Zhao Cuilian, Yang Hualong, Lin Yongsui, Research on Ship Component Management Based on Model Structure Tree , *Computer Aided Engineering*, 2004.9
4. Zhao Cuilian, Yang Hualong, Zhang Hong, Lin Yongsui, 3D Modeling and Simulation of Cabin Based on Assembly Environment, *CAD/CAM*, 2004.5

STUDY ON 3D SOLID RECONSTRUCTION FROM 2D VIEWS BASED ON INTELLIGENT UNDERSTANDING OF MECHANICAL ENGINEERING DRAWINGS

Jianping Liu[1,2], Bangyan Ye[1], Xiaohong Wu[3], Miaoan Ouyang[2]
[1]College of Mechanical Engineering, South China Univesity of Technology, Guangzhou, 510640, China; Email: ljp68@tom.com. [2] Department of Mechatronics, Foshan Polytechnic College, Foshan, 528237, China; [3] College of Information, South China Agricultural Univesity, Guangzhou, 510640, China.

Abstract: This study presents a new algorithm of 3D reconstruction to obtain automatically 3D solid from mechanical engineering drawings. Main steps of the algorithm are as follows: first, reconstruct multiple views from dxf-type files; second, recognize primitives feature and combined relationship between primitives by engineering semantics understanding and loops searching on views; third, partition and reconstruct body primitives; last, all reconstructed primitives are assembled into a goal solid by Boolean operation. The algorithm adopts integrated reconstruction techniques to enlarge the figuration domain of reconstructed objects. And it uses case-based reasoning to set up a self-learning mechanism of feature recognition. It is implemented at the desktop of AutoCAD 2004 by using Object ARX 2002 and VC++6.0. The results show that utility scope and reconstruction speed are improved when using the new algorithm.

Key words: 3D reconstruction; mechanical engineering drawings; primitive; feature recognition; engineering semantics understanding.

1. INTRODUCTION

3D reconstruction from mechanical engineering drawings is a method of automatic constructing 3D solid objects. It is the pivot part of the system converting 2D paper-based engineering drawings into 3D CAD model files. It can provide 3D CAD model for application of advanced manufacturing

This project is supported by the Key Item of Science and Technology of Guangdong, China under the grant No.2002A1020103 and No.2005C1030038.

Please use the following format when citing this chapter:
Liu, Jianping, Ye, Bangyan, Wu, Xiaohong, Ouyang, Miaoan, 2006, in International ederation for Information Processing (IFIP), Volume 207, Knowledge Enterprise: Intelligent Strategies In Product Design, Manufacturing, and Management, eds. K. Wang, Kovacs G., Wozny M., Fang M., (Boston: Springer), pp. 704-709.

technology such as CAM, CAPP, CAE, PDM, CIMS, VM, RP&M, *etc.* Its application can not only save tremendous time and labor but also broaden the domain of computer vision from perspective images understanding to projective graphics understanding.

Idesawa first presented this issue in his paper "A system to generate a solid figure from three view" in 1973 [1]. Various algorithms starting from three views have been published in the literature since then. Among them, boundary representations (B-rep), constructive solid geometry (CSG) and the approach-based algorithm are typical algorithms and most of other algorithms are improvements of the typical ones [2]. Main steps of the B-Rep algorithm [3] are as follows: a wire-frame that is composed of 3D vertices and 3D edges is first recognized from the input orthographic views. Then, candidate faces are found from all these 3D edges. Pseudo elements are deleted by certain criteria and all true faces are assembled to form a final solid. The CSG oriented approach [4] assumes that each 3D object can be built from certain primitives in a hierarchical manner. It selects a 2D loop as a base and generates the 3D primitives by matching their corresponding 2D patterns in each view, then assembles reconstructed primitives to form a goal solid. The approach-based algorithm [5] is similar to cutting mechanical part. Its main steps are as fellows: First, a reference solid is generated by translational–sweeping and Boolean intersection operation. Then, the difference solid is reconstructed by comparing the views between the reference solid and the result solid. Finally, by subtracting the difference solid from the reference solid one by one and testing the residual solid by views, the goal solid can be finally obtained. Although 3D reconstruction from 2D orthographic views has been studied for over three decades, the present algorithms are still far from the desired goal of a general solution and practical application [6]. The main problems of geometric reconstruction are as follows: 1) Reconstructed objects are usually three orthographic views, where there are no centerlines, no dimension, no cutaway views, no sectional views and no assistant views and so on. 2) For B-Rep, its reconstruction efficiency is low and it is easy to generate pathological cases; For CSG, it is difficult to deal with complex figuration because limited pre-defined primitives patterns restrict its applicability; For the approach-based algorithm, there are a great deal of operations and complicated calculation in 3D space. 3) Using single reconstruction technique leads to limited figuration domain of reconstructed objects. 4) Present algorithms of primitive feature recognition are difficult to deal with intersection and complex primitive features.

Researchers think limits of traditional algorithms based on geometric reconstruction cause the less progress on 3D reconstruction research. In order to solve existing problems and seek a new breakthrough, they attempt to use the technique of engineering semantics understanding to recognize primitive feature. In 1984, Yashiura [7] first uses the information conveyed by dimension annotation to recognize primitives. In 1992, Joseph [8] makes feature recognition easier by analyzing the semantics of sectional views. In China, Lu [9] first uses the semantics information conveyed by semi-section or full section views to reconstruct 3D solid. In order to technique practicability of 3D reconstruction, we take practical mechanical engineering drawings which contain section views, local views and have dimensions as reconstructed objects and present a new algorithm based on engineering semantics understanding.

2. NEW ALGORITHM OF 3D RECONSTRUCTION

The key problem of engineering drawings understanding is how to recovery the losing information of geometry and topology at the projective depth direction. Experienced engineers often understand projections by the method of body analysis. Main steps of the method are as follows: 1) A complex body is divided into parts by decomposing view-loops. 2) Corresponding loops selected from three views are validated their coherence according to rules of orthographic projection; 3) Corresponding 3D model are generated from typical patterns according to view loops and multiple interpretations are eliminated by views testing.

2.1 Main idea and whole framework of the new algorithm

This paper simulates the method of body analysis and presents a new algorithm of 3D reconstruction. Reconstruction process of a goal solid is similar to that of machine assembly. Figure 1 shows the whole framework of the new algorithm. The main steps are as follows: First, multiple views are separated and graphics feature information is extracted from dxf-type files. Second, engineering semantics and possible a *priori* knowledge are used to recognize components or primitives and the combined relationship between those. Third, sweeping representation is firstly used for components reconstruction. If unsuccessful, the approach algorithm is then arranged for 3D reconstruction. Provided that the two methods are not fit for this component, the component can be reconstructed by program module A based on primitives decomposition. Fourth, each of decomposed primitives is supposed to be among of elementary feature, sweeping feature, volume-cutting feature and boundary feature. And it is reconstructed by using a corresponding algorithm in order of precedence, i.e. CSG → sweeping representation→the approach algorithm→B-Rep. Every reconstructed 3D component or primitive is validated by its projective views. Finally, all reconstructed components and primitives are assembled into a goal solid by Boolean operation and 3D editing operations.

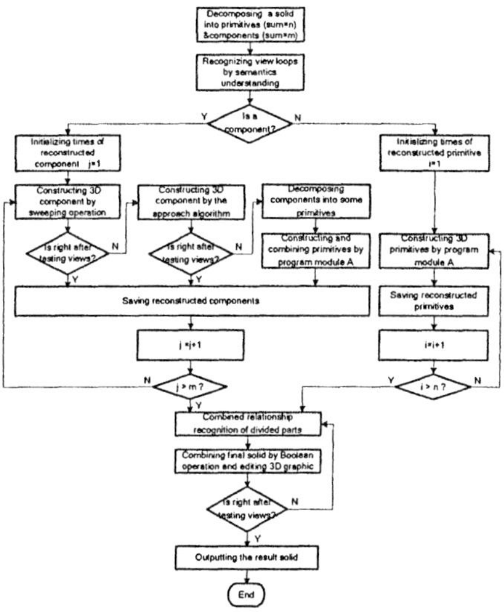

Figure 1. Whole framework of the new algorithm

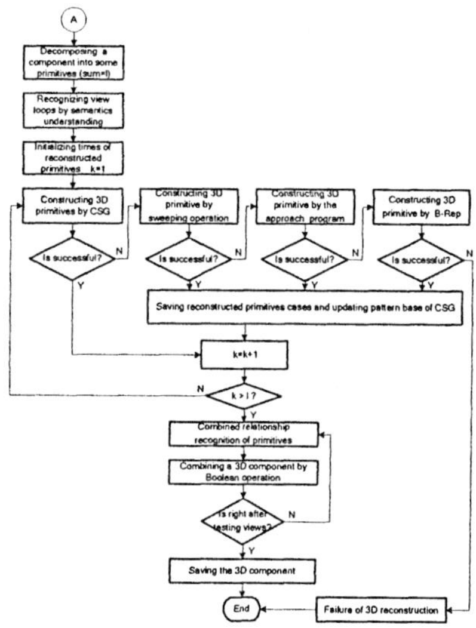

Figure 2. Framework graph of part reconstruction

2.2 Partition and reconstruction of components /primitives

In order to partition components and primitives, we take full advantage of the semantics of engineering drawings to accelerate primitive recognition. According to the process of the projection, the semantics are divided into expression mode semantics, projection relation semantics and dimension constraint semantics. Furthermore, every level is classified to explicit semantics and hidden semantics. Expression semantics understanding is the precondition of engineering drawings understanding; projection semantics understanding is the key process; understanding of dimension constraint semantics plays an assistant role. For example, the modeling information of sweeping-feature primitive, such as sectional loops and the sweeping height or path, is recognized through analyzing the explicit semantics and digging the hidden semantics. When multiple interpretations exist, views testing and visual reasoning are activated to pick out the most acceptable interpretation. Sectional views could also be involved to eliminate the unsuitable interpretations of projection views. From Fig.2, we can see that the new algorithm sets up a mechanism of supposing and ascertaining to reconstruct parts of the goal solid. It uses integrated reconstruction techniques. Different kinds of reconstruction algorithms are arranged to reconstruct different feature primitives according to their reconstruction efficiency from high to low. CSG is first used to reconstruct 3D primitives due to its highest reconstruction efficiency. But limited pre-defined primitive patterns restrict its applicability. So, every new reconstructed primitive is saved as a new pattern to enlarge the pattern base of CSG for feature recognition.

3. IMPLEMENTATION

The combination of ObjectARX 2002 and VC++6.0 is used to implement the new algorithm at the platform of AutoCAD 2004. Fig.3 shows the reconstructed solid of a mechanical part and its reconstructed parts. From it, we can see that it is divided into two components and two primitives by engineering semantics understanding. Fig.4 illustrates the reconstruction process of components and primitives reconstructed by different algorithms.

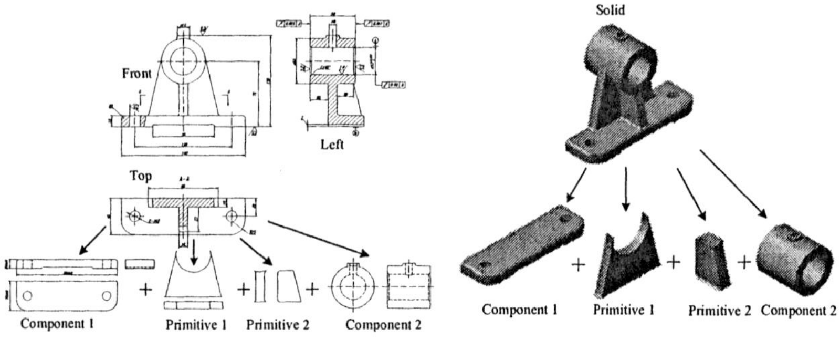

Figure 3. Original engineering drawings, its processed views of decomposed parts and reconstructed solid

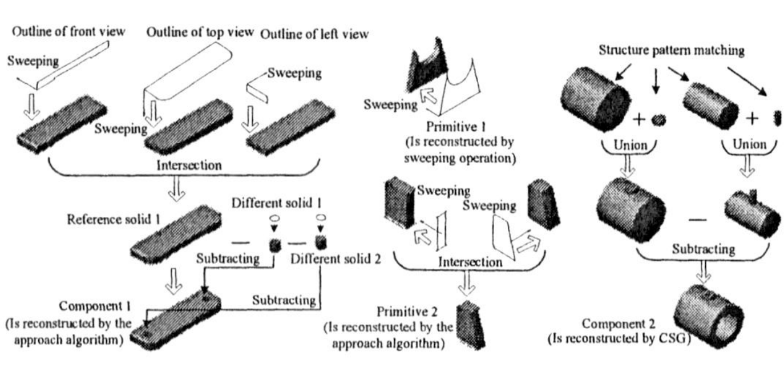

Figure 4. Diagram of reconstruction process of components and primitives

4. CONCLUSION

In order to enhance the practicability of 3D reconstruction, this paper presents a new algorithm based on intelligent understanding engineering drawings. The characteristic of the algorithm is as follows: 1) complex body is divided into different kinds of components and primitives and engineering semantics is used to recognizing them and their combined relationships; 2) hypothesis and views testing are adopted to reconstruct and to validate the result solid; 3) integrated reconstruction techniques are used to enlarge the figuration domain of reconstructed objects; 4) case-based reasoning is used to establish a self-learning mechanism of feature recognition.

5. REFERENCES

1. Idesawa, M., (1973), A Systems to Generate a Solid Figure from Three View, *Bulletin Of the JSME*, Vol. 16(92), pp. 216-225.
2. Nagendra, I.V., and Gujar, V.G., (1988), 3D Objects from 2D Orthographic Views—A Survey, *Computers & Graphics*, Vol. 12(1), pp. 111-114.
3. Preiss, K., (1984), Constructing the a Solid Representation from Engineering Projections, *Computer and Graphics*, Vol.8, pp. 381-389.
4. Aldefeld, B., (1983), On Automatic Recognition of 3-D Structures from 2D Representations, *Computer-Aided Design*, Vol. 15(2), pp. 59-64.
5. Chen, Z., and Perng, D.B., (1988), Automatic Reconstruction of 3D Solid Objects from 2D Orthographic Views, *Pattern Recognition*, Vol. 21(5), pp. 439-449.
6. Wang, W.D., and Grinstein, G.G., (1993), A Survey of 3D Solid Reconstruction from 2D Projection Line Drawings, *Computer Graphics Forum*, Vol. 12(2), pp.137-158.
7. Yoshiura, H., Fujimura, K. and Kunii, T.L., (1984), Top-down Construction of 3-D Mechanical Object Shape from Engineering Drawing, *IEEE Computer Magazine*, Vol.(12), pp. 32-40.
8. Joseph, S.H., and Pridmore, T.P., (1992), *Knowledge-Directed Interpretation of Mechanical Engineering, IEEE Trans Pattern Analysis & Machine Intelligence*, Vol. 14(9), pp. 928-940.
9. Lu, G.D., and Peng, Q.S., (2000), *Research on Primitive Relation Recognition Based on the Engineering Drawing Semantics, Journal of Computer Aided Design and Computer Graphics*, Vol. 12(9), pp. 700-704 (in Chinese).

BONE TISSUE ENGINEERING USING Β-TRICALCIUM PHOSPHATE SCAFFOLDS FABRICATED VIA SELECTIVE LASER SINTERING

Baigong Ma, Liulan Lin, Xianxu Huang, Qingxi Hu, Minglun Fang
Rapid Manufacturing Engineering Center Shanghai University, Shanghai University, China;Email:huqingxi@mail.shu.edu.cn

Abstract: β - tricalcium phosphate (β -TCP) is a biodegradable ceramic with potential application for bone replacement. In this work, porous β -TCP scaffolds were fabricated via selective laser sintering, which is a rapid prototyping technique. β -TCP and glue composite was developed as a powder material to product porous scaffolds, using this computer-controlled sintering and deposition process. The β -TCP scaffolds were produced with the strut size 800 μm and max porosity 55.4%. Compressive modulus ranged from 13MPa to 18Mpa, and shrinkage was about 26%. Analysis of the measured data shows a high correlation between the scaffold porosity and the compressive properties based on SLS process relationship.

Key words: bone tissue engineering, rapid prototyping, β - tricalcium phosphate, bionic scaffolds, porosity

1. INTRODUCTION

Scaffold-guided tissue engineering (TE) has been developed to regenerate specific and functional human tissues or organs [1, 2]. In view of bone tissue engineering, the bone substitute must have proper pore and porosity for bone repair so as to accelerate bone regeneration [3]. TE scaffolds should facilitate the colonization of cells and possess characteristics that enhance cell attachment, proliferation, migration and expression of native phenotypes. Scaffold characteristics and properties such as porosity, surface area to volume ratio, pore size, pore interconnectivity, structural strength, shape (or overall geometry) and biocompatibility are often considered to be critical factors in their design and fabrication [4]. It may be best achieved by

This project is supported by the Shanghai Education Foundation of China under the grant No. 05A281.

Please use the following format when citing this chapter:
Ma, Baigong, Lin, Liulan, Huang, Xianxu, Hu, Qingxi, Fang, Minglun, 2006, in nternational Federation for Information Processing (IFIP), Volume 207, Knowledge Enterprise: Intelligent Strategies In Product Design, Manufacturing, and Management, eds. K. Wang, Kovacs G., Wozny M., Fang M., (Boston: Springer), pp. 710-716.

combining image-based computational design techniques and solid free-form fabrication (SFF) methods.

Selective laser sintering (SLS) is one of SFF method. SLS constructs scaffolds from 3-D digital data by sequentially fusing regions in a powder bed, layer by layer, and then the laser beam is selectively scanned over the powder, causes the particles to be fused together to form a solid mass [5, 6]. SLS is a well-adapted technology, this makes it fast and cost effective to fabricate tissue engineering scaffolds [7, 8]. We have employed SLS to create β-TCP scaffolds.

β- TCP is an important biodegradable ceramic material and has been subjected to intensive studies, which has potential application for bone replacement. But this material cannot be sintered directly by the laser, here epoxy and nylon was used binder. The work presented here is part of results on the manufacture and characterization of β-TCP scaffolds by SLS.

First, various types of structures with different mechanical properties were constructed with a computer-aided design. Then β-TCP matrix was produced on a selective laser sintering technique. After high temperature treatment, the ceramic scaffolds were obtained with designed microporosity. The result of β-TCP scaffolds with 800 μm pore diameter and fully interconnected pore morphology were discussed.

2. MATERIALS AND FABRICATION

2.1 Material preparation

The bioceramic can be degraded in different degrees under the biological environment and be absorbed by organization. β-TCP powders used in this work were from shanghai tissue engineering research center. Its average particle size was 100μm. IT was certified that β -TCP scaffolds repair defective bone very well with compounding bone marrow matrix stem cell.

The mixture powders of epoxy resin and nylon were used as the adhesion of β-TCP powder. Epoxy resin and nylon's glass transition temperature is respectively about 65-75℃ and 110-120℃. Particle average size of the blending powders is 200μm.

2.2 Scaffold design

Rectangular block porous scaffolds (21mm length, 21 mm width, 5mm height), with three-dimensional orthogonal periodic porous architectures, 800μm in strut diameter, were designed using Unigraphics NX3 solid modeling software (UGS PLM Solutions, Plano, TX). An example was shown in Figure1.

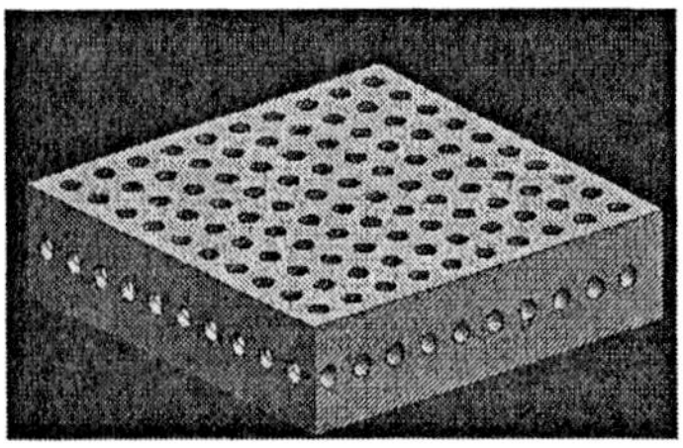

Fig. 1. Virtual structure designed with CAD

2.3 Scaffold fabrication

The settings for the SLS process parameters were kept at their default values except for laser power and scanning speed. Laser power was set at 10W. Then scaffolds were built layer-by-layer using a powder layer thickness of 0.1mm. It was found that part was constructed well without preheating. Different ratio mixture was subjected to laser sintering at different laser power for the optimal setting. By varying the process parameter, three groups of specimens comprising of different ratio of β-TCP/ epoxy resin and nylon were tested, the ratio was 2:1:1, 5:3:2 and 5:2:3. The scaffold surface quality of ratio 5:2:3 was the best, and its strength was more advanced. The specimen is shown in Figure 2(a).

Then this specimen was put into a stove (SX2-1013, Zufa Ltd. China). During the high temperature treatment, the temperature should rise step by step. It may lead to thermal expansion mismatch and a breaking structure if temperature was changed so fast. It took about 40min to elevate the temperature from room temperature to 200℃. Then it took 800min to achieve 600℃. Above 600 ℃, epoxy resin and nylon were completely burnt and the sintering temperature (1100 ℃) could be reached during 250min. Then kept the temperature for 3 hours. A piece of the scaffold treated by high temperature was shown as figure2 (b).

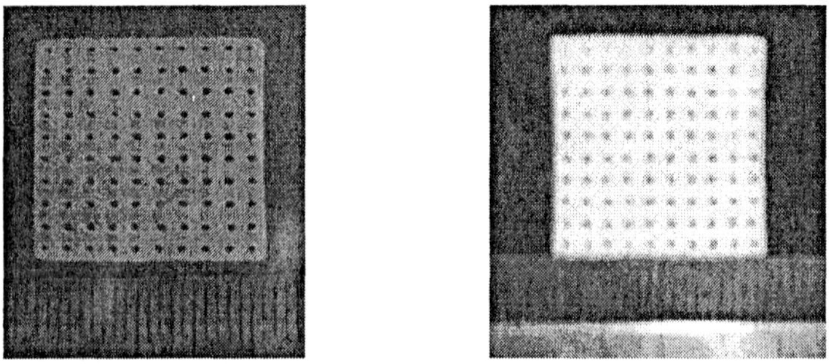

a) before high temperature treatment b) after high temperature treatment

Fig.2 the scaffold of laser power 10w with ratio 5:2:3, each grid under the scaffold was 1 mm

3. RESULTS

3.1 Porosity and volume density of sintered specimen

The theoretical porosity P (%) was estimated based on the ratio, which the volume of the pores in the scaffold divided the whole volume of the scaffold. According to static equilibrium method in liquid each porosity value was determined as:

$$P=(V_1-V_3)/(V_2-V_3)\times 100\% \qquad (1)$$

Where P (%) was the porosity ratio of specimen, V_1 (ml) was the volume of ethanol in the measuring cylinder, V_2 (ml) was the volume of ethanol after the specimen saturated by the ethanol, V_3 (ml) was the volume of ethanol after taking out the specimen.

Table 1 the porosity of scaffold

Parameters	Specimen 1	Specimen 2	Specimen 3
W (g)	0.2	0.2	0.1
V_1 (cm^3)	38.25	13.92	20.00
V_2 (cm^3)	33.90	9.71	16.00
V_3 (cm^3)	33.82	9.60	15.90
P (%)	55.375	39.27	41.00

From table 1 it was found that the porosity of scaffold ranged from 39.27% to 55.375% with laser power changing. It is to say, more higher porosity can be arrived with changing the process characters.

3.2 Microstructure analysis

Microstructure characterization of the SLS-fabricated β-TCP scaffold specimens were carried out using a scanning electron microscope (SEM, JEOL JSM-6700F, Japan) to analyze the surface morphology and microstructure of the sintered specimen.

Figure 3(a) shows that the bigger pores of the β-TCP specimen are surrounded by innumerable micro-pores. And the diameter of the macro-pores is about 700μm, it is designed by CAD; the diameter of the micro-pores is ranging from 1 to 200μm. The internal pores are mainly distributed uniform. The specimen is porosity and the pores are interconnectivity.

Figure 3(b) describes the internal structure of the sintered specimen after high temperature treatment. The crystalloid particles have nearly consistent size, and these particles conglutinate with each other, which form a near-net-shape. The near-net-shape has advantages for porosity and interconnectivity. The diameter is smaller than the diameter without high temperature treatment, its average size is about 20 to 100μm.

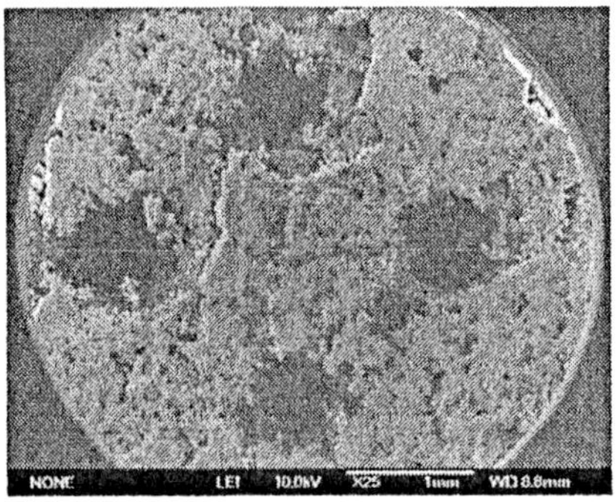

a before high temperature treatment

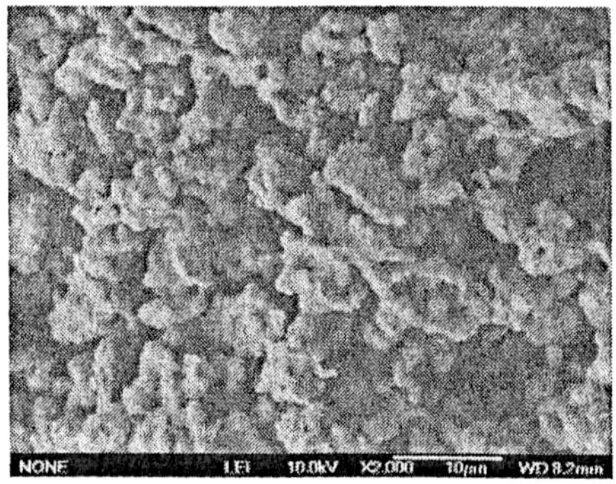

b after high temperature treatment

Fig. 3 SEM micrograph of scaffold

3.3 Shrinkage assessment

The conception of multihole bioceramic shrinkage was the ratio of volume change after sintering under high temperature to original volume. The relationship can be derived as:

$$S=(V_1-V_2)/V_1\times 100\% \qquad (2)$$

Where S is the shrinkage of the scaffold, V_1 (mm^3) is the volume of original specimen, and V_2 (mm^3) is the volume of specimen after high temperature treatment.

Table 2 the shrinkage of scaffold

High temperature treatment		Specimen 1	Specimen 2	Specimen 3
Before	L (mm)	20	20	20
	W (mm)	20	20	20
	H (mm)	2.5	4	2
	V (mm^3)	1000	1600	800
After sinter	L (mm)	17.78	17.88	18.08
	W (mm)	17.56	17.56	18.74
	H (mm)	2.36	3.8	1.96
	V (mm^3)	736.83	1193.10	664.09
S (%)		26.32	25.43	16.99

As the Table 2 shows that the shrinkage of the specimens is ranging from 16.99% to 26.32%. While shrinkage usually occurred in the debinding step due to the collapse of voids created in the part as the polymer binders burned out. As a result, the degree of shrinkage was mainly determined by the amount of binder used. However shrinkage is a very complicated phenomenon dependent on a variety of parameters. The particle diameter has a strong effect on the shrinkage. The smaller the particles, the greater the shrinkage. Sintering time and sintering temperature are also important parameters. And degree of shrinkage is also dependent on the packing density, the particle shape and particle size distribution.

4. CELL SEEDING

The work of cell seeding was carried on Shanghai tissue engineering research center. Several specimens with an orthogonal interconnected pore design were manufactured as described previously and seeded with cells (figure 5). Firstly the specimens were moistened in ethanol. After the sterilization in an autoclave, the specimens were washed with Phosphate Buffered Solution (PBS). Then the specimens were cultured in Dulbecco's Modified Eagle Medium (DMEM) for one night. After the specimens sipped up, they were seeded with the bone mesenchymal cells and cultured in DMEM. The density of the cells is 2-5 million/ml. Cells were cultured in an incubator at 37℃ under humid atmosphere in 5% CO_2.Culture medium was changed every 3 days [10].

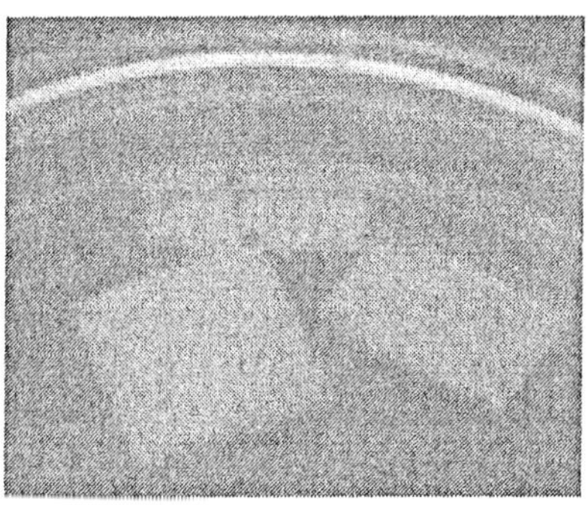

Fig.4 Scaffolds seeded with cells

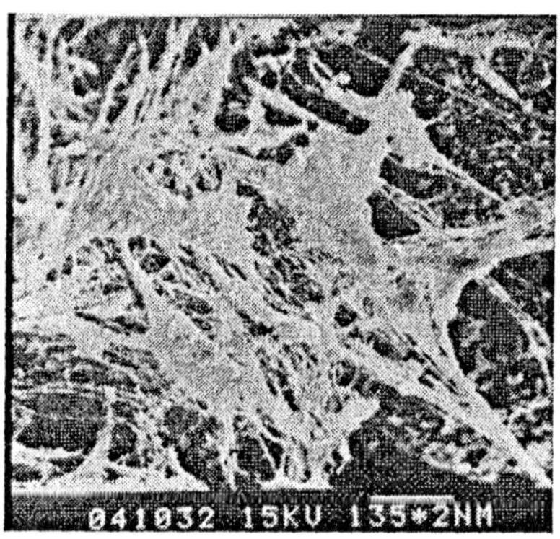

Fig.5 SEM micrograph of scaffold with attached bone mesenchymal cells

Figure 4 shows the scaffolds seeded with cells are cultured in DMEM. Figure 6 is the SEM micrograph of a scaffold specimen, which was covered with bone mesenchymal cells. Figure 5 shows that bone mesenchymal cells adhere and feel well on the specimen, that is to say, the scaffolds fabricated on SLS work well with cell.

5. CONCLUSION

The advantages of using SLS for bone tissue engineering scaffolds lie in the ability to control pore structure for biogenesis through the control of polymer contents and the ability to construct complex three-dimensional structure for bone tissue engineering applications. β -TCP scaffolds fabricated via SLS show great potential for replacement of bone tissues. The porosity of the specimens in our experiment achieved to 55%. It's already content to the requirement of the bone tissue engineering. And with the growth of the new bone tissue, β-TCP scaffold, which was embedded in human body, will be biodegraded. While the shrinkage that ranging from 16.99% to 26.32% is high, so we have to offset these shrinkage in CAD modeling. As the cause of the shrinkage is complex, so it's a difficult point in our following work. In vivo results show that SLS-fabricated β-TCP scaffolds are propitious to the growth of bone mesenchymal cells.

6. REFERENCES

1. Patrick Jr CW, Mikos AG, Mcintire LV. Prospectus of tissue engineering. In: Patrick Jr CW, Mikos AG, Mcintire LV, editors. *Frontiers in Tissue Engineering*, New York: Pergamon, 1998; 3 - 5.
2. Hutmacher DW. Scaffolds in tissue engineering bone and cartilage, *Biomaterials*, 2000;21:2529 - 43.
3. Zhongzhong Chen, Dichen Li, Bingheng Lu, Yiping Tang, Minglin Sun, Songfeng Xu. Fabrication of osteo-structure analogous scaffolds via fused deposition modeling, *Scripta Materialia*, 2005;52:157 - 161.
4. K.H. Tan, C.K. Chua, K.F. Leong, C.M. Cheah, P. Cheang, M.S. Abu Bakar, S.W. Cha. Scaffold development using selective laser sintering of polyetheretherketone - hydroxyapatite biocomposite blends. *Biomaterials*, 2003;24:3115 - 3123.
5. Deckard CR. Part Generation by Layerwise Selective Sintering, MS thesis, Department of Mechanical Engineering, University of Texas at Austin; 1986.
6. Deckard, CR. Selective Laser Sintering, PhD dissertation, Department of Mechanical Engineering, University of Texas at Austin; 1988.
7. Das S, Hollister SJ, Flanagan C, Adewunmi A, Bark K, Chen C, Ramaswamy K, Rose D, Widjaja E. Freeform fabrication of nylon-6 tissue engineering scaffolds. *Rapid Prototyping* 2003;9(1):43– 9.
8. Tan KH, Chua CK, Leong KF, Cheah CM, Cheang P, Abu Bakar MS, Cha SW. Scaffold development using selective laser sintering of polyetheretherketone-hydroxyapatite biocomposite blends. *Biomaterials*, 2003;24(18):3115– 23.
9. Ernst Magnus Noah, Jingsong. Chen, Xiangyang Jiao, and Ingo Heschel, Impact of sterilization on the porous design and cell behavior in collagen sponges prepared for tissue engineering, *Biomaterials*, 2002;23: 2855–2861.)

LOGIC PROGRAMMING FOR MACHINE TOOLS

Reggie Davidrajuh
Department of Electrical & Computer Engineering, University of Stavanger, Norway. Email:reggie.davidrajuh@uis.no

Abstract: Logic programming is a very important issue in programming for machine tools, especially for decision-making for machine operations. First, this paper presents a survey on some of the mathematical approaches for logic programming. Second, it presents a case study on logic programming a flexible manufacturing cell.

Key words: Logic programming, mathematical approach, machine tools

1. INTRODUCTION

Logic functions are very important parts of programs for machine tools as the logic functions take decisions during the execution of the program. Making wrong or bad decisions can be very costly.

The aim of logic programming for machine tools is to develop a logic model of the system so that decisions can be made out of the model. Generally, the logic models and the decision making operations run on them are based on one of the following mathematical approaches:

1) Algebraic operations on Cartesian axes,
2) Searching,
3) Mapping,
4) Geometry.

Structure of this paper: Section-2 presents the mathematical approaches for logic programming. Section-3 presents a case study on logic programming a flexible manufacturing cell.

Please use the following format when citing this chapter:
Davidrajuh, Reggie, 2006, in International Federation for Information Processing (IFIP), Volume 207, Knowledge Enterprise: Intelligent Strategies In Product Design, Manufacturing, and Management, eds. K. Wang, Kovacs G., Wozny M., Fang M., (Boston: Springer), pp. 717-722.

2. MATHEMATICAL REASONING APPROACHES

2.1 Algebraic approaches using Cartesian Axes

Propositional logic and predicate logic falls under this category. Prepositional logic uses simple Boolean connectives (like 'not', 'and', etc.), and lacks quantifiers like 'all', 'among', 'only', 'at least one', etc. Predicate logic has these quantifiers. ***Mathematical reasoning approach***: First, the fundamental logic variables are identified and each logic variable is assigned an axis, creating the whole universe of discourse. Then the logic variables are connected into premises, creating subspaces. Finally, the premises are combined to form the logic system, connecting subspaces spanned by the premises. By connection, spaces that do not satisfy the constraints are removed, leaving a smaller space that represents the feasible solution. ***Advantages and disadvantages of this approach***: This approach is useful in providing formal proofs as it offers clarity. Since a Cartesian axis is assigned to each logic variable, generating subspaces spanning all possible states of all the variables, a *complete model* is obtained. The serious shortcoming of propositional logic and predicate logic is the *exponential growth* (also known as 'combinatorial explosion') of the logic model with increasing number of variables. For example, for M Boolean logic variables, the resulting space of M axes will contain M^2 subspaces. Decision-making is slower due to this exponential growth of the subspaces with increasing number of variables.

2.2 Searching

Production rules use searching as the basic inference mechanism. A production rule has the following form: IF (condition) THEN (conclusion). ***Mathematical reasoning approach***: The reasoning approach starts with a set of facts and look for those rules in which the IF clause matches the facts; if such rules are found ('hit'), then it proceeds to the THEN clause. This reasoning is known as 'forward reasoning'. In 'backward-reasoning', searching starts with a set of desired goals and to look for those rules in which the THEN clause (conclusion) matches the goals. ***Advantages and disadvantages***: The main problem associated with the searching approach is that the decision-making is slow. In addition to this shortcoming, production rules also have a weakness: a logic system is evaluated with a couple of 'if-then' statements; thus, for M multi-valued logic variables with N values M^N 'if-then' statements are needed to span all combinations of the variables. For a system of many logic variables, it will be impossible to write so many if-then statements to take care of all possible combinations of variables; thus, creating a complete model is not easy and prone to errors. The main advantage of production rules is that the rules are simple, easy to understand, modify, and extend.

2.3 Mapping

Fuzzy logic is a popular technology that uses mapping with mathematical approximation for decision-making. ***Mathematical reasoning approach***: Decision-making in fuzzy logic is based on fuzzy rules that connect input and output parameters (fuzzy rule base), and the membership functions for input and output parameters. Inference mechanism in Fuzzy logic is implemented in three phases: 1) Fuzzification, which converts crisp input value into fuzzy value, 2) Inference, which computes fuzzy output value using fuzzy rules base, and 3) Defuzzification, which converts fuzzy output value into crisp value. ***Advantages and disadvantages***: The first limitation is that fuzzy logic does not guarantee completeness; it is up to the designer to include all the fuzzy rules connecting all possible combinations between the input and output parameters. The second limitation is the difficulty in generating fuzzy rule base. The fuzzy rules generated for an application must properly adhere to the process dynamics with no contradictions between the rules; this demands deep understanding of the process dynamics.

2.4 Geometry

Array-based logic uses geometry (topology of connections between the fundamental components of a system) for creating a system model and for running operations on the model. A system consists of three fundamental components, namely *elements*, *connections*, and *sources*. Elements carry all the physical properties of the system; thus, elements are the fundamental building blocks of a system. Connections reflect how elements in a system influence each other; thus, connections represent the structure of a system. Finally, sources are the environment's influence on a system. ***Mathematical reasoning approach***: The inference mechanism consists of three phases (Davidrajuh, 2000): 1) Identifying the primitive system, 2) Making the connected system, and 3) Applying the sources, and solving the connected system. ***Advantages and disadvantages***: The previous subsections state that a complete model of M multi-valued logic variables with a domain of N values contains M^N subspaces. Array-Based Logic avoids this exponential problem by compressing M^N subspaces into $M \times N$ linear representation (Møller, 1995). Array-based logic also provides mechanisms for operations to operate on the compressed representation in linear time.

2.5 Summary of the approaches

Given below are the important characteristics of the different types of mathematical approaches and the sample logic types that fall under the approaches: ***Inference cycle time***: Due to the combinatorial explosion, algebraic methods are slower. By nature, searching approaches are also slow, unless the implementations use some special data structures like 'hash coding'. Mapping by fuzzy logic is fast. Finally, array-based logic is fast due to the compact linear model and the linear-time operation it provides. ***Complete system***: Both the algebraic approaches by propositional logic and predicate logic, and the geometrical approach by array-based logic provides complete logic models. ***Implementation languages***: Most general purpose programming languages can be used for implementing propositional logic statements. Prolog can be used to implement predicate logic (Emden and Kowalski, 1979). Prolog, C-LIPS and Jess programming languages can be used for implementing production rules (Wagner, 2002). There are also numerous commercial packages for fuzzy logic, e.g. fuzzy logic toolbox for MATLAB platform. SABL is a toolbox of functions for implementing array-based logic (Davidrajuh, 2000).

3. CASE STUDY

The case study is about logic programming a flexible manufacturing cell (FMC) consisting of a vertical machining center (VMC), a horizontal machining center (HMC), a robot, and a conveyor belt. Here are the simplified operational specifications of the FMC: Step-1: To start a cycle, a raw part must be available on the incoming conveyor belt, and the robot is also available. Step-2: Robot moves a raw part from conveyor and loads it at HMC. Step-3: Milling operations are performed at HMC while robot backs off (returns). Step-4: Robot unloads semi-finished part from HMC, loads it to VMC and returns. Step-5: Drilling operations are performed at VMC; simultaneously robot perform step 2. Step-6: Robot unloads the finished part from VMC, deposits it on the outgoing conveyor belt and returns. In steady state steps 2-6 are repeated.

3.1 Vertical and Horizontal machining centers

Components of the VMC are shown in figure-1. Three sensors (micro-switches) *B*, *M*, and *E* control the operation of the VMC: Sensor *B* indicates that the boring spindle is at the rear position; Sensor *M* indicates that the boring spindle has reached the feeding position; Sensor *E* indicates that the boring spindle has reached the final destination, and ready for backward motion.

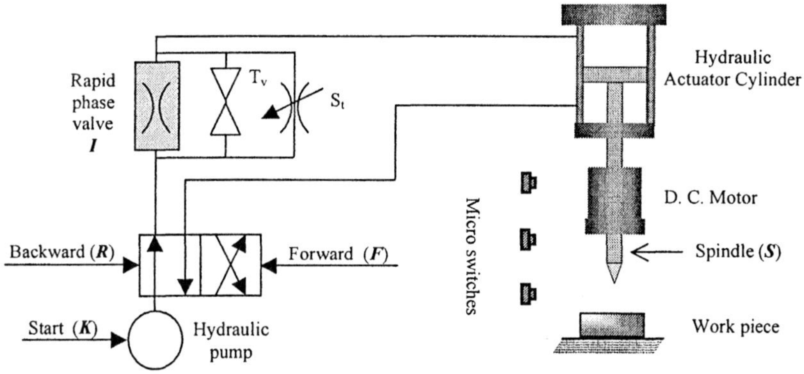

Figure 1. The Vertical Machining Center (VMC)

The spindle is at the rear position initially, and the operator switches on the system by a very short signal *K*. VMC then will go through three modes of operation: Mode-1: Starting from the initial state, hydraulic circuit will open rapid phase valve *I*, and the spindle will go forward by the opening the valve *F*. Mode-2: At position *M*, the rapid phase valve *I* will be switched off to start a controlled feed forward motion; This motion is regulated manually by the servo valve S_t. At position *M*, the spindle motor will also be switched on by a signal *S*. Mode-3: At position *E*, the backward motion *R* will begin. Simultaneously, the rapid phase valve *I* will be switched on.

Components and operations of the HMC are very similar to that of the VMC. The only difference is the orientation of the machining operations.

3.2 Logic programming the FMC

Logic programming is divided into two tasks: 1) Programming the machine centers VMC and HMC, and 2) Programming the robot.

Logic programming the machine centers VMC and HMC: Descriptions about the VMC/HMC reveal that both the VMC and the HMC have 4 inputs (K, B, M, and E) and 4 outputs (F, S, I, and R). Thus, the logic model has 8 Boolean variables, making 256 possible combinations. This means, 1) a *faster inference mechanism* is needed due to the large size of the model. In addition, 2) the logic model must be *complete model* as well, as one would minimize costly mistakes due to operating on a incomplete model. Only array-based logic based on geometry fulfills these two criteria. If speed is not that important (assuming that the VMC/HMC machines are slower) then propositional logic can also be used as they satisfy the complete modeling requirement; Though predicate logic can also be used, but it is not necessary, as the operational specifications seemingly do not demand for logic quantifiers.

Logic programming the robot: Programming an industrial robot is done at three different levels (Nilsson, 1996): 1) The motion control level, 2) The application-specific level, and 3) The end-user level. Only the end-user programming is done by customers, using specialized robot programming tools (before delivering robots to customers, robot manufacturers do the programming at the other two levels as it requires detailed knowledge of the robotic system dynamics). Usage of specialized robot programming tools suffers from a number of drawbacks (Gustavo et al, 2005): high learning curve of the specialized languages, non-reusable code, and limited portability across different robots. Most importantly, unlike the higher-level languages, these specialized languages are *opaque* meaning they reveal no details about how the logic models are created by the compilers and how the logic variables are manipulated. ***Thus, thinking about the mathematical approaches for logic modeling is not applicable in for robot programming.***

4. CONCLUSION

This paper presents 4 fundamental mathematical approaches for logic programming for machine tools. Choosing an appropriate mathematical approach for logic programming is very important as different approaches provide different characteristics such as complete models, fast decision-making, choice of implementation languages, etc. However, as the case study proves, taking a mathematical view is not always applicable for logic programming for machine tools.

5. REFERENCES

Davidrajuh, R. (2000). Automating Supplier Selection Procedures. PhD dissertation, Norwegian University of Science and Technology (NTNU).

Emden, M. and Kowalski, R. (1979). The semantics of Predicate Logic as a Programming Language. *Journal of the Association for Computing Machinery*, Vol. 23, No. 4, October 1976, pp. 733-742

Møller, G. (1995) On the Technology of Array-based Logic. Ph.D. dissertation, Technical University of Denmark

Nilsson, K (1996). Industrial Robot Programming. PhD thesis, Department of Automatic Control, Lund Institute of Technology

Wagner, G. (2002) How to Design a General Rule Markup Language. The workshop on XML Technologies for the Semantic Web (XSW2002). Humbolt University, Berlin, Germany, June 2002

ANALYZE AND RESEARCH OF CORROSION RESISTANCE OF LASER CLADDING LAYER ON THE ANTI-ACID STAINLESS STEEL SURFACE

Geyan Fu[1] , Jianjun Shi[2], Tuo Shi[3]
[1]*Soochow University, China; Email:Fugeyan@suda.edu.cn.* [2]*Soochow University, China.* [3] *Dongnan University, China.*

Abstract: The effect of coating, which is formed with laser cladding and plasma spray welding on the base metal of anti-acid stainless steel, on anti-corrosion has been analyzed and studied. A 5KW laser with CO2 flow transverse is used for cladding Co-base alloy or Ni-base alloy powder replaced on the base metal. Comparing with the plasma spray welding layers, the laser cladding layers have lower defect rate and higher product rate. The result of test shows that the laser cladding layers have higher corrosion resistance properties.

Key words: laser cladding, plasma spray welding, corrosion resistance.

1. FOREWORD

In petrochemical industry, reactor and nuclear power plants, varieties of anti-acid stainless steel valves are widely used. The working mediums in these sectors are corrosive and even radioactive. The results of the corrosion not only destroy the sealing face and shorten the life of the valves, but also can cause the laying off or production halts because of the leakage of the mediums and may pollute the environment and even cause the vicious accidents. The quality of the sealing face is the important index checking the valves' fundamental characteristics. For the stainless steel valves, the requirements are even higher. The way of strengthening the sealing face of the high parameter stainless steel valves is by bead overlaying welding on the valves directly, but not inserting[1,2].In this article, the laser cladding layer on the surface of 20HJ63-20P stainless valve and methylamine-pump

Please use the following format when citing this chapter:
Fu, Geyan, Shi, Jianjun, Shi, Tuo, 2006, in International Federation for Information Processing (IFIP), Volume 207, Knowledge Enterprise: Intelligent Strategies In Product Design, Manufacturing, and Management, eds. K. Wang, Kovacs G., Wozny M., Fang M., (Boston: Springer), pp. 723-728.

valve in urea production line, and comparied in property of anti-corrosion with traditional plasma spray welding coating.

2. EXPERIMENT CONDITIONS

The valve used in this experiment is 20HJ63-20P cut-off valve used in nuclear power station and 3W-2BJ1 in-outlet valve of methylamine-pump used in urea production line. The materials are both austenite stainless steel, and the chemical constitutions (Wt%) are: 0.08%C, 1%Si, 2%Mn, 12%Ni, 18%Cr, 0.7%Ti, 3%Mo. There are two kinds of specimens: square and annulus. Cladding layer is made of a HGL-90 5kW laser with CO_2 flow transverse. The alloy powder of Co-base or Ni-base is used on the coat of sealing face, which is glued by 2123 PF resins. The thickness of the presetting layer is from 1 to 3.5mm according to different requirements. The laser cladding is done on the 2-dimension numeric linked platform. The square specimens are done using multiple tracks scanning and the lap of splice is half of the laser spot's diameter. The annulus specimens are done on the numeric gyrated platform using single track scanning. The process parameters are: laser power $P = 2000 \sim 3000$ kW, scanning speed $v = 8 \sim 12$ mm/s, spot dimension $\phi = 5$ mm. To compare with the traditional process, other specimens which are machined by plasma spray welding. Alloy powder and thickness of the coating are the same as the laser machining. The pattern of the welding machine is DP-500, and the blowpipe is LFH, using flour handing.

After the experiment observe and analysis, the microscopic structure of the laser cladding layer, grade the grains, measure the thickness of the layer and the width of the micro hardness and heat-affected zone and analysis the defects of the layer. We used MEF3 optical metallographic microscope to observe in microscopic structure of the layer. We also observe and analysis the pattern of the melting zone and basal body using the JSM35C made in Japan and S650 and SEM, and assay the component of the site and appraisal the rate of dilution of the laser cladding layer using EDX-9100 energy spectrometer.

3. DETECT AND ANALYSIS COMPARISON

Fig.1 and fig.2 show the metallurgical structure of the cladding layer and indentation photo of microhardness test of the laser cladding and the plasma spray welding. It can be divided into 3 areas: melting zone, bonding zone and basal body along the vertical direction to the machined surface. The texture of the laser melting zone is minute and homogeneous, the measuring grain degree is grade 11~12, and the width of the bonding zone is 5~20um; while the texture of the plasma spray welding is rougher, the grain degree is grade 9~10, and the width of the bonding zone about 80~120 μm. The quality of the cladding layer is measured among the 100 dissected specimens. The yield of non-defect of the laser cladding is above 95%. While there are many defects such as crack, voids, foreign materials in plasma spray welding coating. Seen in fig.3 and fig.4, the yield is about 65%. While we test the micro-hardness along the crucible zone of the top middle and down layer in lengthwise direction and from the edge to center in cross direction, the area of indentation can be seen in fig.1 and fig.2. The result shows that the mean micro-hardness of laser cladding is $HV_{200g} = 581$ kg/mm^2, and it is 20%~40% higher than the spurt welding of the same flour, and the uniformity is also higher than the spurt welding in both lengthwise and cross direction.

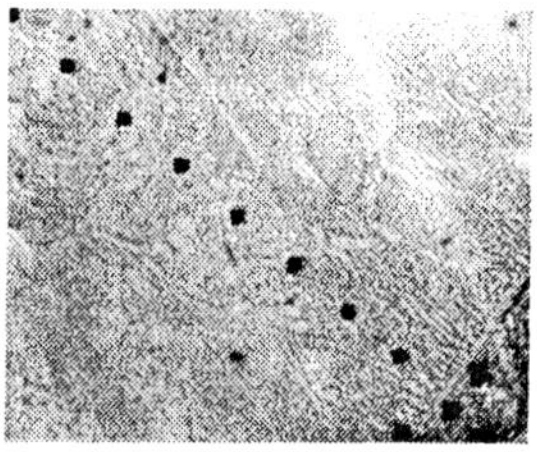

Figure 1. laser clad layer and its combination region

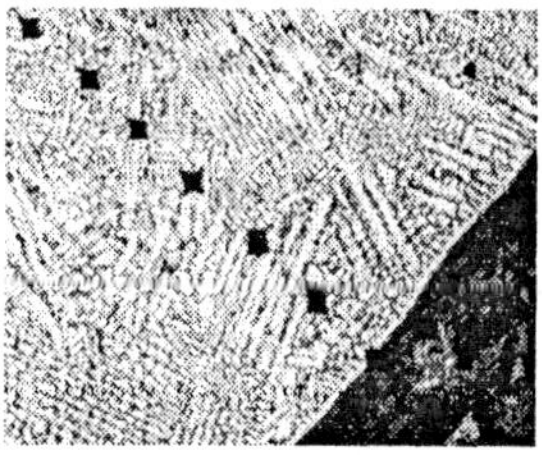

Figure 2. plasma-spray welding layer and its combination region

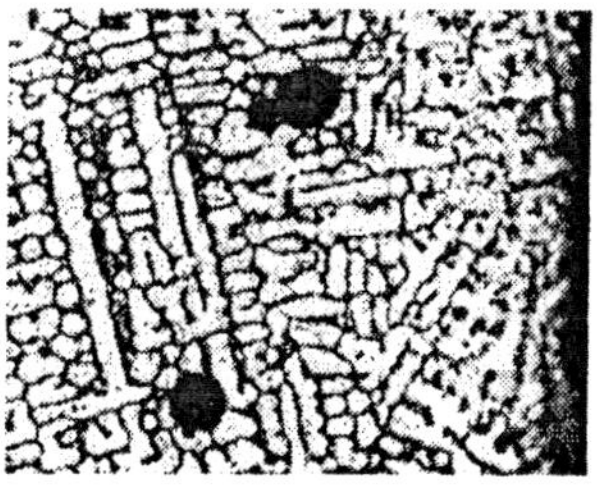

Figure 3. gas holes in the lower part of plasma-spray welding layer

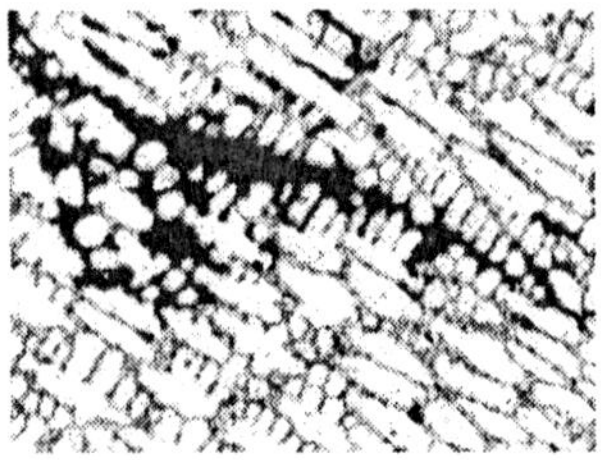

Figure 4. impurity in arborescent crystal of plasma welding layer

Observed by the energy spectrometer, the mean content of the Ti, Fe and Ni in the laser cladding layer are diluted to 0.075%wt, 8.58%wt and 1.385%wt by the basal body. While in the plasma spray welding layer the mean content of the 3 elements are 0.83%wt, 16.19%wt and 2.06%wt. It can be seen that the rate of dilution to the 3 elements of the plasma spray welding are 10 times, 2 times and 1.5 times higher than the laser cladding. As for the loss factor of the Co, Cr and W diffuse to the heat-affected zone and basal body, plasma spray welding coating is more serious than laser cladding layer.

According to the test results, the analysis can be gotten below:

(1) The basic difference between the two kinds of technology is the different heat source. Comparing the high-energy laser beam with the plasma arc, the former have the characteristics of quick heating-up, high temperature, short actuation duration and concentrated heat source. The laser cladding can get a high speed of 10^5~10^6C/s of the heating-up and cooling velocity of the crucible zone. The laser beam can melt the flour layer instantly, and at the same time melt the surface of the basal body and make it combine with the crucible zone to form firm metallurgy layer. The high-speed heating-up and cooling shorten the actuation duration, decrease the melting percentage of the basal body, reduce the diffusion of the elements between the crucible zone and the basal body and minish the dilution rate. Because of the smaller energy density of the plasma spray welding, the heat-affected duration is longer so that the diffusion between the crucible and the basal body is aggravated and the width of heat-affected zone become lager so that increase the pollution of the crucible zone. Because of the high-speed heating-up and cooling of the laser cladding, the metallurgy elements in the melting bath can form many kinds of chemical compounds quickly so that the number of the non-self crystal nucleuses increase and it greatly improves the nucleation rate so that many small and homogeneity microscopic structure form. The minute structure can improve the bonding force at the grain boundary and strengthen the work piece's strength and ductility. The minute structure not only reduces the impurity content at every unit of grain boundary, but also

monishes the components' aliquation degree during the quick cooling so that monishes the affection of corrosion generated by the galvanic cell effect.

(2) In laser machining, the laser beam can generate convection mass transfer effect in the melting bath[3,4]. It can agitate the melting bath fully and the bubbles in the melting bath rise up with the foreign materials to form the dense layer. This improves the quality of the crucible zone. While the plasma spray welding uses plasma to melt the flour and then accelerate to a certain speed and eject to the surface of the basal body through the air. During this process, many bubbles and contaminant may distribute among the large arborescent crystal at the cladding layer and interface, see fig3 and fig4. Electrochemical inhomogeneity generated by the non-metal foreign material, the inhomogeneity of chemical constitution because of the dilution of the crucible zone and the inhomogeneity of the metal texture reduce the corrosion resistance.

4. THE CORROSION RESISTANCE OF THE CRUCIBLE ZONE

We use the method of linear cutting to get two kinds of layer as test piece, and one is the laser cladding layer with the same materials as the basal body, the other is the plasma spray welding layer, and 4 pieces are needed for each. The corrosion experiment is done by throw them into different solutions. The solutions are placed in the water channel of 60℃ to keep its temperature, and the test pieces of two different technologies will be corrupted in the four different solutions for 8 hours, 24 hours, 48 hours and 72 hours. Before weighing the loss weight with rigorous analysis balance, clean the test pieces. The test results and analysis data are shown in table 1. P in column 2 table 1 stands for the test piece of plasma spray welding, while L stands for laser cladding. The unit of corrosion speed in the table is $g/m^2 \cdot h$, i.e. the loss weight of the test piece by corrosion in every 1 hour and every 1 square. From the data, it is concluded that the corrosion speed of the laser cladding test pieces in the four solutions (H_2SO_4, HNO_3, NaOH and urea) are all lower than the speed of plasma spray welding test pieces, so it has a better corrosion resistance, especially in the solutions of H_2SO_4 and HNO_3.

Table 1. the test results of corrosion

the corrosion medium	test piece	the corrosion speed($g/m^2 \cdot h$)			
		8h	24h	48h	72h
10%H_2SO_4	P	139.98	133.96	79.28	
	L	5.28	3.53	2.55	
speed ratio	P/L	26.5	37.95	31.1	

		the corrosion speed($g/m^2 \cdot h$)			
10%HNO_3	P	629.85	233.19	127.22	
	L	1.83	0.67	0.46	
speed ratio	P/L	344.2	348	276.6	
20%NaOH	P	0.043	0.070	0.047	0.043
	L	0.039	0.066	0.036	0.028
speed ratio	P/L	1.10	1.06	1.30	1.53
30%urea	P	0.664	0.415	0.37	0.345
	L	0.55	0.349	0.0246	0.186
speed ratio	P/L	1.21	1.19	1.50	1.85

5. CONCLUSIONS

(1)By comparing the two kinds of surface strength technologies on anti-acid stainless steel surface, it is found that the laser cladding is better than plasma spray welding. Because of the high speed melting and cooling generated by the energetic laser beam scanning, the melting bath is formed on the basal body surface and at the effect of convection mass transfer, the minute and dense layer with no voids and foreign materials is formed. It can realize the metallurgical bonding between the coating layer and the basal body. It reduces the dilution rate and diffusion rate of the components of the crucible zone, and improves the hardness and strength and corrosion resistance. The carbon content in the flour can reduces properly, so that the corrosion resistance will be improved.

(2) The corrosion speed of the laser cladding test pieces in the many kinds of solutions is lower than the speed of plasma spray welding test pieces, so it has better corrosion resistance, especially in the solutions of H_2SO_4 and HNO_3.

6. REFERENCES

1. Gao Qing-bao., (1994), The bead weld technology of valve, *Mechanical Industry Press*, pp. 305–416.
2. Yang Yuan-yuan., (1995), Hand book of valves design, *Mechanical Industry Press*, pp. 398–440.
3. Wang Jia-jin., (1992), The laser machining technology, *China Measurement Press*, pp. 274–279.
4. Zhengzhong Gguang., (1998), handbook of laser processing technique, *Measurement Press of China*, pp. 255–264.

MATERIAL REMOVAL MECHANISMS ANALYSIS IN THE FINISHING MACHINING OF ENGINEERING CERAMICS

Changhe Li[1,2] , Guangqi Cai[1]
[1]School of Mechanical Engineering & Automation, Northeastern University, China; Email: sy_lichanghe@163.com. [2] Inner Mongolia university for Nationalities, China

Abstract: According to the critical size ratio for the characteristic particle size to film thickness between grinding wheel and workpiece, the machining mechanisms in abrasive jet precision finishing with grinding wheel as restraint of advanced engineering ceramic can be categorized into three states, namely, two-body lapping, three-body polishing, abrasive flow jet machining. The critical condition of two-body lapping to three-body polishing transition was analyzed. The single particle material removal modes of two-body lapping, three-body polishing, abrasive flow jet machining were proposed. Experiments were conduced in the refitted surface grinder for academic modes verification. It was found that experimental results were agreement with theoretical modes and the modes validity were verified.

Key words: Material Removal Mechanisms, Free Abrasive, Grinding Wheel as Restraint, Two-body Machining, Three-body Polishing Process, Engineering Ceramic Machining

1. INTRODUCTION

The conventional grinding of engineering ceramic makes surface defects such as micro-crack, tensile residual stresses and burnout because of high grinding zone temperatures. Authors invent a newly abrasive jet precision finishing method, which inject slurry of abrasive and liquid solvent to grinding zone under haven't cut-deep infeed condition when accomplished workpiece grinding. The abrasive particles with grinding wheel as restraint are driven and energized by the rotating grinding wheel and liquid hydrodynamic pressure and increased slurry speed between grinding wheel

This project is supported by the National Natural Science Foundation of China under the grant No.50475052.

Please use the following format when citing this chapter:
Li, Changhe, Cai, Guangqi, 2006, in International Federation for Information Processing (IFIP), Volume 207, Knowledge Enterprise: Intelligent Strategies In Product Design, Manufacturing, and Management, eds. K. Wang, Kovacs G., Wozny M., Fang M., (Boston: Springer), pp. 729-734.

and work surface. The method can either require high form accuracy or attain free damage finishing surface. The goal of this project is establish the foundationally academic system of abrasive jet precision finishing by theoretical and experimental investigation[1,2].

2. THE MACHINING THEOREM

2.1 Two-body lapping

The abrasive powder is mixed with a fluid carrier to form a slurry which is turned into the gaps between wheel and workpiece. A schematic of the process is showed in Fig.1. In the abrasive polishing begin, because of the grinding wheel and workpiece clearance is too small, the abrasives between wheel and workpiece are forced into wheel surface pore under liquid hydrodynamic pressure and grinding wheel drag. Because the two force will not be collinear, form a couple tending to totate the particle[3,4]. The sequence of events is illustrated in Fig.2(a) to (c). The particle continues to rotate until the blunt corner makes contact with the work Fig.2(d). In the condition, there is no impetus to cause further rotation and so the particle will tend to remain at the inclination similar to abrasive two-body machining[5].

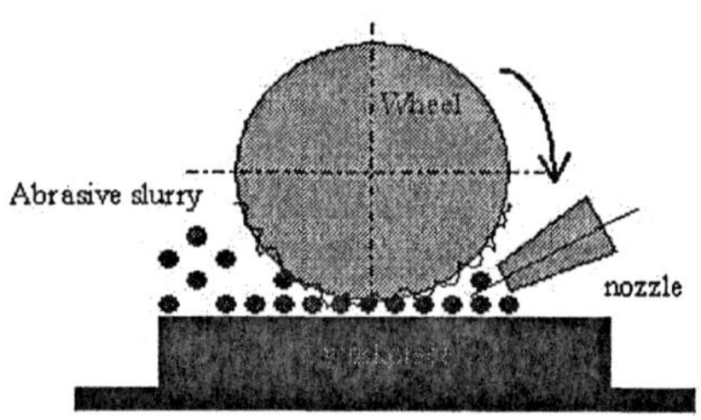

*Figure 1.*The schematic of abrasive jet finishing.

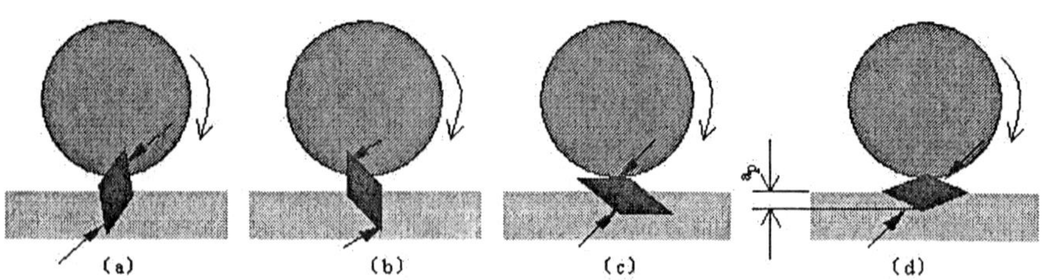

Figure 2. Abrasive movement analyse in two-body process.

2.2 Three-body polishing

With polishing circles increasing, abrasive size gradually diminished and work surface micro-protrusion was removed. The film thickness is increasing lead to the size rate that the characteristic particle size to film thickness is reduce. The particle may rotate sufficiently until lose contact with the work surface, Fig.3.The machining mode is three-body process in which particles are able to tumble and rotate between the wheel and piecework.

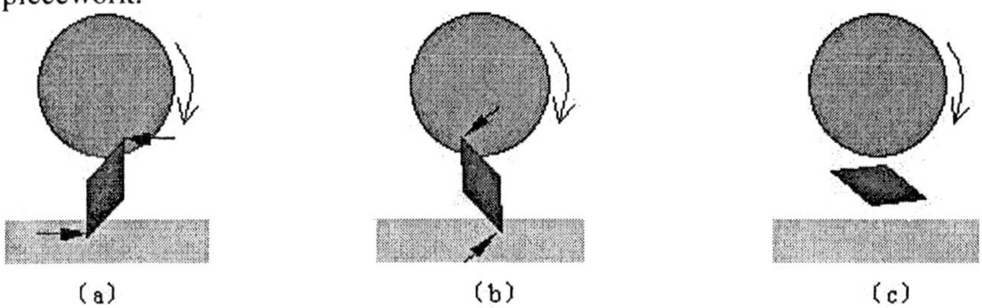

Figure 3. Abrasive movement analyse in three-body process

2.3 The transition critical condition

Since the abrasives are mixed with a fluid carrier to form a slurry and uniformly suspended in the slurry fluid, it is possible that there are several thousand particles present in the gap at any moment. They are likely to be several tens of particle diameters apart, and it is reasonable to think of them acting independently of each other. Therefore, we developed a transition mode from two body lapping to three body polishing, to consider the mechanics of a single representative particle. In the study, the abrasive particle as a long prism with a cross-section which is rhombic in shape. The shortest diagonal of the cross-section is D and h_g is the smallest film thickness. The rate of D and h_g is defined as the critical size rate. when the critical size rate is larger than 1, the machining mode is two-body lapping. On the other hand, the machining mode is three-body polishing[6].

3. EXPERIMENTAL SETUP

The machining system of the experiments is composed of a rotating grinding wheel and a work between which the slurry was filled in by the nozzle, shown in Fig. 1. The wheel type is WA60J8V, which have diameter 240mm and 26mm width. The grinding wheel main axis revolution are 2880rpm and the nozzle supply slurry pressure are 3×10^3 Pa. The slurry is a mixture of abrasive particles and liquid solvent. The abrasive particle is W7 Al_2O_3 with primary grain size of 5～7μm. Three types weight percentage grain to slurry are used as the solvents, which are the 5%; 10%; 15%. The workpiece material is alumina ceramic which is machined up-grinding from 10 circles to 40 circles in the experiments. The specimens have dimensions of $50\times25\times15$mm.The work surface roughness values are measured with a Talysurf instrument[7].

4. RESULTS AND DISCUSSION

4.1 SEM examinations

With machining circles increasing, workpiece surface micrographs changing as Fig.4 showed. As might be expected, the work surface was found to be the most aggressive of the abrasives, in that it produced the greatest removal volumes and directionality groove(Fig.4a) at machining circles is 10 (workpiece move to-and-fro one defined as one circles). At the same time, the critical size rate is larger than 1, therefore, the machining mode is two-body lapping; When the machining circles increased to 20, the workpiece surface not only produced uniformity groove, but with very little sign of random pits. As the discussed above, the machining mode is mixed of two-body lapping and three-body polishing(Fig.4b). When machining to 30 circles, the characteristic particle size to film thickness rate is lower than 1 because of abrasive worn and workpiece surface micro-protrusion removed. The surface produced randomly discontinuous micro-pit absolutely(Fig.4c) and the phenomenon is defined as three-body polishing machining.

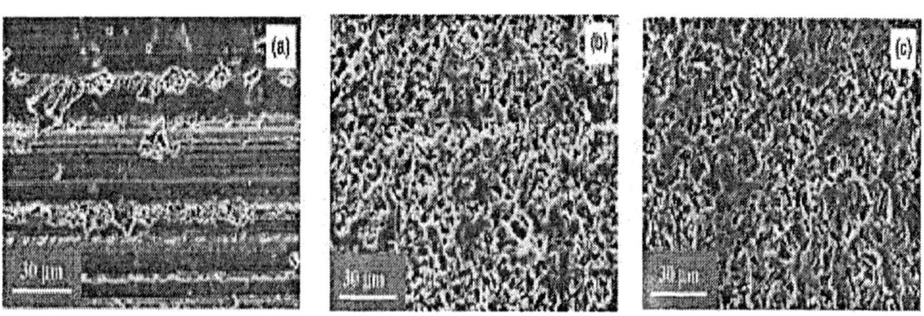

Figure 4. SEM images of abrasive jet process on different machining circles

4.2 Surface roughness

The values of surface roughness of the polishing surface are plotted in Fig.5. It is seen that for all the three percentage concentration the surface roughness decrease with the machining circles increasing. Before 20 cycles, the roughness improvement is low because the machining mode is two-body machining and workpiece surface with continuous and parallel micro-groove. On the other hand, the workpiece surface roughness value obviously diminish after 20 cycles to all abrasive concentration. From the Fig.5, it can be inferred that 15% concentration of abrasive roughness value is larger than 10% concentration of abrasive. With higher concentration, more abrasive grains come into contact with the workpiece resulting in more abrasion, hence higher Ra value.

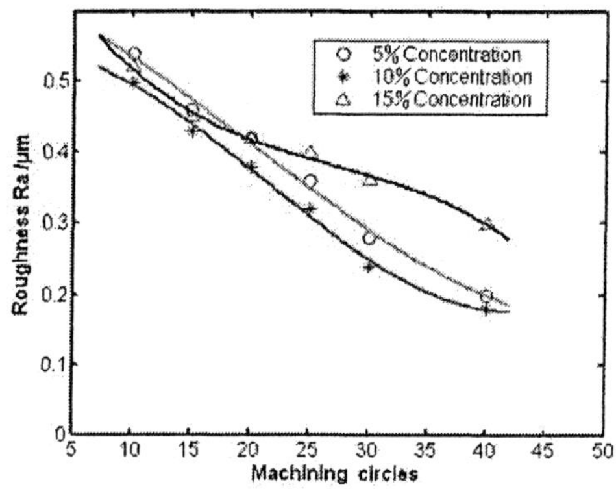

Figure 5. Effect of different abrasive concentration on surface roughness values

5. CONCLUSION

The material removal mechanisms are investigated according to the critical size rate. The machining mechanisms are combined with two-body lapping and three-body polishing. Experiments were performed in the refited surface grinder for academic modes verify. It was found that workpiece surface roughness gradually diminish with machining cycles increasing, but the roughness decrease is low before 20 cycles. The micrograph showed, the machined surface from micro-groove to random pits transition with machining process. experimental results were agree with academic modes and the modes validity were verified.

6. REFERENCES

1. LI Chang-he, Cai Guang-qi and LI Qi,Generating mechanism of surface morphology finished by abrasive jet with grinding wheel as restraint, *Journal of Northeastern University (natural science),* 2005,vol.26(6), pp.578-581.
2. C.H.Li, G.Q.Cai and S.C.Xiu,Material removal model and experimental verification for abrasive jet precision finishing with wheel as restraint, *Key Engineering Materlials*, 2006, vols. 304-305, pp. 555–559.
3. Chang Y P, Hashimura M, Dornfeid D A,An investigation of material removal mechanisms in lapping with grain size transition, *Journal of Manufacturing Science and Engineering,* 2000,vol.122, pp.413-419
4. Chauhan R, Ahn Y, Chandrasekar S,Mechanisms of abrasive wear in lubricated contacts, *Wear,* 1993, vols.162-164, pp.246-257
5. Buijs M, Korpel-van Houten K,Three-body abrasion of brittle materials as studied by lapping, *Wear,* 1993, vol.166, pp.237-245
6. Li Changhe, Cai Guangqi and Yuan Suoxian, Material removal model for abrasive jet precision finishing restricted by abrasive wheel, *Transactions of the Chinese Society for Agricultural Machinery,* 2005,vol.36(11),pp.132-135.
7. Li Changhe,Cai Guangqi,Li Qi,et al. Investigation on material removal mechanism for abrasive jet precision finishing with grinding wheel as restraint, *China Mechanical Engineering*, 2005,vol.16(23), pp.2116-2119

INVESTIGATION OF EXPANSION CHARACTERISTICS OF CORONARY SLOT STENTS USING FINITE ELEMENT ANALYSIS

Xuanzhi Wang, Syed H. Masood
Industrial Research Institute Swinburne, Faculty of Engineering and Industrial Sciences, Swinburne University of Technology, Hawthorn, Victoria, Australia 3122, xuanzwang@yahoo.com.au

Abstract: Coronary stents are metallic tubes inserted in coronary arteries to treat cardiovascular diseases. The Palmaz-Schatz® Stent is the most representative member of the family of slot stents. In this paper, the impact of the geometry of the stent on its expansion behaviours was investigated using the ANSYS 9.0 finite element analysis (FEA) software. Several design cases, with the presence of a certain stenosed coronary artery, were selected for our research investigation. Results of the analysis, including expansion and radial displacement, radial recoil percentage, foreshortening percentage and von Mises stress distribution of various stents have been compared and discussed. This revealed significant potential relationships between the geometry of the stent and its expansion behaviours.

Key words: Angioplasty; Expansion characteristics; Finite Element Analysis; Simulation; Stent.

1. INTRODUCTION

Cardiovascular diseases constitute one of the major human health problems of modern times. Recently, the implantation of metallic coronary stents in the coronary artery has become an integral part of most interventional procedures for percutaneous revascularization[1]. The Palmaz-Schatz® stent (Johnson and Johnson Company, Warren, NJ, USA) is the most representative member of the family of slot stents. Much work[2-5] has been done to investigate different expansion behaviours of stents with

Please use the following format when citing this chapter:
Wang, Xuanzhi, Masood, Syed, H., 2006, in International Federation for Information Processing (IFIP), Volume 207, Knowledge Enterprise: Intelligent Strategies In Product Design, Manufacturing, and Management, eds. K. Wang, Kovacs G., Wozny M., Fang M., (Boston: Springer), pp. 735-742.

various geometries. To date, however, very little research work has investigated expansion characteristics of stents with various geometries with the presence of a stenosed artery. A methodology for studying expansion characteristics of stents with various geometries and the presence of a stenosed artery by the FEA method using ANSYS 9.0 will be presented in detail. Characteristics of stents with various geometries and the presence of a stenosed artery by the FEA method using ANSYS 9.0 will be presented in detail.

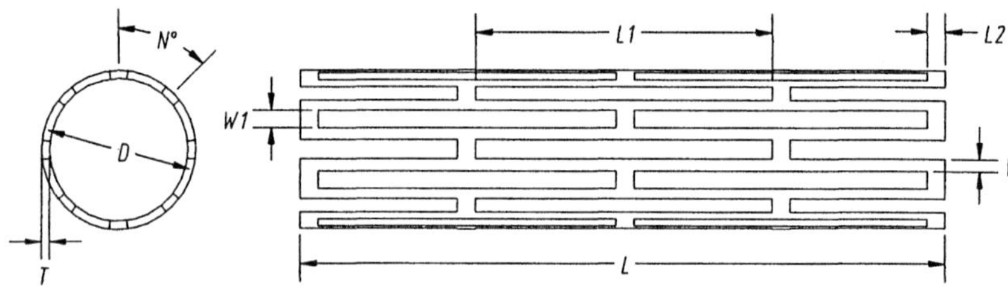

Figure 1. Geometric parameters of stent

2. FINITE ELEMENT ANALYSIS OF STENTS

In this investigation, seven varying designs of stent were modelled. It was assumed that all stents were equal in whole length L, thickness T and mean diameter D. Figure 1 illustrates the geometric parameters of the stent: L_1 and W_1 are the slot length and slot width, L_2 and W_2 are the strut length and strut width, D is stent thickness, N is sector angle. Case 1 is the stent with standard geometry. Cases 2 and 3 are the stents with same stent width W_1 and strut width W_2 as Case1. However, Case 2 has shorter slots in which the slots were modelled to be 5% shorter than in Case 1. Slots of Case 3, however, are 5% longer than those of Case 1. The slot length L_1 and strut length L_2 of Cases 4 and 5 are equal to those of Case 1. The slot width of Case 4, however, was decreased by 10%, while that of Case 5 was increased by 10%. The number of slots of Cases 1 – 5 is 24, while the number of slots of Cases 6 and 7 are 18 and 30, respectively, due to varying the sector angle N. Table 1 summarises the values of geometric parameters.

According to a typical environment and structure of a stenosed coronary artery[6], a three-dimensional model of a stent, a balloon, a fibrous cap, a lipid core and an arterial wall was created. To take advantage of these symmetric characteristics, a basic sector (see Figure 2) of it was used in the analysis using ANSYS 9.0 finite element analysis software. Thus, in the axial direction, a half sector of the model was selected, and in the circumferential direction, a sector of the model was selected.

Table 1. Values of geometric parameters in seven cases

Stent design case	Slot length L_1 (mm)	Slot width W_1 (mm)	Strut length L_2 (mm)	Strut width W_2 (mm)	Stent thickness D (mm)	Sector angle N (°)	Number of slots
1	3.6200	0.2200	0.2200	0.1400	0.1000	45	24
2	3.4390	0.2200	0.3407	0.1400	0.1000	45	24
3	3.8010	0.2200	0.0993	0.1400	0.1000	45	24
4	3.6200	0.1980	0.2200	0.1620	0.1000	45	24
5	3.6200	0.2420	0.2200	0.1180	0.1000	45	24
6	3.6200	0.2933	0.2200	0.1867	0.1000	60	24
7	3.6200	0.1760	0.2200	0.1120	0.1000	36	24

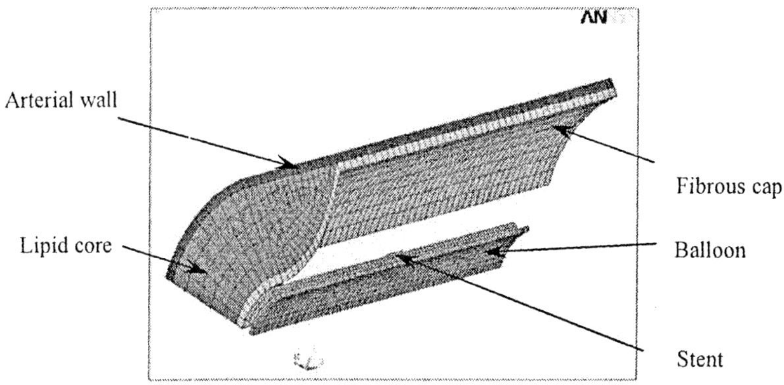

Figure 2. Finite Element Model

All five components were meshed using three dimensional 8-node solid brick elements. The material of the stent is stainless steel 316L. The material properties[7-11] of five components in the model are defined as Table 2.

Table 2. Material properties of components in model

Component	Material model	Young's modulus (MPa)	Poisson's ratio	Yield stress (MPa)	Tangent modulus (MPa)
Balloon	Linear isotropic	6.9×10^2	0.3	–	–
Stent	Bilinear isotropic	2.03×10^5	0.35	3.5×10^2	9.15×10^2
Fibrous cap	Linear isotropic	2.19	0.4999	–	–
Lipid core	Linear isotropic	1.74×10^{-2}	0.4999	–	–
Arterial wall	Linear isotropic	1.74	0.4999	–	–

Three consecutive displacement ramped loads were applied in order on the inner surface of the balloon. From 0 s to 1.1 s, a ramped radial displacement load was increased gradually from 0 mm to 1.1 mm. From 1.1 s to 1.6 s, a constant radial displacement load was maintained at 1.1 mm. From 1.6 s to 2.7 s, a ramped radial displacement load was decreased gradually from 1.1 mm to 0 mm.

3. RESULTS AND DISCUSSION

3.1 Radial displacement

Figure 3 illustrates the radial displacement of various stents at the distal end at the corresponding simulation time obtained from the FEA. The radial displacement of the balloon is also shown as a reference.

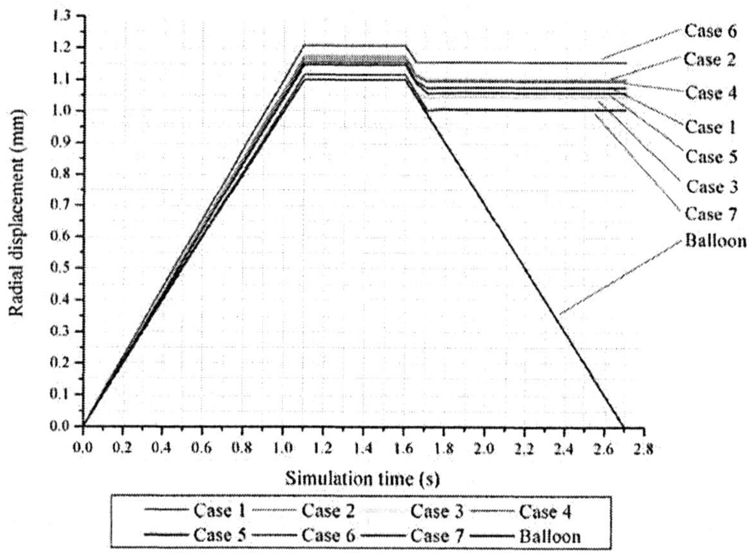

Figure 3. Radial displacement of various stents and balloon

In Case 1, opposite radial displacement characteristics were observed compared to Cases 2 and 3. The magnitude of the ultimate radial displacement of Case 2 was more than that of Case 1, while the magnitude of ultimate radial displacement of Case 3 was less than that of Case 1. It was revealed that decreasing the slot length enlarges the ultimate radial displacement, while increasing the slot length results in the reduction of the radial displacement.

Similar conclusions were observed in the impact of the changes of the slot width and slot length upon the radial displacement. It was shown that the change of the slot width has an opposite influence on the change of the ultimate radial displacement. The wider the slot was, the less the ultimate radial displacement.

Similar conclusions were also revealed in the influence of the number of slots on the radial displacement characteristic. The less the number of slots was, i.e. the larger the angle of the basic sector of the stent, the better radial displacement characteristic the stent had. The number of slots has a much more significant influence on radial displacement than other factors beyond the same displacement loading of the balloon.

3.2 Radial recoil percentage

By comparison of Cases 1, 2 and 3, increasing or decreasing the slot length has the same impact trend on the magnitude of the recoil percentage, which resulted in reduction of the recoil. This revealed that the radial recoil percentage of Case 1 was near the extremum with the corresponding slot length. It was also concluded that the impact of the slot length on the recoil was less notable than other factors, because the slot length was the measurement along the axial direction, and the deformation and stress of the axial direction were far less than that of the radial direction.

From Cases 1, 4 and 5, it was seen that increase of the slot width enhances the recoil of the stent, while decrease of the slot has negative impact on the recoil. Because the slot width was a circumferential measurement, it played a more crucial role in the recoil characteristic than the slot length.

From Cases 1, 6 and 7, it was observed that the number of slots was an important factor in the recoil characteristic of the stent. The recoil phenomenon in Case 7 was stronger than in Cases 1 and 6. More slots lead to small plastic deformation of the stent in a single slot under the same diameter of expansion, and as the major part of its deformation is elastic, so the stent was able to recover from the deformation easily.

3.3 Foreshortening

Figure 4 illustrates the radial displacement, i.e. foreshortening, of various stents at the distal end at the corresponding simulation time obtained from the FEA. The radial displacement of the balloon is also shown as a reference.

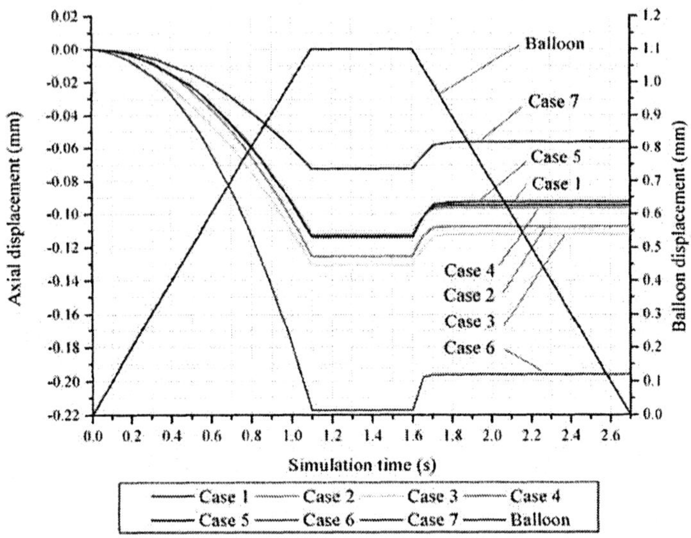

Figure 4. Axial displacement of various stents and balloon

By comparison of Cases 1, 2 and 3, both increasing and decreasing the slot length enhanced the foreshortening of the stent. This revealed that the value of the foreshortening of Case 1 was near the extremum with the corresponding slot length. Because the slot length was an axial measurement, it played a more crucial role in the recoil characteristic than the slot width.

From Cases 1, 4 and 5, it was seen that variety of the slot width has no significant impact on the foreshortening of the stent, because the slot width is the measurement along the conferential direction.

From Cases 1, 6 and 7, it was observed that the number of slots was a crucial factor in the foreshortening of the stent. The foreshortening phenomenon of Case 6 was stronger than in Cases 1 and 7, because fewer slots resulted in more axial deformation of the stent in a single slot under the same diameter of expansion.

3.4 von Mises stress distribution

As shown in Figure 5 (a) (b) (c), from Cases 1, 2 and 3, it was revealed that there are more high stress areas in the stent with the longer slot. The low stress area in Case 2 was more than that of Case 1, especially at the end of the stent. The minimum stress of Case 2 was about half as much as that of Case 1. In Case 2, the slot was shorter and the strut was longer, which decreases the stress concentration and was propitious to decentralization of the stress. In Case 3, the high stress area occurs at the end of the stent as well.

The minimum stress of Case 3 was in excess of twice of that of Case 1. The longer slots and shorter strut meant that much stress was focused at the end and middle of the slots in Case 3. The results revealed that reducing the slot length, i.e. increasing the strut length, was able to significantly eliminate the concentration of stress and lessen its value.

As shown in Figure 5 (a) (d) (e), from Cases 1, 4 and 5, lower stress was found in the stent with a wider slot. In Case 4, with decreasing the slot width, the minimum value of stress was increasing notably. Narrower slots and a wider strut made the stent stiffer, which required more energy to deploy the stent. In Case 5, by increasing the slot width, both minimum and maximum values of stress were reduced because of the improvement of flexibility of the stent. Even so the high stress area in the middle of the slots was enlarged to some extent. The results revealed that increasing the slot width was able to reduce the value of stress.

As shown in Figure 5 (a) (f) (g), from Cases 1, 6 and 7, the stress was dramatically decreased by adding the number of slots. In Case 6, both the maximum and minimum values of stress were the largest in all cases, respectively. It required much more deformation for the stent with fewer slots, which resulted in the appearance of higher stress during the deformation. On the contrary, in Case 7, both the maximum and minimum values of stress were smallest in all cases, respectively. The results revealed that increasing the number of slots was able to reduce the value of stress.

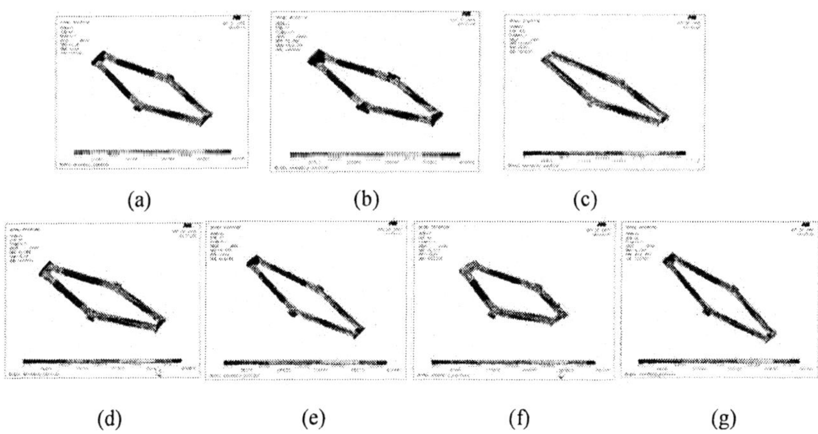

(a) (b) (c)

(d) (e) (f) (g)

Figure 5. Distribution of von Mises stress of Stent

4. SUMMARY

In this paper, a parametric model of a certain Palmaz-Schatz® stent has been developed by varying the slot length, slot width and number of slots. The expansion characteristics have been investigated by the FEA. Comparisons and discussions of these characteristics have revealed potential relationships between the geometry of the stent and its expansion characteristics. This study will provide better understanding of the structural characteristics of metallic stents from design and deployment points of view.

5. REFERENCES

1. A. Colombo, G. Stankovic and J. W. Moses, *Selection of coronary stents*, J. Am. Coll. Cardiol. **40**(6), 1021–1033 (2002).
2. F. D. Whitcher, Simulation of in vivo loading conditions of nitinol vascular stent structures, *Comp. & Struct.* **64**(5), 1005–1011 (1997).
3. C. Dumoulin, and B. Cochelin, *Mechanical behaviour modelling of balloon-expandable stents*, J. Biomech. **33**(11), 1461–1470 (2000).
4. S. N. D. Chua, B. J. MacDonald and M. S. J. Hashmi, *Finite-element simulation of stent expansion,* J. Mater. Process. Tech. **120**(1), 335–340 (2002).
5. S. N. D. Chua, B. J. MacDonald and M. S. J. Hashmi, *Finite element simulation of stent and balloon interaction,* J. Mater. Process. Tech. **143–144**, 591–597 (2003).
6. P. Libby, *Atheroma: more than mush,* Lancet. **348**(Suppl. 1), s4–s7 (1996).
7. P. H. Grewe, K. M. Müler, T. Deneke, E. Harrer, A. Germing, A. Müge, and B. Lemke, Stents: material, surface texture and design, in theory and practice, Minim. Invasive. Ther. Allied. Technol. **11**(4), 157–163 (2002).
8. K. Mori and T. Saito, *Effects of stent structure on stent flexibility measurements*, Annals. Biomed. Eng. **33**(6), 733–742 (2005).
9. C. Rogers, D. Y. Tseng, J. C. Squire and E. R. Edelman, Balloon-artery interactions during stent placement: a finite element analysis approach to pressure, compliance, and stent design as contributors to vascular injury, *Circ. Res.* **84**(4), 378–383 (1999).
10. A. I. Veress, D. G. Vince, P. M. Anderson, J. F. Cornhill, E. E. Herderick, J. D. Klingensmith, B. D. Kuban, N. L. Greenberg and J. D. Thomas, *Vascular mechanics of the coronary artery*, Z. Kardiol. **89**(Suppl. 2), 92–100 (2000).
11. R. A. Baldewsing, C. L. de Korte, J. A. Schaar, F. Mastik and A. F. van der Steen, Finite element modeling and intravascular ultrasound elastography of vulnerable plaques: parameter variation, *Ultrasonics*. **42**(1–9), 723–729 (2004).

STUDY ON APPLICATION OF HIGH SPEED MILLING IN DIES MANUFACTURING FOR PLATE HEAT EXCHANGERS

Xiaoyang Wang[1], Limin Li[2]

[1]*School of Mechatronics Engineering & Automation, Shanghai University, China; Email:* wangxiaoyang1997@126.com [2]*CIMS & Robot Center of Shanghai University, Shanghai University, China.*

Abstract: This paper studies the application strategies of high speed milling (HSM) technology in dies manufacturing for plate heat exchangers (PHE) according to the features of dies for plate heat exchangers (DPHE). It mainly analyzes the safety strategy, the using strategy of cutting tools, the technics, the virtual reality and simulation and so on under HSM situation, which is very significance for promoting deep application of HSM technology in industrial enterprises and improving aggregate level of our manufacturing industry.

Key words: High Speed Milling; HSM strategy; Dies for Plate Heat Exchangers; Technics.

1. INTRODUCTION

High speed milling (HSM) technology is one kind of emerging advanced manufacturing technologies (AMT), which results from the integrated application and development of many new technologies, such as new material technology, computer technology, control technology, precision manufacturing and so on. HSM technology refers to a system, as shown in Fig. 1. HSM technology has more superiority than low speed milling, for example, greatly increasing the material cutting ratio per unit time, getting to much smooth surface quality, successfully machining high rigidity materials. Since the early 1990s, with the maturing of high-speed numerical control machine tools technique and cutting tools technique, HSM Technics has increasingly become the key technique to make successfully high speed milling. In the recent past, application of HSM in difficult-to-machine

Please use the following format when citing this chapter:

Wang, Xiaoyang, Li, Limin, 2006, in International Federation for Information Processing (IFIP), Volume 207, Knowledge Enterprise: Intelligent Strategies In Product Design, Manufacturing, and Management, eds. K. Wang, Kovacs G., Wozny M., Fang M., (Boston: Springer), pp. 743-748.

materials and complicated shapes is more and more extensive. And dies for plate heat exchangers (DPHE) are one of the complicated shapes. Due to the big size and complicated shape of plate and present low velocity machining which results in poor surface quality only to be improved by hand-grinding, the whole time is very long which limits the development of plate heat exchangers. Hence, it is very significant to shorten the machining time and improve its surface quality by adopting HSM technology in dies manufacturing for plate heat exchangers (PHE).

Based on the key technology of HSM, aimed at the features of DPHE, the paper studies the technics strategy of HSM in dies manufacturing for PHE. The practice indicates adopting high speed milling is not only the effective method to dies manufacturing for PHE but also the best choice to large and difficult-to-machine work piece.

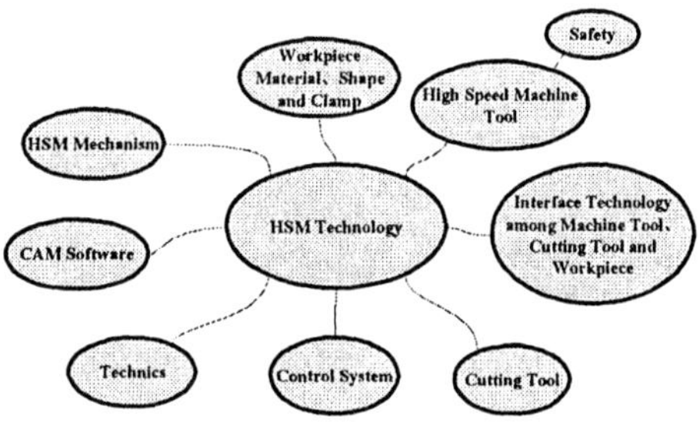

Figure 1. High Speed Milling System

2. SUMMARIZING OF HSM TECHNOLOGY

HSM refers to milling in much high speed than conventional milling speed. Therefore, sometimes it is called Ultra-High Speed Milling.

Through ultra-high speed simulation test, Dr. Carl Salomon (Germany's cutting physicist) found that cutting temperature increases with the increasing cutting speed within conventional cutting speed. There exists a speed scope for every work piece material in which cutting is impossible because cutting temperature is too high to withstand for any cutting tool. However, cutting temperature will reduce when cutting speed continues increasing and exceeds such a speed scope. At the same time, cutting force also drops greatly. In other words, there will be lower cutting temperature and lower cutting force within some high cutting speed scope. We can not only make ultra-high speed cutting with existing tools which will reduce cutting time sharply and improve producing efficiency of machine tool greatly, but also bring a series of fine characteristics into cutting process [1].

During the early 1990s, application of Motor-spindles technique and Linear-motor technique in machine tools, it has realized high speed some up to 60000～100000r/min, reached high feed rate and high fast-stroke speed up to 60～100m/min and obtained high acceleration/deceleration of spindle and table, from past 0.1g～0.2g to now1g～8g, which makes speed transition finished instantly.

HSM technics technology is one of the key technologies to make successfully high speed milling, including selecting different milling methods aiming at different materials, optimizing machining parameters and selecting appropriate cutting tool materials and cutting tool parameters.

3. FEATURES OF DPHE

PHE is usually composed of corrugating and concavo-convex metallic plates touched each other. It can exchange heat sufficiently because of no bypass in the heat transferring. Corrugation makes fluid turbulently flow. So its heat exchange coefficient is very high. DPHE is the key equipment to produce plates and is one kind of high-precision dies. Our study concentrates in the cores' milling. The core material of DPHE is the 40Cr, which corrugates the stainless steel plates with thickness 0.6 ~ 0.8mm, as shown in Fig. 2. DPHE mainly has the following features:

Figure 2. Dies Cores of Forming Plate

(1) Big outline dimensions: The die core area is generally above $1m^2$. Some even achieve above $1.5m^2$.

(2) Small contact point sizes: Plate often adopts small sizes in need of heat exchanging and intensity. Minimal end size is generally between 2.5mm and 6mm.

(3) Changeful corrugation angles: There are three angles around sealed groove between degree 25 to 45. So different angle surfaces mutually pass through that is difficult to milling.

(4) Small corrugation depth: The corrugation is usually about 3.5mm.

(5) High request for size precision: Dies corrugates the stainless steel plates with thickness 0.6 to 0.8mm. After corrugating, shape depth deviation and sealed groove size deviation does not surpass ±0.10mm.

(6) High surface quality: Obvious tool tracks can't be on the formation surfaces of die cores because the mediums that plates transmit are corrosive that request corrugating plates have much high surface precision which is not allowed depth 0.05mm upwards pits, scratches or indentation on surfaces.

4. HSM STRATEGY OF DPHE

The particular structure and the special request decide that only HSM can obtain the high production efficiency and the high surface quality. The cores of DPHE have so massive light-sized machining surfaces and the neighboring incline angles are so different that it is necessary to adopt small diameter cutting tools and very high spindle rotational speed. But then, it is possible to select small cutting depth value because the total depth is only 3.5mm. The key is optimizing the cutting tool path and other machining parameters to avoid that cutting quantity changes sharply when milling work piece. In practice, it should mainly use the following machining strategies:

(1) Safety strategy

Most of cutting tools in HSM are end milling tools and face milling tools whose centrifugal forces surpass greatly cutting force when the tools are in high speed rotating. So the first important thing when adopting HSM is safety because accident would bring great injury and loss to person and machine tools. The safety strategy is that the machining parameters never exceed the rotation speed and correlative characteristic parameters that cutting tools allow.

(2) Using strategy of cutting tools

Using coated cemented carbide is the best choice because the cutting is not steady especially when machining the head plate die cores. Cutting tools should be high wear-resisting, high toughness and high anti- impact. Coated cemented carbide tools have high heat-resist and resistance to wear, as well as good toughness and anti- impact.

(3) Rough machining technics

Surface quality and outline precision is not especially important during rough machining. This moment, tools attrition is most important because it is very difficult to achieve the stable cutting condition. Considering machining efficiency and cutting tool life, it is much important to keep even cutting quantity, constant cutting direction and linear speed, good transient equilibrium. Simultaneously maintain climbing by any possibility.

From the above, when rough machining: (a) maintain stable cutting depth and quite small undulation scope cutting width; (b) add filleting transition at tool path corner; (c) feed tool along spiral line or maximal angle 5° oblique line when machining the interior part of work piece.

(4) Half-finish machining technics

Half-finish machining is mainly to retain a certain working allowance for finish machining. The main task of this process is to smooth the sharp corner and get rid of surplus material. At present majority CAM software can automatically identify rest material and create cleaner tool path. Therefore, this process mainly confirms the results of clearing corner and cutting allowance to retain even milling allowance for finish machining.

(5) Finish machining technics

Finish machining is the last step of procedure. Its goal is to meet the requirements for best surface quality and outline precision. Too high cutting speed, too small federate per tooth (fz) or unstable allowance can cause coating layer burning and destroy the material. Therefore, when finish machining: (a) avoiding turning sharply; (b) reach next cutting depth directly from contour in order to avoid outside feeding or retracting operation. If feeding or retracting outside, then feed or retract along spiral line or oblique line. (c) fz should be suit for radial cutting depth (ae); (d) select stable fz to get requesting surface quality and maximum tool life.

(6) Virtual reality and simulation

Small sizes of corrugation contact points and consecutive concave-convex shapes cause frequently over-cut and no-cut phenomenon. It is important to make full use of the simulation function of advanced CAM software and inspect carefully tool path.

5. COMPARISON OF HSM AND CONVENTION MILLING

It takes about four to seven months with convention milling method to machine DPHE. Moreover the surface quality does not satisfy the product requirement only to be improved by hand grinding. Big outline size and small concave-convex shapes make hand-grinding very difficult and taking time. However, there will be the following superiorities if using HSM:

(1) With cutting speed increasing greatly, feed speed improves about five to ten times and material cutting ratio per unit time enhances three to six

times than conventional method. It reduces greatly idle motion time of no-cut operation because its speed increases a lot. As a result, the efficiency of machine tools increases about four times.

(2) The exciting frequency is so high in HSM that is far away from the inherent frequency of machine-tool-part. So machine working is steady with light vibration and surface is the same as grinding surface.

(3) Cutting force may reduce above 30%, and about 95% ~98% cutting heat is taken away by chip before it is transmitted to work piece. Therefore, work piece can basically maintain cold state all milling long, no distortion, no case hardening.

(4) HSM can cut many kinds of difficult-to-machine materials. It can mill the blade shapes of bordering-ornament blanking dies for PHE after quencher, which can avoid the distortion that heat treatment brings and which also resolves the problem of blade sharpening.

(5) High speed machine tools can bring quick return on its investment by shortening time of delivery and reducing floor space and the number of staff.

6. CONCLUSION

HSM technology would have become a kind of general and foundation technology of advanced manufacturing technology. It represents the development direction to cutting technology and has widespread application prospect. HSM can substitute for hand grinding in dies manufacturing for PHE. Its successful application will improve the production efficiency and surface quality, shorten machining period, and enhance the level of DPHE technology and the ability of quick reaction to marketplace. It is important to impel application of HSM in difficult-to-machine structure and shape. It is also important to develop our country HSM technology and improve application level of high speed machining and develop advanced manufacturing technologies. Realizing HSM needs to integrate and optimize many foundation subjects and implement systems engineering technology.

7. REFERENCES

1. Bailin Zhang. *High Speed Machining Technology and Application*. Beijing: Machine Press.2002.9
2. Eing-Jer Wei, Ming-Chang Lin. Study on general analytical method for CNC machining the free-form surfaces. *Journal of Materials Processing Technology* 168 (2005) 408–413

THE APPLICATION OF LEAST SQUARE IN PRECISION CYLINDRICAL GRINDING

Dawei Fu, Li Su,Yang Wang
School of Mechanical Engineering,Hebei Unibersity of Technlolgy, China; Email: ffddww@163.com.

Abstract: In this paper, PMAC(Programmable Multi-Axis Controller) is used as machine tool CNC system to control linear motor's micro-movement and to achieve on-line measurement of part circularity error. The part circularity is calculated through least square method by measured circularity error. The part circularity error can be decreased by PID control of PMAC, so precision cylindrical grinding can be achieved.

Key words: PMAC(Programmable Multi-Axis Controller); PID control ; precision grinding, LSM(Least Square Method).

1. INTRODUCTION

As high-speed and super high-speed grinding can greatly improve grinding efficiency and machine-part manufacturing quality, it is a kind of technology of powerful vitality and vast range of prospects. It is devoted much attention by the advanced industry countries of the world. In this paper, a LSM is proposed to fulfill high-speed precision grinding of cylindrical parts based on PMAC system. The linear motor's micro-movement and electric spindle of machine tool are controlled by open controller PMAC. Also, a set of numerical control system of grinder is integrated. A good experimental result is gotten in this grinder to grind cylindrical parts.

This project is supported by Doctor foundation of Hebei education bureau under grant B2003205.

Please use the following format when citing this chapter:
Fu, Dawei, Su, Li, Wang, Yang, 2006, in International Federation for Information Processing (IFIP), Volume 207, Knowledge Enterprise: Intelligent Strategies In Product Design, Manufacturing, and Management, eds. K. Wang, Kovacs G., Wozny M., Fang M., (Boston: Springer), pp. 749-754.

2. SYSTEM INTEGRATION OF CNC GRINDER

As shown in Fig.1,this machine tool adopts two-axis PMAC to control linear motor and servo motor. At the same time electric spindle is cooled by circulatory water.

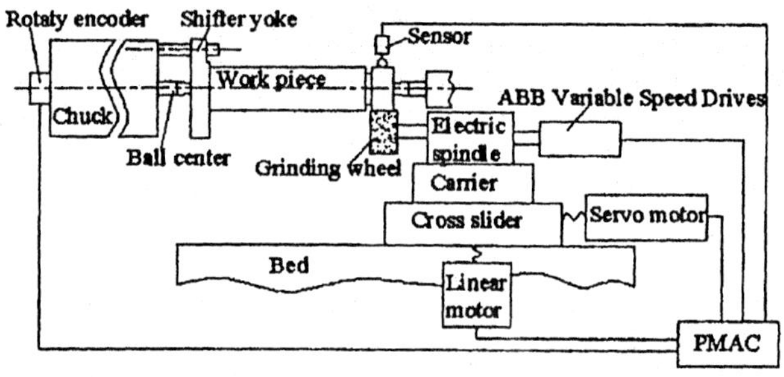

Fig.1 The principle of high-speed precision cylindrical grinding

3. THE MATHEMATICAL METHOD OF LEAST SQUARES CIRCLE OF CIRCULARITY ERROR

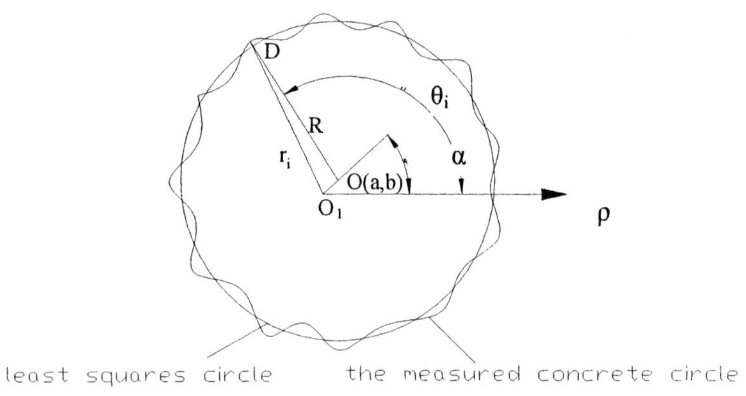

Figure2. The mathematical method of least squares circle of circle of circularity error.

As shown in fig.2, took the rotary center O_1 of the measured section as the center of a circle and divided the measured circle into m equal angles, radii $r_i(i=1,2....m)$from the point O_1 can be obtained. The points of intersection between these m radii r_i and the measured circle are P_i. At the pole coordinate of O_1 as the origin, let the center of least squares circle as O, the radius an R, the offset of least squares circle and rotary center as e. The radial offset from each discrete sampling point Pi to least squares circle as ϵ_i (i=1,2 m), the angle of line O_1P_i from each equal sampling point Pi to rotary center O_1 and the axis of pole coordinate as θ_i(i=1,2 m). The point of the intersection between line OP_i and the least squares circle is D. The radial error Δr_i from the center O_1 is collected by the sensor. From the Fig.2, it is obtained OD=R, OO_1=e, P_iD= ϵ_i , O_1P_i=r_i, $\angle P_iO_1\rho$ =θ_i, $\angle OO_1\rho$ =α, a=ecosα, b=esinα

Also:

$$\begin{aligned} r_i &= e\cos(\theta_i - \alpha) + [(R+\varepsilon_i)^2 - e^2\sin^2(\theta_i - \alpha)]^{1/2} \\ &= e\cos\theta_i\cos\alpha + e\sin\theta_i\sin\alpha + [(R+\varepsilon_i)^2 - e^2\sin^2(\theta_i - \alpha)]^{1/2} \end{aligned} \tag{1}$$

Make the non-linear function $[(R+\varepsilon_i)^2 - e^2\sin^2(\theta_i - \alpha)]^{1/2}$ linear.

Let $f(e) = [(R+\varepsilon_i)^2 - e^2\sin^2(\theta_i - \alpha)]^{1/2}$

By means of Maclaurin formula: $f(x) = \sum_{n=0}^{\infty} \frac{1}{n!} f^{(n)}(0)X^n$

$$f(e) = R + \varepsilon_i - \frac{e^2}{2(R+\varepsilon_i)}\sin^2(\theta_i - \alpha) - \frac{e^4}{8(R+\varepsilon_i)^3}\sin^4(\theta_i - \alpha) \tag{2}$$

If e<<R, and $\sin^2(\theta_i - \alpha)$ <<1

Therefore $f(e) \approx R + \varepsilon_i$ where the maximum error caused is $\frac{e^2}{2(R+\varepsilon_i)}$.

Accordingly: $r_i = R + a\cos\theta_i + b\sin a\theta_i + \varepsilon_i$ i=1,2 .m (3)

The structural matrix of data is:

$$\theta = \begin{bmatrix} 1 & 1 & \ldots\ldots & 1 \\ \cos\theta_1 & \cos\theta_2 & \ldots\ldots & \cos\theta_m \\ \sin\theta_1 & \sin\theta_2 & \ldots\ldots & \sin\theta_m \end{bmatrix}^T$$

Coefficient matrix of normal equations is:

$$A = \theta^T\theta = \theta = \begin{bmatrix} M & \sum_i \cos\theta_i & \sum_i \sin\theta_i \\ & \sum_i \cos^2\theta_i & \sum_i \cos\theta_i \sin\theta_i \\ & & \sum_i \sin^2\theta_i \end{bmatrix}$$

As m is even number

Therefore $A = diag[m, {}^{m}\!/_{2}, {}^{m}\!/_{2}]$

$$A^{-1} = diag[{}^{1}\!/_{m}, {}^{2}\!/_{m}, {}^{2}\!/_{m}]$$

Let $D = [r_1, \ r_2, \ldots\ldots, \ r_m]^T$

So constant matrix of normal equations is:

$$B = \theta^T D = [\sum r_i \ \sum r_i \cos\theta_i \ \sum r_i \sin\theta_i]^T$$

Let $P = [R, \ a, \ b]^T$

So $P = A^{-1}B = [\frac{1}{m}\sum r_i, \frac{2}{m}\sum r_i \cos\theta_i, \frac{2}{m}\sum r_i \sin\theta_i,]^T$

That is: $R = {\sum_i r_i}\big/ m = r_0 + \Delta R$, where $\Delta R = \frac{1}{m}\sum \Delta r_i$

$$a = \frac{2}{m}\sum_i r_i \cos\theta_i = \frac{2}{m}\sum \Delta r_i \cos\theta_i$$

similarly: $b = \frac{2}{m}\sum \Delta r_i \sin\theta_i$

So $\varepsilon_i = r_i - R - a\cos\theta_i - b\sin\theta_i$ i=1,2, ,m

Or $\varepsilon_i = \Delta r_i - \Delta R - a\cos\theta_i - b\sin\theta_i$

So the circularity error of workpiece is:

$$f_o = \max\{\varepsilon_i\} - \min\{\varepsilon_i\}$$

$$= \max\{\Delta r_i - a\cos\theta_i - b\sin\theta_i\} - \min\{\Delta r_i - a\cos\theta_i - b\sin\theta_i\}$$

i=1,2, ,m

4. THE PID CONTROL PRINCIPLE OF PMAC

This numerical control grinder adopts two-axis PMAC with which we can control Z-phase servomotor and X-phase linear motor's feed movement. The sensor collects workpiece circularity error signals by A/D channel on the PMAC. We can attain the error on a certain intersection of workpiece. Comparing this error with tolerance of workpiece, we can figure out deviation and decide the linear motor's feed in accordance with this deviation. Finally we can attain precise machining accuracy of workpiece by many times circle count (as shown in Fig.3). Different speeds of rotation of electric-spindle are achieved by D/A channel on PMAC. The travel limits switch and the cooling water for electric-spindle are controlled by I/O channel on the PMAC.

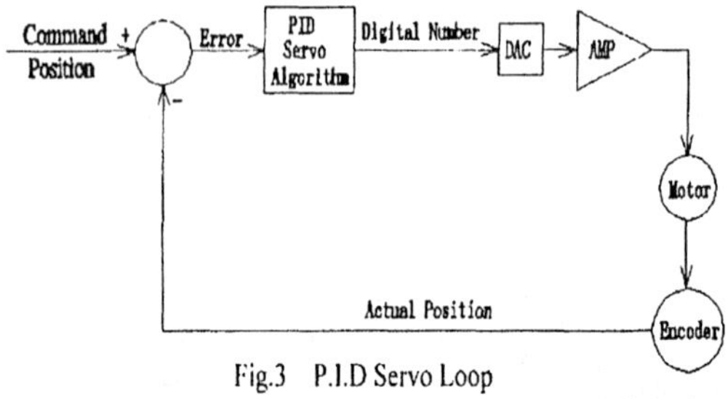

Fig.3 P.I.D Servo Loop

5. CONCLUSIONS

This paper adopts PMAC to accomplish the precision machining for grinding. The feature of PMAC is that it can control two different types of motors to accomplish coordinate motion of two axis, and it can realize some special functions such as precise linear and circle interpolator, "S-Curve" acceleration and deceleration. Thus it can fulfill high precision grinding for cylindrical, elliptical, taper contours and other kinds of irregular machine ones.

6. REFERENCES

1. Zhang Yu: Computer Aided Precise Measurement.(Northeastern University Press, China,1993).
2. Francis.T.Farago: *Abrasive Methods Engineering*. (Industrial Press, Americal,1995).
3. Wang,YongZhang etc: Numerical Control for Machine Tool.(Harbin Institute of Technology Press,China,1995).
4. Pmac User'S Manual.(DELTA TAU Data System, Inc.1998).

THE APPLICATION OF THE GENETIC ALGORITHM TO THE NUMERICAL SIMULATION IN SHEET METAL FORMING

Jingjing Xu, Changdong Li, Yimin Wu, Wei Huang
[1]*School of Mechanical Engineering, Shanghai University, China; Email: xujingjing1125@yahoo.com.cn*

Abstract: This study presents a genetic algorithm GA adopted for recognizing the material parameters with the nonlinear relation in the sheet metal forming. Firstly the nonlinear regression model is established for the parameters estimation. Then based on the material tensile test the parameters estimation is finished by GA and the least square method LS. At last GA is applied for recognizing the material model parameters and the numerical simulation is finished in the V-shape sheet metal forming.

Key words: Genetic Algorithm, Numerical Simulation, Parameter Recognition, Finite Element Method, Nonlinear Material Model.

1. INTRODUCTION

The accurate input of the material model parameters is highly important for the numerical simulation in the sheet metal forming, and the parameters recognition of the material model is a complex combinatorial optimization problem. Frequently the material model is so complicated that it is difficult to get the optimum model parameters with conventional methods such as direct search method, Hoole-Jeeves method, and Gauss-Newton method etc. [1] As a probabilistic search technique the genetic algorithm GA base on the principles of biological evolution, which a biological organism evolves to more fully adapt to its environment, GA becomes a powerful analytical tool to solve an array of constrained and combinatorial optimization problems rapidly.[2-6] Now GA is less applied for the field of the numerical simulation. In this paper GA is applied to deal with the optimization problem of the parameters recognition by using propagation, crossover and mutation

This project is supported by the Ford-China Research and Development Foundation under the grant No.9716214.

Please use the following format when citing this chapter:
Xu, Jingjing, Li, Changdong, Wu, Yimin, Huang, Wei, 2006, in International Federation for Information Processing (IFIP), Volume 207, Knowledge Enterprise: Intelligent Strategies In Product Design, Manufacturing, and Management, eds. K. Wang, Kovacs G., Wozny M., Fang M., (Boston: Springer), pp. 755-760.

operation. Firstly based on the material tensile data the parameters estimation is finished by GA and the least square method LS, and the compared results show that GA is better than the result of LS. Then the gotten model parameters are used for the numerical simulation in the V-shape sheet metal forming.[7, 8]

2. PROBLEM DESCRIPTION FOR PARAMETERS RECOGNITION OF THE MATERIAL MODEL

The material parameters are where the shoe pinches for the numerical simulation in the sheet metal forming, and the material model is nonlinear elastic-plastic material. In this paper two material models are adopted. The material model 1 expresses the elastic-plastic behavior with a power exponent relationship, and the yield stress function is shown as equation (1)

$$\sigma_y = k\varepsilon^m \dot{\varepsilon}^n \qquad (1)$$

Where, k is strength coefficient, m is hardening index, n is strain rate sensitivity coefficient.

The material model 2 provides the elastic-plastic behavior with isotropic hardening and has a power law constitutive relationship, which includes the Cowper-Symonds multiplier to account for strain rate as equation (2).

$$\sigma_y = \left[1 + \left(\frac{\varepsilon^{\cdot}}{c}\right)^{\frac{1}{p}}\right] k\varepsilon^n \qquad (2)$$

Where σ_y is yield stress, $\dot{\varepsilon}$ is strain rate, ε is strain, c, p are Cowper-Symonds strain rate coefficient, k is strength coefficient, n is hardening index.

The nonlinear regression method is used for the parameters estimation of the nonlinear function. Based on two material models the nonlinear regression model can be expressed as follows:

$$\begin{cases} \sigma_y = f(\varepsilon, \varepsilon^{\cdot}, \theta) + e \\ e \sim N(0, \sigma^2) \end{cases} \qquad (3)$$

Where θ respectively express the parameters k, m, n in equation (1) and the parameters c, p, k, n in equation (2), $f(\varepsilon_i, \dot{\varepsilon}_i, \hat{\theta})$ is the nonlinear function, and error e yield to the standard normal distribution function, and the value $\{(\varepsilon_i, \dot{\varepsilon}_i, \sigma_{yi}),\ i=1,2,\ldots,n\}$ can be gotten by the material tensile test data, now the nonlinear parameters recognition is means that how to get the estimation parameters $\hat{\theta}$ to satisfy the formula (4) for any value θ.

$$f(\hat{\theta}) \le f(\theta)$$

$$f(\theta) = \sum_{i=1}^{n} [\sigma_{y_i} - f(\varepsilon_i, \dot{\varepsilon}, \theta)]^2 \quad (4)$$

3. MATERIAL PARAMETERS ESTIMATION

It is necessary to build the nonlinear elastic-plastic material model for the process simulation in the sheet metal forming, so the unilateral tensile test of the thin steel plate is finished by ZWICK tensile machine at various speed, and the data of true stress *ss*, true strain *sn* and strain rate *snr* are gotten and shown as table.1.

Table 1. True stress *ss* (MPa) , true strain *sn* ($x10^{-6}$) and strain rate *snr* ($x10^{-4}$/s) data

snr1	sn1	ss1	snr2	sn2	ss2	snr3	sn3	ss3	snr4	sn4	ss4
3	0.06	120	3	0.06	143	3	0.06	153	3	0.06	147
3	0.10	199	3	0.10	222	3	0.10	229	3	0.10	219
12	0.50	312	15	0.50	315	13	0.51	312	543	0.58	337
24	1.01	329	30	1.01	326	27	1.00	327	561	1.51	357
21	2.00	350	47	2.01	351	30	1.50	341	532	2.59	378
20	4.00	379	63	4.01	386	58	3.50	380	527	4.60	406
17	6.00	399	85	6.01	409	74	5.51	403	556	6.52	423
16	10	426	114	10	439	115	9.51	433	540	10.6	449
13	11	431	113	11	446	124	10.5	439	529	11.5	454
8	12	435	112	12	451	128	11.5	444	508	12.6	458
176	15	461	116	13	456	138	12.5	449	471	13.6	462
264	18.5	475	324	20	479	392	21.5	481	386	23.1	473

Confirming the GA essential points is necessary before the genetic code is finished by program language VC++6.0.

GA code is a combination result of the code scheme, strategy selection, and genetic arithmetic operators. In the code scheme the real number encoding is selected, and the standard fitness function is applied for the measuring the fitness of the population. The original fitness function is transferred into the maximal standard $fit(\theta)$ measure and the value of fitness always is not less than zero. According to the formula (4) the original fitness function estimation is shown as follows.

$$fit(\hat{\theta}) = \sum_{i=1}^{n} [\sigma_{yi} - f(\varepsilon_i, \dot{\varepsilon}_i, \hat{\theta})]^2 \quad (5)$$

Where, $(\varepsilon_i, \dot{\varepsilon}_i, \sigma_{yi}), i = 1,2...n)$ is the value of the strain, strain rate and yield stress at i point in the tensile test, $\hat{\theta}$ is parameters estimate, and $f(\varepsilon_i, \dot{\varepsilon}_i, \hat{\theta})$ is the yield stress estimation of the material model as equation (1) and equation (2).

Different strategy selection will produce the different selection pressure. The convergent speed of the high pressure is faster than the speed of the low pressure in the genetic algorithm. Here, the linear ranking selection is adopted; the selection probability p_i is confirmed by the sort of the individual fitness value in population as $p_i=2*(i+1) / N / (N+1)$. Where, N is the population number, and i is the selected number from 1 to N. based on the select probability p_i a random number r between 0 and 1 is generated, and if the equation, $p_0+p_1+ \quad +p_{i-1}<r\leq p_1+p_2+ \quad p_i$, is satisfied by the roulette wheel selection the individual i is gained.

The control parameters consist of the population number $N=50$, the maximal iterative number $gen=1000$, and the probability of the genetic operations such as propagation probability $p_r=0.1$, crossover probability $p_c=0.9$, mutation probability $p_m=0.15$.

The global correlation coefficient r is used for the estimate the results of the recognition parameters as equation (6), and the meanings of the other symbol in the equation are as before.

$$r = \sqrt{1 - \frac{\sum_{i=1}^{n}(\sigma_{yi} - f(\varepsilon, \dot{\varepsilon}, \hat{\theta}))^2}{\sum_{i=1}^{n}(\sigma_{yi} - \frac{1}{N}\sum_{i=1}^{N}\sigma_{yi})^2}} \tag{6}$$

According the test data, the nonlinear model parameters is estimated by GA and the estimated result is compared with the result of LS method, and the compared results are shown as table 2.

Table 2. The identified parameters of the material model 1 by using GA and PLS

method	k (MPa)	m	n	r
LS	609.7	0.1196	0.0133	0.9975
GA	609	0.1197	0.0131	0.999855

Table 2 shows that two methods have the same results, but the value of the correlation coefficient shows that the genetic algorithm has the better precision. Using LS method can't identify the model parameters of the material model 2, so GA program is applied for the parameters estimate as table 3. Based on the stress, strain and strain rate data in table 1, the parameters of the material model 2 are identified by GA code.

Table 3. The parameters of the material model 2 by GA

Method	C	P	k (MPa)	n	r
GA	0.062	46.5	300	0.13	0.9973

Table.3 shows that using GA the correlation coefficient is highly at 0.99. Subsequently the gotten accurate parameters can input the finite element model in the sheet metal forming, and the numerical simulation can be finished.

4. NUMERICAL SIMULATION OF V-SHAPE PART

Now the forming process of the V-shape part in the sheet metal forming with the 800mm/min punch speed is simulated by finite element method. The process simulation is presented by ANSYS/LS DYNA software. The sheet metal material is tinned steel, the punch and die material is 45# steel, the material model 2 is adopted, and the other input parameters of the finite element model are shown as table 4.

Table 4. The input parameters of the finite element model

Punch density ρ (kg/m^3)	Punch elastic modul E (MPa)	Punch Poisson rate μ	Plate thickness t (mm)	Plate width b (mm)	Plate length l (mm)	Pressing length lp (mm)
7850	204000	0.271	0.34	25	75	64
Die density ρ (kg/m^3)	Die elastic modul E (MPa)	Die Poisson rate μ	Punch radius Rt (mm)	Die radius Rb (mm)	Stoke S (mm)	Pressing force P (kgf)
7820	207000	0.29	2	5	43	3.968

Based on the geometry shape and loading properties the global model is symmetry structure, so the finite element model is only built 1/4. During the calculation the four-node shell element is selected with Belytschkl-Leriathan element algorithm. The symmetrical constrains are applied to the symmetry boundary of the V-shape part, and the interface friction is deal with the various dynamic and static behavior. The punch speed is alterable virtual speed in order to reduce the calculation time, and the die is fixed, and the simulation forming time is the stop time when the trip is achieved. After the forming phase the springback simulation use the implicit program, and the explicit element is transferred into the implicit element type, and the material type is becoming the linear elastic material. The original structure is the last step shape after the finishing of the explicit calculation. At the same time, the die is moved and the stress status is kept. The symmetrical constrains are loaded, the structural big displacement behavior is considered, and the springback simulation is finished. The finite element model of the V-shape part and the simulation result is shown as Fig.1.

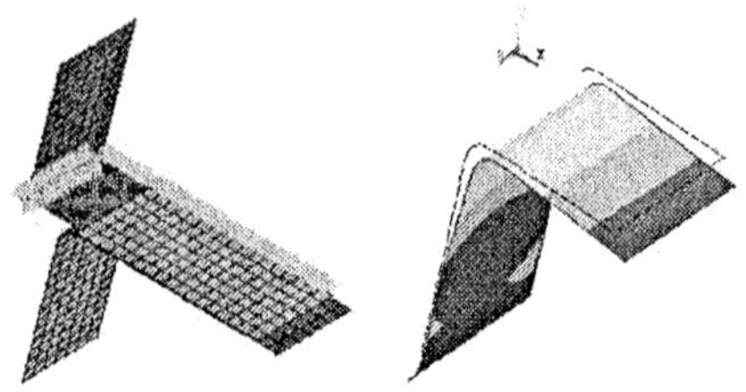

Figure 1. The finite element model and simulation result

5. CONCLUSION

1) The genetic algorithm using the real number encoding, using the standard fitness function to describe the fitness of nonlinear material model, through the propagation, crossover and mutation operation the accurate material parameters can be gotten. Comparing with the least square method, the genetic algorithm is simple and flexible and less constrained, and it is more adapted to deal with complicated model parameters with nonlinear material behavior.

2) The genetic algorithm provides a new tool for forming and springback simulation in the sheet metal forming, the use of genetic algorithms to better design and produce reliable products of high quality, and a new approach is achieved for the complicated parameters input in finite element calculation.

6. REFERENCE

1. Nash J. C. and Walker Smith M., *Nonlinear Parameter Estimation*, Marcel Dekker Inc., New York,Basel,1987
2. Buckles B. P. and Perry F. E., *Genetic Algorithms*, IEEE Computer Society Press. Los Alamitos,CA, 1992
3. Galletly J. E., (1992), An Overview of Genetic Algorithms, *Kybernetics*, 21(6), pp.26-30,
4. Angeline P.J, Reynolds R.G., (Eds.), Proc.of the 6th Int'l Conf. On Evolutionary Programming (EP'97), *Lecture Notes in Computer Science*, NO.1213, Springer-Verlag, 1997
5. Srinivas M. and Patnaik L.M., , Genetic Algorithms: a Survey, Computer, June, pp.17-26,1994
6. Pan Zhengjun etc., *Evolution Algorithm*, Qinghua University Press,2000.
7. *LS-DYNA Keyword User's Manual*, , Livemore Software Technology Corporation ,1999
8. *ANSYS/LS DYNA Basic Algorithm and Using Method,* (2000), USA ANSYS LTD.

THE COVERAGE FORM OF SURFACE COATING AND CRACK OF BASAL BODY UNDER MULTI-IMPACT LOADS

Shihong Shi [1], Keyun Di[2], Anjun Wang[2]

[1]*College of Mechanical Electronic Engineering, Soochow University, China; Email: Shishihong@suda.edu.cn.* [2]*Soochow University, China.*

Abstract: During the experiment for thick test piece with laser coating under multi-impact loads. We discover that the margin of stress force with the form of Total covering is lower than that with the form of partial covering. However, the life of the former is longer than the latter when they are under the same multi-impact stress force action. There will be an intensely deformative hardening and material floating phenomena on the lower part of the edge of covered coating on the basal body of the latter. And the same time, flaw is more easily to come into being and spread at this place, then cause abscission of coat.

Key words: Laser coat, cladding structure, repeated-impact crack.

1. INTRODUCTION

Covering of a coat of excellence performance functional material on the surface of parts. The surface function and operation performance of the part will be improved greatly .Various technology of surface coating prolongs the life of the machine parts working under execrable and complex condition. Then it improves security reliability of the whole machine or the system.

The coating structural member has various failure form under different external loads and working environment. During the experiment we discover that the thick test piece with different way of coating on it's basal body have the different failure method. The current research adopt the laser melting and cladding methods to make two coat on the basal body, one of which has the same width with the basal body and the other does not(figure 1). This article also concentrates in the research of metamorphose and crack behavior of the

Please use the following format when citing this chapter:

Shi, Shihong, Di, Keyun, Wang, Anjun, 2006, in International Federation for Information Processing (IFIP), Volume 207, Knowledge Enterprise: Intelligent Strategies In Product Design, Manufacturing, and Management, eds. K. Wang, Kovacs G., Wozny M., Fang M., (Boston: Springer), pp. 761-766.

basal body with the two structural forms under multi-impact loads which are modal in engineering.

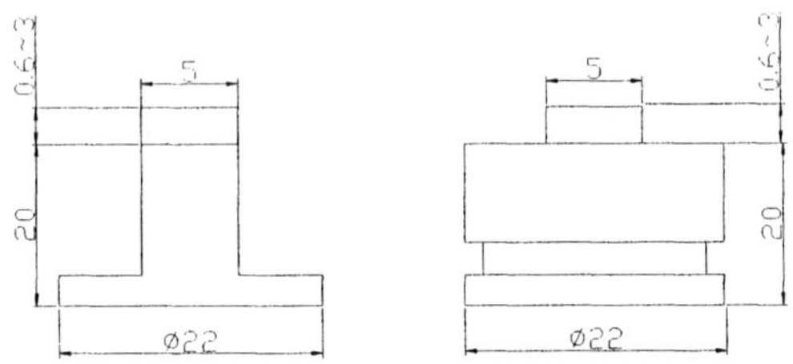

(a)The basal body is totally covered by coat (b)The basal body is partially covered by coat
Figure 1. schematic of test piece

2. EXPERIMENT CONDITION

Test piece material: basal body is medium carbon steel (45#) or austenitic stainless steel (1Cri8Ni9Ti), coating is Ni-base self -fluxing alloy.

Test piece preparation: The Coat is prepared by TJ-HL-T5000 laser device, which use the powder transport method to process the test piece. The thickness of the powder is 1-3.5mm. The parameter of melting and cladding technique: Power of the laser: 3-4.5kw, Diameter of light spot φ: 5mm, Scan Speed V: 5-10 mm/s, Laser mode: multiple-moulding: Test piece post process anneal temperature: 200℃-400℃. All of the coat are inspected. The surface of coat is finely grind by surface grinding machine. And the surface roughness is Ra 0.16mm. The test piece is in figure 1.

Impact & Collision experiment have been done in multi-impact test equipment. And the frequency of impact is tuned to 4 Hz.

Adopt OLYMPUS GX51 optical microscope to observe the section structure, use CL-YD piezoelectric crystal impact force sensor and YE5852 charge amplifier, TDS3000 digital fluorescence memorable oscillo graph to measure, record and display the wave form of the impact force.

3. EXPERIMENT RESULT AND ANALYSIS

In theory, the limit loads and deformation of the coat and basal body can be approximately analysed by applied theory of plane deformation of the ideal rigid-perfectly plastic material and the theory of slippage line field(It has ignore elastic deformation and plastic deformation strengthening) [2、3].

Figure 1 show that basal body is totally covered by coat, and the coat is the same width with the basal body. In figure 2 (σ is always negative in current experiment, The direction is oppositive to the showing).Applying uniform impacting force on the coating layer, The direction of trace of main stress force is perpendicularity downward to coat layer, The Load limitation is $P_t = \sigma_1 = 2K$ ($K = \tau_{max}$, is limit shearing strength) meanwhile $\sigma_2 = \sigma_1 / 2$ (perpendicular to paper plane), $\sigma_3 = 0$ (flank is free surface). Slippage lines are two families of straight lines which are in quadrature with each other. They are should be 45°angle to main stress trace or the surface of test piece.

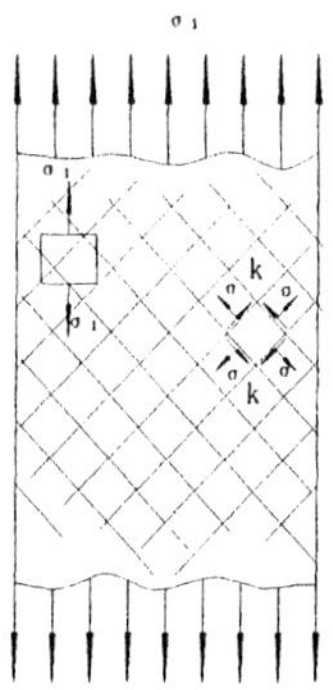

Figure 2. Uniformity stress force slippage line field

Figure 1(b) show that it is partially covered, of which the limit load and slippage of field of lines are the same with the former. The further detail information is indicated by figure 3(a). Provided HBAE is the basal body surface, of which p is impact force, which is working in extreme status, applying on the basal body surface through the coat layer and along the AB

And then the question is equal to like wedge compressed in single side under such condition that it take A,B as vertex and vertex angle is $2\gamma = \pi$. Then we can construct the two connected slippage field of lines under the surface of the basal body, which is indicated by figure 3(a). By way of illustration of right side, the Slippage field of lines is composed of stress field OCA, ADE and center sector stress field ACD .Then we can get the solution that the extreme pressure $P_t = K(2+\pi)$. Then the impact pressure force that is perpendicular to paper plane is $P = 2b \cdot K(2+\pi)$, the figure 3(b) show that the circumstances which appear after the right part flow (solid line). It is easy to get conclusion that OCDE are flow down to O'C'D'E' , Point A is singular point, Area OCA and ADE rigidity translation, Area ACD happens internal translation along the slip line.

Figure4 is the right half part of the graph that is the section deformation of the piece that is partial covered by coat. In which the deformation of coat layer are uniformity deformation. A little pane in Figure 4 is the orthogonal chiseled line which had been drawn on the section before the test. We can get conclusion that basal body covered by coat has been forced to move downwards, because of the deformation of the little pane. The material of basal body outside of covering area has been squeezed and come into bulge, And the deformation in lower edge of coat where is the light region is the most fierce. The deformation is according to deformation of field of slip translation that shown in figure 3(b).

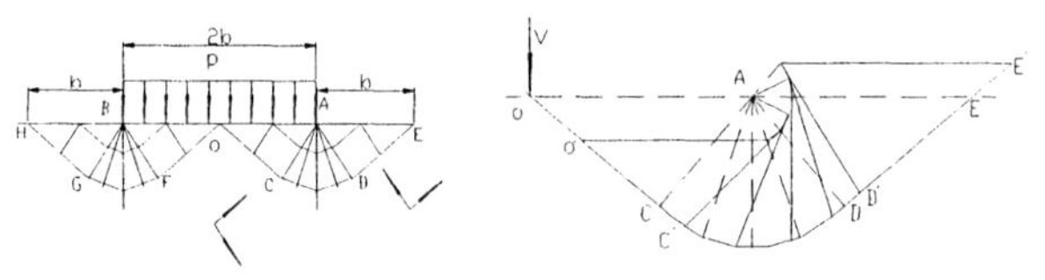

(a)Connected field of slip line (b)Field of flow in right part material

Figure 3. The field of slip line of the basal body with the coating layer of rigidity flat punch

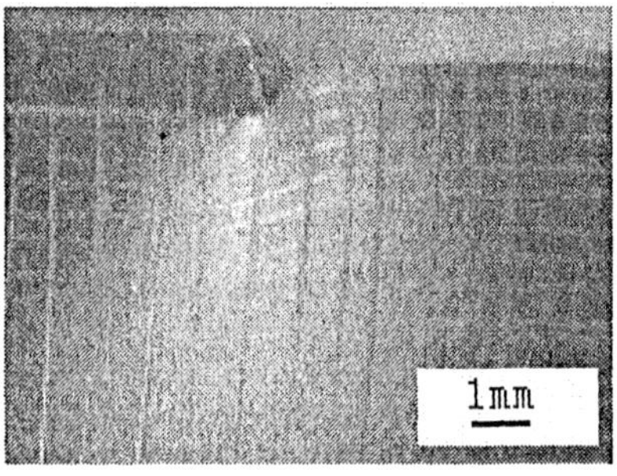

Base: 1Cr18Ni9Ti; Impact stress : 158MPa; Times of Impact 1.28×10^6

Figure 4. The deformation of basal body which totally covered by coat

On the part of the basal body with partial covered , A majority of test piece have came into emergence of crack which shown in figure5(a) and (b) .According to figure 3 we can learn that the deformation in singular point A and b are the most violent and they are the start point of crack . The cracks are spread along with slip line AC or BF which are 45 degree angle to the surface. And in the second period it is perpendicular to the lines of forces gradually because of the alternatively impact force. So point A and point B have come into being cracks and meet in the middle of the basal body and then make the coat fall off.

During the test experiment, Most of the test pieces which is partial covered by coat layer have appeared similar crack, However we don't find any test piece that is totally covered has the same phenomena.

The deformation of the basal body shown in figure5(b) is small, that because the material in the surface of 45 carbon steel basal body get quenching harden as quickly hot and cooled when the laser melt and cover the material. And the material 1Cr18Ni9Ti shown in figure4 basically can not be quenched, then the ability of the plastic flow are more intensive.

The experiment discover that the basal body partially covered by coat appear fewer deformation and crack under strike of lower impact stress time after time. And it's life is longer and accord to the plastic theory[2]. It is because that basal body partially covered with coat is wider than coat and then it can share the pressure force. However when the multi-impact load get more powerful, the material that close to edge part of the area covered by coat in the basal body of the partially covered test piece will produce fierce squeeze deformation and harden phenomenon . The character is that the deformation form of soft basal body is mainly squeeze deformation (figure 4); But in the same place, the rigidity basal body and basal body that can not get rid of rest tension stress will be easily to produce crack and spread into basal body along with 45°slip line figure5 (b), And then it is likely to be failure and it's life is far shorter than the structure of totally covered by coat. As the result of the experiment, we can draw a conclusion that on the part of form of covering of coat on the basal body, the structure of totally covered by coat is better under high impact stress.

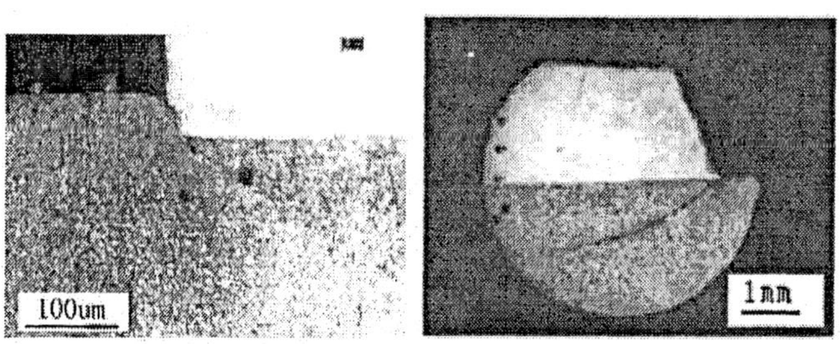

(a)Base: 1Cr18Ni9Ti
Impact stress: 158MPa
Times of Impact: 5.5×10^4

(b) Base: carbon steel 45#
Impact stress: 261MPa
Times of shock: 46×10^4

Figure 5. The crack on the base and its spread direction

4. SUMMING-UP

(1) Under multi-impact loads, we can apply the theory of the field of slip line and the surface deformation on ideal perfectly rigid-plastic material to analyse the extreme load and deformation of coat and basal body approximately, And mechanics model inosculate with the slip deformation of the basal body of the test piece.

(2) Under multi-impact loads, if the impact force is smaller, the test piece whose basal body is totally covered by the coat allow lower limit stress force than that whose basal body is partially covered by the coat. Then the latter has longer life than the former. However when the multi-impact load get more powerful, the material that close to edge part of the area covered by coat in the basal body of the partially covered test piece will produce fierce squeeze deformation and harden phenomenon . The character is that the deformation form of soft basal body is mainly squeeze deformation (figure 4); But in the same place, The rigidity basal body and basal body that can not get rid of rest tension stress will be easily to produce crack and spread into basal body along with 45°slip line figure5 (b), And then it is likely to be failure and it's life is far shorter than the structure of totally covered by coat.

5. REFERENCES

1. Hu Chuanxi., (2000), Theory and Applications of Coating Technology, *Press of Chemical Industry.*
2. Yan Zongda., (1998), Plastic Mechanics, *Tianjin University Press.*
3. Xu binye., (1981), Concise tutorial of plastic theory, *Beijing,Qinghua University Press.*

EFFECT OF SPECIMEN SIZE ON STRENGTH RELIABILITY OF FIBER REINFORCED COMPOSITE

Yun Lu[1], Hideharu Fukunaga[2]

[1]Faculty of Engineering, Chiba University, 1-33, Yayoi-cho, Inage-ku, Chiba, 263-8522, Japan; E-mail: luyun@faculty.chiba-u.jp, [2]Kure College of Technology, Institute of National Colleges, Japan, 2-2-11, Agaminami, Kure, Hiroshima, 737-8506, Japan

Abstract: A tensile test was carried out for continuous SiC fiber reinforced aluminum matrix composites (SiC/Al), which have different specimen width and length. A Monte Carlo simulation using finite element method was also performed with changing the width and length of composite model. The size effect of strength reliability was examined and discussed. The results showed that composite strength does not appear to have a large change but its reliability is improved with increasing in width of the composite specimen. Also, the composite strength decreases, but Weibull shape parameter has an increase tendency with increasing the gauge length. Effect of specimen width and length on composite strength and its reliability is dependent on the both factors, that is, the incidence of accumulating fracture process, the mean fiber strength and the strength scatter of weaker fibers. The accumulation of the fracture process is related to the fiber stress distribution in the composite.

Key words: fiber reinforced metal, SiC/Al composite, tensile test, Monte Carlo simulation, specimen width, gauge length, strength reliability, size effect.

1. INTRODUCTION

It was clarified that both strength and its reliability of single or bundle ceramic fiber have an obvious size effect[1-4]. According to these studies, the strength decreases and its scatter becomes wide with the gauge length of single ceramic fiber[3]. In the case of ceramic fiber bundle, the strength and its

Please use the following format when citing this chapter:
Lu, Yun, Fukunaga, Hideharu, 2006, in International Federation for Information Processing (IFIP), Volume 207, Knowledge Enterprise: Intelligent Strategies In Product Design, Manufacturing, and Management, eds. K. Wang, Kovacs G., Wozny M., Fang M., (Boston: Springer), pp. 767-779.

scatter go down and finally the both converge to the value from the Coleman theory with increase of the fiber number in a bundle [1,4]. Therefore, strength and its reliability of the fiber reinforced metal composite fabricated with use of ceramic fibers and metal matrix can be understood to have the size effect. It is very important to investigate the size effect of composite strength and its reliability from the view point of application for structural materials. Recently, it was reported that the strength has large size effect for fiber reinforced plastic composites[5-6].

In this study, a unidirectional SiC/Al composite having a constant fiber volume fraction was produced by hot press method. A tensile test was carried out for SiC/Al composite specimens with different width or gauge length. The size effect of composite strength and its reliability was examined. Meanwhile, a Monte Carlo simulation using finite element method was performed. The size effect was discussed taking fracture process or stress distribution in the composite into consideration.

2. SPECIMEN AND EXPERIMENTAL

A1050 pure aluminum plate with a thickness of 0.3mm was used as the matrix material. SiC fiber manufactured by a chemical vapor deposition method, obtained from Textron Specialty Materials Company (type: SCS-2, diameter: 140μm), was used as the reinforcement. SiC/Al composite was fabricated by hot press process method. Firstly an aluminum plate was cut into $30\times70mm^2$ pieces which were then annealed at 573K for 30min in air. In order to arrange the fibers on the aluminum pieces orderly, U-shaped grooves with an interspace of 0.5mm were made on one side of the aluminum pieces by pressing the stainless steel wires (with a diameter of 140μm) into the piece. Secondly, SiC fibers, which were cut into a length of 70mm from the obtained successive SiC fiber, were put into the grooves on the surface of each aluminum piece, which was then overlapped by another aluminum piece. Finally, this one layer SiC/Al perform was hot pressed into the composite by the process of temperature 893K, under the pressure of 56MPa and holding time 40min in air. The prepared SiC/Al composites have a dimension about 70mm length and 0.55mm thickness. Fiber volume fraction is about 4.7%. In the case of changing the specimen width, tensile specimens were cut out from SiC/Al composite, changing the specimen width and making the fibers to be being 10, 15, 20 and 30 in the cross section. Tabs of aluminum plate were set to both the ends and the gauge length was a constant, 20mm for all the composite specimens. In the case of changing the gauge length, tensile specimens with gauge lengths of 5, 10, 20, 30 and 40mm, and a constant fiber number, 10 in the cross section were prepared by locating the tabs of aluminum plate.

In addition, to understand the fiber strength in the SiC/Al composite, the fibers were extracted from SiC/Al composite by dissolving the aluminum matrix with 10%NaOH water solution, and then a tensile test was carried out for the fibers. Tabs of aluminum plate were set to both the ends of all tensile specimens.

The tensile test was completed in a screw driven tester (Shimadzu, AG-5000ES). All tests were done at room temperature with a constant crosshead speed of 0.5mm/min. The load was recorded as a function of time using a load-time recorder. For each specimen width and gauge length, the numbers of tensile specimens for SiC/Al composite were over 40, but extracted SiC fiber were over 50, respectively. The strength of the composite and the extracted SiC fiber was evaluated by using 2 parameters Weibull distribution function. Figure 1 shows a SEM image of cross section microstructure of SiC/Al composite produced in the present study. It can be seen that the SiC fibers arrange in almost the same interspace orderly and a good compound between the fiber and the matrix is obtained.

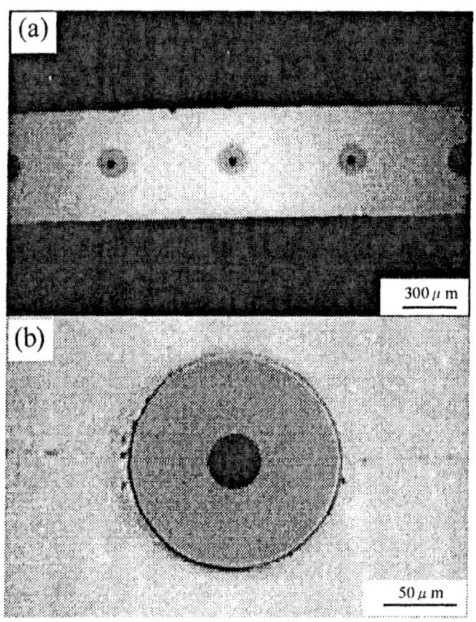

Figure 1. SEM image of cross section microstructure of SiC/Al composite.

3. MODEL AND TECHNIQUE OF THE SIMULATION

A Monte-Carlo simulation using elastic-plastic finite element method for evaluating the tensile strength and its scatter of the unidirectional SiC/Al composite was performed. The finite element model and mesh in the present study are shown as in Fig.2. This model is built on a monolayer plate of unidirectional fiber reinforced metal including 3, 5, 10, 15, 20 and 25 fibers. The two one-dimensional line elements, which represent the reinforcing fibers, are incorporated into two sides along the y-axis of a 4-node isoparametric element (plane stress), which represents the metal matrix. The

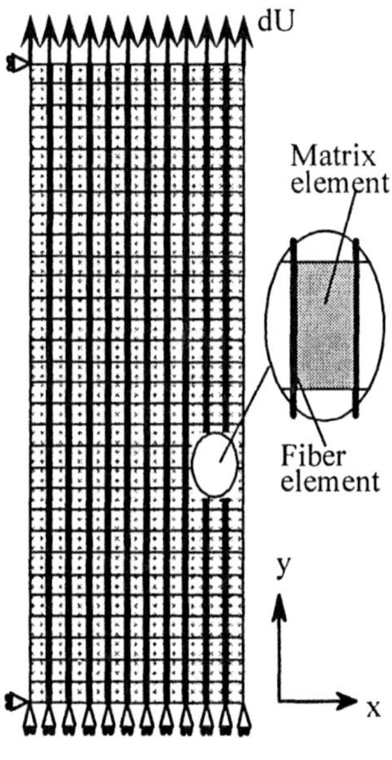

Figure 2. Finite element model (including 10 fibers).

model is divided into 30 elements in y-axis, and has $N_f \times 30$ fiber elements, $(N_f+1)\times 30$ matrix elements and $(N_f+2)\times 31$ nodes. Where, N_f shows fiber number in the model. y-axis of nodes in the lower side is fixed, but displacement increment is given along y-axis of nodes in the upper side. The strength, σ_{fi} of fiber elements is given by Weibull random number, using the inverse function of Weibull distribution as equation 1.

$$\sigma_{fi} = \sigma_0 \{ (L_0/\Delta X) ln[1/(1-Z)]\}^{1/mf} \tag{1}$$

Where, σ_0 and m_f show Weibull scale and shape parameters of the fiber strength, respectively. L_0 is a standard length. ΔX corresponds to the element length along the y-axis. Z is random number in (0, 1) which was produced by the multiplicative congruential method. In the present study, σ_0 and m_f were substituted with ones of the SiC fiber, which were obtained by a tensile test of the extracted fiber from SiC/Al composite. Also, L_0 was substituted with 10mm, which is the gauge length of a tensile test for the extracted fiber. Where, the fiber elastic modulus is maker's data. The parameters obtained by a tensile test of A1050 pure aluminum were used as mechanical property parameters of the matrix in this simulation. The parameters of the SiC fiber and the matrix used in the present simulation are shown in Table 1.

Table 1. Material parameters for the present simulation.

Fiber	
Diameter (mm)	0.14
Elastic module (GPa)	421.7
Strength (GPa)	4.15
Weibull scale parameter (GPa)	4.15
Weibull shape parameter	4.42
Matrix	
Elastic module (GPa)	70
Yield stress (MPa)	23.2
Work hardening coefficient (MPa)	176
Work hardening exponent	0.27
Poisson's ratio	0.33
Composite	
Gauge length (mm)	20, 4-100
Element length (mm)	0.67, 2.0
Thickness (mm)	0.6
Distance between fibers (mm)	0.5
Fiber number	3-25
Fiber volume fraction (%)	4.7

Furthermore, for exactly estimating reasonably each element stress increment necessary to the fiber breaking or the matrix yielding, the r_{min} method is applied to the simulation procedure[7]. The fibers are regarded as elastic state but the matrix is assumed to be elastic-plastic. Fiber element is considered to break when the stress, which is along the axial direction of the fiber, reaches the strength of that fiber element. After a fiber element breaks, the stiffness is changed to zero. The axial load, which was loading on that fiber element, is decreased to zero by dividing the load into 20 times, which was decided by pre-simulation. When composite stress decreases below 80% of the maximum stress, the composite is considered to be failure, and then the simulation is finished. This simulation was carried out 50 times for supposed each specimen width, respectively. The flow chart of the present simulation is shown in Fig.3. For the composite with different gauge length, the above simulation was carried out by changing the gauge length from 4mm to 100mm, but using a model with a constant fiber number of 10 fibers and the same parameters as listed in Table 1.

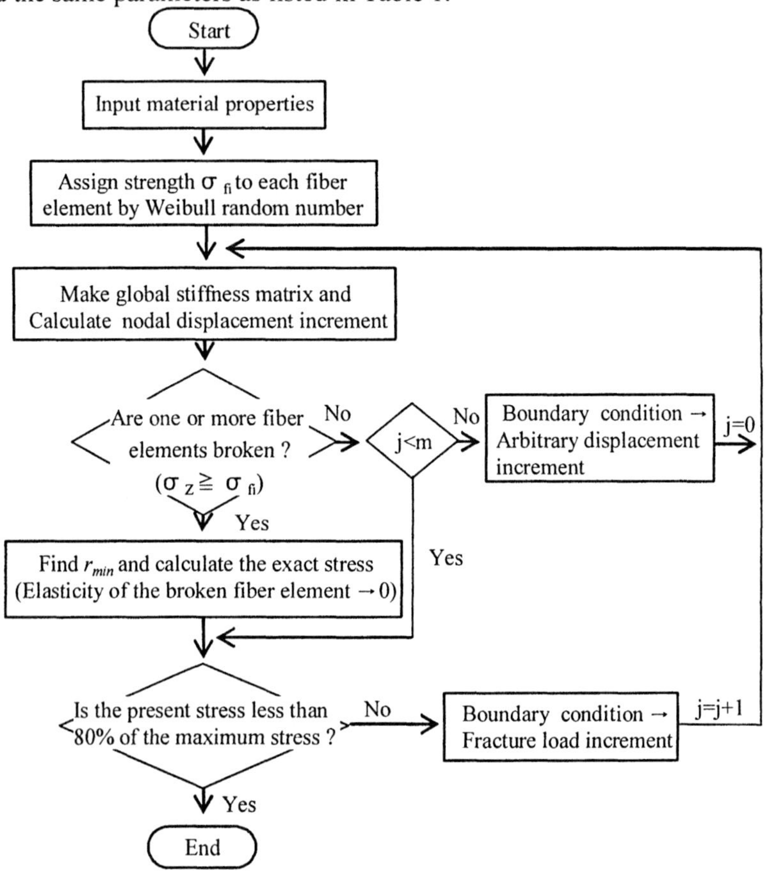

Figure 3. Flow chart of the present simulation.

4. RESULTS AND DISCUSSION

Tensile stress-strain curves obtained from tensile test and the present simulation for SiC/Al composite are shown in Fig.4. It can be seen that the slope of the curves has a change owing to yielding of the matrix. Till the maximum stress (composite strength), there are abrupt drops of the stress at some locations on the curves. The abrupt stress drops are considered to originate in the fibers breaking in the composite. Also, it is understood from the strain-stress curves to be possible to simulate the tensile behavior of the composite by the present simulation. Relationship between specimen width against composite strength and Weibull shape parameter is shown in Fig. 5. The specimen width was represented by the fiber number in the specimen. Composite strength almost does not change with increasing the specimen width (see Fig.5 (a)). But Weibull shape parameter *m* goes up; that is, the reliability is improved with increasing the specimen width (see Fig.5 (b)).

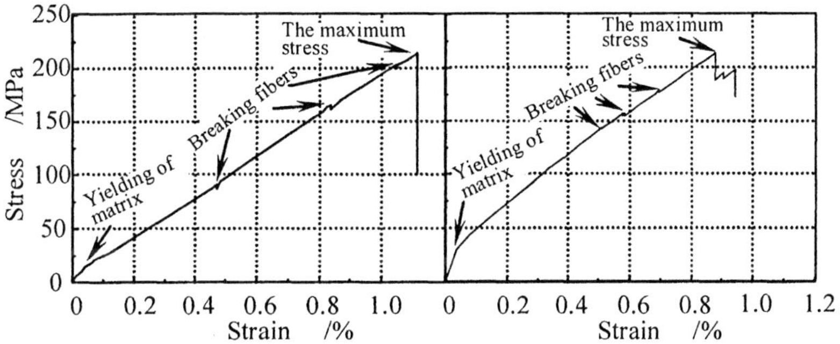

Figure 4. Stress-strain curves of SiC/Al composite with 15 fibers and 20mm gauge length

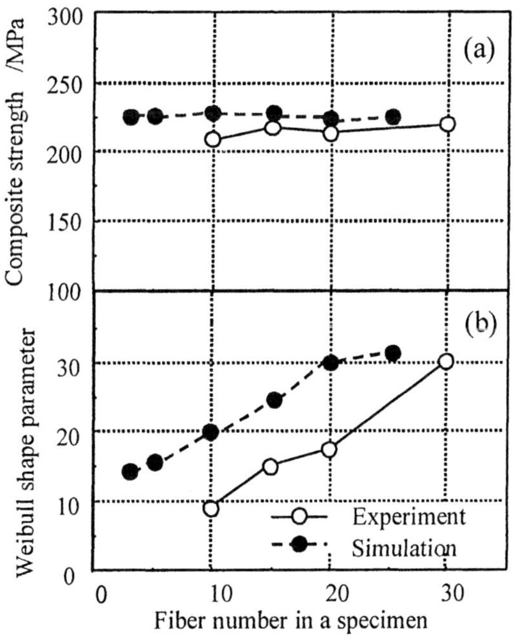

Figure 5. Relationship between specimen width against composite strength and Weibull shape parameter.

Generally, when a composite is loaded, the relatively weak fibers break at first. Till the maximum stress, fiber breaking occurs many times as shown in Fig.4. To understand the relationship between the composite strength, its reliability and the fracture process, the accumulated broken fiber was counted by reading out the number of breaking fiber element till the maximum stress during loading the composites in the present simulation. Relationship between the average count of broken fiber against composite strength and Weibull shape parameter is shown in Fig. 6. The strength does not display any change, but Weibull shape parameter increases with the average count. It hints that the reliability is related to fracture process of the composite.

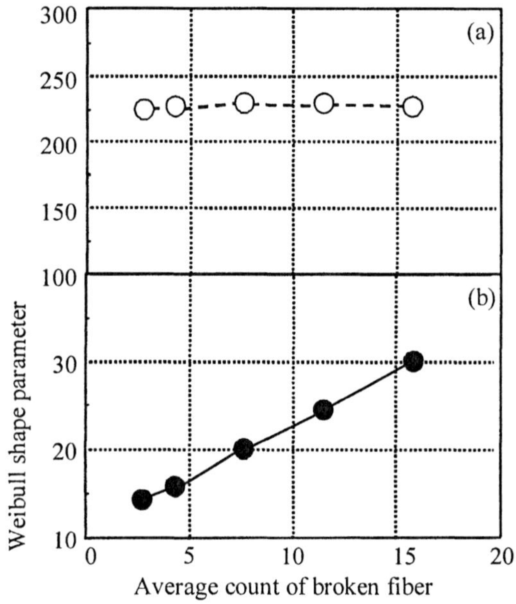

Figure 6. Relationship between average count against composite strength and Weibull shape parameter from the simulation.

Furthermore, in order to discuss the effect of the accumulation of breaking fiber on composite strength and its reliability, the fiber stress in the composite was read out in the order of broken fiber on every simulation for each the specimen width. Then average fiber stress and Weibull shape parameter were calculated by Weibull distribution for each fiber breaking. Figure 7 shows relationship between the order number of broken fiber against the fiber stress and Weibull shape parameter. The fiber stress and Weibull shape parameter show a increase tendency with the order number for each width. That is, the more the accumulating fiber fracture is, the higher fiber stress needed for the fracture is, and the lower the fiber stress scatter is. Therefore, the more the accumulating fiber fracture is, it becomes more obvious that composite strength is controlled by the fiber, which has a high strength and low strength scatter.

On the other hand, it is easy to understand that number of weaker fiber increases with increasing the width, that is, fiber number in the specimen. Figure 8 shows relationship between fiber number in SiC/Al composite against fiber strength and Weibull shape parameter of the weakest fiber element which were the first broken fiber element shown as order number 1 in Fig.7. The weakest fiber strength goes down extensively but Weibull shape parameter just only decreases gently. From above results, it can be understood that effect of specimen width on composite strength and its reliability is dependent on both factors, that is, the one is the accumulating of fracture process and the other is fiber strength and its scatter of the weaker fibers in the composite. It can be explained from both the above factors why composite strength almost does not change, but Weibull shape parameter *m* of the strength goes up with increasing the specimen width in the present study.

Finally, in order to investigate the relationship between stress distribution in the composite specimen and specimen width, stress distribution was calculated when making only the center fiber element to break. In the calculation model, it is supposed that there are 5, 9, 19 and 31

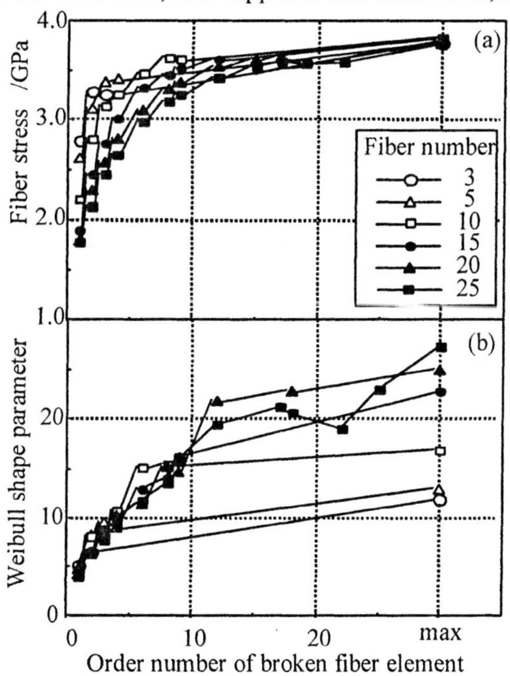

Figure 7. Order number of broken fiber vs fiber stress and Weibull shape parameter from the simulation.

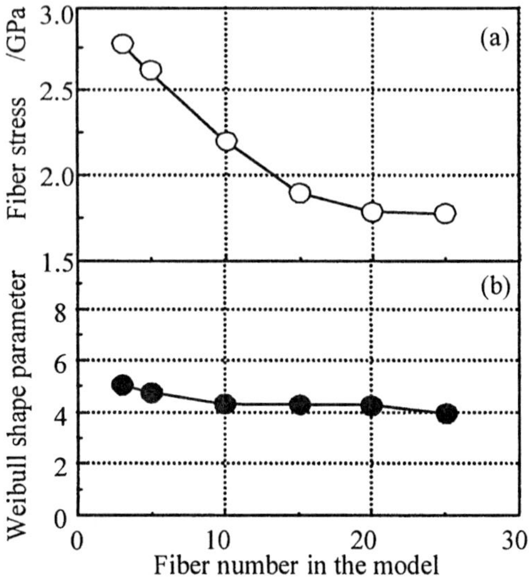

Figure 8. Fiber number in the composite model vs stress and Weibull shape parameter of the weakest fiber from the simulation.

fibers. To calculate the stress distribution after fiber breaking accurately, the extracted fiber strength, 4.15GPa shown in Table 1 was given to the central fiber element, and enough large strength value (100GPa) was given to the other fiber elements in this model. Thus, only the central fiber element breaks but the other fiber elements do not break when the composite is loaded. The fiber stress distribution along the broken fiber and the vertical direction of the fiber were calculated. Where the fiber stress distribution is represented by stress concentration factor, which is calculated out by dividing stresses of all fiber element in the fiber stress at the lowest fiber elements in the model. The results are illustrated in Fig. 9. From this figure, stress concentration factor along broken fiber almost is not different for each case of the width. However, stress concentration factor along the vertical direction of the fiber goes up with decreasing the width. Thus, the fiber breaking is spread or continued easily to other fibers due to the large stress concentration factor when the width is small. It decreases the accumulating of the fracture process in the case of narrow composite specimen.

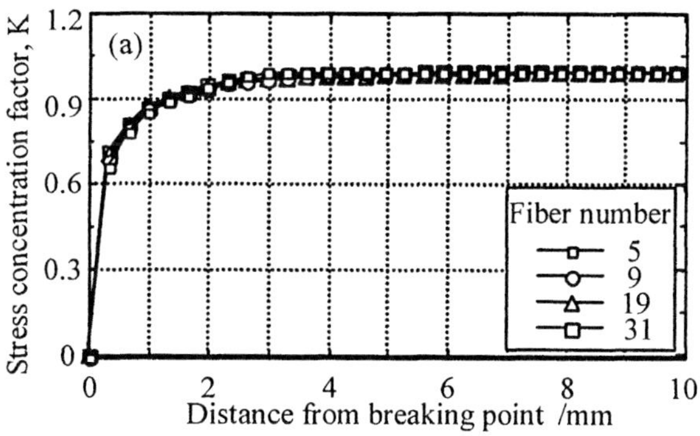

(a) Distribution of fiber stress along broken fiber from breaking point.

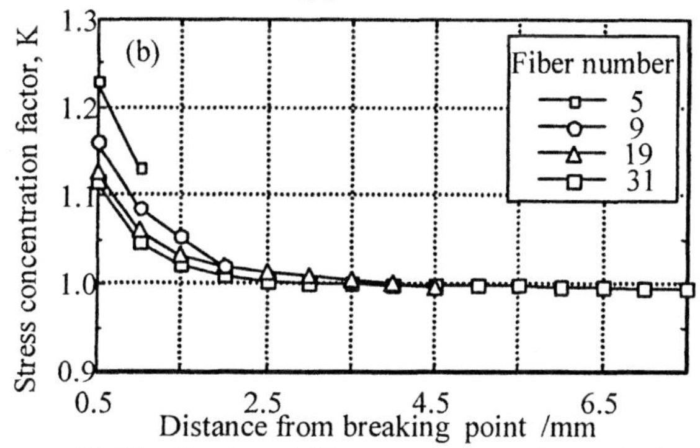

(b) Distribution of fiber stress along the vertical direction of the fiber from breaking point.

Figure 9. Distribution of fiber stress along the broken fiber and vertical direction of the fibers for the composite with different width.

Figure 10 shows strength, Weibull shape parameter and average count of broken fiber for the composite with different gauge length from the present experiment and simulation. The composite strength decreases, but Weibull shape parameter has an increase tendency with increasing the gauge length. It is understood to relate with decrease of reinforced fiber strength with the specimen length and the accumulation of breaking fiber in the composite showed as in Fig.10(c). Also, the fiber stress distribution along the broken fiber and the vertical direction of the fiber were calculated for the composite with different gauge length as the case of the different width. The results are illustrated in Fig. 11. From this figure, stress concentration factor along broken fiber almost is not different for each case of the gauge length. However, stress concentration factor along the vertical direction of the fiber goes up with increasing the gauge length to 22mm, but then thereafter, does not have change. That may be related to the end effect when the gauge length is shorter than 22mm.

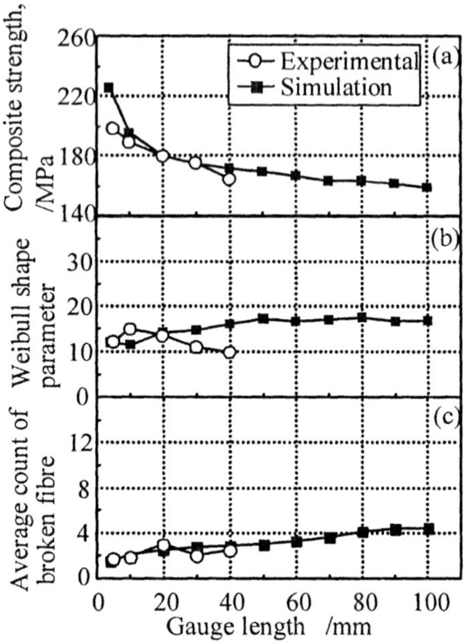

Figure 10. Relationships between gauge length and (a) composite strength, (b) Weibull shape parameter of the strength, and (c) average count of broken fiber for the case of different gauge length.

5. CONCLUSION

1) Composite strength almost does not change, but Weibull shape parameter of the strength goes up with increasing the specimen width. It indicates that the reliability is improved with increasing the specimen width. In the case of the composite with different gauge length, the composite strength decreases, but Weibull shape parameter has an increase tendency with increasing the gauge length.
2) Effect of specimen width on composite strength and its reliability is dependent on both factors, that is, (a) the accumulating of the fracture process and (b) strength and its scatter of the weaker fibers in the composite.
3) The accumulating of the fracture process is related to the fiber stress distribution in the composite.

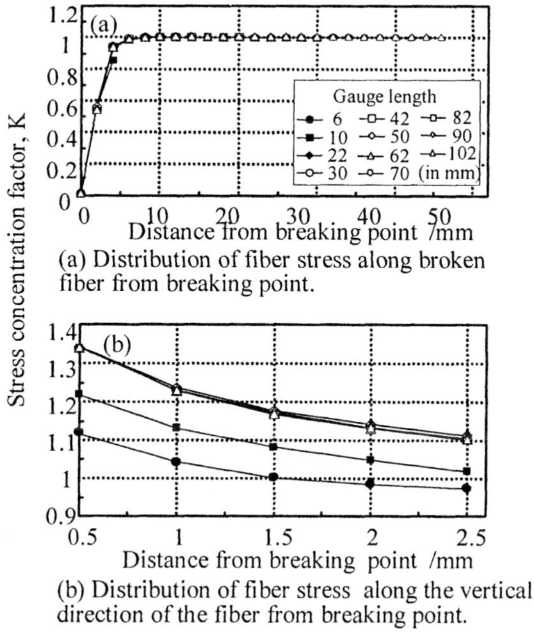

Figure 11. Distribution of fiber stress along the broken fiber and vertical direction of the fibers for the composite with different gauge length.

6. REFERENCES

1. B. D. Coleman, (1958), *J. Mech. Phys. Solids*, Vol.7, 60-70.
2. M. G. Bader and A. M. Priest, (1982), *Progress in science and engineering of composite,* T. Hayashi, K. Kawata & S. Umekawa, Ed., ICCM-IV, Tokyo, pp.1129-1136.
3. Yun LU, M. Hirohashi and O. Yashima, (1996),*Proceeding of the 73rd JSME spring annual meeting*, Vol.II, pp.280-281.
4. Yun LU, M. Hirohashi T. Nomiya and Y. Araoka, (1995),*Proceeding of the 72nd JSME spring annual meeting*, Vol.II, pp.250-251.
5. Pickering K. L and Murray T. L., (1999), *Composites Part A*, Vol.30A, No.8, 1017-1021.
6. Wisnom M. R., (1999), Compos. *Sci. Technol.*, Vol.59, No.13, 1937-1957.
7. K. Washizu, H. Miyamoto, Y. Yamata, Y. Yamamoto and T. Kawai, (1992), *Handbook of Finite Element Method*, Baifukan Ltd., Japan.

A HYBRID INTELLIGENT APPROACH FOR OPTIMAL CONTROL OF SEED CLEANER

Jin Yuan[1,2], Tao Yu[1], Kesheng Wang[3]
[1]CIMS & Robot Center of Shanghai University, Shanghai University, China; Email: yuanjin@graduate.shu.edu.cn. [2]School of Mechanical & Electronic Engineering, Shandong Agricultural University, China; [3]Department of Production and Quality Engineering, Norwegian University of Science and Technology, Norway. E-mail:kesheng.wang@ntnu.no

Abstract: This study presents a hybrid intelligent approach to model and optimize a seed cleaning/separating process, which separates seeds varying with different types, varieties and lots in a seed cleaner system (5XZW-1.5). In this controlling scheme, the nonlinear system is modeled by Artificial Neural Network, and optimized by Genetic Algorithms. The uniform design is used to sample datasets. Control results suggest that the intelligent approach is useful for the optimization of such a complex process.

Key words: Artificial Neural Networks, Genetic Algorithms, seed cleaner, uniform design.

1. INTRODUCTION

In recent years, intelligent control techniques, such as Artificial Neural Networks (ANNs) and Genetic Algorithms (GAs), have been extensively developed in the control of complex processes, to which mathematical approaches are not easily applied [1][2]. ANNs are capable of generating complex mapping between the input and the output space, and thus, arbitrarily complex nonlinear decision boundaries can be formed by these networks. On the other hand, GAs is one of the combinatorial optimization techniques. It searches for an optimal value of a complex objective function by simulating the biological evolutionary process, based on crossover and mutation in genetics [3]. GAs needs only the fitness values to guide its search. However, it is difficult (sometimes impossible) and time consuming to obtain fitness values for some control problems as it is impractical to run a

This project is supported by the National Natural Science Foundation of China under the grant No.50375090.

Please use the following format when citing this chapter:
Yuan, Jin, Yu, Tao, Wang, Kesheng, 2006, in International Federation for Information Processing (IFIP), Volume 207, Knowledge Enterprise: Intelligent Strategies In Product Design, Manufacturing, and Management, eds. K. Wang, Kovacs G., Wozny M., Fang M., (Boston: Springer), pp. 780-785.

real system for each combination of control parameters. ANNs as a fitness estimator would be an alternative. A well trained ANNs would be used as an objective function for the GAs optimization process and generate the output value based on the input value.

This paper proposes a hybrid intelligent control system integrating ANNs and GAs, which is applied to the optimal control of the gravity seed cleaner. Section 2 describes the optimal control problem. The methods of control system, ANNs modeling, sampling strategy and design of GAs, are presented in section 3. Finally, the results obtained show the usefulness of the hybrid intelligent system.

2. OPTIMAL CONTROL PROBLEM

Seed cleaner uses a combination of air, gravity, and screens to separate seed based on different motion principle of seeds on the screen. Such motion principle determined from the size, shape, and density of seeds and the control parameters: velocity of airflow, oscillation frequency, longitudinal pitch, and transverse pitch of screens. Four fractions of seeds evaluated by kernel weight of seeds outflow from four slots installed at gravity seed cleaner. The kernel weight of each fraction is measured online as the controlled variable of the system.

The control parameters should be adjusted at suitable setpoints before the seed cleaner starts. During the process of separating the seeds, the control parameters are unchangeable. Obviously, such a system is a static, nonlinear and decoupling control system. With the suitable setpoints, the seed cleaner can obtain excellent results. However, the setpoints are very difficult to tune with many control parameters and variable varieties and lots of seeds. Liu *et al* [4] used the good parameters, which are obtained from experiment with familiar breed and saved in control system, and adjusted the seed cleaner. But it is not flexible for variable breeds and batches of seeds. For finding out the near-optimal values for different breeds and batches of seeds, we propose a new optimal control system.

3. METHODOLGY FOR OPTIMAL CONTROL

3.1 Design of an optimal control system

One of the important objects for optimal control is how to deal with the search process for near-optimal parameters to control nonlinear system. GAs is the successful option to optimize various control parameters. However, it is very time consuming to obtain fitness values for some control problems as it is impractical to run a real system for each combination of control parameters. Due to good generalization abilities, ANNs are often used to model and approximate a real control system. So the ANNs could be used in fitness evaluation rather than the real control system [5]. Fig. 1 shows the block diagram of the hybrid intelligent optimal control system.

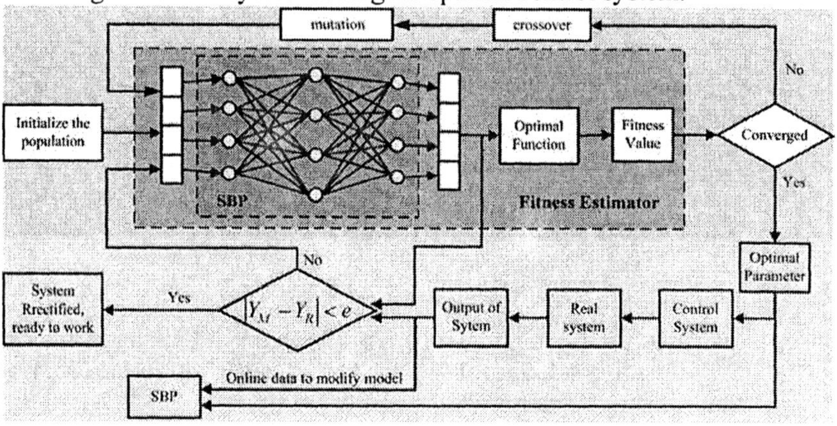

Figure 1. Block diagram of hybrid intelligent optimal control system.

The optimal control is realized through two procedures: identification of seed cleaner system using ANNs and the search for a near-optimal value through simulation of the identified model using GAs.

Because of an optimal value is obtained from a model simulation, the high accuracy of model approximation is badly needed. Therefore, once the approximate error between model and real system exceed the threshold, the ANNs model should be updated with new sample, meanwhile, the search of GAs should be started again. The optimization is accomplished by applying the optimal values to the setpoints in the feedback control system.

3.2 Sampling based on uniform design

The training set influences modeling accuracy of ANNs. In theory, an infinite training set would provide an exact measure of test accuracy and discourage overfitting. Hence, it is desirable to incorporate a training set into the learning phase as large as possible. However, training on very large data set is problematic, as training time tends to increase. In particular, sampling large data sets is impractical at this case. A sampling approach based on uniform design has shown to decrease data sets, but reserving the underlying distribution by sampling.

Uniform Design (UD) is an experimental design method proposed by Fang [6]. UD is equivalent to generating a relative small set of design points that are uniformly scattered on experiment domain, which reflects self-contained information of the system. Data sets have more generalization performance. Its main theory is using certain method to design the uniformly scattered point sets instead of Monte Carlo random point sets in problem space.

UD is often expressed as a table, denoted by $U_n(q^s)$: a matrix of q rows and s columns, which means s factors with q levels for each of the factors in the problem space. Generally, n presented as the number of experiment is equal to q, namely the number of factor's level. If the level is different among factors, the UD table should be designed as mixed level UD table, for example, table 1 used in our experiment. Its first factor has 12 levels, but the other 3 factors have 6 levels. For detail UD, reference Fang [6].

Table 1. Mixed Level Uniform Design table $U_{12}(12\times 6^3)$

No. of Experiment	Oscillation Frequency (450~1000 Hz)	Longitudinal Obliquity (1.5° ~9°)	Screen Airflow Velocity (1~6 m/s)	Transverse Obliquity (1° ~6°)
1	1(450)	2(3)	6(6)	2(2)
2	2(500)	3(4.5)	5(5)	4(4)
3	3(550)	5(7.5)	4(4)	6(6)
4	4(600)	6(9)	3(3)	2(2)
5	5(650)	1(1.5)	2(2)	4(4)
6	6(700)	3(4.5)	1(1)	6(6)
7	7(750)	4(6)	6(6)	1(1)
8	8(800)	6(9)	5(5)	3(3)
9	9(850)	1(1.5)	4(4)	5(5)
10	10(900)	2(3)	3(3)	1(1)
11	11(950)	4(6)	2(2)	3(3)
12	12(1000)	5(7.5)	1(1)	5(5)

3.3 Artificial Neural Network model

The establishment of the model using SBP (Standard Back-Propagation) is realized through two steps. In the first step, the relative exact approximation of SBP model is identified fast with less sample sets which sampled offline. Next, the approximation of SBP will be trained more accurately using the feedforward modeling approach.

The nonlinear model identified by SBP of seed cleaner system can be expressed as follows:

$$G = F(X,W) \quad (1)$$

Where, the input vector is defined $X = [\alpha\ \beta\ f\ v]^T$, the output vector is defined $G = [g_1\ g_2\ g_3\ g_4]^T$, where α denotes longitudinal obliquity, β denotes transverse obliquity, f denotes oscillation frequency and v denotes screen airflow velocity. g_i represents the kernel weight of fraction i. **W** is the weight matrix that is evaluated in the network training process.

Due to the relative less sample sets, the leave-one-out cross validation was used in training process of SBP, and the architecture of SBP was built up by *NEUframe* from Neuscience, UK.[7]

3.4 GAs design

GAs refers to a population-based stochastic search algorithms that are developed from ideas and principles of natural evolution, which is useful for dealing with large complex problems. It does not depend on gradient information and thus are quite suitable for problems where such information is unavailable or very costly to obtain or estimate. Here, the searching process for near-optimal setpoints is using GAs within the nonlinear model identified by ANN. This part just focuses on the fitness value computation.

3.4.1 Optimal objective function

There was variability between lots in the percentage of seed contained in each fraction. Generally, the percentages of fractions are given based on the experiment of germination percentage according to the lots, thus, the optimal objective function is determined:

$$\max f(g_1, g_2, g_3, g_4) = \frac{1}{((\lambda_1 gk - g_1 k_1)^2 + (\lambda_1 gk - g_1 k_1)^2 + (\lambda_1 gk - g_1 k_1)^2 + (\lambda_1 gk - g_1 k_1)^2)} \quad (2)$$

Where g_i, λ_i, k_i is the kernel weight, presumed seeds percentage and grain number of fraction i, $k = k_1 + k_2 + k_3 + k_4$.

3.4.2 Fitness function of GAs

The fitness value could be computed by combining the formulation (1) and (2). The difference from general GAs is that the variables expressed by chromosome is not directly independent variable of optimal function, but independent variable of optimal function via nonlinear mapping of ANNs.

4. RESULTS

The test was done on 5XZW-1.5 type seed cleaner control system with 3 lots of seeds. And the separating result used this optimal approach (O) was compared with that of mechanical adjustment (M).

Table 2. The comparison of testing results for various approaches

Seed lot	Fraction 1(%)		Fraction 2(%)		Fraction 3(%)		Fraction 4(%)		Max f(g1, g2, g3, g4)		Percent germination	
	M	O	M	O	M	O	M	O	M	O	M	O
A	40	48	27	30	21	15	12	7	91.5	95.7	91.3	95.2
B	35	42	29	27	17	18	19	13	93	97.2	90.8	94.8
C	42	49	24	25	22	15	12	11	90.5	96.8	92.9	97.3

Quality of seed fractions was defined finally by determining the percent germination of first two fractions. The percent germination in table 2 shows that the optimal control approach is consistently higher than the adjustment with control parameters saved before.

5. CONCLUSION

In this paper, a hybrid intelligent technique based on ANNs and GAs was applied to the optimal control. First, seed cleaner system was identified using SBP. For minimizing approximate error, uniform design plans sampling the sample sets of SBP, and leave-one-out cross validation was used in training process of SBP. Such a trained network becomes a specialised process simulation tool which is valuable in prediction and optimization of control system. Prediction of process output can be used to determine whether some separated result of seeds will be satisfactory, given the input conditions and process parameters. Next, near-optimal setpoints of the seed cleaner, which maximize the objective function, were sought from among the numerous values of the objective function obtained from the simulation. Thus, this hybrid intelligent approach is suitable for the optimal control of complex systems, e.g. seed cleaner system.

6. REFERENCES

1. Wang, K., (2005), *Applied Computational Intelligence to Intelligence Manufacturing Systems,* Advanced Knowledge International (ISBN 0-9751004-9-1).
2. Tarik C., Mehmet B. Y., and Mehmet B. (2005), A neuro-genetic approach to design and planning of a manufacturing cell, *J of Intelligent Manufacturing,* vol. 16, pp. 255–274.
3. Holland, J. H., (1989), Searching nonlinear functions for high values, *Applied Mathematics and Computation*, vol. 32, pp. 255–274.
4. Liu, C. L. et al., (1997), Study on the design and adaptive control system of the gravity seed cleaner, *Trans. the Chinese Society of Agricultural Machinery*, v28,no. 1, p74-77.
5. Yao, X., (1999), Evolving artificial neural networks, *Proceedings of the IEEE*, 87(9):1423-1447.
6. Fang, K. T. and Wang, Y., (1994), *Number-theoretic Methods in Statistics.* Chapman & Hall.
7. NEUframe User's Guide, www.neusciences.com

NUMERICAL SIMULATION OF TRANSIENT TEMPERATURE FIELD FOR LASER DIRECT METAL SHAPING

Risheng Long [1, 2] , Weijun Liu [1] , Xiaofeng Shang [1]
[1]*Advanced Manufacture Lab, Shenyang Institute of Automation, CSA, Shenyang, Liaoning Province,China;E-mail:* longrisheng@sia.cn. [2]*Graduate School, Chinese Academy of Sciences, Beijing, China.*

Abstract: In order to improve the mechanical properties of the parts fabricated by Laser Direct Metal Shaping (LDMS), it is of great significance to understand the distribution regularities of transient temperature field during LDMS process. Based on the "element birth and death" technique of finite element method, a three-dimensional multi-track and multi-layer model for the transient temperature field analysis of LDMS is developed by ANSYS Parametric Design Language (APDL) for the first time. In the fabricated model, X-direction parallel reciprocating scanning paths is introduced. Using the same process parameters, the results of simulation show good agreement with the microstructure features of samples which fabricated by LDMS.

Key words: Laser Direct Metal Shaping, Transient temperature field, Numerical simulation, laser cladding.

1. INTRODUCTION

Laser Direct Metal Shaping is a new additive layered manufacturing technology for building components from a computer-aided design model without the use of forming dies, tooling or machining, and it is the further development of rapid prototyping (RP). Compared with conventional manufacturing processes, LDMS has the advantages of lower cost, shorter cycle time, and higher manufacturing precision and flexibility. However, due to highly localized laser heating feature and the cooling effect of substrate, high temperature gradients in Z-direction are produced during LDMS

Please use the following format when citing this chapter:
Long, Risheng, Liu, Weijun, Shang, Xiaofeng, 2006, in International Federation for Information Processing (IFIP), Volume 207, Knowledge Enterprise: Intelligent Strategies In Product Design, Manufacturing, and Management, eds. K. Wang, Kovacs G., Wozny M., Fang M., (Boston: Springer), pp. 786-796.

process. Not only does it lead to the directional selection of microstructure in the parts fabricated by LDMS, but also results in large thermal stresses in sample and substrate.

Over the past 10 years, due to the complexity of the physical processes involved in LDMS, the research of transient temperature field in LDMS process is quite deficient. Simple mathematical solutions cannot address the practical manufacturing processes, and it is also impossible for any experimental technique to obtain a complete mapping of the temperature distribution in sample. Computational simulation thus plays an indispensable role in the analysis of such process. Especially, with recent advancements in computational power, FEA simulation of transient temperature in LDMS has gradually become feasible. Recently, 3D laser forming simulations are conducted using commercial FEA software or manual programming. Xi Mingzhe[1], Zhang Yongzhong and Shi Likai developed a 3D single-track FEM model to simulate the transient temperature field in the real process of the metal thin wall fabricated by laser direct deposition. L. Zhanga[2], E.W. Reutzelb and P.Michalerisc established the minimum requirements for discretization of 3D 3nite element models for laser forming by examining the convergence of the numerical solutions with increased discretization. AnK Kyrsanidi[3], ThB Kermanidis and SpG Pantelakis carried out numerical and experimental investigation of the laser forming process using finite element method. K. Dai and L. Shaw[4] developed a 3D thermo-mechanical finite element model including the effect of the powder-to-solid transition. H.Hsiehshen[5] and Lin Jehnming carried out thermal-mechanical analysis on transient deformation during pulsed laser forming.

In the present paper, according to the "element birth and death"[6] technique, a three-dimensional multi-track and multi-layer finite element model for the transient temperature field of LDMS is developed with APDL for the first time. The distribution regularities of transient temperature field during LDMS process are discussed in detail, and the results of simulation are verified by the samples fabricated by LDMS under the same conditions with simulation.

2. COMPUTATIONAL FORMALTATIONS

2.1 Theoretical formulations

The heat transfer behavior during LDMS process is a complex, nonlinear, transient process and it can be described with the classical heat conduction equation in spatial coordinates(x, y, z):

$$k_{e1}\left(\frac{\partial^2 T}{\partial x^2}+\frac{\partial^2 T}{\partial y^2}+\frac{\partial^2 T}{\partial z^2}\right)+k_{e2}\left(\frac{\partial^2 T}{\partial x^2}+\frac{\partial^2 T}{\partial y^2}+\frac{\partial^2 T}{\partial z^2}\right)$$
$$+q=\rho_1\cdot c_1\frac{\partial T}{\partial t}+\rho_2\cdot c_2\frac{\partial T}{\partial t} \tag{1}$$

Where ρ_1 and ρ_2 are density of the metal powder and substrate, c_1 and c_2 are specific of the metal powder and substrate, k_{e1} and k_{e2} are the thermal conductivity of the metal powder and substrate, respectively, and q is the total input energy.

Considering the natural convection and radiation, and assuming $k_e = k_{e1} + k_{e2}$, $c = c_1 + c_2$, $\rho = \rho_1 + \rho_2$ express (1) becomes:

$$k_e\left(\frac{\partial^2 T}{\partial x^2}+\frac{\partial^2 T}{\partial y^2}+\frac{\partial^2 T}{\partial z^2}\right)+q+h(T-T_h)+\sigma\varepsilon(T^4-T_h^4)$$
$$=\rho\cdot c\frac{\partial T}{\partial t} \tag{2}$$

Where h is natural convection exchange coefficient, σ is radiation constant of black body, T_h is room temperature, ε is actual radiation coefficient of object.

The equation (2) must be solved with prescribed boundary and initial conditions. The initial conditions can be written as:

$$T(x,y,z,0)=T_h \tag{3}$$

On the boundary and the upside surface of the sample, convection and radiation in heat transfer are responsible for heat loss to the ambient. So, the boundary conditions of them can be written as:

$$-k_e\frac{\partial T}{\partial z}=h(T-T_h)+\sigma\varepsilon(T^4-T_h^4) \tag{4}$$

The underside of substrate can be treat as adiabatic, and its boundary condition can be written as:

$$-k_e\frac{\partial T}{\partial z}=0 \tag{5}$$

2.2 "Element birth and death" technique

"Element birth and death" technique does not actually remove "killed" elements from model. Instead, it deactivates them by multiplying their stiffness or conductivity by a severe reduction factor. This factor is set to 1.0E-6 by default, but can be given other values. Element loads associated with deactivated elements are zeroed out of the load vector, however, they still appear in element-load lists. Similarly, mass, damping, specific heat, and other such effects are set to zero for deactivated elements.

In like manner, when elements are "born", they are not actually added to the model; they are simply reactivated. So, you must create all elements, including those to be born in later stages of your analysis, while in preprocessor. When an element is reactivated, its stiffness, mass, element loads, etc. return to their full original values, but there is no record of strain history (or heat storage, etc.) for it.

2.3 Heat source and latent heat of phase change

The heat flux to system is put in by a highly focused area on the molten pool in LDMS process, and it is assumed that the heat flux, q(r), follows Gauss distribution in the radial direction, and has the following form:

$$q(r) = \frac{3Q}{\pi R^2} \exp(-\frac{3r^2}{R^2}) \tag{6}$$

Where R is the laser spot radius, r is the distance of a point away from the centre of the laser beam, and Q is the total heat input energy.

Simultaneously, phase change, such as meltage and solidification, are existed during LDMS process, and the latent heat of phase change can't be neglected for the crystal materials, such as metal. Phase change problems are usually solved with finite difference or finite element methods, and the most common method used is the enthalpy method.

3. GEOMETRY CONFIGURE

To accurately reflect of the characteristics during LDMS process, according to the former theories and "element birth and death" technique, a multi-track and multi-layer FEM model is developed by APDL, in which, X-direction reciprocating scanning paths are also introduced into the model.

Fig.1 demonstrates the FEM model for temperature field simulation and the Location of nodes on the longitudinal section. As shown in Fig.1a, the dimension of substrate is 100mm×25mm×10mm, and the size of sample is 50mm×4mm×6mm. To decrease the large computational time and still maintain reasonable accuracy, a fine mesh is used in sample which is divided to six layers, each layer 1mm, and the coarse mesh is adopted in the substrate.

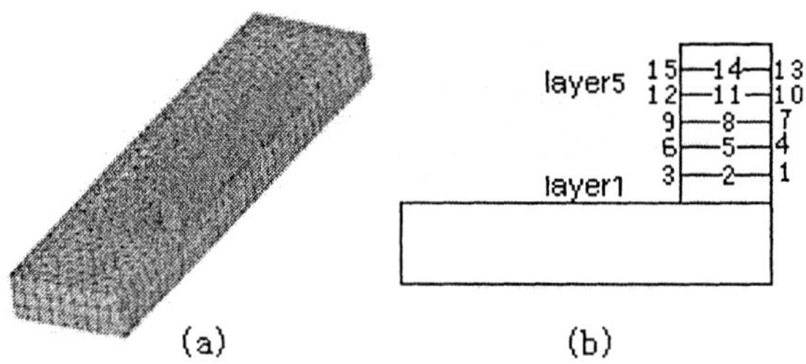

Figure 1. Schematic of model (a) the FEM model of temperature field, (b) the location of nodes on the longitudinal section.

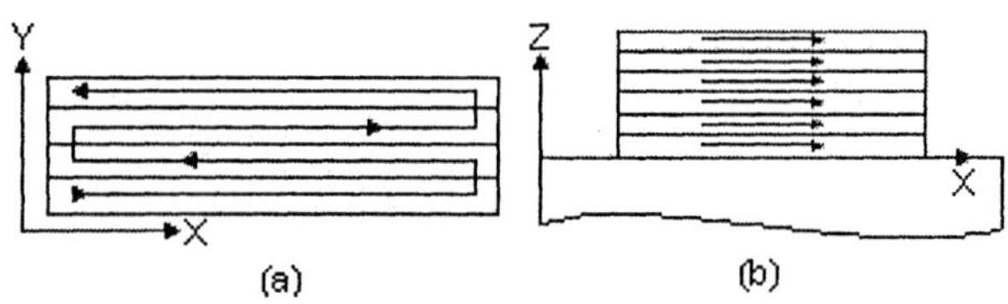

Figure 2. Schematic of scanning paths in simulation. (a) in XY plane, (b) in XZ plane.

Moreover, Fig.2 displays the X-direction parallel reciprocating scanning paths, in which Fig.2a and Fig.2b show the schematic of scanning paths in XY plane and XZ plane, respectively. As shown in the figure, after one layer is finished, the laser firstly moves from the end of current layer to the start, then lift a little distance along Z direction, which is equal to the height of single layer.

4. RESULTS AND DISCUSSION

In simulation, the laser power equals 600 W, spot diameter equals 1mm, scanning velocity equals 5mm/s, room temperature equals 20℃.The material of metal powder used in simulation is Ni60A, its material properties are listed in Table.1. The material of substrate is 45 steel whose material properties are taken from *Complex Alloy Steel Manual* [7].

Table 1. Parameters list of Ni60A

Parameter name	value
Specific heat of Ni60A (J/mg·℃)	0.000464
Latent heat of Ni60A (J/mg)	0.248
Density of Ni60A (kg/m^3)	8522
Melting point of Ni60A (℃)	950~1000

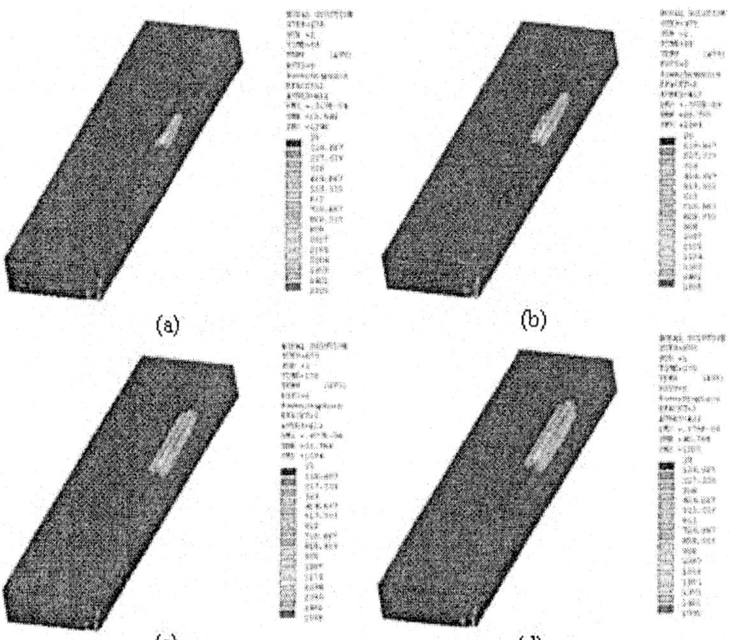

Figure 3. Temperature field contours at different time

Under the condition of Table.1, the temperature field contours of model at different time are demonstrated in Fig.3. As shown in it, Fig.3a shows the contour of temperature field at the 55th second, on this occasion, the second track of the second layer is being processed; Fig.3b shows the contour of temperature field at the 95th second, on this occasion, the second track of the third layer is being processed; Fig.3c shows the contour of temperature field at the 135th second, on this occasion, the second track of the fourth layer is being processed; Fig.3d shows the contour of temperature field at the 175th second, on this occasion, the second track of the fifth layer is being processed.

In addition, as seen from Fig.3, the distribution of temperature field in model takes on motion state with the movement of laser heat source, and the high temperature regions enlarge gradually with the increment of layers. The reason is because that the fabricated layers of sample are heated again and again with the increment of layers, the heat accumulation effect existed in them decreases the cooling velocity of the new layer gradually, and it resulted in the highest temperature of molten pool becoming higher and higher.

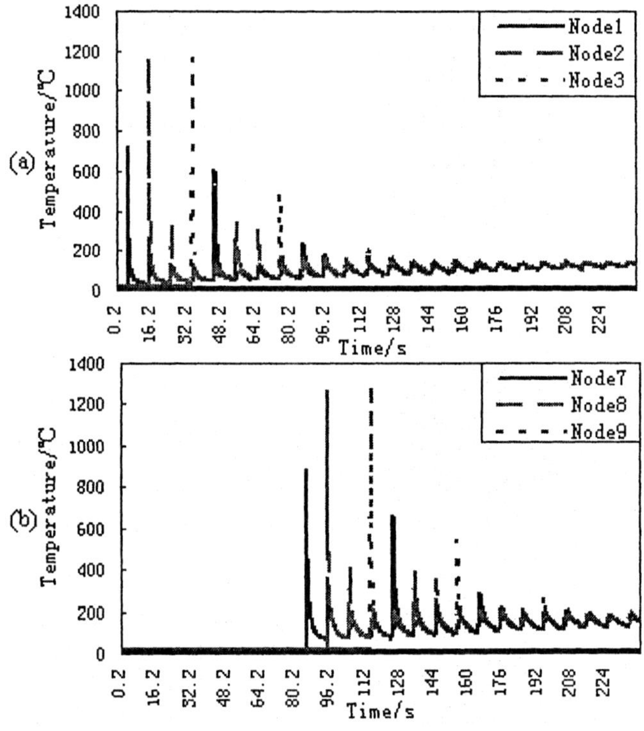

Figure 4. Variations of temperature at locations of different nodes on the longitudinal section. (a) of node 1, 2 and 3; (b) of node 7, 8 and 9.

Fig.4 displays the variations of temperature at locations of different nodes on the longitudinal section (see Fig.1b). As shown in Fig.4a, which compares the temperature history curves of node 1, 2 and 3 in the first layer of sample, the temperature of node 1 is 20℃ before the 5th second, just the room temperature that we set, and it rises to 638℃ quickly at the 5th second. The reason is because node 1 is under "death" condition before the 5th second, and there is no temperature variation record for it. But at the 5th second, the laser heat source just moves to the upside of the element which node 1 belonged to, and at the same time the element are reactivated with node1. Once the node1 is reactivated, the temperature of it rises quickly. In the same way, node 2 is under "death" condition before the 15th second. But at the 15th second, the laser heat source just moves to the upside of the element which node 2 belonged to, node 2 is reactivated, the temperature of it rises from room temperature to 1168 ℃ quickly. Meanwhile, because of the near distance between node1 and node 2, the temperature variation of node1 will be affected by the heat source synchronously. The temperature variation of node 3 is similar to node 1 and 2. In the following simulation, node 1, 2 and 3 are heated and cooled down again and again and are tend to the same temperature gradually. The reason is because when the following layer's process moves to the upside of node 1, 2 and 3, they will be heated by the energy conducted from molten pool and neighboring areas, but the range of temperature variation of them will become smaller and smaller with the increment of layers.

The temperature variation regularities of the node 7, 8 and 9 are similar to the regularities of the node 1, 2 and 3, such as variation period etc., besides the difference of reactivated time, as shown in Fig.4b.

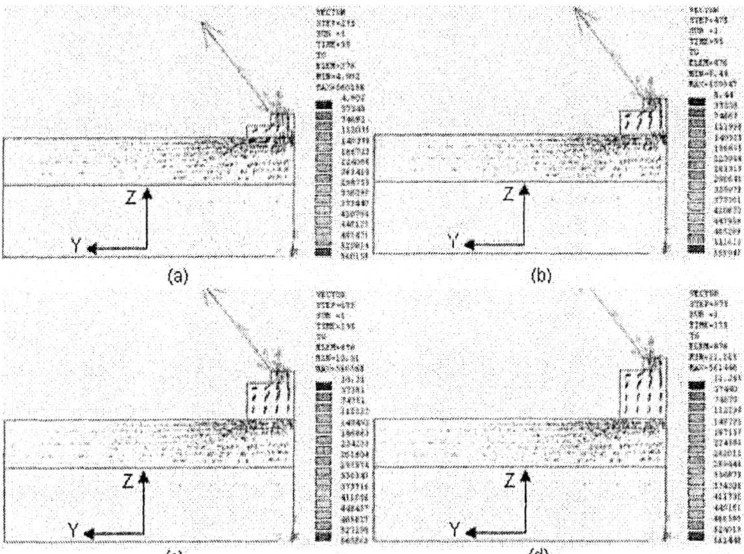

Figure 5. Temperature gradient vector diagrams in simulation at different time. (a) at 55s, (b) at 95s, (c) at 135s, (d) at 175s

Fig.5 shows the temperature gradient vector diagrams in simulation at different time. From it, there are several features can be gotten and they are listed as the following: (1) The distribution of temperature gradient in sample is non-uniform. It is strong and concentrated at the district which subjected to the laser energy influence, and it is small and dispersive at the district which far away from the laser energy influence. The temperature gradients in molten pool and its nearby district are quite big. (2) The temperature gradient in sample is mainly along the Z direction, the perpendicular way of substrate, and it is small in other directions. (3) The distribution of temperature gradient in substrate is non-uniform. It is strong and concentrated at the district which near the sample, and it is small and dispersive at the district which far away from the sample. (4) The temperature gradient in substrate is mainly along the direction which parallels the substrate plane, having obvious layering phenomenon.

5. LDMS EXPERIMENT AND SEM ANALYSIS

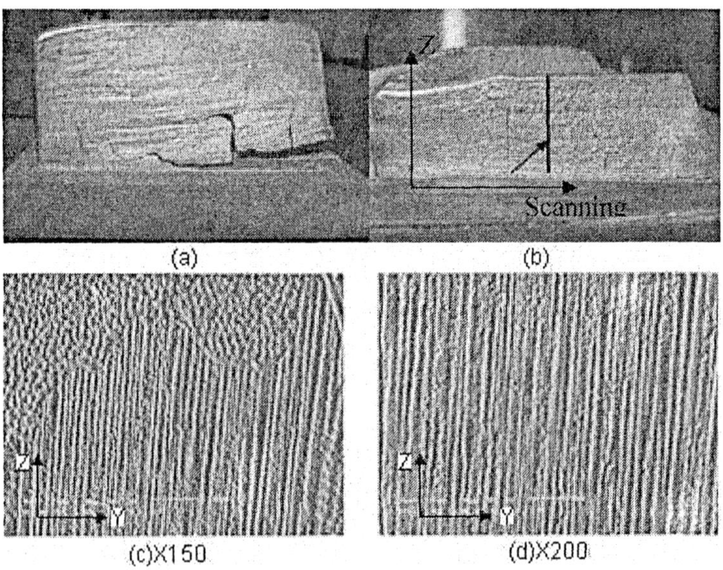

Figure 6. Photos of the samples fabricated by LDMS and SEM macrographs of the longitudinal section of sample along Z-direction. (in the same cladding layer).(a) at the top of same layer; (b) at the middle of same layer

In the case of 800W laser power, 1mm spot diameter, 5mm/s scanning velocity, 20℃room temperature, as shown in Fig.6, some nickel-based superalloy components are fabricated through the LMDS system which has been developed successfully by Chinese Academy of Sciences Shenyang Institute of Automation (see Fig.6a and Fig.6b). Fig.6c and Fig.6d exhibit the Z direction SEM micrograph of the longitudinal section which marked in Fig.6b.

As shown in Fig.6c and Fig.6d, equiaxed grains appear on the top of the fine dendrites at the top of one layer. Similarly, the solidification microstructure at the middle of same layer is composed of parallel dendrites whose growing direction is nearly parallel to the positive Z direction. The reason is because the temperature gradient in the positive Z-direction is remarkably dominant in the process of cooling solidification, and heat mostly dissipates in the negative Z-direction. Therefore, under the action of the highest temperature gradient and solidification rate in the Z-direction, grains grow with the directional selection, thus forming the dendrites which are almost parallel to Z-direction.

6. CONCLUSIONS

In this paper, based on the FEM technique of "life and death element", a three-dimensional multi-track and multi-layer model for the transient temperature field analysis of LDMS is developed with APDL for the first time. Through the model, the transient temperature field's distribution regularities are discussed detailedly. Under the same conditions in simulation, the results from simulation show good agreement with the microstructure features of the samples which fabricated by LDMS.

7. REFERENCES

1. Xi Mingzhe, Zhang Yongzhong and Shi Likai, (2003), Numerical simulation of 3D transient temperature field in thin-wall metal parts fabricated by laser direct deposition, *The Chinese Journal of Nonferrous Metals*, vol.13, pp.887~892.
2. L. Zhanga, E.W. Reutzelb and P.Michalerisc, (2004), Finite element modeling discretization requirements for the laser forming process. *International Journal of Mechanical Sciences* 46, pp.623 - 637.
3. AnK Kyrsanidi, ThB Kermanidis and SpG Pantelakis, (1999), Numerical and experimental investigation of the laser forming process, *Journal of Material Process Technology*, vol.872, pp.281 - 290.
4. K. Dai, L. Shaw, (2004), Thermal and mechanical finite element modeling of laser forming from metal and ceramic powders, *Acta Materialia 52*, 69 80.

5. H.Hsiehshen, Lin Jehnming, (2004), Thermal-mechanical analysis on transient deformation during pulsed laser forming, *International Journal of Machine Tools & Manufacture*, vol.44, pp.191-199.
6. ANSYS Theory Manual, (2003), Release 8.0, ANSYS Inc., USA.
7. Chinese Iron-Steel Research Institue, (1992), *Complex Alloy Steel Manual: lower Volume*, Metallurgical Industry Press, Beijing, pp.3-45.

FAULT TOLERANT SYSTEM DESIGN BASED FUZZY OBSERVER

Yihu Huang[1], Guangxin Zhang[2], Ximei Jia[3]
[1]College of Automation and Electron Engineering, Qingdao University of Science and Technology,China; Email: hyhuzi@163.com. [2]Control Science and Engineering Department, Zhejiang University, China. [3]College of Electromechanical Engineering, Qingdao Science and Technology, China

Abstract: Crude oil in the pipeline evaporation is one of hidden safety trouble for directly-fired kiln, which maybe be caused by severe flame deflecting. According to redundancy between technology parameters of furnace, redundancy select majority strategy was proposed, and used as the inference rule of fuzzy observer, the output of the fuzzy observer was tendency of furnace working condition, then it compared with deviation of other local input parameters to diagnose effectively fire deflect and sensors fault. Thus safety and fault-tolerant capacity of furnace are improved, and stronger robustness is also obtained.

Key words: Furnace, Fuzzy Observer, Fault-Tolerant System

1. INTRODUCTION

Large long-distance pipeline transferring system is main mode of crude oil storage and transportation in China, direct furnace is one of three main parts of it. Crude oil pipeline is laid down along two side of furnace. So there are separately two crude oil temperature of inlet and outlet. This means there are several incipient faults during furnace operation.

Incipient fault one: Ejection gun is fixed in front of furnace by bolts. It has rather shock during working. After long running, the fixed bolt probably shock loose; this can make severe blaze deflection, and cause oil gasification accident. Some protective measures was designed to avoiding the accident, such as alarm or shutdown furnace, if crude oil temperature on the left or

Please use the following format when citing this chapter:
Huang, Yihu, Zhang, Guangxin, Jia, Ximei, 2006, in International Federation for Information Processing (IFIP), Volume 207, Knowledge Enterprise: Intelligent Strategies In Product Design, Manufacturing, and Management, eds. K. Wang, Kovacs G., Wozny M., Fang M., (Boston: Springer), pp. 797-802.

right side of furnace is high(>70℃); Alarm or shutdown furnace if crude oil pipeline temperature in the left or right side of furnace is high(>350℃), and so on. But because the furnace is a severe inertia system, above protective measures can not absolutely prevent oil gasification accident.

Incipient fault two: Sensor fault happen in industry field engineering; electromagnetism disturbs impact accuracy of acquisition data either, they all probably cause furnace fluctuating or shutting down.

2. FUZZY OBSERVER DESIGN

Furnace is a long time-lag non-linear system, which its multi-input and multi-output are strong-coupling. Those characteristic make automatic control difficulties. We also can use it to design a fuzzy observer.

Furnace meets the necessity of heat transfer theory and dynamic thermal balance principle. Flux diversification of fuel or combustion-supporting air can change working condition of furnace, cause a serious of movement of technic index, which include hearth temperature, pipeline temperature on the left and right side of furnace, outlet temperature on the left and right side of furnace. It means those parameters contain redundancy information each other. When sensors are not faulty and there are not disturb in acquisition data, above 7 measure parameters all reflect hearth combustion in different level, they are local parameter of working condition of furnace.

Base on above analyzing, a fuzzy observer was designed to solve above incipient faults and realize fault-tolerant control.

2.1 Fuzzy Interface

Fuzzy interface change real measured value to fuzzy vector. In the industry system, fuzzification gradation is not suitable to divide meticulous, otherwise some information would be lose, and the advantage of fuzzy value could not be utilized (King, 1975). Because the major aid of fuzzy observer is to judge trend of furnace condition, fuzzy input and out variable e is divided by:

$$e = \{\text{Negative, Zero, Positive}\} = \{\text{N, Zo, P}\}$$

Because flow control valve of fuel are one-to-one correspondence with fuel flow, fuel valve can substitute fuel flow to be an input parameter of fuzzy observer.

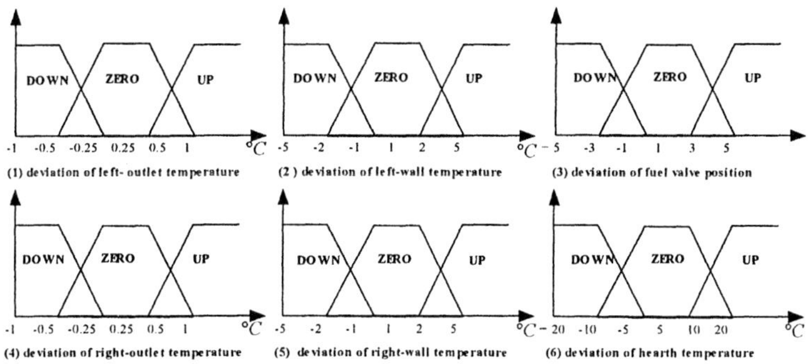

Figure 1. Membership function of input parameters

2.2 DB-Data Base

The design chooses trapezoidal membership function as input parameter function of fuzzy observer, and stores them in data base. The data base provides data when system solves equation of fuzzy relation (Lee, 1990).

Fig.1 shows 6 input parameters of membership function. Valves of the nodal point are defined by multi-trial. Definite principles are as follows:

① Considering measured precision of temperature sensor and valve transducer to prevent misjudging;

② Considering output precision of fuzzy observer to prevent under-reporting.

2.3 Rule of Fuzzy Observer

Base on expert knowledge, rule of the fuzzy observer is a kind language form matching instinct inference. It is usually connected by a series of relative, such as if-then, else, also, end, and. All fuzzy rules store in rule base. Rule base and data base compose knowledge base of fuzzy observer.

2.4 Inference and Defuzzy-Interface

Consider the process of fuzzy observation as follow:

According to input fuzzy value, rule of fuzzy observer fulfill fuzzy inference, then solve equation of fuzzy relation, and acquire fuzzy observation output. Inference and the part that provides conversion function are called defuzzy-interface （Zhujin, 1995） . Output parameter is defined as tendency of working condition of furnace.

Routine Zadeh approximate forward reference （Zedeh,1965） is:

$$R^i: \text{If } x_1 \text{ is } A_1^i \text{ and } x_2 \text{ is } A_2^i \ldots\ldots \text{and } x_n \text{ is } A_n^i \text{ then } y \text{ is } B^i \quad (1)$$

If several sensors are faulty, the integral tendency from (1) is incorrect. Inference rule of select majority based redundancy of input parameters was put forward as follows:

R^i : If x_1 is A_1^i, then N_1^i=1;

x_2 is A_2^i, then N_2^i=1;

……;

x_6 is A_6^i, then N_6^i=1.

$$\text{Where, } i = \begin{cases} -1 & \text{when } A_j^i = -1 \\ 0 & \text{when } A_j^i = 0 \\ 1 & \text{when } A_j^i = 1 \end{cases}, \quad j = 1, 2\ldots, 6;$$

$$\text{If } \sum_{j=1}^{6} N_j^{-1} \geq 4 \text{ then y=-1}$$

$$\text{If } \sum_{j=1}^{6} N_j^{0} \geq 4 \text{ then y=0} \quad (2)$$

$$\text{If } \sum_{j=1}^{6} N_j^{1} \geq 4 \text{ then y=1}$$

Where $x_1,\ldots,x_6$ are input parameters of fuzzy observer, A_j^i is fuzzy value of number j input parameter, y is output result of fuzzy observer.

Above inference principle of select majority based redundancy of input parameters firstly accumulate separately numbers of input parameter fuzzy value equal to −1,0 and 1, if numbers of one fuzzy value >=4, then fuzzy observer output this fuzzy value. This means even if one or two of 6 input parameters is faulty, accurate tendency of working condition of furnace can be obtained by redundancy select majority principle.

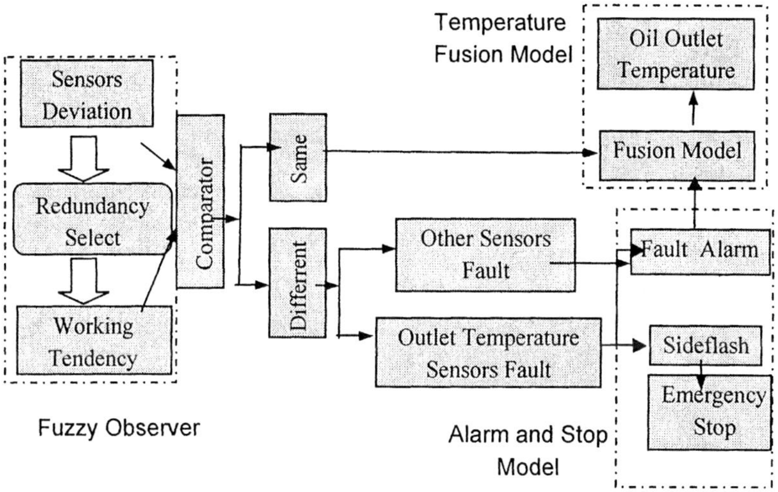

Figure 2. Process Chart of Fault Diagnosis

3. FAULT DIAGNOSIS PROCESS

Fig.2 shows implementation procedure of fault diagnosis. Fault Diagnosis include three parts, part 1 is Fuzzy Observer, which deduce the working tendency of furnace; part 2 is Alarm and Stop Furnace Model, which alarm and stop furnace when sensors is faulty; part 3 is oil outlet temperature fusion model, which include dual fusion model and simple fusion model.

In the comparator, 6 input parameter deviation separately compare with tendency of working condition of furnace, if tendencies are identical, then to go to model of double-fusion, to calculate crude oil temperature of furnace outlet (Huang, *et al.*, 2004); otherwise judge whether the parameter is outlet crude oil temperature on the left(or right) side of furnace, if answer is no, judge that the sensor is faulty, and alarm; if answer is yes, then judge whether tendency of crude oil temperature on outlet is same with crude oil pipeline on the same side, if trend is same, define n=1; Otherwise n=0, then to judge temperature sensor of furnace outlet is faulty, and at same time, to go to simple temperature fusion model of furnace outlet. If n=3, then to judge flame deflection, and to stop furnace urgently.

4. CONCLUSION

The field test shows the method can not only deal with conflict information gathered from different sensors and information of fault sensors, which ensures reliability of furnace, but also overcome safe incipient fault two. In the simple fusion model, the faulty sensor is abandoned, so that accuracy of outlet temperature is assured, and fault-tolerant capacity is improved. Because the method based on the dynamic characteristic of system, the method can abroad apply to other dynamic system, which the parameters are strongly-couple.

5. REFERENCES

1. YiHu Huang and ZeKui Zhou, Other (2004). Application of Deviation Parameter Fuzzy Fusion Method in Furnace. In: *3rd International Symposium on Instrumentation Science and Technology*, 224-227, Harbin Institute of Technology Press, Harbin.
2. WANG Fang, LIU Yan, SHAO Jun-peng. (2001), The Application of Muti-sensor Combination Technology, *JOURAL HARBIN UNIV. SCI. & TECH*, **1**(6): 81-84
3. Lee C C (1990), Fuzzy Logic in Control Systems: Fuzzy Logic Controller. In *Part 1. IEEE Trans. SMC.* **20(2)**: 404-418
4. LUO Zhang-liang, FANG Qing-cheng, ZHANG Qian-jin. (2002). Method of Data Fusion for Multi-Sensor and its Application, *Journal of Transducer Technology*, **2** (21): 27-29
5. King P J and Other (1975). The Application of Fuzzy Control System to Industrial Processes. In *IFAC world Congress. MIT. Boston.* 132-127.
6. Zadeh L A. (1965). Fuzzy Sets. In *Information and Control*, **8**: 338-353.
7. Zhujin, et （1995）, *Fuzzy Control Principle and Application.* Publishing Company of Engineering Industry.

RELIABILITY ANALYSIS FOR MANUFACTURING GRID

Qiming Zou, Tao Yu, Haiyang Sun
CIMS & Robot Center of Shanghai University, Shanghai University, China; Email: kim@staff.shu.edu.cn

Abstract: Manufacturing Grid (MG) emerges as a new manufacturing model to provide a general platform for networked-manufacturing. The reliability is a significant and complex issue in MG system as resources in it are geographically distributed, heterogeneous and information dynamic in nature, owned by different enterprises or organizations with their own policies. In order to make MG work efficiently, the reliability for MG had been analyzed; a hierarchical evaluation model was proposed. In the end, the use of the model was also presented in system recovery policies.

Key words: Manufacturing Grid (MG); reliability analysis; evaluation model; system recovery

1. INTRODUCTION

Based on the research of Networked Manufacturing, Manufacturing Grid 1 utilize not only the advanced concept and idea of Grid but also the open architecture and universal specification and tools to realize the resource sharing and cooperation for the distributed, dynamic manufacturing resources to promote the development of networked manufacturing towards openness and specification.

MG system is a distributed and heterogeneous system, in which resources are highly heterogeneous, diversity and distributed, ranging from information resources, service resources and human resources to design and manufacturing software. With MG technologies, it is possible to realize integration and optimization of all kinds of resources, provide the

This work is supported by the E-institute of Shanghai Education committee-Campus Grid Foundation of Shanghai under grant #200302.

Please use the following format when citing this chapter:
Zou, Qiming, Yu, Tao, Sun, Haiyang, 2006, in International Federation for Information Processing (IFIP), Volume 207, Knowledge Enterprise: Intelligent Strategies In Product Design, Manufacturing, and Management, eds. K. Wang, Kovacs G., Wozny M., Fang M., (Boston: Springer), pp. 803-811.

environment for commerce collaboration, design collaboration, manufacturing collaboration and supply chain collaboration among different enterprises, enables manufacture high quality product meet the market's requirement by lowest cost and shortest development period. However, due to the diverse failures and error conditions in the MG environments, developing, deploying and executing such activities is a challenge. Dependability is a key factor for MG. Reliability analysis and some techniques for fault tolerance are the solutions for improving the MG dependability.

This paper presented reliability hierarchical evaluation model for MG., a simple application case had been given out. Also, we drew the conclusion and gave out future work.

2. RELIABILITY ANALYSIS MODEL OF MG

2.1 The definition of MG system reliability

Generally speaking, the MG system reliability should include the following three aspects:

1) System Operational Reliability (R_O): the reliability when system is running, is a synthetic index;

2) System Inherent Reliability (R_I): the attributes are determined at the beginning of system is produced, there are many factors such as used material, component, design schema, hardware and software structure and manufacturing process are related to R_I;;

3) System Uses Reliability (R_U): is a dynamic attribute, is related to the system's environment and human resources.

The three kinds of reliability mentioned above exists a relationship, which can be expressed with the following formula:

$$R_o \approx R_u \cdot R_I \qquad (1)$$

This expression explains that Ro is decided by R_U and R_I;. To a certain system, we can acquire higher R_O index by improving usage environment, operation and maintenance level. For the MG system, the most important reliability index is R_O. So we define MG system reliability as follows:

The ability of completing specified functions consistently under various conditional factors during the whole running process.

2.2 Reliability Analysis Model of MG

2.2.1 The running model of MG

Now we present our running model of MG. As depicted in Figure.1, MG system is constructed on the basis of grid platform Globus. User can access the sharing resources and services through MG portal, which is the bridge between user and MG system. The portal provides many functions including information service, interface service and user authentication. The various middleware of MG system is running on the database server and application server, its function mainly provide support of system and data. The sharing resources in manufacturing industry are encapsulated as service node according to the rule and standard of grid service; the resource management is responsible for autonomy management; the resource monitor transfer dynamic status to MG system.

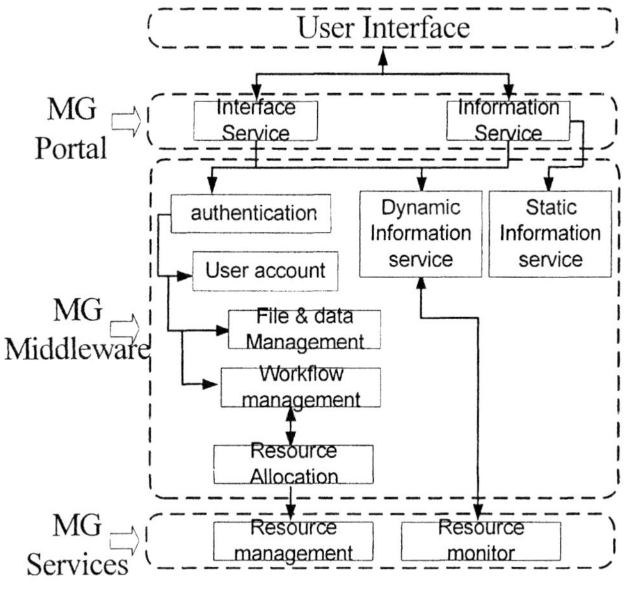

Figure 1. The running model of MG

2.2.2 The reliability block diagram (RBD) of MG

The reliability block diagram analysis is a kind of analysis method of studying the logic relation of system and parts. The RBD of MG system has taken the consideration of failure or expiration of each job element.

From the running model of MG, we knew that MG portal is the entrance which user access the MG, is also the bottleneck of MG, so it' s important to guarantee the MG operating persistently and steadily. Therefore, at the time of carrying on monitoring and management, we adopt hardware redundancy method to raise its reliability. The relation of two portals is parallel. Every kind of application server and database server in system enable system properly running. If there is a server breakdown, then the whole system will breakdown, so the relationship of two kinds of server is series. The service nodes in MG are distributed and independent, so their relation is parallel. The RBD of MG is illustrated in Figure. 2.

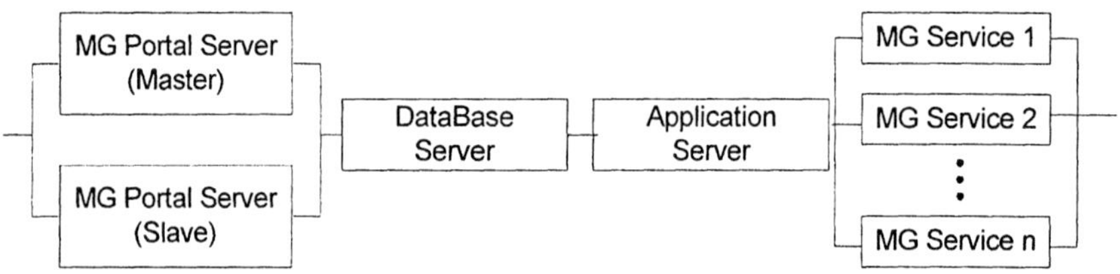

Figure 2. The RBD of MG

The reliability of MG can be expressed with the following formula:

$$R_{MG} = R_{SYS} \cdot R_{Service} = (R_{Portal} \cdot R_{Sys}) \cdot R_{Service} \qquad (2)$$

Where R_{sys} denotes system resource reliability and $R_{service}$ denotes reliability of each element in MG.

2.2.3 Hierarchical Reliability Evaluation Model of MG

As mentioned above, MG is a complex system, from the view of reliability analysis, can be divided into four levels, as depicted in Figure.3. The four levers are system level, sub-system level, child-system level and element level.

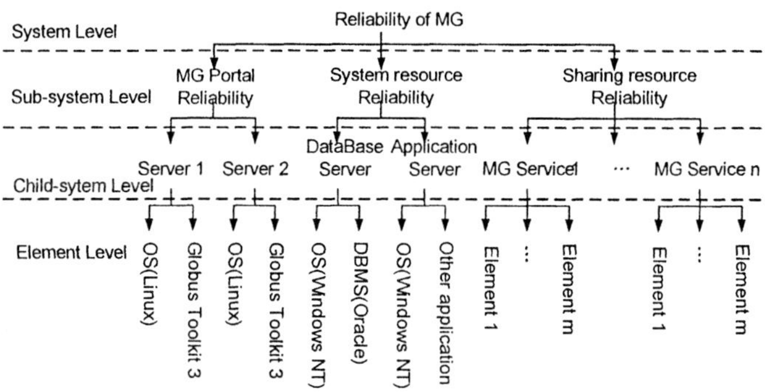

Figure 3. Hierarchical reliability evaluation model of MG

The child system, which is one element in MG, is independent to some degree, also is a integration as a whole for MG. First of all, the reliability for child system is very important to reliability analysis for MG. An element's reliability evaluation model (REM) can be expressed with

$$REM = (V, VS, \Gamma_v) \qquad (3)$$

Where $V = \{v_1, v_2, \cdots, v_n\}$ is a group of variables that stand for object running status; $VS = \{0, 1, 2, 3\}$ is variable status space; while Γv is a reflection relation from V to VS.

The MG reliability evaluation model (MGREM) is based on element's REM, so it can be formally expressed with $MGREM = f(rem_1, rem_2, \cdots, rem_n)$, where rem_i denotes *i-th* element's reliability. The influence degree of various element is different, so must consider the weight index λ_i ($i = 1, 2, \cdots, n$), and $\sum_{i=1}^{n} \lambda_i = 1$, $\lambda = (\lambda_1, \lambda_2, \cdots \lambda_m)$ is the volume of reliability evaluation. So the MGREM can be computed as:

$$MGREM = f(rem_1, rem_2, \cdots, rem_n) = \sum_{i=1}^{n} \lambda_i rem_i \Big/ n \qquad (4)$$

In the reliability analysis for single element process, status parameters are real-time variables from running element, while performance parameters can trace running status of element. There is a mapping relation to be decided between above two parameters. For different element, their parameters values are different and special. When the performance parameters of element are used to reliability analysis, the result can be used in the super level reliability analysis.

In the following example, we adapt reliability level and average life to illustrate the MGREM. Suppose that k ($k = 1,2,\cdots,n$) is element number, $i \in VS$ ($i = 0,1,2,3$) is the status number; v_k ($v_k \in \{0,1,2,3\}$) is the status variable of element in MG; y ($y \in \{0,1,2,3\}$) is the status variable of MG system.

From Figure.3, we knew that MG system is made up of three sub-systems. Its status is decided by the worst system's status $y = \min\{v_1, v_2, v_3\}$.

Suppose that $p(t,i)$ and $p_k(t,i)$ denotes the probability located in state i and time t of MG respectively, where $i = 0,1,2,3,4$, $t \in [0,\infty]$; the whole MG system and *k-th* element performance reliability are $R(t,i)$ and $R_k(t,i)$, can be computed as:

$$R(t,i) = p\{y(t) \ge i, t \in [0,\infty)\} = \sum_{j=i}^{m} p(t,j) \qquad (5)$$

$$R_k(t,i) = \sum_{j=i}^{m} p_k(t,j) \qquad (6)$$

where $\sum_{i=0}^{m} p(t,i) = 1$, $\sum_{i=0}^{m} p_k(t,i) = 1$, $1 = R(t,0) \ge R(t,1) \ge R(t,2) \ge \cdots \ge R(t,m)$.
When $i = m$, then the equation $p(t,m) = R(t,m)$ is correct. According to the definition of $p(t,i)$ and $R(t,i)$, the MG system average occupied time $M_s(i)$ and average life $\theta(i)$ ($i = [i,m]$) at the state i can be computed respectively as follows:

$$M_s(i) = \int_0^{\infty} p(t,i)dt \qquad (7)$$

$$\theta(i) = \int_0^\infty R(t,i)dt = \int_0^\infty [\sum_{j=i}^{m} p(t,i)]dt = \sum_{j=i}^{m} M_s(j), i = 1,2,\cdots,m \qquad (8)$$

3. USES OF RELIABILITY ANALYSIS MODEL

3.1 Reliability Analysis Process for MG

As shown in Figure.4, there are five stages in the MG reliability analysis process: data collection, data storage & management, data analysis and event generation, storage and handling.

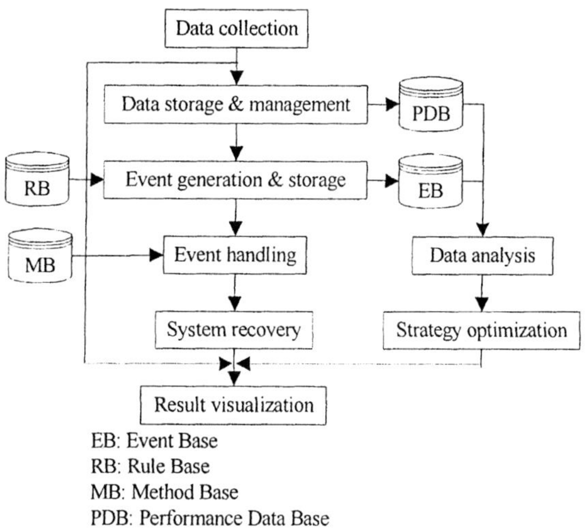

Figure 4. Reliability analysis process

The data collection module collects performance data from the managed resources, which is used for reliability evaluation. After data cleaning and storage, the event management starts to determine if there are any alert or warning events. When some events are generated, warning messages are sent to the system administrator, or corresponding recovery actions are taken if they are defined in the method base (MB). At the same time, the data analysis module extracts data from PDB and EB, and performs further analysis to provide proposals for optimization of the managerial strategy or direct system recovery. The process of reliability management is a closed loop. The task of monitoring is performed from data collection to result visualization. Then, performance optimization is accomplished from the strategy optimization to data collection. The optimization results and performance problems are fed back after a new monitor process is started. The iterative process is a basic workflow of the reliability management.

3.2 Recovery policy of MG

3.2.1 Recovery system of MG

As shown in Figure. 5, there are key four agents in the recovery system 2,3. They are reliability manager, Mobile Agent (MA) server, MA generator and MA.

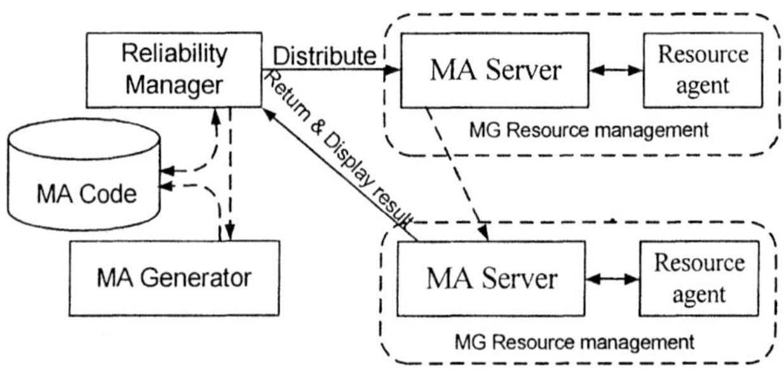

Figure 5. Recovery system of MG

G reliability manager with GUI moderates monitor and control strategy about MG resources. Manager can discover the agent process in working, maintain and update agent list.

User access MA through MA server module, which is installed on each managing resources. Logically, MA server located in the upper layer of resource agent, create effective environment for receiving, instantiation, running and scheduling MA object. It's very import that integration with resource management. There are four components in the MA server: Mobile Agent Listener, Security Facility Component, Service Facility Component and Mobile Facility Component.

Mobile Agent Generator (MAG) is an automatic tool for generating MA code in nature, can customize the combination of MA according to service requirement. The code is stored in the MA base. So MA can extend frame function dynamically, transmit initialization state of MA server.

MA is a Java object with only identifier can be transmitted among hosts, and executing particular management task in independent process. There is a routine table and a data folder in MA, the function of the folder is store and collects management information and method that interact with polling resources.

3.2.2 Implementation process of recovery system

One of the important tasks for reliability management is recover data after system had breakdown. So the prerequisite preparation should include backup data, storage changed data and trace required resources. When the events that influence integrality, consistency and validity of system data occurs, MG reliability manager get the recovery method from event handling method base according to the monitor rule, create mobile agent, then dispatch it to fault resource node, put it into event handling buffer list, wait for executing auto recovery operation or notify the related resource node to execute corresponding operation.

4. CONCLUSIONS AND FUTURE WORK

The generic, heterogeneous, dynamic, and high-latency nature of the MG makes it fragile, and how to measure the reliability is a key issue to provide high availability services. This paper presents a general reliability analysis model for MG, and uses of the model are also discussed. In the future, more use cases will be considered for the model. In this paper, we do not consider the dynamic attribute parameters of manufacturing resources in MG, such factors will be considered in our future model and application.

5. REFERENCES

1. Lilan Liu, TaoYu, Zhanbei Shi, et al, Self organization manufacturing grid and its task scheduling algorithm, *Computer Integrated Manufacturing System*, 2003, 9 (6):449-455.
2. D.Gavalas, D.Greenwood, Advanced Network Monitoring Applications Based on Mobile/Intelligent Agent Technology, *Computer Communications*, 2000, (23):720-730.
3. Andre Mello Barotto, Anddriano de Souza, Carlos Becker Westphall, Distributed Network Management Using SNMP, Java, WWW and CORBA, Journal of Network and Systems Management, 2000, 8(4):483-497.

CONSTRUCTION OF FOREWARNING RISK INDEX SYSTEMS OF VENTURE CAPITAL BASED ON ARTIFICIAL NEURAL NETWORK

Guozheng Zhang, Yun Chen, Dengfeng Hu
School of Public Economy Administration, Shanghai University of Finance & Economics, P.R.china, 200433, Email:guozhengzhang@gmail.com

Abstract: The risk evaluation is a very important step of decision process of venture capital. This article constructs forewarning risk index systems of venture capital, and applies Artificial Neural Network (ANN) to forecast the latent risk level of venture capital. Empirical analysis indicates that this model could predict the latent risk of venture capital very well.

Key words: Artificial neural network, Forewarning risk index system, Venture capital

1. INTRODUCTION

Venture capital is a kind of equity capital that the professional venture capitalist puts into venture enterprises [1, 2]. Presently, Risk forewarning system based on neural network has been applied in many fields such as Credit risk evaluation, Project management and so on [7, 8, 6, 10]. Kuldeep Kumar, John D Hayneshave provided that the ANN is better than statistical techniques on forecasting credit risk level because risk has non-linear relationship with its affected factors[9]. None of current literatures about venture capital has constructed forewarn system based on neural network to forecast the investment risk. There are two questions: (1) How to construct forewarning risk index systems; (2) How to apply neural network to predict risk level? The evaluation target is very complicated. It is a multifactor

The project is supported by the Shanghai Shuguang Project of China under the grant No: 05SG38

Please use the following format when citing this chapter:
Zhang, Guozheng, Chen, Yun, Hu, Dengfeng, 2006, in International Federation for Information Processing (IFIP), Volume 207, Knowledge Enterprise: Intelligent Strategies In Product Design, Manufacturing, and Management, eds. K. Wang, Kovacs G., Wozny M., Fang M., (Boston: Springer), pp. 812-817.

comprehensive decision question [3]. This article effectively evaluates the risk constructing appropriate forewarning risk index system and effectively applying artificial neural network technology.

2. FOREWARNING RISK INDEX SYSTEM OF VENTURE CAPITAL

2.1 Contents of forewarning risk index system

The research of venture capital decision has interested people since the 1980s abroad. Tyebjee and Bruno has listed the evaluation index from qualitative angle, using questionnaire investigation and factor analytic approach to construct the U.S.A evaluation models of venture capital at first[1]. Vance H Fried and Robert D Hisrich extended this model and indicated that such as managerial ability and resistance ability of environment threaten, etc. has a great influence on expected risk of venture capital[3]. While carrying on venture capital decision, venture capitalist must consider every factor impacting on risk synthetically [2, 4]. According to the characteristic of operating venture capital, this article carries on systemic analysis on every factor which forms the risk and attempts to construct a forewarning risk index system including managerial ability, resistive ability to environmental threat, market risk, technological risk, production risk, withdraw channel, etc. 6 first class index and relevant 36 sub index.(Table.1)

Table 1. Forewarning risk index system of venture capital

Forewarning Index Systems of the Venture Enterprise (U)			
First grade index	Managerial ability	resist ability to environment threatens	Technological ability
(U_i)	(U_1)	(U_2)	(U_3)
Sub grade index	Administrator's ethical quality	National political economic environment	substitutability of Technology
(U_{ij})	Academic level	policies and regulations	Advancement of technology
	Experience of management	Change of the bank rate	Suitability of technology
	Ability of association	Investor's income of the market	Dependability of technology
	Environmental adaptive capacity	Potential rival	Intellectual property right

Forewarning Index Systems of the Venture Enterprise (U)			
First grade index	Market risk	Production risk	Withdrawing channel
(U_i)	(U_4)	(U_5)	(U_6)
Sub grade index	Market	Return on net assets	IPO
(U_{ij})	Market scale	equipment utilization degree	the way of Selling
	products competence	Producer's state	the way of Withdrawing
	Service level	Materials supplies	Liquidating
	Marketing ability	Additional investment	Payback period of investment

3. APPLYING ARTIFICIAL NEURAL NETWORK (ANN) TO FORECAST THE RISK LEVEL

3.1 Basic principle of BP neural network

Figure.1 shows a three layer BP neural topological network. Little circle means neuron unit, and the line connecting each neuron units means weight coefficient W. The neural network imports X, and exports Y after the weight coefficient W is dealt with. According to certain algorithm, ANN confirms the right weight coefficient W of each line through studying sample data. After right weight coefficient W is confirmed and the structure is steady, ANN can deal with the new data [5, 7, 8, 6, 10].

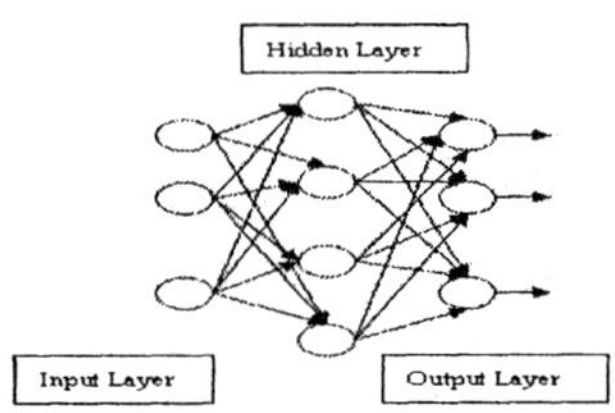

Figure 1. ANN structure

3.2 BP ANN forewarning models

(1) Input and Output. This article uses all sub index as input vector quantity X = $(u_{11}, u_{12}, u_{12}, \ldots, u_{64}\ \ u_{65})$, and regards forewarning risk grade as output value Y. Input vector quantity includes quantitative and qualitative index. Qualitative index ought to be converted to quantitative index by that experts give grade on them. This article uses interval law to define

forewarning risk grade according to corresponding interval as table.2 shows. C is the highest forewarning risk grade; AAA is the lowest forewarning risk grade.

(2) Neural network structure. There aren't perfect methods to confirm the number of neuron units of hidden layer at present, so this article adopts experience rule. If the study samples is N, the number of neurons unit of hidden layer is ($2\sqrt{n}$) [5]. In this way, the basic structure of the three layers ANN network is a type of 30-12-1 BP ANN.

(3) Neuron network function, error function and studying algorithm. Input layer neuron network function is f (x) = X, Function of hidden layer and output layer is Sigmoid function: f(x) =1/(1 +exp (- x)). Initialization of all the neuron units is 0. Weight coefficient $W_k = (w_{ij}^k)_{n_{k-1} \quad k}$, among them, n_k means the number of neuron units of layer k, while w_{ij} means weight coefficient between neuron unit i of layer k-1 and neuron unit j of layer k. The receiving value of neuron unit of layer k is $t_{jk} = \sum_i^{n_{k-1}} w_{ij} x_i$, j =1,2 ,..., n_k. The output value from layer k to layer k +1 is $y_j = f(t_{ij})$, j=1,2, ..., n_k. Error function is $E = \frac{1}{2}\sum_1^n |y_k - \hat{y}_k|^2$, among them, y_k is the expected output of layer k, $\hat{y}_k$ is the actual output of layer k.

Table 2. Forewarning risk grade

C	CC	CCC	B	BB
(0 ,0.1]	(0. 1,0. 2]	(0. 2,0. 3]	(0.3 ,0. 4]	(0. 4,0. 5]
BBB	A	AA	AAA	
(0. 5,0. 6]	(0.6 ,0. 7]	(0. 7,0. 8]	(0. 8,0. 9]	

3.3 Empiric analysis

The research object is 43 venture capital projects. At first, Initialize vector X of each venture capital project to gain the importing vector group X_i, i =1, 2, ..., 43. Then refer to table.2 and through expert's giving grade of 43 venture capital projects, get the expect exporting vector group Y_i, i =1,2,...,43. Use (X_i, Y_i) as the study samples, among them, i =1, 2, ..., 40. Use (X_j, Y_j) as the evaluation target, among them j =41, 42, 43. The purpose of the experiment is to compare the output evaluation result of ANN with the expert's evaluation conclusion, proving ANN risk evaluation is valid on venture capital.

The structure of ANN is 30 -12-1 according to the order of input layer, hidden layer and output layer. Initialization value of each neuron unit is random in [0, 1].Study precision ε =0. 005. The increment coefficient of excellent fast algorithm of dropping and seeking η =0.5. In order to prove the ANN risk evaluation model is valid, I have done the experiment for 10 times, contrasting expected output value Y_{41} =0.75, Y_{42} =0.55, Y_{43} =0.25 (Forewarning risk grade is AA, BBB, CCC) with corresponding actual output risk value $Y_j^{'}$ (j=41, 42, 43). The experimental result is shown in Table.3, Table.4 and Tab.5.

Table 3. Contrast of network output $Y_{41}^{'}$ and expected output Y_{41}

times of experiment	1	2	3	4	5
network Output	0.753	0.764	0.755	0.786	0.793
expected output	0.75	0.75	0.75	0.75	0.75
Result of evaluation	AA	AA	AA	AA	AA
Result of expectation	AA	AA	AA	AA	AA
times of experiment	6	7	8	9	10
network Output	0.731	0.722	0.684	0.773	0.729
expected output	0.75	0.75	0.75	0.75	0.75
Result of evaluation	AA	AA	A	AA	AA
Result of expectation	AA	AA	AA	AA	AA

Table 4. Contrast of network output $Y_{42}^{'}$ and expectation output Y_{42}

times of experiment	1	2	3	4	5
network Output	0. 534	0.533	0.554	0.620	0.589
expected output	0.55	0.55	0.55	0.55	0.55
Result of evaluation	BBB	BBB	BBB	A	BBB
Result of expectation	BBB	BBB	BBB	BBB	BBB
times of experiment	6	7	8	9	10
network Output	0.528	0.584	0.509	0.540	0.558
expected output	0.55	0.55	0.55	0.55	0.55
Result of evaluation	BBB	BBB	BBB	BBB	BBB
Result of expectation	BBB	BBB	BBB	BBB	BBB

Table 5. Contrast of network output $Y_{43}^{'}$ and expected output Y_{43}

times of experiment	1	2	3	4	5
network Output	0.233	0.243	0.256	0.282	0.246
expected output	0.25	0.25	0.25	0.25	0.25
Result of evaluation	CCC	CCC	CCC	CCC	CCC
Result of expectation	CCC	CCC	CCC	CCC	CCC
times of experiment	6	7	8	9	10
network Output	0.297	0.311	0.271	0.245	0.318
expected output	0.25	0.25	0.25	0.25	0.25
Result of evaluation	B	CCC	CCC	CCC	B
Result of expectation	CCC	CCC	CCC	CCC	CCC

4. CONCLUSIONS

Empiric analysis indicates that forewarning index risk system based on ANN could export comparatively accurate forewarning risk grade of venture capital. In this article the forewarning risk index system and ANN forewarning model are based on the visual angle of venture capital forms. The forewarning risk index system based on ANN has solved the problem of comprehensive evaluation of venture capital very well.

We should pay attention to that the study speed of neural network is influenced by the neuron unit number of each layer, neuron network function and study algorithm, etc[5]. We should apply ANN forewarning risk model according to the actual conditions. Because the research of risk evaluation system and risk evaluation method is beginning, here remain many problems to discuss thoroughly further. I hope this paper can play a role in casting a brick to attract jade.

5. REFERENCES

1. T T Tyebjee and A V Bruno, A model of venture capitalist investment activity, *Management Science*, 1984 (9):300-330
2. Masako Ueda, Banks versus Venture Capital: Project Evaluation, Screening, and Expropriation, *Journal of finance*, 2004(4): 604-621
3. Vance H Fried and Robert D Hisrich, Toward a model of venture capital investment decision making, *Financial Management*, 1994 (3):212-230
4. Alex Proimos, Sue Wright, A pilot study of venture capital investment appraisal in Australia, *Journal of Financial Services Marketing*, 2005(9):272 - 286
5. Narendra K S, Parthasarathy K, Identification and control of dynamical systems using neural network, *IEEE Trans Neural Networks* ,1999(1):4--27.
6. Mohamed Attalla and Tarek Hegazy, Predicting Cost Deviation in Reconstruction Projects: Artificial Neural Networks versus Regression, *Journal of construction engineering and management*, 2003(5): 405-510
7. Patrick L. Brockett, William W. Cooper, Linda L. Golden, Utai Pitaktong, A Neural Network Method for Obtaining an Early Warning of Insurer Insolvency, *The Journal of Risk and Insurance*, 1994(3): 402-424
8. Bart Baesens, et al, Using Neural Network Rule Extraction and Decision Tables for Credit-Risk Evaluation, *Management Science*, 2003 (5): 312–329
9. Kuldeep Kumar, John D Haynes, Forecasting credit ratings using an ann and statistical techniques, *International journal of business studies*, 2003(6): 91-108
10. Dariusz Skorupka, Neural Networks in the Risk Management of a Project, *2004 AACE International Transactions[C]* ,RISK.15

RESEARCH ON PERFORMANCE TEST AND EVALUATION METHOD OF ASSEMBLY PACKAGING AMMUNITION

Wenzhao Li, Wei Wang, Min Gao, Junbo Wang
Ordnance Engineering College, Shijiazhuang, China; Email:sharpshaw@sina.com

Abstract: In order to evaluate accurately the comprehensive performance of assembly packaging ammunition using small samples, a new test and evaluate method is presented. Firstly carry out the consistency test on the performance parameters of electronic parts of ammunition before and after the experiment, which are liable to the influence of various environmental factors, if the result of test is consistent, then firing test is carried out in order to check the reliability of whole ammunition. In the course of the consistency test, according to the difference of electric parameters rank sum test and T-test is used, and the Bayes method is applied in the whole ammunition's reliability evaluation, and an example is demonstrated in order to verify the validity of this test and evaluation method.

Key words: Ammunition packaging, Small samples, Rank sum test, T-test, Bayes method.

1. INTRODUCTION

For testing the quality of inner ammunition in general, all kinds of environmental test on assembly packaging ammunition should be carried out according to the demand of ammunition support. By virtue of the difference of experiment's method, packaging test can be divided into two kinds, one is static test, such as stack test, pressure test, temperature test, humidity test etc, the other is dynamic test, such as falling test, impacting test, collision test and others. After doing traditional tests, firing test must be put up to analysis the protecting ability of the packaging material and the rationality of structure. But the test samples of new armed ammunition which is expensive

Please use the following format when citing this chapter:
Li, Wenzhao, Wang, Wei, Gao, Min, Wang, Junbo, 2006, in International Federation for Information Processing (IFIP), Volume 207, Knowledge Enterprise: Intelligent Strategies In Product Design, Manufacturing, and Management, eds. K. Wang, Kovacs G., Wozny M., Fang M., (Boston: Springer), pp. 818-824.

is small, it is very difficult to get exactly the performance of packaging with common methods[1,2].

From the result of research on the design and production process, in expensive ammunition we can see the quality of inner electronic parts are liable to the influence of various environmental factors compared with others in the same condition. The states of most electronic parts can be checked through special test window, therefore we can get the data of performance before and after traditional packaging test experiment easily. If the difference between two results is distinct, we can draw a conclusion that packaging is not reasonable, if the results are consistent, then firing test is carried out to test the reliability of whole ammunition in order to judge whether it reaches an appointed technical index.

2. EXPERIMENT SCHEME

At first we select n ammunition samples from a batch of accepted products at random and test electrical parameters, and get the test results, which is denoted as: $X=(X_1, X_2, \cdots, X_n)$. Then the environmental stress test is carried out which includes stack test, pressure test, temperature test, humidity test, falling test, impacting test, collision test etc. At last we take out the ammunition tested with stress experiment from package box, and test each one's electrical parameter again, which is denoted as $Y=(Y_1, Y_2, \cdots, Y_n)$.

Analyze the consistency of X and Y with the statistical method, if the difference of the two results is distinct, the quality of the packaging is not reasonable, otherwise if the result is compatible, then the firing test is carried out in order to verify the quality of the whole ammunition further. Through the firing test, the result function (s, f) is obtained, the variable s indicates the number of success, f indicates the number of failure. Given the degree of confidence γ, we can figure out the lower level of the reliability R_L with the formula (s, f), and compare R_L with the ammunition's reliability technical index R_L'. By above way, we can judge the level of the ammunition's reliability is remarkable or not, and the conclusion of the packaging test is got. The experiment scheme is shown in figure 1.

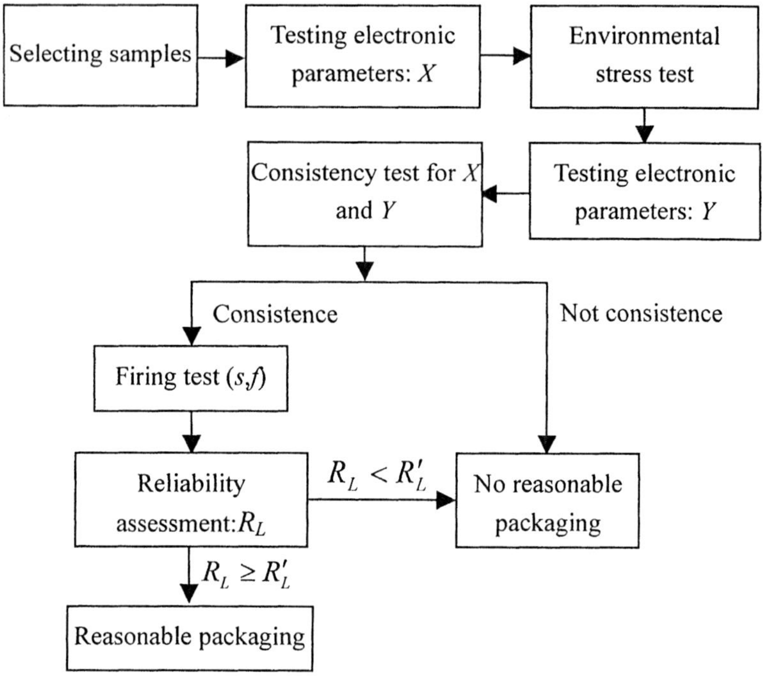

Figure 1 Experiment scheme

3. STATISTIC ANALYSIS METHOD OF DATA

3.1 Test method of consistency

3.1.1 T-test of two collectivities

Given $X_1, X_2, \cdots, X_n$ and $Y_1, Y_2, \cdots, Y_m$ are two different samples selected from continuous collectivity $F(x) \sim N(\mu_1, \sigma^2)$ and $G(x) \sim N(\mu_2, \sigma^2)$. For checking the consistency of *X* and *Y*, that is, the hypothesis is as follows: $H_0 : \mu_1 = \mu_2$; $H_1 : \mu_1 \neq \mu_2$.

we sign:

$$S = \left\{ (n+m)\frac{\sum_{i=1}^{n}(X_i - \overline{X})^2 + \sum_{i=1}^{m}(Y_i - \overline{Y})^2}{(n+m-2)nm} \right\}^{1/2} \quad (1)$$

Given a check level α , the formula (2) is set up.

$$\begin{cases} \left|\overline{X} - \overline{Y}\right| > St_{n+m-2}(\alpha/2) \Rightarrow \text{Reject;} \\ \left|\overline{X} - \overline{Y}\right| < St_{n+m-2}(\alpha/2) \Rightarrow \text{Accept.} \end{cases} \quad (2)$$

3.1.2 Rank sum test

Given $X_1, X_2, \cdots, X_n$ and $Y_1, Y_2, \cdots, Y_m$ are two different samples selected from continuous collectivity $F(x)$ and $G(x)$. For checking the consistency of X and Y, that is, the hypothesis is as follows: $H_0 : F(x) = G(x)$.

Put X and Y together, and arrange these data in order from small to big. The serial number of each data is called rank. Sum the rank of small samples compared with X and Y, denoted as T. If the hypothesis H_0 is true, the difference of two collectivities $F(x)$ and $G(x)$ is no distinct, and the value of T will be suitable.

Lookup the rank sum table of Wilcoxon[3,4], we can get the critical value T_1 and T_2 under the test level α.

$$P\{T_1 < T < T_2\} = 1 - \alpha \quad (3)$$

If $T_1 < T < T_2$, we say the hypothesis H_0 is true, and the difference between $F(x)$ and $G(x)$ is not distinct. If $T \le T_1$ or $T \ge T_2$, we say the hypothesis H_0 is false, and the difference between $F(x)$ and $G(x)$ is distinct. If the value of n and m is larger than 10, the statistical variable T belongs to the normal distribution, this is:

$$T \sim N\left(\frac{n(n+m+1)}{2}, \frac{nm(n+m+1)}{12}\right) \quad (4)$$

Therefore we can use U check method, and select statistical variable is shown bellow.

$$U = \frac{T - \frac{n(n+m+1)}{2}}{\sqrt{\frac{nm(n+m+1)}{12}}} \sim N(0,1) \tag{5}$$

Under the check level α, the rejection region W is as follows.

$$W = \{|u| > u_{1-\alpha/2}\} \tag{6}$$

We can use software such as SAS, GAUSS, SPSS, MATLAB etc to help us calculate the number of T easily. In this paper, we use the rank sum function [p,h]=ranksum(x,y,alpha) in MATLAB to do this work, in which the variable P is shown below.

$$p = P\{T_1 < T < T_2\} \tag{7}$$

If the value of P is larger than the given level, we reject H_0 and draw a conclusion that the difference of two collectivities is distinct, otherwise we accept H_0 and draw a consistent conclusion.

3.2 Method of reliability assessment

The result of the reliability assessment is explained with the lower limit of the reliability R_L, that is, given the degree of confidence r the equation $P\{R \geq R_L\} = r$ is true, in which the variable R is the reliability of system.

If the collectivity belong to the distribution $B(1,R)$, the test result is false or true only, we select n products as test samples at random, and the number of false result X meets the equation (8).

$$P\{X = k\} = C_n^k R^{n-k} (1-R)^k \quad k = 0,1,,\cdots,n \tag{8}$$

If we select Bayes method to assess the reliability of system, the distribution $Beta(a,b)$ is adopted, in which it's density function is shown as follows.

$$\pi(R) = \frac{\Gamma(a+b)}{\Gamma(a)\Gamma(b)} R^{a-1}(1-R)^{b-1} \quad 0 \leq R \leq 1 \tag{9}$$

In above function the variable a and b is the super parameters of distribution before testing. After we get $Beta(a,b)$ and the test samples (s,f), by the Bayes theory calculate the distribution after test $\pi(R \mid s,f) = Beta(a+s,b+f)$. So the lower limit R_{LB} under the level of γ can be gotten by the formula (10).

$$\int_0^{R_{LB}} \pi(R \mid s,f)dR = I(R_{LB}; a+s, b+f) = 1-\gamma \tag{10}$$

Parameters a and b is confirmed with the information before test. Usually for the ammunition product, (a,b) is set $(1/2,1/2)$[5,6].

4. APPLICATION EXAMPLE

In order to test the packaging quality of one type missile, according to the test and accepting rules of ammunition, select eight products as test samples in qualified products. By special measurement instrument, 28 different photoelectricity parameters of each product were attained, in which 16 parameters belong to the distribution of $B(1,R)$ and they need not test. Among all parameters only 12 connect with the quality of ammunition closely, in which there are 8 parameters belong to the normal distribution and we use T-test to check their consistency before and after experiment, use rank sun test to check other 4 parameters which the distribution is unknown. If one of 12 items checked parameters is distinct, it is shown the quality of packaging will not be qualified. Table 1 is the test data of winding's voltage of OK_1, which belong to the normal distribution and Table 2 is the test data of EBT resistance values. The analysis procedure is shown as follows.

Table 1 The Voltage of OK_1 (V)

Results before test X	0.11	-0.09	0.16	0.27	0.20	0.08	0.22	0.03
Results after test Y	-0.10	0.10	0.28	-0.16	0.08	0.19	-0.20	-0.39

Under the check level of $\alpha = 0.05$, we get $\overline{X} - \overline{Y} = 0.1475$ and $S = 0.96981$ with the formula (1). Lookup the T-test table, we get $t_{14}(0.975) = 2.14479$, because of $|\overline{X} - \overline{Y}| < St_{14}(0.975)$, the difference before and after test is not distinct.

Table 2 The resistance values of EBT (Ω)

Results before test X	3.5	3.0	3.3	2.7	1.5	4.3	2.2	1.3
Results after test Y	2.9	4.5	2.0	3.6	1.9	1.8	1.6	2.6

Under the check level of $\alpha = 0.05$, we get the value of variable T is 70. Lookup the Rank sum test table, we get $T_1 = 52$ and $T_2 = 84$. Because $T_1 < T < T_2$, the difference before and after test is not distinct.

According to above method, under the level of $\alpha = 0.05$, 12 test parameters are all not distinct. Then the firing test is carried out, and the result is: the success number s is 8 and the failure number f is 0, given the degree of confidence γ is 0.8, we calculate R_L with the formula (10), $R_L = 0.90532$. Compare with the designed index $R'_L = 0.9$ ($\gamma = 0.8$), we can get the conclusion that the quality of ammunition is good and the protecting ability of ammunition packaging is eligible.

5. CONCLUSIONS

Because of the particularity of ammunition, it is not easy thing very much for assessing and testing the integral performance of the assembly packaging ammunition, so this paper presents a problem of testing the performance of the assembly packaging ammunition under the condition of small samples. About this problem, a method of connecting checking and assessment is introduced, and an example is demonstrated in order to verify the validity of this method, which can evaluate the integral performance of ammunition packaging accurately and offer a new mathematics theory method and path for testing the integral performance of ammunition.

6. REFERENCES

1. Jinhuai Zhang. *Ordnance Equipment Small Sample Test Analysis and Assessment.* Defence industrial press, 2001
2. Efron.B.. Bootstrap Methods: Another Look at Jackknife. *The Annals of Statistical,* 1979, 7: 1-26
3. Xiru Chen. Introduction of Mathematic Statistics. Science and technology press, 1997
4. Fengqin Xu. A Summary of Chain Sampling Plans(Part I). *Mathematic Statistics and Management*, 1999, 18(2): 34-39
5. Fengqin Xu. A Summary of Chain Sampling Plans(Part II). Mathematic *Statistics and Management*, 1999, 18(4): 38-42
6. Shisong Mao. *Bayes Mathematic Statistics.* China Statistic Press, 1999.

SUPPORT VECTOR REGRESSION FOR FINANCIAL TIME SERIES FORECASTING

Wei Hao, Songnian Yu
School of Computer Engineering and Science, Shanghai University, Shanghai, China; Email: haowei@graduate.shu.edu.cn

Abstract: Recently, Support Vector Regression (SVR) has been a popular tool in financial time series forecasting. This study deals with the application of Support Vector Regression in stock composite index forecasting. A preprocessing method for accelerating support vector regression training is presented in this paper. Then we propose a method of support vector regression by modifying the regularized risk function. A data set from Shanghai Stock Exchange is used for the experiments to test the validity of our methods.

Key words: Support Vector Regression, Financial Time Series forecasting

1. INTRODUCTION

Stock market prediction is regarded as one of the most challenging applications of modern time series forecasting. Over the past decade, neural networks have been successfully used for modeling financial time series [1,2]. A large number of successful applications have shown that ANN can be a very useful tool for time-series modeling and forecasting [3]. The early days of these studies focused on application of ANNs to stock market prediction [4,5]. Recent research tends to hybridize several artificial intelligence techniques [6,7]. However, ANN has a difficulty in explaining the prediction results due to the lack of explanatory power, and suffers from difficulties with generalization because of overfitting.

Based on Statistical Learning Theory, Support Vector Machine (SVM), was first developed by Vapnik [8,9]. It has become a hot topic of intensive

Please use the following format when citing this chapter:
Hoa, Wei, Yu, Songnian, 2006, in International Federation for Information Processing (IFIP), Volume 207, Knowledge Enterprise: Intelligent Strategies In Product Design, Manufacturing, and Management, eds. K. Wang, Kovacs G., Wozny M., Fang M., (Boston: Springer), pp. 825-830.

study due to its successful application in classification tasks[10,11] and regression tasks[12,13], specially on time series prediction[14] and financial related applications[15]. This paper focuses on the application of Support Vector Machine in regression tasks to make a new attempt to stock composite index forecasting. A novel method for accelerating support vector regression (SVR) training based on a measurement of similarity among samples is presented in this paper. Then, based on the prior knowledge that in the non-stationary financial time series the dependency between input variables and output variable gradually changes over the time, specifically, the recent past data could provide more important information than the distant past data. We modified the regularized risk function in support vector regression, whereby the recent ε -insensitive errors are penalized more heavily than the distant ε -insensitive errors.

2. PREPROCESSING TO REDUCE TRAINING DATA

From the implementation point of view, training SVM is equivalent to solving a linearly constrained quadratic programming (QP) problem with the number of variables equal to the number of training data points. However, with the great increment of training data in size, the memory space for storing the kernel matrix will increase with the level $O(N^2)$, where N is the number of the training data. Hence, designing effective SVM training algorithms for large data sets will be practically significant.

A novel method based on the similarity measurement, for reducing training data to accelerate support vector machines training, is proposed here. The notion of similarity, as a measurement of the approaching degree between two samples, is introduced to select new training data. The similarity function S is defined as:

$$S(x,y) = f\left(\frac{1}{\|x-y\|^2}\right) = f\left(\frac{1}{\sqrt{\sum_{i=1}^{M}(x_i - y_i)^2}}\right) \tag{2.1}$$

those data points, which are close to some referenced data point according to the similarity function, will be discarded because they can be replaced by the special data point without great influence on the final solution[16]. The division of data groups will affect generalization performance and it can be achieved by clustering. In order to reduce possible errors from clustering phase, we designed a novel clustering algorithm based on K-means and t-test, in which after finding the nearest cluster for a data point, we add a hypothesis testing to verify the statistical significance of correlation between them.

3. THE MODIFIED SUPPORT VECTOR REGRESSION

In the field of financial time series forecasting, numerous studies show that the relationship between input variables and output variable gradually changes over time, and recent data could provide more information than distant data. Therefore, it is advantageous to give more weights on the information provided by the recent data than that of the distant data based on this prior knowledge[17,18]. In the light of this characteristic, we introduced a novel approach whereby more weights are given to the recent ε-insensitive errors than the distant ε-insensitive errors in the regularized risk function. The regularized term in the regularized risk function is retained, regardless of the empirical error.

In Support Vector Regression, the empirical risk function has equal weight C to all the ε-insensitive errors between the predicted and actual values. The regularization constant C determines the trade-off between the empirical risk and the regularized term. Increasing the value of C, the relative importance of the empirical risk with respect to the regularized term grows. For illustration, the empirical risk function is expressed as:

$$E_{svr} = C\sum_{i=1}^{l}(\zeta_i + \zeta_i^*) \tag{3.1}$$

In our approach, instead of a constant value, the regularization constant C adopts a weight function:

$$E_{svr} = \sum_{i=1}^{l} C_i(\zeta_i + \zeta_i^*) \tag{3.2}$$

$$C_i = w(i)C \tag{3.3}$$

Where $w(i)$ is the weight function satisfying $w(i) > w(i-1), i = 2,...,n.$ As the weights will incline from the distant training data points to the rent training data points, C_i will give more weights on the more recent training data points.

4. EXPERIMENTS

Experiments have been conducted to illustrate effectiveness of the presented method. A data set from Shanghai Stock Exchange was used for the experiments. We selected daily closing prices of stock composite index of shanghai stock exchange between Jan.21, 2002 and Sep. 27, 2002. There are totally 140 data points. 100 data points in front of the data series may be used as training data sets and the rest 40 data points as testing data. The experiments will emphasize on verifying the performance of the preprocessing method and evaluating the effectiveness of the modified SVR.

The experiments first emphasize on verifying the effectiveness of preprocessing method for reduction training data. Table 1 summarizes the training and testing results. In practical applications, a suitable similarity threshold may be selected according to specific requirement of speed and accuracy. The prediction performance is evaluated using two standard errors; mean absolute error (MAE) and root mean squared error (RMSE), there are defined as the following, respectively:

$$MAE = \frac{1}{l} * \sum_{i=1}^{l} abs(y_i - d_i), \tag{4.1}$$

$$RMSE = \sqrt{\frac{1}{l} \sum_{i=1}^{l} (y_i - d_i)^2}, \tag{4.2}$$

Where, y_i, d_i denote the predicted result and the measured value, respectively.

Table 1. The training and testing results for different similarity thresholds

S	CPU time (Second)	No. of training vectors	No. of Support vectors	MAE	RMSE
0.2	45.3	83	23	0.8281	0.8791
0.1	18.6	79	20	0.9194	1.0275
0.05	5.4	51	18	1.6259	2.1851
0.03	2.8	37	19	1.5343	1.9146
0.02	1.9	22	16	3.0517	3.6579
0.01	0.5	10	10	32.5634	38.1064

We use general RBF as the kernel function. There are two parameters while using RBF kernels: C and γ. It is not known beforehand which C and γ are the best for one problem; consequently some kind of model selection (parameter search) must be done. The goal is to identify them so that the classifier can accurately predict unknown data (i.e., testing data). Note that it may not be useful to achieve high training accuracy (i.e., classifiers accurately predict training data whose class labels are indeed known). Therefore, a common way is to separate training data to two parts of which one is considered unknown in training the classifier. Then the prediction accuracy on this set can more precisely reflect the performance on classifying unknown data. An improved version of this procedure is cross-validation.

We use a grid-search on C and γ using cross-validation. Basically pairs of (C, γ) are tried and the one with the best cross-validation accuracy is picked. We found that trying exponentially growing sequences of C and γ is a practical method to identify good parameters (for example, $C = 2^{-5},...,2^{15}$, $\gamma = 2^{-15}, 2^{-13},...,2^{3}$).

Figure 1 illustrates the predicted and actual values. The dotted line is the actual value. The smoothing line is the predicted value of our method. From the figure, it can be observed that the method forecast more closely to the actual values in most of the time period. The experimental results show that this method is more effective and efficient in forecasting stock composite index than the standard SVR and traditional time series forecasting models.

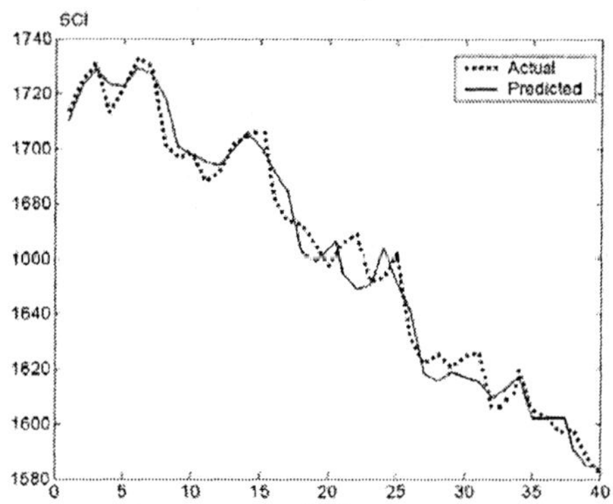

Figure 1. Forecasting result comparison

5. CONCLUSION

In this research, a preprocessing method for accelerating SVR training and a modified SVR are proposed. The objective of this research is also to examine the feasibility of the methods in stock composite index forecasting and to improve the accuracy of SVR in terms of parameters selection. By using our methods, high prediction performance is achieved.

6. REFERENCES

1. W. Cheng, L. Wanger, Forecasting the 30-year US treasury bond with a system of neural networks. *Journal of Computational Intelligence in Finance*, 4 (1996),pp.10 16.
2. R. Sharda, R. Patil. A connectionist approach to time series prediction: an empirical test. *Neural Networks in Finance and Investing*, Chicago: Probus Publishing ,1994, pp.451 464.
3. G. Zhang, B.E. Patuwo, M.Y. Hu, Forecasting with artificial neural networks: the state of the art, *Int.J. Forecasting* 14 (1998) ,pp. 35 62.
4. H. Ahmadi, Testability of the arbitrage pricing theory by neural networks, *Proceedings of the International Conference on Neural Networks*, San Diego, CA, 1990, pp.1385 1393.
5. J.H. Choi, M.K. Lee, M.W. Rhee, Trading S& P 500 stock index futures using a neural network, *Proceedings of the Annual International Conference on Artificial Intelligence Applications on Wall Street*, New York, 1995, pp.63 72.
6. Y. Hiemstra, Modeling structured nonlinear knowledge to predict stock market returns, *Chaos & Nonlinear Dynamics in the Financial Markets: Theory, Evidence and Applications*, Irwin, Chicago, IL, 1995, pp.163 175.
7. I.H. Witten, E. Frank, Data Mining: Practical Machine Learning Tools and Techniques with Java Implementations, *Morgan Kaufmann Publishers*, San Francisco, CA, pp.1999.
8. V.N.Vapnik. The Nature of Statistical Learning Theory. *Springer*, New York, pp.1995.
9. V.N.Vapnik. Statistical Learning Theory. Wiley, New York, 1998.
10. Edgar Osuma and Robert Freund and Federico Girosi. Support Vector Machines: Training and Applications. *AIM-1602*, MIT, 38, 1997.
11. J.C.Burges. A Tutorial on Support Vector Machines for Pattern Recognition. *Data Mining and Knowledge Discovery*, 2(2),pp.121-167, 1998.
12. V.N.Vapink, S.Golowich and A.Smola. Support vector method for function approximation, *regression estimation and signal processing*.
13. A.Smola and B.Scholkopf. A Tutorial on Support Vector Regression. 1998, *Technical Report NeuroCOLT NC-TR-98-030*.
14. K.R.Muller, A.Smola, G.Ratsch, B.Scholkopf, J.Kohlmorgen and V.N.Vapnik. Predicting time series with support vector machines. *ICANN*, 999-1004, 1997.
15. T.B.Trafalis and H.Ince. Support vector machine for regression and applications to financial forecasting. *IJCNN 2000*, 348-353.
16. W.Wang, Z.Xu. A heuristic training for support vector regression. *Neurocomputing* 61 (2004) 259-275.
17. Francis E.H.Tay, L.J.Cao. Modified support vector machines in financial time series forecasting. *Neurocomputing* 48 (2002) 847-861.
18. N.D. Freitas, M. Milo, P. Clarkson, Sequential support vector machines, *Proceedings of the IEEE Signal Processing Society Workshop*, Madison, WI, USA, 1999, 31 40.

QOS MANAGEMENT IN MANUFACTURING GRID

Haiyang Sun, Tao Yu, Lilan Liu, Yu'an He
CIMS & Robot Center of Shanghai University, Shanghai University, China; Email: sunhaiyang@263.net.

Abstract: In order to apply Manufacturing Grid technology to manufacturing field and provide resource-sharing and collaborating for the distributed manufacturing companies on a common platform, Manufacturing Grid system should resolve the problem brought by non-technical factors in manufacturing. Extending the requirement of product quality and regarding it as a kind of provided service which is required in the process of manufacturing. Then importing the idea of QoS (Quality of Service) to Manufacturing Grid and merging it into the manufacturing's characteristics. Based on it, making QoS management as the key research contents of Manufacturing Grid and create Manufacturing Grid-QoS (MG-QoS) system. Furthermore, proposing a kind of preliminary solution based on QoS management to Manufacturing Grid's practical application. It provides quality guarantee for the application of Manufacturing Grid and resolves the issue of prerequisite for Manufacturing Grid's practice from theoretical research.

Key words: Manufacturing Grid, Quality of Service (QoS), QoS Management

1. INTRODUCTION

The term "the Grid" was coined in the mid 1990s [1], and the nature of the Grid is to share the distributed resources. The main aim of the Grid's application is to improve the resources' utilization ratio, reduce cost and complete nontrivial tasks those can't be finished without the Grid [2].

In fact, resources are often leisure in manufacture enterprises because the utilization rations of the resources are low. On the contrary, many enterprises must collaborate with other enterprises because the resources of their own are limited.

This paper is supported by "SEC E-Institute: Shanghai High Institutions Grid" project.

Please use the following format when citing this chapter:
Sun, Haiyang, Yu, Tao, Liu, Lilan, He, Yu'an, 2006, in International Federation for Information Processing (IFIP), Volume 207, Knowledge Enterprise: Intelligent Strategies In Product Design, Manufacturing, and Management, eds. K. Wang, Kovacs G., Wozny M., Fang M., (Boston: Springer), pp. 831-839.

There is a contradiction and it becomes more and more obvious for scarcities of effective collaboration, implementing and management. With the development of information process in manufacturing, people want to change this. Its keys are sharing resources and collaboration. In manufacturing field, cooperation frequently takes place in one enterprise or among them, and it can be regarded as various services. Furthermore, we extend the quality control in the manufacturing enterprise and build mechanism that can be used to evaluate the service provided by enterprise, and then establish Manufacturing Grid (MG) [3]. The Grid's core is Quality of Service (QoS) and its framework is Open Grid Service Architecture (OGSA) [4].

The application of Manufacturing Grid not only introduces the Grid technology into manufacturing, but also to promote the development of manufacturing and accelerate information process in manufacturing. Manufacturing Grid can achieve the aim of manufacturing to share distributed resources and realize coordinated manufacturing by Virtual Organization (VO) [5].

This paper is organized as follows. In section 2, we import the conception of QoS into Manufacturing Grid, and then introduce Manufacturing Grid-Quality of Service (MG-QoS) and QoS management in Manufacturing Grid's application. In section 3, we analyze manufacturing resource management, task disassembling and manufacturing resource allocation. In section 4, we introduce Service Level Agreement (SLA) based on QoS and SLA management. In section 5, we summarize our research and conclude the paper.

2. THE RESEARCH CONTENTS OF QOS IN MANUFACTURING GRID

2.1 QoS management in Manufacturing Grid

The conception of Quality of Service was coined in information field. It refers to the comprehensive result of the service performance which decides the user's satisfaction. That is to say, QoS is a quality agreement about information transmission between the Service User (namely User) and Service Provider [6], and between the clients and the integrated service network used to transfer information. However, Manufacturing Grid, as a new application of grid technology in manufacturing, provides a standard expandable network information platform by Grid technology and implements manufacturing resource's sharing and collaborating.

Manufacturing Grid contains both characteristics of general grid and some unique differences from general computing grid. As an application of service grid in manufacturing, MG has some remarkable differences compared with other Grids, especially in resources, task and task's requirement. We can't import research of QoS from information field into manufacturing field directly. And in manufacturing, quality control impenetrate the whole process of manufacturing and contains almost requirements of products and reference points. So, we regard Manufacturing Grid-Quality of Service as the core of Manufacturing Grid platform system, and emphasize MG-QoS application management based on manufacturing process's characteristics [7].

2.2 Manufacturing Grid-Quality of Service

We regard MG-QoS management as the core of Manufacturing Grid system. According to the ideas of Grid, we can say MG-QoS management is in the application layer of QoS management in manufacturing field. Its primary task is to deal with the logical affairs between user and service providers, and the interaction among different function modules. Just as illustrated in Figure.1.

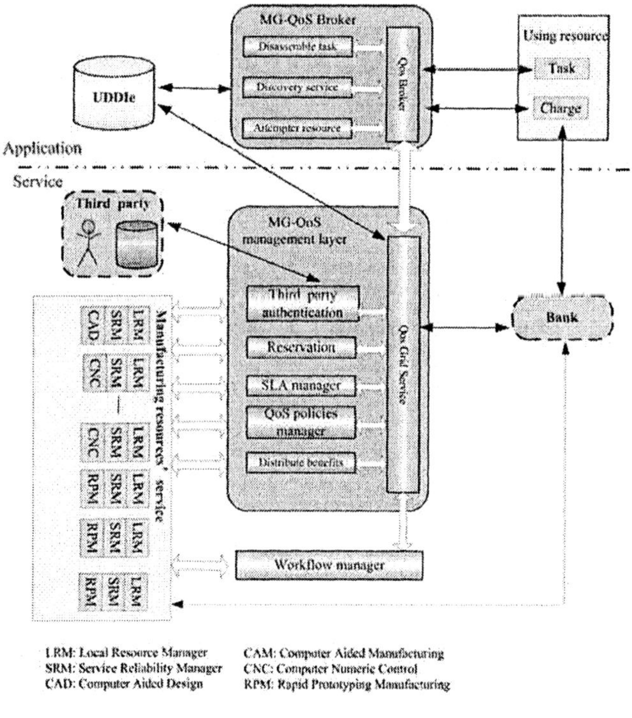

Figure 1. MG-QoS Architecture

2.3 QoS management in Manufacturing Grid application

Then we introduce how to submit a task, and organize manufacturing, then finish task. From the whole process, we show how QoS management works.

When user has enrolled in Manufacturing Grid platform, he can submit task to platform [8]. The description and requirement of task are the contents of MG-QoS management. The manufacturing process in Manufacturing Grid likes the real process in manufacturing enterprise. Then, Manufacturing Grid system analyzes the task submitted by user and judge if the platform can complete user's task by the registered resources in platform. If platform ensures it can complete the user's task, then it disassemble user's task and contract with user. According to user's requirement of QoS and result of disassembling, Manufacturing Grid platform chooses optimal resource from all the registered resources to each subtask by some arithmetic. Following, the platform negotiate the details of subtask with resources' providers. If the provider accepts the subtask, then he can sign SLA agreement with the Manufacturing Grid platform. After signing, workflow management organizes manufacturing and executes workflow according to allocation of subtasks and requirements of QoS until the task is finished, and sends feedback to QoS management. While workflow management works, it can monitor the violation that happens in manufacturing process. If violation happens, workflow management will inform QoS management, and QoS management can executes punishment.

Following, we will introduce two modules which have close connections with QoS management.

3. MANUFACTURING RESOURCES MANAGEMENT, TASKS DISASSEMBLING AND MANUFACTURING RESOURCES ALLOCATION

3.1 Establishing model of resource's QoS properties in Manufacturing Grid

The resources in manufacturing field are different to those in information sciences, and furthermore, different manufacturing resource has different characteristics. So, according to these different characteristics, how to feasibly describe these characteristics' differences is the key to establish models of manufacturing resources' QoS properties [9].

First, we classify the resources of Manufacturing Grid as follows.

1) Technology resource

Technology is a kind of material methods for carrying out yielding, and it manifests the characteristic of manufacturing. It includes design resource, emulation resource, manufacturing resource and data exchange resource.

2) Information resource

Information resources include various manufacturing information. The resources provide necessary condition for managing and utilizing the manufacturing resources. Market information, management information, sharing information and cooperation information compose information resources.

3) Personnel resource

People as organizers or executants participate in almost all manufacturing process, so they play very important roles in the process. Personnel resources include manager, technical adviser, producer and service people.

4) Service resource

Service resources mainly include after service and training.

From the classifying of manufacturing resources, we can see that the manufacturing resources are heterogeneous, distributed, complicated and multifarious and their characteristics show the differences of QoS properties. So, when we establish the model of QoS properties, we must consider these properties' characteristics. Furthermore, in order to easily encapsulate the resources, build up resource nodes, manage resources and allocate resources, we should describe and define the resources' dynamic properties and static properties.

3.2 Describe the user's task based on QoS

If the user wants to take use of the convenience provided by Manufacturing Grid platform, first of all, he should submit his task to the platform. The platform will provide means for user to describe his task and the requirements of QoS. And we regard the requirement of product's quality and additive demands as the services that Manufacturing Grid platform should provide. Namely, user requires the platform to provide satisfying services and ensure good qualities. We introduce several dominating parameters to describe the task. The parameters include delivery time (T), product quality (Q), manufacturing cost (C) and service level (S).

3.3 Disassembling task and allocating resources based on QoS

When user submitted task on the platform, the platform should analyze the whole task and disassemble it into subtasks, and then ensure if it can complete the user's task. The platform's analyzing includes two steps. Firstly, the platform adopts different Global Process Planning (GPP) template to disassemble the task simply. Secondly, the expert belonged to Manufacturing Grid platform will check up the disassembling result.

When the platform decides to accept the task based on the disassembling result, the expert will perfect the result, and synchronously, he should decide the manufacturing method and process. Then, he will rake all the resources they are capable of completing one subtask, and choose the optimal resource. We adopt same method to describe the resources, and then we can distinguish the different resources. We also introduce several dominating parameters to describe the different resources in detail. And the parameters also include delivery time (T), product quality (Q), manufacturing cost(C) and service level(S) [10, 11]. The description of resource is similar to the description of task, so we can easily rake resources and choose optimal one.

4. ERVICE LEVEL AGREEMENT (SLA) MANAGEMENT

4.1 General SLA management and Manufacturing Grid's SLA management

In information business, using information resource is a series of business activities. In order to ensure the rights and the obligations of the user and the provider, the two parties will establish agreements. SLA (Service Level Agreement) is the exact statement of all expectations and obligations in the business relationship between the provider and the user. It is built on the agreement formed by contract between user and service provider, and its purpose is guaranteeing the interest of both parties. In information field, SLA management includes a kind of regular document and can automatically collect useful information and evaluate the parameters in SLA agreement.

Because of the characteristics of manufacturing, SLA management in Manufacturing Grid is different from the SLA management in other fields. The main research of SLA includes negotiation among user, provider and the platform, signing SLA and conserving SLA. Furthermore, the SLA management can provide proof for violation of SLA, analyze breach and execute punishment.

4.2 SLA management's research and its framework in Manufacturing Grid

The main research of SLA management includes the following.

1) Designing SLA management module and interfaces;

2) Establishing SLA contract repository and creating multifarious regulations;

3) Monitoring violation, such as communicating with workflow management, analyzing violation's reason, resuming the interrupted task and recording the violation information;

4) Executing punishment, including deciding the punishment's grade and renewing punishment contents in punished resource's property;

5) Conserving the signed SLA documents.

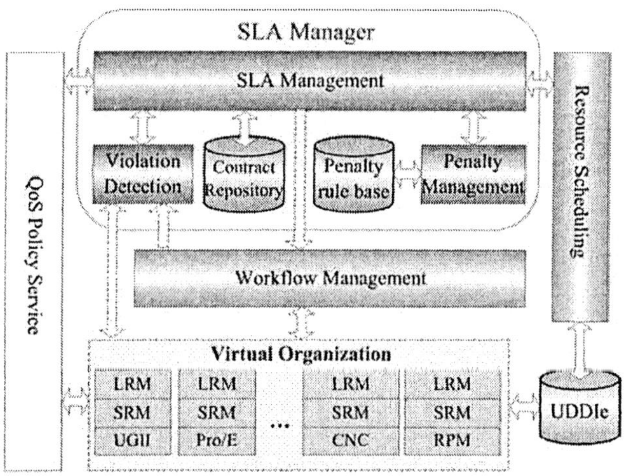

CNC: Computer Numeric Control RPM: Rapid Prototyping Manufacturing

Figure 2. SLA manager

Figure.2 illustrates the framework of SLA Management module. From Figure.2, we can see SLA management module mainly includes SLA Management, Violation Detection, Contract Repository, Penalty Management, Penalty Rule Base and QoS Policy.

4.3 SLA management

SLA management is the dominating content of SLA management module, and it can interact with other functions of the module and manage the module. The main of SLA management includes following.

1) User negotiates with platform about QoS; Platform negotiates with provider about QoS;

2) Create SLA document;

3) Submit workflow to workflow management;

4) When Violation Detection finds a case happens, it will evaluate the violation and decide the punishment, and then penalty management will execute punishment;

5) SLA Management rectifies or resumes the allocation of resources based on the different violations in order to complete the task;

6) Terminate SLA dialogue and renew properties of resources which have provided service.

4.4 SLA template

Because of complexity of manufacturing resources and diversity of functions, the descriptions of their services are different. In order to simplify the platform's management, we design different templates to different services.

SLA templates provide criterion for SLA document. If create a SLA agreement, first, people choose a template from the template repository according the service of the provider. Second, people make sure the requirement of this SLA. Then, finish the details of SLA and create relevant document.

4.5 SLA executing

When finish SLA's creating, SLA management will submit formal workflow to workflow management and the management will execute and manage the whole workflow. It includes informing provider's obligation, monitoring violation and starting punishment when violation happens. The whole process should run until SLA finish.

5. CONCLUSION

Grids are rapidly emerging as a practical means by which to perform new science and new applications. Manufacturing Grid helps the enterprises with similar goals and interests form loosely coupled virtual organizations for collaboration and resource sharing as one of grid applications. In this paper we mainly discuss the QoS management. We extend the requirements of product's quality to the requirements of service provided by provider, and bring forward a method of Manufacturing Grid's application based on QoS management. We introduce the process of Manufacturing Grid's application in detail, and propose the appropriate method to solving practical issues. The implementation in the project of Manufacturing Grid Platform supported by "SEC E-Institute: Shanghai High Institutions Grid" shows that the research is valuable and the solution is effective.

6. REFERENCES

1. Global Grid Forum, http://www.gridforum.org.
2. L. Pearlman, V. Welch, I. Foster, and C. Kesselman,"A Community Authorization Service for Group Collaboration", *Proceedings of the IEEE 3rd International*
3. LIU Lilan, YU Tao, SHI Zhanbei. Research of Self-Organization Manufacturing Grid and its Task Scheduling Algorithm . *Computer integrated manufacturing system*, 2003, 9 (6),pp.449-455
4. Foster, C. Kesselman, J. Nick, S. Tuecke , "The Physiology of the Grid: An Open Grid Services Architecture for Distributed Systems Integration," *Open Grid Service Infrastructure WG*, Global Grid Forum, June 22, 2002.
5. Ian Foster, Carl Kesselman, Steven Tuecke . The Anatomy of the Grid: Enabling Scalable Virtual Organizations . *Intl J.Supercomputer Applications*, 2001, 15 (3),pp.1-21.
6. I.Foster, C.Kesselman, J.M.Nick and S. Tuecke . Grid Service for Distributed System Integratiopn .*IEEE Computer*, 2002, 35 (6),pp.37-46.
7. SHI Zhanbei. *Research of OGSA-based Manufacturing Grid QoS System* . Shanghai: Shanghai University, 2004.
8. Shi Zhanbei, Yu Tao, Liu Lilan, "Service Registry and Discovery in Rapid Manufacturing Grid", *Computer Applications*, 2003, pp.85-87
9. Liu Lilan, Yu Tao, Shi Zhanbei, "Research on Rapid Manufacturing Grid and Its Service Nodes", *Machine Design and Research*, 2003, pp.57-59.
10. ZHAO Nai-yan, FAN Yu-shun. A Product–Structure–Based Solution to Select Collaborating Partners of Agile Virtual Enterprise , *Computer Integrated Manufacturing Systems*, 2002, 8 (2),pp.99-104.

A NEW METHOD BASED ON IMMUNE ALGORITHM TO SOLVE THE UNIT COMMITMENT PROBLEM

Wei Li, Deren Sheng, Jianhong Chen, Zhenfu Yuan, Kefa Cen
College of Mechanical and Energy Engineering, Zhejiang University, 310027 China, Email:energy@cmee.zju.edu.cn

Abstract: A new method based on immune algorithm to solve the thermal unit commitment problem(UCP) is proposed. Through analysis of mathematical model of UCP, application method of immune algorithm was discussed in detail. In the algorithm, the objective function was deemed antigen and the solutions were deemed antibodies. If an antibody fits the antigen best, this antibody was deemed the optimum solution. Encoding the continuous operating time shortened the code length and the searching speed of algorithm was improved greatly. The proposed method was applied in a ten-unit system for a 24 hours period. The calculation results showed immune algorithm had well global searching performance and it was efficient to use this algorithm to solve UCP.

Key words: Unit Commitment Problem(UCP), Immune Algorithm, Optimization, Thermal Unit

1. INTRODUCTION

The thermal unit commitment problem (UCP) is a non-linear combinatorial optimization problem. The task of UCP is to find optimum units combination to meet the requirements of the power system. According to references, the efficiency can be improved by 1% to 2.5% through optimization of unit commitment according to scheduling experience(Pei et al., 2001). So researchers try to solve the unit commitment problem using all kinds of intelligent algorithms for long time.

Please use the following format when citing this chapter:
Li, Wei, Sheng, Deren, Chen, Jianhong, Yuan, Zhenfu, Cen, Kefa, 2006, in International Federation for Information Processing (IFIP), Volume 207, Knowledge Enterprise: Intelligent Strategies In Product Design, Manufacturing, and Management, eds. K. Wang, Kovacs G., Wozny M., Fang M., (Boston: Springer), pp. 840-846.

There are several conventional methods to solve the unit commitment problem, such as priority list, dynamic programming, intege programming, linear programming, Lagrangian relaxation method and so on. Because of the complexity of UCP and the numerous constraints, the optimum solution can't be achieved efficiently using these methods (Cai et al., 1997). Recently genetic algorithm is applied to tackle UCP. However genetic algorithm is easy to suffer from local optimum solution for its poor diversity and premature convergence. A new method based on immune algorithm is proposed to handle UCP. Immune algorithm is a global search method and it has memorization capabilities and good diversity. So this algorithm has good convergence and can search global optimum solution effectively.

2. UCP FORMULATIONS

1. Objective functions (Cai et al., 1997)

The task of unit commitment is to minimize the total fuel consumption of the generating units including operation consumption and start-up consumption. The objective functions can be expressed as below:

$$\min F(U_{it}, P_{it}) = \sum_{t=1}^{24} \sum_{i=1}^{G} [U_{it} F_i(P_{it}) + U_{it}(1 - U_{i(t-1)}) S_i] \quad (1)$$

where $F(U_{it}, P_{it})$ is the total fuel consumption function, U_{it} is the on/off status of unit i at hour t, $U_{it}=1$ if unit i is on and $U_{it}=0$ if unit i is off, P_{it} is the generation output of unit i at hour t, $F_i(P_{it})$ is the fuel consumption of unit i at hour t, G is the total number of units, S_i is the start-up cost of unit i.

2. Constraints

Six constraints should be satisfied. There are electric power balance equation, output limits of generating units, spinning reserve considerations, limits of ramp rates, start-up and down frequency, minimum up-time and minimum down-time.

3. INTRODUCTION OF IMMUNE ALGORITHM

3.1 Summary of immune algorithm

The immune algorithm is inspired by the human immune system that protects our bodies from attack of foreign substance (called antigen). In the algorithm, the objective functions and constraints were deemed antigens. The feasible solutions were deemed antibodies. A computation of affinity was embedded for determining the promotion or suppression of antibodies production. Therefore the diversity of antibodies were kept by the boost or restriction of antibody generations. If an antibody has maximum affinity, this antibody is deemed the optimum solution to the optimization problem.

Immune algorithm has some distinct characteristics as follows:

- It operates on the feasible solutions that guarantee good convergence.
- The diversity is kept by boost or restriction of antibody generations.

3.2 Computation of immune algorithm

1) Computation of affinity (Gao, 2001)

If the antibody pool is composed of N antibodies, each of which has M genes. The entropy $H_j(N)$ of the jth gene can be expressed as follows:

$$H_j(N) = \sum_{i=1}^{N} - p_{ij} \log p_{ij} \tag{2}$$

where p_{ij} is the probability that the ith allele comes out of the jth gene. Then the average information entropy can be expressed as below:

$$H(N) = (\sum_{j=1}^{M} H_j(N)) / M \tag{3}$$

The affinity between antibodies can evaluate the diversity of antibodies. If antibodies are more similar, the affinity between antibodies is higher. The computing formulation can be represented as below:

$$\alpha_{v,w} = 1/(1 + H(2)) \tag{4}$$

where $H(2)$ is the information entropy between antibody v and w.

2) Computation of antibody concentration

The antibody concentration is the proportion of some similar antibodies to the whole population. It can be expressed as follows:

$$c_v = \frac{\sum_{w=1}^{N} S_{v,w}}{N}, S_{v,w} = \begin{cases} 1, & 当 \alpha_{v,w} \geq T_{ac1} \\ 0, & 当 \alpha_{v,w} < T_{ac1} \end{cases} \tag{5}$$

where N is the total number of antibody population, T_{ac1} is the threshold of antibody concentration.

3) Computation of antibody viability

$$e_v = A_v \prod_{s=1}^{S} \frac{1 - L_{v,s}^k}{c_v \sum_{i=1}^{N} A_i}, L_{v,s} = \begin{cases} \alpha_{v,s}, & \alpha_{v,s} \geq T_{ac2} \\ 0, & \alpha_{v,s} < T_{ac2} \end{cases} \quad (6)$$

where A_i is the sufficiency value of antibody, S is the total number of the suppressor cells, k is the suppressor index, T_{ac2} is the viability threshold.

4) Computation process of immune algorithm

Step 1: Initialization.

Some parameters of immune algorithm are set, such as the population size, the total number of memory cells, the concentration threshold, crossover rate, mutation rate and so on.

Step 2: Antigen recognition.

In the immune algorithm, the objective function of unit commitment problem and constraints are deemed as antigen.

Step 3: Generation of initial antibody.

Initial antibody generates according to the encoding method based on binary code. A group of genes form an antibody. Each antibody represents a feasible solution to the optimization problem.

Step 4: Computation of affinity.

Step 5: Computation of antibodies concentration.

Step 6: Differentiation of memory cells.

If some antibody's concentration is over threshold, the memory cell generates to memorize the local optimum solution.

Step 7: Boost or restriction of antibodies.

The antibody with low concentration will most possibly survive to next generation, whereas the suppressor cells will restrict the antibodies with high concentration. Thus the diversity of antibody is kept.

Step 8: Generation of new antibodies.

The new antibodies generate after crossover operation, mutation operation and immune complementarity. Immune complementarity means that the antibody has low viability will be replaced with a new antibody generates randomly.

Step 9: Termination judgment.

The calculation will stop if the termination condition is satisfied. Otherwise the calculation will execute from step 4 to step 8.

Step10: Output optimum solution.

4. IMPLEMENTATION OF IMMUNE ALGORITHM

4.1 Encoding method of antibody

If the operating status of one unit which in single time-interval is represented by one-bit binary code, for G units and 24 hours, the length of an antibody is L=24*G bits. Since the code length is long, the code string that has better mode is easily destroyed under crossover and mutation operation. Considering the constraint of daily start-up and down frequency, some researchers proposed to encode the variable position of units operating status. There are 24 intervals per day and 23 variable positions altogether. One variable position can be represented by 5-bits binary code. It only needs 15 bits binary code if the maximum start-up and down frequency is three.

In addition, generating units should also meet the constraint of minimum up-time and minimum down-time. A new encoding method is proposed, viz. to encode the continuous operating time of each unit. For the maximum continuous operating time is 24, it can be represented by a 5-bits binary code. If the maximum start-up and down frequency is three, then there are four status continuous time altogether. We only need to encode the former three continuous times. Suppose the initial status is on, the encoding method of antibody is given as below:

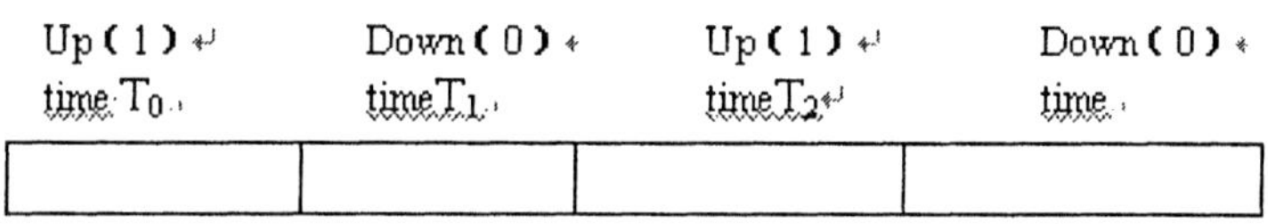

*Figure 1.*encoding figure

In this way, 15-bits binary code is needed to express operating status for 24 hours of each unit. For example, assuming the initial status of one unit is 1, antibody code is 00101 00111 00110, that denotes the operating status in the 1st-5th hour for this unit is 1, in the 6th-12th hour is 0, in the 13th-18th hour is 1, in the 19th-24th hour is 0. Encoding this way not only shortens the code length, but also satisfies the constraint of the start-up and down frequency and the constraint of the minimum continuous operating time so as to quicken the searching speed of immune algorithm.

4.2 An example

The proposed method was tested on a system consist of ten units. Their characteristic data and the data of load are listed in Ref. Han et al., 1994.

And tab.2 tabulated the optimum unit commitment for a 24h period. In immune algorithm, the population size is 100, crossover rate is 0.75, mutation rate is 0.05, the concentration threshold is 0.85, the viability threshold is 0.8, and selection rate is 0.8.

The calculation result using immune algorithm was compared with that using other optimization methods in tab.1. The computing total fuel consumption using the proposed method was 78970.3 tons coals. It was 0.48% better than the calculation result using dynamic programming method, and less than the calculation result using modified genetic algorithm by 27.7 tons. From calculation process, the optimum solution could be achieved through less than twenty generations. The calculation result showed that this method had good convergence and global searching performance and it is effective and efficient to solve the unit commitment problem.

Table 1 calculation results using different optimization methods

Optimization methods	Total fuel consumption (tons)
Lagrangian relaxation	80766
Heuristic genetic algorithm	79807
Dynamic programming	79349
Genetic algorithm in ref.Shi et al., 2002	79004
Improved genetic algorithm in ref.Han et al., 1994	81245.4
Modified genetic algorithm in ref.Pei et al., 2001	78998
Immune algorithm	78970.3

5. CONCLUSIONS

As a novel optimization method, immune algorithm has some distinct features, such as good diversity, boost or restriction mechanism between antibodies and global searching ability and so on. Using immune algorithm to optimize unit commitment can achieve the optimum solution efficiently. The calculation results showed that it is promising to use immune algorithm to solve UCP and other optimization problem in thermal power plant.

6. REFERENCES

1. Cai Chao-hao, Cai Yuan-yu. 1997. Optimization of unit commitment by genetic algorithm. *Power System Technology,* 21(1):44-47,51.(in Chinese)
2. Gao Jie. 2001. the Application of the immune algorithm for power network planning. *Systems Engineering-Theory & Practice,* 5:119-123.(in Chinese)

3. Han Xue-shan, Liu Zhuo. 1994. Optimal unit commitment considering unit's ramp-rate limits. *Power System Technology,* 6:11-16.(in Chinese)
4. Sun Yongzhi, Wei Wei. 2002. Solution of optimal power flow problem based on artificial immune algorithm. *Automation of Electric Power Systems,* 26(12):30-34. (in Chinese)
5. Pei Jin-yong, Lai Yi-fei, Chen Pin et al. 2001.Modified genetic algorithm for optimizing combination of units. Engineering *Journal of Wuhan University,* 34(1):72-76.(in Chinese)
6. Shi Qi, Li Cheng-jun, Wang Jin-wen. 2002. the Application of genetic algorithm to short-term generation scheduling. *Proceedings of the EPSA,* 14(2):56-59. (in Chinese)
7. Shyh-Jier Huang. 1999. Enhancement of thermal unit commitment using immune algorithms based optimization approaches. *Electrical Power and Energy Systems,* 21:245-252.
8. S.O. Orero, M.R.Irving. 1997. A combination of the genetic algorithm and Lagrangian relaxation decomposition techniques for the ge*neration unit commitment problem. Electric Power Systems Research,* 43:149-156.

ENTERPRISE INFORMATION INTEGRATION: STATE OF THE ART AND TECHNICAL CHALLENGES

Jingtao Zhou, Mingwei Wang, Han Zhao
The Key Laboratory of Contemporary Design and Integrated Manufacturing Technology, Ministry of Education, Northwestern Polytechnical University, Xi'an, China, 710072.
zhou.jingtao@gmail.com

Abstract: To gain insight into increasingly intricate business, and deal with highly complex problem situations, an enterprise needs a more generic, standardized, pervasive and scalable infrastructure to fully leverage the information from different data sources, applications, and environments at both system and semantic level. In this context, we first discuss the state of the art of current approaches and solutions of EII (Enterprise Information Integration). Then, we outline the grand challenges of EII from technical perspective based on the analysis of framework, range, scale and performance.

Key words: Enterprise Information Integration; Scalability; Horizontal Integration; Vertical Integration; Centralized Integration; Semantic Interoperability

1. INTRODUCTION

The global changes in economic, shift of the competitive edge and the advent of new technology continually alter the manufacturing and business environment in which enterprises operate. Manufacturing and business management has been extending outside an enterprise in a distributed form geographically or according to business logic spreading across multiple enterprises.[1] The trend of long- or short-time close collaboration across the whole vertical and horizontal manufacturing industries has been illustrated by the implementation of e-commerce, e-business, or virtual enterprise.[2]

Please use the following format when citing this chapter:
Zhou, Jingtao, Wang, Mingwei, Zhao, Han, 2006, in International Federation for Information Processing (IFIP), Volume 207, Knowledge Enterprise: Intelligent Strategies In Product Design, Manufacturing, and Management, eds. K. Wang, Kovacs G., Wozny M., Fang M., (Boston: Springer), pp. 847-852.

Surviving in such an increasing globalization and flexibility environment requires an extremely flexible, self-adaptive IT infrastructure capable of integrating and coordinating any involved information from any heterogeneous data sources, applications, and environments at both system and semantic level to facilitate interoperation and collaboration over large-scale computer networks.

However, in spite of extensive R&D and successful pilots, traditional enterprise information infrastructure is poorly suited for dealing with the strategic, long-term barriers to efficient information sharing across enterprise internal and external boundaries at both system and semantic or knowledge level. The lack of suitable basic framework leads to many information solutions that have to make overmuch tradeoff between long-term adaptability and short-term applicability, broad interoperability and tailored function for very specific purposes. They can not reap the full potential benefits of information, and ultimately fail to the pursuit of setting and realizing corporate strategic and tactical goals. One underlying problem has remained unsolved yet: data resides in thousands of incompatible formats and cannot be systematically and understandably managed, integrated, and reused. As a result, there is mounting pressure from enterprise itself and outside for a direct move away from disparate information systems operating in parallel towards a more common and fundamental shared architecture for semantic information interoperation.

2. STATE OF THE ART

Enterprises have long recognized the value of data integration. Efforts can be roughly classified into two categories: application centric integration (ACI) and data centric integration (DCI).

ACI, such as point to point integration and enterprise application integration (EAI) integrates relative data by linking applications through custom-coding or integration broker that acts as a hub to route messages between connected applications. Connection at the programmatic level, difficulty of metadata reuse, N square problem at data layer, multiple vendors for multiple systems, lack of common protocols, and tight-coupling of technology and systems in the end, make these solutions difficultly suitable to an open, dynamic information interoperation environment of businesses and operations in or across enterprises. In fact, ACI provides little data integration because it operates at the business-process level rather than data level.

DCI can be implemented by creating either a centralized repository for data access and analysis, such as data warehouses, or a data integrating layer over a set of distinct and autonomous data sources, such as federated information systems.

Data warehouses are not effective in meeting ad hoc or opportunistic data integration needs. They fall short where an enterprise needs to address one need today and incrementally add additional requirements over time, or where there is a great deal of change in business requirements.[3]

Federated information integration approaches can be divided into two classes, including tightly-coupled and loosely-coupled systems. Tightly-coupled system integrates all data sources by creating logical mappings between them and a single global schema (a single model or ontology) which may be derived from the actual data sources themselves, or the modeling of current and even future business and operation aspects of information. The main drawbacks when using a global schema are the difficulty of creating a single sufficient global schema to represent a large-scale data sources, maintain of changing and evolution caused both from data sources and the global schema itself. Hence, tightly-coupled approach is only adapted to small-scale integration with little independently changing and evolving. By contrast, the loosely-coupled approach coordinates autonomous component data sources without a single global schema, instead, with a set of federated schemas. In the context of real-time information integration, loosely-coupled federation system is more effective and has advantages than other approaches. However, early loosely-coupled federated systems are not broadly applied in real enterprise environment because of the using of private protocol and data model, low performance, laborious process, critical implementation conditions, immature technology and the lack of reliable infrastructure. Several key problems such as discovery of relevant data, semantic conflicts and violation between the autonomy and privacy hinder its further application, too.

With recent advent of new technologies such as web service, XML and SOA, and new drivers behind e-commerce, e-business and e-enterprise, a representative approach for enterprise information integration (EII) is proposed, which intends to integrate any form data in enterprises, and aims to provide a uniform interface for information access, manipulate, and integrate across multiple data sources. Renaissance of data integration begins with the forming of EII industry. Broadly speaking, the architectures underlying most current new EII approaches (e.g. IBM DB2 II,[4] BEA Liquid,[5] MetaMatrix,[6] etc.) are still based on similar principles of loosely-coupled federated systems although they may support broad type data sources, use new XML model, speak with common protocols and publish integrated results with web services. Therefore, some traditional problems are inherited from loosely-coupled federated information integration system, such as scalable and semantic problem. Furthermore, current EII solutions may encounter their own fierce challenges when they need to integrate all involved information across the whole vertical and horizontal management logic, from both intra-enterprise and inter-enterprise on a highly complex and dynamic networked environment in a timely fashion. The circumstances seem to be too rigorous but are actual in real-world manufacturing and business environment.

3. CHALLENGES

A resent survey[7] has addressed four challenges of EII including scaleup/performance, horizontal or vertical growth, integration with EAI, and metadata management/ semantic heterogeneity from a more general point of view. In this section, we will take a further discussion from a technical perspective.

3.1 Scalability

The framework of most current EII systems is constructed as a hierarchy framework, in general, with (fix) multi-layer, which can be considered as a variation of the traditional five-level federated architecture,[8] and may consist of local schema, export schema (possible wrapped by web service), federated or mediator schema (logic data view), and extern schema (data view for specific application and user, possible wrapped by web service). Generally, approaches relying on a priori creation of federated views do not scale-up efficiently given the complexity involved in constructing and maintaining a shared schema for a large number of, possibly independently managed and evolving, sources.[9] A solution to this problem is to represent federated schemas by a hierarchy of small finely granular schemas, such as business entity schema used in BEA Liquid[5,10] and MetaMatrix[6]. However, this approach needs more endeavors from specialists and relies on their strong intimate knowledge of the desired business entities to be created and deep understanding of the data, underlying schema, and relationships across the various data sources (e.g. BEA Liquid[5,10] and MetaMatrix[6]), which becomes a drawback for scalability when this knowledge grows and changes as more sources join the system and when sources are changing.[9]

3.2 Horizontal vs. vertical integration

From the view of business and manufacturing, data integration requirements come from both the vertical and horizontal logic level. The framework of most current EII system is more appropriate for horizontal dimension of data integration rather than vertical dimension. Although the using of web service in some systems (e.g. data services in BEA liquid[5,10]) has shifted the focus on vertical data integration by service composition, the connection of vertical and horizontal integration has been artificially dissevered by the isolated definition of business concepts, which are dynamically composed only according to the specific application logic, only involving sources that have direct complementary data described by the corresponding federated schema, and ignoring the nature semantic relationship between themselves. As a consequence, it can not consider the potential relative data in other data sources even though the relationship is implied by other federated schema which is not directly involved in an integration process. It does not consider the probable incompleteness of some sources, either. In fact, these scenarios can be avoided by establishing the network of relationships among federated schemas as well as data sources to connect the horizontal and vertical data integration.

3.3 Centralized integration

Most implementations of EII generally resolve queries from dedicated servers that house federated metadata. Not only do these dedicated servers generally form a bottleneck in terms of scaling performance, but the centralized computing model can be a scalability bottleneck in terms of administration[11]. Federated schema design must be done globally; any changes of the federated schema can be only made by the central administrator. This can be especially challenging when data is owned and managed by numerous heterogeneous groups with different needs, and when integrate data across organizations.

3.4 Semantic

Even for information that is carried by web service and represented using XML, a serious and expensive barrier to dramatic improvement exists: a severe lack of explicit knowledge about associated corporate information. The thorny question of locating and understanding the data to be integrated still remains[7]. The assumption of most of the existing approaches that the relevant data is either already known or identified is not reasonable when integration involves multiple domains, multiple boundaries, or large-scale data sources. Key technologies of data semantics discovering, understanding, and semantic conflicts detecting and resolving are still in their infancy stage.

Achieving an information sharing environment enabling the feasible semantic interoperability becomes really a significant challenge.

4. CONCLUSIONS

Clearly, to overcome the challenges of EII and create a more flexibly and dynamically semantic interoperation environment for enterprise information over a large-scale computer networks, we believe there is a need for a new class of data sharing infrastructure. Such infrastructure is fundamentally distributed and dynamic networked, supports highly flexible sharing relationships, ranging from client-server to peer-to-peer, addresses scalability as well as resource control, and achieves interoperability at not only system level but semantic or knowledge level.

5. REFERENCE

1. Q. Wang , L. Y. Kai, and H. Wai, A hierarchical multi-view modeling for Networked Joint Manufacturing System, *Computers in Industry* **53**(1), 59–73 (2004).
2. Y.-E. Nahm and H. Ishikawa, A hybrid multi-agent system architecture for enterprise integration using computer networks, *Robotics and Computer-Integrated Manufacturing* **21**(3), 217-234(2005).
3. Nimble Technology, Next-Generation Data Integration: Harnessing Data for Business Advantage (14 January, 2002); http://uk.builder.com/whitepapers/0,39026692,60087734p-39001077q,00.htm.
4. B. Paolo, A. Francis, B. Amanda, et al., *Data Federation with IBM DB2 Information Integrator V8.1* (IBM Redbook, October 2003).
5. R. B. Vinayak, Liquid Data for WebLogic: Integrating Enterprise Data and Services, in: *ACM SIGMOD International Conference on Management of Data*, edited by W. Gerhard , C. K. Arnd, D. Stefan (ACM, Paris, 2004), pp.917-918.
6. H. Randall, M. Alex, and C. Rob, Information Intelligence: Metadata for Information Discovery, Access, and Integration, in: *ACM SIGMOD International Conference on Management of Data*, edited by O. Fatma (ACM, Baltimore, 2005), pp. 793-798.
7. H. Y. Alon, A. Naveen, B. Dina, C. Michael , et al. Enterprise Information Integration: Successes, Challenges and Controversies, in: *ACM SIGMOD International Conference on Management of Data*, edited by O. Fatma (ACM, Baltimore, 2005), pp. 778-787.
8. A.P. Sheth and J.A. Larson, Federated database systems for managing distributed, heterogeneous, and autonomous databases, *ACM Computing Surveys* **22**(3), 183-236 (1990).
9. C. Nazli, M. E. Stuart, M. Allen, et al., Information Integration for Counter Terrorism Activities: The Requirement for Context Mediation, Working Paper CISL# 2003-09.
10. BEA, BEA AquaLogic Services Platform: Concepts Guide, BEA, 2005.
11. G. I. Zachary, *Efficient Query Processing for Data Integration*, Phd thesis, university of Washington, 2002.

RESOURCE SHARING TECHNOLOGY IN THE AUTOMOBILE PARTS MANUFACTURE GRID

Yufeng Ding[1,2], Zhongling Wei[1], Buyun Sheng[1,2]
[1]*School of Mechanical and Electrical Engineering; Email: dingyf@mail.whut.edu.cn;* [2]*Hubei Province Digit Manufacturing Key Laboratory, Wuhan University of technology, P.R. China.*

Abstract: The manufacture resource classification model is described by using UML class graph in the manufacture grid system of automobile parts industry. Manufacture grid service system structure and resource agent structure is put forward. Manufacture resource attribute is defined using XML standard and the unification standard is provided for manufacture resources description using manufacture resources semantics dictionary. Referencing the manufacture resources semantics dictionary interface naming standard, manufacture resource service interface is described by using WSDL standard. A manufacture resource grid service prototype system is developed on the basis of GT3 development toolkit on the Windows platform, at last the product development process is described in the manufacture grid system by regarding automobile steering product as instance.

Key words: Manufacture grid, Resource sharing, XML, WSDL, PDM.

1. INTRODUCTION

Grid referring to a group of sharing resource integration, also called super virtual computer, is a technology which connect wide-area distributed computer resources by internet or special-purpose network. Ian Foster puts forward the most influential definition of grid in 2001: Dynamic multi-mechanism virtual organization resource sharing and cooperating problem-solved[1,2].From this definition we can see that manufacture grid has three characters: resource sharing, cooperating and dynamic forming virtual organization. At present many domestic and foreign companies and national development facilities have studied the grid technology penetratingly, BOEING, FORD, BMW and other big companies are all attempting to apply

This paper is supported by Hubei province digital manufacturing key lab opening fund project (SZ0406), National Natural Science Foundation of China (50335020) and Wuhan science and technology key plan (20055102021).

Please use the following format when citing this chapter:
Ding, Yufeng, Wei, Zhongling, Sheng, Buyun, 2006, in International Federation for Information Processing (IFIP), Volume 207, Knowledge Enterprise: Intelligent Strategies In Product Design, Manufacturing, and Management, eds. K. Wang, Kovacs G., Wozny M., Fang M., (Boston: Springer), pp. 853-862.

grid computation to the design and simulation of complex products, emphasis on solving the multitude computation problem of complex products such as airplane, automobile products. Ford company built a network bazaar by using grid technology, through which Ford can connect with more than 30,000 Suppliers.

The application of grid technology in manufacture field is just initial stage, manufacture grid which integrated design resources, manufacture resources and flow information in the product life cycle, is the development platform and supporting environment of modern integration manufacture system. The manufacture grid provide enterprises with integrated support service platform for application development and the system function, the platform is open, easy to maintain, and can be reconstructed in the process of enterprise's implementing advancing manufacture technology. The application of grid technology can provide a public manufacture grid platform for the manufacturing industry, can realize resources sharing, cooperating design and cooperating manufacture, so as to achieve the purpose of reducing the production cost, raising the resources utility and speeding up product's going onto market. The existing manufacture resources sharing plan, such as the Internet,B2B,ASP,Java,CORBA and so on, can not satisfy the virtual organization need completely[3]. The superiority of grid technology is to solve the problem of resources sharing and cooperating work among dynamic virtual organizations. Zhou Guanghui et al. proposed resource encapsulation and integration platform and frame[4]. Yang Jingquan et al. studied non-interaction computation resources issuing automatic encapsulation tool[5]. Because manufacture resources are so many that constructing manufacture grid facing all manufacturing industries products is not realistic. It is possible to build an industry manufacture grid to solve resource sharing problem.

2. AUTOMOBILE PARTS MANUFACTURE GRID SERVICE SYSTEM STRUCTURE

2.1 Manufacture resources classification

The manufacture resource includes enterprise equipment, material, staff and the total sum of the hardware and the software of product life cycle involving. The human resources are the industry experts which have the different special skills live in different regions. The on-line experts carry out the product design and make production preparation information by using the sharing tool (CAD/ CAPP /CAM /CAE and so on), then save product information to the resource database. The equipment resources and special technologies include the numerical control machine, the processing center, the special processing equipment (for example nanometer ultra precisely process machine, laser processing machine, rapid prototype manufacture, digitized scanner, casting machine) and so on in the manufacture network. For grid system the most important task is how to select manufacture equipment and processing enterprise can be used in the different regions, it is decided by the product manufacture information and other factors. The last processing enterprises are determined by competing or assigning mode within all selectable enterprise. The corresponding production preparation information is got from network in the resources database by matching the part number and attribute information. The part can be further revised and consummated for procedure, the data, the cutting instrument and the measuring instrument information until the product processing completion based on the network. In the manufacture grid, manufacture resources can be divided into nine classifications according to the resources attribute(physical characteristic, geographical position, dynamics characteristic, sensitivity, function and so on), the customer demand (time, quality, price, service and so on), used way (discovery, proxy, monitor, diagnosis and so on), manufacture resources is described by UML class graph is shown in Figure 1.

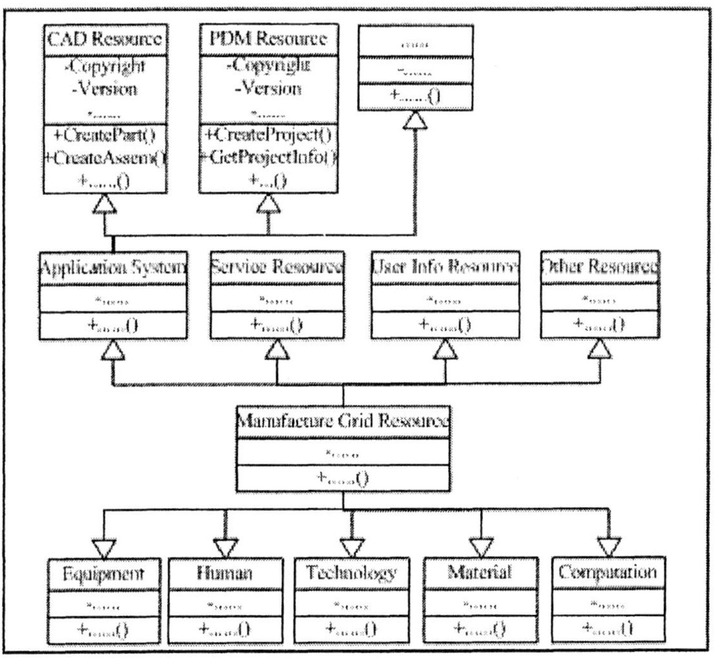

Figure 1. Manufacture resource UML class graph

2.2 Automobile parts industry manufacture grid service system structure

It needs to build up a register service and a handle mapping service in order to share resource in the manufacture grid. Manufacture resources can be encapsulated into services and the service can be issued into grid UDDI (Universal Description and Integration) registered center which may offer registering service. Manufacture grid user may visit registered center through resource agent so as to get all registered information of manufacture resources service factory and the existing service instance information. The user (include resource supplier and resource user) can create an instance by making a request from specified service factory, and send the data to the instance according to service interface standard. Then the user may visit manufacture grid service instance and use the function which the service offers. The process is shown as Figure 2. Resource description is the precondition of building manufacture grid and realizing resource sharing. When a manufacture resource supplier joins manufacture grid, manufacture grid will record resource function, using interface, period of validity and safe coefficient etc. When resource user needs to use a resource, they need put forward a clear function description, and then the resource agent of manufacture grid may seek suitable manufacture according to the user need. It has four kinds of major data type in UDDI registered center. They are BusinessEntity, businessService, bindingTemplate and tModel. To help to issue and seek WSDL service description in UDDI registered center, WSDL document has been divided into two types: Service interface and service implementation.

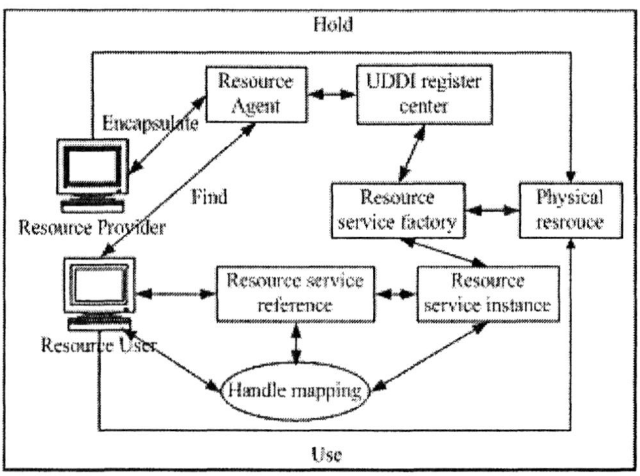

Figure 2. Manufacture grid resource service system

2.3 The structure of resource agent

The structure of resource agent is shown in Figure 3. It includes perception modular, reaction modular, action modular, decision-making modular, communication modular and knowledge base. Resource agents accepts or queries forwardly the resource request form resource user by perception modular, then send different information to corresponding modular after analysis resource description document. If the perception Information is the simple or urgent, it can be directly delivered to reaction modular which can send the order and movement instruction to the action modular, then the action modular carries out action and gives feedback to the user. Otherwise the information is delivered to the decision-making modular which makes decision by reasoning, then send order and movement instruction to the action modular, the action module carries out the action.

(1) Perception module. It constantly catches external incident request information by perceiving message. Every incident has an identification and the property of beginning time B(e), termination time E(e) and occurring period D(e), the 'e' is the name of incident. If resource agent perceives two or many incidents at the same time, it can handle with different method according to the specific condition. The situation includes simultaneity occurring and choice occurring and incorporation occurring, coterminous occurring and nothing occurring etc.

(2) Reaction module. When agent needs to make rapid response to urgent or simple situation, reaction module will be started. Basically it does not carry on reasoning in the reaction module, but directly maps to action by perceive information.

(3) Decision-making module. Agent usually transmits the information from perception information to the decision-making module except the urgent or simple situation. When decision-making module receives it, firstly it carries out information matching retrieval from knowledge library, then uses the reasoning mechanism to infer the information, finally transmit the action order produced from inference to action module.

(4) Action module. Action module carries out some kind of motion based on instruction of reaction module and decision -making module.

(5) Communication module. Agent transmits the information through communication module among themselves..

(6) Knowledge library. There are two kind of knowledge in knowledge library. One is the rule, another is the knowledge block. Knowledge in the knowledge library is the foundation of making decision. The knowledge in the knowledge library is difference based on agent function.

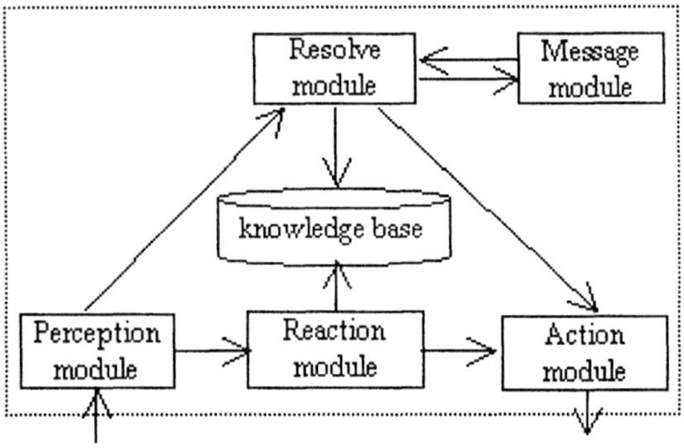

Figure3. The structure of resource agent

3. MANUFACTURE RESOURCE ENCAPSULATION

3.1 Web Services and WSDL standard

Web Service is a software application which can be recognized by URI(Universal resource identification). Its interface and binding can be described and discovered by XML (eXtensible Markup Language). It can communicate with other software application based on XML information by protocol based on the Internet. WSDL (Web Service Description Language) standard is a criterion to realize the Web Service. WSDL can be used to register and discover the description of interface definition and terminal realization, as well as dynamically produce proxy based on specific interface. WSDL supports dynamical discovery and organization of service in the heterogeneous dynamical grid environment through the standard mechanism which separates definition of interface and specific binding.

3.2 Description and encapsulation of manufacture resource

Using WSDL standard to describe and encapsulate manufacture resources usually includes two steps:

Abstract static attribute and dynamic attribute from the manufacture resource, and describe them as service data. It needs to provide a universal standard for attribute definition. Service description will not have misunderstanding between service and user, it also provides safeguard for service querying and discovering. The definition of HUAZHONG NC Machine tool CK6132/750 is shown in Figure 4. NC Machine_CK6132/750 is used to represent equipment resource of NC Machine tool in the XML code, and it can not represent other manufacture resources. The manufacture resources semantics dictionary provides the universal standard for manufacture resource description.

According to manufacture resource semantics dictionary, static attribute and dynamic attribute of each kind of equipment or software application can be ascertained. Static attribute includes equipment name, equipment type, machining size, machining accuracy, document format etc. Dynamic attribute includes machining charge, current equipment working condition, task completion percentage etc. The attribute of HUAZHONG NC Lathe resource is described in Figure 4.

2) Abstract service interface and operation, then describes it as WSDL document, and realize interface. WSDL standard is used to describe interface of manufacture resource service, including the definition of interface type, operation, information, port and related service by referring to naming standard of interface in manufacture resource semantics dictionary. After the WSDL document is defined, the WSDL2Java method can be used to map the WSDL document into a Java file, and interface, operation, binding is realized in Java, and then is compiled into class file.

4. THE DEVELOPMENT AND CASE OF MANUFACTURE GRID SYSTEM

At present most of domestic and foreign grid projects are developed on the basis of GT3 (Globus toolkit 3.0), GT3[6] is a toolkit which can be used to develop the grid application. We construct the manufacture grid in the Windows 2000 operating systems platform. At present in the GT3 toolkit, only GT3 Core may be installed in the Windows platform. In order to construct manufacture grid by using GT3, firstly some supports tool including J2SE1.3.1, Jakarta ant 1.5.2 and Gt3-core-src.tar must be installed, these tools can be downloaded from correlative java website.

GT3 programming is based on distributed computing programming model, also is called proxy-stub mode, It includes service end programming and customer end programming. The service end and customer end is loosely-coupled, their relation is completed through the WSDL service description document. Manufacture resource service provider must provide the related WSDL service description document when develop service. Service interface, method of transferring service and binding condition of transferring service and the bottom layer communication agreement is describe in the document. After the manufacture resource service user obtained WSDL service description document, the stub of transferring service will be produced according to WSDL document, then transfer the service through this stub. The manufacture resource provider may provide many kinds of methods in which transferring service binding with the bottom layer communication agreement. The customer end may choose one kind of methods to transfer service.

```
< Manufacture Resource Semantic Dictionary>
< Properties >
< Name > NC Lathe </Name >
< ID> NCMachine_CK6132/750</ID>
< Manufacture> NC Machine </ID>
< Description > Equipment using NC technology to
process components or product
</Description >
<Parameter>
<machine type> CK6132/750</ machine type >
  < Biggest workpiece length > 750mm </ Biggest
  workpiece length >
  < Biggest turning length > 600mm </ Biggest turning
  length >
  <location accuracy> 0.01/300 </location accuracy>
  ......
<NC system> HNC-21T huazhong" century star"</NC
system>
</Parameter>
</Properties >
</Manufacture Resource Semantic Dictionary >
```

Figure 4. The definition of HUAZHONG NC Lathe

As a case, an automobile steering gear enterprise needs to develop some series of steering gear products, but they developed product mainly by using two-dimensional CAD before. Presently they need to establish the three-dimensional CAD product model, and analyze the stress of steering gear. But they do not want to purchase the expensive three dimensional CAD systems and finite element analysis software for the moment. Therefore the enterprise plans to develop this product using the manufacture resources of manufacture grid system. The description of using sharing resources in the manufacture grid system to design and manufacture the steering screw of circulating ball-type steering gear is as follows:

(1) Designers login the manufacture grid system under the authorized situation using unified Web browser interface, and search the CAD software and collaborative design tool in the manufacture grid system, then they carry out the product collaborative design and the craft collaborative design under the support of these tools.

(2) Corresponding initial condition, machining request, dispatching path and address of aimed mainframe is inputted according to machining task demand. The initial condition and the parameter are got from three-dimensional product model in the local or the long-distance database, or are obtained through visiting local document. The NC code can be produced by

using CAM software automatically. After inspection is finished, if the needed equipment is free, machining code can be transmitted to the aimed computer through clicking the submit button based on the manufacture grid dispatch mechanism. The corresponding encapsulation task will be dispatched to the aimed computer.

(3)The aimed computer server receives dispatched encapsulation task, and analyzes its NC code, then directly transmits it to the local NC machining equipment to carry on the machining process, In the course of process, user may query machining condition of equipments through transmitting inquiry instruction.

(4)The PDM provided by the manufacture grid system will manage all product data in the design and manufacture process.

(5)After the mission is finished, the enterprise will pay for the submitting task through the manufacture grid balance system

The process interface of resources Packaging, resources searching and resources searching transferring is shown in the Figure 5 for PDM software WinCAN of manufacture grid. Encapsulation and transferring process of other software resources and NC equipments is similar to this. Some models of HUAZHONG NC lathe of CK series and NC milling machine of YHM series have already been realized sharing in this manufacture grid platform.

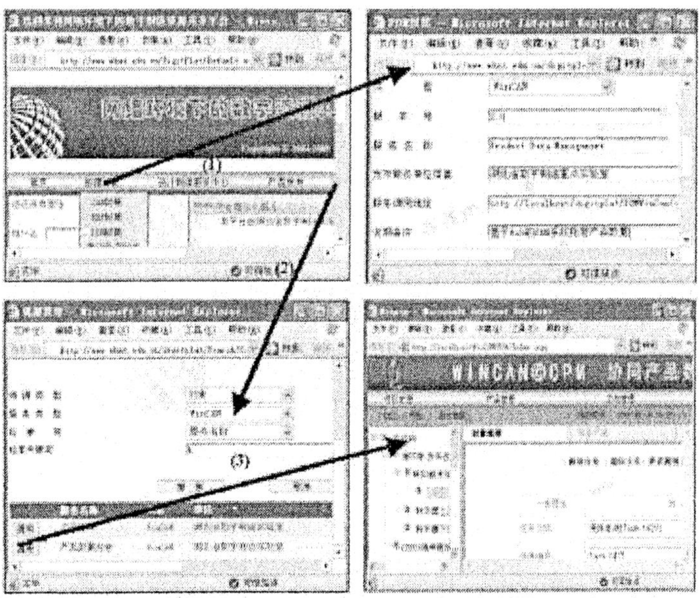

(1) resource Packaging (2)resource searching (3)resource transferring

Figure 5. PDM software in the Manufacture grid platform

5. CONCLUSION

The solution of constructing manufacture grid system of automobile parts industry is put forward through applying the grid technology into automobile parts industry, and solves the resources sharing question in automobile parts industry on the basis of analyzing the shortage of traditional manufacture resources sharing solution. With unceasing maturating of grid technology research and thorough applying in manufacture domain, it is possible to construct sharing platform of manufacture grid resources facing industry application. With the aid of grid technology's superiority, and through constructing the sharing platform of manufacture grid resources, it may fully meet each kind of manufacture resources sharing need among the virtual enterprises, thus provide a new solution way of manufacture resources sharing among manufacture enterprises.

6. REFERENCES

1. I. Foster, C. Kesselman, S.Tuecke(2001),The anatomy of the grid. *Intl. Jour. Supercomputer Applications,* 15, vol.3,pp.1 -21.
2. I. Foster, C. Kesselman, J. Nick et al(2002),The physiology of the grid: an open grid services architecture for distributed systems integration, [EB/OL]. http://www.globus.org/ *research/ papers/ ogsa.pdf*
3. Zhang Qing, Fan Yu-shun(2003), Grid Technology and Its Application in Manufacturing. *Aeronautical Manufacturing Technology,* 2, pp.32-37.
4. Zhou Guang-hui, Jiang Ping-yu(2002),Encapsulation and Integration for Networked Manufacturing Resources Based on Mobile Agents. *Computer Integrated Manufacturing Systems*, 9, pp.28-32
5. Yang Jing-quan, Dai Ya-fei, Li Wei-gang(1998), Fang Bing-xing. Encapsulation technology of Computing resources web issuing". *Computer Engineering and Applications,* 1998, 4, pp.52-54.
6. http://www.globus.org/[EB/OL], Globus Project official website

ANT COLONY OPTIMIZATION FOR MANUFACTURING RESOURCE SCHEDULING PROBLEM

Wang Su, Meng Bo
Computer School, Wuhan University, Hubei Province, China; Email: zzwangsu@163.com.

Abstract: Effective scheduling has a significant influence on efficiency of resources and manufacturing systems. In this paper, we propose a model of resource scheduling based on matching mechanism and apply Ant Colony Optimization to solve this problem. During the process of seeking, ants can change the size of colony adaptively and incline to select the resources which cost less. Finally, we examine the performance of optimization in a test case scenario in container terminals and obtain satisfactory results that suggest that ACO is well suited for application in practice.

Key words: Ant Colony Optimization, Manufacturing Resource Scheduling, Pheromone.

1. INTRODUCTION

Information Technology not only creates new opportunities for manufacturing industry but also gives rise to new challenges and changes. The most prominent is the requirement for higher productivity in the whole manufacturing process. To deal with the challenge, an effective scheduling system that can improve its performance is essential. Scheduling is no longer a static optimization problem. Most practical scheduling problems are NP-hard, dynamic and unpredictable. These manufacturing characteristics limit the effectiveness of conventional scheduling approaches. At present, a key factor of successful scheduling is heuristic optimization.

There are many efforts in the field of optimization of manufacturing resource scheduling. S.G.Ponnambalam[1] adopted Tabu Search (TS) to resolve Flexible Scheduling Problem. Gehlsen[2] proposed a framework for discrete event simulation and applied Genetic Algorithm (GA) to get optima.

This project is supported by the Natural Science Foundation of Hubei Province of China under the grant No.2001ABB058 and the Post Doctoral Science Foundation of China.

Please use the following format when citing this chapter:
Su, Wang, Bo, Meng, 2006, in International Federation for Information Processing (IFIP), Volume 207, Knowledge Enterprise: Intelligent Strategies In Product Design, Manufacturing, and Management, eds. K. Wang, Kovacs G., Wozny M., Fang M., (Boston: Springer), pp. 863-868.

Snke Hartmann[3] used GA for resource-constrained scheduling problem. Chia Jim Tong[4] resolved berth scheduling problem in container terminals based on Ant Colony Optimization (ACO).

In this paper, we introduce a general scheduling model based on matching mechanism. Consider that ACO Algorithm is a good meta-heuristic approach to tackle hard combinatorial optimization problems and it is parallel, robust and positive feedback, we adopt ACO Algorithm to resolve this kind of problem.

2. MANUFACTURING SCHEDULING PROBLEM BASED ON MATCHING

The manufacturing resource scheduling problem based on matching mechanism is described as follows: Resources are scheduled by multiple tasks. Each task can schedule more resources, but one resource can only be scheduled by one task. Resources have different costs. The goal is to determine a matching solution between tasks and resources so that the total cost is minimized on the premise of all tasks are completed. Tasks are defined as $(T_1, T_2, T_3, \ldots, T_n)$. Resources are defined as $(R_1, R_2, R_3, \ldots, R_m)$. t_i is the need of task T_i. c_j is the cost of scheduling resource R_j.

Minimize: $\min \sum_{i=1}^{n} \sum_{j=1}^{m} x_{ij} c_j$

Subject to: : $\sum_{j=1}^{m} x_{ij} c_j \geq t_i$

$$x_{ij} = \begin{cases} 1 & T_i \textit{ schedules } R_j \\ 0 & \textit{otherwise} \end{cases}$$

$$\sum_{i=1}^{n} \sum_{j=1}^{m} x_{ij} \leq m$$

3. ANT COLONY OPTIMIZATION

ACO is an evolutionary approach introduced by Dorigo[5-6]. In ACO several generations of artificial ants search for good solutions. Every ant of a generation builds up a solution step by step. Ants that find a good solution mark their paths through the decision space by putting some amount of pheromone on the path. The following ants of the next generation are attracted by the pheromone so that they will search in the solution space near good solutions.

For scheduling problem in this paper, each ant is assigned to one task and each note represents one resource. T_i schedules resources, then ant k_i deposits pheromone on the path. $\tau_{ij}(t)$ is the pheromone of path l_{ij}, which represents the cost of scheduling R_j after R_i. For example, there are 10 available resources $(n_1, n_2, n_3, ..., n_{10})$ in Figure 1. All ants set out from the initial note n_0 at the beginning. Ants establish four resource unions $\{R_1,R_3,R_4\}$, $\{R_2,R_8,R_{10}\}$, $\{R_6,R_9\}$, $\{R_7\}$ for four tasks and R_5 is not scheduled. The union $\{R_1,R_3,R_4\}$ schedules R_1, R_3 and R_4, and ants deposits pheromone on path l_{01}, l_{13}, l_{34}.

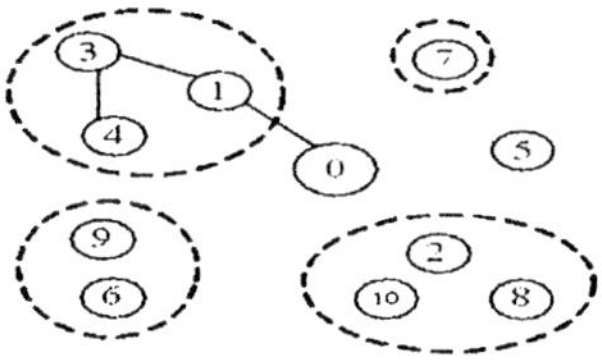

Figure 1. Ten resource scheduled by four tasks

The node transition rule is as follows. Ant k in R_i chooses R_j following the rule:

$$s = \begin{cases} \arg\max\limits_{j \notin tabu_k} \left\{ \tau_{ij}^{\ \alpha} \bullet \eta_{ij}^{\ \beta} \right\} & q \leq q_0 \\ S & q > q_0 \end{cases} \tag{1}$$

q is a random variable between [0,1], q_0 is a tuneable parameter between [0,1]. For the selection of a resource the ant uses pheromone information denoted by τ_{ij} and heuristic information denoted by η_{ij}. The heuristic value is generated by some problem such as resource ability. With probability q_0, the ant chooses R_j from the set of resources that have not been scheduled so far which maximizes $\tau^{\alpha}_{ij} \cdot \eta^{\beta}_{ij}$, where α and β are constants that determine the relative influence of pheromone values and heuristic values. With the probability $1-q_0$ the next resource is chosen according to S determined by p^k_{ij}.

$$p_{ij}^{k} = \begin{cases} \dfrac{\tau_{ij}^{\ \alpha} \bullet \eta_{ij}^{\ \beta}}{\sum\limits_{u \notin tabu_k} \tau_{iu}^{\ \alpha} \bullet \eta_{iu}^{\ \beta}} & j \in allow\ k \\ 0 & otherwise \end{cases} \tag{2}$$

Unlike Traveling Salesman Problem (TSP), ants here seek route for forming some resource-unions for tasks. The ant will stop seeking when it arrives at a certain resource note and finds that the current resource union can accomplish task T_i. It's unnecessary to visit the other resource note. So every time the ant arrives at a resource note, it needs to accumulate the current resource ability and judge whether the combinational ability (ra_i) meets the task demand (t_i). If ra_i is higher than t_i, the ant k_i will stop seeking, otherwise continue. When all the tasks complete, all ants stop and one circle finishes. If not all of the tasks can be accomplished after ants visits all resource notes, the system will make some tasks wait.

In ACO, all ants are allowed to deposit pheromone after completing their tours. The updating rule is as follows:

$$\tau_{ij}^{new} = \rho\tau_{ij}^{old} + (1-\rho)\Delta\tau_{ij} \qquad (3)$$

$$\Delta\tau_{ij} = \begin{cases} \dfrac{t_k}{Q_k} & \textit{ant k select note i and note j} \\ 0 & \textit{otherwise} \end{cases} \qquad (4)$$

(i,j) is one path of the best tour, ρ is a parameter governing pheromone decay. The reason for this is that old pheromone should not have a too strong influence on the future. Q_k is the real cost of completing T_k, which is always higher than t_k.

The algorithm stops when some stopping criterion is met, e.g. a certain number of generations have been done or the best found solution has not changed for several generations.

Using ACO Algorithm to solve resource scheduling problem, the pseudocode is as follows:

```
Step1: Initialization.
Step2: while (available resource) and (uncompleted task)
          for (k=1;k<=n;k++)
            if (task k has completed) Continue;
            else Ants select next paths according to formulation (1), (2);
Step3: All ants complete their searching process;
          if (total cost of current solution < total cost of best solution)
            Update pheromone according to formulation (3), (4);
Step4: NC++;
          if (NC<NCmax ) or (no evolving solution) Goto Step2;
         else Output the solution with minimal cost;
               End the whole program.
```

4. SIMULATION EXPERIMENTS

In this section we describe a test case scenario of scheduling problem in container terminals using ACO Algorithm. Faced with the increasing growth of container throughput and more large ships, the demands on ports of upload/load task with maximum efficiency become more important. Tugboat scheduling is one important scheduling in the whole task.

When ships arrive at a port, they can not enter into the berth directly and need to be tugged by the tugboats. Moreover, the moving between two berths and the department of vessels also need to be tugged. Vessels with different sizes and types should schedule tugboats with different horsepower, avoiding big tugboat for small vessel and small tugboat for big vessel. On the other hand, one tugboat is only scheduled by one vessel simultaneously. To make sure scheduling can be done as rapidly as possible with least cost; tugboats necessary for this procedure must be scheduled at an optimum level.

We make simulation experiments of tugboat scheduling. Table 1 shows the allocation of tugboat at some Chinese ports. Table 2 shows some records of arriving vessel we use in this simulation.

Table 1. Allocation of Tugboat

Horsepower	1200ps	2600ps	3200ps	3400ps	4000ps
Port 1	1	7	2	1	2
Port 2	1	7	2	2	3
Port 3	1	7	4	1	2
Port 4	1	7	2	1	4

Table 1. Records of Arriving Vessels

Name	Country	Time	Length(m)	Width(m)
YuHe	China	03-05-20 10:00	200	28
Aoleta	Poland	03-05-20 10:00	229	32
BuleStar	Britain	03-05-20 10:00	155	25
		……		
Wasalian	Singapore	03-05-22 14:20	67	14
Ankejian	Singapore	03-05-21 14:00	135	31
JiaoCheng	China	03-05-21 14:00	148	22
		……		

The results of simulation without ACO Algorithm are described in Hu Wenbin[7]. We use ACO to optimize simulation results. In the real discreet scheduling problem, the number of ants is a variable parameter which can be changed according to the number of tasks. Other parameters are: τ_0=0.5, α=1, β=3, ρ=0.8, NC_{max}=5000. The results are the average of ten experiments and showed in Table 3: (1) is tugboat utilization rate, (2) is average waiting time of vessel (minute), (3) is max waiting queue length.

Table 3. Comparison of Simulation Results

Method	Simulation Without ACO			Simulation With ACO		
Evaluation	(1)	(2)	(3)	(1)	(2)	(3)
Port 1	23.6%	22.9	9	51.6%	5.4	4
Port 2	20.5%	5.1	8	49.3%	1.8	4
Port 3	18.9%	11.5	9	47.5%	2.2	4
Port 4	19.5%	4.6	8	48.9%	1.6	4

5. CONCLUSION

In this paper we propose a general model for scheduling based on match mechanism and adopt ACO Algorithm to solve this problem.

In the algorithm, ants can change the size of colony adaptively and incline to select the resources which cost less. The results of simulation show that compared with only simulation, applying ACO to optimize the scheduling process can improve the utilization rate of resources and the efficiency of tasks.

Our feature work is to research ACO Algorithm deeply to improve the convergence speed and avoid local optima.

6. REFERENCES

1. S.G.Ponnambalam, V.Ganapathy, S.Saravana Sankar, R.Karthikeyan, (2002), Scheduling flexible manufacturing system using tabu search method, *IEEE International Conference on Industrial Technology*, pp. 1043-1048.
2. Björn Gehlsen, Bernd Page, (2001), A framework for distributed simulation optimization, *Proceedings of the 2001 Winter Simulation Conference*, pp. 508-514.
3. Snke Hartmann., (2002), A self-adapting genetic algorithm for project acheduling under resource constraints. *Naval Research Logistics*.
4. Chia Jim Tong, Hoong Chuin Lau, and Andrew Lim., (1999) Ant colony optimization for the ship berthing problem, *The 5th Asian Computing Science Conference*, LNCS 1742, pp. 59-370.
5. Colorni A., Dorigo M., Maniezzo V., (1991), Distributed optimization by ant colonies, *First Europ. Conf. Artificial Life*, pp. 134-142.
6. Dorigo M., Gambardella L. M., (1997), Ant colonies for the traveling salesman problem, *BioSystems*, vol. 43, pp. 73-81.
7. Hu Wenbin, (2004), Research on the key technology of distributed intelligent group decision supported system based on multi-agent, *PHD Dissertation*, Wuhan University of Technology.

MACHINE WORK PLANNING USING WORKPIECE AGENTS IN AN AUTONOMOUS DECENTRALIZED MANUFACTURING SYSTEM

For High-mix Very-low Volume Production

Michiko Matsuda, Satoshi Utsumi and Yasuaki Ishikawa

Kanagawa Instistute of Technology, Atsugu-shi, Kanagawa, 243-0292 Japan; Email:, matsuda@ic.kanagawa-it.ac.jp

Abstract: In this paper, a workpiece agent is proposed for the configuration of an autonomous decentralized manufacturing system which flexibly responds to high-mix, very-low volume production. When a blank workpiece is put on to the production line, the workpiece agent is dynamically generated with the product model. According to the state of each machine tool on the shop floor, the agent plans the machining processes by itself. Then the workpiece agent determines the selection of the machine tools for each machining step and obtains a schedule in real time through the negotiation among two or more workpiece agents aiming to increase efficiency of the whole manufacturing system. Trial implementation of a workpiece agent has been done.

Key words: multi agent system; workpiece agent; agent negotiation; machine work planning; event driven manufacturing system

1. INTRODUCTION

In recent years, the requirement for high-mix very-low-volume production including flexible order-made production is increasing. Constructing autonomous decentralized manufacturing systems is one of the solutions for the realization of high-mix very-low-volume production. In an autonomous decentralized manufacturing system, a manufacturing activity unit such as machine tool, robot AGV, and/or manufacturing cell has autonomous functionalities. On the other hands, multi agent technology is recently being well applied to construct autonomous systems. The elements

Please use the following format when citing this chapter:

Matsuda, Michiko, Utsumi, Satoshi, Ishikawa, Yasuaki, 2006, in International Federation for Information Processing (IFIP), Volume 207, Knowledge Enterprise: Intelligent Strategies In Product Design, Manufacturing, and Management, eds. K. Wang, Kovacs G., Wozny M., Fang M., (Boston: Springer), pp. 869-874.

in an autonomous system are configured as an agent. Multi agent technology is also applied to the constructions of the autonomous decentralized manufacturing system[1]. Manufacturing activity units, especially manufacturing devices are configured as agents. In these cases, system structure is a device oriented structure. To obtain more flexibility and autonomy, an event driven system structure is useful. For the manufacturing system, the occurrence of an event means that a workpiece is input to the manufacturing shop floor. This paper proposes to represent the workpiece as a dynamic agent. This agent is called a "workpiece agent."

2. MANUFACTURING SYSTEM BASED ON THE WORKPIECE AGENT

A workpiece agent is an intelligent and autonomous software unit attached to a workpiece. This workpiece agent plans the processes and allocates the manufacturing machine for executing the processes required to generate the product. This means that the workpiece agent takes an initiative in the manufacturing system. The conceptual structure of the autonomous decentralized manufacturing system based on a workpiece agent is shown in Figure 1. The workpiece agent is a dynamic agent. It is created when the workpiece is input to the shop floor. The workpiece agent is deleted when the product is completed. At first, after its creation, the workpiece agent generates a process plan. Then the workpiece agent negotiates with other workpiece agents which are already on the shop floor to obtain the manufacturing machine allocation, and orders an AGV for the workpiece transfer. Later on we will discuss the machining system based on workpiece agents.

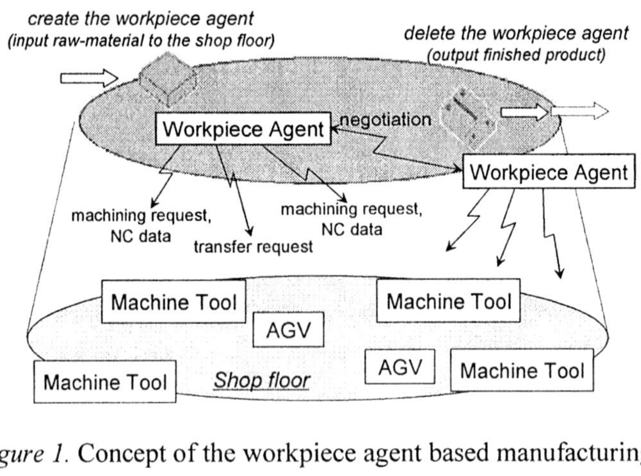

Figure 1. Concept of the workpiece agent based manufacturing system.

3. STRUCTURE OF THE WORKPIECE AGENT

When a workpiece agent is created, it has machining knowledge and product data which describes the finished product. The structure of the workpiece agent is shown in Figure 2. The workpiece agent has a machine work planning function and a machining/transfer requesting function.

At first, several possible machining process plans are generated in the workpiece agent by referring to the product model, machining knowledge and machine tool data which models the available machine tools on the shop floor. Then, workpiece agent selects an executable plan by referring the current machining job allocation schedule, and negotiates with the other workpiece agents by requesting schedule changes according to the selected plan. After a successful negotiation the workpiece agent requests an AGV for a transfer and asks the machine tool for machining by verifying the machining job allocation schedule on the information board of the shop floor.

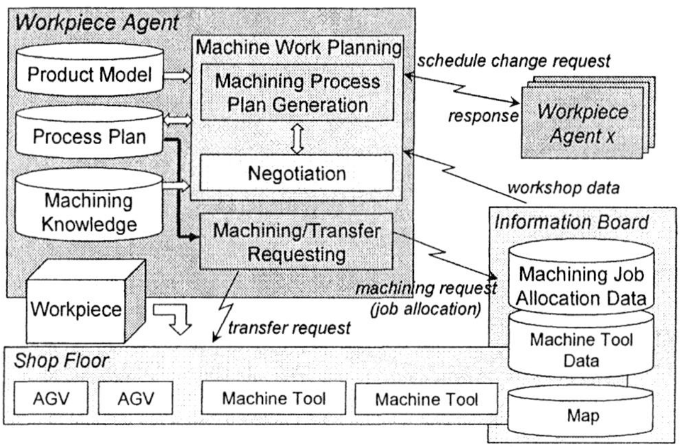

Figure 2. Structure of the workpiece agent.

4. MACHINE WORK PLANNING

4.1 Machining process planning

The first step in the machining process planning procedure is extracting machining features from the finished product model. The second step is lining up candidates of machining processes depending on features using the Feature-Machining Method Table. In this table, possible machining methods for producing corresponding features are described. Usually, one feature type has several possible machining methods. The contents of this table is machining knowledge. The third step is the narrowing down of candidates using machine tool data on the information board. Machining tool data shows available machine tools and machine tool status on the shop floor.

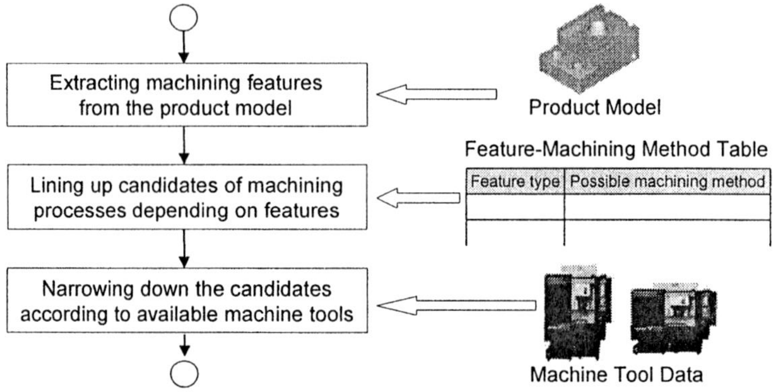

Figure 3. Machining process plan generation in the workpiece agent.

4.2 Machine work scheduling by one-to-one negotiation

When the workpiece is input to the machining shop floor with some urgency machine work scheduling is done by a one-to-one agent negotiation after process planning as shown in Figure 4. In Figure 4, Workpieces A, B and D are already on the shop floor, and Workpiece D is input with some urgency. Machine A and B have the same specification. It is recognized that Machine B has a time margin by referring to the current schedule. Therefore the process plan for Workpiece D which Machine B is able to execute is selected and a candidate for am interrupt schedule is generated. According to this interrupt schedule, Agent D negotiates with Agent C to concede the Machine B allocation. If Agent C accepts this offer, Agent C refers the new current schedule and negotiates with Agent B. Agent B then repeats as same.

4.3 Machine work scheduling by one-to-many negotiation

One-to-one agent negotiation provides a locally optimized machine work plan. To get totally optimized machine work plan, one-to-many agent negotiation is proposed. Figure 5 shows the scheduling process for one-to-many agent negotiation. The case of Figure 5 is the same as in Figure 4. Product D is input to the shop floor. Before the negotiation, an optimized schedule is generated by applying the genetic algorism[2]. One individual

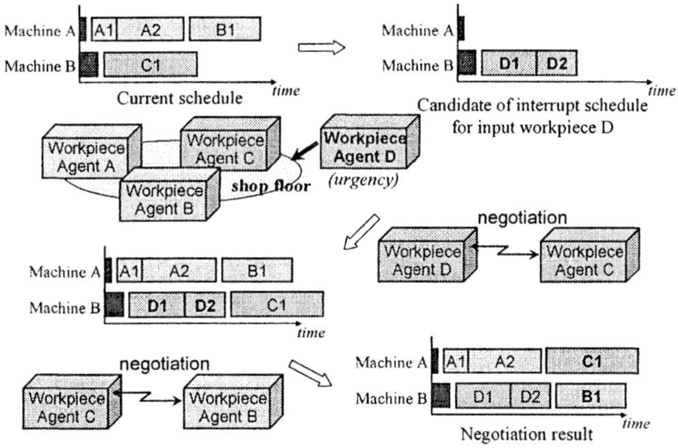

Figure 4. Machine work scheduling by one-to-one negotiation.

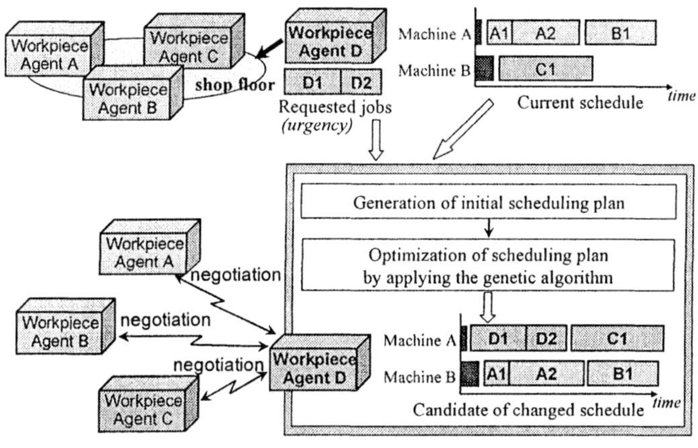

Figure 5. Machine work scheduling by one-to-many negotiation.

represents one scheduling plan. Initial individuals are generated by setting required processes for Workpiece D into the current schedule according to a setting rule. Then, genetic procedure is applied to the initial individuals. The most optimized schedule is the first candidate for the changed schedule. According to this schedule, Agent D negotiates with Agent A, B and C at same time. If negotiation is not successful, another processes for Workpiece D is selected and the procedure is repeated.

4.4 Planning simulation using workpiece agents

A Planning Simulation using workpiece agents is now performed. Figure 6 shows the result of the example simulation. There are three types of products: A, B and C. A product of type D interrupts the line. There are four milling machines on the shop floor. The specifications for the milling machines are different. A1-1 represents product type, lot number and process number. In Figure 6, the upper chart is the result for the case of one-to-one negotiation and the lower chart is for the case of one-to-many negotiation. One-to-many negotiation provides a more even and efficient schedule.

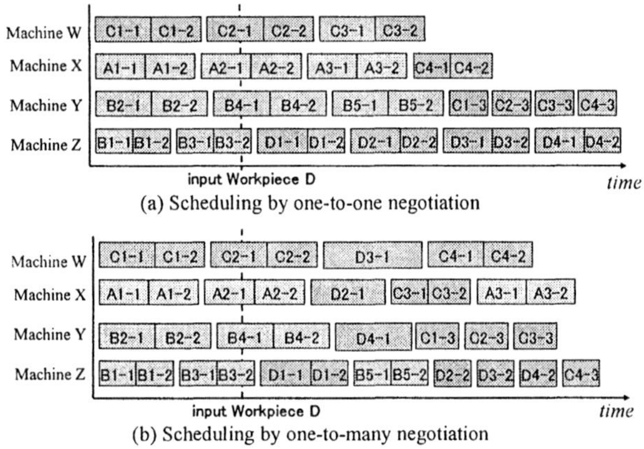

Figure 6. Example results of machine work planning simulation.

5. CONCLUSIONS

Workpiece agents and the configuration of autonomous machining systems using workpiece agents are discussed in this paper. The concepts of a workpiece agent is that a workpiece has the initiative in the manufacturing process and could autonomously decide on the manufacturing process and the allocation of manufacturing machines for itself. Trial implementation of the workpiece agents and simulation of machine work planning using workpiece agents show the effectiveness of event driven configurations on the manufacturing system.

6. REFERENCES

1. N. Sugimura, R. Shrestha, J.Inoue, Integrated process planning and scheduling in holonic manufacturing systems -Optimization based on shop time and machining cost-, Proc. of the 2003 IEEE International symposium on Assembly and task planning (ISATP2003, 2003), pp. 36-41.
2. Y. Tanimizu, T. Sakaguchi, K. Harada, N. Sugimura, Genetic Algorithm Based Reactive Scheduling in Multi-layered Manufacturing System, Proc. of 2002 Japan-USA Symposium on Flexible Automation, 2, (2002), pp.951-958.

GENETIC LOCAL SEARCH ALGORITHMS FOR SINGLE MACHINE SCHEDULING PROBLEMS WITH RELEASE TIME

Jihchang Hsieh[1], Peichann Chang[2], Shihhsin Chen[2]

[1]*Department of Industrial Management, Vanung University, Chung-Li 32061, Tao-Yuan, Taiwan, R.O.C., Email:* hjc@vnu.edu.tw. [2]*Department of Industrial Engineering and Management, Yuan-Ze University, Ne-Li, Tao-Yuan, Taiwan, R.O.C..*

Abstract: Single machine scheduling problems with release time are the prototypes for other complicated scheduling systems. Genetic local search algorithms are proposed to minimize total weighted completion time. Numerical experiments to evaluate the three embedding local search techniques are done. The results indicate that the $2r_i+t_i$ rule proposed by Liu and MacCathy (1991) outperform ERT and WSPT.

Key words: Genetic Local Search Algorithm, Single Machine Scheduling, Release Time.

1. INTRODUCTION

Production scheduling is to arrange job sequences under some manufacturing resources. In this paper, a single machine scheduling with release time is discussed. Although single machine describes one of the most basic production environments, the results can be extended in many other complicated production systems. Therefore it is the most applicable. However, the single machine scheduling problem with release time does not receive much attention in the contrast of the prosperous developments of other scheduling problems.

The traditional ideas developed to solve single machines scheduling with release time is to search optimum solutions. Branch-and-bounds algorithms were ever proposed by Dessouky and Deogun (1981), Deogun (1983), Hariri and Potts (1983), Chu (1992), Chand et al. (1996b), Chang et al. (2004). Rinnooy Kan (1976) has shown that single machine scheduling with release

Please use the following format when citing this chapter:
Hsieh, Jihchang, Chang, Peichann, Chen, Shihhsin, 2006, in International Federation for Information Processing (IFIP), Volume 207, Knowledge Enterprise: Intelligent Strategies In Product Design, Manufacturing, and Management, eds. K. Wang, Kovacs G., Wozny M., Fang M., (Boston: Springer), pp. 875-880.

time is NP-hard. Therefore to search optimum solutions is not efficient. To achieve efficiency, Liu and MacCarthy (1991), Reeves (1995), Chand et al. (1996a) proposed heuristic methods. Those heuristics work for some specified assumptions. Once one of the assumptions is violated, the performance is not guaranteed. In order to sustain stable and satisfactory performance, genetic algorithms are applied in this paper.

Genetic algorithms have widely applied in the optimization problems, of course in scheduling problems as well. However, in the open literature, genetic algorithms only have few applications on single machine scheduling with release time including Liu and Tang (1999), Madureira et al. (2001). In contrast to the rich applications of genetic algorithms on the other fields, it is expected that genetic algorithms should have more contributions in solving the single machine scheduling problems with release time. Therefore a genetic algorithm embedding local search techniques is presented.

2. PROBLEM DEFINITION

A single machine scheduling problem investigates the possible arrangements for jobs that need to be processed on only one machine under one or some specified constraints. Baker (1974) indicated that the research is very important because it may provide good ideas for complex systems. Most of the prior research commonly assumed that jobs are available at time zero. Such an assumption simplifies the complexity of scheduling. But this is not practical since not all jobs are available at time zero in the real-world cases. Therefore, to consider release time is meaningful in scheduling problems.

The following assumptions are used to characterize the scenario of this research.

1. There is a set of independent, single-operation jobs.
2. Release times are positive and not all zeros.
3. Setup times are independent of sequence and included in processing time.
4. The attributes of each job, i.e. release time, processing time and weight, are known in advance.
5. Only one machine is continuously available.
6. No job preemption.

In the shop floor, it is important to keep the work-in-process low in order to control the holding cost. Completion time is an index to show the level of work-in-process. Moreover, each job has its priority to be completed. The priority can be represented in terms of weights. Therefore the total weighted completion time is a suitable objective. In order to minimize the weighted completion time, a genetic local search algorithm is proposed.

3. GENETIC LOCAL SEARCH ALGORITHMS

A genetic local search algorithm is developed to solve the single machine scheduling problem with release time and the detailed procedure is briefly described as follows:

Encoding: Integer encoding.

Generating initial population: The chromosomes in the initial population are generated by random number generation.

Fitness function: Since this problem focuses on minimizing the total completion time, the fitness function needs to be transformed as the following form:

$$fit(x) = \text{MAX} - f(x)$$

To prevent the fitness from being zero, the MAX is set as:

$$\text{MAX} = f_{max}(\Psi_t) \times 1.1$$

Where $f_{max}(\Psi_t)$ means the maximum objective value in population Ψ_t.

Reproduction: Roulette wheel method (Goldberg, 1989).

Crossover: Two-point crossover (Goldberg, 1989).

Mutation: Shift mutation (Goldberg, 1989).

Local Search: Three heuristic-based local search rules are proposed in this research.

1. $2r_i+t_i$: Liu and MacCathy (1991) proposed a heuristic rule to sequence the jobs by $2r_i+t_i$, where r_i and t_i respectively denote the release time and processing time of job i. It is proven that $2r_i+t_i$ rule obtains optimum solutions when there are only two jobs. Therefore, this property is applied to local search. It is expected to improve the solution quality through $2r_i+t_i$ rule.
2. ERT: Earliest release time (ERT) rule is applied to local search.
3. WSPT: Weighted shortest processing time (WSPT) rule is applied to local search.

Any two consecutive jobs are randomly selected in each sequence after mutation. Then the jobs are checked if the heuristic-based local search rule is met. If the rule is not met then the positions of the two jobs are interchanged, else check the next sequence till all the sequences are checked off.

Replacement: An elite preserve strategy is applied. The top 10%

sequences are preserved for the next generation.

Termination: Once the maximum number of generations reaches a prespecified number, the algorithm is terminated.

4. NUMERICAL EXPERIMENTS

To compare the performance of each local search, numerical experiments are performed. Test instances are generated according to the following steps:

(i) Number of jobs: 20 and 40.

(ii) The processing time, weight, and release time follow uniform distribution. The parameters of the uniform distributions are reported in Table 1.

Table 1. Parameters of the Uniform Distributions to Generate Test Instances

Processing time	Weight	Release time
U (1,100)	U (1,10)	U (0, 50.5×n×R) where R equals 0.4

*R is a tardy factor that controls the release times to be tight or loose.

The parameters used in genetic local search algorithm including population size, maximum number of generations, crossover rate, and mutation rate are reported in Table 2.

Table 2. Parameters Used in Genetic Local Search Algorithm

Population size	50
Maximum number of generations	20000
Crossover rate	0.9
Mutation rate	0.1

For each test instance case, 10 test instances are generated and the average is used to be the result for each case. In Table 3, the results of solving 20 jobs are reported. On average, the $2r_i$+t_i rule proposed by Liu and MacCathy (1991) outperform ERT and WSPT.

Table 3. ATWCT of Each Local Search (20 Jobs)

Local Search	$2r_i$+t_i	ERT	WSPT
ATWCT	13449.2	13551.1	13646.56

* **ATWCT**: Average Total Weighted Completion Time

In the 40 jobs instances, the performance is the same as Table 3. $2r_i+t_i$ rule is the most effective, ERT is worse, and WSPT is the worst.

Table 4. ATWCT of Each Local Search (40 Jobs)

Local Search	$2r_i+t_i$	ERT	WSPT
ATWCT	59299.66	60518.1	60847.1

* **ATWCT**: Average Total Weighted Completion Time

Two kinds of number of jobs are used to evaluate the performance of local search techniques. As shown in Figure 1, the result is consistent that $2r_i+t_i$ rule has the minimum ATWCT.

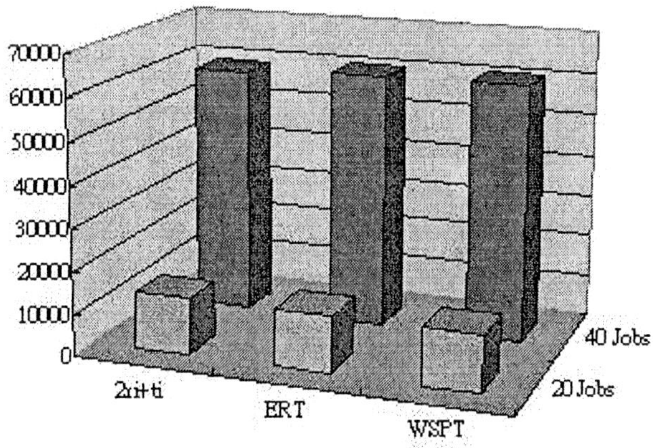

Figure 1. The ATWCT for Solving 20 and 40 Jobs

5. CONCLUSION

Single machine scheduling problems with release time are the prototypes for other complex scheduling systems. The results can be extended to many practical cases. However, this topic does not receive much attention. In this paper, genetic local search algorithms are proposed for this problem. Numerical experiments are done. The results show that the $2r_i+t_i$ rule proposed by Liu and MacCathy (1991) outperform ERT and WSPT.

This paper is a pilot study for single machine scheduling problems with release time. In the future, we may include more test instance cases to further verify the generalized performance of this idea. Moreover, we may extend this result to further extensive application areas.

6. REFERENCES

1. Baker, K., (1974), *Introduction to Sequencing and Scheduling*, New York: Wiley.
2. Chand, S., Traub, R., Uzsoy, R., (1996a), An iterative heuristic for the single machine dynamic total completion time scheduling problem, *Computers and Operations Research*, Vol. 23, pp.641-651.
3. Chand, S., Traub, R., Uzsoy, R., (1996b), Single-machine scheduling with dynamic arrivals: decomposition results and an improved algorithm, *Naval Research Logistics*, Vol. 43, pp.709-719.
4. Chang, P.C., Chung, Y.K., Hsieh, J.C., (2004), On single-machine scheduling with release times to minimize total weighted completion time, *Journal of the Chinese Institute of Industrial Engineers*, Vol. 21, pp.567-575.
5. Chu, C., (1992), A branch-and-bound algorithm to minimize total flow time with unequal release dates, *Naval Research Logistics*, Vol. 39, pp.859-875.
6. Deogun, J.S., (1983), On scheduling with ready times to minimize mean flow time, *Computer Journal*, Vol. 26, pp.320-328.
7. Dessouky, M.I., Deogun, J.S., (1981), Sequencing jobs with unequal ready times to minimize mean flow time, *SIAM Journal of Computing*, Vol. 10, pp.192-202.
8. Goldberg, D.E., (1989), *Genetic Algorithms in Search, Optimization, and Machine Learning*, Addison-Wesley Publishing Company, Inc.
9. Hariri, A.M.A., Potts, C.N., (1983), An algorithm for single machine sequencing with release dates to minimize total weighted completion time, *Discrete Applied Mathematics*, Vol. 5, pp.99-109.
10. Liu, J., MacCarthy, B.L., (1991), Effective heuristics for single machine sequencing problem with ready times, *International Journal of Production Research*, Vol. 29, pp.1521-1533.
11. Liu, J., Tang, L. (1999), A modified genetic algorithm for single machine scheduling, *Computers and Industrial Engineering*, Vol. 37, pp.43-46.
12. Madureira, A., Ramos, C., do Carmo Silva, S., (2001), A GA based scheduling system for dynamic single machine problem, *Proceedings of the 4th IEEE International Symposium on Assembly and Task Planning*, Soft Research Park, Fukuoka, Japan, May 28-29, pp.262-267.
13. Reeves, C., (1995), Heuristics for scheduling a single machine subject to unequal job release times, *European Journal of Operational Research*, Vol. 80, pp.397-403.
14. Rinnooy Kan, A.H.G., (1976), *Machine Scheduling Problem: Classification, Complexity, and Computations*, The Hague: Martinus Nijhoff.

CONTROLLING POWDER DEPOSITION PROCESS USING FUZZY LOGIC SYSTEMS

Kesheng Wang[1] , Paul Akangah[2], Qingfeng Yuan[3]

[1]Department of Production and Quality Engineering, Norwegian University of Science and Technology, Norway; Email:kesheng.wang@ntnu.no. [2]Department of Mechanical Engineering, Kwame Nkrumah University of Science and Technology, Ghana. [3]Knowledge Discovery and Management Laboratory, Shanghai University, China.

Abstract: This paper presents an application of Fuzzy Logic Systems (FLS) to control a powder deposition process, which is used for depositing coatings on the surface of products. Four input variables: work piece rotation speed, gun to work piece distance, powder volume pressure and gun current and one output variable, the first-pass transfer efficiency was chosen. Fuzzy rules have been introduced in this study. Results showed that FLS can be easily implemented in a company; and has the capability to increase first-pass transfer efficiency, minimize waste, reduce the time involved in selecting the best combination of application variables for coating jobs and improves savings.

Key words: Fuzzy Logic Systems; Powder Coating; Rule-Based Inference.

1. INTRODUCTION

Electrostatic powder coating process works on the principle that opposite charges attraction. The powder composed of resins and pigments is pneumatically fed from a reservoir through a spray gun where the powder gains low amperage, high voltage positive charge. Cleaned parts that are to be painted are electrically grounded so that the positively charged powder particles are strongly attracted to the parts' surface. The powder coated part is then pulled through an oven where the powder melts and fuses into a smooth coating. Powder coating can be done manually or through an automatic system.

This technology is used in many applications such as clear coating on brass fixtures and on aluminum cans (Gill, 1990). In powder coating

This project is supported by the National Natural Science Foundation of China under the grant No.50375090.

Please use the following format when citing this chapter:
Wang, Kesheng, Akangah, Paul, Yuan, Qingfeng, 2006, in International Federation for Information Processing (IFIP), Volume 207, Knowledge Enterprise: Intelligent Strategies In Product Design, Manufacturing, and Management, eds. K. Wang, Kovacs G., Wozny M., Fang M., (Boston: Springer), pp. 881-886.

processes, the most crucial questions are how to improve efficiency of powder coating and how to reduce overspray and maximize transfer efficiency while achieving high coating quality. Previous studies showed that parameters related to the powder coating process are adjusted manually and randomly according to the experiences of operators, and it is very difficult to establish a mathematical formula to model the coating process.

This paper presents the design and development of a Fuzzy Logic System (FLS), via a development tool: NEUframe (NEUframe User's Guide) for modelling and controlling the parameters of coating processes. The result obtained shows FLS is an excellent tool for modeling and controlling complex systems that are difficult to model using mathematical approach.

2. PROBLEM DESCRIPTION

The efficiency of a powder application process is indicated by two common terms: (1) System utilization efficiency (STE): ratio of powder applied to powder used (including reclaim) (2) First-pass transfer efficiency (FPTE): the ratio of powder applied to powder sprayed (without reclaim). We are focusing FPTE problem because is the main efficiency used to measure a powder coating system. FPTE describes the productivity of the powder application process.

$$\text{FPTE}(\%) = \frac{\text{amount of powder deposited on part}}{\text{amount of powder sprayed}} * 100$$

Optimizing powder deposited on the work piece is important in achieving higher FPTE, which result in reduced waste and increased savings.

3. FUZZY LOGIC SYSTEM

Fuzzy logic systems have been increasingly applied to wide range of engineering problems since introduced by Zadeh in 1965. Fuzzy Logic modeling is a heuristic method and is suited for systems which require the ability to handle uncertainties and /or approximate reasoning problems. FLS consists of three important procedures: (1) fuzzification; (2) fuzzy inference; and (3) Defuzzification. The system structure is shown in Figure 1. (Wang, 2005; Hopgood, 1998):

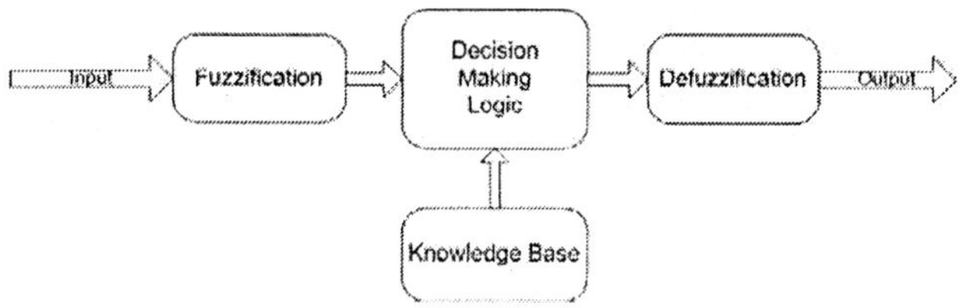

Figure 1. Structure of Fuzzy Logic Systems

4. MEMBERSHIP FUNCTIONS

There are many factors which can affect FPTE. For the problem under review, the four input variables that principally affect the FPTE were selected as inputs: (1) PVP: Powder Volume Pressure (psi); (2) WRS: Workpiece Rotation Speed (r/sec); (3) GPD: Gun to Part Distance (mm); (4) GC: Gun Current (μA). The output variable is PD: Powder Deposition on the work piece. FPTE is easily obtained by dividing Powder Deposition by the amount of powder sprayed. The input variables were divided into three linguistic values: low-, medium- and high-values. To get a more precise result, the output variable was divided into five linguistic values: too low, low, medium, high and too high. All inputs and output investigated in this model use triangular membership functions shown in Table 1.

Table 1. Fuzzy Values for all Variables

Variables	Linguistic Value	Fuzzy Value
PV	Low, Medium, High	(17, 17, 20), (17, 20, 23), (20, 23, 23)
WRS	Low, Medium, High	(1.5, 1.5, 2.25), (1.5, 2.25, 3), (2.25, 3, 3)
GPD	Low, Medium, High	(6, 6, 9), (6, 9, 12), (9, 12, 12)
GC	Low, Medium, High	(40, 40, 50), (40, 50, 60), (50, 60, 60)
PD	Too Low, Low, Medium, High, Too High	(0.82, 0.82, 0.84), (0.82, 0.84, 0.86), (0.84, 0.86, 0.88), (0.86, 0.88, 0.9), (0.88, 0.9, 0.9)

5. DATA ANALYSIS

The work pieces used for the study were 3-piece aerosol containers manufactured from large pre-decorated flat tinplate sheets. A black epoxy polyester gloss was chosen. This is able to produce 60-70 microns coating thickness (Conesa et al, 2004), and when cured gives a glossy surface finish. The 3-piece tinplate aerosol containers were powder coated in the pilot equipment, which utilizes corona charge. During the experiments, 30 cans were sent through the powder booth at various gun to product positions varying from 9 to 23 mm. In all, 135 cans were taken as samples for data analysis. The weights of powder deposited on cans were obtained by measuring the weight of the cans using a microbalance before and after coating. The variations of the four input variables to the output are shown in Figure 2.

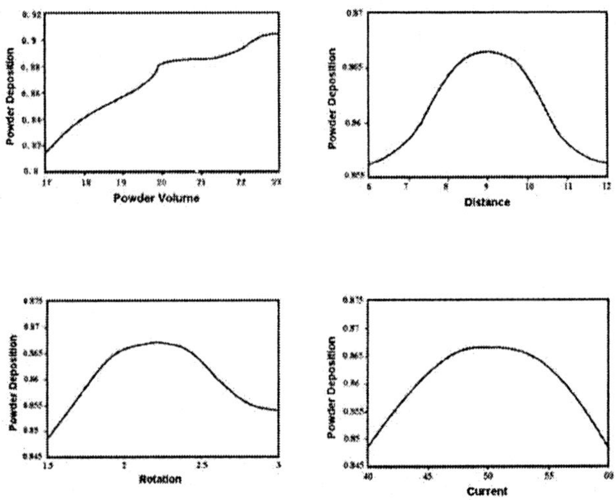

Figure 2. One-dimensional surface view of the four fuzzy input variables

6. DEVELOPING THE DATABASE

The rules for the fuzzy logic were obtained by extracting from the behavior of the process. This approach is preferable because it has the advantage of modeling the behavior of the fuzzy logic model to resemble data diagram of the process. The rule base was modified after a series of testing and simulation to improve the performance of the fuzzy logic system. As the four input variables have three states, there are a total of 3*3*3*3 = 81 rules in the rule base. An example of the rule used to manipulate variables is as follows:

IF Powder volume is High **AND** Work piece rotation speed is Medium **AND** Gun-to-part distance is Medium **AND** Gun current is Low **THEN** Powder deposition on the work piece is Low.

7. RESULTS AND DISCUSSION

To demonstrate the accuracy of this FLS, a sample experimental data input set was chosen and the fuzzy logic model run NEUframe (Figure. 3). The results from the model were then compared with the results obtained from the experiments, which has the same input variables. The outputs predicted by the model matched closely with the plots obtained from the data of experiments at same condition with an average accuracy of 98%. Consequently, the fuzzy logic model performed very well under the conditions investigated.

A three-dimensional surface view indicating the shape of the response variable, powder deposition, given two inputs, rotation speed and powder-volume pressure (psi) has been developed. An optimal powder deposition above 9.0 grams can be achieved if powder-volume pressure (psi) is set at 23 psi and workpiece rotation speed is at 2.25 turns while maintaining gun-to-part distance and gun current at the present levels. The fuzzy-logic based powder coating system thus has the capacity significantly to improve first-pass transfer efficiency by maximizing powder deposition.

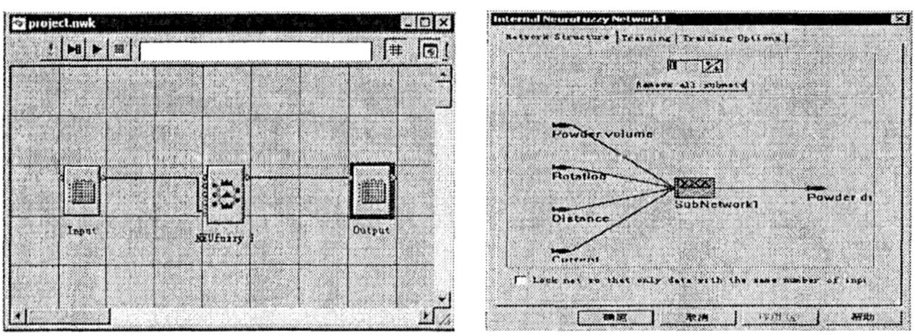

Figure 3. Running Windows of the Development Tool NEUframe

8. CONCLUSIONS

So far, a fuzzy logic model of powder coating system was presented. In powder coating operations, there is the regular need to determine the input variables settings that optimize powder deposition. While each input variable is important in optimizing powder deposition efficiency, it is important to understand how the combination of parameter settings affects coating jobs. A successful model can reduce energy consumption; reduce waste generated and consequently, waste disposal costs; reduce labour and maintenance costs; increase production rates and therefore, reduce reject rates; and increase first-pass transfer efficiency. Because the results derived from the model and results from experiments correlate well, it can be concluded that the model is efficient, and therefore can be used in practice

9. REFERENCES

1. Conesa, C., Saleh, K., Thomas, A., Guigon, A., and Guillot, N., (2004), *Characterization of Flow Properties of Powder coatings used in the Automotive Industry*, Hosokawa Powder Technology Foundation pp 22-106.
2. Gill D. E., (1990), *Powder coatings and their uses, Powder Coating Applications*, Dawson S. and Reddy V., eds., Society of Manufacturing Engineers, Dearborn, Michgam, pp. 52-53
3. Hopgood, A. A., Phillips, H. J., Picton, P. D., and Braithwaite, N. St. J., (1998), Fuzzy logic in a Blackboard System for Controlling Plasma Deposition Processes, *Applied Intelligence in Engineering*, Vol. 12, pp. 253-260.
4. NEUframe User's Guide, www.neusciences.com
5. Wang, K., (2005), *Applied Computational Intelligence to Intelligence Manufacturing Systems,* Advanced Knowledge International (ISBN 0-9751004-9-1).
6. Zadeh, L. A., (1965), Fuzzy Sets, *Information and Control*, Vol. 8, pp. 338-353.

FLEXIBLE DYNAMIC SCHEDULING BASED ON IMMUNE ALGORITHM

Jianjun Yu, Shudong Sun, Jinghui Hao
School of Mechatronic Engineering, Northwestern Polytechnical University, China; Email: npu_yjj@163.com

Abstract: The Job-Shop Scheduling Problem is simplified from the shop production though it is NP-hard, and the actual Scheduling is more complicated than it. The complex flexible manufacturing cell dynamic scheduling (FMCDS) which takes much disturb into account is researched in. The machine may suddenly turn into malfunction during production, and the jobs may go into or be canceled from scheduling for some reason. Each kind of part can be disintegrated into a lot of batch which includes some parts, and batch operation can choice one from some machines. Furthermore, several objects are optimized, which include time, cost and machine use rate. The FMCDS model is set up, and a dynamic evaluation based immune algorithm and rolling-disturb hybrid rescheduling policy are initially brought forward to solve FMCDS. The proposed algorithm, policy and method are applied to Xi'an Aviation Manufacture Corporation, and can fulfill its FMCDS.

Key words: Flexible manufacturing cell dynamic scheduling, Immune algorithm, Temporary solution set, Rolling-disturb hybrid rescheduling policy.

1. INTRODUCTION

Most of the scheduling methods idealize the actual shop condition to reduce the calculation difficulty. However, flexible manufacturing cell dynamic scheduling（FMCDS）is closer to the production practice, thus the research for it is of more theory significance and engineering value, but FMCDS has many characteristics, such as complex modeling, complicated calculation, multi-constraint, multi-objective and dynamic disturb, and belongs to the combinatorial optimization category, and is proved to be the typical NP-hard problem[1,2]. In the paper, the immunity idea is imported into FMCDS, and a novel immune algorithm is put forward to solve the FMCDS.

Please use the following format when citing this chapter:
Yu, Jianjun, Sun, Shudong, Hao, Jinghui, 2006, in International Federation for Information Processing (IFIP), Volume 207, Knowledge Enterprise: Intelligent Strategies In Product Design, Manufacturing, and Management, eds. K. Wang, Kovacs G., Wozny M., Fang M., (Boston: Springer), pp. 887-895.

2. THE MODEL OF FMCDS

In the flexible manufacturing cell, all operations may be completed on several choice machines, and the processing tasks can randomly be added or canceled at any moment during the processing, and the machines may encounter the unexpected breakdown. FMCDS can be described as follows:

The flexible manufacturing cell has H machines, M_1, M_2, ⋯, M_H, and may process N types of work pieces, L_1, L_2, ⋯, L_N, and these types have $N_{p,1}$, $N_{p,2}$, ⋯ $N_{p,N}$ batches respectively which are respectively made up of J_1, J_2, ⋯, J_N constraint operations with the technologic constraint. Supposing that the batch number and the batch size of N type of work pieces are ($L_{i,j}$, $P_{i,j}$), ⋯, ($L_{t,N}$, $P_{t,N}$) respectively, where, $L_{i,j}$ is the j-th batch of i-th work piece and $P_{i,j}$ is the batch size of $L_{i,j}$.

The objective functions include time, cost and machine utilization ratio, etc, and they may be embodied with two sub-objective functions as follows:

① Minimize the penalty due to the earliness/tardiness

$$F_1 = \sum_{i=1}^{N_p} \sum_{j=1}^{N_{p,i}} [b_{i,j} \times \max(0, D_{i,j} - T_{f,i,j}) + a_{i,j} \times \max(0, T_{f,i,j} - D_{i,j})] \tag{1}$$

② Maximize the machine utilization ratio

$$F_2 = \frac{\sum_{m=1}^{H} \sum_{i=1}^{N_p} \sum_{j=1}^{N_{p,i}} \sum_{k=1}^{J_i} T_{i,j,k} \times z_{m,i,j,k} \times c_m}{\sum_{m=1}^{H} (T_{f,m} - T_{s,m}) \times c_m} \tag{2}$$

Supposing that w_1 and w_2 are the weight coefficients, then the chief objective function is defined as follows

$$F = \min(w_1 F_1 + w_2 F_2) \tag{3}$$

The constraint conditions:

① resource constraint: The next batch operation may start on the same machine until current batch operation has finished

$$\begin{cases} T_{s,i,j,k} \geq T_{f,p,q,r} \\ z_{m,i,j,k} = z_{m,p,q,r} = 1 \end{cases} \text{or} \begin{cases} T_{s,p,q,r} \geq T_{f,i,j,k} \\ z_{m,i,j,k} = z_{m,p,q,r} = 1 \end{cases} \tag{4}$$

② technical constraint: The next batch operation of same batch may start until current batch operation has finished

$$T_{s,i,j,k+1} \geq T_{f,i,j,k} \tag{5}$$

$$z_{m,i,j,k} = \begin{cases} 1 & k\text{-th operation of } j\text{-th batch of } i\text{-th type of work piece is allocated on } m\text{-th machine} \\ 0 & \text{otherwise} \end{cases} \tag{6}$$

$$T_{s,m} = \min(T_{s,i,j,k} \in \forall o_{i,j,k} \wedge z_{m,i,j,k} = 1) \tag{7}$$

$$T_{f,m} = \max(T_{f,i,j,k} \in \forall o_{i,j,k} \wedge z_{m,i,j,k} = 1) \tag{8}$$

$o_{i,j,k}$——k-th batch operation of j-th batch of i-th type of work piece

$T_{s,i,j,k}$——the start time of $o_{i,j,k}$

$T_{f,i,j,k}$——the completion time of $o_{i,j,k}$

$T_{i,j,k}$——the processing time of $o_{i,j,k}$

H——the total number of the machines

N_p——the type number of the work pieces

$N_{p,i}$——the batch number of i-th work piece

J_i——the operation number of every batch of i-th work piece

$a_{i,j}$——the penalty coefficient due to tardiness of j-th batch of i-th type of work piece

$b_{i,j}$——the penalty coefficient due to earliness of j-th batch of i-th type of work piece

c_m——the importance coefficient of m-th machine

3. IMMUNE ALGORITHM

Immune algorithm is a new intelligent algorithm enlightened by the biologic immunity system[3, 4]. In order to avoid the algorithm plunging the local optimization, the concept of the temporary solution set is imported into the algorithm to accelerate the algorithm to evolve rapidly for the preponderant direction. The mechanism of the vaccine extraction and inoculation is adopted to lead the algorithm search. The flow of the algorithm is show in Fig.1. The detailed steps are as follow:

step 1 Antigen identification and antibody coding

Every type of work piece is divided into several batches and every operation of every batch may choose one from several machines, therefore new code method showed in Fig.2 is developed.

Step 2 Create solution colony

N_m excellent antibodies are picked up from the memory to form the new generation antibody colony along with the created antibodies.

Step 3 Antibody evaluation

The antibody concentration is adopted to restrain the antibodies which are of large scale but aren't the optimal solution, and the information entropy is regarded as the index to measure the antibody affinity, and the expectant reproduction ratio is applied as the standard of evaluating the antibody.

Supposing that there are N antibodies and every antibody has M genes, and every gene may choose one from S symbols, $k_1, k_2, \cdots, k_s$. The information entropy of N antibodies is defined as follows

$$H(N) = \frac{1}{M}\sum_{i=1}^{M} H_i(N) \tag{9}$$

Where, $H_i(N) = \sum_{j=1}^{S} -p_{ij} \log p_{ij}$, $H_i(N)$ is information entropy of i-th gene of N antibodies, and p_{ij} is the probability by which i-th genes of N antibodies are the symbol k_j.

The antibody affinity: indicating the comparability between two antibodies. The antibody affinity between v-th antibody and w-th antibody is defined as follows

$$a_{v,w} = \frac{1}{1+H(2)} \tag{10}$$

Where, $H(2)$ is information entropy of v-th antibody and w-th antibody. Similarly, the affinity between antigen and antibody is defined as follows

$$a_v = \frac{1}{(1+d_v)} \tag{11}$$

Where, d_v is the difference degree between the antigen and the antibody. The antibody concentration C_v of v-th antibody is defined as follows

$$C_v = \frac{1}{N}\sum_{w=1}^{N} b_{v,w} \tag{12}$$

Where, $b_{v,w} = \begin{cases} 1 & \alpha \cdot a_v \le a_w \le \beta \cdot a_v \\ 0 & \text{otherwise} \end{cases}$, α and β are the adjustment coefficients.

Finally the expectant reproduction ratio e_v is defined as follows

$$e_v = \frac{a_v \prod_{w=1}^{N} (1 - h_{v,w})}{\sqrt{C_v}} \tag{13}$$

Where, $h_{v,w} = \begin{cases} a_{v,w} & \alpha \cdot a_v \le a_w \le \beta \cdot a_v \\ 0 & \text{otherwise} \end{cases}$

Step 4 Vaccine extraction

p_{ij} is the probability by which i-th genes are the symbol k_j in the colony.

$$p_{i,j} = \frac{1}{N}\sum_{j=1}^{N} a_j \tag{14}$$

Where, $a_j = \begin{cases} 1 & g(i) = k_j \\ 0 & \text{otherwise} \end{cases}$, $g(i)$ is the symbol of i-th genes in the colony, and N is the antibody quantity of the colony.

Let the symbol k_j in the allele, whose probability is bigger than the set valve value and is the biggest, as the vaccine segment of the allele, and the vaccine extracted finally is $B=(b_1, b_2, \cdots, b_n)$, where, $b_j = \begin{cases} k_j & \max(p_{i,j}) \geq T_a \\ 0 & \text{otherwise} \end{cases}$, and T_a is the valve value.

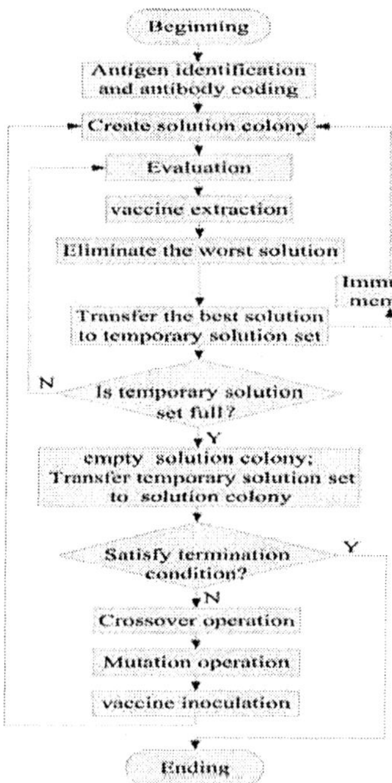

Figure 1. Algorithm flow.

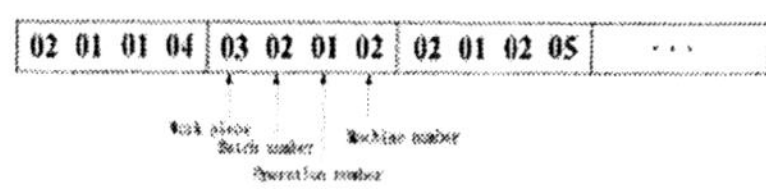

Figure 2. Antibody code.

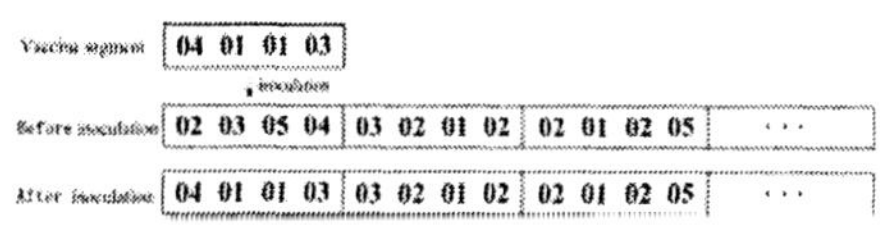

Figure 3. Vaccine inoculation.

Step 5 Eliminate the worst solution and transfer the best solution

Solution colony is ranked according to e_v, then the antibody with least e_v will be eliminated, and at the same time the antibody with biggest e_v will be transferred to the temporary solution set as the temporary optimal solution.

Step 6 Judge whether the temporary solution set has been full

If the temporary solution set has been full, then continue to carry out next step, else its size doesn't reach N_t, therefore go to step 3, and reevaluate the rest solution colony.

Step 7 Empty solution colony and transfer temporary solution set to colony

Abandon the rest solution colony, and transfer temporary solution set to solution colony. At the same time, N_m excellent antibodies are picked from the temporary solution set to be saved in the memory. Finally, the temporary solution set is eliminated.

Step 8 Judge whether the termination condition is satisfied

If satisfied, then terminate, else continue.

Step 9 Crossover operation and mutation operation

The crossover operation doesn't change the machine allocation of the batch operation, but only change the order of the batch operation. The mutation operation is used to change the machine allocation.

Step 10 Vaccine inoculation

The vaccine inoculation operation is shown in Fig.3.

4. FLEXIBLE DYNAMIC SCHEDULING STRATEGY

In the flexible dynamic scheduling, the processing tasks are likely to be inserted or canceled in the midway, and the machine breakdown is likely to appear unexpectedly, so it's necessary to import the flexible dynamic scheduling strategy [5].

4.1 Rolling window technology

The rolling window technology is put forward according to the rolling optimization principle of the forecast control. It isn't the optimization through whole course and isn't optimization only once, but the rolling optimization in the finite period of time. At every sampling time, the optimization performance indexes only involve the finite time from the beginning to the future. When next sampling time appears, the optimization time processes forward accordingly, and this performs repeatedly.

4.2 Scheduling task pool

The scheduling task pool is divided into three consecutive sub-pools. The pending scheduling task pool keeps the work pieces which have been planned but not been scheduled. The window task pool keeps the work pieces which need to be been scheduled and optimized. The completion task pool keeps the work pieces which are completed. The flow course of the work pieces in scheduling pool is shown as Fig.4.

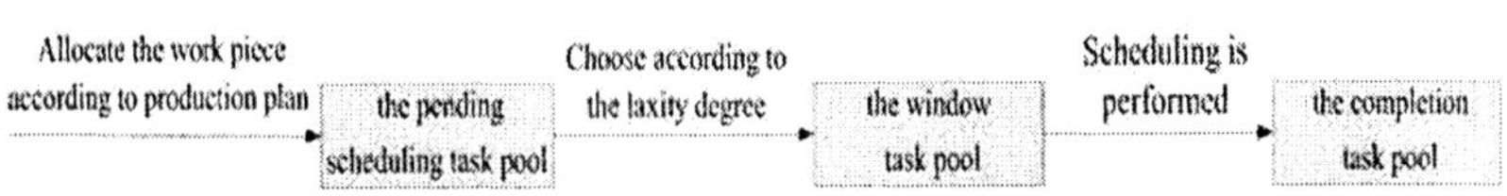

Figure 4. Flow course of the work pieces in scheduling pool.

4.3 Rolling-disturbance hybrid rescheduling

Because the flexible and dynamic characteristic of the system, rolling-disturbance hybrid rescheduling strategy is put forward. If the disturbance doesn't appear in the scheduling system, then the rescheduling will be rolled periodically. However, if disturbance appears unexpectedly, the rescheduling must be carried out immediately to respond to the events.

5. THE SCHEDULING INSTANCE

In the manufacturing cell of Xi'an Aviation Engine Manufacturing Corporation, there are 15 machines and 10 types of work pieces, and every type of work piece includes several batches. Machines, work pieces and batches are numbered as table 1. The technical information is as table 2.

The scheduling of 20 work piece batches is carried out by the immune algorithm put forward, and the optimized scheduling solution is gained, and the evolution curve of the algorithm is shown in Fig.5. It's revealed from the evolution curve that the constringency ability of the algorithm for the scheduling problem is strong and the evolution efficiency is high. The Gantt chart of the first optimized scheduling solution is shown in Fig.6.

The disturbances will appear unexpectedly during the production. Work piece 1 is canceled at time 16, a new batch of work piece type 10 is added at time 18, and the breakdown appears on machine 13 at time 20. Consequently rescheduling is carried out, and the Gantt chart of the whole dynamic scheduling is shown in Fig.7.

Table 1. Work piece number in the manufacturing cell

Work piece type	Batch quantity	Work piece number
1	2	1
		2
2	2	3
		4
3	4	5
		6
		7
		8
4	3	9
		10
		11
5	1	12
6	1	13
7	3	14
		15
		16
8		17
9	2	18
		19
10	1	20

Table 2. Technical information in the manufacturing cell

Work piece type	Batch quantity	Due date	Operation number	Choice machine	Processing time
1	2	38	1	1, 2, 11, 12	3
			2	3, 4, 5	8
			3	6, 8, 9, 10	4
			4	7, 13, 4	6
			5	11, 12	2
			6	15	5
			7	11, 12, 13, 14	7
2	2	28	1	1, 2, 11, 12	2
			2	8, 9, 10	3
			3	11, 12	4
			4	7, 13, 14	7
3	4	26	1	3, 4, 5	5
			2	7, 11, 12	4
			3	7	8
4	3	18	1	3, 4, 5	9
			2	6, 8, 9, 10	5
5	1	14	1	1, 2	6
			2	13, 14, 15	2
			3	7, 10	5
6	1	20	1	8, 9, 10	2
			2	6	6
			3	11, 12, 15	3
			4	3, 4, 5	1
			5	10	4
7	3	19	1	2	5
			2	11, 12, 13, 14	3
8	1	28	1	1	4
			2	11, 12	1
			3	15	6
			4	13, 14	2
			5	3, 4, 5	5
9	2	17	1	3, 4, 5	3
			2	2	5
			3	11, 12	2
10	1	25	1	13, 14	4
			2	2	2
			3	11, 12	6
			4	15	3
			5	3, 4, 5	3
			6	8, 9, 10	8

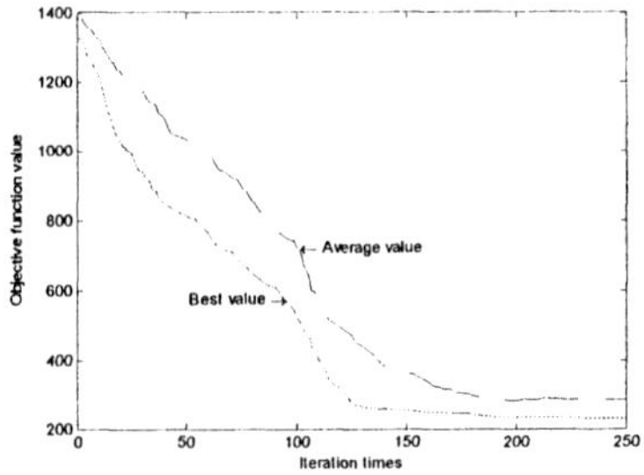

Figure 5. The evolution curve of the algorithm.

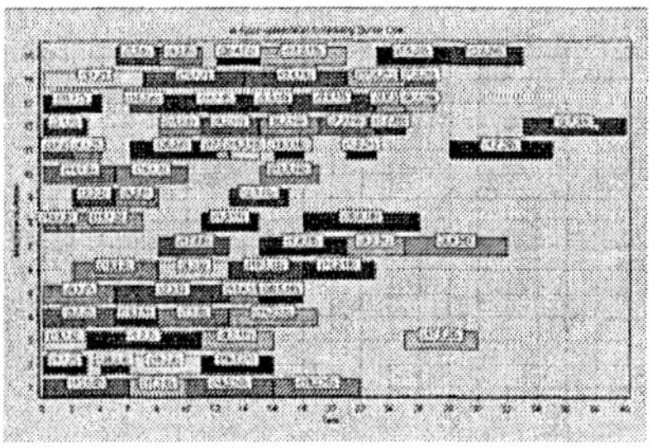

Figure 6. The scheduling Gantt chart.

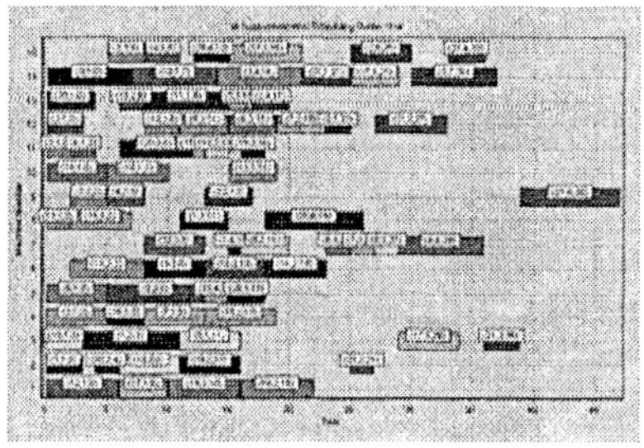

Figure 7. The rescheduling Gantt chart.

6. CONCLUSION

The complex dynamic scheduling of the flexible manufacturing cell is researched, which considers various disturbance factors. The operation has flexible technical routes, and multi-objective optimization is considered. The model of FMCDS is set up, and a new immune algorithm is put forward for FMCDS. The concepts of the rolling window technology and the scheduling mask pool are introduced and the rolling-disturbance hybrid rescheduling strategy is developed. The practice application of FMCDS in Xi'an Aviation Engine Manufacturing Corporation indicates that the algorithm, the strategy and the model can satisfy the requirement of FMCDS.

7. REFERENCES

1. Jain, A.S., Meeran, S.,(1998), A state-of-the-art review of job-shop scheduling techniques, Technical Report, Department of Physics, University of Dundee, Scotland, pp.130–140.
2. Tapan, P., Bagchi, (1999), *MultiObjective Scheduling by Genetic Algorithms*, Kluwer Academic Publishers, NewYork (ISBN 0-7923-8561-6).
3. Leandro, N. C., Jonathan, T., (2002), *Artificial Immnue System: A New Computational Intelligence Approach*, Springer Verlang, Great Britain (ISBN 1-8523-594-7).
4. JIAO, L. DU, H., (2003), Development and prospect of the artificial immune system, *Acta Electronica Sinica*, 31(10), pp.1540–1548.
5. Church, L., Uzsoy. R, (1992), Analy of periodic and event-driven rescheduling policies in dynamic shops, *International Journal of Computer Integrated Manufacturing*, 5(3), pp.153–163.

FUZZY DECISION ON TRAVERSE SPEED FOR ABRASIVE SUSPENSION JET

Wenbin Wang, Tao Yu, Xiaojian Liu, Tan Liu
CIMS & Robot Center, Shanghai University, Shanghai, PR China,Email: wenbin_wang@126.com

Abstract: The traverse speed is very important parameters of cutting process, But the cutting process of the ASJ (abrasive suspension jet) is a complex system with many interactions and influences among process parameters. It is hard to value. The paper presents a fuzzy expert system to get appropriate value of traverse speed. A CAPP system based on the fuzzy expert system for ASJ is also proposed in the paper. At last the application of the fuzzy expert system presents evidence of the usefulness of the system.

Key words: Traverse speed; Abrasive suspension jet; Fuzzy expert system; CAPP;

1. INTRODUCTION

Abrasive suspension jet (ASJ) is a new kind of AWJ (abrasive water jet). Shanghai University has put forward a new slurry and high pressure medium circle isolate device for ASJ (Patent No CN1597253). The device solves the problem of pollution between slurry and high pressure medium. And makes the slurry and high pressure medium recycle and reuse easily. It greatly reduces the user machining cost, and will help ASJ be more popular in mechanical machining [1].

There are two main stages in the cutting process. First stage is the preparative stage. The drawing will be transferred to G-code considering the special cutting process about abrasive suspension jet. The second is cutting stage. The numerical controller will control the movement of the nozzle according to the G-code for cutting. But the cutting process depends on a large number of parameters of an interconnected system. Many parameters show multi-causal behavior and cannot be optimized individually.

Please use the following format when citing this chapter:
Wang, Wenbin, Yu, Tao, Liu, Xiaojian, Liu, Tan, 2006, in International Federation for Information Processing (IFIP), Volume 207, Knowledge Enterprise: Intelligent Strategies In Product Design, Manufacturing, and Management, eds. K. Wang, Kovacs G., Wozny M., Fang M., (Boston: Springer), pp. 896-901.

Furthermore. There is nonlinear relationship among these factors, and it is hard to give traditional mathematic formula. Though Hashish [2] has published an abrasive jet cutting model for predicting some process parameters for brittle materials. Kovacevic [3] used regress equation to predict relationship between traverse speed and cut depth. An error will always exist between the calculated and the actual results because of the simplification and assumption of the modeling. The parameter of traverse speed also encounters the same difficulty without exception, which is usually valued by experience. And the following problems often appear in the cutting stage:

First if the value of the traverse speed is too high, the jet will not cut through the material, it will spay and destroy the material;

Second if the value of the traverse speed is too low, the cost of cutting will be higher.

Third because the lag of jet, if the drawing is made up of complex curve, the unreasonable speed value will cause furrow in the rear of material during the jet turning the small corner, and lead to bad cutting quality.

So the paper presents an effective fuzzy decision method to get appropriate value of traverse speed. It includes five main sections. Section 2 discusses the fuzzy rule of constraints. Section 3 emphasizes on how to use fuzzy expert system to make the appropriate value of traverse speed automatically instead of the experience and expertise of the operator. Then the CAPP based on fuzzy expert system for ASJ has also been put forward in section 4. Finally, deficiency about this system and conclusion are given in section 5.

2. FUZZY RULE OF CONSTRAINTS

From above, some factors are imprecise and ambiguous such as the description about pressure, cutting quality and cost. These factors could not be described by the precise threshold, so it is necessary to fuzzify the traditional production rule. Thus the assumption has been made as follows :

- All rules should come up with multi-premise but single conclusion;
- The premise of rule admits the conjunction "AND" other than "OR". If the premise of one item include K times of "OR", the item of this rule must be considered as K+1 items of single conclusion rule.
- The premise fuzzification: The fuzzy predication and fuzzy state measure word have been inducted in the premise. And the value of fuzzy membership degree also has been considered during fuzzy match.

- Set the rule confidence CT ($0<CT<1$), so as to arrange the rule according to its CT value.
- The conclusion fuzzification.

3. FUZZY EXPERT SYSTEM

3.1 Knowledge base

The knowledge base has four components, as shown in Fig.2. The history database is used to store the appropriate traverse speed under different effect factors. So the history database will grow bigger and bigger with each time new factor has been concerned. The static database is used mainly to store the data of the traverse speed. The dynamic database is used to store all kinds of intermediate information acquired in the inferring process. The rule database represents the expert knowledge and theoretical knowledge abrasive suspension jet cutting. To set up the rule database the paper uses the method of fuzzy orthogonal test [4]. Thus various factors' effect to traverse speed can be evaluated more easily. Then the main domino effect has been used to calculate various membership degrees under different factors. So the rule's confidence can be arranged according the number of the fuzzy membership degree.

There are two kinds of representation of knowledge in the system.

Representation of facts—Predicate calculus is one of the methods used to represent knowledge. It is employed here to represent knowledge. Here the determination of the traverse speed is a combination of a set of facts. It can be represented by the conjunction of the predicates as follows:

Traverse speed $\wedge$ Material $\wedge$ Pressure $\wedge$ Nozzle parameters $\wedge$ ...

Representation of reasoning knowledge—In the system production rules are used to represent reasoning knowledge. The general format is as follows:

IF (fact 1 exists)
(fact 2 exists)
...
(fact n exists)
Then (conclusion P)

3.2 Inference system

In developing the inference mechanism for expert systems, one of the inference strategies is to use both forward and backward chaining [5]. The system mainly uses forward chaining. The machining drawing gains from the CAD parts module or Image vectorization module. With the machining drawing has been split into series of lines and circles in sequence. Then the expert system will search history database first for the appropriate speed value, which matches the premise. If cannot find the answer in the history database, the exert system will infer the speed. And refer to the static database the appropriate value of traverse speed can be made. The appropriate value of traverse speed also has been recorded in the history database for using later. After pretreatment above the appropriate traverse speed of G-code can be made automatically. The whole workflow process is shown in Fig.1.

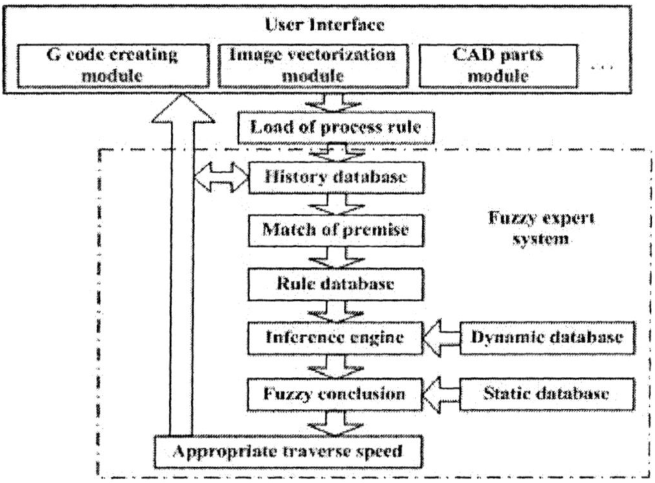

Figure 1. The workflow process of fuzzy expert system.

"Load of process rule" in the workflow means pick up each rule from the fuzzy knowledge base. Then use the conjunction "AND" to deal with multi premise rules.

4. CAPP REALIZATION

As is shown in Fig.2, First we get a AutoCAD design or image information, second we must specify known process parameters such as material parameters, nozzle parameters, slurry parameters, and etc, Then according to the fuzzy expert system the appropriate traverse speed is inferred. Third we specify working procedure such as original point, technical points, start point, end point, cutting path, cutting sequence and etc. thus the whole machining trace has been generated. Finally because some parts using a special technology need to be considered particularly, such as parts motion and rotation, cutter compensation and etc. The machining trace has been modified and the final trace code, G code, has been build by CAPP system.

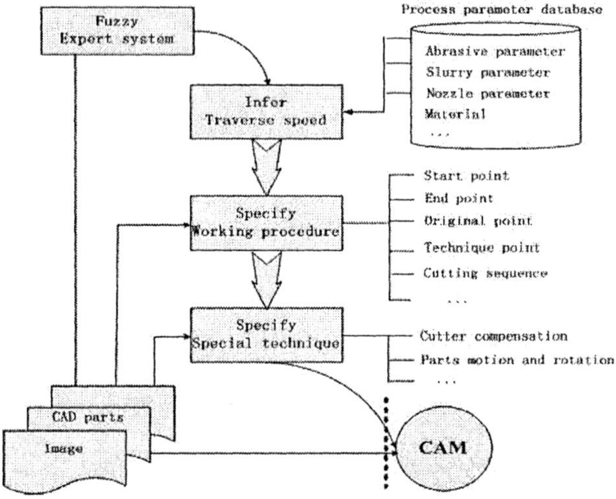

Figure 2. Flowchart of CAPP.

In this paragraph, we show how CAPP parts module makes man-machine interaction. In the left of Fig.3, drawing can be designed or modified in that area. And in the right, it is a CAPP plug in module, where points, lines, layers, dimensions, direction and some process parameters can be easily got. Also can interact with fuzzy expert system via this module to tell the process parameters such as the material, cutting pressure, the kind of abrasive and etc. Then fuzzy expert system generates appropriate traverse speed background. In the top right of Fig.3, it is a frame where G-code with the appropriate traverse speed has been made automatically.

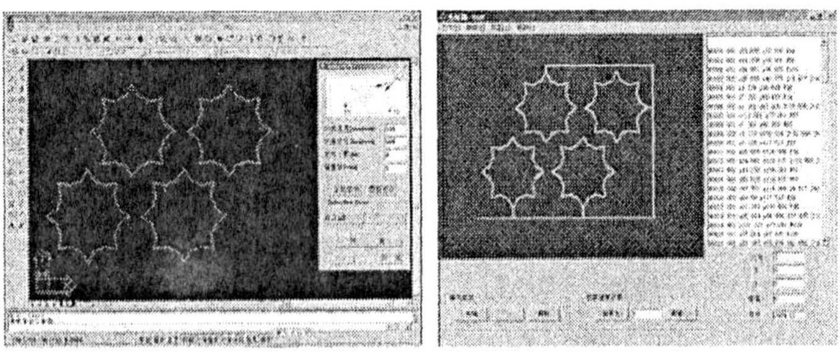

Figure 3. a). CAD parts module interface b) Realization interface

5. CONCLUSION

The paper sets out from the application of the fuzzy expert system in the CAPP field for abrasive suspension jet. And emphasizes on how to use fuzzy expert system to make the appropriate value of traverse speed automatically. The running result show it can better simulate the expert thought in the field, realize the appropriate value of cutting speed and create G-code for working. It will be in favor of popularizing abrasive suspension jet in market.

Although this system has been designed and developed successfully, there is still no clearly estimate standard about cut quality of water-jet, so the expert system may be limited to some special field.

6. REFERENCES

1. Liu Xiaojian, Yu Tao, Liu Linsheng Wang Wenbin "The study on abrasive suspension jet cutting technology", Lubrication Engineering, no.169, pp. 31–33, 2005
2. Kovacevic,R.,et al., "Rehabilitation of the concrete pavements with abrasive waterjets", The 11th International conference on jet cutting Technology, BRHA, st. Andrews Scotland, Sept, 8-10,1992;
3. Chakravarthy, P. Sitarama, Ramesh Babu, N. Source, "New approach for selection of optimal process parameters in Abrasive Water Jet cutting" Materials and Manufacturing Processes, v 14, n 4, Jul, 1999, p 581-600;
4. Honghua Su, Zhengjun Yao, "Fuzzy Analysis Method for Muti-Index Orthogonal Test", Journal of Nanjing University of Aeronautics & Astronautics, Vol.36, No.1, Feb, 2004;
5. Lawrence O. Hall, "Rule Chaining in Fuzzy Expert Systems", IEEE Transactions on Fuzzy Systems, Vol. 9, No. 6, Dec, 2001

RESEARCH ON MANUFACTURING RESOURCE DISCOVERY AND SCHEDULING IN MANUFACTURING GRID

Yu'an He[1,2], Tao Yu[1], Lilan Liu[1], Haiyang Sun [1]
1.Cims & Robot Center, Shanghai University, 200072 China. 2.Shanghai Institute of Technology, China 200233 Email: hya@graduate.shu.edu.cn

Abstract: The significant goal of Manufacturing Grid (MG) is to realize resource sharing and collaboration working among the heterogeneous and geographical distributed manufacturing resources. A TQPS (Time, Quality, Price, Service) - based resource discovery and scheduling algorithm is proposed in the paper. By the four processes of resource discovery, resource scheduling, dynamic negotiation and resource reservation, the optimal resource can be discovered and matched to the task/subtask. Under the precondition of assuring the Quality of Service (QoS), the searching and scheduling algorithm facilitates the realization for the resource discovery and scheduling.

Key words: Manufacturing Grid (MG); Resource management; Resource discovery; Resource scheduling; Quality of Service (QoS).

1. INTRODUCTION

The goal of Manufacturing Grid (MG)[1] is to realize manufacturing resource sharing and collaboration working. When MG receives a manufacturing task, one of the significant functions is to search and schedule appropriate manufacturing resources across the different geographical distributed zones and enterprises, and effectively make these heterogeneous manufacturing resources sharing and collaboration. Because of Manufacturing resources' complexity and diversity, there are many differences between MG and computing grid in resource discovery and scheduling method, the resource management and scheduling methods in MG are more complicated than those in computing grid.

This paper is Supported by "SEC E-Institute: Shanghai High Institutions Grid" project & "SEC project: No. 03HK07.

Please use the following format when citing this chapter:
He, Yu'an, Yu, Tao, Liu, Lilan, Sun, Haiyang, 2006, in International Federation for Information Processing (IFIP), Volume 207, Knowledge Enterprise: Intelligent Strategies In Product Design, Manufacturing, and Management, eds. K. Wang, Kovacs G., Wozny M., Fang M., (Boston: Springer), pp. 902-907.

The resource discovery and scheduling is a user-level middleware in fact. Document 2 proposed that MG system was a four-layer framework: 1) MG Fabric: this consists of all the globally distributed manufacturing resources that are accessible from anywhere on the Internet. 2) MG Core Middleware: this offers core services provided by GT4, such as security management, resource management, etc. 3) MG User-Level Middleware: this includes performance evaluating and monitoring tools, Manufacturing Grid Resource Scheduler (MGRS), etc. 4) MG Applications: MG applications are typically developed using Java language, e.g., MG Portal[3].

Document 4 has ever proposed a QoS-based resource scheduling method, but when MG deals with complicated manufacturing tasks, resource consumers put forward more requirements for manufacturing service level and manufacturing resources, the search and scheduling method can not realize easily. ***In manufacturing, TQPS (Time, Quality, Price or Cost, Service) can describe a task's/subtask's requirements and a resource's characteristics well.*** So a new TQPS -based resource searching and scheduling algorithm is designed in the paper, this algorithm assures the resource consumers that they can obtain manufacturing services of quality assurance in MG.

2. THE PROCESS OF RESOURCE DISCOVERED AND SCHEDULING

A manufacturing task or subtask decomposed by decomposing module must allocate an optimal resource which meets its QoS requirements. The process of searching and scheduling is shown in Fig.3.

Firstly, the matching resources are searched according to the task's QoS constraints, which include all the TQPS conditions, i.e., delivery time T, task's quality Q, task's price P and service requirements S. Comparing the task's TQPS constraints with resources' TQCS properties, i.e., resource spare time T, resource service quality Q, resource cost C and resource service S, a resource list can be obtained if there are matching resources discovered. Then an optimal resource can be matched to the task according to Optimal Evaluation Model (OEM) [4].

Secondly, considering the particularities of manufacturing resources, for the time T, resource providers can compensate the using time for working overtime; for the price P, resource consumers can consult or negotiate with the resource providers. So if all the TQPS constraints do not meet all the resource TQCS properties, the task's Q and S constraints are compared with resources' Q and S properties. If there are matching resources discovered, then T and P constraints are matched with T and C properties successively, or consumers can consult or negotiate with resource providers, then a resource list is also gained.

Thirdly, a Service Level Agreement (SLA) is subscribed by the MG-QoS manager, resources reservation is settled, and workflow model is formed.

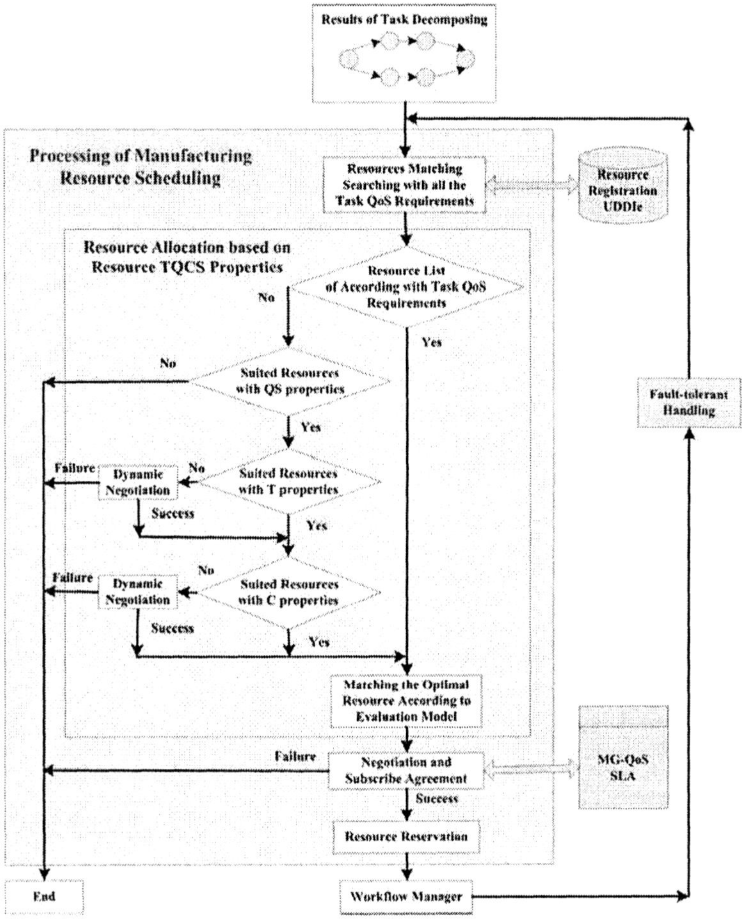

Figure 1. The process of resource discovery and scheduling

In addition, sometimes the SLA can not be carried out for some reasons, e.g., the reason of the resource fault, this time the matching resource must be searched and scheduled again through fault-tolerant handling.

In other word, there are four main steps in this process: TQPS-based Resource Discovery, TQPS-based Resource Scheduling, Dynamic Negotiation and Resource Reservation.

Document 5 presented three types discovery methods: 1) Gold level: The value of resources' QoS attributes should not less than that the customer specified task's QoS constraints. 2) Silver level: the value of resources' QoS attributes could be varied within a range according to the task's QoS constraints. 3) Bronze level: customers just care about one or some QoS constraints, such as delivery time, product quality, cost and service, etc. According to the manufacturing particularities, resource discovery strategy in MG is always bronze level, so the discovery method can not operate and realize easily.

We presented a new algorithm using all TQCS properties or part of them, i.e., if no one resource is discovered using all task's TQPS constraints, then using Q and S constrains searching resources, if a resource list is gained, for the T and P constrains, the resource consumer can negotiate dynamically with resource providers. This method not only accords with the manufacturing facts, but also operates and realizes easily.

3. TQPS-BASED ALGORITHM OF RESOURCE DISCOVERY AND SCHEDULING

The searching process takes resource static QoS attributes (using mainly the TQPS) as restriction conditions, i.e., resource spare time T, service quality Q, cost C and attributes related to service S, to search and discover matching resources from MGIS (Manufacturing Resource Information System). Since UDDIe (Universal Description, Discovery and Integration Extension) provides the Application Programming Interface (API, i.e., the find_service) can realize resource discoveries which are different QoS attributes.

STEP1: constructing the task's subsets and QoS constraints. For a task submitted to MG is $T=\{t_i\}$, $(i=1,2,\cdots n)$, t_i is one of the subtasks which task T is decomposed. Its QoS constraints are $QoS_{ti}=\{T_{si},T_{ei}Q_i,P_i,S_i\}$, T_{si},T_{ei} are the start and end time of the task respectively; Q_i is the quality requirements of the task; P_i is the product price; S_i is the service requirements of the task.

STEP2: constructing the resource subsets and QoS attributes. The resource subsets $R=\{r_j\}$ $(j=1,2,\cdots m)$ are the resources which accord with the requirements of the task and come from all the resource nodes' MGIS. Its QoS attributes are $RQoS_j=\{RT_{sj},RT_{ej},RQ_j,RC_j,RS_j\}$, RT_{sj},RT_{ej} are the start and end spare time of resources respectively; RQ_j is the attributes related to quality; RS_j is the attributes related to service.

STEP3: searching resources according to all the task's QoS constraints. The matching resources $SR = \{R_k\}$, $(k = 1,2,\cdots r)$ must meet the following equation:

$$\begin{cases} T_{si} \geq RT_{sj}, T_{ei} \leq RT_{ej} \\ Q_i \leq RQ_j \\ P_i < RC_j \\ S_i \leq RS_j \end{cases} \quad (1)$$

Where: the start time T_{si} of subtask t_i is later than the start spare time RT_{sj} of the resource R_k, and the end time T_{ei} of subtask t_i is earlier than the end spare time RT_{ej} of the resource R_k. The restriction conditions Q_i of the subtask t_i are less than or equal to the quality levels RQ_j which the resource R_k can provide. The price restriction condition P_i of the subtask t_i is less than using cost RC_j of the resource R_k. The service restriction conditions S_i of the subtask t_i are less than or equal to the service attributes' conditions RS_j of the resource R_k.

STEP4: if there are resources discovered after STEP3's searching, then turning to STEP7 to begin resource scheduling, otherwise just using Q_i and S_i constraints of the subtask to match resources. That is, the matching resources $SR = \{R_k\}$ $(k = 1,2,\cdots r)$ must meet the following equation:

$$\begin{cases} Q_i \leq RQ_j \\ S_i \leq RS_j \end{cases} \quad (2)$$

The meanings of the variables are the same as equation (1).

STEP5: if there are still no resources discovered meeting Q_i and S_i constraints, then returning to the end. It is implied that no resource can undertake the subtask at present, otherwise turning to the STEP6.

STEP6: considering the manufacturing facts, resource providers can solve the problem of resource using time by working overtime, and consumers can negotiate with resource providers about production price. So if there are resources discovered meeting Q_i and S_i constraints of the subtask, then negotiating about the resource using time and production price successively. If the time and price are negotiated successfully, the resource R_k is gained, otherwise returning to the end.

STEP7: according to the matching resources $\{R_k\}$ to choose the optimal resource with OEM. Its application process is as below:

Through this algorithm, manufacturing resources discovered can provide the best manufacturing services for the task in MG.

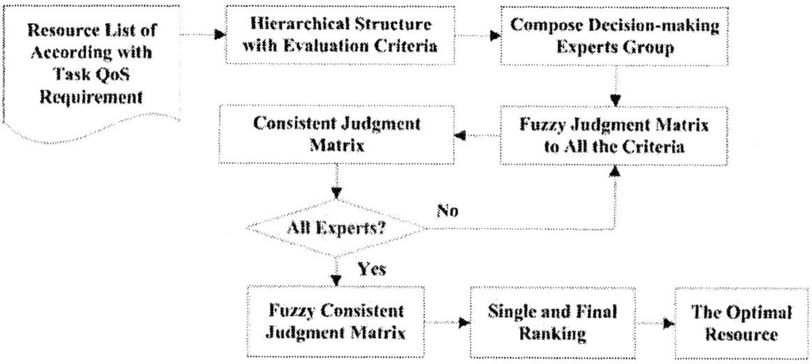

Figure 2. Application process of OEM

4. CONCLUSION

The discovery and scheduling of manufacturing resources were significant functions in MG. As the main requirements of manufacturing tasks and the properties of manufacturing resources, TQPS can describe a task's/subtask's constraints and a resource's characteristics well. In order to realize resource sharing and collaboration working among distributed manufacturing resources. A TQPS-based resource discovery and scheduling algorithm including dynamic negotiation and resource reservation can facilitate the realization for the resource discovery and scheduling.

5. REFERENCES

1. Liu Lilan, Yu Tao, Shi Zhanbei. Research of Self-Organization Manufacturing Grid and its Task Scheduling Algorithm, Computer integrated manufacturing system (CIMS), 2003, 9 (6): 449-455
2. He Yu'an, Yu Tao, Liu Lilan, et al. A QoS-based Task Decomposing Middleware Research for Manufacturing Grid. Proceedings of the First International Conference on Semantics, Knowledge and Grid, Beijing, China, 2005, 513-519
3. I. Foster, J. Frey, S. Graham, et al. "Open Grid Services Infrastructure (OGSI) Version 1.0", 2003, http://www.globus.org/research/papers.html.
4. Liu Lilan, Yu Tao, He Yu'an. A QoS-based GFAM Scheduling Approach for Manufacturing Grid. Proceedings of the Fifth International Conference on Computer and Information Technology (CIT'05). Shanghai, China, 2005,334-338
5. Liu Lilan, Yu Tao, Shi Zhanbei. Research on QoS-based Resource Scheduling in Manufacturing Grid, Computer Integrated Manufacturing System (CIMS). 2005, 11(4): pp 475-480

RESEARCH ON WSRF-BASED RESOURCE MANAGEMENT MODEL AND ENCAPSULATION IN MANUFACTURING GRID

Yu'an He[1,2] , Tao Yu[1] , Bin Shen[1]
1.Cims & robot Center, Shanghai University, China 200072 2.Shanghai Institute of Technology, China 200233 Email: hya@graduate.shu.edu.cn

Abstract: In order to realize resource sharing and collaboration effectively among the heterogeneous and geographical distributed manufacturing resources, a Concept Model of resource management, a WSRF (Web Service Resource Framework)-based Resource Management System Model in Manufacturing Grid (MG) and the resource encapsulation method were presented in this paper. By the two models, the description, encapsulation and publication of manufacturing resources are implemented conveniently, and the manufacturing resource can be encapsulated to a WS (Web Service)-Resource structural MG-Resource. The encapsulation process shields complexity and diversity of manufacturing resources and implements manufacturing resource sharing and collaboration in MG effectively.

Key words: WSRF; Manufacturing Grid (MG); Resource management; Web service; Resource encapsulation.

1. INTRODUCTION

Manufacturing Grid (MG)[1] is the application of Grid Technology in manufacturing, the core idea of MG is manufacturing resource sharing and collaboration. Manufacturing resources mean the equipments, materials, personnel and all the hardware and software related to the product lifecycle[2]. Manufacturing resources have the characteristics of complexity, diversity and distribution, and manufacturing tasks have many characteristics which are different from those in computing grid. Thus there are many differences between MG and computing grid in resource management mode, the

This paper is Supported by "SEC E-Institute: Shanghai High Institutions Grid" project & "SEC project: No. 03HK07.

Please use the following format when citing this chapter:
He, Yu'an, Yu, Tao, Shen, Bin, 2006, in International Federation for Information Processing (IFIP), Volume 207, Knowledge Enterprise: Intelligent Strategies In Product Design, Manufacturing, and Management, eds. K. Wang, Kovacs G., Wozny M., Fang M., (Boston: Springer), pp. 908-913.

resource management method in MG is more complicated than that in computing grid.

An effective Manufacturing Grid Resource Management System (MG-RMS) is crucial. Document 3 has ever proposed a MG-RMS which can solve the resource sharing and collaboration well under the heterogeneous and distributed environment. But when MG deals with complex manufacturing tasks, resource consumers put forward more requirements for manufacturing service level and manufacturing resources, the MG-RMS can not support sufficiently. So a WSRF (Web Service Resource Framework)[4]-based Resource Management System is proposed in this paper. WSRF is the open criterion of the next generation of grid technology, and it defines the sharing mechanism and management mechanism of grid software and hardware resources.

2. WSRF-BASE MG-RMS

2.1 Concept Model

Concept Model is the abstract of design, i.e., it is the visualization representation of design idea. According to function roles, there are three main roles related to resource management in MG: resource providers, resource consumers and Manufacturing Grid Information System (MGIS). The relations between the three roles in concept model are shown in Fig.1. Their definitions are as follows:

Resource providers: they are the principal bodies controlling manufacturing resources, and provide some resources to outside in the service manner.

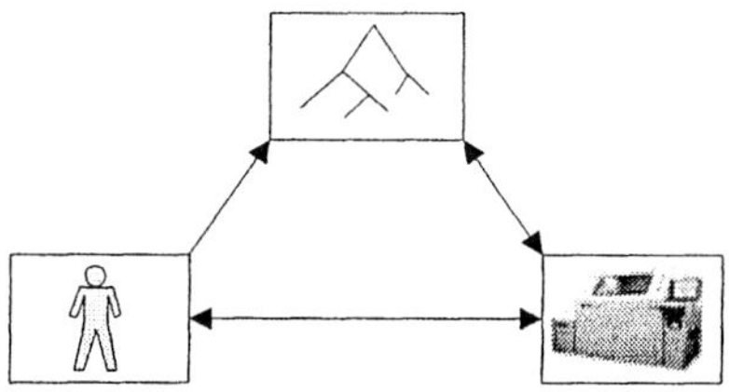

Figure 1. Concept model of MG resource management

MGIS: MGIS is responsible for receiving, sending, indexing and archiving the attributes related to MG-Resources, and dealing with the state information.

Resource consumers: they are the users, usually the members of Visual Organization (VO) or customers using resources.

In MG environment, MG-Resource is the model of manufacturing resources encapsulated in MG. In order to realize resource sharing and collaboration among the heterogeneous and geographical distributed manufacturing resources, manufacturing resource must be encapsulated with uniform criterion, and deployed to MGIS.

2.2 System Model

There are varieties kinds of manufacturing resources, which have great differences among them. In order to solve the geographical distributed sharing problem of manufacturing resources, MG-RMS is developed.

The system model of MG-RMS is shown in Fig.2. It is mainly composed of WSRF encapsulation module, MG-Resource, Manufacturing Grid Information System (MGIS) and task manager in a resource node, and its application module includes security management, task decomposing, Manufacturing Grid Resource Scheduler (MGRS), workflow manager, MG portal and MG-QoS (Quality of Service)manager.

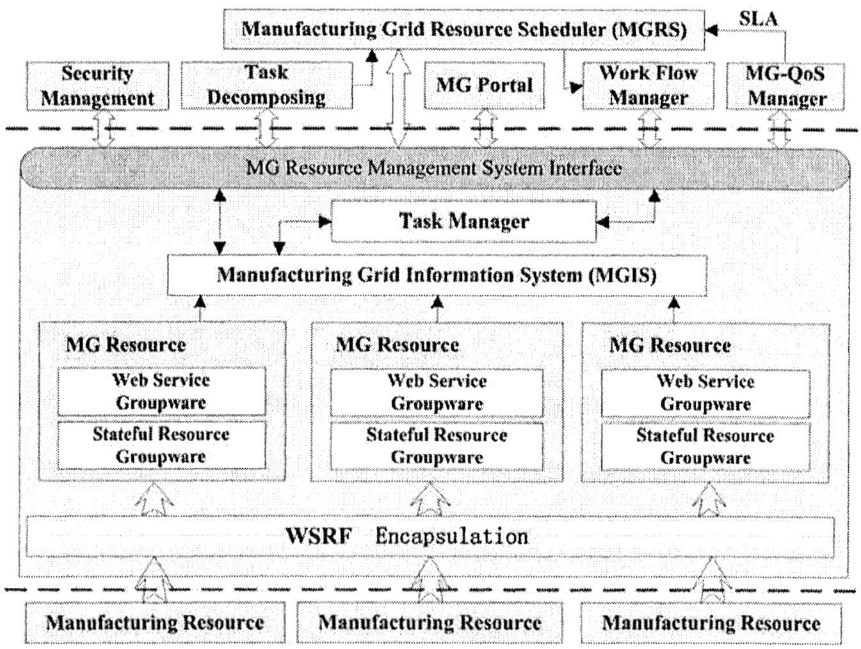

Figure 2. The system model of MG-RMS

1) WSRF encapsulation module: According to the requirements of MG-RMS, WSRF encapsulation module uses WSRF criterion to encapsulate manufacturing resources into MG-Resources, publishes them to the MG platform so as to be invoked by consumers who need these resources.

2) MG-Resource: MG-Resource is the mapping and representation of actual manufacturing resource in MG. All the resources which hope to share in MG need to be encapsulated into MG-Resources and published on the MG platform. Manufacturing resources can be all the software and hardware related to the product lifecycle, e.g., a machine tool, a set of CAD software, even an expert or machining personnel.

3) MGIS: MGRIS transfers attribute information of manufacturing resources in MG. It collects the attribute information (including dynamic and static attribute) then publishes them to consumers according to the requirements, so as to choose a proper manufacturing resource to fulfill the manufacturing task. There are several modes to publish information, and resource providers can choose the mode which they need.

4) MG-RMS interface: the application modules, e.g., security management, MGRS, workflow manager and QoS manager, exchange information through the interface. The interface is by means of Web service to exchange information with these applications.

5) Task manager: Task manger receives the task from work flow manager, chooses manufacturing resources to work in MG-RMS according to task's description, and submits to work flow manager after the task is accomplished.

6) Application modules

MGRS: its functions are mainly composed of QoS-based resource and scheduling, dynamic negotiation, resource allocation and fault-tolerant handling.

MG portal: it publishes MG-Resource attribute information to consumers by HTML (HyperText Markup Language) pages.

Workflow module: it invokes a certain MG-Resource and implements the manufacturing task in appropriate time, and submits the task to workflow after the task is accomplished.

Security module: it monitors the reliability and dynamic attribute of MG-Resource by invoking web service groupware.

Task decomposing module: it is in charge of decomposing a complicated task into many subtasks, so as to search and match resources[5].

MG-QoS manager: it is the management module of MG Quality of Service (QoS), and its functions include management of Service Level Agreement (SLA), resource reservation and management of QoS strategy.

The WSRF encapsulation module is the crucial module, so the paper mainly presents the encapsulation method.

3. RESOURCE-ENCAPSULATION MOTHOD

The WSRF encapsulation module in MG-RMS encapsulates the actual manufacturing resources into MG-Resources by using WSRF criterion.

MG-Resource is a kind of WS-Resource-structure resource, it includes two modules: web service and stateful resource. Web service can realize the function of MG-Resource, provide uniform and standard external invocation interfaces, and make customers invoke MG-Resources by sending standard web service. The stateful resources can save the attribute information about manufacturing resources, publish the information to customers, and make them discover manufacturing resources which are suitable for machining the manufacturing task[6].

Thus, encapsulating manufacturing resources to MG-Resources includes two parts: web service encapsulation and statful resource encapsulation. A WSRF-based MG resource encapsulation structure is shown in Fig.3.

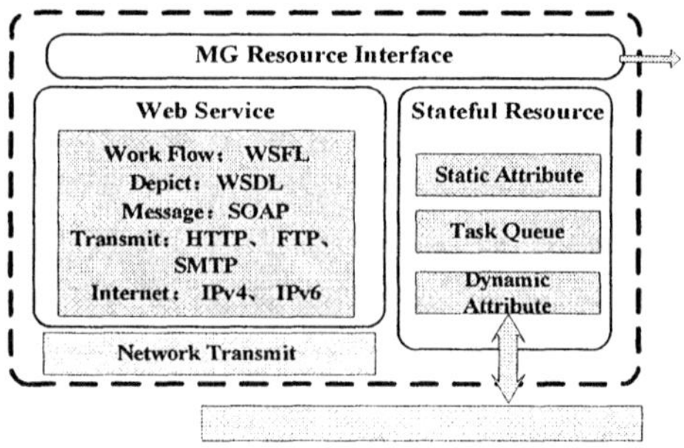

Figure 3. The resource encapsulation model

3.1 The Encapsulation of Stateful Resource

According to the characteristics of attribute information, manufacturing resource information can be divided into three types of information, that is, static attribute information, dynamic attribute information and task queue information, which are crucial for the resource evaluation and scheduling.

Since the quantities of manufacturing resources published in a local MG-RMS are limited, the stateful resources are implemented to records in the Microsoft Access database.

3.2 The Encapsulation of Web Service

The web service module of MG-Resource takes charge of the realization of the functions of manufacturing resources, and management of attribute information of manufacturing resources. Web service is a group of objects deployed on web, provides standard external interface format by WSDL (Web Service Description Language) document. Applicants may send standard SOAP (Simple Object Access Protocol) message to web service so as to invoke these methods.

The encapsulation of web service may carry out according to following steps: 1) Defining interfaces of web service. 2) Creating WSDL document according to the defining interface. 3) Realization of interface document. 4) Compiling the deploying document and publishing the web service on the Axis Server.

4. CONCLUSION

As has been noted, the WSRF-based MG-RMS model constructed by using WSRF criterion can describe, encapsulate and publish manufacturing resources, and the resource is encapsulated to a WS (Web Service)-Resource structural MG-Resource. The MG-RMS can solve resource sharing problem among the heterogeneous and geographical distributed manufacturing resources, make consumers invoke the resources through the standard interface conveniently, as if the manufacturing resources are in local place.

5. REFERENCES

1. Liu Lilan, Yu Tao, Shi Zhanbei. Research of Self-Organization Manufacturing Grid and its Task Scheduling Algorithm, *Computer integrated manufacturing system (CIMS)*, 2003, 9 (6): 449-455
2. Zhou Guanghui, Jiang Pingyu. The Encapsulation and Integration of Networked Manufacturing Resource Based on Mobile Agent. *Computer Integrated Manufacturing System (CIMS)*. 2002, 9(8): pp 530-538
3. LIU Lilan, YU Tao, SHI Zhan. Research on QoS-based Resource Scheduling in Manufacturing Grid, *Computer Integrated Manufacturing System (CIMS)*. 2005, 11(4): pp 475-480
4. The WS-Resource Framework. Available: http://www.globus.org/wsrf/specs/ws-wsrf.pdf
5. He Yu'an, Yu Tao, Liu Lilan, et al. A QoS-based Task Decomposing Middleware Research for Manufacturing Grid. *Proceedings of the First International Conference on Semantics, Knowledge and Grid(SKG05)*, Beijing, China, 2005, 513-519
6. He Yu'an, Yu Tao, Liu Lilan, et al. A WSRF-based Resource Management System of Manufacturing Grid. *Proceedings of 6th IEEE International Symposium on Cluster Computing and the Grid (CCGrid06)*, Singapore, 2006

THE PERFORMANCE ANALYSIS OF A MULTI-OBJECTIVE IMMUNE GENETIC ALGORITHM FOR FLEXIBLE JOB SHOP SCHEDULING

XiuLi Wu , ShuDong Sun, GangGang Niu, YinNi Zhai
Departerment of Mechanical and Electronic Engineering, Northwestern Polytechnical University, Xi'An, 710072, China; Email: henanwxl@hotmail.com

Abstract: First, a multi-objective immune genetic algorithm integrating immune algorithm and genetic algorithm for flexible job shop scheduling is designed. Second, Markov chain is used to analyze quantitatively its convergence. Third, a simulation experiment of the flexible job shop scheduling is carried out. Running results show that the proposed algorithm can converge to the Pareto frontier quickly and distribute evenly along the Pareto frontier.

Key words: flexible job shop scheduling, multi-objective genetic algorithm, convergence, Markov chain.

1. INTRODUCTION

Flexible job shop multi-objective scheduling problem (FJSP) is typical NP hard, which is not easy to solve with a polynomial algorithm. Therefore, evolutionary algorithms, such as genetic algorithm (GA) and immune algorithm (IA), are the efficient procedures. Many literatures presented different multi-objective optimization methods. But the theoretical analysis of the algorithm itself is sparse. Those focusing on the subject mostly concentrate on GA with binary encoding [1], while research on GA with decimal encoding is little. Rudolph's research [2, 3, 4], who is the most contributing scientist, mainly focused on the effect of the elitist strategy on the convergence. The convergence of the hybrid GA taking into more factors, such as the immune technology, has not been discussed in the publicly published literatures. Therefore, a multi-objective immune genetic algorithm (MOIGA) for FJSP is presented and its performance is analyzed.

This project is by the National High-Tech. R&D Program for CIMS, China under the grant No 2003AA411110 and by Doctoral Fund of Ministry of Education of China under the grant No 20040699025

Please use the following format when citing this chapter:
Wu, XiuLi, Sun, ShuDong, Niu, GangGang, Zhai, YinNi, 2006, in International Federation for Information Processing (IFIP), Volume 207, Knowledge Enterprise: Intelligent Strategies In Product Design, Manufacturing, and Management, eds. K. Wang, Kovacs G., Wozny M., Fang M., (Boston: Springer), pp. 914-919.

2. A MULTI-OBJECTIVE IMMUNE GENETIC ALGORITHM FOR SCHEDULING

Let < **P(t),** $\prec$ > be the non-dominated set of the set P(t), P(t) the set of the population, Q(t) the Pareto set of P(t), and S_i the sharing degree of the ith chromosome when processed by the niche technology. The procedure of MOIGA is as follows.

```
P(0) is randomly generated;
Q(0) is the set for Pareto set of P(0);
t=1;
repeat
  if t>=T then end repeat;
  f(X_j)= f(X_j)/S_j;
  get bacterin from P(t-1);
  P(t) = generate(P(t-1));
  crossover(P(t));
  mutation(P(t));
  vaccination(P(t));
  If ∃a ∈ Q(t−1) and ∃x ∈ P(t) and f(x) ≺ f(a) then
  Q(t) ← x, P(t) ← a;
  else if ⟨P(t), ≺⟩ ⊆ Q(t−1) then
  Q(t)=Q(t−1);
  t=t+1;
end.
```

3. ANALYSIS OF THE CONVERGENCE

Now let us describe the proposed MOIGA in terms of Markov Chain. Let $\boldsymbol{\Omega}$ be the finite state space and X ($X \in \boldsymbol{\Omega}$) the individual. All populations cover the whole state space $\boldsymbol{G}$, in which a population represents a state. In the operation-based encoding, encoding character set $\boldsymbol{\lambda}$ is composed of the job numbers. The number of the element C in the set $\boldsymbol{\lambda}$ equals the total number N of the jobs, i.e. $C = N$. Let the length of each chromosome be L, then $L = N \times \max\{n_j\}$. If the population size P is given and finite, then the dimension of $\boldsymbol{\Omega}$ and $\boldsymbol{G}$ is finite and $|\boldsymbol{G}| = |\boldsymbol{\Omega}|^P$. The evolutionary procedure of GA is repeating, i.e. the population transfers from one state to another state after the selection, crossover, mutation and vaccination operation. According to [6], we know the process is a homogeneous Markov Chain with a finite state space.

1) **The crossover operation**

MOIGA adopts linear order crossover operation. Let the transition matrix determined by crossover operation be $\boldsymbol{C}=(c_{ij})_{|G|\times|G|}$, and c_{ij} the probability from the state i to j after the crossover operation. Because one state is transferred to another state at all events, there exists $\sum_{j=1}^{|G|} c_{ij}=1$. According to [6], we know the matrix $\boldsymbol{C}$ is random.

2) **The mutation operation**

The mutation of MOIGA adopts exchanging operation method. Let the transition matrix determined by the mutation operation be $\boldsymbol{T}=(t_{ij})_{|G|\times|G|}$ and t_{ij} the probability from the state i to j after the mutation operation. Let the hamming distance between the state i and j be H_k, k=1, 2..., P. For example, in a FJSP problem with 3 jobs and 2 operations for each, its two states, i.e. the state i and j, are as follows:

s_i= {(1 2 3 3 2 1), (2 1 3 1 2 3), (3 2 1 1 2 3)}

s_j= {(2 1 3 3 2 1), (2 1 2 1 3 3), (3 2 1 3 2 1)}

Then the hamming distance between them is $H_1=H_2=...=H_P=2$.

Each gene of a chromosome has the equal mutation probability p_m ($0<p_m<1$). Therefore

$$t_{ij}=\prod_{i=1}^{P}\left(\frac{p_m}{C-1}\right)^{H_i}(1-p_m)^{L-H_i}=\left(\frac{1}{N-1}\right)^{2P}p_m^{2P}(1-p_m)^{NP\times\max\{n_j\}-2P}>0$$

According to [6], the matrix $\boldsymbol{T}$ is positive and random.

3) **The vaccination operation**

The vaccination of MOIGA is another mutation operation. Let the transition matrix determined by the vaccination operation be $\boldsymbol{T}=(t_{ij})_{|G|\times|G|}$ and t_{ij} the probability from the state i to j after the mutation operation. Each chromosome has the equal vaccination probability p_h ($0<p_h<1$). Therefore

$$h_{ij}=\prod_{i=1}^{P}\left(\frac{p_h}{C-1}\right)^{H_i}(1-p_h)^{L-H_i}=\prod_{i=1}^{P}\left(\frac{p_h}{N-1}\right)^{H_i}(1-p_h)^{N\times\max\{n_j\}-H_i}>0$$

According to [6], the matrix $\boldsymbol{T}$ is positive and random. Therefore, $\boldsymbol{Z=TH}$ is strictly positive.

4) **The selection operation**

Before selection, MOIGA adopts the elitist strategy, i.e. the Pareto optimization solutions will enter into the next generation directly. Recall the probability transition matrix of this operation as $\boldsymbol{E}=(e_{ij})_{|G|\times|G|}$ and the kth chromosome as π_k. Put the Pareto set of the current population in the front of the next generation population and recall them as $\pi_1, \pi_2, ..., \pi_r$. Therefore, the state space size extends to $|\boldsymbol{\Omega}|^r$ times, i.e. $|\boldsymbol{G}|=|\boldsymbol{\Omega}|^{P+r}$. If there is at least a chromosome in the set $P(t+1)$ which can dominate one (or some) in the set $Q(t)$, then replace it (or them). Therefore, the transition probability from the state *i* to *j* is:

$$e_{ij} = \begin{cases} 1 & \text{if } \exists a \in Q(t) \text{ and } \exists x \in P(t+1) \text{ and } f(x) \prec f(a) \\ 1 & \text{if } \langle P(t+1), \prec \rangle \subseteq Q(t) \\ 0 & \text{else} \end{cases}$$

Order the states with the same Pareto set as the original status and order other states according to the order of the Pareto set. Therefore, the matrix $\boldsymbol{E}$ is under-triangle matrix, the element in the matrix $\boldsymbol{E}$ is zero or one, there is only a one in each row, and $\boldsymbol{E}_{11}$ is an identity matrix.

$$E = \begin{bmatrix} E_{11} & & & \\ E_{21} & E_{22} & & \\ \vdots & \vdots & \ddots & \\ E_{|\Omega|^r 1} & E_{|\Omega|^r 2} & \cdots & E_{|\Omega|^r |\Omega|^r} \end{bmatrix}$$

According to [6], we know that the matrix $\boldsymbol{E}$ is a random matrix.

After pre-selection, roulette selection is used to select the outstanding chromosomes to the next generation from the set $P(t)$. Recall the transition matrix of the selection operation as $\boldsymbol{S} = (s_{ij})_{|G|\times|G|}$ and s_{ij} is the probability from the state *i* to *j*. In the roulette selection, there exists

$$s_{ij} = \prod_{j=1}^{P} \left(\frac{f(X_j)/S_j}{\sum_{k=1}^{P} f(X_k)/S_k} \right) > 0$$

According to [6], we know that $\boldsymbol{S} = (s_{ij})_{|G|\times|G|}$ is a column allowable random matrix and that $\boldsymbol{CZS} > \boldsymbol{0}$ .After the elitist strategy, the state space extends to $|\boldsymbol{\Omega}|^r$ times, the transition matrices of the crossover, the mutation and the selection change into the matrices $\boldsymbol{C}^+$, $\boldsymbol{T}^+$ and $\boldsymbol{S}^+$, individually.

$$C^+ = \begin{bmatrix} C & & & \\ & C & & \\ & & \ddots & \\ & & & C \end{bmatrix}, \quad Z^+ = \begin{bmatrix} Z & & & \\ & Z & & \\ & & \ddots & \\ & & & Z \end{bmatrix}, \quad S^+ = \begin{bmatrix} S & & & \\ & S & & \\ & & \ddots & \\ & & & S \end{bmatrix}$$

Therefore, the transition probability matrix is as follows according to the evolutionary process of MOIGA and the Chapman-Kolmogorov equation.

$$P^{+} = C^{+}Z^{+}ES^{+}$$

$$= \begin{bmatrix} C & & & \\ & C & & \\ & & \ddots & \\ & & & C \end{bmatrix} \begin{bmatrix} Z & & & \\ & Z & & \\ & & \ddots & \\ & & & Z \end{bmatrix} \begin{bmatrix} E_{11} & & & \\ E_{21} & E_{22} & & \\ \vdots & \vdots & \ddots & \\ E_{|\Omega|^r 1} & E_{|\Omega|^r 2} & \cdots & E_{|\Omega|^r|\Omega|^r} \end{bmatrix} \begin{bmatrix} S & & & \\ & S & & \\ & & \ddots & \\ & & & S \end{bmatrix}$$

$$= \begin{bmatrix} CZE_{11}S & & & & & \\ CZE_{21}S & CZE_{22}S & & & & \\ \vdots & \vdots & & \ddots & & \\ CZE_{i1}S & \cdots & CZE_{ij}S & \cdots & CZE_{ii}S & \\ \vdots & & \vdots & & \vdots & \ddots \\ CZE_{|\Omega|^r 1}S & CZE_{|\Omega|^r 2}S & & \cdots & & CZE_{|\Omega|^r|\Omega|^r}S \end{bmatrix}$$

Let

$$R = \begin{bmatrix} CZE_{21}S \\ \vdots \\ CZE_{i1}S \\ \vdots \\ CZE_{|\Omega|^r 1}S \end{bmatrix} \quad W = \begin{bmatrix} CZE_{22}S & & & \\ \vdots & \ddots & & \\ CZE_{i2}S \cdots & CZE_{ik}S \cdots & CZE_{ii}S & \\ \vdots & \vdots & \vdots & \ddots \\ CZE_{|\Omega|^r 2}S & & \cdots & CZE_{|\Omega|^r|\Omega|^r}S \end{bmatrix}$$

Because the product of CZS is positive, and E_{11} is an identity matrix, it can be concluded that $P = CTE_{11}S$ is positive with the state space size of $|\Omega|^r \times |\Omega|^r$. Then $P^{+} = \begin{bmatrix} P & O \\ R & W \end{bmatrix}$.

Hereinto, there is at least a nonzero in each row of the matrix R. According to [6], those states without the inclusion of optimal chromosome have zero probability in the limited distribution of Markov Chain; while the limited distribution sum of those states with the optimal chromosome equals to 1. Thus it can be seen that MOIGA can converge to the true Pareto optimal set with the probability 1.

4. AN EXPERIMENT

To prove the performance of the proposed MOIGA, a practical scheduling problem in an aeronautic company with two objectives to be optimized, i.e. production time and production cost. The encoding method is operation-based way. The experiment result is as follows: the two objectives converge to the stable distribution with the evolutionary process (Fig 1). In figure 1, the Pareto frontier is shown with red asterisks and the convergence and diversity are assured perfectly.

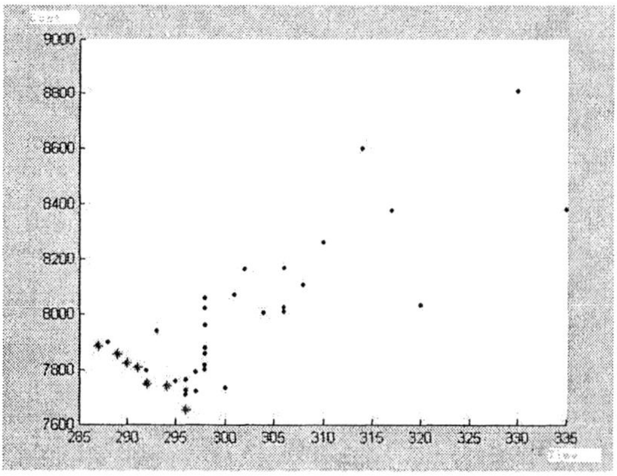

Figure 1 The printed Pareto frontier plot

5. CONCLUSION

The paper analyzed a multi-objective immune genetic algorithm and proofed that it can converge to the Pareto optimization set with the probability one. The running result of a practical problem shows the effectiveness and efficiency of the proposed algorithm.

6. REFERENCES

1. Laumanns M, Thiele L, Zitzler E, et al, (2002), Archiving with Guaranteed Convergence and Diversity in Multi-Objective Optimization, *CECCO*, pp. 439-447.
2. Rudolph G, Agapie A, (2000), Convergence Properties of Some Multi-Objective Evolutionary Algorithms. Piscataway, *Proceedings of the Congress on Evolutionary Computation*, pp. 1010-1016.
3. Rudolph G, (1994), Convergence of non-elitist strategie, *NJ. IEEE Press: Proceedings of the First IEEE Conference on Evolutionary Computation*, pp. 63-66 .
4. Rudolph G, (1998), Evolutionary search for minimal elements in partially ordered finite sets. Piscataway, Springer, Berlin: *Proceedings of the Seventh Annual Conference on Evolutionary Programming,* pp. 25-27.
5. Zhou Ming, Sun Shudong, (2002), *Genetic algorithm and its application,* Beijing: the National Defense Publishers, pp. 75-76.
6. Xing WenXun, Xie JinXing, (1999), *Modern optimization computing algorithms.* BeiJing: QingHua University Publishers, pp. :156-158

TOWARDS SERVICE ORIENTED ENTERPRISE

Sodki Chaari[1,2], Frédérique Biennier[1], Chokri Ben amar[2], Joël Favrel[1]
[1]PRISMa, INSA de Lyon, France; Email: sodki.chaari@insa-lyon.fr. [2]REGIM, Ecole Nationale d'Ingénieurs de Sfax, Tunisia

Abstract: To fit economical constraints, enterprises are more and more involved in collaborative strategies and require agile organization. To support such agile organizations, an efficient and agile IT system should be set. This can be partially achieved thanks to Service Oriented Architectures (SOA) that are able to implement opened and agile information system. The collaborative model we built on this approach is based first on the services exposed by each potential partner and then on a convenient assembly process used to build a cooperation process, designed as a services composition. This architecture impacts the enterprise own organization, leading to a service oriented enterprise (SOE). After describing the context we present the SOE framework and its components. Then the assembly process is described.

Key words: Service oriented architecture, collaborative business infrastructure, interoperability, semantic, policies, ontology

1. INTRODUCTION

Due to the economic pressure and to fit the agility level required by the changing environment, enterprises have to adapt their internal organization and to develop new business opportunities, often leading to Collaborative Business. These collaborative organizations, that can be seen as adaptive and on-demand virtual enterprises (Kwon et al. 2003) involve that enterprises integrate their business processes into a macro cooperative process to achieve common business goals in a more flexible and effective way. This trend has placed an increased emphasis on extending enterprise internal efficiency across their boundaries through inter-enterprise cooperation that is able to answer dynamically to user demands and deal with complex market

---

Please use the following format when citing this chapter:
Chaari, Sodki, Biennier, Frédérique, Ben amar, Chokri, Favrel, Joël, 2006, in International Federation for Information Processing (IFIP), Volume 207, Knowledge Enterprise: Intelligent Strategies In Product Design, Manufacturing, and Management, eds. K. Wang, Kovacs G., Wozny M., Fang M., (Boston: Springer), pp. 920-925.

requirements. These collaborative organizations involve designing agile and inter-operable information systems. To support the necessary openness, one can use Service Oriented Architecture (SOA) to implement efficiently the information systems. Coupled to ontology management, dedicated infrastructures can be built by combining services. Nevertheless, these approaches are often fragmentary so that services combination is taken into account only according to a technical point of view.

In this paper, after describing the current context, we propose a multi-level framework that can be used to improve service combination. Our approach, based on SOA, consists in a set of dynamically combined autonomous services that collectively provide a value-added service to users leading to a Service Oriented Enterprise organization (SOE).

2. CONTEXT

Setting a collaborative infrastructure to support C-Business involves paying a particular attention on inter-operability constraints. This leads to adapt the traditional information system organization to support the necessary openness. Emerging technologies such as the Service Oriented Architecture(SOA) (Baglietto et al., 2005), ontology (Yang et al., 2005) are generally perceived as core technologies to successfully deal with these challenges.

Introducing SOA for enterprises information system can reduce dramatically the funds required to start up a business. Services are readily available for integration and orchestration. Moreover, while each service has a machine understandable representation, discovering the convenient service could be achieved in a dynamic and automated way. Hence, enterprises will be able to form on-the-fly, project-driven and on demand alliances. As the number of services increase, a particular attention must be paid on an automated service combination.

Nevertheless, the service definition varies from one company to another as well as the granularity level of service, which can be associated to a whole workflow made of several tasks, or to a simple elementary task. This means that enterprise must "re-think" their business process organization to link them with these services. To reach this goal, enterprises can use an object oriented approach, modelling and decomposing recursively their business processes into a set of business objects that are turned into appropriate services either by adding them a convenient interface or by associating them to composed services. By this way, the IT support and the associated services are "prepared" both for an internal and an external use.

Moreover, bringing the necessary inter-operability involves also taking semantic constraints into account. This can be solved by using ontology. Basically, ontology is defined as an "explicit specification of a conceptualization" (Hu and Du, 2004). It clarifies a terminology shared by one community. Consequently, ontology can be worthy used to organise services and to allow an effective services discovery starting from requests and ensure a good management of available services area.

After discovering the convenient services, consistency controls must be implemented while orchestrating these services. Academic and industrial research has provided several approaches to reach this goal: (Medjahed et al., 2003) presents a framework for orchestrating atomic services semantically described (using non-functional properties such as function, category and quality) while (Fu et al., 2002) proposes a 3-step process including model checking, a service composition model (Mealy machines) and an automatic composition based on finite state machines. Of course, the automatic services composition can be seen as an "ultimate" goal, taking execution models into account. While (Berardi et al., 2005) presents a framework describing services' behaviour as an external execution tree translated into finite state machines, (Hamadi and Benatallah, 2003) associates a Petri net to each service in order to describe its behaviour and defines some composition operators like sequence, choice, iteration, parallel with communication, discriminator, selection, and refinement.

Nevertheless, these approaches focus on particular technical points. To support dynamic service binding, leading to dynamic virtual organizations, a more complex framework is required.

3. SERVICE ORIENTED ENTERPRISE ARCHITECTURE

As we introduced the Service Oriented Enterprise as a set of on demand well-combined services, a particular attention must be paid on the service selection and combination to set a final cooperation process fitting SOE's goals. Many research works have focused on defining compatibility rules between services. Most of them define syntactic compatibility of input and output parameters and also compatibility rules based on semantic (Medjahed and Bouguettaya, 2005), on behaviour (Bordeaux et al., 2004) or on quality of service requirements (Cardoso et al., 2004). Distribution, dynamicity and heterogeneity of the services can prevent any attempt to understand their semantic. Thus, allowing efficient service sharing needs an adapted infrastructure that must provide flexible tools to organise discovery services, communication facilities, content description, and reconciliation of incompatible syntax and semantics. Combining ontology with services will empower services with semantics and let us having an appropriate organization of our service area. The core of our model is the organization of service area into different service groups called *domain* (according to their functionalities) which are associated to dedicated ontology and their value-added. A domain is defined by a 3-tuple: *domain= (Id, Member, VS)* where *Id* is the domain identifier, *Member* is the set of services suppliers whose publish their services in a particular domain and $VS = \{sv_1 \ldots sv_n\}$ is the set of virtual services published in one domain. Then, each group is associated to ontology. As service can be related to different domains, ontology, relationships are set to improve the service searching process.

To fit these requirements, we propose to organise the services into a multi-level framework so that different compatibility rules can be set to control the service combination, taking into account both the enterprise policy, semantic constraints and operational assembly rules.

3.1 Framework organization

The framework we propose is based on 3 levels (figure 1):

- **The cooperative process level**: is associated to the enterprise policy description and is used to select the suitable services for the SOE creation. Different types of policies related to each service and global process will be also identified. We call services and policies sets the process schema (PS), described with a 3-tuple: *PS= (SD, PSD, PGC)* where *SD* is the set of service domains. *PSD* is the set of policies related to every service domain. PGC represent the set of global policies associated to the SOE process.This PS schema supports the service selection and combination.
- **The virtual services level** is designed to improve the concrete service searching process, taking into account the different semantic domains and the related ontology. More precisely, this service is in charge of checking the compatibility of services belonging to different levels (process cooperation and basic concrete services). Virtual service is defined as: *vs = {In, Out, Function, description}* where: *In* and *Out* are input and output variables respectively, *Function= (Functionality, synonyms)* where *functionality* is used to define what the service can do and *synonyms* presents the set of synonym functionalities, *description* indicates the set of descriptors which can be associated to this service.
- **The concrete services level** is associated to the real IT services. While virtual services are associated to "interface" components (as defined in

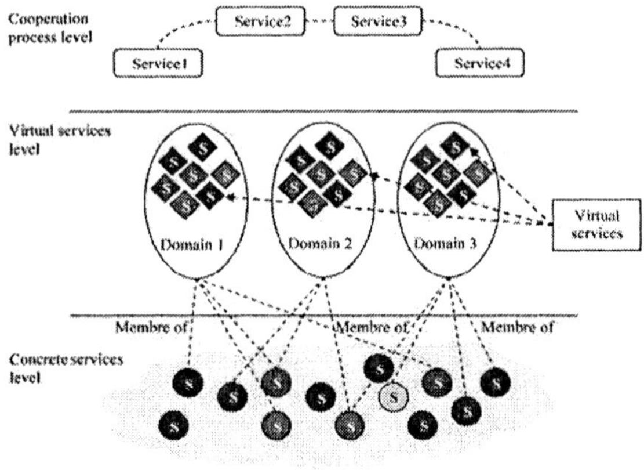

Figure 1. The three level of a service oriented enterprise

the SOA organization), we still have to select the convenient concrete service implementation. This checking process takes both rules and protocols to guide how basic services can be combined to achieve a complex service, fitting the interface requirements defined by the virtual service level.

3.2 Service combination control

Based on this 3 level architecture, our service combination process consists in the following steps:

- *Step 1*: identify the set of virtual services that will be chosen based on the process schema description of the service oriented enterprise and the domain ontology.
- *Step 2*: policies of each service domain extracted from the process schema are associated to the whole selected virtual services sets.
- *Step 3*: based on the set of virtual services defined previously, we try to match each virtual service with concrete services. This task is improved thanks to the different service domains classification. Consequently, the search and match process use syntactic, semantic, and policies compatibility rules. This process implements "vertical" compatibility rules since it verifies inter-levels compatibility and it provides a set of adapted concrete services associated to the virtual services.
- *Step 4*: based on last step output we check whether if the selected concrete services can be composed according to static, semantic and policy rules. We obtain numerous service sequences associated to execution plans. We call this compatibility checking "horizontal compatibilities".

- *Step 5:* having several execution plans, this last step is used to select only one execution plan according to several criteria as global policy, QoS requirements...

4. CONCLUSION

In this paper, we present how a service oriented enterprise can support inter-enterprise cooperation. This architecture, based on the IT SOA, is organised into different levels (business objects used to model enterprise process are associated to virtual services and finally to concrete services) so that the global infrastructure is closely related to the enterprise business strategy and exhibit convenient agility abilities. We propose to add vertical and horizontal combination rules so that the orchestration process provides a consistent service combination.

5. REFERENCES

Baglietto, P., Maresca, M., Parodi, A., and Zingirian, N., 2005, Stepwise deployment methodology of a service oriented architecture for business communities, *IST* **47**:427-436.

Berardi, D., Calvanese, D., and Mecella, M., 2005, Composition of Services with Nondeterministic Observable Behavior, in: *Proceedings of the 3rd conference on Service-Oriented Computing*, B. Benatallah, ed., Springer-Verlag, Amsterdam, pp. 43-58.

Bordeaux, L., Salaun, G., Berardi, D., and Mecella, M., 2004, When are two web services Compatible, in: *Proceedings of the 5th conference on Technologies for E-Services*, M. C. Shan and U. Dayal, eds., Springer-Verlag, Toronto, pp. 15-28.

Cardoso, J., Sheth, A., Miller, J., Arnold, J., and Kochut, K., 2004, Quality of service for workflows and web service processes, *WS* **1**(3),pp.281-308.

Fu, X., Bultan, T., and Su, J., 2002, Formal verification of e-services and workflows, in: *Proceedings of the workshop on Web Services, E-Business, and the Semantic*, C. Bussler, R. Hull, and S. McIlraith, eds., Springer-Verlag, Toronto, pp. 188-202.

Hamadi, R., and Benatallah, B., 2003, A Petri net-based model for web service composition, in: *Proceedings of the 14th Australasian Database Conference*, Adelaide, pp. 191-200.

Hu, H., and Du, X., 2004, Adopting Ontologies and Rules in Web Searching Services, in: *Proceedings of 1st symposium on Computational and Information Science*, J. Zhang, J. He, and Y. Fu, eds., Springer-Verlag, Shanghai, pp. 1047-1053.

Kwon, Y., Lee, H. K., Lee, S., and Lee, J., 2003, The virtual enterprise: redefining the concept, *Lecture Notes in Computer Science*, volume 2713, Springer-Verlag, pp. 249-258.

Medjahed, B., and Bouguettaya, A., 2005, A multilevel composability model for semantic web services, *TKDE* **17**(7),pp. 954-968.

Medjahed, B., Bouguettaya A. and Elmagarmid, A.K., 2003, Composing Web services on the Semantic Web, *VLDB* **12**(4),pp.333-351.

Yang, H., Li, Z., Chen, J., and Xia, H., 2005, Web services composition based on ontology and workflow, in: *Proceedings of the 6th conference on Advances in Web-Age Information Management*, W. Fan, Z. Wu, and J. Yag, eds., Springer-Verlag, Hangzhou, pp. 297-307.

MANUFACTURING RESOURCE MANAGEMENT FOR COLLABORATIVE PROCESS PLANNING

Ziqiang Zhou [1], Mujun Li[2], Liguan Shen [2]
[1]*Mechanical Department, Changshu Institute of Technology, Changshu, China; Email:zzq_hefei@163.com.* [2]*Department of Precision Machinery and Precision Intrumentation, University of Science & Technology of China, Hefei, China.*

Abstract: A framework of collaborative process planning (CPP) platform is introduced. For this purpose, manufacturing resource management system should be constructed. The main modular of them are discussed, which include equipment information management, fixture information management, cutting tool information management, process parameter and NC code information management. In the end, the application of how to work on the process planning in virtual enterprise with manufacturing resource based on the CPP is introduced.

Key words: Manufacturing resource, Collaborative, Process planning, Simulation.

1. INTRODUCTION

With competition in global markets, more and more enterprises seek to make up virtual enterprise or cooperate with each other in the manufacturing process in order to reduce cost of product and increase competitive ability. For this reason, some of the key technologies are developing in the field of collaborative work between enterprise, besides collaborative design, the collaborative process planning (CPP) also very important among them [1-3]. With CPP platform engineers whom belong to the different enterprise may be work more efficiently than before, and the manufacturing resource of virtual enterprise will be utilized more optimized. So a distributed manufacturing resource management system is need in the CPP platform.

Please use the following format when citing this chapter:
Zhou, Ziqiang, Li, Mujun, Shen, Liguan, 2006, in International Federation for Information Processing (IFIP), Volume 207, Knowledge Enterprise: Intelligent Strategies In Product Design, Manufacturing, and Management, eds. K. Wang, Kovacs G., Wozny M., Fang M., (Boston: Springer), pp. 926-931.

2. FRAMEWORK OF COLLABORATIVE PROCESS PLANNING PLATFORM

The framework of CPP platform shown in the Figure.1, which is a multi-agent, based collaborative system. Each partner of the virtual enterprise has a manufacturing resource agent which includes equipments information, cutting tools information, material information and etc. On the sever side, a multi-user collaborative sever program is running for the entire request from the client side. It composed of collaborative process planning modular, collaborative NC simulation modular and collaborative discussion modular. The engineers are operating on the client side to do process planning work, they can collaborative discuss the virtual entity of the part, make label on the 2D drawing of the part, and select suited equipment or cutting tools from the manufacturing resource agent. In the fixture design process, engineers can query from the manufacturing resource agent and get needed part or assembly. Each manufacturing resource agent can transform the data to others automatically, and the new work of engineer will be uploading to the manufacturing resource agent after checkup. Finally, the process engineers can make optimized manufacturing process of the part. With NC collaborative simulation modular, engineers can discuss and evaluate the NC program on the Internet collaboratively. With CPP platform the young engineer will be advantaged from the system and other professional engineer, on the other hand, the partners of the virtual enterprise will be benefited from the sharing manufacturing resource.

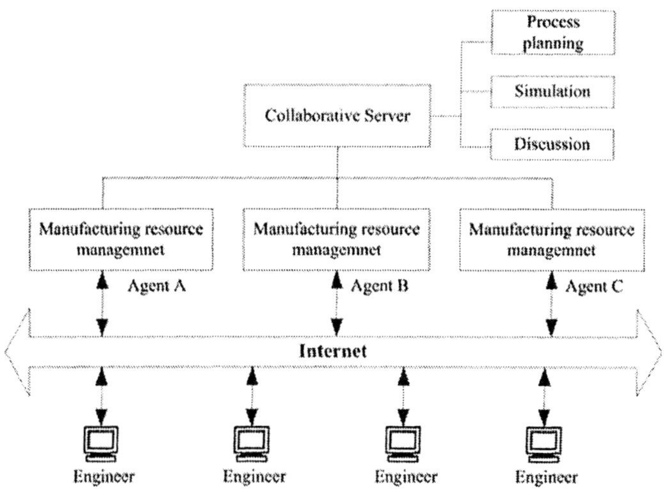

Figure 1. Framework of collaborative process planning platform

3. STRUCTURE OF MANUFACTURING RESOURCE AGENT

The function of manufacturing resource agent is supply rational information for process engineers on the CPP. Four main modular is include: equipment information management, fixture information management, cutting tool information management, process parameter and NC code information management. The resource information is stored with structured data, rule file and 2D or 3D entity of the part. The structured data is stored in the database, the rule of the manufacturing knowledge is stored with a special defined rule language file and the 2D or 3D entity of the part of other component is stored with standard format such STEP or STL file.

Because the structure of manufacturing resource management is based on the multi-agent, so a controlling modular is used to receive outer assignment and supply information for engineer or other agent. In this modular a universal data interface is defined for data transform.

3.1 Equipments information management

All the manufacturing equipments information of the partners within virtual enterprise should be managed in the resource agent. It includes type parameter, precision parameter, working range, and cutting speed of machine tool. All this data is stored in the table of database. The quantity and cost of machine also should be considered. Only by this means, the process engineers can choose cutting machine with high efficiency and low cost. On the other hand, the 3D models of the machine are also created for the collaborative NC simulation.

3.2 Fixtures information management

Fixtures design and manufacture is very important factor in the manufacturing process. The engineers usually start with three options: permanent, modular, and general-purpose fixture. Each of these options has many clamping and locating options of its own. The more standard locating and clamping devices that a designer is familiar with, the more creative he can be. So there are many standardized components in the agent include clamps, locators, supports, studs, nuts, pins, and a host of other elements. The engineers can choose suited components from a intelligent wizard and then download the 3D model from the agent, by this means, the efficiency of fixture design process could be increased, with an electronic white board, engineers can even collaborative discuss the fixture design on the Internet.

3.3 Cutting tools information management

The focus of this modular lies in representing information about cutting tools. That will adequately fulfill the needs of process planning with cutting machine in the virtual enterprise. The flowing data should be contained.

- Parameters of tool components, complete assemblies and tool kits, include dimension, material, and specification, etc.
- Cutting data based on the machining handbook.
- 3D solids with standard format such as STEP, SAT or STL format.

Metal cutting fluids are very important in the process planning, because its characteristics are widely changing, and the cutting quality of the metal part are mostly depend on the cutting fluids. Therefore, the information of cutting fluid also should be contained in the database.

3.4 Process parameters and NC code information management

This is the core segment of manufacturing resource management, all the process engineers will benefited from the process parameters information in CPP during their work. In traditional enterprise, the process planning work mostly depend on the individual knowledge of process engineer, so quality of process planning parameters can not be assured rationally. The goal of this modular is describe all the process information and process parameters with standardized format, and stored them in database. The flowing three aspects information should be stored.

- Universal knowledge or rule such as machining sequence, clamping method and so on. With this rules and instructions, process engineer can fix their attention on the deciding rational process of the part of product, and needn't to do great deal of labors consuming work.
- Process information of typical part or product, which will help to create new manufacturing process of the part. The engineer can find a similar part from the database and edit some of the parameter. Moreover, all the process information created by the CPP will be stored in the system automatically.
- All the machining parameter such as rotate speed of spindle, feeding speed, and so on. The parameter of heat treatment also stored in the database. Engineers can acquire optimized parameter with reasoning algorithms.

NC code information of the part also is managed in the system. It include 3D entity file of the part, project information, version information and related process information, etc. the engineer can edit and simulation an NC code collaboratively with partners in other enterprise. The validated NC code can be sent to the CNC machine immediately.

4. APPLICATION OF MANUFACTURING RESOURCE FOR CPP

With manufacturing resource of CPP platform, process engineers can work collaboratively on the Internet. Firstly the 2D or 3D model of the part will be loaded on the electronic white board for collaborative discussion, in this course, the blue print of manufacturing process will be discussed, and the benchmark of the part, clamping mode, etc. will all be discussed and confirmed. The figure 2 shows the client side of electronic white board. In the next step, the detailed process planning will be confirmed, which include fixture design, process parameter choosing, and NC code programming. Engineers can design fixture based the blue print from first step on the CAD software, in the working process, they can also transport the drawing of the fixture to the electronic white board with DWG or DXF format just for collaborative discussion. In the process parameter planning choosing, the cutting speed, feeding speed, tool type will all be determined. Then engineers can create the NC code with CAM software, after this step, the NC code is loaded in the collaborative NC simulation modular (show in the figure 3), the simulation modular will load the cutting tool chose from the manufacturing resource and simulate the cutting process collaborative. If there are some mistakes found in the simulation. The NC code will be edit or rework.

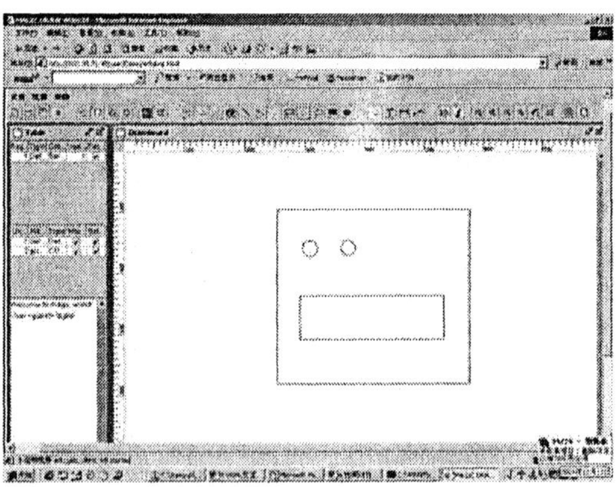

Figure 2. Collaborative discussion on the drawing of part with white board

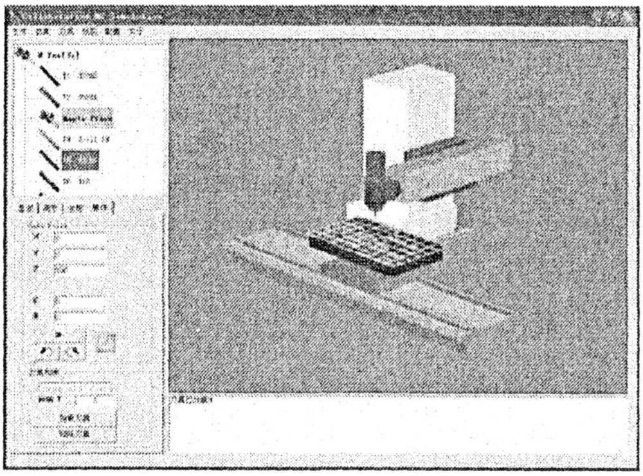

Figure 3. Snapshot of collaborative NC simulation client-side

5. CONCLUSIONS

A multi-agent based manufacturing resource management system is constructed for CPP platform. The contents and functions of its main modular are discussed. With CPP platform, engineers can utilize manufacturing resource of the virtual enterprise sufficiently during the process planning. Thus product can be manufactured with better quality and low cost.

6. REFERENCES

1. Ziqiang Zhou, Xiaoguang Li , Lianguan Shen, Research of Web-based collaborative modeling using layered feature language, *Proceedings of IE&EM'2002 & IceCE'2002.* Sept.19-21, 2002. Beijing, P.R.China.
2. George Q., Huang. Web-based support for collaborative product design review. *Computers in Industry*, 2002, 48(1): 71~88
3. Jan Kempenaers, Jos Pinte, A collaborative process planning and scheduling system, *Advances in Engineering Software* 25(1996):3~8
4. Ziqiang Zhou, Xiaoguang Li, etal. The role of Web-based virtual design in the mass customization, *Proceedings of the Third International Conference on Electronic Commerce Engineering.* Hangzhou, PRC. pp.601-605.
5. Bernhard Jung, Mathias Nousch. Design and Configuration of Furniture Using Internet-based Virtual Reality Techniques. http://www-is.informatik.uni-oldenburg.de /~sauer/ puk2000/ papers/jung.pdf
6. Q. Peng, F.R. Hall, P.M. Lister, Application and evaluation of VR-based CAPP system, *Journal of Materials Processing Technology*, 107 (2000),pp.153~159

THE UNIFORM KNOWLEDGE REPRESENTATION FOR ON-LINE PRODUCT SEMANTIC RECONSTRUCTIONS OF VIRTUAL ORGANIZATION

Chengfeng Jian
Software College at ZheJiang University of Technology,HangZhou, P.R. China; Email: jiancf@zjut.edu.cn.

Abstract: In order to realize the on-line product semantic reconstruction in the virtual organization, the uniform knowledge representation method is put forward including the product data in STEP and the product document data in SGML at first. Second a Semantic Base Cell model of Fractal Hierarchy –Linear Hyperlink is presented. At last according to this model, build the semantic association by the two description layers: the XML template for syntax layer and the OWL template for semantic layer.

Key words: VO, STEP, OWL, Semantic Base Cell model of Fractal Hierarchy –Linear Hyperlink.

1. INTRODUCTION

In the Virtual Organization(VO), knowledge sharing and exchange of product information over the Internet becomes a focus of research, which is different from the virtual enterprise. Because of the characteristics of the autonomy and heterogeneity among the distributed enterprises, the sharing and exchange of product semantic information should meet the requirement of dynamic and agility[1]. With the development of the semantic web and semantic grid, it is possible to realize the on-line semantic reconstruction on the product semantic information. And then the uniform knowledge representation is the prerequisite for the realization of the on-line semantic reconstruction over the Internet.

This project is supported by the National Natural Science Foundation of China under the grant No. 50305033.

Please use the following format when citing this chapter:
Jian, Chengfeng, 2006, in International Federation for Information Processing (IFIP), Volume 207, Knowledge Enterprise: Intelligent Strategies In Product Design, Manufacturing, and Management, eds. K. Wang, Kovacs G., Wozny M., Fang M., (Boston: Springer), pp. 932-937.

In this paper, in order to solve the prerequisite question for the realization of the on-line semantic reconstruction over the Internet, the uniform knowledge representation method is put forward at first. Second we should build the semantic relationship on the basis of the uniform knowledge representation. So a Semantic Base Cell model of Fractal Hierarchy –Linear Hyperlink is presented. And then we build the semantic association by the two description layers: the XML template for syntax layer and the OWL template for semantic layer. According to the XML template we can solve the syntax description for the heterogeneous production information and according to the OWL template we can build the semantic association with the XML-based production information. By means of the dynamic operation and combination among the semantic base cells, we can realize the on-line product semantic reconstruction.

2. THE REALIZATION PROCESS OF THE UNIFORM KNOWLEDGE REPRESENTATION

The control parameters should be adjusted at suitable setpoints before the seed cleaner starts.Figure1 shows the process. It mainly concludes:

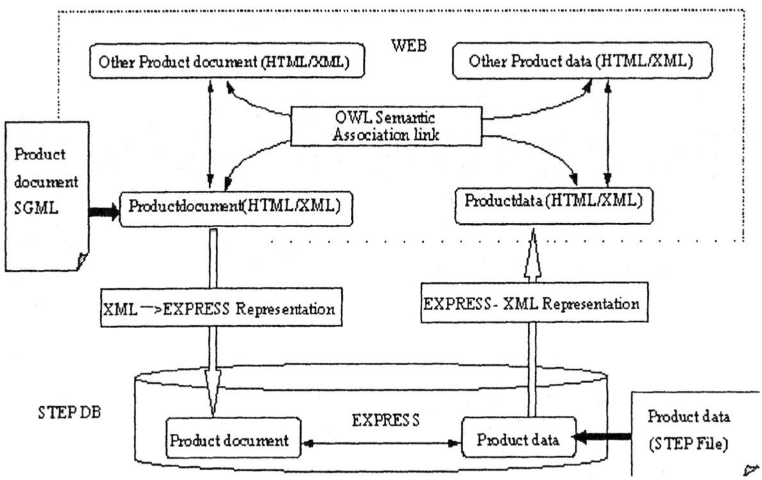

Figure 1. [the realization process of the uniform knowledge representation]

1) Write the product data of the physical STEP file into STEP database

2) Publish the product document information(SGML) to the HTML/XML over the Web

3) Convert the STEP data into the XML representation

4) Publish the product data which represented by XML over the Web

5) By means of a Semantic Base Cell model of Fractal Hierarchy-Extended Hyperlink, build the semantic link to realize the dynamic link among the heterogeneous information.

6) The same kinds information can be directly linked such as among the product document information.

7) In order to realize the on-line semantic reconstruction, the information can over the Web be revised and reflect to the correspondent STEP record in the STEP database or the SGML document.

From the above process, the realization of the XML representation of the product data and the OWL semantic association link are the primary key. The XML representation provides the uniform information representation over the Web. The OWL semantic association link provides the paths of building the correspondent dynamic semantic link among the heterogeneous information.

3. EXPRESS-XML-OWL :THE KNOWLEDGE REPRESENTATION BETWEEN STEP AND SGML

In the Internet, VO contains the two kinds standard information. On the one hand the information on the WWW is represented by the HTML/XML which belongs to SGML. As the product document information, it is necessary to vary with the product data. On the other hand the product data represented by the STEP need the readable understand in the Internet. But it is hard to be understood to the EXPRESS language except the product data structure and the data defined by EXPRESS can't be identified in the WWW browser. EXPRESS is a powerful language for the definition of the data structure, it provides for defining the types of object and their properties and the constraints are far richer than those provided by the XML. On the other hand, XML is far richer than EXPRESS in the data transmission and identification. XML is easily interpretable by humans but EXPRESS not. OWL is also powerful for the semantic description and syntax express uses XML. In order to realize the representation, according to the OO conception, Table1 shows the comparison [2] [3] [4] [5]:

Table 1. the comparison among EXPRESS-XML-OWL

OO	EXPRESS	XML	OWL
Object	Entity	Element type	Class
Object instance	Entity instance	Element	Instances
Object property	Entity attribute	Element	ObjectProperty DatatypeProperty
OO declaration	Entity Schema	DTD	Ontologies

OO	EXPRESS	XML	OWL
OO relationship	Complex Constraint	hiberarchy	Complex Constraint

4. THE DYNAMIC SEMANTIC LINK OVER WEB BETWEEN PRODUCT INFORMATION

After realize the XML representation for the product data represented by STEP, we can build the uniform semantic link between product data and product document information over the WWW.

4.1 A Semantic Base Cell model of Fractal Hierarchy – Linear Hyperlink

In the Virtual organization, the information can decompose the tree structure includes the product data. and the each leaf node has the same structure. We take it as the fractal hierarchy to describe the existing the information. At the same time, in the Internet, XML provides the capability of the data access base the hierarchy tree such as the context query and navigation and so on. So in order to realize the dynamic link among the heterogeneous information, we build a semantic base cell model of fractal hierarchy-linear hyperlink (XML syntax representation).it shows as Figure2:

```
<!ELEMENT cell-element( element-description , context-related-element )
<!ATTLIST cell-element
            id-code             ID      # REQUIRED
            element-name        CDATA   # REQUIRED
            element-reference   IDREFS  # IMPLIED           <!– linear hyperlink-->
>
<!ELEMENT element-description (#PCDATA|empth| content of the element-description)* >
<!ELEMENT content of the element-description (#PCDATA|empth)* >
<!ELEMENT context-related-element (#PCDATA|empth| other cell-element)* >
                                                  <! - - fractal hierarchy- - >
```

Figure 2. [A Semantic Base Cell model of Fractal Hierarchy –Linear Hyperlink]

A cell element is correspondent to the product information object. We can represent the fractal hierarchy by means of the ELEMENT "context-related-element" and the linear hyperlink by means of the ATTLIST of "element-reference". So far as the XML representation of the STEP, we use the STEP data ID as the ATTLIST "id-code", use the ELEMENT "context-related-element" to express the structure of the product object. As the XML representation of the product document, it is a tree of elements and data characters and elements can have the attributes. So we can build the link with the SGML document by means of the ATTLIST of "element-reference".

4.2 The XML template definition of product information

According to the semantic base cell model, we can define the uniform semantic template according to the product information classification for multi VOUs. It includes the two kinds of the template: one is the XML DTD/XML Schema template for syntax layer, the other is OWL template for semantic layer. Each product information record is divided into two parts: product_description and product_associations. Product_description contains the inner information and product_association contains the outer related information. The ways of each product information record is divided into four parts: Product element, Product type element, Production factory element, Location element. Product element mainly contains the detail information about product such as product code, product name, technical parameter, condition, price and so on . Product type element mainly contains the recursive structure of product type such as every son of product type , the son's flag of product type ,the code of product type and so on. Production factory element mainly contains the detail information about factory such as the factory code, the main products about the factory, the flag of VO and so on. Location element mainly contains the recursive structure of the location about the product or factory.

```
<!DOCTYPE ProductInformation>[
    <!ELEMENT ProductItem(Product_Description, Product_Associations ) >
    <!ATTLIST ProductItem   Id ID  #REQUIRED  Nm CDATA  #REQUIRED ref  IDREFS  #IMPLIED >
        <!ELEMENT Product_Description (Product_BusinessInfo, Produc_TechInfo ) >
        ...                          <!-- Product Business Info. and Product Technology Info. -->
        <!ELEMENT Product_Associations (Inherit_elements, Path_elements ) >
                        <!-- context-related-element : Inherit and path -- >
         <!ELEMENT Inherit _elements (#PCDATA/empth/ PartsItem )* >
                        <!-- context-related-element :parts name-- >
        <!ATTLIST Inherit _elements Id ID  #REQUIRED  Nm CDATA  #REQUIRED ref IDREFS  #IMPLIED >
        ...                                 <!--XML Link parts data-- >
         <!ELEMENT Path _elements (#PCDATA/empth/ PathsItem )* >
                                            <!-- context-related-element .paths name-- >
        <!ATTLIST Path _elements Id ID  #REQUIRED  Nm CDATA  #REQUIRED ref IDREFS  #IMPLIED >
        ...                                 <!--XML Link the structure of Product data-- >
         <!ELEMENT PathsItem (MemberItem, ProductTypeItem, LocationItem...) >
                        <!-- VOU Member cell, Product type cell, Location cell... -- >
    <!ELEMENT MemberItem (Member_Description, Member_Associations ) >
    <!ATTLIST MemberItem Id ID  #REQUIRED  Nm CDATA  #REQUIRED ref IDREFS  #IMPLIED >   ...
    <!ELEMENT ProductTypeItem (ProductType_Description, ProductType _Associations ) >
    <!ATTLIST ProductTypeItem Id ID  #REQUIRED  Nm CDATA  #REQUIRED ref IDREFS  #IMPLIED >
    ...
    <!ELEMENT LocationItem (Location_Description, Location _Associations ) >
    <!ATTLIST LocationItem Id ID  #REQUIRED  Nm CDATA  #REQUIRED ref IDREFS  #IMPLIED >
    ...
>]
```

Figure 3. [the XML template for the syntax layer]

The XML template for the syntax layer shows as Figure3, The OWL template for the semantic layer shows as follows: (display in the Semantic works2006 tools.)

Figure 4. the OWL template for the semantic layer

5. CONCLUSION

we deal with the uniform knowledge representation among the heterogeneous information including STEP and SGML. according to the semantic base cell model of fractal hierarchy–linear hyperlink, we build the semantic association by the two description layers: the XML template for syntax layer and the OWL template for semantic layer. By means of the dynamic operation and combination among the semantic base cells, we can realize the on-line product semantic reconstruction.

6. REFERENCES

1. S. Agarwal, S. Handschuh, S. Staab,(2003), Surfing the Service Web, *in Proceedings of the 2nd International Semantic Web Conference (ISWC 2003).*
2. ISO 10303 Industrial automation systems and integration –Product data representation and exchange Part11:Description methods: The EXPRESS language reference manual.
3. W3C, *Extensible Markup Language (XML) 1.0,* http://www.w3.org/TR/1998/REC-xml-19980210.
4. W3C, *OWL Web Ontology Language*, http://www.w3.org/TR/2004/REC-owl-features-20040210/.
5. Chengfeng Jian, Jianrong Tan,(2001),description and identification of XML-based STEP production data, Journal of Computer-Aided Design & Computer Graphics, vol. 13,pp.983-990.

ACCESS CONTROL MODEL OF MANUFACTURING GRID

Hongxia Cai, Tao Yu, Minglun Fang
CIMS&Robot Center, College of Mechatronics Engineering and Automation, Shanghai University, Shanghai, China; Email: hxcai@mail.shu.edu.cn.

Abstract: The overall motivation for Manufacturing Grid is to realize resource sharing and collaboration in the field of Manufacturing. In order to protect the shared resources and services, the access control guarantees in the MG is necessary and important. Due to inherent heterogeneity, multi-domain characteristic and highly dynamic nature of Manufacturing Grid, we propose a new Manufacturing Grid Access Control Model to support the dynamic restricted rights management based on the current Role-based Access Control Model. We employ a distributed access control mechanism to realize the authorization from the local service nodes according to the global and local authorization policy. The implementation within the project of the Shanghai High Institutions Grid proves the authorization architecture and model are workable and our work is valuable.

Key words: Manufacturing Grid, security, Role-based Access Control Model

1. INTRODUCTION

Based on the technology of Grid, we developed a Manufacturing Grid system to realize resource sharing and collaboration in the field of Manufacturing. However, in the Grid environment, the access control challenges are not fully addressed by existing approaches. It is still one of the major remaining obstacles to the wide-spread adoption of the Manufacturing Grid. Due to inherent heterogeneity, multi-domain characteristic and highly dynamic nature of Manufacturing Grid, we propose a new Manufacturing Grid Access Control (MGAC) Model which complements the classic Role-based access control (RBAC) [1], while

Please use the following format when citing this chapter:
Cai, Hongxia, Yu, Tao, Fang, Minglun, 2006, in International Federation for Information Processing (IFIP), Volume 207, Knowledge Enterprise: Intelligent Strategies In Product Design, Manufacturing, and Management, eds. K. Wang, Kovacs G., Wozny M., Fang M., (Boston: Springer), pp. 938-943.

retaining its advantages (i.e. ability to define and manage complex security policies). The model dynamically adjusts Permission Assignments based on global project state. The MGAC model is scalable to combine the coarse-grain authorization policy at local service node with the global fine-grain authorization policy.

The rest of this paper is organized as follows. In section 2, we present the MGAC model. In section 3, we introduce its implementation within the project of the Manufacturing Grid, which is the part of Shanghai High Institutions Grid. A brief summary is given in section 4 along with concluding remarks.

2. ACCESS CONTROL MODEL OF MG

We will present an access control approach to the Manufacturing Grid which allows for the unified and centralized management of authorization. The MGAC model complements the current Role-based Access Control Model to support the dynamic authorization.

In RBAC model, the user has the privilege when he is assigned the role. In the MGAC, the privilege of the user is not only related with static role, but also subjected to the project state. The traditional RBAC model could not satisfy the distributed Manufacturing Grid architecture. In the MGAC, the user accesses the local service node by mapping the global role and to the local role.

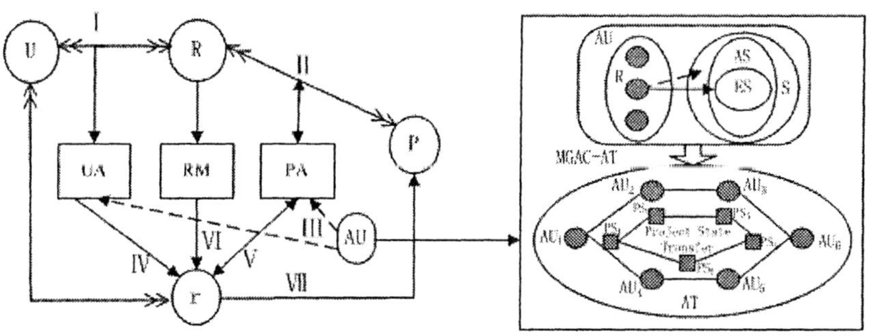

Figure 1. MG Access Control Model

Figure 1 is the Manufacturing Grid access control model. In the MGAC model, I , II shows the principle of the traditional RBAC which illustrates the relationship of the user-to-role-to-privilege assignment. IV, V ,VI, reflects the consistency between the local and global access control management.III shows the dynamic authorization mechanism which is introduced in the

MGAC-AT. In the MGAC-AT, AU is the basic Authorization Unit which comprises the enabled-permissions set to the dynamic role. The AU is transited to confine the enabled permissions of the role according to the project state. We define the transition of the AU as AT. AT is the sequence of the AU: AT= {AU1, AU2 ..... AUn }

We formally define the MGAC model as follows:

$$MGAC=(U,R,r,P,AT) \quad (1)$$

U is the user, R is the global role, r is the local role, P is the privilege, AT is the authorization transition.

We define the following relationship: User assignment (UA), Permission Assignment (PA), Role Mapping (RM), Authorization Unit(AU):

- User assignment(UA), $UA \subset U \times R$, a many-to-many mapping user-to-role assignment relation;
- Permission Assignment(PA),$PA \subset P \times R$, a many-to-many mapping, permission-to-role assignment relation.
- Role Mapping (RM), $RM \subset R \times r$, Global Role-to-Local role assignment relation.
- Authorization Unit (AU), $AU \subset r \times AS$, Local Role-to-Authorized Service assignment relation.

We assign the user the global role to have static privilege:

$$UA(R)=\{ua(u_i, R)|ua(u_i,R) \in UA, i=1,2,......n\}; \quad (2)$$

$$PA(R)=\{pa(R, p_i)|pa(R, p_i) \in PA, i=1,2,......n\}; \quad (3)$$

The authorization of the dynamic role can be transmitted according to the project state:

$$AU(R)=\{au(R, as_i)|au(R, as_i) \in AU, i=1,2,......n\}; \quad (4)$$

We map the global role to local role:

$$RM(r)=\{rm(r, R_i)|rm(r, R_i) \in RM, i=1,2,......n\}; \quad (5)$$

Whether the user can access the local service, it depends on:

$$Authorize(u,s) \rightarrow \exists R(u \in UA(R) \cap p \in PA(R) \cap AS \in AU(R) \cap R \in RM(r)) \quad (6)$$

So in the MGAC model, the authorization is dynamic convergence and consistency.

3. IMPLEMENTATION

The project of Shanghai High Institutions Grid sponsored by the Shanghai Technical Committee, has goals similar to those of the Grid, focusing on establishing a platform for resource sharing and collaboration in the field of rapid manufacturing and so on.

We have tested the scenario on our platform. In our platform, it allows for the unified and centralized management of access control rules to a set of services and resources that may be widely distributed and highly heterogeneous. The global policy repositories determine the global access rules in which the fine-grain policy is defined by the community about the role and privilege assignment, while the dynamic policy is related with the project state. The local policy repositories grant coarse-grain access to the virtual organization of MG. The distributed services are managed locally, while the privileges of the user are authorized at local service node.

Figure 2 is Access Control Working Flow in the MG. The MGAC Global Broker is responsible for authenticating the requesting Grid user, authorizing their request and determining whether the user has the global right to access the services or not. The authorization depends on the fine-grain policy from the global policy repositories and dynamic policy related with the condition of the project. We use the SAML [6] (Security Assertion Markup Language) for exchanging authentication and authorization information. The MGAC Local Broker checks if the global policy statement authorizes the request and the privilege is authorized to the community. The MGAC Local Broker maps the Global Role to Local Role according to the role mapping list. Policy enforcement is carried out by local operating enforcement mechanisms.

In our scenarios, according to the central fine-grain access control policy, the roles of rapid designer and project manager in the project group can access the task file and its attachment. The dynamic policy only allows the role of the rapid designer and so on to update the drawing when the project is in the state of design. According to the local course-grained polices, for example, at the one of the rapid design service node, the drawings are open to the community of MG. Of course, the local access control policy could forbid the blacklist accessing the design drawings. When the provider of the rapid prototyping began to machine, the rapid designer could not access files any more. As soon as the project finished, the global roles assigned to the users are withdrawn, and the users could not access resources any more.

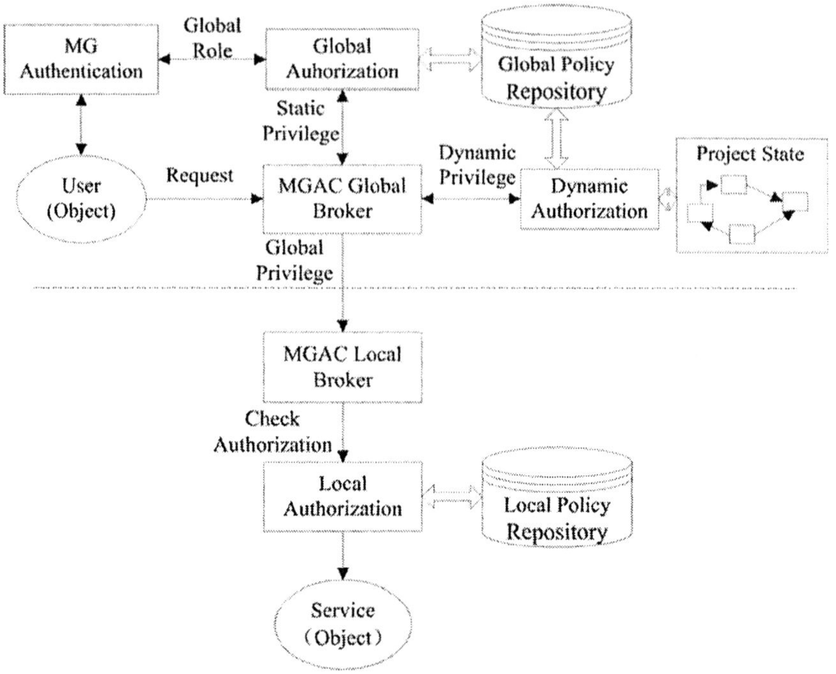

Figure 2. Access Control Working Flow in the MG

4. CONCLUSION

In this paper, we present a new access control model for the Manufacturing Grid. Compared with the traditional Role-based access control model, the Manufacturing Grid Access Control model is more scalable, more applicable to the distributed and dynamic Manufacturing Grid environment. The MGAC model solves the access control issues and satisfies the access control requirement in Manufacturing Grid. The implementation in the project shows that the research is valuable and the MGAC model is workable.

5. REFERENCES

1. R. Sandhu, E. Coyne, H. Feinstein, and C. Youman. "Role-based Access Control Models," *Proceedings of the 5th ACM Workshop on Role-Based Access Control,* 2000, pp. 38- 47.

2. Liu Lilan, Yu Tao, Shi Zhanbei, "Research on Rapid Manufacturing Grid and Its Service Nodes", *Machine Design and Research*, 2003, pp.57-59.
3. I. Foster, C. Kesselman, J. Nick, and S. Tuecke, "The Physiology of the Grid: An Open Grid Services Architecture for Distributed Systems Integration," *Open Grid Service Infrastructure WG (GGF)*, June 2002.
4. L. Pearlman, V. Welch, I. Foster, and C. Kesselman, "A Community Authorization Service for Group Collaboration," *Proceedings of the IEEE 3rd International Workshop on Policies for Distributed Systems and Networks*, 2002.
5. P. Hallam-Baker et al, "Assertions and Protocol for the OASIS Security Assertion Markup Language (SAML)", *Oasis Standard*, November 5th, 2002.
6. S. Godik et al, "eXtensible Access Control Markup Language (XACML) Vers. 1.1", *OASIS Standard*, July,2003

RESEARCH ON STRATEGIC ARCHITECTURE OF COLLABORATIVE BUSINESS SYSTEM

Zhiwei Mi [1], Wenwu Mao [2], Minglun Fang [1]
[1]*CIMS & Robot Center of Shanghai University, Shanghai University, China; Email:* *zwmi@mail.shu.edu.cn.* [2]*College of Engieering, Shanghai Fisheries University, China.*

Abstract: Collaborative business system is perfect combination of the best business practices with information technology. Collaborative business system is an innovation and enabled changes to traditional rules for doing business. The aim of this paper is: firstly to provide an overview of collaborative business by outlining the essence and the framework; secondly, to put forward to the architecture of collaborative layer, the most important part of the framework.

Key words: Collaborative business system, theoretical framework.

1. INTRODUCTION

Even in today's New Economy, traditional rules for doing business still hold true, but the Internet has changed the way business works, linking manufacturers, suppliers, distributors, customers, partners, and employees across entire sectors, by establishing collaborative business networks (virtual communities). Creating value by collaboration, empowerment, and integration in the Internet economy depends on the ability of businesses to work together with its customers, suppliers, and the like across traditional enterprise boundaries in virtual global networks.

The following evolution has recently taken place in the business world. Originally, companies were looking at cost reduction and efficiency by integrating internal business processes. So they use 'ERP functionality' to achieve this. ERP offer enterprise-centric functionality (management of Human Resource, Finance, Material) to integrate internal core processes.

Then, companies realized they must establish relationships with their suppliers and customers, to share information to reduce costs and time within their supply chain network. So the next generation of new dimension

This project is Supported by China Shanghai Municipal Education Commission under the grant No.03AK80.

Please use the following format when citing this chapter:
Mi, Zhiwei, Mao, Wenwu, Fang, Minglun, 2006, in International Federation for Information Processing (IFIP), Volume 207, Knowledge Enterprise: Intelligent Strategies In Product Design, Manufacturing, and Management, eds. K. Wang, Kovacs G., Wozny M., Fang M., (Boston: Springer), pp. 944-949.

software products appeared, extending ERP, taking functionality out of the company, to bring value through **cooperation along the supply chain.**

Currently, the Internet has enabled a collaborative environment where value is created through **collaboration within business communities.** Collaborative business system builds on ERP functionality and also extends the following:

(1) Offering collaborative functions.

(2) Generating value through collaborative possibilities.

(3) Seamlessly integrating heterogeneous platforms.

The term of collaborative business was mainly introduced in 2000 by the world largest commercial software vendors, for instance, SAP, EDS, based on the combination of their best business practices with information technology. Since the introduction of the term, it has received growing interested and research both in literature as well as from industrial practice[1~7].

The aims of this paper are to: firstly provide an overview of collaborative business system by outlining the essence and the framework; secondly explain in greater detail architecture of collaborative layer, the most important part of the framework.

2. FRAMEWORK OF COLLABORATIVE BUSINESS SYSTEM

So far, there is no formal and official definition of collaborative business system. The description of collaborative business system proposed here is only trying to extract the essence:

(1) Collaborative business system is complex system engineering. It regards business community as an entire system in which the parts are dynamic and interrelated, and as well business community and their environment are also regarded as a whole, enterprises should have social responsibilities.

(2) Collaborative business system is an innovation and enabling changes to traditional operating rules for doing business.

(3) Collaborative business system is synthetic approaches and tools using various theories, technologies and methodologies.

(4) Information technology enables collaborative business. Collaborative business system is perfect combination of the best business practices with information technology.

(5) Collaborative business is the task of creating value and benefits for each partner (so called win-win) through collaboration within business community with the aim of improving the competitiveness as a whole, as well fulfilling requirement of customers.

(6) The collaboration and integration are critical in the collaborative business.

According to the description above, the paper puts forward to theoretical framework of collaborative business system as figure 1. As you can see from the figure 1, the theoretical framework consists of three layers:

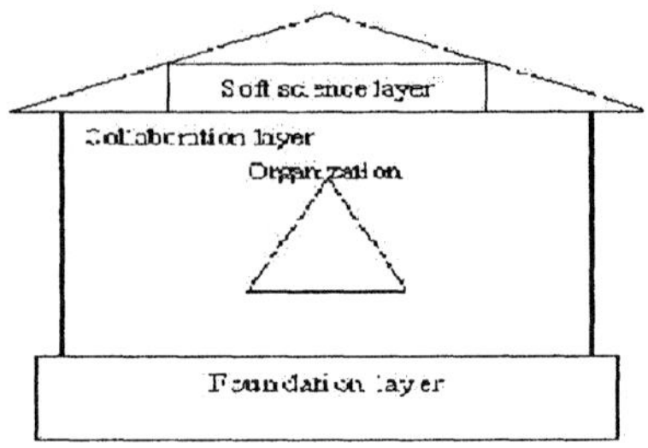

Figure 1. Theoretical framework of collaborative business system

(1) Foundation layer: this layer is the physical foundation of business community running, including various soft and hard resources, for example, distribution centers, ports, plant, other locations, transportation lane, material, tools, flexible manufacturing system, material handling system, computer; standards and protocols.

(2) Collaboration layer: this layer is the core of theoretical framework. This layer brings all together within business communities, coordinates the business operation, keep everything operating at peak efficiency. It consists of three parts, collaborative organization, collaborative business processes, integration and application platform. This layer will be described in great detail below.

(3) Soft science layer: this layer is comprehensive implementation solution available to streamline and facilitate collaborative business system. It consists of science theories, regulations, visions, methodologies, tools, services and training, etc.

3. THE CORE OF THEORETICAL FRAMEWORK

Compared with linear, sequential and separated business operation of traditional SCM solutions, the one of most important objectives of collaborative layer of theoretical framework is to transition today's fragmented sequential supply chain to synchronizing and coordinating and adaptive supply chain network.

The paper divides collaborative layer into three levels of collaboration.

3.1 Collaborative Organization

One of the first steps in configuring the Collaborative business system is mapping and modeling the collaborative organizational structure in accordance with strategic business requirement of collaborative business using the organizational units.

From the business practices, organizational forms within collaborative organizational structure includes a wide array ranging from long-term purchasing agreements to co-marketing and licensing agreements, to R&D collaboration teams, to joint ventures. So, a collaborative organizational structure should create and build a close, medium-long term, mutually beneficial partnerships between two or more organizational units during all phases of business relationships. Resources, knowledge, and capabilities within collaborative organizational structure are shared with the objective of enhancing the competitive position of each unit. Centralization and decentralization are the two extremes in the collaborative organizational structure.

Despite the strategic alliances between enterprises, for example, are growing at ever increasing rate, many alliances fail with low success rates. Lack of communication and mistrust and even cheating are the main causes.

3.2 Collaborative business processes

Based on analysis of operational data from the transactional enterprise resources planning, collaborative business process supports continuously redesigning, coordination and optimization of business processes for gaining competitive advantage.

Collaborative business processes consists of the following contents as figure 2.

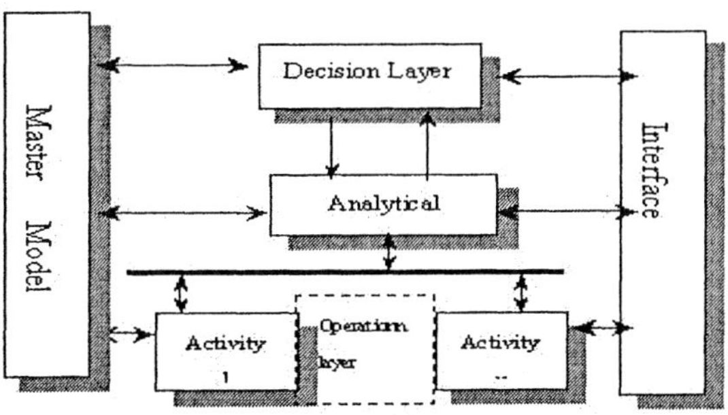

Figure 2. The Structure of Collaborative business

3.3 Integration and application platform

The evolution of IT infrastructure from mainframe to client/server, to web service architecture reflects the ongoing adaptability and flexibility. Open, flexible IT infrastructure for collaboration across companies boundaries, e.g. built on the web service concept and environment, together with the integration of Internet, enables today's collaborative business system.

The paper considers the integration and application platform as the technical foundation for Collaborative business system. It delivers a complete, open and flexible infrastructure that easily integrate the three areas as belows:

(1) Integrate all participating people involved in the business process, whether within an enterprise or outside of it (internal and external, all the relevant roles).

(2) Manage all the relevant information (structured and unstructured) for the whole business processes through data warehouse and knowledge management, for example.

(3) Integrate all systems required for the business processes integration through integration technology, e.g. database oriented technologies, message oriented technologies, distributed object technologies, interfaced oriented technologies, etc, regardless of whether they are internal or external systems.

4. CONCLUSION

Collaborative business system is advanced management thoughts, methodologies and tools through which customers, suppliers, partners, employees, organizational units, business processes and information are brought together to form one collaborative unit within business community.

The foundation for achieving collaboration is as followings:

(1) Modern information and communication technology.

(2) Collaborative business processes, especially for advanced planning, optimization and controlling.

(3) Strategic alliance between organizational units.

5. REFERENCES

1. Escudero, L.F., Galindo, E., Garcia, G., Gomez, E., Sabau, V., 1999. Schumann, a modeling framework for supply chain management under uncertainty. *European Journal of Operational Research* 119, 14 - 34.

2. Fawcett, S.E., Myers, M.B., 2001. Product and employee development in advanced manufacturing: Implementation and impact. *International Journal of Production Research* 39, 65 - 79.
3. Fox, M.S., Barbuceanu, M., Teigen, R., 2000. Agent oriented supply chain management. *International Journal of Flexible Manufacturing Systems* 12, 165 - 188.
4. G. Gunnarsson, S. Jonsson, Charge the relationships and gain loyalty e.ects: Turning the supply link alert to IT opportunities, *European Journal of Operational Research.* 144 (2003) 257 - 269.
5. G. Stefansson, Business-to-business data sharing: A source for integration of supply chains, *International Journal of Production Economics* 75 (2002) 135 - 146.
6. Cooper, M.C., Lambert, D.M., Pagh, J.D., 1997. Supply chain management: More than a new name for logistics. *International Journal of Logistics Management* 8 (1), 1 - 13.

A WORKFLOW MANAGEMENT SYSTEM FOR DYNAMIC ALLIANCE

Zhiqiang Gu[1], Gangyan Li[2], Kunpeng Wang[2]
1Science and Technology Office, Wuhan University of Technology, Wuhan, P.R. China; Email: gxkwhut@mail.whut.edu.cn. 2School of Electronic and Mechanical Engineering, Wuhan University of Technology; Email: gangyanli@mail.whut.edu.cn, robbie82@sohu.com.

Abstract: In order to support operation recombination of enterprise, based on analyzing some important problems, including flow start-up, flow navigation, resource subsequent control, activity performance and task management, a workflow system support dynamic alliance was designed. A resource conflict resolution strategy based on queue and request in the process of workflow was introduced. Strategies of occupancy and release resource in inaction were introduced. The main module and structure of database of the workflow management system were designed.

Key words: Dynamic Alliance, Workflow Management System, Control, Database.

1. INTRODUCTION

Workflow management is a kind of rising technology span subjects. It refers to many subject fields such as information, management, psychology, and so on. At the same time, it can be applied in many fields such as e-business, product design and remote-distance education and so on. Researching the technology of workflow management thoroughly is significant to improve informational degree, efficiency and competitive abilities of enterprise and government department. Based on analyzing some important problems, including flow start-up, flow navigation, resource subsequent control, activity performance and task management, a workflow

This research is funded and supported by the National Natural Science Foundation of China (NO.E0550335020), the National Key Technologies R&D Program of China (NO.2003BA103C) and the Key Technologies R&D Program of Wuhan City (NO. 20043007077-02).

Please use the following format when citing this chapter:
Gu, Zhiqiang, Li, Gangyan, Wang, Kunpeng, 2006, in International Federation for Information Processing (IFIP), Volume 207, Knowledge Enterprise: Intelligent Strategies In Product Design, Manufacturing, and Management, eds. K. Wang, Kovacs G., Wozny M., Fang M., (Boston: Springer), pp. 950-959.

system support dynamic alliance was designed. A resource conflict resolution strategy based on queue and request in process of workflow was introduced. Strategies of occupancy and release resource in inaction were introduced. The main module and structure of database of the workflow management system support were designed.

2. OBJECT DESIGN OF SYSTEM

The Object-oriented method (OO) is applied to encapsulate all the data in the form of object. All services offer their exterior interfaces in the form of objects.

2.1 Object of system

The objects in the system are as follows: workflow definition, node definition, resource instance, user instance, role instance, activity instance, application data and mission.

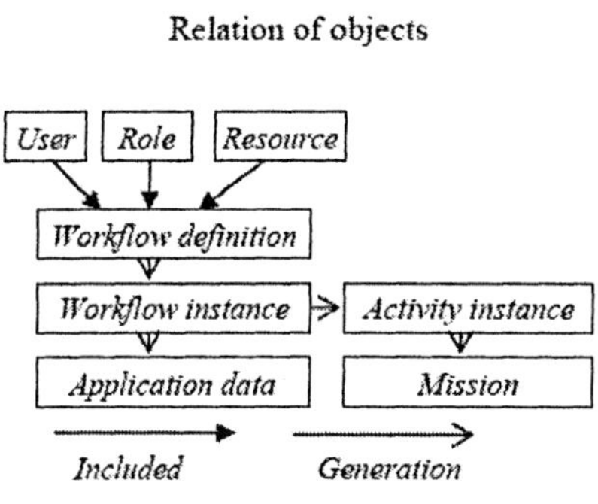

Figure 1. Mapping relation

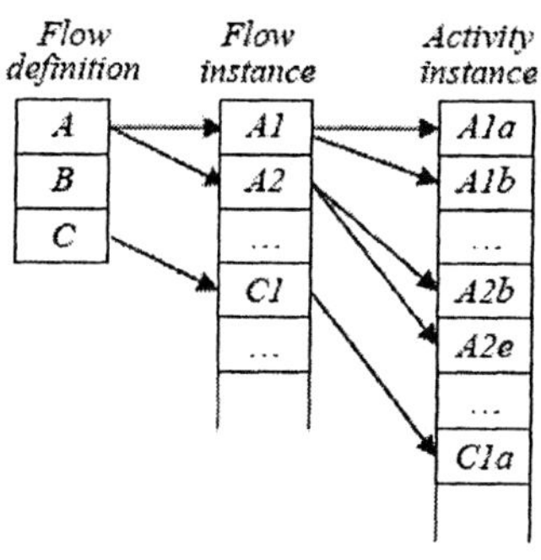

Figure 2. Mapping relation

Relations between all data objects in the system are illustrated in Figure 1. Making the work flow run normally according to the defined operation model is the main aim of workflow management system. A great deal of activity instances can be derived from a flow instance. (There are several nodes in the workflow model). At the same time, a great deal of flow instances can be derived from a workflow instance so that an enterprise can deal with several same businesses at the same time.

Figure 2 shows the mapping relation between flow definition, flow instance and activity instance.

3. DATABASE DESIGN

The relational database management system is applied to save the data running in workflow. Relational database is based on a solid relation theory foundation. There is a series of design methods in relational database design. Furthermore, there are various commercial database softwares suitable for developing various software systems. Figure 3 illustrates a logic model of system database based on above discussion.

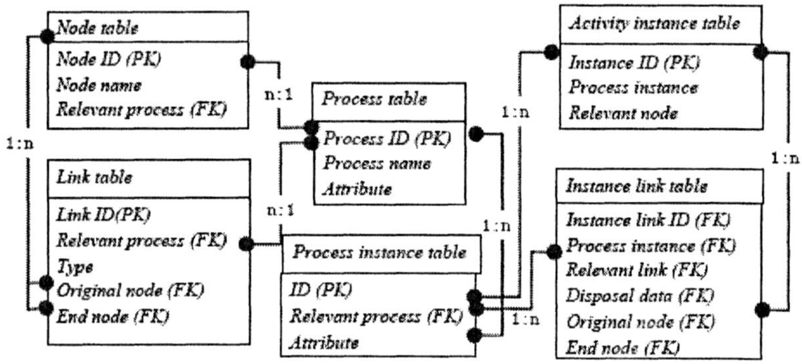

Figure 3. Logic model of database.

In order to realize that data are separated from program, database is applied to save data. A mass of running information are saved in database according to the designed relation. The design information of main table of the database is as follows:

Process table: process table includes workflow ID, workflow name, workflow content and other attributes.

Process instance. Process instance table logs the flow running in system. Instance information of flow is saved in database. A workflow definition can include several instances.

Node table.

Activity instance table. It includes instance ID, instance name, Input data.

Variable table. It logs all variable of all nodes. Figure 4 shows data model of system.

Variable instance table. Variable instance is the object encapsulated from control data produced while flows are running. In this system, data ID, relevant flow, relevant activity, data content are encapsulated to a variable instance.

Variable type table. Control data are used to navigate and assign. The types of control data are defined in advance. We can choose the types from the table when we design the variables of nodes.

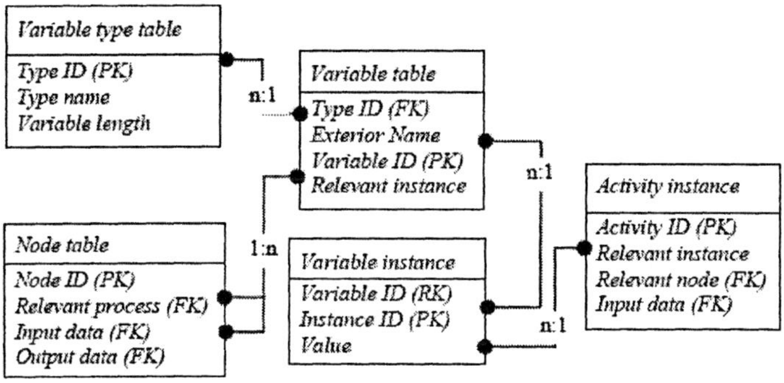

Figure 4. Logic model of data.

Role. Role which includes someone's rights and function is a set of a group right. Role is an object which is used to cooperate among department. Member should have different roles in different projects. A task is assigned to a role while a workflow is defined.

Resource. Resource is composed of equipments, workshop and other hardware equipments. It should include human resource especially.

People. People include all employees of enterprise. Attributes of personnel are as follows: name, age, length of service, department, strong suit and so on. Each employee has an ID as only identifier.

Task. Task is a command to relevant employee sent by activity instance running. Employee can get task information from task manager after login. Task is generated from activity instance.

Alliance table. The table is used to memorize information of alliance in that enterprise joined. Workflow engine is associated with enterprise's name according to the data in the table. Local enterprise chooses enterprise to communicate with by name of enterprise. The attributes of the table include name of enterprise, address of enterprise, and type of enterprise, about enterprise and so on.

4. DESCRIPTION OF WORKFLOW ARITHMETIC

4.1 Design of system class

It is necessary to design the detail part of class. In order to decrease length of code and reuse existent code, the design function of Object Oriented is introduced to encapsulate function models of system to classes.

Running process of system is as follow:

The chief control workflow system marks out flow and decomposes task. Then, it assigns task unit to relevant enterprise workflow system.

Initializing run environment of alliance, synchronizing all subsystem workflow system, preparing for communication among system.

Administrator starts up a work flow.

System flow begins using activity navigation service. Activity data object is initialized. Process control parameters are modified. Automatic flow nodes are executed. Activity is transferred to another workflow system according to assignation. Then, next activity is navigated.

Manpower atom activity execution: all data of activity are initialized. All need resources are configured. Task notice is created and added in task list.

Sub-process execution: Flow is dynamic decomposed and initialized. Then, flow begins.

Workflow system watches flow run, activity execution, and task submission and so on.

4.2 The start up of flow

There are three modes to start flow up: administrator start, user start and automatic start. Flow is started up by flow watching model. Watching model initializes flow instance according to the flow definition (see figure 5.).

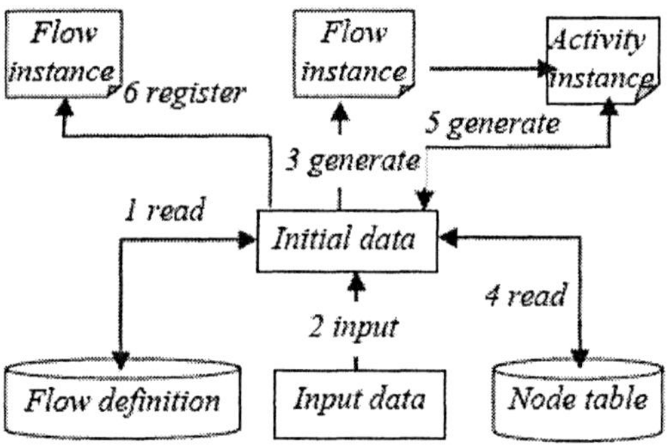

Figure 5. Initializing flow.

The arithmetic of flow Initialization is described as follows (see figure 6.):

Choose a workflow definition from workflow table. If finished, next.

Create a flow instance according to workflow definition and provide input interface. Users must input dynamic attributes of the instance.

Initialize activity instance according to node definition and prepare data boxes and variable instance.

If all nodes and links are initialized successfully, run state of flow will be configured and the instance will be registered in global information.

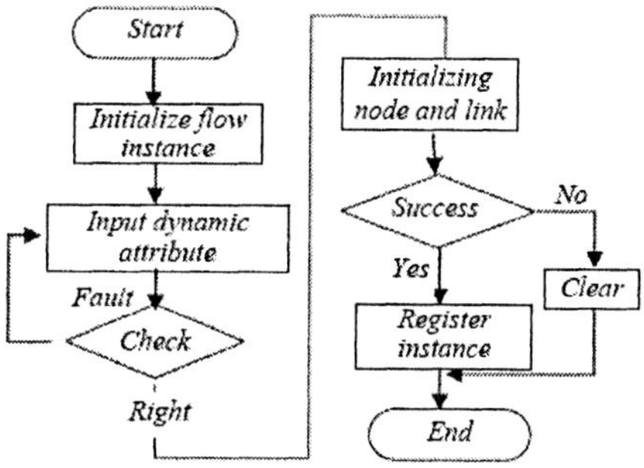

Figure 6. Process of initialization.

4.3 Flow navigation

System needs to confirm the next execution activity according to data, instance and link running in flow when an activity finished. Activity assigning engine answers for navigation of activity (see figure 7.).

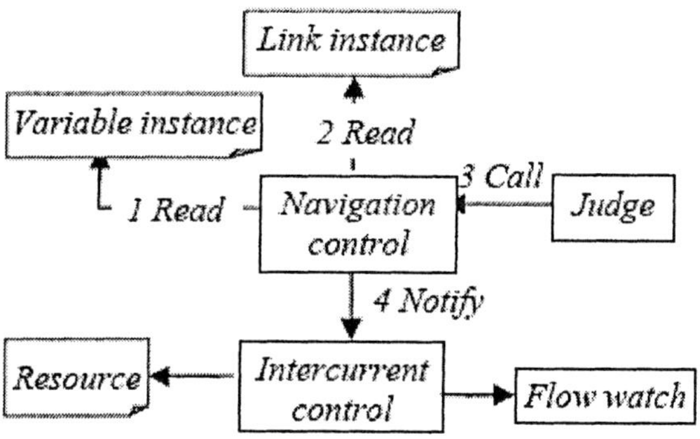

Figure 7. Flow navigation.

The arithmetic of navigation flow is described as follows:

Read next link, variable and judgment expression according to finished activity.

Get next activity calling judgment model.

Endow the activity with dynamic attributes and examine intercurrent control.

Intercurrent control model examines activities which are able to begin according to assignation rules. Flow watch model is notified to begin to execute the activity.

4.4 Intercurrent control

Some instances of a model or different instances of different models maybe exist at the same time while system is running. Furthermore, there will be competition for resources among instances. Intercurrent control model is used to deal with this problem.

4.4.1 Intercurrent resolution

We introduced the preemptive resolution. Each resource has an activity competition queue. When the priority of activity in queue changed, the queue must be adjusted. Activities which have the same priority hold the resource according to the order. Figure 8 shows an activity wait queue of resource.

In order to avoid deadlock (a state that a suspended activity remains hold the resource) while activity wait and execute, we introduced a method to adjust activity. The method is as follows:

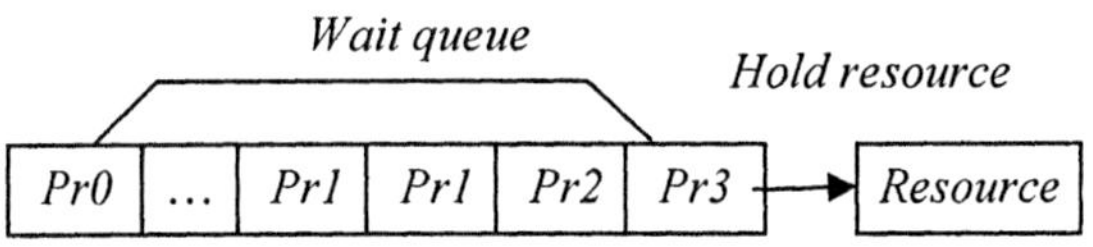

Figure 8. Activity resource wait queue.

Priority modulus (Pr) is set 0 when the activity is suspended. Pr=0.

Priority modulus must be counted and reset when an activity starts or resumes. Pr=wait modulus+ hold modulus +import modulus

System count waiting modulus of activity timely and adjust the queue.

According to above discussion, intercurrent control model should count wait modulus of activity in resource wait queue when the activity requests resource. Executing activity will be deprived of resource he held for lessening of wait modulus caused by holding resource long time excessively. Figure 9 shows the expressions of wait modulus.

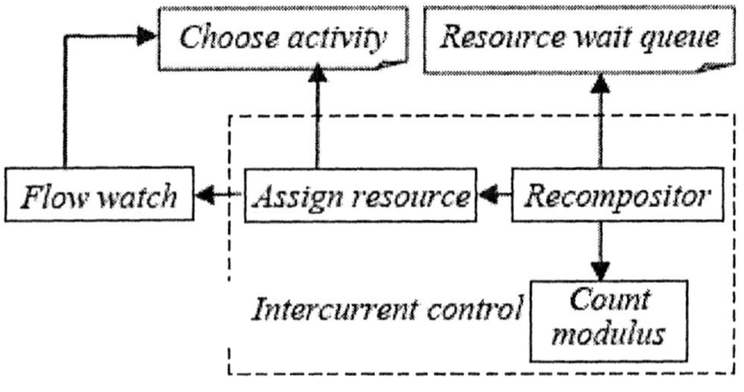

Figure 9. Process of Intercurrent control.

4.4.2 The arithmetic of intercurrent control

The arithmetic of intercurrent control is described as follows:
Count waits modulus of activities in resource wait queue.
Count waits modulus of activity hold the resource.
Choose an activity that has the most wait modulus.
Assign resource to the resource and notify watch model that the activity begins.

4.5 Execution control of activity

An instance of activity has six states: initialization, execution, ready, wait, suspended and achieved. Initialization state means period between creating instance and executing it. Ready state means that the instance has gotten all need parameters and other conditions. Wait state means that an activity is waiting for a time or resources to execute. Suspended state means that instance is suspended when there is suddenness. Figure 10 shows Activity instance state transition.

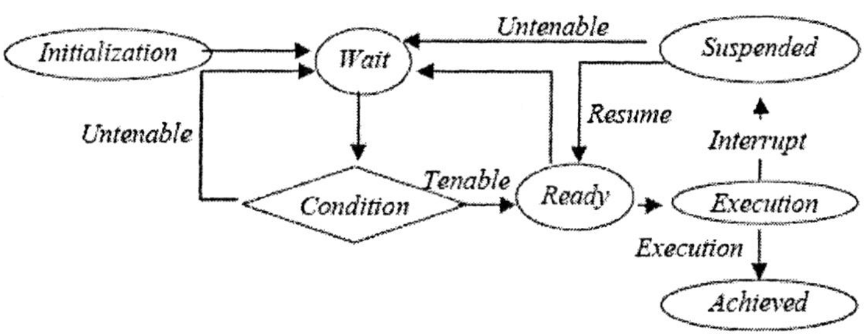

Figure 10. Activity instance state transition.

Activity may need various resources while it is executing. So it is state transition based on multi-resource and preemptive model.
The arithmetic of activity state control is described as follows:
Search resource that activity needs and put the activity in wait queue of the resource.
Execute the activity which is ready and set right state of activity according to its execution state.
Examine the suspended activity whether it hold resource. If it is deprived of resource, its state will be set wait state.
Examine the ready activity whether it hold resource. If it is deprived of resource, its state will be set wait state.

4.6 Task management

As a manpower activity, system needs to translate activity to customization form and gives notice to user as task message. Task management model provides of function which manages task list of user. The function gives notice to flow watch model and instance manager to run flow control.

Task is oriented user. Generalized user contains some resources needed in activity execution. A manpower activity means that the activity needs people (or resource) to execute it. So Task list is manpower resource wait queue.

Figure 11 shows four submodel of task management model and relation among them. Execution submodel executes manpower activity wholly. The process is as follows:

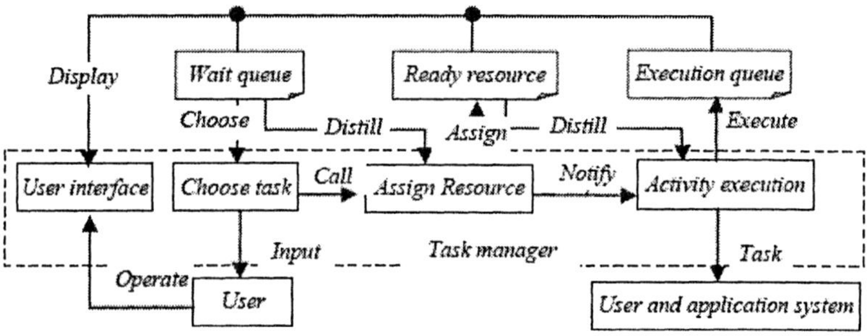

Figure 11. Activity instance state transition.

Distill an activity of manpower which do not begin or suspend.

Search other resources needed.

Choose an activity which hold all resources from flow ready queue to execute if it needs user's input data, user inputs data through user interface provided by task management.

If activity needs one more different type resources, resource wait of activity is very complex. The activity may get all resources at the same time for ever, which is called deadlock of resource.

In order to solve the deadlock of resource, we designed a resolution based on resource message. It is means that user can modify the position of activities in resource wait queue, force-deprive of held resource of activity. The mechanism of solution is defined as follows:

The deprivation is unconditional. But when a resource is held by other activity in un-deprivation pattern, user can not hold the resource until the activity achieves. Un-deprivation hold pattern is designed for activity which is not allowed to be interrupted while is executing.

In situation that activity needs several people to join in, user can send messages to other people of the activity. The activity begins after all people replied.

In situation that activity needs various hardware resource (all resources except people), user cannot operate for no user interface. System runs intercurrent control in background. The arithmetic is used to avoid deadlock. System force-assigns other resource to activity when the activity hold a resource.

5. CONCLUSION

A novel workflow system support dynamitic alliance was designed. The realization method of the system was introduced. The memory formats in database of workflow definition and workflow instance were designed. Intercurrent conflict resolution and harmony strategy for several resources request were introduce.

6. REFERENCES

1. Guangtai Min, Tao Feng, *Dynamic Flexible Workflow Management System Based on Middleware Technology*, Computer Integrated Manufacturing System, 2002, 8(8), pp.655-659 (in Chinese).
2. Tao Feng, Qian Wang, *An Approach to the Flexible Workflow Management Based on Doc*, Computer Application, 2002, 22(2), pp.20-23 (in Chinese).
3. Tong Lin, Zaizhong Zhang, Hongchen Li, *Process Models in Workflow Management System*, Computer Application, 2002, 22(6), pp.13-16 (in Chinese).
4. Yongquan Fan, Zhujun Wang, Yujun Luo, *A Workflow Management System Based on relation Database*, Computer Application, 2002, 22(7), pp.75-77 (in Chinese).
5. Bussler C., User mobility in workflow-management systems, *Proc Telecommunications In formation Networking Conference*, Melbourne, 1995.

A REGION GROWING ALGORITHM FOR SOLVING CVE PARTITIONING PROBLEM

Xiaomei Hu, Zhengjun Zhai , Xiaobin Cai
School of Computer, No.127 West Youyi Road, Northwestern Polytechnical University, Xi'an, China 710072,sufeimasohxm@163.com

Abstract: Collaborative Virtual Environment (CVE) systems offer multi-user support, in which many users share a virtual world and work collaboratively, and in which multiple-server architectures are widely used and multiple servers deployed, with each one handling a part of the virtual world. One of the key issues in the design of scalable CVE systems based on multiple-server architecture is the partitioning problem. In this paper, a new quality function of partitioning is used to evaluate the partitioning method and a region growing (RG) algorithm is proposed for solving the partitioning problem in CVE systems. The RG algorithm regards the partitioning problem as a dynamic assignment of cells to the servers. Some cells are chosen as growing points, and their neighbor cells are merged into the same region. Users in the same region are connected with the same server until the server is overloaded. Through the experiment, we discover that our method can provide better partitioning solutions with shorter execution time than the methods proposed in the previous literature of large-scale CVE systems. Therefore, the proposed implementation of RG algorithm improves scalability and efficiency of CVE systems.

Key words: collaborative virtual environment; partitioning; region growing.

1. INTRODUCTION

Collaborative virtual environment (CVE) supports users working on different computers to share a virtual environment and accomplish collaborative works. A scalable CVE system usually adopts the multiple-server architecture so that more users can participate in the system. A CVE is partitioned into several parts and each server is in charge of a part of the virtual world. When a new user logs in the CVE system, it will select one of

---

Please use the following format when citing this chapter:
Hu, Xiaomei, Zhai, Zhengjun, Cai, Xiaobin, 2006, in International Federation for Information Processing (IFIP), Volume 207, Knowledge Enterprise: Intelligent Strategies In Product Design, Manufacturing, and Management, eds. K. Wang, Kovacs G., Wozny M., Fang M., (Boston: Springer), pp. 960-965.

the servers in the system to connect with, which makes the server side increase the computation workload. In addition, when the user modifies the state of system, it sends the update message to its server, which in turn propagates this message to other servers or clients to maintain the consistency of the system, which makes the server side increase the communication cost. There are two kinds of communication messages in the CVE systems: Inter-server ones and Intra-server ones as shown in Figure 1.

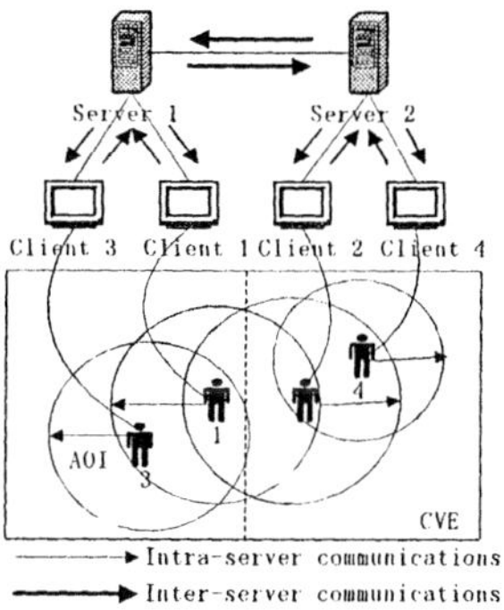

Figure 1. Two kinds of communication messages in CVE

Lui and Chan[1] defined a quality function C_p consisting of the computing workload C_p^W and overall inter-server communication cost C_p^L to evaluate the partitioning algorithm and used Linear Optimization Technique (LOT) algorithm to minimize the value of function C_p. The LOT algorithm is periodically executed for adapting the partition to the current state of the CVE system, which may take much execution time.

P. Morillo et. al. [2,3,4,5] put forward several heuristic search methods and presented a comparison study of several heuristics methods for solving the partitioning problem in CVE systems. These methods improve the performance of the partitioning algorithm by providing better C_p values and requiring shorter execution time than LOT method. The characterization study of CVE system[6] shows that the partitioning problem seriously affects the overall performance of the CVE system, since it determines not only the workload each server is assigned to, but also the inter-server communication requirements. According to the character of the CVE system, the effective partitioning algorithm should not only balance the workload among the servers in the system in such a way that none of them reaches its saturation point to improve the system throughput, but also assign the users doing the same collaborative task to the same server to minimize the inter-server messages in order to reduce the network traffic.

2. A NEW PARTITIONING APPROACH

In this section, a new quality function of partitioning is introduced and a region growing algorithm for solving the partitioning problem is proposed.

2.1 Quality function of partitioning

Average system response time is the optimal parameter to evaluate the partitioning algorithm. But it is hard to get the actual data before a real CVE system is constructed. Through the analysis of the CVE system response character[6], we propose the quality function of partitioning R_p.Assuming that CVE where the load is W is partitioned into p regions and the network traffic of inter-server communication of region i and region j is denoted as $D_{i,j}$, the inter-server communication is defined as

$$C_p^L = \sum_{j=1}^{p} \sum_{i=1}^{p} D_{i,j} \quad (1)$$

Let the threshold δ be the maximum number of clients that a server can handle smoothly. Assuming the workload of the server i is W_i and the threshold of the server i is δ_i, the CPU utilization of the server i can be approximated by:

$$L_i = \frac{W_i}{\delta_i} \quad (2)$$

The quality function R_p is defined as

$$R_p = C_p^L \times \max\{1, L_1, L_2, \cdots L_p\} \quad (3)$$

The best partitioning strategy $P*$ should make the quality function R_p* minimum. In order to get R_p*, the inter-communication should be minimized and the workload of servers should be lower δ. If the workload of servers does not balance, some servers can easily reach to the saturation point to make them overloaded. If servers are overloaded seriously, namely the value of L_i is very high, the value of R_p will increase rapidly. Therefore, R_p become a near optimal quality function to evaluate the partitioning algorithm.

2.2 A RG algorithm for partitioning the CVE

CVE is divided into $N \times M$ cells and the size of cell is the average value of the users' vision domain radius. The neighborhood of a cell is the adjacent cells sharing the same border of the cell. The users only communicate with the other users in the neighborhood of its residing cells. So communication cost between two partitions can be simplified to the case that we only need to consider the users of the adjacent cells that are divided into different partitions. Suppose each cell c_i in CVE is labeled with the number of users that it contains $L(c_i)$ and the assigned server $S(c_i)$, the communication cost in the system is considered the sum of users in the cells c_i whose adjacent cells have different value of $S(c_i)$. In order to find the optimal partitioning, the communication cost should be reduced as the same time each server should not exceed δ .Therefore, the cells which contain fewer users can be considered as the border of partitions.

RG algorithm first initializes growing points which is near the general idea of the graph theoretic method. Assuming that there are *k* servers in the system, CVE is initially separated into *k* regions. Then all the cells in each region are sorted by their $L(c_i)$ value, and the first *k* cells called mass-center (mc_k) that are located in different regions are chosen as growing point of each region and assigned to different servers. At this time, the workload of each server is the value of $L(mc_k)$. Next is to assign the other cells to servers. The neighborhood of the cell mc_k is defined as $N(mc_k)$ and the cell which has the maximum value $L(c_j)$ ($c_j \in N(mc_k)$) and is not assigned to servers is merged into the cell mc_k and assigned to the server *k* if the workload of the server *k* is not beyond the threshold. This process is cycled until all the cells are assigned to servers and the pseudo-code is shown in Algorithm 1:

- Algorithm 1. Region Growing algorithm

```
Begin
  Divide CVE into k regions
  let S(c_i)=0;
  For i=1 to k do
      Choose the biggest element mc_i of the region i;
      S(mc_i)= i;
      n--;
      Workload(Server_i)=L(mc_i);
  Endfor
  While(n<the number of the cells)
     For i=1 to k do
       List=obtain(N(mc_i));
      Sorted-list=Quicksort(List);
      c_j =The first element in Sorted-list;
      If (S(c_j)=0&&L(c_j)+Workload(Server_i)< δ)
         Merge(c_j,mc_i);
         S(c_j)=i;
         n--;
         Workload(Server_i)+=L(c_j);
      Endif
```

Endfor
Endwhile
End.

According to the RG algorithm, partitioning problem is the dynamic assignment of cells to the servers. In this process, the workload of server i increases gradually until server i reaches to a saturation point δ . The CVE systems are extremely changing environments since users can join or leave the system at any time and they can also freely move throughout the virtual world. The RG algorithm should be executed periodically. With this algorithm, the servers cannot be overloaded because new cells are not assigned to the server when the number of users in the region managed by the server is close to the threshold δ . Since the cells with high $L(c_i)$ value in the neighborhood of mc are first assigned to the same region with mc , the cells with low $L(c_i)$ value become the border of regions, which reduces the inter-server communication.

3. EXPERIMENTS AND RESULTS

In this section, we present the performance evaluation of the RG algorithm for solving the partitioning problem in CVE systems. We use the hardware platform of 2.4 GHz Pentium IV with 512 Mbytes of RAM and develop a simulator in C++ running on windows2000 to model a CVE system based on the hardware platform. The virtual environment is divided into 609 cells. There are six servers in the CVE system and the threshold δ of each server is 100. Users are randomly distributed in the CVE in the initialization. During the simulation, they move to six hotspots and finally show clustering distributions in the CVE.

The value of the quality function R_p and execution time T/ms is used to evaluate partitioning algorithm. According to [2], the ACS search method currently provides the best heuristic method for solving the CVE partitioning problem. Table 1 shows the results of two partitioning algorithms. It proves that RG algorithm provides a lower value of R_p and requires a shorter execution time than the ACS method. Thus we learn that the ACS method is not the best partitioning solution because it does not provide the lowest R_p value though it can provide the lowest C_p value. In addition, the execution time of ACS method increases more quickly with the users increasing in the CVE because the complexity of calculation in ACS algorithm is $O(B)$ where B is the number of users in the border of regions which increases with the number of users in the system while the complexity of calculation in RG algorithm is $O(n)$ where n is the number of the cells in the CVE system which is a fixed value.

Table 1. The results of partitioning CVE

user	RG		ACS	
	R_p	*T(ms)*	R_p	*T(ms)*
300	36	613	45	1413
400	42	688	49	1920
500	53	725	57	2491
600	64	852	66	3214

4. CONCLUSIONS AND FUTURE WORK

The CVE systems are becoming an important research area in parallel and distributed processing field. In this paper, we discussed a new region growing algorithm for solving the partitioning problem in scalable CVE systems. A new quality function of partitioning is given to evaluate the partitioning algorithm for a large CVE system with clustering distributions of the existing users in the systems. The experiments prove that the region growing algorithm takes lower execution time and provides better solutions to the partitioning problem than current approaches. As future work to be done, one should consider use of grouping strategy in the partition management in CVE systems with multiple-server architecture to reduce the network traffic.

5. REFERENCES

1. John C.S.Lui and M.F.Chan,(2002), An Efficient Partitioning Algorithm for Distributed Virtual Environment Systems, *Parallel and Distributed Systems*, Vol.13, pp.193-211.
2. P. Morillo, M. Fernandez and J.M. Orduna, (2003), A comparison study of modern Heuristics for Solving the Partitioning Problem in Distributed Virtual Environment Systems, In: *proceedings of the International Conference on Computational Science and its Applications*, Montreal, Canada, Springer, pp. 458-467.
3. P. Morillo and M. Fernandez,(2003),A GRASP-based algorithm for solving CVE partitioning problem, In: *Proceedings of, Parallel and Distributed Processing Symposium*, Nice, France, IEEE Computer Society, pp.60.2
4. P. Morillo, M. Fernandez and J. M. Orduna, (2003), An ACS-Based Partitioning Method for Distributed Virtual Environment Systems, In: *Proceedings of, Parallel and Distributed Processing Symposium*, Nice, France, IEEE Computer Society, p.148.2
5. P. Morillo, P. Lopez , J. M. Orduna, and M. Fernandez, (2003), A New Genetic Approach for the Partitioning Problem in Distributed Virtual Environment Systems, In: *Proceeding of the 10th Conference of the Spanish Association for Artificial Intelligence*, Spain, Springer-Verlag GmbH, pp.76-85.
6. P. Morillo, J. M. Orduna, M. Fernandez and J. Duato, (2003), On the Characterization of Distributed Virtual Environment Systems, In: *proceedings of Euro-Par2003 Parallel Processing*, Austria, Computer Science 2790 Springer, 2003, 1190 – 1198.

VISUAL STUDIES ON RICE BRAN EXTRUSION

Wenhua Zhu [1], Huiyuan Yao [2]
[1]*CIMS & Robot Center of Shanghai University, Shanghai University, China; Email: toney_wh_zhu@staff.shu.edu.cn.* [2]*Food School, Southern Yantze University, China.*

Abstract: This paper presents to make use of visualization technique to study the process of extrusion and extrusion assembly, the visual phenomena indicate that material transportation underlying the extrusion course is a combination of two transportations, one is positive displacement pumping transportation in the intermeshing region, and the other is partially filled transportation in the translation region. Virtual assembly technique is used to design the extrusion assembly, and interference is checked, and animations are created to demonstrate assembly sequence and path. This product has been designed successfully by computer in virtual environment.

Key words: Visualization technique, Virtual assembly, Rice bran, Extrusion.

1. INTRODUCTION

Nowadays visualization technique is a valid and useful mean to research the complicated problem of engineering that have not been reached and researched intensively. Owing to the complexity of process, extrude process is called as "Black Box" for many years, as is shown in the fig.1[1], we can see clearly that process equipment is a black box between the input and output. Visualization technique is adopted to research the order of rice bran extrusion, so that to provide the academic basis that designs the rice bran extruder and establishes theory model of rice bran extrusion, hence it will be made the theory model match the fact better.

This project is supported by the 10th Five-year Plan Key Technologies R&D Program of China under the grant No.2001BA501A03.

Please use the following format when citing this chapter:
Zhu, Wenhua, Yao, Huiyuan, 2006, in International Federation for Information Processing (IFIP), Volume 207, Knowledge Enterprise: Intelligent Strategies In Product Design, Manufacturing, and Management, eds. K. Wang, Kovacs G., Wozny M., Fang M., (Boston: Springer), pp. 966-971.

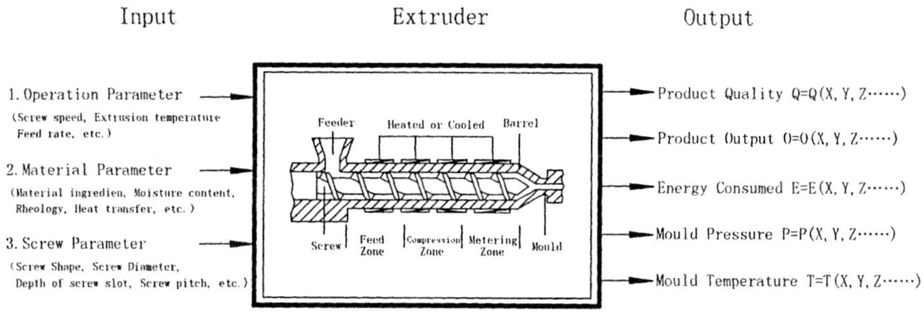

Figure 1. Black box model of food extruder

Rice bran is a sort of nutritious and functional food material, In order to develop rice bran resources, the first thing is how to stabilize rice bran. For the stabilization research of rice bran, there are many different approaches, J.R. Loeb etc.[2] used refrigerating method to handle the rice bran, M. Williams and the S. Baer[3] first time adopted extrusion to stabilize rice bran in 1965, J.V. Prabhakar etc.[4] stabilized rice bran with the chemistry method, J. Tao etc.[5] utilized microwave to stabilize rice bran, Fernando etc.[6] stabilized rice bran in the fluidized bed. It was latest reported by American RiceX company[7], extrusion with high temperature and high pressure and high shearing is an economical and effective method, also become unique engineering method available. All these above researches do not mentioned visualization method, and this paper will study rice bran extrusion with visualization technique.

2. VISUAL PHENOMENA AND ANALYSIS OF RICE BRAN EXTRUSION[8]

2.1 Positive displacement pumping transportation in the intermeshing region

When rice bran is added to the barrel from feeder, one part of materials is piled up in the intermeshing region and transported forward along the length direction of screw, and another part of materials is overflowed to the under intermeshing region and also transported forward. This phenomenon is called positive displacement pumping transportation in the intermeshing region, fig.2(a) is the illustration and fig.2(b) is the visual picture

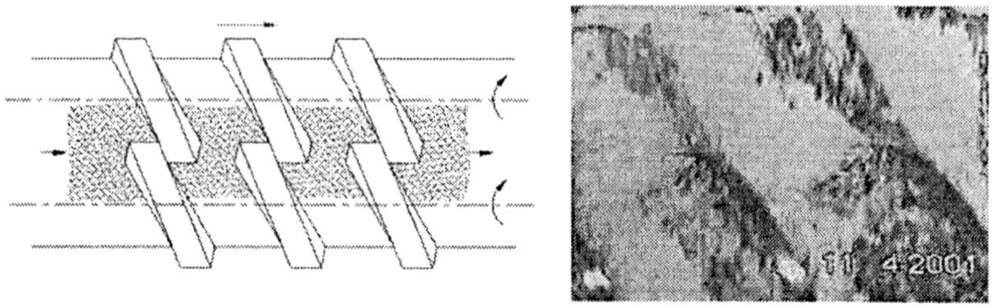

(a) Illustration of positive displacement. (b) Visual picture of positive displacement.

Figure 2. Positive displacement pumping transportation in the intermeshing region

2.2 Partially filled transportation in the translation region

When rice bran is added to the barrel from feeder, the other part of materials is transported in the side screw slot, rice bran entered side slot is moved upward into the upper intermeshing region by fiction drag, thus the consecutive movement of rice bran is formed the "∞" shape, but the screw slot isn't fully filled rice bran. This phenomenon is called partially filled transportation in the translation region, fig.3(a) is the illustration, fig.3(b) is the visual picture.

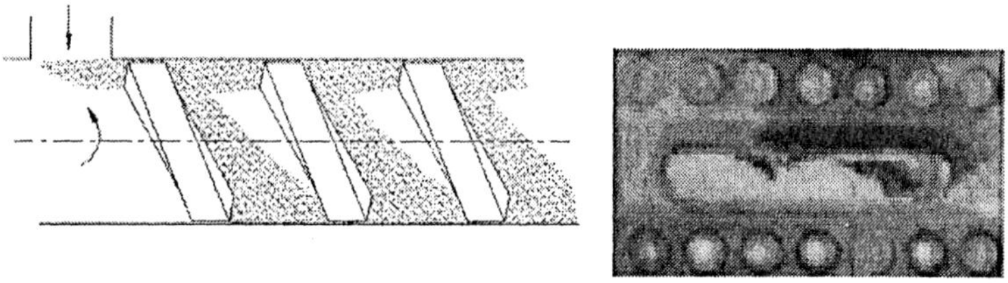

(a) Illustration *(b) Visual picture*

Figure 3. Partially filled transportation in the translation region

As a result, material transportation underlying the extrusion course was a combination of two transportation, one was positive displacement pumping transportation in the intermeshing region, and the other was partially filled transportation in the translation region。

3. VIRTUAL ASSEMBLY[9]

Virtual assembly is a kind of technique to take into the comprehensive usage of CAD technique, visualization technique, and simulated technique. Assembly of virtual prototype is worked out in the virtual environment without the real physical model and real process of assembly.

3.1 Process of virtual assembly

In the process of virtual assembly, assembly plan must be carried out first, it includes the enough information to support the process choice decision, and acquire the more detailed process definition, consider the relationship of hierarchy and constraint between parts and subassemblies, estimate the feasibility of assembly, choose the good assembling sequence and path. At the same time, the moving path of each parts need to be recorded in the assembling space environment, thus these paths become the soft districts, need to carry out the method to establish, save and review these soft districts.

The assembling description in virtual environment is reflected to the real assembling process. Virtual assembly will produce some useful information, e.g. assembly interference, sequence and path etc. and the information should be sent to the CAD system to modify the 3D model further. fig.4 is the process of virtual assembly.

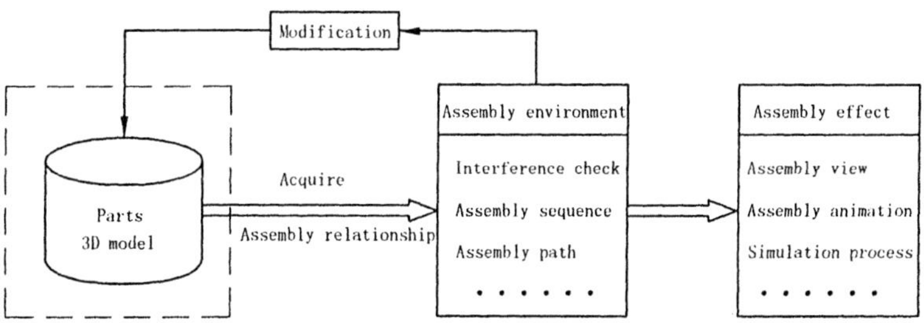

Figure 4. Process of virtual assembly

3. 2 Interference check in virtual assembly environment

Taking Teamcenter Visualization software as the virtual environment, rice bran extruder is designed with 3D software, then these 3D model is transferred into the Teamcenter Visualization to assemble visually.

Interference check needs to set the check scope first, you can select the shown parts on the screen or all parts of prototype. Moreover clearance needs to setup, and run the interference check program, environment will show you the interference between every two parts which have been assembled each other, and the report is in a matrix list. Such as the fig.5 show, the diagram is the interference circumstance of the mould and the barrel head, the manifestation of interference is penetration, this is highlighted with the red color on the entity, and the section view shows the section circumstance clearly. This interference can be removed by the further modification.

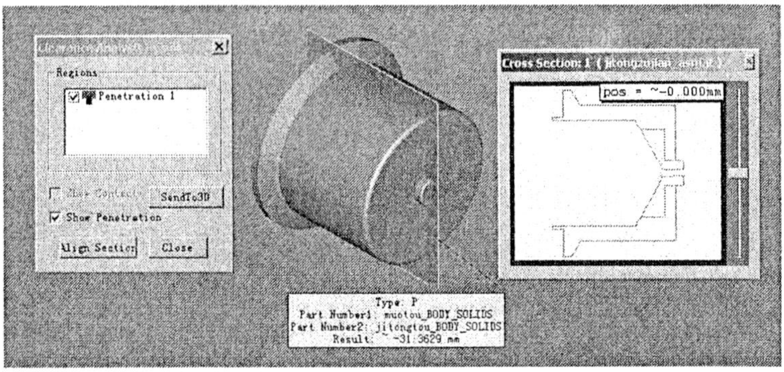

Figure 5. Interference check in virtual assembly environment

3. 3 Virtual assembly sequence and path

To create the assembly sequence and path should disassemble the assembly model artificially, base on the supposition which assembly is disassembly available, we can acquire the sequence and path by tracking parts reversely. Such as fig.6 show, the diagram is the assembly sequence and path of screw subassembly, small balls indicate the assembly paths, the sequence of these paths also means the assembly sequence.

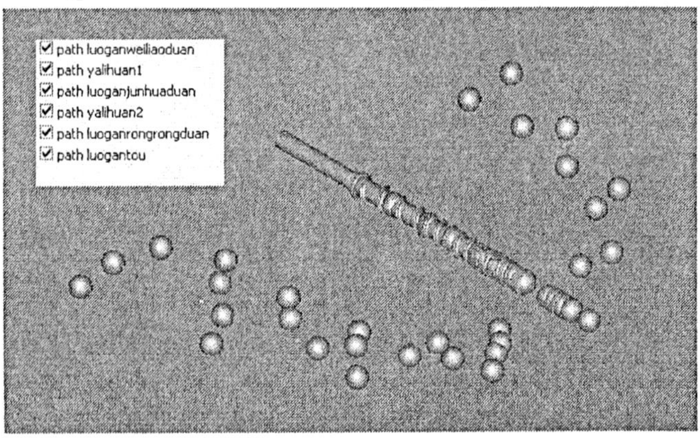

Figure 6. Assembly sequence and path layout

Assembly animation can be established to demonstrate an assembly sequence or some other series actions, firstly we specify the animation properties to control the whole speed of animation and keep on time, then add a series of events, these events include assembly path, and Organize actions into events, thus broadcast the animation to validate the assembly process, further edit the animation and create the movie document of AVI. Finally the animation of virtual assembly is completed.

4. CONCLUSION

Visualization technique is modern mean to research the complicated problem of engineering. We make use of visualization technique to open the "Black Box" that it is hard to understand at former time.

Visual experiments has been done on the process of rice bran extrusion, the phenomena indicate that material transportation underlying the extrusion course is a combination of two transportation, one is positive displacement pumping transportation in the intermeshing region, and the other is partially filled transportation in the translation region; extrusion of rice bran is a process of cooking at high temperature and under high pressure。

Virtual assembly technique has been used to design the extrusion assembly, and interference is checked, and animations are created to demonstrate assembly sequence and path. This product has been designed successfully by computer in virtual environment.

5. REFERENCES

1. Connor, Colm, (1987), *Extrusion Technology For the Food Industry,* Great Britain: Galliard Ltd, Great Yarmouth.
2. Loeb, J.R., et al., (1949), Storage of the Bran as It Affects Hydrolysis of the Oil, *Journal of the American Oil Chemists' Society,* vol.27, pp738 743.
3. Williams, M., and Baer, S., (1965), The expansion and extraction of rice bran, *Journal of the American Oil Chemists' Society,* vol.42, pp151 155.
4. Prabhakar, J.V., et al., (1986), A Simple Chemical Method for Stabilization of Rice Bran, *Journal of the American Oil Chemists' Society,* vol.63(5), pp644 648.
5. Tao, J., et al., (1993), Microwave Heating for Rice Bran Stabilization, *Journal of Microwave Power Electromagnetic Energy,* vol.28(3), pp156 159.
6. Fernando, W.J.N., et al., (1993), Effect of Fluidized Bed Drying on Stabilization of Rice Bran, *Drying Technology,* vol.11(5), pp1115 1119.
7. The RiceX Company--Feeding the World, (2000), [EB/OL] www.RiceX.com
8. Zhu Wen-hua, (2001), *Studies on Stabilization of Rice bran by Extrusion Cooking,* PhD. Dissertation of Southern Yangtze University, (in Chinese).
9. Zhu Wen-hua, Ma Deng-zhe, (2004). Application Research of Virtual Assembly Technology, *Journal of Design and Research,* vol.94(6), pp47 48, (in Chinese).

IMPLEMENTATION OF WEB RESOURCE SERVICE TO PRODUCT DESIGN

Baoli Dong[1,2], Sumei Liu[1]
[1]Department of Mechnical Engieering, Taiyuan University of Technology and Science, China; Email: tydbl@hotmail.com; [2]Department of Mechnical Engieering, Zhejiang Univ., China.

Abstract: Manufacturing enterprises needs the wider external resources during the product design phrase. A typical web resource service approach is presented in this paper. It realizes Web-based manufacturing resource discovery, process and application in product design. This pattern addresses on the differences of service ways considering different types of resources. The relevant enabling technologies are put forward. Manufacturing resource information is achieved through Web search engine and processed with semantic Web. Meanwhile, the Web resource repositories such as Web parts library and distributed knowledge bases are set up preliminarily. Finally, the two application cases of Internet-based collaborative product design are introduced briefly.

Key words: Manufacturing Service, Web Resource, Product Design.

1. INTRODUCTION

Manufacturing enterprises must respond speedily to the market and customer needs[1,2]. The increasing product complexity and the globalization manufacturing further necessitate a wide-ranging distributed manufacturing resource demand involved in the product development stage. Product development more and more requires that manufacturing enterprises possess of the higher design level, advanced manufacturing equipments, better network communication environment[3]. However, the distribution of manufacturing resources is unbalanced. Most enterprises are incapable of satisfying these conditions and unable to implement product development alone. As a possible solution, the Web sharing of qualified manufacturing resources can help them utilize the external resources adequately. For this

This project is supported by the Key Project of Ministry of Education of China under the grant No.211CERS12.

Please use the following format when citing this chapter:
Dong, Baoli, Liu, Sumei, 2006, in International Federation for Information Processing (IFIP), Volume 207, Knowledge Enterprise: Intelligent Strategies In Product Design, Manufacturing, and Management, eds. K. Wang, Kovacs G., Wozny M., Fang M., (Boston: Springer), pp. 972-977.

purpose, the enabling technologies of Web-oriented manufacturing resource discovery, integration and sharing services for product design are analyzed in the following sections.

2. RESOURCE SERVICE AND KEY TECHNOLOGIES

Product development is viewed as integration and sharing process of design and manufacturing resources[4]. Manufacturing resources available on the Web are much more extensive and the application ways are of diversity. Manufacturing resources can be divided into hardware, software, data and knowledge[5]. The unified data exchange standards like STEP, XML and semantic description like RDF, Ontology realize the in-depth data sharing and semantic services. The interlinking and interoperability through Internet directly for most software and hardware resources are difficult at present due to the technical and management factors. So Web resource services can help the enterprises to get external service resources and implement collaborative design and manufacturing. Considering the actual enterprise needs and product design process, we address to the following key technologies relating to Web manufacturing resource services.

- **Web resource discovery**: The enterprises need to acquire the external existing resources timely, apply them to enterprise decision-making and design activities fully. So resource discovery affects the succeeding resource integration and sharing process. We employ Web search, information processing and data mining applied to enterprise intelligence collection, knowledge management and decision analysis relating to product design activities. The Web manufacturing information and knowledge are achieved through focus crawling, information extraction and data mining, thus provide service contents such as resource navigation to enterprises.
- **Web resource repository**: Shared manufacturing resources determine the service quantity, range and quality. Web resource repositories realize resource integration, sharing and reuse. They can be divided into two types: one is information base and resource portal through automatic search from the Web, the other is that enterprises issue and provide such resource as distributed parts bases and knowledge ones. Both provide the general resource retrieval services, parts downloading and knowledge management.
- **Web sharing supportive environment**: Web resource applications need a communication, coordination and cooperation network supportive environment. Considering resource difference, the sharing tools consist of network conference, white board, RPC and etc, which conduce to cooperative product development among users.

3. APPLICATION CASE

3.1 Manufacturing information and resource discovery

The enterprise portals and domain sites contain abundant of Web manufacturing resources and information. The general search engines like Google and Baidu can't satisfy the search requirement of enterprise resources and the precision ratio is very low. The active and open Web resource discovery aiming at manufacturing field is an effective approach to search specialized resources. It can extend the acquiring range of shared resources and improve the search accuracy. Fig. 1 specifies the overall process of Web resource discovery.

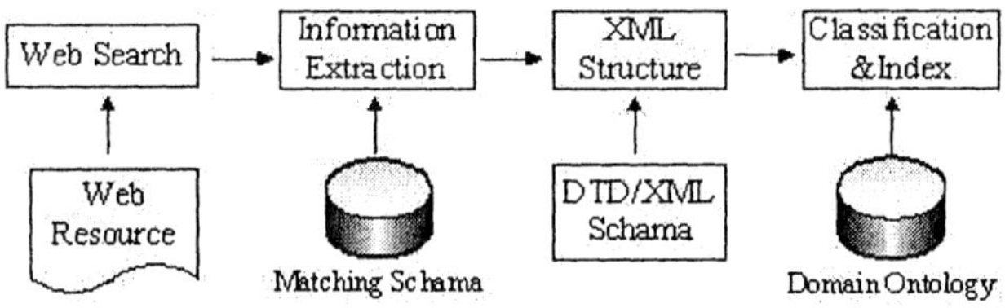

Figure 1. Web discovery and semantic transaction of manufacturing resource

Focus crawling can guarantee the search accuracy to specialized resources on the Web[6]. To a certain type of topic enterprise site, website topic feature modeling is a critical issue for focus crawling. The content and structure characteristics of different topic sites are obvious distinct. The semantic of feature text reflect its topic. The semantic text information extracted from the site is mapped to HVSM-based topic feature vector[7]. Then the topic classifier using support vector machine algorithm (SVM) determines its topic. The search engine only rambles topic-specific sites and gathers relevant Web pages.

The wrapper deals with parsing page contents and extracting manufacturing information. Web pages are of semi-structured documents, and manufacturing information are mixed with HTML tags and other useless contents. They are normalized the well-structured ones firstly, and converted to DOM (Document Object Tree) tree with DOM API of W3C. DOM tree determines the locating tags and extraction path of HTML document[8]. Considering the existing structure forms of HTML documents are very complicated, the extraction models are achieved by the Extraction Rule Learning Algorithm (ERLA)[8]. According to the definite pattern matching rules, the manufacturing information is extracted through the prefix and suffix forms of table attribute terms like product name, specification and price, etc. The extraction models include the locating tags, keywords, etc.

Finally, the finely granular information is extracted and structured through these processes above mentioned, and saved to the resource database of server. Therefore, the service system can gather specialized resource information automatically and update timely.

To enhance the accuracy of resource retrieval, Resource Description Framework (RDF) is employed to describe, categorize and index extracted information. As a unified metadata description norm based on XML syntax, RDF strengthens the semantic retrieval ability[9]. It is independent of specific fields, so RDF Schema (RDFS) defines the domain taxonomy of resource description, and constructs the hierarchy model of manufacturing resource entity types such as enterprise, service, product and function. This strengths the resource description capacity of RDF and provides the semantic support for resource in-depth application. For example, the service information about mould design that Institute A provides, whose URL address is "http://www.amtcn.org/", can be described with RDF/XML as following.

```
< ? Xml version="1.0" encoding="UTF-8" ? >
< rdf:RDF xmlns:rdf="http://www.w3.org/1999-02-22-rdf-syntax-ns#"
      xmlns:s="http://description.org/schema/" >
    < rdf:Description about=" http://www.amtcn.org/" >
     < rdf:type rdf:resource=" http:// description.org / term#Service"/ >
        < s:Item >automobile mould< /s:Item >
        < s:Creator > Institute A< /s:Creator >
    < /rdf:Description >
< /rdf:RDF >
```

The extracted resource metadata information is processed with RDF/XML. While the user send the resource search request, the server parses the RDF-based resource documents with Jena, converts retrieval-matching results to XML form and offers the semantic search service. The prototype system is developed with Java. The search objects aim at the manufacturing equipments, software, products and suppliers, etc. Fig. 2 shows a resource discovery example. The users can get various enterprise and service information results related to querying contents.

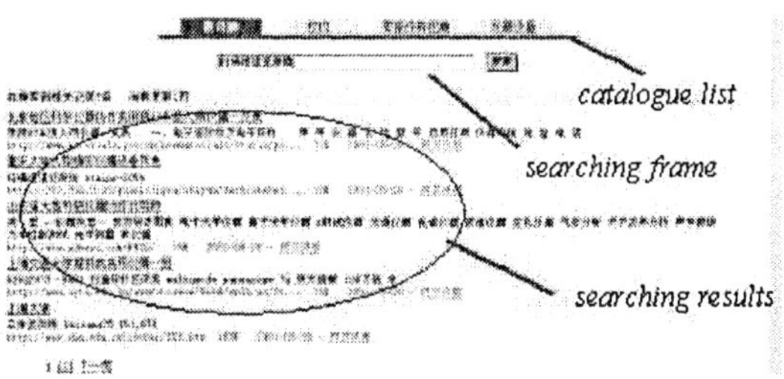

Figure 2. Resource search results of mould enterprises and organizations

3.2 Web parts library

Web parts library is one of design resource bases for rapid product development. The enterprise is able to issue its product models and technical files to Web parts library. Other enterprises lacking design capacities can retrieval, preview, download and reuse the needed product models from it directly. The system architecture of Web parts library is shown as Fig.3. The service information contain the parts geometric models and other relevant contents relating to parts like supplier information.

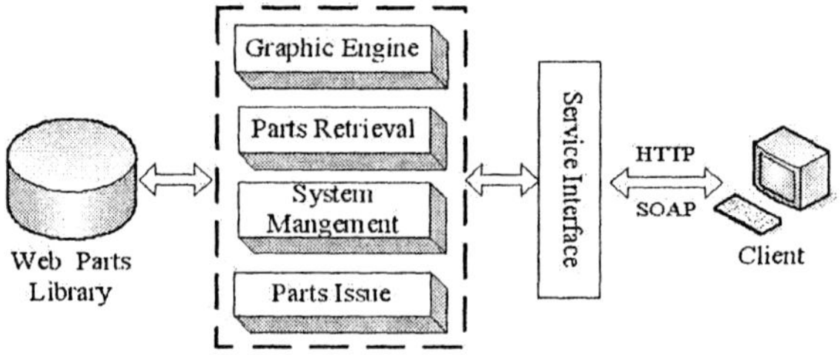

Figure 3. Architecture of Web parts library

As to parts retrieval, the modeling of parts family adopts the international standard ISO13584. This standard defines the product data representation and exchange format[10]. Parts description information includes two parts: parts data dictionary and its concrete contents. The former defines the parts type and relevant attributes, provides the parts classification identification standard. The latter is represented with XML syntax according to the parts classification standard and attribute elements by the data dictionary. The enterprise uploads the parts models by ISO13584 to Web parts library. The users retrieval parts information in terms of parts family and parts function characteristics. Fig.4 shows that the users can retrieval, browse and download parts information of tire mould through Web parts library.

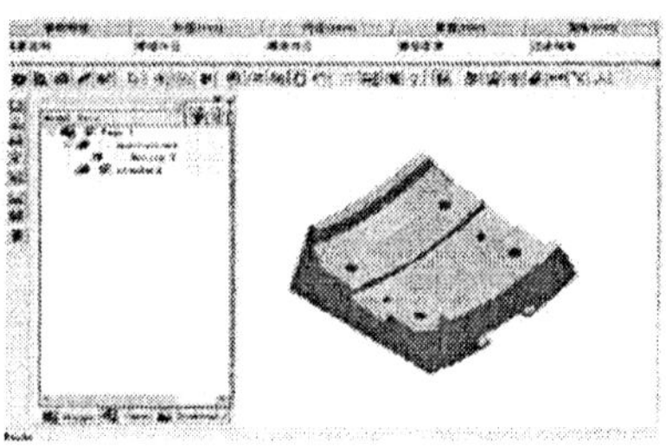

Figure 4. Browsing and downloading service of parts model

Web parts library provides a successful sharing environment and a seamless interactive channel through Internet. It integrates with the existing commercial design system and enterprise operational activities, allows distributed users to exchange and share product visualized information and meets collaborative work requirements.

4. CONCLUSION

In this paper, the Web-oriented resource discovery and sharing service pattern is introduced, and the key implementation technologies are discussed in detail. These service ways meets the enterprise demand on quality specialized resources from the Web. We have built a prototype system towards the in-depth and overall sharing of network manufacturing resources for production development. Relevant services realize resource discovery and reuse in the whole product development cycle. Up till now, they have applied to the diversified application service objects for various product development objects covering mould, automobile parts, pump, clothing, medicine, etc. This resource-sharing service method has certain of instruction signification for rapid collaborative product development of manufacturing enterprises.

5. REFERENCES

1. Xie, S., Huang, H., Tu, Y., (2002), A WWW-based information management system for rapid and integrated mould product development, *Advanced Manufacturing Technology* **20**(1), 50-57.
2. Huang, G., (2002), Web-based support for collaborative product design review, *Computers in Industry* **48**(1), 71-88.
3. Huang, G., Lee, S. and Mak, K. L., (2001), Synchronized web applications for product development in the 21st century, *Advanced Manufacturing Technology* **18**(8), 605-613.
4. Zhou, S., Chin, K., Xie, Y., et al, (2003), Internet-based distributive knowledge integrated system for product design, *Computer in Industry* **50**(2), 195-205.
5. Yang, T., Wang, Y., Xiao, T., et al, (2003), Resource discovery in context of Internet-based manufacturing, *Computer integrated Manufacturing System* **9**(1), 47-51.
6. Chakrabarti, S., Dom, B., Berg, M., (1999), Focused crawling: a new approach to topic specific web resource discovery, *Computer Networks* **31**(11), 1623-1640.
7. Dong, B., Qi, G., Gu, X., (2005), Domain-specific website recognition using hybrid vector space model, *Lecture Notes of Computer Science* 3739, 840-845.
8. Li, X., Gu, Y., (2002), DOM-based information extraction for the web sources, *Chinese Journal of Computers* **25**(2), 526-533.
9. Cai, M., Lin, L., Chen, G., et al, (2003), Intelligent resource discovery in Internet-based manufacturing, *Computer Integrated Manufacturing System* **9**(7), 589-594.
10. Yang, Z., Qi, G., Gu, X., (2005), Research on web parts library for mass customization, *Computer Integrated Manufacturing System* **11**(4), 541-547.

ENGINEERING TESTING PROCESS MANAGEMENT SYSTEM BASED ON .NET FRAMEWORK

Liang Jin, Gandi Xie, CaiXing Lin
College of Mechanicals Engineering and Automation, Shanghai University, Shanghai 200072, China.Email:jackyking2004@163.com

Abstract: This paper raises the idea that the traditional way to manage engineering testing documents by hands is not competent for Business Process Reengineering after the author has analyzed the status quo of the engineering testing process management in the auto production industries. The author raises Engineering Testing Process Management (ETPM) system software based on ASP.NET technology to solve the practical problems. Additionally, the architecture of the software and the realization of the system are discussed subsequently. The basic model of the ETPM system software and the prime modules which have been developed are introduced.

Key words: Engineering Test Process Management, ASP.NET, .NET architecture

1. INTRODUCTION

With the development of economy, enterprises faced with greater competition than ever. That demands enterprise to shorten the production cycle. It is proved in practice that Business Process Reengineering (BPR) is the most useful way to enhance the ability of the long-term competition for enterprise strategically.

The new product development in motor manufacturing is composed of design testing and trial-manufacture. There are so many processes in every aspect of them that caused the great complication of the engineering testing

Please use the following format when citing this chapter:
Jin, Liang, Xie, Gandi, Lin, CaiXing, 2006, in International Federation for Information Processing (IFIP), Volume 207, Knowledge Enterprise: Intelligent Strategies In Product Design, Manufacturing, and Management, eds. K. Wang, Kovacs G., Wozny M., Fang M., (Boston: Springer), pp. 978-983.

process management. So how to improve the engineering testing process management in BPR is becoming one of the crucial factors for the large-scale manufacturing enterprise to enhance its ability of competition.

The author raised the ETPM system to solve these problems. The prime modules have been developed successfully, and the software is begun to be applied in practical use.

2. ANALYSIS OF THE STATUS QUO OF EPTM

Engineering testing is indispensability in product developing. However, there are many processes and departments involved in this course. The testing documents have to be signed and be delivered by hand under the traditional way. That naturally prolonged the cycle of file delivery.

Obviously, the biggest problem is that all the documents (include test proxy, test program, test project, work order, test monitor data file, etc.) of processes are all finished by hands. That truly causes the low efficiency. Furthermore, some issues which have to be solved in engineering can hardly be realized such as version management, track of process design and so on under the traditional way. Accordingly, that absolutely becomes the bottleneck of the value chain in lean production.

On the other hand, if Chinese economy needs greater development, enterprises must think much of product innovation. Engineering testing is the important course and means in product innovation.

In a word, the effective management in engineering testing will vigorously ameliorate and optimize productivity.

3. THE MODEL OF ETPM SYSTEM

According to practical demands, the ETPM system is a full set of computer aided engineering testing process management system.

The design aims of the ETPM system are as follow, (1) It can collect the affairs which are fussy and hard to quantized to one uniform platform for users to deal with. (2) It can supervise the engineering testing documents (especially test proxy, test program, test project, work order, test monitor data file, etc.)systematically and gather those data to form a knowledge base which can supply historical data and data support for new product development in the future. (3) Different authorized user of engineering testing department can finish their respective work in this platform in cooperative way. (4) Authorized user can share the resources on the net according to the different purview. Figure 1 shows the systemic management model.

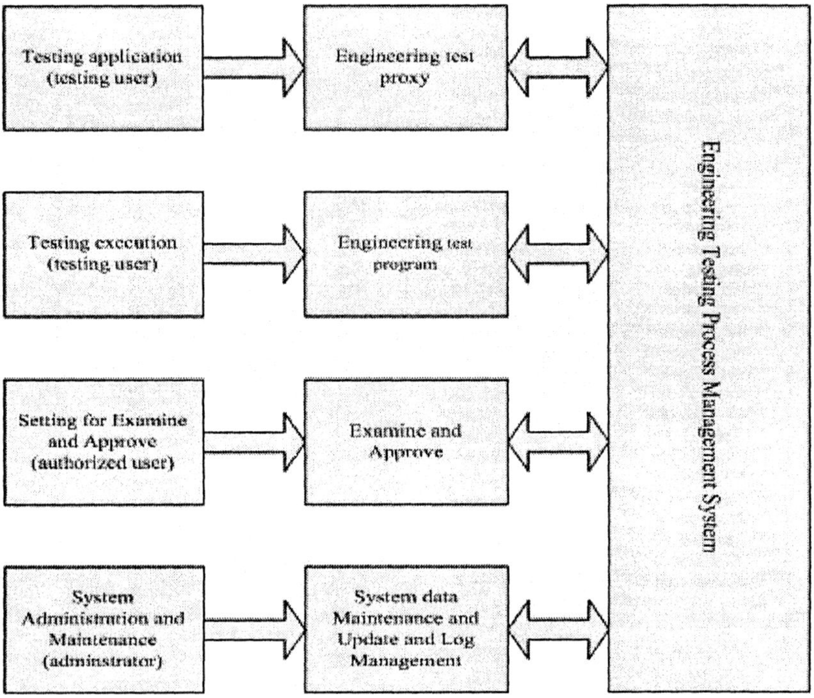

Figure 1: the Systemic Management Model

4. FUNCTIONAL DESIGN OF ETPM SYSTEM

The ETPM system is composed of 5 modules. They are test document management module, work project management module, system management module, security management module and examine and approve management module. (See figure 2)

4.1 Test document management module

Test document management module is used to realize the foundation and management of the documents, such as test proxy, test program, test project, work order, test monitor data file and so on.

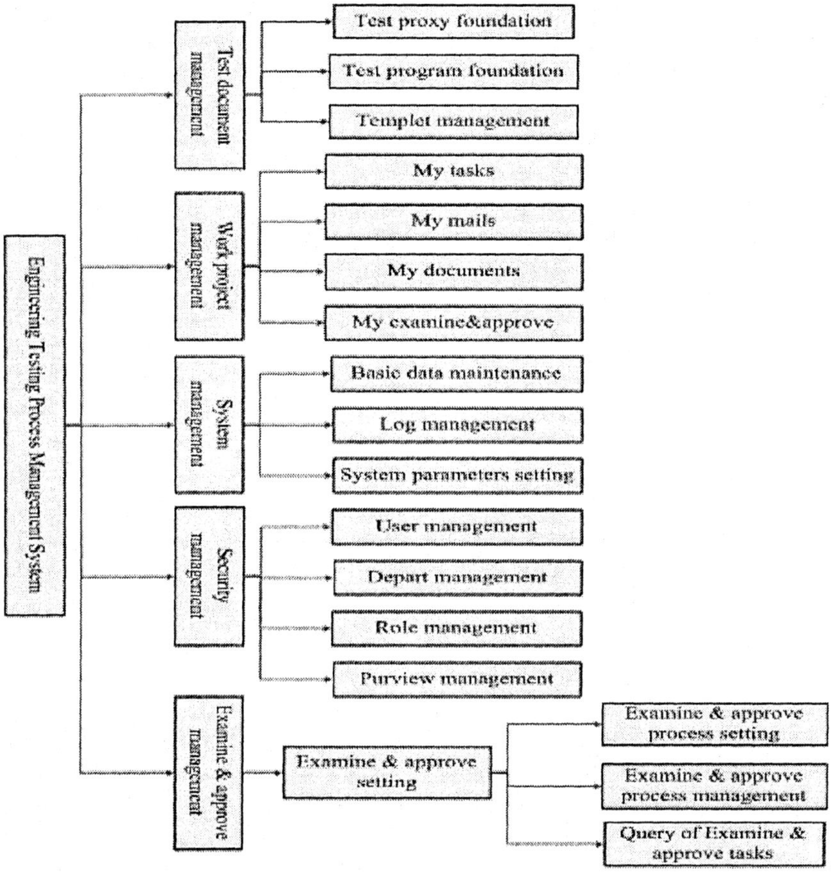

Figure 2: the Composition of the ETPM system

Test users who apply for the testing task are able to create new test proxy and test program by this module. User can pick contents from a knowledge base to create documents easily. In addition, user can use the templates stored in the system to create documents since some documents will be used frequently. And all the documents can be appended accessories to make the information of the document intact during the documents foundation.

4.2 Work project management module

Work project management module helps user to manage and arrange individual work scheme.

Authorized user can query their task and scheme or execute daily job. And it can help user to finish the tasks in time, such as test application, send and receive mails, examine and approve etc.

On the other hand, authorized user can arrange personal task according to own schedule. At the same time, the leaders can master the employee's schedule. The management can be bidirectional. The module also provide special task reminder assistant as an auxiliary function for user. So the on-line work management is realized. All the users can query the interior sharing resource and browse the real-time news.

4.3 System management module

System management module helps system administrator to maintain and update the data of the ETPM system.

The administrator can use the relevant functions which are provided by the ETPM system to fulfill the following tasks. First, the administrator can append, modify, delete and backup both the data of the document knowledge base and the basic information. Second, the administrator can manage the enterprise news and the messages. Third, the administrator can manage the logs which memorize all users' operation records. Fourth, the administrator can fulfill the task of setting system parameters.

4.4 Security management module

Security management module also helps system administrator to supervise user, department and role.

This module has the fundamental function of registration and modification for user, department and role. The administrator can use the retrieve tool which is provided by the module to search the basic information of the specified user, role or department quickly, and then the administrator can finish the update operations such as modify and delete. In addition, the system administrator can operator the purview management of user, department and role.

4.5 Examine and approve management module

Examine and approve management module provides users the function of foundation of examine and approve task toward the documents applied for authorized user.

After a testing user has submitted the test application documents, the specified user who has the purview to set the examine-and-approve task to create a new examine-and-approve task by this module. The user can arrange different group of people to finish the work of examine and approve according to practical demand. The examine-and-approve types can be parallel or serial. The examine-and-approve task really begins after it is completely created. At the same time, a new work task has been produced in the work schedule of the each person of the examine-and-approve group.

5. TECHNOLOGY REALIZATION OF ETPM

The ETPM system is based on distributed architecture. The user in the interior LAN can store, dispose, query and supervisor the engineering test documents. The software adopts the B/S structure. It is developed by Microsoft ASP.NET technology. All the maintenance and update will be just fulfilled on the server. That makes the cost minimization. The presentation layer is the web platform, the operation toward database and workflow controls are in the middle layer. The relational database, MS SQL Server2000, which lies in the data link layer supplies fundamental database platform.

6. CONCLUSION

The ETPM system which this paper discussed is more than software; it is a kind of resolve scheme. It totally meets the demand of the engineering testing process management in the auto production industries. Tough the ETPM system still need to be improved in some aspects, such as the foundation and the management of work process for complex system etc. The ETPM system has realized the foundation, submission, examine and approve, query, monitor etc on the net by the means of advanced platform technology and design structure. It not only supplies a convenient friendly interactive environment for users, but also supports the requirements of distributed application in engineering effectively.

7. REFERENCE

1. WANG Zhaixiao LV Lintao YAN Wenyao, Online Endorsement System Based on ASP.NET and Workflow Model, *Computer Engineering*, vol.30, No.17, 2004, p83-85.
2. Liu Haichuan Chen Peijiu, Design and Implementation of Workflow Management System Based on XML and ASP.NET, *Computer Engineering and Application*, 2004.15, p214-217.
3. Tu Yan, On Workflow Integration And Engineering Management Information Systems, *Journal of Beijing Electronic Science and Technology Institute*, vol.12, No.4, 2004, p64-68.

RESEARCH ON THE PICTURE COLLECTED BY MICROMINIATURE AIRCRAFT TO CONSTRUCT DIGITAL ELEVATION MODEL

Likang Shao[1], Guangxing Wei[2] ,Qiyu Yang [3], Xisheng Wu[2]
[1]*Machinery Staff Room* Artillery Academy *,HeFei ,China ;Email: mailto:yang_baojiang@163.com.* [2]*School of information technology, Southern Yangtze university,Wuxi, China; Email: weigx407@126.com .* [3] *Precision Machinery Department, Shanghai University, Shanghai China, Email: .shaolikang@163.com*

Abstract: It is the primary step of the "digital city" that construct tridimensional digital map[1]. We consider that construct it by computer vision base on the characteristic of the image microminiature aircraft collected[2]. But the quantity of the pairs of the characteristic point cannot meet the request to construct Digital Elevation Model. This paper provide a latest correlate matching method base on Epipolar Constrain after introduce the characteristic point extract and matching. the paper adopt a precise Fundamental Matrix estimate algorithm to calculated epipolar equation and increase the matching point's quantity effectively. It has met the real time 's request by experimental verification.

Key words: Epipolar Constrain Computer Vision Digital , Elevation Model Fundamental Matrix.

1. FOREWORD

Satellite remote sensing and navigate photography adapt to the data source of big region cartography. But the costs of conversion is too large, the cycle of the data acquisition is too long. It is feasible and practicable that that method of construct tridemensional map which use the pictures from the low-level microminiature aerocraft contrapose the current technology and cost. It suitable for the request of construct partly clearly tridimensional digital map because of its spending , flight level and mobility.

This project is supported by "863" project under the grant No.2001AA422160 and No. 2004AA420110

Please use the following format when citing this chapter:
Shao, Likang, Wei, Guangxing, Yang, Qiyu, Wu, Xisheng, 2006, in International Federation for Information Processing (IFIP), Volume 207, Knowledge Enterprise: Intelligent Strategies In Product Design, Manufacturing, and Management, eds. K. Wang, Kovacs G., Wozny M., Fang M., (Boston: Springer), pp. 984-989.

This research based on the nonstandard aerophotogrammetry picture construct the Digital Elevation Model (DEM). It needs enough right pairs of matching characteristic point to calculate the position's space coordinate. It can calculate the space coordinate by the pairs of matching point's image coordinate and camera's inside and outside parameter. And it also can construct DEM by those space coordinates.

2. CHACTERISTIC POINT EXTRACT AND MATCHING AND THEIR EXPERIMENT

2.1 Characteristic Point Extract

Make the corner of the image be the character[3]. Harris arithmetic operators differential coefficient's formula:

$$f_x(x,y)=\frac{\partial f}{\partial x}=f(x+1,y)-f(x-1,y) \quad f_y(x,y)=\frac{\partial f}{\partial y}=f(x,y+1)-f(x,y-1) \quad (1)$$

and it introduce the gauss smoothness strain in Harris operator. So in fact the characteristic matrix $\tilde{Q}$:

$$\tilde{Q}=\begin{pmatrix} \Sigma \tilde{f}_x^2 & \Sigma \widetilde{f_x f_y} \\ \Sigma \widetilde{f_x f_y} & \Sigma \tilde{f}_y^2 \end{pmatrix} \quad (2)$$

Replace the $\tilde{Q}$ for M, the last characteristic matrix $\tilde{Q}$can be:

$$M=\begin{pmatrix} A & C \\ C & B \end{pmatrix} \quad (3)$$

Harris define the estimate strength factor of the corner:

$$H(x,y)=det(M)-k*trace(M)^2 \quad (4)$$

Ahead of schedule it must define the strength criterion H_{thr}. It confirm it be the corner when $H(x,y) > H_{thr}$, So it can select a comparatively lower H_{thr} to get enough corner and sort it by the $H(x,y)$ and select definite quantity.

2.2 Characteristic point matching

Find out the characteristic point which the same scene in different image project named characteristic point matching. There are many matching method adapt to different condition[4]. The matching means base on the regiondepent make a presumption that the corresponding pixel has similitude brightness. We must consider several neighbor pixel's brightness. for example, make a hypothesis that $\boldsymbol{u}_i = (u,v)^T$ be a point of the first image, and its grey scale value be $h_1(u,v)$. $\boldsymbol{u}'_j = (u',v')^T$ should be any point of the second image, and its grey scale value should be $h_2(u',v')$ .Then define that the gradation similarity of the point $\boldsymbol{u}_i$ and $\boldsymbol{u}'_j$ as the gradation correlation value of the neighborhood Ω :

$$\sigma_{ij} = \frac{1}{N} \sum_{k,l \in \Omega} \left(h_1(u-k, v-l) - \bar{h}_1 \right) \left(h_2(u'-k, v'-l) - \bar{h}_2 \right) \tag{5}$$

If the characteristic point $\boldsymbol{u}_i$ in first image match to the characteristic point $\boldsymbol{u}'_m$ in the second image ,the $\boldsymbol{u}'_m$ must be close to the position of $\boldsymbol{u}_i$ in the second image because that the camera's displacement and rolling angle will not be very large. If the distance of the two point larger than the radius which define first then make the corresponding correlation coefficient be 0. It can be consider as a possible matching if the distance of the two point less than this radius and its correlation coefficient less than the defined threshold. it is the mostfrequentlyused means that the regional correlation coefficient(LACC) to determine whether the two point matching in practice : $c_{ij} = \dfrac{\sigma_{ij}}{\sqrt{\sigma_1^2 \sigma_2^2}}$.

It is necessary to presumption that it is similar that the characteristic excursion vector in a small neighborhood for the sake of improving the matching effect. So it used weighting function to calculate the matching INTENSIVE of the point multitude which got from the region gradation correlation method .We can presume that there is a pair of matching point

u_i and u'_j. If there is a point u_k nearby u_i in the first image then the relation between the point u'_l match to u_k in the second image similar to the relation between u_i and u_k. It can express as:

$$\mu_{(ij,kl)} = \frac{(u_k - u_i) \bullet (u'_l - u'_j)}{max(\|u_k - u_i\|, \|u'_l - u'_j\|)^2} \tag{6}$$

We call it as discommender when $\mu_{(ij,kl)} < 0$. p_n express the percent of those discommenders in those matching candidate .It define matching intensity as:

$$s_{ij} = c_{ij}(\sum_{k\in\Omega} \max_{l\in\Omega}(c_{kl}\mu_{(ij,kl)}f_{ik}) + \sum_{l\in\Omega} \max_{k\in\Omega}(c_{kl}\mu_{(ij,kl)}f_{jl}))(1-p_n) \tag{7}$$

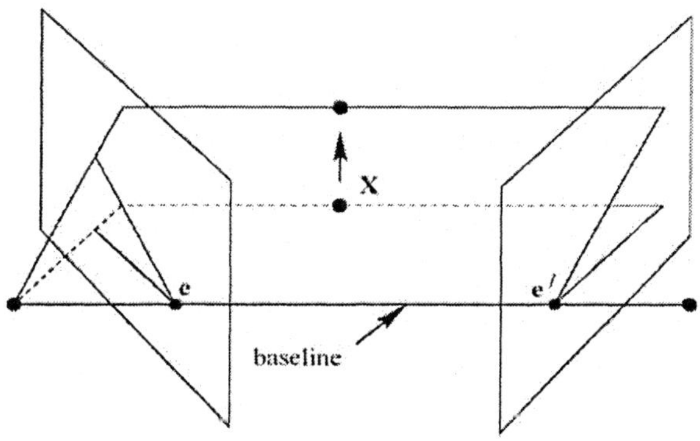

Figure 1. the geometry relation of the two picture

3. FUNDAMENTAL MATRIX CALCULATE AND ITS EXPERIMENT

As*Figure* 1 noted ,it take two picture from two view point. The plane which overpast the two viewpoint intersect with the two picture can get two line (base line). Different base line intersect with each other on one point(base point) in one picture. The corresponding point of Any image point must in it's base line of other picture. It can be recount by Fundamental matrix in algebra. The Fundamental matrix is a three row three rang matrix , and it is homogeneous and Its rank is two ,so it has seven free degree. Any pair of corresponding points X, X' and Fundamental Matrix satisfied:

$$\mathbf{x}'^T\mathbf{F}\mathbf{x} = 0 \tag{8}$$

Given the Fundamental Matrix, the corresponding base line of the image point can be ascertain as:

$$\mathbf{l}' = \mathbf{F}\mathbf{x}, \quad \mathbf{l} = \mathbf{F}^T\mathbf{x}' \tag{9}$$

The Fundamental Matrix can be estimate from the corresponding point. For the given points (x, y) and (x', y'), it establish linear equation as:

$$[x'x \quad x'y \quad x' \quad y'x \quad y'y \quad y' \quad x \quad y \quad 1]\mathbf{f} = 0 \tag{10}$$

So it can get ascertain 8 independence parameter of the Fundamental Matrix from 8 pairs of corresponding point. If given the pair of corresponding point **X**, and the Fundamental Matrix, it can get the he pair of corresponding point $\mathbf{x}'$.

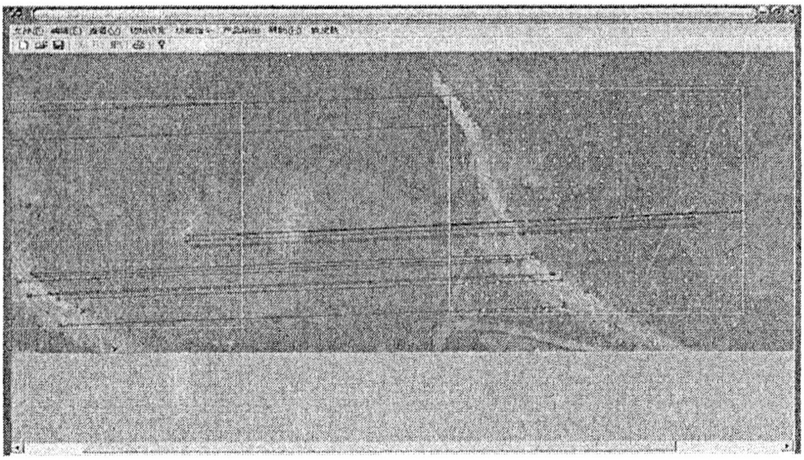

Figure 2. Increase more character base on Epipolar

4. CONSTRUCT DEM AND DISPLAY

It should write the elevation correspond to the net-point to a file follow the format which VirtuoZo System submitted ,named DEM(Digital Elevation Model) (See Figure 3).

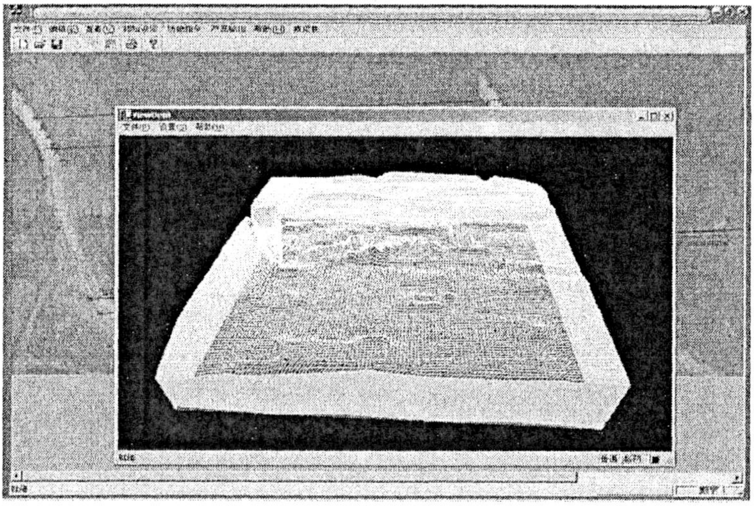

Figure 3. display Dem

5. SUMMARIZE

It is the primary step of the "digital city" that construct tridimensional digital map, We complete the step of increase characteristic point by computer vision because of the specialty of the image collected by microminiature aircraft. It also disposed reference point in fieldwork and counted relevant parameter in office work. It did a favorable support to the next work of construct DEM. The experiment proved the effectiveness and feasibility of this algorithm. And the speed of algorithm's perform meet the request of real time.

6. REFERENCES

1. Liu, Lin Huang, Ying, (2004.12), How to make the development strategies for digital city, *Journal of Harbin Institute of Technology (New Series)*, vol. 11, pp. 649-653
2. Besl, Paul J. (1998.4.21), Geometric Modeling and Computer Vision, *Research Publication - General Motors Research Laboratories*, n 6248, Apr 21, 1988,
3. Dong, Mingli Zhou, Xiaogang; Zhu, Lianqing; Lu, Naiguang; Sun, Yunan,(2005,12), Automated image matching with coded points in stereovision measurement, *Chinese Journal of Mechanical Engineering (English Edition)*, vol:18, n 3 ,pp: 453-456
4. Zhong, Hui-Xiang Feng, Yue-Ping; Pang, Yun-Jie. "Robust estimation of the fundamental matrix based on an error model" 2005 *International Conference on Machine Learning and Cybernetics, ICMLC 2005*, 2005, p 5082-5087

ANALYSIS ON PRODUCT TECHNICAL RISK WITH BAYESIAN BELIEF NETWORKS

Ming Chen[1], Yun Chen[2], Bingsen Chen[1], Qun Wang[1]
[1]Sino-German College of Applied Sciences, Tongji University, China; Email: cdhawcm@yahoo.com.cn. [2]School of Public Economy Administration, Shanghai University of finance & economics, China.

Abstract: With the development of the science and technology, products are more and more complex, e.g. more functions integrated into one product, abilities of products are stronger and stronger, degrees of automatization and intelligentizing are higher and higher. In order to develop products successful, we must analyze product technical risk systematically and comprehensively. All products' behaviors that don't satisfy their functions are function failures, i.e. product technical risk. The sources of causing risk are the products' structures. We anticipate the behaviors of the target product, seek the potential function failures, and analyze the causes, and then set up BBN model. With this methodology, we have analyzed the product technical risk of a micro-electronic equipment.

Key words: Product developing, Product technical risk, Bayesian belief networks.

1. INTRODUCTION

With the development of the science and technology, people have been improving their living and working situations constantly. The requirements of products that serve people are more and more, e.g. multiple functions are integrated into a product, abilities of system are more and more powerful, degrees of automatization and intelligentizing are very highly. It makes required products more and more complex.

Frequently, Design indexes of are diversity and some indexes are not consistent with the others. The target product consists of massive elements.

Please use the following format when citing this chapter:
Chen, Ming, Chen, Yun, Chen, Bingsen, Wang, Qun, 2006, in International Federation for Information Processing (IFIP), Volume 207, Knowledge Enterprise: Intelligent Strategies In Product Design, Manufacturing, and Management, eds. K. Wang, Kovacs G., Wozny M., Fang M., (Boston: Springer), pp. 990-995.

And the product's interfaces are multitudinous and complex. Most products are involved into many fields, e.g. mechanical engineering, electronic engineering, optical engineering etc. Designers lack of the experiences of usage of the advanced technology that are used in product developing.

In order to develop products successful, we must analyze product technical risk.

There are some ways to assess the risk, for example, fuzzy mathematics and neuron network, but they can't solve the essential problem of risk assessment: reasoning integrated with experts' knowledge based on uncertain information.

Bayesian Belief Network is a kind of expert system that can describe uncertain information. It can reflect continuity and accumulation of risk assessment. The consistency of time can't be realized with non-memorable methods, for example, methods based on rules and neuron network. Because of reliability in mathematics, Bayesian Belief Network can be changed into standard model which can express human thinking and reasoning process, while the traditional expert system is entirely oriented to tasks, and different tasks have their own models and they lake of catholicity.

The Product Technical Risk is modeled with Bayesian Belief Networks, constitutes assessing arithmetic, simulates and analyzes the technical risk.

2. FUNCTION, STRUCTURE AND BEHAVIOR

2.1 Product system

Products are entities that provide customers utilities and interests. Large-scale products are systems composed of controls, units and interactions

Controls in product can be on product's level, on sub-systems' level or components' level. Units are non-control elements in product. Interactions are expressed that product's elements affect one other.

Any system has its own functions and shows its behaviors outside. System's behavior is process of realize relevant function, system's function is goal, results or constraints of behaviors. System's structure is total relations of elements in some system; it is a carrier of functions and behaviors.

2.2 Method of analysis on product technical risk

Owing to the complexity of systems in large-scale products, we must start analysis from the sources of product development: customers (Figure 1). All products' behaviors that don't satisfy their functions are function failure, i.e. product technical risk. The sources of causing risk are the products' structures. From the customers' requirements, the products' functions are acquired. According the functions, the structures are designed and after developing, the target product is manufactured. Analyzing the product's behaviors, we can judge whether or not the product satisfies the functions. We anticipate its behaviors, seek the potential function failure, i.e. product technical risk, and analyze the causes, i.e. its structure, and then set up Cause-and-Effect Chain, i.e. BBN model.

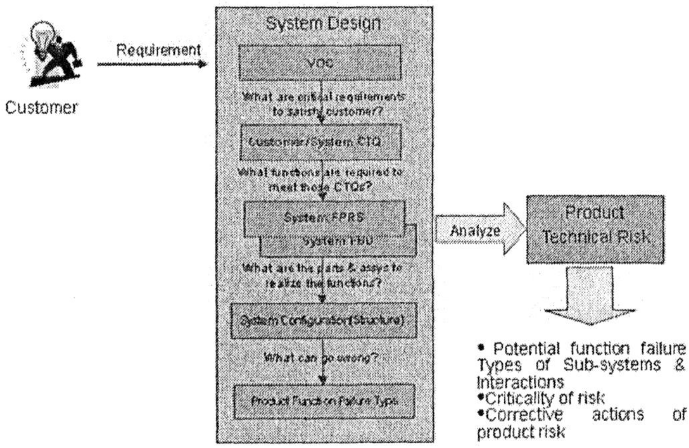

Figure 1. Method of product technical risk analysis

3. MODEL OF PRODUCT TECHNICAL RISK

3.1 VOC, CTQ, FPRS and FBD

The voice of the customer (VOC) is a process used to capture the requirements/feedback from the customer (internal or external) to provide the customers with the best in service/product quality. This process is all about being proactive and constantly innovative to capture the changing requirements of the customers with time.

Customer critical to quality (CTQ) is the key measurable characteristic of top level product whose performance standards or specification limits must be met in order to satisfy the customer. It aligns improvement or design efforts with customer requirements.

Starting with the customer's problem as outlined in the Customer/System CTQ, the Functional Performance Requirement Spec (FPRS) defines a product's functionality, usability, reliability, performance, supportability, availability, localizability, and other requirements. It describes the functions and performance levels that the proposed product should have to successfully address that problem.

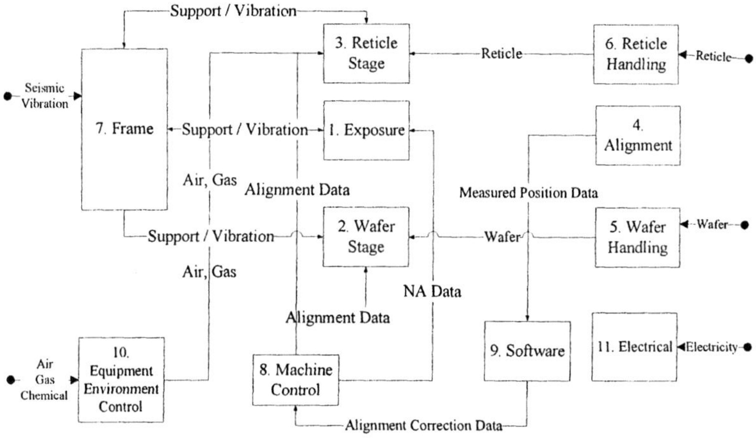

Figure 2. FBD of Step & Scanner

Functional Block Diagram (FBD) is used to show how the different sub-systems / parts interact with one another to verify the critical path. We prepare FBD in 3 different levels: system level FBD, sub-system level FBD and unit FBD. Figure 2 is a sample which shows the System Level of 300mm, 193 ArF, 130nm Step & Scanner.

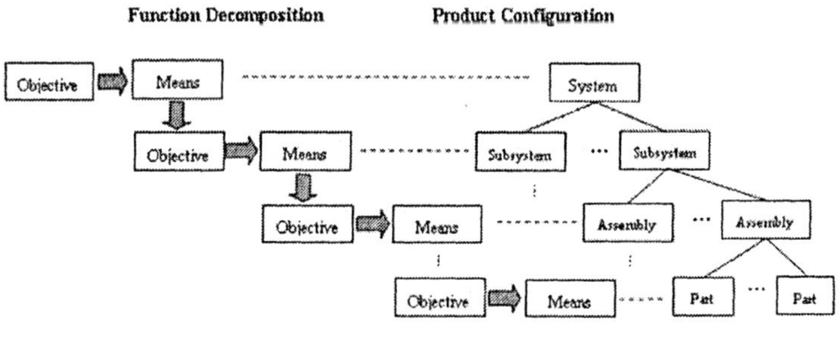

Figure 3. Function decomposition and product configuration

3.2 Product System Configuration

The functions of the target product decompose some objectives and means on the system level. We decide which structures on this level according to the means. These means on this level are the objectives on the next level. We break them to some means. Then the structures of subsystem are designed. In this way, all the system is developed (Figure 3).

3.3 BBN Model

A BBN model is a directed acyclic graph (DAG), which is defined to consist of the qualitative and quantitative relationships. Qualitative Relationships include a set of random or deterministic variables and a set of directed edges or arcs. Quantitative Relationships include a set of root and conditional probabilities.

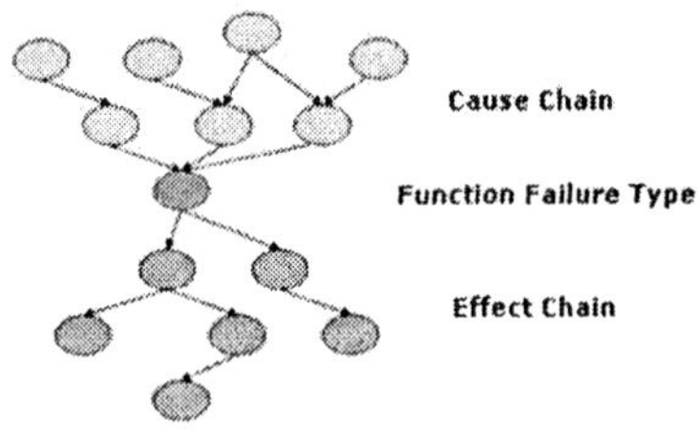

Figure 4. BBN model

Cause-and-Effect chain is a BBN model of the product risk analysis. Every function failure has its own model. It's easy to find causes that result in the function failure, and effect that the failure gives.

First we list the function failure, then list the phenomena occurs directly if the function failure appears, analyze the further effects and get effect chain. After that, we analyze the causes that result in the function failure, and analyze the further causes and get cause chain (Figure 4).

4. CASE STUDY

300mm, 193 ArF, 130nm Step & Scanner is a kind of large-scale complex product which is related to mechanical engineering, electronic engineering, optical engineering, control engineering etc. We must analyze its technical risk during developing. With this methodology, we have analyzed the potential function failures. There are 20 function failure type, for example, CD resolution failure, CD uniformity failure etc. With the BBN model, we calculated the subsystem technical risk (Figure 5). We took some measures to improve the design and the risk has been reduced.

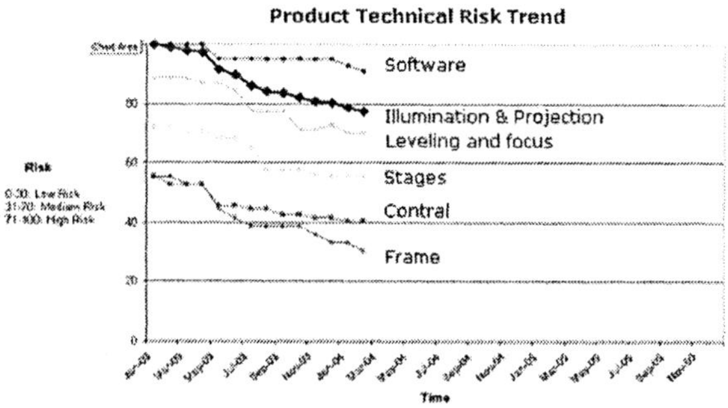

Figure 5. The product technical risk trend of the step & scanner

5. CONCLUSION

In order to develop large-scale complex products successful, we must analyze product technical risk systematically. Product technical risk is its function failure. We analyze the behaviors of the target product and anticipate the potential function, i.e. failure product technical risk, and then analyze the causes, i.e. structure. In this way, its BBN model will be set up. We calculate the risk, and then it can guide us to take corrective actions in order.

6. REFERENCES

1. Chen Ming, Chen Yun, Chen Bingsen. Management of technical risk in the development of complex products. *Manufacturing Automation* 2005; 10:pp. 6-8.
2. Wang Guohua, Chen Ming, Chen Yun. Fuzzy Synthetized Estimating of Technical Risk of Complex Products. *Mechatronic* 2006; 1:
3. Burton Hoyt Lee. *Failure modes and effects analysis with Bayesian belief networks: bridging the design-diagnosis modeling gap*. Dissertation of Stanford University 2002.
4. Elaine M.Hall. *Managing Risk——Methods for Software Systems Development*. Bosten: Addison-Wesley 1998
5. Wang Lianchen. *Engineering System Theory*. Beijing: China Astronautics publishing House, 2002
6. Eigenworx and ConneXwave. *Advanced FMEA*. SMEE 2003
7. ConneXwave . *Managing Complex Projects Plan*. SMEE 2003

HANDLING VARIABILITY IN MASS CUSTOMIZATION OF SOFTWARE FAMILY PRODUCT

Jianhong Ma [1] , Runhua Tan [2]
[1]*School of Computer Science and Software Engineering, Hebei University of Technology, TianJin, 300130, China, jh_ma@eyou.com;* [2]*School of Mechanical Engineering, Hebei University of Technology, TianJin, 300130, China*

Abstract: Similar to mass customization in traditional industry, managing variability of software family products is also the key technology in software product line. In this paper, variability model and variability lifecycle are discussed. Typical approaches of modeling variability are investigated and compared in terms of the process of identifying, managing variability, model expression. The development trend of mass customization in software produce is talked about.

Key words: variability model; variability lifecycle; software product line; mass customization

1. INTRODUCTION

Similar to mass customization in traditional industry, it is the goal to develop and evolve software systems with minimal development effort and time-to-market and maximum quality in software industry. Modeling and managing variability is the key technology in software product line. Method of handling and managing variability has a direct and important effect on software reusability. Many approaches have been presented in literatures. Each of them paid more attention on some aspects but ignored others.

This project is supported by the Project of Natural Science Foundation of China under Grant No.50375045, the Project of Natural Science Foundation of Tianjin under Grant No.043802211 and the Project of Educational Foundation of HeiBei Province under Grant No.2004412

Please use the following format when citing this chapter:
Ma, Jianhong, Tan, Runhua, 2006, in International Federation for Information Processing (IFIP), Volume 207, Knowledge Enterprise: Intelligent Strategies In Product Design, Manufacturing, and Management, eds. K. Wang, Kovacs G., Wozny M., Fang M., (Boston: Springer), pp. 996-1001.

First, the variability lifecycle and variability model in each lifecycle phase are discussed. According to the variability lifecycle, typical approaches are compared in identifying, representing and implementing the variability. The shortage and development trend of handling variability in software are talked about at last.

2. VARIABILITY IN SOFTWARE PRODUCT LINE

Software variability is the ability of a software system or artifact to be changed, customized or configured for use in a particular context.

2.1 The lifecycle of variability in product line

Software product line includes two processes: application engineering and domain engineering. Variability exists throughout the whole software produce process in the product line.

1) in domain engineering phase, variability is identified, designed and implemented for reuse. It includes: variability identification, variability analysis and design and variability implementation.

2) in application engineering phase, the variability in product line is tailored and configured during developing a new member of family products. It includes: variability customization, variability configuration and variability binding.

2.2 Variability model

Variability in a software product line can be modeled at the software requirements level and the software design level. The different variability models with different roles are created at each lifecycle phase.

Variability requirement model: identifies and expresses variability abstracted from the existent family products or predicted from the future products.

Variability design model: shows how to deal with this variability in coding stage for system developer and coding engineers to implement the variability. The model is created based on variability requirement model.

Variability reuse model: directs the reuser to custom, configure or bind the variability when reuse the core asserts. There is not explicitly distinct between variability design model and reuse model, commonly, they are emerged together as design model.

3. COMPARING THE APPROACHES OF MODELING THE VARIABLITY

The approaches to model and manage the variability are classified into three main branches: feature-oriented method, objected-oriented method and integration method.

3.1 Modeling variability with object-oriented method

1992, Hassan Gomaa presented the object-oriented domain analysis and modeling method[1]. It includes two major steps: first is to complete the kernel domain model, then to analyze the variations of the domain. There is no distinction between each phase of variability lifecycle. The variability is expressed in object models based on kernel model. The objects may be kernel, optional or variant objects. The objects relationships may be described as: aggregation hierarchy, supporting the IS-PART-OF relationship; generalization/specialization hierarchy, supporting the IS-A relationship; and feature/object dependencies: showing the objects required to support each feature of the domain.

The advantage of this method is that the notation of models is same as UML notation. It is easy to describe the models with UML tools. The limitation is that the reusable domain object is too difficult to be identified in practice and results in the low reusability. Inheritance mechanism makes the software product more flexibility in some ways, but it reduces the dependency between the objects or modules. There is no explicit variability reuse model, it is implicated in object models. They are expressed with natural language.

3.2 Modeling variability with feature-oriented approach

Feature means the attribute and character belonging to the system. It is the bridge that connects users and application developers. It characterizes one product from other family products. The classical feature-based method is Feature Oriented Domain Analyses(FODA) proposed in 1990 by Kang[2].

3.2.1 Variability identified in FODA

In FODA, the variability requirement model includes feature model, function model and ER entity. Variability is identified and described mainly in feature model. The features in feature model are classified into following categories: mandatory feature,optional0n feature and alternative feature.

3.2.2 Variability analyses and design in FODA

Based on variability requirement model, variability is analyzed, the architectural model(also known as a design reference model) is created. That is variability design model. It provides detailed design and component construction. It focuses on identifying concurrent processes and domain-oriented common modules, allocating the features, functions, and data objects defined in the domain models to the processes and modules. Some features have impact on the final architecture such as when to "bind" or “fix” the value of a feature, features are grouped into three classes according to the binding features such as: compile-time features, load-time features and runtime features.

In 1998, the feature-oriented reuse method(FORM) takes many non-function features into account, it concentrates on analyzing and modeling the variability in terms of capability, operating environment, domain technology, and implementation technique features.

3.2.3 Variability reuse model in FODA

The constraint relationships between variable features are used to validate whether the constraints are satisfied after tailoring and customizing the feature models for special application. The relationships between features are represented as: composition or generalization relationships, dependency relationship, exclusive and non-exclusive alternatives and mutual exclusion relationship. These relationships are modeled with UML notations.

3.3 Integrated method

Aiming at improving both feature-oriented method and object-oriented method, many approaches have been presented to integrate them together.

3.3.1 FeaturRBSE method

In 2000, FeatureRBSE[4] integrated the feature modeling of Feature-Oriented Domain Analysis (FODA) into the approach of the Reuse-Driven Software Engineering Business (RSEB). The RSEB is a use-case driven systematic reuse process: architecture and reusable subsystems are first described by use cases and then transformed into object models that are traceable to these use cases. Variability in the RSEB is captured by structuring use case and object models using explicit variation points and variants. Also, based on the notion of domain-oriented, it emphasizes a group of closely related applications in a domain rather than a single application, first it creates domain use case model.

Comparing with FODA, besides the compose-of relationship, mandatory or optional attributes are considered in featurRBSE, there are some extended feature relationships:

1) the alternative relationship: variation and variant features. A feature can act as a variation point (called variation point feature or vp-feature) in the model, while other features play the role of its possible variants (called variant features).

2) the binding time: attribute of vp-features. Vp-features can be bound at reuse time, i.e. when the reuser accesses the domain infrastructure to configure reusable assets for his development.

3.3.2 Extended FORM method

In 2000, kang extended FORM[3] with integrating object-oriented into FORM to make the gap closely between variability identification and variability analysis and design. It presents a systematic way to identify reusable objects from feature model by linking feature categories to object categories. Variant points are described in object model that consists of objects and their relationships: aggregation, generalization and association. The relationships of the features in feature model are traced to variant points described in object model. The relationships between features (e.g., composed-of, generalization, and implemented-by) are mapped into relationships (e.g., aggregation and generalization) between objects.

3.3.3 FODM method

The feature-oriented domain modeling method (FODM)[5] was presented by Hong Mei at 2003. This method explicitly absorbs and evolves the advantages of FODA method and FeatureRBSE method. It presents more feasibly method for reusing the requirements with use case model and feature model, the feature model is first proposed from five aspects, namely, basic structure, variation, representation mechanism, variation binding time, variation constraint mechanism and quality feature analysis. Variability constraints are described in logic, it is a guideline for reuser to tailor or bind the variability from the requirement model.

The disadvantage of this method is that there need a special tool to support these models.

4. CONCLUSION

The emphasis work of managing variability must be shifted from how to develop the core asserts to how to tailor the core asserts and how to validate its validity. There are some important questions in software product line:

1) How to identify the variability and design variability with the uniform notation in each model.

2) One variant often is reflected in different view, how to trace the variant in different model.

3) How to evaluate and validate the reusability, cohesion, independence of component or object with variant points.

5. REFERENCES

1. Hassan Gomaa, *An Object-Oriented domain analysis and modeling method for software reuse.* 0073-1129-1/92 1992 IEEE.
2. Kang KC, Cohen SG, Hess JA, Novak WE, Peterson AS. *Feature-Oriented domain analysis feasibility study. Technique Report,* SEI-90-TR-21, Pittsburgh: Software Engineering Institute, 1990.
3. Kwanwoo Lee1_, Kyo C. Kang1, Wonsuk Chae1 and ByoungWook Choi2.Feature-based approach to object-oriented engineering of applications for reuse. *Softw. Pract. Exper.* 2000; 30:1025 1046.
4. Griss ML, Favaro J, d'Alessandro M. Integrating feature modeling with the RSEB. Proc. of the 5th Int'l Conf. on Software Reuse. *IEEE Computer Society,* 1998. 76−85.
5. Hong Mei, Wei Zhang, Fang Gu. A Feature Oriented Approach to Modeling and Reusing Requirements of Software Product Lines. *Proceedings of the 27th Annual International Computer Software and Applications Conference* 2003 IEEE.

A PROCESS MODEL OF PRODUCT PLATFORM BUILDING FOR THE PRODUCTS WITH THE SAME FUNCTION STRUCTURE

Huangao Zhang [1], Wenyan Zhao [2,3], Guoping Li [1,4], Runhua Tan [1]
[1]School of Mechanical Engineering, Hebei University of Technology, Tianjin, 300130, PRC ; Email: zhgzwy@eyou.com [2]School of Management, Tianjin University, Tianjin, 300072, PRC [3]School of Management, Hebei University of Technology, Tianjin, 300130, PRC, [4]Zhangjiakou Coal Mining Machinery Co. Ltd, Zhangjiakou,075025, PRC;

Abstract: Product platform is a good way to achieve Mass Customization. The ability of deriving variants of a product platform depends on its configuration, which ultimately is determined by voices of customer. The nature of product platform building is to identify the common parts from the object product family. According to the similarity of a set of products the product platform divided into two cases: similar function with close parameters and same function with different parameters. Based on E-QFD a process model of product platform building for the last case is put forward that start from analyzing customers' demands. And the steps according to the process model are discussed. As an illustration the model is applied to build the product platform for armed face conveyer.

Key words: product platform; HOQ; customers' demands; technical measure; module; armed face conveyer(AFC)

1. INTRODUCTION

Product platform is a good way to achieve Mass Customization. A well-defined product platform can succeed in the purpose that the company can provide as much variety as possible for market with as little variety as possible between products.(Mayer, 1997)

Considering the similarity of a set of products the product platform can include two cases:

Please use the following format when citing this chapter:
Zhang, Huangao, Zhao, Wenyan, Li, Guoping, Tan, Runhua, 2006, in International Federation for Information Processing (IFIP), Volume 207, Knowledge Enterprise: Intelligent Strategies In Product Design, Manufacturing, and Management, eds. K. Wang, Kovacs G., Wozny M., Fang M., (Boston: Springer), pp. 1002-1009.

- Product platform for a set of products with similar function and close parameters. Similarity of function is that a set of products vary in functions but share several items such as input and output in their function structure chart, for example the truck and the bus has the same input and output and then their function is similar. (McAdams, 1997) Just those common parts make it possible for the objective products to build a product platform, because the common function model can be implement by the same concept solution, furthermore, the close parameters of the function give more chances to use the same physical model to implement the common concept solution. The main problem of this case for product platform building is to identify the common function models.
- Product platform for a set of products with same function and different parameters. The same function means a set of products have the same function structure chart while their performance parameters are different. At this case the objective products used to be classified into family tree according to their main parameters. (Javier, 2000) There are good chances in each branch to build a platform. And there always have chance to build a platform which can cover several branch. The main problem of this case is to identify the main parameters and common parts of each branch. Next parts a process model will be discussed for product platform building in this case.

2. A PROCESS MODEL FOR PRODUCT PLATFORM BUILDING

Good product platforms must be carefully planned, designed, and developed. This requires intimate knowledge of customer requirements and a thorough understanding of the market. (Simpson, 1998) As shown in figure 1, a process model for product platform building is developed that start from the customers' demands and end with platform configuration. The model extends QFD (E-QFD) to identify the common parts of a set of product with the same function.

According to the model the general steps will be included as follows:

Step1. Identify and Analyze the Demands of Customers

Customers' demands can be collected by market research. The demands should be classified into two types: basic demands and special demands.

- Basic demands are the demands that the customers desire for the main function and primary performance of the products. Generally, basic demands include power, kinetic and dynamic index, essential function, reliability, life, important dimension and structure parameters and so on.

- Special demands are the demands other than basic demands. Special demands usually include three aspects: additional function desired by customers, additional requirements for adapt to different object and environment, other demands result in irregular change in parameter of performance and architecture. If the products should attract customers and carry out their function perfectly, the special demands must be met.

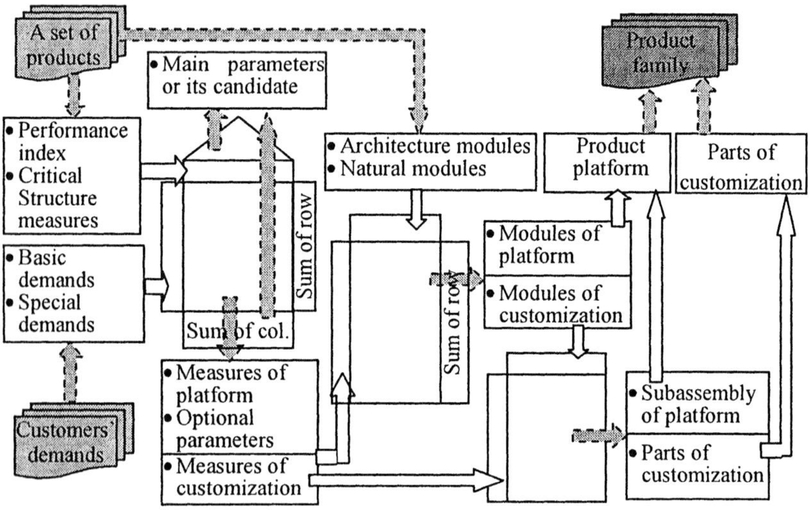

*Figure 1.*a process model for product platform building

Step2. Select Technical Measures of the Object Product Family

According to the product object the performance measures and critical structure measures could be identified. The performance measures represent the function and capability of the product, such as Power, velocity etc. And the critical structure measures are items of measure that must be decided before the detail design begin. The measures rest with the interior and exterior demands, such as capability, environment restriction and so on.

Step3. Identify the Measures Respectively Belong to Platform and Customization

The HOQ is developed to relate the customers' demands and the technical measures. (Jack, 1997) The sums are calculated for each column and each row in the relationship matrix. An item of customers' demands with higher sum value of each row has more influence on the technical measures than other items. The weights that represent relative importance of each demand can be given according as the sum of the row. The technical measures with higher sum value of each column could be select to be main parameters candidate according to which the product family tree is plotted.

Those technical measures that relate to basic demands while didn't relate to special demands are ranged in the measures of product platform. Some other technical measures whose value can be selected from a few of constant value are classified into the optional parameters. The rest of technical measures should are defined as the measures of customization, which will influence on the structure and parameters of the parts of customization.

The roof of the HOQ is the correlation matrix that describes the relationships between technical measures. The items that correlate to most of the others should be main parameter candidates.

Step4. Model Analysis for Function and Architecture

First, define the general function according to customers' demands. Second, implement function analysis and obtain function configuration chart. Third, identify the function module according to three-flow-principle. (Stone, 1997) Fourth, implement architecture analysis according to architecture of the object product family and function module. At last the architecture modules of the object product family can be identified. Note that the components can be used as the natural modules of the object product family. The architecture modules identified can be used as the initial module of modularization design for product platform.

Step5. Identify the Modules Belong to Platform

The second HOQ is developed to relate the technical measures and architecture modules of the object product family. The sums are calculated for each row in the relationship matrix. Those technical measures with higher value of sum of each row are main parameter candidates.

Those modules having no relation to measures of customization will be ranged in modules of product platform. Some of them that associate with optional parameters are modules belong to multi-platform.

The modules associated with the measures of customization are modules of customization, two cases would include:

- If the module is a part, the part is module of customization.
- If the module is a component or assembly, the component or assembly should be analyzed further for the further processing.

Step6. Identify the Parts or Sub-assembly of Platform from Modules of Customization

First, the component or assembly of customization is disassembled into parts. Second, the HOQs are developed to relate the measures of customization and the parts for each component or assembly of customization. Third, identify the parts out of product platform. The parts have relation to measures of customization are parts out of platform that must be designed for each product. Others can be assembled into sub-assembly and add to product platform.

Step7. Integrate and Redesign Modules

The modules of platform identified in step5 and sub-assembles or parts of platform identified in step6 combined into the product platform of the object product family. All elements are prescribed by synthesized attributes that include measure, shape, material and so on and integrated into a database of product platform. The parts of customization are redesigned according to the method of broad-sense-modularization so that the change is the least when the parts are used as prototypes for redesign. According to existing products the interfaces between modules are redesigned for standardization.

3. AN EXAMPLE

Armored face conveyer (AFC) is typical product for customization. Most AFC design is variants design that the variant use an existing product as a prototype. There are many common parts between different variant. AFC is a type of product with same function, and its variants can be compartmentalized in series. The above-mentioned process model is applied to build the product platform for AFC.

As shown in figure 2, the customers' demands and technical measures are integrated in HOQ1. The demands in the shadow are special demands, and other demands are basic demands. The A1-matrix is evaluated according to the strength of relationship between customers' demands and technical measures: (Jack, 1997)

- Strong=5. The strength is strong when the value of the technical measure is determined directly by the customer's demand.
- Medium=3. The strength is medium when the customer's demand influences the value of the technical measure indirectly.
- Weak=1. The strength is weak when the technical measure has some relation with the customer's demands and the value of the measure can be selected in a broad range.
- Blank=0. An item of the A1-matrix is blank, which represent the technical measure has no relation with the customer's demand.

In the roof of HOQ1, " + " represents that the two technical intersected at the position correlate to each other.

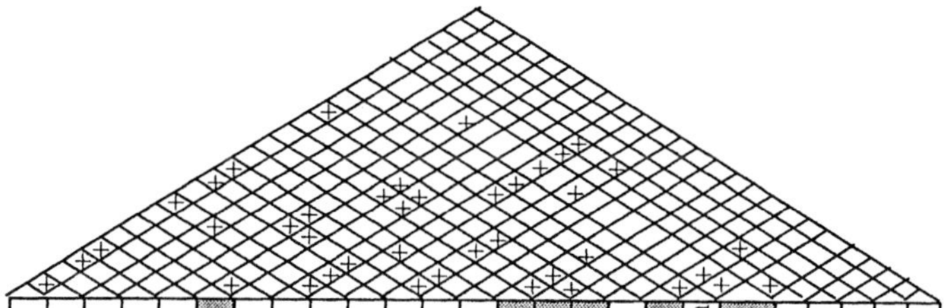

		Total power	Disposal of motor	Flight bar type	Chain speed	Chain size	Height of discharge	Method of discharge	Connector type	Breadth of pans	Length of line pan	Height of line pan	Design length	Inspection structure	Length of reroute	Obliquity of ramp pan	Obliquity of connection pan	Type of chain track	Fixing dimension of track	Dimension of cable Trough	Type and dimensions of pushing ear	Process of plate	Dimension of spade plate	Type of chain tension	Reliability of parts	Antiskid
Basic demands	capacity	5	3	5	5	5			3	5	3	5	3			1	1	5	5	1	5	3	1	1	3	
	conveyance distance	5	5	3	3	5			3	3	5	5	5		3	5	5	1	1	1	1		3	1	3	
	reliability	5	3	3	3	5			5	3				1	1	3	3	3			5	3	3	3	5	3
	working life	5	3	5	5	5			3	5	3	5	3			1	1	5	5	1	5	5	1	1	5	
	flexible								5	3	3	3	3							3	3	3	1		3	
	easy to maintain								5		1		1	5				3	1	1	5	3	1		3	
	moveable						3	3	5	3	3	3	3							3	5	3	1		3	
	easy to assemble								5		3		1	5				3		3	3	3		5		
	cleaning floor									3											3	3	5			
Special demands	Skid resistance																									5
	demands for special parts					5				3		3										5		3	5	
Fit shearer	appropriate track									5	3	3						5			3	3			5	
	appropriate span					3				5		5						1	5		3	3	5			
	no interference		5			3	3	3	5	5	3	3	3		5	5	5		5	5	3	3	5			
	protect cable									3	5	3							5	5	3					
Fit support	Appropriate link	3	5							5	5	5							3	3	5	5	1		5	
	no interference		5					3	3	1	3	5							1	5	5	1			1	
Fit stage loader	unload absolutely			3	5	3	5	5		5		3				1	1									
	no interference		5			3	5	5	1	3	3	3			3	3										

*Figure2.*a process model for product platform building

In the technical measures associated with the special demands the height of discharge, the length of re-route, obliquity of ramp pan, obliquity of connection pan, fixing dimension of track, the type and dimensions of pushing ear can not be fixed. The other technical measures associated with the special demands are the optional parameters.

As shown in table 1, the technical measures and module of AFC are integrated in HOQ2. The modules in the house are components for field installation. From the table we can see that the breadth of pan correlate to most modules, which confirm the conclusion that the breadth is the main parameters for AFC family tree analysis. If ignore some fringe demands, the transmission unit of delivery end and return end can add to platform directly. In view of the seat of track and spilling plate are joined by weld, the track assembly can add to platform directly too. Ignore the weak relation the chain assembly that includes chain, flight bar assembly and chain connector can also add to platform directly. The other module should be analyzed further to find invariant parts for platform. As an example the line pane is analyzed further into four parts: spade plate, spilling plate, deck plate and bottom plate. Only spilling plate relate with the measure of customization, then the other three parts integrate into a sub-assembly and add to the product platform.

Table 1. the HOQ2 that relate the technical measures to module of AFC

	Transmission Unit of delivery end	Transmission Unit of return end	Pushing unit of delivery end	Pushing unit of return end	Ramp pan of delivery end	Ramp pan of return end	Connection pan	Re-route pan	Inspection pan	Line pan	Track assembly	Connector assembly	Chain and flight bar	Cable trough	Other parts
Total power	5	5	3	3	3	3	3	3	3	3	1	3	5		
Disposal of motor	5	5	3	3											
Flight bar type	5	5			5	5	5	5	5	5	1	3	5		
Chain speed	3	3	1	1	1	1	1	1	1	1			5		
Chain size	5	5	5	5	5	5	5	5	5	5	3	3	5	3	1
Height of discharge			5		5		5	3							
Method of discharge	5	5	5		3		3	3							
Connector type	5	5			5	5	5	5	5	5	1	5	1		
Breadth of pans	5	5	5	5	5	5	5	5	5	5	3	3	3	3	1
Length of line pan					3	3	5	5	5	5		3	3	5	
Height of line pan	3	3			5	5	5	5	5	5	3	3	5	3	
Design length	5	5			5	5		3						5	
Inspection structure									5						
Length of reroute					5	5	3	5							
Obliquity of ramp pan					5	5	5	5					1		
Obliquity of connection pan					5	5	5	3					1		
Type of chain track											5				

Fixing dimension of track					3	3		5			5		1	3	
Dimension of cable Trough											3		3	5	
Type and dimensions of pushing ear			5	5	5	5	5	5	5	5		1	1	1	
Process of plate							5	5	5	5	3	3	3		
Dimension of spade plate					3	3	3	3	5	5			1		
Type of chain tension	5	5		5		5									
Reliability of parts	5	5	3	3							3	3	5		
Antiskid					3	3	3	3	5	5					

4. SUMMARY

Product platform is a set of modules or parts shared by a set of products. The nature of product platform building is a process that the common items are identified by some means. A process model of building product platform has been presented. This model takes into account voice of customers and restrictions of environment, and is propitious to build product platform for the products with the same function structure.

The process model shown in this paper has been applied to a set of AFC. The elements of platform are identified. It will help many enterprises to realize mass customization.

5. ACKNOWLEDGMENT

The research is supported in part by Natural Science Foundation of China under Grant Numbers 50375045 and TianJin Natural Science Foundation Key Projects under Grant Number 043802211.

6. REFERENCES

1. Meyer M H, Lehnerd A P, 1997, *The power of product platforms*, USA, Free Press, New York
2. Simpson T. W., 1998, *A Concept Exploration Method for Product Family Design*, Ph. D Dissertation, Georgia Institute of Technology
3. McAdams D A, Stone R B, 1998, Wood K L. Understanding Product Similarity Using Customer Needs, *ASME Design Engineering Technical Conference*, DETC98/DTM-5660
4. Javier P. Gonzalez-Zugasti, 2000, Kevin N. Otto, John D. Baker, A Method for Architecting Product platforms, 12, *Research in Engineering Design*, pp.61-72
5. Stone R. B., 1997, *Towards a Theory of Modular Design*, Ph. D Dissertation, University of Texas
6. Jack B. Revelle, John W. Moran, Charles A. Cox, 1998, *The QFD Handbook*, USA: John Wiley & Sons, Inc., New York, pp.3-21.

A TOP-DOWN APPROACH TO AN ARMOURED FACE CONVEYOR FAMILY DESIGN

Guoping Li[1,2], Huangao Zhang[1], Wenyan Zhao[3], Haixian Zhao[1], Runhua Tan[1]

[1]School of Mechanical Engineering, Hebei University of Technology, Tianjin, 300130, PRC ; Email: zhgzwy@sina.com [2]Zhangjiakou Coal Mining Machinery Co. Ltd, Zhangjiakou,075025, PRC; [3]School of Management, Hebei University of Technology, Tianjin, 300130, PRC

Abstract: The Armored Face Conveyer is typical product for customization. To achieve Mass customization, a top-down approach to product family design is introduced. The steps and tools of the approach are discussed when it is used to design AFC family. The key of the approach includes three aspects: the object product is classified in series according to the main parameters of the product; each series build a platform as the design frame; a database is build to deploy variants. After the platforms are built, software is developed to implement the process of variant deployment. With the help of the software, the design and manufacture periods are much shortened.

Key words: Armed Face Conveyer (AFC); product platform; product family; variant; variant deployment, software

1. INTRODUCTION

The Armored Face Conveyor (AFC) is one of important conveying equipment used at coal mining face under ground. Many restrictions for AFC design couldn't predefine and a number of uncertain elements influence the product design and manufacture, then the AFC is typical product for customization. In order to obtain new products more efficiently and effectively, it is essential for the AFC to implement Mass Customization. In most case the AFC design is variants design that the variant use an existing product as a prototype. There are many common parts between different variant. It is feasible to implement Mass Customization.

Please use the following format when citing this chapter:

Li, Guoping, Zhang, Huangao, Zhao, Wenyan, Zhao, Haixian, Tan, Runhua, 2006, in International Federation for Information Processing (IFIP), Volume 207, Knowledge Enterprise: Intelligent Strategies In Product Design, Manufacturing, and Management, eds. K. Wang, Kovacs G., Wozny M., Fang M., (Boston: Springer), pp. 1010-1015.

A good way to achieve Mass Customization is to build one or more effective product platforms. As Mayer and Lehnerd defined "A product platform is a set of subsystems and interfaces that form a common structure from which a stream of derivative products can be efficiently developed and produced". (Mayer and Lehnerd, 1997) Every product derived from the platform is a "variant" and all the variants form a product family.

2. A TOP-DOWN APPROACH TO PRODUCT FAMILY DESIGN

Simpson classifies the approaches of product family design into two types: bottom-up and top-down approaches.(Simpson,1997) The two approaches are different in the way of product platform building. In the former the product platform is built by the way that a group of distinct products were redesigned or consolidated to realize components standardization without significantly compromising product quality and performance. In the latter the subsystems of all existing products in a company are integrated into one or more product platforms and the new product development must surround that or those platforms. Because the capital investment and redesign costs can be significant when the bottom-up approach is carried out, most companies would rather adopt the top-down approach to design their product family.According to the character of AFC, a top-down approach to design AFC family is put forward, which steps was shown in figure 1.

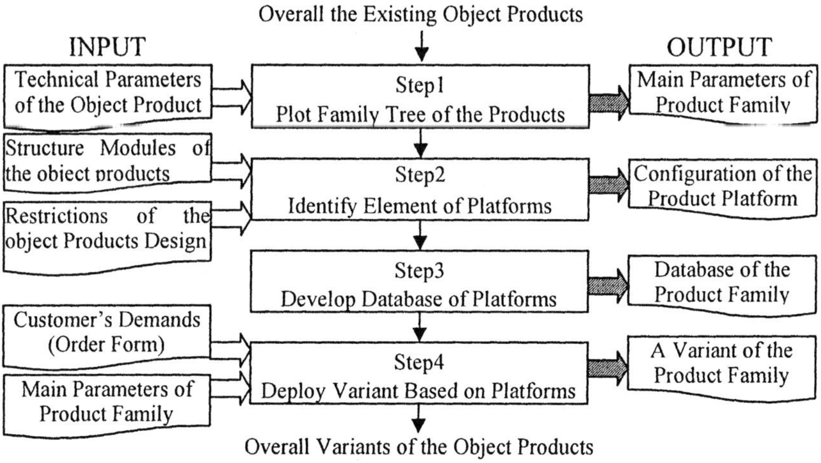

Figure1. the steps of a top-down approach to design product family

The four steps of the approach show four key problems to obtain mass customizations. The method to solve this four problem will be discussed together with AFC in next section.

3. AFC FAMILY DESIGN

AFC family is designed according to above-mentioned approach.

3.1 AFC Family Tree Plotting

Because the requirement of output and the applied environment vary greatly, each individuals of AFC vary greatly from each other in the performance parameter and structure measures. But if it is well planed the variants have chance to be compartmentalized in series. Individuals in same series should be same in the main parameters and most parts, close in performance parameters and architecture parameters. Then each series should build a platform as a common frame of the AFC design. All the platforms with their variant can integrate into a product family tree.

The key to plot product family tree is to identify the main parameters by which one group is distinguished from others. The main parameters can be one or more performance index or structure measures. Many methods can be used to identify the main parameters for a set of product, such as AHP, Pareto diagram, HOQ etc.

The main function of AFC is to carry coal, so output capacity is a primary technical parameter. While output capacity depend on the width of the line pan and the chain speed. And the range of the chain speed design is very small. Then the width of the line pan represents the ability of the conveyor and can be used as a main parameter.

Considering all the AFC with the same width of pans, they differ greatly in the measure of power, the flight-bar type and chain size and design length. The Σ type of pan, power, design length both relate to chain size and the flight-bar type are same in most part. After analyze the structures of the AFC, the individuals with the same width of pan, chain size and the flight-bar type only differ in a few parts. Therefore the main parameters include the width of pan, chain size and the flight-bar type. According to those three parameters the existing AFC is classified into 26 series. Then 26 platforms should be built for each series, and each platform is a branch of the AFC family tree.

3.2 Product Platform Building for AFC

As shown by figure 2, a variant derived from a platform compose of two parts: parts from platform and parts of customization. (Zhang,2003)

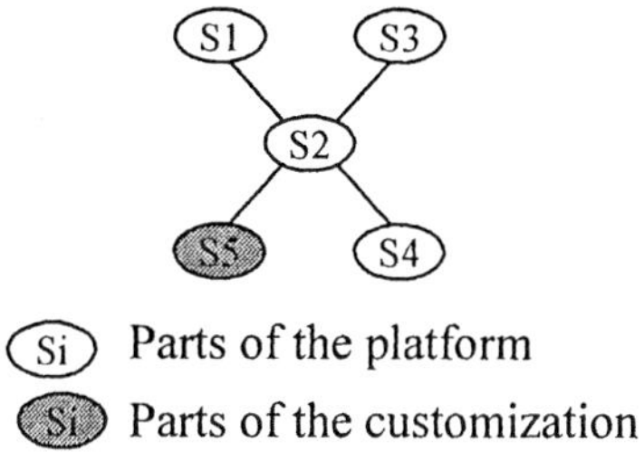

Figure 2. The configuration of a variant of a platform

From the definition of product platform the platform can be seen as the common items shared by a set of products. Therefore, the essential of product platform building can be seen as a process that the common items are identified by some means. Considering in reverse, if the parts of customization are identified, the rest parts could integrate into a platform. Because all AFC have similar configuration, once a platform for a series is built, other platforms are the same in configuration. One series were used to study object to build a platform. By identifying the parts of customization, the process of product platform design should include four steps (as shown in figure 3).

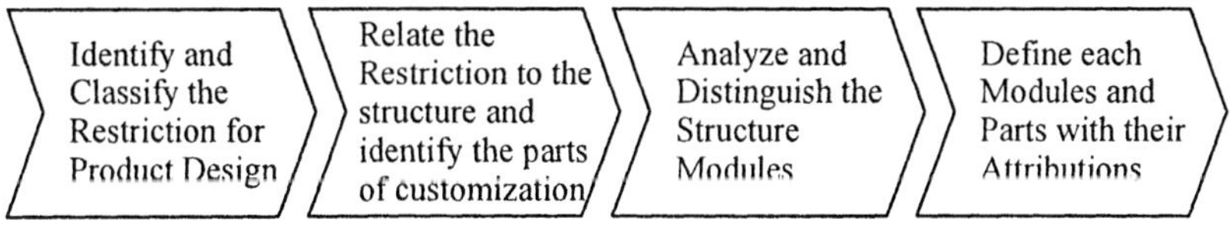

Figure 3. Four steps of product platform building

The structure change result from the change of design restriction that can be extracted from the demands of customers. First, the customers' demands should be studied carefully to find out the items that changed irregularly as restrictions for variants design. Second, the restrictions for variants design should relate to the structure of the product and find out the parts that relate to those restrictions. (McAdams, 1998) Third, the object product is decomposed into modules according to its function structure and its physical configurations. The parts identified from the step3 should depart from others and work as independent modules. Last, the modules should be described by attributions for later deployment. In the process QFD can be used to analyze the relationship between two items (Jack, 1998) and the method of modular design can help to identify the modules (Stone, 1997).

The restrictions for variant design of AFC mainly come from the environment, which include the geological situation, the shear, the stage conveyer and the self-advanced hydraulic supports. If well planed, the influence of these restrictions on the structure of AFC can be limited in some parts, such as pushing ear of spilling plate, pushing unit of delivery and return ends, seat of track and so on. Considering the customers sometimes will emphasize on some parts. The module of the platform should be decomposed into parts and define each part or subsystem with some attributions.

3.3 The Database of the AFC Family Building and Variants Deployment

After all the platforms are built, each module of a platform should be described by some attributions to distinguish from the same module of the other platform. Some complex modules should be decomposed into smaller subsystems in case that some special requirements result in the change of the modules. All data of AFC of a company has been input the e-database of a PDM system, and the software for variant design has been developed and added to the PDM system (As shown in figure 4).

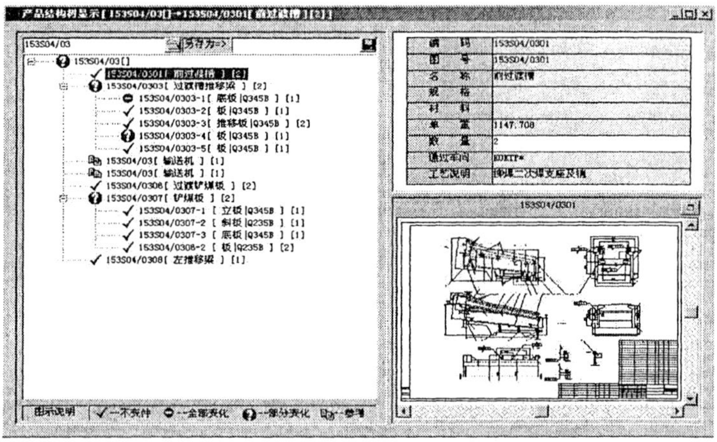

Figure 4. A window of the software for variant deployment

When the order forms are arrived, the engineers input the main structure parameters and then get a product platform. On the basis of the product platform, the software can deploy a new variant by giving some value of the modules attributions. The parts of the platform can be selected promptly and one or more prototype similar to the parts of the customization can also be found. With the help of the software, the design and manufacture periods are much shortened.

4. SUMMARY

A well-defined product platform can succeed in the purpose that the company can provide as much variety as possible for market with as little variety as possible between products.

The Armored Face Conveyer is typical product for customization. To achieve Mass customization, a top-down approach to product family design is introduced. The steps and tools of the approach are discussed when it is used to design AFC family. After the platforms are built, software is developed to implement the process of variant deployment. With the help of the software, the design and manufacture periods are much shortened.

5. ACKNOWLEDGMENT

The research is support by Zhangjiakou Coal Mining Machinery Co. Ltd. and also is supported in part by Natural Science Foundation of China (50375045) and Tianjin Natural Science Foundation Key Projects (043802211)

6. REFERENCES

1. Meyer M H, Lehnerd A P, 1997, *The Power of Product Platforms, USA*, Free Press, New York
2. Simpson T. W., 1998, *A Concept Exploration Method for Product Family Design, Ph. D Dissertation, Georgia Institute of Technology*
3. McAdams D A, Stone R B, 1998, Wood K L. Understanding Product Similarity Using Customer Needs, *ASME Design Engineering Technical Conference*, DETC98/DTM-5660
4. Zhang Huangao, Zhao Wenyan, etc., 2003, A New Model of the Design Process Using Product Platform/QFD/TRIZ, Proceeding of 2003 *International Conference of Integrated Design and Process Technology*,IDPT2003
5. Jack B. Revelle, John W. Moran, Charles A. Cox, 1998, *The QFD Handbook*, USA: John Wiley & Sons, Inc., New York, pp.3-21
6. Stone R. B., 1997,*Towards a Theory of Modular Design*, Ph. D Dissertation, University of Texas.

STUDY ON THE RESPONSE CHARACTERISTICS OF AUTOMATIC WELDING FILTER

Huifang Liu[1], Shiliang Jin[1], Huiting Xiao[1], Yue Hong[1], Longji Shen[2]
[1]*School of Mecanical Engineering, Shanghai University, China; Email:liudf-163@163.com.*
[2]*Department of Teaching, Shanghai University, China.*

Abstract: To facilitate the application and development of the automatic welding filter, it is necessary to study its response characteristics. On the basis of the analysis and research about luminous transmittance and the scale number of the filter, this article proposes a solution to the specific definition of the response and holding time. By comparing different calculation methods in different standards, a felicitous way is chosen to describe the response characteristics accurately.

Key words: luminous transmittance; scale number; response time; holding time

1. INTRODUCTION

There are various hazards during the welding process, such as ultraviolet, infrared, organic gas, and so on. Until now, the welders usually use ordinary welding filter, it is harmful to their eyes. In China, lots of welders have ophthalmia every year. In this situation, the use of the automatic welding filter is extremely necessary.

The automatic welding filter has been used more and more extensively in modern society. It is a protective filter which automatically switches its scale number from a lower value (light state scale number) to a higher value (dark state scale number) when the welding arc is ignited, and protects the worker's eyes effectively so that the welding could be formed more efficiently and good quality is guaranteed [3].

Please use the following format when citing this chapter:
Liu, Huifang, Jin, Shiliang, Xiao, Huiting, Hong, Yue, Shen, Longji, 2006, in International Federation for Information Processing (IFIP), Volume 207, Knowledge Enterprise: Intelligent Strategies In Product Design, Manufacturing, and Management, eds. K. Wang, Kovacs G., Wozny M., Fang M., (Boston: Springer), pp. 1016-1021.

The structure of the automatic welding filter is as follows:

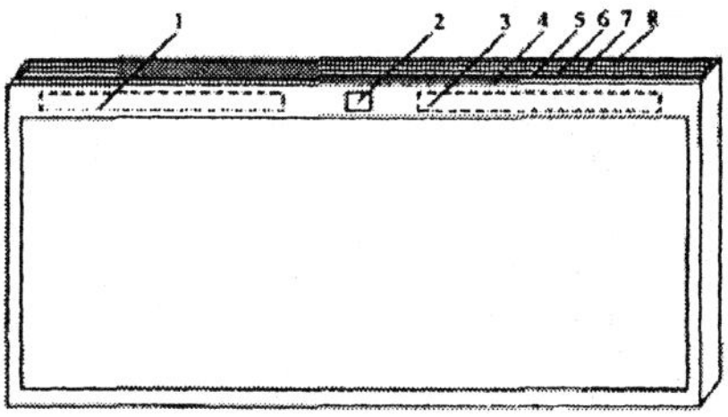

Figure 1. Structure of the automatic welding filter: 1—control circuit, 2—light detector, 3—electrical source, 4—frame, 5—outer protector, 6—ultraviolet and infrared filter, 7—light valve, 8—inner protector.

2. LUMINOUS TRANSMITTANCE AND THE SCALE NUMBER

The luminous transmittance of a filter is the ratio of the transmitted spectral radiant flux ϕ_τ to the incident spectral radiant flux ϕ_i. In other words, $\tau = \frac{\phi_\tau}{\phi_i}$. To describe the response characteristics clearly, another parameter called scale number is introduced. The scale number consists of a coding number and a shade number: the coding number indicates the type of filter, with the exception of "welding filter without coding number"; the shade number is defined by the following expression:

$$N = 1 + \frac{7}{3} \cdot \log_{10}\left(\frac{1}{\tau}\right) \tag{1}$$

Here τ is the luminous transmittance. The shade number is the scale number for the filter that is studied in this article [5].

To describe the state of the automatic welding filter in detail, it is better to divide the scale number into three parts, that is light state scale number, dark state scale number and darkest state scale number.

Light state scale number is the scale number corresponding to the maximum value of luminous transmittance τ_1, dark state scale number is the scale number corresponding to the value of the luminous transmittance τ_d reached after the ignition of the welding arc by an automatic welding filter, and darkest state scale number is the highest scale number corresponding to the minimum value of luminous transmittance τ_2 of an automatic welding filter claimed by the manufacture (see Figure 2).

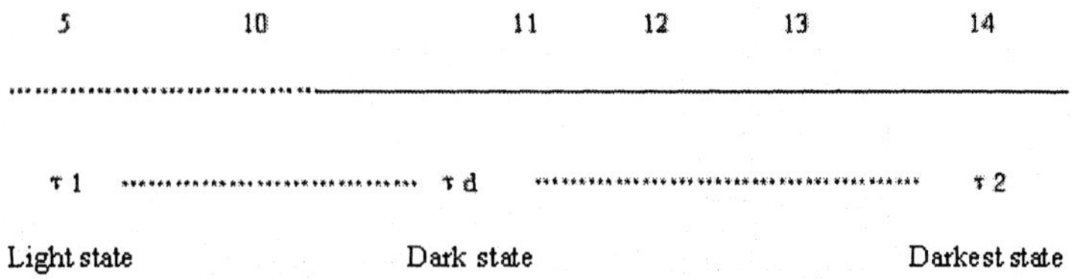

Figure 2. Example showing the meaning of the terms in the case of an automatic welding filter with a light state scale number of 5 and dark state scale numbers between 10 and 14.

The luminous transmittance requirements for filters used in welding and related techniques are given in Table 1.

Table 1. Luminous transmittance requirements

Scale number	Luminous transmittance τ		Scale number	Luminous transmittance τ	
	Maximum %	Minimum %		Maximum %	Minimum %
1.2	100	74.4	8	0.16	0.061
1.4	74.4	58.1	9	0.061	0.023
1.7	58.1	43.2	10	0.023	0.0085
2	43.2	29.1	11	0.0085	0.0032
2.5	29.1	17.8	12	0.0032	0.0012
3	17.8	8.5	13	0.0012	0.00044
4	8.5	3.2	14	0.00044	0.00016
5	3.2	1.2	15	0.00016	0.000061
6	1.2	0.44	16	0.000061	0.000023
7	0.44	0.16			

3. RESPONSE TIME AND HOLDING TIME

There are two critical parameters for the automatic welding filter, one is the response time and the other is holding time. These two parameters determine whether the automatic welding filter being used is good or not.

Response time is the time from the light detector receiving the arc signal to the automatic welding filter turning to dark during the filter tested. It is not only related to the time in which the welders suffer from strong visible light, and it is also related to whether the welders can observe clearly during the welding process.

Holding time is the time from the light detector stopping the arc signal to the automatic welding filter turning back to its original state when the filter is in the dark state. During the welding process, it will happen 10 to 15 times instant short circuit per second, so the filter will change its state frequently if the holding time is too short. Then, the welders will be exposed to the strong visible light, this is bad to their health. Choosing a proper holding time is propitious to the continuous observation and the protection of the welder.

At present, based on the research provided above, different countries establish different standard definitions about the response characteristics of the automatic welding filter.

The European standard EN 379:2003 and the American standard ANSI Z87.1-2003 regulate that the response time of automatic welding filter is related to the filter's luminous transmittance and the scale number.

In these standards, the response time t_s of an automatic welding filter is defined by the following integral:

$$t_s = \frac{1}{\tau_1} \int_{t=0}^{t=t(\tau(t)=3\tau_2)} \tau(t)dt \tag{2}$$

Here: τ_1 is the light state luminous transmittance; τ_2 is the darkest state luminous transmittance; $t = 0$ is the time at which the arc ignites; $\tau(t)$ is luminous transmittance at a time t after the ignition of the welding arc; $t=t(\tau(t)=3\tau_2)$ is the time at which the luminous transmittance falls to 3 times the luminous transmittance in the darkest state.

But the Australian and New Zealand standard AS/NZS 1338. 1:1992 specifies the definition of response time and holding time in a different way. It records the response time as the time from the filter triggering as the shuttered light starts to rise in intensity, to the point where the filter reaches 90%of the dark state. And it records the holding time as the time from the filter triggering as the shuttered light starts to fall in intensity, to the point where the filter drops below 90% of the dark. In this standard, it stipulates that the minimum of the holding time should reach 20 ms [4].

If we use luminous transmittance to define the holding time, the holding time will be the time from stopping the arc signal to the luminous transmittance reaches 90% of the light state.

From the above different definitions of the response characteristics we can get that the start point of the response time is uniform. Every standard begins the record with the light state (the luminous transmittance is maximum) when the filter receives the arc signal. And the start point of the holding time is the same too. They begin the record with the dark state (the luminous transmittance is minimum) when the filter eliminates the arc signal. However, there are differences among them, and they are the confirmation of the end point and the calculation method.

According to the analysis about response time and holding time above, we can easily get the conclusion: for the Australian and New Zealand standard, the definition of response and holding time is just related to the dark state intensity. They have nothing to do with the light state intensity. Furthermore, this definition just considers the beginning state and the ending state of the welding process, but doesn't consider the change during the test. It can't reflect the response characteristics timely, and the result won't be very precise. But for the European and American standards, the definition of response time not only has something to do with the light state luminous transmittance and dark state luminous transmittance of the automatic welding filter, but also reflects the change of the luminous transmittance during the welding. So, this definition can express the protection performance of the filter directly. Thus, the author suggests adopting the European and American definition to measure the response time. That is the formula (2).

As described above, it is better to adopt the time from stopping the arc signal to the luminous transmittance reaches 90% of the light state to measure the holding time. And also, for the health of the welder, the holding time should not be too short. It is recommended that the holding time should be more than 20 ms, since the intensity of light is greatly reduced after the 20 ms of the extinguishments of the welding arc.

Different filters have different light state scale number and dark state scale number, so their response time is different. But for all of them, the response time given in Table 2 shall not be exceeded.

Table 2. Maximum response time

Dark state scale number	Light state scale number						
	1.7	2	2.5	3	4	5	6
	Response time (ms)						
7	300	400	500	700	1000	-	-
8	100	150	200	300	500	1000	-
9	40	50	70	100	200	400	700
10	20	20	30	40	70	100	300
11	6	7	10	15	30	50	100
12	2	3	4	5	10	20	40
13	0.8	1	1.5	2	4	7	10
14	0.3	0.4	0.5	0.7	1	3	5
15	0.10	0.15	0.2	0.3	0.5	1	2
16	0.04	0.05	0.07	0.1	0.2	0.4	0.7

This is an example of the measurement. The range of the filter's scale number is 5 to 11. It's response characteristics are described by the curve given in Figure 3.

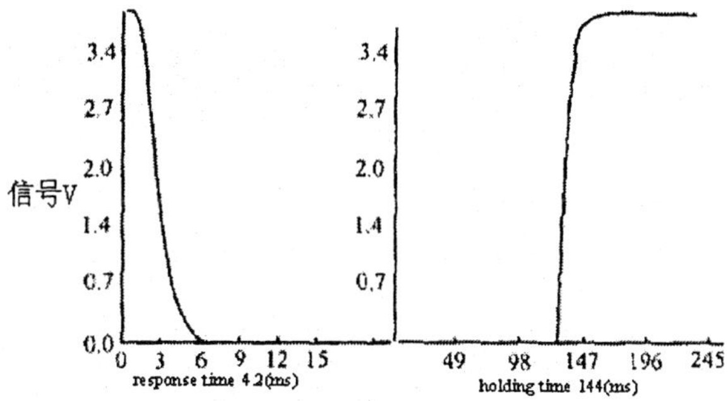

Figure 3. The curve of response characteristics.

4. CONCLUSIONS

According to the compare and research on the response characteristics of automatic welding filters in different countries, this article provides some new requirements and thoughts: the response characteristics have something to do with the luminous transmittance of the filter itself; the response characteristics should reflect the change of the state during the welding process, not just the start point and the end point; the response time is defined as formula (2), and the holding time should be more than 20 ms. This research has great significance to the development and enlargement of new products, and it is very useful to evaluate the quality of new products too.

5. REFERENCES

1. Jeng Jeng-Ywan, Mau Tzuoh-Fei, Leu Shyeu-Ming. Prediction of laser butt joint welding parameters using back propagation and learning vector quantization networks. *Journal of Materials Processing Technology* Volume: 99, Issue: 1-3, March 1, 2000, pp. 207-218
2. Hsieh J.C., Liu C.J., Ju Y.H. Response characteristics of lead phthalocyanine gas sensor: effects of film thickness and crystal morphology. *Thin Solid Films* Volume: 322, Issue: 1-2, June 8, 1998, pp. 98-103
3. EN379: 2003, *Personal Eye-protection-Automatic welding filters* [S], 2003
4. AS/NZS1338.1: 1992, *Filters for eye protectors* [S], 1992
5. ANSI Z87.1-2003 [S], 2003
6. EN 165:1995, *Personal eye-protection-Vocabulary* [S], 1995
7. ISO 4850-1979, *Personal eye-protectors for welding and related techniques- Filters-Utilisation and transmittance requirements* [S], 1979

RESEARCH ON MODULE-BASED VARIANT DESIGN FOR MASS CUSTOMIZATION

Junjian Ding, Shili Tan, Haihong Zhang, Xiaofeng Song
School of Mechanical&Electronic Engineering and Automation, Shanghai University, China; Email: ddajian@sohu.com

Abstract: The mass customization can make customized products with the mass products efficiency, which will become mainstream manufacturing mode in the 21 century. The design for mass customization is the most important factor in the mass customization, on the base of extracting the excellences of the modulization design and the variant design, the article brings forward an idea of module-based variant design. Combining with the technologies of module coding, database, case-based design and CAD, The idea has been used in the customized design of the equipment of pushing and measuring automobile chair track, some equipments have been designed and manufactured successfully.

Key words: Mass Customization; Modulization Design; Variant Design; Module-Based Variant Design

1. INTRODUCTION

The mass customization can make customized products with the mass products efficiency. Like the status of the mass products in the earlier of 20 century, the mass customization which is the coming reformation will take great changes in manufacturing. In order to realize the strategic target in manufacturing, some developed countries have applied mass customization to small amount or single product customization, and the ideas and methods of

This project is supported by the scientific development foundation of Shanghai Automobile Industry under the grant No. 0212.

Please use the following format when citing this chapter:
Ding, Junjian, Tan, Shili, Zhang, Haihong, Song, Xiaofeng, 2006, in International Federation for Information Processing (IFIP), Volume 207, Knowledge Enterprise: Intelligent Strategies In Product Design, Manufacturing, and Management, eds. K. Wang, Kovacs G., Wozny M., Fang M., (Boston: Springer), pp. 1022-1029.

mass customization have been looked as the important measure, at the same time, the design for mass customization is the most important factor.[8,9]

Each of modulization design and variant design is one of the key methods of the mass customization. However when the transverse and longitudinal variants are occurred in the mass customized products, the modulization or variant design can not accomplish the design singly. The transverse variant is that the function, form and structure of a product family are changeless, while the size of the component is changeable. The longitudinal variant is that the function of a product family is changeless, while the form and structure are changeable. So the article brings forward a method------module-based variant design, the method has been used in the designs and equipments to make the automobile chair tracks.

2. MODULIZATION DESIGN

Modulization design is a design method facing the whole product family. On the base of forecasting products market and analyzing products function, the product family is been divided into a series of general modules. According to the customers' demands, modulization design can select and reconfigure different modules, and get a new product, which function, performance and specification are different with the old ones.

Modulization design is one of the key factors of the mass customization, it can make good use of the design resource effectively, reduce the management and assembly cost, simplify the design process and renew the product, at the same time it is good at managing and maintain the product during its life cycle. However using this method, the design system can become complicated, the product structure can not be harmony, the assembly fit can not be very well.

In generally, the model of the product family includes three parts[3], 1) function model, it expresses the function, elements of the function and performance owned by the product. 2) principle model, it expresses the method used in mapping function model to structure model, and the method includes some knowledge, such as design constraint, calculation relation and the control of design process. 3) structure model, it expresses the assembly level of the product family and the possible structures of all modules.

According to the components of the product family, it is easy to get the map from the function model to the structure model, and modules of the product family.

3. VARIANT DESIGN

Variant design is another design method, the new product can be got by using it, which has the same principle, similar structure comparing with the old products, while partial structure and/or design parameter of the old products have been changed. At present, most of mechanical designs belong to variant design, few of them is innovation design[2]. In innovation design, some existed products or components can not been reused, but they can enlighten the designer. In an equipment designed by innovation design, the reused components accounts for about 40%, the ratio of the components amended from the existed ones is about 40%, there is about 20% is the new component. Variant design suits for longitudinal change of the mass customization product mainly[1].

Case-based design (CBD) is one of the common methods of variant design[4,5,6,7,8], it is an application of CBR (Case-based reasoning) technology in the field of CAD. The CBR is one of Artificial Intelligence technologies, so the CBD can well present the characters of variant design, such as generalization, process and individuation. CBR includes questioning, exploring, extracting, evaluating, and storing[6]. According to customers demands, The CBD extracts the similar or same case from the existed database, then analyzes the case's structure and parameter, if they meets the design requests, they can look as the object of the design, or if they can partly meet, CBD will amend them until they are well.

4. MODULE-BASED VARIANT DESIGN

In order to realize the transverse and longitudinal variant of the mass customization, combining modulization design with variant design gets the method of module-based variant deign, which has been applied in the some designs. Figure 1 shows the processes of the module-based variant design.

In fact, module-based variant design is the method applying the idea of modulization design into the variant design, it has the flexibility of variant design and overcomes the rigidity of modulization design. Using it can shorten the time of design and manufacture, reduce the cost and improve the quality, the products also have great advantages to response the market needs for many kinds and small batch products.

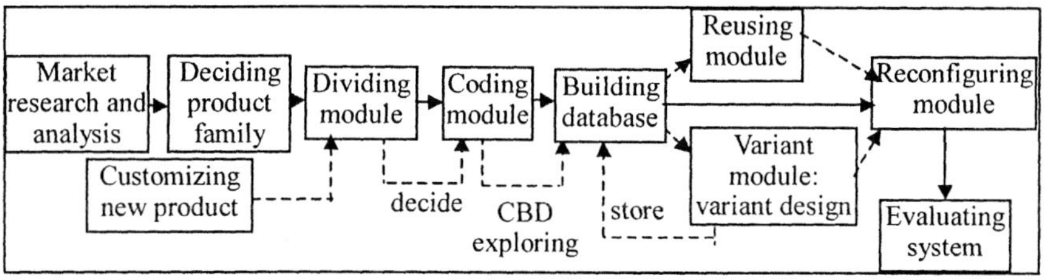

Figure 1. the processes of the module-base variant design

5. CASE

With the development of automobile industry, the people's needs take on individual trend, and the kinds of automobile chair become more and more (such as, different chairs can move in two, four, six, or eight direction, and their fixing modes are different), while they have the same character, the track is their basic safe component, so the product lines of the automobile chairs are similar. At present, the automobile chair manufacturing line can be looked as the kind of such products which are designed and made individually.

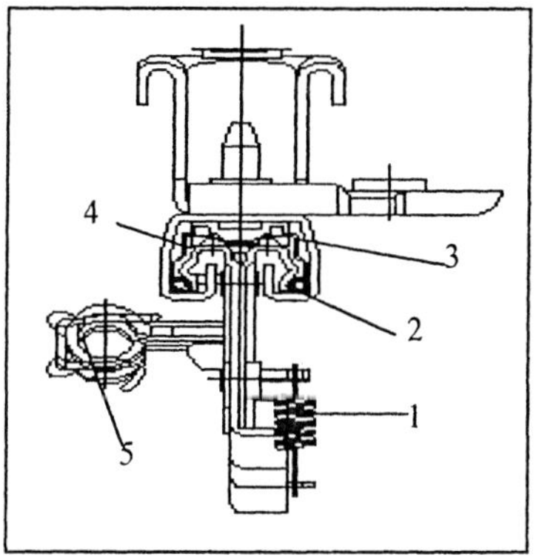

Figure 2. side-glance of the chair track

1. lock spring 2. upper slotware 3.bearing parts 4. lower slotware 5.lock structure

The automobile chair track mainly includes: upper slotware, lower soltware and front and behind bearing parts(such as figure 2). The lower slotware is fixed on the automobile, the upper slotware can move along the lower. The bearing part has some steel balls and a bracket, which is been thrust into the track and between the upper and the lower slotware, when the steel balls rolled, the relative movement between the upper and the lower can be occurred. When the bearing part is thrust into the track, the upper, the lower and the bearing part

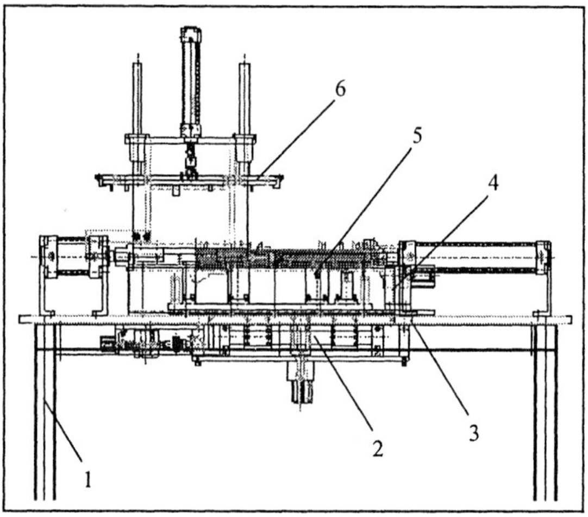

Figure 3. Hcar rover p32e-- the equipment of pushing and measuring track

1.support shelf 2. measure part 3.worktable 4.clamp part 5.orientation part 6.load/push part

integrate tightly, in order to measure the force of track's relative movement, at first, the track must been pushed and become flexible, then the force can be measured precisely.

The equipment of pushing and measuring track can attain the former task, and it is one of the key equipments in the automobile chair track manufacturing line. Figure 3 shows its structures.

After researching and analyzing the market of the equipments of pushing and measuring track, the equipments can be divided into some modules, figure 4 shows the map of function–structure modules of the equipments. The load module can simulate the real environment to load the vertical force on the track. The unlock module can disable the lock structure in the track, so the upper and lower slotware can move relatively. In figure 4, the basic module and auxiliary module can form the prototype equipment, the special module and additional module can extend the function of the equipment.

Analyzing all the equipments, each module includes some different contents:

Orientation module: pin orientation, block orientation.

Unlock module: upper unlock module, lower unlock module.

Table 1. the sum table of different equipment of pushing and measuring track and different tracks

Track style	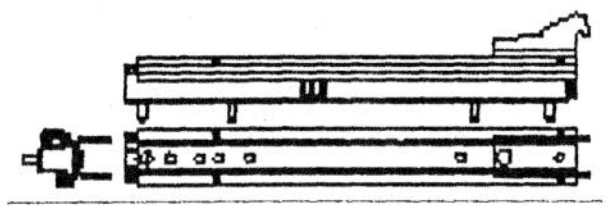	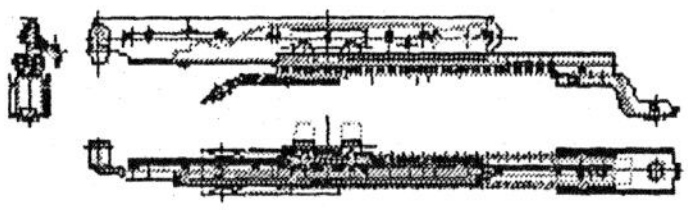
Track name	Hcar track	TT track
Equipment component	Orientation, Clamp, Push, Measure, Data display and store module	Orientation, Clamp, Push module
Track style	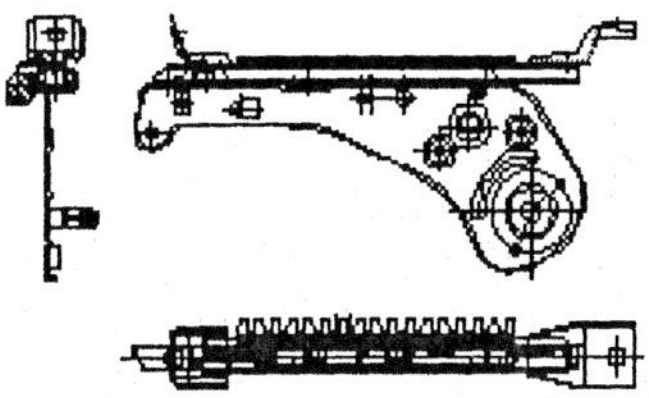	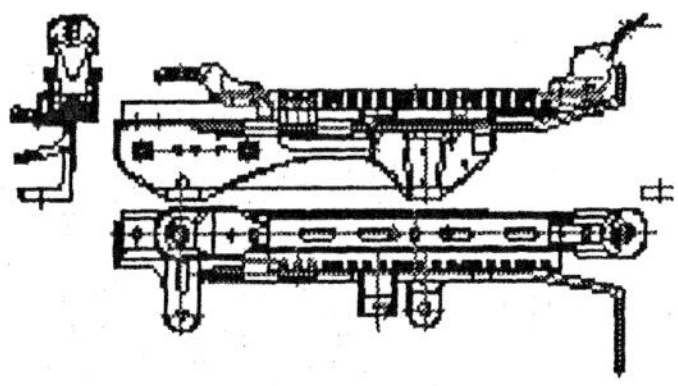
Track name	Spark track	Leli track
Equipment component	Orientation, Clamp, Push, Measure, Data display module	Orientation, Clamp, Push, Measure, Data display module
Track style		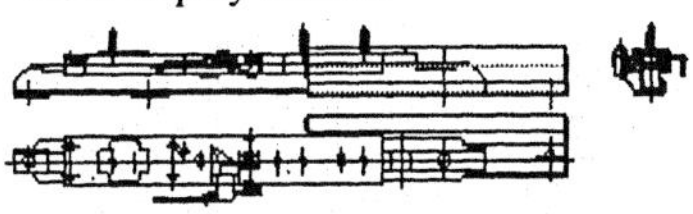
Track name	Mingfang trac	Quanshun track
Equipment component	Orientation, Clamp, Push, Measure module	Orientation, Clamp, Push, Measure, Unlock module
Track style	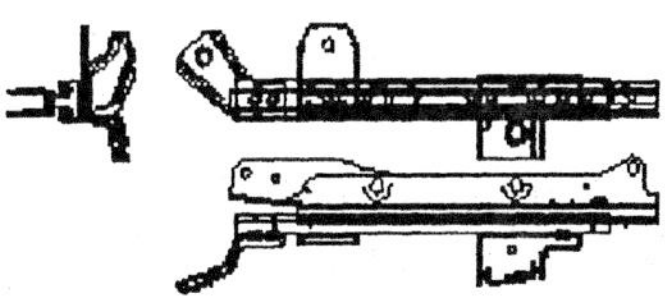	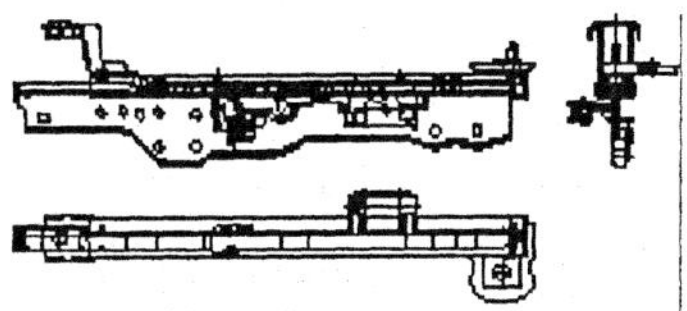
Track name	A21 track	LIANA track
Equipment component	Orientation, Clamp, Push, Measure module	Orientation, Clamp, Push, Measure, Data display module

Clamp module: across clamp module, axis-direction clamp module.
Push module: driving push, pushing push, pulling push.
Load module: single load, pushing-loading load.
Measure, data display, and data store module are the common module in all the equipments.

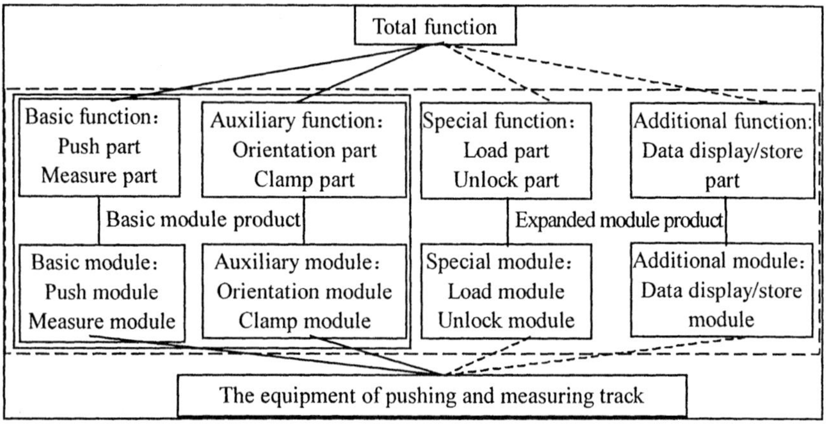

Figure 4. the map of function-structure modules of the equipment

Table 1 includes the most styles of the existing chair tracks. When designing the equipment of pushing and measuring track, the modules is been divided, the codes is been get, the database is been built. If there is a new customized demand, the designer will divide the customized equipment into different modules, code them and explore them in the database according to CBD rules. When the requested module is in the database, it will be used right away, if the requested module is not, the system will find a similar module, so the designer can do variant design on the base of the similar module, at last, the new module will be stored in the database.

6. SUMMARY AND PROSPECT

By using the module-base variant design, all the equipments in the table 1 have been designed and employed, a general database which can meet the current customized demands for the equipment of pushing and measuring track has been built, at the same time, it can been extended. The module-based variant design can be applied into other mass customization products which have the characters of many kinds and small batch.

Some evaluations are needed in the module-based variant design process, they can not only evaluate the design result, but also supply reasons to improve the product, and the evaluation is a key tool to get knowledge. So the evaluation system, which includes evaluation index, evaluation module, analysis method and so on, will become a future research emphasis.

7. REFERENCES

1. Dobrescu G, Reich Y, (2003), Progressive sharing of modules among product variants, *Computer-Aided Design*, vol. 3, pp.791-806
2. Zhong Tingxiu, (2004), Theory and method of rapid response product design, *Hydraulics Pneumatics&Seals*, vol.4, pp.4-9
3. Zeng Fengfang, You Xiao, Yan Xiaoguang, (2004), Technology research on product family modeling under mass customization, *Computer and Digital Engineering*, vol.2, pp.48-50
4. Jiang Li, Sun Shouqian, (1997), An intelligent variant-design supporting system model, *KonstruKtion*, vol.3, pp.14-19
5. Luo Haiyu, (2001), Design strategy for rapid variant design of mechanical products, *Mechanical Research&Applicatin*, vol.12, pp.69-71
6. Zhao Jiyun, Zhong Tingxiu, (1999), The key techniques and application of CBD system to the variant design, *Mechanical Design and Research*, vol.2, pp.36-40
7. *Mechanical Design Handbook* edition two(China Machine Press, Beijing, 2003.6)
8. Hwai-En Tseng, Chien-Chen Chang, (2005), Applying case-based reasoning for product configuration in mass customization environments, *Expert Systems with Applications*, vol.29, pp.913-925
9. Jari Partanen, Harri Haapasalo, (2004), Fast production for order fulfillment: Implementing mass customization in electronics industry, *Production Economics*, vol.90, pp.213-222

APPLIED RESEARCH ON TECHNOLOGIES OF DESIGN FOR MASS CUSTOMISATION

Qiaoxiang Gu, Guoning Qi, Shaohui Su, Yujun Lu
Institute of Contemporary Manufacturing Eng., Zhejiang Univ., Hangzhou, China ; Email: guqiaoxiang@163.com

Abstract: Utilizing the rotor and cylinder of the steam turbine as examples, the key technologies of design for mass customization (DFMC) have been researched. Out of this research, an applied system of design for mass customisation has been assembled. This system consists of several modules, including the system integrated Smart Team PDM, with a coding module and tabular layouts of article characteristics module (both developed by the research team), and Solid Edge CAD software. The system provides facilities for designing customisation products that satisfy individual customers' needs, and has been applied in the design of the steam turbine.

Key words: Mass customisation; product master model; product master structure

1. INTRODUCTION

Mass customization has recently received much attention from both industry and academia. It aims to deliver an increasing product variety to satisfy diverse customer needs while maintaining near mass production efficiency[1,2].

Du et al. describe how an architecture of product families contributes to the generation of families of products efficiently[4]. Jianxin Jiao et al. propose a triple-view scheme for modelling single products, including function view, technical view and structural view[5]. Sehyun et al. describe the parametric modelling process for machine tools, and propose a framework which parametrically models a machine tool assembly based on a design expert system[6].

The research team's framework of technology for mass customisation consists of four parts, of which design for mass customisation (DFMC) is the

This project is supported by the National Natural Science Foundation of China under the grant No. 50575204 and 60374057

Please use the following format when citing this chapter:
Gu, Qiaoxiang, Qi, Guoning, Su, Shaohui, Lu, Yujun, 2006, in International Federation for Information Processing (IFIP), Volume 207, Knowledge Enterprise: Intelligent Strategies In Product Design, Manufacturing, and Management, eds. K. Wang, Kovacs G., Wozny M., Fang M., (Boston: Springer), pp. 1030-1036.

key part[3]. Under the direction of the technologies for modularisation and tabular layouts of article characteristics (TLAC module), the system constructs the product master models (PMM) of variant products and establishes the product master structure (PMS). On the basis of the PMM and PMS, using the configuration design and variant design, the engineers can design the individual product to satisfy customer needs.

2. FRAMEWORK OF THE SYSTEM FOR DFMC

The framework of the system for DFMC is a three-layer architecture, consisting of the user layer, logic layer and database (see Figure 1). In the system, Smart Team PDM was chosen as the user platform, and integrated with the coding module, tabular layouts of article characteristics module and the Solid Edge CAD system. The members of the project developed the coding module and tabular layouts of article characteristics module. The coding module is the core and base of DFMC. Although Smart Team PDM includes a code generation function, this does not meet the requirements of the DFMC. Therefore, we developed the coding module for DFMC.

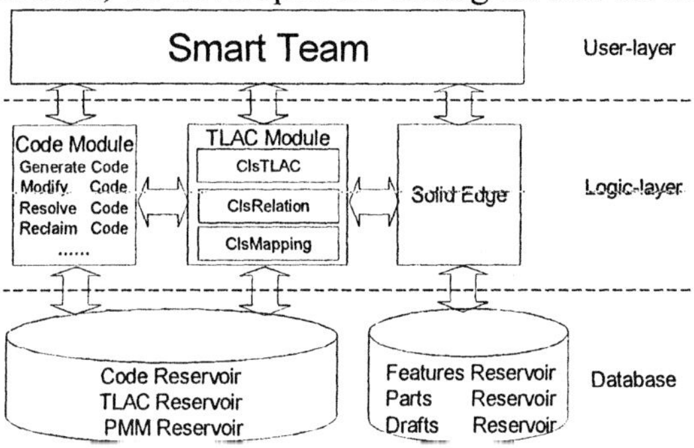

Figure 1. The framework of the system for DFMC

Using the tabular layouts of article characteristics module, the designers can construct product master models of variant products and generate variant products, which meet customer requirements. The module comprises the class of TLAC (ClsTLAC), the class of relations of the TLACs (ClsRelation) and the class of mapping relations between TLAC and CAD (ClsMapping). This module provides the facility to establish the TLAC of the product master models and to create the variant products that meet customer requirements.

3. THE LOGIC STRUCTURE OF THE DATABASE

The database of DFMC consists of code reservoir, features reservoir, PMM reservoir, TLAC reservoir, parts reservoir, drafts reservoir, etc. The system uses the same code module in order to ensure the consistency of data in the development of the product. In this way, all products data and documents are searched and utilized effectively. The logic structure of the database is illustrated in Figure 2.

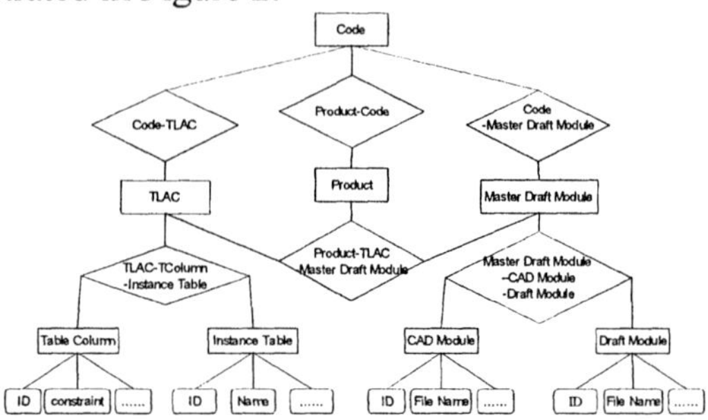

Figure 2. The logic structure of the database

4. THE KEY TECHNOLOGIES OF DFMC

DFMC includes development for mass customization and design for mass customization [3]. The key technologies of DFMC consist of code for DFMC, modularisation, product modelling, product master structure and the product configuration, and so on. Connected with the design for the cylinder and rotor of the steam turbine, how to apply the key technologies will be researched in the next paragraphs.

4.1 The code for DFMC

Figure 3. The code for DFMC

The code module is the base and core of the system for DFMC. The code for DFMC is composed of the classification code, identification code and view code (Figure 3). These can be used in combination.

The classification code (TLAC code) is used to manage the part family in the product family. It can adopt the membership code or classification code.

The identification code is unique in the life cycle of the parts, and it adopts a sequential code.

4.2 Modularization technology

On the basis of overall analysis and research into customer needs, using the modularisation technology, the designers developed an independent function module, which can be used to assemble the complete product family. This module is the key to product configuration design and variant design. It can reduce the internal diversification of products and enhance the exterior diversification of products.

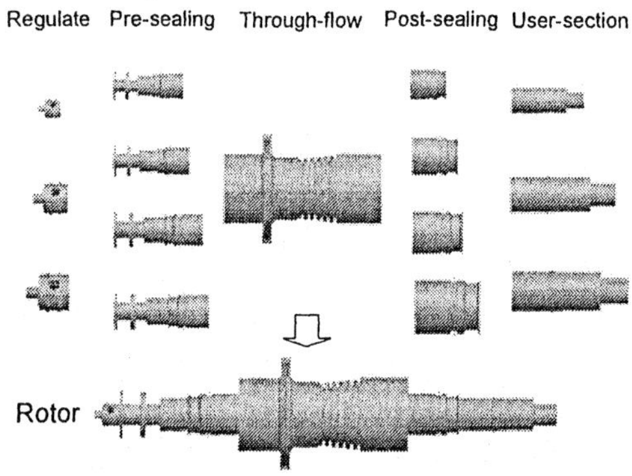

Figure 4. Theory for module design the rotor of industrial steam turbine

Under the direction of the principles of modularization, the rotor of the steam turbine can be divided into five sections in logic, which are regulate, pre-sealing, through-flow, post-sealing and user sections. With the exception of the through-flow section, the sections are designed for standardisation modules, which are in a numerical system. According to customer needs, the engineers design the model of the through-flow section. The designers can then choose the corresponding models from the series models of the others sections, and assemble the rotor model quickly.

4.3 Product master model

The product master model contains the geometric and non-geometric information of product, which is used to for the full life cycle of the product. The system induces the TLAC to create the product master model [7,8]. Inputting a set of parameters that reflect customer needs into the TLAC of the PMM, we can generate the variant products automatically.

Depending on the product, there are two kinds of variant, dimension variant and structure variant. Gu et al. present how to construct the TLAC of the dimension variant [8]. We now discuss how to construct the structure variant part's TLAC.

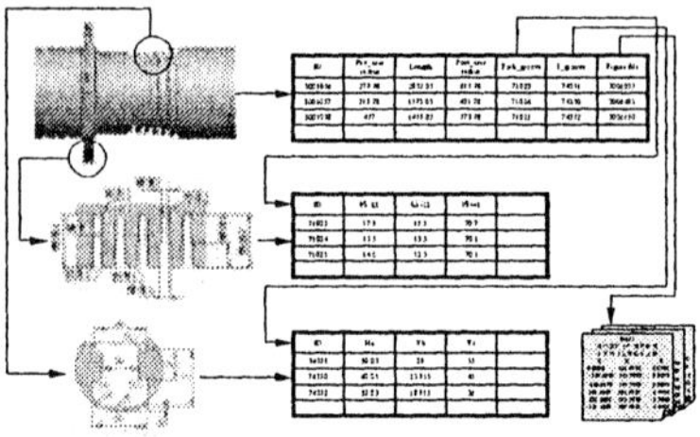

Figure 5. TLAC of through-flow section

The through-flow section of the steam turbine is a typical structure variant part. From customer needs, the designers determine the model parameters. The through-flow section model includes the steam-seal groove and lamina-foot groove. Except for the location dimension of the model, the other dimensions are structure variant. The structure variant items consist of type of steam-seal groove and location, type of lamina-foot groove and location. These are saved in data files. The TLAC of the model saves the index of the data files (Figure 5).

4.4 Product master structure

The product master structure (PMS) describes the configurable product family, which contains the standardization modules. Based on the customer needs, the designer can determine the customisation product's structure from the PMS.

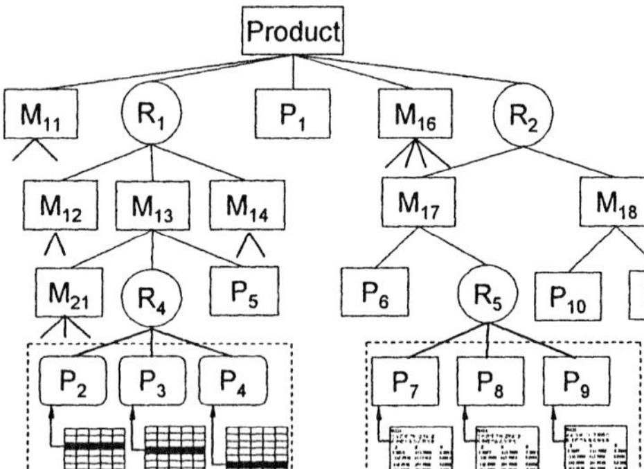

Figure 6. Product Master Structure

When we construct the PMS, the hierarchical structure of the basic modules should be constructed, and then added to the selected modules (Figure 6). In the PMS for product family, it used the rules to choose the selected modules. Here the configuration rules were described by Event-Condition-Action rule. In the PMS, it includes the dimension and structure variant module, which were mentioned in subsection 4.3.

Definition: PMS is a pair, PMS::=({Modules}, {ECA rules}),Where:

{Modules} represents a set of modules. {Module}::= (ID, Name, Parameter, Description), the elements of set describe the identification number, name, feature-parameter and description of the module.

{ECA rules} is a set of configuration rules. R::=(ID, Event, Condition, Action). Where the elements of set represent the identification number, trigger, precondition and conclusion of the configuration rules.

4.5 Product configuration design

According to the customer needs, the product structure and parts of customisation product can be gained from the Product Configuration Design. We can use Solutions=F (PMS, Requirements) to describe the process of configuration. Where F is the solution process; PMS describes product master structure; Requirements represents customer needs; Solutions describes the configuration scheme, which satisfied with customers.

To customize the steam turbine, customer needs include the type of the steam turbine, the pressure and temperature of inlet steam, the pressure and temperature of exhaust steam, the output power and rotate speed of working machine, and so on. Surely, it is an iterative process to solve of the product configuration. To the steam turbine, which is ETO, it is impossible to depend on the computer to accomplish the configuration completely, and sometimes need the designer to join.

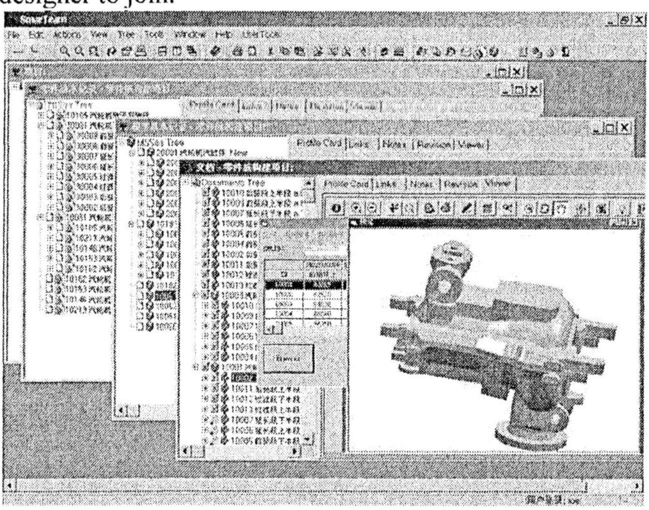

Figure 7. Application Instance

5. CASE STUDY

The steam turbine is customised completely by customer needs. If designers induce DFMC to design it, it will reduce parts number obviously and enhance the design efficiency. The cylinder of steam turbine includes pre-section, decompression-section, transition-section, prolong-section and exhaust-steam-section. According to the through-flow section, these modules can be assembled to the cylinder by rules.

Figure 7 provides a rotor instance. The cylinder includes of pre-section, transition-section, prolong-section and exhaust-steam-section.

6. CONCLUSIONS

Using the system for DFMC to develop the steam turbine, it not only enhances the design efficiency, but also reduces the cost of production. Further, in order to exploit the effect of DFMC, it should be combined with management for mass customization and manufacture for mass customization.

7. REFERENCES

1. Pine II, B.J.,(1993),*Mass Customization: The New Frontier in Business Competition,* Harvard Business School Press（ISBN: 0875849466）.
2. Tseng, M. M., Jiao, J.,(1996),Design for mass customization,*Annals of the CIRP, 45,* pp. 153-156.
3. Qi Guoning, Gu Xinjian, Tan Janrong,(2003),*Mass Customization Technologies and Development,*china machine press（ISBN: 7-111-13110-x）.
4. Xuehong Du, Mitchell M.Tseng, Jianxin Jiao, (2001),Product Families for Mass Customization-Understanding the architecture,*The Customer Centric Enterprise - Advances in Mass Customization and Personalization,* Springer-Verlag Berlin Heidelberg New York.（ISBN: 3-540-02492-1）
5. Jianxin Jiao, Mitchell M.Tseng, Vincent G. Duffy, (1998), Product Family Modeling for Mass Customization, *Computers ind. Engng* ,Vol.35, Nos 3-4, pp.495-498.
6. Sehyun Myung, Soonhung Han, (2001),Knowledge-based parametric design mechanical products based on configuration design method,*Expert Systems with Applications 21.* pp.99-107.
7. Lu Yunjun,Yu Junhe, Qi Guoning, (2003), Product Variant Design Based on Tabular Layouts of Article Characteristics,*Computer Integrated Manufacturing Systems,* Vol. 9 No. 10, pp.840-844.
8. Gu Qiaoxiang, Su Shaohui, Yu Junhe, (2004),Research on Product Variant Design Based on the Tabular Layouts of Article Characteristics, China Mechanical Engineering, Vol.15 No.19, pp.1713-1716.

HIDDEN PATTERN OF INNOVATION

Innovation of Knowledge Enterprises by WOIS

Hansjürgen Linde, Gunther Herr, Andreas Rehklau
WOIS Innovaiton School Coburg, Germany, Email: linde@fh-coburg.de

Abstract: This paper will give an overview on the key elements of the innovation technology of WOIS that can be used for generating a sustainable decision basis for future development directions.

Key words: knowledge enterprise, innovation, WOIS, knowledge management.

1. INTRODUCTION

Especially in critical economic situations, safeguarding company's future requires strengthening the innovation power. Consequently companies necessitate a clear picture of the company's competitive situation in future as well as reliable innovation processes for heading towards this future in a focused way. In particular fundamental decisions throughout the earliest phases of innovation processes define the achievable output later on. Therefore, it is crucial to generate an as strong argumentation basis for early strategic decisions as possible, even though these need to be taken in diffuse situations always. This requires severe innovation knowledge, the availability of strategic orientation tools and an innovation technology that enables efficient and effective innovation management process. Efficient in a manner that not all employees that need to be involved in innovation processes can be trained to become innovation experts first. Consequently, processes need to be in place that provides a proven innovation technology. Even more important, this innovation technology needs to foster an effective innovation philosophy that focuses on generating future benefits and therefore competitive advantages and differentiation potentials. Accordingly three essential key elements exist for mastering strategic innovations:

Please use the following format when citing this chapter:
Linde, Hansjürgen, Herr, Gunther, Rehklau, Andreas, 2006, in International Federation for Information Processing (IFIP), Volume 207, Knowledge Enterprise: Intelligent Strategies In Product Design, Manufacturing, and Management, eds. K. Wang, Kovacs G., Wozny M., Fang M., (Boston: Springer), pp. 1037-1041.

- Innovation Knowledge for strengthening the innovation power.
- Direction Finding Models for increasing the prognosis effectiveness-
- Innovation Technology for supporting the realisation efficiency.

The innovation strategy WOIS combines these elements in a unique Contradiction Oriented Innovation Strategy by integrating key elements of known successful methodologies such as TRIZ with new aspects of encouraging a challenging innovation culture, designing highly competitive value creation chains and realising successful innovations in known and new created markets. Partners that have already applied this innovation theory successfully are companies such as BMW, Bosch, Braun, Linde, Melitta, Nestlé Schöller, Ergoline, Grohe, Hilit, Siemens, Tesa and many others.

Especially for long term sucess company's requires a clear picture of the future competitive situation as well as reliable innovation processes for heading towards this future. In particular "weak signals" become important as it is likely that they end up having "strong impacts". The key questions are as the following.

2. WHERE TO LOOK AND HOW TO INSPIRE NEW DEVELOPMENT DIRECTIONS?

For building a sustainable picture it is important to consider how the society, technology, resources and the thinking culture will develop. All four aspects will develop, including their interdependencies, no matter whether "our company" contributes to the future development or not.

Consequently the challenge will become to influence this development in a way that we design and meet future expectations of the markets. Facing this task from a business point of view.

3. WHO IS RESPONSIBLE FOR TURNING THE SWITCHES TOWARDS THE RIGHT TRACKS?

Is it only the management of the company? Frequently the top management faces a lack of detailed information. Therefore the decision making basis is not as accurate as desirable. On the other hand it could be asked if employees with too little decision making empowerment and too little overview should take the decisions?

Throughout the industry detailed descriptions exist for all kinds of business processes. Often, future perspectives are developed based on 'roadmaps' that result from evaluating spontaneous ideas, former project results and preliminary concept proposals.

But this is not sufficient for developing a clear picture for the future of the business.

It can be observed, that many companies are confronted with a threatening gap especially within the early, strategic phases of development processes. The importance and impact of defining repeatable and reliable approaches for initiating strategic innovations is frequently underestimated.

One might think that companies that fail in defining sustainable strategic approaches will suffer considerable disadvantages - but due to similar behaviours of the competitors this is not the case.

Nevertheless, significant advantages can be gained through implementing systems that permanently allow the flexible adaptation of the development program, without loosing the clear focus for medium and long term targets.

4. BUT HOW TO FILL THE STRATEGIC GAP? WHAT CAN BE THE SCIENTIFIC BASIS FOR DOING SO?

In times of multi-criteria problems, company over spanning development networks and ever faster development processes, it is essential for companies to ground significant decisions on a reliable sustainable decision basis.

Especially 'stock market governed' company's belief in 'management by financial target agreements', but fail in implementing the ability for developing future perspectives. They are dead-locked in efficiency programs, unable to develop and implement strategies for future wealth creation.

Future orient companies require a system that consists of

- A growth oriented economical model,
- An energy focusing philosophy,
- A culture of a commonly shared understanding concerning future challenges and
- Knowledge about hidden pattern of competition.

The innovation strategy WOIS integrates these aspects into one powerful strategy.

5. IS THERE A NEED FOR ANOTHER APPROACH?

When studying the literature hundreds of tools, tactics, methodologies exist for a wide range of situations. But there is no approach that would support business, marketing, product, process, organisation and resource innovations from the early strategic definition phases to its market implementation.

It is the intention of WOIS to provide a flexible mental model and innovation technology for innovation leadership, by supporting the processes required for recognising and realising innovative shortcuts in a more focused way. This is an exiting intension.

6. WHAT IS THE STRATEGY BEHIND THE INTENSION OF SHORTCUTTING THE SPIRAL OF EVOLUTION?

In general, the decision making processes can be described as a process of three distinct phases:

- An orientation,
- Decision making and
- Innovation-finding phase.

Knowing the phases is good, but not yet helpful for guiding a project team through a major innovation project.

7. WHICH ELEMENTS ARE NEEDED TO RUN SUCCESSFUL INNOVATION PROJECTS?

For being able to allow chaotic association phases throughout the project it is essential to be able to rely on the governing overlaying process. Therefore, models exist for each of the phases that support project work with

- Strategic orientation tools,
- Distinct analysis perspectives,
- Inspiration resources and
- Models for deriving key findings.

When trying to consider developments of the society, technology, natural resources and global competition networks as well as there interdependencies:

8. FROM AN ABSTRACT POINT OF VIEW, WHAT ARE KEY SUCCESS FACTORS FOR INNOVATION?

Managing Professional Strategic Innovations needs to integrate three Leading Core Competences for Innovation:

- Knowledge on Innovation Sciences as a theoretical basis
- Models for finding High Potential Future Directions as common process basis and
- Innovation Technologies for identifying key barriers and innovative perspectives as process support.

It is only the combination of these pillars that sustains competitive leadership by generating the required continues growth.

The following companies use the scientific background of WOIS and its Innovation Technology to innovate their business: Braun, Siemens, Linde, BMW, Grohe, Mondi, Hilti, Netzsch, SaintGobain, Pierburg, Tenneco, Schmitz, Tesa, Vorwerk, Mann+Hummel, Siemens VDO, Kapp, Brose, Hailo, Knorr, BSH, Bosch, Berker, Stöbich, Nestlé Schöller, Fischer, Demang, VOSS, Dahle, Kermi, Winkhaus, AquaRotter, EBS, Melitta, Keramag, Winkhaus, Viking, Luk and others.

9. REFERENCE

WOIS INSTITUTE, Am Hofbrauhaus 1, D-96450 Coburg, GERMANY
Tel.: +49 9561 8363 240, www.wois-innovation.de, info@wois-innovation.de

Author Index